Lecture Notes in Computer Science 6942

Commenced Publication in 1973
Founding and Former Series Editors:
Gerhard Goos, Juris Hartmanis, and Jan van Leeuwen

Advanced Research in Computing and Software Science

Subline of Lectures Notes in Computer Science

D1697060

Camil Demetrescu
Magnús M. Halldórsson (Eds.)

Algorithms – ESA 2011

19th Annual European Symposium
Saarbrücken, Germany, September 5-9, 2011
Proceedings

 Springer

Volume Editors

Camil Demetrescu
University of Rome
"La Sapienza"
Dipartimento di Informatica e Sistemistica
Via Ariosto 25, 00185 Roma, Italy
E-mail: demetres@dis.uniroma1.it

Magnús M. Halldórsson
Reykjavik University
School of Computer Science
Menntavegur 1, 101 Reykjavik, Iceland
E-mail: mmh@ru.is

ISSN 0302-9743   e-ISSN 1611-3349
ISBN 978-3-642-23718-8   e-ISBN 978-3-642-23719-5
DOI 10.1007/978-3-642-23719-5
Springer Heidelberg Dordrecht London New York

Library of Congress Control Number: 2011935220

CR Subject Classification (1998): F.2, I.3.5, C.2, E.1, G.2, D.2, F.1

LNCS Sublibrary: SL 1 – Theoretical Computer Science and General Issues

Typesetting: Camera-ready by author, data conversion by Scientific Publishing Services, Chennai, India

Printed on acid-free paper

Springer is part of Springer Science+Business Media (www.springer.com)

Preface

This volume contains the papers presented at the 19th Annual European Symposium on Algorithms (ESA 2011) held at the Max Planck Institute for Informatics, Saarbrücken, during September 5–7, 2011. ESA 2011 was organized as part of ALGO 2011, which also included the Workshop on Algorithms for Bioinformatics (WABI), the International Symposium on Parameterized and Exact Computation (IPEC), the Workshop on Approximation and Online Algorithms (WAOA), the International Symposium on Algorithms for Sensor Systems, Wireless Ad Hoc Networks and Autonomous Mobile Entities (ALGOSENSORS), and the Workshop on Algorithmic Approaches for Transportation Modeling, Optimization, and Systems (ATMOS). The preceding symposia were held in Liverpool (2010), Copenhagen (2009), Karlsruhe (2008), Eilat (2007), Zürich (2006), Palma de Mallorca (2005), Bergen (2004), Budapest (2003), Rome (2002), Aarhus (2001), Saarbrücken (2000), Prague (1999), Venice (1998), Graz (1997), Barcelona (1996), Corfu (1995), Utrecht (1994), and Bad Honnef (1993).

The ESA symposia are devoted to fostering and disseminating the results of high-quality research on the design and evaluation of algorithms and data structures. The forum seeks original algorithmic contributions for problems with relevant theoretical and/or practical applications and aims at bringing together researchers in the computer science and operations research communities. Papers were solicited in all areas of algorithmic research, both theoretical and experimental, and were evaluated by two Program Committees (PC). The PC of Track A (Design and Analysis) selected contributions with a strong emphasis on the theoretical analysis of algorithms. The PC of Track B (Engineering and Applications) evaluated papers reporting on the results of experimental evaluations and on algorithm engineering contributions for practical applications.

The conference received 255 submissions from 39 countries. Each submission was reviewed by at least three PC members, and carefully evaluated on quality, originality, and relevance to the conference. Overall, the PCs wrote 760 reviews with the help of 355 trusted external referees. Based on an extensive electronic discussion, the committees selected 67 papers (55 out of 209 in Track A and 12 out of 46 in Track B), leading to an acceptance rate of 26%. In addition to the accepted contributions, the symposium featured two distinguished plenary lectures by Berthold Vöcking (RWTH Aachen University) and Lars Arge (University of Aarhus and MADALGO).

The European Association for Theoretical Computer Science (EATCS) sponsored a best paper award and a best student paper award. The former went to Nikhil Bansal and Joel Spencer for their contribution titled "Deterministic Discrepancy Minimization." The best student paper prize was awarded to Pawel Gawrychowski for his paper titled "Pattern Matching in Lempel-Ziv Compressed

Strings: Fast, Simple, and Deterministic." Our warmest congratulations to them for these achievements.

We would like to thank all the authors who responded to the call for papers, the invited speakers, the members of the PCs, as well as the external referees and the Organizing Committee members. We also gratefully acknowledge the developers and maintainers of the EasyChair conference management system, which provided invaluable support throughout the selection process and the preparation of these proceedings. We hope that the readers will enjoy the papers published in this volume, sparking their intellectual curiosity and providing inspiration for their work.

July 2011   Camil Demetrescu
                                                               Magnús M. Halldórsson

Conference Organization

Program Committees

Track A (Design and Analysis)

Therese Biedl	University of Waterloo, Canada
Philip Bille	Technical University of Denmark
Allan Borodin	University of Toronto, Canada
Amr Elmasry	University of Copenhagen, Denmark
Sándor Fekete	TU Braunschweig, Germany
Martin Fürer	Pennsylvania State University, USA
Cyril Gavoille	LaBRI, University of Bordeaux, France
Fabrizio Grandoni	University of Rome Tor Vergata, Italy
Magnús M. Halldórsson	Reykjavik University, Iceland
Tao Jiang	UC Riverside, USA
Dieter Kratsch	University Paul Verlaine – Metz, France
James R. Lee	University of Washington, USA
Daniel Lokshtanov	University of Bergen, Norway
Pradipta Mitra	Reykjavik University, Iceland
Yoshio Okamoto	Japan Advanced Institute of Science and Technology
Guido Schäfer	CWI and VU University Amsterdam, The Netherlands
Danny Segev	University of Haifa, Israel
Hadas Shachnai	Technion, Israel
Cliff Stein	Columbia University, USA
Ileana Streinu	Smith College, USA
Takeshi Tokuyama	Tohoku University, Japan

Track B (Engineering and Applications)

Umut Acar	Max Planck Institute for Software Systems, Germany
Michael Bender	SUNY Stony Brook, USA
Alberto Caprara	University of Bologna, Italy
Carlos Castillo	Yahoo! Research, Spain
Graham Cormode	AT&T Labs–Research, USA
Daniel Delling	Microsoft Research–Silicon Valley, USA
Camil Demetrescu	Sapienza University of Rome, Italy
Thomas Erlebach	University of Leicester, UK
Rudolf Fleischer	Fudan University, China
Monika Henzinger	University of Vienna, Austria

Michael Jünger   University of Cologne, Germany
Ian Munro   University of Waterloo, Canada
Sotiris Nikoletseas   University of Patras and CTI, Greece
Rasmus Pagh   IT University of Copenhagen, Denmark
Bettina Speckmann   TU Eindhoven, The Netherlands

Organizing Committee

Christina Fries   Max Planck Institute for Informatics, Germany
Tomasz Jurkiewicz   Max Planck Institute for Informatics, Germany
Kurt Mehlhorn (Chair)   Max Planck Institute for Informatics, Germany

External Reviewers

Aardal, Karen   Carmi, Paz
Abraham, Ittai   Carrère, Frédérique
Albagli-Kim, Sivan   Censor-Hillel, Keren
Andoni, Alex   Charguéraud, Arthur
Arge, Lars   Chekuri, Chandra
Arroyuelo, Diego   Chen, Xi
Asahiro, Yuichi   Chowdhury, Rezaul
Asano, Tetsuo   Christ, Tobias
Ásgeirsson, Eyjólfur Ingi   Christodoulou, Giorgos
Aziz, Haris   Chrobak, Marek
Bacon, Dave   Chun, Jinhee
Bae, Sang Won   Codenotti, Paolo
Bansal, Nikhil   Cormode, Graham
Becchetti, Luca   Costello, Kevin
Bein, Wolfgang   Cotter, Andrew
Benkoczi, Robert   Daly, Elizabeth M.
Berger, André   Daskalakis, Constantinos
Blelloch, Guy   Devanur, Nikhil
Bodlaender, Hans L.   Ding, Jian
Bonifaci, Vincenzo   Dinitz, Michael
Bonsma, Paul   Dobzinski, Shahar
Brandenburg, Franz   Durocher, Stephane
Braverman, Mark   Dutot, Pierre-François
Briest, Patrick   Dwork, Cynthia
Buchbinder, Niv   Ebrahimi, Roozbeh
Buchin, Maike   Edmonds, Jeff
Buhler, Jeremy   Efrat, Alon
Cacchiani, Valentina   Eickmeyer, Kord
Caminiti, Saverio   Elbassioni, Khaled

Elberfeld, Michael
Englert, Matthias
Eppstein, David
Epstein, Leah
Erickson, Jeff
Eyraud-Dubois, Lionel
Fagerberg, Rolf
Farach-Colton, Martin
Farzan, Arash
Favrholdt, Lene M.
Feldman, Dan
Feldman, Michal
Fiat, Amos
Finocchi, Irene
Fischer, Johannes
Fomin, Fedor V.
Friedetzky, Tom
Fujito, Toshihiro
Fukasawa, Ricardo
Fusco, Emanuele Guido
Fusy, Eric
Gagie, Travis
Galli, Laura
Gamzu, Iftah
Ganguly, Sumit
Gao, Jie
Gaspers, Serge
Georgiou, Konstantinos
Ghasemi, Golnaz
Gionis, Aristides
Gkatzelis, Vasilis
Goldwasser, Michael
Golovach, Petr
Goussevskaia, Olga
Goyal, Vineet
Grant, Elyot
Green Larsen, Kasper
Gutin, Gregory
Gutwenger, Carsten
Hajiaghayi, MohammadTaghi
Hallgren, Sean
Hanczar, Blaise
Hanrot, Guillaume
Hanusse, Nicolas
Har-Peled, Sariel

Harks, Tobias
Harrow, Aram
Hartvigsen, David
Hassin, Refael
Haverkort, Herman
He, Meng
Heggernes, Pinar
Hermelin, Danny
Hertli, Timon
Hochstättler, Winfried
Hoefer, Martin
Hu, Yifan
Huang, Chien-Chung
Iacono, John
Immorlica, Nicole
Ivan, Ioana
Iwama, Kazuo
Jacobs, Tobias
Jansen, Bart
Jansen, Klaus
Jeannot, Emmanuel
Jerrum, Mark
Johnson, Aaron
Joy, Joseph
Kakugawa, Hirotsugu
Kamiyama, Naoyuki
Kanj, Iyad
Kanté, Mamadou
Kaplan, Haim
Karloff, Howard
Karpinski, Marek
Kash, Ian
Katajainen, Jyrki
Kavitha, Telikepalli
Kawachi, Akinori
Kawamura, Akitoshi
de Keijzer, Bart
Kellerer, Hans
Kern, Walter
Kesselheim, Thomas
Khandekar, Rohit
Kijima, Shuji
Kim, Yoo-Ah
King, James
King, Valerie

Kirkpatrick, David
Kiyomi, Masashi
Kleinberg, Jon
Klimm, Max
Knauer, Christian
Könemann, Jochen
Koivisto, Mikko
Kontogiannis, Spyros
Korman, Matias
Koster, Arie
Král', Daniel
Kratsch, Stefan
Kreutzer, Stephan
van Kreveld, Marc
Kriegel, Klaus
Krishnaswamy, Ravishankar
Kröller, Alexander
Kulik, Ariel
Labourel, Arnaud
Landa-Silva, Dario
Langer, Alexander
Langerman, Stefan
Larsen, Kim S.
Laszlo, Michael
Lattanzi, Silvio
Laurent, Monique
van Leeuwen, Erik Jan
Letchford, Adam
Levin, Asaf
Li, Jian
Li, Yanbo
Liberty, Edo
Liedloff, Mathieu
Ljubic, Ivana
Lotker, Zvi
Lu, Hsueh-I
Lubiw, Anna
Lucier, Brendan
Lübbecke, Marco
Mahmoud, Hosam
Makowsky, Johann
Makris, Christos
Manlove, David
Marchetti-Spaccamela, Alberto
Martin, Barnaby

Marx, Dániel
Matsliah, Arie
Mazoit, Frédéric
Medina, Moti
Medjedovic, Dzejla
Mehta, Aranyak
Meijer, Henk
Meister, Daniel
Mestre, Julián
Meyer auf der Heide, Friedhelm
Meyerhenke, Henning
Mihal'ák, Matúš
Mirrokni, Vahab
Misra, Neeldhara
Mitchell, Joseph
Mitzenmacher, Michael
Miyamoto, Yuichiro
Miyazaki, Shuichi
Molloy, Michael
Montes, Pablo
Morin, Pat
Moruz, Gabriel
Moseley, Benjamin
Mucha, Marcin
Müller, Haiko
Mulzer, Wolfgang
Munro, Ian
Muthukrishnan, S.
Mutzel, Petra
Mäkinen, Erkki
Mäkinen, Veli
Nagamochi, Hiroshi
Nagano, Kiyohito
Nagarajan, Viswanath
Nakano, Shin-Ichi
Nekrich, Yakov
Nelson, Jelani
Neumann, Frank
Newman, Alantha
Newman, Ilan
Niedermeier, Rolf
Nikolova, Evdokia
Nishimura, Naomi
Nobili, Paolo
Nussbaum, Yahav

Nöllenburg, Martin
Onak, Krzysztof
Ono, Hirotaka
Orecchia, Lorenzo
Otachi, Yota
Oudot, Steve
Ozkok, Umut
Pagh, Rasmus
Paluch, Katarzyna
Panagiotou, Konstantinos
Panigrahi, Debmalya
Pardella, Gregor
Pasquale, Francesco
Paul, Christophe
Pearce, David
Peis, Britta
Perkovic, Ljubomir
Pferschy, Ulrich
Piazza, Carla
Pritchard, David
Procaccia, Ariel
Pruhs, Kirk
Rabani, Yuval
Rahman, M. Sohel
Raman, Rajiv
Raptopoulos, Christoforos
Rawitz, Dror
Rendl, Franz
Reyzin, Lev
Ricca, Federica
Rizzi, Romeo
Ron, Dana
Rossmanith, Peter
Rothvoss, Thomas
Röglin, Heiko
Sadakane, Kunihiko
Saha, Barna
Salvy, Bruno
Sanders, Peter
Sanità, Laura
Satti, Srinivasa Rao
Sau, Ignasi
Saurabh, Saket
Schapira, Michael
Schieber, Baruch

Schmidt, Christiane
Schmidt, Daniel R.
Schmutzer, Andreas
Schudy, Warren
Schultz, Rüdiger
Schwartz, Roy
Severs, Christopher
Shashidhar, K.C.
Sherette, Jessica
Shioura, Akiyoshi
Shmoys, David
Sidiropoulos, Anastasios
Silveira, Rodrigo
Sitters, Rene
Sivan, Balasubramanian
Skutella, Martin
Smyth, William F.
Sorenson, Jonathan
Soto, Jose A.
Souza, Alexander
Srivastava, Nikhil
van Stee, Rob
Stehle, Damien
Stiller, Sebastian
Suomela, Jukka
Svensson, Ola
Ta, Vinh-Thong
Tamir, Tami
Tangwongsan, Kanat
Tanigawa, Shin-Ichi
Tazari, Siamak
Telikepalli, Kavitha
Thaler, Justin
Thilikos, Dimitrios
Ting, Hing-Fung
Tiwary, Hans Raj
Todinca, Ioan
Toma, Laura
Tsichlas, Kostas
Tsigaridas, Elias
Tsourakakis, Charalampos
Uchoa, Eduardo
Uehara, Ryuhei
Uetz, Marc
Ukkonen, Antti

Úlfarsson, Henning A.
Uno, Takeaki
Varadarajan, Kasturi
Végh, László
Verbeek, Kevin
Verbin, Elad
Viennot, Laurent
Villanger, Yngve
Vöcking, Berthold
Vondrák, Jan
Ward, Justin
Wei, Fang
Weimann, Oren
Weller, Andreas
Werneck, Renato
Westermann, Matthias

Wiese, Andreas
Williamson, David
Woelfel, Philipp
Wulff-Nilsen, Christian
Yan, Li
Yi, Ke
Yin, Yitong
Yousefi, Arman
Yu, Gexin
Yuster, Raphael
Zaroliagis, Christos
Zeh, Norbert
Zelikovsky, Alex
Zenklusen, Rico
Zhang, Yong

Table of Contents

Graph Algorithms I

Stable Matchings and Auctions

Optimization

Online Algorithms I

Exponential-Time Algorithms

Online Algorithms I

Parameterized Algorithms

Best Paper Session

Graph Algorithms I

Computational Geometry II

Scheduling

Data Structures

Approximation Algorithms I

Graphs and Games

Distributed Computing and Networking

Strings and Sorting

Local Search and Set Systems

Polynomial-Time Approximation Schemes for Maximizing Gross Substitutes Utility under Budget Constraints

Akiyoshi Shioura

GSIS, Tohoku University, Sendai 980-8579, Japan
shioura@dais.is.tohoku.ac.jp

Abstract. We consider the maximization of a gross substitutes utility function under budget constraints. This problem naturally arises in applications such as exchange economies in mathematical economics and combinatorial auctions in (algorithmic) game theory. We show that this problem admits a polynomial-time approximation scheme (PTAS). More generally, we present a PTAS for maximizing a discrete concave function called an $\mathrm{M}^{\natural}$-concave function under budget constraints. Our PTAS is based on rounding an optimal solution of a continuous relaxation problem, which is shown to be solvable in polynomial time by the ellipsoid method. We also consider the maximization of the sum of two $\mathrm{M}^{\natural}$-concave functions under a single budget constraint. This problem is a generalization of the budgeted max-weight matroid intersection problem to the one with a nonlinear objective function. We show that this problem also admits a PTAS.

1 Introduction

We consider the problem of maximizing a nonlinear utility function under a constant number of budget (or knapsack) constraints, which is formulated as

$$\text{Maximize } f(X) \quad \text{subject to } X \in 2^N,\ c_i(X) \le B_i\ (1 \le i \le k), \qquad (1)$$

where N is a set of n items, $f : 2^N \to \mathbb{R}$ is a nonlinear utility function[1] of a consumer (or buyer) with $f(\emptyset) = 0$, k is a constant positive integer, and $c_i \in \mathbb{R}_+^N$, $B_i \in \mathbb{R}_+$ $(i = 1, 2, \ldots, k)$. For a vector $a \in \mathbb{R}^N$ and a set $X \subseteq N$, we denote $a(X) = \sum_{v \in X} a(v)$. The problem (1) is a natural generalization of budgeted combinatorial optimization problems ([20,21,38], etc.), and naturally arises in applications such as exchange economies with indivisible objects in mathematical economics ([18,19], etc.) and combinatorial auctions in (algorithmic) game theory ([4,7,22], etc.).

The problem (1) with a submodular objective function f is extensively discussed in the literature of combinatorial optimization, and constant-factor approximation algorithms have been proposed. Wolsey [39] considered the problem

---

[1] Monotonicity of f is not assumed throughout this paper, although utility functions are often assumed to be monotone.

C. Demetrescu and M.M. Halldórsson (Eds.): ESA 2011, LNCS 6942, pp. 1–12, 2011.

(1) with a monotone submodular f and $k = 1$, and proposed the first constant-factor approximation algorithm with the ratio $1 - e^{-\beta} \simeq 0.35$, where β satisfies $e^{\beta} = 2 - \beta$. Later, Sviridenko [38] improved the approximation ratio to $1 - 1/e$, which is the best possible under the assumption that P $\neq$ NP [10]. For the case of a monotone submodular f and a general constant $k \geq 1$, Kulik et al. [20] proposed a $(1 - 1/e)$-approximation algorithm by using the approach of Calinescu et al. [5] for the submodular function maximization under a matroid constraint. For a non-monotone submodular f and a general constant $k \geq 1$, a $(1/5 - \varepsilon)$-approximation local-search algorithm was given by Lee et al. [21].

Submodularity for set functions is known to be equivalent to the concept of decreasing marginal utility in mathematical economics. In this paper, we focus on a more specific subclass of decreasing marginal utilities, called *gross substitutes utilities*, and show that the problem (1) admits a polynomial-time approximation scheme (PTAS) if f is a gross substitutes utility.

A *gross substitutes utility* (*GS utility*, for short) function is defined as a function $f : 2^N \to \mathbb{R}$ satisfying the following condition:

$\forall p, q \in \mathbb{R}^N$ with $p \leq q$, $\forall X \in \arg\max_{U \subseteq N}\{f(U) - p(U)\}$,
$\exists Y \in \arg\max_{U \subseteq N}\{f(U) - q(U)\}$ s.t. $\{v \in X \mid p(v) = q(v)\} \subseteq Y$,

where p and q represent price vectors. This condition means that a consumer still wants to get items that do not change in price after the prices on other items increase. The concept of GS utility is introduced in Kelso and Crawford [19], where the existence of a Walrasian (or competitive) equilibrium is shown in a fairly general two-sided matching model. Since then, this concept plays a central role in mathematical economics and in auction theory, and is widely used in various models such as matching, housing, and labor market (see, e.g., [1,3,4,7,15,18,22]).

Various characterizations of gross substitutes utilities are given in the literature of mathematical economics [1,15,18]. Among them, Fujishige and Yang [15] revealed the relationship between GS utilities and discrete concave functions called $M^{\natural}$-*concave functions*, which is a function on matroid independent sets. It is known that a family $\mathcal{F} \subseteq 2^N$ of matroid independent sets satisfies the following property [30]:

(B$^{\natural}$-EXC) $\forall X, Y \in \mathcal{F}$, $\forall u \in X \setminus Y$, at least one of (i) $X - u, Y + u \in \mathcal{F}$, and (ii) $\exists v \in Y \setminus X$: $X - u + v, Y + u - v \in \mathcal{F}$, holds,

where $X - u + v$ is a short-hand notation for $X \setminus \{u\} \cup \{v\}$. We consider a function $f : \mathcal{F} \to \mathbb{R}$ defined on matroid independent sets $\mathcal{F}$. A function f is said to be $M^{\natural}$-*concave* [30] (read "M-natural-concave") if it satisfies the following:[2]

(M$^{\natural}$-EXC) $\forall X, Y \in \mathcal{F}$, $\forall u \in X \setminus Y$, at least one of (i) $X - u, Y + u \in \mathcal{F}$ and $f(X) + f(Y) \leq f(X - u) + f(Y + u)$, and (ii) $\exists v \in Y \setminus X$: $X - u + v$, $Y + u - v \in \mathcal{F}$ and $f(X) + f(Y) \leq f(X - u + v) + f(Y + u - v)$, holds.

---

[2] The concept of M$^{\natural}$-concavity is originally introduced for functions defined on generalized (poly)matroids (see [30]). In this paper we mainly consider a restricted class of M$^{\natural}$-concave functions.

The concept of M$^\natural$-concave function is introduced by Murota and Shioura [30] as a class of discrete concave functions (independently of gross substitutes utilities). It is an extension of the concept of M-concave function introduced by Murota [25,27]. The concepts of M$^\natural$-concavity/M-concavity play primary roles in the theory of discrete convex analysis [28], which provides a framework for well-solved nonlinear discrete optimization problems.

It is shown by Fujishige and Yang [15] that gross substitutes utilities constitute a subclass of M$^\natural$-concave functions.

Theorem 1.1. *A function* $f : 2^N \to \mathbb{R}$ *defined on* 2^N *is a gross substitutes utility if and only if* f *is an* M$^\natural$*-concave function.*

This result initiated a strong interaction between discrete convex analysis and mathematical economics; the results obtained in discrete convex analysis are used in mathematical economics ([3,22], etc.), while mathematical economics provides interesting applications in discrete convex analysis ([32,33], etc.).

In this paper, we mainly consider the *k-budgeted M$^\natural$-concave maximization problem*:

$(k\mathbf{BM^\natural M})$ Maximize $f(X)$ subject to $X \in \mathcal{F}$, $c_i(X) \le B_i$ $(1 \le i \le k)$,

which is (slightly) more general than the problem (1) with a gross-substitutes utility. Here, $f : \mathcal{F} \to \mathbb{R}$ is an M$^\natural$-concave function with $f(\emptyset) = 0$ defined on matroid independent sets $\mathcal{F}$, and k, c_i, and B_i are as in (1). We assume that the objective function f is given by a constant-time oracle which, given a subset $X \in 2^N$, checks if $X \in \mathcal{F}$ or not, and if $X \in \mathcal{F}$ then returns the value $f(X)$. The class of M$^\natural$-concave functions includes, as its subclass, linear functions on matroid independent sets. Hence, the problem $(k\mathbf{BM^\natural M})$ is a nonlinear generalization of the max-weight matroid independent set problem with budget constraints, for which Grandoni and Zenklusen [16] have recently proposed a conceptually simple, deterministic PTAS using the polyhedral structure of matroids.

Our Main Result In this paper, we propose a PTAS for $(k\mathbf{BM^\natural M})$ by extending the approach of Grandoni and Zenklusen [16]. We show the following property, where OPT denotes the optimal value of $(k\mathbf{BM^\natural M})$.

Theorem 1.2. *A feasible solution* $\tilde{X} \in 2^N$ *of* $(k\mathbf{BM^\natural M})$ *satisfying* $f(\tilde{X}) \ge$ OPT $- 2k \max_{v \in N} f(\{v\})$ *can be computed deterministically in polynomial time.*

The algorithm used in Theorem 1.2 can be converted into a PTAS by using a standard technique called *partial enumeration*, which reduces the original problem to a family of problems with "small" elements, which is done by guessing a constant number of "large" elements contained in an optimal solution (see, e.g., [2,16,20,34]). Hence, we obtain the following:

Theorem 1.3. *For every fixed* $\varepsilon > 0$, *a* $(1-\varepsilon)$-*approximate solution of* $(k\mathbf{BM^\natural M})$ *can be computed deterministically in polynomial time.*

To prove Theorem 1.2, we use the following algorithm, which is a natural extension of the one in [16]:

STEP 1: Construct a continuous relaxation problem (CR) of (kBM$^\natural$M).
STEP 2: Compute a vertex optimal solution $\hat{x} \in [0, 1]^N$ of (CR).
STEP 3: Round down the non-integral components of the optimal solution $\hat{x}$.

In [16], LP relaxation is used as a continuous relaxation, and it is shown that a vertex optimal solution (i.e., an optimal solution which is a vertex of the feasible region) of the resulting LP is nearly integral. Since the LP relaxation problem can be solved in polynomial time by the ellipsoid method, rounding down a vertex optimal solution yields a near-optimal solution of the budgeted max-weight matroid independent set problem.

These techniques in [16], however, cannot be applied directly since the objective function in (kBM$^\natural$M) is nonlinear. In particular, our continuous relaxation problem (CR) is a nonlinear programming problem formulated as

$$\textbf{(CR)} \quad \text{Maximize } \overline{f}(x) \quad \text{subject to } x \in \overline{\mathcal{F}}, \ c_i^\top x \le B_i \ (1 \le i \le k). \qquad (2)$$

Here, $\overline{\mathcal{F}}$ is a matroid polytope of the matroid $(N, \mathcal{F})$ and $\overline{f}$ is a concave closure of the function f (see §3 for definitions).

To extend the approach in [16], we firstly modify the definition of vertex optimal solution appropriately since there may be no optimal solution which is a vertex of the feasible region if the objective function is nonlinear. Under the new definition, we show that a vertex optimal solution of (CR) is nearly integral by using the polyhedral structure of M$^\natural$-concave functions.

We then show that if f is an M$^\natural$-concave function, then (CR) can be solved (almost) optimally in polynomial time by the ellipsoid method [17]. Note that the function $\overline{f}$ is given implicitly, and the evaluation of the function value is still a nontrivial task; even if f is a monotone submodular function, the evaluation of $\overline{f}(x)$ is NP-hard [5]. To solve (CR) we use the following new algorithmic property concerning the concave closure of M$^\natural$-concave functions, which is proven by making full use of conjugacy results of M$^\natural$-concave functions in the theory of discrete convex analysis.

Lemma 1.1. *Let $x \in \overline{\mathcal{F}}$.*

(i) *For every $\delta > 0$, we can compute in polynomial time $p \in \mathbb{Q}^N$ and $\beta \in \mathbb{Q}$ satisfying $\overline{f}(y) - \overline{f}(x) \le p^\top (y - x) + \delta \ (\forall y \in \overline{\mathcal{F}})$ and $\overline{f}(x) \le \beta \le \overline{f}(x) + \delta$.*
(ii) *If f is an integer-valued function, then we can compute in polynomial time $p \in \mathbb{Q}^N$ with $\overline{f}(y) - \overline{f}(x) \le p^\top (y - x) \ (\forall y \in \overline{\mathcal{F}})$ and the value $\overline{f}(x)$.*

Our Second Result. We also consider another type of budgeted optimization problem, which we call the *budgeted M$^\natural$-concave intersection problem*:

$$\textbf{(1BM}^\natural\textbf{I)} \quad \text{Maximize } f_1(X) + f_2(X) \quad \text{subject to } X \in \mathcal{F}_1 \cap \mathcal{F}_2, \ c(X) \le B,$$

where $f_j : \mathcal{F}_j \to \mathbb{R} \ (j = 1, 2)$ are M$^\natural$-concave functions with $f_j(\emptyset) = 0$ defined on matroid independent sets $\mathcal{F}_j$, $c \in \mathbb{R}_+^N$ and $B \in \mathbb{R}_+$. This is a nonlinear generalization of the budgeted max-weight matroid intersection problem. Indeed, if each f_j is a linear function on matroid independent sets $\mathcal{F}_j$, then the problem (1BM$^\natural$I) is nothing but the budgeted max-weight matroid intersection problem, for which Berger et al. [2] proposed a PTAS using Lagrangian relaxation and a novel patching operation. We show that the approach can be extended to (1BM$^\natural$I).

Theorem 1.4. *For every fixed $\varepsilon > 0$, a $(1-\varepsilon)$-approximate solution of* (1BM$^\natural$I) *can be computed deterministically in polynomial time.*

The following is the key property to prove Theorem 1.4, where OPT denotes the optimal value of (1BM$^\natural$I). We may assume that $\{v\} \in \mathcal{F}_1 \cap \mathcal{F}_2$ and $f_1(\{v\}) + f_2(\{v\}) > 0$ hold for all $v \in N$.

Theorem 1.5. *For* (1BM$^\natural$I), *there exists a polynomial-time algorithm which computes a set $\tilde{X} \in \mathcal{F}_1 \cap \mathcal{F}_2$ satisfying $f_1(\tilde{X}) + f_2(\tilde{X}) \geq$ OPT$-2\cdot\max_{v \in N}\{f_1(\{v\})+f_2(\{v\})\}$ and $c(\tilde{X}) \leq B + \max_{v \in N} c(v)$.*

To extend the approach in [2], we use techniques in Murota [26] developed for M$^\natural$-concave intersection problem *without* budget constraints. An important tool for the algorithm and its analysis is a *weighted* auxiliary graph defined by local information around the current solution, while an *unweighted* auxiliary graph is used in [2]. This makes it possible, in particular, to analyze how much amount the value of the objective function changes after updating a solution.

Both of our PTASes for (kBM$^\natural$M) and (1BM$^\natural$I) are based on novel approaches in Grandoni and Zenklusen [16] and in Berger et al. [2], respectively. The adaption of these approaches in the present settings, however, are not trivial as they involve nonlinear discrete concave objective functions. The main technical contribution of this paper is to show that those previous techniques for budgeted *linear* maximization problems can be extended to budgeted *nonlinear* maximization problems by using some results in the theory of discrete convex analysis.

In the following, we omit some of the proofs due to the page limit.

2   Gross Substitutes Utility and M$^\natural$-Concave Functions

We give some examples of gross substitutes utility and M$^\natural$-concave functions and explain some known results. Recall the notation $a(X) = \sum_{v \in X} a(v)$ for $a \in \mathbb{R}^N$ and $X \subseteq N$.

A simple example of M$^\natural$-concave function is a linear function $f(X) = a(X)$ $(X \in \mathcal{F})$ defined on a family $\mathcal{F} \subseteq 2^N$ of matroid independent sets, where $a \in \mathbb{R}^N$. In particular, if $\mathcal{F} = 2^N$ then f is a GS utility function. Below we give some nontrivial examples. See [28,29] for more examples of M$^\natural$-concave functions.

Example 1. (Weighted rank functions) Let $\mathcal{I} \subseteq 2^N$ be the family of independent sets of a matroid, and $w \in \mathbb{R}_+^N$. Define a function $f : 2^N \to \mathbb{R}_+$ by $f(X) = \max\{w(Y) \mid Y \subseteq X, Y \in \mathcal{I}\}$ $(X \in 2^N)$, which is called the *weighted rank function* [6]. If $w(v) = 1$ $(v \in N)$, then f is an ordinary rank function of the matroid $(N, \mathcal{I})$. Every weighted rank function is a GS utility function [5].   □

Example 2. (Laminar concave functions) Let $\mathcal{T} \subseteq 2^N$ be a laminar family, i.e., $X \cap Y = \emptyset$ or $X \subseteq Y$ or $X \supseteq Y$ holds for every $X, Y \in \mathcal{T}$. For $Y \in \mathcal{T}$, let $\varphi_Y : \mathbb{Z}_+ \to \mathbb{R}$ be a univariate concave function. Define a function $f : 2^N \to \mathbb{R}$ by

$f(X) = \sum_{Y \in \mathcal{T}} f_Y(|X \cap Y|) \ (X \in 2^N)$, which is called a *laminar concave function* [28, §6.3] (also called an *S-valuation* in [3]). Every laminar concave function is a GS utility function.   □

Example 3. (Maximum-weight bipartite matching) Consider a bipartite graph G with two vertex sets N, J and an edge set $E \ (\subseteq N \times J)$, where N and J correspond to workers and jobs, respectively. Every $(u, v) \in E$ means that worker $u \in N$ has ability to process job $v \in J$, and profit $p(u, v) \in \mathbb{R}_+$ can be obtained by assigning worker u to job v. Consider a matching between workers and jobs which maximizes the total profit, and define $\mathcal{F} \subseteq 2^N$ by $\mathcal{F} = \{X \subseteq N \mid \exists M :$ matching in G s.t. $\partial_N M = X\}$, where $\partial_N M$ denotes the set of vertices in N covered by edges in M. It is well known that $\mathcal{F}$ is a family of independent sets in a transversal matroid. Define $f : \mathcal{F} \to \mathbb{R}$ by

$$f(X) = \max\{ \sum_{(u,v) \in M} p(u, v) \mid \exists M : \text{matching in } G \text{ s.t. } \partial_N M = X\} \ (X \in \mathcal{F}).$$

Then, f is an M$^\natural$-concave function [29, §11.4.2]. In particular, if G is a complete bipartite graph, then $\mathcal{F} = 2^N$ holds, and therefore f is a GS utility function.   □

GS utility is a sufficient condition for the existence of a Walrasian equilibrium [19]; it is also a necessary condition in some sense [18]. GS utility is also related to desirable properties in the auction design (see [4,7]); for example, an ascending item-price auction gives an approximate equilibrium, while an exact equilibrium can be computed in polynomial time (see, e.g., [22, §5]).

M$^\natural$-concave functions have various desirable properties as discrete concavity. Global optimality is characterized by local optimality, which implies the validity of a greedy algorithm for M$^\natural$-concave function maximization. Maximization of the sum of two M$^\natural$-concave functions is a nonlinear generalization of the max-weight matroid intersection problem, and can be solved in polynomial time as well. A budget constraint with uniform cost is equivalent to a cardinality constraint. Hence, (kBM$^\natural$M) and (1BM$^\natural$I) with uniform cost can be solved in polynomial time as well. The maximization of a single M$^\natural$-concave function under a general matroid constraint can be also solved exactly in polynomial time, while the corresponding problem for the sum of two M$^\natural$-concave functions is NP-hard (see [28]).

3   PTAS for k-Budgeted M$^\natural$-Concave Maximization

We prove Theorem 1.2, a key property to show the existence of a PTAS for (kBM$^\natural$M). Due to the page limitation, we mainly consider the case where f is an integer-valued function, and a more complicated proof for the general case is omitted; the proof for the integer-valued case is much simpler, but gives an idea of our algorithm for the general case.

Continuous Relaxation. Our continuous relaxation of (kBM$^\natural$M) is given by (2), where $\overline{\mathcal{F}}$ and $\overline{f}$ are defined as follows. For $X \subseteq N$ the characteristic vector of X is denoted by $\chi_X \in \{0, 1\}^N$. We denote by $\overline{\mathcal{F}} \subseteq [0, 1]^N$ the convex hull of vectors $\{\chi_X \mid X \in \mathcal{F}\}$, which is called a *matroid polytope*. The *concave closure* $\overline{f} : \overline{\mathcal{F}} \to \mathbb{R}$ of function f is given by

$$\overline{f}(x) = \max \Big\{ \sum_{X \in \mathcal{F}} \lambda_X f(X) \mid \sum_{X \in \mathcal{F}} \lambda_X \chi_X = x,$$
$$\sum_{X \in \mathcal{F}} \lambda_X = 1, \ \lambda_X \geq 0 \ (X \in \mathcal{F}) \Big\} \quad (x \in \overline{\mathcal{F}}).$$

Note that for a general (not necessarily $M^\natural$-concave) f, the concave closure $\overline{f}$ is a polyhedral concave function satisfying $\overline{f}(\chi_X) = f(X)$ for all $X \in \mathcal{F}$. Let $S \subseteq [0, 1]^N$ denote the set of feasible solutions to (CR), which is a polyhedron.

The set $P = \{(x, \alpha) \in [0, 1]^N \times \mathbb{R} \mid x \in S, \alpha \leq \overline{f}(x)\}$ is a polyhedron. We say that x is a *vertex feasible solution* of (CR) if $(x, \overline{f}(x))$ is a vertex of the polyhedron P. There always exists an optimal solution of (CR) which is a vertex feasible solution, and we call such a solution a *vertex optimal solution*. Note that a vertex optimal solution does not correspond to a vertex of S in general.

Solving Continuous Relaxation. We show that if f is an integer-valued function, then (CR) can be solved exactly in polynomial time by using the ellipsoid method. Similar approach is used in Shioura [37] for the problem with a monotone $M^\natural$-concave function. We here extend the approach to the case of non-monotone $M^\natural$-concave function.

The ellipsoid method finds a vertex optimal solution of (CR) in time polynomial in n and in $\log \max_{X \in \mathcal{F}} |f(X)|$ if the following oracles are available [17]:

(O-1) polynomial-time strong separation oracle for the set S,
(O-2) polynomial-time oracle for computing a subgradient of $\overline{f}$.

The oracle (O-1) can be realized as follows. Let $x \in [0, 1]^N$ be a vector. We firstly check whether the inequalities $c_i^\top x \leq B_i$ are satisfied or not. If not, then the corresponding inequality can be used as a separating hyperplane of S. We next check whether $x \in \overline{\mathcal{F}}$ or not. Recall that for a given subset $X \subseteq N$, we have an oracle to check $X \in \mathcal{F}$ in constant time. This enables us to compute the rank function $\rho : 2^N \to \mathbb{Z}_+$ of the matroid $(N, \mathcal{F})$ in polynomial time (see, e.g., [14,28]). We have $\overline{\mathcal{F}} = \{y \in [0, 1]^N \mid y(X) \leq \rho(X) \ (\forall X \in 2^N)\}$. Hence, the membership in $\overline{\mathcal{F}}$ can be checked by solving the problem $\min_{X \in 2^N} \{\rho(X) - x(X)\}$, which is a submodular function minimization and can be done in polynomial time [8,14,17]. Let $X_* \in \arg\min_{X \in 2^N} \{\rho(X) - x(X)\}$. If $\rho(X_*) \geq x(X_*)$, then $x \in \overline{\mathcal{F}}$ holds; otherwise, $\rho(X_*) \geq x(X_*)$ gives a separating hyperplane of S.

We then consider the oracle (O-2). A vector $p \in \mathbb{R}^N$ is called a *subgradient* of $\overline{f}$ at $x \in \overline{\mathcal{F}}$ if it satisfies $\overline{f}(y) - \overline{f}(x) \leq p^\top(y - x)$ for all $y \in \overline{\mathcal{F}}$. Lemma 1.1 (ii) states that if f is an integer-valued function, then a subgradient of $\overline{f}$ can be computed in polynomial time, i.e., the oracle (O-2) is available.

We give a proof of Lemma 1.1 (ii) by using conjugacy results of $M^\natural$-concave functions. We define a function $\overline{g} : \mathbb{R}^N \to \mathbb{R}$ by $\overline{g}(p) = \inf\{p^\top y - \overline{f}(y) \mid y \in \overline{\mathcal{F}}\}$ ($p \in \mathbb{R}^N$). Note that $\inf\{p^\top y - \overline{f}(y) \mid y \in \overline{\mathcal{F}}\} = \inf\{p(Y) - f(Y) \mid Y \in \mathcal{F}\}$ holds, and therefore the evaluation of the function value of $\overline{g}$ can be done in polynomial time by using an $M^\natural$-concave function maximization algorithm [36].

It is well known in the theory of convex analysis (see, e.g., [35]) that $p \in \mathbb{R}^N$ is a subgradient of $\overline{f}$ at $x \in \overline{\mathcal{F}}$ if and only if $p \in \arg\max\{\overline{g}(q) - q^\top x \mid q \in \mathbb{R}^N\}$. The next lemma shows that the maximum in $\max\{\overline{g}(q) - q^\top x \mid q \in \mathbb{R}^N\}$ can be achieved by an integral vector in a finite set.

Lemma 3.1. *For every $x \in \overline{\mathcal{F}}$, there exists a subgradient p of $\overline{f}$ at x such that $p \in \mathbb{Z}^N$ and $|p(v)| \leq 2n \max_{X \in \mathcal{F}} |f(X)|$ for all $v \in N$.*

The discussion above and Lemma 3.1 imply that it suffices to compute an optimal solution of the problem $\max\{\overline{g}(q) - q^\top x \mid q \in \mathbb{Z}^N, \; |p(v)| \leq 2n \max_{X \in \mathcal{F}} |f(X)| \; (v \in N)\}$. The function $\overline{g}$ has a nice combinatorial structure called *L-concavity* [27,28], and this problem can be solved exactly in time polynomial in n and in $\log \max_{X \in \mathcal{F}} |f(X)|$. Hence, we obtain the following property:

Lemma 3.2. *If f is an integer-valued function, then a vertex optimal solution of (CR) can be computed in polynomial time.*

Rounding of Continuous Solution. It is shown that there exists an optimal solution of (CR) which is nearly integral.

Lemma 3.3. *Let $\hat{x} \in [0,1]^N$ be a vertex optimal solution of (CR). Then, $\hat{x}$ has at most $2k$ non-integral components.*

This generalizes a corresponding result in [16] for the budgeted matroid independent set problem. Below we give a proof of Lemma 3.3.

In the proof we use the concept of g-polymatroids. A *g-polymatroid* [13] is a polyhedron $Q = \{x \in \mathbb{R}^N \mid \mu(X) \leq x(X) \leq \rho(X) \; (X \in 2^N)\}$ given by a pair of submodular/supermodular functions $\rho : 2^N \to \mathbb{R} \cup \{+\infty\}$, $\mu : 2^N \to \mathbb{R} \cup \{-\infty\}$ satisfying the inequality $\rho(X) - \mu(Y) \geq \rho(X \setminus Y) - \mu(Y \setminus X) \; (X, Y \in 2^N)$. If ρ and μ are integer-valued, then Q is an integral polyhedron; in such a case, we say that Q is an *integral g-polymatroid*.

Let $\hat{x} \in [0,1]^N$ be a vertex optimal solution of (CR). Then, $\hat{x}$ is a vertex of a polyhedron given as the intersection of a set $Q = \arg\max\{\overline{f}(x) - p^\top x \mid x \in \overline{\mathcal{F}}\}$ for some $p \in \mathbb{R}^N$ and the set $K = \{x \in [0,1]^N \mid c_i^\top x \leq B_i \; (i = 1, \ldots, k)\}$. Since f is an M$^\natural$-concave function, Q is an integral g-polymatroid [31, §6]. Hence, $\hat{x}$ is contained in a d-dimensional face F of Q with $d \leq k$. The next lemma is a generalization of [16, Th. 3].

Lemma 3.4. *Let $Q \subseteq \mathbb{R}^N$ be an integral g-polymatroid and let $F \subseteq Q$ be a face of dimension d. Then, every $x \in F$ has at most $2d$ non-integral components.*

By Lemma 3.4, the number of non-integral components in $\hat{x}$ is at most $2d \leq 2k$. This concludes the proof of Lemma 3.3.

Lemma 3.3 implies the following property, stating that a solution obtained by rounding down non-integral components of a vertex optimal solution satisfies the condition in Theorem 1.2.

Lemma 3.5. *The set $\tilde{X} = \{v \in N \mid \hat{x}(v) = 1\}$ is a feasible solution to (kBM$^\natural$M) satisfying $f(\tilde{X}) \geq \mathrm{OPT} - 2k \max_{v \in N} f(\{v\})$.*

This, together with Lemma 3.2, implies Theorem 1.2 for integer-valued functions.

Algorithm for General Case. If the function f is not integer-valued, then it is difficult to compute a vertex optimal solution of (CR). Instead, we compute the set $\tilde{X}$ in Lemma 3.5 directly, without computing a vertex optimal solution, by using Lemma 3.3 and Lemma 3.6 below.

Lemma 3.6. *For every fixed $\varepsilon > 0$, we can compute a feasible solution $x \in [0,1]^N$ of (CR) with $\overline{f}(x) \geq (1 - \varepsilon)\overline{\text{OPT}}$ in polynomial time, where $\overline{\text{OPT}}$ is the optimal value of (CR).*

4   PTAS for 1-Budgeted M$^\natural$-Concave Intersection

We give a proof of Theorem 1.5 for (1BM$^\natural$I). With a parameter $\lambda \in \mathbb{R}_+$, the Lagrangian relaxation of (1BM$^\natural$I) is given by

(LR(λ)) Maximize $f_1(X) + f_2(X) + \lambda\{B - c(X)\}$ subject to $X \in \mathcal{F}_1 \cap \mathcal{F}_2$.

The problem (LR(λ)) is an instance of the M$^\natural$-concave intersection problem *without* budget constraint, which is essentially equivalent to the valuated matroid intersection problem discussed in [25]. Therefore, the theorems and algorithms in [25] can be used to (LR(λ)) with slight modification. In particular, (LR(λ)) can be solved in polynomial time.

Below we explain how to compute a set $\tilde{X} \in \mathcal{F}_1 \cap \mathcal{F}_2$ satisfying the condition in Theorem 1.5. We firstly compute the value $\lambda = \lambda_*$ minimizing the optimal value of (LR(λ)), together with two optimal solutions X_*, Y_* of (LR(λ_*)) satisfying $c(X_*) \leq B \leq c(Y_*)$. This can be done by Megiddo's parametric search technique (see [24]; see also [2,34]). Note that the inequality $c(X_*) \leq B \leq c(Y_*)$ implies $f_1(X_*) + f_2(X_*) \leq \text{OPT} \leq f_1(Y_*) + f_2(Y_*)$, where OPT is the optimal value of the original problem (1BM$^\natural$I). Hence, if $X_* = Y_*$ then we have $c(X_*) = B$ and $f(X_*) = \text{OPT}$, implying that $\tilde{X} = X_*$ satisfies the condition in Theorem 1.5. Otherwise (i.e., $X_* \neq Y_*$), "patching" operations are applied to X_* and Y_* to construct a better approximate solution.

The patching operations are done by using cycles in a weighted auxiliary graph. In the following, we assume that $|X_*| = |Y_*|$ holds, since in this case the description of the algorithm can be simplified (and does not lose the generality so much). We define an *auxiliary graph* $G_{X_*}^{Y_*} = (V, A)$ with arc weight $\omega : A \to \mathbb{R}$ associated with X_* and Y_* by $V = (X_* \setminus Y_*) \cup (Y_* \setminus X_*)$, $A = E_1 \cup E_2$, and

$$E_1 = \{(u,v) \mid u \in X_* \setminus Y_*, \ v \in Y_* \setminus X_*, \ X_* - u + v \in \mathcal{F}_1\},$$
$$\omega(u,v) = f_1(X_* - u + v) - f_1(X_*) + \lambda_*\{c(u) - c(v)\} \quad ((u,v) \in E_1),$$
$$E_2 = \{(v,u) \mid v \in Y_* \setminus X_*, \ u \in X_* \setminus Y_*, \ X_* + v - u \in \mathcal{F}_2\},$$
$$\omega(v,u) = f_2(X_* + v - u) - f_2(X_*) \quad ((v,u) \in E_2).$$

A cycle in $G_{X_*}^{Y_*}$ is a directed closed path which visits each vertex at most once. In every cycle in $G_{X_*}^{Y_*}$, arcs in E_1 and arcs in E_2 appear alternately, and every cycle contains an even number of arcs.

For a cycle C in the graph $G_{X_*}^{Y_*}$, we define a set $X_* \oplus C \ (\subseteq N)$ by

$$X_* \oplus C = X_* \setminus \{u \in X_* \setminus Y_* \mid (u,v) \in C \cap E_1\} \cup \{v \in Y_* \setminus X_* \mid (u,v) \in C \cap E_1\}.$$

Lemma 4.1. (i) *A maximum-weight cycle in $G_{X_*}^{Y_*}$ is a zero-weight cycle.*
(ii) *Let C be a zero-weight cycle in $G_{X_*}^{Y_*}$ with the minimum number of arcs. Then, $X_* \oplus C$ is an optimal solution of $(\mathrm{LR}(\lambda_*))$ with $X_* \oplus C \neq X_*$.*

Lemma 4.1 (i) implies that a zero-weight cycle C in $G_{X_*}^{Y_*}$ with the minimum number of arcs can be computed by using a shortest-path algorithm. The set $X' = X_* \oplus C$ is an optimal solution of $(\mathrm{LR}(\lambda_*))$ by Lemma 4.1 (ii). If $X' = Y_*$, then an additional patching operation explained below is applied. If $c(X') = B$, then we stop since X' satisfies the condition in Theorem 1.5. If $c(X') < B$ then we replace X_* with X'; otherwise (i.e., $c(X') > B$), we replace Y_* with X'; in both cases, we repeat the same patching operations.

We explain the additional patching operation in the case where $X_* \oplus C = Y_*$. In this case, C contains all vertices in the graph $G_{X_*}^{Y_*}$. Let $a_1, a_2, \ldots, a_{2h} \in A$ be a sequence of arcs in the cycle C, where $2h$ is the number of arcs in C. We may assume that $a_j \in E_1$ if j is odd and $a_j \in E_2$ if j is even. For $j = 1, 2, \ldots, h$, let $\alpha_j = \omega(a_{2j-1}) + \omega(a_{2j})$. Since C is a zero-weight cycle, we have $\sum_{j=1}^{h} \alpha_j = 0$.

Lemma 4.2 (Gasoline Lemma (cf. [23])). *Let $\alpha_1, \alpha_2, \ldots, \alpha_h$ be real numbers satisfying $\sum_{j=1}^{h} \alpha_j = 0$. Then, there exists some $t \in \{1, \ldots, h\}$ such that $\sum_{j=t}^{t+i} \alpha_{j \pmod h} \geq 0$ $(i = 0, 1, \ldots, h-1)$, where $\alpha_0 = \alpha_h$.*

By Lemma 4.2, we may assume that $\sum_{j=1}^{i} \alpha_j \geq 0$ for all $i = 1, 2, \ldots, h$. For $j = 1, 2, \ldots, h$, we denote $a_{2j-1} = (u_j, v_j)$, and let $\eta_j = c(v_j) - c(u_j)$. Then, $c(Y_*) = c(X_*) + \sum_{j=1}^{h} \eta_j$ holds. Let $t \in \{1, 2, \ldots, h\}$ be the minimum integer such that $c(X_*) + \sum_{j=1}^{t} \eta_j > B$. Since $c(X_*) < B$, we have $t \geq 1$. In addition, the choice of t implies that $c(X_*) + \sum_{j=1}^{t-1} \eta_j \leq B$. With the arc set $C' = \{a_1, a_2, \ldots, a_{2t-1}, a_{2t}\}$, we define $\tilde{X} \subseteq N$ by

$$\tilde{X} = X_* \setminus \{u \in X_* \mid (u,v) \in C' \cap E_1 \text{ or } u = u_{t+1}\} \cup \{v \in N \setminus X_* \mid (u,v) \in C' \cap E_1\}.$$

We show that the set $\tilde{X}$ satisfies the desired condition in Theorem 1.5.
We have

$$\mathrm{OPT} \leq f_1(X_*) + f_2(X_*) + \lambda_*\{B - c(X_*)\} + \sum_{j=1}^{t} \alpha_j$$
$$\leq [f_1(X_*) + \sum_{j=1}^{t}\{f_1(X_* - u_j + v_j) - f_1(X_*)\}]$$
$$+ [f_2(X_*) + \sum_{j=1}^{t}\{f_1(X_* - u_{j+1} + v_j) - f_1(X_*)\}].$$

We define $\tilde{X}_1 = \tilde{X} \cup \{u_{t+1}\}$ and $\tilde{X}_2 = \tilde{X} \cup \{u_1\}$. Note that $\tilde{X}_1 \cap \tilde{X}_2 = \tilde{X}$. By using the fact that C' is a subpath of a zero-weight cycle with the smallest number of arcs, we can show the following:

Lemma 4.3. *We have $\tilde{X}_1 \in \mathcal{F}_1$, $\tilde{X}_2 \in \mathcal{F}_2$, $f_1(\tilde{X}_1) = f_1(X_*) + \sum_{j=1}^{t}\{f_1(X_* - u_j + v_j) - f_1(X_*)\}$, and $f_2(\tilde{X}_2) = f_2(X_*) + \sum_{j=1}^{t}\{f_1(X_* - u_{j+1} + v_j) - f_1(X_*)\}$.*

Hence, we obtain $f_1(\tilde{X}_1) + f_2(\tilde{X}_2) \geq \text{OPT}$. $M^\natural$-concavity of f_1 and f_2 implies

$$f_1(\tilde{X}) + f_2(\tilde{X}) \geq f_1(\tilde{X}_1) - \{f_1(\tilde{X}_1) - f_1(\tilde{X})\} + f_2(\tilde{X}_2) - \{f_2(\tilde{X}_2) - f_2(\tilde{X})\}$$
$$\geq f_1(\tilde{X}_1) - \{f_1(\{u_{t+1}\}) - f_1(\emptyset)\} + f_2(\tilde{X}_2) - \{f_2(\{u_1\}) - f_2(\emptyset)\}$$
$$\geq \text{OPT} - 2 \cdot \max_{v \in N}\{f_1(\{v\}) + f_2(\{v\})\},$$

from which the former inequality in Theorem 1.5 follows. The latter inequality in Theorem 1.5 can be shown as follows:

$$c(\tilde{X}) = c(X_*) + \sum_{j=1}^{t} \eta_j - c(u_{t+1})$$
$$\leq \{c(X_*) + \sum_{j=1}^{t-1} \eta_j\} + \eta_t \leq B + \eta_t \leq B + \max_{v \in N} c(v).$$

References

1. Ausubel, L.M., Milgrom, P.: Ascending auctions with package bidding. Front. Theor. Econ. 1, Article 1 (2002)
2. Berger, A., Bonifaci, V., Grandoni, F., Schäfer, G.: Budgeted matching and budgeted matroid intersection via the gasoline puzzle. Math. Programming (2009) (to appear)
3. Bing, M., Lehmann, D., Milgrom, P.: Presentation and structure of substitutes valuations. In: Proc. EC 2004, pp. 238–239 (2004)
4. Blumrosen, L., Nisan, N.: Combinatorial auction. In: Nisan, N., Roughgarden, T., Tardos, É., Vazirani, V.V. (eds.) Algorithmic Game Theory, pp. 267–299. Cambridge Univ. Press, Cambridge (2007)
5. Calinescu, G., Chekuri, C., Pál, M., Vondrák, J.: Maximizing a submodular set function subject to a matroid constraint (Extended abstract). In: Fischetti, M., Williamson, D.P. (eds.) IPCO 2007. LNCS, vol. 4513, pp. 182–196. Springer, Heidelberg (2007)
6. Calinescu, G., Chekuri, C., Pál, M., Vondrák, J.: Maximizing a submodular set function subject to a matroid constraint. SIAM J. Comput. (to appear)
7. Cramton, P., Shoham, Y., Steinberg, R.: Combinatorial Auctions. MIT Press, Cambridge (2006)
8. Cunningham, W.H.: On submodular function minimization. Combinatorica 5, 185–192 (1985)
9. Dress, A.W.M., Wenzel, W.: Valuated matroids. Adv. Math. 93, 214–250 (1992)
10. Feige, U.: A threshold of $\ln n$ for approximating set cover. J. ACM 45, 634–652 (1998)
11. Feige, U.: On maximizing welfare when utility functions are subadditive. SIAM J. Comput. 39, 122–142 (2009)
12. Feige, U., Mirrokni, V.S., Vondrák, J.: Maximizing non-monotone submodular functions. In: Proc. FOCS 2007, pp. 461–471 (2007)
13. Frank, A., Tardos, É.: Generalized polymatroids and submodular flows. Math. Programming 42, 489–563 (1988)
14. Fujishige, S.: Submodular Functions and Optimization, 2nd edn. Elsevier, Amsterdam (2005)
15. Fujishige, S., Yang, Z.: A note on Kelso and Crawford's gross substitutes condition. Math. Oper. Res. 28, 463–469 (2003)

16. Grandoni, F., Zenklusen, R.: Approximation schemes for multi-budgeted independence systems. In: de Berg, M., Meyer, U. (eds.) ESA 2010. LNCS, vol. 6346, pp. 536–548. Springer, Heidelberg (2010)
17. Grötschel, M., Lovász, L., Schrijver, A.: Geometric algorithms and combinatorial optimization, 2nd edn. Springer, Heidelberg (1993)
18. Gul, F., Stacchetti, E.: Walrasian equilibrium with gross substitutes. J. Econ. Theory 87, 95–124 (1999)
19. Kelso, A.S., Crawford, V.P.: Job matching, coalition formation and gross substitutes. Econometrica 50, 1483–1504 (1982)
20. Kulik, A., Shachnai, H., Tamir, T.: Maximizing submodular set functions subject to multiple linear constraints. In: Proc. SODA 2009, pp. 545–554 (2009)
21. Lee, J., Mirrokni, V.S., Nagarajan, V., Sviridenko, M.: Maximizing nonmonotone submodular functions under matroid or knapsack constraints. SIAM J. Discrete Math. 23, 2053–2078 (2010)
22. Lehmann, B., Lehmann, D., Nisan, N.: Combinatorial auctions with decreasing marginal utilities. Games Econom. Behav. 55, 270–296 (2006)
23. Lin, S., Kernighan, B.W.: An effective heuristic algorithm for the traveling salesman problem. Oper. Res. 21, 498–516 (1973)
24. Megiddo, N.: Combinatorial optimization with rational objective functions. Math. Oper. Res. 4, 414–424 (1979)
25. Murota, K.: Convexity and Steinitz's exchange property. Adv. Math. 124, 272–311 (1996)
26. Murota, K.: Valuated matroid intersection, I: optimality criteria, II: algorithms. SIAM J. Discrete Math. 9, 545–561, 562–576 (1996)
27. Murota, K.: Discrete convex analysis. Math. Programming 83, 313–371 (1998)
28. Murota, K.: Discrete Convex Analysis. SIAM, Philadelphia (2003)
29. Murota, K.: Recent developments in discrete convex analysis. In: Cook, W.J., Lovász, J., Vygen, J. (eds.) Research Trends in Combinatorial Optimization, pp. 219–260. Springer, Heidelberg (2009)
30. Murota, K., Shioura, A.: M-convex function on generalized polymatroid. Math. Oper. Res. 24, 95–105 (1999)
31. Murota, K., Shioura, A.: Extension of M-convexity and L-convexity to polyhedral convex functions. Adv. in Appl. Math. 25, 352–427 (2000)
32. Murota, K., Tamura, A.: Application of M-convex submodular flow problem to mathematical economics. Japan J. Indust. Appl. Math. 20, 257–277 (2003)
33. Murota, K., Tamura, A.: New characterizations of M-convex functions and their applications to economic equilibrium models with indivisibilities. Discrete Appl. Math. 131, 495–512 (2003)
34. Ravi, R., Goemans, M.X.: The constrained minimum spanning tree problem. In: Karlsson, R., Lingas, A. (eds.) SWAT 1996. LNCS, vol. 1097, pp. 66–75. Springer, Heidelberg (1996)
35. Rockafellar, R.T.: Convex Analysis. Princeton University Press, Princeton (1970)
36. Shioura, A.: Minimization of an M-convex function. Discrete Appl. Math. 84, 215–220 (1998)
37. Shioura, A.: On the pipage rounding algorithm for submodular function maximization: a view from discrete convex analysis. Discrete Math. Algorithms Appl. 1, 1–23 (2009)
38. Sviridenko, M.: A note on maximizing a submodular set function subject to a knapsack constraint. Oper. Res. Lett. 32, 41–43 (2004)
39. Wolsey, L.A.: Maximising real-valued submodular functions: primal and dual heuristics for location problems. Math. Oper. Res. 7, 410–425 (1982)

Approximating the Smallest 2-Vertex Connected Spanning Subgraph of a Directed Graph*

Loukas Georgiadis

Department of Informatics and Telecommunications Engineering,
University of Western Macedonia, Greece
lgeorg@uowm.gr

Abstract. We consider the problem of approximating the smallest 2-vertex connected spanning subgraph (2-VCSS) of a 2-vertex connected directed graph, and explore the efficiency of fast heuristics. First, we present a linear-time heuristic that gives a 3-approximation of the smallest 2-VCSS. Then we show that this heuristic can be combined with an algorithm of Cheriyan and Thurimella that achieves a $(1 + 1/k)$-approximation of the smallest k-VCSS. The combined algorithm preserves the 1.5 approximation guarantee of the Cheriyan-Thurimella algorithm for $k = 2$ and improves its running time from $O(m^2)$ to $O(m\sqrt{n} + n^2)$, for a digraph with n vertices and m arcs. Finally, we present an experimental evaluation of the above algorithms for a variety of input data. The experimental results show that our linear-time heuristic achieves in practice a much better approximation ratio than 3, suggesting that a tighter analysis may be possible. Furthermore, the experiments show that the combined algorithm not only improves the running time of the Cheriyan-Thurimella algorithm, but it may also compute a smaller 2-VCSS.

1 Introduction

A directed (undirected) graph is *k-vertex connected* if it has at least $k+1$ vertices and the removal of any set of at most $k-1$ vertices leaves the graph strongly connected (connected). The computation of a smallest (i.e., with minimum number of edges) k-vertex connected spanning subgraph (k-VCSS) of a given k-vertex connected graph is a fundamental problem in network design with many practical applications [11]. This problem is NP-complete for $k \geq 2$ for undirected graphs, and for $k \geq 1$ for directed graphs [10]. Recently, the more general problem of approximating minimum-cost subgraphs that satisfy certain connectivity requirements has also received a lot of attention. See, e.g., [8,21].

Here we consider the problem of approximating the smallest k-VCSS of a directed graph (digraph) for $k = 2$. The key contribution of this work is a linear-time heuristic that provides a 3-approximation of the smallest 2-VCSS. Our new

* Research funded by the John S. Latsis Public Benefit Foundation. The sole responsibility for the content of this paper lies with its authors.

C. Demetrescu and M.M. Halldórsson (Eds.): ESA 2011, LNCS 6942, pp. 13–24, 2011.
© Springer-Verlag Berlin Heidelberg 2011

heuristic is based on recent results on independent spanning trees [14], testing 2-vertex connectivity [12], and computing strong articulation points [18] in digraphs. The approximation bound of the new heuristic is by a factor of two worse compared to the currently best known bound of 1.5. The latter is achieved by the algorithm of Cheriyan and Thurimella [4], which computes a $(1 + 1/k)$-approximation of the smallest k-VCSS and runs in $O(km^2)$ time for a k-vertex connected digraph with m arcs. However, our experimental results show that in practice the linear-time heuristic performs much better than the 3 approximation bound suggests, which implies that a tighter analysis may be possible. Furthermore, we provide a simplified version of the Cheriyan-Thurimella algorithm for $k = 2$, and present a combination of our linear-time heuristic with this algorithm. The combined algorithm achieves the same approximation guarantee of 1.5 as the Cheriyan-Thurimella algorithm, and it runs in $O(m\sqrt{n} + n^2)$ time. Finally, we present some experimental results for all the above algorithms on artificial and synthetic graphs. The experiments show that the combined algorithm, besides improving the running time of the Cheriyan-Thurimella algorithm, can also compute a smaller 2-VCSS.

1.1 Definitions and Notation

We denote the vertex set and the arc (edge) set of a directed (undirected) graph G by $V(G)$ and $A(G)$ ($E(G)$), respectively.

For an undirected graph G and a subset of edges $M \subseteq E(G)$, we let $\deg_M(v)$ denote the degree of vertex v in M, i.e., the number of edges in M adjacent to v. The subset M is a *matching* if for all vertices v, $\deg_M(v) \leq 1$. A vertex v is *free* (with respect to M) if $\deg_M(v) = 0$. If for all vertices v, $\deg_M(v) \geq k$, then we call M a $(\geq k)$-*matching*. Given a function $b : V \mapsto \mathbf{Z}$ the *b-matching* problem is to compute a maximum-size subgraph $G' = (V, E(G'))$ of G such that $\deg_{E(G')}(v) \leq b(v)$ for all v. This problem is also referred to as the *degree-constrained subgraph problem* [9].

For a directed graph (digraph) G and a subset of arcs $M \subseteq A(G)$, we let $\deg_M^-(v)$ denote the in-degree of vertex v in M, i.e., the number of arcs in M entering v, and let $\deg_M^+(v)$ denote the out-degree of vertex v in M, i.e., the number of arcs in M leaving v. If for all vertices v, $\deg_M^+(v) \geq k$ and $\deg_M^-(v) \geq k$, then we call M a $(\geq k)$-*matching*.

2 Approximation Algorithms

In this section we describe the three main algorithms and their combinations that we consider in our experimental study. First, we describe a simple heuristic that computes a minimal k-VCSS (Section 2.1). It considers one arc of the digraph at a time and decides whether to include it in the computed k-VCSS or not. This heuristic computes a 2-approximation of the smallest k-VCSS. Next we consider the algorithm of Cheriyan and Thurimella that computes a $(1 + 1/k)$-approximation of the smallest k-VCSS, and present a simplified version of this algorithm for the $k = 2$ case (Section 2.2). Despite the vast literature on connectivity and network design problems, the above two algorithms are, to the best of our knowledge, the only previously known algorithms for approximating the 2-VCSS of a digraph. (We refer to [4] for results on the undirected or edge-connected versions of the problem.) Then, we present our linear-time heuristic (Section 2.3) and two hybrid algorithms that are derived by combining the linear-time heuristic with either the minimal k-VCSS or the Cheriyan-Thurimella algorithm (Section 2.4).

The quality of the k-VCSS computed by any algorithm can be bounded by the following simple fact: In any k-vertex (or arc) connected digraph each vertex has outdegree $\geq k$. Thus any k-vertex (arc) connected digraph has at least kn arcs. We use this to bound the approximation ratio of the fast heuristic of Section 2.3. We note, however, that the approximation ratio achieved by the algorithm of Cheriyan and Thurimella [4] is based on deeper lower bounds on the number of arcs in k-connected digraphs.

2.1   Minimal k-VCSS

Results of Edmonds [7] and Mader [20] imply that every minimal k-VCSS has at most $2kn$ arcs [4]. This fact implies that the following simple heuristic guarantees a 2-approximation of the smallest k-VCSS. We process the arcs in an arbitrary order, and while doing so we maintain a current subgraph $\widehat{G}$ of G. Initially $\widehat{G} = G$. When we process an arc (x, y) we test if $\widehat{G}$ contains at least $k+1$ vertex-disjoint paths from x to y (including the arc (x, y)). If this is the case then we can set $\widehat{G} \leftarrow \widehat{G} - (x, y)$. At the end of this procedure $\widehat{G}$ is a minimal k-VCSS of G, i.e., for any arc $(x, y) \in A(\widehat{G})$, $\widehat{G} - (x, y)$ is not k-vertex connected.

Testing if a digraph G has $k + 1$ vertex-disjoint paths from x to y can be carried out efficiently, for constant k, by computing $k + 1$ arc-disjoint paths in a derived digraph G_{der} using a flow-augmenting algorithm [1]. The vertex set $V(G_{\mathrm{der}})$ contains a pair of vertices v_- and v_+ for each vertex $v \in V(G)$. The arc set $A(G_{\mathrm{der}})$ contains the arcs (v_-, v_+) corresponding to all $v \in V(G)$. Also, for each arc (v, w) in $A(G)$, $A(G_{\mathrm{der}})$ contains the arc (v_+, w_-). A single flow-augmentation step in G_{der} takes $O(m)$ time, therefore the $k + 1$ arc-disjoint x-y paths are constructed in $O(km)$, and the overall running time of the algorithm is $O(km^2)$.

In the $k = 2$ case, if we wish to avoid the construction of the derived graph, we can replace the flow-augmenting algorithm with an algorithm that computes dominators in $G - (x, y)$ with x as the source vertex. The arc (x, y) can be removed if and only if y is not dominated by any vertex other than x and y. (See Section 2.3 and also [14].) We refer to the resulting algorithm as MINIMAL.

2.2   The Cheriyan-Thurimella Algorithm

Cheriyan and Thurimella [4] proposed an elegant algorithm that achieves an $(1 + 1/k)$-approximation of the smallest k-VCSS. The algorithm consists of the following two phases:

Phase 1. This phase computes a minimum $(\geq k - 1)$−matching M of the input digraph G. This is transformed to a b-matching problem on a bipartite graph

B associated with G. The bipartite graph is constructed as follows. For each vertex $v \in V$ there is a pair of vertices v_- and v_+ in $V(B)$. For each arc (v, w) in $A(G)$ there is an edge $\{v_+, w_-\}$ in $E(B)$. The problem of computing M in G is equivalent to computing a minimum $(\geq k - 1)$-matching M_B in B. This, in turn, is equivalent to computing a maximum b-matching M'_B in B, where $b(v) = \deg_{E(B)}(v) - (k-1)$, since $M_B = E(B) \setminus M'_B$. A b-matching problem on a graph with n vertices and m edges can be solved in $O(\sqrt{m}\alpha(m, m) \log m \ m \log m)$ time [9].

Phase 2. The second phase runs the minimal k-VCSS algorithm of Section 2.1 but only for the arcs in $A(G) \setminus M$. The algorithm maintains a current graph $\widehat{G}$. Initially we set $A(\widehat{G}) = A$, and at the end of this phase $A(\widehat{G}) = M \cup F$, where $F \subseteq A(G) \setminus M$ is a minimal subset of arcs such that $\widehat{G}$ is k-vertex connected. Using the algorithm of Section 2.1, phase 2 takes $O(k|E|^2)$ time, which is also the asymptotic running time of the whole algorithm.

Simplification for $k = 2$. Although the Cheriyan-Thurimella algorithm is conceptually simple (but its analysis is intricate), it is challenging to provide an efficient implementation, especially for phase 1. For the case $k = 2$ we propose the following simpler implementation of phase 1. First we compute a maximum matching M in B. Then we augment this matching to a set M_2 by adding an edge incident to each free vertex. Next we show that M_2 is indeed a minimum (≥ 1)-matching in B.

Lemma 1. M_2 *is a minimum (≥ 1)-matching in B.*

Proof. Clearly, M_2 is a (≥ 1)-matching in B, so it remains to show that it is minimum. Let M' be a minimum (≥ 1)-matching in B. Let $X = \{x \in V(B) \mid \deg_{M'}(x) > 1\}$. If $X = \emptyset$ then $\deg_{M'}(x) = 1$ for all $x \in V(B)$. In this case M' is a perfect matching, hence $|M'| = |M_2|$.

Consider now $X \neq \emptyset$. Let x be any vertex in X. Then, for any edge $\{x, y\}$ in M', $\deg_{M'}(y) = 1$. Otherwise, $M' - \{x, y\}$ is a (≥ 1)-matching in B which contradicts the fact that M' is minimum. Therefore, there is no edge $\{x, y\} \in M'$ such that both x and y are in X. Let N be a subset of M' that is left after removing $\deg_{M'}(x) - 1$ edges for each $x \in X$. Suppose that ℓ edges are removed from M' to form N. Then $|N| = |M'| - \ell$ and B has ℓ free vertices with respect to N. We show that N is a maximum matching in B. Suppose, for contradiction, that it is not. Let M be a maximum matching. Then, for some $\ell' \geq 1$ we have $|M| = |N| + \ell' = |M'| + (\ell' - \ell)$. Next note that there are $\ell - 2\ell'$ free vertices with respect to M. Therefore $|M_2| \leq |M| + (\ell - 2\ell') = |M'| + \ell' - \ell + \ell - 2\ell' = |M'| - \ell' < |M'|$, a contradiction. So $|M| = |N|$ which implies $|M_2| = |M'|$. □

We refer to the above implementation of the Cheriyan-Thurimella algorithm, for $k = 2$, as CT.

2.3   Linear-Time Algorithm

Now we present a fast heuristic which is based on recent work on independent spanning trees [14], testing 2-vertex connectivity [12], and computing strong articulation points [18] in digraphs. Our algorithm uses the concept of dominators and semi-dominators. Semi-dominators were introduced by Lengauer and Tarjan [19] in their fast algorithm for computing dominators in flowgraphs.

A flowgraph $G(s) = (V, A, s)$ is a directed graph with a distinguished source vertex $s \in V$ such that every vertex is reachable from s. A vertex w *dominates* a vertex v if every path from s to v includes w. By Menger's theorem (see, e.g., [6]), it follows that if G is 2-vertex connected then, for any $x \in V$, $G(x)$ has only trivial dominators, meaning that any vertex in $G(x)$ is dominated only by itself and x.

Let D be a depth-first search tree of $G(s)$, rooted at s. We assign to each vertex a preorder number with respect to D and identify the vertices by their preorder numbers. Then, $u < v$ means that u was visited before v during the depth-first search. A path $P = (u = v_0, v_1, \ldots, v_{k-1}, v_k = v)$ is a *semi-dominator path* (abbreviated sdom-path) if $v < v_i$ for $1 \leq i \leq k - 1$. The *semi-dominator* of vertex v is defined by

$$sdom(v) = \min\{u \mid \text{there is an sdom-path from } u \text{ to } v\}.$$

From the properties of depth-first search it follows that, for every $v \neq s$, $sdom(v)$ is a proper ancestor of v in D [19].

For any vertex $v \neq s$, we define $t(v)$ to be a predecessor of v that belongs to an sdom-path from $sdom(v)$ to v. Such vertices can be found easily during the computation of semi-dominators [14]. Therefore, by [3], we can compute $t(v)$ for all $v \neq s$ in linear time. Next, we define the *minimal equivalent sub-flowgraph* of $G(s)$, $\min(G(s)) = (V, A(\min(G(s))), s)$, where

$$A(\min(G(s))) = \big\{(p(v), v) \mid v \in V - s\big\} \cup \big\{(t(v), v) \mid v \in V - s\big\}.$$

The following lemma from [14] shows that $G(s)$ and $\min(G(s))$ are equivalent with respect to the dominance relation.

Lemma 2 ([14]). *The flowgraphs $G(s)$ and $\min(G(s))$ have the same dominators.*

Let $G_{\not s}$ denote the graph that remains after deleting s from G. Since G is 2-vertex connected then $G_{\not s}$ is strongly connected. For any digraph H, let H^r denote the digraph that is formed after reversing all arc directions in H; we apply the same notation for arc sets. In particular, the notation $\min(G^r(s))$ refers to the minimal equivalent sub-flowgraph of $G^r(s)$.

The algorithm constructs a subgraph $G^* = (V, A^*)$ of G as follows. First, it chooses two arbitrary vertices $s, s' \in V$, where $s' \neq s$. Then, it computes $\min(G(s))$, $\min(G^r(s))$, a spanning out-tree $T_{\not s}$ of $G_{\not s}(s')$ (e.g., a depth-first search tree of $G_{\not s}$, rooted at s') and a spanning out-tree $T'_{\not s}$ of $G^r_{\not s}(s')$. Then, it sets

$$A^* = A(\min(G(s))) \cup A^r(\min(G^r(s))) \cup A(T_{\not s}) \cup A^r(T'_{\not s}).$$

Theorem 1. G^* *is 2-vertex connected.*

Proof. From [12,18], we have that a digraph H is 2-vertex connected if and only if it satisfies the following property: $H(s)$ and $H^r(s)$ have trivial dominators only, and H_s is strongly connected, where $s \in V(H)$ is arbitrary. Lemma 2, implies that G^* satisfies the above property by construction. Therefore, G^* is 2-vertex connected. □

An alternative way to express the construction of G^* is via independent spanning trees [14]. Two spanning trees, T_1 and T_2 of a flowgraph $G(s)$ are independent if, for all vertices $x \neq s$, the paths from s to x in T_1 and T_2 have only s and x in common. In [14] it is shown that $\min(G(s))$ is formed by two independent spanning trees of $G(s)$. Therefore, G^* is formed by two independent spanning trees of $G(s)$, two independent spanning trees of $G^r(s)$ (with their arcs reversed), a spanning tree of $G_s(s')$, and a spanning tree of $G_s^r(s')$ (with its arcs reversed).

Theorem 2. G^* *is a 3-approximation of the smallest 2VCSS.*

Proof. Since G^* is formed by 4 spanning trees on n vertices and 2 spanning trees on $n - 1$ vertices, we have $|A(G^*)| \leq 4(n-1) + 2(n-2)$. (The inequality holds because the spanning trees may have some arcs in common.) In any 2-vertex connected digraph each vertex has outdegree at least 2, so a 2-vertex connected digraph has at least $2n$ arcs. These facts imply an approximation ratio of at most $3 - \frac{4}{n}$. □

Practical Implementation. The experimental results presented in Section 3 show that in fact our algorithm performs much better than the bound of Theorem 2 suggests, especially after taking the following measures. First, when we have computed $\min(G(s))$ and $\min(G^r(s))$ the algorithms checks if $A(\min(G(s))) \cup A^r(\min(G^r(s)))$ already forms a 2-vertex connected digraph. If this is the case then it returns this graph instead of G^*, therefore achieving a 2-approximation of the smallest 2-VCSS. Another idea that helps in practice is to run the algorithm from several different source vertices and return the best result. For our experiments we used five sources, chosen at random. In all of our experiments, the returned 2-VCSS was formed only by the set $A(\min(G(s))) \cup A^r(\min(G^r(s)))$ for one of the five randomly chosen sources s. (We note that for some choices of s the above set did not suffice to produce a 2-VCSS.) We leave as open question whether there is a tighter bound on the approximation ratio of our heuristic, or whether some variant of the heuristic can achieve better than a 2-approximation.

The above algorithm runs in $O(m + n)$ time using a linear-time algorithm to compute semi-dominators [2,3,13], or in $O(m\alpha(m, n))$ time using the Lengauer-Tarjan algorithm [19]. (Here $\alpha(m, n)$ is a very slow-growing functional inverse of Ackermann's function.) Previous experimental studies (e.g., [15]) have shown that in practice the simple version of the Lengauer-Tarjan algorithm, with $O(m \log n)$ worst-case running time, outperforms the $O(m\alpha(m, n))$-time version. In view of these experimental results, our implementation of the above heuristic, referred to as FAST, was based on the simple version of Lengauer-Tarjan.

2.4   Hybrid Algorithms

We can get various hybrid algorithms by combining FAST with either MINIMAL or CT. A first idea is to run MINIMAL with the 2-VCSS returned by FAST as input. We call this algorithm FASTMINIMAL. Since this algorithm also computes a minimal 2-VCSS, it also achieves a 2-approximation. The running time is improved from $O(m^2)$ to $O(n^2)$.

The above idea does not work for CT, since FAST may filter out the wrong arcs and, therefore, the final 2-VCSS may not be a 1.5-approximation. To fix this problem we propose the following combination of FAST and CT, referred to as FASTCT, which runs FAST between the two phases of CT. Following the computation of the (≥ 1)-matching M, we run FAST with the initial digraph G as input. This returns a 2-VCSS G^* of G. The input to the second phase of CT is the digraph with arc set $A(G^*) \cup M$.

Theorem 3. *Algorithm* FASTCT *computes a 1.5-approximation of the smallest 2-VCSS in $O(m\sqrt{n} + n^2)$ time.*

Proof. After having computed a minimum (≥ 1)−matching M of the input digraph G, the Cheriyan-Thurimella algorithm can process the arcs in $A(G) \setminus M$ in an arbitrary order. So if it processes the arcs in $A(G) \setminus \big(A(G^*) \cup M\big)$ first, all these arcs will be removed from the current graph $\widehat{G}$ that is maintained during the second phase. Thus, the approximation guarantee of the Cheriyan-Thurimella algorithm is preserved.

Regarding the running time, by Lemma 1 we have that a minimum (≥ 1)−matching can be computed in $O(m\sqrt{n})$ time using the Hopcroft-Karp maximum bipartite matching algorithm [17]. Also, by Section 2.3, the computation of $A(G^*)$ takes linear time. Then, we are left with a 2-VCSS with $O(n)$ arcs, so the last phase of the algorithm runs in $O(n^2)$ time.   □

3   Experimental Results

3.1   Implementation and Experimental Setup

For the first phase of CT and FASTCT, we used an implementation of the push-relabel maximum-flow algorithm of Goldberg and Tarjan [16] from [5]. This implementation does not use a dynamic tree data structure [22], which means that the worst-case bound for the first phase of these algorithms is $O(n^3)$. However, as we confirmed experimentally, the push-relabel algorithm runs very fast in practice. For the implementation of FAST, as well as for the last phase of CT, MINIMAL, and FASTCT, we adapted the implementation of the simple version of the Lengauer-Tarjan dominators algorithm [19] from [15]. The source code was compiled using g++ v. 3.2.2 with full optimization (flag -O4). All tests were conducted on an Intel Xeon E5520 at 2.27GHz with 8192KB cache, running Ubuntu 9.04 Server Edition.

The main focus of the experiments in this paper is the quality of the 2-VCSS returned by each algorithm. Therefore we did not put much effort in optimizing our source code. (Note, however, that the implementations in [5] and [15] are highly optimized.) For completeness we do mention the running times for a subset of the experiments, but leave a thorough experimental study of the running times for the full version of the paper.

3.2   Instances

We considered three types of instances: bicliques, internet peer-to-peer networks taken from the Stanford Large Network Dataset Collection [23], and random digraphs. Some of these graphs are not 2-vertex connected to begin with, so we need to apply a suitable transformation to make them 2-vertex connected. We tested two transformations: adding a bidirectional Hamiltonian cycle (indicated by the letter H in the tables below), and removing strong articulation points (indicated by the letter A in the tables below).

The term bidirectional Hamiltonian cycle refers to two oppositely directed Hamiltonian cycles. We create these cycles by taking a random permutation of the vertices of the input digraph, say $v_0, v_1, \ldots, v_{n-1}$, and adding the arcs (v_i, v_{i+1}) and (v_{i+1}, v_i), where the addition is computed mod n. Note that a bidirectional Hamiltonian cycle is by itself a minimum 2-VCSS with exactly $2n$ arcs, so in this case it is easy to assess how close to optimal are the 2-VCSS produced by the tested algorithms.

A strong articulation point of a strongly connected digraph is a vertex whose removal increases the number of strongly connected components [18]. To apply the second transformation method, we first introduce a directed Hamiltonian cycle to the input digraph if it is not strongly connected. Then we compute the strong articulation points of the resulting digraph and add appropriate arcs so that the final digraph is 2-vertex connected. More details are given below.

Bicliques. Graphs that consist of a complete bipartite graph (biclique) together with a Hamiltonian cycle were suggested in [4] as instances for which getting a better than a 2-approximation of the smallest k-VCSS may be hard. We let bcH(n_1, n_2) denote a graph that consists of a biclique K_{n_1, n_2} plus a Hamiltonian cycle; we convert this graph to a digraph by making all edges bidirectional, i.e. replace each edge $\{x, y\}$ with the arcs (x, y) and (y, x).

We remark that the order in which the arcs are read from the input file can affect the quality of the returned 2-VCSS significantly. To that end we considered the following example using bcH(1000, 2) as input. If the arcs of the Hamiltonian cycle are read first then FAST, running from an appropriately chosen source vertex, returns a 2-VCSS with 2004 arcs, which is the optimal value. On the other hand, MINIMAL returns 4000 arcs. The situation is reversed if the arcs of the Hamiltonian cycle are read last, i.e., FAST returns 4000 arcs and MINIMAL returns 2004 arcs. The cause of this effect is that FAST favors the arcs that are explored first during a depth-first search, while MINIMAL favors the arcs that are processed last. The number of arcs that FASTCT returns is 2005 for both graphs. Also, FASTMINIMAL returns 2004 arcs for the first graph and 2006 for the second.

In view of the above effect, we created several instances of each digraph bcH(n_1, n_2) used in our experiments, by permuting randomly its arcs. The results are shown in Table 1.

Table 1. Bicliques with embedded bidirectional Hamiltonian cycle. Average number of arcs and standard deviation in the computed 2-VCSS for 20 seeds. The best result in each row is marked in bold.

INSTANCE	CT	FASTCT	FAST	MINIMAL	FASTMINIMAL
bcH($1000, 2$)	2664.75	**2045.65**	2458.00	2791.65	2061.90
	16.12	38.29	95.95	17.54	41.13
bcH($1000, 10$)	2925.90	**2132.40**	2771.75	3558.50	2199.50
	7.85	37.77	57.65	14.78	36.27
bcH($1000, 20$)	2977.00	**2159.55**	2916.80	3752.90	2230.90
	5.30	20.44	19.90	12.34	27.90
bcH($1000, 40$)	3020.00	**2227.00**	3048.50	3871.40	2315.50
	4.27	20.25	23.91	11.07	14.97

Internet Peer-to-Peer Networks. We tested digraphs of the family p2p-Gnutella taken from the Stanford Large Network Dataset Collection [23], after performing some preprocessing. These graphs are internet peer-to-peer networks, and therefore relevant for problems concerning the design of resilient communication networks. Also, their size is small enough so that they could be processed within a few hours by the slower algorithms in our study.

The goal of preprocessing was to make these digraphs 2-vertex connected. (Most of these graphs are not strongly connected, and some are not even weakly connected.) To that end we applied the two transformations mentioned earlier, indicated in Table 2 by the letters H and A. The former indicates that a bidirectional Hamiltonian cycle was added to the digraph. The latter transformation first adds a unidirectional Hamiltonian cycle, producing a strongly connected digraph, say G'. Then it chooses two random source vertices r_1 and r_2 and computes strong articulation points (cut vertices) in G' using dominator and post-dominator (i.e., dominator in the reverse digraph) computations with r_1 and r_2 as roots. (We refer to [12,18] for details.) A strong articulation point x is removed by adding one of the arcs (r_1, x), (x, r_1), (r_2, x) or (x, r_2) depending on which dominators calculation revealed the specific articulation point.

Random Graphs. The last type of digraphs we consider are random directed graphs. We denote by rand(n, m) a random directed graph with parameters n and m; n is the number of vertices and m is the expected number of arcs, i.e., each of the $n(n-1)$ possible arcs is chosen with probability $\frac{m}{n(n-1)}$. Since we need to construct 2-vertex connected digraphs, we have to choose large values for m. In order to get a view of the situation for smaller values of m we also include the family of graphs randH(n, m), which contain a bidirectional Hamiltonian cycle. The results are shown in Table 3. Table 4 gives us an indication of the corresponding running times.

Table 2. Internet peer-to-peer networks; the number of vertices n and the average number of arcs $\overline{m}$ for each type of instances is shown. Average number of arcs and standard deviation in the computed 2-VCSS for 20 seeds. The best result in each row is marked in bold.

INSTANCE	CT	FASTCT	FAST	MINIMAL	FASTMINIMAL
p2p-Gnutella04-H	23461.80	23340.10	30024.10	23488.10	**23254.55**
($n = 10876, \overline{m} = 61692$)	23.54	17.37	35.18	21.74	29.86
p2p-Gnutella04-A	28203.35	**27929.60**	33920.60	28131.05	28055.95
($n = 10876, \overline{m} = 63882.50$)	304.09	38.64	414.36	288.35	102.18
p2p-Gnutella08-H	13561.30	13459.60	17048.15	13584.95	**13395.55**
($n = 6301, \overline{m} = 33137$)	14.23	15.65	35.00	14.00	15.45
p2p-Gnutella08-A	16858.20	**16727.85**	19357.65	16859.20	16778.45
($n = 6301, \overline{m} = 33913.35$)	140.13	15.82	198.89	153.26	36.96
p2p-Gnutella09-H	17410.75	17321.40	21922.30	17449.70	**17232.60**
$n = 8114, \overline{m} = 41977$	14.49	18.28	35.55	21.28	20.95
p2p-Gnutella09-A	21740.70	**21606.25**	25076.60	21714.80	21679.15
$n = 8114, \overline{m} = 44081.70$	141.43	14.44	293.33	134.11	42.52

Table 3. Random directed graphs. Average number of arcs and standard deviation in the computed 2-VCSS for 20 seeds. The best result in each row is marked in bold.

INSTANCE	CT	FASTCT	FAST	MINIMAL	FASTMINIMAL
rand(1000, 20000)	2301.40	2167.25	2991.65	2222.65	**2164.00**
	7.41	9.57	13.03	11.09	10.23
rand(1000, 40000)	2336.85	2167.95	2997.10	2224.70	**2163.55**
	10.58	9.46	11.63	7.73	9.20
randH(1000, 2000)	2165.65	2149.75	2694.10	2167.00	**2145.95**
	7.04	7.80	12.94	9.22	5.35
randH(1000, 4000)	2205.80	**2159.25**	2851.45	2200.15	2159.30
	8.13	8.55	13.46	8.52	7.79

Table 4. Random directed graphs. Average running time (in seconds) and standard deviation for 20 seeds.

INSTANCE	CT	FASTCT	FAST	MINIMAL	FASTMINIMAL
rand(1000, 20000)	14.17	0.79	0.02	14.63	4.54
	0.25	0.02	0.01	0.19	0.09
rand(1000, 40000)	52.59	0.77	0.02	54.52	8.89
	0.63	0.02	0.00	0.77	0.14
randH(1000, 2000)	0.81	0.70	0.01	1.00	1.00
	0.03	0.02	0.00	0.03	0.03
randH(1000, 4000)	1.65	0.77	0.01	1.79	1.45
	0.04	0.02	0.00	0.04	0.03

3.3   Evaluation

The above experimental results show that although our fast heuristic returned a larger 2-VCSS than the other algorithms in all tested cases (with the exception of bicliques), the approximation ratio of 3 seems pessimistic, at least in practice. Moreover, the hybrid algorithms FASTCT and FASTMINIMAL had the best performance in all of our tests. The sizes of the subgraphs returned by these two algorithms were very close to each other, with FASTCT being slightly better for bicliques, the A-instances of the peer-to-peer networks, and the larger H-instance of the random digraphs.

The comparison between FASTCT and FASTMINIMAL favors the former, since it guarantees a better approximation ratio, and, as Table 4 shows, it can run significantly faster. The reason for this difference in the running times is that the maximum-flow computation (used for bipartite maximum matching) is completed a lot faster than the time it takes FASTMINIMAL to process the arcs of the matching that are excluded from the corresponding processing during the last phase of FASTCT. We will investigate thoroughly the relative running times of different implementations in the full version of the paper.

Acknowledgements. We would like to thank Renato Werneck for providing the implementation of the push-relabel maximum-flow algorithm from [5] and for several helpful discussions, and the anonymous referees for some useful comments.

References

1. Ahuja, R.K., Magnanti, T.L., Orlin, J.B.: Network flows: theory, algorithms, and applications. Prentice-Hall, Inc., Upper Saddle River (1993)
2. Alstrup, S., Harel, D., Lauridsen, P.W., Thorup, M.: Dominators in linear time. SIAM Journal on Computing 28(6), 2117–2132 (1999)
3. Buchsbaum, A.L., Georgiadis, L., Kaplan, H., Rogers, A., Tarjan, R.E., Westbrook, J.R.: Linear-time algorithms for dominators and other path-evaluation problems. SIAM Journal on Computing 38(4), 1533–1573 (2008)
4. Cheriyan, J., Thurimella, R.: Approximating minimum-size k-connected spanning subgraphs via matching. SIAM J. Comput. 30(2), 528–560 (2000)
5. Delling, D., Goldberg, A.V., Razenshteyn, I., Werneck, R.F.: Graph partitioning with natural cuts. In: 25th International Parallel and Distributed Processing Symposium, IPDPS 2011 (2011)
6. Diestel, R.: Graph Theory, 2nd edn. Springer, New York (2000)
7. Edmonds, J.: Edge-disjoint branchings. Combinatorial Algorithms, 91–96 (1972)
8. Gabow, H.N., Gallagher, S.: Iterated rounding algorithms for the smallest k-edge connected spanning subgraph. In: Proceedings of the Nineteenth Annual ACM-SIAM Symposium on Discrete Algorithms, pp. 550–559 (2008)
9. Gabow, H.N., Tarjan, R.E.: Faster scaling algorithms for general graph matching problems. J. ACM 38, 815–853 (1991)
10. Garey, M.R., Johnson, D.S.: Computers and Intractability: A Guide to the Theory of NP-Completeness. W. H. Freeman & Co., New York (1979)

11. Garg, N., Santosh, V.S., Singla, A.: Improved approximation algorithms for biconnected subgraphs via better lower bounding techniques. In: Proc. 4th ACM-SIAM Symp. on Discrete Algorithms, pp. 103–111 (1993)
12. Georgiadis, L.: Testing 2-vertex connectivity and computing pairs of vertex-disjoint s-t paths in digraphs. In: Proc. 37th Int'l. Coll. on Automata, Languages, and Programming, pp. 738–749 (2010)
13. Georgiadis, L., Tarjan, R.E.: Finding dominators revisited. In: Proc. 15th ACM-SIAM Symp. on Discrete Algorithms, pp. 862–871 (2004)
14. Georgiadis, L., Tarjan, R.E.: Dominator tree verification and vertex-disjoint paths. In: Proc. 16th ACM-SIAM Symp. on Discrete Algorithms, pp. 433–442 (2005)
15. Georgiadis, L., Tarjan, R.E., Werneck, R.F.: Finding dominators in practice. Journal of Graph Algorithms and Applications (JGAA) 10(1), 69–94 (2006)
16. Goldberg, A.V., Tarjan, R.E.: A new approach to the maximum-flow problem. Journal of the ACM 35, 921–940 (1988)
17. Hopcroft, J.E., Karp, R.M.: An $n^{5/2}$ algorithm for maximum matchings in bipartite graphs. SIAM Journal on Computing 2, 225–231 (1973)
18. Italiano, G., Laura, L., Santaroni, F.: Finding strong bridges and strong articulation points in linear time. In: Wu, W., Daescu, O. (eds.) COCOA 2010, Part I. LNCS, vol. 6508, pp. 157–169. Springer, Heidelberg (2010)
19. Lengauer, T., Tarjan, R.E.: A fast algorithm for finding dominators in a flowgraph. ACM Transactions on Programming Languages and Systems 1(1), 121–141 (1979)
20. Mader, W.: Minimal n-fach zusammenhängende digraphen. Journal of Combinatorial Theory, Series B 38(2), 102–117 (1985)
21. Nutov, Z.: An almost $O(\log k)$-approximation for k-connected subgraphs. In: Proc. 20th ACM-SIAM Symp. on Discrete Algorithms, SODA 2009, pp. 912–921 (2009)
22. Sleator, D.D., Tarjan, R.E.: A data structure for dynamic trees. Journal of Computer and System Sciences 26, 362–391 (1983)
23. Stanford network analysis platform (snap), http://snap.stanford.edu/

Improved Approximation Algorithms for Bipartite Correlation Clustering

Nir Ailon[1,*], Noa Avigdor-Elgrabli[1], Edo Liberty[2], and Anke van Zuylen[3]

[1] Technion, Haifa, Israel
{nailon,noaelg}@cs.technion.ac.il
[2] Yahoo! Research, Haifa, Israel
edo.liberty@ymail.com
[3] Max-Planck Institut für Informatik, Saarbrücken, Germany
anke@mpi-inf.mpg.de

Abstract. In this work we study the problem of Bipartite Correlation Clustering (BCC), a natural bipartite counterpart of the well studied Correlation Clustering (CC) problem. Given a bipartite graph, the objective of BCC is to generate a set of vertex-disjoint bi-cliques (clusters) which minimizes the symmetric difference to it. The best known approximation algorithm for BCC due to Amit (2004) guarantees an 11-approximation ratio.[1]

In this paper we present two algorithms. The first is an improved 4-approximation algorithm. However, like the previous approximation algorithm, it requires solving a large convex problem which becomes prohibitive even for modestly sized tasks.

The second algorithm, and our main contribution, is a simple randomized combinatorial algorithm. It also achieves an expected 4-approximation factor, it is trivial to implement and highly scalable. The analysis extends a method developed by Ailon, Charikar and Newman in 2008, where a randomized pivoting algorithm was analyzed for obtaining a 3-approximation algorithm for CC. For analyzing our algorithm for BCC, considerably more sophisticated arguments are required in order to take advantage of the bipartite structure.

Whether it is possible to achieve (or beat) the 4-approximation factor using a scalable *and* deterministic algorithm remains an open problem.

1   Introduction

The analysis of large bipartite graphs is becoming of increased practical importance. Recommendation systems, for example, take as input a large dataset of bipartite relations between users and objects (e.g. movies, goods) and analyze its structure for the purpose of predicting future relations [2]. Other examples may include images vs. user generated tags and search engine queries vs. search results. Bipartite clustering is also studied in the context of gene expression data

---

[*] Supported in part by the Yahoo! Faculty Research and Engagement Program.
[1] A previously claimed 4-approximation algorithm [1] is erroneous; due to space constraints, details of the counterexample are omitted from this version.

C. Demetrescu and M.M. Halldórsson (Eds.): ESA 2011, LNCS 6942, pp. 25–36, 2011.
© Springer-Verlag Berlin Heidelberg 2011

analysis (see e.g. [3][4][5] and references therein). In spite of the extreme practical importance of bipartite clustering, far less is known about it than standard (non-bipartite) clustering. Many notions of clustering bipartite data exist. Some aim at finding the best cluster, according to some definition of 'best'. Others require that the entire data (graph) be represented as clusters. Moreover, data points (nodes) may either be required to belong to only one cluster or allowed to belong to different overlapping clusters. Here the goal is to obtain non-overlapping (vertex disjoint) clusters covering the entire input vertex set. Hence, one may think of our problem as *bipartite graph partitioning*.

In Bipartite Correlation Clustering (BCC) we are given a bipartite graph as input, and output a set of disjoint clusters covering the graph nodes. Clusters may contain nodes from either side of the graph, but they may possibly contain nodes from only one side. We think of a cluster as a bi-clique connecting all the elements from its left and right counterparts. An output clustering is hence a union of bi-cliques covering the input node set. The cost of the solution is the symmetric difference between the input and output edge sets. Equivalently, any pair of vertices, one on the left and one of the right, will incur a unit cost if either (1) they are connected by an input edge but the output clustering separates them into distinct clusters, or (2) they are not connected by an input edge but the output clustering assigns them to the same cluster. The objective is to minimize this cost. This problem formulation is the bipartite counterpart of the more well known Correlation Clustering (CC), introduced by Bansal, Blum and Chawla [6], where the objective is to cover the node set of a (non-bipartite) graph with disjoint cliques (clusters) minimizing the symmetric difference with the given edge set. One advantage of this objective is in alleviating the need to specify the number of output clusters, as often needed in clustering settings such as k-means or k-median. Another advantage lies in the objective function, which naturally corresponds to some models about noise in the data. Examples of applications include [7], where a reduction from *consensus clustering* to our problem is introduced, and [8] for an application of a related problem to large scale document-term relation analysis.

Bansal et al. [6] gave a $c \approx 10^4$ factor for approximating CC running in time $O(n^2)$ where n is the number of nodes in the graph. Later, Demaine, Emanuel, Fiat and Immorlica [9] gave a $O(\log(n))$ approximation algorithm for an *incomplete* version of CC, relying on solving an LP and rounding its solution by employing a region growing procedure. By incomplete we mean that only a subset of the node pairs participate in the symmetric difference cost calculation.[2] BCC is, in fact, a special case of incomplete CC, in which the non-participating node pairs lie on the same side of the graph. Charikar, Guruswami and Wirth [10] provide a 4-approximation algorithm for CC, and another $O(\log n)$-approximation algorithm for the incomplete case. Later, Ailon, Charikar and Newman [11] provided a 2.5-approximation algorithm for CC based on rounding an LP. They also provide a simpler 3-approximation algorithm, QuickCluster, which runs in time

---

[2] In some of the literature, CC refers to the much harder incomplete version, and "CC in complete graphs" is used for the version we have described here.

linear in the number of edges of the graph. In [12] it was argued that QuickCluster runs in expected time $O(n + cost(OPT))$ over the algorithm randomness.

Van Zuylen and Williamson [13] provided de-randomization for the algorithms presented in [11] with no compromise in the approximation guarantees. Giotis and Guruswami [14] gave a PTAS for the CC case in which the number of clusters is constant. Later, (using other techniques) Karpinski and Schudy [15] improved the runtime.

Amit [3] was the first to address BCC directly. She proved its NP-hardness and gave a constant 11-approximation algorithm based on rounding a linear programming in the spirit of Charikar et al.'s [10] algorithm for CC.

It is worth noting that in [1] a 4-approximation algorithm for BCC was presented and analyzed. The presented algorithm is incorrect (we give a counter example in the paper) but their attempt to use arguments from [11] is an excellent one. We will show how to achieve the claimed guarantee with an extension of the method in [11].

1.1   Our Results

We first describe a deterministic 4-approximation algorithm for BCC (Section 2). It starts by solving a Linear Program in order to convert the problem to a non bipartite instance (CC) and then uses the pivoting algorithm [13] to construct a clustering. The algorithm is similar to the one in [11] where nodes from the graph are chosen randomly as 'pivots' or 'centers' and clusters are generated from their neighbor sets. Arguments from [13] derandomize this choice and give us a deterministic 4-approximation algorithm. This algorithm, unfortunately, becomes impractical for large graphs. The LP solved in the first step needs to enforce the transitivity of the clustering property for all sets of three nodes and thus contains $\Omega(n^3)$ constraints.

Our main contribution is an extremely simple combinatorial algorithm called PivotBiCluster which achieves the same approximation guarantee. The algorithm is straightforward to implement and terminates in $O(|E|)$ operations (the number of edges in the graph). We omit the simple proof of the running time since it is immediate given the algorithm's description, see Section 3.2. A disadvantage of PivotBiCluster is the fact that it is randomized and achieves the approximation guarantee only in expectation. However, a standard Markov inequality argument shows that taking the best solution obtained from independent repetitions of the algorithm achieves an approximation guarantee of $4 + \varepsilon$ for any constant $\varepsilon > 0$.

While the algorithm itself is simple, its proof is rather involved and requires a significant extension of previously developed techniques. To explain the main intuition behind our approach, we recall the method of Ailon et al. [11]. The algorithm for CC presented there (the unweighted case) is as follows: choose a random vertex, form a cluster including it and its neighbors, remove the cluster from the graph, and repeat until the graph is empty. This random-greedy algorithm returns a solution with cost at most 3 times that of the optimal solution, in expectation. The key to the analysis is the observation that each part of the cost of the algorithm's solution can be naturally related to a certain *minimal contradicting structure*, for CC, an induced subgraph of 3 vertices and exactly 2 edges. Notice that in any such structure, at least one vertex pair must be violated. A vertex pair being violated means it contributes to the symmetric difference between the graph and the clustering. In other words, the vertex pairs that a clustering violates must *hit* the set of minimal contradicting structures. A corresponding hitting set LP lower bounding the optimal solution was defined to capture this simple observation. The analysis of the random-greedy solution constructs a dual feasible solution to this LP, using probabilities arising in the algorithm's probability space.

It is tempting here to consider the corresponding minimal contradicting structure for BCC, namely a set of 4 vertices, 2 on each side, with exactly 3 edges between them. Unfortunately, this idea turned out to be evasive. A proposed solution attempting this [1] has a counter example; due to space constraints, details are omitted from this version. Our attempts to follow this approach have also failed. In our analysis we resorted to contradicting structures of unbounded size. Such a structure consists of two vertices ℓ_1, ℓ_2 of the left side and two sets of vertices N_1, N_2 on the right hand side such that N_i is contained in the neighborhood of ℓ_i for $i = 1, 2$, $N_1 \cap N_2 \neq \emptyset$ and $N_1 \neq N_2$. We define a hitting LP as we did earlier, this time of possibly exponential size, and analyze its dual in tandem with a carefully constructed random-greedy algorithm, PivotBiCluster. At each round PivotBiCluster chooses a random pivot vertex on the left, constructs a cluster with its right hand side neighbors, and for each other vertex on the left randomly decides whether to join the new cluster or not. The new cluster is removed and the process is repeated until the graph is exhausted. The main challenge is to find joining probabilities of left nodes to new clusters which can be matched to a feasible solution to the dual LP.

1.2 Paper Structure

We first present a deterministic LP rounding based algorithm in Section 2. Our main algorithm in given in Section 3. We start with notations and definitions in Section 3.1, followed by the algorithm's description and our main theorem in Section 3.2. The algorithm's analysis is logically partitioned between Sections 3.3, 3.4, and 3.5. Finally, we propose future research and conjectures in Section 4.

2 A Deterministic LP Rounding Algorithm

We start with a deterministic algorithm with a 4-approximation guarantee by directly rounding an optimal solution to a linear programming relaxation $\mathrm{LP}_{\mathrm{det}}$ of BCC. Let the input graph be $G = (L, R, E)$ where L and R are the sets of left and right nodes and E be a subset of $L \times R$. For notational purposes, we define the following constants given our input graph: for each edge $(i, j) \in E$ we define $w_{ij}^+ = 1, w_{ij}^- = 0$ and for each non-edge $(i, j) \notin E$ we define $w_{ij}^+ = 0, w_{ij}^- = 1$. Our integer program has an indicator variable y_{ij}^+ which equals 1 iff i and j are placed in the same cluster. The variable is defined for each pair of vertices, *and not only for pairs* (ℓ, r) with $\ell \in L, r \in R$. Hence, in a certain sense, this approach forgets about bipartiteness. For ease of notation we define $y_{ij}^- = 1 - y_{ij}^+$. The objective function becomes $\sum_{(i,j)} (w_{ij}^+ y_{ij}^- + w_{ij}^- y_{ij}^+)$. The clustering consistency constraint is given as $y_{ij}^- + y_{jk}^- + y_{ik}^+ \geq 1$ for all (ordered) sets of three vertices $i, j, k \in V$, where $V = L \cup R$. The relaxed LP is given by:

$$\mathrm{LP_{det}} = \min \sum_{(i,j)} (w_{ij}^+ y_{ij}^- + w_{ij}^- y_{ij}^+)$$

$$\text{s.t } \forall i, j, k \in V : y_{ij}^- + y_{jk}^- + y_{ik}^+ \geq 1, y_{ij}^+ + y_{ij}^- = 1, y_{ij}^+, y_{ij}^- \in [0,1].$$

Given an optimal solution to $\mathrm{LP_{det}}$, we partition the pairs of distinct vertices into two sets E^+ and E^-, where $e \in E^+$ if $y_e^+ \geq \frac{1}{2}$ and $e \in E^-$ otherwise. Since each distinct pair is in either E^+ or E^-, we have an instance of CC which can then be clustered using the algorithm of Van Zuylen et al. [13]. The algorithm is a derandomization of Ailon et al.'s [11] randomized QuickCluster for CC. QuickCluster recursively constructs a clustering simply by iteratively choosing a pivot vertex i at random, forming a cluster C that contains i and all vertices j such that $(i, j) \in E^+$, removing them from the graph and repeating. Van Zuylen et al. [13] replace the random choice of pivot by a deterministic one, and show conditions under which the resulting algorithm output is a constant factor approximation with respect to the LP objective function. To describe their choice of pivot, we need the notion of a "bad triplet" [11]. We will call a triplet (i, j, k) a bad triplet if exactly two of the pairs among $\{(i, j), (j, k), (k, i)\}$ are edges in E^+. Consider the pairs of vertices on which the output of QuickCluster disagrees with E^+ and E^-, i.e., pairs $(i, j) \in E^-$ that are in the same cluster, and pairs $(i, j) \in E^+$ that are not in the same cluster in the output clustering. It is not hard to see that in both cases, there was some call to QuickCluster in which (i, j, k) formed a bad triplet with the pivot vertex k. The pivot chosen by Van Zuylen et al. [13] is the pivot that minimizes the ratio of the weight of the edges that are in a bad triplet with the pivot and their LP contribution.

Given an optimal solution y to $\mathrm{LP_{det}}$ we let $c_{ij} = w_{ij}^+ y_{ij}^- + w_{ij}^- y_{ij}^+$. Recall that E^+, E^- are also defined by the optimal solution. We are now ready to present the deterministic LP rounding algorithm:

Theorem 1. *[From Van Zuylen et al. [13]] Algorithm QuickCluster* (V, E^+, E^-) *from [11] returns a solution with cost at most* 4 *times the cost of the optimal solution to* $\mathrm{LP_{det}}$ *if in each iteration a pivot vertex is chosen that minimizes:*

$$F(k) = \left(\sum_{(i,j) \in E^+ : (i,j,k) \in \mathcal{B}} w_{ij}^+ + \sum_{(i,j) \in E^- : (i,j,k) \in \mathcal{B}} w_{ij}^- \right) / \sum_{(i,j,k) \in \mathcal{B}} c_{ij},$$

where $\mathcal{B}$ is the set of bad triplets on vertices that haven't been removed from the graph in previous steps.

Due to space constraints, the proof of Theorem 1 is omitted from this version.

3   The Combinatorial 4-Approximation Algorithm

3.1   Notation

Before describing the framework we give some general facts and notations. Let the input graph again be $G = (L, R, E)$ where L and R are the sets of left and right nodes and E be a subset of $L \times R$. Each element $(\ell, r) \in L \times R$ will be referred to as a *pair*.

A solution to our combinatorial problem is a clustering $C_1, C_2, \ldots, C_m$ of the set $L \cup R$. We identify such a clustering with a bipartite graph $B = (L, R, E_B)$ for which $(\ell, r) \in E_B$ if and only if $\ell \in L$ and $r \in R$ are in the same cluster C_i for some i. Note that given B, we are unable to identify clusters contained exclusively in L (or R), but this will not affect the cost. We therefore take the harmless decision that single-side clusters are always singletons.

We will say that a pair $e = (\ell, r)$ is violated if $e \in (E \setminus E_B) \cup (E_B \setminus E)$. For convenience, let $x_{G,B}$ be the indicator function for the violated pair set, i.e., $x_{G,B}(e) = 1$ if e is violated and 0 otherwise. We will also simply use $x(e)$ when it is obvious to which graph G and clustering B it refers. The cost of a clustering solution is defined to be $\mathrm{cost}_G(B) = \sum_{e \in L \times R} x_{G,B}(e)$. Similarly, we will use $\mathrm{cost}(B) = \sum_{e \in L \times R} x(e)$ when G is clear from the context, Let $N(\ell) = \{r | (\ell, r) \in E\}$ be the set of all right nodes adjacent to ℓ.

It will be convenient for what follows to define a *tuple*. We define a tuple T to be $(\ell_1^T, \ell_2^T, R_1^T, R_{1,2}^T, R_2^T)$ where $\ell_1^T, \ell_2^T \in L$, $\ell_1^T \neq \ell_2^T$, $R_1^T \subseteq N(\ell_1^T) \setminus N(\ell_2^T)$, $R_2^T \subseteq N(\ell_2^T) \setminus N(\ell_1^T)$ and $R_{1,2}^T \subseteq N(\ell_2^T) \cap N(\ell_1^T)$. In what follows, we may omit the superscript of T. Given a tuple $T = (\ell_1^T, \ell_2^T, R_1^T, R_{1,2}^T, R_2^T)$, we define the *conjugate tuple* $\bar{T} = (\ell_1^{\bar{T}}, \ell_2^{\bar{T}}, R_1^{\bar{T}}, R_{1,2}^{\bar{T}}, R_2^{\bar{T}}) = (\ell_2^T, \ell_1^T, R_2^T, R_{1,2}^T, R_1^T)$. Note that $\bar{\bar{T}} = T$.

3.2   Algorithm Description

We now describe PivotBiCluster. The algorithm runs in rounds. In every round it creates one cluster and possibly many singletons, all of which are removed from the graph before continuing to the next round. Abusing notation, by $N(\ell)$ we mean, in the algorithm description, all the neighbors of $\ell \in L$ which have not yet been removed from the graph.

Every such round performs two phases. In the first phase, PivotBiCluster picks a node on the left side uniformly at random, ℓ_1, and forms a new cluster $C = \{\ell_1\} \cup N(\ell_1)$. This will be referred to as the ℓ_1-phase and ℓ_1 will be referred to as the left center of the cluster. In the second phase the algorithm iterates over all other remaining left nodes, ℓ_2, and decides either to (1) append them to C, (2) turn them into singletons, or (3) do nothing. This will be denoted as the ℓ_2-sub-phase corresponding to the ℓ_1-phase. We now explain how to make this decision. let $R_1 = N(\ell_1) \setminus N(\ell_2)$, $R_2 = N(\ell_2) \setminus N(\ell_1)$ and $R_{1,2} = N(\ell_1) \cap N(\ell_2)$. With probability $\min\{\frac{|R_{1,2}|}{|R_2|}, 1\}$ do one of two things: (1) If $|R_{1,2}| \geq |R_1|$ append ℓ_2 to C, and otherwise (2) (if $|R_{1,2}| < |R_1|$), turn ℓ_2 into a singleton. In the remaining probability, (3) do nothing for ℓ_2, leaving it in the graph for future

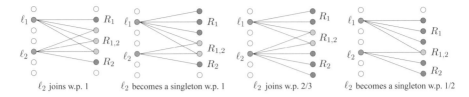

Fig. 1. Four example cases in which ℓ_2 either joins the cluster created by ℓ_1 or becomes a singleton. In the two right most examples, with the remaining probability nothing is decided about ℓ_2.

iterations. Examples for cases the algorithm encounters for different ratios of R_1, $R_{1,2}$, and R_2 are given in Figure 1.

Theorem 2. *Algorithm PivotBiCluster returns a solution with expected cost at most 4 times that of the optimal solution.*

3.3   Algorithm Analysis

We start by describing *bad events*. This will help us relate the expected cost of the algorithm to a sum of event probabilities and expected consequent costs.

Definition 1. *We say that a bad event, X_T, happens to the tuple $T = (\ell_1^T, \ell_2^T, R_1^T, R_{1,2}^T, R_2^T)$ if during the execution of PivotBiCluster, ℓ_1^T was chosen to be a left center while ℓ_2^T was still in the graph, and at that moment, $R_1^T = N(\ell_1^T)\setminus N(\ell_2^T)$, $R_{1,2}^T = N(\ell_1^T)\cap N(\ell_2^T)$, and $R_2^T = N(\ell_2^T)\setminus N(\ell_1^T)$. (We refer by $N(\cdot)$ here to the neighborhood function in a particular moment of the algorithm execution.)*

If a bad event X_T happens to tuple T, we *color* the following pairs with color T: (1) $\{(\ell_2^T, r) \;:\; r \in R_1^T \cup R_{1,2}^T\}$, (2) $\{(\ell_2^T, r) \;:\; r \in R_2^T\}$. We color the latter pairs only if we decide to associate ℓ_2^T to ℓ_1^T's cluster, or if we decide to make ℓ_2^T a singleton during the ℓ_2-sub-phase corresponding to the ℓ_1-phase. Notice that these pairs are the remaining pairs (in the beginning of event X_T) from ℓ_2^T that after the ℓ_2^T-sub-phase will be removed from the graph. We also denote by $X_{e,T}$ the event that the edge e is colored with color T.

Lemma 1. *During the execution of PivotBiCluster each pair $(\ell, r) \in L \times R$ is colored at most once, and each violated pair is colored exactly once.*

Proof. For the first part, we show that pairs are colored at most once. A pair (ℓ, r) can only be colored during an ℓ_2-sub-phases with respect to some ℓ_1-phase, if $\ell = \ell_2$. Clearly, this will only happen in one ℓ_1-phase, as every time a pair is labeled either ℓ_2 or r are removed from the graph. Indeed, either $r \in R_1 \cup R_{1,2}$ in which case r is removed, or $r \in R_2$, but then ℓ is removed since it either joins the cluster created by ℓ_1 or becomes a singleton. For the second part, note that during each ℓ_1-phase the only pairs removed from the graph not colored are between left centers, ℓ_1, and right nodes in the graph at that time. All of these pairs are clearly not violated. $\square$

We denote by q_T the probability that event X_T occurs and by $\text{cost}(T)$ the number of violated pairs that are colored by X_T. From Lemma 1, we get:

Corollary 1. *Letting random variable* COST *denote* $\text{cost}(PivotBiCluster)$:

$$\mathbb{E}[\text{COST}] = \mathbb{E}\left[\sum_{e \in L \times R} x(e)\right] = \mathbb{E}\left[\sum_T \text{cost}(T)\right] = \sum_T q_T \cdot \mathbb{E}[\text{cost}(T)|X_T] .$$

3.4 Contradicting Structures

We now identify bad structures in the graph for which every output must incur some cost, and use them to construct an LP relaxation for our problem. In the case of BCC the minimal such structures are "bad squares": A set of four nodes, two on each side, between which there are only three edges. We make the trivial observation that any clustering B must make at least one mistake on any such bad square, s (we think of s as the set of 4 pairs connecting its two left nodes and two right nodes). Any clustering solution's violating pair set must hit these squares. Let S denote the set of all bad squares in the input graph G.

It will not be enough for our purposes to concentrate on squares in our analysis. Indeed, at an ℓ_2-sub-phase, decisions are made based on the intersection pattern of the current neighborhoods of ℓ_2 and ℓ_1 - a possibly unbounded structure. The *tuples* now come in handy.

Consider tuple $T = (\ell_1^T, \ell_2^T, R_1^T, R_{1,2}^T, R_2^T)$ for which $|R_{1,2}^T| > 0$ and $|R_2^T| > 0$. Notice that for every selection of $r_2 \in R_2^T$, and $r_{1,2} \in R_{1,2}^T$ the tuple contains the bad square induced by $\{\ell_1, r_2, \ell_2, r_{1,2}\}$. Note that there may also be bad squares $\{\ell_2, r_1, \ell_1, r_{1,2}\}$ for every $r_1 \in R_1^T$ and $r_{1,2} \in R_{1,2}^T$ but these will be associated to the *conjugate tuple* $\bar{T} = (\ell_2^T, \ell_1^T, R_2^T, R_{1,2}^T, R_1^T)$.

For each tuple we can write a corresponding linear constraint for the vector $\{x(e) : e \in L \times R\}$, indicating, as we explained above, the pairs the algorithm violates. A tuple constraint is the sum of the constraints of all bad squares it is associated with, where a constraint for square s is simply defined as $\sum_{e \in s} x(e) \geq 1$. The purpose of this constraint is to encode that we must violate at least one pair in a bad square. Since each tuple corresponds to $|R_2^T| \cdot |R_{1,2}^T|$ bad squares, we get the following constraint:

$$\forall\, T: \sum_{r_2 \in R_2^T, r_{1,2} \in R_{1,2}^T} \left(x_{\ell_1^T, r_2} + x_{\ell_1^T, r_{1,2}} + x_{\ell_2^T, r_2} + x_{\ell_2^T, r_{1,2}} \right) =$$

$$\sum_{r_2 \in R_2^T} |R_{1,2}^T| \cdot (x_{\ell_1^T, r_2} + x_{\ell_2^T, r_2}) + \sum_{r_{1,2} \in R_{1,2}^T} |R_2^T| \cdot (x_{\ell_1^T, r_{1,2}} + x_{\ell_2^T, r_{1,2}}) \geq |R_2^T| \cdot |R_{1,2}^T|$$

The following linear program hence provides a lower bound for the optimal solution: $LP = \min \sum_{e \in L \times R} x(e)$

$$\text{s.t. } \forall T \quad \frac{1}{|R_2^T|} \sum_{r_2 \in R_2^T} (x_{\ell_1^T, r_2} + x_{\ell_2^T, r_2}) + \frac{1}{|R_{1,2}^T|} \sum_{r_{1,2} \in R_{1,2}^T} (x_{\ell_1^T, r_{1,2}} + x_{\ell_2^T, r_{1,2}}) \geq 1$$

The dual program is as follows: $DP = max \sum_T \beta(T)$

s.t. $\forall (\ell, r) \in E :$

$$\sum_{T: \ell_2^T = \ell, r \in R_2^T} \frac{\beta(T)}{|R_2^T|} + \sum_{T: \ell_1^T = \ell, r \in R_{1,2}^T} \frac{\beta(T)}{|R_{1,2}^T|} + \sum_{T: \ell_2^T = \ell, r \in R_{1,2}^T} \frac{\beta(T)}{|R_{1,2}^T|} \le 1 \quad (1)$$

and $\forall (\ell, r) \notin E :$ $\sum_{T: \ell_1^T = \ell, r \in R_2^T} \frac{1}{|R_2^T|} \beta(T) \le 1$ (2)

3.5   Obtaining the Competitive Analysis

We now relate the expected cost of the algorithm on each tuple to a feasible solution to DP. We remind the reader that q_T denotes the probability of event X_T corresponding to tuple T.

Lemma 2. *The solution* $\beta(T) = \alpha_T \cdot q_T \cdot \min\{|R_{1,2}^T|, |R_2^T|\}$ *is a feasible solution to* DP, *when* $\alpha_T = \min\left\{1, \frac{|R_{1,2}^T|}{\min\{|R_{1,2}^T|, |R_1^T|\} + \min\{|R_{1,2}^T|, |R_2^T|\}}\right\}$.

Proof. First, notice that given a pair $e = (\ell, r) \in E$ each tuple T can appear in at most one of the sums in the LHS of the DP constraints (1) (as $R_1^T, R_{1,2}^T, R_2^T$ are disjoint). We distinguish between two cases.

1. Consider T appearing in the first sum of the LHS of (1), meaning that $\ell_2^T = \ell$ and $r \in R_2^T$. e is colored with color T if ℓ_2^T joined the cluster of ℓ_1^T or if ℓ_2 was turned into a singleton. Both cases happen, conditioned on X_T, with probability $\Pr[X_{e,T}|X_T] = \min\left\{\frac{|R_{1,2}^T|}{|R_2^T|}, 1\right\}$. Thus, we can bound the contribution of T to the sum as follows:

$$\frac{1}{|R_2^T|}\beta(T) = \frac{1}{|R_2^T|}\alpha_T \cdot q_T \cdot \min\{|R_{1,2}^T|, |R_2^T|\} \le q_T \cdot \min\left\{\frac{|R_{1,2}^T|}{|R_2^T|}, 1\right\}$$

$$= \Pr[X_T]\Pr[X_{e,T}|X_T] = \Pr[X_{e,T}].$$

   (The inequality is simply because $\alpha_T \le 1$.)
2. T contributes to the second or third sum in the LHS of (1). By definition of the conjugate $\bar{T}$, the following holds:

$$\sum_{T \text{ s.t } \ell_1^T = \ell, r \in R_{1,2}^T} \frac{\beta(T)}{|R_{1,2}^T|} + \sum_{T \text{ s.t } \ell_2^T = \ell, r \in R_{1,2}^T} \frac{\beta(T)}{|R_{1,2}^T|} = \sum_{T \text{ s.t } \ell_1^T = \ell, r \in R_{1,2}^T} \frac{(\beta(T) + \beta(\bar{T}))}{|R_{1,2}^T|}.$$

It is therefore sufficient to bound the contribution of each T to the RHS of the latter equality. We henceforth focus on tuples T for which $\ell = \ell_1^T$ and $r \in R_{1,2}^T$. Consider a moment in the algorithm execution in which both ℓ_1^T and ℓ_2^T were still present in the graph, $R_1^T = N(\ell_1^T) \setminus N(\ell_2^T)$, $R_{1,2}^T = N(\ell_1^T) \cap N(\ell_2^T)$, $R_2^T = N(\ell_2^T) \setminus N(\ell_1^T)$ and one of ℓ_1^T, ℓ_2^T was chosen to be a left center.[3] Either one of ℓ_1^T and ℓ_2^T had the same probability to be chosen. In other words, $\Pr[X_T | X_T \cup X_{\bar{T}}] = \Pr[X_{\bar{T}} | X_T \cup X_{\bar{T}}]$, and hence, $q_T = q_{\bar{T}}$. Further, notice that $e = (\ell, r)$ is never colored with color T, and if event $X_{\bar{T}}$ happens then e is colored with color $\bar{T}$ with probability 1. Therefore:

$$\frac{1}{|R_{1,2}^T|}\left(\beta(T) + \beta(\bar{T})\right) = \frac{1}{|R_{1,2}^T|} \cdot q_T \cdot \min\left\{1, \frac{|R_{1,2}^T|}{\min\{|R_{1,2}^T|,|R_1^T|\} + \min\{|R_{1,2}^T|,|R_2^T|\}}\right\} \cdot$$
$$\left(\min\{|R_{1,2}^T|,|R_2^T|\} + \min\{|R_{1,2}^{\bar{T}}|,|R_2^{\bar{T}}|\}\right)$$
$$\leq q_T = q_{\bar{T}} = \Pr[X_{\bar{T}}] = \Pr[X_{e,\bar{T}}] + \Pr[X_{e,T}].$$

Summing this all together, for every edge $e \in E$:

$$\sum_{T \ s.t \ \ell_2^T = \ell, r \in R_2^T} \frac{\beta(T)}{|R_2^T|} + \sum_{T \ s.t \ \ell_1^T = \ell, r \in R_{1,2}^T} \frac{\beta(T)}{|R_{1,2}^T|} + \sum_{T \ s.t \ \ell_2^T = \ell, r \in R_{1,2}^T} \frac{\beta(T)}{|R_{1,2}^T|} \leq \sum_T \Pr[X_{e,T}].$$

By the first part of Lemma 1 we know that $\sum_T \Pr[X_{e,T}]$ is exactly the probability of the edge e to be colored (the sum is over probabilities of disjoint events), therefore it is at most 1, as required to satisfy (1).

Now consider a pair $e = (\ell, r) \notin E$. A tuple T contributes to (2) if $\ell_1^T = \ell$ and $r \in R_2^T$. Since, as before, $q_T = q_{\bar{T}}$ and since $\Pr[X_{e,\bar{T}} | X_{\bar{T}}] = 1$ (this follows from the first coloring rule described in the beginning of Section 3.3) we obtain the following:

$$\sum_{T \ s.t \ \ell_1^T = \ell, r \in R_2^T} \frac{1}{|R_2^T|} \beta(T) = \sum_{T \ s.t \ \ell_1^T = \ell, r \in R_2^T} \frac{1}{|R_2^T|} \cdot \alpha_T \cdot q_T \cdot \min\{|R_{1,2}^T|,|R_2^T|\}$$
$$\leq \sum_{T \ s.t \ \ell_1^T = \ell, r \in R_2^T} q_T = \sum_{\bar{T} \ s.t \ \ell_2^{\bar{T}} = \ell, r \in R_1^{\bar{T}}} q_{\bar{T}} = \sum_{\bar{T} \ s.t \ \ell_2^{\bar{T}} = \ell, r \in R_1^{\bar{T}}} \Pr[X_{\bar{T}}]$$
$$= \sum_{\bar{T} \ s.t \ \ell_2^{\bar{T}} = \ell, r \in R_1^{\bar{T}}} \Pr[X_{e,\bar{T}}] = \sum_T \Pr[X_{e,T}].$$

From the same reason as before, this is at most 1, as required for (2).   □

After presenting the feasible solution to our dual program, we have left to prove that the expected cost of PivotBiCluster is at most 4 times the DP value of this solution. For this we need the following:

Lemma 3. *For any tuple T,*
$$q_T \cdot \mathbb{E}[\mathrm{cost}(T)|X_T] + q_{\bar{T}} \cdot \mathbb{E}[\mathrm{cost}(\bar{T})|X_{\bar{T}}] \leq 4 \cdot \left(\beta(T) + \beta(\bar{T})\right).$$

Proof. We consider three cases, according to the structure of T.
Case 1. $|R_1^T| \leq |R_{1,2}^T|$ and $|R_2^T| \leq |R_{1,2}^T|$ (equivalently $|R_1^{\bar{T}}| \leq |R_{1,2}^{\bar{T}}|$ and $|R_2^{\bar{T}}| \leq |R_{1,2}^{\bar{T}}|$): For this case, $\alpha_T = \alpha_{\bar{T}} = \min\left\{1, \frac{|R_{1,2}^T|}{|R_1^T| + |R_2^T|}\right\}$, and we get that

$$\beta(T) + \beta(\bar{T}) = \alpha_T \cdot q_T \cdot \left(\min\{|R_{1,2}^T|,|R_2^T|\} + \min\{|R_{1,2}^T|,|R_1^T|\}\right)$$
$$= q_T \cdot \min\{(|R_2^T| + |R_1^T|),|R_{1,2}^T|\} \geq \frac{1}{2} \cdot q_T \cdot (|R_2^T| + |R_1^T|).$$

---
[3] Recall that $N(\cdot)$ depends on the "current" state of the graph at that moment, after removing previously created clusters.

Since $|R_1^T| \leq |R_{1,2}^T|$, if event X_T happens PivotBiCluster adds ℓ_2^T to ℓ_1^T's cluster with probability $\min\left\{\frac{|R_{1,2}^T|}{|R_2^T|}, 1\right\} = 1$. Therefore the pairs colored with color T that PivotBiCluster violates are all the edges from ℓ_2^T to R_2^T and all the non-edges from ℓ_2^T to R_1^T, namely, $|R_2^T| + |R_1^T|$ edges. The same happens in the event $X_{\bar{T}}$ as the conditions on $|R_1^{\bar{T}}|$, $|R_{1,2}^{\bar{T}}|$, and $|R_2^{\bar{T}}|$ are the same, and since $|R_2^{\bar{T}}| + |R_1^{\bar{T}}| = |R_1^T| + |R_2^T|$. Thus,

$$q_T \cdot \left(\mathbb{E}[\text{cost}(T|X_T)] + \mathbb{E}[\text{cost}(\bar{T}|X_{\bar{T}})]\right) = q_T \left(2\left(|R_2^T| + |R_1^T|\right)\right) \leq 4 \cdot \left(\beta(T) + \beta(\bar{T})\right).$$

Case 2. $|R_1^T| < |R_{1,2}^T| < |R_2^T|$ (equivalently $|R_1^{\bar{T}}| > |R_{1,2}^{\bar{T}}| > |R_2^{\bar{T}}|$)[4]: The details for this case are omitted from this version due to lack of space.

Case 3. $|R_{1,2}^T| < |R_1^T|$ and $|R_{1,2}^T| < |R_2^T|$ (equivalently, $|R_{1,2}^{\bar{T}}| < |R_2^{\bar{T}}|$ and $|R_{1,2}^{\bar{T}}| < |R_1^{\bar{T}}|$): The details for this case are omitted from this version due to lack of space. $\square$

By Corollary 1: $\mathbb{E}[\text{cost}(PivotBiCluster)] = \sum_T \Pr[X_T] \cdot \mathbb{E}[\text{cost}(T)|X_T]$

$$= \frac{1}{2} \sum_T \left(\Pr[X_T] \cdot \mathbb{E}[\text{cost}(T)|X_T] + \Pr[X_{\bar{T}}] \cdot \mathbb{E}[\text{cost}(\bar{T})|X_{\bar{T}}]\right).$$

By Lemma 3 the above RHS is at most $2 \cdot \sum_T (\beta(T) + \beta(\bar{T})) = 4 \cdot \sum_T \beta(T)$. We conclude that $\mathbb{E}[\text{cost}(PivotBiCluster)] \leq 4 \cdot \sum_T \beta(T) \leq 4 \cdot OPT$. This proves our main Theorem 2.

4 Future Work

The main open problem is that of improving the factor 4 approximation ratio. We believe that it should be possible by using both symmetry and bipartiteness simultaneously. Indeed, our LP rounding algorithm in Section 2 is symmetric with respect to the left and right sides of the graph. However, in a sense, it "forgets" about bipartiteness altogether. On the other hand, our combinatorial algorithm in Section 3 uses bipartiteness in a very strong way but is asymmetric which is counterintuitive.

References

1. Guo, J., Hüffner, F., Komusiewicz, C., Zhang, Y.: Improved Algorithms for Bicluster Editing. In: Agrawal, M., Du, D.-Z., Duan, Z., Li, A. (eds.) TAMC 2008. LNCS, vol. 4978, pp. 445–456. Springer, Heidelberg (2008)
2. Symeonidis, P., Nanopoulos, A., Papadopoulos, A., Manolopoulos, Y.: Nearest-biclusters collaborative filtering (2006)

---

[4] From symmetry reasons (between T and $\bar{T}$) here we also deal with the case $|R_2^T| < |R_{1,2}^T| < |R_1^T|$.

3. Amit, N.: The bicluster graph editing problem (2004)
4. Madeira, S.C., Oliveira, A.L.: Biclustering algorithms for biological data analysis: A survey. IEEE/ACM Trans. Comput. Biol. Bioinformatics 1, 24–45 (2004)
5. Cheng, Y., Church, G.M.: Biclustering of expression data. In: Proceedings of the Eighth International Conference on Intelligent Systems for Molecular Biology, pp. 93–103. AAAI Press, Menlo Park (2000)
6. Bansal, N., Blum, A., Chawla, S.: Correlation clustering. Machine Learning 56, 89–113 (2004)
7. Fern, X.Z., Brodley, C.E.: Solving cluster ensemble problems by bipartite graph partitioning. In: Proceedings of the Twenty-First International Conference on Machine Learning, ICML 2004, p. 36. ACM, New York (2004)
8. Zha, H., He, X., Ding, C., Simon, H., Gu, M.: Bipartite graph partitioning and data clustering. In: Proceedings of the Tenth International Conference on Information and Knowledge Management, CIKM 2001, pp. 25–32. ACM, New York (2001)
9. Demaine, E.D., Emanuel, D., Fiat, A., Immorlica, N.: Correlation clustering in general weighted graphs. Theoretical Computer Science (2006)
10. Charikar, M., Guruswami, V., Wirth, A.: Clustering with qualitative information. J. Comput. Syst. Sci. 71(3), 360–383 (2005)
11. Ailon, N., Charikar, M., Newman, A.: Aggregating inconsistent information: Ranking and clustering. J. ACM 55(5), 1–27 (2008)
12. Ailon, N., Liberty, E.: Correlation Clustering Revisited: The True Cost of Error Minimization Problems. In: Albers, S., Marchetti-Spaccamela, A., Matias, Y., Nikoletseas, S., Thomas, W. (eds.) ICALP 2009. LNCS, vol. 5555, pp. 24–36. Springer, Heidelberg (2009)
13. van Zuylen, A., Williamson, D.P.: Deterministic pivoting algorithms for constrained ranking and clustering problems. Math. Oper. Res. 34(3), 594–620 (2009); Preliminary version appeared in SODA 2007 (with Rajneesh Hegde and Kamal Jain)
14. Giotis, I., Guruswami, V.: Correlation clustering with a fixed number of clusters. In: Proceedings of the Seventeenth Annual ACM-SIAM Symposium on Discrete Algorithms (SODA), pp. 1167–1176. ACM, New York (2006)
15. Karpinski, M., Schudy, W.: Linear time approximation schemes for the gale-berlekamp game and related minimization problems. CoRR, abs/0811.3244 (2008)

Bounds on Greedy Algorithms for MAX SAT*

Matthias Poloczek

Institute of Computer Science, University of Frankfurt, Frankfurt, Germany
matthias@thi.cs.uni-frankfurt.de

Abstract. We study adaptive priority algorithms for MAX SAT and show that no such deterministic algorithm can reach approximation ratio $\frac{3}{4}$, assuming an appropriate model of data items. As a consequence we obtain that the Slack–Algorithm of [13] cannot be derandomized. Moreover, we present a significantly simpler version of the Slack–Algorithm and also simplify its analysis. Additionally, we show that the algorithm achieves a ratio of $\frac{3}{4}$ even if we compare its score with the optimal *fractional* score.

1 Introduction

In the maximum satisfiability problem (MAX SAT) we are given a collection of clauses and their (nonnegative) weights. Our goal is to find an assignment that satisfies clauses of maximum total weight.

The currently best approximation ratio of 0.797 for MAX SAT is achieved by an algorithm due to Avidor, Berkovitch and Zwick [3] that uses semidefinite programming. Moreover, they give an additional algorithm with a conjectured performance of 0.843 (see their paper for further details).

The first greedy algorithm for MAX SAT is due to Johnson [11] and is usually referred to as *Johnson's algorithm*. Chen, Friesen and Zheng [7] showed that Johnson's algorithm guarantees a $\frac{2}{3}$ approximation for any variable ordering and asked whether a random variable ordering gives a better approximation. Recently, Costello, Shapira, and Tetali [8] gave a positive answer by proving that the approximation ratio is actually improved to $\frac{2}{3} + c$ for some constant $c > 0$. However, Poloczek and Schnitger [13] showed the approximation ratio can be as bad as $2\sqrt{15} - 7 \approx 0.746 < \frac{3}{4}$. Instead they proposed the Slack–Algorithm [13], a greedy algorithm that achieves an expected approximation ratio of $\frac{3}{4}$ without sophisticated techniques such as linear programming. This result even holds if variables are given in a worst case fashion, i.e. the Slack–Algorithm can be applied in an online scenario.

To study the limits of greedy algorithms for MAX SAT, we employ the concept of priority algorithms, a model for *greedy–like* algorithms introduced in a seminal paper by Borodin, Nielsen and Rackoff [6]. The model was extended to algorithms for graph problems in [9,5]. Angelopoulos and Borodin [2] studied randomized priority algorithms for facility location and makespan scheduling.

---

* Partially supported by DFG SCHN 503/5-1.

C. Demetrescu and M.M. Halldórsson (Eds.): ESA 2011, LNCS 6942, pp. 37–48, 2011.

MAX SAT was first examined by Alekhnovich et al. [1] in the more powerful model of priority branching–trees. Independently of our work, Yung [16] gave a $\frac{5}{6}$ inapproximability result for adaptive priority algorithms.

1.1   Priority Algorithms for MAX SAT

The framework of priority algorithms covers a broad variety of greedy–like algorithms. The concept crucially relies on the notion of *data items* from which input instances are built. Our data items contain the name of a variable, say x, to be processed and a list of clauses x appears in. For each clause c the following information is revealed:

- the sign of variable x in c,
- the weight of c,
- a list of the names of still unfixed variables appearing in c. No sign is revealed.

(Identical clauses are merged by adding their weights.)

Thus, we restrict access to information that, at least at first sight, is helpful only in revealing global structure of the formula.

A *fixed* priority algorithm specifies an ordering of all possible data items in advance and receives in each step the currently smallest data item of the actual input. In our situation, upon receiving the data item of variable x the algorithm is forced to fix x irrevocably. An *adaptive* priority algorithm may submit a new ordering in each step and otherwise behaves identically.

The variable order is an influential parameter for greedy algorithms. A randomized ordering was shown to improve the approximation ratio of Johnson's algorithm by some additive constant [8]. Moreover, for every 2CNF formulae there exists a variable ordering such that Johnson's algorithm determines an optimal solution [13] – and in fact our lower bounds utilize clauses of length at most two. An adaptive priority algorithm may reorder data items at will, using all the knowledge about existing and non–existing data items collected so far.

Observe that priority algorithms are not resource bounded. Hence any lower bound utilizes "the nature and the inherent limitations of the algorithmic paradigm, rather than resource constraints" [2].

Johnson's algorithm as well as the Slack–Algorithm can be implemented with our data type, even assuming a worst case ordering. However, no deterministic greedy algorithm with approximation ratio $\frac{3}{4}$ is known. For this data type, is randomization more powerful than determinism? We show in Sect. 2:

Theorem 1. *For any $\varepsilon > 0$, no* adaptive *algorithm can approximate MAX SAT within $\frac{\sqrt{33}+3}{12} + \varepsilon$ (using our data type). Thus, the Slack–Algorithm cannot be derandomized.*

We compare our work to Yung [16]: Utilizing a stronger data type that reveals all signs of neighboring variables, Yung considers the version of MAX SAT where each clause contains exactly two literals and shows that no adaptive algorithm can obtain an approximation ratio better than $\frac{5}{6}$. Note that Johnson's algorithm guarantees a $\frac{3}{4}$ approximation in this case.

The result of Alekhnovich et al. [1] (for priority branching trees, using the stronger data type) implies a lower bound of $\frac{21}{22} + \varepsilon$ for *fixed* priority algorithms that must stick with an a priori ordering of potential data items.

1.2 The Slack–Algorithm

The Slack–Algorithm works in an online scenario where variables are revealed one by one together with the clauses they appear in. The algorithm has to decide on the spot: assume that the variable x is to be fixed. Let fanin (fanout) be the weight of all clauses of length at least two that contain the literal x (resp. $\overline{x}$). For the weights $w_x, w_{\overline{x}}$ of the unit clauses $x, \overline{x}$ and $\Delta = \text{fanin} + 2w_x + \text{fanout} + 2w_{\overline{x}}$ we define

$$q_0 := \text{prob}[x = 0] = \frac{2w_{\overline{x}} + \text{fanout}}{\Delta} \quad \text{and} \quad q_1 := \text{prob}[x = 1] = \frac{2w_x + \text{fanin}}{\Delta}$$

as the canonical assignment probabilities. (Notice that we double the weight of unit clauses but keep the weight of longer clauses unchanged.) However these probabilities do not yield an expected $\frac{3}{4}$ approximation if the slack of the decision is small compared to the weights of the unit clauses [13]. In this case the Slack–Algorithm modifies the assignment probabilities suitably.

Since carefully adjusted assignment probabilities are essential for this algorithm, is it possible that the approximation ratio can be improved by some elaborate fine-tuning of its probabilities?

Theorem 2. *For any $\varepsilon > 0$, no randomized priority algorithm achieves a $\frac{3}{4} + \varepsilon$ approximation for Online MAX SAT whp (using our data type). Hence, the Slack–Algorithm is an optimal online algorithm.*

In this proceedings Azar, Gamzu and Roth [4] present a randomized $\frac{2}{3}$ approximation algorithm for submodular MAX SAT that also works in an online fashion. They prove their algorithm optimal by showing that for Online MAX SAT no algorithm can achieve a $\frac{2}{3} + \varepsilon$ approximation for any $\varepsilon > 0$, using a different data type. Note that this result does not contradict Theorem 2, since their data items do not reveal the length of a clause and thus the Slack–Algorithm cannot be applied.

In Sect. 3 we present a significantly simpler version of the Slack–Algorithm and also simplify its analysis. A preview of the refined algorithm is given below. Moreover, we show that the Slack–Algorithm satisfies clauses with an expected weight of at least $\frac{3}{4}$ of the score of an optimal *fractional* assignment. The score of the fractional optimum can be larger by a factor of $\frac{4}{3}$ than the integral optimum – as for the clauses $(x \vee y), (\overline{x} \vee y), (x \vee \overline{y}), (\overline{x} \vee \overline{y})$ – and this bound is tight [10].

Independently of our work, van Zuylen [14] gave an alternative randomized approximation algorithm that also achieves $\frac{3}{4}$ of the objective value of the LP relaxation. The new algorithm can be implemented using our data type and processes variables in an arbitrary order, hence the optimality of its assignment probabilities is implied by Theorem 2. Moreover, van Zuylen proposed a novel deterministic rounding scheme for the LP relaxation that gives a performance guarantee of $\frac{3}{4}$.

---

procedure SLACK–ALGORITHM(Formula ϕ) ▷ the formula ϕ has variable set V
    for all $x \in V$ **do**   ▷ process variables in the given order
        $q_0 \leftarrow \frac{2w_{\overline{x}}+\text{fanout}}{\Delta}$, $q_1 \leftarrow \frac{2w_x+\text{fanin}}{\Delta}$
        if $q_0 \leq q_1$ **then**
            $\varepsilon^* \leftarrow$ the smallest $\varepsilon \geq 0$ with
            $(q_1 - q_0) \cdot \Delta \left((q_1 - q_0) + 2\varepsilon\right) - (q_1 - q_0)(w_x + w_{\overline{x}}) + \varepsilon(\text{fanin} + \text{fanout}) \geq 0$
            $p_0 \leftarrow q_0 - \varepsilon^*$, $p_1 \leftarrow q_1 + \varepsilon^*$
        else
            $\varepsilon^* \leftarrow$ the smallest $\varepsilon \geq 0$ with
            $(q_0 - q_1) \cdot \Delta \left((q_0 - q_1) + 2\varepsilon\right) - (q_0 - q_1)(w_x + w_{\overline{x}}) + \varepsilon(\text{fanin} + \text{fanout}) \geq 0$
            $p_0 \leftarrow q_0 + \varepsilon^*$, $p_1 \leftarrow q_1 - \varepsilon^*$
        assign $x = 0$ with probability p_0 and $x = 1$ with probability p_1

Fig. 1. The Slack–Algorithm

In the full version of the paper [12] we also give an improved bound on Johnson's algorithm that relates to the optimal fractional score and thereby demonstrate that our technique is also useful for the analysis of other greedy algorithms.

2   Limits of Priority Algorithms for MAX SAT

In the next section we investigate the optimality of the Slack Algorithm for Online MAX SAT, before we give a bound on adaptive priority algorithms in Sect. 2.2.

2.1   Optimal Bounds for Online MAX SAT

In Online MAX SAT the adversary reveals data items successively. The algorithm must fix the corresponding variable instantly and irrevocably. We do not impose restrictions on the decision process: the algorithm may use all information deducible from previous data items and may decide probabilistically.

In [13] the Slack–Algorithm was shown $\frac{3}{4}$ approximative under these conditions. Can one come up with a better algorithm?

Proof of Theorem 2. Let X_b and Y_b for $b \in \{0, 1\}$ denote sets of n variables each. The clause set is constructed as follows: For each variable pair in $X_b \times Y_b$ there is an equivalence and for each pair in $X_b \times Y_{1-b}$ an inequivalence. It is crucial that the algorithm cannot distinguish both constraint types.

The bound follows by an application of Yao's principle [15]. Since we are dealing with an online setting, the adversary selects a *random* permutation π on the data items of $X_0 \cup X_1$. When given the data items according to π, the (deterministic) algorithm cannot decide whether the current variable belongs to X_0 or X_1, since the sets of clauses look identical for all variables in $X_0 \cup X_1$ according to our type of data items. Hence the algorithm ends up with at least $2n^2 - o(n^2)$ violated (in-)equivalences whp. The bound on the approximation ratio follows, since $2n^2 - o(n^2)$ clauses are not satisfied out of $8n^2$ obtained by the optimum. □

Remark. We point out that all variables in $X_0 \cup X_1$ must be fixed before the first y–variable is decided. If the algorithm was allowed to fix a single y–variable in advance, it could easily come up with an optimal assignment.

Note that randomized online algorithms are strictly more powerful than deterministic ones: assume that the adversary presents the clauses $(\overline{x} \vee y), (x \vee \overline{y})$ for variable x. If the algorithm fixes $x = 1$, the formula is completed by adding unit clause $(\overline{y})$ and (y) otherwise. Thus, no deterministic online algorithm is better than $\frac{2}{3}$ approximative – and Johnson's algorithm is optimal in its class. Observe that the (randomized) Slack–Algorithm achieves approximation ratio $\frac{3}{4}$.

2.2 Adaptive Priority Algorithms for MAX SAT

In this section we study adaptive priority algorithms that decide deterministically. In particular, an adaptive priority algorithm may choose the order in which variables are to be processed by utilizing the previously obtained information on the structure of the formula.

In the lower bound we study the Adaptive Priority Game which is played between an adversary and the (adaptive priority) algorithm.

1. The game starts with the adversary announcing the number n of variables as well as their names.
2. In each round the algorithm submits an ordering on the data items and the adversary picks a smallest data item d according to this ordering. In particular, no unseen data item less than d appears in the input.

Using our data type, we show that the adversary is able to force any algorithm into an approximation ratio of at most $\frac{\sqrt{33}+3}{12} + o(1) \approx 0.729 < \frac{3}{4}$.

Outline of the Proof of Theorem 1. The adversary constructs a 2CNF formula ϕ with n variables. At any time there is an equivalence or an inequivalence of weight 1 for every pair of still unfixed variables. The adversary utilizes that our data items do not allow the algorithm to distinguish equivalences from inequivalences. Thus, data items differ only in the name of the respective variable and in the weights of their two unit clauses; the initial weights of unit clauses will always belong to the set $\{0, 1, 2, \ldots, G\}$.

The game is divided into two phases. Phase (I) lasts for $G < n$ steps, where G is determined later. We set a trap for the algorithm by introducing contradictions for the variables of phase (II) in each round.

In round $k \in \{1, 2, \ldots, G\}$ the adversary allows data items for the remaining variables only if the weights of both unit clause are at least $k-1$ and at most G. In particular, the combinations $(G, 0), (0, G)$ are feasible at the beginning of round 1. A fact that will be used for the variables of phase (II).

Phase (II) consists of $(n - G)$ rounds. Here we make sure that the algorithm does not recover. At the beginning of phase (II), in round $G + 1$, the adversary allows only variables with weight combination (G, G) for unit clauses. (It turns out that the algorithm can be forced to "increase" the initial weight combination $(G, 0)$ or $(0, G)$ to (G, G).) From now on, the adversary observes two consecutive steps of the algorithm and then adapts the formula ϕ appropriately. W.l.o.g.

assume that $(n - G)$ is even. In round $G + 2m$ the combinations $(G+m, G+m)$ and in round $G+2m+1$ the combinations $(G+m+1, G+m), (G+m, G+m+1)$ are allowed, where $m \in \{0, 1, \ldots, \frac{n-G}{2} - 1\}$ holds. Observe that, if the adversary is actually able to achieve this goal, the algorithm has absolutely no advantage whatsoever in determining the "optimal" value of the current variable.

Without knowing ϕ exactly, we can give an upper bound on the total weight satisfied by the algorithm. The algorithm may always satisfy the unit clause of larger weight and exactly one clause of each equivalence, resp. inequivalence. Thus, the weight of satisfied clauses of length two equals

$$\sum_{k=1}^{n} n - k = \sum_{k=1}^{n-1} k = \frac{n^2 - n}{2}. \tag{1}$$

The second clause of an (in-)equivalence loses one literal and hence appears again as a unit clause. It will be counted only if the algorithm satisfies the unit clause at some later time.

The weight of unit clauses satisfied in phase (I) is at most G^2, since G is the maximal weight for a unit clause. In phase (II) units of weight at most

$$\sum_{m=0}^{\frac{n-G}{2}-1} (G+m) + (G+m+1) = (2G+1) \cdot \frac{n-G}{2} + \frac{2}{2} \cdot \frac{n-G}{2} \cdot \left(\frac{n-G}{2} - 1 \right)$$

$$= \frac{n^2}{4} + \frac{1}{2}nG - \frac{3}{4}G^2$$

are satisfied. Thus, the total weight of satisfied clauses is at most

$$\text{Sat} = \frac{n^2 - n}{2} + G^2 + \frac{n^2}{4} + \frac{1}{2}nG - \frac{3}{4}G^2$$

$$= \frac{3}{4}n^2 + \frac{1}{2}nG + \frac{1}{4}G^2 - \frac{n}{2}. \tag{2}$$

We have to show that the adversary is able to construct an almost satisfiable formula.

Proof. We begin by describing the adversary in phase (I). Assume that the algorithm has already fixed variables in Z_b to b and that it processes variable x in round k. If it sets x to 0, the adversary introduces equivalences between x and all variables in Z_0 and inequivalences between x and all variables in Z_1. These insertions are consistent with the data items presented in the previous rounds, since the algorithm cannot distinguish between equivalences and inequivalences. The insertions are also consistent with the current data item of x, where falsified (in-)equivalences appear again as unit clauses. Thus, the weight of unit clause (x) is unchanged, whereas the weight of $(\overline{x})$ increases by $k - 1$. The adversary may drop any data item. Therefore, it can enforce that the weight of (x) is at least $k - 1$, but at most G, and additionally guarantee that the weight of $(\overline{x})$ is bounded by G. If x is set to 1, the argumentation is analogous.

The assignment of the adversary is obtained by flipping the assignment of the algorithm. As a consequence, the adversary also satisfies all (in-)equivalences between variables fixed in phase (I).

Now we deal with the variables of phase (II). How does the adversary justify that the weights of unit clauses at the beginning of phase (II), i.e. in round $G + 1$, always equal G? The adversary now drops any data item with an initial weight combination different from $(G, 0), (0, G)$. For any variable with weight combination $(G, 0)$ for (x) and $(\overline{x})$ resp. the adversary inserts equivalences with all variables in Z_0 and inequivalences with all variables in Z_1. Observe that the adversary never drops such a variable in phase (I), since its weight combination equals $(G, k - 1)$ in round k. In particular, the final weight after phase (I) is G for both unit clauses. As before, the case of weights 0 for (x) and G for $(\overline{x})$ is treated analogously.

During phase (II) the adversary observes two consecutive assignments x, y made by the algorithm. Let us assume inductively, that the adversary only allows variables with weight combination $(G + m, G + m)$ at the beginning of round $G + 2m$. Assume w.l.o.g. that the algorithm fixes variable x to 1.

Case 1: Variable y has weight combination $(G+m+1, G+m)$ and the algorithm fixes variable y to 0. The adversary sets both variables to 1.

Observe that y has unit weights $(G+m, G+m)$ at the beginning of round $G + 2m$ and the weight of unit clause (y) increases by 1 according to the case assumption. The adversary inserts an equivalence between x and y. Hence, y indeed has weight combination $(G + m + 1, G + m)$ at the beginning of round $G + 2m + 1$. Moreover, the adversary inserts equivalences between x, y and all variables of phase (II) that it set to 1 and inequivalences between x, y and all variables of phase (II) that it set to 0.

Contrary to the adversary the algorithm treated x and y differently. Any untouched variable z is later connected with x and y via equivalences only or inequivalences only. Hence both unit weights of z increase by 1 and we can conclude the inductive argument.

Case 2: Variable y has weight combination $(G+m+1, G+m)$ and the algorithm fixes variable y to 1. The adversary sets x to 1 and y to 0.

Observe that y has unit weights $(G+m, G+m)$ at the beginning of round $G + 2m$ and the weight of unit clause (y) increases by 1 according to the case assumption. Again the adversary inserts an equivalence between x and y, but in this case satisfies only one out of two clauses. Hence, y indeed has weight combination $(G + m + 1, G + m)$ at the beginning of round $G + 2m + 1$. This time, the adversary inserts equivalences between $x, \overline{y}$ and all variables of phase (II) that it set to 1 and inequivalences between $x, \overline{y}$ and all variables of phase (II) that it set to 0.

Contrary to the adversary the algorithm treated x and y identically. Any untouched variable z is later connected with x and $\overline{y}$ via equivalences only or inequivalences only. Hence both unit weights of z increase by 1. And we can conclude the inductive argument in this case as well.

The cases that variable y has weight combination $(G + m, G + m + 1)$ in round $G + 2m + 1$ is treated analogously.

What is the total weight satisfied by the assignment of the adversary? Only at most $n - G$ (in-)equivalences of phase (II) are falsified and hence the adversary satisfies at least $\binom{n}{2} - (n - G)$ (in-)equivalences, i.e. at least $n \cdot (n - 1) - (n - G)$ 2-clauses of weight 1.

For every variable of phase (II) the unit clause of initial weight G is satisfied: the adversary satisfies units of phase (II) variables with a combined weight of $(n - G) \cdot G$.

Now consider the variable fixed in round k of phase (I). Remember that for a variable x set to 0 by the algorithm, the adversary inserts equivalences between x and all variables in Z_0 as well as inequivalences between x and all variables in Z_1. Thus, only the weight of $(\bar{x})$ has been increased by the assignments the algorithm made so far in phase (I). Moreover, remember that the adversary drops all variables with a unit weight smaller than $k - 1$ at the beginning of round k, thus the unit clause (x) has initial weight at least $k - 1$. We obtained the assignment of the adversary by flipping the assignment of the algorithm and hence the adversary satisfies (x). The case that the algorithm sets variable x to 1 is treated analogously. Thus, the weight of satisfied unit clauses for phase (I) variables is at least $\sum_{k=1}^{G} k - 1 = \sum_{k=1}^{G-1} k = \frac{G \cdot (G-1)}{2}$.

The assignment presented by the adversary satisfies clauses of total weight at least

$$n \cdot (n - 1) - (n - G) + (n - G) \cdot G + \frac{G \cdot (G - 1)}{2}$$
$$= n^2 + nG - \frac{1}{2}G^2 - 2n + \frac{1}{2}G. \tag{3}$$

Combining (2) and (3) and then choosing $G := \alpha \cdot n$ yields

$$\frac{\text{Sat}}{\text{Opt}} \leq \frac{\frac{3}{4}n^2 + \frac{1}{2}nG + \frac{1}{4}G^2 - \frac{n}{2}}{n^2 + nG - \frac{1}{2}G^2 - 2n + \frac{1}{2}G} = \frac{\frac{3}{4} + \frac{1}{2}\alpha + \frac{1}{4}\alpha^2 - o(1)}{1 + \alpha - \frac{1}{2}\alpha^2 - o(1)}$$

as an upper bound on the approximation ratio. A simple calculation shows that the ratio is minimized for $\alpha = \frac{\sqrt{33}-5}{4}$ and that the approximation ratio converges to at most $\frac{\sqrt{33}+3}{12}$. □

3   A Refined Analysis of the Slack–Algorithm

In this section we present a refinement of the Slack–Algorithm [13]. In what follows we assume that a CNF formula ϕ as well as weights w_k for clauses $k \in \phi$ are given. Let $L_+(k)$ be the set of positive and $L_-(k)$ be the set of negative literals of clause k. If π is a *fractional* assignment, then

$$\pi(k) = \sum_{l \in L_+(k)} \pi_l + \sum_{l \in L_-(k)} (1 - \pi_{\bar{l}})$$

is the score of π on k and $\sum_{k \in \phi} w_k \cdot \min\{1, \pi(k)\}$ is the score of π on formula ϕ. We call a fractional assignment optimal for ϕ iff its score is optimal among fractional assignments and denote its score by Opt.

Theorem 3. *Let W be the combined weight of all clauses with optimal (fractional) score Opt. Then*

$$\mathbb{E}[\mathrm{Sat}] \geq \frac{2\mathrm{Opt} + W}{4} \geq \frac{3}{4}\mathrm{Opt},$$

where $\mathbb{E}[\mathrm{Sat}]$ is the expected clause weight satisfied by the Slack–Algorithm.

Since Opt is at least as large as the score of an integral assignment, the approximation ratio of the Slack–Algorithm is at least $\frac{3}{4}$. Also, better approximation ratios are achievable for a highly contradictory formula, i.e., if Opt is small in comparison to the total weight W.

The algorithm processes the variables in arbitrary order and decides instantly. Remember that if the variable x is to be decided, we denote by fanin (fanout) the weight of all clauses of length at least two that contain the literal x (resp. $\overline{x}$). For the weights $w_x, w_{\overline{x}}$ of the unit clauses $x, \overline{x}$ and $\Delta = \mathrm{fanin} + 2w_x + \mathrm{fanout} + 2w_{\overline{x}}$ we define

$$q_0 := \mathrm{prob}[x = 0] = \frac{2w_{\overline{x}} + \mathrm{fanout}}{\Delta} \quad \text{and} \quad q_1 := \mathrm{prob}[x = 1] = \frac{2w_x + \mathrm{fanin}}{\Delta}$$

as the canonical assignment probabilities.

An Overview of the Analysis. We aim for an approximation ratio of $\frac{2\mathrm{Opt}+W}{4}$, where Opt is the weight of all clauses satisfied by some fixed fractional optimal assignment π and W is the total clause weight. Goemans and Williamson [10] showed that their LP based algorithm satisfies each clause k with probability at least $\frac{3}{4}\pi(k)$. The Slack–Algorithm, however, cannot give such a guarantee for individual clauses. For an example consider the following formula $(x \lor y), (\overline{x})$ where the first clause has a weight of w and the unit clause of 1. The Slack–Algorithm satisfies the unit clause with probability $\frac{2}{2+w}$ only, whereas its fractional score is 1. Our analysis follows the different approach of [13].

Let "Sat" and "Unsat" be the random variables indicating the total weight of satisfied and falsified clauses after fixing *all* variables. Then the algorithm achieves the desired performance if $\mathbb{E}[\mathrm{Sat}] \geq \frac{2\mathrm{Opt}+W}{4}$ or equivalently if $\mathbb{E}[\mathrm{Sat}] + 3\mathbb{E}[\mathrm{Sat}] - 2\mathrm{Opt} - W \geq 0$. But

$$\mathbb{E}[\mathrm{Sat}] + 3\mathbb{E}[\mathrm{Sat}] - 2\mathrm{Opt} - W = \mathbb{E}[\mathrm{Sat}] - 3 \cdot (W - \mathbb{E}[\mathrm{Sat}]) - 2 \cdot (\mathrm{Opt} - W)$$
$$= \mathbb{E}[\mathrm{Sat}] - 3\mathbb{E}[\mathrm{Unsat}] - 2 \cdot (\mathrm{Opt} - W) \qquad (4)$$

holds and we have to guarantee that the right hand side is nonnegative. We call $W - \mathrm{Opt}$ the *initial contradiction*; observe that $W - \mathrm{Opt}$ coincides with the weight of all clauses multiplied with the fraction to what extend a clause is *falsified* by the optimal fractional assignment π.

We perform a step–by–step analysis of the algorithm. Imagine the algorithm processing a variable x. After assigning a binary value to x, some contradictions are resolved by terminally falsifying or satisfying clauses. However new contradictions may be introduced. In particular, let the random variable c be the partially contradictory weight of all clauses *before* fixing x and c' be the same quantity *after* fixing x.

We show that the right hand side of (4) is nonnegative by establishing an invariant which holds whenever fixing a variable. In particular, if "sat" and "unsat" are the random variables expressing the total weight of clauses satisfied, resp. falsified when fixing x, then we choose assignment probabilities such that the invariant

$$\mathbb{E}[\text{sat} - 3 \cdot \text{unsat} - 2(c' - c)] \geq 0. \tag{5}$$

holds. Where is the problem in determining appropriate assignment probabilities $p_0 = \text{prob}[x = 0]$, $p_1 = \text{prob}[x = 1]$ such that (5) holds? We have no problem computing $\mathbb{E}[\text{sat}]$ and $\mathbb{E}[\text{unsat}]$, since

$$\mathbb{E}[\text{sat}] = p_0(w_{\overline{x}} + \text{fanout}) + p_1(w_x + \text{fanin}), \quad \mathbb{E}[\text{unsat}] = p_0 w_x + p_1 w_{\overline{x}} \tag{6}$$

holds. However computing $\mathbb{E}[c' - c]$ will in general be infeasible. But we may bound the unknown quantity $\mathbb{E}[c' - c]$, assuming that we know π_x, the fractional value of x in the optimum π.

Lemma 1. *Assume that the assignment probabilities are given by $p_0 = \text{prob}[x = 0]$, $p_1 = \text{prob}[x = 1]$ and that $\pi_x \in [0; 1]$ is the optimal fractional assignment of variable x. The increase in contradiction is bounded as follows:*

$$\mathbb{E}[c' - c] \leq \pi_x \cdot (p_0 \text{fanin} - w_{\overline{x}}) + (1 - \pi_x)(p_1 \text{fanout} - w_x).$$

Proof. Let k be a clause containing the literal x, so k is a clause contributing to fanin. If k is satisfied by fixing x, which happens with probability p_1, then no new contradictions are introduced.

In case k is (temporarily) falsified, however, the contradiction is increased by at most $\pi_x \cdot w_k$, where w_k is the weight of clause k. Hence, the increase related to fanin is at most $p_0 \cdot \pi_x \cdot \text{fanin}$. The expected gain in contradiction related to fanout follows analogously.

Before x is fixed, the unit clauses $x, \overline{x}$ contribute $(1 - \pi_x) \cdot w_x$ and $\pi_x \cdot w_{\overline{x}}$ to all contradictions. Irrespective of the particular assignment, both unit clauses are removed from the set of clauses, and hence these contradictions are resolved. □

How to Adjust the Assignment Probabilities? W.l.o.g. we assume $q_0 \leq q_1$ in the following and hence Johnson's algorithm would assign $x = 1$. The Slack–Algorithm increases the majority probability slightly, i.e., replaces q_1 by $p_1 = q_1 + \varepsilon$ and q_0 by $p_0 = q_0 - \varepsilon$. How can the algorithm determine ε? We evaluate invariant (5) after we replaced $\mathbb{E}[c - c']$ by the bound supplied in Lemma 1.

Lemma 2. *Assume that $q_0 \leq q_1$. For assignment probabilities $p_0 = q_0 - \varepsilon$, $p_1 = q_1 + \varepsilon$ we get*

$$\mathbb{E}[\,\mathrm{sat} - 3 \cdot \mathrm{unsat} - 2(c' - c)]$$
$$\geq (q_1 - q_0) \cdot ((q_1 - q_0) \cdot \Delta + (1 - 2\pi_x)(w_x + w_{\bar{x}}))$$
$$+ \varepsilon \cdot (2(q_1 - q_0) \cdot \Delta - (1 - 2\pi_x)(\mathrm{fanin} + \mathrm{fanout})). \qquad (7)$$

We give the proof in [12]. The algorithm has to find an ε^* for which (7) is nonnegative, regardless of the value π_x. As we show in the proof of the following lemma, the algorithm may set $\pi_x = 1$ (resp. $\pi_x = 0$ for $q_0 > q_1$) and compute the smallest ε for which the right side of (7) is nonnegative. Observe that the probabilities are adjusted only if the slack of the decision is small compared to the weights of the unit clauses, i.e. $(q_1 - q_0) \cdot \Delta < w_x + w_{\bar{x}}$.

Lemma 3. *Let ε^* denote the adjustment determined by the algorithm.*
(a) Then the right side of (7) is nonnegative regardless of the value of π_x and hence local invariant (5) holds in each step.
(b) The assignment probabilities are well defined, i.e. $0 \leq \varepsilon^* \leq q_0$.

The proof is given in [12]. This concludes the proof of Theorem 3.

Acknowledgement. The author would like to thank Georg Schnitger for his helpful comments.

References

1. Alekhnovich, M., Borodin, A., Buresh-Oppenheim, J., Impagliazzo, R., Magen, A.: Toward a model for backtracking and dynamic programming. Electronic Colloquium on Computational Complexity (ECCC) 16, 38 (2009)
2. Angelopoulos, S., Borodin, A.: Randomized priority algorithms. Theor. Comput. Sci. 411(26-28), 2542–2558 (2010)
3. Avidor, A., Berkovitch, I., Zwick, U.: Improved approximation algorithms for MAX NAE-SAT and MAX SAT. In: Erlebach, T., Persinao, G. (eds.) WAOA 2005. LNCS, vol. 3879, pp. 27–40. Springer, Heidelberg (2006)
4. Azar, Y., Gamzu, I., Roth, R.: Submodular Max-SAT. In: Demetrescu, C., Halldórsson, M.M. (Eds.): ESA 2011. LNCS, vol. 6942, pp. 323–334. Springer, Heidelberg (2011)
5. Borodin, A., Boyar, J., Larsen, K.S., Mirmohammadi, N.: Priority algorithms for graph optimization problems. Theor. Comput. Sci. 411(1), 239–258 (2010)
6. Borodin, A., Nielsen, M.N., Rackoff, C.: (Incremental) priority algorithms. Algorithmica 37(4), 295–326 (2003)
7. Chen, J., Friesen, D.K., Zheng, H.: Tight bound on Johnson's algorithm for maximum satisfiability. J. Comput. Syst. Sci. 58(3), 622–640 (1999)
8. Costello, K.P., Shapira, A., Tetali, P.: Randomized greedy: new variants of some classic approximation algorithms. In: SODA, pp. 647–655 (2011)
9. Davis, S., Impagliazzo, R.: Models of greedy algorithms for graph problems. Algorithmica 54(3), 269–317 (2009)
10. Goemans, M.X., Williamson, D.P.: New 3/4-approximation algorithms for the maximum satisfiability problem. SIAM J. Discrete Math. 7(4), 656–666 (1994)
11. Johnson, D.S.: Approximation algorithms for combinatorial problems. J. Comput. Syst. Sci. 9(3), 256–278 (1974)
12. Poloczek, M.: Bounds on greedy algorithms for MAX SAT,
    http://www.thi.cs.uni-frankfurt.de/poloczek/maxsatesa11.pdf

13. Poloczek, M., Schnitger, G.: Randomized variants of Johnson's algorithm for MAX SAT. In: SODA, pp. 656–663 (2011)
14. van Zuylen, A.: Simpler 3/4 approximation algorithms for MAX SAT (submitted)
15. Yao, A.C.-C.: Lower bounds by probabilistic arguments. In: FOCS, pp. 420–428. IEEE, Los Alamitos (1983)
16. Yung, C.K.: Inapproximation result for exact Max-2-SAT (unpublished manuscript)

Approximating Minimum Manhattan Networks in Higher Dimensions

Aparna Das[1], Emden R. Gansner[2], Michael Kaufmann[3], Stephen Kobourov[1], Joachim Spoerhase[4], and Alexander Wolff[4]

[1] Dept. of Comp. Sci., University of Arizona, Tucson, AZ, U.S.A.
[2] AT&T Labs Research, Florham Park, NJ, U.S.A.
[3] Wilhelm-Schickard-Institut für Informatik, Universität Tübingen, Germany
[4] Institut für Informatik, Universität Würzburg, Germany

Abstract. We consider the minimum Manhattan network problem, which is defined as follows. Given a set of points called *terminals* in $\mathbb{R}^d$, find a minimum-length network such that each pair of terminals is connected by a set of axis-parallel line segments whose total length is equal to the pair's Manhattan (that is, L_1-) distance. The problem is NP-hard in 2D and there is no PTAS for 3D (unless $\mathcal{P} = \mathcal{NP}$). Approximation algorithms are known for 2D, but not for 3D.

We present, for any fixed dimension d and any $\varepsilon > 0$, an $O(n^\varepsilon)$-approximation. For 3D, we also give a $4(k-1)$-approximation for the case that the terminals are contained in the union of $k \geq 2$ parallel planes.

1 Introduction

In a typical network construction problem, one is given a set of objects to be interconnected such that some constraints regarding the connections are fulfilled. Additionally, the network must be cheap. For example, if the objects are points in Euclidean space and the constraints say that, for some fixed $t > 1$, each pair of points must be connected by a path whose length is bounded by t times the Euclidean distance of the points, then the solution is a so-called *Euclidean t-spanner*. Concerning cost, one usually requires that the total length of the network is proportional to the length of a Euclidean minimum spanning tree of the points. Such cheap spanners can be constructed efficiently [2].

In this paper, we are interested in constructing 1-spanners, with respect to the Manhattan (or L_1-) metric. Our aim is to minimize the total length (or *weight*) of the network. Note that the Euclidean 1-spanner of a set of points is simply the complete graph (if no three points are collinear) and hence, its weight is completely determined. Manhattan 1-spanners, in contrast, have many degrees of freedom and vastly different weights.

More formally, given two points p and q in d-dimensional space $\mathbb{R}^d$, a *Manhattan path* connecting p and q (a p–q *M-path*, for short) is a sequence of axis-parallel line segments connecting p and q whose total length equals the Manhattan distance of p and q. Thus an M-path is a monotone rectilinear path. For our purposes, a set

C. Demetrescu and M.M. Halldórsson (Eds.): ESA 2011, LNCS 6942, pp. 49–60, 2011.

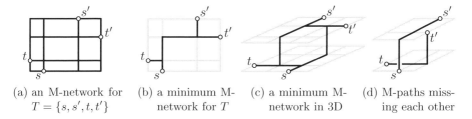

(a) an M-network for $T = \{s, s', t, t'\}$

(b) a minimum M-network for T

(c) a minimum M-network in 3D

(d) M-paths missing each other

Fig. 1. Examples of M-networks in 2D and 3D

of axis-parallel line segments is a *network*. Given a network N, its *weight* $\|N\|$ is the sum over the lengths of its line segments. A network N *Manhattan-connects* (or *M-connects*) two given points p and q if it "contains" a p–q M-path π. Note that we slightly abuse the notation here: we mean pointwise containment, that is, we require $\bigcup \pi \subseteq \bigcup N$. Given a set T of points—called *terminals*—in $\mathbb{R}^d$, a network N is a *Manhattan network* (or *M-network*) for T if N M-connects every pair of terminals in T. The *minimum Manhattan network problem* (MMN) consists of finding, for a given set T of terminals, a minimum-weight M-network. For examples, see Fig. 1.

M-networks have important applications in computational biology; Lam et al. [13] use them in gene alignment in order to reduce the size of the search space of the Viterbi algorithm for pair hidden Markov models.

Previous work. 2D-MMN, the 2D-version of the problem, was introduced by Gudmundsson et al. [10]. They gave an 8- and a 4-approximation algorithm. Later, the approximation ratio was improved to 3 [3,9] and then to 2, which is the currently best result. It was achieved in three different ways: via linear programming [5], using the primal–dual scheme [16] and with purely geometric arguments [11]. The last two algorithms run in $O(n \log n)$ time. A ratio of 1.5 was claimed [17], but apparently the proof is incomplete [9]. Chin et al. [6] finally settled the complexity of 2D-MMN by proving it NP-hard.

A little earlier, Muñoz et al. [15] considered 3D-MMN. They showed that the problem is NP-hard and NP-hard to approximate beyond a factor of 1.00002. For the special case of 3D-MMN where any cuboid spanned by two terminals contains other terminals or is a rectangle, they gave a 2α-approximation, where α denotes the best approximation ratio for 2D-MMN. They posed the design of approximations algorithms for general 3D-MMN as an open problem.

Related problems. In d-dimensional MMN (dD-MMN) we consider the dimension d fixed. As we observe in the full version of our paper [7], dD-MMN is a special case of the *directed Steiner forest problem* (DSF). More precisely, an instance of MMN can be decomposed into a constant number of DSF instances. The input of DSF is an edge-weighted directed graph G and a set of vertex pairs. The goal is to find a minimum-cost subgraph of G (not necessarily a forest!) that connects all given vertex pairs. Recently, Feldman et al. [8] reported an $O(n^{4/5+\varepsilon})$-approximation for DSF. This bound carries over to dD-MMN.

An important special case of DSF is the *directed Steiner* tree *problem* (DST). Here, the input instance specifies a digraph G, a *root* vertex r, and a subset S of the vertices of G that must be connected to r. An optimum solution for DST is an r-rooted subtree of G spanning S. DST admits an $O(n^\varepsilon)$-approximation for any $\varepsilon > 0$ [4].

A *geometric* optimization problem that resembles MMN is the *rectilinear Steiner arborescence problem* (RSA). Given a set of points in $\mathbb{R}^d$ with non-negative coordinates, a rectilinear Steiner arborescence is a spanning tree that connects all points with M-paths to the origin. As for MMN, the aim is to find a minimum-weight network. For 2D-RSA, there is a polynomial-time approximation scheme (PTAS) [14] based on Arora's technique [1]. It is not known whether 2D-MMN admits a PTAS. Arora's technique does not seem to work here as M-paths between terminals forbid detours and thus may not respect portals. RSA is a special case of DST; see the full version of this paper [7].

Our contribution. We give, for any $\varepsilon > 0$, an $O(n^\varepsilon)$-approximation algorithm for 3D-MMN; see Section 3. In the full version of this paper [7], we extend this result to arbitrary but constant dimension. We also present a $4(k-1)$-approximation algorithm for the special case of 3D-MMN where the given terminals are contained in $k \geq 2$ planes parallel to the x–y plane; see Section 2.

Our $O(n^\varepsilon)$-approximation for 3D-MMN constitutes a significant improvement upon the best known ratio of $O(n^{4/5+\varepsilon})$ for (general) directed Steiner forest [8]. We obtain this result by exploiting the geometric structure of the problem. To underline the relevance of our result, we remark that the bound of $O(n^\varepsilon)$ is the best known result also for other directed Steiner-type problems such as DST [4] or even acyclic DST [19].

Our $O(k)$-approximation for the k-plane case strongly relies on a deep result that Soto and Telha [18] achieved recently. They showed that, given a set of red and blue points in the plane, one can determine efficiently a minimum-cardinality set of points that together *pierce* all rectangles having a red point in the lower left corner and a blue point in the upper right corner. Their algorithm, together with an approximation algorithm for 2D-MMN, basically yields an approximation for the 2-plane case. We show how to generalize this idea to k planes.

Intuitively, what makes 3D-MMN more difficult than 2D-MMN is the fact that in 2D, if the bounding box of terminals s and s' and the bounding box of t and t' cross (as in Fig. 1b), then any s–s' M-path will intersect any t–t' M-path, which yields s–t' and t–s' M-paths for free (if s and t are the lower left corners of their respective boxes). A similar statement for 3D does not hold; M-paths can "miss" each other—even if their bounding cuboids cross; see Fig. 1d.

Let us formalize this observation. Given a set T of terminals, a set Z of pairs of terminals is a *generating set* [12] if any network that M-connects the pairs in Z in fact M-connects *all* pairs of terminals. In 2D, any MMN instance has a generating set of linear size [12]. This fact is exploited by most approximation algorithms for 2D-MMN. In the full version of this paper [7], we show that there are 3D instances where any generating set has quadratic size.

Notation and an observation. Given a point $p \in \mathbb{R}^3$, we denote the x-, y- and z-coordinate of p by $x(p)$, $y(p)$, and $z(p)$, respectively. Given two points a and c in $\mathbb{R}^2$, let $R(a,c) = \{b \in \mathbb{R}^2 \mid x(a) \leq x(b) \leq x(c), y(a) \leq y(b) \leq y(c)\}$ be the *rectangle spanned by a and c*. If a line segment is parallel to the x-, y-, or z-axis, we say that it is x-, y-, or z-*aligned*.

Gudmundsson et al. [10] observed that any instance of MMN has a solution that is contained in the *Hanan grid*, the grid induced by the terminals.

2 k-Plane Case

In this section, we consider 3D-MMN under the assumption that the set T of terminals is contained in the union of $k \geq 2$ planes $E_1, \ldots, E_k$ that are parallel to the x–y plane. Of course, this assumption always holds for some $k \leq n$. We present a $4(k-1)$-approximation algorithm, which outperforms our algorithm for the general case in Section 3 if $k \in o(n^\varepsilon)$.

Let N_{opt} be some fixed minimum M-network for T, let $N_{\mathrm{opt}}^{\mathrm{hor}}$ be the set of all x-aligned and all y-aligned segments in N_{opt}, and let $N_{\mathrm{opt}}^{\mathrm{ver}}$ be the set of all z-aligned segments in N_{opt}. Let OPT denote the weight of N_{opt}. Of course, OPT does not depend on the specific choice of N_{opt}; the weights of $N_{\mathrm{opt}}^{\mathrm{hor}}$ and $N_{\mathrm{opt}}^{\mathrm{ver}}$, however, may depend on N_{opt}. Further, let T_{xy} be the projection of T onto the x–y plane. For $i \in \{1, \ldots, k\}$, let $T_i = T \cap E_i$ be the set of terminals in plane E_i.

Our algorithm consists of two phases. Phase I computes a set N^{hor} of horizontal (that is, x- and y-aligned) line segments, phase II computes a set N^{ver} of vertical line segments. Finally, the algorithm returns the set $N = N^{\mathrm{hor}} \cup N^{\mathrm{ver}}$.

Phase I is simple; we compute a 2-approximate M-network N_{xy} for T_{xy} (using the algorithm of Guo et al. [11]) and project N_{xy} onto each of the planes $E_1, \ldots, E_k$. Let N^{hor} be the union of these projections. Note that N^{hor} M-connects any pair of terminals that lie in the same plane.

Observation 1. $\|N^{\mathrm{hor}}\| \leq 2k\|N_{\mathrm{opt}}^{\mathrm{hor}}\|$.

Proof. The projection of $N_{\mathrm{opt}}^{\mathrm{hor}}$ to the x–y plane is an M-network for T_{xy}. Hence, $\|N_{xy}\| \leq 2N_{\mathrm{opt}}^{\mathrm{hor}}$. Adding up over the k planes yields the claim.   □

In what follows, we describe phase II, which computes a set N^{ver} of vertical line segments, so-called *pillars*, of total cost at most $4(k-1)\|N_{\mathrm{opt}}^{\mathrm{ver}}\|$. This yields an overall approximation factor of $4(k-1)$ since $\|N^{\mathrm{hor}} \cup N^{\mathrm{ver}}\| \leq 2k\|N_{\mathrm{opt}}^{\mathrm{hor}}\| + 4(k-1)\|N_{\mathrm{opt}}^{\mathrm{ver}}\| \leq 4(k-1)(\|N_{\mathrm{opt}}^{\mathrm{hor}}\| + \|N_{\mathrm{opt}}^{\mathrm{ver}}\|) \leq 4(k-1)\mathrm{OPT}$.

For simplicity, we restrict ourselves to the *directional* subproblem of M-connecting all terminal pairs (t, t') such that t *dominates* t', that is, $x(t) \leq x(t'), y(t) \leq y(t'), z(t) \leq z(t')$ and $t \neq t'$. We call such terminal pairs *relevant*.

One may think of this directional subproblem as solving the part of the problem that runs in direction north-east (NE) in the x–y plane (with increasing z-coordinates). We construct a pillar network $N_{\mathrm{dir}}^{\mathrm{ver}}$ of weight at most $(k-1)\|N_{\mathrm{opt}}^{\mathrm{ver}}\|$ that, together with N^{hor}, M-connects all relevant pairs. We solve the analogous subproblems for the directions NW, SE, and SW symmetrically. Then N^{ver} is the union of the four partial solutions and has weight at most $4(k-1)\|N_{\mathrm{opt}}^{\mathrm{ver}}\|$ as desired.

Our directional subproblem is closely linked to the *(directional) bichromatic rectangle piercing problem* (BRP), which is defined as follows. Let R and B be sets of red and blue points in $\mathbb{R}^2$, respectively, and let $\mathcal{R}(R,B)$ denote the set of axis-aligned rectangles each of which is spanned by a red point in its SW-corner and a blue point in its NE-corner. Then the aim of BRP is to find a minimum-cardinality set $P \subset \mathbb{R}^2$ such that every rectangle in $\mathcal{R}(R,B)$ is *pierced*, that is, contains at least one point in P. The points in P are called *piercing points*.

The problem dual to BRP is the *(directional) bichromatic independent set of rectangles problem* (BIS) where the goal is to find the maximum number of pairwise disjoint rectangles in $\mathcal{R}(R,B)$, given the sets R and B.

Recently, Soto and Telha [18] proved a beautiful min–max theorem saying that, for $\mathcal{R}(R,B)$, the minimum number of piercing points always *equals* the maximum number of independent rectangles. This enabled them to give efficient exact algorithms for BRP and BIS running in $\tilde{O}(n^{2.5})$ worst-case time or $\tilde{O}(n^{\gamma})$ expected time, where the $\tilde{O}$-notation ignores polylogarithmic factors, $\gamma < 2.4$ is the exponent for fast matrix multiplication, and $n = |R| + |B|$ is the input size.

2.1 Two Planes

Our phase-II algorithm for two planes is very simple. We sketch it in order to give a smooth introduction to the k-plane case. Imagine the terminals in T_1 to be red and those in T_2 to be blue. Ignore the z-coordinates of the terminals. Then the relevant red–blue point pairs span exactly the rectangles in $\mathcal{R}(T_1, T_2)$, which we call relevant, too.

Our algorithm consists of two steps. First, we compute a minimum piercing P of $\mathcal{R}(T_1, T_2)$ using the algorithm of Soto and Telha [18]. Second, we move each piercing point $p \in P$ to a new position $\hat{p}$—a nearby junction of N_{xy}—and erect, at $\hat{p}$, a pillar connecting the two planes. Let $\hat{P}$ be the set of piercing points after the move, and let $N_{\mathrm{dir}}^{\mathrm{ver}}$ be the corresponding set of pillars.

Lemma 1. *It holds that $\|N_{\mathrm{dir}}^{\mathrm{ver}}\| \leq \|N_{\mathrm{opt}}^{\mathrm{ver}}\|$.*

Proof. Clearly, $|\hat{P}| = |P|$. Integrating over the distance d of the two planes yields $\|N_{\mathrm{dir}}^{\mathrm{ver}}\| = |\hat{P}| \cdot d = |P| \cdot d \leq \|N_{\mathrm{opt}}^{\mathrm{ver}}\|$. The last inequality is due to the fact that P is a *minimum* piercing of $\mathcal{R}(T_1, T_2)$ and that the pillars in $N_{\mathrm{opt}}^{\mathrm{ver}}$ pierce $\mathcal{R}(T_1, T_2)$—otherwise N_{opt} would not be feasible. $\square$

Now we turn to feasibility. We first detail how we move each piercing point p to its new position $\hat{p}$. For the sake of brevity, we identify terminals with their projections to the x–y plane. Our description assumes that we have at our disposal some network M (such as N_{xy}) connecting the relevant pairs in T_{xy}.

Given a piercing point $p \in P$, let A_p be the intersection of the relevant rectangles pierced by p. Clearly, $p \in A_p$. Refer to Fig. 2a; note that the bottom and left sides of A_p are determined by terminals t_W and t_S to the west and south of A_p, respectively. Symmetrically, the top and right sides of A_p are determined

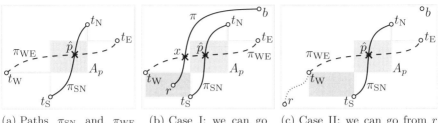

(a) Paths π_{SN} and π_{WE} meet in a point $\hat{p}$ in A_p.   (b) Case I: we can go from r via x to $\hat{p}$.   (c) Case II: we can go from r via t_{W} or t_{S} to $\hat{p}$.

Fig. 2. Sketches for the proof of Lemma 2

by terminals t_{E} and t_{N} to the east and north of A_p, respectively. Note that t_{W} and t_{S} may coincide, and so may t_{E} and t_{N}. Clearly, the network M contains an M-path π_{SN} connecting t_{S} and t_{N} and an M-path π_{WE} connecting t_{W} and t_{E}. The path π_{SN} goes through the bottom and top sides of A_p and π_{WE} goes through the left and right sides. Hence, the two paths intersect in a point $\hat{p} \in A_p$. This is where we move the original piercing point p.

Since $\hat{p} \in A_p$, the point $\hat{p}$ pierces the same relevant rectangles as p, and the set $\hat{P} = \{\hat{p} \mid p \in P\}$ is a (minimum) piercing for the set of relevant rectangles.

Lemma 2. *Let $\mathcal{R}(R, B)$ be an instance of* BRP *and let M be a network that M-connects every relevant red–blue point pair. Then we can efficiently compute a minimum piercing of $\mathcal{R}(R, B)$ such that M contains, for every relevant red–blue point pair (r, b) in $R \times B$, an r–b M-path that contains a piercing point.*

Proof. We use the algorithm of Soto and Telha [18] to compute a minimum piercing P of $\mathcal{R}(R, B)$. Then, as we have seen above, $\hat{P}$ is a minimum piercing of $\mathcal{R}(R, B)$, too. Now let (r, b) be a relevant red–blue pair in $R \times B$, and let $\hat{p} \in \hat{P}$ be a piercing point of $R(r, b)$.

Then, by the definition of $A_{\hat{p}}$, the point r lies to the SW of $A_{\hat{p}}$ and b to the NE. We prove that M contains an r–$\hat{p}$ M-path; a symmetric argument proves that M also contains a $\hat{p}$–b M-path. Concatenating these two M-paths yields the desired r–b M-path since b lies to the NE of $\hat{p}$ and $\hat{p}$ lies to the NE of r. Recall that $\hat{p}$ lies on the intersection of the t_{W}–t_{E} M-path π_{WE} and the t_{S}–t_{N} M-path π_{SN}, where t_{W}, t_{E}, t_{S}, t_{N} are the terminals that determine the extensions of $A_{\hat{p}}$; see Fig. 2a. To show that M M-connects r and $\hat{p}$, we consider two cases.

Case I: $r \in R(t_{\mathrm{W}}, t_{\mathrm{S}})$ (dark shaded in Fig. 2b).

According to our assumption, M contains *some* r–b M-path π. It is not hard to see that π must intersect π_{WE} or π_{SN} in some point x to the SW of $\hat{p}$. Thus, we can go, in a monotone fashion, on π from r to x and then on π_{WE} or π_{SN} from x to $\hat{p}$. This is the desired r–$\hat{p}$ M-path.

Case II: r lies to the SW of t_{W} or t_{S}; see Fig. 2c.

Clearly, M contains M-paths from r to t_{W} and to t_{S}. If r lies to the SW of t_{W}, we can go, again in a monotone fashion, from r to t_{W} and then on π_{WE} from t_{W}

to $\hat{p}$. Otherwise, if r lies to the SW of t_S, we can go from r to t_S and then on π_{SN} from t_S to $\hat{p}$. $\qquad\square$

Lemmas 1 and 2 (with $R = T_1$, $B = T_2$, and $M = N_{xy}$) yield the following.

Theorem 1. *We can efficiently compute a 4-approximation for the 2-plane case.*

2.2 Recursive Algorithm for k Planes

Now we extend our 2-plane algorithm to the k-plane case.

Theorem 2. *There is a $4(k-1)$-approximation algorithm for 3D-MMN where the terminals lie in the union of $k \geq 2$ planes parallel to the x–y plane.*

In the remainder of this section, we restrict ourselves to the directional subproblem of M-connecting all relevant terminal pairs. Specifically, we construct a pillar network $N_{\mathrm{dir}}^{\mathrm{ver}}$ of weight at most $(k-1)\|N_{\mathrm{opt}}^{\mathrm{ver}}\|$. As we have argued at the beginning of Section 2, this suffices to prove Theorem 2.

Our pillar-placement algorithm is as follows. Let $i \in \{1, \ldots, k-1\}$. We construct an instance $\mathcal{I}_i$ of BRP where we two-color T_{xy} such that each point corresponding to a terminal of some plane E_j with $j \leq i$ is colored red and each point corresponding to a terminal of some plane $E_{j'}$ with $j' \geq i+1$ is colored blue. For $\mathcal{I}_i$, we compute a minimum piercing $\hat{P}_i$ according to Lemma 2 with $M = N_{xy}$. In other words, for any relevant pair $(t_j, t_{j'})$, there is some M-path in N_{xy} that contains a piercing point of $\hat{P}_i$. We choose $i^\star \in \{1, \ldots, k-1\}$ such that $\hat{P}_{i^\star}$ has minimum cardinality. This is crucial for our analysis. At the piercing points of $\hat{P}_{i^\star}$, we erect pillars spanning all planes $E_1, \ldots, E_k$. Let $\hat{N}_{i^\star}$ be the set of these pillars. We first investigate feasibility of the resulting network.

Lemma 3. *The network $N^{\mathrm{hor}} \cup \hat{N}_{i^\star}$ M-connects any relevant terminal pair $(t_j, t_{j'})$ with $j \leq i^\star$ and $j' \geq i^\star + 1$.*

Proof. Consider a pair $(t_j, t_{j'})$ as in the statement. We construct an M-path from t_j to $t_{j'}$ as follows. We know that there is an M-path π that connects the projections of t_j and $t_{j'}$ in N_{xy} and contains a piercing point p of $\hat{P}_{i^\star}$. So we start in t_j and follow the projection of π onto plane E_j until we arrive in p. Then we use the corresponding pillar in $\hat{N}_{i^\star}$ to reach plane $E_{j'}$ where we follow the projection of π (onto that plane) until we reach $t_{j'}$. $\qquad\square$

In order to also M-connect terminals in planes $E_j, E_{j'}$ where either $j, j' \leq i^\star$ or $j, j' \geq i^\star + 1$, we simply apply the pillar-placement algorithm recursively to the sets $\{E_1, \ldots, E_{i^\star}\}$ and $\{E_{i^\star+1}, \ldots, E_k\}$. This yields the desired pillar network $N_{\mathrm{dir}}^{\mathrm{ver}}$. By Lemma 3, $N_{\mathrm{dir}}^{\mathrm{ver}} \cup N^{\mathrm{hor}}$ is feasible. Next, we bound $\|\hat{N}_{i^\star}\|$.

Lemma 4. *Let M be an arbitrary directional Manhattan network for T, and let M^{ver} be the set of vertical segments in M. Then the pillar network $\hat{N}_{i^\star}$ has weight at most $\|M^{\mathrm{ver}}\|$.*

Proof. Without loss of generality, we assume that M is a subnetwork of the Hanan grid [10]. We may also assume that any segment of M^{ver} spans only consecutive planes. For $1 \le i \le j \le k$, let $M_{i,j}$ denote the subnetwork of M^{ver} lying between planes E_i and E_j. Let $d_{i,j}$ be the vertical distance of planes E_i and E_j.

We start with the observation that, for any $j = 1, \ldots, k$, the network $M_{j,j+1}$ is a set of pillars that forms a valid piercing of the piercing instance $\mathcal{I}_j$. Hence, $|M_{j,j+1}| \ge |\hat{P}_j| \ge |\hat{P}_{i^\star}|$, which implies the claim of the lemma as follows:

$$\|M^{\mathrm{ver}}\| = \sum_{j=1}^{k-1} \|M_{j,j+1}\| = \sum_{j=1}^{k-1} |M_{j,j+1}| \cdot d_{j,j+1} \ge \sum_{j=1}^{k-1} |P_{i^\star}| \cdot d_{j,j+1} = \|P_{i^\star}\|.$$

$\square$

It is crucial that the pillars constructed recursively span either $E_1, \ldots, E_{i^\star}$ or $E_{i^\star+1}, \ldots, E_k$ but not all planes. For $1 \le j \le j' \le k$, let $\mathrm{weight}_z(j, j')$ denote the weight of the vertical part of the network produced by the above pillar-placement algorithm when applied to planes $E_j, \ldots, E_{j'}$ recursively. Assume that $j < j'$ and that the algorithm splits at plane $E_{i'}$ with $j \le i' < j'$ when planes $E_j, \ldots, E_{j'}$ are processed. By means of Lemma 4, we derive the recursion

$$\mathrm{weight}_z(j, j') \le \|M_{j,j'}\| + \mathrm{weight}_z(j, i') + \mathrm{weight}_z(i' + 1, j'),$$

which holds for any M-network M for T. We now claim that

$$\mathrm{weight}_z(j, j') \le (j' - j)\|M_{j,j'}\|.$$

Our proof is by induction on the number of planes processed by the algorithm. By the inductive hypothesis, we have that $\mathrm{weight}_z(j, i') \le (i' - j)\|M_{j,i'}\|$ and $\mathrm{weight}_z(i' + 1, j') \le (j' - i' - 1)\|M_{i'+1,j'}\|$. Since $\|M_{j,i'}\| + \|M_{i'+1,j'}\| \le \|M_{j,j'}\|$ and $\mathrm{weight}_z(l, l) = 0$ for any $l \in \{1, \ldots, k\}$, the claim follows.

We conclude that the weight of the solution produced by the algorithm when applied to all planes $E_1, \ldots, E_k$ is bounded by $\mathrm{weight}_z(1, k) \le (k - 1)\|M_{1,k}\| = (k - 1)\|M^{\mathrm{ver}}\|$. This finishes the proof of Theorem 2.

3   General Case

In this section, we present an approximation algorithm, the *grid algorithm*, for general 3D-MMN. Our main result is the following.

Theorem 3. *For any $\varepsilon > 0$, there is an $O(n^\varepsilon)$-approximation algorithm for 3D-MMN that runs in time $n^{O(1/\varepsilon)}$.*

Our approach extends to higher dimensions; see the full version of our paper [7].

For technical reasons we assume that the terminals are in general position, that is, any two terminals differ in all three coordinates. As in the k-plane case it suffices to describe and analyze the algorithm for the directional subproblem.

Algorithm. We start with a high-level summary. To solve the directional subproblem, we place a 3D-grid that partitions the instance into a constant number of cuboids. Cuboids that differ in only two coordinates form *slabs*. We connect terminals from different slabs by M-connecting each terminal to the corners of its cuboid and by using the edges of the grid to connect the corners. We connect terminals from the same slab by recursively applying our algorithm to the slabs.

Step 1: Partitioning into cuboids and slabs. Consider the bounding cuboid C of T and set $c = 3^{1/\varepsilon}$. Partition C by $3(c-1)$ separating planes into $c \times c \times c$ axis-aligned subcuboids C_{ijk} with $i, j, k \in \{1, \ldots, c\}$. The indices are such that larger indices mean larger coordinates. Place the separating planes such that the number of terminals between two consecutive planes is at most n/c. This can be accomplished by a simple plane-sweep for each direction x, y, z, and placing separating planes after n/c terminals. Here we exploit our general-position assumption. The edges of the resulting subcuboids—except the edges on the boundary of C which we do not need—induce a three-dimensional grid $\mathcal{G}$ of axis-aligned line segments. We insert $\mathcal{G}$ into the solution.

For each $i \in \{1, \ldots, c\}$, define the x-aligned *slab*, C_i^x, to be the union of all cuboids C_{ijk} with $j, k \in \{1, \ldots, c\}$. Define y-aligned and z-aligned slabs C_j^y, C_k^z analogously; see the full version of this paper [7] for figures.

Step 2: Add M-paths between different slabs. Consider two cuboids C_{ijk} and $C_{i'j'k'}$ with $i < i'$, $j < j'$, and $k < k'$. Any terminal pair $(t, t') \in C_{ijk} \times C_{i'j'k'}$ can be M-connected using the edges of $\mathcal{G}$ as long as t and t' are connected to the appropriate corners of their cuboids; see the full version [7] for a figure. To this end, we use the following *patching* procedure.

Call a cuboid C_{ijk} *relevant* if there is a non-empty cuboid $C_{i'j'k'}$ with $i < i'$, $j < j'$, and $k < k'$. For each relevant cuboid C_{ijk}, let $\hat{p}_{ijk}$ denote a corner that is dominated by all terminals inside C_{ijk}. By *Up-patching* C_{ijk} we mean to M-connect every terminal in C_{ijk} to $\hat{p}_{ijk}$. We up-patch C_{ijk} by solving (approximately) an instance of 3D-RSA with the terminals in C_{ijk} as points and $\hat{p}_{ijk}$ as origin. We define *down-patching* symmetrically; cuboid C_{ijk} is relevant if there is a non-empty cuboid $C_{i'j'k'}$ with $i > i'$, $j > j'$, $k > k'$ and $\check{p}_{ijk}$ is the corner that dominates all terminals in C_{ijk}.

We insert the up- and down-patches of all relevant cuboids into the solution.

Step 3: Add M-paths within slabs. To M-connect relevant terminal pairs that lie in the same slab, we apply the grid algorithm (steps 1–3) recursively to each slab C_i^x, C_j^y, and C_k^z with $i, j, k \in \{1, \ldots, c\}$.

Analysis. We first show that the output of the grid algorithm is feasible, then we establish its approximation ratio of $O(n^\varepsilon)$ and its running time of $n^{O(1/\varepsilon)}$ for any $\varepsilon > 0$. In this section, OPT denotes the weight of a minimum M-network (*not* the cost of an optimal solution to the directional subproblem).

Lemma 5. *The grid algorithm M-connects all relevant terminal pairs.*

Proof. Let (t, t') be a relevant terminal pair.

First, suppose that t and t' lie in cuboids of different slabs. Thus, there are $i < i', j < j', k < k'$ such that $t \in C_{ijk}$ and $t' \in C_{i'j'k'}$. Furthermore, C_{ijk} and $C_{i'j'k'}$ are relevant for up- and down-patching, respectively. When up-patching, we solve an instance of RSA connecting all terminals in C_{ijk} to $\hat{p}_{ijk}$. Similarly, down-patching M-connects t' to $\check{p}_{i'j'k'}$. The claim follows as $\mathcal{G}$ M-connects $\hat{p}_{ijk}$ and $\check{p}_{i'j'k'}$.

Now, suppose that t and t' lie in the same slab. As the algorithm is applied recursively to each slab, there is a recursion step where t and t' lie in cuboids in different slabs. Here, we need our general-position assumption. Applying the argument above to that particular recursive step completes the proof. $\square$

Next, we turn to the performance of our algorithm. Let $r(n)$ be its approximation ratio, where n is the number of terminals in T. The total weight of the output is the sum of $\|\mathcal{G}\|$, the cost of patching, and the cost for the recursive treatment of the slabs. We analyze each of the three costs separately.

The grid $\mathcal{G}$ consists of all edges induced by the c^3 subcuboids except the edges on the boundary of C. Let ℓ denote the length of the longest side of C. The weight of $\mathcal{G}$ is at most $3(c-1)^2\ell$, which is bounded by $3c^2\mathrm{OPT}$ as $\mathrm{OPT} \geq \ell$.

Let $r_{\mathrm{patch}}(n)$ denote the cost of patching all relevant cuboids in step 2. Lemma 6 (given below) proves that $r_{\mathrm{patch}}(n) = O(n^\varepsilon)\mathrm{OPT}$.

Now consider the recursive application of the algorithm to all slabs. Recall that N_{opt} is a fixed minimum M-network for T. For $i \in 1, \ldots, c$, let OPT_i^x be the optimum cost for M-connecting *all* (not only relevant) terminal pairs in slab C_i^x. Costs OPT_i^y and OPT_i^z are defined analogously.

Slightly abusing of notation, we write $N_{\mathrm{opt}} \cap C_i^x$ for the set of line segments of N_{opt} that are completely contained in slab C_i^x. Observe that $N_{\mathrm{opt}} \cap C_i^x$ forms a feasible solution for C_i^x. Thus, $\mathrm{OPT}_i^x \leq \|N_{\mathrm{opt}} \cap C_i^x\|$. By construction, any slab contains at most n/c terminals. Hence, the total cost of the solutions for slabs $C_1^x, \ldots, C_c^x$ is at most

$$\sum_{i=1}^{c} r\left(\frac{n}{c}\right) \mathrm{OPT}_i^x \leq r\left(\frac{n}{c}\right) \sum_{i=1}^{c} \|N_{\mathrm{opt}} \cap C_i^x\| \leq r\left(\frac{n}{c}\right) \mathrm{OPT}.$$

Clearly, the solutions for the y- and z-slabs have the same bound.

Summing up all three types of costs, we obtain the recursive equation

$$r(n)\mathrm{OPT} \leq 3c^2\mathrm{OPT} + 3r\left(\frac{n}{c}\right) \mathrm{OPT} + r_{\mathrm{patch}}(n)\mathrm{OPT}.$$

Hence, $r(n) = O(n^{\max\{\varepsilon, \log_c 3\}})$. Plugging in $c = 3^{1/\varepsilon}$ yields $r(n) = O(n^\varepsilon)$, which proves the approximation ratio claimed in Theorem 3.

Lemma 6. *Patching all relevant cuboids costs $r_{\mathrm{patch}}(n) \in O(n^\varepsilon)\mathrm{OPT}$.*

Proof. It suffices to consider up-patching; down-patching is symmetric.

Lemma 7 shows that by reducing the patching problem to 3D-RSA, we can find such a network of cost $O(\rho)\mathrm{OPT}$, where ρ is the approximation factor of 3D-RSA.

In the full version of our paper [7], we argue that there is an approximation preserving reduction from 3D-RSA to DST. DST, in turn, admits an $O(n^\varepsilon)$-approximation for any $\varepsilon > 0$ [4]. Hence, the cost of up-patching is indeed bounded by $O(n^\varepsilon)$OPT. □

The following lemma is proved in the full version [7] of this paper.

Lemma 7. *Given an efficient ρ-approximation of 3D-RSA, we can efficiently up-patch all relevant cuboids at cost no more than $12(c^2 + c)\rho$OPT.*

Finally, we analyze the running time. Let $T(n)$ denote the running time of the algorithm applied to a set of n terminals. The running time is dominated by patching and the recursive slab treatment. Using the DST algorithm of Charikar et al. [4], patching cuboid C_i requires time $n_i^{O(1/\varepsilon)}$, where n_i is the number of terminals in C_i. As each cuboid is patched at most twice and there are c^3 cuboids, patching takes $O(c^3)n^{O(1/\varepsilon)} = n^{O(1/\varepsilon)}$ time. The algorithm is applied recursively to $3c$ slabs. This yields the recurrence $T(n) = 3cT(n/c) + n^{O(1/\varepsilon)}$, which leads to the claimed running time.

4   Open Problems

We have presented a grid-based $O(n^\varepsilon)$-approximation for dD-MMN. This is a significant improvement over the ratio of $O(n^{4/5+\varepsilon})$ which is achieved by reducing the problem to DSF [8]. For 3D, we have described a $4(k-1)$-approximation for the case when the terminals lie on $k \geq 2$ parallel planes. This outperforms our grid-based algorithm when $k \in o(n^\varepsilon)$. Whereas 2D-MMN admits a 2-approximation [5,11,16], it remains open whether $O(1)$- or $O(\log n)$-approximations can be found in higher dimensions.

Our $O(n^\varepsilon)$-approximation algorithm for dD-MMN solves instances of dD-RSA for the subproblem of patching. We conjecture that dD-RSA admits a PTAS, which is known for 2D-RSA [14]. While this is an interesting open question, a positive result would still not be enough to improve our approximation ratio, which is dominated by the cost of finding M-paths inside slabs.

The complexity of the *undirectional* bichromatic rectangle piercing problem (see Section 2) is still unknown. Currently, the best approximation has a ratio of 4, which is (trivially) implied by the result of Soto and Telha [18]. Any progress would immediately improve the approximation ratio of our algorithm for the k-plane case of 3D-MMN (for any $k > 2$).

Acknowledgments. This work was started at the 2009 Bertinoro Workshop on Graph Drawing. We thank the organizers Beppe Liotta and Walter Didimo for creating an inspiring atmosphere. We also thank Steve Wismath, Henk Meijer, Jan Kratochvíl, and Pankaj Agarwal for discussions. We are indebted to Stefan Felsner for pointing us to Soto and Telha's work [18].

The research of Aparna Das and Stephen Kobourov was funded in part by NSF grants CCF-0545743 and CCF-1115971.

References

1. Arora, S.: Approximation schemes for NP-hard geometric optimization problems: A survey. Math. Program. 97(1–2), 43–69 (2003)
2. Arya, S., Das, G., Mount, D.M., Salowe, J.S., Smid, M.: Euclidean spanners: Short, thin, and lanky. In: Proc. 27th Annu. ACM Symp. Theory Comput (STOC), pp. 489–498. ACM Press, New York (1995)
3. Benkert, M., Wolff, A., Widmann, F., Shirabe, T.: The minimum Manhattan network problem: Approximations and exact solutions. Comput. Geom. Theory Appl. 35(3), 188–208 (2006)
4. Charikar, M., Chekuri, C., Cheung, T.Y., Dai, Z., Goel, A., Guha, S., Li, M.: Approximation algorithms for directed Steiner problems. In: Proc. 9th ACM-SIAM Sympos. Discrete Algorithms (SODA), pp. 192–200 (1998)
5. Chepoi, V., Nouioua, K., Vaxès, Y.: A rounding algorithm for approximating minimum Manhattan networks. Theor. Comput. Sci. 390(1), 56–69 (2008)
6. Chin, F., Guo, Z., Sun, H.: Minimum Manhattan network is NP-complete. Discrete Comput. Geom. 45, 701–722 (2011)
7. Das, A., Gansner, E.R., Kaufmann, M., Kobourov, S., Spoerhase, J., Wolff, A.: Approximating minimum Manhattan networks in higher dimensions. ArXiv e-print abs/1107.0901 (2011)
8. Feldman, M., Kortsarz, G., Nutov, Z.: Improved approximating algorithms for directed Steiner forest. In: Proc. 20th ACM-SIAM Sympos. Discrete Algorithms (SODA), pp. 922–931 (2009)
9. Fuchs, B., Schulze, A.: A simple 3-approximation of minimum Manhattan networks. Tech. Rep. 570, Zentrum für Angewandte Informatik Köln (2008)
10. Gudmundsson, J., Levcopoulos, C., Narasimhan, G.: Approximating a minimum Manhattan network. Nordic J. Comput. 8, 219–232 (2001)
11. Guo, Z., Sun, H., Zhu, H.: Greedy construction of 2-approximation minimum Manhattan network. In: Hong, S., Nagamochi, H., Fukunaga, T. (eds.) ISAAC 2008. LNCS, vol. 5369, pp. 4–15. Springer, Heidelberg (2008)
12. Kato, R., Imai, K., Asano, T.: An improved algorithm for the minimum Manhattan network problem. In: Bose, P., Morin, P. (eds.) ISAAC 2002. LNCS, vol. 2518, pp. 344–356. Springer, Heidelberg (2002)
13. Lam, F., Alexandersson, M., Pachter, L.: Picking alignments from (Steiner) trees. J. Comput. Biol. 10, 509–520 (2003)
14. Lu, B., Ruan, L.: Polynomial time approximation scheme for the rectilinear Steiner arborescence problem. J. Comb. Optim. 4(3), 357–363 (2000)
15. Muñoz, X., Seibert, S., Unger, W.: The minimal Manhattan network problem in three dimensions. In: Das, S., Uehara, R. (eds.) WALCOM 2009. LNCS, vol. 5431, pp. 369–380. Springer, Heidelberg (2009)
16. Nouioua, K.: Enveloppes de Pareto et Réseaux de Manhattan: Caractérisations et Algorithmes. Ph.D. thesis, Université de la Méditerranée (2005)
17. Seibert, S., Unger, W.: A 1.5-approximation of the minimal Manhattan network problem. In: Deng, X., Du, D. (eds.) ISAAC 2005. LNCS, vol. 3827, pp. 246–255. Springer, Heidelberg (2005)
18. Soto, J.A., Telha, C.: Jump number of two-directional orthogonal ray graphs. In: Gülük, O., Woeginger, G. (eds.) IPCO 2011. LNCS, vol. 6655, pp. 389–403. Springer, Heidelberg (2011)
19. Zelikovsky, A.: A series of approximation algorithms for the acyclic directed Steiner tree problem. Algorithmica 18(1), 99–110 (1997)

On Isolating Points Using Disks[*]

Matt Gibson[1], Gaurav Kanade[2], and Kasturi Varadarajan[2]

[1] Department of Electrical and Computer Engineering
The University of Iowa
Iowa City, IA 52242
matthew-gibson@uiowa.edu
[2] Department of Computer Science
The University of Iowa
Iowa City, IA 52242
gaurav-kanade@uiowa.edu, kvaradar@iowa.uiowa.edu

Abstract. In this paper, we consider the problem of choosing disks (that we can think of as corresponding to wireless sensors) so that given a set of input points in the plane, there exists no path between any pair of these points that is not intercepted by some disk. We try to achieve this separation using a minimum number of a given set of unit disks. We show that a constant factor approximation to this problem can be found in polynomial time using a greedy algorithm. To the best of our knowledge we are the first to study this optimization problem.

1   Introduction

Wireless sensors are being extensively used in applications to provide barriers as a defense mechanism against intruders at important buildings, estates, national borders etc. Monitoring the area of interest by this type of coverage is called *barrier* coverage [16]. Such sensors are also being used to detect and track moving objects such as animals in national parks, enemies in a battlefield, forest fires, crop diseases etc. In such applications it might be prohibitively expensive to attain blanket coverage but sufficient to ensure that the object under consideration cannot travel too far before it is detected. Such a coverage is called *trap* coverage [4,19].

Inspired by such applications, we consider the problem of isolating a set of points by a minimum-size subset of a given set of unit radius disks. A unit disk crudely models the region sensed by a sensor, and the work reported here readily generalizes to disks of arbitrary, different radii.

Problem Formulation. The input to our problem is a set I of n unit disks, and a set P of k points such that I *separates* P, that is, for any two points $p, q \in P$, every path between p and q intersects at least one disk in I. The goal is to find a minimum cardinality subset of I that separates P. See Figure 1 for an illustration of this notion of separation.

---

[*] This material is based upon work supported by the National Science Foundation under Grant No. 0915543.

C. Demetrescu and M.M. Halldórsson (Eds.): ESA 2011, LNCS 6942, pp. 61–69, 2011.

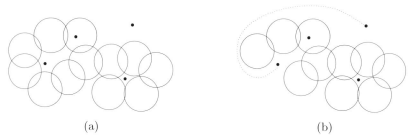

(a)   (b)

Fig. 1. (a) This set of disks separates the points because every path connecting any two points must intersect a disk. (b) This set of disks does not separate the points.

There has been a lot of recent interest on geometric variants of well-known NP-hard combinatorial optimization problems, and our work should be seen in this context. For several variants of the geometric set cover problem, for example, approximation algorithms have been designed [10,3,17] that improve upon the best guarantees for the combinatorial set cover problem. For the problem of covering points by the smallest subset of a given set of unit disks, we have approximation algorithms that guarantee an $O(1)$ approximation and even a PTAS [6,17]. These results hold even for disks of arbitrary radii. Our problem can be viewed as a set cover problem where the elements that need to be covered are not points, but paths. However, known results only imply a trivial $O(n)$ approximation when viewed through this set cover lens.

Another example of a problem that has received such attention is the *independent set* problem. For many geometric variants [8,9,12], approximation ratios that are better than that for the combinatorial case are known.

Our problem is similar to the node multi-terminal cut problem in graphs [13]. Here, we are given a graph $G = (V, E)$ with costs on the vertices and a subset $U \subseteq V$ of k vertices, and our goal is to compute a minimum cost subset of vertices whose removal disconnects every pair of vertices in U. This problem admits a poly-time algorithm that guarantees an $O(1)$ approximation. We note however that the problem we consider does not seem to be a special case of the multi-terminal cut problem.

Contribution and Related Work. Our main result is a polynomial time algorithm that guarantees an $O(1)$ approximation for the problem. To the best of our knowledge, this is the first non-trivial approximation algorithm for this problem. Our algorithm is simple and combinatorial and is in fact a greedy algorithm. We first present an $O(1)$ approximation for the following two-point separation problem. We are given a set of unit disks G, and two points s and t, and we wish to find the smallest subset $B \subseteq G$ so that B separates s and t.

Our greedy algorithm to the overall problem applies the two-point separation algorithm to find the cheapest subset B of I that separates some pair of points in P. Suppose that P is partitioned into sets $P_1, P_2, \ldots, P_\tau$ where each P_i is the subset of points in the same "face" with respect to B. The algorithm then recursively finds a separator for each of the P_i, and returns the union of these and B.

The analysis to show that this algorithm has the $O(1)$ approximation guarantee relies on the combinatorial complexity of the boundary of the union of disks. It uses a subtle and global argument to bound the total size of all the separators B computed in each of the recursive calls.[1]

Our approximation algorithm for the two-point separation problem, which is a subroutine we use in the overall algorithm, is similar to fast algorithms for finding minimum s-t cuts in undirected planar graphs, see for example [18]. Our overall greedy algorithm has some resemblance to the algorithm of Erickson and Har-Peled [11] employed in the context of approximating the minimum cut graph of a polyhedral manifold. The details of the our algorithm and the analysis, however, are quite different from these papers since we do not have an embedded graph but rather a system of unit disks. Sankararaman et al. [19] investigate a notion of coverage which they call *weak coverage*. Given a region $\mathcal{R}$ of interest (which they take to be a square in the plane) and a set I of unit disks (sensors), the region is said to be k-weakly covered if each connected component of $\mathcal{R} - \bigcup_{d \in I} d$ has diameter at most k. They consider the situation when a given set I of unit disks *completely* covers $\mathcal{R}$, and address the problem of partitioning I into as many subsets as possible so that $\mathcal{R}$ is k-weakly covered by every subset. Their work differs in flavor from ours mainly due to the assumption that I completely covers $\mathcal{R}$.

Bereg and Kirkpatrick [5] consider a problem that loosely resembles the two-point separation problem that is used as a subroutine here: Given a set of unit disks and two points s and t, find a path from s to t that intersects the smallest number of disks. In recent work that is simultaneous with and independent of ours, Alt et al. [2] address the two-point separation problem for a set of line segments. They show that the problem in fact admits a polynomial-time exact algorithm. They also consider the problem addressed by Bereg and Kirkpatrick [5], but in the context of line segments.

Organization. In Section 2, we discuss standard notions we require, and then reduce our problem to the case where none of the points in input P are contained in any of the input disks. In Section 3, we present our approximation algorithm for separating two points. In Section 4, we describe our main result, the constant factor approximation algorithm for separating P. We conclude in Section 5 with some remarks.

2 Preliminaries

We will refer to the standard notions of vertices, edges, and faces in arrangements of circles [1]. In particular, for a set R of m disks, we are interested in the faces

---

[1] In an earlier version of this paper [14], a similar algorithm was analyzed in a more "local" fashion. The basic observation was that the very first separator B that is computed has size $O(|OPT|/k)$, where OPT is the optimal solution for the problem. Subsequent separators computed in the recursive calls may be more expensive, but it was shown that the overall size is $O((\log k) \cdot |OPT|)$. In contrast, the present analysis does not try to bound the size of the individual separators, but just the sum of their sizes. As a consequence, the analysis also turns out to be technically simpler.

in the complement of the union of the disks in R. These are the connected components of the set $\Re^2 - \bigcup_{d \in R} d$. We also need the combinatorial result that the number of these faces is $O(m)$. Furthermore, the total number of vertices and edges on the boundaries of all these faces, that is, the combinatorial complexity of the boundary of the union of disks in R, is $O(m)$ [15]. We make standard general position assumptions about the input set I of disks in this article. This helps simplify the exposition and is without loss of generality.

Lemma 1. *Let R be a set of disks in the plane, and Q a set of points so that (a) no point from Q is contained in any disk from R, and (b) no face in the complement of the union of the disks in R contains more than one point of Q. Then $|R| = \Omega(|Q|)$.*

Proof. The number of faces in the in the complement of the union of the disks in R is $O(|R|)$.   □

Covering vs. Separating. The input to our problem is a set I of n unit disks, and P a set of k points such that I separates P. Let $P_c \subseteq P$ denote those points contained in some disk in I; and P_s denote the remaining points. We compute an α-approximation to the smallest subset of I that covers P_c using a traditional set-cover algorithm; there are several poly-time algorithms that guarantee that $\alpha = O(1)$ [6,17]. We compute a β-approximation to the smallest subset of I that separates P_s, using the algorithm developed in the rest of this article. We argue below that the combination of the two solutions is an $O(\alpha + \beta)$ approximation to the smallest subset of I that separates P.

Let $OPT \subseteq I$ denote an optimal subset that separates P. Suppose that OPT covers k_1 of the points in P_c and let $k_2 = |P_c| - k_1$. By Lemma 1, $|OPT| = \Omega(k_2)$.

Now, by picking one disk to cover each of the k_2 points of P_c not covered by OPT, we see that there is a cover of P_c of size at most $|OPT| + k_2 = O(|OPT|)$. Thus, our α-approximation has size $O(\alpha) \cdot |OPT|$. Since OPT also separates P_s, our β-approximation has size $O(\beta) \cdot |OPT|$. Thus the combined solution has size $O(\alpha + \beta) \cdot |OPT|$.

In the rest of the article, we abuse notation and assume that no point in the input set P is contained in any disk in I, and describe a poly-time algorithm that computes an $O(1)$-approximation to the optimal subset of I that separates P.

3   Separating Two Points

Let s and t be two points in the plane, and G a set of disks such that no disk in G contains either s or t, but G separates s and t. See Figure 2. Our goal is to find a minimum-cardinality subset B of G that separates s and t. We describe below a polynomial-time algorithm that returns a constant factor approximation to this problem.

Without loss of generality, we may assume that the intersection graph of G is connected. (Otherwise, we apply the algorithm to each connected component for which the disks in the component separate s and t. We return the best solution

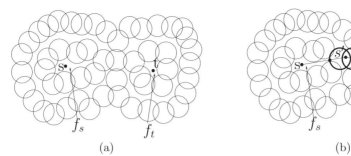

Fig. 2. (a) The figure shows faces f_s and f_t (b) This figure shows the sequence of disks in σ (their boundaries are bold) and the path π

obtained.) Let f_s and f_t denote the faces containing s and t, respectively, in the arrangement of G. We augment the intersection graph of G with vertices corresponding to s and t, and add an edge from s to each disk that contributes an edge to the boundary of the face f_s, and an edge from t to each disk that contributes an edge to the boundary of the face f_t. We assign a cost of 0 to s, t, and a cost of 1 to each disk in G. We then find a shortest path from s to t in this graph, where the length of a path is the number of the *vertices* on it that correspond to disks in G. Let σ denote the sequence of disks on this shortest path. Note that any two disks that are not consecutive in σ do not intersect.

Using σ, we compute a path π in the plane, as described below, from s to t so that (a) there are points s' and t' on π so that the portion of π from s to s' is in f_s, and the portion from t' to t is in f_t; (b) every point on π from s' to t' is contained in some disk from σ; (c) the intersection of π with each disk in σ is connected. See Figure 2.

Suppose that the sequence of disks in σ is $d_1, \ldots, d_{|\sigma|}$. Let s' (resp. t') be a point in d_1 (resp. d_σ) that lies on the boundary of f_s (resp. f_t). For $1 \le i \le |\sigma|-1$, choose x_i to denote an arbitrary point in the intersection of d_i and d_{i+1}. The path π is constructed as follows: Take an arbitrary path from s to s' that lies within f_s, followed by the line segments $\overline{s'x_1}, \overline{x_1 x_2}, \ldots, \overline{x_{|\sigma|-2} x_{|\sigma|-1}}, \overline{x_{|\sigma|-1} t'}$, followed by an arbitrary path from t' to t that lies within f_t.

Properties (a) and (b) hold for π by construction. Property (c) is seen to follow from the fact that disks that are not consecutive in σ do not overlap.

Notice that π "cuts" each disk in σ into two pieces. (Formally, the removal of π from any disk in σ yields two connected sets.) The path π may also intersect other disks and cut them into two or more pieces, and we refer to these pieces as disk pieces. For a disk that π does not intersect, there is only one disk piece, which is the disk itself.

We consider the intersection graph H of the disk pieces that come from disks in G. Observe that a disk piece does not have points on π, since π is removed; so two disk pieces intersecting means there is a point outside π that lies in both of them. In this graph, each disk piece has a cost of 1.

In this graph H, we compute, for each disk $d \in \sigma$, the shortest path between the two pieces corresponding to d. Suppose $d' \in \sigma$ yields the overall shortest path

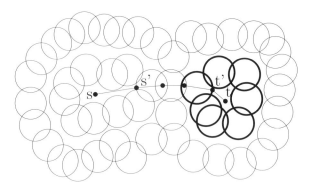

Fig. 3. This figure continues with the example of Figure 2. The disks with bold boundary are the set D computed by our algorithm. The only disk from G with bold boundary has two disk pieces, and the shortest path between them in graph H yields D.

σ'; let D denote the set of disks that contribute a disk piece to this shortest path. Our algorithm returns D as its computed solution. See Figure 3.

We note that D separates s and t – in particular, the union of the disk pieces in σ' and the set $\pi \cap d'$ contains a cycle in the plane that intersects the path π between s and t exactly once.

3.1 Bounding the Size of the Output

Let B^* denote the smallest subset of G that separates s and t. We will show that $|D| = O(|B^*|)$. Let f^* denote the face containing s in the arrangement of B^*. Due to the optimality of B^*, we may assume that the boundary of f^* has only one component. Let a (resp. b) denote the first (resp. last) point on path π where π leaves f^*. It is possible that $a = b$. We find a minimum cardinality contiguous subsequence $\overline{\sigma}$ of σ that contains the subpath of π from a to b; let d_a and d_b denote the first and last disks in $\overline{\sigma}$. See Figure 4.

We claim that $|\overline{\sigma}| \leq |B^*| + 2$; if this inequality does not hold, then we obtain a contradiction to the optimality of σ by replacing the disks in the $\overline{\sigma} \setminus \{d_a, d_b\}$ by B^*.

Consider the face f containing s in the arrangement with $B^* \cup \overline{\sigma}$. Each edge that bounds this face comes from a single disk piece, except for one edge corresponding to d_a that may come from two disk pieces. (This follows from the fact that the portion of π between a and b is covered by the disks in $\overline{\sigma}$.) These disk pieces induce a path in H in between the two pieces from d_a, and their cost therefore upper bounds the cost of D. We may bound the cost of these disk pieces by the number of edges on the boundary of f (with respect to $B^* \cup \overline{\sigma}$). The number of such edges is $O(|B^*| + |\overline{\sigma}|) = O(|B^*|)$.

Theorem 1. *Let s and t be two points in the plane, and G a set of disks such that no disk in G contains either s or t, but G separates s and t. There is a polynomial time algorithm that takes such G, s, and t as input, and outputs a subset $B \subseteq G$ that separates s and t; the size of B is at most a multiplicative constant of the size of the smallest subset $B^* \subseteq G$ that separates s and t.*

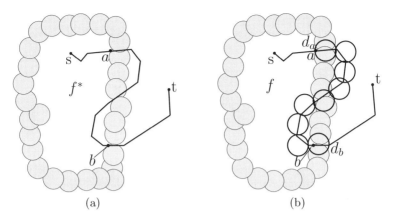

Fig. 4. The shaded disks are in B^*.(a) This figure shows points a and b where π leaves f^* for the first and last time, respectively. (b) This figure shows the face f containing s in the arrangement with $B^* \cup \bar{\sigma}$.

4   Separating Multiple Points

We now present a polynomial time algorithm that yields an $O(1)$ approximation to the problem of finding a minimum subset of I that separates the set P of points. The algorithm is obtained by calling recSep(P), where recSep(Q), for any $Q \subseteq P$ is the following recursive procedure:

1. If $|Q| \leq 1$, return $\emptyset$.
2. For every pair of points $s, t \in Q$, invoke the algorithm of Theorem 1 (with $G \leftarrow I$) to find a subset $B_{s,t} \subseteq I$ such that $B_{s,t}$ separates s and t.
3. Let B denote the minimum size subset $B_{s,t}$ over all pairs s and t considered.
4. Consider the partition of Q into subsets so that each subset corresponds to points in the same face (with respect to B). Suppose $Q_1, \ldots, Q_\tau$ are the subsets in this partition. Note that $\tau \geq 2$, since B separates some pair of points in Q.
5. Return $B \cup \bigcup_{j=1}^{\tau} \text{recSep}(Q_j)$.

Clearly, recSep(P) yields a separator for P. To bound the size of this separator, let us define a set $\mathcal{Q}$ that contains as its element any $Q \subseteq P$ such that $|Q| \geq 2$ and recSep(Q) is called somewhere within the call to recSep(P). For any $Q \in \mathcal{Q}$, define B_Q to be the set B that is computed in the body of the call to recSep(Q). Notice that recSep(P) returns $\cup_{Q \in \mathcal{Q}} B_Q$.

Now we "charge" each such B_Q to an arbitrary point within $p_Q \in Q$ in such a way that no point in P is charged more than once. A moment's thought reveals that this is indeed possible. (In a tree where each interval node has degree at least 2, the number of leaves is greater than the number of internal nodes.)

Let OPT denote the optimal separator for P and let $F_Q \subseteq OPT$ denote the disks that contribute to the boundary of the face (in the arrangement of OPT)

containing p_Q. We claim that $|B_Q| = O(|F_Q|)$; indeed F_Q separates $p_Q \in Q$ from any point in P, and thus any point in Q. Thus for any $t \in Q \setminus \{p_Q\}$, we have $|B_Q| \leq B_{p_Q, t} = O(|F_Q|)$.

We thus have

$$\bigcup_{Q \in \mathcal{Q}} |B_Q| \leq \sum_{Q \in \mathcal{Q}} O(|F_Q|) = O(|OPT|),$$

where the last equality follows from union complexity.

We have derived the main result of this paper:

Theorem 2. *Let I be a set of n unit disks and P a set of k points such that I separates P. There is a polynomial time algorithm that takes as input such I and P, and returns a subset $O \subseteq I$ of disks that also separates P, with the guarantee that $|O|$ is within a multiplicative $O(1)$ of the smallest subset of I that separates P.*

5   Conclusions

We have a presented a polynomial-time $O(1)$-approximation algorithm for finding the minimum subset of a given set I of disks that separates a given set of points P. One way to understand our contribution is as follows. Suppose we had at our disposal an efficient algorithm that (nearly) optimally separates a single point $p \in P$ from every other point in P. Then applying this algorithm for each point in P, we get a separator for P. That the size of this separator is within $O(1)$ of the optimal is an easy consequence of union complexity. However, we only have at our disposal an efficient algorithm for a weaker task: that of approximately separating two given points in P. What we have shown is that even this suffices for the task of obtaining an $O(1)$ approximation to the overall problem.

It is easy to see that our algorithm and the approximation guarantee generalize, for example, to the case when the disks have arbitrary and different radii.

We have not really attempted to establish the NP-hardness of the problem considered here. Interestingly, the two-point separation problem appears to admit a polynomial-time algorithm [7], via the same approach used in the context of line segments [2].

Acknowledgements. We thank Alon Efrat for discussions that led to the formulation of the problem, and Sariel Har-Peled for discussions that led to the algorithm described here.

References

1. Agarwal, P.K., Sharir, M.: Davenport-Schinzel Sequences and Their Geometric Applications. Cambridge University Press, Cambridge (1998)
2. Alt, H., Cabello, S., Giannopoulos, P., Knauer, C.: Minimum cell connection and separation in line segment arrangements (2011)(manuscript),
   http://arxiv.org/abs/1104.4618

3. Aronov, B., Ezra, E., Sharir, M.: Small-size epsilon-nets for axis-parallel rectangles and boxes. SIAM J. Comput. 39(7), 3248–3282 (2010)
4. Balister, P., Zheng, Z., Kumar, S., Sinha, P.: Trap coverage: Allowing coverage holes of bounded diameter in wireless sensor networks. In: Proc. of IEEE INFOCOM, Rio de Janeiro (2009)
5. Bereg, S., Kirkpatrick, D.: Approximating Barrier Resilience in Wireless Sensor Networks. In: Dolev, S. (ed.) ALGOSENSORS 2009. LNCS, vol. 5804, pp. 29–40. Springer, Heidelberg (2009)
6. Brönnimann, H., Goodrich, M.T.: Almost optimal set covers in finite VC-dimension. Discrete & Computational Geometry 14(4), 463–479 (1995)
7. Cabello, S.: Personal Communication (May 2011)
8. Chalermsook, P., Chuzhoy, J.: Maximum independent set of rectangles. In: Proceedings of the Twentieth Annual ACM-SIAM Symposium on Discrete Algorithms, SODA 2009, pp. 892–901. Society for Industrial and Applied Mathematics, Philadelphia (2009)
9. Chan, T.M., Har- Peled, S.: Approximation algorithms for maximum independent set of pseudo-disks. In: Proc. Symposium on Computational Geometry, SCG 2009, pp. 333–340 (2009)
10. Clarkson, K.L., Varadarajan, K.: Improved approximation algorithms for geometric set cover. In: Proc. Symposium on Computational Geometry, SCG 2005, pp. 135–141 (2005)
11. Erickson, J., Har- Peled, S.: Optimally cutting a surface into a disk. Discrete & Computational Geometry 31(1), 37–59 (2004)
12. Fox, J., Pach, J.: Computing the independence number of intersection graphs. In: Proceedings of the ACM-SIAM Symposium on Discrete Algorithms, SODA 2011 (2011)
13. Garg, N., Vazirani, V.V., Yannakakis, M.: Multiway cuts in node weighted graphs. Journal of Algorithms 50(1), 49–61 (2004)
14. Gibson, M., Kanade, G., Varadarajan, K.: On isolating points using disks (2011) (manuscript), http://arxiv.org/abs/1104.5043v1
15. Kedem, K., Livne, R., Pach, J., Sharir, M.: On the union of Jordan regions and collision free translational motion amidst polygonal obstacles. Discrete Comput. Geom. 1, 59–71 (1986)
16. Kumar, S., Lai, T. H., Arora, A.: Barrier coverage with wireless sensors. In: MobiCom 2005: Proceedings of the 11th Annual International Conference on Mobile Computing and Networking, pp. 284–298. ACM, New York (2005)
17. Mustafa, N.H., Ray, S.: PTAS for geometric hitting set problems via local search. In: Proc. Symposium on Computational Geometry, SCG 2009, pp. 17–22 (2009)
18. Reif, J.: Minimum s-t cut of a planar undirected network in $o(n \log^2 n)$ time. SIAM Journal on Computing 12, 71–81 (1983)
19. Sankararaman, S., Efrat, A., Ramasubramanian, S., Taheri, J.: Scheduling sensors for guaranteed sparse coverage (2009) (manuscript), http://arxiv.org/abs/0911.4332

An Output-Sensitive Approach for the L_1/L_∞ k-Nearest-Neighbor Voronoi Diagram*

Chih-Hung Liu[1], Evanthia Papadopoulou[2], and D.T. Lee[1,3]

[1] Research Center for Information Technology Innovation, Academia Sinica, Taiwan
chliu_10@citi.sinica.edu.tw, dtlee@iis.sinica.edu.tw
[2] Faculty of Informatics, University of Lugano, Lugano, Switzerland
evanthia.papadopoulou@usi.ch
[3] Institute of Information Science, Academia Sinica, Taipei, Taiwan

Abstract. This paper revisits the k-nearest-neighbor (k-NN) Voronoi diagram and presents the first output-sensitive paradigm for its construction. It introduces the k-NN Delaunay graph, which corresponds to the graph theoretic dual of the k-NN Voronoi diagram, and uses it as a base to directly compute the k-NN Voronoi diagram in R^2. In the L_1, L_∞ metrics this results in $O((n + m) \log n)$ time algorithm, using segment-dragging queries, where m is the structural complexity (size) of the k-NN Voronoi diagram of n point sites in the plane. The paper also gives a tighter bound on the structural complexity of the k-NN Voronoi diagram in the L_∞ (equiv. L_1) metric, which is shown to be $O(\min\{k(n - k), (n - k)^2\})$.

1 Introduction

Given a set S of n point sites $\in R^d$ and an integer k, $1 \leq k < n$, the *k-nearest-neighbor Voronoi Diagram*, abbreviated as *k-NN Voronoi diagram* and denoted as $V_k(S)$, is a subdivision of R^d into regions, called *k-NN Voronoi regions*, each of which is the locus of points closer to a subset H of k sites, $H \subset S$, called a *k-subset*, than to any other k-subset of S, and is denoted as $V_k(H, S)$. The distance between a point p and a point set H is $d(p, H) = \max\{d(p, q), \forall q \in H\}$, where $d(p, q)$ denotes the distance between two points p and q.

A k-NN Voronoi region $V_k(H, S)$ is a polytope in R^d. The common face between two neighboring k-NN Voronoi regions, $V_k(H_1, S)$ and $V_k(H_2, S)$, is portion of the bisector $B(H_1, H_2) = \{r \mid d(r, H_1) = d(r, H_2), r \in R^d\}$. In R^2, the boundary between two neighboring k-NN Voronoi regions is a *k-NN Voronoi edge*, and the intersection point among more than two neighboring k-NN Voronoi

---

* This work was performed while the first and third authors visited University of Lugano in September/October 2010. It was supported in part by the University of Lugano during the visit, by the Swiss National Science Foundation under grant SNF-200021-127137, and by the National Science Council, Taiwan under grants No. NSC-98-2221-E-001-007-MY3, No. NSC-98-2221-E-001-008-MY3, and No. NSC-99-2918-I-001-009.

C. Demetrescu and M.M. Halldórsson (Eds.): ESA 2011, LNCS 6942, pp. 70–81, 2011.

regions is a *k-NN Voronoi vertex*. In the L_p-metric the distance $d(s,t)$ between two points s,t is $d_p(s,t) = (|x_{1_s} - x_{1_t}|^p + |x_{2_s} - x_{2_t}|^p + \cdots + |x_{d_s} - x_{d_t}|^p)^{1/p}$ for $1 \le p < \infty$, and $d_\infty(s,t) = \max(|x_{1_s} - x_{1_t}|, |x_{2_s} - x_{2_t}|, \ldots, |x_{d_s} - x_{d_t}|)$.

Lee [14] showed that the *structural complexity*, i.e., the size, of the k-NN Voronoi diagram in the plane is $O(k(n-k))$, and proposed an iterative algorithm to construct the diagram in $O(k^2 n \log n)$ time and $O(k^2(n-k))$ space. Agarwal et al. [3] improved the time complexity to be $O(nk^2 + n \log n)$. Based on the notions of geometric duality and arrangements, Chazelle and Edelsbrunner [8] developed two versions of an algorithm, which take $O(n^2 \log n + k(n - k) \log^2 n)$ time and $O(k(n-k))$ space, and $O(n^2 + k(n-k) \log^2 n)$ time and $O(n^2)$ space, respectively. Clarkson [9], Mulmuley [16], and Agarwal et al. [2] developed randomized algorithms, where the expected time complexity of $O(k(n - k) \log n + n \log^3 n)$ in [2] is the best. Boissonnat et al. [6] and Aurenhammer and Schwarzkopf [5] proposed on-line randomized incremental algorithms: the former in expected $O(n \log n + nk^3)$ time and $O(nk^2)$ space, and the latter in expected $O(nk^2 \log n + nk \log^3 n)$ time and $O(k(n - k))$ space. For higher dimensions, Edelsbrunner et al. [11] devised an algorithm to compute all $V_k(S)$ in R^d with the Euclidean metric for $1 \le k < n$, within optimal $O(n^{d+1})$ time and space, and Clarkson and Shor [10] showed the total size of all $V_k(S)$ to be $O(k^{\lceil (d+1)/2 \rceil} n^{\lfloor (d+1)/2 \rfloor})$.

All the above-mentioned algorithms [2,5,6,8,9,14,16] focus on the Euclidean metric. However, the computationally simpler, piecewise linear, L_1 and L_∞ metrics are very well suited for practical applications. For example, L_∞ higher-order Voronoi diagrams have been shown to have several practical applications in VLSI design e.g., [17,18,19]. Most existing algorithms compute k-NN Voronoi diagrams using reductions to arrangements or geometric duality [2,5,8,9,16] which are not directly applicable to the L_1, L_∞ metrics. Furthermore, none of the existing deterministic algorithms is output-sensitive, i.e., their time complexity does not only depend on the actual size of the k-NN Voronoi diagram. For example, the iterative algorithm in [14] needs to generate $V_1(S), V_2(S), \ldots, V_k(S)$, and thus has a lower bound of time complexity $\Omega(nk^2)$. The algorithm in [8] needs to generate $\Theta(n^2)$ bisectors, while not all the bisectors appear in $V_k(S)$.

In this paper, we revisit the k-NN Voronoi diagram and propose the first direct output-sensitive approach to compute the L_∞ (equiv. L_1) k-NN Voronoi diagram. We first formulate the k-NN Delaunay graph (Section 2), which is the graph-theoretic dual of the k-NN Voronoi diagram. Note that the k-NN Delaunay graph is a graph-theoretic structure, different from the k-Delaunay graph in [1,12]. We then develop a traversal-based paradigm to directly compute the k-NN Delaunay graph of point sites in the plane (Section 3). In the L_∞ metric we implement our paradigm by applying segment-dragging techniques (Section 4), resulting in an $O((m + n) \log n)$-time algorithm for the L_∞ planar k-NN Delaunay graph of size m. As a by-product, we also derive a tighter bound on the structural complexity m of the L_∞ k-NN Voronoi diagram, which is shown to be $O(\min\{k(n-k), (n-k)^2\})$. Since the L_1 metric is equivalent to L_∞ under rotation, these results are also applicable to the L_1 metric. Due to the limit of space, we remove most proofs except Theorem 2.

2   The k-Nearest-Neighbor Delaunay Graph

Given a set S of n point sites in R^d, we define the k-NN Delaunay graph following the notion of the Delaunay Tessellation [4]. Let a *sphere* denote the boundary of an L_p-ball, and the interior of a sphere denote the interior of the corresponding L_p-ball. A sphere is said to *pass through* a set H of sites, $H \subset S$, if and only if its boundary passes through at least one site in H and it contains in its interior all the remaining sites in H. A sphere *contains* a set H of sites if and only if its interior contains all the sites in H. Given a sphere, sites located on its boundary, in its interior, and in its exterior are called *boundary sites*, *interior sites*, and *exterior sites*, respectively. A k-element subset of S is called a *k-subset*.

Definition 1. *Given a set S of n point sites $\in R^d$ and an L_p metric, a k-subset H, $H \subset S$, is called* valid *if there exists a sphere that contains H but does not contain any other k-subset; H is called* invalid, *otherwise. A valid k-subset H is represented as a graph-theoretic node, called a k-NN Delaunay node.*

A k-NN Delaunay node H can be embedded in R^d as a point in $V_k(H, S)$ i.e., as the center of a sphere that contains H but no other k-subset. A k-NN Delaunay node is a graph-theoretic node, however, it also has a geometric interpretation as it corresponds to the k-subset of sites it uniquely represents.

Definition 2. *Two k-Delaunay nodes, H_1 and H_2, are connected with a k-NN Delaunay edge (H_1, H_2) if and only if there exists a sphere that passes through both H_1 and H_2 but does not pass through any other k-subset. The graph $G(V, E)$, where V is the set of all the k-NN-Delaunay nodes, and E is the set of all k-NN Delaunay edges, is called a k-NN Delaunay graph (under the corresponding L_p metric).*

Lemma 1. *Given a set S of point sites $\in R^d$, two k-NN Delaunay nodes H_1 and H_2, are joined by a k-NN Delaunay edge if and only if (1) $|H_1 \cap H_2| = k-1$, and (2) There exists a sphere whose boundary passes through exactly two sites, $p \in H_1 \setminus H_2$ and $q \in H_2 \setminus H_1$, and whose interior contains $H_1 \cap H_2$ but does not contain any site $r \in S \setminus (H_1 \cup H_2)$.*

Let $H_1 \oplus H_2 = H_1 \setminus H_2 \cup H_2 \setminus H_1$. Following Lemma 1, a k-NN Delaunay edge (H_1, H_2), corresponds to a collection of spheres each of which passes through exactly two sites, p and $q \in H_1 \oplus H_2$, and contains exactly $k-1$ sites in $H_1 \cap H_2$ in its interior (Note that under the general position assumption, $|H_1 \oplus H_2| = 2$).

Theorem 1. *Given a set S of point sites $\in R^d$, the k-NN Delaunay graph of S is the graph-theoretic dual of the k-NN Voronoi diagram of S.*

3   Paradigm for the k-NN Delaunay Graph in R^2

In this section we present a paradigm to directly compute a k-NN Delaunay graph for a set S of n point sites $\in R^2$. We make the general position assumption that no more than three sites are located on the same circle, where a circle is a sphere in R^2. Under the general position assumption, the k-NN Delaunay graph is a planar triangulated graph, in which any chordless cycle is a triangle, and we call it a k-NN Delaunay triangulation. This assumption is removed in Section 3.2.

Our paradigm consists of the following two steps:

1. (Section 3.1) Compute the *k-NN Delaunay hull* and all the *extreme k-NN Delaunay circuits*, defined in Section 3.1 (Definition 4).
2. (Section 3.2) For each extreme k-NN Delaunay circuit find a *k-NN Delaunay triangle* of a *k-NN Delaunay component*, defined in Section 3.1 (Definition 4), traverse from the k-NN Delaunay triangle to its adjacent k-NN Delaunay triangles, and repeatedly perform the traversal operation until all triangles of the k-NN Delaunay component have been traversed.

3.1 k-NN Delaunay Triangles, Circuits, Components and Hull

A *k-NN Delaunay triangle*, denoted as $T(H_1, H_2, H_3)$, is a triangle connecting three k-NN Delaunay nodes, H_1, H_2, and H_3, by three k-NN Delaunay edges. The circle passing through all the three k-subsets H_1, H_2, and H_3 is called the *circumcircle* of $T(H_1, H_2, H_3)$. Note that the circumcircle of $T(H_1, H_2, H_3)$ is a circle induced by three points $p \in H_1$, $q \in H_2$, and $r \in H_3$.

Lemma 2. *Given a set S of point sites in R^2, three k-NN Delaunay nodes, H_1, H_2, and H_3, form a k-NN Delaunay triangle if and only if (1) $|H_1 \cap H_2 \cap H_3|$ $= k - 1$ or $k - 2$, and (2) there exists a circle that passes through $p \in H_1 \setminus H_2$, $q \in H_2 \setminus H_3$, and $r \in H_3 \setminus H_1$, contains $H_1 \cap H_2 \cap H_3$ in its interior, but does not contain any site $t \in S \setminus H_1 \cup H_2 \cup H_3$ in its interior or boundary. This circle is exactly the unique circumcircle of $T(H_1, H_2, H_3)$.*

Following Lemma 2, a k-NN Delaunay triangle $T(H_1, H_2, H_3)$ is also denoted as $T(p, q, r)$, where $p \in H_1 \setminus H_2$, $q \in H_2 \setminus H_3$, and $r \in H_3 \setminus H_1$ are the boundary sites of its circumcircle.

Definition 3. *An* unbounded circle *is a circle of infinite radius. A k-NN Delaunay node H is called* extreme *if there exists an unbounded circle that contains H but does not contain any other k-subset. A k-NN Delaunay edge is* extreme *if it connects two extreme k-NN Delaunay nodes.*

Definition 4. *The k-NN Delaunay hull is a cycle connecting all the extreme k-NN Delaunay nodes by the extreme k-NN Delaunay edges. An* extreme k-NN *Delaunay circuit is a simple cycle consisting of extreme k-NN Delaunay nodes and extreme k-NN Delaunay edges. A k-NN Delaunay component is a maximal collection of k-NN Delaunay triangles bounded by an extreme k-NN Delaunay circuit.*

Remark 1. *In the L_1, L_∞ metrics a k-NN Delaunay hull may consist of several extreme k-NN Delaunay circuits. In addition, a k-NN Delaunay graph may consist of several k-NN Delaunay components, and some k-NN Delaunay components may share a k-NN Delaunay node, called k-NN Delaunay cut node.*

A triangulated graph TG is *triangularly connected* if for each pair of triangles T_s and $T_t \in$ TG, there exists a sequence of triangles, $T_s = T_1, T_2, \cdots, T_l = T_t \in$ TG, where T_i and T_{i+1} are adjacent to each other (i.e., they share a common edge) for $1 \leq i < l$. A k-NN Delaunay component is triangularly connected.

To compute the k-NN Delaunay hull, we first find an extreme k-NN Delaunay edge, then traverse from it to its adjacent k-NN Delaunay edge, and repeatedly perform the traversal operation until all the extreme k-NN Delaunay edges have been traversed. Lemma 3 implies that there must always exist at least one extreme k-NN Delaunay edge.

Lemma 3. *Consider a set S of point sites in R^2. There exist two k-subsets, H_1 and H_2 and an unbounded circle that passes through H_1 and H_2 but does not pass through any other k-subset.*

Consider two adjacent extreme k-NN Delaunay edges (H_1, H_2) and (H_2, H_3), where $H_1 = H \cup \{p, r\}$, $H_2 = H \cup \{q, r\}$, and $|H| = k - 2$. Since $|H_2 \cap H_3| = k - 1$, by Lemma 1, H_3 is either $H \cup \{r, t\}$ or $H \cup \{q, t\}$ for some site $t \notin H$. If $H_3 = H \cup \{r, t\}$, there exists an unbounded circle that passes through q and t and contains $H \cup \{r\}$ but does not contain any other site. If $H_3 = H \cup \{q, t\}$, there exists an unbounded circle that passes through r and t and contains $H \cup \{q\}$ but does not contain any other site. For the former case, to identify H_3, it is enough to compute a site $t \notin H_1 \cup H_2$ such that the unbounded circle formed by q and t contains $H_1 \cap H_2$ but does not contain any other site. For the latter case, we compute a site $r \in H_1 \cap H_2$ and a site $t \notin H_2$ such that the unbounded circle formed by r and t contains $H_2 \setminus \{q\}$ but does not contain any other site. The details of the hull construction in the L_∞ metric are discussed in Section 4.1.

3.2   Traversal-Based Operation among Triangles

Under the general position assumption, a k-NN Delaunay triangle is dual to a k-NN Voronoi vertex (Theorem 1). A k-NN Delaunay triangle $T = T(H_1, H_2, H_3)$ is categorized as *new* or *old* according to the number of interior sites of its circumcircle as follows: if $|H_1 \cap H_2 \cap H_3| = k - 1$, T is *new*, and if $|H_1 \cap H_2 \cap H_3| = k - 2$, T is *old*. The terms new and old follow the corresponding terms for k-NN Voronoi vertices in [14].

We propose a circular wave propagation to traverse from $T_1 = T(H_1, H_2, H_3) = T(p, q, r)$ to $T_2 = T(H_1, H_2, H_4) = T(p, q, t)$. As mentioned in Section 2, a k-NN Delaunay edge (H_1, H_2) corresponds to a collection of circles whose boundary sites are exactly two sites, p and q, $\in H_1 \oplus H_2$ and whose interior sites are exactly $k - 1$ sites $\in H_1 \cap H_2$. Therefore, traversal from T_1 to T_2 is like having a specific circular wave propagation which begins as the circumcircle of T_1, then follows the collection of circles corresponding to Delaunay edge (H_1, H_2) in a continuous order, and ends as the circumcircle of T_2. During the propagation, the circular wave keeps touching p and q and contains exactly the $k - 1$ sites in $H_1 \cap H_2$ in its interior, while moving its center along $B(H_1, H_2) = B(p, q)$.

If T_1 is new, the circular wave moves along the direction of $B(p, q)$ that excludes r, and if T_1 is old, the circular wave moves along the opposite direction of

$B(p,q)$ to include r. Otherwise, the circular wave would not contain $k-1$ sites in its interior, and thus it would not correspond to the common k-NN Delaunay edge (H_1, H_2). The circular wave terminates when it touches a site $t \notin \{p, q, r\}$. If $t \in H_1 \cap H_2 \cap H_3$, the resulting circle contains $k-2$ sites in its interior and T_2 is old; if $t \notin H_1 \cup H_2 \cup H_3$, the resulting circle contains $k-1$ sites in its interior and T_2 is new.

Using the traversal operation and assuming that we can identify a k-NN Delaunay triangle in an extreme k-NN Delaunay circuit as a starting triangle, we can compute the entire incident k-NN Delaunay component. Since we already have all the extreme k-NN Delaunay edges after the k-NN Delaunay hull construction, we can use an extreme k-NN Delaunay edge to compute its incident k-NN Delaunay triangle and use it as a starting triangle.

If we remove the general position assumption, the dual of a k-NN Voronoi vertex becomes a chordless cycle of the k-NN Delaunay graph, called a *k-NN Delaunay cycle*, and Lemma 2 generalizes to Lemma 4.

Lemma 4. *Given a set S of point sites $\in R^2$, l k-NN Delaunay nodes, $H_1, H_2, \cdots,$ and H_l, form a k-NN Delaunay cycle $(H_1, H_2, \cdots, H_l)$ if and only if (1) $k+1-l \leq |H_1 \cap H_2 \cap \cdots \cap H_l| \leq k-1$ and $|H_i \setminus H_{i+1}| = 1$, for $1 \leq i \leq l$, where $H_{l+1} = H_1$, (2) there exists a circle that passes through $c_1 \in H_1 \setminus H_2$, $c_2 \in H_2 \setminus H_3$, $\cdots$, and $c_l \in H_l \setminus H_1$, and contains $\bigcap_{1 \leq i \leq l} H_i$ but does not contain any site $t \in S \setminus \bigcup_{1 \leq i \leq l} H_i$ in its interior or boundary. This circle is the unique circumcircle of the l k-NN Delaunay nodes.*

We use Fig. 1 to illustrate Lemma 4. Fig. 1(a) shows a circle passing through five sites and containing three sites. According to Lemma 4, the circle corresponds to a k-NN Delaunay cycle, $4 \leq k \leq 7$. Fig. 1(b) shows a 5-NN Delaunay cycle. As shown in Fig. 1(c), in order to traverse from this 5-NN Delaunay cycle to its adjacent 5-NN Delaunay cycle via the 5-NN Delaunay edge (H_1, H_2), the corresponding circular wave will follow $B(H_1, H_2) = B(s_1, s_3)$ to exclude s_4 and s_5 and to include s_2 such that it contains $k-1 = 4$ sites, s_2, s_6, s_7, and s_8.

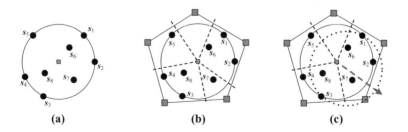

(a)   **(b)**   **(c)**

Fig. 1. (a) A circle passing through five sites (s_1, s_2, s_3, s_4, and s_5) and containing three sites (s_6, s_7, and s_8). (b) A 5-NN Delaunay cycle associated with $H_1 = \{s_1, s_2, s_6, s_7, s_8\}$, $H_2 = \{s_2, s_3, s_6, s_7, s_8\}$, $H_3 = \{s_3, s_4, s_6, s_7, s_8\}$, $H_4 = \{s_4, s_5, s_6, s_7, s_8\}$, and $H_5 = \{s_1, s_5, s_6, s_7, s_8\}$. (c) The circular wave touches two sites (s_1 and s_3) and contains $k-1 = 4$ sites (s_2, s_6, s_7, and s_8).

4   Planar k-NN Delaunay Graph in the L_∞ Metric

We implement our paradigm in the L_∞ metric such that the hull construction takes $O(n \log n)$ time (Section 4.1) and each traversal operation between two triangles takes $O(\log n)$ time (Section 4.2). Since the number of traversal operations is bounded by the number m of k-NN Delaunay edges, we have an $O((n+m) \log n)$-time algorithm to directly compute the L_∞ planar k-NN Delaunay graph. In the L_∞ metric, general position is augmented with the assumption that no two sites are located on the same axis-parallel line.

4.1 L_∞ k-NN Delaunay Hull Computation

To compute the k-NN Delaunay hull, we traverse from one extreme k-NN Delaunay edge to all the others. In the L_∞ metric an unbounded circle passing through two sites is an axis-parallel L-shaped curve. An L-shaped curve partitions the plane into two portions, where one portion is a *quarter-plane*, illustrated shaded in Fig. 2. Therefore, an extreme k-NN Delaunay edge (H_1, H_2) corresponds to an L-shaped curve which passes through exactly two sites, p and q $\in H_1 \oplus H_2$, and whose quarter-plane exactly contains $k-1$ sites $\in H_1 \cap H_2$ in its interior. All the extreme k-NN Delaunay edges can be classified into four categories, $\{\mathrm{NE}, \mathrm{SE}, \mathrm{SW}, \mathrm{NW}\}$, according to the orientation of their corresponding quarter-plane.

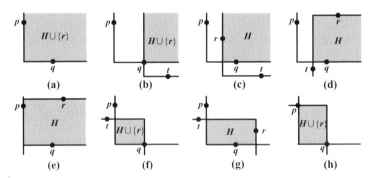

Fig. 2. $e_1 = (H_1, H_2)$ and $e_2 = (H_2, H_3)$, where e_1 is NE. (a) $H_1 = H \cup \{p, r\}$, and $H_2 = H \cup \{q, r\}$. (b)–(c) e_2 is NE. (b) $H_3 = H \cup \{r, t\}$. (c) $H_3 = H \cup \{q, t\}$. (d)–(e) e_2 is SE. (d) $H_3 = H \cup \{q, t\}$. (e) $H_3 = H \cup \{p, q\}$. (f)–(h) e_2 is SW. (f) $H_3 = H \cup \{r, t\}$. (g) $H_3 = H \cup \{q, t\}$. (h) $e_1 = e_2$.

Given an extreme k-NN Delaunay edge $e_1 = (H_1, H_2)$, we propose an approach to find its clockwise adjacent extreme k-NN Delaunay edge $e_2 = (H_2, H_3)$. We only discuss the cases where e_1 is NE, If $H_1 = H \cup \{p, r\}$ and $H_2 = H \cup \{q, r\}$, where $|H| = k-2$, then H_3 is either $H \cup \{r, t\}$ or $H \cup \{q, t\}$, as shown in Fig. 2(b) and Fig. 2(c). Below, we discuss the two cases of H_3 assuming that e_2 is NE, SE, SW, and NW, respectively. The coordinates of p, q, r, and t are denoted as (x_p, y_p), (x_q, y_q), (x_r, y_r), and (x_t, y_t), respectively. Moreover, ε is any value between 0 and the minimum distance among the sites.

1. e_2 is **NE** (Fig. 2(b)–(c)): We first drag a vertical ray, $[(x_p, y_q), (x_p, \infty)]$, right to touch a site $v \in H_2$. Then, we drag a horizontal ray, $[(x_v, y_q), (\infty, y_q)]$, down to touch a site $t \notin H_1 \cup H_2$. If $v = q$, $H_3 = H \cup \{r, t\}$; otherwise, $v = r$ and $H_3 = H \cup \{q, t\}$.

2. e_2 is **SE** (Fig. 2(d)–(e)): If H_3 is $H \cup \{r, t\}$, H_3 cannot exist since there does not exist an SE L-shaped curve which passes through q and a site $t \notin H_2$ and whose quarter half-plane contains $H \cup \{r\}$. Therefore, we first drag a horizontal ray, $[(x_p + \varepsilon, \infty), (\infty, \infty)]$, from infinity downward to touch a site r. Then, we drag a vertical ray, $[(x_p + \varepsilon, y_q), (x_p + \varepsilon, \infty)]$, right to touch a site v. As last, we drag a vertical ray, $[(x_v, y_r), (x_v, -\infty)]$, left to touch a site t. In fact, t is possibly p. Fig. 2(d)–(e) shows the two cases $t \neq p$ and $t = p$.

3. e_2 is **SW** (Fig. 2(f)–(h)): We first drag a vertical ray, $[(\infty, y_q), (\infty, \infty)]$, left to touch a site v. Then, we drag a horizontal ray, $[(x_p + \epsilon, \infty), (\infty, \infty)]$, from infinity downward to touch a site u. At last, we drag a horizontal ray, $[(x_p, y_u), (-\infty, y_u)]$, up to touch a site t. If $v = q$ and $t \neq p$, $H_3 = H \cup \{r, t\}$. If $v \neq q$, $H_3 = H \cup \{q, r\}$. If $v = q$ and $t = p$, $e_2 = e_1$, i.e., e_1 is also SW.

4. e_2 is **NW**: If e_2 is not NE, SE, or SW, e_1 must also be SW, and we consider this case as the case where e_1 is SW.

For each extreme k-NN Delaunay edge, this approach uses a constant number of segment dragging queries to compute its adjacent one. Chazelle [7] proposed an algorithmic technique to answer each orthogonal segment dragging query in $O(\log n)$ time using $O(n \log n)$-time preprocessing and $O(n)$ space.

A sequence of extreme k-NN Delaunay edges in the same category is called *monotonic*. An L_∞ k-NN Delaunay hull consists of at most four monotonic sequences of extreme k-NN Delaunay edges whose clockwise order follows $\{$NE, SE, SW, NW$\}$. Therefore, the number of extreme k-NN Delaunay edges is $O(n - k)$, and it takes $O(n \log n)$ time to compute the k-NN Delaunay hull. During the traversal procedure, once an extreme k-NN Delaunay node has been traversed an even number of times, an extreme k-NN Delaunay circuit has been constructed.

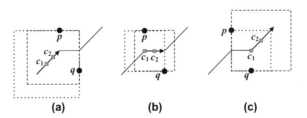

(a)   (b)   (c)

Fig. 3. Square wave propagation along $B(p, q)$. Solid segments are $B(p, q)$, arrow heads are moving directions, and c_1 and c_2 are the centers of dot squares and dash squares, respectively. (a) square contraction. (b) square movement. (c) square expansion.

4.2 L_∞ k-NN Traversal Operation among Triangles

As mentioned in Section 3.2, a traversal operation between triangles corresponds to a circular wave propagation whose center is located on the bisector $B(p, q)$ and which passes through p and q and contains $k-1$ sites in its interior. Since the circular wave propagation will terminate when it touches a site t, the problem reduces to computing the first site to be touched during the circular wave propagation.

In the L_∞ metric, a circle is an axis-parallel square, and a bisector between two points may consist of three parts as shown in Fig. 3. Therefore, a square wave propagation along an L_∞ bisector would consist of three stages: square contraction, square movement, and square expansion. As shown in Fig. 3, square contraction occurs when the center of the square wave moves along ray towards the vertical or horizontal segment, square movement occurs when the center moves along the vertical or horizontal segment, and square expansion occurs when the center moves along a $45°$ or $135°$ ray to infinity.

Below, we discuss the square wave propagation for k-NN Delaunay triangles. Without loss of generality, we only discuss the case that the three boundary sites are located on the left, bottom, and right sides of the corresponding square, respectively. The remaining cases are symmetric.

For a new Delaunay triangle $T(p, q, r)$, the square wave propagation, traversing from $T(p, q, r)$ to its neighbor $T(p, q, t)$, will move along $B(p, q)$ to exclude r until it touches t. If p and q are located on adjacent sides, as shown in Fig. 4(a)–(c), the corresponding portion of bisector $B(p, q)$ may consist of three parts. This square wave propagation will begin as square contraction, then become square movement, and finally turn into square expansion until it first touches a site t. If p and q are located on parallel sides, as shown in Fig. 4(d)–(e), the

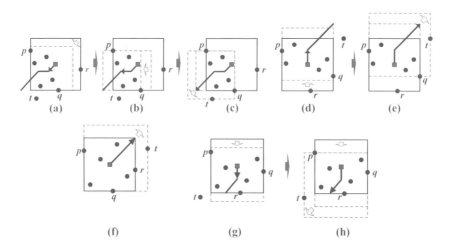

Fig. 4. Square wave propagation for $T(p, q, r)$. (a)–(e) $T(p, q, r)$ is new. (a)–(c) p and q are on adjacent sides. (d)–(e) p and q are on opposite sides. (f)–(h) $T(p, q, r)$ is old. (f) p and q are on adjacent sides. (g)–(h) p and q are on opposite sides.

corresponding portion of bisector $B(p,q)$ may consist of two parts. This square wave propagation will begin as square movement, and finally turn into square expansion until it first touches a site t.

For an old Delaunay triangle $T(p,q,r)$, the square wave propagation, traversing from $T(p,q,r)$ to its neighbor $T(p,q,t)$, will move along $B(p,q)$ to include r until it touches t. If p and q are located on adjacent sides, as shown in Fig. 4(f), the corresponding portion of bisector $B(p,q)$, is just a $45°$ or $135°$ ray, and thus this square wave propagation is just square expansion until it first touches a site t. If p and q are located on parallel sides, as shown in Fig. 4(g)–(h), the corresponding portion of bisector $B(p,q)$ may consist of two parts. This square wave propagation will begin as square movement, and then turn into square expansion until it first touches a site t.

Square contraction is equivalent to dragging two axis-parallel segments perpendicularly and then selecting the closer one of their first touched sites (see Fig. 5(a)). Square movement is similar to square contraction, but the two dragged segments are parallel to each other (see Fig. 5(b)). Since an orthogonal segment-dragging query can be computed in $O(\log n)$ time after $O(n\log n)$-time preprocessing [7], both square contraction and square movement can be answered in $O(\log n)$ time. On the other hand, square expansion is equivalent to four segment-dragging queries (see Fig. 5(c)). However, two of the four segment-dragging queries fall into a new class, in which one endpoint is located on a fixed vertical or horizontal ray, and the other endpoint is located on a fixed $45°$ or $135°$ ray. In [15], Mitchell stated that this class of segment dragging queries can be transformed into a point location query in a specific linear-space subdivision (see Fig. 5(d)), and thus this class of segment dragging queries can be answered in $O(\log n)$ time using $O(n\log n)$-time preprocessing and $O(n)$ space. Therefore, square expansion can also be answered in $O(\log n)$ time.

To conclude, each traversal operation for any k-NN Delaunay triangle takes at most one square contraction, one square movement, and one square expansion, and thus it can be computed in $O(\log n)$ time.

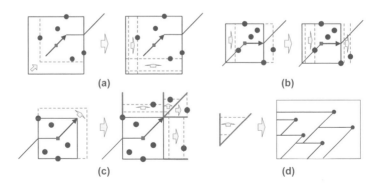

Fig. 5. Square wave propagation and segment-dragging queries. (a) square contraction. (b) square movement. (c) square expansion. (d) the second class of segment-dragging query and the point location query.

4.3   Complexity of L_∞ k-NN Voronoi Diagram

In this section, we use the Hanan grid [13] to derive a tighter bound on the structural complexity of the L_∞ k-NN Voronoi diagram. Given a set S of n point sites in the plane, the Hanan grid is derived by drawing the axis-parallel lines through every point in S.

Given the L_∞ k-NN Delaunay graph, the L_∞ circumcircle of a k-NN Delaunay triangle is a unique square, called k-NN Delaunay square, that passes through three sites (assuming general position i.e., no two points are on the same axis parallel line) and contains $k - 1$ (in case of a new triangle) or $k - 2$ (in case of an old triangle) sites in its interior. The center of a k-NN Delaunay square is exactly a k-NN Voronoi vertex.

Lemma 5. *In the L_∞ metric, a k-NN Delaunay square must have at least two corners on the Hanan grid.*

Theorem 2. *The structural complexity of the L_∞ k-NN Voronoi diagram is $O((n-k)^2)$.*

Proof. Let us number the rows and columns of the Hanan grid 1 to n from right to left and from top to bottom respectively. Let us assume that points are in general position i.e., no two points are on the same axis parallel line. Let p be a point on the Hanan grid such that p is the NW corner of a k-NN Delaunay square D. Square D must enclose exactly $k + 1$ or $k + 2$ sites, including sites on its boundary, and thus no point on the Hanan grid past column $(n - k)$ or below row $(n - k)$ can serve as a NW corner to a k-NN Delaunay square. Hence, there are at most $(n - k)^2$ Hanan grid points that can serve as NW corners of a k-NN Delaunay square. Similarly for all four corner types of a k-NN Delaunay square. In addition, point p can be the NW corner of at most two k-NN Delaunay squares, one containing $k + 1$ sites and the other containing $k + 2$ sites. By Lemma 5, a k-NN Delaunay square must have at least two corners on the Hanan grid. Thus, there can be at most $O((n-k)^2)$ distinct k-NN Delaunay squares, and $O((n - k)^2)$ distinct k-NN Delaunay triangles.

As shown in [14], the size of the k-NN Voronoi diagram is $O(k(n - k))$ and the bound remains valid in the L_P metric. Thus, the following corollary is implied.

Corollary 1. *The structural complexity of the L_∞ k-NN Voronoi diagram, equivalently the k-NN Delaunay graph, is $O(\min\{k(n - k), (n - k)^2\})$.*

5   Conclusion

Based on our proposed traversal-based paradigm, we develop an $O((n+m)\log n)$-time algorithm for the L_∞ k-NN Voronoi diagram of size m using segment-dragging queries. This bound is output-sensitive and it can be considerably smaller than the time complexities of previous methods in the Euclidean metric, $O(nk^2 \log n)$ of [14] and $O(n^2 \log n + k(n - k)\log^2 n)$ or $O(n^2 + k(n - k)\log^2 n)$ of [8]. Since the L_∞ k-NN Voronoi diagram can be computed directly, it is likely that the L_2 k-NN Voronoi diagram can also be computed in a similar manner.

References

1. Abellanas, M., Bose, P., Garcia, J., Hurtado, F., Nicolas, C.M., Ramos, P.A.: On structural and graph theoretic properties of higher order Delaunay graphs. Internat. J. Comput. Geom. Appl. 19(6), 595–615 (2009)
2. Agarwal, P.K., de Berg, M., Matousek, J., Schwarzkopf, I.: Constructing levels in arrangements and higher order Voronoi diagrams. Siam J. on Computing 27(3), 654–667 (1998)
3. Aggarwal, A., Guibas, L.J., Saxe, J., Shor, P.W.: A linear-time algorithm for computing Voronoi diagram of a convex polygon. Discrete and Computational Geometry 4, 591–604 (1984)
4. Aurenhammer, F., Klein, R.: Voronoi Diagrams. In: Handbook of Computational Geometry. Elseiver, Amsterdam (2000)
5. Aurenhammer, F., Schwarzkopf, O.: A simple on-line randomized incremental algorithm for computing higher order Voronoi diagrams. Internat. J. Comput. Geom. Appl. 2, 363–381 (1992)
6. Boissonnat, J.D., Devillers, O., Teillaud, M.: A semidynamic construction of higher-order Voronoi diagrams and its randomized analysis. Algorithmia 9, 329–356 (1993)
7. Chazelle, B.: An algorithm for segment dragging and its implementation. Algorithmica 3, 205–221 (1988)
8. Chazelle, B., Edelsbrunner, H.: An improved algorithm for constructing kth-order Voronoi Diagram. IEEE Trans. on Computers 36(11), 1349–1454 (1987)
9. Clarkson, K.L.: New applications of random sampling in computational geometry. Discrete and Computational Geometry 2, 195–222 (1987)
10. Clarkson, K.L., Shor, P.W.: Applications of random sampling in computational geometry, II. Discrete and Computational Geometry 4, 387–421 (1989)
11. Edelsbrunner, H., O'Rourke, J., Seidel, R.: Constructing arrangements of lines and hyperplanes with applications. SIAM J. on Computing 15, 341–363 (1986)
12. Gudmundsson, J., Hammar, M., van Kreveld, M.: Higher order Delaunay triangulations. Computaional Geometry 23(1), 85–98 (2002)
13. Hanan, M.: On Steiner's problem with rectilinear distance. SIAM J. on Applied Mathematics 14, 255–265 (1966)
14. Lee, D.-T.: On k-nearest neighbor Voronoi Diagrams in the plane. IEEE Trans. on Computers 31(6), 478–487 (1982)
15. Mitchell, J.S.B.: L1 Shortest Paths Among Polygonal Obstacles in the Plane. Algorithmica 8, 55–88 (1992)
16. Mulmuley, K.: On levels in arrangements and Voronoi diagrams. Discrete and Computational Geometry 6, 307–338 (1991)
17. Papadopoulou, E.: Critical Area computation for missing material defects in VLSI circuits. IEEE Trans. on CAD 20(5), 583–597 (2001)
18. Papadopoulou, E.: Net-aware Critical area extraction for opens in VLSI circuits via higher-order Voronoi diagrams. IEEE Trans. on CAD 30(5), 704–716 (2011)
19. Papadopoulou, E., Lee, D.-T.: The L_∞ Voronoi Diagram of Segments and VLSI Applications. Internat. J. Comput. Geom. Appl. 11(5), 503–528 (2001)

Can Nearest Neighbor Searching Be Simple and Always Fast?

Victor Alvarez[1,*], David G. Kirkpatrick[2], and Raimund Seidel[3]

[1] Fachrichtung Informatik, Universität des Saarlandes
`alvarez@cs.uni-saarland.de`
[2] Department of Computer Science, University of British Columbia
`kirk@cs.ubc.ca`
[3] Fachrichtung Informatik, Universität des Saarlandes
`rseidel@cs.uni-saarland.de`

Abstract. Nearest Neighbor Searching, *i.e.* determining from a set S of n sites in the plane the one that is closest to a given query point q, is a classical problem in computational geometry. Fast theoretical solutions are known, e.g. point location in the Voronoi Diagram of S, or specialized structures such as so-called Delaunay hierarchies. However, practitioners tend to deem these solutions as too complicated or computationally too costly to be actually useful.

Recently in ALENEX 2010 Birn *et al.* proposed a simple and practical randomized solution. They reported encouraging experimental results and presented a partial performance analysis. They argued that in many cases their method achieves logarithmic expected query time but they also noted that in some cases linear expected query time is incurred. They raised the question whether some variant of their approach can achieve logarithmic expected query time in all cases.

The approach of Birn *et al.* derives its simplicity mostly from the fact that it applies only one simple type of geometric predicate: which one of two sites in S is closer to the query point q. In this paper we show that any method for planar nearest neighbor searching that relies just on this one type of geometric predicate can be forced to make at least $n-1$ such predicate evaluations during a worst case query.

1 Introduction

Nearest Neighbor Searching is a classical problem in computational geometry. In the simplest non-trivial case it asks to preprocess a set S of n sites in the plane so that for any query point q the site in S that is closest to q can be determined quickly. The problem appears already in Shamos' 1975 seminal paper on "Geometric Complexity" [9]. Already there the problem is reduced to point location in the Voronoi diagram of S, which was then solved in logarithmic time, although using quadratic space, by what is now commonly known as the Dobkin-Lipton slab method. Of course later on various logarithmic time, linear space

---

* Partially supported by CONACYT-DAAD of México.

C. Demetrescu and M.M. Halldórsson (Eds.): ESA 2011, LNCS 6942, pp. 82–92, 2011.

methods for planar point location were proposed, e.g. [6,4], providing a solution to the Nearest Neighbor Searching problem that was completely satisfactory and asymptotically optimal from a theoretical point of view. However, practitioners have considered these solution too complicated or the constants hidden in the asymptotics too large and have frequently employed solutions based on kd-trees [1,8] that do not guarantee optimal worst case performance but appear to be very fast in practice.

Recently Birn *et al.* [2] proposed a new randomized method for Nearest Neighbor Searching that is simple and elegant and appears to be fast in practice in many cases. The method does not rely on a straightforward reduction of Nearest Neighbor Searching to planar point location. Birn *et al.* presented an analysis showing that under certain plausible circumstances their method achieves logarithmic expected query time. However they also noted that in some circumstances their method incurs linear expected query time. They also reported on some attempts to modify their approach to achieve logarithmic query time in all cases, while maintaining simplicity and practicality.

Our paper shows that such attempts must be futile unless additional types of geometric predicates are employed or the input is restricted in some form. The method of Birn *et al.* derives some of its simplicity from the fact that they use only one type of geometric predicate: which of the two sites $a, b \in S$ is closer to the query point q? Or, in other words, where does q lie in relation to the perpendicular bisector $\mathcal{L}_{a,b}$ of sites a and b.

In this paper we consider the problem of Nearest Neighbor searching under the condition that the only geometric predicates used are such tests against perpendicular bisectors between sites, let us call them *bisector tests*. Note that all the fast planar point location methods when applied to Voronoi diagrams use besides such bisector tests (which typically are just tests against Voronoi edges) also some other tests, typically how a query point lies in relation to a vertical or horizontal line through a Voronoi vertex.

Is it possible to achieve fast Nearest Neighbor Searching by just using bisector tests? At first you are inclined to answer no: consider a set S of n sites with one site having a Voronoi region V_a with $n-1$ edges. Testing whether a query point q is contained in this region seems to require a test against each of its $n-1$ edges. But on further thought you notice that V_a can be crossed by many bisectors and containment of q in V_a could perhaps be certified by relatively few tests of q against such bisectors.

In this paper we show that this is in general not the case.

Theorem 1. *There is a set S of n sites in the plane so that any method for Nearest Neighbor Search in S that relies exclusively on bisector tests (i.e. testing the location of the query point against the perpendicular bisector between two sites $a, b \in S$) can be forced to use $n-1$ such tests in the worst case.*

At first sight this theorem seems to convey only a negative message: With bisector tests alone you cannot do guaranteed fast Nearest Neighbor Searching. But there is also a positive take to this, namely advice to the algorithm designer: You want to do guaranteed fast Nearest Neighbor Searching? Then you better use other geometric predicates besides bisector tests, or — your algorithm or its analysis should assume small coordinate representations for the sites and must exploit this assumption. The latter alternative derives from the fact that the set of sites S constructed in the proof of Theorem 1 has exponential spread, i.e. the ratio of smallest and largest inter-site distance in S is exponential in n. This exponential behavior does not seem to be a fluke of our method but in some sense seems inherent in such a lower bound construction.

2 Preliminaries

All of the following will be set in the plane $\mathbb{R}^2$. By *comparing a (query) point q against a (test) line L* we mean determining whether q lies on L or, if not, which of the two open halfplanes bounded by L contains q. We will denote the halfplane that contains q by $L[q]$. We will use the expression "testing q against L" synonymously with "comparing q against L."

Let P be a convex polygon with n edges and let $\mathcal{T}$ be a finite set of lines. We are interested in the computational complexity of methods that ascertain containment of a query point q in polygon P by just comparing q against testlines in $\mathcal{T}$. We will measure this complexity only in terms of number of comparisons of the query point against lines in $\mathcal{T}$. We will completely ignore all other costs of querying, all costs incurred by preprocessing, and also all space issues.

Let us consider a few examples: If $\mathcal{T}$ consists of all lines that support edges of P, then determining whether $q \in P$ can be achieved by comparing q against each line in $\mathcal{T}$. If one of these comparisons is omitted, say the one against the line that supports edge e, then you cannot be sure of actual containment in P since a query point q' close to edge e but outside P would show the same comparison outcomes as a point q close to edge e but inside P. If, on the other hand, $\mathcal{T}$ consists of all lines through pairs of vertices of P then for any query point q only $\log_2 n + O(1)$ comparisons against lines in $\mathcal{T}$ will suffice to determine whether $q \in P$ or not – essentially you just need to perform a binary search of q among all lines in $\mathcal{T}$ that contain the same vertex.

We want to say that a test set $\mathcal{T}$ is *k-bad* for polygon P if no method that determines whether a query point q is contained in P and that employs just comparisons of q against lines in $\mathcal{T}$ can always give correct answers using fewer than k such comparisons. More formally we say that a test set $\mathcal{T}$ is *k-bad* for P if there is a point c so that for every $\mathcal{S} \subset \mathcal{T}$ with $|\mathcal{S}| < k$ we have that $\bigcap\{L[c] | L \in \mathcal{S}\}$ intersects P as well as its complement $\overline{P}$.

Lemma 1. *If a test set $\mathcal{T}$ is k-bad for a polygon P, then any method that determines membership of a query point in P by just using comparisons against lines in $\mathcal{T}$ can be forced to make at least k such comparisons.*

Proof. Let c be the point mentioned in the definition of k-badness of $\mathcal{T}$. We will use an adversary argument with the following strategy: during a query answer all line comparisons as if c were the query point. Assume $k - 1$ tests had been made during a query and $\mathcal{S}$ were the set of lines in $\mathcal{T}$ against which the query point was compared. Let q and q' be points in $\bigcap\{L[c]|L \in \mathcal{S}\}$ with $q \in P$ and $q' \notin P$. Such points must exist by the definition of k-badness. For the query algorithm the points q and q' are indisdinguishable from c, since all three points behave the same on all comparisons against lines in $\mathcal{S}$. But since $q \in P$ and $q' \notin P$ the query needs at least one more comparison against some line in $\mathcal{T}$ in order to produce a correct answer.

Let us call a line L a *shaving line* of polygon P if one of the two closed halfspaces bounded by L contains exactly one edge e of P completely. We say in this case that L *"shaves off e."* Note that every line that contains an edge of P is a shaving line of P.

Lemma 2. *Let P be a convex polygon with $n > 6$ sides and let $\mathcal{T}$ be a set of shaving lines of P. The set $\mathcal{T}$ is $\lfloor n/2 \rfloor$-bad for P.*

Proof. Let $\mathcal{T}$ be such a set of shaving lines. For $L \in \mathcal{T}$ let L^+ be the open halfspace bounded by L that does **not** contain the edge shaved off by L. The halfspace L^+ contains $n - 2 > 4$ vertices. Thus, since $n > 5$ for any three lines $L_1, L_2, L_3 \in \mathcal{T}$ the intersection $L_1^+ \cap L_2^+ \cap L_3^+$ must be non-empty. Thus by Helly's theorem the intersection $\bigcap\{L^+|L \in \mathcal{T}\}$ is non-empty. Let c be a point in this intersection.

Now let $\mathcal{S} \subset \mathcal{T}$ with $|\mathcal{S}| = k - 1 < \lfloor n/2 \rfloor$. There must be two consecutive edges of P so that neither is shaved off by any line in $\mathcal{S}$. Let v be the vertex[1] joining those two edges. Certainly $\bigcap\{L[c]|L \in \mathcal{S}\}$ contains c but also v and also an entire neighborhood of v, and therefore also points that are in P and points that are not in P. Thus $\mathcal{T}$ is $\lfloor n/2 \rfloor$-bad for P.

Lemma 2 allows to prove $\Omega(n)$ lower bounds for anwering point containment queries in an n-sided convex polygon using comparisons against a set of test lines. We want to strengthen this to be able to claim that in the worst case actually at least n such comparisons are needed.

We will strengthen the notion of "shaving" to "closely shaving." For this purpose pick from the edge e of P some point m_e in its relative interior, and we will refer to m_e as the *midstpoint of e* and we will denote the set of all chosen midstpoints, one for each edge, by M_P. We will call a line L that shaves off edge e of P *closely shaving* iff the closed halfplane bounded by L that contains e contains no midstpoint except for m_e, or, in other words, the open halfspace bounded by L that does not contain e contains all midstpoints except for m_e.

Lemma 3. *For $n > 6$ let P be an n-sided convex polygon with set M_P of chosen midstpoint, and let $\mathcal{T}$ be a set of closely shaving test lines.*
    The set $\mathcal{T}$ is n-bad for P.

Proof. Let c be the point as in the proof of Lemma 2, and let $\mathcal{S}$ be a subset of $\mathcal{T}$ containing fewer than n lines. There must be some edge e of P that is not shaved

---

[1] We ignore here the case that P is unbounded and those two edges are unbounded and hence v is a vertex "at infinity." This case can be taken care of either by arguing it separately or by just claiming the slightly weaker ($\lfloor n/2 \rfloor - 1$)-badness.

off by any line in $\mathcal{S}$. Because of the closely shaving property the midstpoint m_e must be contained in the open halfplane $L[c]$ for each $L \in \mathcal{S}$ and the same must be true for an entire neighborhood of m_e. Thus $\bigcap\{L[c]|L \in \mathcal{S}\}$ contains m_e and also an entire neighborhood of m_e, and therfore also points that are in P and points that are not in P. Thus $\mathcal{T}$ is n-bad for P.

3   A Voronoi Diagram Construction

With the results of the preliminary section the strategy for proving the main Theorem 1 should be clear: Construct a set S of n sites, so that in its Voronoi diagram there is a Voronoi cell V_p with $n-1$ edges so that the set $\mathcal{L}_S$ of all perpendicular bisectors between pairs of points in S forms a set of closely shaving lines of V_p (for appropriately chosen midstpoints). Lemma 3 then immediately implies Theorem 1.

For our set S we will choose points from the non-negative part $\mathbb{P}$ of the unit parabola described by $\{u(t)|t \geq 0\}$ with $u(t) = (t, t^2)$. Let $0 \leq t_1 < t_2 < \cdots < t_n$ and let $S = \{u(t_i)|1 \leq i \leq n\}$. It is well known that the structure of the Voronoi diagram of S is completely determined and independent of the actual choices of the t_i's. To see this note that any circle can intersect $\mathbb{P}$ in at most 3 points, which is a consequence of the fact that these intersection points are given by the non-negative roots of a polynomial $(a-t)^2 + (b-t^2)^2 - r^2$, which has coefficient 0 for t^3, but this coefficient is the sum of the four roots, and hence at most 3 of them can be non-negative (except for the uninteresting case that all are 0). If there are 3 intersection points between $\mathbb{P}$ and a circle, then $\mathbb{P}$ must cross the circle in those points and some parts of $\mathbb{P}$ must lie inside the circle. From this, and the fact that for every circle $u(t)$ is outside for sufficiently large t you can characterize which triples of points from S span Delaunay triangles: all triples of the form $(u(t_1), u(t_{i-1}), u(t_i))$ with $3 \leq i \leq n$. This is akin to the well known Gale's evenness condition for the description of the structure of cyclic polytopes [10, page 14].

The structure of the Voronoi diagram of S is now as follow: let V_i denote the Voronoi region of $u(t_i)$; the Voronoi region V_1 neighbors every region $V_2, V_3, \ldots, V_n$ in this counterclockwise order; V_2 neighbors V_1, V_3, for $3 \leq i < n$ the region V_i neighbors V_{i-1}, V_1, V_{i+1}, and V_n neighbors V_1, V_{n-1}. All Voronoi regions are unbounded (since S is in convex position). See Fig. 1 for an example.

For $a, b \geq 0$ consider the perpendicular bisector between the points $u(a)$ and $u(b)$. It is described by the equation

$$y = \frac{-1}{a+b}x + (a^2 + b^2 + 1)/2 \,.$$

Note that its slope is always negative and for a or b sufficiently large the slope can be made arbitrarily close to 0.

Let $\mathcal{L}_{i,j}$ be the perpendicular bisector between $u(t_i)$ and $u(t_j)$. By the discussion above each bisector $\mathcal{L}_{1,j}$ contributes an edge e_j to the Voronoi region $P = V_1$. Our goal will be to show that the t_i's can be chosen so that for each

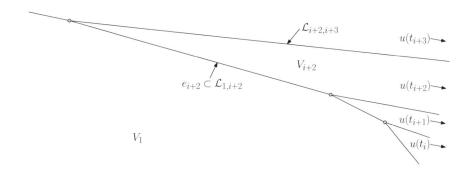

Fig. 1.

$j > 2$ each bisector $\mathcal{L}_{ij}$, with $1 \leq i < j$, closely shaves off edge e_j from polygon P (of course with an appropriate choice of midstpoints m_i on e_i, for $1 < i \leq n$).

We will prove this by induction on n. So assume that for some $n \geq 3$ values $0 = t_1 < t_2 < \cdots < t_n$ have been chosen and for each j with $1 < j \leq n$ a midstpoint m_j on Voronoi edge e_j has been chosen, so that for $1 \leq i < j$ the bisector $\mathcal{L}_{i,j}$ closely shaves off edge e_j from V_1. Now we want to choose $t_{n+1} > t_n$ so that the bisectors $\mathcal{L}_{i,j}$ with $1 \leq i < j \leq n$ remain closely shaving, and the "new" bisectors $\mathcal{L}_{i,n+1}$ with $1 \leq i \leq n$ are closely shaving as well, in particular they closely shave off the new edge e_{n+1} of V_1 contributed by $\mathcal{L}_{1,n+1}$.

Since "closely shaving off" is a local condition, in that it depends only on edge e_j and the midstpoints m_{j-1} and m_{j+1} all bisectors $\mathcal{L}_{i,j}$ with $i < j < n$ definitely remain closely shaving, and the bisectors $\mathcal{L}_{i,n}$ with $i < n$ remain closely shaving, provided the new Voronoi vertex v_n between e_n and e_{n+1} (which is the intersection of $\mathcal{L}_{1,n}$ and $\mathcal{L}_{1,n+1}$) is to the left of midstpoint m_n. It can easily be checked that the x-coordinate of v_n is given by $-(t_n^2 t_{n+1} + t_n t_{n+1}^2)/2$. Thus by making t_{n+1} large enough the Voronoi vertex v_n can be moved as far left on $\mathcal{L}_{1,n}$ as desired.

We further need that all the bisectors $\mathcal{L}_{i,n}$ with $i \leq n$ intersect the new Voronoi edge e_{n+1}. This happens if $\mathcal{L}_{1,n+1}$ has slope closer to 0 than the slopes of all the $\mathcal{L}_{i,n}$'s. Since the slope of $\mathcal{L}_{i,j}$ is given by $-1/(t_i + t_j)$ this can also be achieved by making t_{n+1} suitably large, in particular

$$t_{n+1} \geq t_n + t_{n-1}. \tag{1}$$

The midstpoint m_{n+1} for e_{n+1} can then be chosen as any point on e_{n+1} to the left of all of those intersections $\mathcal{L}_{i,n} \cap \mathcal{L}_{1,n+1}$.

Finally we need to ensure that "new" bisectors $\mathcal{L}_{i,n+1}$ with $1 < i \leq n$ intersect edge e_n between midstpoint m_n and the new Voronoi vertex v_n, moreover their slope should be closer to 0 than the slope of $\mathcal{L}_{1,n+1}$ so that no $\mathcal{L}_{i,n+1}$ can intersect e_{n+1}. Fortunately this slope condition holds automatically, since for $i > 1$ we have $t_i > t_1 = 0$.

Now consider the intersection $\mathcal{L}_{1,n} \cap \mathcal{L}_{i,n+1}$ for some i with $1 < i < n$. Its x-coordinate is given by

$$-\frac{t_n(t_{n+1}^3 + t_{n+1}^2 t_i + t_{n+1}(t_i^2 - t_n^2) + t_i^3 - t_n^2 t_i)}{2(t_i + t_{n+1} - t_n)},$$

which clearly can be made as small as desired by making t_{n+1} sufficiently large. Thus all these intersections can be moved to the left of m_n.

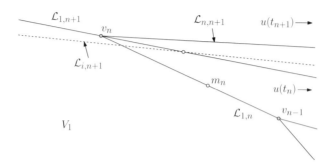

Fig. 2.

It remains to show that all these intersections occur to the right of v_n. Since the slope of $\mathcal{L}_{i,n+1}$ is closer to 0 than the slope of $\mathcal{L}_{1,n+1}$, it suffices to show that the intersections $\mathcal{L}_{1,n+1} \cap \mathcal{L}_{i,n+1}$ happen on $\mathcal{L}_{1,n+1}$ to the right of $v_n = \mathcal{L}_{1,n+1} \cap \mathcal{L}_{n,n+1}$. The x-coordinate of such an intersection is given by

$$-(t_i^2 t_{n+1} + t_i t_{n+1}^2)/2,$$

which, since $t_i < t_n$, is clearly larger than the x-coordinate of $v_n = \mathcal{L}_{1,n+1} \cap \mathcal{L}_{n,n+1}$, which is

$$-(t_n^2 t_{n+1} + t_n t_{n+1}^2)/2.$$

4 Exponentiality of the Construction

The Fibonacci type Inequality (1) implies that t_n, and hence some site coordinate in S, is at least exponential in n. Sites with such large coordinates do not seem to occur naturally in actual inputs, and thus our lower bound construction seems artificial and not really relevant "in practice." We therefore tried to alter our construction in order to avoid such exponential behavior. The most natural approach seemed to be to replace the parabola from which we chose the sites of S by some other curve. We tried several curves, e.g. $(t, t^{1+\varepsilon})$ for $t \geq 0$, or $(t, t \log t)$ for $t \geq 1$, or the hyperbola $(t, 1/t)$ for $t \geq 1$. Each admitted the same inductive construction we used in the proof in the previous section, but in each case we arrived at exponentially large coordinates (particularly bad in the case of the hyperbola). This raised the suspicion that this exponential behavior is inherent in this inductive construction. This turns out to be indeed the case.

Assume we choose S from some curve γ. By translation and rotation[2] we may assume without loss of generality that the curve starts at the origin, lies in the positive quadrant, is monotonically increasing and convex. Let $p_1, p_1, \ldots, p_n$ be the points chosen from the curve in their natural order with p_1 being the origin.

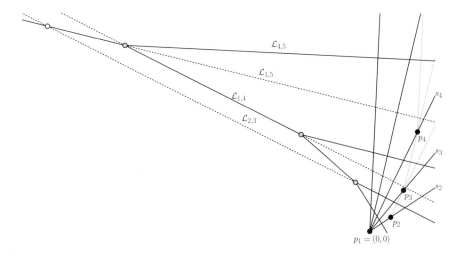

Fig. 3.

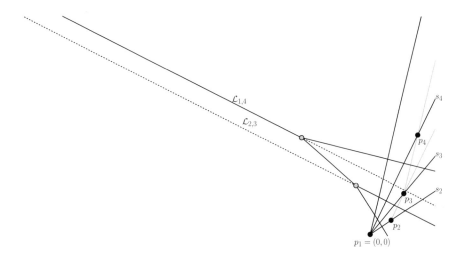

Fig. 4.

---

[2] Such a rotation can be realized with "small" rational numbers in the transformation matrix [3].

The construction of the previous section places a number of constraints on the p_i's. Here we will be interested in only one of them, namely the one that led to inequality (1). It requires that for each i the bisector $\mathcal{L}_{1,i+1}$ between p_1 and p_{i+1} has slope closer to 0 than the bisector $\mathcal{L}_{i-1,i}$ between p_{i-1} and p_i. Since the bisector $\mathcal{L}_{k,\ell}$ is normal to the line through p_ℓ and p_k this implies the following condition: for each i the line through p_{i-1} and p_i must have slope at most as large as the slope of the line between p_1 and p_{i+1}. If we set $p_i = (x_i, s_i \cdot x_i)$, i.e. the slope of the line through $p_1 = (0, 0)$ and p_i is s_i, then this slope condition is expressed algebraically by

$$\frac{s_i x_i - s_{i-1} x_{i-1}}{x_i - x_{i-1}} \leq s_{i+1},$$

which implies that

$$x_i \geq \frac{s_{i+1} - s_{i-1}}{s_{i+1} - s_i} x_{i-1}.$$

This must hold for $3 \leq i < n$. Now let d_i denote the slope difference $s_{i+1} - s_i$. We obtain that

$$x_i \geq \frac{d_i + d_{i-1}}{d_i} x_{i-1} = \left(1 + \frac{d_{i-1}}{d_i}\right) x_{i-1},$$

and putting all these inequalities together we get that

$$x_{n-1} \geq \left(1 + \frac{d_2}{d_3}\right)\left(1 + \frac{d_3}{d_4}\right)\cdots\left(1 + \frac{d_{n-2}}{d_{n-1}}\right) x_2. \tag{2}$$

Below we show that inequality (2) implies that x_{n-1}/x_2 or d_{n-1}/d_2 must be exponentially large. Both can be seen as constant sized rational expressions of the coordinates of p_2, p_3, p_{n-1}, and p_n. If those coordinates are rational numbers then therefore at least one of the involved numerators or denominators must be exponentially large also.

Lemma 4. *Let $D_0, D_1, \ldots, D_N$ be positive numbers and let*

$$P = \left(1 + \frac{D_0}{D_1}\right)\left(1 + \frac{D_1}{D_2}\right)\cdots\left(1 + \frac{D_{N-1}}{D_N}\right).$$

Then $P \geq \varphi^N$ or $D_N/D_0 \geq \varphi^N$, where $\varphi = 1.618...$ is the golden ratio, the larger root of $x^2 = x + 1$.

Proof. It is an easy exercise to show that for $\prod_{1 \leq i \leq N} \rho_i = A$, with $\rho_i > 0$ for all i, the product $\prod_{1 \leq i \leq N}(1 + \rho_i)$ is minimized when all ρ_i are the same, i.e. $\rho_i = A^{1/N}$.

With $\rho_i = D_{i-1}/D_i$ we get that $\prod_{1 \leq i \leq N} \rho_i = D_0/D_N$, and hence

$$P \geq \left(1 + (D_0/D_N)^{1/N}\right)^N.$$

Now let $X = (D_N/D_0)^{1/N}$. We have $P \geq (1 + 1/X)^N$ and $D_N/D_0 = X^N$.
Clearly for $X \geq \varphi$ we have

$$D_N/D_0 \geq \varphi^N$$

.

For $0 < X \leq \varphi$ we have

$$P \geq (1 + 1/X)^N \geq (1 + 1/\varphi)^N = \varphi^N,$$

since by the definition of φ we have $1 + 1/\varphi = \varphi$.

5   Conclusions and Remarks

We have shown that Nearest Neighbor Searching in the plane must be slow in the
worst case if the only type of primitive predicate allowed is comparing the query
point against bisectors of sites. This raises the question which additional type
of predicates must be used in order to facilitate guaranteed fast query time.
Certainly comparing query points against horizontal or vertical lines through
Voronoi points would do the job. However these predicates are a bit more complicated than you would like, since algebraically they are realized by evaluating
the sign of a degree-3 polynomial in the coordinates of the sites and the query
point. Note that in contrast comparing a query point against a site bisector
amounts to evaluating the sign of a degree 2 polynomial. There has been some
work on reducing the degrees of the predicate polynomials that are used in a
nearest neighbor query [5,7], however they apply only to the case where the
query points have integral coordinates.

In the context of this work an interesting question is whether comparing
query points against horizontal or vertical lines through sites could be useful.
Algebraically this is possibly the simplest of all imaginable predicates, since it
amounts to the evaluation of a degree 1 polynomial. Unfortunately at this point
we do not see any way of adapting our construction so that our lower bound also
works if those types of predicates are allowed. We leave it as in interesting open
problem to either prove that such predicates are not useful for nearest neighbor
searching or to devise a method that profitably exploits such predicates.

Our lower bound constructions involve very large numbers. So it is conceiva-
ble that if all numbers involved have small rational representations then just
bisector tests suffice for fast Nearest Neighbor Searching. At this point we do
not see how such a restriction on the coordinates can be profitably exploited.

The results in this paper should be regarded as advice to the algorithm designer. If you want guaranteed fast Nearest Neighbor Search, then using just
bisector tests cannot do the job, unless you make assumptions on your input
coordinates and exploit these assumptions.

References

1. Bentley, J.L.: Multidimensional Binary Search Trees used for Associative Searching. Communications of the ACM 18(9), 509–517 (1975)
2. Birn, M., Holtgrewe, M., Sanders, P., Singler, J.: Simple and Fast Nearest Neighbor Search. In: Proceedings of the Twelfth Workshop on Algorithm Engineering and Experiments (ALENEX). SIAM, Philadelphia (2010)
3. Canny, J.F., Donald, B.R., Ressler, E.K.: A Rational Rotation Method for Robust Geometric Algorithms. In: Proc. of the Eigth ACM Symposium on Computational Geometry (SOCG), pp. 251–260 (1992)
4. Kirkpatrick, D.G.: Optimal Search in Planar Subdivisions. SIAM J. Comput. 12(1), 28–35 (1983)
5. Liotta, G., Preparata, F.P., Tamassia, R.: Robust Proximity Queries: An Illustration of Degree-driven Algorithm Design. SIAM J. Comput. 28(3), 864–889 (1998)
6. Lipton, R.J., Tarjan, R.E.: Applications of a Planar Separator Theorem. SIAM J. Comput. 9(3), 615–627 (1980)
7. Millman, D., Snoeyink, J.: Computing Planar Voronoi Diagrams in Double Precision: A Further Example of Degree-driven Algorithm Design. In: Proceedings of the Annual Symposium on Computational Geometry (SoCG), pp. 386–392 (2010)
8. Mount, D.M., Arya, S.: ANN: A Library for Approximate Nearest Neighbor Searching. In: CGC 2nd Annual Fall Workshop on Computational Geometry (1997)
9. Shamos, M. I.: Geometric Complexity. In: Proceedings of Seventh Annual ACM Symposium on Theory of Computing (STOC), pp. 224–233 (1975)
10. Ziegler, G.M.: Lectures on Polytopes. Springer, Heidelberg (1995)

On the Approximation Performance of Fictitious Play in Finite Games[*]

Paul W. Goldberg[1], Rahul Savani[1], Troels Bjerre Sørensen[2], and Carmine Ventre[1]

[1] Department of Computer Science, University of Liverpool, UK
[2] Department of Computer Science, University of Warwick, UK

Abstract. We study the performance of Fictitious Play, when used as a heuristic for finding an approximate Nash equilibrium of a two-player game. We exhibit a class of two-player games having payoffs in the range $[0, 1]$ that show that Fictitious Play fails to find a solution having an additive approximation guarantee significantly better than $1/2$. Our construction shows that for $n \times n$ games, in the worst case both players may perpetually have mixed strategies whose payoffs fall short of the best response by an additive quantity $1/2 - O(1/n^{1-\delta})$ for arbitrarily small δ. We also show an essentially matching upper bound of $1/2 - O(1/n)$.

1   Introduction

Fictitious Play is a very simple iterative process for computing equilibria of games. A detailed motivation for it is given in [5]. When it converges, it necessarily converges to a Nash equilibrium. For two-player games, it is known to converge for zero-sum games [10], or if one player has just 2 strategies [2]. On the other hand, Shapley exhibited a 3×3 game for which it fails to converge [9,11].

Fictitious Play (FP) works as follows. Suppose that each player has a number of *actions*, or pure strategies. Initially (at iteration 1) each player starts with a single action. Thereafter, at iteration t, each player has a sequence of $t - 1$ actions which is extended with a t-th action chosen as follows. Each player makes a best response to a distribution consisting of the selection of an opponent's strategy uniformly at random from his sequence. (To make the process precise, a tie-breaking rule should also be specified; however, in the games constructed here, there will be no ties.) Thus the process generates a sequence of mixed-strategy profiles (viewing the sequences as probability distributions), and the hope is that they converge to a limiting distribution, which would necessarily be a Nash equilibrium.

The problem of computing *approximate* equilibria was motivated by the apparent intrinsic hardness of computing exact equilibria [8], even in the two-player case [4]. An ϵ-Nash equilibrium is one where each player's strategy has a payoff of at most ϵ less than the best response. Formally, for 2 players with pure strategy sets M, N and payoff functions $u_i : M \times N \to \mathbb{R}$ for $i \in \{1, 2\}$, the mixed strategy σ is an ϵ-*best-response* against the mixed strategy τ, if for any $m \in M$, we have $u_1(\sigma, \tau) \geq u_1(m, \tau) - \epsilon$. A pair of strategies σ, τ is an ϵ-Nash equilibrium if they are ϵ-best responses to each other.

[*] Supported by EPSRC grants EP/G069239/1 and EP/G069034/1 "Efficient Decentralised Approaches in Algorithmic Game Theory."

C. Demetrescu and M.M. Halldórsson (Eds.): ESA 2011, LNCS 6942, pp. 93–105, 2011.

Typically one assumes that the payoffs of a game are rescaled to lie in $[0, 1]$, and then a general question is: for what values of ϵ does some proposed algorithm guarantee to find ϵ-Nash equilibria? Previously, the focus has been on various algorithms that run in polynomial time. Our result for FP applies without any limit on the number of iterations; we show that a kind of cyclical behavior persists.

A recent paper of Conitzer [5] shows that FP obtains an approximation guarantee of $\epsilon = (t + 1)/2t$ for two-player games, where t is the number of FP iterations, and furthermore, if both players have access to infinitely many strategies, then FP cannot do better than this. The intuition behind this upper bound is that an action that appears most recently in a player's sequence has an ϵ-value close to 0 (at most $1/t$); generally a strategy that occurs a fraction γ back in the sequence has an ϵ-value of at most slightly more than γ (it is a best response to slightly less than $1 - \gamma$ of the opponent's distribution), and the ϵ-value of a player's mixed strategy is at most the overall average, i.e., $(t + 1)/2t$, which approaches $1/2$ as t increases.

However, as soon as the number of available pure strategies is exceeded by the number of iterations of FP, various pure strategies must get re-used, and this re-usage means, for example, that every previous occurrence of the most recent action all have ϵ-values of $1/t$. This appears to hold out the hope that FP may ultimately guarantee a significantly better additive approximation. We show that unfortunately that is not what results in the worst case. Our hope is that this result may either guide the design of more "intelligent" dynamics having a better approximation performance, or alternatively generalize to a wider class of related algorithms, e.g., the ones discussed in [6].

In Section 2 we give our main result, the lower bound of $1/2 - O(1/n^{1-\delta})$ for any $\delta > 0$, and in Section 3 we give the corresponding upper bound of $1/2 - O(1/n)$. Proofs omitted due to lack of space may be found in the full version of the paper.

2   Lower Bound

We specify a class of games with parameter n, whose general idea is conveyed in Figure 1, which shows the row player's matrix for $n = 5$; the column player's matrix is its transpose. A blank entry indicates the value zero; let $\alpha = 1 + \frac{1}{n^{1-\delta}}$ and $\beta = 1 - \frac{1}{n^{2(1-\delta)}}$ for $\delta > 0$. Both players start at strategy 1 (top left). Generally, let $\mathcal{G}_n$ be a $4n \times 4n$ game in which the column player's payoff matrix is the transpose of the row player's payoff matrix R, which itself is specified as follows. For $i \in [2 : n]$, $R_{i,i-1} = 1$. If $i \in [n + 1 : 4n]$, $R_{i,i} = 1$. If $i \in [n + 1 : 4n]$, $R_{i,i-1} = \alpha$. Also, $R_{2n+1,4n} = \alpha$. Otherwise, if $i > j$ and $j \leq 2n$, $R_{i,j} = \beta$. Otherwise, if $i > j$ and $i - j \leq n$, $R_{i,j} = \beta$. For $j \in [3n + 1 : 4n]$ and $i \in [2n + 1 : j - n]$, $R_{i,j} = \beta$. Otherwise, we have $R_{ij} = 0$. For ease of presentation we analyze FP on $\mathcal{G}_n$; the obtained results can be seen to apply to a version of $\mathcal{G}_n$ with payoffs rescaled into $[0, 1]$ (cf. the proof of Theorem 2).

Overview. We next give a general overview and intuition on how our main result works, before embarking on the technical details. Number the strategies $1, \ldots 4n$ from top to bottom and left to right, and assume that both players start at strategy 1. Fictitious Play proceeds in a sequence of steps, which we index by positive integer t, so that step t consists of both players adding the t-th element to their sequences of length $t - 1$. It is easy to observe that since the column player's payoff matrix is the transpose of the row

1																			
β	1																		
β	β	1																	
β	β	β	1																
β	β	β	β	α	1														
β	β	β	β	β	α	1													
β	β	β	β	β	β	α	1												
β	β	β	β	β	β	β	α	1											
β	β	β	β	β	β	β	β	α	1										
β	β	β	β	β	β	β	β	β	α	1					β	β	β	β	α
β	β	β	β	β	β	β	β	β	β	α	1				β	β	β	β	
β	β	β	β	β	β	β	β	β	β	β	α	1			β	β	β		
β	β	β	β	β	β	β	β	β	β	β	β	α	1		β	β			
β	β	β	β	β	β	β	β	β	β	β	β	β	α	1	β				
β	β	β	β	β	β	β	β	β	β	β	β	β	β	α	1				
β	β	β	β	β	β	β	β	β	β		β	β	β	β	α	1			
β	β	β	β	β	β	β	β	β	β		β	β	β	β	α	1			
β	β	β	β	β	β	β	β	β	β		β	β	β	β	α	1			
β	β	β	β	β	β	β	β	β	β		β	β	β	β	α	1			

Fig. 1. The game $\mathcal{G}_5$ belonging to the class of games used to prove the lower bound

player's, at every step both players play the same action. This simplifies the analysis since it means we are analyzing a single sequence of numbers (the shared indices of the actions chosen by the players).

A basic insight into the behavior of Fictitious Play on the games in question is provided by Lemma 1, which tells us a great deal about the structure of the players' sequence. Let s_t be the action played at step t. We set $s_1 = 1$.

Lemma 1. *For any t, if $s_t \neq s_{t+1}$ then $s_{t+1} = s_t + 1$ (or $s_{t+1} = 2n + 1$ if $s_t = 4n$).*

Proof. The first n steps are similar to [5]. For step $t > n$, suppose the players play $s_t \neq 4n$ (by the above observation, the two players play the same strategy). s_t is a best response at step t, and since $R_{s_t+1,s_t} > R_{s_t,s_t} > R_{j,s_t}$ ($j \notin \{s_t, s_t + 1\}$), strategy $s_t + 1$ is the only other candidate to become a better response after s_t is played. Thus, if $s_{t+1} \neq s_t$, then $s_{t+1} = s_t + 1$. Similar arguments apply to the case $s_t = 4n$. $\square$

The lemma implies that the sequence consists of a block of consecutive 1's followed by some consecutive 2's, and so on through all the actions in ascending order until we get to a block of consecutive $4n$'s. The blocks of consecutive actions then cycle through the actions $\{2n + 1, \ldots, 4n\}$ in order, and continue to do so repeatedly. As it stands, the lemma makes no promise about the lengths of these blocks, and indeed it does not itself rule out the possibility that one of these blocks is infinitely long (which would end the cycling process described above and cause FP to converge to a pure Nash equilibrium). The subsequent results say more about the lengths of the blocks. They show that in fact the process never converges (it cycles infinitely often) and furthermore, the lengths of the blocks increase in geometric progression. The parameters α and β in $\mathcal{G}_n$ govern the ratio between the lengths of consecutive blocks. We choose a ratio large enough that ensures that the n strategies most recently played, occupy all but an exponentially-small fraction of the sequence. At the same time the ratio is small enough that the corresponding probability distribution does not allocate much probability to any individual strategy.

The proof. We now identify some properties of probabilities assigned to strategies by FP. We let $\ell_t(i)$ be the number of times that strategy i is played by the players until time step t of FP. Let $p_t(i)$ be the corresponding probability assigned by the players to strategy i at step t, also for any subset of actions S we use $p_t(S)$ to denote the total probability of elements of S. So it is immediate to observe that $p_t(i) = \frac{\ell_t(i)}{\sum_{j=1}^{4n} \ell_t(j)} = \frac{\ell_t(i)}{t}$. The next fact easily follows from the FP rule.

Lemma 2. *For all strategies* $i \leq n$, $p_t(i) = \frac{1}{t}$ *and therefore* $\ell_t(i) = 1$ *for any* $t \geq i$.

By Lemma 1, each strategy is played a number of consecutive times, in order, until the strategy $4n$ is played; at this point, this same pattern repeats but only for the strategies in $\{2n+1, \ldots, 4n\}$. We let $t^\star$ be the length of the longest sequence containing all the strategies in ascending order, that is, $t^\star$ is the last step of the first consecutive block of $4n$'s. We also let t_i be the last time step in which i is played during the first $t^\star$ steps, i.e., t_i is such that $\ell_{t_i}(i) = \ell_{t_i-1}(i) + 1$ and $\ell_t(i) = \ell_{t^\star}(i)$ for $t \in \{t_i, \ldots, t^\star\}$.

Lemma 3. *For all strategies* $n+1 \leq i \leq 3n$ *and all* $t \in \{t_i, \ldots, t^\star\}$, *it holds:*

$$\frac{\alpha - \beta}{\alpha - 1} p_t(i-1) \leq p_t(i) \leq \frac{1}{t} + \frac{\alpha - \beta}{\alpha - 1} p_t(i-1)$$

and therefore,

$$\frac{\alpha - \beta}{\alpha - 1} \ell_t(i-1) \leq \ell_t(i) \leq 1 + \frac{\alpha - \beta}{\alpha - 1} \ell_t(i-1).$$

Proof. By definition of t_i, strategy i is played at step t_i. This means that i is a best response for the players given the probability distributions at step $t_i - 1$. In particular, the expected payoff of i is better than the expected payoff of $i+1$, that is,

$$\beta \sum_{j=1}^{i-2} p_{t_i-1}(j) + \alpha p_{t_i-1}(i-1) + p_{t_i-1}(i) \geq \beta \sum_{j=1}^{i-2} p_{t_i-1}(j) + \beta p_{t_i-1}(i-1) + \alpha p_{t_i-1}(i).$$

Since $\alpha > 1$, the above implies that $p_{t_i-1}(i) \leq \frac{\alpha-\beta}{\alpha-1} p_{t_i-1}(i-1)$. By explicitly writing the probabilities, we get

$$\frac{\ell_{t_i-1}(i)}{t_i-1} \leq \frac{\alpha-\beta}{\alpha-1} \frac{\ell_{t_i-1}(i-1)}{t_i-1} \iff \ell_{t_i}(i) - 1 \leq \frac{\alpha-\beta}{\alpha-1} \ell_{t_i}(i-1) \iff \quad (1)$$

$$\frac{\ell_{t_i}(i)}{t_i} \leq \frac{1}{t_i} + \frac{\alpha-\beta}{\alpha-1} \frac{\ell_{t_i}(i-1)}{t_i} \iff p_{t_i}(i) \leq \frac{1}{t_i} + \frac{\alpha-\beta}{\alpha-1} p_{t_i}(i-1). \quad (2)$$

At step $t_i + 1$ strategy i is not a best response to the opponent's strategy. Then, by Lemma 1, $i + 1$ is the unique best response and so the expected payoff of $i + 1$ is better than the expected payoff of i given the probability distributions at step t_i, that is,

$$\beta \sum_{j=1}^{i-2} p_{t_i}(j) + \alpha p_{t_i}(i-1) + p_{t_i}(i) \leq \beta \sum_{j=1}^{i-2} p_{t_i}(j) + \beta p_{t_i}(i-1) + \alpha p_{t_i}(i).$$

Since $\alpha > 1$, the above implies that

$$p_{t_i}(i) \geq \frac{\alpha - \beta}{\alpha - 1} p_{t_i}(i-1), \text{ and then that } \ell_{t_i}(i) \geq \frac{\alpha - \beta}{\alpha - 1} \ell_{t_i}(i-1). \tag{3}$$

By definition of t_i action i will not be played anymore until time step t^*. Similarly, Lemma 1 shows that $i - 1$ will not be a best response twice in the time interval $[1, t^*]$ and so will not be played until step t^*. So, the claim follows from (1), (2) and (3). $\qquad \square$

Using similar arguments, we can prove a similar result on the actions $3n+1, \ldots, 4n-1$.

Lemma 4. *For $i \in \{3n + 1, \ldots, 4n - 1\}$ and $t \in \{t_i, \ldots, t^*\}$, it holds:*

$$\frac{\alpha - \beta}{\alpha - 1} p_t(i-1) \leq p_t(i) \leq \frac{1}{t} + \frac{\alpha - \beta}{\alpha - 1} p_t(i-1) + \frac{\beta}{\alpha - 1} p_t(i-n)$$

and therefore,

$$\frac{\alpha - \beta}{\alpha - 1} \ell_t(i-1) \leq \ell_t(i) \leq 1 + \frac{\alpha - \beta}{\alpha - 1} \ell_t(i-1) + \frac{\beta}{\alpha - 1} \ell_t(i-n).$$

Exploiting the properties given by Lemmata 2, 3 and 4, the next lemma shows that we can "forget" about the first $2n$ actions at the cost of paying an exponentially small addend in the payoff function.

Lemma 5. *For any $\delta > 0$, $\alpha = 1 + \frac{1}{n^{1-\delta}}$ and $\beta = 1 - \frac{1}{n^{2(1-\delta)}}$, $\sum_{j=1}^{2n} p_{t^*}(j) \leq 2^{-n^\delta}$.*

The theorem below generalizes the above arguments to the cycles that FP visits in the last block of the game, i.e., the block which comprises strategies $S = \{2n+1, \ldots, 4n\}$. Since we focus on this part of the game, to ease the presentation, our notation uses circular arithmetic on the elements of S. For example, the action $j + 2$ will denote action $2n+2$ for $j = 4n$ and the action $j - n$ will be the strategy $3n+1$ for $j = 2n+1$. Note that under this notation $j - 2n = j + 2n = j$ for each $j \in S$.

Theorem 1. *For any $\delta > 0$, $\alpha = 1 + \frac{1}{n^{1-\delta}}$ and $\beta = 1 - \frac{1}{n^{2(1-\delta)}}$, n sufficiently large, any $t \geq t^*$ we have*

$$\frac{p_t(i)}{p_t(i-1)} \geq 1 + \frac{1}{n^{1-\delta}} \; \forall i \in S \text{ with } i \neq s_t, s_t + 1, \text{ and } \frac{p_t(i)}{p_t(i-1)} \leq 1 + \frac{3}{n^{1-\delta}} \; \forall i \in S.$$

Proof. The proof is by induction on t.

Base. For the base of the induction, consider $t = t^\star$ and note that at that point $s_{t^\star} = 4n$ and $s_{t^\star} + 1 = 2n + 1$. Therefore we need to show the lower bound for any strategy $i \in \{2n + 2, \ldots, 4n - 1\}$. From Lemmata 3 and 4 we note that for $i \neq 4n, 2n + 1$,

$$\frac{p_{t^\star}(i)}{p_{t^\star}(i-1)} \geq \frac{\alpha - \beta}{\alpha - 1} = 1 + \frac{1}{n^{1-\delta}}.$$

As for the upper bound, we first consider the case of $i \neq 4n, 2n + 1$. Lemma 3 implies that for $i = 2n + 2, \ldots, 3n$,

$$\frac{p_{t^\star}(i)}{p_{t^\star}(i-1)} \leq \frac{1}{t^\star} + \frac{\alpha - \beta}{\alpha - 1},$$

while Lemma 4 implies that for $i = 3n + 1, \ldots, 4n - 1$,

$$\frac{p_{t^\star}(i)}{p_{t^\star}(i-1)} \leq \frac{1}{t^\star} + \frac{\alpha - \beta}{\alpha - 1} + \frac{\beta}{\alpha - 1}\frac{p_{t^\star}(i-n)}{p_{t^\star}(i-1)} = \frac{1}{t^\star} + \frac{\alpha - \beta}{\alpha - 1} + \frac{\beta}{\alpha - 1}\frac{\ell_{t^\star}(i-n)}{\ell_{t^\star}(i-1)}.$$

To give a unique upper bound for both cases, we only focus on the above (weaker) upper bound and next are going to focus on the ratio $\frac{\ell_{t^\star}(i-n)}{\ell_{t^\star}(i-1)}$. We use Lemmata 3 and 4 and get

$$\ell_{t^\star}(i-1) \geq \frac{\alpha - \beta}{\alpha - 1}\ell_{t^\star}(i-2) \geq \left(\frac{\alpha - \beta}{\alpha - 1}\right)^2 \ell_{t^\star}(i-3) \geq \left(\frac{\alpha - \beta}{\alpha - 1}\right)^{n-1}\ell_{t^\star}(i-n).$$

By setting α and β as in the hypothesis and noticing that $t^\star \geq n \geq n^{1-\delta}$ we then obtain that

$$\frac{p_{t^\star}(i)}{p_{t^\star}(i-1)} \leq 1 + \frac{2}{n^{1-\delta}} + \left(1 + \frac{1}{n^{1-\delta}}\right)^{1-n}\frac{n^{2(1-\delta)} - 1}{n^{1-\delta}}.$$

We end this part of the proof by showing that the last addend on the right-hand side of the above expression is upper bounded by $\frac{1}{4n^{1-\delta}}$. To do so we need to prove

$$\left(1 + \frac{1}{n^{1-\delta}}\right)^{1-n} \leq \frac{1}{4n^{1-\delta}}\frac{n^{1-\delta}}{n^{2(1-\delta)} - 1}, \tag{4}$$

which is equivalent to $\left(\left(1 + \frac{1}{n^{1-\delta}}\right)^{n^{1-\delta}}\right)^{n^{\delta} - \frac{1}{n^{1-\delta}}} \geq 4(n^{2(1-\delta)} - 1)$. We now lower bound the left-hand side of the latter inequality: $\left(\left(1 + \frac{1}{n^{1-\delta}}\right)^{n^{1-\delta}}\right)^{n^{\delta} - \frac{1}{n^{1-\delta}}} > \frac{2^{n^{\delta}}}{2^{\frac{1}{n^{1-\delta}}}} > \frac{2^{n^{\delta}}}{2}$, where the first inequality follows from the fact that the function $(1 + 1/x)^x$ is greater than 2 for $x > 2$ and the second one follows from the fact that $2^{\frac{1}{n^{1-\delta}}} < 2$ for $n^{1-\delta} > 1$. Then, since for $n \geq {}^{2(1-\delta)}\!\!\sqrt[]{4}$, $5n^{2(1-\delta)} \geq 4(n^{2(1-\delta)} - 1)$, to prove (4) is enough to show $2^{n^{\delta}} \geq 2(5n^{2(1-\delta)}) \iff n^{\delta} \geq 2(1 - \delta)\log_2(10n)$. To prove the latter, since $\delta > 0$, it is enough to observe that the function n^{δ} is certainly bigger than the function $2\log_2(10n) > 2(1 - \delta)\log_2(10n)$ for n large enough (e.g., for $\delta = 1/2$, this is true for $n > 639$). Similarly to the proof of Lemma 4, we can prove that the upper bound holds at time step $t^\star$ for $i = 4n, 2n + 1$. This concludes the proof of the base of the induction.

Inductive step. Now we assume the claim is true until time step $t - 1$ and we show it for time step t. By inductive hypothesis, the following is true, with $j \neq s_{t-1}, s_{t-1} + 1$

$$1 + \frac{1}{n^{1-\delta}} \le \frac{p_{t-1}(j)}{p_{t-1}(j-1)} \le 1 + \frac{3}{n^{1-\delta}}, \tag{5}$$

$$\frac{p_{t-1}(s_{t-1}+1)}{p_{t-1}(s_{t-1})} \le 1 + \frac{3}{n^{1-\delta}}. \tag{6}$$

We first consider the case in which $s_t \neq s_{t-1}$. By Lemma 1, the strategy played at time t is $s_{t-1}+1$, i.e., $s_t = s_{t-1}+1$. Let $s_{t-1} = i$ and then we have $s_t = i+1$. By inductive hypothesis, for all the actions $j \neq i, i+1, i+2$ we have

$$\frac{\alpha - \beta}{\alpha - 1} = 1 + \frac{1}{n^{1-\delta}} \le \frac{p_t(j)}{p_t(j-1)} \le 1 + \frac{3}{n^{1-\delta}}. \tag{7}$$

Indeed, for these actions j, $\ell_{t-1}(j) = \ell_t(j)$ and $\ell_{t-1}(j-1) = \ell_t(j-1)$. Therefore the probabilities of j and $j-1$ at time t are simply those at time $t-1$ rescaled by the same amount and the claim follows from (5). The upper bound on the ratio $\frac{p_t(i+2)}{p_t(i+1)}$ easily follows from the upper bound in (5) as $\ell_{t-1}(i+2) = \ell_t(i+2)$ and $\ell_{t-1}(i+1) < \ell_t(i+1) = \ell_{t-1}(i+1) + 1$. However, as $s_t = i+1$ here we need to prove lower and upper bound also for the ratio $\frac{p_t(i)}{p_t(i-1)}$ and the upper bound for the ratio $\frac{p_t(i+1)}{p_t(i)}$ (this proof can be found in the full version of the paper).

Claim. It holds $1 + \frac{1}{n^{1-\delta}} \le \frac{p_t(i)}{p_t(i-1)} \le 1 + \frac{3}{n^{1-\delta}}$.

Proof. To prove the claim we first focus on the last block of the game, i.e., the block in which players have strategies in $\{2n+1, \ldots, 4n\}$. Recall that our notation uses circular arithmetic on the number of actions of the block.

The fact that action $i+1$ is better than action i after $t-1$ time steps implies that

$$p_{t-1}(i) + \alpha p_{t-1}(i-1) + \beta p_{t-1}(i-n) \le \alpha p_{t-1}(i) + p_{t-1}(i+1) + \beta p_{t-1}(i-1)$$

and then since $\alpha > 1$

$$\frac{p_{t-1}(i)}{p_{t-1}(i-1)} \ge \frac{\alpha - \beta}{\alpha - 1} + \frac{\beta}{\alpha - 1}\frac{p_{t-1}(i-n)}{p_{t-1}(i-1)} - \frac{1}{\alpha - 1}\frac{p_{t-1}(i-2n+1)}{p_{t-1}(i-1)}. \tag{8}$$

We next show that $\frac{\beta p_{t-1}(i-n) - p_{t-1}(i-2n+1)}{(\alpha-1)p_{t-1}(i-1)} \ge -\frac{1}{4n^{1-\delta}}$ or equivalently that $\frac{p_{t-1}(i-n)}{p_{t-1}(i-2n+1)} \ge \frac{1}{\beta} - \frac{(\alpha-1)p_{t-1}(i-1)}{4\beta n^{1-\delta}p_{t-1}(i-2n+1)}$. To prove this it is enough to show that $\frac{p_{t-1}(i-n)}{p_{t-1}(i-2n+1)} \ge \frac{1}{\beta}$. We observe that

$$\frac{p_{t-1}(i-n)}{p_{t-1}(i-2n+1)} = \frac{p_{t-1}(i-n)}{p_{t-1}(i-n-1)} \cdots \frac{p_{t-1}(i-2n+2)}{p_{t-1}(i-2n+1)} \ge \left(\frac{\alpha-\beta}{\alpha-1}\right)^{n-1} \ge \frac{1}{\beta},$$

where the first inequality follows from inductive hypothesis (we can use the inductive hypothesis as all the actions involved above are different from i and $i + 1$) and the second inequality follows from the aforementioned observation that, for sufficiently large n, $n^\delta \geq 2 \log_2(2n)$. Then to summarize, for α and β as in the hypothesis, (8) implies that

$$\frac{p_t(i)}{p_t(i-1)} = \frac{p_{t-1}(i)}{p_{t-1}(i-1)} \geq 1 + \frac{1}{n^{1-\delta}} - \frac{1}{4n^{1-\delta}},$$

where the first equality follows from $\ell_{t-1}(i) = \ell_t(i)$ and $\ell_{t-1}(i-1) = \ell_t(i-1)$, which are true because $s_t = i + 1$.

Since action $i + 1$ is worse than strategy i at time step $t - 1$ we have that

$$p_{t-1}(i) + \alpha p_{t-1}(i-1) + \beta p_{t-1}(i-n) \geq \alpha p_{t-1}(i) + p_{t-1}(i+1) + \beta p_{t-1}(i-1)$$

and then since $\alpha > 1$, $\frac{p_{t-1}(i)}{p_{t-1}(i-1)} \leq \frac{\alpha-\beta}{\alpha-1} + \frac{\beta}{\alpha-1} \frac{p_{t-1}(i-n)}{p_{t-1}(i-1)} - \frac{1}{\alpha-1} \frac{p_{t-1}(i-2n+1)}{p_{t-1}(i-1)}$. Similarly to the proof of Lemma 3 above this can be shown to imply

$$\frac{p_t(i)}{p_t(i-1)} \leq \frac{1}{t} + \frac{\alpha-\beta}{\alpha-1} + \frac{\beta}{\alpha-1} \frac{p_t(i-n)}{p_t(i-1)} - \frac{1}{\alpha-1} \frac{p_t(i-2n+1)}{p_t(i-1)}$$

$$\leq \frac{1}{t} + \frac{\alpha-\beta}{\alpha-1} + \frac{\beta}{\alpha-1} \frac{p_t(i-n)}{p_t(i-1)}. \tag{9}$$

We now upper bound the ratio $\frac{\beta}{\alpha-1} \frac{p_t(i-n)}{p_t(i-1)}$. By repeatedly using the inductive hypothesis (7) we have that

$$p_t(i-1) \geq \frac{\alpha-\beta}{\alpha-1} p_t(i-2) \geq \left(\frac{\alpha-\beta}{\alpha-1}\right)^2 p_t(i-3) \geq \left(\frac{\alpha-\beta}{\alpha-1}\right)^{n-1} p_t(i-n).$$

(Note again that we can use the inductive hypothesis as none of the actions above is i or $i + 1$.) This yields

$$\frac{\beta}{\alpha-1} \frac{p_t(i-n)}{p_t(i-1)} \leq \frac{\beta}{\alpha-1} \left(\frac{\alpha-1}{\alpha-\beta}\right)^{n-1} \leq \frac{1}{4n^{1-\delta}},$$

where the last inequality is proved above (see (4)). Therefore, since $t \geq n^{1-\delta}$, (9) implies the following

$$\frac{p_t(i)}{p_t(i-1)} \leq 1 + \frac{2}{n^{1-\delta}} + \frac{1}{4n^{1-\delta}}.$$

To conclude the proof we must now consider the contribution to the payoffs of the actions $1, \dots, 2n$ that are not in the last block. However, Lemma 5 shows that all those actions are played with probability $1/2^{n^\delta}$ at time $t^\star$. Since we prove above (see Lemma 1) that these actions are not played anymore after time step $t^\star$ this implies that $\sum_{j=1}^{2n} p_t(j) \leq \sum_{j=1}^{2n} p_{t^\star}(j) \leq 2^{-n^\delta}$. Thus the overall contribution of these strategies is upper bounded by $\frac{1}{2^{n^\delta}}(\alpha - \beta) \leq \frac{1}{2^{n^\delta}} \leq \frac{1}{4n^{1-\delta}}$ where the last bound follows from the aforementioned fact that, for n sufficiently large, $n^\delta \geq (1-\delta) \log_2(4n)$. This concludes the proof of this claim.   □

Finally, we consider the case in which $s_{t-1} = s_t$. In this case, for the actions $j \neq s_t, s_t + 1$ it holds $\ell_{t-1}(j) = \ell_t(j)$ and $\ell_{t-1}(j-1) = \ell_t(j-1)$. Therefore, similarly to the above, for these actions j the claim follows from (5). The upper bound for the ratio $\frac{p_t(s_t+1)}{p_t(s_t)}$ easily follows from (6) as $\ell_{t-1}(s_t+1) = \ell_t(s_t+1)$ and $\ell_{t-1}(s_t) < \ell_t(s_t) = \ell_{t-1}(s_t) + 1$. The remaining case to analyze is the upper bound on the ratio $\frac{p_t(s_t)}{p_t(s_t-1)}$. To prove this we can use *mutatis mutandis* the proof of the upper bound on the ratio $\frac{p_t(i)}{p_t(i-1)}$ for the case $s_{t-1} \neq s_t$ (in that proof simply set with $s_t = i$).   □

The claimed performance of Fictitious Play, in terms of the approximation to the best response that it computes, follows directly from this theorem.

Theorem 2. *For any value of $\delta > 0$ and any time step t, Fictitious Play returns an ϵ-NE with $\epsilon \geq \frac{1}{2} - O\left(\frac{1}{n^{1-\delta}}\right)$.*

Proof. For $t \leq n$ the result follows since the game is similar to [5]. In details, for $t \leq n$ the payoff associated to the best response, which in this case is $s_t + 1$, is upper bounded by 1. On the other hand, the payoff associated to the current strategy prescribed by FP is lower bounded by $\frac{\beta}{i^2} \sum_{j=0}^{i-1} j$ where $i = s_t$. Therefore, the regret of either player normalized to the $[0, 1]$ interval satisfies: $\epsilon \geq \frac{1}{\alpha} - \frac{\beta}{\alpha} \frac{i-1}{2i}$. Since $\frac{i-1}{2i} < 1/2$, the fact that $1 - \frac{\beta}{2} - \frac{\alpha}{2} + \frac{\alpha}{n^{1-\delta}} \geq 0$ (which is true given the values of α and β) yields the claim. For $t \leq t^\star$ the result follows from Lemmata 3 and 4; while the current strategy s_t (for $t \leq t^\star$) has payoff approximately 1, the players' mixed strategies have nearly all their probability on the recently played strategies, but with no pure strategy having very high probability, so that some player is likely to receive zero payoff; by symmetry each player has payoff approximately $\frac{1}{2}$. This is made precise below, where it is applied in more detail to the case of $t > t^\star$.

We now focus on the case $t > t^\star$. Recall that for a set of strategies S, $p_t(S) = \sum_{i \in S} p_t(i)$. Let S_t be the set $\{2n+1, \ldots, s_t\} \cup \{s_t + n, \ldots, 4n\}$ if $s_t \leq 3n$, or the set $\{s_t - n, \ldots, s_t\}$ in the case that $s_t > 3n$. Let $S_t' = \{2n+1, \ldots, 4n\} \setminus S_t$. Also, let $s_t^{\max} = \arg\max_{i \in \{2n+1, \ldots, 4n\}} (p_t(i))$; note that by Theorem 1, $s_t^{\max}$ is equal to either s_t or s_t^-, where $s_t^- = s_t - 1$ if $s_t > 2n$, or $4n$ if $s_t = 2n$. We start by establishing the following claim:

Claim. For sufficiently large n, $p_t(S_t) \geq 1 - \frac{2n-1}{2^{n\delta}}$.

Proof. To see this, note that for all $x \in S_t'$, by $p_t(s_t^{\max}) \geq p_t(s_t^{\max} - 1)$ and Theorem 1 we have

$$\frac{p_t(s_t^{\max})}{p_t(x)} = \frac{p_t(s_t^{\max})}{p_t(s_t^{\max} - 1)} \frac{p_t(s_t^{\max} - 1)}{p_t(s_t^{\max} - 2)} \cdots \frac{p_t(x+1)}{p_t(x)} \geq \left(1 + \frac{1}{n^{1-\delta}}\right)^{k-1},$$

where k is the number of factors on the right-hand side of the equality above, i.e., the number of strategies between x and $s_t^{\max}$. Thus, as $k \geq n$,

$$p_t(x) \leq \frac{p_t(s_t^{\max})}{\left(1 + \frac{1}{n^{1-\delta}}\right)^{k-1}} \leq \left(1 + \frac{1}{n^{1-\delta}}\right)^{1-k} \leq \left(1 + \frac{1}{n^{1-\delta}}\right)^{1-n} \leq 4^{(1-n)/(n^{1-\delta})}.$$

Hence $p_t(S'_t) \leq (2n)4^{(1-n)/(n^{1-\delta})} = \frac{n4^{1/n^{1-\delta}}}{2^{n^\delta}} < \frac{2n}{2^{n^\delta}}$, where the last inequality follows from the fact that, for large n, $4^{1/n^{1-\delta}} < 2$. Then $p_t(S_t) \geq 1 - p_t(S'_t) - p_t(\{1, \ldots, 2n\})$, which establishes the claim, since Lemma 5 establishes a strong enough upper bound on $p_t(\{1, \ldots, 2n\})$.   □

Claim. The current best response at time t, s_t, has payoff at least $\beta\left(1 - \frac{2n-1}{2^{n^\delta}}\right)$.

Proof. The action s_t receive a payoff of at least β when the opponent plays any strategy from S_t; the claim follows using Claim 2.   □

Let E_t denote the expected payoff to either player that would result if they both select a strategy from the mixed distribution that allocates to each strategy x, the probability $p_t(x)$. The result will follow from the following claim:

Claim. For sufficiently large n, $E_t \leq \frac{\alpha}{2} + \frac{6}{n^{1-\delta}} + \alpha\frac{2n}{2^{n^\delta}}$.

Proof. The contribution to E_t from strategies in $\{1, \ldots, n\}$, together with strategies in S'_t, may be upper-bounded by α times the probability that any of that strategies get played. This probability is by Lemma 5 and the claim above exponentially small, namely $2n/2^{n^\delta}$.

Suppose instead that both players play from S_t. If they play different strategies, their total payoff will be at most α, since one player receives payoff 0. If they play the same strategy, they both receive payoff 1. We continue by upper-bounding the probability that they both play the same strategy. This is upper-bounded by the largest probability assigned to any single strategy, namely $p_t(s_t^{\max})$.

Suppose for contradiction that $p_t(s_t^{\max}) > 6/n^{1-\delta}$. At this point, note that by Theorem 1, for any strategy $s \in S_t$, we have $\frac{p_t(s_t^{\max})}{p_t(s)} \leq \left(1 + \frac{3}{n^{1-\delta}}\right)^k$, where k is the distance between s and $s_t^{\max}$. Therefore, denoting $r = \left(1 + \frac{3}{n^{1-\delta}}\right)^{-1}$, we obtain

$$p_t(S_t) = \sum_{s \in S_t} p_t(s) = p_t(s_t) + \sum_{i=s_t-n}^{s_t-1} p_t(i) \geq p_t(s_t^{\max}) \sum_{k=0}^{n-1} r^k.$$

Applying the standard formula for the partial sum of a geometric series we have $p_t(S_t) \geq \frac{6}{n^{1-\delta}}\left(\frac{1-r^n}{1-r}\right)$. Noting that $1 - r^n > \frac{1}{2}$ we have $p_t(S_t) > \frac{6}{n^{1-\delta}} \cdot \left(\frac{1}{2}\right) \cdot \left(\frac{n^{1-\delta}}{3}\right)$ which is greater than 1, a contradiction.

The expected payoff E_t to either player, is, by symmetry, half the expected total payoff: we have $E_t \leq (1 - \frac{2n}{2^{n^\delta}} - \frac{6}{n^{1-\delta}})\frac{\alpha}{2} + \frac{6}{n^{1-\delta}} + \frac{2n}{2^{n^\delta}}\alpha$ which yields the claim.   □

We now show that Fictitious Play never achieves an ϵ-value better than $\frac{1}{2} - O\left(\frac{1}{n^{1-\delta}}\right)$. From the last two claims the regret of either player normalized to $[0, 1]$ satisfies:

$$\epsilon \geq \frac{\beta}{\alpha}\left(1 - \frac{2n-1}{2^{n^\delta}}\right) - \frac{1}{2} - \frac{6}{\alpha n^{1-\delta}} - \frac{2n}{2^{n^\delta}} = \frac{1}{2} - O\left(\frac{1}{n^{1-\delta}}\right).$$

This concludes the proof.   □

3 Upper Bound

In this section, n denotes the number of pure strategies of both players. Let a, b denote the FP sequences of pure strategies of length t, for players 1 and 2 respectively. Let $a_{[k:\ell]}$ denote the subsequence $a_k, \ldots, a_\ell$. We overload notation and use a to also denote the mixed strategy that is uniform on the corresponding sequence.

Let m^* be a best response against b, and let ϵ denote the smallest ϵ' for which a is an ϵ'-best-response against b. To derive a bound on ϵ, we use the most recent occurrence of pure strategy in a. For $k \in \{1, \ldots, t\}$, let $f(k)$ denote the last occurrence of a_k in the sequence a, that is, $f(k) := \max_{\ell \in \{1,\ldots,t\}, \, a_\ell = a_k} \ell$. It is not hard to show that

$$\epsilon \leq 1 + \frac{1}{t} - \frac{1}{t^2} \sum_{i=1}^{t} f(i). \tag{10}$$

To provide a guarantee on the performance of FP, we find the sequence a that maximizes the right hand-side of (10), i.e., that minimizes $\sum_{i=1}^{t} f(i)$.

Definition 1. *For a FP sequence a, let $S(a) := \sum_{i=1}^{t} f(a_i)$ and let $\hat{a} = \arg\min_a S(a)$.*

The following three lemmata allow to characterize $\hat{a}$, the sequence that minimizes $S(a)$.

Lemma 6. *The entries of $\hat{a}$ take on exactly n distinct values.*

We now define a transformation of an FP sequence a into a new sequence a' so that $S(a') < S(a)$ if $a \neq a'$.

Definition 2. *Suppose the entries of a take on d distinct values. We define $x_1, \ldots, x_d$ to be the last occurrences, $\{f(a_i) \mid i \in [t]\}$, in ascending order. Formally, let $x_d := a_t$ and for $k < d$ let $x_k := a_i$ be such that $i := \arg\max_{j=1,\ldots,t} a_j \notin \{x_{k+1}, \ldots, x_d\}$. For $i = 1, \ldots, d$, let $\#(x_i) := |\{a_j \mid j \in [t], a_j = x_i\}|$, which is the number of occurrences of x_i in a. Define a' as $a' := \underbrace{x_1, \ldots, x_1}_{\#(x_1)}, \underbrace{x_2, \ldots, x_2}_{\#(x_2)}, \cdots, \underbrace{x_d, \ldots, x_d}_{\#(x_d)}$.*

Lemma 7. *For any FP sequence a, let a' be as in Definition 2. If $a' \neq a$ then $S(a') < S(a)$.*

Lemma 8. *Let $n, t \in \mathbb{N}$ be such that $n \mid t$. Let a be a sequence of length t of the form $a = 1, \ldots, 1, 2, \ldots, 2, \cdots, n, \ldots, n$, where the blocks of consecutive actions have length $c_1, c_2, \ldots, c_n$, respectively. Then $S(a)$ is minimized if and only if $c_1 = \cdots = c_n = t/n$.*

Proof. We refer to the maximal length subsequence of entries with value $u \in \{1, \ldots n\}$ as *block u*. Consider two adjacent blocks u and $u + 1$, where block u starts at i and block $u + 1$ starts at j and finishes at k. The contribution of these two blocks to $S(a)$ is $\sum_{i}^{j-1}(j-1) + \sum_{j}^{k} k = j^2 - (k+i)j + (i+k)$. If $k + i$ is even, this contribution is minimized when $j = \frac{k+i}{2}$. If $k + i$ is odd, this contribution is minimized for both values $j = \lfloor \frac{k+i}{2} \rfloor$ and $j = \lceil \frac{k+i}{2} \rceil$.

Now suppose for the sake of contradiction that $S(a)$ is minimized when $c_1 = \cdots = c_n = t/n$ does not hold. There are two possibilities. Either there are two adjacent blocks whose lengths differ by more than one, in which case we immediately have a contradiction. If not, then it must be the case that all pairs of adjacent blocks differ in length by at most one. In particular, there must be a block of length $t/n + 1$ and another of length $t/n - 1$ with all blocks in between of length t/n. Flipping the leftmost of these blocks with its right neighbor will not change the sum $S(a)$. Repeatedly doing this until the blocks of lengths $t/n + 1$ and $t/n - 1$ are adjacent, does not change $S(a)$. Then we have two adjacent blocks that differ in length by more than one, which contradicts the fact that $S(a)$ was minimized. $\square$

Theorem 3. *If $n|t$, the FP strategies (a, b) are a $\left(\frac{1}{2} + \frac{1}{t} - \frac{1}{2n}\right)$-equilibrium.*

Proof. By symmetry, it suffices to show that $\hat{a}$ is a $\left(\frac{1}{2} + \frac{1}{t} - \frac{1}{2n}\right)$-best-response against b. Applying Lemma 6, Lemma 7 and Lemma 8, we have that $\hat{a} = m_1, \ldots, m_1, m_2, \ldots,$ $m_2, \cdots, m_n, \ldots, m_n$, where $m_1, \ldots, m_n$ is an arbitrary labeling of player 1's pure strategies and where each block of actions has length t/n. Using (10), we have that

$$\epsilon \leq 1 + \frac{1}{t} - \frac{1}{t^2} \sum_{i=1}^{t} f(\hat{a}_i) = 1 + \frac{1}{t} - \frac{1}{t^2} \frac{t}{n} \sum_{i=1}^{n} \left(\frac{i \cdot t}{n}\right) = 1 + \frac{1}{t} - \frac{n+1}{2n} = \frac{1}{2} + \frac{1}{t} - \frac{1}{2n}.$$

This concludes the proof. $\square$

4   Discussion

Daskalakis et al. [7] gave a very simple algorithm that achieves an approximation guarantee of $\frac{1}{2}$; subsequent algorithms e.g. [3,12] improved on this, but at the expense of being more complex and centralized, commonly solving one or more derived LPs from the game. Our result suggests that further work on the topic might address the question of whether $\frac{1}{2}$ is a fundamental limit to the approximation performance obtainable by certain types of algorithms that are in some sense simple or decentralized. The question of specifying appropriate classes of algorithms is itself challenging, and is also considered in [6] in the context of algorithms that provably fail to find Nash equilibria without computational complexity theoretic assumptions.

References

1. Brandt, F., Fischer, F., Harrenstein, P.: On the Rate of Convergence of Fictitious Play. In: 3rd Symposium on Algorithmic Game Theory, pp. 102–113 (2010)
2. Berger, U.: Fictitious play in 2 × n games. Journal of Economic Theory 120, 139–154 (2005)
3. Bosse, H., Byrka, J., Markakis, E.: New Algorithms for Approximate Nash Equilibria in Bimatrix Games. In: Proceedings of the 3rd International Workshop on Internet and Network Economics, pp. 17–29 (2007)
4. Chen, X., Deng, X., Teng, S.-H.: Settling the complexity of computing two-player Nash equilibria. Journal of the ACM 56(3), 1–57 (2009)
5. Conitzer, V.: Approximation Guarantees for Fictitious Play. In: Procs of 47th Annual Allerton Conference on Communication, Control, and Computing, pp. 636–643 (2009)

6. Daskalakis, C., Frongillo, R., Papadimitriou, C.H., Pierrakos, G., Valiant, G.: On Learning Algorithms for Nash Equilibria. In: Kontogiannis, S., Koutsoupias, E., Spirakis, P.G. (eds.) AGT 2010. LNCS, vol. 6386, pp. 114–125. Springer, Heidelberg (2010)
7. Daskalakis, C., Mehta, A., Papadimitriou, C.H.: A Note on Approximate Nash Equilibria. Theoretical Computer Science 410(17), 1581–1588 (2009)
8. Daskalakis, C., Goldberg, P.W., Papadimitriou, C.H.: The Complexity of Computing a Nash Equilibrium. SIAM Journal on Computing 39(1), 195–259 (2009)
9. Fudenberg, D., Levine, D.K.: The Theory of Learning in Games. MIT Press, Cambridge (1998)
10. Robinson, J.: An Iterative Method of Solving a Game. Annals of Mathematics 54(2), 296–301 (1951)
11. Shapley, L.: Some topics in two-person games. In: Advances in Game Theory. Annals of Mathematics Studies, vol. 52, Princeton University Press, Princeton (1964)
12. Tsaknakis, H., Spirakis, P.G.: An Optimization Approach for Approximate Nash Equilibria. Internet Mathematics 5(4), 365–382 (2008)

How Profitable Are Strategic Behaviors in a Market?

Ning Chen[1], Xiaotie Deng[2], and Jie Zhang[3]

[1] Division of Mathematical Sciences, School of Physical and Mathematical Sciences,
Nanyang Technological University, Singapore
ningc@ntu.edu.sg
[2] Department of Computer Science, University of Liverpool, UK
xiaotie@liv.ac.uk
[3] Department of Computer Science, City University of Hong Kong, Hong Kong
csjiezhang@gmail.com

Abstract. It is common wisdom that individuals behave strategically in economic environments. We consider Fisher markets with Leontief utilities and study strategic behaviors of individual buyers in market equilibria. While simple examples illustrate that buyers do get larger utilities when behaving strategically, we show that the benefits can be quite limited: We introduce the concept of *incentive ratio* to capture the extent to which utility can be increased by strategic behaviors of an individual, and show that the incentive ratio of Leontief markets is less than 2. We also reveal that the incentive ratios are insensitive to market sizes. Potentially, the concept incentive ratio can have applications in other strategic settings as well.

1 Introduction

Market equilibrium is a vital notion in classical economic theory. Understanding its properties and computation has been one of the central questions in Algorithmic Game Theory. For the Fisher market model [7], we consider pricing and allocation of m products to n buyers, each with an initially endowed amount of cash e_i and with a non-decreasing concave utility function. At a market equilibrium, all products are sold out, all cash is spent, and, most importantly, the set of products purchased by each buyer maximizes his utility function for the given equilibrium price vector constrained by his initial endowment. It is well-known that a market equilibrium always exists given mild assumptions on the utility functions [5,7].

A critical assumption in studying and analyzing marketplaces is that all individuals are rational and would like to seek their utility maximized. In a market equilibrium, while it is true that every buyer gets his maximum utility through an equilibrium allocation, his utility function indeed takes an effect on generating equilibrium prices, which in turn affect his utility in the equilibrium allocation. In other words, (strategic) buyers may affect market equilibria and their own utilities through the information that the market collects.

C. Demetrescu and M.M. Halldórsson (Eds.): ESA 2011, LNCS 6942, pp. 106–118, 2011.

From a different angle, market equilibrium maps input information from buyers to prices and allocations; this defines a natural *market equilibrium mechanism*, in which buyers report to the market their utility functions (their strategies consist of all possible functions in a given domain) and the market generates an equilibrium output for the reported functions. Adsul et al. [1] recently observed that bidding truthfully is not a dominant strategy in the market equilibrium mechanism for linear utility functions. That is, buyers can behave strategically in a linear market to increase their utilities. This phenomenon is not restricted to the linear utility functions. The following example suggests that it may happen for other utility functions as well, e.g., Leontief functions [9,10,16] — the utility to buyer i is given by $u_i = \min_j \left\{ \frac{x_{ij}}{a_{ij}} \right\}$, where x_{ij} is the allocation of item j to i and a_{ij}'s are parameters associated with buyer i.

Example 1. There are two buyers i_1 and i_2 with endowment $e_1 = 2/3$ and $e_2 = 1/3$ respectively. There are two items j_1 and j_2 both with unit quantity. Buyer i_1's utility is $u_1 = \min \left\{ \frac{x_{11}}{\frac{2}{3}}, \frac{x_{12}}{\frac{1}{3}} \right\}$, and buyer i_2's utility is $u_2 = \min \left\{ \frac{x_{21}}{\frac{1}{3}}, \frac{x_{22}}{\frac{2}{3}} \right\}$. Then in an equilibrium output[1] with price vector $\mathbf{p} = (1,0)$ and allocations $\mathbf{x}_1 = (\frac{2}{3}, \frac{1}{3})$ and $\mathbf{x}_2 = (\frac{1}{3}, \frac{2}{3})$, the utilities are $u_1 = 1$ and $u_2 = 1$. Now if buyer i_1 strategizes in the market and reports $u_1' = \min \left\{ \frac{x_{11}}{\frac{5}{9}}, \frac{x_{12}}{\frac{4}{9}} \right\}$, price vector $\mathbf{p}' = (0,1)$ and allocations $\mathbf{x}_1' = (\frac{5}{6}, \frac{2}{3})$ and $\mathbf{x}_2' = (\frac{1}{6}, \frac{1}{3})$ give an equilibrium for the new setting. Now the utilities are $u_1' = \frac{5}{4}$ and $u_2' = \frac{1}{2}$, where buyer i_1 gets a strictly larger utility.

These intriguing phenomena motivate us to consider the following questions: How much can utility be increased with strategic behaviors in a market? How does the answer to the question depend on the domains of utility functions? The answers to these questions would help us to understand the effects of strategic behaviors in marketplaces on their equilibria and utilities of individuals. In this paper, we take a first step towards answering these questions by considering Leontief utility functions. Leontief utilities are a special case of well-studied Constant Elasticity of Substitution (CES) functions [4] and represent perfect complementarity preferences of buyers. They allow us to model a wide range of realistic preferences of buyers, illustrating a variety of different phenomena of markets.

We study this problem by introducing a quantity called *incentive ratio*, which characterizes the extent to which utilities can be increased by strategic behaviors of individuals. Formally, for any fixed bids of other buyers in a given market, the incentive ratio of any given buyer is defined to be his maximum possible utility by behaving strategically divided by his utility when bidding truthfully, given any fixed bids of all other buyers. Note that the definition of incentive ratio can be generalized to any mechanisms. Indeed, if a mechanism is incentive

---

[1] In a Leontief market, equilibrium allocation may not be unique, but the utility of every buyer in all equilibria, evaluated using the utility functions reported, is unique [24]. Therefore, the selection of output equilibria will not affect the utility of any buyer.

compatible, its incentive ratio is one. A mechanism/market with a small incentive ratio implies that the "invisible hand" in the market is strong in the sense that no one can benefit much from (complicated) strategic considerations, even if one has complete information of the others. Incentive ratio therefore characterizes robustness of incentives in a mechanism/market, and has potential applications in other incentive scenarios.

A related concept, *price of anarchy* [19], together with several variants (e.g., *price of stability* [2]), is becoming one of the most important solution concepts in our understanding of Algorithmic Game Theory. While the price of anarchy considers the ratio of the social welfare achieved in the worst case Nash equilibrium versus the optimum social welfare, our concept of incentive ratio does not deal with social welfare but that of individual optimality, for everyone. It is close to the approximation market equilibrium price introduced in [11] where every individual achieves a solution within a constant factor from its own optimum under the given price. In both concepts, individuals do not achieve their own optimum but bounded by a constant factor away. However, in [11], the consideration is about computational difficulty, but, in our model, we consider an exact equilibrium mechanism when market participants may play strategically.

In this paper, we show that the incentive ratio of Leontief markets is less than 2. Our proof lies on revealing interconnections of the incentive ratios of markets with different sizes. In particular, we prove that the incentive ratio is independent of the number of buyers (Theorem 5), by showing a reduction from any n-buyer market to a 2-buyer market, and vice versa. A similar result holds for items as well, i.e., the incentive ratio is independent of the number of items (Theorem 6). These results are of independent interests and imply that the incentive ratio is insensitive to the size of a market. In other words, the size of a market is not a factor to affect the largest possible utility that a strategic individual can obtain. Given these properties, we therefore focus on a market with 2 buyers and 2 items to bound the incentive ratio. Our proof involves in best response and dominant strategy analysis for Leontief markets.

1.1 Related Work

Market equilibrium, especially its algorithmic perspective, has received extensive attention in the last decade [11,17,10,25,12]. For Fisher markets, Devanur et al. [13] developed a polynomial primal-dual algorithm with linear utility functions. Codenotti and Varadarajan [9] modeled the Leontief market equilibrium problem as a concave maximization problem; Garg et al. [16] gave a polynomial time algorithm to compute an equilibrium for Leontief utilities. Other algorithmic results of computing a market equilibrium in Fisher markets including, e.g., Cobb-Douglas utilities [14] and logarithmic utilities [8].

Roberts and Postlewaite [21] observed that as the number of buyers becomes large, the Walrasian allocation rule is asymptotically incentive compatible. This has a similar flavor to our results that the incentive ratio does not increase as the size of the market enlarges. In a recent paper, Adsul et al. [1] studied strategic behaviors in a Fisher market with linear utilities; they showed that all Nash equilibria in their setting are conflict-free, i.e., buyers have no conflict on allocations that maximize their utilities. They also showed that a symmetric strategy profile, i.e., all buyers report the same utility function, is a Nash equilibrium if and only if it is conflict-free. Our paper however has completely different focuses as we study Leontief rather than linear utilities, and we consider strategic behaviors for each individual "locally" whereas [1] studied Nash equilibrium for all the buyers.

2   Preliminaries

In a Fisher market M, we are given a set of n buyers and a set of m divisible items. Without loss of generality, assume that all items are of unit quantity supply. Each buyer i has an initial endowment $e_i > 0$, which is normalized to be $\sum_i e_i = 1$, and a utility function $u_i(\mathbf{x}_i)$, where $\mathbf{x}_i = (x_{i1}, \ldots, x_{im})$ is a vector denoting the amount of items that i receives.

One of the most important classes of utility functions is that of Constant Elasticity of Substitution (CES) functions [23]: $u_i(\mathbf{x_i}) = \left(\sum_{j=1}^{m} a_{ij} x_{ij}^{\rho} \right)^{\frac{1}{\rho}}$, for $-\infty < \rho < 1$ and $\rho \neq 0$, where $\mathbf{a_i} = (a_{i1}, \ldots, a_{im})$ is a given vector associated with each buyer; its elements $a_{ij} > 0$. Let $a_i^{\max} = \max_j \{a_{ij}\}$ and $a_i^{\min} = \min_j \{a_{ij}\}$. CES functions allow us to model a wide range of realistic preferences of buyers, and have been shown to derive, in the limit, special cases including, e.g., linear, Leontief, and Cobb-Douglas utility functions [4].

The output of a market is given by a tuple $(\mathbf{p}, \mathbf{x})$, where $\mathbf{p} = (p_1, \ldots, p_m)$ is a price vector and $\mathbf{x} = (\mathbf{x}_1, \ldots, \mathbf{x}_n)$ is an allocation vector. We say $(\mathbf{p}, \mathbf{x})$ is a *market equilibrium* (and $\mathbf{p}$ equilibrium price and $\mathbf{x}$ equilibrium allocation respectively) if the following conditions hold:

- The allocation $\mathbf{x}_i$ maximizes the utility of each buyer i given his endowment e_i and price vector $\mathbf{p}$.
- The market is clear, i.e., all items are sold out and all endowments are exhausted. That is, for any item j, $\sum_i x_{ij} = 1$; and for any buyer i, $\sum_j p_j x_{ij} = e_i$. (Note that this implies that $\sum_j p_j = \sum_i e_i = 1$.)

For CES functions, the equilibrium allocation can be captured by the seminal Eisenberg-Gale convex program [15].

In this paper, we will focus on *Leontief utility functions*: The utility of every buyer is given by $u_i(\mathbf{x_i}) = \min_j \left\{ \frac{x_{ij}}{a_{ij}} \right\}$, where $a_{ij} > 0$. Leontief utility function indicates perfect complementarity between different items and is the case when $\rho \to -\infty$ in the CES functions. Codenotti and Varadarajan [9] gave the following convex program to encode a Leontief market:

$$\max \quad \sum_{i=1}^{n} e_i \log u_i \tag{1}$$

$$s.t. \quad \sum_{i=1}^{n} a_{ij} u_i \leq 1, \quad \forall \, j = 1, \ldots, m$$

$$u_i \geq 0, \quad \forall \, i = 1, \ldots, n$$

Recall that the KKT conditions [6] are necessary for a feasible solution of a nonlinear program to be optimal, provided that some regularity conditions are satisfied. If the objective function and the inequality constraints are continuously differentiable convex functions and the equality constraints are affine functions, the KKT necessary conditions are also sufficient for optimality. Moreover, if the objective function is strictly convex, then the optimal solution is unique [6]. Note that (1) satisfies all the necessary conditions; thus, it possesses all these properties. In particular, this implies that the utility of every buyer is unique in all market equilibria.

Applying the KKT necessary conditions on (1), we get

$$
\begin{cases}
-\dfrac{e_i}{u_i} + \displaystyle\sum_{j=1}^{m} p_j a_{ij} - \mu_i = 0, \ \ \forall \, i \\[4mm]
\mu_i u_i = 0, \ \ \forall \, i \\[4mm]
p_j \left(\displaystyle\sum_{i=1}^{n} a_{ij} u_i - 1 \right) = 0, \ \ \forall \, j
\end{cases}
\tag{2}
$$

where the Lagrangian variable p_j is actually an equilibrium price of the market. Since $u_i > 0$, we have $\mu_i = 0$ and $u_i = \frac{e_i}{\sum_{j=1}^{m} p_j a_{ij}}$.

We have the following simple characterization on the uniqueness of allocation in a Leontief market.

Lemma 1. *The equilibrium allocation is unique if and only if the first constraint of the convex program (1) is tight. That is, $\sum_{i=1}^{n} a_{ij} u_i = 1$, for $j = 1, \ldots, m$.*

3   Market Equilibrium Mechanism

Market equilibrium provides a natural mirror that maps input information of a market (i.e., endowments and utility functions) to an output (i.e., prices and allocations). The participation of every buyer in the market is to seek for a utility-maximized allocation. That is, for any given price vector **p**, every buyer would like to get an allocation that maximizes his utility subject to the budget constraint. This fact is indeed captured by the market equilibrium solution concept.

If we consider rational behaviors of buyers, however, as observed in Example 1, they may actually behave strategically to increase their utilities. This suggests a natural *market equilibrium mechanism*, in which buyers report to the market their utility functions[2] (their strategy space consists of all possible functions in a given domain, i.e., Leontief functions considered in this paper), and then the market generates an equilibrium output based on the reported functions.

---

[2] We assume that buyers do not play with their budget and always report the true value e_i. This assumption is without loss of generality as we are able to show that for any reported utility functions, bidding the true budget is always a dominant strategy.

Formally, each buyer i has a private vector $\mathbf{a_i} = (a_{i1}, \ldots, a_{im})$, which denotes his true Leontief utility function, and may bid an arbitrary vector $\mathbf{b_i} = (b_{i1}, \ldots, b_{im})$ where each $b_{ij} > 0$ to the mechanism. Upon receiving the reported vectors $\mathbf{b_1}, \ldots, \mathbf{b_n}$ (as well as endowments) from all buyers, the mechanism outputs prices and allocations of the items; the true utility that buyer i obtains from the output is denoted by $u_i(\mathbf{b_1}, \ldots, \mathbf{b_n})$.

In this section, we will establish a few properties regarding best response and dominant strategy for the market equilibrium mechanism. Since bidding truthfully is not necessarily a dominant strategy, for any given fixed reported functions of other buyers, every individual buyer will report a function that maximizes his own utility; such a bidding function is called a *best response*. Before considering best response strategies, we first notice that since equilibrium allocations may not be unique for given reported utility functions, the real utility that a strategic buyer obtains depends on different allocations, which in turn affects his strategic behaviors. The following claims, however, indicate that there is always a best response such that the resulting equilibrium allocation is unique. In the following we denote $b_i^{\max} = \max_j\{b_{ij}\}, b_i^{\min} = \min_j\{b_{ij}\}, \forall i$.

Theorem 1. *For any given market with two buyers i_1, i_2 and any reported bid $\mathbf{b_2}$ of i_2, there is a best response strategy $\mathbf{b_1}$ of i_1 such that the equilibrium allocation with $(\mathbf{b_1}, \mathbf{b_2})$ is unique. Further, this best response strategy $\mathbf{b_1}$ is given by*

$$
b_{1j} = \begin{cases} e_1 & \text{if } j \in \arg\max\{b_{2j}\} \\ 1 - \frac{b_{2j}}{b_2^{\max}} e_2 & \text{otherwise} \end{cases}
$$

where the maximum utility of i_1 is $u_1(\mathbf{b_1}, \mathbf{b_2}) = \min_j \left\{ \frac{1 - \frac{b_{2j}}{b_2^{\max}} e_2}{a_{1j}} \right\}$.

For a market with n buyers, we have the following result which follows from Theorem 1 directly.

Theorem 2. *For any given market with n buyers and reported bids $\mathbf{b_2}, \ldots, \mathbf{b_n}$ of $i_2, \ldots, i_n$, respectively, let $S = \{j \mid b_{ij} = b_i^{\max}, i \neq i_1\}$. If $S \neq \emptyset$, then a best response strategy $\mathbf{b_1}$ of i_1 is given by*

$$
b_{1j} = \begin{cases} e_1 & \text{if } j \in S \\ 1 - \sum_{i \neq i_1} \frac{b_{ij}}{b_i^{\max}} e_i & \text{otherwise} \end{cases}
$$

where the maximum utility of i_1 is $u_1(\mathbf{b_1}, \mathbf{b_2}, \ldots, \mathbf{b_n}) = \min_j \left\{ \frac{1 - \sum_{i \neq i_1} \frac{b_{ij}}{b_i^{\max}} e_i}{a_{1j}} \right\}$, and the resulting equilibrium allocation is unique.

The following theorem is a sufficient condition for being a dominant strategy for all buyers to bid truthfully, i.e., $\mathbf{b_i} = \mathbf{a_i}$.

Theorem 3. *For any fixed bid $\mathbf{b_{-i}}$ of all the buyers except i, if there exists an item j such that $a_{ij} = a_i^{max}$, and $b_{i'j} = b_{i'}^{max}$ for all $i' \neq i$, then bidding truthfully is dominant strategy for i.*

The above theorem implies that if there is a common item on which all buyers have the largest weight, then that item "dominates" the utilities and it is impossible to increase one's allocation.

4   Incentive Ratio

In this section, we will present the definition of incentive ratio and our main results.

Definition 1 (Incentive Ratio). *For a given market M and any fixed bids $\mathbf{b}_{-i}$ of other buyers, let $u_i(\mathbf{a_i}, \mathbf{b_{-i}})$ be the utility of i when he bids truthfully, and $\max_{\mathbf{b_i}} u_i(\mathbf{b_i}, \mathbf{b_{-i}})$ be the largest possible utility of i when he unilaterally changes his bid.[3] Define*

$$R_i^M = \max_{\mathbf{b}_{-i}} \frac{\max_{\mathbf{b_i}} u_i(\mathbf{b_i}, \mathbf{b_{-i}})}{u_i(\mathbf{a_i}, \mathbf{b_{-i}})}$$

to be the incentive ratio *of i in the market M.*

Incentive ratio quantifies the benefit of strategic behaviors of each individual buyer. Let $R^M \triangleq \max_i R_i^M$ denote the largest individual incentive ratio of M. Our main result is the following.

Theorem 4 (main). *For any given Leontief market M and a buyer $i \in M$, his incentive ratio is smaller than 2, i.e., $R_i^M < 2$. Thus, $R^M < 2$.*

The ratio given by the theorem is tight; the following example shows that R^M can take on a value arbitrarily close to 2.

Example 2 (Tight example). There are 2 buyers and 2 items with $\mathbf{a_1} = (1 - \epsilon, \epsilon)$ and $\mathbf{a_2} = (\frac{1}{2} - \epsilon, \frac{1}{2} + \epsilon)$. Their budgets are $e_1 = 4\epsilon - 4\epsilon^2 + \epsilon^3$ and $e_2 = 1 - 4\epsilon + 4\epsilon^2 - \epsilon^3$, where $\epsilon > 0$ is an arbitrarily small number. Assume $\mathbf{b_2} = \mathbf{a_2}$. When i_1 bids truthfully, his utility is $u_1 = \frac{4\epsilon - 4\epsilon^2 + \epsilon^3}{1 - \epsilon}$. If i_1 strategically bids

$$\mathbf{b_1} = \left(\frac{8\epsilon - 12\epsilon^2 + 9\epsilon^3 - 2\epsilon^4}{1 + 2\epsilon}, 4\epsilon - 4\epsilon^2 + \epsilon^3 \right)$$

then his utility will be

$$u_1' = \frac{8\epsilon - 12\epsilon^2 + 9\epsilon^3 - 2\epsilon^4}{(1 + 2\epsilon)(1 - \epsilon)}$$

and the incentive ratio is

$$\frac{8 - 12\epsilon + 9\epsilon^2 - 2\epsilon^3}{4 + 4\epsilon - 7\epsilon^2 + 2\epsilon^3}$$

which converges to 2 as ϵ approaches 0.

---

[3] As mentioned earlier, equilibrium allocation may not be unique for different bids of i, which may lead to different true utilities for him. Our definition of incentive ratio is the strongest in the sense that it bounds the largest possible utility in all possible equilibrium allocations, which include, of course, a best possible allocation.

Before proving Theorem 4, we first establish a connection of the incentive ratios of different markets in terms of the number of buyers and items. These properties are of independent interests to reveal the effect of market sizes on strategic behaviors. We will prove the theorem at the end of the section.

4.1   Incentive Ratio Is Independent of the Number of Buyers

In this subsection, we will show that the incentive ratio of Leontief markets is independent of the number of buyers.

Theorem 5. *Incentive Ratio is independent of the number of buyers. That is, if $R_n = \max \left\{ R^{M_n} \mid M_n \text{ is a market with } n \text{ buyers} \right\}$, then $R_n = R_{n'}$ for any $n > n' \geq 2$. (Note that $R_1 = 1$.)*

The proof is by showing that $R_n = R_2$ for any $n \geq 3$; it consists of two directions of reductions: one is from R_n to R_2 and the other is from R_2 to R_n.

Reduction from R_n to R_2. We construct a reduction from any market with n buyers to a 2-buyer market, as the following lemma states. The reduction is by unifying $n - 1$ buyers in the n-buyer market, and comparing the incentive ratio of the buyer whose utility functions are the same in both of the markets.

Lemma 2. *For any n-buyer market M_n, there is a 2-buyer market M_2 such that $R^{M_2} \geq R^{M_n}$. This implies that $R_2 \geq R_n$.*

Reduction from R_2 to R_n. We can have a similar reduction from any market with 2 buyers to an n-buyer market, as the following lemma states. (Note that such a reduction is necessary for Leontief utility functions, since we cannot simply add dummy buyers and items as they may affect the allocations of existing items.)

Lemma 3. *For any 2-buyer market M_2, there is an n-buyer market M_n such that $R^{M_n} \geq R^{M_2}$. This implies that $R_n \geq R_2$.*

4.2   Incentive Ratio Is Independent of the Number of Items

The following claim shows that the incentive ratio does not depend on the number of items as well.

Theorem 6. *Incentive ratio is independent of the number of items.*

4.3   Proof of Theorem 4

In this subsection, we will prove our main theorem. Given the properties of the incentive ratio established in the above subsections, it suffices to bound the incentive ratio for a 2-buyer 2-item market.

Let M be a market with 2 buyers i_1, i_2 and 2 items j_1, j_2 with true utility vectors $\mathbf{a_1} = (a_{11}, a_{12})$ and $\mathbf{a_2} = (a_{21}, a_{22})$. We will only analyze the incentive ratio of i_1; the same argument applies to i_2. Suppose the reported utility vector of i_2 is $\mathbf{b_2} = (b_{21}, b_{22})$, and assume without loss of generality that $b_{21} < b_{22}$. According to Theorem 3, we can assume that $a_{11} > a_{12}$ (otherwise truthful strategy is dominant strategy for i_1 and his incentive ratio is 1).

Let $\mathbf{b_1} = (b_{11}, b_{12})$ be i_1's best response for i_2's bid $\mathbf{b_2}$. Let $u_1 = u_1(\mathbf{a_1}, \mathbf{b_2})$, $u_1' = (\mathbf{b_1}, \mathbf{b_2})$ and $u_2 = u_2(\mathbf{a_1}, \mathbf{b_2})$. Hence, the incentive ratio of i_1 is

$$R_1^M = \frac{u_1'}{u_1} = \frac{\min\limits_j \left\{ \frac{1 - \frac{b_{2j}}{b_2^{\max}} e_2}{a_{1j}} \right\}}{u_1} = \frac{\min \left\{ \frac{1 - \frac{b_{21}}{b_{22}} e_2}{a_{11}}, \frac{e_1}{a_{12}} \right\}}{u_1}$$

There are the following two cases.

- $\frac{b_{21}}{b_{22}} > e_2$, or equivalently, $e_1 > 1 - \frac{b_{21}}{b_{22}}$. For this case, we will construct a new market $\tilde{M}$ where a_{12} is replaced by $\tilde{a}_{12} = \min \left\{ a_{12}, (1 - \frac{b_{22}}{b_{21}} e_2) \frac{a_{12}}{e_1} - \epsilon \right\}$, and all other parameters remain unchanged. Let $\tilde{u}_1 = \tilde{u}_1(\tilde{\mathbf{a}}_1, \mathbf{b_2})$, where $\tilde{\mathbf{a}}_1 = (a_{11}, \tilde{a}_{12})$.

Recall that $\tilde{u}_1 \leq \frac{e_1}{\tilde{a}_1^{\min}} = \frac{e_1}{\tilde{a}_{12}}, u_2 \leq \frac{e_2}{b_2^{\min}} = \frac{e_2}{b_{21}}$, and by definition $\tilde{a}_{12} < (1 - \frac{b_{22}}{b_{21}} e_2) \frac{a_{12}}{e_1}$, we can verify that when i_1 bids truthfully in market $\tilde{M}$, for item j_2 we have

$$a_{12} u_1 + b_{22} u_2 - 1 = \tilde{a}_{12} u_1 + b_{22} u_2 - 1$$
$$< \left(1 - \frac{b_{22}}{b_{21}} e_2 \right) \frac{a_{12}}{e_1} \cdot \frac{e_1}{a_{12}} + b_{22} \frac{e_2}{b_{21}} - 1 = 0$$

According to linear complementary conditions, $p_2 = 0$ and $p_1 > 0$. Recall that $p_1 + p_2 = e_1 + e_2 = 1$; we get $p_1 = 1$. Hence, $\tilde{u}_1 = \frac{e_1}{p_1 a_{11} + p_2 \tilde{a}_{12}} = \frac{e_1}{a_{11}}, u_2 = \frac{e_2}{b_{21}}$. Therefore, $\tilde{u}_1 = \frac{e_1}{a_{11}} = \frac{e_1}{a_1^{\max}} \leq u_1$. Hence,

$$R_1^M = \frac{\min\limits_j \left\{ \frac{1 - \frac{b_{2j}}{b_2^{\max}} e_2}{a_{1j}} \right\}}{u_1} \leq \frac{\min \left\{ \frac{1 - \frac{b_{21}}{b_{22}} e_2}{a_{11}}, \frac{e_1}{a_{12}} \right\}}{\tilde{u}_1} \leq \frac{\min \left\{ \frac{1 - \frac{b_{21}}{b_{22}} e_2}{a_{11}}, \frac{e_1}{\tilde{a}_{12}} \right\}}{\frac{e_1}{a_{11}}} \leq R_1^{\tilde{M}}$$

Now we can get the ratio of $\tilde{M}$. For item j_2, we have

$$a_{1j} u_1 + b_{2j} u_2 - 1 = \tilde{a}_{12} u_1 + b_{22} u_2 - 1 = \tilde{a}_{12} \frac{e_1}{a_{11}} + b_{22} \frac{e_2}{b_{21}} - 1 < 0$$

Thus, $e_1 > \frac{a_{11}(b_{22} - b_{21})}{a_{11} b_{22} - \tilde{a}_{12} b_{21}}$. Recall that we have $\frac{b_{21}}{b_{22}} > e_2$, which is equivalent to $e_1 > 1 - \frac{b_{21}}{b_{22}}$. In addition, $\frac{a_{11}(b_{22} - b_{21})}{a_{11} b_{22} - \tilde{a}_{12} a b_{21}} > 1 - \frac{b_{21}}{b_{22}}$, we conclude the constraint for e_1 satisfies

$$e_1 > \frac{a_{11}(b_{22} - b_{21})}{a_{11} b_{22} - \tilde{a}_{12} b_{21}}$$

Now comparing the two terms of buyer i_1's utility when he uses best response strategy, we have

$$
\begin{aligned}
\frac{1 - \frac{b_{21}}{b_{22}} e_2}{a_{11}} - \frac{e_1}{\tilde{a}_{12}} &= \frac{1}{a_{11}} \left[\left(1 - \frac{b_{21}}{b_{22}} \right) - \frac{a_{11} b_{22} - \tilde{a}_{12} b_{21}}{\tilde{a}_{12} b_{22}} e_1 \right] \\
&< \frac{1}{a_{11}} \left[\left(1 - \frac{b_{21}}{b_{22}} \right) - \frac{a_{11} b_{22} - \tilde{a}_{12} b_{21}}{\tilde{a}_{12} b_{22}} \cdot \frac{a_{11}(b_{22} - b_{21})}{a_{11} b_{22} - \tilde{a}_{12} b_{21}} \right] \\
&= \frac{1}{a_{11}} \left(1 - \frac{b_{21}}{b_{22}} \right) \left(1 - \frac{a_{11}}{\tilde{a}_{12}} \right) < 0
\end{aligned}
$$

Therefore,

$$
R = \frac{u_1'}{u_1} = \frac{\min\limits_{j} \left\{ \frac{1 - \frac{b_{2j}}{b_2^{\max}} e_2}{a_{1j}} \right\}}{u_1} \leq \frac{\frac{1 - \frac{b_{21}}{b_{22}} e_2}{a_{11}}}{\frac{e_1}{a_{11}}} = \frac{1 - \frac{b_{21}}{b_{22}} e_2}{e_1} = \frac{b_{22} - b_{21}}{b_{22}} \frac{1}{e_1} + \frac{b_{21}}{b_{22}}
$$

$$
< \frac{b_{22} - b_{21}}{b_{22}} \cdot \frac{a_{11} b_{22} - \tilde{a}_{12} b_{21}}{a_{11}(b_{22} - b_{21})} + \frac{b_{21}}{b_{22}} = 1 + \frac{b_{21}}{b_{22}} \left(1 - \frac{\tilde{a}_{12}}{a_{11}} \right) < 1 + 1 = 2
$$

where the last inequality is due to $0 < b_{22} - b_{21} < \epsilon$ and $0 < \tilde{a}_{12} < \epsilon$, as ϵ approaches 0.

$-\frac{b_{21}}{b_{22}} \leq e_2$, or equivalently, $e_1 \leq 1 - \frac{b_{21}}{b_{22}}$. Since $\frac{b_{21}}{b_{22}} \leq e_2$, we know that when buyer i_1 bids truthfully, $u_2 < \frac{e_2}{b_{21}}$. Otherwise if $u_2 = \frac{e_2}{b_{21}}$, then for item j_2, $a_{12} u_1 + b_{22} \frac{e_2}{b_{21}} - 1 \geq a_{12} u_1 + 1 - 1 > 0$, which is a contradiction. On the other hand, if $u_2 = \frac{e_2}{b_{22}}$, then buyer i_1 cannot improve his utility any further and his incentive ratio is 1. Therefore, the utilities satisfy

$$
\frac{e_1}{a_{11}} < u_1 < \frac{e_1}{a_{12}}, \quad \frac{e_2}{b_{22}} < u_2 < \frac{e_2}{b_{21}} \tag{3}
$$

According to the KKT conditions, we have

$$
0 < p_1 < 1, \quad 0 < p_2 < 1
$$

$$
\begin{cases}
a_{11} u_1 + b_{21} u_2 - 1 = 0 \\
a_{12} u_1 + b_{22} u_2 - 1 = 0
\end{cases}
$$

Thus,

$$
u_1 = \frac{b_{22} - b_{21}}{a_{11} b_{22} - a_{12} b_{21}}, \quad u_2 = \frac{a_{11} - a_{12}}{a_{11} b_{22} - a_{12} b_{21}}
$$

Plug it into (3), we have

$$
\frac{a_{12}(b_{22} - b_{21})}{a_{11} b_{22} - a_{12} b_{21}} < e_1 < \frac{a_{11}(b_{22} - b_{21})}{a_{11} b_{22} - a_{12} b_{21}} \tag{4}
$$

Since $1 - \frac{b_{21}}{b_{22}} < \frac{a_{11}(b_{22} - b_{21})}{a_{11} b_{22} - a_{12} b_{21}}$, we conclude the constraint for e_1 is,

$$
\frac{a_{12}(b_{22} - b_{21})}{a_{11} b_{22} - a_{12} b_{21}} < e_1 \leq 1 - \frac{b_{21}}{b_{22}} \tag{5}
$$

Thus,

$$\frac{1 - \frac{b_{21}}{b_{22}} e_2}{a_{11}} - \frac{e_1}{a_{12}} = \frac{1}{a_{11}} \left[\left(1 - \frac{b_{21}}{b_{22}} \right) - \frac{a_{11} b_{22} - a_{12} b_{21}}{a_{12} b_{22}} e_1 \right]$$

$$< \frac{1}{a_{11}} \left[\left(1 - \frac{b_{21}}{b_{22}} \right) - \frac{a_{11} b_{22} - a_{12} b_{21}}{a_{12} b_{22}} \cdot \frac{a_{12} (b_{22} - b_{21})}{a_{11} b_{22} - a_{12} b_{21}} \right]$$

$$= \frac{1}{a_{11}} \left[\left(1 - \frac{b_{21}}{b_{22}} \right) - \left(1 - \frac{b_{21}}{b_{22}} \right) \right] = 0$$

Therefore,

$$R_1^M = \frac{u_1'}{u_1} = \frac{\min\limits_j \left\{ \frac{1 - \frac{b_{2j}}{b_2^{\max}} e_2}{a_{1j}} \right\}}{u_1} = \frac{\frac{1 - \frac{b_{21}}{b_{22}} e_2}{a_{11}}}{\frac{b_{22} - b_{21}}{a_{11} b_{22} - a_{12} b_{21}}}$$

$$= \frac{a_{11} b_{22} - a_{12} b_{21}}{a_{11} (b_{22} - b_{21})} \left(1 - \frac{b_{21}}{b_{22}} + \frac{b_{21}}{b_{22}} e_1 \right)$$

$$\leq \frac{a_{11} b_{22} - a_{12} b_{21}}{a_{11} (b_{22} - b_{21})} \left(\frac{b_{22} - b_{21}}{b_{22}} + \frac{b_{21}}{b_{22}} \cdot \frac{b_{22} - b_{21}}{b_{22}} \right)$$

$$= (1 - \frac{a_{12}}{a_{11}} \cdot \frac{b_{21}}{b_{22}})(1 + \frac{b_{21}}{b_{22}}) < (1 - 0)(1 + 1) = 2$$

where the last inequality is due to $0 < b_{22} - b_{21} < \epsilon$ and $0 < a_{12} < \epsilon$, as ϵ approaches 0.

This completes the proof.

5   Conclusions

We introduce the concept of incentive ratio to characterize the extent to which utilities can be increased by strategic behaviors of individuals in a marketplace. It would be interesting to study the incentive ratio for other market models, e.g., linear, Cobb-Douglas, or general, CES utility functions, as well as Arrow-Debreu markets. The definition of incentive ratio can be generalized to other mechanism design settings. For example, if a mechanism is incentive compatible, its incentive ratio is one. A notion similar to incentive ratio is approximate truthfulness, which has been considered in, e.g., [22,20,3,18]. The concept of incentive ratio focuses on individual participant rather than the worst case analysis of the whole market. In particular, for markets with asymmetric information, the incentive ratios of different individuals could be very different, depending on their knowledge and unbalanced situation in the market. The incentive ratio defined in our paper characterizes robustness of incentives for individuals, and has potential applications in other settings.

References

1. Adsul, B., Babu, C. S., Garg, J., Mehta, R., Sohoni, M.: Nash Equilibria in Fisher Market. In: Kontogiannis, S., Koutsoupias, E., Spirakis, P.G. (eds.) AGT 2010. LNCS, vol. 6386, pp. 30–41. Springer, Heidelberg (2010)
2. Anshelevich, E., Dasgupta, A., Kleinberg, J.M., Tardos, É., Wexler, T., Roughgarden, T.: The Price of Stability for Network Design with Fair Cost Allocation. SIAM Journal on Computing 38(4), 1602–1623 (2008)
3. Archer, A., Papadimitriou, C., Talwar, K., Tardos, E.: An Approximate Truthful Mechanism for Combinatorial Auctions with Single Parameter Agents. In: SODA 2003, pp. 205–214 (2003)
4. Arrow, K.J., Chenery, H., Minhas, B., Solow, R.: Capital-Labor Substitution and Economic Efficiency. The Review of Economics and Statistics 43(3), 225–250 (1961)
5. Arrow, K., Debreu, G.: Existence of an Equilibrium for a Competitive Economy. Econometrica 22, 265–290 (1954)
6. Bazaraa, M.S., Sherali, H.D., Shetty, C.M.: Nonlinear Programming: Theory and Algorithms. John Wiley & Sons, Chichester (2006)
7. Brainard, W., Scarf, H.E.: How to Compute Equilibrium Prices in 1891. Cowles Foundation Discussion Paper (2000)
8. Chen, N., Deng, X., Sun, X., Yao, A.C.: Fisher Equilibrium Price with a Class of Concave Utility Functions. In: Albers, S., Radzik, T. (eds.) ESA 2004. LNCS, vol. 3221, pp. 169–179. Springer, Heidelberg (2004)
9. Codenotti, B., Varadarajan, K.R.: Efficient Computation of Equilibrium Prices for Markets with Leontief Utilities. In: Díaz, J., Karhumäki, J., Lepistö, A., Sannella, D. (eds.) ICALP 2004. LNCS, vol. 3142, pp. 371–382. Springer, Heidelberg (2004)
10. Codenotti, B., Saberi, A., Varadarajan, K., Ye, Y.: Leontief Economies Encode Nonzero Sum Two-Player Games. In: SODA 2006, pp. 659–667 (2006)
11. Deng, X., Papadimitriou, C., Safra, S.: On the Complexity of Equilibria. In: STOC 2002, pp. 67–71 (2002)
12. Devanur, N., Kannan, R.: Market Equilibria in Polynomial Time for Fixed Number of Goods or Agents. In: FOCS 2008, pp. 45–53 (2008)
13. Devanur, N., Papadimitriou, C., Saberi, A., Vazirani, V.: Market Equilibrium via a Primal-Dual Algorithm for a Convex Program. JACM 55(5) (2008)
14. Eaves, B.C.: Finite Solution of Pure Trade Markets with Cobb-Douglas Utilities. Mathematical Programming Study 23, 226–239 (1985)
15. Eisenberg, E., Gale, D.: Consensus of Subjective Probabilities: The Pari-Mutuel Method. Annals Of Mathematical Statistics 30, 165–168 (1959)
16. Garg, D., Jain, K., Talwar, K., Vazirani, V.: A Primal-Dual Algorithm for Computing Fisher Equilibrium in the Absence of Gross Substitutability Property. Theoretical Computer Science 378, 143–152 (2007)
17. Jain, K.: A polynomial Time Algorithm for Computing an Arrow-Debreu Market Equilibrium for Linear Utilities. SIAM Journal on Computing 37(1), 303–318 (2007)
18. Kothari, A., Parkes, D., Suri, S.: Approximately-Strategyproof and Tractable Multiunit Auctions. Decision Support Systems 39(1), 105–121 (2005)
19. Koutsoupias, E., Papadimitriou, C.H.: Worst-Case Equilibria. In: Meinel, C., Tison, S. (eds.) STACS 1999. LNCS, vol. 1563, pp. 404–413. Springer, Heidelberg (1999)
20. Lehmann, D., O'Callaghan, L., Shoham, Y.: Truth Revelation in Approximately Efficient Combinatorial Auctions. JACM 49(5), 577–602 (2002)

21. Roberts, D., Postlewaite, A.: The Incentives for Price-Taking Behavior in Large Exchange Economies. Econometrica 44, 113–127 (1976)
22. Schummer, J.: Almost Dominant Strategy Implementation, MEDS Department, Northwestern University, Discussion Papers 1278 (1999)
23. Solov, R.: A Contribution to the Theory of Economic Growth. Quarterly Journal of Economics 70, 65–94 (1956)
24. Varian, H.: Microeconomic Analysis. W. W. Norton & Company, New York (1992)
25. Ye, Y.: A Path to the Arrow-Debreu Competitive Market Equilibrium. Mathematical Programming 111(1-2), 315–348 (2008)

Improving the Price of Anarchy for Selfish Routing via Coordination Mechanisms

George Christodoulou[1], Kurt Mehlhorn[2], and Evangelia Pyrga[3]

[1] University of Liverpool, United Kingdom
gchristo@liv.ac.uk
[2] Max-Planck-Institut für Informatik, Saarbrücken, Germany
mehlhorn@mpi-inf.mpg.de
[3] Technische Universität München, Germany
pyrga@in.tum.de

Abstract. We reconsider the well-studied Selfish Routing game with affine latency functions. The Price of Anarchy for this class of games takes maximum value 4/3; this maximum is attained already for a simple network of two parallel links, known as Pigou's network. We improve upon the value 4/3 by means of Coordination Mechanisms.

We increase the latency functions of the edges in the network, i.e., if $\ell_e(x)$ is the latency function of an edge e, we replace it by $\hat{\ell}_e(x)$ with $\ell_e(x) \leq \hat{\ell}_e(x)$ for all x. Then an adversary fixes a demand rate as input. The *engineered Price of Anarchy* of the mechanism is defined as the worst-case ratio of the Nash social cost in the modified network over the optimal social cost in the original network. Formally, if $\hat{C}_N(r)$ denotes the cost of the worst Nash flow in the modified network for rate r and $C_{opt}(r)$ denotes the cost of the optimal flow in the original network for the same rate then

$$ePoA = \max_{r \geq 0} \frac{\hat{C}_N(r)}{C_{opt}(r)}.$$

We first exhibit a simple coordination mechanism that achieves for any network of parallel links an engineered Price of Anarchy strictly less than 4/3. For the case of two parallel links our basic mechanism gives 5/4 = 1.25. Then, for the case of two parallel links, we describe an *optimal* mechanism; its engineered Price of Anarchy lies between 1.191 and 1.192.

1 Introduction

We consider single-commodity congestion games on networks, defined by a directed graph $G = (V, E)$, designated nodes $s, t \in V$, and a set $\ell = (\ell_e)_{e \in E}$ of non-decreasing non-negative functions; ℓ_e is the latency function of edge $e \in E$. Let P be the set of all paths from s to t, and let $f(r)$ be a feasible s,t-flow routing r units of flow. For any $p \in P$, let $f_p(r)$ denote the amount of flow that $f(r)$ routes via path p. For ease of notation, when r is fixed and clear from context, we will write simply f, f_p instead of $f(r), f_p(r)$. By definition, $\sum_{p \in P} f_p = r$. Similarly, for any edge $e \in E$, let f_e be the amount of flow going through e. We define the latency of p under flow f as $\ell_p(f) = \sum_{e \in p} \ell_e(f_e)$ and the

C. Demetrescu and M.M. Halldórsson (Eds.): ESA 2011, LNCS 6942, pp. 119–130, 2011.

cost of flow f as $C(f) = \sum_{e \in E} f_e \cdot \ell_e(f_e)$ and use $C_{opt}(r)$ to denote the minimum cost of any flow of rate r. We will refer to such a minimum cost flow as an *optimal* flow (Opt). A feasible flow f that routes r units of flow from s to t is at *Nash (or Wardrop [28]) Equilibrium*[1] if for $p_1, p_2 \in P$ with $f_{p_1} > 0$, $\ell_{p_1}(f) \le \ell_{p_2}(f)$. We use $C_N(r)$ to denote the maximum cost of a Nash flow for rate r. The *Price of Anarchy (PoA)* [22] (for demand r) is defined as

$$PoA(r) = \frac{C_N(r)}{C_{opt}(r)} \quad \text{and} \quad PoA = \max_{r > 0} PoA(r).$$

PoA is bounded by $4/3$ in the case of affine latency functions $\ell_e(x) = a_e x + b_e$ with $a_e \ge 0$ and $b_e \ge 0$; see [27,12]. The worst-case is already assumed for a simple network of two parallel links, known as Pigou's network; see Figure 1.

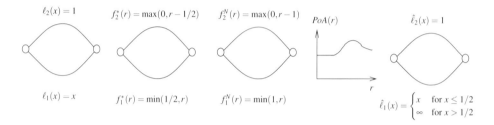

Fig. 1. Pigou's network: We show the original network, the optimal flow and the Nash flow as a function of the rate r, respectively, the Price of Anarchy as a function of the rate ($PoA(r)$ is 1 for $r \le 1/2$, then starts to grow until it reaches its maximum of $4/3$ at $r = 1$, and then decreases again and approaches 1 as r goes to infinity), and finally the modified latency functions. We obtain $ePoA(r) = 1$ for all r in the case of Pigou's network

A *Coordination Mechanism*[2] replaces the cost functions $(\ell_e)_{e \in E}$ by functions[3] $\hat{\ell} = (\hat{\ell}_e)_{e \in E}$ such that $\hat{\ell}_e(x) \ge \ell_e(x)$ for all $x \ge 0$. Let $\hat{C}(f)$ be the cost of flow f when for each edge $e \in E$, $\hat{\ell}_e$ is used instead of ℓ_e and let $\hat{C}_N(r)$ be the maximum cost of a Nash flow of rate r for the modified latency functions. We define the *engineered Price of Anarchy* (for demand r) as

$$ePoA(r) = \frac{\hat{C}_N(r)}{C_{opt}(r)} \quad \text{and} \quad ePoA = \max_{r > 0} ePoA(r).$$

We stress that the optimal cost refers to the original latency functions ℓ.

Non-continuous Latency Functions: In the previous definition, as it will become clear in Section 2, it is important to allow non-continuous modified latencies. However, when

---

[1] This assumes continuity and monotonicity of the latency functions. For non-continuous functions, see the discussion later in this section.

[2] Technically, we consider *symmetric* coordination mechanisms in this work, as defined in [8] i.e., the latency modifications affect the users in a symmetric fashion.

[3] One can interpret the difference $\hat{\ell}_e - \ell_e$ as a flow-dependent toll imposed on the edge e.

we move from continuous to non-continuous latency functions, Wardrop equilibria do not always exist. Non-continuous functions have been studied by transport economists to model the effects of step-function congestion tolls and traffic lights. Several notions of equilibrium that handle discontinuities have been proposed in the literature[4]. The ones that are closer in spirit to Nash equilibria, are those proposed by Dafermos[5] [14] and Berstein and Smith [4]. According to the Dafermos' [14] definition of *user optimization*, a flow is in equilibrium if no *sufficiently small* fraction of the users on any path, can decrease the latency they experience by switching to another path[6]. Berstein and Smith [4] introduced the concept of *User Equilibrium*, weakening further the Dafermos equilibrium, taking the fraction of the users to the limit approaching 0. The main idea of their definition is to capture the notion of the *individual commuter* of the users, that was implicit in Wardrop's definition for continuous functions. The Dafermos equilibrium on the other hand is a stronger concept that captures the notion of coordinated deviations by *groups of commuters*.

We adopt the concept of User Equilibrium. Formally, we say that a feasible flow f that routes r units of flow from s to t is a User Equilibrium, iff for all $p_1, p_2 \in P$ with $f_{p_1} > 0$,

$$\ell_{p_1}(f) \leq \liminf_{\varepsilon \downarrow 0} \ell_{p_2}(f + \varepsilon \mathbf{1}_{p_2} - \varepsilon \mathbf{1}_{p_1}),$$

where $\mathbf{1}_p$ denotes the flow where only one unit passes along a path p.

Note that for continuous functions the above definition is identical to the Wardrop Equilibrium. One has to be careful when designing a Coordination Mechanism with discontinuous functions, because the existence of equilibria is not always guaranteed[7]. It is important to emphasize, that all the mechanisms that we suggest in this paper use both lower semicontinuous and regular[8] latencies, and therefore User Equilibrium existence is guaranteed due to the theorem of [4]. Moreover, since our modified latencies are non-decreasing, all User Equilibria are also Dafermos-Sparrow equilibria. Finally, the lower bounds that we provide, do not rely on any monotonicity assumptions, and hold for general coordination mechanisms as defined above. From now on, we refer to the User Equilibria as Nash Equilibria, or simply Nash flows.

Our Contribution: We demonstrate the possibility of reducing the Price of Anarchy for Selfish Routing via Coordination Mechanisms. We obtain the following results for networks of k parallel links.

– if original and modified latency functions are continuous, no improvement is possible, i.e., *ePoA* ≥ *PoA*; see Section 2.

---

[4] See [25,23] for an excellent exposure of the relevant concepts, the relation among them, as well as for conditions that guarantee their existence.

[5] In [14], Dafermos weakened the orginal definition by [13]to make it closer to the concept of Nash Equilibrium.

[6] See Section 5 for a formal definition.

[7] See for example [15,4] for examples where equilibria do not exist even for the simplest case of two parallel links and non-decreasing functions.

[8] See [4] for a definition of regular functions.

– for the case of affine cost functions, we describe a simple coordination mechanism that achieves an engineered Price of Anarchy strictly less than 4/3; see Section 3. The functions $\hat{\ell}_e$ are of the form

$$\hat{\ell}_e(x) = \begin{cases} \ell_e(x) & \text{for } x \leq r_e \\ \infty & \text{for } x > r_e. \end{cases} \tag{1}$$

For the case of two parallel links, the mechanism gives 5/4 (see Section 3.1), for Pigou's network it gives 1, see Figure 1.

– For the case of two parallel links with affine cost functions, we describe an *optimal*[9] mechanism; its engineered Price of Anarchy lies between 1.191 and 1.192 (see Sections 4 and 5). It uses modified cost functions of the form

$$\hat{\ell}_e(x) = \begin{cases} \ell_e(x) & \text{for } x \leq r_e \text{ and } x \geq u_e \\ \ell_e(u_e) & \text{for } r_e < x < u_e. \end{cases} \tag{2}$$

The Price of Anarchy is a standard measure to quantify the effect of selfish behavior. There is a vast literature studying the Price of Anarchy for various models of selfish routing and scheduling problems (see [24]). We show that simple coordination mechanisms can reduce the Price of Anarchy for selfish routing games below the 4/3 worst case for networks of parallel links and affine cost functions.

We believe that our arguments extend to more general cost functions, e.g., polynomial cost functions. However, the restriction to parallel links is crucial for our proof. We leave it as a major open problem to prove results for general networks or at least more general networks, e.g., series-parallel networks.

Due to space limitations, some of the proofs are omitted and deferred to the full version of this paper.

Implementation: We discuss the realization of the modified cost function in a simple traffic scenario where the driving speed on a link is a decreasing function of the flow on the link and hence the transit time is an increasing function. The step function in (2) can be realized by setting a speed limit corresponding to transit time $\ell_e(u_e)$ once the flow is above r_e. The functions in (1) can be approximately realized by access control. In any time unit only r_e items are allowed to enter the link. If the usage rate of the link is above r_e, the queue in front of the link will grow indefinitely and hence transit time will go to infinity.

Related Work: The concept of Coordination Mechanisms was introduced in (the conference version of) [8]. Coordination Mechanisms have been used to improve the Price of Anarchy in scheduling problems for parallel and related machines[8,18,21] as well as for unrelated machines [3,6]; the objective is makespan minimization. Very recently,

---

[9] The lower bound that we provide in Section 5 holds for all deterministic coordination mechanisms with respect to the Dafermos-Sparrow[13] definition of equilibria. However, the arguments of our proof work for all deterministic coordination mechanisms that use *non-decreasing modified latencies* even for the weaker definition of User Equilibrium. Therefore the mechanism of Section 4 is optimal for these two classes of mechanisms.

[9] considered as objective the weighted sum of completion times. Truthful coordination mechanisms have been studied in [1,7,2].

Another very well-studied attempt to cope with selfish behavior is the introduction of taxes (tolls) on the edges of the network in selfish routing games [10,17,19,20,16,5]. The disutility of a player is modified and equals her latency plus some toll for every edge that is used in her path. It is well known (see for example [10,17,19,20]) that so-called marginal cost tolls, i.e., $\hat{\ell}_e(x) = \ell_e(x) + x\ell'_e(x)$, result in a Nash flow that is equal to the optimum flow for the original cost functions.[10] Roughgarden [26] seeks a sub-network of a given network that has optimal Price of Anarchy for a given demand. [11] studies the question whether tolls can reduce the cost of a Nash equilibrium. They show that for networks with affine latencies, marginal cost pricing does not improve the cost of a flow at Nash equilibrium, as well as that the maximum possible benefit that one can get is no more than that of edge removal.

Discussion: The results of this paper are similar in spirit to the results discussed in the previous paragraph, but also very different. The above papers assume that taxes or tolls are determined with full knowledge of the demand rate r. Our coordination mechanisms must *a priori* decide on the modified latency functions *without knowledge of the demand*; it must determine the modified functions $\hat{\ell}$ and then an adversary selects the input rate r. More importantly, our target objectives are different; we want to minimize the ratio of the modified cost (taking into account the increase of the latencies) over the *original* optimal cost. Our simple strategy presented in Section 3 can be viewed as a generalization of link removal. Removal of a link reduces the capacity of the edge to zero, our simple strategy reduces the capacity to a threshold r_e.

2   Continuous Latency Functions Yield No Improvement

The network in this section consists of k parallel links connecting s to t and the original latency functions are assumed to be continuous and non-decreasing. We show that substituting them by continuous functions brings no improvement.

Lemma 1. *Assume that the original functions ℓ_e are continuous and non-decreasing. Consider some modified latency functions $\hat{\ell}$ and some rate r for which there is a Nash Equilibrium flow $\hat{f}$ such that the latency function $\hat{\ell}_i$ is continuous at $\hat{f}_i(r)$ for all $1 \leq i \leq k$. Then $ePoA(r) \geq PoA(r)$.*

3   A Simple Coordination Mechanism

Let $\ell_i(x) = a_i x + b_i = (x + \gamma_i)/\lambda_i$ be the latency function of the i-th link, $1 \leq i \leq k$. We call λ_i the *efficiency* of the link. We order the links in order of increasing b-value and

---

[10] It is important to observe that although the Nash flow is equal to the optimum flow, its cost with respect to the marginal cost function can be twice as large as its cost with respect to the original cost function. For Pigou's network, the marginal costs are $\hat{\ell}_1(x) = 2x$ and $\hat{\ell}_2(x) = 1$. The cost of a Nash flow of rate r with $r \leq 1/2$ is $2r^2$ with respect to marginal costs; the cost of the same flow with respect to the original cost functions is r^2.

assume $b_1 < b_2 < \ldots < b_k$ as two links with the same b-value may be combined (by adding their efficiencies). We may also assume $a_i > 0$ for $i < k$; if $a_i = 0$, links $i+1$ and higher will never be used. We say that a link is *used* if it carries positive flow. The following theorem summarizes some basic facts about optimal flows and Nash flows; it is proved by straightforward calculations.[11]

Theorem 1. *Let* $\Lambda_j = \sum_{i \leq j} \lambda_i$ *and* $\Gamma_j = \sum_{i \leq j} \gamma_j$. *Consider a fixed rate r and let f_i^* and f_i^N, $1 \leq i \leq k$ be the optimal flow and the Nash flow for rate r respectively. Let*

$$r_j = \sum_{1 \leq i < j} (b_{i+1} - b_i)\Lambda_i.$$

Then Nash uses link j for $r > r_j$ and Opt uses link j for $r > r_j/2$. If Opt uses exactly j links at rate r then

$$f_i^* = \frac{r\lambda_i}{\Lambda_j} + \delta_i/2, \quad where \ \delta_i = \frac{\Gamma_j \lambda_i}{\Lambda_j} - \gamma_i$$

and

$$C_{opt}(r) = \frac{1}{\Lambda_j}\left(r^2 + \Gamma_j r\right) - \sum_{i \leq j}\frac{\delta_i^2}{4\lambda_i} = \frac{1}{\Lambda_j}\left(r^2 + \Gamma_j r\right) - C_j, where \ C_j = \left(\sum_{i \leq h \leq j}(b_h - b_i)^2\lambda_h\lambda_i\right)/(4\Lambda_j).$$

If Nash uses exactly j links at rate r then

$$f_i^N = \frac{r\lambda_i}{\Lambda_j} + \delta_i \quad and \quad C_N(r) = \frac{1}{\Lambda_j}\left(r^2 + \Gamma_j r\right).$$

If $s < r$ and Opt uses exactly j links at s and r then

$$C_{opt}(r) = C_{opt}(s) + \frac{1}{\Lambda_j}\left((r-s)^2 + (\Gamma_j + 2s)(r-s)\right).$$

If $s < r$ and Nash uses exactly j links at s and r then

$$C_N(r) = C_N(s) + \frac{1}{\Lambda_j}\left((r-s)^2 + (\Gamma_j + 2s)(r-s)\right).$$

Finally, $\Gamma_j + r_j = b_j\Lambda_j$ *and* $\Gamma_{j-1} + r_j = b_j\Lambda_{j-1}$.

We next define our simple coordination mechanism. It is governed by parameters R_1, $R_2, \ldots, R_{k-1}$; $R_i \geq 2$ for all i. We call the j-th link *super-efficient* if $\lambda_j > R_{j-1}\Lambda_{j-1}$. In Pigou's network, the second link is super-efficient for any choice of R_1 since $\lambda_2 = \infty$ and $\lambda_1 = 1$. Super-efficient links are the cause of high Price of Anarchy. Observe that Opt

---

[11] In a Nash flow all used links have the same latency. Thus, if j links are used at rate r and f_i^N is the flow on the i-th link, then $a_1 f_1^N + b_1 = \ldots = a_j f_j^N + b_j \leq b_{j+1}$ and $r = f_1^N + \ldots + f_j^N$. The values for r_j and f_i^N follow from this. Similarly, in an optimal flow all used links have the same marginal costs.

starts using the j-th link at rate $r_j/2$ and Nash starts using it at rate r_j. If the j-th link is super-efficient, Opt will send a significant fraction of the total flow across the j-th link and this will result in a high Price of Anarchy. Our coordination mechanism induces the Nash flow to use super-efficient links earlier. The latency functions $\hat{\ell}_i$ are defined as follows: $\hat{\ell}_i = \ell_i$ if there is no super-efficient link $j > i$; in particular the latency function of the highest link (= link k) is unchanged. Otherwise, we choose a threshold value T_i (see below) and set $\hat{\ell}_i(x) = \ell_i(x)$ for $x \leq T_i$ and $\hat{\ell}(x) = \infty$ for $x > T_i$. The threshold values are chosen so that the following behavior results. We call this behavior *modified Nash (MN)*.

Assume that Opt uses h links, i.e., $r_h/2 \leq r \leq r_{h+1}/2$. If $\lambda_{i+1} \leq R_i \Lambda_i$ for all i, $1 \leq i < h$, MN behaves like Nash. Otherwise, let j be minimal such that link $j+1$ is super-efficient; MN changes its behavior at rate $r_{j+1}/2$. More precisely, it freezes the flow across the first j links at their current values when the total flow is equal to $r_{j+1}/2$ and routes any additional flow across links $j+1$ to k. The thresholds for the lower links are chosen in such a way that this freezing effect takes place. The additional flow is routed by using the strategy recursively. In other words, let $j_1 + 1, \ldots, j_t + 1$ be the indices of the super-efficient links. Then MN changes behavior at rates $r_{j_i+1}/2$. At this rate the flow across links 1 to j_i is frozen and additional flow is routed across the higher links.

We use $C_{MN}(r) = \hat{C}_N^{R_1,\ldots,R_{k-1}}(r)$ to denote the cost of MN at rate r when operated with parameters R_1 to R_{k-1}. Then $ePoA(r) = C_{MN}(r)/C_{opt}(r)$. For the analysis of MN we use the following strategy. We first investigate the benign case when there is no super-efficient link. In the benign case, MN behaves like Nash and the worst case bound of 4/3 on the PoA can never be attained. More precisely, we will exhibit a function $B(R_1,\ldots,R_{k-1})$ which is smaller than $4/3$ for all choices of the R_i's and will prove $C_{MN}(r) \leq B(R_1,\ldots,R_{k-1})C_{opt}(r)$. We then investigate the non-benign case. We will derive a recurrence relation for

$$ePoA(R_1,\ldots,R_{k-1}) = \max_r \frac{\hat{C}_N^{R_1,\ldots,R_{k-1}}(r)}{C_{opt}(r)}$$

In the case of a single link, i.e., $k = 1$, MN behaves like Nash which in turn is equal to Opt. Thus $ePoA() = 1$. The coming subsections are devoted to the analysis of two links and more than two links, respectively.

3.1 Two Links

The modified algorithm is determined by a parameter $R > 1$. If $\lambda_2 \leq R\lambda_1$, modified Nash is identical to Nash. If $\lambda_2 > R\lambda_1$, the modified algorithm freezes the flow across the first link at $r_2/2$ once it reaches this level. In Pigou's network we have $\ell_1(x) = x$ and $\ell_2(x) = 1$. Thus $\lambda_2 = \infty$. The modified cost functions are $\hat{\ell}_2(x) = \ell_2(x)$ and $\hat{\ell}_1(x) = x$ for $x \leq r_2/2 = 1/2$ and $\hat{\ell}_1(x) = \infty$ for $x > 1/2$. The Nash flow with respect to the modified cost function is identical to the optimum flow in the original network and $\hat{C}_N(f^*) = C(f^*)$. Thus $ePoA = 1$ for Pigou's network.

Theorem 2. *For the case of two links, $ePoA \leq \max(1 + 1/R, (4 + 4R)/(4 + 3R))$. In particular $ePoA = 5/4$ for $R = 4$.*

3.2   Many Links

As already mentioned, we distinguish cases. We first study the benign case $\lambda_{i+1} \leq R_i \Lambda_i$ for all i, $1 \leq i < k$, and then deal with the non-benign case.

The Benign Case: We assume $\lambda_{i+1} \leq R_i \Lambda_i$ for all i, $1 \leq i < k$. Then MN behaves like Nash. We will show $ePoA \leq B(R_1, \dots, R_{k-1}) < 4/3$; here B stands for benign case or base case. Our proof strategy is as follows; we will first show (Lemma 2) that for the i-th link the ratio of Nash flow to optimal flow is bounded by $2\Lambda_k/(\Lambda_i + \Lambda_k)$. This ratio is never more than two; in the benign case, it is bounded away from two. We will then use this fact to derive a bound on the Price of Anarchy (Lemma 4).

Lemma 2. *Let h be the number of links that Opt is using. Then*

$$\frac{f_i^N}{f_i^*} \leq \frac{2\Lambda_h}{\Lambda_i + \Lambda_h}$$

for $i \leq h$. If $\lambda_{j+1} \leq R_j \Lambda_j$ for all j, then

$$\frac{2\Lambda_h}{\Lambda_i + \Lambda_h} \leq \frac{2P}{P+1},$$

where $P := R_1 \cdot \prod_{1 < i < k}(1 + R_i)$.

Proof (Sketch). For $i > j$, the Nash flow on the i-th link is zero and the claim is obvious. For $i \leq j$, we can write the Nash and the optimal flow through link i as

$$f_i^N = r\lambda_i/\Lambda_j + (\Gamma_j \lambda_i/\Lambda_j - \gamma_i) \quad \text{and} \quad f_i^* = r\lambda_i/\Lambda_h + (\Gamma_h \lambda_i/\Lambda_h - \gamma_i)/2$$

Therefore their ratio as a function of r is

$$F(r) = \frac{f_i^N}{f_i^*} = \frac{\Lambda_h}{\Lambda_j} \cdot \frac{2r + 2\Gamma_j - 2b_i\Lambda_j}{2r + \Gamma_h - b_i\Lambda_h}$$

A tedious calculation shows that $F(r)$ is bounded by $2\Lambda_h/(\Lambda_i + \Lambda_h)$.

If $\lambda_{j+1} \leq R_j \Lambda_j$ for all j, then $\Lambda_{j+1} = \lambda_{j+1} + \Lambda_j \leq (1 + R_j)\Lambda_j$ for all j and hence $\Lambda_h \leq P\Lambda_1$. □

Lemma 3. *For any reals μ, α, and β with $1 \leq \mu \leq 2$ and $\alpha/\beta \leq \mu$, $\beta\alpha \leq \frac{\mu-1}{\mu^2}\alpha^2 + \beta^2$.*

Proof. We may assume $\beta \geq 0$. If $\beta = 0$, there is nothing to show. So assume $\beta > 0$ and let $\alpha/\beta = \delta\mu$ for some $\delta \leq 1$. We need to show (divide the target inequality by β^2) $\delta\mu \leq (\mu - 1)\delta^2 + 1$ or equivalently $\mu\delta(1 - \delta) \leq (1 - \delta)(1 + \delta)$. This inequality holds for $\delta \leq 1$ and $\mu \leq 2$. □

Lemma 4. *If $f_i^N/f_i^* \leq \mu \leq 2$ for all i, then $PoA \leq \mu^2/(\mu^2 - \mu + 1)$. If $\lambda_j \leq R_j \Lambda_j$ for all j, then*

$$PoA \leq B(R_1, \dots, R_{k-1}) := \frac{4P^2}{3P^2 + 1},$$

where $P = R_1 \cdot \prod_{1 < j < k}(1 + R_j)$.

Proof. Assume that Nash uses j links and let L be the common latency of the links used by Nash. Then $L = a_i f_i^N + b_i$ for $i \leq j$ and $L \leq b_i = a_i f_i^N + b_i$ for $i > j$. Thus

$$C_N(r) = Lr = \sum_i L f_i^* \leq \sum_i (a_i f_i^N + b_i) f_i^* \leq \frac{\mu-1}{\mu^2} \sum_i a_i (f_i^N)^2 + \sum_i (a_i (f_i^*)^2 + b_i f_i)$$

$$\leq \frac{\mu-1}{\mu^2} C_N(r) + C_{opt}(r)$$

and hence $PoA \leq \mu^2/(\mu^2 - \mu + 1)$. If $\lambda_j \leq R_j \Lambda_j$ for all j, we may use $\mu = 2P/(P+1)$ and obtain $PoA \leq 4P^2/(3P^2 + 1)$.   □

The General Case: We come to the case where $\lambda_{i+1} \geq R_i \Lambda_i$ for some i. Let j be the smallest such i. For $r \leq r_{j+1}/2$, MN and Opt use only links 1 to j and we are in the benign case. Hence *ePoA* is bounded by $B(R_1, \ldots, R_{j-1}) < 4/3$. MN routes the flow exceeding $r_{j+1}/2$ exclusively on higher links.

Lemma 5. *MN does not use links before Opt.*

Proof. Consider any $h > j+1$. MN starts to use link h at $s_h = r_{j+1}/2 + \sum_{j+1 \leq i < h} (b_{i+1} - b_i)(\Lambda_i - \Lambda_j)$ and Opt starts to use it at $r_h/2 = r_{j+1}/2 + \sum_{j+1 \leq i < h} (b_{i+1} - b_i)\Lambda_i/2$. We have $s_h \geq r_h/2$ since $\Lambda_i - \Lambda_j \geq \Lambda_i/2$ for $i > j$.   □

We need to bound the cost of MN in terms of the cost of Opt. In order to do so, we introduce an intermediate flow Mopt (modified optimum) that we can readily relate to MN and to Opt. Mopt uses links 1 to j to route $r_{j+1}/2$ and routes $f = r - r_{j+1}/2$ optimally across links $j+1$ to k. Let f_i^* and f_i^m be the optimal flows and the flows of Mopt, respectively, at rate r. Let $r_s = \sum_{i \leq j} f_i^* \geq r_{j+1}/2$ be the total flow routed across the first j links in the optimal flow (the subscript s stands for small) and let

$$t = \frac{r - r_{j+1}/2}{r - r_s}$$

We will show $t \leq 1 + 1/R_j$ below. We next relate the cost of Mopt on links $j+1$ to k to the cost of Opt on these links. To this end we scale the optimal flow on these links by a factor of t, i.e., we consider the following flow across links $j+1$ to k: on link i, $j+1 \leq i \leq k$, it routes $t \cdot f_i^*$. The total flow on the high links (= links $j+1$ to k) is $r - r_{j+1}/2$ and hence Mopt incurs at most the cost of this flow on its high links. Thus

$$\sum_{i>j} \ell_i(f_i^m) f_i^m \leq \sum_{i>j} \ell_i(t f_i^*) t f_i^* \leq t^2 \left(\sum_{i>j} \ell_i(f_i^*) f_i^* \right).$$

The cost of MN on the high links is at most $ePoA(R_{j+1},\ldots,R_{k-1})$ times this cost by the induction hypothesis. We can now bound the cost of MN as follows:

$$C_{MN}(r) = C_N(r_{j+1}/2) + C_{MN}(\text{flow } f \text{ across links } j+1 \text{ to } k)$$

$$\leq B(R_1,\ldots,R_{j-1})C_{opt}(r_{j+1}/2) + t^2 ePoA(R_{j+1},\ldots,R_{k-1})\left(\sum_{i>j}\ell_i(f_i^*)f_i^*\right)$$

$$\leq B(R_1,\ldots,R_{j-1})C_{opt}(r_s) + t^2 ePoA(R_{j+1},\ldots,R_{k-1})\left(\sum_{i>j}\ell_i(f_i^*)f_i^*\right)$$

$$\leq \max(B(R_1,\ldots,R_{j-1}), t^2 ePoA(R_{j+1},\ldots,R_{k-1}))C_{opt}(r)$$

Lemma 6. $t \leq 1 + 1/R_j$.

We summarize the discussion.

Lemma 7. *For every k and every j with $1 \leq j < k$. If $\lambda_{j+1} > R_j\Lambda_j$ and $\lambda_i \leq R_i\Lambda_{i+1}$ for $i < j$ then*

$$ePoA(R_1,\ldots,R_{k-1}) \leq \max\left(B(R_1,\ldots,R_{j-1}), \left(1+\frac{1}{R_j}\right)^2 ePoA(R_{j+1},\ldots,R_{k-1})\right).$$

We are now ready for our main theorem.

Theorem 3. *For any k, there is a choice of the parameters R_1 to R_{k-1} such that the engineered Price of Anarchy with these parameters is strictly less than 4/3.*

Proof. We define R_{k-1}, then R_{k-2}, and so on. We set $R_{k-1} = 8$. Then $ePoA(R_{k-1}) = 5/4$ and $(1 + 1/R_{k-1})^2 ePoA() = (9/8)^2 < 4/3$. Assume now that we have defined R_{k-1} down to R_{i+1} so that $ePoA(R_{i+1},\ldots,R_{k-1}) < 4/3$ and $(1+1/R_j)^2 ePoA(R_{j+1},\ldots,R_{k-1}) < 4/3$ for $j \geq i+1$. We next define R_i. We have

$$ePoA(R_i,\ldots,R_{k-1}) \leq \max\left(\begin{array}{l} B(R_i,\ldots,R_{k-1}), \\ \max_{j;i\leq j<k}\left(B(R_i,\ldots,R_{j-1}), \left(1+\frac{1}{R_j}\right)^2 ePoA(R_{j+1},\ldots,R_{k-1})\right) \end{array}\right),$$

where the first line covers the benign case and the second line covers the non-benign case. We choose R_i such that $(1+1/R_i)^2 ePoA(R_{i+1},\ldots,R_{k-1}) < 4/3$. Then, $B(R_i,\ldots,R_k) < 4/3$ and $B(R_i,\ldots,j-1) < 4/3$ by Lemma 4 and the induction step is complete. $\square$

4   An Improved Mechanism for the Case of Two Links

In this section we present a mechanism which achieves $ePoA = 1.192$ for a network that consists of two parallel links. The ratio $C_N(r)/C_{opt}(r)$ is maximal for $r = r_2$. At this rate Nash still uses only the first link and Opt uses both links. In order to avoid this maximum ratio (if larger than 1.192), we force MN to use the second link earlier by increasing the latency of the first link after some rate x_1, $r_2/2 \leq x_1 \leq r_2$ to a value above b_2. In the preceding section, we increased the latency to ∞. In this way, we avoided a bad ratio at r_2, but paid a price for very large rates. The idea for the improved construction, is to increase the latency to a finite value. This will avoid the bad ratio, but also allows MN to use both links for large rates. In particular, we obtain the following result.

Theorem 4. *There is a mechanism for a network of two parallel links that achieves* $ePoA = 1.192$.

5   A Lower Bound for the Case of Two Links

We prove that the construction of the previous section is optimal among the following class of deterministic mechanisms; mechanisms that consider Dafermos-Sparrow equilibria, and mechanisms that use non-decreasing[12] latency functions *even for the weaker notion of User Equilibria*. For the above mechanisms we show that $ePoA \geq 1.191$. The proof of the following theorem does not make any assumptions on the monotonicity of the modified latencies.

Theorem 5. *The construction of Section 4 is optimal and $ePoA \geq 1.191$.*

6   Open Problems

Clearly the ultimate goal is to design coordination mechanisms for general networks. Our mechanism approaches $4/3$, as k grows. Can we improve the upper bound for the case of k parallel links? A possible approach could be to use the ideas of Section 4. Another approach would be to define the benign case more restrictively. Assuming $R_i = 8$ for all i, we would call the following latencies benign: $\ell_1(x) = x$, and $\ell_i(x) = 1 + \varepsilon \cdot i + x/8^i$ for $i > 1$ and small positive ε. However, Opt starts using the k-th link shortly after $1/2$ and hence uses an extremely efficient link for small rates. What can be said about atomic (weighted or unweighted) scheduling games and for games with polynomial latencies? What is the exact value of *ePoA* for the case of two parallel links? We conjecture that a reinspection of Sections 4 and 5 settles this question.

Acknowledgments. We would like to thank Elias Koutsoupias and Spyros Angelopoulos for many fruitful discussions.

References

1. Angel, E., Bampis, E., Pascual, F.: Truthful algorithms for scheduling selfish tasks on parallel machines. Theor. Comput. Sci. 369(1-3), 157–168 (2006)
2. Angel, E., Bampis, E., Pascual, F., Tchetgnia, A.-A.: On truthfulness and approximation for scheduling selfish tasks. J. Scheduling 12(5), 437–445 (2009)
3. Azar, Y., Jain, K., Mirrokni, V.S.: (Almost) optimal coordination mechanisms for unrelated machine scheduling. In: SODA, pp. 323–332 (2008)
4. Bernstein, D., Smith, T.E.: Equilibria for networks with lower semicontinuous costs: With an application to congestion pricing. Transportation Science 28(3), 221–235 (1994)

---

[12] It remains open whether similar arguments can be applied for showing the lower bound for non-monotone mechanisms with respect to User Equilibria.

5. Bonifaci, V., Salek, M., Schäfer, G.: On the efficiency of restricted tolls in network routing games. In: SAGT (2011) (to appear)
6. Caragiannis, I.: Efficient coordination mechanisms for unrelated machines scheduling. In: SODA, pp. 815–824 (2009)
7. Christodoulou, G., Gourvès, L., Pascual, F.: Scheduling selfish tasks: About the performance of truthful algorithms. In: Lin, G. (ed.) COCOON 2007. LNCS, vol. 4598, pp. 187–197. Springer, Heidelberg (2007)
8. Christodoulou, G., Koutsoupias, E., Nanavati, A.: Coordination mechanisms. Theor. Comput. Sci. 410(36), 3327–3336 (2009)
9. Cole, R., Correa, J.R., Gkatzelis, V., Mirrokni, V., Olver, N.: Inner product spaces for minsum coordination mechanisms. In: STOC (2011)
10. Cole, R., Dodis, Y., Roughgarden, T.: Pricing network edges for heterogeneous selfish users. In: STOC, pp. 521–530 (2003)
11. Cole, R., Dodis, Y., Roughgarden, T.: How much can taxes help selfish routing? J. Comput. Syst. Sci. 72(3), 444–467 (2006)
12. Correa, J.R., Schulz, A.S., Stier-Moses, N.E.: A geometric approach to the Price of Anarchy in nonatomic congestion games. Games and Economic Behavior 64, 457–469 (2008)
13. Dafermos, S.C., Sparrow, F.T.: The traffic assignment problem for a general network. Journal of Research of the National Bureau of Standards, Series B 73B(2), 91–118 (1969)
14. Dafermos, S.: An extended traffic assignment model with applications to two-way traffic. Transportation Science 5, 366–389 (1971)
15. de Palma, A., Nesterov, Y.: Optimization formulations and static equilibrium in congested transportation networks. In: CORE Discussion Paper 9861, pp. 12–17. Université Catholique de Louvain, Louvain-la-Neuve (1998)
16. Fleischer, L.: Linear tolls suffice: New bounds and algorithms for tolls in single source networks. Theor. Comput. Sci. 348(2-3), 217–225 (2005)
17. Fleischer, L., Jain, K., Mahdian, M.: Tolls for heterogeneous selfish users in multicommodity networks and generalized congestion games. In: FOCS, pp. 277–285 (2004)
18. Immorlica, N., Li, L., Mirrokni, V.S., Schulz, A.: Coordination mechanisms for selfish scheduling. In: Deng, X., Ye, Y. (eds.) WINE 2005. LNCS, vol. 3828, pp. 55–69. Springer, Heidelberg (2005)
19. Karakostas, G., Kolliopoulos, S.G.: Edge pricing of multicommodity networks for heterogeneous selfish users. In: FOCS 2004, pp. 268–276 (2004)
20. Karakostas, G., Kolliopoulos, S.G.: The efficiency of optimal taxes. In: López-Ortiz, A., Hamel, A.M. (eds.) CAAN 2004. LNCS, vol. 3405, pp. 3–12. Springer, Heidelberg (2005)
21. Kollias, K.: Non-preemptive coordination mechanisms for identical machine scheduling games. In: Shvartsman, A.A., Felber, P. (eds.) SIROCCO 2008. LNCS, vol. 5058, pp. 197–208. Springer, Heidelberg (2008)
22. Koutsoupias, E., Papadimitriou, C.: Worst-case equilibria. Computer Science Review 3(2), 65–69 (2009)
23. Marcotte, P., Patriksson, M.: Traffic equilibrium. In: Transportation, Handbooks in Operations Research and Management Science. ch.10, vol. 14, pp. 623–713. North-Holland, Amsterdam (2007)
24. Nisan, N., Roughgarden, T., Tardos, E., Vazirani, V.V.: Algorithmic Game Theory. Cambridge University Press, Cambridge (2007)
25. Patriksson, M.: The Traffic Assignment Problem: Models and Methods. V.S.P. Intl. Science (1994)
26. Roughgarden, T.: Designing networks for selfish users is hard. In: FOCS (2001)
27. Roughgarden, T., Tardos, É.: How bad is selfish routing? J.ACM 49, 236–259 (2002)
28. Wardrop, J.G.: Some theoretical aspects of road traffic research. In: Proceedings of the Institute of Civil Engineers, Part II, vol. 1, pp. 325–378 (1952)

Algorithms for Finding a Maximum
Non-k-linked Graph

Yusuke Kobayashi[1,*] and Yuichi Yoshida[2,**]

[1] Department of Mathematical Informatics, Graduate School of Information Science
and Technology, University of Tokyo
kobayashi@mist.i.u-tokyo.ac.jp
[2] School of Informatics, Kyoto University, and Preferred Infrastructure, Inc.
yyoshida@lab2.kuis.kyoto-u.ac.jp

Abstract. A graph with at least $2k$ vertices is said to be k-linked if for any ordered k-tuples $(s_1, \ldots, s_k)$ and $(t_1, \ldots, t_k)$ of $2k$ distinct vertices, there exist pairwise vertex-disjoint paths $P_1, \ldots, P_k$ such that P_i connects s_i and t_i for $i = 1, \ldots, k$. For a given graph G, we consider the problem of finding a maximum induced subgraph of G that is not k-linked. This problem is a common generalization of computing the vertex-connectivity and testing the k-linkedness of G, and it is closely related to the concept of H-linkedness. In this paper, we give the first polynomial-time algorithm for the case of $k = 2$, whereas a similar problem that finds a maximum induced subgraph without 2-vertex-disjoint paths connecting fixed terminal pairs is NP-hard. For the case of general k, we give an $(8k - 2)$-additive approximation algorithm. We also investigate the computational complexities of the edge-disjoint case and the directed case.

Keywords: k-linkedness, H-linkedness, disjoint paths, connectivity.

1 Introduction

A graph is said to be k-*linked* if it has at least $2k$ vertices and for any ordered k-tuples $(s_1, \ldots, s_k)$ and $(t_1, \ldots, t_k)$ of $2k$ distinct vertices, there exist pairwise vertex-disjoint paths $P_1, \ldots, P_k$ such that P_i connects s_i and t_i for $i = 1, \ldots, k$. The k-linkedness has been well-studied by many graph theorists, and there are many results on relationships between the k-linkedness and the vertex-connectivity of graphs [1,8,10,14,20]. From the algorithmic point of view, the k-linkedness has attracted attention because of similarities with the vertex-disjoint paths problem, which is one of the most important problems in computer science and algorithmic graph theory. In the *vertex-disjoint paths problem*, we

---

* Supported by Grant-in-Aid for Scientific Research and by Global COE Program
"The research and training center for new development in mathematics", MEXT,
Japan.
** Supported by MSRA Fellowship 2010.

C. Demetrescu and M.M. Halldórsson (Eds.): ESA 2011, LNCS 6942, pp. 131–142, 2011.

are given a graph G and $2k$ distinct vertices $s_1, \ldots, s_k, t_1, \ldots, t_k$ called *termi-nals*, and the objective is to find pairwise vertex-disjoint paths $P_1, \ldots, P_k$ such that P_i connects s_i and t_i for $i = 1, \ldots, k$. With the terminology of the vertex-disjoint paths problem, a graph is k-linked if and only if the vertex-disjoint paths problem has a solution for any choice of $2k$ terminals. In this paper, we consider the problem of finding a minimum number of vertices whose removal makes the graph non-k-linked, which can be stated as follows.

Max Non-k-Linked Induced Subgraph

Input. A graph $G = (V, E)$.
Problem. Find a vertex set $V_0 \subseteq V$ with maximum cardinality such that $G[V_0]$ (the subgraph induced by V_0) is not k-linked.

We mainly discuss the case of $k = 2$, which is interesting because of its relation to the problem of finding a maximum planar induced subgraph. By a classical result on the 2-vertex-disjoint paths problem [18], it is well-known that the graph is not 2-linked if and only if it cannot be embedded in a plane up to "3-separations" (see Theorem 6 for the precise statement). That is, the non-2-linkedness is a similar concept to the planarity. The problem of finding a maximum planar induced subgraph is an important problem in theoretical computer science, because it amounts to computing a measure for non-planarity of graphs (see e.g. [2,15]). Max Non-2-Linked Induced Subgraph can also be regarded as a problem of computing a measure for non-planarity of graphs, which is one of our motivations for studying Max Non-k-Linked Induced Subgraph. As we will describe later, we show that Max Non-2-Linked Induced Subgraph can be solved in poly-nomial time (Theorem 5). This result is surprising because most of all natural problems of computing measures for non-planarity, such as finding a maximum planar (induced) subgraph or computing the minimum number of crossings in an embedding in a plane, are known to be NP-hard (see [15]).

Max Non-k-Linked Induced Subgraph is motivated also by the concept of H-linkedness that has been studied [4,6,11,12,13] as a common generalization of the graph connectivity and the k-linkedness. For a multigraph H, an H-*subdivision* in a graph G is a pair of mappings $f : V(H) \to V(G)$ and $g : E(H) \to \mathcal{P}$, where $\mathcal{P}$ is the set of paths in G, such that:

1. $f(u) \neq f(v)$ for all distinct $u, v \in V(H)$,
2. $g(uv)$ is a path connecting $f(u)$ and $f(v)$ in G for $uv \in E(H)$, and the paths are internally disjoint.

For a multigraph H, a graph G is H-*linked* if every injective mapping $f : V(H) \to V(G)$ can be extended to an H-subdivision in G. This is a generalization of the notions of k-linkedness and k-connectivity, because the H-linkedness is equiva-lent to the k-linkedness when H is a matching with k edges, and it is equivalent to the k-connectivity when H consists of $k + 1$ vertices and one edge.

For a multigraph H (or an integer k, respectively), determining whether a given graph G is H-linked (resp. k-linked) or not is a natural algorithmic problem. When a multigraph H (resp. an integer k) is fixed, Robertson and Seymour [16] gave a polynomial-time algorithm for this problem based on their seminal work on graph minor project, which spans 23 papers and gives several deep and profound results in discrete mathematics. On the other hand, when H or k is a part of the input, no polynomial-time algorithm is known for the problem. Thus, determining the H-linkedness for non-fixed multigraphs H is an interesting open problem.

Our second motivation for Max Non-k-Linked Induced Subgraph comes from the fact that it is a special case of the problem of determining the H-linkedness. More precisely, Max Non-k-Linked Induced Subgraph is equivalent to the case when H is a union of a matching of size k and l distinct vertices, i.e., H has $2k + l$ vertices and k edges. Let us emphasize that k and/or l are a part of the input throughout this paper, and so this problem setting is completely different from the case when H is fixed. Note that when $k = 1$, determining the H-linkedness is equivalent to testing the vertex-connectivity of an input graph. On the other hand, the polynomial-time solvability of the case of $k = 2$, which corresponds to Max Non-2-Linked Induced Subgraph, is non-trivial.

We can also consider the edge-disjoint version. We say that a graph G is *weakly k-linked* if for any ordered k-tuples $(s_1, \ldots, s_k)$ and $(t_1, \ldots, t_k)$ of $2k$ vertices (not necessarily distinct), there exist pairwise edge-disjoint paths $P_1, \ldots, P_k$ such that P_i connects s_i and t_i for $i = 1, \ldots, k$.

Max Weakly Non-k-Linked Subgraph

Input. A graph $G = (V, E)$.
Problem. Find an edge set $E_0 \subseteq E$ with maximum cardinality such that the subgraph $G_0 = (V, E_0)$ is not weakly k-linked.

Note that we find an edge set E_0 in this problem, whereas we find a vertex set V_0 in Max Non-k-Linked Induced Subgraph. This problem setting is natural because the weakly k-linkedness is closely related to the edge-connectivity rather than the vertex-connectivity. When $k = 1$, Max Weakly Non-k-Linked Subgraph is equivalent to computing the edge-connectivity of an input graph, that is, the edge-connectivity is c if and only if the optimal value is $|E| - c$. Thus, this problem is a generalization of computing the edge-connectivity and testing the weakly k-linkedness. We also note that in the same way as the relationship between Max Non-k-Linked Induced Subgraph and the H-linkedness, Max Weakly Non-k-Linked Subgraph is related to the concept of H-*immersion* studied in [3,17].

Related work: Many graph theorists are interested in how much connectivity is necessary to ensure that a graph is k-linked [1,8,10,14,20]. It is shown (implicitly) in [20] that, every $10k$-connected graph is k-linked, which is the currently best bound. Similar results are known for the edge-disjoint case, that is, it is shown in [7] that every $(k + 2)$-edge-connected graph is weakly k-linked. In the same way as the k-linkedness, the main interest on the H-linkedness goes to sufficient conditions for graphs to be H-linked [4,6,11,12,13].

The k-linkedness is closely related to the vertex-disjoint paths problem with k terminal pairs (k-vertex-disjoint paths problem). When k is a part of the

Table 1. Our results on the problems

Problems	Vertex-disjoint case	Edge-disjoint case
Max Non-2-Linked Induced Subgraph	P (Thm. 5)	P (Cor. 1)
Max 2-VDP-free Induced Subgraph	NP-hard (Thm. 7)	P (Thm. 4)
Max Non-k-Linked Induced Subgraph	$\text{OPT} - 8k + 2$	$\text{OPT} - 2$

input of the problem, this is one of Karp's NP-complete problems [9]. In 1980, it was shown that the 2-vertex-disjoint paths problem is solvable in polynomial time [18,19,21]. In particular, the following characterization is shown for the existence of the 2-vertex-disjoint paths.

Theorem 1 ([18]). *Let $G = (V, E)$ be a graph and let s_1, t_1, s_2, t_2 be distinct terminals. The 2-vertex-disjoint paths problem has no solution if and only if there exists a partition $U, A_1, \ldots, A_l$ $(l \geq 0)$ of V with $s_1, t_1, s_2, t_2 \in U$ such that*

(1) *for $1 \leq i, j \leq l$ with $i \neq j$, $N(A_i) \cap A_j = \emptyset$,*
(2) *for $1 \leq i \leq l$, $|N(A_i)| \leq 3$ and $G[A_i]$ is connected, and*
(3) *if G' is the graph obtained from G by deleting A_i and adding new edges joining every pair of distinct vertices in $N(A_i)$ for every i, then G' can be embedded in a plane so that s_1, s_2, t_1, t_2 are on the outer boundary of G' in this order.*

The characterization will be used in our argument. On the other hand, the 2-vertex-disjoint paths problem (or the 2-edge-disjoint paths problem) in digraphs, in which we find directed paths P_1, P_2 such that P_i is from s_i to t_i for $i = 1, 2$, was shown to be NP-hard [5].

For fixed k, Robertson and Seymour [16] gave a polynomial-time algorithm for the k-vertex-disjoint (edge-disjoint) paths problem based on their graph minor theory. When k is fixed, by solving the k-vertex-disjoint (or edge-disjoint) paths problem for every choice of the terminals, the k-linkedness (resp. the weakly k-linkedness) of an input graph can be tested in polynomial time. Similarly, for any fixed multigraph H, we can determine whether an input graph is H-linked or not in polynomial time. We emphasize here that this algorithm runs in polynomial time only when H is fixed.

Our contributions: In this paper, we consider algorithms for Max Non-k-Linked Induced Subgraph and the corresponding edge-disjoint version (Max Weakly Non-k-Linked Subgraph). We summarize our results in Table 1.

First, we show that Max Non-2-Linked Induced Subgraph can be solved in polynomial time (Theorem 5), which is one of the main results in this paper. This problem corresponds to the H-linkedness, where H consists of two edges and some isolated vertices. This is the first non-trivial case in which the H-linkedness can be determined in polynomial time when H is not fixed. We also give a polynomial-time algorithm for Max Weakly Non-2-Linked Subgraph (Corollary 1).

We now give some remarks on proof techniques for Theorem 5 and Corollary 1. A natural approach to solve Max Non-2-Linked Induced Subgraph is to consider the problem of finding a maximum vertex set whose inducing subgraph contains no two vertex-disjoint paths connecting fixed terminal pairs. We call it Max 2-VDP-free Induced Subgraph, whose formal description is as follows.

Max 2-VDP-free Induced Subgraph

Input. A graph $G = (V, E)$ and distinct terminals $s_1, t_1, s_2, t_2 \in V$.

Problem. Find a vertex set $V_0 \subseteq V$ with maximum cardinality such that $\{s_1, t_1, s_2, t_2\} \subseteq V_0$ and the vertex-disjoint paths problem with terminal pairs (s_1, t_1) and (s_2, t_2) has no solution in $G[V_0]$.

We can easily see that by solving Max 2-VDP-free Induced Subgraph for every choice of the terminals s_1, t_1, s_2, t_2, we obtain a solution of Max Non-2-Linked Induced Subgraph. However, we show that Max 2-VDP-free Induced Subgraph is NP-hard (Theorem 7), which suggests that this reduction does not work for solving Max Non-2-Linked Induced Subgraph. Therefore, we need another approach to Max Non-2-Linked Induced Subgraph.

In the same way as Max 2-VDP-free Induced Subgraph, we consider the edge-disjoint version of the problem, which we call Max 2-EDP-free Subgraph.

Max 2-EDP-free Subgraph

Input. A graph $G = (V, E)$ and terminals $s_1, t_1, s_2, t_2 \in V$.

Problem. Find an edge set $E_0 \subseteq E$ with maximum cardinality such that the edge-disjoint paths problem with terminal pairs (s_1, t_1) and (s_2, t_2) has no solution in the subgraph $G_0 = (V, E_0)$.

We give a polynomial-time algorithm for this problem (Theorem 4), and consequently, we show the polynomial solvability of Max Weakly Non-2-Linked Subgraph (Corollary 1). Since most problems on vertex-disjoint paths and their corresponding edge-disjoint versions are equivalent with respect to their polynomial solvability, it is interesting to note that Max 2-EDP-free Subgraph is solvable in polynomial time, whereas Max 2-VDP-free Induced Subgraph is NP-hard.

Second, for general k, we show that there exists an $(8k - 2)$-additive approximation algorithm for Max Non-k-Linked Induced Subgraph by using a known sufficient condition for a graph to be k-linked, that is, our algorithm finds a feasible solution V_0 whose cardinality is at least the optimum value minus $8k - 2$. Similarly, for general k, we give a 2-additive approximation algorithm for Max Weakly Non-k-Linked Subgraph.

Finally, we also consider the directed versions of these problems. It is well-known that the directed versions of Max 2-VDP-free Induced Subgraph and Max 2-EDP-free Subgraph are NP-hard [5]. By observing that the weakly k-linkedness is equivalent to the k-connectivity for digraphs, we see that Directed Max Weakly Non-k-Linked Subgraph can be solved in polynomial time for general k. On the other hand, based on the arguments in [22], we show that Directed Max Non-k-Linked Induced Subgraph is NP-hard even when $k = 2$.

Notation: In this paper, we use the following notations. Let $G = (V, E)$ be a graph with a vertex set V and an edge set E. For a vertex set $X \subseteq V$, let

$\delta_G(X)$ be the set of edges between X and $V \setminus X$, and such an edge set is called a *cut*. Let $N_G(X)$ denote the set of vertices in $V \setminus X$ that are adjacent to X. We simply denote $\delta(X)$ and $N(X)$ if no confusion may arise. For $X \subseteq V$, the subgraph induced by X is denoted by $G[X]$, and the graph $G[V \setminus X]$ is denoted by $G - X$. For an edge set $F \subseteq E$, let $G - F = (V, E \setminus F)$ and let G/F denote the graph obtained from G by contracting all edges in F. For $s, t \in V$, a cut $\delta(X)$ is called an *s-t cut* if exactly one of s and t is contained in X, and a vertex set X is called an *s-t vertex cut* if s and t are contained in different connected components of $G - X$.

Organization: The rest of this paper is organized as follows. In Section 2, we deal with the edge-disjoint case and give polynomial-time algorithms for Max 2-EDP-free Subgraph and Max Weakly Non-2-Linked Subgraph. In Section 3, we give a polynomial-time algorithm for Max Non-2-Linked Induced Subgraph, which is the main part of this paper. In Section 4, we show the NP-hardness of Max 2-VDP-free Induced Subgraph. Due to space constraints, some cases are omitted and dealt with in the full version.

2   Weakly 2-Linkedness of Graphs

In this section, we show that Max 2-EDP-free Subgraph and Max Weakly Non-2-Linked Subgraph can be solved in polynomial time.

First, we consider Max 2-EDP-free Subgraph. Let $G = (V, E)$ be a graph, and fix terminal pairs (s_1, t_1) and (s_2, t_2). In this section, 2 edge-disjoint paths mean 2 edge-disjoint paths with respect to those terminals, and a graph is said to be *2-EDP-free* if it does not have 2 edge-disjoint paths.

We now give an upper bound on the number of edges we must remove to make G 2-EDP-free. Let C_1, C_2 be minimum cuts separating s_1 from t_1 and s_2 from t_2, respectively. It is easy to see that G becomes 2-EDP-free after removing C_1 or C_2. Let C_{12}, C'_{12} be minimum cuts separating $\{s_1, s_2\}$ from $\{t_1, t_2\}$ and $\{s_1, t_2\}$ from $\{t_1, s_2\}$, respectively. Let F_{12} (resp., F'_{12}) be an edge set obtained from C_{12} (resp., C'_{12}) by discarding an edge. It is also easy to see that G becomes 2-EDP-free after removing F_{12} or F'_{12}. This motivates us to define $c = \min\{|C_1|, |C_2|, |F_{12}|, |F'_{12}|\}$, and let F be the edge set that attains the minimum. When $c = 0$, we say that G violates the *cut condition*. Clearly, c is the upper bound on the number of edges we must remove to make G 2-EDP-free. Now, we show that the converse also holds when $c \geq 4$.

Theorem 2. *If $c \geq 4$, $G \setminus F$ is an optimal solution for Max 2-EDP-free Subgraph.*

To prove Theorem 2, We use the following theorem by Seymour [18] on the feasibility of the 2-edge-disjoint paths problem.

Theorem 3 ([18]). *Let $G = (V, E)$ be a graph and let s_1, t_1, s_2, t_2 be terminals. Then, the 2-edge-disjoint paths problem has no solution if and only if it violates the cut condition or there exists an edge set $E' \subseteq E$ such that*

(1) *each vertex of G/E' has degree at most 3,*
(2) *terminals of G/E' are distinct, and have degree at most 2,*
(3) *G/E' can be embedded in a plane so that s_1, s_2, t_1, t_2 are on the outer boundary of G/E' in this order.*

Proof (of Theorem 2). Since we have seen that $G - F$ is a solution for Max 2-EDP-free Subgraph, it suffices to show that it is optimal. Assume that there exists an edge set $F' \subseteq E$ with $|F'| \leq c - 1$ such that $G - F'$ is 2-EDP-free for terminal pairs (s_1, t_1) and (s_2, t_2).

Since $G - F'$ satisfies the cut condition, there exists an edge set $E' \subseteq E \setminus F'$ satisfying the three conditions of Theorem 3. Let $G' = (G - F')/E'$, and let S_1, T_1, S_2, T_2 be connected subgraphs of G that correspond to s_1, t_1, s_2, t_2 in G', respectively. Note that, since s_1, t_1, s_2, t_2 are distinct in G', the subgraphs S_1, T_1, S_2, T_2 are disjoint. By the condition (2), we have $\delta_{G-F'}(S_1) \leq 2$, $\delta_{G-F'}(T_1) \leq 2$, $\delta_{G-F'}(S_2) \leq 2$, and $\delta_{G-F'}(T_2) \leq 2$. On the other hand, since the size of a minimum s_i-t_i edge cut is at least c for $i = 1, 2$, we have $\delta_G(S_1) \geq c$, $\delta_G(T_1) \geq c$, $\delta_G(S_2) \geq c$, and $\delta_G(T_2) \geq c$. Observing that removing $|F'|$ edges decreases the value $\delta_G(S_1) + \delta_G(T_1) + \delta_G(S_2) + \delta_G(T_2)$ by at most $2|F'|$, we have $4c - 8 \leq 2|F'|$. This contradicts that $|F'| \leq c - 1$ and $c \geq 4$. $\square$

Theorem 4. *Max 2-EDP-free Subgraph is solvable in polynomial time.*

Proof. First, we compute c and F using any polynomial-time algorithm for the minimum cut problem. If $c \geq 4$, by Theorem 2, $G - F$ is an optimal solution. If $c \leq 3$, for every set F' of at most c edges, we test whether $G - F'$ is 2-EDP-free, which can be done in polynomial time. $\square$

Corollary 1. *Max Weakly Non-2-Linked Subgraph is solvable in polynomial time.*

Proof. We solve Max 2-EDP-free Subgraph for each terminal pairs (s_1, t_1) and (s_2, t_2), and we take the maximum of them. $\square$

3   2-Linkedness of Graphs

This section is devoted to proving the following theorem.

Theorem 5. *Max Non-2-Linked Induced Subgraph is solvable in polynomial time.*

Since we consider vertex-disjoint paths in this section, we assume that all graphs are simple. By Theorem 1, we immediately see the following theorem.

Theorem 6. *A graph $G = (V, E)$ is not 2-linked if and only if there exists a partition $U, A_1, \ldots, A_l$ ($l \geq 0$) of V such that*

(1) *for $1 \leq i, j \leq l$ with $i \neq j$, $N(A_i) \cap A_j = \emptyset$,*
(2) *for $1 \leq i \leq l$, $|N(A_i)| \leq 3$ and $G[A_i]$ is connected, and*
(3) *if G' is the graph obtained from G by deleting A_i and adding new edges joining every pair of distinct vertices in $N(A_i)$ for every $i = 1, \ldots, l$, then G' can be embedded in a plane so that the boundary of some face contains at least four vertices.*

We note that, from the conditions (1) and (2), $A_1, \ldots, A_l$ must induce connected components in $G - U$.

First, we observe that if a graph G' is not 3-connected, there exist two vertices x and y separated by a vertex cut $\{v_1, v_2\}$ of size two. Then, we can easily see that G' is not 2-linked by setting $U = \{x, y, v_1, v_2\}$ in Theorem 6. Suppose that $|V| \geq 4$ and the vertex-connectivity of the input graph G is $c \geq 2$. By the above observation, we can make G non-2-linked by removing at most $c - 2$ vertices. Thus, it suffices to give algorithms for finding a vertex set $X \subseteq V$ with $|X| \leq c - 3$ such that $G - X$ is not 2-linked. We consider the cases $|X| \leq c - 4$ and $|X| = c - 3$ in Sections 3.1 and 3.2, separately (see Propositions 1 and 2).

We note that when c is bounded by a fixed constant, the problem can be solved in polynomial time by enumerating all possible vertex sets X. Thus, in what follows, we suppose that c is sufficiently large (e.g. $c \geq 50$).

3.1   Finding a Vertex Set X with $|X| \leq c - 4$

Suppose that $G - X$ is not 2-linked for some vertex set $X \subseteq V$ with $|X| \leq c - 4$. In this case, $G - X$ is 4-connected, and hence by Theorem 6, $G - X$ is not 2-linked if and only if $G - X$ can be embedded in a plane so that the boundary of some face contains at least four vertices.

We observe that $G - X$ contains a vertex v of degree at most five when $G - X$ is planar, which implies that $d_G(v) \leq c + 1$. Recall that we have assumed that G is simpe. On the other hand, since G is c-connected, the degree of v in G is at least c, and $|N_G(v) \cap X| \geq c - 5$. With this observation, we can find all possible vertex sets with at most $c - 4$ vertices by executing the following procedure:

> For every $v \in V$ with degree at most $c + 1$, and for every vertex set X with $|X| \leq c - 4$ and $|N_G(v) \cap X| \geq c - 5$, test the 2-linkedness of $G - X$.

Since the number of the choices of X is at most $n \cdot \binom{c+1}{c-5} \cdot n = O(n^8)$, this procedure can be done in polynomial time, and we have the following proposition.

Proposition 1. *We can enumerate all vertex sets X such that $G - X$ is not 2-linked and $|X| \leq c - 4$ in polynomial time.*

3.2   Finding a Vertex Set X with $|X| = c - 3$

In this subsection, we give an algorithm for finding a vertex set X such that $|X| = c - 3$ and $G - X$ is not 2-linked. If there exists a vertex set X with $|X| \leq c - 3$ such that $G - X$ can be embedded in a plane so that the boundary of some face contains at least four vertices, then such a set X can be found in polynomial time in the same way as Section 3.1. Thus, it suffices to consider the case when there exist a *positive* integer l and a partition $U, A_1, \ldots, A_l$ of $V \setminus X$ satisfying the conditions of Theorem 6. Note that, since $|X| = c - 3$, we must have $N_{G-X}(A_i) = 3$ for every i. The main idea behind our algorithm is to guess a vertex $r \in A_1$ and a set of three vertices $W = N_{G-X}(A_1)$ and then check whether a required partition exists under this condition.

Case 1: When $|U| \leq 6$

First, we find a vertex set X and a partition $U, A_1, \ldots, A_l$ of $V \setminus X$ with $|U| \leq 6$. We note that an algorithm in this part can be applied even if $|U|$ is more than six but bounded by a fixed constant. In order to find such vertex sets, we consider the following subproblem.

Problem A

Input. A c-connected graph $G = (V, E)$, a vertex set U, a vertex $r \in V \setminus U$, and a set of three vertices $W \subseteq U$.

Problem. Find a vertex set $X \subseteq V \setminus (U \cup \{r\})$ with $|X| = c - 3$ satisfying the following conditions: $V \setminus X$ can be partitioned into $U, A_1, \ldots, A_l$ ($l \geq 1$) such that they satisfy the conditions of Theorem 6, $r \in A_1$, and $N_{G-X}(A_1) = W$.

Lemma 1. *Problem A is solvable in polynomial time.*

Proof. If $|U| \leq 3$, then the graph G' defined as in the condition (3) in Theorem 6 has at most three vertices, which violates the condition (3). Hence, a desired set X obviously does not exist. Suppose that $|U| \geq 4$. We compute a minimum vertex cut separating $U \setminus W$ and r in $G - W$. Among them, let $S \subseteq V$ be the minimum vertex cut such that the connected component of $G - W - S$ containing r is maximum. Let A_1^S be the vertex set of the connected component containing r. If $|S| \geq c - 2$, then we can conclude that the required X does not exist. Thus, since G is c-connected, we may assume that $|S| = c - 3$ and $S \cup W$ is a minimum vertex cut of G.

In this case, let $X = S$ and $A_1 = A_1^S$, and define $A_2, \ldots, A_l$ as the vertex sets of the connected components of $G - S - U - A_1^S$. Let $\mathcal{P}$ be the partition $U, A_1^S, A_2, \ldots, A_l$ of $V \setminus S$. If $\mathcal{P}$ satisfies the conditions of Theorem 6, then $X = S$ is a desired set. Now we show the following claim, which says that we do not have to consider other sets.

Claim. If $X = S$ is not a solution of Problem A, then there exists no solution.

Proof. Assume that S is not a solution of Problem A, but $X' \subseteq V \setminus (U \cup \{r\})$ is a solution. Since G is c-connected and $|X'| = c - 3$, X' is a minimum vertex cut separating $U \setminus W$ and r. Let A_1' be the vertex set of the connected component of $G - W - X'$ containing r, and define $A_2', \ldots, A_{l'}'$ as the vertex sets of the connected components of $G - X' - U - A_1'$. Let $\mathcal{P}'$ be the partition $U, A_1', A_2', \ldots, A_{l'}'$ of $V \setminus X'$. Since X' is a solution of Problem A, $\mathcal{P}'$ satisfies the conditions of Theorem 6. Note that, by the maximality of A_1, $G - X' - U - A_1' = G[A_2' \cup \cdots \cup A_{l'}']$ contains $G - S - U - A_1^S = G[A_2 \cup \cdots \cup A_l]$ as a subgraph (see Fig. 1).

Since S is not a solution, the partition $\mathcal{P}$ violates the conditions of Theorem 6. Assume that $|N_{G-S}(A_i)| \geq 4$ for some $i = 2, 3, \ldots, l$. Since A_i is contained in A_j' for some $j \in \{2, 3, \ldots, l'\}$, we have $|N_{G-X'}(A_j')| \geq |N_{G-S}(A_i)| \geq 4$, which contradicts that the partition $\mathcal{P}'$ satisfies the conditions of Theorem 6.

Therefore, $|N_{G-S}(A_i)| = 3$ for every $i = 2, 3, \ldots, l$, and $\mathcal{P}$ violates the condition (3) if we apply Theorem 6 to $G - S$. That is, the graph G^S obtained by the operations in the condition (3) is either a non-planar graph or a planar graph

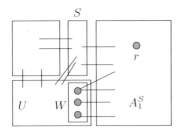 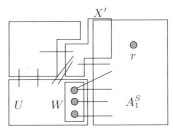

Fig. 1. An example of Case 1

whose every face has three vertices. On the other hand, for any $i = 2, 3, \ldots, l$, $N_{G-S}(A_i) \subseteq N_{G-X'}(A'_j)$ holds for some $j \in \{2, 3, \ldots, l'\}$. Thus, if we apply Theorem 6 to $G - X'$, then either the partition $\mathcal{P}'$ violates the condition (2), or the graph G' defined as in the condition (3) contains G^S as a subgraph and $V(G') = U = V(G^S)$. This means that the partition $\mathcal{P}'$ violates the conditions of Theorem 6, which contradicts the assumption.   □

By this claim, in order to solve Problem A, it suffices to test whether $X = S$ is a desired set or not, which can be done in polynomial time.   □

We can find a solution X and a partition $U, A_1, \ldots, A_l$ of $V \setminus X$ with $|U| \leq 6$ by solving Problem A for every choice of U, r and W, which can be done in polynomial time by Lemma 1.

Case 2: When $|U| \geq 7$

Second, we find a solution X and a partition $U, A_1, \ldots, A_l$ of $V \setminus X$ with $|U| \geq 7$. Due to the space constraints, this case is dealt with in the full version.

Thus, by Cases 1 and 2, we have the following proposition.

Proposition 2. *We can find a vertex set X such that $G - X$ is not 2-linked and $|X| = c - 3$ in polynomial time (if one exists).*

4   NP-Hardness of **Max 2-VDP-free Induced Subgraph**

Theorem 7. *Max 2-VDP-free Induced Subgraph is NP-hard.*

Proof. We show that Vertex Cover can be reduced to Max 2-VDP-free Induced Subgraph. Note that Vertex Cover is an NP-hard problem, in which we are given a simple graph $G = (V, E)$ and the objective is to find a vertex set $X \subseteq V$ with minimum cardinality such that every edge in E is incident to at least one vertex in X. Suppose that we are given an instance $G = (V, E)$ of Vertex Cover, where $V = \{1, \ldots, n\}$. We construct a new graph $G' = (V', E')$ as follows (see Fig. 2):

$$V' = \{s_1, t_1, s_2, t_2\} \cup \{p_{i,j}, q_{i,j} \mid i, j \in \{1, \ldots, n\}\},$$

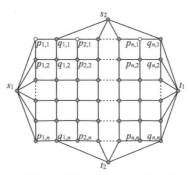

Fig. 2. The graph G' obtained from G by the reduction. For each edge uv in G, there is an edge $p_{u,1}p_{v,1}$ in G' (edges may be added among empty vertices).

$$E' = \{p_{i,j}q_{i,j} \mid 1 \leq i \leq n, 1 \leq j \leq n\} \cup \{p_{i+1,j}q_{i,j} \mid 1 \leq i \leq n-1, 1 \leq j \leq n\}$$
$$\cup \{p_{i,j}p_{i,j+1}, q_{i,j}q_{i,j+1} \mid 1 \leq i \leq n, 1 \leq j \leq n-1\}$$
$$\cup \{s_1p_{1,j}, t_1q_{n,j} \mid 1 \leq j \leq n\} \cup \{s_2q_{i,1}, q_{i,n}t_2, \mid 1 \leq i \leq n\}$$
$$\cup \{p_{u,1}p_{v,1} \mid u, v \in V, uv \in E\}.$$

We show that, for a positive integer $k \leq n-1$, G has a vertex cover of size k if and only if Max 2-VDP-free Induced Subgraph in G' has a feasible solution of size $|V'| - k$.

Suppose that $G = (V, E)$ has a vertex cover X with $|X| = k$. If we remove the vertex set $X' = \{p_{v,1} \mid v \in X\} \subseteq V'$ from G', then the obtained graph is 2-VDP-free (i.e., it does not have 2 vertex-disjoint paths) by the planarity of $G' - X'$. This means that $V' \setminus X'$ is a feasible solution of Max 2-VDP-free Induced Subgraph whose size is $|V'| - k$.

Conversely, suppose that Max 2-VDP-free Induced Subgraph in G' has a feasible solution of size $|V'| - k$, that is, there exists a vertex set $X' \subseteq V'$ with $|X'| = k$ such that $G' - X'$ is 2-VDP-free. First, we note that $|X'| \leq n-1$ since the graph obtained from G' by removing $p_{1,1}, \ldots, p_{n-1,1}$ clearly avoids 2 vertex-disjoint paths. Define $X \subseteq V$ by

$$X = \{v \in V \mid X' \cap \{p_{v,1}, \ldots, p_{v,n}, q_{v,1}, \ldots, q_{v,n}\} \neq \emptyset\}.$$

We now show that X is a vertex cover of G. In order to derive a contradiction, assume that there exist $u, v \in V$ such that $u < v$, $uv \in E$, $X' \cap \{p_{u,1}, \ldots, p_{u,n}, q_{u,1}, \ldots, q_{u,n}\} = \emptyset$ and $X' \cap \{p_{v,1}, \ldots, p_{v,n}, q_{v,1}, \ldots, q_{v,n}\} = \emptyset$. Since $|X'| \leq n-1$, $G' - X'$ contains a path P_s from s_1 to $p_{u,1}$ and a path P_t from $p_{v,1}$ to t_1 which are both not intersecting with $\{q_{u,1}, \ldots q_{u,n}\}$. This means that G' contains two vertex-disjoint paths $P_1 = P_s \cup \{p_{u,1}q_{v,1}\} \cup P_t$ and $P_2 = (s_2, q_{u,1}, \ldots q_{u,n}, t_2)$, which contradicts the definition of X'. By the above argument, G contains a vertex cover X of size at most k.

Therefore, Vertex Cover in G can be reduced to Max 2-VDP-free Induced Subgraph in G', which shows the NP-hardness of Max 2-VDP-free Induced Subgraph.   □

References

1. Bollobás, B., Thomason, A.: Highly linked graphs. Combinatorica 16, 313–320 (1996)
2. Faria, L., de Figueiredo, C.M.H., Gravier, S., Mendonç, C.F., Stolfi, J.: Nonplanar vertex deletion: maximum degree thresholds for NP/Max SNP-hardness and a $\frac{3}{4}$-approximation for finding maximum planar induced subgraphs. Electronic Notes in Discrete Mathematics 18, 121–126 (2004)
3. Ferrara, M., Gould, R.J., Tansey, G., Whalen, T.: On H-immersions. Journal of Graph Theory 57, 245–254 (2008)
4. Ferrara, M., Gould, R., Tansey, G., Whalen, T.: On H-linked graphs. Graphs and Combinatorics 22, 217–224 (2006)
5. Fortune, S., Hopcroft, J., Wyllie, J.: The directed subgraph homeomorphism problem. Theoretical Computer Science 10, 111–121 (1980)
6. Gould, R.J., Kostochka, A., Yu, G.: On minimum degree implying that a graph is H-linked. SIAM J. Discret. Math. 20, 829–840 (2006)
7. Huck, A.: A sufficient condition for graphs to be weakly k-linked. Graphs and Combinatorics 7, 323–351 (1991)
8. Jung, H.: Eine Verallgemeinerung des n-fachen Zusammenhangs für Graphen. Mathematische Annalen 187, 95–103 (1970)
9. Karp, R.M.: On the computational complexity of combinatorial problems. Networks 5, 45–68 (1975)
10. Kawarabayashi, K., Kostochka, A., Yu, G.: On sufficient degree conditions for a graph to be k-linked. Comb. Probab. Comput. 15, 685–694 (2006)
11. Kostochka, A., Yu, G.: An extremal problem for H-linked graphs. J. Graph Theory 50, 321–339 (2005)
12. Kostochka, A., Yu, G.: Minimum degree conditions for H-linked graphs. Discrete Appl. Math. 156, 1542–1548 (2008)
13. Kostochka, A.V., Yu, G.: Ore-type degree conditions for a graph to be H-linked. J. Graph Theory 58, 14–26 (2008)
14. Larman, D., Mani, P.: On the existence of certain configurations within graphs and the 1-skeletons of polytopes. Proc. of the London Mathematical Society 20, 144–160 (1970)
15. Liebers, A.: Planarizing graphs – a survey and annotated bibliography. J. Graph Algorithms Appl. 5, 1–74 (2001)
16. Robertson, N., Seymour, P.D.: Graph minors. XIII. the disjoint paths problem. Journal of Combinatorial Theory B63, 65–110 (1995)
17. Robertson, N., Seymour, P.D.: Graph minors XXIII. Nash-Williams' immersion conjecture. Journal of Combinatorial Theory, Series B 100, 181–205 (2010)
18. Seymour, P.D.: Disjoint paths in graphs. Discrete Mathematics 29, 293–309 (1980)
19. Shiloach, Y.: A polynomial solution to the undirected two paths problem. Journal of the ACM 27(3), 445–456 (1980)
20. Thomas, R., Wollan, P.: An improved linear edge bound for graph linkages. European Journal of Combinatorics 26, 309–324 (2005)
21. Thomassen, C.: 2-linked graphs. European Journal of Combinatorics 1, 371–378 (1980)
22. Thomassen, C.: Highly connected non-2-linked digraphs. Combinatorica 11, 393–395 (1991)

An $\mathcal{O}(n^4)$ Time Algorithm to Compute the Bisection Width of Solid Grid Graphs*

Andreas Emil Feldmann and Peter Widmayer

Institute of Theoretical Computer Science, ETH Zürich, Switzerland
{feldmann,widmayer}@inf.ethz.ch

Abstract. The *bisection problem* asks for a partition of the n vertices of a graph into two sets of size at most $\lceil n/2 \rceil$, so that the number of edges connecting the two sets is minimised. A *grid graph* is a finite connected subgraph of the infinite two-dimensional grid. It is called *solid* if it has no holes. Papadimitriou and Sideri [8] gave an $\mathcal{O}(n^5)$ time algorithm to solve the bisection problem on solid grid graphs. We propose a novel approach that exploits structural properties of optimal cuts within a dynamic program. We show that our new technique leads to an $\mathcal{O}(n^4)$ time algorithm.

1 The Problem and Our New Approach

The problem of partitioning a graph into pieces of roughly equal sizes by cutting edges continues for decades to be of genuine theoretical interest and practical relevance, since it is usable in many divide-and-conquer algorithms. In our case the application stems from load distribution in parallel finite element simulations, where the input graph is a huge 3D-grid. The aim is to cut it into near equal sized pieces that each is scheduled onto a different machine. At the same time the interprocessor communication, modelled by the number of edges connecting the pieces, is to be minimised since this constitutes a bottleneck in parallel computations.

Of the above problem class we study a variant in which the graph is to be partitioned into *two* equal sized pieces. Our goal is to understand the nature of this problem from a theoretical viewpoint so as to lay the ground for further investigations of other problems of this type. A partition of the n vertices of a graph into two sets of size at most $\lceil n/2 \rceil$ each is called a *bisection*, and the number of edges connecting the two sets is its *cut-size*. The *bisection problem* asks for a bisection of minimum cut-size, the *bisection width*. For general graphs, it is NP-hard [6], and it can be approximated with ratio $\mathcal{O}(\log n)$ [10]. Even though the problem is weakly NP-hard [9] for planar graphs with vertex weights, the complexity is unknown for unweighted planar graphs. However a PTAS has been proposed [2]. For other graph classes such as trees and hypercubes the problem can be solved to optimality in polynomial time [1,7]. We are interested in

---

* We gratefully acknowledge discussions with Peter Arbenz, and the support of this work through the Swiss National Science Foundation under Grant No. 200021_125201/1.

grid graphs, defined as finite, connected subgraphs of the infinite two-
dimensional grid. We assume that a
grid graph is given together with its
natural embedding in the plane, where
each vertex is a coordinate in $\mathbb{N}^2$. We
call a grid graph *solid* if it has no
holes, i.e. it has no interior face sur-
rounded by more than four edges (Figure 1). There is a polynomial time reduction [8] from planar graphs to grid

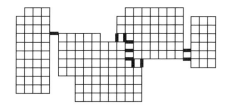

Fig. 1. An optimal bisection in a solid grid
graph. Note that the sets of the cut out
partition are not connected.

graphs with holes, and to determine the complexity for planar graphs has been
an open problem for many years. Hence restricting ourselves to solid grid graphs
is a natural way to gain insights about the more general problem. Motivated by a
VLSI layout problem, Papadimitriou and Sideri [8] presented an algorithm that
computes the bisection width for a solid grid graph in time $\mathcal{O}(n^5)$. More than
15 years ago, they asked whether this runtime can be improved. This problem
appeared to be difficult ever since, because the approach by Papadimitriou and
Sideri [8] was the only known one, and appeared not to be open to modifications
that might lead to faster runtimes. In this paper we show that with a novel
approach that makes use of some structural properties of a solution, a better
runtime is indeed possible, and we present an $\mathcal{O}(n^4)$ time algorithm.

A Bird's Eye View of Our Technique. Our approach is based on two
key findings. The first one limits the shapes of the pieces that are cut off from
the solid grid, and it bounds the number of cut edges in an optimum cut. The
second one uses these limits in a dynamic program.

The shape of a piece cut off from a solid grid is determined by the edges
being cut, or alternatively by the sequence of faces (grid cells) leading from the
exterior face through the grid and back to the exterior face (it is easy to see that
the exterior face needs to be present for the cut to be optimum; see also [8]). It
will be convenient to view this sequence of faces (including the exterior face) of
the grid graph G as a simple cycle in the dual graph of G. The dual graph is
defined as the (multi-)graph whose vertices are faces of G and whose edges are
between vertices of adjacent faces. The edge set in G corresponding to such a
simple cycle is called a *segment*. The optimum cut will in general cut the graph
into more than two pieces (more than two connected components). A cut in
a grid graph therefore corresponds to a set of simple cycles in its dual graph.
Hence the segments of a grid graph can be seen as building blocks of which a
cut consists.

We recall [8] that it is enough for an optimum cut to limit the segments to
the shape of a straight line, a corner, a stair, a clamp, or a square (Figure 2),
with one small variation: for the sake of being able to cut off exactly the desired
number of vertices, a side-step by one grid cell can be present that we call a *break*
and define precisely later. Furthermore, we will show that a single stair, clamp,
or square segment (with or without break) is enough in an optimum cut. That

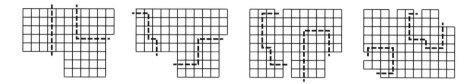

Fig. 2. The types of segments occurring in optimal k-cuts. In this order from left to right: straight and corner segments, stair segments with and without breaks, clamp segments with and without breaks, and square segments without and with breaks.

is, all others can be straight and corner segments. In addition, we can bound the number of edges of any segment in an optimum cut to $\mathcal{O}(\sqrt{n})$, by recalling [3] that the bisection width of a solid grid on n vertices is $\mathcal{O}(\sqrt{n})$.

We make use of these limitations on segments by explicitly considering all possible stairs, clamps, and squares, without and with a break. For each of both parts into which such a segment cuts the grid, we only consider straight and corner segments that cut away exactly the desired number of vertices. We are able to compute the optimal way to cut out any desired number of vertices in each part using only straight and corner segments inductively in a dynamic program. The efficiency of our approach rests on the fact that there are only $\mathcal{O}(n^2)$ segments to be considered, since each segment is defined by three parameters: first, one of the corners in its shape (at one of at most n positions); second, the distance to a (suitable) neighbouring corner of its shape (at most $\mathcal{O}(\sqrt{n})$); and third, the potential position of a break (at most $\mathcal{O}(\sqrt{n})$ possible ones). We will show that only an additional multiplicative term of $\mathcal{O}(n^2)$ is needed to compute the optimal bisection. This proves the claimed runtime of $\mathcal{O}(n^4)$.

2 Properties of Optimal k-Cuts

For our dynamic program we need to generalise the bisection problem to considering k-*cuts*. These are sets of segments that, when removed from the graph, leave a spanning subgraph that contains a set of connected components including exactly k vertices. We say that a k-cut S *cuts out* the k vertices of these connected components, and we call the two sets of vertices of size k and $n-k$ the *parts* cut out by S. Given a k-cut S we call the number of edges $\sum_{s \in S} |s|$ in S its *cut-size*, and we call a k-cut that minimises the cut-size over all k-cuts *optimal*. Notice that some edges may be counted several times in the sum. However, edges that appear more than once in different segments can be removed from a k-cut. This is why this generalisation does not change the optimal solution.

In order to prove our claimed results we need to analyse the types of segments that may occur in an optimal k-cut. We first recall [8] the result that these only include so called straight, corner, stair, clamp, and square segments. Thereafter we will prove that at most one of the segments in an optimal k-cut is not a straight or corner segment. We begin by formally defining the above types of segments (Figure 2) with the help of the dual (multi-)graph D of a solid grid graph G. In the following any face refers to a face of G, i.e. a vertex of D. Let

s be a segment in G such that the simple cycle p corresponding to s in the dual D includes the exterior face. An interior face f in D lying on p is called a *bend* of s if f touches two edges $e_1, e_2 \in s$ such that e_1 and e_2 share a vertex. We say that the bend f *points* in two *directions*: the directions are *up* and *right* if e_1 and e_2 lie above and to the right of f respectively, and analogously they can be *down* or *left* if the edges lie appropriately. Two bends of s are said to *point in opposing directions* if they do not share any direction in which they point. If they share at least one direction they are said to *point in a common direction*. A *break* of s is an edge $e \in s$ such that e touches two bends of s that point in opposing directions. Let q be a sub-path of p such that q starts and ends in two faces f_1, f_2 of D and f_1 and f_2 are bends of s or equal the exterior face f_∞. The subset b of s corresponding to q is called a *bar* of s if no face on q between f_1 and f_2 is a bend of s or equal to f_∞. The bar b is said to *end at* the two faces f_1 and f_2. The subset b is called a *broken bar* of s if b can be partitioned into three bars of s and the one ending neither at f_1 nor at f_2, i.e. the middle one, is a break of s. Also the broken bar b is said to *end at* the two faces f_1 and f_2. Two bends of s are called *consecutive* if there is a bar of s ending at them.

Definition 1. *The segment s is called (Figure 2)*

- *a* straight *segment if it has no bend.*
- *a* corner *segment if it has exactly one bend.*
- *a* stair *segment if any consecutive bends of s point in opposing directions. Additionally, if s has no break then it has exactly two bends. Otherwise it contains a broken bar b such that $s \setminus b$ constitutes at most two bars each of which ends at f_∞.*
- *a* clamp *segment if there are two bends f_1, f_2 of s pointing in a common direction. Additionally s can be partitioned into (1) a bar or broken bar b ending at f_1 and f_2, and (2) two bars that both end at f_∞.*
- *a* square *segment if there are three bends f_1, f_2, f_3 of s such that s can be partitioned into (1) a bar or broken bar b ending at f_1 and f_2, (2) a bar b' ending at f_2 and f_3, and (3) two bars ending at f_∞, and f_1 or f_3 respectively. Additionally each of the pairs f_1, f_2 and f_2, f_3 point in a common direction, and if b is a broken bar then f_1 and its consecutive bend of s touching the break, point in a common direction. Moreover $|b'| \leq \beta \leq |b'| + 1$, where β is the length of b without counting its breaks. That is, $\beta = |b|$ if b is a bar, and $\beta = |b| - 1$ if b is a broken bar.*

If any of the above segments contains a broken bar we say that s has a break, and we refer to the corresponding break of the broken bar as the break of s. For a clamp or square segment s let $v \in V$ be the vertex of the grid graph that is shared by the edges touching the bend f_1 of s. The part cut out by s including v is referred to as convex, *and the other part as* concave.

The next lemma, which states that there exists an optimal k-cut containing only the types of segments in the above definition, follows from Lemmas 3 and 4 in [8]: Lemma 3 therein can be used to convert the segments described by Lemma 4 in [8] in order to derive the particular shape required by the above definition.

Lemma 2 (follows from [8]). *There is an optimal k-cut containing only straight, corner, stair, clamp, and square segments.*

Furthermore at most one segment in an optimal k-cut is not a straight or corner segment. We prove this by shifting pieces of cut out areas from one segment to another. This is done so that the overall cut out area stays the same while the cut-size does not increase. The proof of the following theorem can be found in the full version of the paper [5].

Theorem 3. *There is an optimal k-cut that contains only straight and corner segments except at most one which is either a stair, clamp, or square segment.*

As a consequence of the above theorem the obvious way to proceed at this point would be to consider each stair, clamp, and square segment explicitly in the algorithm, as described in the introduction. However the runtime would in this case be larger than claimed since, for instance, there are $\Theta(n^3)$ stair segments in the worst case. However not all of these will appear in an optimal k-cut since some of them are too large. This follows from the fact that the maximum degree of a grid graph is 4, and a result from [3] where it was shown that the bisection width of any planar graph of maximum degree Δ is $\mathcal{O}(\sqrt{\Delta n})$. Hence by further restricting some of the segments to such ones that contain at most $\mathcal{O}(\sqrt{n})$ edges we are able to reduce the runtime.

Theorem 4 (follows from [3]). *The bisection width of a grid graph is $\mathcal{O}(\sqrt{n})$.*

3 Computing Optimal k-Cuts

In this section we will present an algorithm to compute optimal k-cuts in solid grid graphs. We do this by assuming that we are given a solid grid graph G and the set $\mathcal{S}$ of straight, corner, stair, clamp, and square segments in G. Some of the segments in $\mathcal{S}$ will have a length of at most $\mathcal{O}(\sqrt{n})$. We will specify exactly which ones will be short in the next section where the runtime of our algorithm will be determined. According to Lemma 2 and Theorem 4 it suffices to compute an optimal k-cut that only uses segments from $\mathcal{S}$. More formally, for any set of segments $\widetilde{\mathcal{S}}$ we say that any k-cut S is $\widetilde{\mathcal{S}}$-restricted if $S \subseteq \widetilde{\mathcal{S}}$, and our goal is to compute an optimal non-crossing $\mathcal{S}$-restricted k-cut. Additionally we assume that we are given the set $\mathcal{K} \subseteq \mathcal{S}$ of straight and corner segments in G. According to Theorem 3 we know that any optimal $\mathcal{S}$-restricted k-cut contains at most one segment that is not from $\mathcal{K}$.

One crucial observation needed to construct the dynamic program is that we can assume that no segments cross in the optimal k-cut: observe that a simple cycle in the dual of a planar graph corresponds to a closed curve in the embedding of the dual graph in the plane. Hence the cycle divides the plane into an interior and an exterior area. We say that a pair of cycles *cross* if the corresponding closed curve of one of them both contains points belonging to the interior and the exterior area into which the other cycle divides the plane. Note that any pair of simple cycles that cross can be seen as a (different) pair

of simple cycles that do not cross
(Figure 3). Hence we may limit ourselves to cuts in which no segments
cross and we call these *non-crossing*.

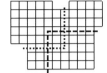 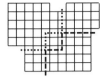

The idea of the algorithm is to
guess a stair, clamp, or square seg-
ment $s \in \mathcal{S} \setminus \mathcal{K}$ from which we know
that it is contained in the optimal
solution and all other segments are
straight and corner segments from $\mathcal{K}$.
The case when the optimum only con-

Fig. 3. A crossing and an equivalent non-
crossing k-cut containing two segments.
The segments are indicated by a dashed and
a dotted line.

tains segments from $\mathcal{K}$ is dealt with separately. We split the graph into the two
parts V_s^1 and V_s^2 cut out by s. If the optimum $\mathcal{K}$-restricted cuts in these two
parts are known, then these can be used to compute the optimum containing s.
That is, we can compute the cut-size $C_s(k)$ of an optimal k-cut that contains
s and only segments from $\mathcal{K}$ that do not cross s. Let for $i \in \{1,2\}$ the set $\mathcal{K}_s^i$
include every segment $t \in \mathcal{K}$ that cuts out a part $V_t \subseteq V_s^i$. Let also $C_s^i(k)$ denote
the cut-size of an optimal $\mathcal{K}_s^i$-restricted k-cut. We define the value of $C_s^i(k)$ to
be infinite if no such cut exists. Using $C_s^1(\cdot)$ and $C_s^2(\cdot)$ we compute $C_s(k)$ as
follows. The corresponding k-cut cuts out some number k' of the vertices from
V_s^1. Since the computed solution should contain s as a segment the number of
vertices cut out from V_s^2 is $|V_s^2| - (k - k')$. Thus the optimal cut-size is

$$C_s(k) = \min\{|s| + C_s^1(k') + C_s^2(|V_s^2| - (k - k')) \mid k' \in \{0, ..., k\}\}. \qquad (1)$$

Since the optimal solution contains at most one segment from $\mathcal{S} \setminus \mathcal{K}$ and all
others from $\mathcal{K}$, taking the minimum over all $s \in \mathcal{S} \setminus \mathcal{K}$ of all computed values
$C_s(k)$ correctly computes the optimal k-cut if it contains a segment from $\mathcal{S} \setminus \mathcal{K}$.
To handle the case when the optimum only contains segments from $\mathcal{K}$, we define
$C_s^i(k)$ and $\mathcal{K}_s^i$ accordingly for any $s \in \mathcal{K}$. Notice that then $s \in \mathcal{K}_s^i$. We treat
this special case by also taking the cut-size $C_s^i(k)$ of a segment $s \in \mathcal{K}$ for which
$\mathcal{K}_s^i = \mathcal{K}$ into account in the final step. It is easy to see that such a segment
always exists since G is a solid grid graph and $\mathcal{K}$ contains all straight and corner
segments. Hence also taking the corresponding value $C_s^i(k)$ into account will
correctly find the optimal solution for the given solid grid graph. Note that
given the functions $C_s^i(\cdot)$, for a fixed k the algorithm will take $\mathcal{O}(\sum_{s \in \mathcal{S} \setminus \mathcal{K}} n)$
time according to Equation (1) to compute the optimum k-cut. This is because
for each segment $s \in \mathcal{S} \setminus \mathcal{K}$ it needs to consider all possible values for k'.

It remains to be shown how the $\mathcal{K}$-restricted optima $C_s^1(\cdot)$ and $C_s^2(\cdot)$ are
computed. The main inspiration for this part of our algorithm is taken from the
corresponding algorithm for trees [7]. In a tree the segments correspond to single
edges and a dynamic program is used to compute an optimal k-cut bottom-up
from the leaves to the root. For each edge e of the tree the algorithm computes
the optimal solution for the subtree at e. It will decide whether to include e in
the solution by considering the optimal cuts in the subtrees immediately below

e. Combining the cuts in the subtrees in order to compute the optimum up to e is easy since they do not interfere with one another.

For our case we proceed in a similar way as for trees by inductively computing the cut-size $C_s^i(k)$ of an optimal $\mathcal{K}_s^i$-restricted k-cut for every $k \in \{0, ..., n\}$ in each part V_s^i, $i \in \{1, 2\}$, cut out by a segment $s \in \mathcal{S} \setminus \mathcal{K}$. To compute the $\mathcal{K}_s^i$-restricted solutions in the two parts we will also need the solutions for any part cut out in V_s^1 and V_s^2 by a segment from $\mathcal{K}$. Hence we show how to compute the optima for the parts cut out by any segment from $\mathcal{S}$, i.e. not excluding the straight and corner segments. Fix one of the parts cut out by $s \in \mathcal{S}$ and call it V_s. We will decide whether to include s into the solution for part V_s by considering the cuts computed for segments cutting out parts from V_s. However these solutions do interfere with one another since the parts can overlap. In order to circumvent this problem the idea is to guess where V_s has to be split so that each segment of the non-crossing optimum is contained in one of the resulting pieces of V_s. We will use segments cutting out parts in V_s for splitting. To find the correct way to split a part we give the following definition.

Definition 5. *Let V_s be a part cut out by a segment s and let S denote a set of segments such that each $t \in S$ cuts out a part $V_t \subset V_s$. The set S is called an interference-free set (IFS) in V_s if $V_t \cap V_{t'} = \emptyset$ for each pair $t \neq t'$ from S. Let $\mathcal{K}_S$ contain all segments $u \in \mathcal{K}$ that cut out a part $V_u \subseteq V_t$ for some $t \in S$.*

Note that s itself can not be contained in an IFS in V_s. In order to find the cut-size of the optimal $\mathcal{K}$-restricted k-cut in V_s we will split V_s according to each IFS from a small predefined set of IFSs in V_s. We will need one such predefined set for each part cut out by a segment $s \in \mathcal{S}$ and hence call them $\mathcal{I}_s^1$ and $\mathcal{I}_s^2$. In the next section we will show that for each part we can find such a set that is small enough in order to guarantee the claimed runtime. The IFSs in the set include segments from $\mathcal{S}$ and together have the property that they cover all IFSs including segments from $\mathcal{K}$ in V_s^i, in the following sense.

Definition 6. *Let $s \in \mathcal{S}$ cut out a part V_s. An IFS covering set $\mathcal{I} \subseteq 2^{\mathcal{S}}$ (w.r.t. $\mathcal{K}$) includes IFSs in V_s such that for any IFS $S \subseteq \mathcal{K}$ also in V_s there is a set $S^* \in \mathcal{I}$ for which $S \subseteq \mathcal{K}_{S^*}$.*

Fix an IFS covering set $\mathcal{I}$ for the cut out part V_s we are considering. For any IFS S in V_s let $C_S(k)$ denote the cut-size of the optimal $\mathcal{K}_S$-restricted k-cut. To compute the cut-size of the optimal $\mathcal{K}$-restricted k-cut in V_s we can split V_s according to each IFS $S^* \in \mathcal{I}$ and make use of the functions $C_{S^*}(\cdot)$. To see this we show that any non-crossing $\mathcal{K}$-restricted k-cut S in V_s is $\mathcal{K}_{S^*}$-restricted for some $S^* \in \mathcal{I}$. Consider the set S' of segments containing any $t \in S \setminus \{s\}$ that cuts out a part $V_t \subset V_s$ such that there is no other segment $t' \in S \setminus \{s\}$ that cuts out a superset $V_{t'} \subset V_s$ of V_t. Since S is non-crossing the set S' is an IFS. Since also $S' \subseteq \mathcal{K}$, by the definition of an IFS covering set this means that $S \setminus \{s\}$ is $\mathcal{K}_{S^*}$-restricted for some $S^* \in \mathcal{I}$. Hence also the optimal non-crossing $\mathcal{K}$-restricted k-cut in V_s is $(\mathcal{K}_{S^*} \cup \{s\})$-restricted for some $S^* \in \mathcal{I}$. Thus we need only consider the sets $S^* \in \mathcal{I}$ and pick the solution that has the minimum cut-size according to the functions $C_{S^*}(\cdot)$ to compute the optimum in V_s. If $s \in \mathcal{K}$

we also need to consider the case when s is included in the solution. In this case the number of cut-out vertices from V_s is $|V_s| - k$.

Hence for any $i \in \{1, 2\}$ and $s \in \mathcal{S}$, if the IFS covering set $\mathcal{I}_s^i$ for the cut out part V_s^i is non-empty then $C_s^i(k)$ can be computed by

$$C_s^i(k) = \begin{cases} \min\{C_{S^*}(k) \mid S^* \in \mathcal{I}_s^i\} & \text{if } s \in \mathcal{S} \setminus \mathcal{K}, \\ \min\{C_{S^*}(k), |s| + C_{S^*}(|V_s^i| - k) \mid S^* \in \mathcal{I}_s^i\} & \text{if } s \in \mathcal{K}. \end{cases} \quad (2)$$

In case $\mathcal{I}_s^i$ is empty there are no segments that cut out a subset of V_s^i. Hence then $C_s^i(0) = 0$ and if $s \in \mathcal{K}$ also $C_s^i(|V_s^i|) = |s|$. All other values of $C_s^i(\cdot)$ are infinite. Computing the table containing all values of the function $C_s^i(\cdot)$ clearly takes $\mathcal{O}(n \cdot \sum_{s \in \mathcal{S}} |\mathcal{I}_s^1 \cup \mathcal{I}_s^2|)$ steps if all values of the functions $C_{S^*}(\cdot)$ of corresponding IFSs S^* are given.

The last missing part of this section is to show how a function $C_S(\cdot)$ for a non-empty IFS S in a cut out part V_s of a segment $s \in \mathcal{S}$ can be computed. In order to find the cut size $C_S(k)$ of an optimal $\mathcal{K}_S$-restricted k-cut, the algorithm will combine the solutions computed for the segments in S in the same way the solutions for subtrees were combined in the algorithm for

1. **for all $s \in \mathcal{S}$ and all $S^* \in \bigcup_{i,s} \mathcal{I}_s^i$ do**
   – compute C_s^i using C_{T^*}, where $T^* \in \mathcal{I}_s^i$
   – compute C_{S^*} using C_t^i, where $t \in S^*$
2. **for all $s \in \mathcal{S} \setminus \mathcal{K}$ do**
   – compute C_s using C_s^1 and C_s^2
3. **return**
   $\min\{C_s(k), C_t^i(k) \mid s \in \mathcal{S} \wedge \mathcal{K}_t^i = \mathcal{K}\}$

Fig. 4. The overall structure of the algorithm to compute an optimum k-cut

trees in [7]. If S contains only a single segment t then obviously $C_S(k) = C_t^i(k)$, where $i \in \{1, 2\}$ such that V_t^i is the part cut out by t from V_s. In case S contains more than one segment, the value of $C_S(k)$ can, for any fixed $t \in S$, be recursively computed using the solutions to the IFS $S \setminus \{t\}$ and the solution for the part $V_t^i \subset V_s$. An optimal $\mathcal{K}_S$-restricted k-cut must cut out some number k' of the k vertices from V_t^i. The remaining $k - k'$ vertices are taken from the parts cut out by the segments in $S \setminus \{t\}$. Thus finding the minimum cut-size among all possible values of k' will find the optimal solution. Hence the following equation is correct:

$$C_S(k) = \min\{C_t^i(k') + C_{S \setminus \{t\}}(k - k') \mid k' \in \{0, ..., k\}\}. \quad (3)$$

To evaluate the right hand side of Equation (3) we need only consider values of k' that are at most the number n_t of vertices in V_t^i since all other values of $C_t^i(\cdot)$ are infinite. This means that we can amortise the runtime needed to compute all values of $C_S(\cdot)$ for a particular IFS S to $\mathcal{O}(\sum_{t \in S} n \cdot n_t) = \mathcal{O}(n^2)$. This is true because we need to consider at most all the $n + 1$ possible values of k for each $t \in S$ while the parts cut out by the segments in S are disjoint. To compute all values of $C_s^i(\cdot)$ for all $s \in \mathcal{S}$ we need to compute all values of $C_{S^*}(\cdot)$ for all $S^* \in \mathcal{I}_s^i$ and all $s \in \mathcal{S}$. Hence computing the whole table for all values of $C_{S^*}(\cdot)$ takes $\mathcal{O}(n^2 \cdot \sum_{s \in \mathcal{S}} |\mathcal{I}_s^1 \cup \mathcal{I}_s^2|)$ time. Therefore the runtime of the algorithm

(Figure 4) is dominated by the time needed to compute the table containing the values of the functions $C_{S^*}(\cdot)$.

4   Counting Segments and IFS Covering Sets

In Section 3 we have seen that we can efficiently compute optimal k-cuts for solid grids if the number of considered segments and IFS covering sets is small. Hence we need to identify a small IFS covering set $\mathcal{I}_s^i$ for each considered segment s and each cut out part V_s^i, where $i \in \{1, 2\}$. In this section we will prove that the runtime of the given algorithm is $\mathcal{O}(n^4)$ as claimed, by counting the number of segments and the sizes of the IFS covering sets. In order for the involved sets not to be too large, the set $\mathcal{S}$ includes only straight, and corner segments, together with the stair, clamp, and square segments without breaks. It also contains all stair segments that consist of only a single broken bar. Additionally $\mathcal{S}$ contains all stair, clamp, and square segments with breaks that have at most $c\sqrt{n}$ edges, for some constant c according to Theorem 4. The latter theorem together with Lemma 2 guarantees that these sets $\mathcal{K}$ and $\mathcal{S}$ suffice in order to compute an optimal k-cut using the algorithm in Section 3.

According to the results in Section 3 we need to show that $\sum_{s \in \mathcal{S}} |\mathcal{I}_s^1 \cup \mathcal{I}_s^2| \in \mathcal{O}(n^2)$ and that all required segments and IFS covering sets can be found efficiently, in order to guarantee a total runtime of $\mathcal{O}(n^4)$. It was already shown in [4] that the number of straight and corner segments is $\mathcal{O}(n)$, and the number of IFSs in each set $\mathcal{I}_s^1$ and $\mathcal{I}_s^2$ for any such segment s also is $\mathcal{O}(n)$. Additionally these segments and their IFS covering sets can also be listed [4] in time $\mathcal{O}(n^2)$. Hence we only need to prove similar results for the stair, square, and clamp segments. We start by counting the number of segments for each such type.

Lemma 7. *There are $\mathcal{O}(n^2)$ stair, clamp, and square segments without breaks, and $\mathcal{O}(n^2)$ of these segments with breaks having a length of at most $c\sqrt{n}$. Also there are $\mathcal{O}(n)$ stair segments that consist of only a single broken bar. Furthermore all of these segments can be listed in time $\mathcal{O}(n^2)$.*

Proof. There are $\mathcal{O}(n^2)$ stair, clamp, and square segments without breaks since according to Definition 1 each such segment can be identified with the respective bend f_2, the directions in which f_2 points (the directions in which the other bends point is determined by this), and a distance to the consecutive bends of f_2 (which in the case of a square segment amounts to choosing a length for the bar b). Since there are $\mathcal{O}(n)$ faces in the grid graph that can be used for f_2, four directions in which to point, and the distance can be at most n, the result follows.

For segments with breaks each choice of the above three parameters also leaves the choice of a position of the break along the length of the respective broken bar. If the segment is a stair segment consisting of only a single broken bar then the only choice is its break and the direction in which the bends point. Thus there are only $\mathcal{O}(n)$ many such segments. For all other segments with breaks the length of the segment is assumed to be at most $c\sqrt{n}$. Therefore there are only $\mathcal{O}(\sqrt{n})$ choices for the distance between the bends. This also means that

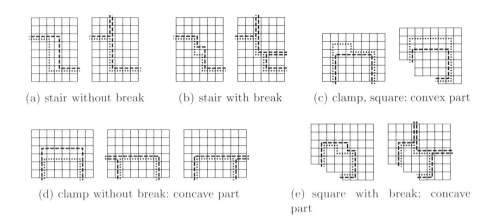

(a) stair without break   (b) stair with break   (c) clamp, square: convex part

(d) clamp without break: concave part   (e) square with break: concave part

Fig. 5. The IFS covering sets for all different types of segments s. The segment s is indicated by the dotted line and the segments in an IFS by the dashed lines.

there are $\mathcal{O}(\sqrt{n})$ possible positions for the break, and hence the total number of segments with breaks is again $\mathcal{O}(n^2)$.

The above counting arguments clearly give a straightforward way of listing all these segments too. Hence they can be found in time $\mathcal{O}(n^2)$.   □

Finally we need to identify the IFS covering sets used for each of the segments $s \in \mathcal{S} \setminus \mathcal{K}$ and each $i \in \{1, 2\}$. We will prove that there is an IFS covering set of constant size for each such segment. The given construction of the sets can be used to compute them in a preprocessing step of the algorithm in $\mathcal{O}(n^2)$ total time.

In all but the case of clamp or square segments with breaks we will ignore the fact that some of the segments in the claimed IFS covering set may partially lie outside of the grid graph. The reason why we can ignore these special cases is that if any segment, except for a clamp or square segment with a break, is split at any point (by the border of the grid graph) then the resulting segments that lie inside the grid graph are all segments of types again included in $\mathcal{S}$. Hence these can be used for the IFS. On the other hand, splitting a clamp or square segment s with a break may not result in a set of segments from $\mathcal{S}$. We will therefore need to handle these two cases separately.

Theorem 8. *For any stair, clamp, or square segment $s \in \mathcal{S} \setminus \mathcal{K}$ in a solid grid graph G and any $i \in \{1, 2\}$ there is an IFS covering set $\mathcal{I}_s^i$ (w.r.t. $\mathcal{K}$) containing at most three IFSs. Moreover all sets $\mathcal{I}_s^i$ for segments in $s \in \mathcal{S} \setminus \mathcal{K}$ can be listed in time $\mathcal{O}(n^2)$.*

Proof. We will show the statement for every type of segment separately by giving a construction of an IFS covering set $\mathcal{I}_s^i$. According to Definition 6, for each case we need to show that for any IFS $S \subseteq \mathcal{K}_s^i$ there exists a set $S^* \in \mathcal{I}_s^i$ for which $S \subseteq \mathcal{K}_{S^*}$. In all cases, if $\mathcal{K}_s^i = \emptyset$ then $\mathcal{I}_s^i = \emptyset$ is an IFS covering set. Hence throughout this proof we assume that $\mathcal{K}_s^i \neq \emptyset$. To prove the following results we need to give the respective segments an orientation in the grid G. Therefore we define *horizontal* respectively *vertical* edges of G to be those for which the incident vertices have the same y- respectively x-coordinates. A *horizontal* bar of a segment contains only vertical edges, and a *vertical* bar only horizontal edges. A *horizontal* respectively *vertical* broken bar of a segment contains two horizontal respectively vertical bars.

Consider the case when s is a stair segment. If s has no break let b be the bar of s ending at its two bends. Otherwise b is the broken bar of s. We assume w.l.o.g. that any bend of s points down and left, or up and right. Furthermore b is assumed to be vertical, and the vertices that are the upper and right incident vertices of the vertical and horizontal edges of s, respectively, are those belonging to V_s^i.

If s has no break (Figure 5(a)) it has two bends and hence any straight segment including edges from b would in this case cut out vertices not contained in V_s^i. Thus any segment from $\mathcal{K}_s^i$ that includes edges from b must be a corner segment. Let $T \subseteq \mathcal{K}_s^i$ contain all these corner segments. There is a corner or stair segment without break $t \in \mathcal{S}$ that cuts out all vertices from V_s^i except those that are incident to the edges in b. For the IFS $S_1^* = \{t\}$ in V_s^i it holds that $\mathcal{K}_{S_1^*} = \mathcal{K}_s^i \setminus T$ since the only edges that can not be used to form a segment from $\mathcal{K}_{S_1^*}$ are those in b. Hence for any IFS $S \subseteq \mathcal{K}_s^i$ which does not contain a segment from T it holds that $S \subseteq \mathcal{K}_{S_1^*}$. Let $u \in \mathcal{S}$ be the corner segment that cuts out a part V_u from V_s^i and has the same bend as s, say f_1, pointing up and right. The vertices in $V_s^i \setminus V_u$ are cut out by the corner segment $u' \in \mathcal{S}$ having the other bend f_2 of s as its bend which for u' however points up and left. The only segments from $\mathcal{K}_s^i$ that are not included in $\mathcal{K}_{S_2^*}$, where $S_2^* = \{u, u'\}$, are those that have a horizontal bar crossing the vertical bars of u and u' above the bend f_2 of s. The vertical bar of any $r \in T$ is included in the vertical bar of u. Hence if an IFS $S \subseteq \mathcal{K}_s^i$ contains a corner segment $r \in T$ then $S \subseteq \mathcal{K}_{S_2^*}$ since S is non-crossing. In conclusion the set $\mathcal{I}_s^i$, for any $i \in \{1,2\}$, is an IFS covering set if it contains the IFSs S_1^* and S_2^* in case s is a stair segment without break.

All other cases can be shown in a similar way. The complete proof can be found in the full version of the paper [5].   □

5 Discussion

We presented a novel approach for solid grid bisection that runs in time $\mathcal{O}(n^4)$, in contrast to the best earlier algorithm whose runtime was $\mathcal{O}(n^5)$. The new algorithm is based on structural properties of the cut, and on the use of these properties within a dynamic program. Just as in the original algorithm by Papadimitriou and Sideri [8] we not only compute the optimal bisection but the optimum k-cut for every $k \in \{0, \ldots, n\}$. This means that also objective functions depending on the cut size and the balance of the cut can be optimised in time $\mathcal{O}(n^4)$. For instance in load distribution applications, such as ours, the cut size may be more important than the balance.

One might ask whether the structural properties could also have been used with the original algorithm [8]. The answer is negative. The reason is that the original algorithm recursively computes an optimal k-cut for each pair of boundary edges of the grid by splitting the boundary in clockwise direction from one edge of the pair to the other. A limitation like ours on the useful segments does not positively affect the runtime of this approach. It remains to be seen whether these structural properties of the cut, cast into an algorithm, have the potential to lead to improved solutions for more general classes of graphs (such as planar graphs) and for more general cuts (such as partitions into any given number of parts). We remark that for some immediate generalisations from solid grid graphs the presented observations on the structure of the segments do not hold. For instance giving weights to the vertices or edges makes the shapes of the segments in a bisection much more complex. Also it follows from [8] that grid graphs with holes are essentially as hard to bisect as planar graphs, and determining the complexity of the latter is a long standing open problem. However we believe that generalisations to graphs with a very regular structure are possible. For instance if the interior faces of the considered (unweighted) graph constitute a tessellation of the plane, then similar observations on the structure of segments and their IFS covering sets as the ones given for solid grid graphs should be achievable. One motivation for such instances is that many 2D finite element models use triangles as tessellations.

References

1. Díaz, J., Petit, J., Serna, M.J.: A survey of graph layout problems. ACM Comput. Surv. 34(3), 313–356 (2002)
2. Díaz, J., Serna, M.J., Torán, J.: Parallel approximation schemes for problems on planar graphs. Acta Informatica 33(4), 387–408 (1996)
3. Diks, K., Djidjev, H.N., Sykora, O., Vrto, I.: Edge separators of planar and outerplanar graphs with applications. Journal of Algorithms 14(2), 258–279 (1993)
4. Feldmann, A.E., Das, S., Widmayer, P.: Simple cuts are fast and good: Optimum right-angled cuts in solid grids. In: Wu, W., Daescu, O. (eds.) COCOA 2010, Part I. LNCS, vol. 6508, pp. 11–20. Springer, Heidelberg (2010)
5. Feldmann, A.E., Widmayer, P.: An $O(n^4)$ time algorithm to compute the bisection width of solid grid graphs. Technical Report 730, Institute of Theoretical Computer Science, ETH Zürich (July 2011)
6. Garey, M.R., Johnson, D.S., Stockmeyer, L.: Some simplified NP-complete graph problems. Theoretical Computer Science 1(3), 237–267 (1976)
7. MacGregor, R.M.: On partitioning a graph: a theoretical and empirical study. PhD thesis, University of California, Berkeley (1978)
8. Papadimitriou, C., Sideri, M.: The bisection width of grid graphs. Theory of Computing Systems 29, 97–110 (1996)
9. Park, J.K., Phillips, C.A.: Finding minimum-quotient cuts in planar graphs. In: Proceedings of the Twenty-Fifth Annual ACM Symposium on Theory of Computing, STOC 1993, pp. 766–775. ACM, New York (1993)
10. Räcke, H.: Optimal hierarchical decompositions for congestion minimization in networks. In: Proceedings of the 40th Annual ACM Symposium on Theory of Computing (2008)

Min-Cuts and Shortest Cycles in Planar Graphs in $O(n \log \log n)$ Time*

Jakub Łącki and Piotr Sankowski

Institute of Informatics
University of Warsaw
Warsaw, Poland

Abstract. We present a deterministic $O(n \log \log n)$ time algorithm for finding shortest cycles and minimum cuts in planar graphs. The algorithm improves the previously known fastest algorithm by Italiano *et al.* in STOC'11 by a factor of $\log n$. This speedup is obtained through the use of dense distance graphs combined with a divide-and-conquer approach. Extending this approach we are able to show an $O(n^{5/6} \log^{5/2} n)$ time dynamic algorithm al well.

1   Introduction

In this paper we study the minimum cut and shortest cycle problems in planar graphs. The *minimum cut problem* is to find the cut with minimum capacity, whereas the *shortest cycle problem* is to find the cycle with minimum total length. These two problems are actually equivalent, since a shortest cycle corresponds to a minimum cut in the dual graph. Moreover, the size of the minimum cut is equal to the weighted edge-connectivity of the graph. In this paper when presenting the algorithms we only talk about the problem of finding shortest cycle keeping in mind that the min-cut problem can be solved using this reduction.

In general graphs the minimum cut can be found in $O(m \log^3 n)$ randomized time as shown by Karger [11], or in $O(mn + n^2 \log n)$ deterministic time as given by Nagamochi and Ibraki [13]. On the other hand, the shortest odd cycle can be found in $O(nm)$ time [9], whereas the shortest even length cycle can be found in $O(n^2)$ time [15].

In the case of planar graphs these two problems have attracted considerable attention in recent years. Even in the case of the unweighted graphs these problems are interesting. However, one needs to keep in mind that the duals of unweighted graphs are no longer unweighted. Eppstein [4] was the first one to show how to find cycles of constant weight in $O(n)$ time. This result was later on improved by Alon, Yuster and Zwick [1], who have shown an $O(n)$ time algorithm for finding shortest cycles of length ≤ 5. On the other hand, the fastest

---

* This work has been partially supported by the Polish Ministry of Science, Grant N N206 355636 and by the ERC StG Project PAAl no. 259515.

C. Demetrescu and M.M. Halldórsson (Eds.): ESA 2011, LNCS 6942, pp. 155–166, 2011.

algorithm for finding the shortest cycle in unweighted graph (also called the *girth* of the graph), was given by Chang [8]. This very recent algorithm works in $O(n)$ time.

In the case of weighted planar graphs the fastest algorithm was proposed in this year and works in $O(n \log n \log \log n)$ time [10]. This algorithm is obtained by simply joining the $O(n \log^2 n)$ time divide-and-conquer approach of Chalermsook *et al.* [3] with a faster $O(n \log \log n)$ time max-flow algorithm given in [10]. Here we show an even further improvement by showing an $O(n \log \log n)$ time algorithm for both the minimum cut and the shortest cycle problems in weighted planar graphs.

The minimum cut problem is related to *minimum st-cut problem*, where we need to find minimum cut that separates s from t. Up until the paper of Italiano *et al.* [10] the fastest known algorithm worked in $O(n \log n)$ time [6]. You may notice that the approach of Chalermsook *et al.* [3] results in a $\log n$ complexity gap between min-cut and min st-cut problems. In this paper we actually show how to close this gap, and we believe that the techniques introduced here will be useful as well when faster min st-cut algorithms are developed in the future.

In addition to the static results, we give a dynamic algorithm for computing the minimum cut size and the shortest cycle length in planar graphs. It processes updates and answers queries in $O(n^{5/6} \log^{5/3} n)$ time and is the first known dynamic result that is able to handle weighted edges and arbitrary edge connectivity. The only previously known exact dynamic algorithm with update time sublinear in n was able to maintain the information about polylogarithmic edge connectivity in $O(\sqrt{n})$ time per update [14]. For the history of this problem we refer you to the description in [14].

This paper is organized as follows. In next section we give a summary of the techniques developed in the previous papers that we use. In Section 3 we recall the Chalermsook *et al.* [3] algorithm. Our first algorithm that works in $O(n \log n)$ time is given in Section 4. This algorithm actually builds a part of the main result of this paper that is given in Section 5. The dynamic algorithm is given in the final section of this paper.

2   Preliminaries

For a graph $G = (V, E)$, we denote the set of its vertices and edges by $V(G)$ and $E(G)$ respectively. Additionally, if G is planar, $F(G)$ denotes the set of its faces. If C is a cycle in a planar graph, we define its interior and exterior (denoted by $int(C)$ and $ext(C)$) to be the subgraphs embedded inside and outside the cycle, both containing the cycle itself.

We use n and m to denote the number and vertices and edges in the graph. Throughout the paper all graphs are connected, undirected and have nonnegative edge weights.

Simplifying Assumptions. We will assume that the graph we work on is both triangulated and has a constant degree. This can be easily achieved in $O(n)$ time by first triangulating the dual graph with zero weight edges and then triangulating the primal graph with infinite length edges using zigzag triangulations [2].

Graph r-division. Define a *piece* $P = (V_P, E_P)$ of G to be the subgraph of G induced by a subset E_P of E. In G, the vertices of V_P adjacent to vertices in $V \setminus V_P$ are the *border vertices* of P. We will denote the set of border vertices of piece P by ∂P.

We define an *r-division* of G, to be a partition of (the edges of) G into $O(n/r)$ pieces each containing $O(r)$ vertices and $O(\sqrt{r})$ border vertices. A *hole* is a finite face of P which is not a face of G. The following theorem was shown by Italiano *et al.* [10].

Theorem 1. *For a plane n-vertex graph, an r-division in which each piece has $O(1)$ holes can be found in $O(n \log r + (n/\sqrt{r}) \log n)$ time.*

In this paper, when we talk about an r-division, we shall assume that it has the form as in Theorem 1.

For an r-division $\mathcal{P}$, define a *skeleton graph* $G_{\mathcal{P}} = (\partial \mathcal{P}, E_{\mathcal{P}})$ to be a graph over the set of border vertices in $\mathcal{P}$. The edge set $E_{\mathcal{P}}$ is composed of infinite length edges connecting *consecutive* border vertices of each hole.

Dense Distance Graphs. We use the r-division to define a representation for shortest paths in a graph that has a similar number of edges, but fewer vertices. In order to achieve this, we use the notion of dense distance graphs. If G is edge-weighted, we define the *dense distance graph* of a piece P to be the complete graph on the set of border vertices of P where each edge (u, v) has weight equal to the shortest path distance (w.r.t. the edge weights) in P between u and v. A dense distance graph for an r-division is a set of dense distance graphs of all its pieces. Observe that it contains $O(\frac{n}{\sqrt{r}})$ nodes and $O(n)$ edges.

Italiano *et al.* [10] have used an algorithm by Klein [12] to compute a dense distance graph for any division.

Lemma 1. *Given an r-division $\mathcal{P}$, its dense distance graph can be computed in $O(n \log r)$ time.*

Fast Dijkstra The dense distance graphs can be used to speed up shortest path computations using Dijkstra's algorithm. It was shown by Fakcharoenphol and Rao ([5], Section 3.2.2) that a Dijkstra-like algorithm can be executed on a dense distance graph for a piece P in $O(|\partial P| \log^2 |\partial P|)$ time. Having constructed the dense distance graphs, we can run Dijkstra in time almost proportional to the number of vertices (rather than to the number of edges, as in standard Dijkstra). We use this algorithm in graphs composed of dense distance graphs and a subset E' of edges of the original graph $G = (V, E)$:

Corollary 1. *Dijkstra can be implemented in $O(|E'| \log |V| + \sum_i |\partial G_i| \log^2 |\partial G_i|)$ time on a graph composed of a set of dense distance graphs G_i and a set of edges E' over the vertex set V.*

Proof. In order to achieve this running time we use Fakcharoenphol and Rao [5] data structure for each G_i. Moreover, minimum distance vertices from each G_i and all endpoints of edges in E' are kept in a global heap.

Max Flow Queries. Italiano *et al.* [10] have shown an $O(n^{2/3} \log^{8/3} n)$ time dynamic algorithm for computing max-flow values in planar graphs. More generally speaking, they have presented an algorithm that allows the following tradeoffs between preprocessing, update and query times.

Theorem 2. *There exists a data structure that after $O(n \log r + \frac{n}{\sqrt{r}} \log n)$ preprocessing time, supports: edge insertions and edge deletions in $O((r + \frac{n}{\sqrt{r}}) \log^2 n)$ time; s to t distance queries in $O((r + \frac{n}{\sqrt{r}}) \log^2 n)$ time; max st-flow queries in $O((r + \frac{n}{\sqrt{r}}) \log^3 n)$ time, where $r \in [1, \ldots, n]$.*

The only information maintained by the algorithm is an r-division together with dense distance graphs for all pieces. If we set $r = \log^8 n$, then the initialization takes $O(n \log \log n)$ time and the query time becomes $O((\log^8 n + \frac{n}{\log^4 n}) \log^3 n) = O(\frac{n}{\log n})$. Hence, we obtain the following static algorithm that will be very useful for us:

Corollary 2. *There exists a data structure that after $O(n \log \log n)$ preprocessing time, can compute a max st-flow value in $O(\frac{n}{\log n})$ time.*

3   Chalermsook *et al.* Algorithm

Chalermsook, Fakcharoenphol and Nanongkai [3] have shown an $O(n \log^2 n)$ algorithm (we call it CFN from now on) for finding minimum cuts in undirected, weighted planar graphs. It uses a divide and conquer approach.

Dividing Step. The algorithm computes a shortest paths tree using a linear time algorithm by Henzinger *et al.* [7]. Then it finds a cycle C that divides the graph into two parts, both containing a constant fraction of all faces. The cycle C consists of two shortest subpaths Q_{ab} and Q_{ac}, that belong to the tree, and a path bc, which goes along a boundary of a face. After that, the algorithm recursively computes the length of the shortest cycles inside and outside C. In addition, a shortest cycle that crosses C is computed.

Conquering Step. Let $Q = q_1 q_2 \ldots q_k$ be some shortest path in G, and let F_i be any face incident to q_i, for $i = 1$ and $i = k$. Let F_e be a face adjacent to bc that lies inside C and let F_a be a face from outside of C that is adjacent to the first edge of Q_{ab}. The the following lemma was stated by Chalermsook *et al.* [3].

Lemma 2. *The length of the shortest cycle is equal to the minimum of the length of (1) the shortest cycle which which separates which separates F_a from F_e, (2) the shortest cycle contained within $int(C)$ and (3) the shortest cycle contained within $ext(C)$.*

By duality of shortest cycles and minimum cuts, we can find such a shortest cycle using a single maximum flow computation. This can be done in $O(n \log \log n)$ time using a recent algorithm by Italiano *et al.* [10].

Reducing Step. In the reducing step, we remove vertices of degree 2 by merging their incident edges. As a result, all vertices have degree at least 3 and, by Euler's formula, the number of vertices is at most twice the number of faces. Moreover, each dividing step adds at most one new face, so the total number of faces in every recursion level is bounded by $O(n)$. The same bound holds for the number of vertices. There are $O(\log n)$ recursion levels and each requires $O(n \log \log n)$ time. Hence, the overall running time is $O(n \log n \log \log n)$.

4   An $O(n \log n)$ Time Algorithm

In this section we show how to obtain a faster algorithm by a simple modification of the CFN algorithm. We present an improved version of the CFN algorithm that still has $O(\log n)$ levels of recursion, but each of them will require $O(n)$ amortized time.

We now run the recursion as follows. Every $\log \log n$ levels of the recursion, in every branch of the recursion tree we reinitialize the maximum flow algorithm from Corollary 2. Over all levels of the recursion this takes $O(\frac{\log n}{\log \log n} n \log \log n) = O(n \log n)$ time. Within $\log \log n$ levels following the initialization, we issue at most $2^{\log \log n} = \log n$ maximum flow queries to each maximum flow structure. As we have observed, this requires only linear time. Hence, all maximum flow computations require $O(n \log n)$ time.

The data structure is not recomputed in every step, but only from time to time. Hence, a query for a shortest cycle separating two faces is answered using the structure for larger part of the graph. However, this does not affect the final result. When the graph has some additional vertices and edges, the length of the shortest cycle can only decrease. Moreover, every cycle we find is a valid cycle in the original graph, so the length of the shortest cycle in the whole graph is computed correctly.

The CFN algorithm runs in $O(n \log n)$ time if we exclude the time needed for maximum flow computations. Here, we have shown how to perform all maximum flow computations in $O(n \log n)$ time, thus reducing the running time of the whole algorithm from $O(n \log n \log \log n)$ to $O(n \log n)$.

5   An $O(n \log \log n)$ Time Algorithm

We show that the algorithm of Chalermsook *et al.* [3] can be implemented on *dense distance graphs*. Instead of recursing on the subgraphs, we use the skeleton graph. The dense distance graphs are kept in a global memory and referred when needed. We follow the structure of Section 3 and describe how to implement all three steps of the algorithm. However, here we stop the recursion when the

subgraph for recursion contains less than r nodes. Hence, we require a terminal
step that handles such small subgraphs at the end of the algorithm.

The first step in the CFN algorithm is building a shortest paths tree. We
also start by computing a shortest paths tree T in a dense distance graph of an
r-division. However, we require T to be *noncrossing*, which means that its every
edge can be mapped to an underlying shortest path inside one piece in such a way,
that the paths do not cross [1]. We use the linear time algorithm by Henzinger
et al. [7] for finding shortest paths in a planar graph with nonnegative edge
weights. Then for each piece of the decomposition we map the shortest subpaths
connecting border vertices to their corresponding edges in the dense distance
graph. The resulting tree is noncrossing.

The main part of the algorithm is based on divide and conquer technique.
We start by building an r-division and a skeleton graph for the given graph. We
define a *recursion graph*. This is a planar graph that is used to represent the
parts of the entire r-division that are considered in recursive calls. Initially this
is a skeleton graph. In every step the graph is divided by intersecting it with an
interior and exterior of some cycle. We represent this division in the recursion
graph by mapping every part of the cycle contained within one piece into an edge
connecting two border vertices of the piece. Then we insert each of the resulting
edges into the recursion graph. Those edges are called *division edges*. We never
add vertices to the recursion graph, so it has $O(\frac{n}{\sqrt{r}})$ vertices all the time.

A *region* is a subgraph of the recursion graph bounded with division edges.
Every region represents a part of the graph that is processed in one recursive
call (see Fig. 1).

Regions contain faces of two kinds. Some faces contain parts of the graph that
are represented by this region. We call those faces *interior faces*. The rest of faces

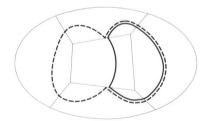 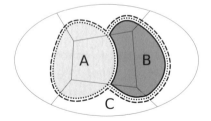

Fig. 1. The recursion graph (in the right picture) after dividing the graph along the
cycles from the left picture. Note that parts of the cycles are drawn close to each other
to mark that they overlap. The recursion graph has three regions A, B and C. The
interiors of regions are colored light gray, gray and white, respectively. Each edge is
marked with two or three strokes to represent boundaries of regions going through
this edge. In particular, the path along the boundary of region B belongs to all three
regions. This is because for any cycle C' both $int(C')$ and $ext(C')$ contain C'.

---

[1] Note that if a piece has holes, this can happen even if edges of T can be embedded
in the plane without crossing.

contains parts of the graph, that are to be processed in other branches of the recursion (e.g. the face of region A which contains region B or the outer face of A). Every face of the graph belongs to exactly one region, but edges and vertices can be in multiple regions.

Note that we can insert a division edge connecting some pair of border vertices multiple times. This means that for some region R only the shortest path connecting those two vertices belongs to R. Since we want the recursion graph to be simple, if there are multiple division edges connecting a pair of border vertices within the same piece, we merge them into a single edge. Such edges can belong to many regions.

Whenever we need to extract the distance between two border vertices from one piece, we check whether they belong to the same interior face of the region or if they are connected with a division edge (or a super edge, which is defined later). If this is the case, we return length of the appropriate edge from the dense distance graph. Otherwise, the distance is infinite.

5.1   Dividing Step

The dividing step of the CFN algorithm finds a cycle that splits the graph into two subgraphs, both containing at most $\frac{2}{3}$ of all faces. It is a fundamental cycle determined by one edge in a shortest paths tree. This is what we do as well, but we use the noncrossing shortest paths tree to do it more efficiently.

We find a cycle that is composed of two subpaths $Q_{ab'}$ and $Q_{ac'}$ of the shortest paths tree T, two shortest paths $Q'_{b'b}$ and $Q'_{c'c}$ (in the original graph) fully contained within one piece of the decomposition and an edge bc. Those paths form a cycle which cuts the original graph into two pieces, each containing a constant fraction of faces.

This is done in two phases. First, we use the skeleton graph to find the approximate location of the edge bc, i.e., the piece it belongs to. Then, we plant the piece to the corresponding face of the skeleton graph and use a similar procedure to find the exact location of the edge bc. The details follow.

We build a planar graph F by taking the union of edges of the skeleton graph $G_{\mathcal{P}}$ and T. F can contain multiple edges between some pairs of vertices. A spanning tree S_F of its dual can be constructed by taking edges corresponding to edges from the skeleton graph.

We assign weights to vertices of S_F. A vertex v of S_F corresponds to a face f from F. We define its weight as the number of faces of G which are inside f in the common embedding of G and F. Those values can be computed together with building the shortest paths tree in the beginning.

Denote by W the total weight of all vertices. The goal is to find a vertex w in S_F such that the total weight of its subtree is at least $\frac{W}{2}$ and the weight of its every child subtree is smaller than $\frac{W}{2}$. This can be achieved in linear time by walking down the tree (starting from the root) and always choosing the heaviest subtree.

Let us denote by P the piece containing the face corresponding to w. We now plant the piece P in the corresponding face of the skeleton graph. More formally speaking, we build a graph F_P in the following way. By running Dijkstra's algorithm in piece P, we extend the shortest paths tree T with paths to all vertices of P, thus obtaining a tree T_P. This requires $O(r \log r)$ time. The graph F_P consists of the skeleton graph G_P, the tree T_P and all edges of P, which do not belong to T_P.

Again, we build a spanning tree S_{F_P} of the dual graph, by taking edges that do not belong to T_P. The weight of a vertex $v \in V(S_{F_P})$ is the number of faces of G embedded inside v. Thus, total weight of S_{F_P} is equal to the total weight of S_F. In fact, S_{F_P} can be obtained from S_F by splitting some vertices.

Then we find a vertex w in the tree S_{F_P}, such that the total weight of its subtree is at least $\frac{W}{2}$ and the weight of subtrees rooted in all its children is smaller than $\frac{W}{2}$. If we apply the routine describe earlier and break ties the same way as previously, we find a vertex w, which corresponds to a face belonging to piece P. This means that the weight of w is equal to 1 and its degree is 3. It is easy to observe that by removing an appropriate edge e' incident to w, we can split the tree into two parts which contain a constant fraction of the total weight.

The edge e which is a primal edge corresponding to e' determines a fundamental cycle C_e in T_P, which splits the graph into two pieces, both containing a constant fraction of all faces of the original graph. We now want to cut the graph with this cycle and recurse into two smaller subgraphs.

It remains to show how to carry out the cutting. We describe how to do it for pieces that contain no holes. Due to space limitations the description of how to handle the general case will be given in the full version of this paper.

The cycle C_e consists of shortest subpaths between border vertices (except for the piece P). Hence, for any piece P' other than P, shortest paths both in $P' \cap int(C_e)$ and $P' \cap ext(C_e)$ do not cross C_e. This implies that the distances in both parts are preserved, so it suffices to insert one division edge that connects two respective border vertices to the recursion graph.

To cut the piece P, we have to use a different approach. We use Klein's algorithm [12] to rebuild the dense distance graphs of $P \cap int(C_e)$ and $P \cap ext(C_e)$ in $O(r \log r)$ time, as given by Corollary 1. A division edge is inserted between the border vertices of P connected by C_e.

The last step is to cut the noncrossing shortest paths tree T. Since the cut C_e does not cross any edge of T, suffices to divide it into $int(C_e) \cap T$ and $ext(C_e) \cap T$. Note that the shortest path tree $ext(C_e) \cap T$ is rooted in the vertex a, which belongs to C_e.

5.2   Conquering Step

In the previous section we have shown how to find a cycle C_e, which divides the graph into two subgraphs of roughly the same size. The cycle C_e can be mapped to a fundamental cycle determined by a single edge e in the shortest paths tree of the original graph. This means that it consists of two shortest subpaths Q_{ab} and Q_{ac} and the edge $e = bc$. By Lemma 2, to find the shortest cycle crossing C_e we need a single maximum-flow computation. Note that the cycle C_e might not be simple, i.e. Q_{ab} and Q_{ac} can share some edges in the beginning, but the lemma still holds. As discussed in the previous section, we have the r-division and dense distance graphs ready in each recursive call. This allows us to use the algorithm from Theorem 2 to answer max st-flow queries in $O((r + \frac{n}{\sqrt{r}}) \log^3 n)$ time.

5.3 Reducing Step

In the CFN algorithm in the reduction step we have removed degree 2 vertices. This was done in order to bound the total number of vertices in all branches on each level of the recursion. Here, we would like to do the same.

The total number of vertices in one level of the recursion is the total number of vertices in all regions. Some vertices can be counted many times, if they belong to multiple regions. Each face, on the other hand, is in exactly one region.

Consider a region R. Some of its vertices are incident to the faces of the recursion graph, which are interior faces of R. It follows that the number of such vertices is equal to the sum of sizes of all faces from the interior of R, so there are $O(\frac{n}{\sqrt{r}})$ such vertices in total.

R can also contain some vertices that are not incident to any face from its interior. From among those, we find vertices incident to exactly two division edges. If we take the part of the original graph corresponding to R, those vertices also have degree two. Therefore, we remove each such vertex and replace its incident edges e_1 and e_2 with a *super edge*, whose length is the sum of lengths of e_1 and e_2. Note that there can be no vertices incident to 0 or 1 division edge in R. For the analysis, let us also get rid of all other vertices of degree 2 with a a similar procedure. We obtain a planar graph R' with vertices of degree at least 3.

The degree bound implies that $|V(R')| = O(|F(R')|)$. Removing vertices of degree two does not increase the number of faces, so $|V(R')| = O(|F(R)|)$ [2]. In every recursion step, we divide one region into two regions with a single cycle. This increases the total number of faces in all regions by a constant. Since there are $O(\frac{n}{\sqrt{r}})$ recursion steps, we conclude that $|V(R')| = O(\frac{n}{\sqrt{r}})$, so there are $O(\frac{n}{\sqrt{r}})$ vertices not incident to an interior face in any region.

Corollary 3. *The total number of vertices in all regions in every recursion level is* $O(\frac{n}{\sqrt{r}})$.

Observe that the same argument allows us to bound the number of super edges with $O(\frac{n}{\sqrt{r}})$.

5.4 Terminal Step

Each recursive call for region with k vertices consists of the following steps:

1. Find a cycle C_e that, when mapped to the original graph, divides it into two parts containing constant fraction of faces. This requires $O(k + r \log r)$ time.

---

[2] Here $F(R)$ denotes the set of all faces of R, not only those from its interior.

2. Compute the length of the shortest cycle separating two given faces. This step runs in $O((r+k)\log^3 k)$.
3. Insert a division edges corresponding to C_e to the recursion graph and divide the shortest path tree.

Hence, the total running time of each call is $O((r+k)\log^3 k)$. We run the recursion as long as $r \le k$. This implies that the total cost of each step would be dominated by the summand depending on k. Since we start with a graph with $O(\frac{n}{\sqrt{r}})$ vertices, the recursion takes $O(\frac{n}{\sqrt{r}}\log^4 n)$ total time.

If in any recursive call $r > k$, we abandon the recursion and use a different approach for finding the shortest cycle within the current region R. We will refer to such recursive calls as *terminal* recursive calls.

For each region R in from a terminal recursive call we need to compute the part of the original graph, which corresponds to it. We find a graph G_R, which represents the part corresponding to R, in a compressed way. Namely, some paths composed of vertices of degree 2 are replaced with one edge. The process of finding graphs G_R is rather technical and due to space limitations will be described in the full version of this paper.

Lemma 3. *All graphs G_R can be computed in $O(n \log r)$ time. The sum of $|V(G_R)|$ over all regions R is $O(n)$.*

It remains to find the length of the shortest cycle within each G_R. In order to do that, we use the $O(n \log n)$ algorithm for computing the girth from section 4. The following lemma is necessary to bound the running time of the terminal calls.

Lemma 4. *For each R, $|V(G_R)| = O(r^2)$.*

Proof. By the definition of the terminal recursive call, $|V(R)| < r$, so G_R can contain at most r vertices that are border vertices of some pieces. All other vertices lie inside the interior faces of R. Each such face is fully contained within one piece of the r-division, so it contains at most $O(r)$ vertices. Since there are at most $O(r)$ faces in R, we conclude that there are $O(r^2)$ vertices in G_R in total.   □

From the above lemmas, it follows that the running time of the $O(n \log n)$ algorithm on all G_R is $\sum_R O(|V(G_R)| \log |V(G_R)|) \le \sum_R O(|V(G_R)| \log r^2) = O(\log r) \sum_R |V(G_R)| = O(n \log r)$.

In this way we obtain a pair of faces s and t, such that the shortest cycle separating s and t is the globally shortest cycle in the graph. In the end, we run the $O(n \log \log n)$ algorithm by Italiano *et al.* [10] to find the shortest cycle separating s and t, which computes the final result.

Thus, the whole algorithm requires $O(\frac{n}{\sqrt{r}}\log^4 n) + O(n \log r) + O(n \log \log n)$ time. Setting $r = \log^8 n$ yields an $O(n \log \log n)$ algorithm.

6 Dynamic Shortest Cycle

In this section we show how to use the ideas introduced in the previous section to construct a dynamic algorithm for finding minimum cuts in planar graphs. We show how to maintain a planar graph G with positive edge weights under an intermixed sequence of the following operations: $insert(x, y, c)$ add to G an edge of weight c between vertex x and vertex y, provided that the embedding of G does not change; $delete(x, y)$ delete from G the edge between vertex x and vertex y; $shortest\text{-}cycle$ return the length of the shortest cycle in G.

6.1 Shortest Cycles through Given Edge

In our algorithm we use a dynamic data structure that supports the following operations: $insert(x, y, c)$ add to G an edge of weight c between vertex x and vertex y, provided that the embedding of G does not change; $delete(x, y)$ delete from G the edge between vertex x and vertex y; $shortest\text{-}cycle(x, y)$ return the length of the shortest cycle that includes edge (x, y).

The existence of such dynamic algorithm is implied by Theorem 2.

Lemma 5. *Given a planar graph G with positive edge weights, we can insert edges, delete edges and report the length of a shortest cycle that includes given edge in $O(n^{2/3} \log^{5/3} n)$ worst-case time per operation.*

Proof. For supporting updates we simply use Theorem 2 for $r = n^{2/3} \log^{2/3} n$. When we need to answer a query for an edge (x, y) we: delete edge (x, y); ask for the shortest distance from x to y – this plus the length of (x, y) is the shortest cycle length; reinsert edge (x, y). Hence, the answer to a $shortest\text{-}cycle(x, y)$ query can be computed using three operations of the original data structure. □

6.2 Data Structures and Updates

In our dynamic algorithm we maintain two data structures from Theorem 2: structure A for $r = n^{1/3}$; structure B for $r = n^{2/3} \log^{2/3} n$.

Initialization of both structures requires $O(n \log n)$ time. Additionally, in the beginning we compute the length of the shortest cycle fully contained within each piece used by structure A. This also runs in $O(n \log n)$ time.

The edge updates are handled using Theorem 2, i.e., for each piece in the decomposition we maintain the dense distance graph. Additionally, in structure A we find a shortest cycle contained fully inside the piece, which the inserted or deleted edge belongs to. This does not increase running time of the algorithm even when we use the $O(n \log n)$ time algorithm from Section 4. The update time is then $O(n^{5/6} \log^2 n)$.

Our algorithm for answering queries follows the lines of the $O(n \log \log n)$ algorithm. However, instead of using the r-division for $r = \log^8 n$ we use the r-division given by structure A for $r = n^{1/3}$. In previous section we have essentially shown that for polylogarithmic r we obtained polylogarithmic speed up for the

CFN algorithm. By taking polynomial r we are able to obtain a polynomial speed up of CFN. For $r = n^{1/3}$ the speed up will be by a factor of $n^{1/6}$ and the running time of our algorithm will be $O(n^{5/6} \log^3 n)$. Nevertheless there are some technical differences between the static and dynamic algorithm which will be given in the full version of this paper. The main difference is that we might need to divide the graph using cycles defined by several non-tree edges. In order to find shortest cycles crossing many non-tree edges we use Lemma 5 applied to structure B.

References

1. Alon, N., Yuster, R., Zwick, U.: Finding and counting given length cycles. Algorithmica 17, 209–223 (1997)
2. Borradaile, G., Sankowski, P., Wulff-Nilsen, C.: Min st-cut oracle for planar graphs with near-linear preprocessing time. In: FOCS, pp. 601–610. IEEE Computer Society, Los Alamitos (2010)
3. Chalermsook, P., Fakcharoenphol, J., Nanongkai, D.: A deterministic near-linear time algorithm for finding minimum cuts in planar graphs. In: Ian Munro, J. (ed.) SODA, pp. 828–829. SIAM, Philadelphia (2004)
4. Eppstein, D.: Subgraph isomorphism in planar graphs and related problems. Journal of Graph Algorithms and Applications 3(3), 1–27 (1999)
5. Fakcharoenphol, J., Rao, S.: Planar graphs, negative weight edges, shortest paths, and near linear time. J. Comput. Syst. Sci. 72(5), 868–889 (2006)
6. Frederickson, G.N.: Fast algorithms for shortest paths in planar graphs, with applications. SIAM J. Comput. 16(6), 1004–1022 (1987)
7. Henzinger, M.R., Klein, P.N., Rao, S., Subramanian, S.: Faster shortest-path algorithms for planar graphs. J. Comput. Syst. Sci. 55(1), 3–23 (1997)
8. Lu, H.-I., Chang, H.-C.: Subquadratic algorithm for dynamic shortest distances. In: COCOON (to appear, 2011)
9. Itai, A., Rodeh, M.: Finding a minimum circuit in a graph. SIAM J. Comput. 7(4), 413–423 (1978)
10. Italiano, G.F., Nussbaum, Y., Sankowski, P., Wulff-Nilsen, C.: Improved Min Cuts and Max Flows in Undirected Planar Graphs. In: STOC (2011)
11. Karger, D.R.: Minimum cuts in near-linear time. Journal of the ACM (JACM) 47(1), 46–76 (2000)
12. Klein, P.N.: Multiple-source shortest paths in planar graphs. In: SODA, pp. 146–155. SIAM, Philadelphia (2005)
13. Nagamochi, H., Ibaraki, T.: Computing edge-connectivity in multiple and capacitated graphs. In: Asano, T., Imai, H., Ibaraki, T., Nishizeki, T. (eds.) SIGAL 1990. LNCS, vol. 450, pp. 12–20. Springer, Heidelberg (1990)
14. Thorup, M.: Fully-dynamic min-cut. In: Proceedings of the Thirty-Third Annual ACM Symposium on Theory of Computing, STOC 2001, pp. 224–230. ACM, New York (2001)
15. Yuster, R., Zwick, U.: Finding even cycles even faster. SIAM J. Discrete Math. 10(2), 209–222 (1997)

Near-Popular Matchings in the Roommates Problem

Chien-Chung Huang[1],[*] and Telikepalli Kavitha[2]

[1] Humboldt-Universität zu Berlin, Germany
`villars@informatik.hu-berlin.de`
[2] Tata Institute of Fundamental Research, India
`kavitha@tcs.tifr.res.in`

Abstract. Our input is a graph $G = (V, E)$ where each vertex ranks its neighbors in a strict order of preference. The problem is to compute a matching in G that captures the preferences of the vertices in a *popular* way. Matching M is more popular than matching M' if the number of vertices that prefer M to M' is more than those that prefer M' to M. The *unpopularity factor* of M measures by what factor any matching can be more popular than M. We show that G always admits a matching whose unpopularity factor is $O(\log |V|)$ and such a matching can be computed in linear time. In our problem the optimal matching would be a *least* unpopularity factor matching - we show that computing such a matching is NP-hard. In fact, for any $\epsilon > 0$, it is NP-hard to compute a matching whose unpopularity factor is at most $4/3 - \epsilon$ of the optimal.

1 Introduction

Our input is a graph $G = (V, E)$ where each $u \in V$ ranks its neighbors in a strict order of preference. Preference lists can be incomplete, i.e., a vertex need not be adjacent to all the other vertices. Each vertex seeks to be matched to a neighbor and for any $u \in V$, if u ranks v higher than v' in its preference list, then u prefers v as its partner to v'. This is the same as an instance of the *roommates problem* with incomplete lists, which is a generalization of the *stable marriage problem*.

In the stable marriage problem, the input graph is bipartite and the problem is to compute a *stable* matching M, i.e., one where there is no edge (a, b) where both a and b prefer each other to their respective assignments in M. While every bipartite graph admits a stable matching, it is easy to come up with roommates instances that admit no stable matchings. Our problem is to compute a matching in G that captures the preferences of the vertices *optimally*, where we need to come up with a definition of optimality so that such a matching always exists.

Popular Matchings. Popularity is a weaker notion than stability, in other words, every stable matching is popular but not vice-versa [2]. We define popular matchings below. For any two matchings M and M', we say that vertex u prefers

---

[*] Supported by an Alexander von Humboldt fellowship.

C. Demetrescu and M.M. Halldórsson (Eds.): ESA 2011, LNCS 6942, pp. 167–179, 2011.
© Springer-Verlag Berlin Heidelberg 2011

M to M' if u is *better off* in M than in M' (i.e., u is either matched in M and unmatched in M' or matched in both and prefers $M(u)$ to $M'(u)$).

Let $\phi(M, M')$ = the number of vertices that prefer M to M'. We say that M' is more popular than M if $\phi(M', M) > \phi(M, M')$.

Definition 1. *A matching M is popular if there is no matching that is more popular than M, i.e., $\phi(M, M') \geq \phi(M', M)$ for all matchings M' in G.*

Since there is no matching where more vertices are better-off than in a popular matching, a popular matching is a natural candidate for an optimal matching. A simple instance from [2] that admits no stable matching but has popular matchings is the following: consider the complete graph H on 4 vertices $\{a_0, a_1, a_2, a_3\}$ where for $i = 0, 1, 2$, a_i prefers a_{i+1} to a_{i+2} (all indices are mod 3). The fourth vertex a_3 is at the tail of the preference lists of a_0, a_1, and a_2. The vertex a_3's preference list is a_0 followed by a_1, and then a_2. It is easy to check that this instance has no stable matching, however it admits 2 popular matchings: $M_1 = \{(a_0, a_3), (a_1, a_2)\}$ and $M_2 = \{(a_1, a_3), (a_0, a_2)\}$. But popular matchings also do not always exist: the above instance H with the fourth vertex a_3 removed, has no popular matching. In any instance G, let us measure by what factor one matching (say, M_1) can be more popular than another (say, M_0) as follows:

$$\Delta(M_0, M_1) = \begin{cases} \frac{\phi(M_1, M_0)}{\phi(M_0, M_1)} & \text{if } \phi(M_0, M_1) \neq 0 \\ \infty & \text{otherwise.} \end{cases}$$

Let $\mathcal{M}$ denote the set of all matchings in G. The *unpopularity factor* of M is $u(M) = \max_{M' \in \mathcal{M} \setminus \{M\}} \Delta(M, M')$. A matching M with a low value of $u(M)$ can be considered a good matching because $\phi(M', M) \leq u(M) \cdot \phi(M, M')$ for all M'; hence when comparing M with any matching M', the number of vertices that prefer M' to M cannot be larger than the number of other vertices by a factor of more than $u(M)$. The matching M with the least value of $u(M)$ among all the matchings in G is called a *least unpopularity factor* matching.

When G admits popular matchings, it is easy to see that every popular matching M is a least unpopularity factor matching. However unlike popular matchings, least unpopularity factor matchings always exist. Hence, a least unpopularity matching is a promising candidate for an optimal matching in G. However finding such a matching is APX-hard, as shown by our following result.

Theorem 1. *It is NP-hard to find a least unpopularity factor matching in a roommates instance $G = (V, E)$. In fact, for any $\epsilon > 0$, it is NP-hard to compute a matching whose unpopularity factor is at most $4/3 - \epsilon$ of the optimal. These hardness results hold even in the special case when G is a complete graph.*

Nevertheless, there is always a matching whose unpopularity factor is $O(\log n)$ and this can be computed efficiently, as shown by our following result.

Theorem 2. *Let G be a roommates instance on n vertices and m edges. Then G always admits a matching whose unpopularity factor is at most $4 \log n + O(1)$ and such a matching can be computed in $O(m + n)$ time.*

Thus we propose a solution that always exists and is efficiently computable for the problem of finding a good matching in a roommates instance. We also show an instance $G = (V, E)$ where every matching has unpopularity factor $\Omega(\log |V|)$, hence the upper bound in Theorem 2 cannot be improved to $o(\log n)$.

Popular matchings have been well-studied during the last few years [1, 6, 8, 9, 10, 11, 12] in bipartite graphs where only vertices on one side (called *applicants*) have preferences while vertices on the other side (called *jobs*) have no preferences. So when we compare two matchings wrt popularity, it is only applicants that cast their votes. This is called the one-sided preference lists model and popular matchings need not always exist here. Also, there exist simple instances here where every matching has unpopularity factor $\Omega(n)$: for instance, $A = \{a_1, \ldots, a_n\}$ and $B = \{b_1, \ldots, b_n\}$ where each $a \in A$ has the same preference list, which is, b_1 followed by b_2 followed by b_3 and so on till b_n as the last choice - every perfect matching here has unpopularity factor $n - 1$ (and non-perfect matchings have unpopularity factor ∞). Thus the existence of an $O(\log n)$ unpopularity factor matching in the roommates problem is surprising. Section 2 has this result.

Gale and Shapley [3] introduced the stable marriage problem and Irving [7] showed an efficient algorithm for determining if a roommates problem admits a stable matching or not and if so, to compute one. Gärdenfors [4] introduced the notion of popularity in the stable marriage problem. When ties are allowed in preference lists here, it has been shown by Biró, Irving, and Manlove [2] that the problem of computing an arbitrary popular matching is NP-hard. For one-sided preference lists, there are efficient algorithms known for determining if the input instance admits a popular matching or not, and if so, to compute one [1]. However, McCutchen [11] showed that the problem of computing a least unpopularity factor matching in the domain of one-sided preference lists is NP-hard. In bipartite graphs with two-sided strict preference lists, this problem becomes easy due to the existence of stable matchings. We show that the hardness returns when we generalize to non-bipartite graphs. Section 3 has this result.

2   A Low Unpopularity Factor Matching

Preliminaries. Let $G = (V, E)$ be an instance of the roommates problem. No matter whether G admits a stable matching or not, Tan [13] showed that G always admits a *stable partition*, which generalizes the notion of a stable matching (which can be regarded as a partition into pairs and perhaps, singletons).

A stable partition is a partition $\{A_1, \ldots, A_k\}$ of the vertex set V, where each A_i is an *ordered set* $\langle a_i^0, \ldots, a_i^\ell \rangle$. We call a_i^{j-1} the *predecessor* of a_i^j and for every i and j, if $a_i^{j+1} \neq a_i^{j-1}$, then a_i^j prefers a_i^{j+1} (its *successor*) to a_i^{j-1}, where all superscripts are modulo $|A_i|$. When $A_i = \langle a_i^0 \rangle$, i.e., if A_i is a singleton set, let the predecessor of a_i^0 be $\emptyset$. Note that any cyclic permutation of A_i is also an ordered set. For every edge (a, b) in G, the following stable condition holds:

(∗) If a prefers b to its predecessor, then b does not prefer a to its predecessor.

Note that by this condition, if $A_i = \langle a_i^0 \rangle$, then no vertex adjacent to a_i^0 can belong to a singleton set as such a vertex prefers its predecessor to a_i^0. By this observation, for any $(a, b) \in E$, either a is b's predecessor/vice-versa or one of $\{a, b\}$ strictly prefers its predecessor to the other. A stable partition for the graph H on $\{a_0, \ldots, a_3\}$ described in Section 1 is $\{\langle a_0, a_1, a_2 \rangle, \langle a_3 \rangle\}$.

We will use the following notation: for any vertex u and neighbors v and w of u, $\mathsf{vote}_u(v, w)$ is 1 if u prefers v to w, it is -1 if u prefers w to v, and it is 0 otherwise (i.e., $v = w$). Also, if matching M leaves u unmatched, then $\mathsf{vote}_u(v, M(u)) = 1$ where v is any neighbor of u.

2.1 Our Algorithm

We present an algorithm for finding an $O(\log|V|)$ unpopularity factor matching in $G = (V, E)$. Input is the graph $G_0 = G$ (so the vertex set V_0 refers to V).

1. Let $\mathcal{P}_0$ be a stable partition $\{A_1, \ldots, A_k\}$ of G_0.
   1.1 Set $X_0 = \cup_{i=1}^{k} \{a_i^{2t-1} : t = 1, \ldots, \lceil |A_i|/2 \rceil\}$, that is, X_0 is the set of all *odd* indexed vertices in all the ordered sets in the partition $\mathcal{P}_0$.
       {*recall that vertices in the ordered set A_i are indexed $\langle a_i^0, a_i^1, \ldots, a_i^{|A_i|-1} \rangle$*}
   1.2 Run the Gale-Shapley algorithm on $(X_0, V_0 \setminus X_0)$.
       {*vertices of X_0 propose to those in $V_0 \setminus X_0$ and vertices of $V_0 \setminus X_0$ dispose*}
       Let M_0 denote the resulting matching and let Y_0 denote the set of matched vertices in $V_0 \setminus X_0$.
2. Let V_1 denote the set of *unmatched* vertices in $V_0 \setminus X_0$. In other words, $V_1 = V_0 \setminus (X_0 \cup Y_0)$.
   2.1 If $V_1 = \emptyset$ or it is an independent set, then return $S = M_0$.
   2.2 Let G_1 be the induced subgraph on V_1. Let S' be the matching returned by running our algorithm recursively on G_1. Return $S = M_0 \cup S'$.

Note that the Gale-Shapley algorithm is described in the proof of Lemma 1. In our algorithm above, just as V_0 was partitioned into X_0, Y_0, V_1, in the recursive call for G_1, the vertex set V_1 gets partitioned into X_1, Y_1, V_2, and in the recursive call for G_2 (the induced subgraph on V_2), the vertex set V_2 gets partitioned into X_2, Y_2, V_3, and so on till V_{r+1} is empty or an independent set, where r is the recursion depth in our algorithm. Let $X = X_0 \cup X_1 \cup \cdots \cup X_r$ and let $Y = Y_0 \cup Y_1 \cup \cdots \cup Y_r \cup V_{r+1}$. We will show that the following properties hold:

(I) For each $0 \leq i \leq r$, every vertex in X_i is matched to a vertex in Y_i and vice-versa. Every unmatched vertex has to be in V_{r+1}.
(II) If we label each edge $(u, v) \in E \setminus S$ by (α, β) where α is $\mathsf{vote}_u(v, S(u))$ and β is $\mathsf{vote}_v(u, S(v))$, then
    - there is no $(1, 1)$ edge between any two vertices in X, and
    - there is no $(1, 1)$ edge between any $x \in X_i$ and any $y \in Y_i \cup Y_{i+1} \cup \cdots \cup Y_r \cup V_{r+1}$.

Lemma 1. *For each i, every $x \in X_i$ gets matched in S to a vertex that x considers at least as good as its predecessor in the stable partition $\mathcal{P}_i$ of G_i.*

Proof. Recall the Gale-Shapley algorithm on the edge set restricted to $X_i \times (V_i \setminus X_i)$ when vertices of X_i propose and vertices of $V_i \setminus X_i$ dispose. While there exists a vertex $x \in X_i$ that is not yet matched (we assume that all the unmatched vertices of X_i are placed in a queue), x is removed from the queue and it proposes to its most preferred neighbor y in $V_i \setminus X_i$ that has not yet rejected x. If y is unmatched or prefers x to its current neighbor z, then x gets matched to y (in the latter case, z gets rejected by y), else x is rejected by y. The rejected vertex gets added to the queue of unmatched vertices. This goes on till every $x \in X_i$ is either matched or has been rejected by all its neighbors in $V_i \setminus X_i$.

For each $x \in X_i$, let $p_i(x)$ denote the predecessor of x in the stable partition $\mathcal{P}_i$ of G_i. We need to show that each $x \in X_i$ gets matched to a vertex y that x considers at least as good as $p_i(x)$. Suppose not. Let $z \in X_i$ be the first vertex in the Gale-Shapley algorithm that gets rejected by $p_i(z)$. That is, the vertex $p_i(z)$ is matched to a vertex w that $p_i(z)$ prefers to z. Since z was the *first* vertex that is rejected by its predecessor in the Gale-Shapley algorithm, it has to be the case that w prefers $p_i(z)$ to $p_i(w)$. Also, $p_i(z)$ prefers w to its predecessor (since $p_i(z)$ prefers w to its successor z) - this contradicts the stable property $(*)$ stated at the beginning of Section 2. Hence there cannot be any vertex z during the entire course of the Gale-Shapley algorithm that gets rejected by its predecessor. The lemma follows.   □

It follows from Lemma 1 that every $x \in X_i$ is matched in S. For each $0 \leq i \leq r$, let M_i be the matching S restricted to $X_i \cup Y_i$. Since $Y_i \subseteq V \setminus X_i$ is the set of matched vertices in M_i, we can conclude the following corollary.

Corollary 1. *For each $0 \leq i \leq r$, M_i is a perfect matching on $X_i \cup Y_i$.*

Hence it follows that every unmatched vertex has to be in $V \setminus (\cup_{i=0}^r X_i \cup_{i=0}^r Y_i)$, which is V_{r+1}. Thus we have proved property (I). Now we show property (II).

Lemma 2. *With respect to the matching S,*

(1) there is no $(1,1)$ edge between any pair of vertices in X, and
(2) for every $0 \leq i \leq r$, there is no $(1,1)$ edge between a vertex $x \in X_i$ and a vertex $y \in Y_i \cup Y_{i+1} \cup \cdots \cup Y_r \cup V_{r+1}$.

Proof. Lemma 1 tells us that each vertex $x \in X_0$ gets matched to a vertex that is at least as good as its predecessor in $\mathcal{P}_0$. So property $(*)$ of a stable partition implies that there can be no $(1,1)$ edge between any two vertices of X_0. Since we run Gale-Shapley algorithm between $(X_0, V \setminus X_0)$, there can be no $(1,1)$ edge between a vertex of X_0 and a vertex of $V \setminus X_0$. Thus it follows that there can be no $(1,1)$ edge between an $x_0 \in X_0$ and any vertex of V. Hence there is no $(1,1)$ edge between an $x_0 \in X_0$ and any $x \in X$; also there is no $(1,1)$ edge between an $x_0 \in X_0$ and a vertex $y \in Y_0 \cup \cdots \cup Y_r \cup V_{r+1}$.

Applying the same argument in G_1, we see that there is no $(1,1)$ edge between an $x_1 \in X_1$ and any vertex of V_1, i.e., there is no $(1,1)$ edge between an $x_1 \in X_1$

and any vertex of $X_1 \cup \cdots X_r$; also there is no $(1,1)$ edge between an $x_1 \in X_1$ and a vertex $y \in Y_1 \cup \cdots \cup Y_r \cup V_{r+1}$.

Continuing this argument, we see that there is no $(1,1)$ edge between an $x_i \in X_i$ and any vertex of $X_i \cup \cdots X_r$; also there is no $(1,1)$ edge between an $x_i \in X_i$ and a vertex $y \in Y_i \cup \cdots \cup Y_r \cup V_{r+1}$.

Thus there is no $(1,1)$ edge between any pair of vertices in X and for every $0 \le i \le r$, there is no $(1,1)$ edge between a vertex $x \in X_i$ and a vertex $y \in Y_i \cup Y_{i+1} \cup \cdots \cup Y_r \cup V_{r+1}$.   □

The properties of the matching S (as given by Lemmas 1 and 2) enable us to show Lemma 3. We say an alternating path/cycle ρ wrt S has k *consecutive* $(1,1)$ edges if ρ has a subpath v_0-v_1-$S(v_1)$-$v_2 \cdots S(v_{k-1})$-v_k where every unmatched edge (note that there are k of them) is a $(1,1)$ edge.

Lemma 3. *There can be at most $2r+1$ consecutive $(1,1)$ edges in any alternating path/cycle wrt S. No alternating cycle can consist solely of $(1,1)$ edges.*

Proof. Let ρ be an alternating path wrt S and let ρ' be the longest subpath of consecutive $(1,1)$ edges in ρ. Since only edges of $E \setminus S$ get labeled by the votes of their endpoints, we can assume that the first edge of ρ' is an unmatched edge. Let u_0 be an endpoint of ρ'. There are two cases: (i) $u_0 \in X$, (ii) $u_0 \in Y$.

Case (i). Let u_0 be in X_i. Since every unmatched edge of ρ' is marked $(1,1)$, it has to be the case that the vertex that follows u_0 in ρ', call this vertex v_0, has to be in $Y_{i-1} \cup Y_{i-2} \cup \cdots \cup Y_0$, since there are no $(1,1)$ edges in $X \times (X \cup Y_i \cup \cdots \cup V_{r+1})$. Suppose $v_0 \in Y_j$, where $0 \le j \le i-1$. Then $S(v_0) \in X_j$. Let u_1 be the vertex $S(v_0)$. It follows from the same argument that the vertex after u_1 in ρ', call this vertex v_1, has to be in $Y_{j-1} \cup Y_{j-2} \cup \cdots \cup Y_0$. Suppose $v_1 \in Y_k$, where $0 \le k \le j-1$. Then $S(v_1) \in X_k$. However we cannot continue in this manner for more than r edges since we will be at a vertex in X_0 after r such edges and there are no $(1,1)$ edges incident on any vertex in X_0. Thus ρ' has at most r consecutive $(1,1)$ edges in this case.

Case (ii). Let u_0 be in Y_i. The vertex v_0 succeeding u_0 in ρ' can be either in X or in Y. If $v_0 \in Y$, then $u_1 = S(v_0)$ has to be in X and this becomes exactly the same as Case (i) and so there can be at most r consecutive $(1,1)$ edges after u_1. So let us assume that $v_0 \in X$. Since the edge (u_0, v_0) is labeled $(1,1)$ and there are no $(1,1)$ edges between Y_i and $X_i \cup X_{i-1} \cup \cdots \cup X_0$, the vertex v_0 has to be in X_j, where $j \ge i+1$. Hence $u_1 = S(v_0)$ is in Y_j. Again, the vertex v_1 that follows u_1 in ρ' is either in X or in Y. If $v_1 \in Y$, then this again becomes exactly the same as Case (i). Hence we assume that $v_1 \in X$. Since (u_1, v_1) is labeled $(1,1)$, it follows that v_1 has to be in X_k, where $k \ge j+1$. We cannot see more than r such $(1,1)$ edges of $Y \times X$ in ρ', since after seeing r such edges, we reach a vertex in Y_r (call this vertex u_ℓ) and there are no $(1,1)$ edges between any vertex in Y_r and a vertex in X. Hence the vertex v_ℓ that follows u_ℓ is also in Y. Then the vertex $u_{\ell+1} = S(v_\ell)$ is in X and the same argument as in Case (i) goes through, and so we have at most r consecutive $(1,1)$ edges in ρ' after we reach $u_{\ell+1} \in X$. Thus the total number of $(1,1)$ edges in ρ' is at most r (from u_0 to u_ℓ) + 1 (for the edge (u_ℓ, v_ℓ)) + r (from $u_{\ell+1}$ onwards), which is $2r+1$.

Let ρ be an alternating cycle. First, we prove that not every non-matching edge in ρ can be a $(1,1)$ edge. Pick any vertex $u_0 \in X$ as our starting vertex in ρ. The same argument as in Case (i) holds and if we traverse r consecutive $(1,1)$ edges starting from u_0, then we have to be at a vertex u_ℓ in X_0. As there are no $(1,1)$ edges incident on any vertex in X_0, we have to see an edge in $\{(-1,1),(1,-1),(-1,-1)\}$ after reaching u_ℓ. Thus there cannot be an alternating cycle with only $(1,1)$ edges. Let ρ' be a subpath of ρ that consists of only $(1,1)$ edges. Now the same proof as above (when ρ was an alternating path) holds here too: thus the total number of $(1,1)$ edges in ρ' is at most $2r+1$.   □

Lemma 3 leads to Lemma 4 that bounds $u(S)$ in terms of r. Lemma 5 bounds r in terms of $|V|$. The proof of Lemma 4 is given in the full version.

Lemma 4. $u(S) \leq 4r + O(1)$.

Lemma 5. *The recursion depth in our algorithm, i.e., r, is at most $\log |V|$.*

Proof. For any i, where $0 \leq i \leq r$, let $\mathcal{P}_i$ be the stable partition of the graph G_i computed in our algorithm. Let o_i be the number of odd cardinality ordered sets in $\mathcal{P}_i$. Since $|M_i| = 2|X_i|$ where X_i includes exactly $\lfloor |A_j|/2 \rfloor$ vertices from each ordered set A_j in $\mathcal{P}_i$, it follows that the number of unmatched vertices in M_i is o_i. That is, $|V_{i+1}| = o_i$.

The set V_{i+1} includes all vertices that formed singleton sets in $\mathcal{P}_i$, let U_i denote this set of vertices. From the definition of a stable partition, it follows that U_i has to form an independent set in G_i. Thus the size of a minimum cardinality vertex cover in G_{i+1} is at most $o_i - |U_i|$, which is the number of odd cardinality ordered sets of size ≥ 3 in $\mathcal{P}_i$.

Let C_i be a vertex cover of G_i. Since C_i has to include at least 2 vertices from every odd cardinality ordered set of size ≥ 3 in $\mathcal{P}_i$, we have $|C_i| \geq 2(o_i - |U_i|)$. Thus $c_i \geq 2c_{i+1}$, where c_i (similarly, c_{i+1}) is the size of a minimum cardinality vertex cover of G_i (resp., G_{i+1}). This inequality holds for every $0 \leq i \leq r$. Since the edge set of G_r is non-empty, $c_r \geq 1$. Thus we get $r \leq \log c_0 \leq \log |V|$.   □

It follows from Lemmas 4 and 5 that $u(S) \leq 4 \log |V| + O(1)$. We now bound the running time of our algorithm. Gale-Shapley algorithm takes time linear in the size of the graph [3] and the time taken for computing a stable partition is also linear in the graph size [13]. Thus the time taken in the i-th level of recursion to compute $\mathcal{P}_i$, X_i, M_i, Y_i, and come up with the graph G_{i+1} for the next level of recursion is linear in the size of G_i. Before computing $\mathcal{P}_i$, we delete all the isolated vertices in G_i. Hence the total time taken by our algorithm is $O(n + \sum_{i=0}^{r} (\text{number of edges in } G_i))$, which is $O(n + m \log n)$, where $n = |V|$ and $m = |E|$. In the full version, we show how to modify the algorithm so that it takes $O(n + m)$ time This concludes the proof of Theorem 2 stated in Section 1.

In the full version, we also exhibit an example $G = (V, E)$ with $|V| = 3^k$ where every matching has unpopularity factor at least $2k = 2 \log_3 |V|$. Thus it is not possible to have an $o(\log |V|)$ bound in Theorem 2.

3   Least Unpopularity Factor Matching

In this section, we prove Theorem 1 by presenting a reduction from 1-IN-3 SAT. In 1-IN-3 SAT, a formula $\phi = C_1 \wedge C_2 \wedge \cdots \wedge C_m$ in conjunctive normal form is given, where each clause $C_j = x_1^j \vee x_2^j \vee x_3^j$ is a disjunction of three non-negated literals. The formula ϕ is satisfiable iff there exists an assignment where exactly one literal in each clause is set to true. This decision problem is NP-complete [5].

Our reduction will construct a roommates instance G_ϕ such that if ϕ is a *yes* instance for 1-IN-3 SAT, then there exists a matching M with $u(M) \leq 1.5$, and if ϕ is a *no* instance, then every matching M has $u(M) \geq 2$. Also, G_ϕ will be a complete graph.

To avoid confusion, we use the upper case X_i to refer to a variable, while the lower case x_t^j means the t-th literal in clause C_j. For instance, if $C_j = (X_1 \vee X_5 \vee X_{10})$, then $x_1^j = X_1$, $x_2^j = X_5$, and $x_3^j = X_{10}$.

We have two types of gadgets: variable gadget and clause gadget. For each variable X_i, we create 16 vertices that form a variable gadget and for each clause C_j, we create 20 vertices that form a clause gadget. Note that since the preferences are complete and the number of vertices is even, if a matching is not perfect, then its unpopularity factor will be ∞. In the following discussion, we implicitly assume that a matching is perfect.

Variable Gadget. For each variable X_i, we create 16 vertices: $a_1, \ldots, a_7$ and $b_1, \ldots, b_9$. Their preference lists are shown in Table 1. The function $\pi(\cdot)$ is an arbitrary permutation of some vertices from clause gadgets and we will explain who they are later. The "$\cdots$" is an arbitrary permutation of all the remaining vertices.

Table 1. The preference lists of the vertices in the variable gadget for X_i

Vertex	Preference List	Vertex	Preference List
a_1^i	$a_2^i\ b_1^i\ a_7^i\ \cdots$	b_1^i	$b_2^i\ b_9^i\ \pi(b_1^i)\ a_1^i\ \cdots$
a_2^i	$a_3^i\ a_1^i\ \cdots$	b_2^i	$b_3^i\ b_1^i\ \cdots$
a_3^i	$a_4^i\ \pi(a_3^i)\ a_2^i\ \cdots$	b_3^i	$b_4^i\ b_2^i\ \cdots$
a_4^i	$a_5^i\ a_3^i\ \cdots$	b_4^i	$b_5^i\ b_3^i\ \cdots$
a_5^i	$a_6^i\ b_9^i\ a_4^i\ \cdots$	b_5^i	$b_6^i\ b_4^i\ \cdots$
a_6^i	$a_7^i\ a_5^i\ \cdots$	b_6^i	$b_7^i\ b_5^i\ \cdots$
a_7^i	$a_1^i\ \pi(a_7^i)\ a_6^i\ \cdots$	b_7^i	$b_8^i\ b_6^i\ \cdots$
		b_8^i	$b_9^i\ b_7^i\ \cdots$
		b_9^i	$b_1^i\ b_8^i\ \pi(b_9^i)\ a_5^i\ \cdots$

We will define the function $\pi(\cdot)$ in such as way so that the following holds.

Proposition 1. *In any matching M in G_ϕ, if $u(M) < 2$, then for every i:*

(i) b_k^i is not matched to any vertex in $\pi(b_k^i)$ for $k = 1$ and $k = 9$.
(ii) None of the vertices in $\{a_t^i\}_{t=1}^7 \cup \{b_t^i\}_{t=1}^9$ is matched to any vertex in the "$\cdots$" part of their preference lists.

We will show Proposition 1 holds after we finish the description of G_ϕ. For now, we assume it is true and show Lemma 6.

Lemma 6. *If there is a matching M with $u(M) < 2$, then the vertices corresponding to variable X_i can only be matched in one of the following two ways.*

(i) $(a_1^i, b_1^i), (a_2^i, a_3^i), (a_4^i, a_5^i), (a_6^i, a_7^i), (b_2^i, b_3^i), (b_4^i, b_5^i), (b_6^i, b_7^i), (b_8^i, b_9^i) \in M$ —*in this case, we say the variable X_i is set to true.*

(ii) $(a_1^i, a_2^i), (a_3^i, a_4^i), (a_5^i, b_9^i), (a_6^i, a_7^i), (b_1^i, b_2^i), (b_3^i, b_4^i), (b_5^i, b_6^i), (b_7^i, b_8^i) \in M$ —*in this case, we say the variable X_i is set to false.*

Proof. We cannot have M match all the vertices in $\{b_1^i, \ldots, b_9^i\}$ among themselves, since there are an odd number of these vertices. Since $u(M) < 2$, by Proposition 1, it follows that either b_1^i is matched to a_1^i, or b_9^i is matched to a_5^i, but not both, otherwise, some vertices in $\{b_t^i\}_{t=2}^8$ would have to be matched to the vertices in the "$\cdots$" part in their lists, contradicting Proposition 1.

Now if $(a_1^i, b_1^i) \in M$, it is easy to see that (i) is the only possible way to match these vertices so as to maintain the property that $u(M) < 2$. The same applies to (ii) if $(a_5^i, b_9^i) \in M$.   □

Clause Gadget. For each clause $C_j = x_1^j \vee x_2^j \vee x_3^j$, we create 20 vertices: $c_1^j, c_2^j, c_3^j, d_1^j, \ldots, d_{17}^j$. The preference list for 14 of the vertices is given below:

$$d_t^j : \quad d_{t+1}^j \quad d_{t-1}^j \quad \cdots \qquad \text{for } t \in \{2, \ldots, 8\} \cup \{10, \ldots, 16\}.$$

The "$\cdots$" is an arbitrary permutation of those remaining vertices not explicitly listed. The preference lists of the other vertices are shown in Table 2. As before, $\pi(\cdot)$ stands for an arbitrary permutation of some vertices from variable gadgets and we will explain who they are later.

Table 2. The preference lists of six vertices in the clause gadget for $C_j = x_1^j \vee x_2^j \vee x_3^j$

Vertex	Preference List	Vertex	Preference List
c_1^j	$c_2^j \ d_1^j \ c_3^j \ \pi(c_1^j) \ \cdots$	d_1^j	$d_2^j \ d_{17}^j \ \pi(d_1^j) \ c_1^j \ \cdots$
c_2^j	$c_3^j \ d_9^j \ c_1^j \ \pi(c_2^j) \ \cdots$	d_9^j	$d_{10}^j \ d_8^j \ \pi(d_9^j) \ c_2^j \ \cdots$
c_3^j	$c_1^j \ d_{17}^j \ c_2^j \ \pi(c_3^j) \ \cdots$	d_{17}^j	$d_1^j \ d_{16}^j \ \pi(d_{17}^j) \ c_3^j \ \cdots$

We will define the function $\pi(\cdot)$ in such as way so that the following holds.

Proposition 2. *In any matching M in G_ϕ, if $u(M) < 2$, then for every j:*

(i) d_k^j *is not matched to any vertex in $\pi(d_k^j)$ for $k = 1, 9$, and 17.*

(ii) *None of the vertices in $\{c_t^j\}_{t=1}^3 \cup \{d_t^j\}_{t=1}^{17}$ is matched to any vertex in the "$\cdots$" part of their preference lists.*

We will show Proposition 2 holds after we finish the description of G_ϕ. For now, we assume it is true and show Lemma 7.

Lemma 7. *If there is a matching M such that $u(M) < 2$, then the vertices corresponding to clause $C_j = x_1^j \vee x_2^j \vee x_3^j$ can only be matched in one of the following three ways:*

(i) *(c_1^j, d_1^j), (c_2^j, c_3^j), (d_{2k}^j, d_{2k+1}^j), for $1 \leq k \leq 8$, are in M —in this case we say the first literal x_1^j is set to true.*

(ii) *(c_2^j, d_9^j), (c_1^j, c_3^j), (d_{2k-1}^j, d_{2k}^j), for $1 \leq k \leq 4$, (d_{2k}^j, d_{2k+1}^j), for $5 \leq k \leq 8$, are in M —in this case we say the second literal x_2^j is set to true.*

(iii) *(c_3^j, d_{17}^j), (c_1^j, c_2^j), (d_{2k-1}^j, d_{2k}^j), for $1 \leq k \leq 8$, are in M —in this case we say the third literal x_3^j is set to true.*

Proof. If $u(M) < 2$, then by Proposition 2, exactly one of the following three edges can be in M: (c_1^j, d_1^j), (c_2^j, d_9^j), (c_3^j, d_{17}^j), otherwise, some vertices in $\{d_t^j\}_{t=2}^8 \cup \{d_t^j\}_{t=10}^{16}$ would have to be matched to the vertices in the "$\cdots$" part in their lists, contradicting Proposition 2.

Now if $(c_1^j, d_1^j) \in M$, it is easy to see that (i) is the only possible way to match all the vertices so as to maintain the property that $u(M) < 2$. The same applies to (ii) if $(c_2^j, d_9^j) \in M$ and to (iii) if $(c_3^j, d_{17}^j) \in M$. □

How the two types of gadgets interact. We now explain how the two types of gadgets work together by specifying the function π. It may be helpful to first use a simple example to illustrate our ideas. Suppose $C_1 = (X_1 \vee X_5 \vee X_{10})$. Intuitively, we want the following when the derived instance has a matching M with $u(M) < 2$: if the first literal of C_1 is set to true (i.e., $(c_1^1, d_1^1) \in M$—see Lemma 7), then we want to make sure that X_1 is set to true while X_5 and X_{10} are set to false (i.e., we want (a_1^1, b_1^1), (a_5^5, b_9^5), and (a_5^{10}, b_9^{10}) part of M—see Lemma 6.)

Our construction of the function π makes sure if the assignment is "inconsistent", for instance, the first literal $x_1^j = X_1$ of C_1 is set to true but the variable X_1 itself is set to false, i.e., if both (c_1^1, d_1^1) and (a_5^1, b_9^1) are in M, then we can find an alternating cycle with a $(1, 1)$ and two $(1, -1)$ edges, where every edge in $(u, v) \in E \setminus M$ is labeled $(\mathsf{vote}_u(v, M(u)), \mathsf{vote}_v(u, M(v)))$. This would cause M to have unpopularity factor at least 2. Specifically, we define $\pi(\cdot)$ as follows.

1. For all i and j: a_3^i, a_7^i are in $\pi(c_1^j)$, in $\pi(c_2^j)$, and in $\pi(c_3^j)$. Symmetrically, c_1^j, c_2^j, c_3^j are in $\pi(a_3^i)$ and in $\pi(a_7^i)$.
2. For each j, we ensure the following inclusions: suppose $C_j = x_1^j \vee x_2^j \vee x_3^j$ and $X_i = x_1^j$, $X_k = x_2^j$, $X_t = x_3^j$. Then
   (a) b_9^i, b_1^k, b_1^t are in $\pi(d_1^j)$; symmetrically, d_1^j is in $\pi(b_9^i), \pi(b_1^k)$, and $\pi(b_1^t)$.
   (b) b_1^i, b_9^k, b_1^t are in $\pi(d_9^j)$; symmetrically, d_9^j is in $\pi(b_1^i), \pi(b_9^k)$, and $\pi(b_1^t)$.
   (c) b_1^i, b_1^k, b_9^t are in $\pi(d_{17}^j)$; symmetrically, d_{17}^j is in $\pi(b_1^i), \pi(b_1^k)$, and $\pi(b_9^t)$.

[Observe that the function π is symmetrical. If a vertex $c \in \pi(a)$, then $a \in \pi(c)$; if a vertex $b \in \pi(d)$, then $d \in \pi(b)$. Moreover, our construction ensures that β belongs to the "$\cdots$" part of α's list if and only is α belongs to the "$\cdots$" part of β's list.]

To illustrate how the above definitions of $\pi(\cdot)$ help us achieve consistency, consider the above example. Suppose that (c_1^1, d_1^1), (a_5^1, b_9^1) are in M. Consider the alternating cycle $\rho = c_1^1\text{-}d_1^1\text{-}b_9^1\text{-}a_5^1\text{-}a_6^1\text{-}a_7^1\text{-}c_1^1$. $\Delta(M, M \oplus \rho) = 4/2$ since d_1^1, b_9^1, a_5^1, and a_7^1 are better off in $M \oplus \rho$ while a_6^1 and c_1^1 are worse off. Thus $u(M) \geq 2$.

The construction of G_ϕ is complete and we now prove Propositions 1 and 2.

Proofs of Propositions 1 and 2. Let b_t^i be an element in $\pi(d_s^j)$, so d_s^j is also in $\pi(b_t^i)$. Suppose the edge $(b_t^i, d_s^j) \in M$. Then (b_t^i, b_{t-1}^i) and (d_s^j, d_{s-1}^j) are both $(1, 1)$ edges, where b_0^i (similarly, d_0^j) is the same as b_9^i (resp., d_{17}^j).

The matching obtained by augmenting M along the alternating path $\rho = M(b_{t-1}^i)\text{-}b_{t-1}^i\text{-}b_t^i\text{-}d_s^j\text{-}d_{s-1}^j\text{-}M(d_{s-1}^j)$ makes b_t^i, b_{t-1}^i, d_s^j, and d_{s-1}^j better off while $M(b_{t-1}^i)$ and $M(d_{s-1}^j)$ are worse off. Thus $\Delta(M, M \oplus \rho) = 2$. Thus $u(M) \geq 2$ and we have proved (i) in both the propositions.

To prove (ii) in both the propositions, assume that $(y, z) \in M$ and y and z list each other in the "$\cdots$" part of their preference lists.

- for $1 \leq t \leq 7$, if y (or z) is a_t^i, then y' (resp., z') is a_{t-1}^i (where a_0^i is a_7^i);
- for $1 \leq t \leq 9$, if y (or z) is b_t^i, then y' (resp., z') is b_{t-1}^i (where b_0^i is b_9^i);
- for $1 \leq t \leq 3$, if y (or z) is c_t^j, then y' (resp., z') is c_{t-1}^j (where c_0^j is c_3^j);
- for $1 \leq t \leq 17$, if y (or z) is d_t^j, then y' (resp., z') is d_{t-1}^j (where d_0^j is d_{17}^j).

Consider the alternating path $\rho = M(y')\text{-}y'\text{-}y\text{-}z\text{-}z'\text{-}M(z')$. Augmenting M along ρ makes y, y', z, and z' better off while $M(y')$ and $M(z')$ are worse off. Thus $u(M) \geq 2$ and we have proved (ii) in both the propositions. $\qquad\square$

3.1   Correctness of Our Reduction

Lemma 8. *Suppose that there is a matching M with $u(M) < 2$ in G_ϕ. Then there exists a satisfying assignment to ϕ.*

Proof. We construct a truth assignment for ϕ based on M as follows. By Lemma 7, for each clause gadget of $C_j = x_1^j \vee x_2^j \vee x_3^j$, one of its three literals x_t^j is set to true. Set the variable X_i to true if $X_i = x_t^j$. X_i is set to false if it is never set to true. We claim that this yields a satisfying assignment for ϕ.

First note that at least one of the literals in each clause is set to true. So if we do not have a satisfying assignment, it must be the case that some clause $C_j = x_1^j \vee x_2^j \vee x_3^j$ has two (or more) literals being set to true. Without loss of generality, assume that $X_1 = x_1^j$ and $X_2 = x_2^j$ and X_1 is set to true because in the matching M, the first literal of C_j is satisfied, i.e., the edges (c_1^j, d_1^j), (c_2^j, c_3^j), (d_{2k}^j, d_{2k+1}^j), for $1 \leq k \leq 8$, are in M.

Then as X_2 is also set to true, there must exist another clause $C_t \neq C_j$ and $C_t = x_1^t \vee x_2^t \vee x_3^t$ and C_t is satisfied by its, say, first literal. So $x_1^t = X_2$ and (c_1^t, d_1^t), (c_2^t, c_3^t), (d_{2k}^t, d_{2k+1}^t), for $1 \leq k \leq 8$, are in M.

Now by Lemma 6, either (a_1^2, b_1^2) and (a_2^2, a_3^2) are in M, or (a_5^2, b_9^2) and (a_6^2, a_7^2) are in M. In the former case, augmenting M along the alternating cycle $b_1^2\text{-}d_1^t\text{-}c_1^t\text{-}a_3^2\text{-}a_2^2\text{-}a_1^2\text{-}b_1^2$ makes b_1^2, d_1^t, a_3^2, and a_1^2 better off while c_1^t and a_2^2 are worse off; in the latter case, augmenting M along the alternating cycle $b_9^2\text{-}d_1^j\text{-}c_1^j\text{-}a_7^2\text{-}a_6^2\text{-}a_5^2\text{-}b_9^2$ makes b_9^2, d_1^j, a_7^2, and a_5^2 better off while c_1^j and a_6^2 are worse off. In both cases, we have $u(M) \geq 2$, a contradiction.   □

Conversely, suppose ϕ is satisfiable. Then Lemma 9 constructs a matching M in G_ϕ, based on Lemmas 6 and 7, so that $u(M) \leq 1.5$. The proof of Lemma 9 is given in the full version.

Lemma 9. *Suppose that there is a satisfying assignment for ϕ. Then there is a matching M with $u(M) \leq 1.5$ in G_ϕ.*

Using Lemma 8 and 9, we now show Theorem 1 stated in Section 1.

Proof of Theorem 1. Given an input instance ϕ of 1-IN-3 SAT, we build the graph G_ϕ. Let M be a matching in G_ϕ whose unpopularity factor is strictly smaller than $4/3$ of the least unpopularity factor matching.

- If $u(M) < 2$, then ϕ is a *yes* instance (as shown by Lemma 8).
- If $u(M) \geq 2$, then ϕ has to be a *no* instance. Otherwise, by Lemma 9, we know that there is a matching with unpopularity factor at most 1.5. This implies $u(M) < 2$ since $u(M)$ has to be smaller than $4/3$ of the optimal.

Thus the problem of computing a matching whose unpopularity factor is strictly smaller than $4/3$ of the optimal is NP-hard.   □

References

1. Abraham, D.J., Irving, R.W., Kavitha, T., Mehlhorn, K.: Popular matchings. SIAM Journal on Computing 37(4), 1030–1045 (2007)
2. Biró, P., Irving, R.W., Manlove, D.F.: Popular Matchings in the Marriage and Roommates Problems. In: Calamoneri, T., Diaz, J. (eds.) CIAC 2010. LNCS, vol. 6078, pp. 97–108. Springer, Heidelberg (2010)
3. Gale, D., Shapley, L.S.: College admissions and the stability of marriage. American Mathematical Monthly 69, 9–15 (1962)
4. Gärdenfors, P.: Match making: assignments based on bilateral preferences. Behavioural Sciences 20, 166–173 (1975)
5. Garey, M., Johnson, D.: Computers and Intractablility. Freeman, New York (1979)
6. Huang, C.-C., Kavitha, T.: Popular Matchings in the Stable Marriage Problem. In: Aceto, L., Henzinger, M., Sgall, J. (eds.) ICALP 2011. LNCS, vol. 6755, pp. 666–677. Springer, Heidelberg (2011)
7. Irving, R.W.: An Efficient Algorithm for the "Stable Roommates" Problem. Journal of Algorithms 6, 577–595 (1985)
8. Kavitha, T., Mestre, J., Nasre, M.: Popular mixed matchings. In: Albers, S., Marchetti-Spaccamela, A., Matias, Y., Nikoletseas, S., Thomas, W. (eds.) ICALP 2009. LNCS, vol. 5555, pp. 574–584. Springer, Heidelberg (2009)
9. Mahdian, M.: Random popular matchings. In: Proceedings of the ACM EC 2006, pp. 238–242 (2006)
10. Manlove, D.F., Sng, C.T.S.: Popular matchings in the capacitated house allocation problem. In: Azar, Y., Erlebach, T. (eds.) ESA 2006. LNCS, vol. 4168, pp. 492–503. Springer, Heidelberg (2006)

11. McCutchen, R.M.: The least-unpopularity-factor and least-unpopularity-margin criteria for matching problems with one-sided preferences. In: Laber, E.S., Bornstein, C., Nogueira, L.T., Faria, L. (eds.) LATIN 2008. LNCS, vol. 4957, pp. 593–604. Springer, Heidelberg (2008)
12. Mestre, J.: Weighted popular matchings. In: Bugliesi, M., Preneel, B., Sassone, V., Wegener, I. (eds.) ICALP 2006. LNCS, vol. 4051, pp. 715–726. Springer, Heidelberg (2006)
13. Tan, J.J.M.: A necessary and sufficient condition for the existence of a complete stable matching. Journal of Algorithms 12, 154–178 (1991)

The Hospitals/Residents Problem with Quota Lower Bounds

Koki Hamada[1], Kazuo Iwama[2,*], and Shuichi Miyazaki[3,**]

[1] NTT Information Sharing Platform Laboratories, NTT Corporation
hamada.koki@lab.ntt.co.jp
[2] Graduate School of Informatics, Kyoto University
iwama@kuis.kyoto-u.ac.jp
[3] Academic Center for Computing and Media Studies, Kyoto University
shuichi@media.kyoto-u.ac.jp

Abstract. The Hospitals/Residents problem is a many-to-one extension of the stable marriage problem. In its instance, each hospital specifies a quota, i.e., an upper bound on the number of positions it provides. It is well-known that in any instance, there exists at least one stable matching, and finding one can be done in polynomial time. In this paper, we consider an extension in which each hospital specifies not only an upper bound but also a *lower* bound on its number of positions. In this setting, there can be instances that admit no stable matching, but the problem of asking if there is a stable matching is solvable in polynomial time. In case there is no stable matching, we consider the problem of finding a matching that is "as stable as possible", namely, a matching with a minimum number of blocking pairs. We show that this problem is hard to approximate within the ratio of $(|H|+|R|)^{1-\epsilon}$ for any positive constant ϵ where H and R are the sets of hospitals and residents, respectively. We tackle this hardness from two different angles. First, we give an exponential-time exact algorithm for a special case where all the upper bound quotas are one. This algorithm runs in time $O(t^2(|H|(|R|+t))^{t+1})$ for instances whose optimal cost is t. Second, we consider another measure for optimization criteria, i.e., the number of residents who are involved in blocking pairs. We show that this problem is still NP-hard but has a polynomial-time $\sqrt{|R|}$-approximation algorithm.

1 Introduction

In the *stable marriage problem* [10], we are given sets of men and women, and each person's preference list that orders the members of the other sex according to his/her preference. The question is to find a *stable matching*, that is, a matching containing no pair of man and woman who prefer each other to their partners. Such a pair is called a *blocking pair*. Gale and Shapley proved that any instance admits at least one stable matching, and gave an algorithm to find one, known as the Gale-Shapley algorithm.

---

* Supported by KAKENHI 22240001.
** Supported by KAKENHI 20700009.

C. Demetrescu and M.M. Halldórsson (Eds.): ESA 2011, LNCS 6942, pp. 180–191, 2011.

They also proposed a many-to-one extension of the stable marriage problem, which is currently known as the *Hospitals/Residents problem* (*HR* for short) [10]. In HR, the two sets corresponding to men and women are residents and hospitals. Each hospital specifies its *quota*, which means that it can accept at most this number of residents. Hence in a feasible matching, the number of residents assigned to each hospital is up to its quota. Most properties of the stable marriage problem carry over to HR, e.g., any instance admits a stable matching, and we can find one by the appropriately modified Gale-Shapley algorithm. As the name of HR suggests, it has real-world applications in assigning residents to hospitals in many countries, known as NRMP in the U.S. [12], CaRMS in Canada [7], and SFAS in Scotland [17]. Along with these applications and due to special requirements in reality, several useful extensions have been proposed, such as HR with couples [24,23,3,22], and the Student-Project Allocation problem [2].

In this paper, we study another extension of HR where each hospital declares not only an upper bound but also a *lower* bound on the number of residents it accepts. Consequently, a feasible matching must satisfy the condition that the number of residents assigned to each hospital is between its upper and lower bound quotas. This restriction seems quite relevant in several situations. For example, shortage of doctors in hospitals in rural area is a critical issue; it is sometimes natural to guarantee some number of residents for such hospitals in the residents-hospitals matching. Also, when determining supervisors of students in universities, it is quite common to consider that the number of students assigned to each professor should be somehow balanced, which can be achieved again by specifying both upper and lower bounds on the number of students accepted by each professor. We call this problem *HR with Minimum Quota* (*HRMQ* for short).

The notion of minimum quota was first raised in [13] and followed by [5,16] (see "Related Work" below). In this paper, we are interested in a most natural question, i.e., how to obtain "good" matchings in this new setting. In HRMQ, stable matchings do not always exist. However, it is easy to decide whether or not there is a stable matching for a given instance, since in HR the number of students a specific hospital h receives is identical for any stable matching (this is a part of the well-known *Rural Hospitals Theorem* [11]). Namely, if this number satisfies the upper and lower bound conditions of all the hospitals, it is a feasible (and stable) matching, and otherwise, no stable matching exists. In case there is no stable matching, it is natural to seek for a matching "as stable as possible".

Our Contributions. We first consider the problem of minimizing the number of blocking pairs, which is quite popular in the literature (e.g., [20,1,6,15]). As shown in Sec. 2, it seems that the introduction of the quota lower bound intrinsically increases the difficulty of the problem. Actually, we show that this problem is NP-hard and cannot be approximated within a factor of $(|H| + |R|)^{1-\varepsilon}$ for any positive constant ε unless P=NP, where H and R denote the sets of hospitals and residents, respectively. This inapproximability result holds even if all the preference lists are complete, all the hospitals have the same preference list, (e.g., determined by scores of exams and known as the *master list* [18]), and all the hospitals have upper bound quota of one. On the positive side, we give a polynomial-time $(|H| + |R|)$–approximation algorithm, which shows that the above inapproximability result is almost tight.

We then tackle this hardness from two different angles. First, we restrict ourselves to instances where upper bound quotas of all the hospitals are one, which correspond to the marriage case and are still hard to approximate as shown above. We give an exponential-time exact algorithm which runs in time $O(t^2(|H|(|R| + t))^{t+1})$ for instances whose optimal cost is t. Note that this is a polynomial-time algorithm when the optimal cost is constant. Second, we go back to the original many-to-one case, and consider another measure for optimization criteria, i.e., the number of residents who are involved in blocking pairs. We show that this problem is still NP-hard, but give a quadratic improvement, i.e., we give a polynomial-time $\sqrt{|R|}$-approximation algorithm. We also give an instance showing that our analysis is tight up to a constant factor. Furthermore, we show that if our problem has a constant approximation factor, then the Dense k-Subgraph Problem (DkS) has a constant approximation factor also. Note that the best known approximation factor of DkS has long been $|V|^{1/3}$ [21] in spite of extensive studies, and was recently improved to $|V|^{1/4+\epsilon}$ for an arbitrary positive constant ϵ [4]. The reduction is somewhat tricky, which is done through the third problem, called the Minimum Coverage Problem (MinC), and exploits the best approximation algorithm for DkS. MinC is relatively less studied and only the NP-hardness is known for its complexity [25]. As a byproduct, our proof gives a similar hardness for MinC, which is of independent interest.

Because of the space restriction, most of the proofs are omitted. They are included in the full version [14].

Related Work. Biró, et al. [5] also considers HR with quota lower bounds. In contrast to our model, which requires to satisfy the lower bound quota of all the hospitals, their model allows some hospitals to be closed, i.e., to receive no residents. Huang [16] considers *classified stable matchings*, in which not only individual hospitals but also selected sets of hospitals declare quota upper and lower bounds. He proved a dichotomy theorem for the problem of deciding the existence of a stable matching, in terms of the structural property of the family of the sets of hospitals that declare quota bounds.

2 Preliminaries

An instance of the *Hospitals/Residents Problem with Minimum Quota* (HRMQ for short) consists of the set R of residents and the set H of hospitals. Each hospital h has lower and upper bounds of quota, p and q $(p \leq q)$, respectively. We sometimes say that the quota of h is $[p,q]$, or h is a $[p,q]$-hospital. For simplicity, we also write the name of a hospital with its quota bounds, such as $h[p,q]$. Each member (resident or hospital) has a preference list that orders a subset of the members of the other party.

A *matching* is an assignment of residents to hospitals (possibly, leaving some residents unassigned), where matched residents and hospitals are in the preference list of each other. Let $M(r)$ be the hospital to which resident r is assigned under a matching M (if it exists), and $M(h)$ be the set of residents assigned to hospital h. A *feasible matching* is a matching such that $p \leq |M(h)| \leq q$ for each hospital $h[p, q]$. We may sometimes call a feasible matching simply a matching when there is no fear of confusion. For a matching M and a hospital $h[p, q]$, we say that h is *full* if $|M(h)| = q$ and that h is *under-subscribed* if $|M(h)| < q$.

For a matching M, we say that a pair comprising a resident r and a hospital h who include each other in the list forms a *blocking pair* for M if the following two conditions are met: (i) r is either unassigned or prefers h to $M(r)$, and (ii) h is under-subscribed or prefers r to one of the residents in $M(h)$. We say that r is a *blocking resident* for M if r is involved in a blocking pair for M.

Minimum-Blocking-Pair HRMQ (*Min-BP HRMQ* for short) is the problem of finding a feasible matching with the minimum number of blocking pairs. *Min-BP 1ML-HRMQ* ("1ML" standing for "1 Master List") is the restriction of Min-BP HRMQ so that preference lists of all the hospitals are identical. *0-1 Min-BP HRMQ* is the restriction of Min-BP HRMQ where a quota bound of each hospital is either $[0, 1]$ or $[1, 1]$. *0-1 Min-BP 1ML-HRMQ* is Min-BP HRMQ with both "1ML" and "0-1" restrictions. *Minimum-Blocking-Resident HRMQ* (*Min-BR HRMQ* for short) is the problem of finding a feasible matching with the minimum number of blocking residents. *Min-BR 1ML-HRMQ, 0-1 Min-BR HRMQ*, and *0-1 Min-BR 1ML-HRMQ* are defined similarly.

We assume without loss of generality that the number of residents is at least the sum of the lower bound quotas of all the hospitals. Also, in this paper we impose the following restriction $\mathcal{Z}$ to guarantee existence of a feasible solution: every resident's list includes all hospitals with positive quota lower bounds, and such hospitals' lists include all the residents. (We remark in Sec. 5 that allowing arbitrarily incomplete preference lists makes the problem extremely hard.)

We say that an algorithm A is an $r(n)$-approximation algorithm if it satisfies $A(x)/opt(x) \leq r(n)$ for any instance x of size n, where $opt(x)$ and $A(x)$ are the costs (i.e., the number of blocking pairs in the case of Min-BP HRMQ) of the optimal and the algorithm's solutions, respectively.

As a starting example, consider n residents and $n + 1$ hospitals, whose preference lists and quota bounds are as follows. Here, "$\cdots$" in the residents' preference lists denotes an arbitrary order of the remaining hospitals.

$$
\begin{aligned}
&r_1 : h_1 \quad h_{n+1} \cdots \\
&r_2 : h_1 \quad h_2 \quad h_n \cdots && h_1[0, 1] : r_1 \, r_2 \cdots r_n \\
&r_3 : h_2 \quad h_1 \quad h_3 \cdots && h_2[1, 1] : r_1 \, r_2 \cdots r_n \\
&r_4 : h_3 \quad h_1 \quad h_4 \cdots && \vdots \\
&\vdots && h_n[1, 1] : r_1 \, r_2 \cdots r_n \\
&r_i : h_{i-1} \, h_1 \quad h_i \cdots && h_{n+1}[1, 1] : r_1 \, r_2 \cdots r_n \\
&\vdots \\
&r_n : h_{n-1} \, h_1 \quad h_n \cdots
\end{aligned}
$$

Note that we have n $[1,1]$-hospitals all of which have to be filled by the n residents. Therefore, let us modify the instance by removing the $[0,1]$-hospital h_1 and apply the Gale-Shapley algorithm (see e.g., [12] for the Gale-Shapley algorithm; in this paper it is always the residents-oriented version, namely, residents make and hospitals receive proposals). Then the resulting matching is $M_1 = \{(r_1, h_{n+1}), (r_2, h_2), (r_3, h_3), \cdots, (r_n, h_n)\}$, which contains at least n blocking pairs (between h_1 and every resident). However, the matching $M_2 = \{(r_1, h_{n+1}), (r_2, h_n), (r_3, h_2), (r_4, h_3), \ldots, (r_n, h_{n-1})\}$ contains only three blocking pairs (r_1, h_1), (r_2, h_1), and (r_2, h_2).

3   Minimum-Blocking-Pair HRMQ

In this section, we give both approximability and inapproximability results. For the latter, we prove a strong inapproximability result for the restricted subclass, as mentioned in Sec. 1. On the other hand, we can show that this inapproximability result is almost tight by providing an approximation algorithm for the general class.

Theorem 1. *For any positive constant ε, there is no polynomial-time $(|H| + |R|)^{1-\varepsilon}$-approximation algorithm for 0-1 Min-BP 1ML-HRMQ unless P=NP, even if all the preference lists are complete. (Proof is omitted. See [14].)*

Theorem 2. *There is a polynomial-time $(|H| + |R|)$–approximation algorithm for Min-BP HRMQ.*

Proof. Here we give only a sketch. See [14] for the complete proof. The following simple algorithm (Algorithm I) achieves an approximation ratio of $|H| + |R|$: Given an instance I of Min-BP HRMQ, consider it as an instance of HR by ignoring quota lower bounds. Then, apply the Gale-Shapley algorithm to I and obtain a matching M. In M, define a *deficiency of a hospital $h_i[p_i, q_i]$* to be $\max\{p_i - x_i, 0\}$ where x_i is the number of residents assigned to h_i by M. We then move residents arbitrarily from hospitals with surplus to the hospitals with positive deficiencies (but so as not to create new positive deficiency) to fill all the deficiencies. This is possible because of the restriction $\mathcal{Z}$.

   Let k be the sum of the deficiencies over all the hospitals. Then, k residents are moved by the above procedure. We can show that at most $|H| + |R|$ new blocking pairs can arise per resident movement and hence at most $k(|H| + |R|)$ in total. On the other hand, we can prove that if there are k deficiencies in M, an optimal solution contains at least k blocking pairs. $\square$

3.1   Exponential-Time Exact Algorithm

In this section we consider only the cases where quota bounds are $[0,1]$ or $[1,1]$, as in the example given in the previous section. Recall that the problem is still hard to approximate, and our goal here is to design nontrivial exponential-time algorithms by using the parameter t denoting the cost of an optimal solution. Probably, a natural idea is to change some subset H_0 of $[0,1]$-hospitals into

[1,1]-hospitals so that in the optimal solution residents are assigned to hospitals in H_0 plus original [1,1]-hospitals exactly. However, there is no obvious way of selecting H_0 rather than exhaustive search, which will result in blow-ups of the computation time even if t is small. Furthermore, even if we would be able to find a correct H_0, we are still not sure how to assign the residents to these (expanded) [1,1]-hospitals optimally.

Theorem 3. *There is an $O(t^2(|H|(|R| + t))^{t+1})$-time exact algorithm for 0-1 Min-BP HRMQ, where t is the cost of an optimal solution of a given instance.*

Proof. For a given integer $k > 0$, our algorithm $A(k)$ finds a solution (i.e., a matching between residents and hospitals) whose cost is at most k if any. Starting from $k = 1$, we run $A(k)$ until we find a solution, by increasing the value of k. $A(k)$ is quite simple, for which the following informal discussion might be enough.

Classify the blocking pairs as follows: A blocking pair (r, h) such that hospital h is empty is a Type-I blocking pair, and one such that h is non-empty is a Type-II blocking pair. First, guess the numbers k_1 of Type-I blocking pairs and k_2 of Type-II blocking pairs such that $k_1 + k_2 = k$, thus $k + 1$ possibilities.

We next guess, for each resident r_i, the number $b_i(\geq 0)$ of Type-I blocking pairs r_i is involved in, such that $b_1 + b_2 + \cdots + b_{|R|} = k_1$. Note that there are $O((|R| + k_1)^{k_1})$ possibilities. Then, for each r_i such that $b_i > 0$, we again guess the set S_i of b_i hospitals that form Type-I blocking pairs with r_i. Note that there are at most $O(|H|^{k_1})$ different possibilities for selecting such k_1 blocking pairs. Let S be the union of S_i over all r_i. Note that all the hospitals in S are [0, 1]-hospitals since they are supposed to be empty. Now remove these hospitals from the instance. For a resident r_i and hospital $h \in S$, suppose that $h \notin S_i$ in the current guess. Then, we remove all the hospitals lower ranked than h in r_i's list, since if r_i is assigned to one of such hospitals, then r_i forms a Type-I blocking pair with h, contradicting our guess.

Next, we guess k_2 Type-II blocking pairs similarly. Let T be the set of those pairs. Since there are at most $|H||R|$ pairs, there are $O((|H||R|)^{k_2})$ choices of T. For each pair $(r, h) \in T$, we remove h from r's list. Finally, we apply the Gale-Shapley algorithm to the modified instance. If all the [1,1]-hospitals are full, then it is a desired solution, otherwise, we fail and proceed to the next guess.

We show that the algorithm runs correctly. Consider any optimal solution M_{opt} and consider the execution for $k = t$ for which we assume to have made a correct guess of the t blocking pairs of M_{opt}. Then, it is not hard to see that M_{opt} is stable in the above modified instance and makes all the [1,1]-hospitals full. Then by the Rural Hospitals Theorem, any stable matching for this new instance makes all the [1,1]-hospitals full. Hence if we apply the Gale-Shapley algorithm to this instance, we find a matching M in which all the [1,1]-hospitals are full. Note that M has no blocking pair in the modified instance. Then, observe that M has at most k_1 Type-I blocking pairs in the original instance because, when hospitals in S are returned back to the instance, each resident r_i can form Type-I blocking pairs with only the hospitals in S_i. Also, only the pairs in T

can be Type-II blocking pairs of M in the original instance. Therefore, M has at most t blocking pairs in the original instance.

Finally, we bound the time-complexity. For each k, we apply the Gale-Shapley algorithm to at most $O(k) \cdot O((|R| + k_1)^{k_1}) \cdot O(|H|^{k_1}) \cdot O((|H||R|)^{k_2}) = O(k(|H|(|R| + k))^k)$ instances. Therefore, the time-complexity is $O(k(|H|(|R| + k))^{k+1})$ for each k. Since we find a solution when k is at most t, the whole time-complexity is at most $\Sigma_{k=1}^{t} O(k(|H|(|R| + k))^{k+1}) = O(t^2(|H|(|R| + t))^{t+1})$. □

4   Minimum-Blocking-Resident HRMQ

In this section, we consider the problem of minimizing the number of blocking residents. We first show a negative result.

Theorem 4. *0-1 Min-BR 1ML-HRMQ is NP-hard. (Proof is omitted. See [14].)*

For the approximability, we note that Algorithm I in the proof of Theorem 2 does not work. For example, consider the instance introduced in Sec. 2. If we apply the Gale-Shapley algorithm, resident r_i is assigned to h_i for each i, and we need to move r_1 to h_{n+1}. However since h_1 becomes unassigned, all the residents become blocking residents. On the other hand, the optimal cost is 2 as we have seen there. Thus the approximation ratio becomes as bad as $\Omega(|R|)$.

Theorem 5. *There is a polynomial-time $\sqrt{|R|}$–approximation algorithm for Min-BR HRMQ.*

We prove Theorem 5 by giving a $\sqrt{|R|}$–approximation algorithm for 0-1 Min-BR HRMQ (Lemmas 1 and 3). In 0-1 Min-BR HRMQ, the number of residents assigned to each hospital is at most one. Hence, for a matching M, we sometimes abuse the notation $M(h)$ to denote the resident assigned to h (if any) although it is originally defined as the set of residents assigned to h.

Lemma 1. *If there is a polynomial-time α–approximation algorithm for 0-1 Min-BR HRMQ, then there is a polynomial-time α–approximation algorithm for Min-BR HRMQ. (Proof is omitted. See [14].)*

4.1   Our Algorithm

To describe the idea behind our algorithm, recall Algorithm I presented in the proof of Theorem 2: First, apply the Gale-Shapley algorithm to a given instance I and obtain a matching M. Next, move residents arbitrarily from assigned $[0, 1]$-hospitals to unassigned $[1, 1]$-hospitals. Suppose that in the course of the execution of Algorithm I, we move a resident r from a $[0, 1]$-hospital h to an unassigned $[1, 1]$-hospital. Then, of course r creates a blocking pair with h, but some other residents may also create blocking pairs with h because h becomes unassigned. Then, consider the following modification. First, set the upper bound quota of a $[0, 1]$-hospital h to ∞ and apply the Gale-Shapley algorithm. Then, all residents who "wish" to go to h actually go there. Hence, even if we move all such residents to other hospitals, only the moved residents can become blocking residents. By doing this, we can bound the number of blocking residents by the number (given by the function g introduced below) of those moving residents. In the above example, we extended the upper bound quota of only one hospital, but in fact, we may need to select two or more hospitals to select sufficiently many residents to be sent to other hospitals so as to make the matching feasible. However, at the same time, this number should be kept minimum to guarantee the quality of the solution.

We define $g(h, h)$: For an instance I of HR, suppose that we extend the upper bound quota of hospital h to ∞ and find a stable matching of this new instance. Define $g(h, h)$ be the number of residents assigned to h in this stable matching. Recall that this quantity does not depend on the choice of the stable matching by the Rural Hospitals Theorem [11]. Extend $g(h, h)$ to $g(A, B)$ for $A, B \subseteq H$ such that $g(A, B)$ denotes the number of residents assigned to hospitals in A when we change upper bound quotas of all the hospitals in B to ∞.

We now propose Algorithm II for 0-1 Min-BR HRMQ. Let I be a given instance. Define $H_{p,q}$ to be the set of $[p, q]$-hospitals of I. Recall from Sec. 3 that the deficiency of a hospital is the shortage of the assigned residents from its lower bound quota. Now define the *deficiency of the instance I* as the sum of the deficiencies of all the hospitals of I, and denote it $D(I)$. Since we are considering 0-1 Min-BR HRMQ, $D(I)$ is exactly the number of empty $[1, 1]$-hospitals.

Algorithm II

1: Apply the Gale-Shapley algorithm to I by ignoring the lower bound quotas. Let M_s be the obtained matching. Compute the deficiency $D(I)$.

2: $H'_{0,1} := \{h \mid M_s(h) \neq \emptyset, h \in H_{0,1}\}$. (If $M_s(h) = \emptyset$, then residents never come to h in this algorithm.)

3: Compute $g(h, h)$ for each $h \in H'_{0,1}$ by using the Gale-Shapley algorithm.

4: From $H'_{0,1}$, select $D(I)$ hospitals with smallest $g(h, h)$ values (ties are broken arbitrarily). Let S be the set of these hospitals. Extend the quota upper bounds of all hospitals in S to ∞, and run the Gale-Shapley algorithm. Let M_∞ be the obtained matching.

5: In M_∞, move residents who are assigned to hospitals in S arbitrarily to empty hospitals to make the matching feasible. (We first make $[1, 1]$-hospitals full. This is possible because of the restriction $\mathcal{Z}$. If there is a hospital in S still having two or more residents, then send remaining residents arbitrarily to empty $[0, 1]$-hospitals, or simply make them unassigned if there is no $[0, 1]$-hospital to send them.) Output the resulting matching M^*.

We first prove the following property of the original HR problem.

Lemma 2. *Let I_0 be an instance of HR, and h be any hospital. Let I_1 be a modification of I_0 so that only the upper bound quota of h is increased by 1. Let M_i be a stable matching of I_i for each $i \in \{0, 1\}$. Then, (i) $|M_0(h)| \leq |M_1(h)|$, and (ii) $\forall h' \in H \setminus \{h\}, |M_0(h')| \geq |M_1(h')|$.*

Proof. If M_0 is stable for I_1, then we are done, so suppose not. Because M_0 is stable for I_0, if M_0 has blocking pairs for I_1, then all of them involve h. Let r be the resident such that (r, h) is a blocking pair and there is no blocking pair (r', h) such that h prefers r' to r. If we assign r to h, all blocking pairs including h are removed. If no new blocking pairs arise, again, we are done. Otherwise, r must be previously assigned to some hospital, say h', and all the new blocking pairs involve h'. We then choose the resident r', most preferred by h' among all the blocking residents, and assign r' to h'. We continue this operation until there arise no new blocking pairs. This procedure eventually terminates because each iteration improves exactly one resident. By the termination condition, the resulting matching is stable for I_1. Note that by this procedure, only h can gain one more resident, and at most one hospital may lose one resident. This completes the proof.   □

Obviously, Algorithm II runs in polynomial time. We show that Algorithm II runs correctly, namely that the output matching M^* satisfies the quota bounds. To do so, it suffices to show the following: (1) $|H'_{0,1}| \geq D(I)$ so that we can construct S at Step 4, and (2) in M_∞, the total number of residents assigned to hospitals in S is at least the number of empty $[1, 1]$-hospitals, so that Step 5 is executable.

For (1), recall that we have assumed that the number of residents is at least the sum of the lower bound quotas, which is the number of $[1, 1]$-hospitals. Also, we can assume that all the residents are assigned in M_s (since otherwise, we already have a feasible stable matching). Then, by the definition of $D(I)$, we have that $|H'_{0,1}| \geq D(I)$. For (2), it suffices to show that the number N of residents assigned to $S \cup H_{1,1}$ in M_∞ is at least $|H_{1,1}|$. Note that empty hospitals in M_s are also empty in M_∞ by Lemma 2. Therefore, the number $\overline{N}$ of residents assigned to hospitals in $H \setminus (S \cup H_{1,1})$ in M_∞ is at most the number of hospitals in $H'_{0,1} \setminus S$. Thus $\overline{N} \leq |H'_{0,1}| - |S|$ and $N = |R| - \overline{N} \geq |R| - (|H'_{0,1}| - |S|)$. By the definition of $D(I)$, we have that $|H'_{0,1}| + |H_{1,1}| = |R| + D(I)$. Thus, $N \geq |R| - (|R| + D(I) - |H_{1,1}| - |S|) = |H_{1,1}|$ (recall that $|S| = D(I)$).

4.2 Analysis of the Approximation Ratio

Lemma 3. *The approximation ratio of Algorithm II is at most* $\sqrt{|R|}$.

Proof. Let I be a given instance of 0-1 Min-BR HRMQ and let f_{opt} and f_{alg} be the costs of an optimal solution and the solution obtained by Algorithm II, respectively. First, note that only residents moved at Step 5 can be blocking residents. Hence, (1) $f_{alg} \leq g(S, S)$.

We then give a lower bound on the optimal cost. To do so, see the proof of Theorem 2 [14], where it is shown that any optimal solution for instance I of Min-BP HRMQ has at least $D(I)$ blocking pairs. It should be noted that those $D(I)$ blocking pairs do not have any common resident. Thus we have (2) $f_{opt} \geq D(I)$.

Now here is our key lemma to evaluate the approximation ratio.

Lemma 4. *In Step 3 of Algorithm II, there are at least $D(I)$ different hospitals $h \in H'_{0,1}$ such that $g(h, h) \leq f_{opt}$.*

The proof will be given in a moment. By this lemma, we have $g(h,h) \leq f_{opt}$ for any $h \in S$ by the construction of S. This implies that (3) $\sum_{h \in S} g(h,h) \leq D(I)f_{opt}$. Also, by Lemma 2, we have (4) $g(h,S) \leq g(h,h)$ for any $h \in S$. Hence, by (1), (4), (3) and (2), we have

$$f_{alg} \leq g(S,S) = \sum_{h \in S} g(h,S) \leq \sum_{h \in S} g(h,h) \leq D(I)f_{opt} \leq (f_{opt})^2.$$

Therefore, we have that $\sqrt{f_{alg}} \leq f_{opt}$, and hence $\frac{f_{alg}}{f_{opt}} \leq \sqrt{f_{alg}} \leq \sqrt{|R|}$, completing the proof of Lemma 3. □

Proof of Lemma 4. Let h be a hospital satisfying the condition of the lemma. In order to calculate $g(h,h)$ in Step 3, we construct a stable matching, say M_h, for the instance $I_\infty(h)$ in which the upper bound quota of h is changed to ∞. We do not know what kind of matching M_h is, but in the following, we show that there is a stable matching, say M_2, for $I_\infty(h)$ such that $|M_2(h)| \leq f_{opt}$. M_h and M_2 may be different, but we can guarantee that $|M_h(h)| \leq f_{opt}$ by the Rural Hospitals Theorem. A bit trickily, we construct M_2 from an optimal matching.

Let M_{opt} be an optimal solution of I (which of course we do not know). Let R_b and R_n be the sets of blocking residents and non-blocking residents for M_{opt}, respectively. Then $|R_b| = f_{opt}$ by definition. We modify M_{opt} as follows: Take any resident $r \in R_b$. If r is unassigned, we do nothing. Otherwise, force r to be unassigned. Then there may arise new blocking pairs, all of which include the hospital h' to which r was assigned. Among residents who are included in such new blocking pairs, we select the resident in R_n who is most preferred by h' (if any) and assign her to h'. In a similar way as the proof of Lemma 2, we continue to move residents until no new blocking pair arises (but this time, we move only residents in R_n as explained above). We do this for all the residents in R_b, and let M_1 be the resulting matching.

The following (a) and (b) are immediate: (a) There are at least f_{opt} unassigned residents in M_1, since residents in R_b are unassigned in M_1. (b) Residents in R_n are non-blocking for M_1. We prove the following properties: (c) There are at most f_{opt} unassigned $[1,1]$-hospitals in M_1. (d) Define $H' = \{h \mid h \in H'_{0,1}$ and h is unassigned in $M_1\}$. Then $|H'| \geq D(I)$.

(c) In M_{opt}, all the $[1,1]$-hospitals are full. It is easy to see that an unassigned hospital of M_{opt} is also unassigned in M_1. Since at most f_{opt} residents are made to be unassigned by the above procedure, the claim holds.

(d) Let H_1 be the set of hospitals assigned in M_1. By the definition of H', $H' = H'_{0,1} \setminus (H_1 \cap H_{0,1})$. By the definition of $D(I)$, $|H'_{0,1}| = |R| + D(I) - |H_{1,1}|$, and by the properties (a) and (c), $|H_1 \cap H_{0,1}| \leq |R| - |H_{1,1}|$. Then $|H'| \geq |H'_{0,1}| - |H_1 \cap H_{0,1}| \geq (|R| + D(I) - |H_{1,1}|) - (|R| - |H_{1,1}|) = D(I)$.

For any $h \in H'$, we show that $g(h,h) \leq f_{opt}$. Then, this completes the proof of Lemma 4 because $H' \subseteq H'_{0,1}$ and (d) $|H'| \geq D(I)$. Since h is unassigned in M_1, residents in R_n are still non-blocking for M_1 in $I_\infty(h)$ (whose definition is in the beginning of this proof) by the property (b). Now, choose any resident r from R_b, and apply the Gale-Shapley algorithm to $I_\infty(h)$ starting from M_1. This execution starts from the proposal by r, and at the end, nobody in $R_n \cup \{r\}$ is a blocking resident for $I_\infty(h)$. Since hospitals assigned in M_1 never become unassigned, and since unassigned residents in R_n never become assigned, h receives at most one resident. If we do this for all the residents in R_b, the resulting matching M_2 is stable for $I_\infty(h)$, and h is assigned at most $|R_b| = f_{opt}$ residents. As mentioned previously, this implies $g(h,h) \leq f_{opt}$. $\square$

We can show that the analysis of Lemma 3 is tight up to a constant factor: There is an instance of 0-1 Min-BR HRMQ for which Algorithm II produces a solution of cost $|R| - \sqrt{|R|}$ but the optimal cost is at most $2\sqrt{|R|}$ (see [14]).

4.3   Inapproximability of Min-BR HRMQ

For the hardness of Min-BR HRMQ, we have only NP-hardness, but we can give a strong evidence for its inapproximabitily. The *Dense k-Subgraph Problem* (DkS) is the problem of finding, given a graph G and a positive integer k, an induced subgraph of G with k vertices that contains as many edges as possible. This problem is NP-hard because it is a generalization of Max CLIQUE. Its approximability has been studied intensively but there still remains a large gap between approximability and inapproximability: The best known approximation ratio is $|V|^{1/4+\epsilon}$ [4], while there is no PTAS under reasonable assumptions [8,19]. The following Theorem 6 shows that approximating Min-BR HRMQ within a constant ratio implies the same for DkS.

Theorem 6. *If Min-BR 1ML-HRMQ has a polynomial-time c-approximation algorithm, then DkS has a polynomial-time $(1 + \epsilon)c^4$-approximation algorithm for any positive constant ϵ. (Proof is omitted. See [14].)*

5   Concluding Remarks

An obvious future research is to obtain lower bounds on the approximation factor for Min-BR HRMQ (we even do not know its APX-hardness at this moment). Since the problem is harder than DkS, it should be a reasonable challenge. Another direction is to develop an FPT algorithm for Min-BP HRMQ, improving Theorem 3. Finally, we remark on the possibility of generalization of instances: In this paper, we guarantee existence of feasible matchings by the restriction $\mathcal{Z}$ (Sec. 2). However, even if we allow arbitrarily incomplete lists (and even ties), it is decidable in polynomial time if the given instance admits a feasible matching [9]. Thus, it might be interesting to seek approximate solution for instances without restriction $\mathcal{Z}$. Unfortunately, however, we can easily imply its $|R|^{1-\epsilon}$-hardness (see [14]).

References

1. Abraham, D.J., Biró, P., Manlove, D.F.: "Almost stable" matchings in the room-mates problem. In: Erlebach, T., Persinao, G. (eds.) WAOA 2005. LNCS, vol. 3879, pp. 1–14. Springer, Heidelberg (2006)
2. Abraham, D.J., Irving, R.W., Manlove, D.F.: Two algorithms for the Student-Project Allocation problem. J. Discrete Algorithms 5(1), 73–90 (2007)

3. Aldershof, B., Carducci, O.M.: Stable matchings with couples. Discrete Applied Mathematics 68, 203–207 (1996)
4. Bhaskara, A., Charikar, M., Chlamtac, E., Feige, U., Vijayaraghavan, A.: Detecting high log-densities – an $O(n^{1/4})$ approximation for densest k-subgraph. In: Proc. STOC 2010, pp. 201–210 (2010)
5. Biró, P., Fleiner, T., Irving, R.W., Manlove, D.F.: The College Admissions problem with lower and common quotas. Theoretical Computer Science 411(34-36), 3136–3153 (2010)
6. Biró, P., Manlove, D.F., Mittal, S.: Size versus stability in the marriage problem. Theoretical Computer Science 411(16-18), 1828–1841 (2010)
7. Canadian Resident Matching Service (CaRMS), http://www.carms.ca/
8. Feige, U.: Relations between average case complexity and approximation complexity. In: Proc. STOC 2002, pp. 534–543 (2002)
9. Gabow, H.N.: An efficient reduction technique for degree-constrained subgraph and bidirected network flow problems. In: Proc. STOC 1983, pp. 448–456 (1983)
10. Gale, D., Shapley, L.S.: College admissions and the stability of marriage. Amer. Math. Monthly 69, 9–15 (1962)
11. Gale, D., Sotomayor, M.: Some remarks on the stable matching problem. Discrete Applied Mathematics 11, 223–232 (1985)
12. Gusfield, D., Irving, R.W.: The Stable Marriage Problem: Structure and Algorithms. MIT Press, Boston (1989)
13. Hamada, K., Iwama, K., Miyazaki, S.: The hospitals/residents problem with quota lower bounds. In: Proc. MATCH-UP (satellite workshop of ICALP 2008), pp. 55–66 (2008)
14. Hamada, K., Iwama, K., Miyazaki, S.: The hospitals/residents problem with quota lower bounds (manuscript) ,
    http://www.lab2.kuis.kyoto-u.ac.jp/~shuichi/ESA11/esa11-final-long.pdf
15. Hamada, K., Iwama, K., Miyazaki, S.: An improved approximation lower bound for finding almost stable maximum matchings. Information Processing Letters 109(18), 1036–1040 (2009)
16. Huang, C.-C.: Classified stable matching. In: Proc. SODA 2010, pp. 1235–1253 (2010)
17. Irving, R.W., Manlove, D.F., Scott, S.: The hospitals/Residents problem with ties. In: Halldórsson, M.M. (ed.) SWAT 2000. LNCS, vol. 1851, pp. 259–271. Springer, Heidelberg (2000)
18. Irving, R.W., Manlove, D.F., Scott, S.: The stable marriage problem with master preference lists. Discrete Applied Math. 156(15), 2959–2977 (2008)
19. Khot, S.: Ruling out PTAS for graph min-bisection, densest subgraph and bipartite clique. In: Proc. FOCS 2004, pp. 136–145 (2004)
20. Khuller, S., Mitchell, S.G., Vazirani, V.V.: On-Line algorithms for weighted bipartite matching and stable marriages. Theoretical Computer Science 127(2), 255–267 (1994)
21. Feige, U., Kortsarz, G., Peleg, D.: The dense k-subgraph problem. Algorithmica 29, 410–421 (2001)
22. McDermid, E.J., Manlove, D.F.: Keeping partners together: algorithmic results for the hospitals/residents problem with couples. Journal of Combinatorial Optimization 19(3), 279–303 (2010)
23. Ronn, E.: NP-complete stable matching problems. J. Algorithms 11, 285–304 (1990)
24. Roth, A.E.: The evolution of the labor market for medical interns and residents: a case study in game theory. J. Political Economy 92(6), 991–1016 (1984)
25. Vinterbo, S.A.: A stab at approximating minimum subadditive join. In: Dehne, F., Sack, J.-R., Zeh, N. (eds.) WADS 2007. LNCS, vol. 4619, pp. 214–225. Springer, Heidelberg (2007)

Multi-parameter Mechanism Design under Budget and Matroid Constraints[*]

Monika Henzinger and Angelina Vidali

Theory and Applications of Algorithms Research Group
University of Vienna, Austria

Abstract. The design of truthful auctions that approximate the optimal expected revenue is a central problem in algorithmic mechanism design. 30 years after Myerson's characterization of Bayesian optimal auctions in single-parameter domains [8], characterizing but also providing efficient mechanisms for multi-parameter domains still remains a very important unsolved problem. Our work improves upon recent results in this area, introducing new techniques for tackling the problem, while also combining and extending recently introduced tools.

In particular we give the first approximation algorithms for Bayesian auctions with multiple heterogeneous items when bidders have additive valuations, budget constraints and general matroid feasibility constraints.

1 Introduction

Assume n bidders are competing for m items. Each bidder i has a private valuation $v_{ij} \geq 0$ for item j, drawn from a publicly known distribution. Assume further there is either an *individual* matroid $\mathcal{M}_i$ for each bidder i such that each bidder can only receive an independent set of items (the *individual matroid case*) or a global matroid $\mathcal{M}$ for *all* bidders (the *global matroid case*) such that the set of all bidder-item pairs assigned should be an independent set. What is the optimal revenue maximizing auction?

In his seminal paper [8] Myerson gave a complete characterization of the optimal auction for the case $m = 1$ if the distributions of valuations are uncorrelated. Papadimititriou and Pierrakos [9] recently showed that for $n > 2$ bidders with correlated distributions finding the optimal (dominant strategy incentive compatible) deterministic auction is NP-hard, even if $m = 1$. Thus, one of the main open questions in this area is to deal with multiple items, i.e., the case of $m > 1$, when the bidders' distributions are uncorrelated. This is the problem we study in this paper together with matroid and budget constraints.

---

[*] This project has been funded by the Vienna Science and Technology Fund WWTF grant ICT10-002.

C. Demetrescu and M.M. Halldórsson (Eds.): ESA 2011, LNCS 6942, pp. 192–202, 2011.
© Springer-Verlag Berlin Heidelberg 2011

Truthfulness. For any mechanism there are various criteria for evaluation. One criterion is which notion of *truthfulness* or *incentive compatibility* it fulfills. Every definition of truthfulness involves some notion of *profit* or *optimality* for a bidder. In our setting we assume bidder i receives a set S_i and has to pay p_i for it. Then the *profit* of bidder is $\sum_{j \in S_i} v_{ij} - p_i$. An *optimal* outcome for a given bidder is an outcome that maximizes his profit. We distinguish three notions of truthfulness. (1) A mechanism is *dominant strategy incentive compatible (DSIC)* if truth-telling is optimal for the bidder even if he knows the valuations of the other bidders and the random choices made by the mechanism[1]. (2) A mechanism is *truthful in expectation* if revealing the true value maximizes the *expected* profit of every bidder, where the expectation is taken over the internal random coin flips of the mechanism. (3) If a prior distribution of the bidders' valuations is given, then a mechanism is *Bayesian incentive compatible (BIC)* if revealing the true value maximizes the *expected* profit of every bidder, where the expectation is over the internal random coin flips of the mechanism *and* the valuations of the *other* bidders.

Optimal or approximate. The revenue, of a mechanism is the sum of the payments collected by the auctioneer $\sum_i p_i$. If a mechanism returns the maximum revenue out of all mechanisms fulfilling a certain type of constraints (e.g. all BIC mechanisms), it is an *optimal* mechanism. If it returns a fraction k of the revenue of the optimal mechanism, we call it a *k-approximation*.

Value distributions. If a prior distribution on the bidders' valuations is assumed then there are also multiple cases to distinguish. In the *correlated bidders setting* the value distributions of different bidders can be correlated. Except for [6] and [9], all prior work and also our work assumes that the distributions of different bidders are uncorrelated. We further distinguish the case that for each bidder the distributions of different items are uncorrelated (*the independent items case*), and the case that the value distributions of the same bidder for different items to be correlated (*the correlated items case*). There is strong evidence that it is not possible to design an optimal DSIC mechanism for the correlated items case [1]: Even if there is just *one unit-demand* bidder, but his valuations for the items are correlated, the problem of assigning the optimal item to the bidder can be reduced to the problem of unlimited supply *envy-free pricing* with m bidders [7]. For the latter problem the best known mechanism is a logarithmic approximation and there is strong evidence that no better approximation is possible [2].

Running time model. A final criterion to evaluate a mechanism is whether it runs in time polynomial in the input size. Of course this depends on how the input size is measured. We use the model used in [1], where the running time has to be polynomial in n, m, and the support size of the valuation distributions. All the results we list take polynomial time in this model.

---

[1] This is independent of whether a distributions of the valuations are given or not.

Related work. *Correlated bidders*. Dobzinski, Fu, and Kleinberg [6] gave an optimal *truthful-in-expectation* mechanism for $m \geq 1$ in the *correlated* bidders and items setting but *without* any matroid or budget constraints.

Uncorrelated bidders. Chawla et al. [4] studied the case $m \geq 1$ with a universal matroid constraint and general valuation distributions, but with only *unit-demand* bidders *without* budget constraints. For a variety of special matroids, like uniform matroids and graphical matroids, they gave constant factor approximations. Very recently, Chawla et al. [5] gave a constant factor approximation for general matroid constraints *with* budgets, but again only with *unit-demand* bidders. Bhattacharya et al. [1] studied the case of individual uniform matroid constraints and budget constrained bidders. For the correlated items case they presented a BIC mechanism whose revenue is within a factor 4 of the optimal BIC mechanism. Their mechanism is truthful only in expectation. For the independent items case if the valuations additionally fulfill the *monotone hazard rate assumption (MHR)* (see Section 4) they gave a DSIC mechanism that achieves a constant factor of the revenue of the optimal BIC mechanism. For the independent items case where the valuations do *not* fulfill MHR they gave a DSIC mechanism that achieves an $O(\log L)$ approximation of the revenue of the optimal DSIC mechanism and they showed that no better posted-price (defined below) mechanism exists. Here L is the maximum value that any bidder can have for any item. In a very recent work Cai et al. [3] give almost optimal mechanisms for the case when either the items or the bidders are i.i.d. and there exist budget and uniform matroid constraints.

Our results. We use the same model as Bhattacharya et al. [1], i.e., both matroid and (public) budget constraints. We improve upon their work since they studied only individual matroid constraints where the matroid is a uniform matroid. Specifically we show the following results. (1) For the correlated items with individual matroid constraints case we present a BIC mechanism whose revenue is within a factor 2 of the optimal BIC mechanism. In [1] a 4-approximation for uniform matroids was given. (2) For the independent items case we study general matroid constraints, both in the global and the individual setting. Our mechanisms are DSIC sequential posted price mechanisms [2]. Our results are summarized in the following table:

Our results on global matroid constraints are a generalization of the work by Chawla et al. [4,5]. They gave a constant approximation for global uniform matroids and global graphical matroids, and very recently in [5] also for general matroids (in [5] however the authors do not provide a polynomial-time

---

[2] In [1] a *sequential posted-price (spp)* mechanism is defined as follows: The bidders are considered sequentially in arbitrary order and each bidder is offered a subset of the remaining items at a price for this item and bidder; the bidder simply chooses the profit maximizing bundle out of the offered items. These mechanisms have the advantage of being more practical as they do not require from the bidders to report their valuations but only to take or leave items with posted prices. Experimental evidence also suggests that in spp players tend to act more rationally and they are more likely to participate.

	Individual matroids	Global matroid
General matroids	$O(\log L)$	$O(\log L)$
Uniform matroids	(previous: $O(\log L)$ [1])	$O(\log L)$
General matroids & MHR	$O(\log m), O(k)$	$O(\log m), O(k)$
Uniform matroids & MHR	9, (previous: 24 [1])	9
Graphical matroids & MHR	16	64 with budgets and 3 without

Fig. 1. Independent distributions case: A summary of our and earlier results. Multiply the given constants by $8e^2$ to get the approximation ratio.

algorithm), but only for the special case when the bidders are *unit demand* (which is reducible to the single-parameter problem). We give constant approximations for *bidders with arbitrary demands* (i.e. for the general multi-parameter problem) for the case of global uniform and graphical matroids however with the assumption that the valuation distributions fulfill the monotone hazard rate condition. All our results take polynomial time.

Our tools and techniques. The basic idea of [1] and [6] is to solve a linear program to determine prices and assignment probabilities for bidder-item pairs. We use the same general approach but extend the linear programs of [1] by suitable constraints that are (i) "strong enough" to enable us to achieve approximation ratios for *general* matroids, but also (ii) "weak enough" so that they can still be solved in polynomial time using the *Ellipsoid method with a polynomial-time separation oracle*. In the correlated items case the results of this new LP together with a modified mechanism and a careful analysis lead to the *improved approximation factor over [1], even with general matroid constraints.*

In the independent items case Bhattacharya et al. [1] used Markov inequalities to reason that uniform matroid constraints and budget constraints reduce the expected revenue only by a constant factor. This approach, however, exploits certain properties of uniform matroids and cannot be generalized to graphical or general matroids. Thus, we extended ideas from Chawla et al [4] to develop different *techniques to deal with non-uniform matroid constraints*: (1) For graphical matroids we combine a graph partitioning technique and prophet inequalities [10]. (2) For general matroids we use Lemma 3 (see also Theorem 10 in [4]) together with a bucketing technique. The lemma says roughly that if a player is asked the same price for all items then the matroid constraints reduce the expected revenue by at most a factor of 2 in the approximation. As we show it holds both in the global as well as in the individual matroids setting. In combination with a bucketing technique that partitions the items into buckets so that all items in the same bucket have roughly the same price the lemma allows us to tackle general matroid constraints in all of our non-constant approximation algorithms. The generality of the lemma makes it very likely that it is further applicable.

We also develop a new way to deal with *budget constraints* that simplifies the proofs and enables us to *improve the approximation factors*, e.g. for uniform matroids from 24 [1] to 9.

Apart from improving and extending recent results, more importantly our paper sheds light on multi-parameter Bayesian mechanism design and evolves, combines and proposes alternatives for important recent techniques. Thus it represents one further step towards the better understanding of this very important, timely and still wide open problem.

The paper is organized as follows. The next section contains all necessary definitions. Section 3 presents the result for correlated, Section 4 for independent valuations. All missing proofs and details can be found in the full version of the paper.

2   Problem Definition

There are n bidders and a set J of m distinct, indivisible items. Each bidder i has a private valuation $v_{ij} \geq 0$ for each item j drawn from a publicly known distribution $\mathcal{D}_{i,j}$. Additionally each bidder has a budget B_i and cannot be charged more than B_i. If bidder i receives a subset S_i of items and is charged p_i for it then the *profit* of bidder i is $\sum_{j \in S_i} v_{ij} - p_i$. Bidders are *individually rational*, i.e. bidder i only selects S_i if his profit in doing so is non-negative. A bidder is *individually rational in expectation* if his expected profit is non-negative. The goal of the mechanism is to maximize its *revenue* $\sum_i p_i$ under the constraint that $p_i \leq B_i$ for all i, that all bidder are individually rational or at least individually rational in expectation, and that each item can be sold only once. Additionally there are matroid constraints on the items. We analyze two types of matroid constraints: In the *universal matroid constraint* problem there exists *one* matroid $\mathcal{M}$ such that $\cup_i S_i$ has to be an independent set in $\mathcal{M}$. In the *individual matroid constraint* problem there exists one matroid $\mathcal{M}_i$ *for each* bidder i such that S_i has to be an independent set in $\mathcal{M}_i$.

Assumptions. We make the same assumptions as in [1,6] (1) For all i and j the number of valuations with non-zero probability, i.e., the support of $\mathcal{D}_{ij}$ is finite. The running time of our algorithms is polynomial in n, m, and the size of the support of $\mathcal{D}_{ij}$ for all bidders i and items j, i.e., in the *size of the input*. (2) The random variable v_{ij} takes only rational values and that there exists an integer L polynomial in the size of the input such that for all i, j, $1/L \leq v_{ij} \leq L$. (3) For each of the matroids $\mathcal{M}_i$ if given a subset S of J we can in time polynomial in the size of the input compute $rank_{\mathcal{M}_i}(S)$ and determine whether S is independent in $\mathcal{M}_i$ or not.

3   Correlated Item Valuations

Here we study the setting that the distribution of the valuations of a *fixed* bidder for different items can be arbitrarily correlated, while the distributions of different bidders are independent. To model this setting we assume that (a) the valuations of a bidder i are given by its *type*, (b) there is a publicly known probability distribution $f_i(t)$ on the types of bidder i with finite support, and

(c) $(v_{t1}, \ldots, v_{tm})$ is the vector of valuations for item $1, \ldots, m$ for a bidder with type t. Additionally we assume in this section that every probability $f_i(t)$ is a rational number such that $1/L \leq f_i(t) \leq 1$, where L is polynomial in the size of the input. We present a BIC mechanism that gives a 2-approximation of the optimal revenue.

The mechanism works as follows: Based on the distributions $\mathcal{D}_{ij}$ the mechanism solves a linear programming relaxation of the assignment problem, whose objective function is an upper bound on the value achieved by the optimal mechanism. The linear program returns values for variables $y_{iS}(t)$, where S is an independent set in $\mathcal{M}_i$, for "payment" variables $p_i(t)$ for each i and t, and for variables $x_{ij}(t)$ for each i, j, and t. Then the mechanism interprets y_{iS} as the probability that the optimal BIC mechanism assigns S to i and picks an assignment of items to i based on the probability distribution $y_{iS}(t_i)$, where t_i is the type reported by bidder i. The constraints in the linear program guarantee that the mechanism is BIC.

(LP1)   Maximize $\sum_{i,t} f_i(t)p_i(t)$ such that

$$\forall i,j,t \quad m_{ij} - x_{ij}(t) \geq 0 \tag{1}$$

$$\forall j \quad -\sum_i m_{ij} \geq -1 \tag{2}$$

$$\forall i,t,s \quad \sum_j v_{tj} x_{ij}(t) - \sum_j v_{tj} x_{ij}(s) - p_i(t) + p_i(s) \geq 0 \tag{3}$$

$$\forall i,t \quad \sum_{j \in J_i} v_{tj} x_{ij}(t) - p_i(t) \geq 0 \tag{4}$$

$$\forall i,t,j \quad \sum_{\text{independent } S \text{ with } j \in S} y_{iS}(t) - x_{ij}(t) = 0 \tag{5}$$

$$\forall i,t \quad \sum_{\text{independent } S} -y_{iS}(t) \geq -1 \tag{6}$$

$$\forall i,j,t \quad -x_{ij}(t) \geq -1 \tag{7}$$

$$\forall i,t \quad -p_i(t) \geq -B_i \tag{8}$$

$$\forall i,j,t : x_{ij}(t) \geq 0, \forall i,t : p_i(t) \geq 0, \forall i,t,S : y_{iS}(t) \geq 0, \forall i,j : m_{ij} \geq 0 \tag{9}$$

Note that the optimal BIC mechanism is a feasible solution to LP1: Set $x_{ij}(t)$ to the probability that the mechanism assigns item j to bidder i when the bidder reports type t and $y_{iS}(t)$ to the probability that the mechanism assigns set S to bidder i when the bidder reports type t. This assignment fulfills all constraints of LP1, i.e. it gives a feasible solution to LP1. Thus LP1 has a solution and its optimal solution gives an upper bound on the revenue of the optimal BIC mechanism. LP1 has an exponential number of variables but using the fact the dual LP has only a polynomial number of variables and a polynomial time separation oracle, we show that LP1 can be solved in polynomial time.

Lemma 1. *The linear program LP1 can be solved in polynomial time.*

For each bidder i we treat y_{iS} as a probability distribution over the independent sets S of $\mathcal{M}_i$ and pick an independent set T_i according to that probability distribution. We define for all items j, $Z_{0j} = 1$ and for all items j and bidders i let $Z_{ij} = 1 - \sum_{i' < i} x_{i'j}(t_{i'})/2$. Note that $Z_{ij} \geq 1/2$ and thus $1/(2Z_{ij}) \leq 1$. The mechanism assigns the items to bidders as follows:

1. $A = J$
2. For $i = 1, 2, \ldots, n$
   (a) Pick an indep. set T_i using the distribution $y_{S,i}(t_i)$; set $S_i = \emptyset$
   (b) for each $j \in T_i$: if $j \in A$ then with probability $1/(2Z_{ij})$ do:
       i. $S_i = S_i \cup \{j\}$; $A = A - \{j\}$
   (c) Bidder i gets S_i and pays $p_i(t_i)/2$.

One can show that $P(j \in S_i) = x_{ij}(t_i)/2$. Thus (3) and (4) of LP1 together with the fact that $p_i = p_i(t_i)/2$ guarantee incentive compatibility in expectation and individual rationality in expectation.

Theorem 1. *The above mechanism is Bayesian incentive compatible, individually rational in expectation, and its revenue is a 2-approximation to the optimal BIC mechanism.*

4   Independent Item Valuations

In this section we assume that for each bidder i the distributions of v_{ij} for different j are independent. The goal is to achieve for this case a stronger notion of truthfulness, namely a DSIC instead of a BIC mechanism. All mechanisms in this section are sequential-posted-price mechanisms.

For each item j to bidder i let $\mathcal{V}_{ij} := \min\{v_{ij}, B_i/4\}$ and let f_{ij} be its density function, i.e. $\mathcal{V}_{ij} \sim f_{ij}$. We assume that for all i, j and r all values $f_{ij}(r)$ are rational numbers.

Bhattacharya et al. [1] formulated an LP with variables $x_{ij}(r)$, where $x_{ij}(r)$ denotes the expected amount of item j bidder i gets when $\mathcal{V}_{ij} = r$. We modify their LP by generalizing their constraint for uniform matroids to general matroids and call it LP2. As a result the LP has now an exponential number of constraints, but it can still be solved in polynomial time using the fact that is has only a polynomial number of variables.

Lemma 2. *The maximum value of the optimal solution for LP2 achieves at least $1/4$ of the revenue of the optimal BIC mechanism. Additionally there exists an optimal solution for LP2 such that $x_{ij}(r)$ is a monotonically non-decreasing function of r. The solution can be computed in time polynomial in the size of the input.*

The following Lemma is an important tool, that we repeatedly use. It says that by giving the bidder the freedom to choose the independent set he likes, and not assigning him the set with the maximum revenue one will not loose more than a factor of 2 of the maximum possible revenue.

Lemma 3. *Assume that bidder i is offered a set S of items where each item has the same price r_i. Let q_{ij} be the probability, that i is offered item j and picks it. Assume that q_{ij} satisfies that for every subset T of S, $\sum_{j \in T} q_{ij} \leq rank(T)$. Let S_i be the independent set of $\mathcal{M}_{\rangle}$ picked by the individual rational bidder i, and let $R_i = \sum_{j \in S_i} q_{ij} r_i$ be the expected revenue from S_i. Then $\sum_{j \in S \setminus S_i} q_{ij} r_i \leq R_i$, i.e. the expected revenue lost because the items in $S \setminus S_i$ were not picked is at most R_i.*

In the appendix we state the equivalent lemma for the case of universal matroid constraints. Using a bucketing technique we show the following result, which does not make any assumptions on the hazard rate.

Theorem 2. *Assume for all i and j, $v_{ij} \in [1, L]$ follow independent distributions f_{ij}. Then there is a $O(\log L)$ approximation of the revenue of the optimal BIC mechanism through a spp mechanism under any matroid constraint.*

4.1   Valuation Distributions with Monotone Hazard Rate

In the previous section we gave a $\Theta(\log L)$ spp mechanism for general matroids. Since this matches the known lower bound for spp mechanisms and the lower bound is achieved by a distribution that satisfies regularity [1], the natural question to ask is whether we can do better for valuations v_{ij} whose distributions satisfy the monotone hazard rate condition[3] [8].

We modify the LP given by Bhattacharya et al [1] for the special case of uniform, individual matroids to work for both general, individual matroids and general universal matroids. The resulting LP3 has an exponential number of constraints but the same argument as in Lemma 2 shows that the LP can be solved in polynomial time. We then generalize the proof of the two subsequent lemmata to work for the modified LPs.

Lemma 4 ([1]). *If the valuations follow distributions that fullfill the MHR condition, then the revenue of LP3 is at least $\frac{1}{2e^2}$ times the revenue of LP2.*

Lemma 5 ([1]). *The optimal solution to (LP3) satisfies the following property. For all i, j x_{ij} be decomposed in polynomial time in the following way: $x_{ij} = p_{ij} y_{ij} + (1-p_{ij}) z_{ij}$, where $r_{ij}^* + 1 \leq \frac{B_i}{4}$, $r_{ij}^* \geq 1$, $y_{ij}(r) := 0$ for $r < r_{ij}^*$ and else 1, and $z_{ij}(r) := 0$ for $r < r_{ij}^* + 1$ and else 1.*
  Define $R_{ij} := p_{ij} r_{ij}^* P(\mathcal{V}_{ij} \geq r_{ij}^*) + (1 - p_{ij})(1 + r_{ij}^*) P(\mathcal{V}_{ij} \geq r_{ij}^* + 1)$, and $q_{ij} := p_{ij} P(\mathcal{V}_{ij} \geq r_{ij}^*) + (1 - p_{ij}) P(\mathcal{V}_{ij} \geq r_{ij}^* + 1)$.
  Then (a) $\sum_r x_{ij}(r) \phi_{ij}(r) f_{ij}(r) = R_{ij}$, and (b) $\sum_r x_{ij}(r) f_{ij}(r) = q_{ij}$.

---

[3] The function $h(r) = f(r)/(1 - F(r))$ is called the *hazard rate* of f. The probability distribution f_{ij} has a *monotone hazard rate (MHR)* if $1/h_{ij}(r)$ is non-decreasing as a function of r. The function $\phi_{ij}(r) = r - 1/h_{ij}(r)$ is called the *virtual valuation* of player i for item j. The distribution is called *regular* if the virtual valuation is a non-decreasing function of r. Clearly, MHR distributions are regular, but the converse doesn't hold.

The optimal mechanism extracts a revenue of $2kR_{ij}$ for each bidder item pair (i, j), while our mechanism achieves an expected revenue of $R_{ij}/3$, where k is the maximum rank in the matroids $\mathcal{M}_i$.

Theorem 3. *Assume for all i and j, $v_{ij} \in [1, L]$ follow independent distributions f_{ij} and satisfy the monotone hazard rate condition. Then there is an $O(k)$ approximation of the revenue of the optimal BIC mechanism, through a spp mechanism under any matroid constraint.*

To achieve an $O(\log m)$ approximation we bucket for each bidder i all the items above a certain threshold according to their r_{ij}^* value. Note that this is different from the bucketing in the proof of Theorem 2. Then we pick the bucket with the largest expected revenue, assign all items in it the same price, and let the bidder pick an independent set of items from the bucket. We use Lemma 3 to show that the expected revenue lost due to the matroid constraints is only a factor of 2. Here is the mechanism in detail for individual matroids (for global matroids see the full version of the paper):

1. Solve LP3 to get the $x_{ij}(r)$ values, set for all i, j and r, $\tilde{x}_{ij}(r) = x_{ij}(r)/2$. Decompose $\tilde{x}_{ij}(r)$ to get p_{ij} and r_{ij}^* values for all i and j. Set $R_{ij} = p_{ij} r_{ij}^* P(\mathcal{V}_{ij} \geq r_{ij}^*) + (1 - p_{ij})(1 + r_{ij}^*) P(\mathcal{V}_{ij} \geq r_{ij}^* + 1)$.
2. For each bidder i do
   (a) Set $OPT_i = \sum_j R_{ij}$, let $r_i^{max} = \max_j\{r_{ij}^* + 1\}$, and let $r_i^{min} = OPT_i/m^2$.
   (b) For $l = \lfloor \log(r_i^{min}) \rfloor$ to $\lfloor \log(r_i^{max}) \rfloor$ do: Set $\Gamma_l = \emptyset$
   (c) For all items j with $r_{ij}^* \geq r_{min}$ do:
       i. Set $k = \lfloor \log r_{ij}^* \rfloor$, $\Gamma_k = \Gamma_k \cup \{(j, p_{ij} \cdot P(\mathcal{V}_{ij} \geq r_{ij}^*)/P(\mathcal{V}_{ij} \geq 2^k))\}$, set $k' = \lfloor \log(r_{ij}^* + 1) \rfloor$ and $\Gamma_{k'} = \Gamma_{k'} \cup \{(j, (1 - p_{ij}) \cdot P(\mathcal{V}_{ij} \geq r_{ij}^* + 1)/P(\mathcal{V}_{ij} \geq 2^{k'}))\}$
   (d) Set $\mathcal{B}_i := \Gamma_{k_i}$ with $k_i = \operatorname{argmax}_k \sum_{(j,p) \in \Gamma_k} p 2^k P(\mathcal{V}_{ij} \geq 2^k)$.
3. $A = J$ and order the bidders arbitrarily.
4. For $i = 1, 2, \ldots, n$
   (a) Let $S = \emptyset$.
   (b) For every item $j \in A$: If $(j, p) \in \mathcal{B}_i$ then with probability p add item j to S with a price of 2^{k_i}.
   (c) Let the bidder pick an independent subset S_i of S such that $|S_i| 2^{k_i} \leq B_i$. Assign S_i to i at a cost of $|S_i| 2^{k_i}$ and set $A = A \setminus S_j$.

Lemma 6. *Suppose $v_{ij} \in [1, L]$, that follow independent distributions f_{ij} for different (i, j), satisfying the monotone hazard rate condition and that the items allocated are subject to a matroid constraint. Then for every bidder i the above mechanism is an $O(\log(r_i^{max}/r_i^{min}))$ approximation of OPT_i under any matroid constraint.*

We sketch the proof of the lemma: Since all items with r_{ij}^* value below r_i^{min} contribute at most OPT_i/m in total to the revenue collected from bidder i, omitting them decreases the collected revenue at most by a factor of $(1 - 1/m)$.

Since the mechanism picks the bucket $\mathcal{B}_i$ with largest expected revenue and there are only $\log{(r_i^{max}/r_i^{min})}$, the mechanism would collect an expected revenue of at least $OPT_i/(c\log{(r_i^{max}/r_i^{min})})$ for some constant c from the items in $\mathcal{B}_i$ if bidder i was not restricted to picking only an *independent* set. This holds even if it charges the *same* price for all items in $\mathcal{B}_i$, as this "rounding" only affects the constant in the approximation. As Lemma 3 shows in this setting restricting the bidder to pick an *independent* set decreases the expected revenue collected from the bidder at most by another factor of 2.

If we assume $f_{ij}(r) \geq 1/m^c$ then $r_i^{\max} \leq m^{c+1}OPT_i$ and $r_i^{\min} = OPT_i/m^2$. The corollary follows.

Corollary 1. *If $f_{ij}(r) \geq 1/m^c$ for all i, j, and r and some constant c, then our mechanism is an $O(\log m)$ approximation of the optimal revenue.*

4.2   Constant Approximations for Specific Matroids and MHR

If the valuations v_{ij} follow regular distributions and the feasibility constraint is described by a k-uniform matroid, [1] gives a constant approximation algorithm for the optimal expected revenue. We improved the approximation ratio from 24 [1] to 9, by arguing differently (and more simply) about the fulfillment of the budget constraints: *If the expected revenue from a bidder i is at least $3/4B_i$, the mechanism achieved a constant factor of the optimal revenue for i. Otherwise, since $\mathcal{V}_{ij} \leq B_i/4$, the budget constraints did not keep i from taking more items and can be ignored in the future analysis of the revenue collected from i.* The proof also easily extends for the case of a universal matroid constraint.

Theorem 4. *Assume that for all i and j, $v_{ij} \in [1, L]$ follow independent distributions f_{ij} and satisfy the monotone hazard rate condition. There is a constant approximation of the revenue of the optimal BIC mechanism, that achieves a 9-approximation to LP3, through a spp mechanism under global or individual k-uniform matroid constraints.*

In the case of graphical matroids we use a graph decomposition technique employed by [4] for matroids with budgets constraints and prophet inequalities for matroids without budgets contraints to achieve the following results.

Theorem 5. *If for all i and j, $v_{ij} \in [1, L]$ follow independent distributions f_{ij} and satisfy the monotone hazard rate condition, then there is a constant approximation of the revenue of LP3, through a spp mechanism under global or individual graphical matroid constraints.*

Acknowledgments. We would like to thank Tim Roughgarden for proposing the topic and many helpful comments, insights and discussions.

References

1. Bhattacharya, S., Goel, G., Gollapudi, S., Munagala, K.: Budget constrained auctions with heterogeneous items. In: STOC, pp. 379–388 (2010)
2. Briest, P.: Uniform budgets and the envy-free pricing problem. In: Aceto, L., Damgård, I., Goldberg, L.A., Halldórsson, M.M., Ingólfsdóttir, A., Walukiewicz, I. (eds.) ICALP 2008, Part I. LNCS, vol. 5125, pp. 808–819. Springer, Heidelberg (2008)
3. Cai, Y., Daskalakis, C., Matthew Weinberg, S.: On optimal multidimensional mechanism design. SIGecom Exchanges 10(2) (2011)
4. Chawla, S., Hartline, J.D., Malec, D.L., Sivan, B.: Multi-parameter mechanism design and sequential posted pricing. In: STOC, pp. 311–320 (2010)
5. Chawla, S., Malec, D.L., Malekian, A.: Bayesian mechanism design for budget-constrained agents. In: EC 2011 (2011)
6. Dobzinski, S., Fu, H., Kleinberg, R.: Optimal auctions with correlated bidders are easy. In: STOC 2011 (2011)
7. Guruswami, V., Hartline, J.D., Karlin, A., Kempe, D., Kenyon, C., McSherry, F.: On profit-maxmizing envy-free pricing. In: SODA, pp. 1165–1173 (2005)
8. Myerson, R.B.: Optimal auction design. Mathematics of Operations Research 6, 58–73 (1981)
9. Papadimitriou, C.H., Pierrakos, G.: On optimal single-item auctions. In: STOC (2011)
10. Samuel-Cahn, E.: Comparison of threshold stop rules and maximum for independent nonnegative random variables. The Annals of Probability 12(4), 1213–1216 (1984)

Quantified Linear Programs: A Computational Study[*]

Thorsten Ederer[1], Ulf Lorenz[1], Alexander Martin[2], and Jan Wolf[1]

[1]Institute of Mathematics, Technische Universität Darmstadt, Germany
{ederer,lorenz,wolf}@mathematik.tu-darmstadt.de
[2]Institute of Mathematics, Universität Erlangen-Nürnberg, Germany
alexander.martin@math.uni-erlangen.de

Abstract. Quantified linear programs (QLPs) are linear programs with variables being either existentially or universally quantified. The integer variant (QIP) is PSPACE-complete, and the problem is similar to games like chess, where an existential and a universal player have to play a two-person-zero-sum game. At the same time, a QLP with n variables is a variant of a linear program living in $\mathbb{R}^n$, and it has strong similarities with multi-stage stochastic linear programs with variable right-hand side. Our interest in QLPs stems from the fact that they are LP-relaxations of QIPs, which themselves are mighty modeling tools. In order to solve QLPs, we apply a nested decomposition algorithm. In a detailed computational study, we examine, how different structural properties like the number of universal variables, the number of universal variable blocks as well as their positions in the QLP influence the solution process.

1 Introduction

In the 1940s, linear programming arose as a mathematical planning model and rapidly found its daily use in many industries. However, integer programming, which was introduced in 1951, became dominant far later at the beginning the 1990s. Certainly, one reason for the delay of the integer programming success story stems from the fact that linear programming resides in the complexity class P, while integer programming is NP complete. Nowadays, we are able to solve very large mixed integer programs of practical size, but companies observe an increasing danger of disruptions, i.e., events occur which prevent companies from acting as planned. Therefore, there is a need for planning and deciding under uncertainty. Uncertainty, however, often pushes the complexity of traditional optimizations problems, which are in P or NP, to PSPACE. The quantified versions of linear integer programs cover the complexity class PSPACE. The relaxed versions, which we examine in this paper, additionally have remarkable polyhedral properties. The idea of our research is to explore the abilities of linear programming when applied to PSPACE-complete problems, similar as it was applied to NP-complete problems in the 1990s.

1.1 State-of-the-Art

For traditional deterministic optimization one assumes data for a given problem to be fixed and exactly known when the decisions have to be taken. However, data are often afflicted with some kinds of uncertainties, and only estimations, maybe in form of

---
[*] Research partially supported by German Research Foundation (DFG) funded SFB 805.

C. Demetrescu and M.M. Halldórsson (Eds.): ESA 2011, LNCS 6942, pp. 203–214, 2011.

probability distributions, are known. Examples are flight or travel times. Throughput-time, arrival times of externally produced goods, and scrap rate are subject to variations in production planning processes. One possibility to deal with these uncertainties is to aggregate a given probability distribution to a single estimated number. Then, the optimum concerning these estimated input data can be computed with the help of traditional optimization tools. In some fields of application, as e.g. the fleet assignment problem of airlines, this procedure was successfully established. In other fields, like production planning and control, this technique could not be successfully applied, although mathematical models do exist [22]. Some newer approaches also deal with stochastic aspects of special optimization problems [11,16,20,19,28].

Prominent solution paradigms for optimization under uncertainty are Dynamic Programming [3], Sampling [15], the exploration of Markov-Chains [34], Robust Optimization [17], and Stochastic Programming [7,11,27,10]. Markov-Chains and Dynamic Programming are often used for problems from complexity class P, but for more complex problems, other algorithms are often faster, like e.g. the Alphabeta algorithm for two-person zero-sum games [9]. At two-stage Stochastic Programming, a set of initial decisions are taken first, followed by a random event. After this, recourse decisions are taken, which allow to compensate for events that have been observed in the previous stage. The multi-stage problem, accordingly consists of multiple stages, with a random event occurring between each stage. Such a problem can be transformed into a so-called deterministic equivalent program (DEP) and then be solved with the help of linear programming. An appropriate procedure for solving multi-stage stochastic programs are the nested decomposition procedure and its variants [7].

We can interpret a set of expressions, which can encode a PSPACE-complete problem, as a very powerful modeling language: more powerful than necessary to encode any NP-complete problem. The fact that it is not possible to find polynomial time algorithms for all problems that are encoded with the help of such a powerful modeling language, leads to the consequence that research for new solutions must be driven from the application-side or even from the instances-side, as e.g. presented in [16]. Relatively unexplored are the abilities of linear programming extensions in the PSPACE-complete world. In this context, Subramani introduced the notion of quantified linear programs (QLP) [30,31]. While it is known that quantified linear integer programs describe PSPACE, the complexity class of their LP-relaxations, i.e. quantified linear programs is unknown. In [18], we were able to show that the solution space of QLPs forms a polytope, and the problem is in PSPACE.

In Section 2, we formally describe the QLP-problem. In Section 3, we describe how the Benders Decomposition can be recursively applied to solve QLPs, followed by a detailed computational study in Section 4.

2   The Problem Statement: Quantified Linear Programs

Within this paper, we intend to concentrate on quantified linear programs, as they were introduced in [31,30], and in-depth analyzed in [18]. We present the definition of a QLP in form of a decision problem.

Definition 1. *(QLP-Instance) Given some vector of variables* $x = (x_1, \ldots, x_n)^T \in \mathbb{Q}^n$, *upper and lower bounds* $u \in \mathbb{Z}^n$ *and* $l \in \mathbb{Z}^n$ *with* $l_i \leq x_i \leq u_i$, *a matrix* $A \in \mathbb{Q}^{m \times n}$, *a vector* $b \in \mathbb{Q}^m$ *and a quantifier string* $Q = (q_1, \ldots, q_n) \in \{\forall, \exists\}^n$, *where the quantifier* q_i *belongs to the variable* x_i, *for all* $1 \leq i \leq n$. *We denote by* $[Q \circ x : Ax \leq b, l \leq x \leq u]$ *an instance of a quantified linear program (QLP), where* $Q \circ x$ *denotes* $(q_1 x_1, \ldots, q_n x_n)^T$.

A maximal subset of Q, which contains a consecutive sequence of quantifiers of the same type, is called a (variable-) *block*. For ease of explanation, we sometimes denote the universal variables by y and the existential by x and rewrite the system as $[Q \circ \binom{x}{y} : A \binom{x}{y} \leq b]$ for short, ignoring the sequence of the variables if it is clear from the context.

We interpret each QLP-instance as a two-person zero-sum game between an *existential player* setting the $\exists$-variables and a *universal player* setting the $\forall$-variables. Each fixed vector $x \in \mathbb{Q}^n$, that is when the existential player has fixed the existential variables and the universal player has fixed the universal variables, is called a *game*. If x satisfies $Ax \leq b$, we say the existential player wins. The variables are set in consecutive order according to the quantifier string Q. Consequently, we say that a player makes the move $x_k = z$, if he fixes the variable x_k to the value z. At each such move, the corresponding player knows the settings of $x_1, \ldots, x_{i-1}$ before setting x_i.

Definition 2. *(Strategy) Given a QLP-instance* $[Q \circ \binom{x}{y} : A \binom{x}{y} \leq b, l \leq \binom{x}{y} \leq u]$, *a strategy* $(V_x \dot\cup V_y, E, c)$ *for the existential player* S *is a labeled tree of depth* n, *where* V_x *and* V_y *are two disjoint sets of nodes. Nodes from* V_x *represent existential variables, nodes from* V_y *universal variables. Each tree level* i *consists either of only existential nodes or of only universal nodes and is associated with variable* x_i. *Each edge* $e = (v, w) \in E$, *with* v *being a node at level* i *and* w *being a node at level* $i+1$, *represents an assignment of variable* x_i *to the label* c_e, *where* $c \in Q^{|E|}$ *and* $l_i \leq c_e \leq u_i$. *Existential nodes have one successor, universal nodes have two successors, one representing* $y_i = 0$ *and* $y_i = 1$ *the other* [1]. *A strategy is called a* winning *strategy, if all paths from the root to a leaf represent a vector* $\binom{x}{y}$ *such that* $A \binom{x}{y} \leq b$.

Definition 3. *(Policy) Given a QLP-instance* $[Q \circ x : Ax \leq b, l \leq x \leq u]$, *a* policy *is an algorithm that fixes a variable* x_i *with the knowledge, how* $x_1, \ldots, x_{i-1}$ *have been set before. Observe that a policy implements a set of computable functions of the form* $x_i = f_i(x_1, \ldots, x_{i-1})$ *for all existentially quantified variables* x_i. *A policy is called a* winning *policy, if it implements functions* $f_1, \ldots, f_m$, *m being the number of existentially quantified variables, such that* $Ax \leq b$, *independently of the universal-player's actions.*

Definition 4. *(QLP-Problem) Given a QLP-instance* $[Q \circ x : Ax \leq b, l \leq x \leq u]$, *is there a winning policy for the existential player?*

It has been shown that the restricted QLP problem with only one quantifier-change is either in P (when the quantifier string begins with existential quantifiers and ends with universal ones) or is coNP-complete (when the quantifier string begins with universal quantifiers and ends with existential ones) [31].

---

[1] According to the results from [18] it suffices to mention the case $x_i \in 0, 1$.

Moreover, one may distinguish between QIPs (quantified integer programs with $x \in \mathbb{Z}$), QLPs where the variables of the existential player are rational numbers and the universal variables are integer numbers, and QLPs where all variables are rational numbers. However, it was shown [18] that

- If $l_1, ..., l_m$ are the lower bound restrictions to the universal variables $y_1, .., y_m$ of a given QLP, and $u_1, ..., u_m$ the corresponding upper bounds, the existential player has a winning strategy against the universal player who takes his choices $y_i \in \{l_i, ..., u_i\}, i \in 1...m$, if and only if the existential player has a winning policy against the universal player who takes his choices from the corresponding continuous intervals, i.e. $y_i \in [l_i, u_i], i \in 1...m$.
- Whether or not there is a winning strategy for the existential player can be determined with polynomially many bits.

3   How to Solve QLP Problems

Besides the *Quantifier Elimination Algorithm* as described by Subramani [31], which is known to use double exponential space in the worst case, another possibility to solve QLPs is described in the following. The proposed algorithm relies on decomposition techniques that are widely used in the stochastic programming community to solve stochastic programs, which have some interesting similarities to QLPs.

In the following, we motivate out approach by showing similarities between QLPs and stochastic linear programming problems (SLPs). We show how QLPs can be transformed into deterministic equivalent linear programs (DEPs), which can have very large scale, but can be tackled by decomposition techniques due to their special matrix structure. At the end, we describe some details of the implementation of our algorithm.

3.1   Equivalent Linear Program Formulations of SLPs and QLPs

Stochastic programming problems can be essentially divided into two-stage and multi-stage problems (MSSLPs). In the former, a set of initial decisions are taken first, followed by a random event. After this, recourse decisions, which are based on this event, are taken. The multi-stage problem, accordingly consists of multiple stages, with a random event occurring between each stage. We briefly review the latter problem, details can be found in [7].

The *multi-stage stochastic linear programming problem with recourse* can be formulated as

$$\min c^1 x^1 + E_{\xi^2}\left[\min c^2(\omega)x^2(\omega^2) + \cdots + E_{\xi^H}\left[\min c^H(\omega)x^H(\omega^H)\right]\ldots\right]$$
$$s.t. \ W^1 x^1 = h^1,$$
$$T^1(\omega)x^1 + W^2 x^2(\omega^2) = h^2(\omega),$$
$$\ddots$$
$$T^{H-1}(\omega)x^{H-1}(\omega^{H-1}) + W^H + x^H(\omega^H) = h^H(\omega),$$
$$x^1 \geq 0; x^t(\omega^t) \geq 0, t = 2, ..., H$$

where $c^1 \in \mathbb{R}^{n1}$ and $h^1 \in \mathbb{R}^{m1}$ are known vectors and each matrix $W^t \in \mathbb{R}^{m_t \times n_t}$ is known. The random N^t-Vector $\xi^t(\omega)^T = (c^t(\omega)^T, h^t(\omega)^T, T_1^{t-1}, ..., T_{m_t}^{t-1})$ is defined on the probability space (Ω, Σ^t, P) (where $\Sigma^t \subset \Sigma^{t+1}$) for all $t = 2, .., H$. Ξ^t is supposed to be the support of ξ^t, the decision x depends on the history up to time t and is indicated by $x^t(\omega^t)$.

A common strategy to solve a stochastic program is an approximation that assumes a finite time horizon and a discrete probabilistic representation of uncertainty. Usually, it is assumed that the stochastic elements are defined over a discrete probability space $(\Xi, \sigma(\Xi), P)$, where $\Xi = \Xi^1 \otimes \cdots \otimes \Xi^H$ is the support of the random data at stage $\Xi^t = \{\xi_s^t = (T_s^t, W_s^t, h_s^t, c_s^t), s = 1, \ldots, S^t\}$.

Thus, the uncertainty can be represented through a so-called *scenario tree*, which defines the possible sequence of realizations over all time-stages as depicted in Figure 1 a). Nodes in the scenario tree at stage t are decision points where variable allocations must be determined, with respect to all realized data $(\xi^1 \ldots \xi^{t-1})$ of the last $t-1$ stages. Arcs of the tree represent realizations of the random variables. The root node is associated with the first stage decision variables and a path from the root to a leaf is called a *scenario*. Thus, solving a MSSLP can be interpreted as solving a tree of linear programs, and passing information among nodes as depicted in Figure 1 a).

Scenario trees have strong similarities to *decision trees* of the universal player when interpreting QLPs as some sort of game between an existential and a universal player. Finiteness of the time horizon is equivalent to a finite number of quantifier changes, the discrete distribution of the underlying random variables is given because we only need to consider the lower and upper bounds of each universally quantified variable [18]. A path from the root to a leaf is called a *game*. A mapping from a QLP with the quantifier sequence $\exists\forall\exists\forall\exists$ to a decision tree is shown in Figure 1 b).

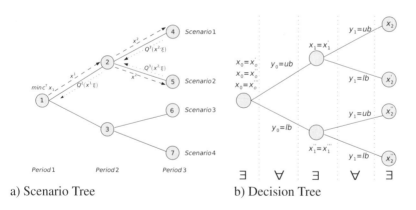

a) Scenario Tree   b) Decision Tree

Fig. 1. MSSLP Scenario Tree and QLP Decision Tree

The information embodied in the scenario tree can be used to formulate a corresponding deterministic equivalent linear program (DEP). Even though, DEPs can have very large scale, since their size increases exponentially with the number of time stages and realizations of the random variables, the resulting matrix structure can be exploited by decomposition methods [26].

The *compact-variable* formulation[24] directly exploits the tree structure by defining a reduced set of decision variables, which implicitly satisfy the *nonanticipativity* property. This means that each period's decision may only depend on information that is available at the time, and thus can be seen as a direct mapping on the scenario tree. The underlying LP has so-called (nested) block-ladder structure.

In a similar way we can create the DEP of a QLP. Instead of encoding the scenario tree, we encode the decision tree into the DEP by replicating the inequalities of the QLP for each possible decision combination of the universal player.

Substituting the current decision of the universal player into the corresponding matrix columns and taking them to the right side, results in a multistage decision problem with variable right-hand side.

3.2 Nested Benders Decomposition

The compact-variable formulation can be exploited by the Benders Decomposition principle [4], which is the basis of many algorithms that have been developed in the stochastic programming community over the past decades. For the two-stage case most proposed methods trace back to the L-shaped method of Van Slyke and Wets [32], which builds on arguments similar to Benders Decomposition. Based on the observation that the nested block ladder structure that results from a multistage problem can be solved by a recursive application of Benders Decomposition, the L-shaped method has been successfully extended to the multistage case in [5].

Today, some of the most efficient solution approaches for stochastic programming problems rely on these basics. Over the past decades many improvements have been suggested in the literature. Serious progress was made by parallelizing the solution process [25,8,5] and also ideas for combining decomposition techniques with ideas from other fields like for example sampling techniques [33] or genetic algorithms were proposed [23,29]. The vast amount of theoretical basics and computational experiences motivated us to adopt decomposition techniques and its extensions to solve QLPs.

Solving QLPs with the Nested Benders Decomposition. In [18] a decomposition based algorithm was presented to determine the *feasibility* of QLPs with at most polynomial space. Nevertheless, for the rest of the paper, we do not care of polynomial space but concentrate on the implementation of an objective function and how to use this objective for a target-driven search process. While in stochastic programming the goal is to minimize or maximize the expected value of all possible realizations of the scenarios, the goal of an QLP is to find the *optimal worst-case* value of an objective function, and thus, can be seen as some sort of robust optimization technique.

Formally, the DEP of a QLP, using the notion of $x^t(\omega^t)$ for states and the assumption that the stages are 1 to H, the DEP can be recursively formulated similar as described in [7] for MSSLPs as follows:

$$\min c^T x^1 + \mathcal{Q}^2(x^1)$$
$$s.t \quad A^1 x^1 \leq b^1,$$
$$x^1 \geq 0$$

where at inner nodes

$$\mathcal{Q}^{t+1}(x^t) = max_{\xi^{t+1}} \left[\mathcal{Q}^{t+1}(x^t, \xi^{t+1}(\omega)) \right]$$

with

$$Q^t(x^{t-1}, \xi^t(\omega)) = min\{c^t x^t + Q^{t+1}(x^t)|$$
$$A^t x^t \leq b^t(\omega) - T^{t-1} x^{t-1}, x^t \geq 0\}.$$

for $t = 2, \ldots, H - 1$, and finally at the terminal nodes

$$Q^H(x^{H-1}, \xi^H(\omega)) = min\{c^H x^H|$$
$$W^H x^H = b^H(\omega) - T^{H-1} x^{H-1}, x^H \geq 0\}.$$

The vector $\xi^t(\omega)^T = (b^t(\omega)^T)$ contains the right-hand side of each event, which results when the desicions of the universal player so far are applied to the corresponding matrix part of the universal player and then subtracted from b^t. This DEP can be solved with the traditional Nested Benders Decomposition but since we are interested in the worst-case solution instead of an expected value, the computation of the bounds only considers the worst-case subsolution at each node, instead of computing a weighted sum as the original algorithm does.

Implementation of the Nested Benders Decomposition. Any implementation of the Nested Benders Decomposition approach has to deal with a number of details, which can substantially effect the overall performance.

Before a subproblems is solved, it has to be loaded into the linear programming solver. As resolving is usually much faster than solving it from scratch, having one LP for each node of the tree might lower the computational time required. However, due to memory constraints, this is not possible for larger problems. Compromising we decided to share LPs for all nodes at the same stage of the tree, since those problems only differ in their right-hand sides, apart from cuts being generated during the solution process. This saves a significant amount of memory and has the advantage that nodes at a stage can benefit from sharing information among each other.

The order in which the tree is traversed is determined by the *sequencing protocol*. We use the *fast-forward* method, which is described in [5,12]. Instead of the more commonly used breath-first search as used in the stochastic programming community, the nodes of the tree are visited depth-first in our implementation. This allows to use advanced memory saving techniques [2]. Moreover, the depth-first search is the most successful approach to traverse a game-tree in two-person zero sum games like chess, othello etc., and the most valuable advantage of the depth-first decomposition approach is that far larger instances can be solved, because of the low memory requirements. In an upcoming version of our algorithm, we plan to combine both techniques with branching schemes from linear mixed integer programming [1,21,14].

For cut generation, we decided to use the *multicut* approach suggested by Birge and Louveaux [6]. In contrast to the traditional approach summing up the weighted dual variables for each node forming one optimality cut, they suggest to disaggregate optimality cuts by placing one cut for each subproblem in the corresponding master. Experiments showed, that multicuts reduce the solution time for large problems [12].

4   Computational Experiments

In the following we present the results of our experimental tests. All algorithms where coded in C++ and run on a quad-core processor AMD Phenom II X4 945 with 3.00GHz and 8GB RAM running Ubuntu 10.04. We resort to the callable library CPLEX 12 as external solver software.

In the first part, we describe our test instances that were generated for the experiments. Afterwards, we examine the effect of different quantifier block positions. It follows a survey of the effect when the number of universally quantified variables increases. We proceed by checking whether there are noticeable differences when the universally quantified variables are distributed over several blocks. We also solve the corresponding DEP with CPLEX in each case and compare the solution times with our algorithm.

4.1   The Data Set

Since no QLP instances do exist in the literature, we decided to use instances from existing test sets and transform them to QLPs by adding universally quantified variables. From the field of linear programming we decided to use the Netlib Library [2] because it is well known and contains many real world applications. Instances from this suite have been successfully converted to stochastic programming problems and used in many experimental papers that deal with implementations of Nested Benders Decomposition variants in the field of stochastic programming [12,8,5,13].

As mentioned at the beginning, our interest in QLPs stems from the fact that they are QLP-relaxations of QIPs. Therefore, we were also interested in creating QIPs from real world integer and mixed integer programs. We decided to use instances from the Miplib2003[3] test set to create QIPs, whose QLP-relaxations were then solved in our experiments.

To convert LP instances to QLPs, we inserted blocks of universally quantified variables into the original LP. The corresponding columns for each new variable $y_i \in \forall$ with $y_i \in [0, 1]$ have a density of 25% of non-zero coefficients from the interval $[-1, 1]$. The coefficients were positioned randomly in the corresponding column. We varied the position of the blocks, the number of blocks, and the number of universal variables in each block in our tests. The corresponding DEPs were computed using the compact-variable representation and stored in the CPLEX LP file format on the harddisk.

4.2   Analysis of Twostage QLPs

First of all we considered twostage QLPs with a single block of universally quantified variables. In order to investigate the effect of different quantifier block positions we solved QLPs with $\forall$-blocks positioned after the first third, in the middle, and before the last third of the quantifier string. To investigate how the solution process is affected when the number of universally quantified variables increases, we took QLPs with five and ten universally quantified variables into account. The QLP sizes range from about twenty variables and constraints right up to several hundreds of both.

Table 1 shows the summed up results for QLPs generated from Netlib and Miplib2003 instances. Column 1 contains the used test set followed by the number of universally quantified variables and the position of the universal quantifier block. It

---

[2] http://www.netlib.org/lp/
[3] http://miplib.zib.de/

Table 1. Twostage QLPs with different quantifier block positions

			TwoStage Optimality Tests					
			CPLEX	Nested Benders Decomposition				
Test Set	∀-Variables	Block	Time (s)	Time (s)	Iterations	Feas Cuts	Opt Cuts	Subproblems
Netlib	5	1st third	2.2	1.5	378	360	2211	2950
	5	centered	1.6	1.6	578	566	2907	4051
	5	2nd third	1.0	1.8	781	772	3636	5191
	10	1st third	1110.4	39.7	687	674	94430	95793
	10	centered	679.2	59.7	1036	1027	131711	133774
	10	2nd third	350.3	50.8	1343	1333	169135	171812
Miplib2003	5	1st third	4.4	21.9	1828	1655	12418	15902
	5	centered	3.7	21.1	2052	1915	15652	19619
	5	2nd third	2.6	15.0	1543	1399	11620	14562
	10	1st third	1660.8	459.1	2149	2002	316729	320880
	10	centered	968.9	467.8	3692	3602	439221	446515
	10	2nd third	516.2	390.8	2629	2513	378275	383417

follows the total solution time when solving the corresponding DEPs with CPLEX. Columns 5 shows the solution time of our algorithm followed by the number of iterations, the number of non-redundant feasibility and optimality cuts, and finally the number of subproblems that were solved during the solution process.

For twostage QLPs the results suggest that our solver outperforms solving the DEP with CPLEX clearly when the number of universally quantified variables increase, independently from the quantifier block position. However, the difference between our solver and the result from CPLEX becomes smaller when the quantifier block is moved towards the end of the quantifier string. This is not suprising, since the resulting DEP becomes much smaller in this case. For the Netlib instances, moving the quantifier block from the left to the right leads to a higher number of iterations, cuts and subproblems that must be solved. We observed that the solution time is at its highest point, when the quantifier block was positioned in the middle. We think that there is a tradeoff between the size of the master and the subproblems. If the block is positioned in the first third, the algorithm benefits from a small master with less variables that must be fixed via cuts. If the block is positioned in the second third, the advantage comes from the fact that the subproblems that must be solved very often become smaller. For the Miplib instances this is not so clear with respect to the solution times, but the number of iterations, cuts and subproblems to be solved seem to confirm this suspicion. Tests where we only checked the instances for feasibility reaffirmed the observations. Again our solver was much faster than checking the DEPs for feasibility.

4.3 Analysis of Multistage QLPs

Another point of interest was the effect when the universally quantified variables are spreaded via several ∀-blocks, resulting in a decision tree with multiple stages. We therefore considered instances with one, two and five blocks in our tests. The results are shown in Table 2 and contain the same columns as in the prior table, except column 3 that shows the number of universal variable blocks instead of the block position.

The results show that for three stages and an increasing number of universally quantified variables our algorithm also performs better in general, but with an increasing number of quantifier blocks, and thus stages in the tree, there is a growing number of instances where solving the DEP with CPLEX is significantly faster. A causal connection to the

Table 2. QLPs with different numbers of quantifier blocks

Test Set	∀-Variables	∀-Blocks	CPLEX Time (s)	Time (s)	Iterations	Feas Cuts	Opt Cuts	Subproblems
			MultiStage Optimality Tests		Nested Benders Decomposition			
Netlib	5	1	1.6	1.6	578	566	2907	4051
	5	2	1.9	5.4	2855	2031	4204	8934
	5	5	2.1	98.1	22754	9370	26448	51615
	10	1	679.2	59.7	1036	1027	131711	133774
	10	2	177.2	104.3	21193	15807	236530	273770
	10	5	125.9	925.2	226051	80629	600441	816297
Miplib2003	5	1	3.7	21.1	2052	1915	15652	19619
	5	2	3.7	299.9	12974	7874	23015	44624
	5	5	2.7	7152.0	110946	13187	172124	263866
	10	1	968.9	467.8	3692	3602	439221	446515
	10	2	907.0	3212.6	58610	34166	875395	1040134
	10	5	468.0	14280.0	407040	68627	1205383	1794417

size of the instances and the depth of the tree could be especially recognized for instances generated from Miplib2003. For an increasing number of ∀-blocks we observed an increasing number of outliers, which destroyed the overall performance of the decomposition approach. This behavoir was strongly weakened when checking for simple feasibility and we observed that the algorithm stays a long time inside the tree when trying to find the optimal solution after feasibility is detected.

In detail, referring to this, we profiled our code and furthermore detected that our algorithm spends most of the time in the deletion routines *IloNumLinTerI::removeMarked, CPXdelrows, IloRecycleBinI::removeFromAll* of CPLEX. Since our algorithm holds one CPLEX instance per stage, which is shared among all nodes of a stage, these routines get called every time when we move down the tree and reach a node, whoose cuts from former iterations at this stage must be removed. We hope to mitigate this drawback by the use of better sequencing protocols to reduce the total amount of iterations, and thus the number of cut deletions. The latter could be furthermore improved by the use of cut sharing techniques.

5   Conclusion and Further Work

This paper discusses QLPs, which are linear programs with variables being either existentially or universally quantified. Their close resemblance to multistage stochastic linear programs, as well as the relationship of their integer variant, called QIPs, to PSPACE-complete problems, were illustrated.

We proposed a decomposition based algorithm to solve QLPs and used its implementation for a detailed computational study to examine how different structural properties of QLPs affect the solution process. We furthermore compared the performance of our algorithm with the approach of solving the deterministic equivalent linear program.

The tests suggest that our QLP-solver is clearly faster for QLP instances with a quantifier string of the form $\exists^k \forall^l \exists^m$. Moreover, the decomposition solver outperforms the approach of solving the corresponding DEP in many instances when we inspect multistage problems with up to five stages. We interpret the fact that DEP approach outperforms our solver in some instances as a hint that the depth-first character of our decomposition might be improvable with the help of other sequencing protocols and cut sharing techniques. The most valuable advantage that we see for the decomposition-based method is its efficient use of memory. Principally by far larger instances can be solved by the decomposition method than by solving the exponentially large DEP.

Further improvements of our solver can be done by implementing other techniques as described in the literature. Especially parallelism seems to be a promising approach but also the combination with sampling techniques might be beneficial. In addition, we are concerned to extend our implementation to cope with QIPs.

References

1. Achterberg, T., Koch, T., Martin, A.: Branching rules revisited. Operations Research Letters 33, 42–54 (2005)
2. Altenstedt, F.: Memory consumption versus computational time in nested benders decomposition for stochastic linear programming. Tech.Rep., Chalmers University, Goteborg (2003)
3. Bellmann, R.: Dynamic programming. Princeton University Press, Princeton (1957)
4. Benders, J.F.: Partitioning procedures for solving mixed-variables programming problems. Numerische Mathematik 4(1), 238–252 (1962)
5. Birge, J., Donohue, C., Holmes, D., Svintsitski, O.: A parallel implementation of the nested decomposition algorithm for multistage stochastic linear programs. Math. Program. 75, 327–352 (1996)
6. Birge, J., Louveaux, F.: A multicut algorithm for two-stage stochastic linear programs. European Journal of Operational Research 34(3), 384–392 (1988)
7. Birge, J., Louveaux, F.: Intro. to Stochastic Programming. Springer, Heidelberg (1997)
8. Dempster, M., Thompson, R.: Parallelization and aggregation ofnested benders decomposition. Annals of Operations Research 81, 163–188 (1998), doi:10.1023/A:1018996821817
9. Donninger, C., Kure, A., Lorenz, U.: Parallel brutus: The first distributed, fpga accelarated chess program. In: Proc. of 18th International Parallel & Distributed Processing Symposium (IPDPS). IEEE Computer Society, Santa Fe (2004)
10. Dyer, M., Stougie, L.: Computational complexity of stochastic programming problems. Math. Program. 106(3), 423–432 (2006)
11. Engell, S., Märkert, A., Sand, G., Schultz, R.: Aggregated scheduling of a multiproduct batch plant by two-stage stochastic integer programming. Optimiz. and Engineering 5 (2004)
12. Gassmann, H.: Mslip: A computer code for the multistage stochastic linear programming problem. Mathematical Programming 47(1-3), 407–423 (1990)
13. Ho, J., Sundarraj, R.P.: Distributed nested decomposition of staircase linear programs. ACM Trans. Math. Softw. 23, 148–173 (1997)
14. Kılınç Karzan, F., Nemhauser, G., Savelsbergh, M.: Information-based branching schemes for binary linear mixed integer problems. Mathematical Programming Computation (January 2010)
15. Kleywegt, A., Shapiro, A., Homem-De-Mello, T.: The sample average approximation method for stochastic discrete optimization. SIAM Jour. of Opt., 479–502 (2001)
16. König, F.G., Lübbecke, M., Möhring, R.H., Schäfer, G., Spenke, I.: Solutions to real-world instances of PSPACE-complete stacking. In: Arge, L., Hoffmann, M., Welzl, E. (eds.) ESA 2007. LNCS, vol. 4698, pp. 729–740. Springer, Heidelberg (2007)
17. Liebchen, C., Lübbecke, M., Möhring, R., Stiller, S.: The concept of recoverable robustness, linear programming recovery, and railway applications. Robust and online large-scale optimization, 1–27 (2009)

18. Lorenz, U., Martin, A., Wolf, J.: Polyhedral and algorithmic properties of quantified linear programs. In: de Berg, M., Meyer, U. (eds.) ESA 2010. LNCS, vol. 6346, pp. 512–523. Springer, Heidelberg (2010)
19. Megow, N., Vredeveld, T.: Approximation in preemptive stochastic online scheduling. In: Azar, Y., Erlebach, T. (eds.) ESA 2006. LNCS, vol. 4168, pp. 516–527. Springer, Heidelberg (2006)
20. Möhring, R., Schulz, A., Uetz, M.: Approximation in stochastic scheduling: The power of lp-based priority schedules. Journal of ACM 46(6), 924–942 (1999)
21. Ostrowski, J.P., Linderoth, J., Rossi, F., Smriglio, S.: Orbital branching. In: Fischetti, M., Williamson, D.P. (eds.) IPCO 2007. LNCS, vol. 4513, pp. 104–118. Springer, Heidelberg (2007)
22. Pochet, Y., Wolsey, L.: Production planning by mixed integer programming. Springer Series in Operations Research and Financial Engineering. Springer, New York (2006)
23. Poojari, C., Beasley, J.: Improving benders decomposition using a genetic algorithm. European Journal of Operational Research 199(1), 89–97 (2009)
24. Rockafellar, R.T., Wets, R.-B.: Scenarios and policy aggregation in optimization under uncertainty. Math. Oper. Res. 16, 119–147 (1991)
25. Ruszczyński, A.: Parallel decomposition of multistage stochastic programming problems. Math. Program. 58, 201–228 (1993)
26. Ruszczyński, A.: Decomposition methods in stochastic programming. Math. Program. 79, 333–353 (1997)
27. Schultz, R.: Stochastic programming with integer variables. Math. Progr. 97, 285–309 (2003)
28. Shmoys, D., Swamy, C.: Stochastic optimization is (almost) as easy as deterministic optimization. In: Proc. FOCS 2004, pp. 228–237 (2004)
29. Sirikum, J., Techanitisawad, A., Kachitvichyanukul, V.: A new efficient GA-benders' decomposition method: For power generation expansion planning with emission controls. IEEE Transactions on Power Systems 22(3), 1092–1100 (2007)
30. Subramani, K.: Analyzing selected quantified integer programs. In: Basin, D., Rusinowitch, M. (eds.) IJCAR 2004. LNCS (LNAI), vol. 3097, pp. 342–356. Springer, Heidelberg (2004)
31. Subramani, K.: On a decision procedure for quantified linear programs. Annals of Mathematics and Artificial Intelligence 51(1), 55–77 (2007)
32. Van Slyke, R.M., Wets, R.: L-Shaped linear programs with applications to optimal control and stochastic programming. SIAM Journal on Applied Mathematics 17(4), 638–663 (1969)
33. Verweij, B., Ahmed, S., Kleywegt, A.J., Nemhauser, G., Shapiro, A.: The sample average approximation method applied to stochastic routing problems: A computational study. Comput. Optim. Appl. 24, 289–333 (2003)
34. Zhang, L., Hermanns, H., Eisenbrand, F., Jansen, D.: Flow faster: Efficient decision algorithms for probabilistic simulations. Logical Methods in Computer Science 4(4) (2008)

Recoverable Robustness by Column Generation

P.C. Bouman, J.M. van den Akker, and J.A. Hoogeveen

Department of Information and Computing Sciences
Utrecht University
Princetonplein 5, 3584 CC Utrecht, The Netherlands
research@pcbouman.nl, J.M.vandenAkker@uu.nl, J.A.Hoogeveen@uu.nl

Abstract. Real-life planning problems are often complicated by the occurrence of disturbances, which imply that the original plan cannot be followed anymore and some recovery action must be taken to cope with the disturbance. In such a situation it is worthwhile to arm yourself against common disturbances. Well-known approaches to create plans that take possible, common disturbances into account are robust optimization and stochastic programming. Recently, a new approach has been developed that combines the best of these two: recoverable robustness. In this paper, we apply the technique of column generation to find solutions to recoverable robustness problems. We consider two types of solution approaches: separate recovery and combined recovery. We show our approach on two example problems: the size robust knapsack problem, in which the knapsack size may get reduced, and the demand robust shortest path problem, in which the sink is uncertain and the cost of edges may increase.

1 Introduction

Most optimization algorithms rely on the assumption that all input data are deterministic and known in advance. However, in many practical optimization problems, such as planning in public transportation or health care, data may be subject to changes. To deal with this uncertainty, different approaches have been developed. In case of *robust optimization* (see [4], [2]) we choose the solution with minimum cost that remains feasible for a given set of disturbances in the parameters. In case of *stochastic programming* [3], we take *first stage decisions* on basis of the current information and, after the value of the unknown data has been revealed, we take the *second stage* or *recourse decisions*. The objective here is to minimize the cost of the first stage decisions plus the expected cost of the recourse decisions. The recourse decision variables may be restricted to a polyhedron through the so-called technology matrix [3]. So robust optimization wants the initial solution to be completely immune for a predefined set of disturbances, while stochastic programming includes a lot of options to postpone decisions to a later stage or change decisions in a later stage.

Recently, the notion of *recoverable robustness* [11] has been developed, which combines robust optimization and second-stage recovery options. Recoverable robust optimization computes solutions, which for a given set of scenarios can

C. Demetrescu and M.M. Halldórsson (Eds.): ESA 2011, LNCS 6942, pp. 215–226, 2011.

be recovered to a feasible solution according to a set of pre-described, fast, and simple recovery algorithms. The main difference between recoverable robustness and stochastic programming is the way in which recourse actions are limited. The property of recoverable robustness that recourse actions must be achieved by applying a simple algorithm instead of being bounded by a polyhedron makes it very suitable for combinatorial problems. As an example, consider the planning of buses and drivers in a large city. We may expect that during rush hour buses may be delayed, and hence may be too late to perform the next trip in their schedule. In case of robust optimization, we can counter this only by making the time between two consecutive trips larger than the maximum delay that we want to take into account. In case of recoverable robustness, we are allowed to change, if necessary, the bus schedule, but this is limited by the choice of the recovery algorithm. For example, we may schedule a given number of stand-by drivers and buses, which can take over the trip of a delayed driver/bus combination. Especially in the area of railway optimization (see e.g. [7] and [8]) recoverable robust optimization methods have gained a lot of attention.

In this paper we present a new approach for solving recoverable robust optimization problems. We use *column generation* for recoverable robust optimization. We will present column generation models for the size robust knapsack problem and for the demand robust shortest path problem. Our approach can be generalized to many other problems. To the best of our knowledge, this is the first paper applying column generation to recoverable robust optimization. Another decomposition approach, namely Benders decomposition, is used in [6] to assess the Price of Recoverability for recoverable robust rolling stock planning in railways.

The remainder of the paper is organized as follows. In Section 2, we define the concept of recoverable robustness. In Section 3, we consider the size robust knapsack problem and in Section 4 the demand robust shortest path problem. In Section 5, we report on computational results. Finally, Section 6 concludes the paper.

2   Recoverable Robustness

In this section we formally define the concept of recoverable robustness. We are given an optimization problem

$$P = \min\{f(x)|x \in F\},$$

where $x \in \mathbb{R}^n$ are the decision variables, f is the objective function and F is the set of feasible solutions.

Disturbances are modeled by a set of scenarios S. We use F_s to denote the set of feasible solutions for scenario $s \in S$, and we denote the decision variables for scenarios s by y^s. The set of algorithms that can be used for recovery are denoted by $\mathcal{A}$, where $A(x, s) \in \mathcal{A}$ determines a feasible solution y^s from a given initial solution x in case of scenario s. In case of planning buses and drivers a scenario corresponds to a set of bus trips that are delayed, and the algorithms in $\mathcal{A}$ decide about the use of standby drivers.

The *recovery robust* optimization problem is now defined as:

$$\mathcal{RRP}_{\mathcal{A}} = \min\{f(x) + g(\{y^s | s \in S\}) | x \in F,\ A \in \mathcal{A},\ \forall_{s \in S} y^s = A(x, s)\}.$$

Here, $g(\{y^s | s \in S\})$ denotes the cost associated with the recovery variables y^s. There are many possible choices for g. A few examples are as follows:

1. g is defined as the all-zero function. This models the situation where our only concern is the feasibility of the recovered solutions.
2. g is equal to the maximal cost of the recovered solutions y^s. This corresponds to minimizing the worst-case cost.
3. g measures the deviation of the solutions y^s from the initial solution x. Note that this deviation may also be limited by the recovery algorithms.
4. Suppose we are given the probabilities p_s of scenarios s. Then g is defined as expected value of the solution after recovery, i.e., $g(\{y^s | s \in S\}) = \sum_{s \in S} p_s g(y^s)$, where $g(y^s)$ is the cost of solution y^s.

Although earlier papers on recoverable robustness (e.g. [11]) consider the latter type of definition of g as two-stage stochastic programming, we think that the requirement of a pre-described easy recovery algorithms makes this definition fit into the framework of recoverable robustness.

3   Size Robust Knapsack Problem

We consider the following knapsack problem. We are given n items, where item j $(j = 1, \ldots, n)$ has revenue c_j and weight a_j. Each item can be selected at most once. The knapsack size is b. We define the *size robust knapsack problem* as the knapsack problem where the knapsack size b is subject to uncertainty. We denote by $b_s < b$ the size of the knapsack in scenario $s \in S$. We assume that the knapsack will keep its original size with probability p_0 and that scenario s will occur with probability p_s. Our objective is to maximize the expected revenue after recovery. We study the situation in which recovery has to be performed by removing items. As soon as it becomes clear which scenario s appears, recovery is performed by removing items in such a way that the remaining items give a knapsack with maximal revenue and size at most b_s. This boils down to solving a knapsack problem were the item set is the set of items selected in the initial solution and the knapsack size is b_s. Hence, our set of recovery algorithms is given by the dynamic programming algorithm for solving these knapsacks. Recently [5] have studied an extended version of our knapsack problem. They show $\mathcal{NP}$-hardness of several variants and develop a polyhedral approach to solve these problems.

We are going to discuss two decomposition approaches for the size robust knapsack problem. In both cases we reformulate the problem such that we have to select one knapsack filling for the initial problem and all scenarios from a given set. The difference consists of the way we deal with the scenarios. In *Separate Recovery Decomposition*, we select an initial knapsack filling and separately we select a knapsack filling for each scenario; the relation that the initial knapsack filling should contain all scenario fillings is enforced by constraints in the master problem. In *Combined Recovery Decomposition*, we select for each scenario a combination of an initial knapsack filling together with the optimal recovery knapsack for that single scenario. We enforce that only one initial knapsack filling will get selected in the master problem.

3.1   Separate Recovery Decomposition

We define $K(b)$ as the set of feasible knapsack fillings with size at most b. For $k \in K(b)$, we denote its revenue by $C_k = \sum_{i \in k} c_i$. In the same way, we denote the revenue of $k \in K(b_s)$ by $C_k^s = \sum_{i \in k} c_i$.

We define decision variables

$$x_k = \begin{cases} 1 \text{ if knapsack } k \in K(b) \text{ is selected,} \\ 0 \text{ otherwise.} \end{cases}$$

and

$$y_k^s = \begin{cases} 1 \text{ if knapsack } k \in K(b_s) \text{ is selected for scenario } s, \\ 0 \text{ otherwise.} \end{cases}$$

The problem can now be formulated as follows. This is called the Master ILP.

$$\max p_0 \sum_{k \in K(b)} C_k x_k + \sum_{s \in S} p_s \sum_{k \in K(b_s)} C_k^s y_k^s$$

subject to

$$\sum_{k \in K(b)} x_k = 1 \tag{1}$$

$$\sum_{k \in K(b_s)} y_k^s = 1 \quad \text{for all } s \in S \tag{2}$$

$$\sum_{k \in K(b)} a_{ik} x_k - \sum_{k \in K(b_s)} a_{ik}^s y_k^s \geq 0 \quad \text{for all } i \in \{1, 2, \ldots, n\}, s \in S \tag{3}$$

$$x_k, \in \{0, 1\} \quad \text{for all } k \in K(b) \tag{4}$$

$$y_k^s, \in \{0, 1\} \quad \text{for all } k \in K(b_s), s \in S, \tag{5}$$

where the index variables a_{ik} and a_{ik}^s are defined as follows:

$$a_{ik} = \begin{cases} 1 \text{ if item } i \text{ is in knapsack } k \in K(b), \\ 0 \text{ otherwise.} \end{cases}$$

and

$$a_{ik}^s = \begin{cases} 1 \text{ if item } i \text{ is in knapsack } k \in K(b_s), \\ 0 \text{ otherwise.} \end{cases}$$

In the above model constraint (1) states that exactly one knapsack is selected for the original situation and constraints (2) that exactly one knapsack is selected for each scenario. Constraints (3) ensures that recovery is done by removing items.

We want to solve this ILP formulation using Branch-and-Price [1]. We relax the integrality constraints (4) and (5) into $x_k \geq 0$ and $y_k^s \geq 0$, and solve this LP-relaxation. To deal with the large number of variables we are going to solve the problem by *column generation*. We start with a limited subset of the variables and solve the LP-relaxation for this subset only; this is called the restricted master LP. Then we solve the pricing problem, i.e., we look for variables that are not yet included in the restricted master LP and can to improve the solution if their value is made positive. If such variables are found, they are added to the restricted master LP, it is solved again, after which pricing is performed etc. If pricing does not find any new variables we know that the master LP has been solved to optimality.

The pricing problem
From the theory of linear programming it is well-known that for a maximization problem increasing the value of a variable will improve the current solution if and only if its reduced cost is positive. The pricing problem then boils down to maximizing the reduced cost.

Let λ, μ_s, and π_{is} be the dual variables of constraints 1, 2, and 3 respectively. Now the reduced cost $c^{\mathrm{red}}(x_k)$ of x_k is given by

$$c^{\mathrm{red}}(x_k) = p_0 \sum_{i \in k} c_i - \lambda - \sum_{i=1}^{n} \sum_{s \in S} a_{ik} \pi_{is}$$
$$= \sum_{i=1}^{n} a_{ik} (p_0 c_i - \sum_{s \in S} \pi_{is}) - \lambda.$$

The pricing problem is to find a feasible knapsack for the original scenario, where the revenue of item i, equals $(p_0 c_i - \sum_{s \in S} \pi_{is})$. This is just the original knapsack problem with modified objective coefficients. Similarly the reduced cost $c^{\mathrm{red}}(y_k^s)$ are given by $c^{\mathrm{red}}(y_k^s) = \sum_{i=1}^{n} a_{ik}^s (p_s c_i + \pi_{is}) - \mu_s$ It follows that the pricing is exactly the knapsack problem with knapsack size b_s and modified objective coefficients. Note that in the pricing problem an item may have a negative revenue. Clearly such items can be discarded.

To find an integral solution, we are going to apply Branch-and-Price. We branch on items that are fractional in the current solution; this is easily combined with column generation.

3.2 Combined Recovery Decomposition

In contrast to the *Separate Recovery Decomposition*, we consider fillings of the initial knapsack in combination with the optimal recovery for *one* scenario. Consequently, we introduce decision variables:

$$z_{kq}^s = \begin{cases} 1 \text{ if the combination of initial solution } k \in K(b) \\ \quad \text{ and recovery solution } q \in K(b_s) \text{ is selected for scenario } s, \\ 0 \text{ otherwise.} \end{cases}$$

Clearly, z_{kq}^s is only defined if q is a subset of k. The ILP model further contains the original variable x_i signaling if item i is contained in the initial knapsack. We can formulate the problem as follows:

$$\max p_0 \sum_{i=1}^n c_i x_i + \sum_{s \in S} p_s \sum_{(k,q) \in K(b) \times K(b_s)} C_q^s z_{kq}^s$$

subject to

$$\sum_{(k,q) \in K(b) \times K(b_s)} z_{kq}^s = 1 \quad \text{for all } s \in S \tag{6}$$

$$x_i - \sum_{(k,q) \in K(b) \times K(b_s)} a_{ik} z_{kq}^s = 0 \quad \text{for all } i \in \{1, 2, \ldots, n\}, s \in S \tag{7}$$

$$x_i, \in \{0, 1\} \quad \text{for all } i \in \{1, 2, \ldots, n\} \tag{8}$$

$$z_{kq}^s, \in \{0, 1\} \quad \text{for all } k \in K(b), q \in K(b_s), s \in S, \tag{9}$$

Constraints (6) enforce that exactly one combination is selected for each scenario; constraints (7) ensure that the same initial knapsack filling is selected for all scenarios.

Again, we are going to solve the LP-relaxation by column generation. We include the variables x_i in the restricted master LP and, hence pricing is only performed for the variables z_{kq}^s. We denote the dual variables of constraints (6) and (7) by ρ_s and σ_{is}, respectively. The reduced cost of z_{kq}^s is now equal to:

$$c^{\text{red}}(z_{kq}^s) = \sum_{i=1}^n a_{iq}^s p_s c_i + \sum_{i=1}^n a_{ik} \sigma_{is} - \rho_s.$$

We solve the pricing problem for each scenario separately. We have to find an initial and recovery solution. This can be solved by dynamic programming. The main observation is that there are three types of items: items included in both the initial and recovery knapsack, items selected for the initial knapsack, but removed by the recovery, and non-selected items. We define state variables $D(i, w_0, w_s)$ as the best value for a combination of an initial and recovery knapsack for scenario s, such that the initial knapsack is a subset of $\{1, 2 \ldots, i\}$, the recovery knapsack is a subset of the initial knapsack, and the initial and recovery knapsack have weight w_0 and w_s, respectively. The recurrence relation is as follows:

$$D(i, 0, 0) = 0 \quad \forall i$$
$$D(0, w_0, w_s) = -\infty \quad \text{for } w_0, w_s > 0$$

$$D(i, w_0, w_s) = \max \begin{cases} D(i-1, w_0, w_s) \\ D(i-1, w_0 - a_i, w_s) + \sigma_{is} \\ D(i-1, w_0 - a_i, w_s - a_i) + \sigma_{is} + p_s c_i \end{cases}$$

4   Demand Robust Shortest Path Problem

The demand robust shortest path problem is an extension of the shortest path problem and has been introduced in [9]. We are given a graph (V, E) with cost c_e on the edges $e \in E$, and a source node $v_{\text{source}} \in V$. The question is to find the cheapest path from source to the sink, but the exact location of the sink is subject to uncertainty. Moreover, the cost of the edges may change over time. More formally, there are multiple scenarios $s \in S$ that each define a sink v_{sink}^s and a factor $f^s > 1$ by which the cost of the edges are scaled.

[12] has studied two variants of this problem. In both cases, the sink is known, but the costs of the edges can vary. Initially, a path has to be chosen. In the first variant, recovery is limited by replacing at most k edges in the chosen path; [12] shows this problem to be $\mathcal{NP}$-hard. In the second variant any new path can be chosen, but you get a discount on already chosen edges; [12] looks at the worst case behavior of a heuristic.

In contrast to [12], we can buy any set of edges in the initial planning phase. In the recovery phase, we have to extend the initial set such that it contains a path from the source to the sink v_{sink}^s, while paying increased cost for the additional edges. Our objective is to minimize the cost of the worst case scenario. Remark that, when the sink gets revealed, the recovery problem can be solved as a shortest path problem, where the edges already bought get zero cost. Hence, the recovery algorithm is a shortest path algorithm.

Observe that the recovery problem has the constraint that the union of the edges selected during recovery and the initially selected edges contains a path from source v_{source} to sink v_{sink}^s. It is not straightforward to express this constraint as a linear inequality, and hence to apply Separate Recovery Decomposition. However, the constraint fits very well into Combined Recovery Decomposition.

Our Combined Recovery Decomposition model contains the variable x_e signaling if edge $e \in E$ is selected initially. Moreover, for each scenario, it contains variables indicating which edges are selected initially and which edges are selected during the recovery:

$$z_{kq}^s = \begin{cases} 1 \text{ if the combination of initial edge set } k \subseteq E \\ \quad \text{and recovery edge set } q \subseteq E \text{ is selected for scenario } s, \\ 0 \text{ otherwise.} \end{cases}$$

Observe that z_{kq}^s is only defined if k and q are feasible, i.e., their intersection is empty and their union contains a path from v_{source} to v_{sink}^s. Finally, it contains $z_{\max}$ defined as the maximal recovery cost.

We can formulate the problem as follows:

$$\min \sum_{e \in E} c_e x_e + z_{\max}$$

subject to

$$\sum_{(k,q) \subseteq E \times E} z_{kq}^s = 1 \quad \text{for all } s \in S \tag{10}$$

$$x_e - \sum_{(k,q) \subseteq E \times E} a_{ek} z_{kq}^s = 0 \quad \text{for all } e \in E, \, s \in S \tag{11}$$

$$z_{\max} - \sum_{e \in E} f^s c_e \sum_{(k,q) \subseteq E \times E} a_{eq}^s z_{kq}^s \geq 0 \quad \text{for all } s \in S \tag{12}$$

$$x_e, \in \{0, 1\} \quad \text{for all } e \in E \tag{13}$$

$$z_{kq}^s, \in \{0, 1\} \quad \text{for all } k \subseteq E, \, q \subseteq E, \, s \in S, \tag{14}$$

where the binary index variables a_{ek} signal if edge e is in edge set k and the binary index variables a_{eq}^s signal if edge e is in edge set q for scenario s.

Constraints (10) ensure that exactly one combination of initial and recovery edges is selected for each scenario; constraints (11) enforces that the same set of initial edges is selected for each scenario. Finally, constraints (12) make sure that $z_{\max}$ represents the cost of the worst case scenario.

Let λ_s, ρ_{es}, and π_s be the dual variables of the constraints (10), (11), and (12) respectively. The reduced cost of z_{kq}^s is now equal to:

$$c^{\mathrm{red}}(z_{kq}^s) = -\lambda_s + \sum_{e \in E} \rho_{es} a_{ek} + \sum_{e \in E} \pi_s f^s c_e a_{eq}^s$$

Since we are dealing with a minimization problem, increasing the value of a variable improves the current LP-solution if and only if the reduced cost of this variable is negative. We have to solve the pricing problem for each scenario separately. For a given scenario s, the pricing problem amounts to minimizing $c^{\mathrm{red}}(z_{kq}^s)$ over all feasible a_{ek} and a_{eq}^s. This means that we have to select a subset of edges that contains a path from v_{source} to v_{sink}. This subset consists of edges which have been bought initially and edges which are attained during recovery. The first type corresponds to $a_{ek} = 1$ and has cost ρ_{es} and the second type to $a_{eq}^s = 1$ and has cost $\pi_s f^s c_e$. The pricing problem is close to a shortest path problem, but we have two binary decision variables for each edge. However, we can apply the following preprocessing steps:

– First, we select all edges with negative cost. From LP theory it follows that all dual variables π_s are nonnegative, and hence, all recovery edges have nonnegative cost. So we only select initial phase edges. From now on, the cost of these edges is considered to be 0.
– The other edges can either be selected in the initial phase or in the recovery phase. To minimize the reduced cost, we have to choose the cheapest option. This means that we can set the cost of an edge equal to $\min(\rho_{es}, \pi_s f^s c_e)$.

The pricing problem now boils down to a *shortest path* problem with nonnegative cost on the edges and hence can be solved by Dijkstra's algorithm.

5 Computational Results

We performed extensive computational experiments with the knapsack problem. The algorithms were implemented in the Java Programming language and the Linear Programs were solved using ILOG CPLEX 11.0. All experiments were run on a PC with an Intel Core^{TM}2 Duo 6400 2.13GHz processor.

Our experiments were performed in three phases. Since we want to focus on difficult instances, in the first phase we tested 12 different instance types to find out which types are the most difficult. Our instance types are based on the instance types in [10], where we have to add the knapsack weight b_s and the probability p_s for each of the scenarios. In the second phase, we tested many different algorithms on relatively small instances. In the third phase we tested the best algorithms from the second phase on larger instances. In this section, we will present the most important issues from the second and third phase. We omit further details for reasons of brevity.

In the second phase we tested 5 instance classes, including subset sum instances. We considered instances with 5, 10, 15 and 25 items and with 2, 4, 6 and 8 scenarios (except for 5 items were we only considered 2 and 4 scenarios). For each combination we generated 100 item sets (20 from each instance class) and for each item set we generated 3 sets of scenarios, with large, middle, and small values of b_s relative to b, respectively. This means that we considered 4200 instances in total.

We report results on the following algorithms:

- Separate Recovery Decomposition with Branch-and-Price, where we branch on the fractional item with largest $\frac{c_j}{a_j}$ ratio and first evaluate the node which includes the item.
- Combined Recovery Decomposition with Branch-and-Price, where we branch in the same way as in Separate Recovery decomposition.
- Branch-and-Bound where we branch on the fractional item with smallest $\frac{c_j}{a_j}$ ratio and first evaluate the node which includes the item.
- Dynamic programming: a generalization of the DP solving the pricing problem in case of Combined Recovery Decomposition.
- Hill Climbing: we apply neighborhood search on the initial knapsack and compute for each initial knapsack the optimal recovery by Dynamic programming. Hill climbing performs 100 restarts.

For the branching algorithms we tested different branching strategies. In the branch-and-price algorithms the difference in performance turned out to be minor and we report on the strategy that performed best in Separate Recovery Decomposition. However, in the Branch-and-Bound algorithm some difference could be observed and we report on the strategy that shows the best performance for this algorithm.

The results of the second phase are given in Table 1. For each instance and each algorithm, we allowed at most 3000 milliseconds of computation time. For each algorithm, we report on the number of instances (out of 4200) that could not be solved within 3000 ms, the average and maximum computation time over the successful instances. For Hill Climbing we give the average and minimal performance ratio and for the branching algorithms the average and maximum number of evaluated nodes. For Hill Climbing 'Failed' means that it was not able to finish all restarts in the given time.

Table 1. Second Phase Results

Algorithm	Failed	avg t(ms)	max t(ms)	avg $\frac{c}{c*}$	min $\frac{c}{c*}$	avg nodes	max nodes
Seperate Recovery	128	107	2563	-	-	3.27	122
Combined Recovery	1407	417	2969	-	-	1.12	17
Branch and Bound	190	111	2906	-	-	1281	33321
DP	2840	347	2984	-	-	-	-
Hill Climbing	0	17.3	422	0.99	0.85	-	-

The results indicate that for this problem Separate Recovery Decomposition outperforms Combined Recovery Decomposition. DP is inferior to Branch-and-Bound and Hill Climbing.

In the third phase we experimented with larger instances for the two best algorithms. We present results for instances with 50 and 100 items and 2, 3, 4, 10, or 20 scenarios. Again, for each combination of number of items, number of scenarios, we generated 100 item sets (20 from each instance class) with each 3 scenario sets. This results in 300 instances per combination of number of items and number of scenarios, where the maximum computation time per instance per algorithm is 4 minutes. The results are depicted in Tables 2 and 3.

Table 2. Third Phase Results for Separate Recovery decomposition

Items	Scenarios	Failed	avg ms	max ms	avg nodes	max nodes
50	2	2	686	56,312	1.56	68
50	3	12	2,724	53,454	1.7	25
50	4	46	3,799	58,688	2.6	35
50	10	125	3,295	53,483	2.29	35
50	20	144	1,473	38,766	1.4	17
100	2	114	1,695	47,531	1.05	5
100	3	173	703	24,781	1.16	11
100	4	176	964	46,172	2.03	59
100	10	213	469	34,547	1.39	25
100	20	210	103	2,703	1.13	13

The results suggest that the computation time of Separate Recovery Decomposition scales very well with the number of scenarios. As may be expected, Hill Climbing shows a significant increase in the computation time when the number of scenarios is increased. Moreover, the small number of nodes indicates that Separate Recovery Decomposition is well-suited for instances with a larger number of scenarios. On average the quality of the solutions form Hill Climbing is very high. However, the minimum performance ratios of about 0.66 show that there is no guarantee of quality. Observe that there is a difference in the notion of

Table 3. Third Phase Results for Hill Climbing

Items	Scenarios	Failed	avg ms	max ms	avg $\frac{c}{c^*}$	min $\frac{c}{c^*}$
50	2	0	104	969	0.98	0.68
50	3	0	173	1,204	0.98	0.84
50	4	0	180	1,203	0.98	0.83
50	10	0	268	1,407	1	0.94
50	20	0	309	1,515	1	0.84
100	2	0	887	19,656	0.98	0.66
100	3	0	1,257	25,578	1	0.86
100	4	0	1,783	32,625	1	0.8
100	10	0	3,546	34,703	1	0.8
100	20	0	4,546	37,312	1	0.94

Failed. For the Separate Recovery Decomposition it means failed to solve to full optimality and for Hill Climbing failed complete the algorithm with 100 restarts.

6   Generalization and Conclusion

In this paper we investigated column generation for recoverable robust optimization. We think that our approach is very promising and that it can be generalized to many other problems.

We presented two methods: Separate Recovery Decomposition and Combined Recovery Decomposition. In the first approach, we work with separate solutions for the initial problem and recovery solutions for the different scenarios; in the second one, we work with combined solutions for the initial problem and the recovery problem for a *single* scenario.

We considered the size robust knapsack problem. We applied Separate Recovery Decomposition and Combined Recovery Decomposition. In the first model, the pricing problem is a knapsack problem for both the initial solution columns and the recovery solution columns. In the second model, the pricing problem is to find an optimal column containing a combination of initial and recovery solution for a single scenario, i.e., recoverable robust optimization for a single scenario case. We implemented branch-and-price algorithms for both models. Our computational experiments revealed that for this problem Separate Recovery Decomposition outperformed Combined Recovery Decomposition and the first method scaled very well with the number of scenarios. If we improve the primal heuristic in the algorithm, it will find a feasible solution faster, which is able to reduce the computation time and in this way the number of Failed instances as reported in Table 2. This is an interesting topic for further research.

We developed a Combined Recovery model for the demand robust shortest path problem. We intend to implement this model in the near future. Interesting issues for further research are restrictions on the recovery solution such as a limited budget for the cost of the recovery solution or an upper bound on the number of edges obtained during recovery.

Finally, we think that our approach can be generalized to solve many other problems. We are currently developing a framework for recoverable robustness by column generation. The application of the presented methods to other problems is a very interesting area for further research.

References

1. Barnhart, C., Johnson, E.L., Nemhauser, G.L., Savelsbergh, M.W.P., Vance, P.H.: Branch-and-price: column generation for solving huge integer programs. Operations Research 46, 316–329 (1998)
2. Ben-Tal, A., El Ghaoui, L., Nemirovski, A.: Robust Optimization. Princeton University Press, Princeton (2009)
3. Birge, J.R., Louveaux, F.: Introduction to Stochastic Programming. Springer, Heidelberg (1997)
4. Bertsimas, D., Sim, M.: The Price of Robustness. Operations Research 52(1), 35–53 (2004)
5. Büsing, C., Koster, A.M.C.A., Kutschka, M.: Recoverable Robust Knapsacks: the Discrete Scenario Case (2010),
   ftp://ftp.math.tu-berlin.de/pub/Preprints/combi/Report-018-2010.pdf
6. Cacchiani, V., Caprara, A., Galli, L., Kroon, L., Maroti, G., Toth, P.: Recoverable Robustness for Railway Rolling Stock Planning. In: 8th Workshop on Algorithmic Approaches for Transportation Modeling, Optimization, and Systems, ATMOS 2008 (2008), http://drops.dagstuhl.de/opus/volltexte/2008/1590/
7. Caprara, A., Galli, L., Kroon, L.G., Maróti, G., Toth, P.: Robust Train Routing and Online Re-scheduling. In: ATMOS (2010),
   http://drops.dagstuhl.de/opus/volltexte/2010/2747/
8. Cicerone, S., D'angelo, G., Di Stefano, G., Frigioni, D., Navarra, A., Schachtebeck, M., Schöbel, A.: Recoverable Robustness in Shunting and Timetabling. In: Ahuja, R.K., Möhring, R.H., Zaroliagis, C.D. (eds.) Robust and Online Large-Scale Optimization. LNCS, vol. 5868, pp. 28–60. Springer, Heidelberg (2009)
9. Dhamdhere, K., Goyal, V., Ravi, R., Singh, M.: How to pay, come what may: Approximation algorithms for demand-robust covering problems. In: Proceedings of the Annual IEEE Symposium on Foundations of Computer Science (FOCS 2005), pp. 367–378. IEEE Computer Society, Los Alamitos (2005)
10. Kellerer, H., Pferschy, U., Pisinger, D.: Knapsack Problems. Springer, Germany (2004)
11. Liebchen, C., Lübbecke, M., Möhring, R., Stiller, S.: Recoverable robustness. In: Ahuja, R.K., Möhring, R.H., Zaroliagis, C.D. (eds.) Robust and Online Large-Scale Optimization. LNCS, vol. 5868, pp. 1–27. Springer, Heidelberg (2009)
12. Puhl, C.: Recoverable robust shortest path problems. Report 034-2008. Mathematics. Technical University, Berlin (2008)

Passenger Flow-Oriented Train Disposition[*]

Annabell Berger[1], Christian Blaar[1], Andreas Gebhardt[1],
Matthias Müller-Hannemann[1], and Mathias Schnee[2]

[1] Department of Computer Science
Martin-Luther-Universität Halle-Wittenberg
{berger,gebhardt,muellerh}@informatik.uni-halle.de,
christian.blaar@student.uni-halle.de
[2] Department of Computer Science
Technische Universität Darmstadt
schnee@algo.informatik.tu-darmstadt.de

Abstract. Disposition management solves the decision problem whether
a train should wait for incoming delayed trains or not. This problem has
a highly dynamic nature due to a steady stream of update information
about delayed trains. A dispatcher has to solve a global optimization
problem since his decisions have an effect on the whole network, but he
takes only local decisions for subnetworks (for few stations and only for
departure events in the near future). In this paper, we introduce a new
model for an optimization tool. Our implementation includes as building blocks (1) routines for the permanent update of our graph model
subject to incoming delay messages, (2) routines for forecasting future
arrival and departure times, (3) the update of passenger flows subject
to several rerouting strategies (including dynamic shortest path queries),
and (4) the simulation of passenger flows. The general objective is the
satisfaction of passengers. We propose three different formalizations of
objective functions to capture this goal. Experiments on test data with
the train schedule of German Railways and real delay messages show
that our disposition tool can compute waiting decisions within a few seconds. In a test with artificial passenger flows it is fast enough to handle
the typical amount of decisions which have to be taken within a period
of 15 minutes in real time.

1 Introduction

Motivation. Delay and disposition management is about waiting decisions of
connecting trains in a delay scenario: Shall a train wait for passengers from a
feeding train or not? Dispatchers work at regional centers and are responsible
for a small number of stations where they have to resolve all waiting conflicts
which occur in the next few minutes. Classic delay management [1] solves a
global and static optimization problem. It thus makes simultaneous decisions

---

[*] This work was supported by the DFG Focus Program Algorithm Engineering, grant
MU 1482/4-2. We wish to thank Deutsche Bahn AG for providing us timetable data
for scientific use.

for all stations and in particular even for several hours ahead. This has two drawbacks: it is hardly realizable in daily operation as it modifies the planned schedule at too many places. Moreover, it works with the fictitious assumption that the future is completely known at the time of decision, whereas additional delays will repeatedly occur. Our approach tries to avoid these shortcomings. The dispatcher is still responsible for local decisions, but he shall now use a global optimization. We obtain a steady stream of real-time updates of the current delay situation of trains which we incorporate into our algorithms. From this update information, we repeatedly have to derive estimates for future departure and arrival times.

A major challenge is to find an appropriate objective function. Our general overall objective is a passenger-friendly disposition. It is, however, not at all clear what this goal means, since the individual benefit of a passenger is in conflict with a system-wide optimum. Classic literature proposes a simple sum-objective and minimizes the lateness of all passengers at their destinations [1]. This view obscures the philosophical question whether it is preferable to delay all passengers in a similar way or to let a few passengers suffer from heavy delays (with compensations). These issues deserve further discussion. In this paper, we take a pragmatic view and propose the following three objective functions: (1) as in classic literature, the overall lateness at the destination, (2) the deviation of realized passenger numbers from original travel plans over all trains, (3) the number of passengers which do not reach their destination (within reasonable time). Forecasts of delays for the next one or two hours are relatively accurate, but they bear an increasing uncertainty when they make estimates for events further in the future [2]. Hence, it is natural to weight earlier events higher.

An essential pre-condition to put a passenger-friendly disposition into work is that dispatchers know about current passenger flows. Up to now, this information is only partially available, train dispatchers mainly work with estimates based on their experience or passenger countings in the past. In this paper, we want to study to which extent we can support disposition management under the assumption that reliable information about passenger flows is available. More precisely, we assume that dispatchers know how many passengers are traveling in which train and to which destination they are traveling. Staff on trains can collect these numbers and desired destinations during ticket checking and send them to the disposition center. We believe that such a model could be realized without too much extra effort. German Railways applies a certain set of standard waiting rules which specify how long a train will wait at most for one of its feeding trains in case of a delay. In general, these waiting time rules are automatically applied. Train dispatchers constantly monitor the current delay development. Whenever an incoming train is delayed by so much that the maximum waiting time is exceeded, i.e. the out-going train would not wait according to the automatic standard policy, the dispatcher steps in. He has now to decide within a short period of time whether he wants to overrule the automatic decision.

Up to now, the dispatcher considers the number of passengers which would miss their connection if the out-going train does not wait. The effect of his decision on other passengers which may miss their transfers because of induced delays are ignored. In this sense, a dispatcher optimizes locally. [1] In periods of high traffic and many delays, it is challenging for the dispatchers to find optimal decisions. It is therefore important to develop supporting tools so that many disposition conflicts can be solved automatically by algorithms, and only a few remain to be solved manually. In our scenarios, there is no information about track capacities available. This is a serious restriction since these capacities play a crucial role for accurate predictions of induced delays [3,4,5]. However, it is a realistic assumption in current practice. Non-discrimination rules in the European Union require that track management and disposition management for the different operating companies sharing the same tracks are strictly separated.

Related Work. Gatto et al. [6,7] study the computational complexity of several variants of delay management. They also developed efficient algorithms for special cases based on dynamic programming or by reduction to a sequence of minimum cut problems. Various integer linear programming models for delay management have been proposed and studied by Schöbel [1] and Schachtebeck and Schöbel [8]. These models assume periodic schedules and have a static view: the complete delay scenario is known. In contrast, we consider an online scenario where the newest delay information is revealed step by step. Schöbel and co-authors evaluate their models on relatively small subnetworks of Germany and the Netherlands. Ginkel and Schöbel consider bicriteria delay management, taking into account both the delay of trains and the number of passengers who miss a connection [9]. Efficient deterministic propagation of initial source delays (primary delays) and the cascade of implied delays (secondary delays) has been done by Müller-Hannemann and Schnee [10]. They demonstrated that even massive delay data streams can be propagated instantly, making this approach feasible for real-time multi-criteria timetable information. This approach has very recently been extended to stochastic delay propagation [2].

Our Contribution. We propose a new model for real-time train disposition aiming at passenger-friendly optimization and discuss a number of modeling issues: How to deal with the fact that only partial information is available? What are appropriate models and data structures to represent passenger flows? Since we have to deal with millions of passengers every day, representing each of them (and their complete travel paths!) individually is not feasible. The development of an appropriate compact model is therefore a crucial part of this work. In essence, we propose to group passengers together with respect to their final destination (but ignore their planned paths). If passengers have to be rerouted, we assume that they prefer alternatives leading them to their destination as quickly as possible. We have built a prototypal implementation composed of complex software components for the following main tasks:

---

[1] The true disposition process in Germany is even more complicated. In addition to the regional operators there is a central disposition unit which monitors (but does not optimize in the mathematical sense) the overall situation. Decisions of regional operators having a global effect have to be coordinated with the central disposition.

1. routines for the permanent update of our graph model subject to incoming delay messages,
2. routines for forecasting future arrival and departure times,
3. the update of passenger flows subject to several rerouting strategies (including dynamic shortest path queries), and
4. the simulation of initial passenger flows.

From an algorithmic point of view, the basic question studied in this paper concerns efficiency: Is it possible to achieve passenger flow updates fast enough to be used within an online decision support system? In our experiments, conducted with the complete German train schedule (about 8800 stations, 40000 trains and one million events per day), relatively large passenger flows (the initial flow is built upon up to 32,000 origin-destination pairs), and several delay scenarios, we achieve update times of several seconds. A bottleneck analysis indicates that the dynamic rerouting is currently the most expensive part of the computation. Using data of a typical day of operation, we determined the dispatching demand, i.e. the number of waiting conflicts which have to be resolved by dispatchers. Simulating such a demand we find that our algorithms can solve these problems fast enough to be applicable in a real-time system.

Overview. In Section 2, we describe in detail our model of the input data, the event graph, objective functions, and passenger flows. Afterwards, we explain our update algorithms for the current delay situation and the following rerouting phase of passenger flows. Computational results are given in Section 4. Finally, we conclude with remarks for future research. A full version of the paper is available as a technical report [11].

2   Model

2.1   Timetable and Events

A *timetable* $TT := (P, S, C)$ consists of a tuple of sets. Let P be the set of trains, S the set of stations and C the set of *elementary connections*, that is $C := \left\{ c = (z, s, s', t_d, t_a) \middle| \begin{array}{l} \text{train } z \in P \text{ leaves station } s \text{ at time } t_d. \\ \text{The next stop of } z \text{ is at station } s' \text{ at time } t_a \end{array} \right\}$. A timetable TT is valid for a number of N traffic days. A *validity function val* : $P \mapsto \{0, 1\}^N$ determines on which traffic days the train operates. We define with respect to the set of elementary connections C a set of departure events Dep and arrival events Arr for all stations $v \in S$. Each event $dep_v := (plannedTime, updatedTime, train) \in Dep$ and $arr_v := (plannedTime, updatedTime, train) \in Arr$ represents exactly one departure or arrival event which consists of the three attributes *plannedTime*, *updatedTime* and *train*. The attribute *updatedTime* at an effective time point represents either the forecast time or the realized time for an event. Hence, if the current time is smaller than the updated timestamp, then this attribute represents a forecast otherwise it has already been realized. Additionally, we build for each departure event dep_v a corresponding *get-in event* $in_v \in Getin$ and for each arrival event $arr_v \in Arr$ a *get-off event* $off_v \in Getoff$

representing starts and ends of passenger journeys. Furthermore, at each station we define exactly one *nirvana event* $nir_v \in Nirvana$ to model the possibility that a passenger does not arrive at his destination (within reasonable time after the planned arrival). In the following, we simply write $event_v$ if we do not specify its type. Staying times at a station v can be lower bounded by *minimum staying times* $minstay(arr_v, dep_v) \in \mathbb{Z}^+$ which have to be respected between different events at v. Minimum staying times arise for two purposes in our model: for two events corresponding to the same train they ensured that passengers have enough time to enter and leave the train, whereas when arrival and departure event belong to different trains they specify the minimum time to transfer from one train (the so-called *feeder train*) to another.

2.2   The Event Graph

We construct a directed, acyclic *event graph* $G = (V, A)$ with respect to a given timetable $TT = (P, S, C)$ as follows. The vertex set V consists of all events, i.e. $V := Dep \cup Arr \cup Getin \cup Getoff \cup Nirvana$. Next, we define different types of arcs. *Transfer arcs* $A_1 := \{(arr_v, dep_v) \mid arr_v \in Arr, dep_v \in Dep, v \in S\}$ represent all possibilities to transfer between trains. Note that this set may contain transfers which are not feasible with respect to minimum staying times $minstay$. The reason is that the timestamps $updatedTime$ of arrival and departure events can be changed due to delays. Hence, we may loose possibilities to transfer or get some new chances. The set A_1 also contains arcs (arr_v, dep_v) with $arr_v(train) = dep_v(train)$ meaning that a passenger stays at station v in his train. Furthermore, there exist so-called *waiting times* $waitingTime : A_1 \mapsto \mathbb{Z}_0^+$ modelling the rule that a train $dep_v(train)$ has to wait for a delayed feeder train $arr_v(train)$ at most $waitingTime(arr_v, dep_v)$ time units. This results in a threshold value $threshold : A_1 \mapsto \mathbb{Z}^+$ for each transfer arc determining the latest point in time $threshold(arr_v, dep_v) = dep_v(plannedTime) + waitingTime(arr_v, dep_v)$ for departure at station v with respect to train $arr_v(train)$.

Our model is designed to compute 'passenger flows' — i.e. the current number of passengers traveling to different destinations on each arc in our event graph. Hence, we have to solve two main problems: (1) the repeated computation of all timestamps $event_v(updatedTime)$ for each event and (2) the repeated computation of all passenger flows on all arcs. For both cases we only have to update a small subset of A_1. Moreover, these subsets are not identical and can change in each update scenario. That means, we can determine two different subgraphs of our event graph — only containing a small set of transfer arcs. Our computations can be done on these subgraphs leading to efficient running times. We define two types of validity functions for transfer arcs with respect to the corresponding scenario. Passengers are only able to transfer from one train to another train, if minimum staying times are respected. For that reason, we define with respect to case (2) a *feasibility function* $Z : A_1 \mapsto \{0, 1\}$ setting $Z((arr_v, dep_v)) = 1$ if $minstay(arr_v, dep_v) \le dep_v(updatedTime) - arr_v(updatedTime)$. Otherwise, we have $Z((arr_v, dep_v)) = 0$. Clearly, in a real world scenario all transfer arcs which are used by passengers are feasible. That means we have $minstay(arr_v, dep_v) \le$

$dep_v(plannedTime) - arr_v(plannedTime)$ for each planned transfer arc. Hence, an arc set for case (2) has to contain all transfer arcs containing passengers and all arcs which are feasible after an update (below we give an exact definition).

Somewhat more complicated is the concept for transfer arcs if we want to compute the timestamps of a departure event in case (1). Such a timestamp $dep_v(updatedTime)$ only depends on the arrival of a given feeder train if the value $threshold(arr_v, dep_v)$ is not exceeded. Hence, a transfer arc is not relevant for the computation of a timestamp, if $threshold(arr_v, dep_v)$ is exceeded. Note that there is no reason to consider transfer arcs with a waiting time of zero (i.e., arcs without waiting rule), because the departure time of such an event does not depend on the corresponding arrival event. Hence, we can exclude such transfer arcs from the delay propagation. We define an *activity function*, $active : A_1 \mapsto \{0, 1\}$ representing the relevance of an arc for the computation of timestamp $dep_v(updatedTime)$, with $active(arr_v, dep_v) = 1$ for relevant arcs only.

As a second type of arcs we define *travel arcs* $A_2 := \{(dep_v, arr_w) \mid dep_v, arr_w$ are contained in an elementary connection $c \in C\}$. The set of all *get-in arcs* we define as $A_3 := \{(in_v, dep_v) \mid in_v$ is the corresponding event of $dep_v \}$, the set of all *get-off arcs* as $A_4 := \{(arr_v, off_v) \mid arr_v$ is the corresponding event of $off_v\}$, and the set of all *nirvana arcs* as $A_5 := \{(arr_v, nir_w) \mid \forall arr_v \in Arr \wedge v \neq w \in S\}$.

2.3   Passenger Flows and Objective Functions

We define for our event graph $G = (V, A)$ a *destination-oriented passenger flow* $flow : A \mapsto \mathbb{N}_0^{|T|}$ with respect to a subset $T \subseteq S$ of destinations. This is a variant of an uncapacitated multi-commodity flow with one commodity for each destination in T. Hence, the tth-component $flow_t(a) = j$ represents the number of passengers j on arc a wishing to travel to destination t. Let $d_G^+(event_v)$ be the number of outgoing arcs for vertex $event_v$ and $d_G^-(event_v)$ the number of ingoing arcs. We denote a destination-oriented passenger flow as *feasible*, if it fulfills for all vertices $event_v \in V$ with $d_G^+(event_v) > 0$ and $d_G^-(event_v) > 0$, i.e. for all vertices which are neither sources nor sinks, the flow conservation condition:

$$\sum_{event_w \in V} flow_t(event_v, event_w) = \sum_{event_{w'} \in V} flow_t(event_{w'}, event_v).$$

Moreover, we require that the *get-in* arcs A_3 carry a prescribed amount of entering passengers. In practice we only need to consider transfer arcs $a \in A_1$ which possess a non-zero passenger flow $flow_t(a) > 0$ for at least one destination $t \in T$. We define two new subsets of transfer arcs A_1 —*the passenger propagation transfer arcs* with

$$A_1' := \{(arr_v, dep_v) \mid arr_v \in Arr \wedge dep_v \in Dep \wedge \exists t \text{ with } flow_t(arr_v, dep_v) > 0\}$$
$$\cup \{(arr_v, dep_v) \mid arr_v \in Arr \wedge dep_v \in Dep \wedge dep_v(train) = arr_v(train)\}$$

and *time propagation transfer arcs* with

$$A_1'' := \{(arr_v, dep_v) \mid arr_v \in Arr \wedge dep_v \in Dep \wedge waitingTime(arr_v, dep_v) > 0\}$$
$$\cup \{(arr_v, dep_v) \mid arr_v \in Arr \wedge dep_v \in Dep \wedge dep_v(train) = arr_v(train)\}.$$

Note that both sets include all 'transfer arcs' which model that a train enters and leaves a station. Hence, we define a new arc set $A' := A_1' \cup A_2 \cup A_3 \cup A_4 \cup A_5$ and define the subgraph $G' = (V, A', Z)$ of event graph G with feasibility function Z and call it *passenger propagation graph*. Furthermore, we define a second subgraph $G'' = (V, A'', active)$ where $A'' := A_1'' \cup A_2 \cup A_3 \cup A_4 \cup A_5$ and call it *time propagation graph*. Each of these subgraphs represents exactly all relevant arcs for a propagation of timestamps/passenger flows.

Let us now consider our objective functions. We denote a new feasible flow after one update as $flow'$. We define for each travel arc $u \in A_2$ the *squared passenger deviation* as $\Delta_{flow}(u) := \sum_{t \in T} (flow_t'(u) - flow_t(u))^2$. Furthermore, we want to weight events with respect to their timestamp because it makes sense to assume that it is more likely that forecasts for the near future will actually realize. We define the *current time* as $t_{current}$ and define a weight function $w_{Dep} : Dep \cup Arr \mapsto [0, 1]$, monotonically increasing with decreasing difference of $t_{current}$ and $event(plannedTime)$, and monotonically decreasing with decreasing importance of the train category. Our objective functions consider only events which have not been realized yet. We build our first objective function

$$f_{Dep}(flow, flow') := \sum_{(dep_w, arr_v) \in A_2} \Delta_{flow}(dep_w, arr_v) \cdot w_{Dep}(dep_w).$$

It measures the strength of the change in passenger flows caused by rerouting passengers. In a second optimization criterion we want to model penalties for passengers non-arriving at their destinations. We define $f_{Nir}(flow, flow') :=$

$$\sum_{(arr_v, nir_w) \in A_5} \left(\sum_{t \in T} (flow_t'(arr_v, nir_w) - flow_t(arr_v, nir_w))^2 \right) \cdot w_{nir}.$$

The weight function $w_{nir} : Nir \mapsto \{1, \dots\}$ represents the strength of penalties for non-arriving at the planned destination. In our last criterion we want to minimize the changes within the schedule weighted by the number of affected passengers. Consider for example a situation where many passengers in a train wait for a small number of passengers of a delayed other train. The passenger deviation will be quite small but many passengers of the waiting train will not be satisfied because they all arrive late at their destination.

We propose as a third optimization criterion $f_{TT}(flow, flow') :=$

$$\sum_{(dep_w, arr_v) \in A_2} w_{TT} \cdot \overline{flow(dep_w, arr_v)}^2 \cdot |updatedTime(arr_v) - plannedTime(arr_v)|$$

where $\overline{flow(dep_w, arr_v)}$ denotes the mean of planned and updated flow on this arc. The weighting parameters w_{TT} can be used to model further influences, for example the dependence on the time horizon. In our experiments, we use a weighted combination of all three objectives.

3   Algorithmic Approach

Overview. In this section we sketch our concept for passenger-flow oriented train disposition. The practical scenario is the following. Every minute we obtain from an external source all information about delayed trains at the current point of time $t_{current}$. We forecast future timestamps of trains under consideration of waiting rules of German Railways. These waiting rules can be overruled by a dispatcher in a short time window. Our system regularly checks whether there is a potential waiting conflict requiring an explicit decision of a dispatcher. Detected conflicts are ordered increasingly by the deadline when they have to be resolved at the latest, and then handled in this order one after another. Each potential outcome of a decision changes the forecast timestamps of the timetable and requires an update of the expected passenger flow. We tentatively calculate the updates of timestamps and passenger flows, and evaluate the prospective objective cost. Based on these evaluations, the dispatcher (or the system) takes a final decision, and proceeds to the next conflict. Next we give some more details.

Repeated updates of the timetable. The timetable update works on the time propagation graph G''. Given external messages about new delays, the current timestamp *updatedTime* for each departure or arrival event is recursively determined from its predecessors under consideration of travel times, minimum transfer times and waiting rules. For details we refer to Müller-Hannemann and Schnee [10] and Schnee [12]. We here reuse their efficient implementation within the tool MOTIS (multi-criteria timetable information system).

The update of passenger flows. After the update of timestamps, some transfer arcs may become infeasible. The passenger flow carried by infeasible arcs must be set to zero. To repair the thereby violated flow conservation conditions, we have to redirect affected passengers. All vertices with flow violations are put into a priority queue which is ordered according to the new timestamps, breaking ties arbitrarily. This order corresponds to a topological order of the passenger propagation graph. We process elements from the queue iteratively in this order, and thereby obtain a one-pass algorithm to update the flow. Two cases may occur. First, less passengers with destination t than planned arrive at vertex v. Then we simply scan the outgoing arcs, and remove the appropriate amount of flow with the same destination from these arcs. As the flow was feasible before the update, this is always possible. Second, and more interesting is the case, that several passengers cannot use their planned transfer and must be redirected. Assume that they are heading towards destination t. One possible strategy would be to determine from the current station an earliest arrival path towards t, and then to choose the very first arc on this path as the next travel arc for these passengers. This *dynamic timetable routing* strategy has the drawback of being computationally relatively expensive. Therefore, we designed a much simpler *standard routing* strategy. At the current station we scan the out-going travel arcs and reroute the passengers to the first feasible alternative, provided that this arc already contains passengers towards t. The rationale behind this rule is the assumption, that if some passengers use a train with destination t, then their path should be also reasonable for other passengers with the same destination. Hence, with priority we use standard routing. Only if this fails (in case there is no suitable arc), we apply dynamic timetable routing. For the latter, we use our tool *Dynamic Timetable Information* on a time-dependent graph as a black box, for details see [13]. In case the dynamic timetable routing does not find a path to destination t arriving there within reasonable time, we apply *"nirvana routing"*, i.e., we send such passengers to a nirvana arc, implying that such passengers are affected severely be the flow change and do not have a reasonable alternative.

4   Experiments

Test Instances and Environment. Our computational study is based on the German train schedule. Our data set contains about 8800 stations, 40000 trains and one million events per day. We use real world status messages from German Railways containing all data of delayed trains at each time. Unfortunately, no passenger data are available to us. Hence, we had the additional task to create a *Passenger Flow Simulator* simulating traveling passengers for a whole day, as described in the following subsection. All experiments were run on a PC (Intel(R) Xeon(R), 2.93GHz, 4MB cache, 47GB main memory under ubuntu linux version 10.04.2 LTS). Only one core has been used by our program. Our code is written in C++ and has been compiled with g++ 4.4.3 and compile option -O3.

Passenger Flow Simulator. Our Passenger Flow Simulator works as follows. We randomly generate a large number of triples (s, t, t_{start}), where $s, t \in S$ are source and target stations and t_{start} denotes the earliest allowed departure time at station s. Source and target stations are chosen with a probability proportional to a measure of importance. Here we simply measure importance of a station by the number of its events. For each selected triple, we use our tool *Dynamic Timetbale Information* (described in [13]) to determine an earliest arrival (s, t)-path. Such a path we denote here as *route*. Next, we add c passengers for destination t to each arc $a \in A$ of this (s, t)-path. The parameter c can be varied, for example depending on the time of departure or the importance of source and target stations.

Experiment 1: Running times for one passenger flow update. In Figure 1, we compare two delay scenarios: 20 delays of 5 minutes and 20 delays of 30 minutes, respectively. We observe that running times do increase substantially with increasing number of routes. Furthermore, they are influenced by the strength of a delay scenario. The most interesting observation is that almost all time is consumed by dynamic timetable routing. In comparison, all other steps are negligible. In Figure 2, we show the number of cases where we have chosen the dynamic timetable routing for a varying number of routes. The number of dynamic timetable routing steps increases with increasing number of routes.

Experiment 2: Real time capability. In a second experimental phase we investigated the functionality of our tool in a real time scenario. Clearly, we are not able to test our tool in the real world. Especially, we cannot make decisions

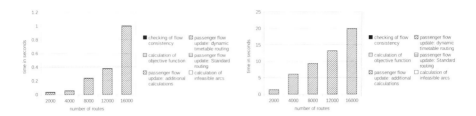

Fig. 1. Comparison of running times dependent on the number of routes for one flow update step for two delay scenarios: 20 delays of 5 (left) and 30 (right) minutes

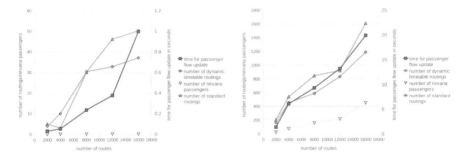

Fig. 2. Comparison of the number of routing steps with standard routing, dynamic timetable routing, and nirvana routing (left axis) and the corresponding running time in seconds (right axis) dependent on the number of routes for two delay scenarios: 20 delays of 5 (left) and 30 (right) minutes.

in the German train network and then observe what happens. Hence, we use real world data but after the first moment where we make a decision we have to simulate further primary delays and cannot use real delay data. For a realistic simulation, we first determined the *dispatching demand* per minute on a typical day. A dispatcher can overrule the waiting times for a departing train by a longer waiting time of at most κ additional minutes. Here, we choose $\kappa := 10$ minutes. Then the dispatching demand is the number of such "potentially exceeded waiting times" on which the dispatcher has to decide. Since, we update the timetable and the passenger flow at a frequency of one minute, we investigated the dispatching demand per minute and accumulated this demand within 15-minute intervals, see Figure 3. We observe a maximum dispatching demand of 8 decisions per minute, but in most cases there are only 2 and rarely 4 decisions to take, if any.

Next we investigated the running times for exactly one optimization step. How do these times depend on the size of the passenger flow (number of routes) or on the dispatching demand? In Figure 4, we analyze several delay scenarios. The scenario on the left-hand side with several new small delays of 5 minutes each can be solved in real time (2 or 3 seconds of computation time). On the right hand-side, we consider a scenario with several large initial delays of 30 minutes. When 20 such delays occur simultaneously, one optimization takes 140 seconds. We clearly observe that the time needed for one optimization step

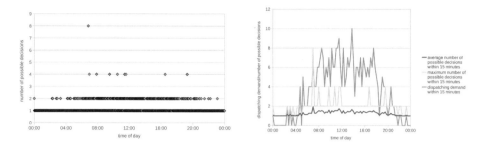

Fig. 3. Dispatching demand for a whole day with real world "delay data" and one passenger flow update per minute. We show the number of dispatching decisions per minute (left) and a cumulated analysis for each interval of 15 minutes (right).

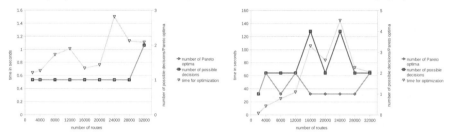

Fig. 4. Comparison of running times dependent on the number of routes for one optimization step for two delay scenarios: 20 delays of 5 (left) and 30 (right) minutes

depends only marginally on the number of routes, it is much stronger correlated with the dispatching demand. Note that the dispatching demand in our simulation scenario is similar to that in the real world data used in Figure 3. In a real decision support system, a dispatcher will need some time to analyze the proposed solution from our algorithm. During this time one has to freeze the actual delay situation in our model. It will usually suffice when the algorithm delivers a solution within 3 minutes. Our simulation shows that our current tool is already capable of meeting these requirements. We would also like to point out that further speed-up is possible, if alternatives are computed in parallel (parallelization of passenger flow updates is relatively easy to achieve). Hence, we conclude that our tool is fast enough to be applied in a real system.

5   Conclusions and Future Work

In this paper we have proposed a new approach for a passenger-friendly train disposition. In experiments with real and artificial delay scenarios we have demonstrated that our approach can evaluate the consequences of single disposition decisions on passenger flows in an efficient way. Our approach can be extended in several ways. In order to make disposition decisions, our current optimization tool requires the knowledge of forecasts of arrival and departure times for future events. For these predictions we use a set of standard waiting rules (of German Railways). This approach hides the problem that the application of our tool overwrites these rules. Thus, to achieve a more realistic model our optimization tool itself has to produce wide-ranging predictions to derive decisions for the near future. Further extensions of our model concern rerouting subject to the booked train category and the incorporation of seat capacities, so that passengers are not guided towards trains operating already at their capacity limit.

References

1. Schöbel, A.: Integer programming approaches for solving the delay management problem. In: Geraets, F., Kroon, L.G., Schoebel, A., Wagner, D., Zaroliagis, C.D. (eds.) Algorithmic Methods for Railway Optimization 2004. LNCS, vol. 4359, pp. 145–170. Springer, Heidelberg (2007)
2. Berger, A., Gebhardt, A., Müller-Hannemann, M., Ostrowski, M.: Stochastic delay prediction in large train networks. Technical Report 2011/1, Institut für Informatik, Martin-Luther-Universität Halle-Wittenberg (2011)
3. Meester, L.E., Muns, S.: Stochastic delay propagation in railway networks and phase-type distributions. Transportation Research Part B 41, 218–230 (2007)
4. Yuan, J.: Stochastic modeling of train delays and delay propagation in stations. PhD thesis. Technische Universiteit Delft, The Netherlands (2006)
5. van der Meer, D., Goverde, R., Hansen, I.: Prediction of train running times and conflicts using track occupation data. In: Proceedings of 12th WCTR (2010)
6. Gatto, M., Glaus, B., Jacob, R., Peeters, L., Widmayer, P.: Railway delay management: Exploring its algorithmic complexity. In: Hagerup, T., Katajainen, J. (eds.) SWAT 2004. LNCS, vol. 3111, pp. 199–211. Springer, Heidelberg (2004)
7. Gatto, M., Jacob, R., Peeters, L., Schöbel, A.: The computational complexity of delay management. In: Kratsch, D. (ed.) WG 2005. LNCS, vol. 3787, pp. 227–238. Springer, Heidelberg (2005)
8. Schachtebeck, M., Schöbel, A.: To wait or not to wait - and who goes first? Delay management with priority decisions. Transportation Science 44, 307–321 (2010)
9. Ginkel, A., Schöbel, A.: To wait or not to wait? The bicriteria delay management problem in public transportation. Transportation Science 41, 527–538 (2007)
10. Müller-Hannemann, M., Schnee, M.: Efficient timetable information in the presence of delays. In: Ahuja, R.K., Möhring, R.H., Zaroliagis, C.D. (eds.) Robust and Online Large-Scale Optimization. LNCS, vol. 5868, pp. 249–272. Springer, Heidelberg (2009)
11. Berger, A., Blaar, C., Gebhardt, A., Müller-Hannemann, M., Schnee, M.: Passenger flow-oriented train disposition. Technical Report 2011/2, Institut für Informatik, Martin-Luther-Universität Halle-Wittenberg (2011)
12. Schnee, M.: Fully realistic multi-criteria timetable information systems. PhD thesis, Fachbereich Informatik, Technische Universität Darmstadt (2009)
13. Berger, A., Grimmer, M., Müller-Hannemann, M.: Fully dynamic speed-up techniques for multi-criteria shortest path searches in time-dependent networks. In: Festa, P. (ed.) SEA 2010. LNCS, vol. 6049, pp. 35–46. Springer, Heidelberg (2010)

One to Rule Them All:
A General Randomized Algorithm for Buffer Management with Bounded Delay

Łukasz Jeż

Institute of Computer Science, University of Wrocław, ul. Joliot-Curie 15, 50-383 Wrocław, Poland*

Abstract. We give a memoryless scale-invariant randomized algorithm MIX-R for *buffer management with bounded delay* that is $e/(e-1)$-competitive against an adaptive adversary, together with better performance guarantees for many restricted variants, including the s-bounded instances. In particular, MIX-R attains the optimum competitive ratio of $4/3$ on 2-bounded instances.

Both MIX-R and its analysis are applicable to a more general problem, called *Item Collection*, in which only the relative order between packets' deadlines is known. MIX-R is the optimal *memoryless* randomized algorithm against adaptive adversary for that problem in a strong sense.

While some of the provided upper bounds were already known, in general, they were attained by several different algorithms.

1 Introduction

In this paper, we consider the problem of *buffer management with bounded delay*, introduced by Kesselman et al. [16]. This problem models the behavior of a single network switch responsible for scheduling packet transmissions along an outgoing link as follows. We assume that time is divided into unit-length steps. At the beginning of a time step, any number of packets may arrive at a switch and be stored in its *buffer*. Each packet has a positive weight, corresponding to the packet's priority, and a deadline, which specifies the latest time when the packet can be transmitted. Only one packet from the buffer can be transmitted in a single step. A packet is removed from the buffer upon transmission or expiration, i.e., reaching its deadline. The goal is to maximize the *gain*, defined as the total weight of the packets transmitted.

We note that *buffer management with bounded delay* is equivalent to a scheduling problem in which packets are represented as jobs of unit length, with given release times, deadlines and weights; release times and deadlines are restricted to integer values. In this setting, the goal is to maximize the total weight of jobs completed before their respective deadlines.

* Supported by MNiSW grants N N206 490638, 2010–2011 and N N206 368839, 2010–2013, and a scholarship co-financed by ESF project *Human Capital*.

C. Demetrescu and M.M. Halldórsson (Eds.): ESA 2011, LNCS 6942, pp. 239–250, 2011.

As the process of managing packet queue is inherently a real-time task, we model it as an *online problem*. This means that the algorithm, when deciding which packets to transmit, has to base its decision solely on the packets which have already arrived at a switch, without the knowledge of the future.

1.1 Competitive Analysis

To measure the performance of an online algorithm, we use the standard notion of *competitive analysis* [6], which, roughly speaking, compares the gain of the algorithm to the gain of the *optimal solution* on the same instance. For any algorithm ALG, we denote its gain on instance I by $\mathcal{G}_{\mathrm{ALG}}(I)$. The optimal offline algorithm is denoted by OPT. We say that a deterministic algorithm ALG is $\mathcal{R}$-competitive if on any instance I it holds that $\mathcal{G}_{\mathrm{ALG}}(I) \geq \frac{1}{\mathcal{R}} \cdot \mathcal{G}_{\mathrm{OPT}}(I)$.

When analyzing the performance of an online algorithm ALG, we view the process as a game between ALG and an *adversary*. The adversary creates the instance on the one hand, i.e., controls what packets are injected into the buffer, and solves the instance optimally on the other, i.e., chooses its packets for transmission. The goal is then to show that the adversary's gain is at most $\mathcal{R}$ times ALG's gain.

If the algorithm is randomized, we consider its expected gain, $\mathbb{E}[\mathcal{G}_{\mathrm{ALG}}(I)]$, where the expectation is taken over all possible random choices made by ALG. However, in the randomized case, the power of the adversary has to be further specified. Following Ben-David et al. [3], we distinguish between an *oblivious* and an *adaptive-online* adversary, which from now on we will call *adaptive*, for short. An oblivious adversary has to construct the whole instance in advance. This instance may depend on ALG but not on the random bits used by ALG during the computation. The expected gain of ALG is compared to the gain of the optimal offline solution on I. In contrast, in case of an adaptive adversary, the choice of packets to be injected into the buffer may depend on the algorithm's behavior up to the given time step. This adversary must also provide an answering entity ADV, which creates a solution in parallel to ALG. This solution may not be changed afterwards. We say that ALG is $\mathcal{R}$-competitive against an adaptive adversary if for any adaptively created instance I and any answering algorithm ADV, it holds that $\mathbb{E}[\mathcal{G}_{\mathrm{ALG}}(I)] \geq \frac{1}{\mathcal{R}} \cdot \mathbb{E}[\mathcal{G}_{\mathrm{ADV}}(I)]$. We note that ADV is (wlog) deterministic, but as ALG is randomized, so is the instance I.

In the literature on online algorithms [6], the definition of the competitive ratio sometimes allows an additive constant, i.e., a deterministic algorithm is then called $\mathcal{R}$-competitive if there exists a constant $\alpha \geq 0$ such that for any instance I it holds that $\mathcal{G}_{\mathrm{ALG}}(I) \geq \frac{1}{\mathcal{R}} \cdot \mathcal{G}_{\mathrm{OPT}}(I) - \alpha$. An analogous definition applies to the randomized case. For our algorithm MIX-R the bound holds for $\alpha = 0$, which is the best possible.

1.2 Basic Definitions

We denote a packet with *weight* w and *relative deadline* d by (w, d), where the relative deadline of a packet is, at any time, the number of steps after which it expires. The packet's *absolute deadline*, on the other hand, is the exact point in time at which the packet expires. A packet that is in the buffer, i.e., has already been released and has neither expired nor been transmitted by an algorithm, is called *pending* for the algorithm. The *lifespan* of a packet is its relative deadline value upon injection, or in other words the difference between its absolute deadline and release time.

The goal is to maximize the weighted throughput, i.e., the total weight of transmitted packets. We assume that time is slotted in the following way. We distinguish between points in time and time intervals, called *steps*. In step t, corresponding to the interval $(t, t+1)$, ADV and the algorithm choose, independently, a packet from their buffers and transmit it. The packet transmitted by the algorithm (ADV) is immediately removed from the buffer and no longer pending. Afterwards, at time $t+1$, the relative deadlines of all remaining packets are decremented by 1, and the packets whose relative deadlines reach 0 expire and are removed from both ADV's and the algorithm's buffers. Next, the adversary injects any set of packets. At this point, we proceed to step $t+1$.

To no surprise, all known algorithms are *scale-invariant*, which means that they make the same decisions if all the weights of packets in an instance are scaled by a positive constant. A class of further restricted algorithms is of special interest for their simplicity. An algorithm is *memoryless* if in every step its decision depends only on the set of packets pending at that step.

1.3   Previous and Related Work, Restricted Variants

The currently best, 1.828-competitive, deterministic algorithm for general instances was given by Englert and Westermann [10]. Their algorithm is scale-invariant, but it is *not* memoryless. However, in the same article Englert and Westermann provide another, 1.893-competitive, deterministic algorithm that is memoryless scale-invariant. The best known randomized algorithm is the 1.582-competitive memoryless scale-invariant RMix, proposed by Chin et al. [7]. For reasons explained in Section 2.1 the original analysis by Chin et al. is only applicable in the oblivious adversary model. However, a refined analysis shows that the algorithm remains 1.582-competitive in the adaptive adversary model [14].

Consider a (memoryless scale-invariant) greedy algorithm that always transmits the heaviest pending packet. It is not hard to observe that it is 2-competitive, and actually no better than that. But for a few years no better deterministic algorithm for the general case was known. This naturally led to a study of many restricted variants. Below we present some of them, together with known results. The most relevant bounds known are summarized in Table 1. Note that the majority of algorithms are memoryless scale-invariant.

For a general overview of techniques and results on buffer management, see the surveys by Azar [2], Epstein and Van Stee [11] and Goldwasser [12].

Uniform Sequences. An instance is *s-uniform* if the lifespan of each packet is exactly s. Such instances have been considered for two reasons. Firstly, there is a certain connection between them and the *FIFO model* of buffer management, also considered by Kesselmann et al. [16]. Secondly, the 2-uniform instances are among the most elementary restrictions that do not make the problem trivial. However, analyzing these sequences is not easy: while a simple deterministic 1.414-competitive algorithm for 2-uniform instances [18] is known to be optimal among memoryless scale-invariant algorithms [7], for unrestricted algorithms a sophisticated analysis shows the optimum competitive ratio is 1.377 [9].

Bounded Sequences. An instance is *s-bounded* if the lifespan of each packet is at most s; therefore every s-uniform instances is also s-bounded. In particular, the strongest lower bounds on the competitive ratio known for the problem employ 2-bounded instances. These are $\phi \approx 1.618$ for deterministic algorithms [1,8,13], 1.25 for randomized algorithms in the oblivious adversary model [8], and 4/3 in the adaptive adversary model [5]. For 2-bounded instances algorithms matching these bounds are known [16,7,5]. A ϕ-competitive deterministic algorithm is also known for 3-bounded instances, but in general the best known algorithms for s-bounded instances, EDF_β, are only known to be $2 - \frac{2}{s} + o(\frac{1}{s})$-competitive [7]; thus the 1.828-competitive algorithm provides better ratio for large s.

Similarly Ordered Sequences. An instance is *similarly ordered* or has *agreeable deadlines* if for every two packets i and j their spanning intervals are not properly contained in one another, i.e., if $r_i < r_j$ implies $d_i \leq d_j$. Note that every 2-bounded instance is similarly ordered, as is every s-uniform instance, for any s. An optimal deterministic ϕ-competitive algorithm [17] and a randomized 4/3-competitive algorithm for the oblivious adversary model [15] are known. With the exception of 3-bounded instances, this is the most general class of instances for which a ϕ-competitive deterministic algorithm is known.

Other restrictions. Among other possible restrictions, let us mention one for which MIX-R provides additional bounds. As various transmission protocols usually specify only several priorities for packets, one might bound the number of different packet weights. In fact, Kesselmann et al. considered deterministic algorithms for instances with only two distinct packet weights [16].

Generalization: Collecting Weighted Items from a Dynamic Queue A generalization of buffer management with bounded delay has been studied, in which the algorithm knows only the relative order between packets' deadlines rather than their exact values [4]; after Bienkowski et al. we dub the generalized problem *Item Collection*. Their paper focuses on deterministic algorithms but it does provide certain lower bounds for memoryless algorithms, matched by our algorithm. See Section 2.3 for details.

1.4   Our Contribution

We consider randomized algorithms against an adaptive adversary, motivated by the following observation. In reality, traffic through a switch is not at all independent of the packet scheduling algorithm. For example, lost packets are typically resent, and throughput through a node affects the choice of routes for data streams in a network. These phenomena can be captured by the adaptive

Table 1. Comparison of known and new results. The results of this paper are shown in boldface; a reference next to such entry means that this particular bound was already known. The results without citations are implied by other entries of the table. An asterisk denotes that the algorithm attaining the bound is memoryless scale-invariant.

		deterministic	(rand.) adaptive	(rand.) oblivious
general	upper	1.828 [10], 1.893* [10]	**1.582*** [14]	1.582* [7]
	lower	1.618	1.333	1.25
s-bounded	upper	$2 - \frac{2}{s} + o(\frac{1}{s})^*$ [7]	$\mathbf{1/\left(1 - (1 - \frac{1}{s})^s\right)^*}$	$1/\left(1 - (1 - \frac{1}{s})^s\right)^*$
	lower	1.618	1.333	1.25
2-bounded	upper	1.618* [16]	**1.333*** [5]	1.25* [7]
	lower	1.618 [1,8,13]	1.333 [5]	1.25 [8]

adversary model but not by the oblivious one. The adaptive adversary model is also of its own theoretical interest and has been studied in other settings [6].

The main contribution of this paper is a simple memoryless scale-invariant algorithm MIX-R. As is reflected in its name, MIX-R is very similar to RMIX, proposed by Chin et al. [7]. In fact, the only, yet crucial, difference between these two algorithms is the probability distribution of transmission over pending packets. RMIX is in fact the EDF_β algorithm, with β chosen randomly in each step. The probability distribution for β is fixed, i.e., it remains the same in every step. It is no longer so in our algorithm MIX-R.

Consequently, our analysis of MIX-R gives an upper bound on its competitive ratio that depends on the maximum number of packets, over all steps, that have positive probability of transmission in the step. Specifically, if we denote that number by N, our upper bound on the competitive ratio of MIX-R is $1/\left(1 - (1 - \frac{1}{N})^N\right)$. Note that $1/\left(1 - (1 - \frac{1}{N})^N\right)$ tends to $e/(e-1)$ from below. The number N can be bounded a priori in certain restricted variants of the problem, thus giving better bounds for them, as we discuss in detail in Section 2.5. For now let us mention that $N \leq s$ in s-bounded instances and instances with at most s different packet weights. The particular upper bound of $4/3$ that we obtain for 2-bounded instances is tight in the adaptive adversary model [5]. We provide new upper bounds for s-bounded instances, where $s > 2$. Our results also imply several other upper bounds that were known previously, but in general were attained by several different algorithms, cf. Table 1.

As is the case with RMIX, both MIX-R and its analysis rely only on the relative order between the packets' deadlines. Therefore our upper bound(s) apply to the *Item Collection* problem [4]. In fact, MIX-R is the optimum randomized memoryless algorithm for that problem in a strong sense, cf. Section 2.3.

2   General Upper Bound

2.1   Analysis Technique

In our analysis, we follow the paradigm of modifying ADV's buffer, introduced by Li et al. [17]. Namely, we assume that in each step MIX-R and ADV have precisely the same pending packets in their buffers. Once they both transmit a packet, we modify ADV's buffer judiciously to make it identical with that of MIX-R. This analysis technique leads to a streamlined and intuitive proof.

When modifying the buffer, we may have to let ADV transmit another packet, inject an extra packet to his buffer, or upgrade one of the packets in its buffer by increasing its weight or deadline. We ensure that these changes are be *advantageous to the adversary* in the following sense: for any adversary strategy ADV, starting with the current step and buffer content, there is an adversary strategy $\overline{\text{ADV}}$ that continues computation with the modified buffer, such that the total gain of $\overline{\text{ADV}}$ from the current step on (inclusive), on any instance, is at least as large as that of ADV.

To prove R-competitiveness, we show that in each step the expected *amortized gain* of ADV is at most R times the expected gain of MIX-R, where the former is the total weight of the packets that ADV eventually transmitted in this step. Both expected values are taken over random choices of MIX-R.

We are going to assume that ADV never transmits a packet a if there is another pending packet b such that transmitting b is always advantageous to ADV. Formally, we introduce a dominance relation among the pending packets and assume that ADV never transmits a dominated packet.

We say that a packet $a = (w_a, d_a)$ is *dominated* by a packet $b = (w_b, d_b)$ at time t if at time t both a and b are pending, $w_a \leq w_b$ and $d_a \geq d_b$. If one of these inequalities is strict, we say that a is *strictly dominated* by b. We say that packet a is (strictly) dominated whenever there exists a packet b that (strictly) dominates it. Then the following fact can be shown by a standard exchange argument.

Fact 1. *For any adversary strategy* ADV, *there is a strategy* $\overline{\text{ADV}}$ *with the following properties:*

1. *the gain of* $\overline{\text{ADV}}$ *on every sequence is at least the gain of* ADV,
2. *in every step* t, $\overline{\text{ADV}}$ *does not transmit a strictly dominated packet at time* t.

Let us stress that Fact 1 holds for the adaptive adversary model. Now we give an example of another simplifying assumption, often assumed in the oblivious adversary model, which seems to break down in the adaptive adversary model. In the oblivious adversary model the instance is fixed in advance by the adversary, so ADV may precompute the optimum schedule to the instance and follow it. Moreover, by standard exchange argument for the *fixed* set of packets to be transmitted, ADV may always send the packet with the smallest deadline from that set—this is usually called the *earliest deadline first* (EDF) property or order. This assumption not only simplifies analyses of algorithms but is often crucial for them to yields desired bounds [7,9,17,15].

In the adaptive adversary model, however, the following phenomenon occurs: as the instance I is randomized, ADV does not know for sure which packets it will transmit in the future. Consequently, deprived of that knowledge, it cannot ensure any specific order of packet transmissions.

2.2   The Algorithm

We describe Mix-R, i.e., its actions in a single step, in Algorithm 1.

Algorithm 1. Mix-R (single step)

1: **if** there are no pending packets **then**
2:     do nothing and proceed to the next step
3: **end if**
4: $m \leftarrow 0$ ▷ counts packets that are not strictly dominated
5: $n \leftarrow 0$ ▷ counts packets with positive probability assigned
6: $r \leftarrow 1$ ▷ unassigned probability
7: $H_0 \leftarrow$ pending packets
8: $h_0 = (w_0, d_0) \leftarrow$ heaviest packet from H_0
9: **while** $H_m \neq \emptyset$ **do**
10:      $m \leftarrow m + 1$
11:      $h_m = (w_m, d_m) \leftarrow$ heaviest not strictly dominated packet from H_{m-1}
12:      $p_{m-1} \leftarrow \min\{1 - \frac{w_m}{w_{m-1}}, \; r\}$
13:      $r \leftarrow r - p_{m-1}$
14:      **if** $r > 0$ **then**
15:          $n \leftarrow n + 1$
16:      **end if**
17:      $H_m \leftarrow \{x \in H_{m-1} \mid x$ is not dominated by $h_m\}$
18: **end while**
19: $p_m \leftarrow r$
20: **transmit** h chosen from $h_1, \ldots, h_n$ with probability distribution $p_1, \ldots, p_n$
21: proceed to the next step

We introduce the packet h_0 to shorten Mix-R's pseudocode. The packet itself is chosen in such a way that $p_0 = 0$, to make it clear that it is not considered for transmission (unless $h_0 = h_1$). The while loop itself could be terminated as soon as $r = 0$, because afterwards Mix-R does not assign positive probability to any packet. However, letting it construct the whole sequence $h_1, h_2, \ldots h_m$ such that $H_m = \emptyset$ simplifies our analysis. Before proceeding with the analysis, we note a few facts about Mix-R.

Fact 2. *The sequence of packets $h_0, h_1, \ldots, h_m$ selected by Mix-R satisfies*

$$w_0 = w_1 > w_2 > \cdots > w_m \; ,$$
$$d_1 > d_2 > \cdots > d_m \; .$$

Furthermore, every pending packet is dominated by one of $h_1, \ldots, h_m$.

Fact 3. *The numbers $p_1, p_2, \ldots, p_m$ form a probability distribution such that*

$$p_i \leq 1 - \frac{w_{i+1}}{w_i} \qquad for \; all \; i < m \; . \tag{1}$$

Furthermore, the bound is tight for $i < n$, while $p_i = 0$ for $i > n$, i.e.,

$$p_i = \begin{cases} 1 - \frac{w_{i+1}}{w_i}, & for \; i < n \\ 0, & for \; i > n \end{cases} \tag{2}$$

Theorem 4. Mix-R *is* $1/\left(1 - (1 - \frac{1}{N})^N\right)$*-competitive against an adaptive adversary, where N is the maximum number of packets, over all steps, that are assigned positive probability in a step.*

Proof. For a given step, we describe the changes to Adv's scheduling decisions and modifications to its buffer that make it the same as Mix-R's buffer. Then, to prove our claim, we will show that

$$\mathbb{E}\left[\mathcal{G}_{\text{Adv}}\right] \leq w_1 \ , \tag{3}$$

$$\mathbb{E}\left[\mathcal{G}_{\text{Mix-R}}\right] \geq w_1 \left(1 - (1 - \frac{1}{n})^n\right) \ , \tag{4}$$

where n is the number of packets assigned positive probability in the step. The theorem follows by summation over all steps.

Recall that, by Fact 1, Adv (wlog) sends a packet that is not strictly dominated. By Fact 2, the packets $h_1, h_2, \ldots h_m$ dominate all pending packets, so the one sent by Adv, say p is (wlog) one of $h_1, h_2, \ldots h_m$: if p is dominated by h_i, but not strictly dominated, then p has the same weight and deadline as h_i.

We begin by describing modifications to Adv's buffer and estimate Adv's amortized gain. Suppose that Adv transmits the packet $h_z = (w_z, d_z)$ while Mix-R transmits $h_f = (w_f, d_f)$. Denote the expected amortized gain of Adv, given that Adv transmits h_z, by $\mathcal{G}_{\text{Adv}}^{(z)}$. The value of $\mathcal{G}_{\text{Adv}}^{(z)}$ depends on the relation between z and f—with f fixed, there are two cases.

Case 1: $d_f \leq d_z$. Then $w_f \leq w_z$, since h_z is not strictly dominated. After both Adv and Mix-R transmit their packets, we replace h_f in the buffer of Adv by a copy of h_z. This way their buffers remain the same afterwards, and the change is advantageous to Adv: this is essentially an upgrade of the packet h_f in its buffer, as both $d_f \leq d_z$ and $w_f \leq w_z$ hold.

Case 2: $d_f > d_z$. After both Adv and Mix-R transmit their packets, we let Adv additionally transmit h_f, and we inject a copy of h_z into its buffer, both of which are clearly advantageous to Adv. This makes the buffers of Adv and Mix-R identical afterwards.

We start by proving (3), the bound on the adversary's expected amortized gain. Note that Adv always gains w_z, and if $d_z < d_f$ ($z > f$), it additionally gains w_f. Thus, when Adv transmits h_z, its expected amortized gain is

$$\mathbb{E}\left[\mathcal{G}_{\text{Adv}}^{(z)}\right] = w_z + \sum_{i < z} p_i w_i \ . \tag{5}$$

As the adversary's expected amortized gain satisfies

$$\mathbb{E}\left[\mathcal{G}_{\text{Adv}}\right] \leq \max_{1 \leq j \leq m} \left\{ \mathbb{E}\left[\mathcal{G}_{\text{Adv}}^{(j)}\right] \right\} \ , \tag{6}$$

to establish (3), we will prove that

$$\max_{1 \leq j \leq m} \left\{ \mathbb{E}\left[\mathcal{G}_{\text{Adv}}^{(j)}\right] \right\} \leq \mathcal{G}_{\text{Adv}}^{(1)} = w_1 \ . \tag{7}$$

The equality in (7) follows trivially from (5). To see that the inequality in (7) holds as well, observe that, by (5), for all $j < m$,

$$\mathbb{E}\left[\mathcal{G}_{\text{ADV}}^{(j)}\right] - \mathbb{E}\left[\mathcal{G}_{\text{ADV}}^{(j+1)}\right] = w_j - w_{j+1} - p_j w_j \geq 0 \ , \tag{8}$$

where the inequality follows from (1).

Now we turn to (4), the bound on the expected gain of MIX-R in a single step. Obviously,

$$\mathbb{E}\left[\mathcal{G}_{\text{MIX-R}}\right] = \sum_{i=1}^{n} p_i w_i \ . \tag{9}$$

By (2), $p_i w_i = w_i - w_{i+1}$ for all $i < n$. Also, $p_n = 1 - \sum_{i<n} p_i$, by Fact 3. Making corresponding substitutions in (9) yields

$$\mathbb{E}\left[\mathcal{G}_{\text{MIX-R}}\right] = \left(\sum_{i=1}^{n-1} (w_i - w_{i+1})\right) + \left(1 - \sum_{i=1}^{n-1} p_i\right) w_n$$

$$= w_1 - w_n \sum_{i=1}^{n-1} p_i \ . \tag{10}$$

As (2) implies $w_i = w_{i-1}(1 - p_{i-1})$ for all $i \leq n$, we can express w_n as

$$w_n = w_1 \prod_{i=1}^{n-1} (1 - p_i) \ . \tag{11}$$

Substituting (11) for w_n in (10), we obtain

$$\mathbb{E}\left[\mathcal{G}_{\text{MIX-R}}\right] = w_1 \left(1 - \prod_{i=1}^{n-1}(1 - p_i) \sum_{i=1}^{n-1} p_i\right) \ . \tag{12}$$

Let $x_i = 1 - p_i$ for $1 \leq i < n$ and let $x_n = \sum_{i=1}^{n-1} p_i$; note that $\sum_{i=1}^{n} x_i = n-1$. The inequality between arithmetic and geometric means for $x_1, \ldots, x_n$ yields

$$\prod_{i=1}^{n-1}(1 - p_i) \sum_{i=1}^{n-1} p_i = \prod_{i=1}^{n} x_i \leq \left(\frac{\sum_{i=1}^{n} x_i}{n}\right)^n = \left(1 - \frac{1}{n}\right)^n \ . \tag{13}$$

Plugging (13) into (12) gives

$$\mathbb{E}\left[\mathcal{G}_{\text{MIX-R}}\right] \geq w_1 \left(1 - (1 - \frac{1}{n})^n\right) \ ,$$

which proves (4), and together with (3), the whole theorem. $\qquad\square$

2.3   Rationale behind the Probability Distribution

Recall that the upper bound on the competitive ratio of MIX-R is

$$\frac{\max_{1\leq z\leq m}\{\mathbb{E}\left[\mathcal{G}_{\text{ADV}}^{(z)}\right]\}}{\mathbb{E}\left[\mathcal{G}_{\text{MIX-R}}\right]} \ , \tag{14}$$

irrespective of the choice of $p_1,\ldots,p_m$.

The particular probability distribution used in MIX-R is chosen to (heuristically) minimize above ratio by maximizing $\mathbb{E}\left[\mathcal{G}_{\text{MIX-R}}\right]$, while keeping (7) satisfied, which, together with (6), implies $\mathbb{E}\left[\mathcal{G}_{\text{ADV}}\right] \leq \mathcal{G}_{\text{ADV}}^{(1)} = w_1$.

The first goal is trivially achieved by setting $p_1 \leftarrow 1$. This however makes $\mathbb{E}\left[\mathcal{G}_{\text{ADV}}^{(z)}\right] > w_1$ for all $z > 1$. Therefore, some of the probability mass is transferred to $p_2, p_3, \ldots$ in the following way. To keep $\mathbb{E}\left[\mathcal{G}_{\text{MIX-R}}\right]$ as large as possible, p_2 is greedily set to its maximum, if there is any unassigned probability left, p_3 is set to its maximum, and so on. As $\mathbb{E}\left[\mathcal{G}_{\text{ADV}}^{(z)}\right]$ does not depend on p_i for $i \geq z$, the values $\mathbb{E}\left[\mathcal{G}_{\text{ADV}}^{(z)}\right]$ can be equalized with w_1 sequentially, with z increasing, until there is no unassigned probability left. Equalizing $\mathbb{E}\left[\mathcal{G}_{\text{ADV}}^{(j)}\right]$ with $\mathbb{E}\left[\mathcal{G}_{\text{ADV}}^{(j-1)}\right]$ consists in setting $p_{j-1} \leftarrow 1 - \frac{w_j}{w_{j-1}}$, as shown in (8). The same inequality shows what is intuitively clear: once there is no probability left and further values $\mathbb{E}\left[\mathcal{G}_{\text{ADV}}^{(z)}\right]$ cannot be equalized, they are only smaller than w_1.

2.4   Optimality for *Item Collection*

In both the algorithm and its analysis it is the respective order of packets' deadlines rather than their exact values that matter. Therefore, our results are also applicable to the *Item Collection* problem [4], briefly described in Section 1.3. MIX-R is optimal among randomized memoryless algorithms for *Item Collection* [4, Theorem 6.3]. In fact, the lower bound construction essentially proves that, among randomized memoryless algorithms, MIX-R is optimal for all values of parameter N—the proof gives an infinite sequence, parametrized by N, of adversary's strategies, such that the N-th strategy forces ratio at least $1/\left(1 - (1 - \frac{1}{N})^N\right)$, while keeping the number of packets pending for the algorithm never exceeds N. In such case, i.e., when the number of pending packets never exceeds N, MIX-R is guaranteed to attain exactly that ratio.

Note that, in particular, this implies that (14) is indeed minimized by the heuristic described in Section 2.3.

2.5   Implications for Restricted Variants

We have already mentioned that for s-bounded instances or those with at most s different packet weights, $N \leq m \leq s$ in Theorem 4, which trivially follows from Fact 2. Thus for either kind of instances MIX-R is $1/\left(1 - (1 - \frac{1}{s})^s\right)$-competitive.

In particular, on 2-bounded instances MIX-R coincides with the previously known optimal 4/3-competitive algorithm RAND [5] for the adaptive adversary model.

Sometimes it may be possible to give more sophisticated bounds on N, and consequently on the competitive ratio for particular variant of the problem, as we now explain. The reason for considering only the packets $h_0, h_1, \ldots, h_m$ is clear: by Fact 1 and Fact 2, ADV (wlog) transmits one of them. Therefore, MIX-R tries to mimic ADV's behavior by adopting a probability distribution over these packets (recall that in the analysis the packets pending for MIX-R and ADV are exactly the same) that keeps the maximum, over ADV's choices, expected amortized gain of ADV and its own expected gain as close as possible. Now, if for whatever reason we know that ADV is going to transmit a packet from some set S, then H_0 can be initialized to S rather than all pending packets, and Theorem 4 will still hold. And as the upper bound it guarantees depends on N, a set S of small cardinality yields improved bound.

While it seems unlikely that bounds for any restricted variant other than s-bounded instances or instances with at most s different packet weights can be obtained this way, there is one interesting example that shows it is possible. For similarly ordered instances (aka instances with agreeable deadlines) and oblivious adversary one can always find such set S of cardinality at most 2 [15, Lemma 2.7]; this was in fact proved already by Li et al. [17], though not explicitly stated therein. Roughly, the set S contains the earliest-deadline and the heaviest packet from any optimal provisional schedule. The latter is the optimal schedule under the assumption that no further packets are ever injected, and as such can be found in any step. The 4/3-competitive algorithm from [15] can be viewed as MIX-R operating on the confined set of packets S.

3   Conclusion and Open Problems

While MIX-R is very simple to analyze, it subsumes almost all previously known randomized algorithms for packet scheduling and provides new bounds for several restricted variants of the problem. One notable exception is the optimum algorithm against oblivious adversary for 2-bounded instances [7]. This exposes that the strength of our analysis, i.e., applicability to adaptive adversary model, is most likely a weakness at the same time. The strongest lower bounds on competitive ratio for oblivious and adaptive adversary differ. And as both are tight for 2-bounded instances, it seems impossible to obtain an upper bound smaller than 4/3 on the competitive ratio of MIX-R for any non-trivial restriction of the problem in the oblivious adversary model.

As explained in Section 2.4, MIX-R is the optimum randomized memoryless algorithm for *Item Collection*. Therefore, to beat either the general bound of $e/(e-1)$, or any of the $1/\left(1 - (1 - \frac{1}{s})^s\right)$ bounds for s-bounded instances for buffer management with bounded delay, one either needs to consider algorithms that are not memoryless scale-invariant, or better utilize the knowledge of exact deadlines—in the analysis at least, if not in the algorithm itself.

References

1. Andelman, N., Mansour, Y., Zhu, A.: Competitive queueing policies for qos switches. In: Proc. of the 14th ACM-SIAM Symp. on Discrete Algorithms (SODA), pp. 761–770 (2003)
2. Azar, Y.: Online packet switching. In: Persiano, G., Solis-Oba, R. (eds.) WAOA 2004. LNCS, vol. 3351, pp. 1–5. Springer, Heidelberg (2005)
3. Ben-David, S., Borodin, A., Karp, R.M., Tardos, G., Wigderson, A.: On the power of randomization in online algorithms. Algorithmica 11(1), 2–14 (1994); also appeared in Proc. of the 22nd STOC, pp. 379–386 (1990)
4. Bienkowski, M., Chrobak, M., Dürr, C., Hurand, M., Jeż, A., Jeż, Ł., Stachowiak, G.: Collecting weighted items from a dynamic queue. In: Proc. of the 20th ACM-SIAM Symp. on Discrete Algorithms (SODA), pp. 1126–1135 (2009)
5. Bienkowski, M., Chrobak, M., Jeż, Ł.: Randomized algorithms for buffer management with 2-bounded delay. In: Bampis, E., Skutella, M. (eds.) WAOA 2008. LNCS, vol. 5426, pp. 92–104. Springer, Heidelberg (2009)
6. Borodin, A., El-Yaniv, R.: Online Computation and Competitive Analysis. Cambridge University Press, Cambridge (1998)
7. Chin, F.Y.L., Chrobak, M., Fung, S.P.Y., Jawor, W., Sgall, J., Tichý, T.: Online competitive algorithms for maximizing weighted throughput of unit jobs. Journal of Discrete Algorithms 4, 255–276 (2006)
8. Chin, F.Y.L., Fung, S.P.Y.: Online scheduling for partial job values: Does time-sharing or randomization help? Algorithmica 37, 149–164 (2003)
9. Chrobak, M., Jawor, W., Sgall, J., Tichý, T.: Improved online algorithms for buffer management in QoS switches. ACM Transactions on Algorithms 3(4) (2007); also appeared in Proc. of the 12th ESA, pp. 204–215 (2004)
10. Englert, M., Westermann, M.: Considering suppressed packets improves buffer management in QoS switches. In: Proc. of the 18th ACM-SIAM Symp. on Discrete Algorithms (SODA), pp. 209–218 (2007)
11. Epstein, L., van Stee, R.: Buffer management problems. Sigact News 35, 58–66 (2004)
12. Goldwasser, M.: A survey of buffer management policies for packet switches. SIGACT News 41(1), 100–128 (2010)
13. Hajek, B.: On the competitiveness of online scheduling of unit-length packets with hard deadlines in slotted time. In: Conference in Information Sciences and Systems, pp. 434–438 (2001)
14. Jeż, Ł.: Randomised buffer management with bounded delay against adaptive adversary. CoRR abs/0907.2050 (2009)
15. Jeż, Ł.: Randomized algorithm for agreeable deadlines packet scheduling. In: Proc. of the 27th Symp. on Theoretical Aspects of Computer Science (STACS), pp. 489–500 (2010)
16. Kesselman, A., Lotker, Z., Mansour, Y., Patt-Shamir, B., Schieber, B., Sviridenko, M.: Buffer overflow management in QoS switches. SIAM Journal on Computing 33(3), 563–583 (2004); also appeared in Proc.of the 33rd STOC, pp 520–529, (2001)
17. Li, F., Sethuraman, J., Stein, C.: An optimal online algorithm for packet scheduling with agreeable deadlines. In: Proc. of the 16th ACM-SIAM Symp. on Discrete Algorithms (SODA), pp. 801–802 (2005)
18. Zhu, A.: Analysis of queueing policies in QoS switches. Journal of Algorithms 53(2), 137–168 (2004)

Better Bounds for Incremental Frequency Allocation in Bipartite Graphs

Marek Chrobak[1], Łukasz Jeż[2], and Jiří Sgall[3]

[1] Department of Computer Science, Univ. of California, Riverside, CA 92521, USA⋆
[2] Institute of Computer Science, University of Wrocław, 50-383 Wrocław, Poland⋆⋆
[3] Dept. of Applied Mathematics, Faculty of Mathematics and Physics, Charles University, Malostranské nám. 25, CZ-11800 Praha 1, Czech Republic⋆⋆⋆

Abstract. We study frequency allocation in wireless networks. A wireless network is modeled by an undirected graph, with vertices corresponding to cells. In each vertex we have a certain number of requests, and each of those requests must be assigned a different frequency. Edges represent conflicts between cells, meaning that frequencies in adjacent vertices must be different as well. The objective is to minimize the total number of used frequencies.

The offline version of the problem is known to be NP-hard. In the incremental version, requests for frequencies arrive over time and the algorithm is required to assign a frequency to a request as soon as it arrives. Competitive incremental algorithms have been studied for several classes of graphs. For paths, the optimal (asymptotic) ratio is known to be $4/3$, while for hexagonal-cell graphs it is between 1.5 and 1.9126. For ξ-colorable graphs, the ratio of $(\xi + 1)/2$ can be achieved.

In this paper, we prove nearly tight bounds on the asymptotic competitive ratio for bipartite graphs, showing that it is between 1.428 and 1.433. This improves the previous lower bound of $4/3$ and upper bound of 1.5. Our proofs are based on reducing the incremental problem to a purely combinatorial (equivalent) problem of constructing set families with certain intersection properties.

1 Introduction

Static frequency allocation. In the frequency allocation problem, we are given a wireless network and a collection of requests for frequencies. The network is modeled by a (possibly infinite) undirected graph G, whose vertices correspond to the network's cells. Each request is associated with a vertex, and requests in the same vertex must be assigned different frequencies. Edges represent conflicts between cells, meaning that frequencies in adjacent vertices must be different as

⋆ Research supported by NSF Grant CCF-0729071.
⋆⋆ Research supported by MNiSW grant N N206 368839, 2010–2013 and a scholarship co-financed by ESF project *Human Capital*.
⋆⋆⋆ Partially supported by Inst. for Theor. Comp. Sci., Prague (project 1M0545 of MŠMT ČR) and grant IAA100190902 of GA AV ČR.

C. Demetrescu and M.M. Halldórsson (Eds.): ESA 2011, LNCS 6942, pp. 251–262, 2011.
© Springer-Verlag Berlin Heidelberg 2011

well. The objective is to minimize the total number of used frequencies. We will refer to this model as *static*, as it corresponds to the scenario where the set of requests in each vertex does not change over time.

A more rigorous formulation of this static frequency allocation problem is as follows: Denote by ℓ_v the *load* (or *demand*) at a vertex v of G, that is the number of frequency requests at v. A frequency allocation is a function that assigns a set L_v of frequencies (represented, say, by positive integers) to each vertex v and satisfies the following two conditions: (i) $|L_v| = \ell_v$ for each vertex v, and (ii) $L_v \cap L_w = \emptyset$ for each edge (v, w). The total number of frequencies used is $|\bigcup_{v \in G} L_v|$, and this is the quantity we wish to minimize. We will use notation $opt(G, \bar{\ell})$ to denote the minimum number of frequencies for a graph G and a demand vector $\bar{\ell}$.

If one request is issued per node, then $opt(G, \bar{\ell})$ is equal to the chromatic number of G, which immediately implies that the frequency allocation problem is NP-hard. In fact, McDiarmid and Reed [7] show that the problem remains NP-hard for the graph representing the network whose cells are regular hexagons in the plane, which is a commonly studied abstraction of wireless networks. (See, for example, the surveys in [8,1]). Polynomial-time $\frac{4}{3}$-approximation algorithms for this case appeared in [7] and [9].

Incremental frequency allocation. In the incremental version of frequency allocation, requests arrive over time and an incremental algorithm must assign frequencies to requests as soon as they arrive. An incremental algorithm $\mathcal{A}$ is called *asymptotically R-competitive* if, for any graph G and load vector $\bar{\ell}$, the total number of frequencies used by $\mathcal{A}$ is at most $R \cdot opt(G, \bar{\ell}) + \lambda$, where λ is a constant independent of $\bar{\ell}$. We allow λ to depend on the class of graphs under consideration, in which case we say that $\mathcal{A}$ is R-competitive for this class. We refer to R as the *asymptotic competitive ratio* of $\mathcal{A}$. As in this paper we focus only on the asymptotic ratio, we will skip the word "asymptotic" (unless ambiguity can arise), and simply use terms "R-competitive" and "competitive ratio" instead. Following the terminology in the literature (see [2,3]), we will say that the competitive ratio is *absolute* when the additive constant λ is equal 0.

Naturally, research in this area is concerned with designing algorithms with small competitive ratios for various classes of graphs, as well as proving lower bounds. For hexagonal-cells graphs, Chan *et al.* [2,3] give an incremental algorithm with competitive ratio 1.9216 and prove that no ratio better than 1.5 is possible. A lower bound of 4/3 for paths was given in [4], and later Chrobak and Sgall [6] gave an incremental algorithm with the same ratio. Paths are in fact the only non-trivial graphs for which tight asymptotic ratios are known. As pointed out earlier, there is a strong connection between frequency allocation and graph coloring, so one would expect that the competitive ratio can be bounded in terms of the chromatic number. Indeed, for ξ-colorable graphs Chan *et al.* [2,3] give an incremental algorithm with competitive ratio of $(\xi + 1)/2$. (This ratio is in fact absolute.) On the other hand, the best known lower bounds on the competitive ratio, 1.5 in the asymptotic and 2 in the absolute case [2,3], hold for hexagonal

-cell graphs (for which $\xi = 3$), but no stronger bounds are known for graphs of higher chromatic number.

Our contribution. We prove nearly tight bounds on the optimal competitive ratio of incremental algorithms for bipartite graphs (that is, for $\xi = 2$), showing that it is between $10/7 \approx 1.428$ and $(18 - \sqrt{5})/11 \approx 1.433$. This improves the lower and upper bounds for this version of frequency allocation. The best previously known lower bound was $4/3$, which holds in fact even for paths [4,6]. The best upper bound of 1.5 was shown in [2,3] and it holds even in the absolute case.

Our proofs are based on reducing the incremental problem to a purely combinatorial (equivalent) problem of constructing set families, which we call *F-systems*, with certain intersection properties. A rather surprising consequence of this reduction is that the optimal competitive ratio can be achieved by an algorithm that is topology-independent; it assigns a frequency to each vertex v based only on the current optimum value, the number of requests to v, and the partition to which v belongs; that is, independently of the frequencies already assigned to the neighbors of v.

To achieve a competitive ratio below 2 for bipartite graphs, we need to use frequencies that are shared between the two partitions of the graph. The challenge is then to assign these shared frequencies to the requests in different partitions so as to avoid collisions—in essence, to break the symmetry. In our construction, we develop a symmetry-breaking method based on the concepts of "collisions with the past" and "collisions with the future", which allows us to derive frequency sets in a systematic fashion.

Our work is motivated by purely theoretical interest, as there is no reason why realistic wireless networks would form a bipartite graph. There is an intuitive connection between the chromatic number of a graph and optimal frequency allocation, and exploring the exact nature of this connection is worthwhile and challenging. Our results constitute a significant progress for the case of two colors, and we believe that some ideas from this paper—the concept of F-systems and our symmetry-breaking method, for example—can be extended to frequency assignment problems on graphs with larger chromatic number.

Other related work. Determining optimal absolute ratios is usually easier than for asymptotic ratios and it has been accomplished for various classes of graphs, including paths [4] and bipartite graphs in general [2,3], and hexagonal-cell graphs and 3-colorable graphs in general [2,3]. The asymptotic ratio model, however, is more relevant to practical scenarios where the number of frequencies is typically very large, so the additive constant can be neglected.

In the *dynamic version* of frequency allocation each request has an arrival and departure time. At each time, any two requests that have already arrived but not departed and are in the same or adjacent nodes must be assigned different frequencies. As before, we wish to minimize the total number of used frequencies. As shown by Chrobak and Sgall [6], this dynamic version is NP-hard even for the special case when the input graph is a path.

It is natural to study the online version of this problem, where we introduce the notion of "time" that progresses in discrete steps, and at each time step some requests may arrive and some previously arrived requests may depart. This corresponds to real-life wireless networks where customers enter and leave a network's cells over time, in an unpredictable fashion. An online algorithm needs to assign frequencies to requests as soon as they arrive. The competitive ratio is defined analogously to the incremental case. (The incremental static version can be thought of as a special case in which all departure times are infinite.) This model has been well studied in the context of job scheduling, where it is sometimes referred to as time-online. Very little is known about this online dynamic case. Even for paths the optimal ratio is not known; it is only known that it is between $\frac{14}{9} \approx 1.571$ [6] and $\frac{5}{3} \approx 1.667$ [4].

2   Preliminaries

For concreteness, we will assume that frequencies are identified by positive integers, although it does not really matter. Recall that we use the number of frequencies as the performance measure. In some literature [4,5,3], authors used the maximum-numbered frequency instead. It is not hard to show (see [6], for example, which does however involve a transformation of the algorithm that makes it *not* topology independent) that these two approaches are equivalent.

For a bipartite graph $G = (A, B, E)$, it is easy to characterize the optimum value. As observed in [4,6], in this case the optimum number of frequencies is

$$opt(G, \bar{\ell}) = \max_{(u,v)\in E} \{\ell_u + \ell_v\}. \tag{1}$$

For completeness, we include a simple proof: Trivially, $opt(G, \bar{\ell}) \geq \ell_u + \ell_v$ for each edge (u, v). On the other hand, denoting by ω the right-hand side of (1), we can assign frequencies to nodes as follows: for $u \in A$, assign to u frequencies $1, 2, \ldots, \ell_u$, and for $u \in B$ assign to u frequencies $\omega - \ell_u + 1, \omega - \ell_u + 2, \ldots, \omega$. This way each vertex u is assigned ℓ_u frequencies and no two adjacent nodes share the same frequency.

Throughout the paper, we will use the convention that if $c \in \{A, B\}$, then c' denotes the partition other than c, that is $\{c, c'\} = \{A, B\}$.

3   Competitive F-Systems

In this section we show that finding an R-competitive incremental algorithm for bipartite graphs can be reduced to an equivalent problem of constructing certain families of sets that we call F-systems.

Suppose that for any $c \in \{A, B\}$ and any integers t, k such that $0 < k \leq t$, we are given a set $F_{t,k}^c$ of positive integers (frequencies). Denote by $\mathcal{F} = \{F_{t,k}^c\}$ the family of those sets. Then $\mathcal{F}$ is called an *F-system* if

(F1) $|F_{t,k}^c| \geq k$ for all c, t, k, and
(F2) $F_{t,k}^A \cap F_{t',k'}^B = \emptyset$ for all k, k', t, t' such that $k + k' \leq \max(t, t')$.

An F-system is called R-*competitive* if for all t we have

$$\left| \bigcup_{\kappa \leq \tau \leq t} (F^A_{\tau,\kappa} \cup F^B_{\tau,\kappa}) \right| \leq R \cdot t + \lambda, \tag{2}$$

where λ is a constant independent of t. The *competititive ratio* of $\mathcal{F}$ is the smallest R for which $\mathcal{F}$ is R-competitive.

Lemma 1. *For any $R \geq 1$, there is an R-competitive incremental algorithm for frequency allocation in bipartite graphs if and only if there is an R-competitive F-system.*

Proof. ($\Rightarrow$) Let $\mathcal{A}$ be an R-competitive incremental algorithm. To prove this implication, we define a "universal" infinite bipartite graph $H = (A, B, E)$ and we will issue requests to this graph. For $c \in \{A, B\}$, the vertices in c have the form $(t, k)_c$, where $k \leq t$. Two vertices $(t, k)_A$ and $(t', k')_B$ are connected by an edge if $k + k' \leq \max(t, t')$.

The requests are issued in phases numbered $t = 1, 2, \ldots$. In phase t, for each node $(t, k)_c$, we issue k requests to this node. Let $F^c_{t,k}$ be the set of frequencies that $\mathcal{A}$ assigns to $(t, k)_c$. After phase t, by the definition of H and by (1), the optimum number of frequencies is t, so $\mathcal{A}$ uses at most $Rt + \lambda$ frequencies, for some λ, implying (2) and proving that $\mathcal{F} = \{F^c_{t,k}\}$ is an R-competitive F-system.

($\Leftarrow$) Let $\mathcal{F}$ be an R-competitive F-system. We use $\mathcal{F}$ to define an incremental algorithm $\mathcal{A}$ that works as follows. Let $G = (A, B, E)$ be the given bipartite graph. Consider one step of the computation in which a new request arrives at a vertex $u \in c$, where $c \in \{A, B\}$. Denote by t the current optimum number of frequencies, that is $t = \max_{(v,w) \in E}(\ell_v + \ell_w)$. Choose any frequency $f \in F^c_{t,k}$, for $k = \ell_u$, that is not yet used on u and assign f to this request. Such f exists, because by property (F1) we have $|F^c_{t,k}| \geq k$ and the number of frequencies assigned so far to u is $k - 1$.

Trivially, all frequencies assigned by $\mathcal{A}$ to one node are different. We claim that adjacent nodes will be assigned different frequencies as well. Consider again a step where a frequency f is assigned to a kth request to a vertex u, when the optimum value is t, as described above. So $k = \ell_u$. Without loss of generality, assume $u \in A$. For an edge $(u, v) \in E$, let $k' = \ell_v$ be the current load at v. If g is a frequency assigned by $\mathcal{A}$ to v then, by the definition of $\mathcal{A}$, we have $g \in F^B_{t',k''}$ for some $t' \leq t$ and $k'' \leq \min(t', k')$. Thus $k + k'' \leq k + k' \leq t$, by the definition of t. Using (F2), we now get that $F^A_{t,k} \cap F^B_{t',k''} = \emptyset$, and therefore $f \neq g$.

Finally, when the optimum is t then any frequency used by $\mathcal{A}$ is from some set $F^c_{\tau,\kappa}$ for $\kappa \leq \tau \leq t$. Therefore $\mathcal{A}$ is R-competitive, by the property (2) of $\mathcal{F}$.

4   An Upper Bound

We now design an R_0-competitive incremental algorithm, for $R_0 = (18-\sqrt{5})/11 \approx 1.433$. Using Lemma 1, it is sufficient to construct an R_0-competitive F-system.

Intuitions. Our construction below may appear rather mysterious, so we begin by gradually introducing its main ideas. We will distinguish between two types

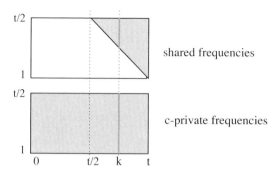

Fig. 1. Frequency sets in the 1.5-competitive algorithm, represented by shaded regions. In this figure, we fix the value of t and show the frequency sets for $k \leq t$. The horizontal axis represents k and the vertical axis represents frequencies. $F_{t,k}^c$ is represented by thick line segments on the vertical line for load k.

of frequencies: private and shared. A-private frequencies will be used only in sets $F_{t,k}^A$, B-private frequencies will be used only in sets $F_{t,k}^B$, while shared frequencies can be used in some sets from both partitions A and B.

The competitive ratio of 2 can be easily achieved using only private frequencies. For each $c \in \{A, B\}$, let P^c be an infinite pool of c-private frequencies, with P^A and P^B disjoint. We simply let $F_{t,k}^c$ consist of the first k frequencies from P^c. Conditions (F1) and (F2) are trivially true. For any given t, the set on the left-hand side of (2) contains $2t$ frequencies, so (2) holds for $R = 2$.

To improve the ratio to 1.5, we introduce an infinite pool S (disjoint with $P^A \cup P^B$) of shared frequencies. To avoid violations of (F2), we need to use these frequencies judiciously. Roughly, each $F_{t,k}^c$ will contain the first $t/2$ c-private frequencies, and if $k > t/2$ then it will also contain $k - t/2$ *last* shared frequencies numbered at most $t/2$, namely those from $t/2 - (k - t/2) + 1 = t - k + 1$ to $t/2$. (See Figure 1.) It is easy to verify that both (F1) and (2) are satisfied with $R = 1.5$. The intuition behind (F2) is that, assuming $t' \leq t$, $F_{t',k'}^B$ uses shared frequencies only when $k' > t'/2$, which together with $k' \leq t - k$ implies that $t'/2 \leq t - k$, so all shared frequencies in $F_{t',k'}^B$ are smaller than those in $F_{t,k}^A$.

To make the above idea more precise, for any real number $x \geq 0$ let

$$S_x = \text{the first } \lfloor x \rfloor \text{ frequencies in } S ,$$
$$P_x^c = \text{the first } \lfloor x \rfloor \text{ frequencies in } P^c, \text{ for } c \in \{A, B\}.$$

We now let $\mathcal{F} = \{F_{t,k}^c\}$, where for $c \in \{A, B\}$ and $k \leq t$ we have

$$F_{t,k}^c = P_{t/2+1}^c \cup (S_{t/2} \setminus S_{t-k}).$$

We claim that $\mathcal{F}$ is a 1.5-competitive F-system. If $k \leq \lfloor t/2 \rfloor + 1$, then $|F_{t,k}^c| \geq k$ is trivial. If $k \geq \lfloor t/2 \rfloor + 2$, then $t - k \leq t - \lfloor t/2 \rfloor - 2 \leq t/2$, so $S_{t-k} \subseteq S_{t/2}$ and thus $|F_{t,k}^c| \geq \lfloor t/2 \rfloor + 1 + (\lfloor t/2 \rfloor - \lfloor t - k \rfloor) \geq k$. So (F1) holds.

To verify (F2), pick any two pairs $k \leq t$ and $k' \leq t'$ with $k + k' \leq \max(t, t')$. Without loss of generality, assume $t' \leq t$. If $k' \leq \lfloor t'/2 \rfloor + 1$, then $F_{t',k'}^B \subseteq P^B$, so

(F2) is trivial. If $k' \geq \lfloor t'/2 \rfloor + 2$, then $t'/2 \leq k' \leq t - k$, so $F^B_{t',k'} \subseteq P^B \cup S_{t'/2} \subseteq P^B \cup S_{t-k}$, which implies (F2) as well.

Finally, for $c \in \{A, B\}$ and $\kappa \leq \tau \leq t$, we have $F^c_{\tau,\kappa} \subseteq P^A_{t/2+1} \cup P^B_{t/2+1} \cup S_{t/2}$, so (2) holds with $R = 1.5$ and $\lambda = 2$, implying that $\mathcal{F}$ is 1.5-competitive.

More intuition. It is useful to think of our constructions, informally, in terms of *collisions*. We designate some shared frequencies as forbidden in $F^c_{t,k}$, because they might be used in some sets $F^{c'}_{t',k'}$ with $k + k' \leq \max(t, t')$. Depending on whether $t' \leq t$ or $t' > t$, we refer to those forbidden frequencies as "collisions with the past" and "collisions with the future", respectively. Figure 1 illustrates this idea for our last construction. For $t' \leq t$ and $k > t/2$, each $F^{c'}_{t',k'}$ uses shared frequencies numbered at most $t'/2 \leq t/2 < t - k$. Thus all shared frequencies that collide with the past are in the region strictly below the line $x = t - k$, which complements the shaded region assigned to $F^c_{t,k}$. This construction does not use collisions with the future.

An R_0-competitive F-system. To achieve a ratio below 1.5 we need to use shared frequencies even when $k < t/2$. For such k near $t/2$, sets $F^A_{t,k}$ and $F^B_{t,k}$ will conflict and each contain shared frequencies, so, unlike before, we need to assign their shared frequencies in a different order—in other words, we need to break symmetry. To achieve this, we introduce A-shared and B-shared frequencies. In sets $F^c_{t,k}$, for fixed c, t and increasing k, we first use c-private frequencies, then, starting at $k = t/\phi^2 \approx 0.382t$ we also use c-shared frequencies, then, starting at $k = t/2$ we add symmetric-shared frequencies, and finally, when k reaches $t/\phi \approx 0.618t$ we "borrow" c'-shared frequencies to include in this set. (See Figure 2.) We remark that symmetric-shared frequencies are still needed; with only private and c-shared frequencies we could only achieve ratio ≈ 1.447.

Now, once the symmetry is broken, we can assign frequencies to $F^c_{t,k}$ more efficiently. The key to this is to consider "collisions with the future", with sets $F^{c'}_{t',k'}$ for $t' > t$. We examine once more the construction for ratio 1.5. For fixed even k, consider $F^A_{t,k}$ as t grows from k to $2k$. For $t = k$, $F^A_{t,k}$ uses shared frequencies $1, 2, ..., k/2$. As t grows, $F^A_{t,k}$ uses higher numbered shared frequencies, even though it needs fewer of them. In particular, for $3k/2 < t < 2k$, $F^A_{t,k}$ is disjoint from $F^A_{k,k}$. This is wasteful, for the following reason: If some $F^B_{t',k'}$ conflicts with this $F^A_{t,k}$ (that is, $k + k' \leq t'$), then it also conflicts with $F^A_{k,k}$, so it cannot use any frequencies in $F^A_{k,k}$. Thus in $F^A_{t,k}$ we can as well reuse, for "free", the shared frequencies from $F^A_{k,k}$.

More generally, if a frequency was used already in $F^c_{t'',k}$ for some $t'' < t$, then using it in $F^c_{t,k}$ cannot create any new collisions in the future. If the sets for all t are represented by the same shape, this means that for every γ, the points on the line $x = \gamma k$ create the same collisions in the future. Thus the natural choice is to avoid frequencies in the the half-plane below the line $x = \gamma k$, for an appropriate γ. Then (in the opposite set) the collisions with the past are represented by the half-plane below the line $x = \gamma(t-k)$ (using the same γ). The optimization of the parameters for all three types of shared frequencies leads to our new algorithm.

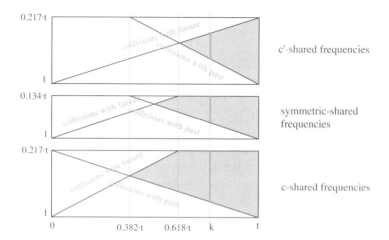

Fig. 2. Frequency sets in the R_0-competitive algorithm. We show only the shared frequencies, represented as in Figure 1.

The pools of c-shared and symmetric-shared frequencies are denoted S^c and Q, respectively. As before, for any real $x \geq 0$ we define

$$S_x^c = \text{the first } \lfloor x \rfloor \text{ frequencies in } S^c, \text{ for } c \in \{A, B\}.$$
$$Q_x = \text{the first } \lfloor x \rfloor \text{ frequencies in } Q.$$

Let $\phi = (\sqrt{5} + 1)/2$ be the golden ratio. Our construction uses three constants

$$\alpha = \frac{2}{\phi + 3} \approx 0.433, \qquad \beta = \frac{1}{\phi + 3} \approx 0.217, \quad \text{and} \quad \rho = \frac{\phi - 1}{\phi + 3} \approx 0.134.$$

We have the following useful identities: $\alpha = R_0 - 1$, $\beta = \alpha/2$, $\rho = \beta/\phi$, and $2\alpha + 2\beta + \rho = R_0$.

We define $\mathcal{F} = \{F_{t,k}^c\}$, where for any $t \geq k \geq 0$ we let

$$F_{t,k}^c = P_{\alpha t+4}^c \cup (S_{\beta \cdot \min(t,\phi k)}^c \setminus S_{\beta(t-k)}^c) \cup (S_{\beta k}^{c'} \setminus S_{\phi\beta(t-k)}^{c'})$$
$$\cup (Q_{\rho \cdot \min(t,\phi k)} \setminus Q_{\phi\rho(t-k)}). \qquad (3)$$

We now show that $\mathcal{F}$ is an R_0-competitive F-system. We start with (2). For $\kappa \leq \tau \leq t$ and $c \in \{A, B\}$ we have

$$F_{\tau,\kappa}^c \subseteq P_{\alpha\tau+4}^c \cup S_{\beta\tau}^c \cup S_{\beta\kappa}^{c'} \cup Q_{\rho\tau} \subseteq P_{\alpha t+4}^c \cup S_{\beta t}^c \cup S_{\beta t}^{c'} \cup Q_{\rho t}$$
$$\subseteq P_{\alpha t+4}^A \cup P_{\alpha t+4}^B \cup S_{\beta t}^A \cup S_{\beta t}^B \cup Q_{\rho t}.$$

This last set has cardinality at most $(2\alpha + 2\beta + \rho)t + 8 = R_0 t + 8$, so (2) holds.

Next, we show (F2). By symmetry, we can assume that $t' \leq t$ in (F2), so $k' \leq t - k$. Then

$$F_{t',k'}^B \subseteq P^B \cup S_{\phi\beta k'}^B \cup S_{\beta k'}^A \cup Q_{\phi\rho k'} \subseteq P^B \cup S_{\phi\beta(t-k)}^B \cup S_{\beta(t-k)}^A \cup Q_{\phi\rho(t-k)},$$

and this set is disjoint with $F_{t,k}^A$ by (3). Thus $F_{t,k}^A \cap F_{t',k'}^B = \emptyset$, as needed.

Finally, we prove (F1), namely that $|F_{t,k}^c| \geq k$. We distinguish two cases.

Case 1: $k > t/\phi$. This implies that $\min(t, \phi k) = t$, so in (3) we have $S_{\beta \cdot \min(t,\phi k)}^c = S_{\beta t}^c$ and $Q_{\rho \cdot \min(t, \phi k)} = Q_{\rho t}$. Thus

$$|F_{t,k}^c| \geq [\alpha t + 3] + [\beta t - \beta(t-k) - 1] + [\beta k - \phi \beta(t-k) - 1] + [\rho t - \phi \rho(t-k) - 1]$$
$$= (\alpha - \phi \beta - (\phi - 1)\rho)t + (2\beta + \phi \beta + \phi \rho)k = k,$$

using the substitutions $\alpha = 2\beta$ and $\rho = \beta/\phi$. Note that this case is asymptotically tight as the algorithm uses all three types of shared frequencies (and the corresponding terms are non-negative).

Case 2: $k \leq t/\phi$. The case condition implies that $\phi k \leq t$, so $S_{\beta \cdot \min(t, \phi k)}^c = S_{\phi \beta k}^c$, $Q_{\rho \cdot \min(t, \phi k)} = Q_{\phi \rho k}$, and $S_{\beta k}^{c'} \setminus S_{\phi \beta(t-k)}^{c'} = \emptyset$. Therefore

$$|F_{t,k}^c| \geq [\alpha t + 3] + [\phi \beta k - \beta(t-k) - 1] + [\phi \rho k - \phi \rho(t-k) - 1]$$
$$= (\alpha - \beta - \phi \rho)t + ((\phi + 1)\beta + 2\phi \rho)k + 1 = k + 1,$$

using $\alpha = 2\beta$ and $\rho = \beta/\phi$ again. Note that this case is (asymptotically) tight only for $k > t/2$ when c-shared and symmetric-shared frequencies are used. For $k \leq t/2$, the corresponding term is negative.

Summarizing, we conclude that $\mathcal{F}$ is indeed an R_0-competitive F-system. Therefore, using Lemma 1, we get our upper bound:

Theorem 1. *There is an R_0-competitive incremental algorithm for frequency allocation on bipartite graphs, where $R_0 = (18 - \sqrt{5})/11 \approx 1.433$.*

5   A Lower Bound

In this section we show that if $R < 10/7$, then there is no R-competitive incremental algorithm for frequency allocation in bipartite graphs. By Lemma 1, it is sufficient to show that there is no R-competitive F-system.

The general intuition behind the proof is that we try to reason about the sets $Z_t = F_{t,\gamma t}^A \cup F_{t,\gamma t}^B$ for a suitable constant γ. These sets should correspond to the symmetric-shared frequencies from our algorithm, for γ such that no c'-shared frequencies are used. If Z_t is too small, then both partitions use mostly different frequencies and this yields a lower bound on the competitive ratio. If Z_t is too large, then for a larger t and suitable k, the frequencies cannot be used for either partition, and hopefully this allows to improve the lower bound. We are not able to do exactly this. Instead, for a variant of Z_t, we show a recurrence essentially saying that if the set is too large, then for some larger t, it must be proportionally even larger, leading to a contradiction.

We now proceed with the proof. For $c \in \{A, B\}$, let $F_t^c = \bigcup_{\kappa \leq \tau \leq t} F_{\tau,\kappa}^c$. Towards contradiction, suppose that an F-system $\mathcal{F}$ is R-competitive for some $R < 10/7$. Then $\mathcal{F}$ satisfies the definition of competitiveness (2) for some positive integer λ. Choose a sufficiently large integer θ for which $R < 10/7 - 1/\theta$.

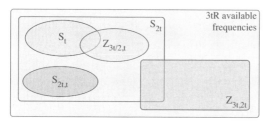

Fig. 3. Illustration of Lemma 4

We first identify shared frequencies in $\mathcal{F}$. Recall that $F_t^c = \bigcup_{\kappa \leq \tau \leq t} F_{\tau,\kappa}^c$, for $c \in \{A, B\}$. Thus the definition of R-competitiveness says that $|F_t^A \cup F_t^B| \leq Rt + \lambda$. The set of *level-$t$ shared* frequencies is defined as $S_t = F_t^A \cap F_t^B$.

Lemma 2. *For any t, we have $|S_t| \geq (2 - R)t - \lambda$.*

Proof. This is quite straightforward. By (F1) we have $|F_t^c| \geq t$ for each c, so $|S_t| = |F_t^A| + |F_t^B| - |F_t^A \cup F_t^B| \geq 2t - (Rt + \lambda) = (2 - R)t - \lambda$.

Now, let $S_{2t,t} = S_{2t} \cap (F_{2t,t}^A \cup F_{2t,t}^B)$ be the level-$2t$ shared frequencies that are used in $F_{2t,t}^A$ or $F_{2t,t}^B$. Each such frequency can only be in one of these sets because $F_{2t,t}^A \cap F_{2t,t}^B = \emptyset$.

Lemma 3. *For any t, we have $|S_{2t,t}| \geq (6 - 4R)t - 2\lambda$.*

Proof. By definition, $F_{2t,t}^A \cup F_{2t,t}^B \cup S_{2t} \subseteq F_{2t}^A \cup F_{2t}^B$, and thus (2) implies

$$
\begin{aligned}
2Rt + \lambda &\geq |F_{2t,t}^A \cup F_{2t,t}^B \cup S_{2t}| \\
&= |F_{2t,t}^A \cup F_{2t,t}^B| + |S_{2t}| - |(F_{2t,t}^A \cup F_{2t,t}^B) \cap S_{2t}| \\
&= |F_{2t,t}^A| + |F_{2t,t}^B| + |S_{2t}| - |S_{2t,t}| \ ,
\end{aligned}
$$

where the identities follow from the inclusion-exclusion principle, disjointness of $F_{2t,t}^A$ and $F_{2t,t}^B$, and the definition of $S_{2t,t}$. Transforming this inequality, we get

$$
|S_{2t,t}| \ \geq \ |F_{2t,t}^A| + |F_{2t,t}^B| + |S_{2t}| - (2Rt + \lambda) \ \geq \ (6 - 4R)t - 2\lambda \ ,
$$

as claimed, by property (F1) and Lemma 2.

For any even t define $Z_{3t/2,t} = F_{3t/2,t}^A \cap F_{3t/2,t}^B$. In the rest of the lower-bound proof we will set up a recurrence relation for the cardinality of sets $S_t \cup Z_{3t/2,t}$. The next step is the following lemma.

Lemma 4. *For any even t, we have $|S_{2t} \setminus Z_{3t,2t}| \geq |S_t \cup Z_{3t/2,t}| + |S_{2t,t}|$.*

Proof. From the definition, the sets $S_t \cup Z_{3t/2,t}$ and $S_{2t,t}$ are disjoint and contained in $S_{2t} - Z_{3t,2t}$ (see Figure 3.) This immediately implies the lemma.

Lemma 5. *For any even t, we have $|Z_{3t,2t}| \geq |S_t \cup Z_{3t/2,t}| - (3R - 4)t - \lambda$.*

Proof. As $F_{3t,2t}^A \cup F_{3t,2t}^B \cup S_t \cup Z_{3t/2,t} \subseteq F_{3t}^A \cup F_{3t}^B$, inequality (2) implies

$$
\begin{aligned}
3Rt + \lambda &\geq |F_{3t,2t}^A \cup F_{3t,2t}^B \cup S_t \cup Z_{3t/2,t}| \\
&= |F_{3t,2t}^A \cup F_{3t,2t}^B| + |S_t \cup Z_{3t/2,t}| \\
&= |F_{3t,2t}^A| + |F_{3t,2t}^B| - |F_{3t,2t}^A \cap F_{3t,2t}^B| + |S_t \cup Z_{3t/2,t}| \\
&= |F_{3t,2t}^A| + |F_{3t,2t}^B| - |Z_{3t,2t}| + |S_t \cup Z_{3t/2,t}| \; ,
\end{aligned}
$$

where the identities follow from the inclusion-exclusion principle, the fact that $F_{3t,2t}^A \cup F_{3t,2t}^B$ and $S_t \cup Z_{3t/2,t}$ are disjoint, and the definition of $Z_{3t,2t}$.

Transforming this inequality, we get

$$
\begin{aligned}
|Z_{3t,2t}| &\geq |F_{3t,2t}^A| + |F_{3t,2t}^B| + |S_t \cup Z_{3t/2,t}| - (3Rt + \lambda) \\
&\geq |S_t \cup Z_{3t/2,t}| - (3R - 4)t - \lambda \; ,
\end{aligned}
$$

as claimed, by property (F1).

We are now ready to derive our recurrence. By adding the inequalities in Lemma 4 and Lemma 5, taking into account that $|S_{2t} \setminus Z_{3t,2t}| + |Z_{3t,2t}| = |S_{2t} \cup Z_{3t,2t}|$, and then applying Lemma 3, for any even t we get

$$
\begin{aligned}
|S_{2t} \cup Z_{3t,2t}| &\geq 2 \cdot |S_t \cup Z_{3t/2,t}| + |S_{2t,t}| - (3R - 4)t - \lambda \\
&\geq 2 \cdot |S_t \cup Z_{3t/2,t}| + (10 - 7R)t - 3\lambda. \quad (4)
\end{aligned}
$$

For $i = 0, 1, \ldots, \theta$, define $t_i = 6\theta\lambda 2^i$ and $\gamma_i = |S_{t_i} \cup Z_{3t_i/2,t_i}|/t_i$. (Note that each t_i is even.) Since $S_{t_i} \cup Z_{3t_i/2,t_i} \subseteq S_{2t_i}$, we have that $\gamma_i \leq |S_{2t_i}|/t_i \leq 2R + 1/(6\theta) < 3$. Dividing recurrence (4) by $t_{i+1} = 2t_i$, we obtain

$$
\gamma_{i+1} \geq \gamma_i + 5 - 7R/2 - 3\lambda/(2t_i) \; \geq \; \gamma_i + 7/(2\theta) - 1/(4\theta) \; \geq \; \gamma_i + 3/\theta,
$$

for $i = 0, 1, \ldots, \theta - 1$. But then we have $\gamma_\theta \geq \gamma_0 + 3 \geq 3$, which contradicts our earlier bound $\gamma_i < 3$, completing the proof. Thus we have proved the following.

Theorem 2. *If $\mathcal{A}$ is an R-competitive incremental algorithm for frequency allocation on bipartite graphs, then $R \geq 10/7 \approx 1.428$.*

As a final remark we observe that our lower bound works even if the additive constant λ is allowed to depend on the actual graph. I.e., for every $R < 10/7$ we can construct a single finite graph G so that no algorithm is R-competitive on this graph. In our lower bound, we can restrict our attention to sets F_{t_i,t_i}^c, F_{2t_i,t_i}^c, $F_{3t_i,2t_i}^c$, and $F_{3t_i/2,t_i}^c$, for $i = 0, 1, \ldots, \theta$ and $c = A, B$. Then, in the construction from the proof of Lemma 1 for the lower bound sequence we obtain a finite graph together with a request sequence. However, for a fixed θ, the graphs for different values of λ are isomorphic, as all the indices scale linearly with λ. So, instead of using different isomorphic graphs, we can use different sequences corresponding to different values of λ on a single graph G.

6   Final Comments

We proved that the competitive ratio for incremental frequency allocation on bipartite graphs is between 1.428 and 1.433, improving the previous bounds of 1.33 and 1.5. Closing the remaining gap, small as it is, remains an intriguing open problem. Besides completing the analysis of this case, the solution is likely to give new insights into the general problem.

Two other obvious directions of study are to prove better bounds for the dynamic case and for k-partite graphs. Our idea of distinguishing "collisions with the past" and "collisions with the future" should be useful to derive upper bounds for these problems. The concept of F-systems extends naturally to k-partite graphs, but with a caveat: For $k \geq 3$ the maximum load on a k-clique is only a lower bound on the optimum (unlike for $k = 2$, where the equality holds). Therefore in Lemma 1 only one direction holds. This lemma is still sufficient to establish upper bounds on the competitive ratio, and it is possible that a lower bound can be proved using graphs where the optimum number of frequencies is equal to the maximum load of a k-clique.

References

1. Aardal, K., van Hoesel, S.P.M., Koster, A.M.C.A., Mannino, C., Sassano, A.: Models and solution techniques for frequency assignment problems. 4OR: Quarterly Journal of the Belgian, French and Italian Operations Research Societies 1, 261–317 (2003)
2. Chan, J.W.-T., Chin, F.Y.L., Ye, D., Zhang, Y.: Online frequency allocation in cellular networks. In: SPAA 2007, pp. 241–249 (2007)
3. Chan, J.W.-T., Chin, F.Y.L., Ye, D., Zhang, Y.: Absolute and asymptotic bounds for online frequency allocation in cellular networks. Algorithmica 58, 498–515 (2010)
4. Chan, J.W.-T., Chin, F.Y.L., Ye, D., Zhang, Y., Zhu, H.: Frequency allocation problems for linear cellular networks. In: Asano, T. (ed.) ISAAC 2006. LNCS, vol. 4288, pp. 61–70. Springer, Heidelberg (2006)
5. Chin, F.Y.L., Zhang, Y., Zhu, H.: A 1-Local 13/9-Competitive Algorithm for Multicoloring Hexagonal Graphs. In: Lin, G. (ed.) COCOON 2007. LNCS, vol. 4598, pp. 526–536. Springer, Heidelberg (2007)
6. Chrobak, M., Sgall, J.: Three results on frequency assignment in linear cellular networks. Theoretical Computer Science 411, 131–137 (2010)
7. McDiarmid, C., Reed, B.: Channel assignment and weighted colouring. Networks 36, 114–117 (2000)
8. Murphey, R.A., Pardalos, P.M., Resende, M.G.C.: Frequency assignment problems. In: Handbook of Combinatorial Optimization, pp. 295–377. Kluwer Academic Publishers, Dordrecht (1999)
9. Narayanan, L., Shende, S.M.: Static frequency assignment in cellular networks. Algorithmica 29, 396–409 (2001)

Two-Bounded-Space Bin Packing Revisited

Marek Chrobak[1], Jiří Sgall[2], and Gerhard J. Woeginger[3]

[1] Department of Computer Science, University of California, Riverside, USA
[2] Department of Applied Mathematics, Charles University, Prague, Czech Republic
[3] Department of Mathematics and Computer Science, TU Eindhoven, Eindhoven,
The Netherlands

Abstract. We analyze approximation algorithms for bounded-space bin packing by comparing them against the optimal bounded-space packing (instead of comparing them against the globally optimal packing that does not necessarily satisfy the bounded-space constraint). For 2-bounded-space bin packing we construct a polynomial time offline approximation algorithm with asymptotic worst case ratio $3/2$, and we show a lower bound of $5/4$ for this scenario. We show that no 2-bounded-space online algorithm can have an asymptotic worst case ratio better than $4/3$.

1 Introduction

An instance of the classical bin packing problem consists of an ordered list $a_1, a_2, \ldots, a_n$ of items with rational sizes $0 \le s(a_i) \le 1$, and the goal is to pack these items into the smallest possible number of bins of unit size. Bin packing is a fundamental problem in combinatorial optimization, and it has been studied extensively since the early 1970s. Since bin packing is NP-hard, one particularly active branch of research has concentrated on approximation algorithms that find near-optimal packings; see for instance [1].

A bin packing algorithm works *offline* if it packs the items with full knowledge of the entire item list, and it works *online* if it packs every item a_i solely on the basis of the preceding items a_j with $1 \le j \le i$, and without any information on subsequent items in the list. A bin packing algorithm uses *k-bounded space*, if for each item a_i the choice of bins in which it can be packed is restricted to a set of k or fewer so-called *active* bins. Once an active bin is closed, it can never become active again. Bounded-space packings arise in many applications, as for instance in packing trucks at a loading dock or in communicating via channels with bounded buffer sizes.

Customarily, the performance of an approximation algorithm A for bin packing is measured by comparing it against the optimal offline packing. Let $\text{OPT}(I)$ denote the number of bins used in an optimal (unconstrained) packing for instance I, and let $A(I)$ denote the number of bins used by algorithm A. Then the *asymptotic worst case ratio* is defined as

$$R^\infty(A) := \lim_{Opt(I) \to \infty} \sup_I A(I)/\text{OPT}(I).$$

C. Demetrescu and M.M. Halldórsson (Eds.): ESA 2011, LNCS 6942, pp. 263–274, 2011.
© Springer-Verlag Berlin Heidelberg 2011

The bin packing literature contains many articles that investigate and evaluate the (absolute and asymptotic) worst case ratios of bounded-space bin packing algorithms; see for instance [2,3,5,6,7]. Csirik & Johnson [3] show that the 2-space-bounded Best Fit algorithm BBF_2 has the asymptotic worst case ratio $R^\infty(BBF_2) = 1.7$. Among all 2-space-bounded algorithms (online and offline) in the bin packing literature, this online algorithm is the champion with respect to worst case ratios. A central result of Lee & Lee [5] designs k-bounded-space online bin packing algorithms whose asymptotic ratios come arbitrarily close to the magic harmonic number $h_\infty \approx 1.69103$, as the space bound k tends to infinity. They also show that every bounded-space online bin packing algorithm A satisfies $R^\infty(A) \geq h_\infty$.

The trouble with the old worst case ratio. The lower bound constructions in [5] use item lists on which *all* bounded-space algorithms must fail, no matter whether they are online or offline and no matter whether they run in polynomial time or in exponential time: In these constructions decent packings can only be reached by using unbounded space, and every bounded-space algorithm has asymptotic worst case ratio at least h_∞.

A cleaner — and more realistic — way of measuring the performance of bounded-space bin packing algorithms is to compare them against optimal offline solutions *that also are subject to the bounded-space constraint*. Hence we let $\text{OPT}_k(I)$ denote the smallest possible number of bins used in a packing produced by a k-bounded-space offline bin packing algorithm for instance I, and define the asymptotic *k-bounded-space ratio* of algorithm A as

$$R_k^\infty(A) := \lim_{Opt_k(I)\to\infty} \sup_I A(I)/\text{OPT}_k(I). \tag{1}$$

Note that the optimal k-bounded-space offline algorithm has a k-bounded-space ratio of 1. Furthermore $\text{OPT}_k(I) \geq \text{OPT}(I)$ implies $R_k^\infty(A) \leq R^\infty(A)$ for any algorithm.

Our contributions. Throughout the paper, we will mainly concentrate on 2-bounded-space scenarios. Our main goal is to beat the 2-bounded-space champion with respect to the classical asymptotic worst case ratio, the Best Fit algorithm BBF_2 of [3] with $R^\infty(BBF_2) = 1.7$. Indeed in Section 3 we construct an offline approximation algorithm A for 2-bounded-space bin packing with $R_2^\infty(A) \leq 3/2 + \varepsilon$. The analysis goes deeply into the structure of 2-bounded-space packings. Some of the techniques involve exhaustive search of instances of size $1/\varepsilon$, thus the running time exponential in ε, but it is linear in n.

Section 4 demonstrates that the k-bounded-space ratio asymptotic worst case ratio in (1) does not allow polynomial time approximation schemes; in fact every polynomial time algorithm A must satisfy $R_2^\infty(A) \geq 5/4$.

In Section 5 we discuss 2-bounded-space bin packing for small instances I with $\text{OPT}_2(I) \leq 4$, and we provide a complete analysis for the absolute performance of polynomial time approximation algorithms for these cases. The problem of finding good packings is closely connected to our offline algorithm.

Finally Section 6 shows that every 2-bounded-space online algorithm must satisfy $R_2^\infty(A) \geq 4/3$. Subsequent to our work, Epstein and Levin improved this lower bound to $3/2$. The 2-bounded-space ratio of 2-space-bounded Best Fit is known to be in $[1.5, 1.7]$. Improving its analysis or designing a new online algorithm matching the lower bound of 1.5 remains an interesting open problem.

2 Technical Preliminaries

Throughout the paper, bins usually have size $\beta = 1$, and items usually have rational sizes in $[0, 1]$. In our lower bound constructions (in Sections 4, 5, and 6) we will work with bins of integer size β and item sizes in $[0, \beta]$. Since bin and item sizes can be scaled by β, both models are computationally equivalent. For a set B of of items (or for the set of items packed into some bin B), we denote by $s(B)$ the sum of the sizes of all the items in B.

For an instance I of k-bounded-space bin packing, we define a *well-ordered* packing $\mathcal{B}$ to be a partition of the items into bins B_{ij} (for $i = 1, \ldots, k$ and $j = 1, \ldots, m_i$), such that for any i and for any $j < j'$, every item in bin B_{ij} precedes every item in bin $B_{ij'}$. We denote by $|\mathcal{B}| = m_1 + \cdots + m_k$ the number of bins in a packing $\mathcal{B}$. A well-ordered packing $\mathcal{B}$ forms a feasible k-bounded-space packing, if every bin B_{ij} satisfies $s(B_{ij}) \leq 1$.

For a given well-ordered packing $\mathcal{B}$, the sequence of bins B_{ij} with $j = 1, \ldots, m_i$ is called *track i*. We will sometimes define a packing $\mathcal{B}$ by specifying an assignment of items to tracks $1, ..., k$ only, rather than to specific bins. The actual bin assignment can be uniquely determined by greedily assigning items to bins, for each track separately.

For the purpose of the design of the offline algorithm, it is convenient to specify when exactly the bins are closed, so that at each time exactly k bins are open. We say that a bin B_{ij} is open from (and including) the moment when the first item in B_{ij} arrives until (not including) the moment when the first item in $B_{i,j+1}$ arrives or until the instance ends. As an exception, the first bin B_{i1} in the track is considered to be open from the beginning of the instance.

3 An Offline Approximation Algorithm

In this section we construct for every $\varepsilon > 0$ an offline approximation algorithm A for 2-bounded-space bin packing whose asymptotic ratio is at most $3/2 + \varepsilon$. The crucial part, which is to convert certain approximate packings in the first 2 or 3 bins into exact packings with one additional bin, is postponed to Section 3.3.

The high-level strategy of our algorithm is as follows. We fix a suitable integer $V \geq 3/\varepsilon$. We cut the given instance I greedily into m sub-lists $I_1, I_2, \ldots, I_m$ with $s(I_j) \leq 2V$ for all j and $V \leq s(I_j)$ for $j < m$. Thus $\text{OPT}_2(I_j) \leq 4V$ and $m \leq 1 + \text{OPT}_2(I)/V$. Since any 2-bounded-space packing of I can be split into 2-bounded-space packings of the sublists I_j $(1 \leq j \leq m)$ with a loss of at most two bins between adjacent sublists, we have

$$\sum_{j=1}^{m} \text{OPT}_2(I_j) \ \leq \ \text{OPT}_2(I) + 2(m-1) \ \leq \ \left(1 + \frac{2}{V}\right) \text{OPT}_2(I).$$

Hence, for getting a good approximation for I it is enough to get a good approximation for every sublist I_j separately.

In Sections 3.1, 3.2 and 3.3, we will show that in polynomial time we can find 2-bounded-space packings of each I_j with at most $\frac{3}{2}\text{OPT}_2(I_j) + 1$ bins. By concatenating these packings, we get the desired approximate 2-bounded-space packing for I with at most

$$\sum_{j=1}^{m} \left(\frac{3}{2}\text{OPT}_2(I_j) + 1\right) \ \leq \ \frac{3}{2}\left(1 + \frac{2}{V}\right) \text{OPT}_2(I) + m \ \leq \ \left(\frac{3}{2} + \varepsilon\right) \text{OPT}_2(I) + 1.$$

This yields the following theorem (and the main result of the paper).

Theorem 3.1. *For every $\varepsilon > 0$ there exists a polynomial time algorithm A for 2-bounded-space bin packing with asymptotic ratio $R_2^{\infty}(A) \leq 3/2 + \varepsilon$.*

3.1 Relaxed Packings

We are left with the following problem: given an instance I with $\text{OPT}_2(I) \leq 4V$, find an approximate 2-bounded-space packing into at most $\frac{3}{2}\text{OPT}_2(I) + 1$ bins. This problem is solved in two steps: first we find a *relaxed* packing of I that may slightly overpack some bins (Lemma 3.3), and then we transform this relaxed packing into an *exact* packing that obeys the bin sizes (Lemma 3.4).

Definition 3.2. *A δ-relaxed (2-bounded-space) packing of an instance I is a well-ordered partition such that for each bin B_{ij} (with $1 \leq i \leq 2$) either*

- $s(B_{ij}) \leq 1$, *in which case we set $D_{ij} = \emptyset$ and $\bar{B}_{ij} = B_{ij}$, or*

- *there exists $d \in B_{ij}$ such that $d \leq \delta$ and $s(B_{ij} \setminus \{d\}) \leq 1$, in which case we fix one such d, call it the special item, and set $D_{ij} = \{d\}$ and $\bar{B}_{ij} = B_{ij} \setminus D_{ij}$.*

In other words, each bin B_{ij} is split into a set $\bar{B}_{ij}$ of size at most 1 and the set D_{ij} with at most 1 item of size at most δ. For the rest of Section 3 we set $\delta = 1/20 = 0.05$.

Lemma 3.3. *For any instance I with $\text{OPT}_2(I) \leq 4V$, we can find in polynomial time a δ-relaxed packing $\mathcal{B}$ into at most $\text{OPT}_2(I)$ bins.*

Proof. We search exhaustively the value $\text{OPT}_2(I)$ and the assignment of all items that are larger than δ. Since the optimal number of bins is at most $4V$ and the number of large items is at most $4V/\delta$, i.e., they are both bounded by a constant, there is a constant number of assignments to try. For each assignment, we pack the items smaller than δ by adding them greedily from the beginning of the instance: Assign each item into the open bin that closes first, until and including the first item that overfills the bin; if the bin is already full then into the other

Two-Bounded-Space Bin Packing Revisited 267

bin. It can be seen that for the optimal assignment of the large items, the number
of bins used is at most $\mathrm{OPT}(I_j)$: Inductively we can verify that if a bin is full
at the time of our assignment of an item, then the total size of this and all the
previous bins is larger than or equal to the size of these bins in the optimum,
thus the other bin cannot be full and we can assign the item there. □

Lemma 3.4. *(Transformation lemma) Given a δ-relaxed packing $\mathcal{B}$ of instance
I, we can construct in polynomial time an exact packing of I with at most $\frac{3}{2}|\mathcal{B}|+1$
bins.*

To prove the transformation lemma, we split the instance again. In each round,
we process the r initial bins of $\mathcal{B}$ for a suitable r and find an exact packing of
them into at most $\frac{3}{2}r$ bins. Then we glue this packing together with a packing
of a slightly modified remaining instance. Some care needs to be taken to ensure
that the overlapping parts of the packings use only one track each. The exact
formulation of the needed building block follows.

Lemma 3.5. *(Initial segment lemma) Given a δ-relaxed packing $\mathcal{B}$ of instance
I with at least 2 bins, we can find in polynomial time r, an instance I' which is
a subsequence of I and an exact packing of $\mathcal{B}'$ of I' such that*

(i) *I' contains all the items from the first r closed bins of $\mathcal{B}$ and possibly some
additional items that arrive before the rth bin of $\mathcal{B}$ is closed;*
(ii) *$\mathcal{B}'$ uses at most $\frac{3}{2}r$ bins;*
(iii) *in $\mathcal{B}'$, all the items that arrive after some item in $I \setminus I'$ are packed into
bins in the same track and each of these bins has size at most $1 - \delta$.*

3.2 Proof of the Transformation from the Initial Segment Lemma

We proceed by induction on the number of bins m in the relaxed packing $\mathcal{B}$. If
I is empty, the statement is trivial. If $\mathcal{B}$ uses a single bin B_{11}, then there is an
exact packing with bins $\bar{B}_{11}$ and D_{11}.

If $\mathcal{B}$ uses $m \geq 2$ bins, then use Lemma 3.5 to find r, I', and $\mathcal{B}'$.

Let C be the set of all items in $I \setminus I'$ that arrive before the last item of I'.
Lemma 3.5 (i) implies that all these items belong to a single bin of $\mathcal{B}$, namely
to the bin that remains open when the rth bin is closed. If C contains a special
item d of that bin, we let $C' = C \setminus \{d\}$; otherwise we let $C' = C$.

Now create the instance I'' as follows. It starts by an item a of size $s(C')$
followed by the items in $(I \setminus I') \setminus C$ in the same order as in I. We claim that
we can construct a δ-relaxed packing of I'' with $m - r$ bins: Take $\mathcal{B}$, remove the
first r bins, remove the items C and put the item a in their bin. If $C = C'$ the
size of the bin does not change. If $C = C' \cup \{d\}$, the size of d is not included
in a, thus the size of the bin drops below 1. In each of these cases, this bin and
whole packing satisfy the condition of δ-relaxed packing.

By the induction assumption there exists an exact packing $\mathcal{B}''$ for I'' with
$\frac{3}{2}(m - r) + 1$ bins.

Now we construct the exact packing $\bar{\mathcal{B}}$ for I. The packing starts by the bins of
$\mathcal{B}'$ put in the same tracks as in $\mathcal{B}'$; Lemma 3.5 (i) implies that there are at most

$\frac{3}{2}r$ bins. If the special item d exists and is in C, Lemma 3.5 (ii) guarantees that the bin in $\mathcal{B}'$ that is open when d arrives has size at most $1 - \delta$; we add d in this bin without violating the size constraint. Next we consider the bin B''_{i1} from $\mathcal{B}''$ that contains a, modified so that we replace a by items C', preserving its size. Lemma 3.5 (ii) guarantees that all the bins in $\mathcal{B}'$ that are open when the items of C' arrive are in a single track, thus we can use the other track for B''_{i1} in $\bar{\mathcal{B}}$. Finally, $\bar{\mathcal{B}}$ contains all the remaining bins from $\mathcal{B}''$. By the definition of C, all the items in these bins arrive after the last item from I'. Thus the only constraint for placing them in tracks is the placement of B''_{i1}: If B''_{i1} is in track i in $\bar{\mathcal{B}}$, all the remaining bins from $\mathcal{B}''$ are in $\bar{\mathcal{B}}$ in the same track as in $\mathcal{B}''$, otherwise they are in the opposite track. Thus $\bar{\mathcal{B}}$ is indeed 2-bounded packing for I. The total number of bins in $\bar{\mathcal{B}}$ is $\frac{3}{2}r + \frac{3}{2}(m - r) + 1 = \frac{3}{2}m + 1$.

3.3   Proof of the Initial Segment Lemma

Without loss of generality, assume that B_{11} is closed no later than B_{21}. Let B_{1t} be the last bin closed before or at the same time when B_{21} is closed.

In the proceed in several cases. The overall idea for finding the packing is always the same. We identify up to two large items and pack them in bins by themselves. These bins are placed in track 1 together with bins $\bar{B}_{1j}$ that are not in an ordering conflict with the large items. The set G of the remaining items is packed greedily in track 2. Since there are no large items among them, this packing is efficient, which is quantified in Claim 3.3. To guarantee Lemma 3.5 (iii), we make sure that the bins containing G have each size at most $1 - \delta$ and that G contains all the items that arrive after the first item in $I \setminus I'$.

In most of the cases the instance I' consists exactly of all items in the first r closed bins, in which case Lemma 3.5 (i) follows immediately. Only in Case 3.1 we use a more complicated I'.

The value of various constants in the proof is somewhat arbitrary. For understanding the idea of the proof, one can assume that δ is set arbitrarily small. Also, the hard case is when $t = 1, 2$ as we can use only one additional bin. With increasing t, there is increasing slack in the construction.

Claim. Suppose that $b \leq 1 - \delta$ is the largest item in an instance G and $s(G) > 1$. Then there exists a 1-bounded packing of G with at most $\lceil (s(G) - 1 - \delta)/(1 - \delta - b) \rceil + 1$ bins, each of size at most $1 - \delta$. In particular for $r \geq 2$, $\delta = 0.05$, and $b \leq 0.75$:

(i) If $s(G) \leq (1.75 - b) + r\delta$ then $\lfloor 1 + r/2 \rfloor$ bins of size $1 - \delta$ suffice for G.
(ii) If $s(G) \leq 0.8 + r\delta$ then $\lfloor r/2 \rfloor$ bins of size $1 - \delta$ are sufficient for G.

Proof. Pack the items greedily from the beginning of the sequence. When it is necessary to close a bin, its size is more than $1 - \delta - b$. Iterating this, and observing that once the volume is below $1 - \delta$, a single bin is sufficient, the general bound follows. To prove that the special cases (i) and (ii) follow, note that they are tight for $r = 3$ and check that this is the tightest case, as for $r = 2$ there is additional slack and whenever r increases by 2, we have 1 additional bin but the additional size is only $2\delta < 1 - \delta - b$.   □

Case 1: All items in B_{21} have size at most 0.75. We set $I' = B_{21} \cup B_{11} \cup \cdots \cup B_{1t}$, thus $r = t+1$ and Lemma 3.5 (i) holds. The packing $\mathcal{B}'$ uses track 1 for t bins $\bar{B}_{1i}$, $i = 1, \ldots, t$. The remaining items $G = B_{21} \cup D_{11} \cup \cdots \cup D_{1t}$ are packed greedily using track 2. No item in G is larger than 0.75 and $s(G) \le 1 + r\delta$. Claim 3.3 (i) implies that $1 + r/2$ bins are sufficient for G. All items from the last open bin, i.e. B_{21}, are included in G, thus Lemma 3.5 (iii) is satisfied. Overall, we have at most $\frac{3}{2}r$ bins as required in Lemma 3.5 (ii).

Case 2: The largest item a in B_{21} has size at least 0.75 and arrives while the bin B_{1j} is open for some $j \le t$. We set $I' = B_{21} \cup B_{11} \cup \cdots \cup B_{1t}$, thus $r = t+1$ and Lemma 3.5 (i) holds. The packing $\mathcal{B}'$ uses track 1 for one bin containing just a and $t-1$ bins $\bar{B}_{1i}$, $i = 1, \ldots, t$, $i \ne j$, in the appropriate order; it also uses at most $1 + r/2$ additional bins in both tracks depending on the two subcases. Altogether we use at most $\frac{3}{2}r$ bins as required.

To satisfy Lemma 3.5 (iii), we will make sure that all items from B_{21} are packed greedily in track 2, with the exception of a. However, a arrives while B_{1j} is open, i.e., before any item in $I \setminus I'$ arrives, thus it does not violate the condition.

Subcase 2.1: The largest item $\bar{a}$ in B_{1j} has size at least 0.5. The packing $\mathcal{B}'$ uses one additional bin containing just $\bar{a}$ in track 1. The remaining items $G = (B_{21} \cup B_{1j} \setminus \{a, \bar{a}\}) \cup D_{11} \cup \cdots \cup D_{1t}$ are packed greedily using track 2. No item in G has size more than 0.5 and $s(G) \le 0.75 + r\delta$. Claim 3.3 (ii) implies that $r/2$ bins are sufficient for G.

Subcase 2.2: No item in B_{1j} has size 0.5 or more. All the remaining items $G = (B_{21} \setminus \{a\}) \cup B_{1j} \cup D_{11} \cup \cdots \cup D_{1t}$ are packed greedily using track 2. No item in G has size more than 0.5 and $s(G) \le 1.25 + r\delta$. Claim 3.3 (i) implies that $1 + r/2$ bins are sufficient for G.

Case 3: The largest item a in B_{21} has size at least 0.75 and arrives while the bin $B_{1,t+1}$ is open. Let C be the set of items in $B_{1,t+1}$ that arrive before the large item a arrives into B_{21}. Since at the time of arrival of a we need to use both tracks (one for a and one for the greedy assignment), we have to include in I' also items in C to guarantee Lemma 3.5 (iii). We distinguish subcases according to $s(C)$.

Subcase 3.1: $s(C) \le 0.5$. We set $I' = B_{21} \cup B_{11} \cup \cdots \cup B_{1t} \cup C$, thus $r = t+1$ and Lemma 3.5 (i) holds because all items in C arrive while B_{21} is open. The packing $\mathcal{B}'$ uses track 1 for t bins B_{1i}, $i = 1, \ldots, t$ and one bin containing just a. The remaining items $G = (B_{21} \setminus \{a\}) \cup D_{11} \cup \cdots \cup D_{1t} \cup C$ are packed greedily using track 2. No item in G has size more than 0.5 (the largest item may be in C) and $s(G) \le 0.75 + r\delta$. Claim 3.3 (ii) implies that we use at most $r/2$ bins for G, a total of $\frac{3}{2}r$ bins.

Subcase 3.2: $s(C) \ge 0.5$. Here we want to include in I' the whole bin $B_{1,t+1}$, so that an additional bin is available to be used for C. This brings more problems, as many bins may close in track 2 before $B_{1,t+1}$ closes, and we have to include them as well to satisfy Lemma 3.5 (i).

Let t' be such that $B_{2t'}$ is the last bin closed while $B_{1,t+1}$ is open. We set
$I' = B_{21} \cup \cdots \cup B_{2t'} \cup B_{11} \cup \cdots \cup B_{1t} \cup B_{1,t+1}$, $r = t + t' + 1$, and Lemma 3.5 (i)
is satisfied now. The packing $\mathcal{B}'$ uses track 1 for t bins $\bar{B}_{1i}$, $i = 1, \ldots, t$, one bin
containing $C \setminus D_{1,t+1}$, one bin containing just a, and $t' - 1$ bins $\bar{B}_{2i}$, $i = 2, \ldots, t'$,
in this order; this is $t + 2 + (t' - 1) = r$ bins total. The remaining items $G =
D_{11} \cup \cdots \cup D_{1t} \cup (B_{1,t+1} \setminus C) \cup (B_{21} \setminus \{a\}) \cup D_{22} \cup \cdots \cup D_{2t'}$ are packed greedily
using track 2. No item in G has size more than $0.5 + \delta$ (the largest item may be
in $B_{1,t+1} \setminus C$) and $s(G) \leq 0.8 + r\delta$. Claim 3.3 (ii) implies that we use at most
$r/2$ bins for G, altogether at most $\frac{3}{2}r$ bins.

4   An Asymptotic Lower Bound for Offline Algorithms

We now prove a lower bound of 1.25 on the asymptotic worst case ratio of any 2-
space-bounded bin packing algorithm. We reduce from the $\mathbb{NP}$-hard PARTITION
problem; see [4]. We consider an instance of PARTITION which is a set $P =
\{q_1, \ldots, q_n\}$ of $n \geq 2$ positive integers that add up to $2Q$. The question is whether
P has an equitable partition, namely a partition into two sets each adding up to
Q. Without loss of generality, we assume that each q_i is a multiple of 3 and is
at most Q, and that Q is large enough, say $Q \geq 100$. Then P has an equitable
partition if and only if it has a partition into two sets each adding up to between
$Q - 2$ and $Q + 2$.

We construct the following bin packing instance. The bins have size $\beta = 2Q+3$.
The instance I consists of m copies of the following sub-list J with $n + 6$ items:
$q_1, \ldots, q_n, Q + 3, Q + 3, Q + 2, Q + 2, Q + 1, Q + 1$.
If P has an equitable partition, we claim that there exists a 2-bounded-space
packing of I into $4m$ bins. It is sufficient to pack J into 4 bins. Partition $q_1, \ldots, q_n$
into two subsets, each adding up to Q, and assign them to different tracks. In
addition, each track has one copy of items $Q + 3$, $Q + 2$, $Q + 1$ and needs only
two bins.

In the rest of Section 4 we show that if P does not have an equitable partition,
then any 2-bounded-space packing of I uses at least $5m$ bins of size $\beta = 2Q + 3$.

We introduce first some terminology. For a given list L, we say that a packing
$\mathcal{B}$ dominates a packing $\mathcal{B}'$ on L if, for any list L' with all items of size at least 3,
$\mathcal{B}$ can be extended to a packing of LL' that uses no more bins that any extension
of $\mathcal{B}'$ to LL'. A set of packings is called dominant on L if any other packing is
dominated by some packing in this set.

By a state we mean a pair of integers $\{u, v\}$ representing the content of the two
open bins. We consider 2-bounded-space packings with arbitrary initial states,
not necessarily $\{0, 0\}$. The concept of dominance extends naturally to such packings. Note that to determine whether $\mathcal{B}$ dominates $\mathcal{B}'$ we only need to know for
each of $\mathcal{B}$ and $\mathcal{B}'$ the final state and the number of bins closed by it. In particular, the following easy observation will be useful in showing that one packing
dominates another.

Claim. Consider two packings $\mathcal{B}$, $\mathcal{B}'$, where $\mathcal{B}$ closes b bins and ends in state
$\{u, v\}$, and $\mathcal{B}'$ closes b' bins and ends in state $\{u', v'\}$, for $u \leq v$ and $u' \leq v'$.

Suppose that either (i) $b \leq b' - 2$, or (ii) $b = b' - 1$ and $u \leq v'$, or (iii) $b = b'$ and $u \leq u'$ and $v \leq v'$. Then $\mathcal{B}$ dominates $\mathcal{B}'$.

We split J into three parts:

$$J_1 \;=\; q_1, \ldots, q_n, Q+3, Q+3, \qquad J_2 \;=\; Q+2, Q+2, \qquad J_3 \;=\; Q+1, Q+1.$$

Figure 1 shows a state diagram, where transitions correspond to packings of sublists J_1, J_2, J_3 from certain states. A transition from $\{u, u'\}$ to $\{v, v'\}$ labeled by J_i / b represents a packing that, starting from state $\{u, u'\}$, packs J_i closing b bins and ending in state $\{v, v'\}$.

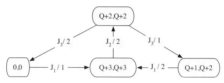

Fig. 1. The state diagram representing dominant packings

We claim that the packings represented in Figure 1 are dominant for the given initial states and lists J_i. Note that for the proof it is not necessary that these packings actually exist, since we only use them to lower-bound the actual packings. Another way to think about these transitions is as "sub-packings", where each item may take less space in the bin than its actual size.

Before proving this claim, we show that it implies the lemma. Indeed, since the packings in Figure 1 are dominant, an optimal packing of $I = (J_1 J_2 J_3)^m$ cannot do better than follow some path in that diagram, closing the numbers of bins indicated on the transitions, and ending either at state $\{0, 0\}$ or $\{Q+1, Q+2\}$. By a straightforward inspection of the cycles in the diagram, if this path ends at $\{0, 0\}$, the optimal packing will close $5m$ bins. If it ends at $\{Q+1, Q+2\}$, it will close $5m - 1$ bins, with two open non-empty bins. Thus the total number of used bins is at least $5m$, and the lemma follows.

To prove our claim, we consider transitions from each state. Since $q_i \geq 3$ for all i, we can assume that all bins with load at least $2Q + 1$ are immediately closed. We also assume that the bins open after packing each J_i that have loads at most $2Q$ remain open until the next item arrives.

Consider a packing of J_1 starting from $\{0, 0\}$. If we put $Q+3$, $Q+3$ in different tracks, then we must close a bin that contains only items from P (because P does not have an equitable partition), so we can as well put them all in one bin, and we reach state $\{Q + 3, Q + 3\}$. If $Q+3$, $Q+3$ are in the same track, then we need to close one bin with the first item $Q + 3$, and the other track will contain items from P that do not fit into that bin, so it will have load $x \geq Q + 3$, by our assumption. Thus the final state is $\{Q + 3, x\}$. In both cases, the packings are dominated by the packing represented by the one in Figure 1.

Consider a packing of J_2 from $\{Q + 3, Q + 3\}$. The items $Q + 2$, $Q + 2$ cannot share a bin and do not fit in the open bins, so any packing must close two bins.

The dominant packing will end up in state with minimum load bins, that is $\{Q + 2, Q + 2\}$, same as the one in Figure 1.

Next, consider a packing from $\{Q + 2, Q + 2\}$ on J_3. If $Q+1, Q+1$ are assigned to different tracks then we close two bins and go to $\{0, 0\}$, otherwise we close one bin and go to $\{Q + 1, Q + 2\}$. These two packings are exactly those from Figure 1.

Finally, consider a packing of J_1 from $\{Q + 1, Q + 2\}$. Items $Q + 3$, $Q + 3$ cannot share a bin and cannot be assigned to already open bins, so we need to close two bins. If $Q + 3, Q + 3$ are assigned to different tracks, the two open bins will close and we will reach a state $\{Q + x, Q + x'\}$ for some $x, x' \geq 3$. If $Q + 3$, $Q + 3$ are on one track, the other track needs to be assigned some items from P, because P does not have an equitable partition. Therefore the new state will be $\{Q + 3, Q + x\}$, for $x \geq 3$. This packing is dominated by the one in Figure 1.

Our reduction yields the main result of this section:

Theorem 4.1. *Any polynomial time algorithm A for 2-bounded-space bin packing has asymptotic ratio $R_2^\infty(A) \geq 1.25$ (unless $\mathbb{P}=\mathbb{NP}$).*   □

5   Absolute Bounds for Offline Algorithms

In this section we discuss 2-bounded-space bin packing for instances that can be packed into a small number b of bins. For $b \geq 1$ we define $\mathcal{I}_b$ as set of all instances I with $\text{OPT}_2(I) = b$. Furthermore, we define $f_2(b)$ as the smallest integer such that for every instance $I \in \mathcal{I}_b$ we can find in polynomial time a 2-bounded-space packing into $f_2(b)$ bins; note that in this definition we assume $\mathbb{P} \neq \mathbb{NP}$. Theorem 4.1 implies $f_2(b) \geq 5b/4$ for large values of b.

It is not hard to see that $f_2(1) = 1$, $f_2(2) = 3$, and $f_2(3) = 4$. A more involved argument shows that $f_2(4) = 6$. Due to the space constraint, we only give the lower bound reduction, omitting both its proof and the upper bound altogether.

Theorem 5.1. *It is $\mathbb{NP}$-hard to distinguish instances in $\mathcal{I}_4$ from those in $\mathcal{I}_6$.*

Proof. We consider an instance P of the $\mathbb{NP}$-hard **PARTITION** problem (see Section 4) that consists of $n \geq 2$ positive integers $q_1, \ldots, q_n$ that add up to $2Q$. We construct a bin packing instance I with bin size $\beta = 2Q + 1$ and the $2n + 4$ items with the following sizes: $q_1, \ldots, q_n, Q + 1, Q + 1, Q + 1, Q + 1, q_1, \ldots, q_n$. It can be verified that (i) if P has an equitable partition then $\text{OPT}_2(I) = 4$, and (ii) if I has a 2-bounded-space packing of I into five bins then P has an equitable partition.   □

6   An Asymptotic Lower Bound for Online Algorithms

Throughout this section we will consider an arbitrary online bin packing algorithm ON that uses 2-bounded space. Our goal is to establish a lower bound of $4/3$ on the performance ratio $R_2^\infty(\text{ON})$. We will make excessive use of the following two straightforward observations. First: if a new item does not fit into any active bin, then it is never wrong to close the fullest active bin. Second: if an active bin cannot accommodate any of the future items, then it can be closed right away.

The proof is done in terms of a standard adversary argument. The adversary works in p phases, where p is a huge integer that can be chosen arbitrarily large. All item sizes in the kth phase ($1 \le k \le p$) depend on a small rational number $\varepsilon_k = 2^{k-p}/100$. Note that these values ε_k form a strictly increasing sequence satisfying $\varepsilon_\ell \ge 2\varepsilon_k$ for all $\ell > k$. Note furthermore that the ε_k are bounded from above by $1/100$. During the kth phase, the adversary may issue items of the following three types: so-called 1-items of size $1 + \varepsilon_k$; so-called 2-items of size $2 - 2\varepsilon_k$; and so-called 3-items of size $3 - \varepsilon_k$.

The bins have size $\beta = 4$. When referring to the load or contents of a bin, we will usually ignore the dependence on the epsilons. Hence the active bins will be bins of load 1, 2 or 3. Bins of load 4 can always be closed right away.

Proposition 6.1. *Consider a partially packed bin that only contains items issued during the first k phases.*

(i) *If the bin has load 1, then any 3-item from phase k or later will fit into it.*
(ii) *If the bin has load 2, then any 2-item from phase k or later will fit into it.*
(iii) *If the bin has load 3, then no item from phase $k + 1$ or later will fit into it.*

Proof. Statements (i) and (ii) are straightforward. For statement (iii) we distinguish three cases: if the bin contains a single 3-item, the free space is at most $1 + \varepsilon_k$. If the bin contains three 1-items, the free space is at most $1 - 3\varepsilon_k$. If the bin contains a 1-item and a 2-item, the free space is at most $1 + 2\varepsilon_k - \varepsilon_k$. All items arriving in phase $k + 1$ or later are of size at least $1 + 2\varepsilon_k$, and hence will not fit into the free space. $\square$

Here is a central ingredient to the adversary's strategy: Whenever ON produces an active bin of load 3, the adversary starts a new phase and increases the value of ε. Then by Proposition 6.1.(iii) ON will never be able to use this bin again, and we may assume without loss of generality that whenever ON produces an active bin of load 3, this bin is closed right away.

The adversary will only close bins of load 4, either by adding a 2-item to an active bin of load 2, or by adding a 3-item to an active bin of load 1. By Proposition 6.1 these items will always fit into their respective bins.

The adversary will always issue one or two elements and make sure it ends in one of the following three states:

ADV$[0, 0]$: Both active bins are empty.

ADV$[0, 1]$: One empty bin; the other bin contains a single 1-item.

ADV$\{1, 1\}$: Two 1-items in the active bins. The adversary may decide later whether these items are in the same bin or in different bins.

Before a new item arrives, the active bins of ON will always have loads $X, Y \in \{0, 1, 2\}$, we may assume $X \le Y$; these six states are denoted by ON$[X, Y]$.

The adversary's strategy is designed to guarantee that the combination of states ADV$\{1, 1\}$ and ON$[X, Y]$ never appears; all other combinations are possible. The following paragraphs provide a full description of the adversary's behavior, in dependence on its current state and the current state of ON.

ADV[0, 0] and any ON[X, Y]: The adversary issues a 1-item and moves to ADV[0, 1]. Depending on its state, ON perhaps closes a bin of load 3. Since all ON states are legal with ADV[0, 1], we end up in a legal combination.

ADV[0, 1] or ADV{1, 1} and ON[X, Y] with $X, Y \in \{0, 2\}$: The adversary issues a 3-item, closes a bin of load 4, and moves to ADV[0, 0] or ADV[0, 1], while ON either closes a bin of load 3, or it closes a bin of load 2 and a bin of load 3. Again, all final state combinations are legal.

ADV{1, 1} and ON[0, 1] or ON[1, 1]: The adversary issues a 2-item, closes a bin of load 4 containing both 1-items, and moves to ADV[0, 0], while ON cannot reach load 4. Perhaps it closes a bin of load 3. All ON states are a legal combination with ADV[0, 0].

ADV[0, 1] and ON[0, 1] or ON[1, 2]: The adversary issues a 1-item and moves to ADV{1, 1}, while ON moves to one of the states ON[1, 1], ON[0, 2], or ON[2, 2], or else it closes a bin of load 3 and moves to ON[0, 1]. In any case, ON final state is not ON[1, 2], so the combination is legal.

ADV[0, 1] and ON[1, 1]: The adversary issues two 2-items, closes a bin with them and remains in ADV[0, 1], while ON closes one or two bins of load 3 and moves to some state. Again, any ON state is a legal combination with ADV[0, 1].

This completes the description of the adversary's strategy, as all the cases are covered. To summarize: the adversary only closes bins of load 4, whereas ON always closes bins of load at most 3. This yields the following theorem.

Theorem 6.2. *Any 2-bounded-space online bin packing algorithm* ON *has performance ratio* $R_2^\infty(\text{ON}) \geq 4/3$.   □

Acknowledgments. This research has been partially supported by NSF Grant CCF-0729071; by Inst. for Theor. Comp. Sci., Prague (project 1M0545 of MŠMT ČR) and grant IAA100190902 of GA AV ČR, by the Netherlands Organization for Scientific Research (NWO), grant 639.033.403; by BSIK grant 03018 (BRICKS: Basic Research in Informatics for Creating the Knowledge Society).

References

1. Coffman Jr., E.G., Csirik, J., Woeginger, G.J.: Approximate solutions to bin packing problems. In: Pardalos, P.M., Resende, M.G.C. (eds.) Handbook of Applied Optimization, pp. 607–615. Oxford University Press, New York (2002)
2. Csirik, J., Imreh, B.: On the worst-case performance of the NkF bin packing heuristic. Acta Cybernetica 9, 89–105 (1989)
3. Csirik, J., Johnson, D.S.: Bounded space on-line bin packing: Best is better than first. Algorithmica 31, 115–138 (2001)
4. Garey, M.R., Johnson, D.S: Computers and Intractability: A Guide to the Theory of NP-Completeness. Freeman, San Francisco (1979)
5. Lee, C.C., Lee, D.T.: A simple on-line bin-packing algorithm. Journal of the ACM 32, 562–572 (1985)
6. Mao, W.: Tight worst-case performance bounds for Next-k-Fit bin packing. SIAM Journal on Computing 22, 46–56 (1993)
7. Woeginger, G.J.: Improved space for bounded-space, on-line bin-packing. SIAM Journal on Discrete Mathematics 6, 575–581 (1993)

Output-Sensitive Listing of Bounded-Size Trees in Undirected Graphs

Rui Ferreira[1], Roberto Grossi[1], and Romeo Rizzi[2]

[1] Università di Pisa
{ferreira,grossi}@di.unipi.it
[2] Università degli Studi di Udine
romeo.rizzi@uniud.it

Abstract. Motivated by the discovery of combinatorial patterns in an undirected graph G with n vertices and m edges, we study the problem of listing all the trees with k vertices that are subgraphs of G. We present the first optimal output-sensitive algorithm, i.e. runs in $O(sk)$ time where s is the number of these trees in G, and uses $O(m)$ space.

1 Introduction

Graphs are employed to model a variety of problems ranging from social to biological networks. Some applications require the extraction of combinatorial patterns [1,5] with the objective of gaining insights into the structure, behavior, and role of the elements in these networks.

Consider an undirected connected graph $G = (V, E)$ of n vertices and m edges. We want to list all the k-trees in G. We define a k-tree T as an edge subset $T \subseteq E$ that is acyclic and connected, and contains k vertices. We denote by s the number of k-trees in G. For example, there are $s = 9$ k-trees in the graph of Fig. 1, where $k = 3$. We present the *first optimal output-sensitive* algorithm for listing all the k-trees in $O(sk)$ time, using $O(m)$ space.

As a special case, our basic problem models also the classical problem of listing the spanning trees in G, which has been largely investigated (here $k = n$ and s is the number of spanning trees in G). The first algorithmic solutions appeared in the 60's [6], and the combinatorial papers even much earlier [7]. Read and Tarjan gave an output-sensitive algorithm in $O(sm)$ time and $O(m)$ space [9]. Gabow and Myers proposed the first algorithm [3] which is optimal when the spanning trees are explicitly listed. When the spanning trees are implicitly enumerated, Kapoor and Ramesh [4] showed that an elegant incremental representation is possible by storing just the $O(1)$ information needed to reconstruct a spanning tree from the previously enumerated one, giving $O(m+s)$ time and $O(mn)$ space [4], later reduced to $O(m)$ space by Shioura et al. [10]. We are not aware of any non-trivial output-sensitive solution for the problem of listing the k-trees in the general case.

We present our solution starting from the well-known binary partition method. (Other known methods are those based on Gray codes and reverse search [2].)

C. Demetrescu and M.M. Halldórsson (Eds.): ESA 2011, LNCS 6942, pp. 275–286, 2011.

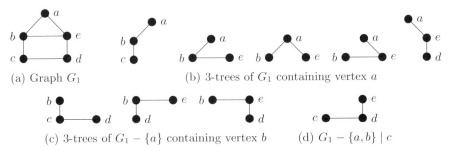

(a) Graph G_1 (b) 3-trees of G_1 containing vertex a

(c) 3-trees of $G_1 - \{a\}$ containing vertex b (d) $G_1 - \{a, b\} \mid c$

Fig. 1. Example graph G_1 and its 3-trees

We divide the problem of listing all the k-trees in two subproblems by choosing an edge $e \in E$: we list the k-trees that contain e and those that do not contain e. We proceed recursively on these subproblems until there is just one k-tree to be listed. This method induces a binary recursion tree, and all the k-trees can be listed when reaching the leaves of the recursion tree.

Although output sensitive, this simple method is not optimal since it takes $O(s(m + n))$ time. One problem is that the adjacency lists of G can be of length $O(n)$ each, but we cannot pay such a cost in each recursive call. Also, we need a *certificate* that should be easily maintained through the recursive calls to guarantee *a priori* that there will be at least one k-tree generated. By exploiting more refined structural properties of the recursion tree, we present our algorithmic ideas until an optimal output-sensitive listing is obtained, i.e. $O(sk)$ time. Our presentation follows an incremental approach to introduce each idea, so as to evaluate its impact in the complexity of the corresponding algorithms.

2   Preliminaries

Given a simple (without self-loops or parallel edges), undirected and connected graph $G = (V, E)$, with $n = |V|$ and $m = |E|$, and an integer $k \in [2, n]$, a *k-tree* T is an acyclic connected subgraph of G with k vertices. We denote the total number of k-trees in G by s, where $sk \geq m \geq n - 1$ since G is connected.

Problem 1 (k-tree listing). Given an input graph G and an integer k, list all the k-trees of G.

We say that an algorithm that solves Problem 1 is *optimal* if it takes $O(sk)$ time, since the latter is proportional to the time taken to explicitly list the output, namely, the $k - 1$ edges in each of the s listed k-trees. We also say that the algorithm has *delay* $t(k)$ if it takes $O(t(k))$ time to list a k-tree after having listed the previous one.

We adopt the standard representation of graphs using adjacency lists $\mathtt{adj}(v)$ for each vertex $v \in V$. We maintain a counter for each $v \in V$, denoted by $|\mathtt{adj}(v)|$, with the number of edges in the adjacency list of v. Additionally, as the graph is undirected, the notations (u, v) and (v, u) represent the same edge.

Let $X \subseteq E$ be a *connected edge* set. We denote by $V[X] \equiv \{u \mid (u, v) \in X\}$ the set of its endpoints, and its *ordered vertex list* $\hat{V}(X)$ recursively as follows: $\hat{V}(\{\cdot, u_0)\}) = \langle u_0 \rangle$ and $\hat{V}(X + (u, v)) = \hat{V}(X) + \langle v \rangle$ where $u \in V[X]$, $v \notin V[X]$, and $+$ denotes list concatenation. We also use the shorthand $E[X] \equiv \{(u, v) \in E \mid u, v \in V[X]\}$ for the induced edges. In general, $G[X] = (V[X], E[X])$ denotes the subgraph of G *induced* by X, which is equivalently defined as the subgraph of G induced by the vertices in $V[X]$.

The *cutset* of X is the set of edges $C(X) \subseteq E$ such that $(u, v) \in C(X)$ if and only if $u \in V[X]$ and $v \in V - V[X]$. Note that when $V[X] = V$, the cutset is empty. Similarly, the *ordered cutlist* $\hat{C}(X)$ contains the edges in $C(X)$ ordered by the rank of their endpoints in $\hat{V}(X)$. If two edges have the same endpoint vertex $v \in \hat{V}(S)$, we use the order as they appear in $\texttt{adj}(v)$ to break the tie.

Throughout the paper we represent an unordered k'-tree $T = \langle e_1, e_2, \ldots, e_{k'} \rangle$ with $k' \leq k$ as an *ordered*, connected and acyclic list of k' edges, where we use a *dummy* edge $e_1 = (\cdot, v_i)$ having a vertex v_i of T as endpoint. The order is the one by which we discover the edges $e_1, e_2, \ldots, e'_k$. Nevertheless, we do not generate two different orderings for the same T.

3   Basic Approach: Recursion Tree

We begin by presenting a simple algorithm that solves Problem 1 in $O(sk^3)$ time, while using $O(mk)$ space. Note that the algorithm is not optimal yet: we will show in Sections 4–5 how to improve it to obtain an optimal solution with $O(m)$ space and delay $t(k) = k^2$.

Top level. We use the standard idea of fixing an ordering of the vertices in $V = \langle v_1, v_2, \ldots, v_n \rangle$. For each $v_i \in V$, we list the k-trees that include v_i and do not include any previous vertex $v_j \in V$ ($j < i$). After reporting the corresponding k-trees, we remove v_i and its incident edges from our graph G. We then repeat the process, as summarized in Algorithm 1. Here, S denotes a k'-tree with $k' \leq k$, and we use the dummy edge $(\cdot, v_i)$ as a start-up point, so that the ordered vertex list is $\hat{V}(S) = \langle v_i \rangle$. Then, we find a k-tree by performing a DFS starting from v_i: when we meet the kth vertex, we are sure that there exists at least one k-tree for v_i and execute the binary partition method with $\texttt{ListTrees}_{v_i}$; otherwise, if there is no such k-tree, we can skip v_i safely. We exploit some properties on the recursion tree and an efficient implementation of the following operations on G:

- $\texttt{del}(u)$ deletes a vertex $u \in V$ and all its incident edges.
- $\texttt{del}(e)$ deletes an edge $e = (u, v) \in E$. The inverse operation is denoted by $\texttt{undel}(e)$. Note that $|\texttt{adj}(v)|$ and $|\texttt{adj}(u)|$ are updated.
- $\texttt{choose}(S)$, for a k'-tree S with $k' \leq k$, returns an edge $e \in C(S)$: e^- the vertex in e that belongs to $V[S]$ and by e^+ the one s.t. $e^+ \in V - V[S]$.
- $\texttt{dfs}_k(S)$ returns the list of the tree edges obtained by a *truncated DFS*, where conceptually S is treated as a *single* (collapsed vertex) source whose adjacency list is the cutset $C(S)$. The DFS is truncated when it finds k tree edges (or less if there are not so many). The resulting list is a k-tree (or

---

Algorithm 1. ListAllTrees($G = (V, E)$, k)

---

1. for $i = 1, 2, \ldots, n - 1$:
   - (a) $S := \langle (\cdot, v_i) \rangle$
   - (b) if $|\mathtt{dfs}_k(S)| = k$ then $\mathtt{ListTrees}_{v_i}(S)$
   - (c) $\mathtt{del}(v_i)$

Algorithm 2. $\mathtt{ListTrees}_{v_i}(S)$

---

1. if $|S| = k$ then:
   - (a) $\mathtt{output}(S)$
   - (b) $\mathtt{return}$
2. $e := \mathtt{choose}(S)$
3. $\mathtt{ListTrees}_{v_i}(S + \langle e \rangle)$
4. $\mathtt{del}(e)$
5. if $|\mathtt{dfs}_k(S)| = k$ then $\mathtt{ListTrees}_{v_i}(S)$
6. $\mathtt{undel}(e)$

---

smaller) that includes all the edges in S. Its purpose is to check if there exists a connected component of size at least k that contains S.

Lemma 2. *Given a graph G and a k'-tree S, we can implement the following operations: $\mathtt{del}(u)$ for a vertex u in time proportional to u's degree; $\mathtt{del}(e)$ and $\mathtt{undel}(e)$ for an edge e in $O(1)$ time; $\mathtt{choose}(S)$ and $\mathtt{dfs}_k(S)$ in $O(k^2)$ time.*

Recursion tree and analysis. The recursive binary partition method in Algorithm 2 is quite simple, and takes a k'-tree S with $k' \leq k$ as input. The purpose is that of listing all k-trees that include all the edges in S (excluding those with endpoints $v_1, v_2, \ldots, v_{i-1}$). The precondition is that we recursively explore S if and only if there is at least a k-tree to be listed. The corresponding recursion tree has some interesting properties that we exploit during the analysis of its complexity. The root of this binary tree is associated with $S = \langle (\cdot, v_i) \rangle$. Let S be the k'-tree associated with a node in the recursion tree. Then, left branching occurs by taking an edge $e \in C(S)$ using $\mathtt{choose}$, so that the *left child* is $S + \langle e \rangle$. Right branching occurs when e is deleted using $\mathtt{del}$, and the *right child* is still S but on the reduced graph $G := (V, E - \{e\})$. Returning from recursion, restore G using $\mathtt{undel}(e)$. Note that we do *not* generate different permutations of the same k'-tree's edges as we either take an edge e as part of S or remove it from the graph by the binary partition method.

Lemma 3 (Correctness). *Algorithm 2 lists each k-tree containing vertex v_i and no vertex v_j with $j < i$, once and only once.*

A closer look at the recursion tree reveals that it is *k-left-bounded*: namely, each root-to-leaf path has exactly $k - 1$ *left branches*. Since there is a one-to-one correspondence between the leaves and the k-trees, we are guaranteed that leftward branching occurs less than k times to output a k-tree.

What if we consider rightward branching? Note that the height of the tree is less than m, so we might have to branch rightward $O(m)$ times in the worst case. Fortunately, we can prove in Lemma 4 that for each internal node S of the recursion tree that has a right child, S has always its left child (which leads to one k-tree). This is subtle but very useful in our analysis in the rest of the paper.

Lemma 4. *At each node S of the recursion tree, if there exists a k-tree (descending from S's right child) that does not include edge e, then there is a k-tree (descending from S's left child) that includes e.*

Note that the symmetric situation for Lemma 4 does not necessarily hold. We can find nodes having just the left child: for these nodes, the chosen edge cannot be removed since this gives rise to a connected component of size smaller than k. We can now state how many nodes there are in the recursion tree.

Corollary 5. *Let s_i be the number of k-trees reported by* ListTrees$_{v_i}$. *Then, its recursion tree is binary and contains s_i leaves and at most $s_i k$ internal nodes. Among the internal nodes, there are $s_i - 1$ of them having two children.*

Lemma 6 (Time and space complexity). *Algorithm 2 takes $O(s_i k^3)$ time and $O(mk)$ space, where s_i is the number of k-trees reported by* ListTrees$_{v_i}$.

Theorem 7. *Algorithm 1 can solve Problem 1 in $O(nk^2 + sk^3) = O(sk^3)$ time and $O(mk)$ space.*

4 Improved Approach: Certificates

A way to improve the running time of ListTrees$_{v_i}$ to $O(s_i k^2)$ is indirectly suggested by Corollary 5. Since there are $O(s_i)$ binary nodes and $O(s_i k)$ unary nodes in the recursion tree, we can pay $O(k^2)$ time for binary nodes and $O(1)$ for unary nodes (i.e. reduce the cost of choose and dfs$_k$ to $O(1)$ time when we are in a *unary* node). This way, the total running time is $O(sk^2)$.

The idea is to maintain a certificate that can tell us if we are in a unary node in $O(1)$ time and that can be updated in $O(1)$ time in such a case, or can be completely rebuilt in $O(k^2)$ time otherwise (i.e. for binary nodes). This will guarantee a total cost of $O(s_i k^2)$ time for ListTrees$_{v_i}$, and lay out the path to the wanted optimal output-sensitive solution of Section 5.

4.1 Introducing Certificates

We impose an "unequivocal behavior" to dfs$_k(S)$, obtaining a variation denoted mdfs$_k(S)$ and called *multi-source truncated DFS*. During its execution, mdfs$_k$ takes the order of the edges in S into account (whereas an order is not strictly necessary in dfs$_k$). Specifically, given a k'-tree $S = \langle e_1, e_2, \ldots, e_{k'} \rangle$, the returned k-tree $D = $ mdfs$_k(S)$ contains S, which is conceptually treated as a collapsed vertex: the main difference is that S's "adjacency list" is now the *ordered cutlist* $\hat{C}(S)$, rather than $C(S)$ employed for dfs$_k$.

Equivalently, since $\hat{C}(S)$ is induced from $C(S)$ by using the ordering in $\hat{V}(S)$, we can see $\mathtt{mdfs}_k(S)$ as the execution of multiple standard DFSes from the vertices in $\hat{V}(S)$, in that order. Also, all the vertices in $V[S]$ are conceptually marked as visited at the beginning of $\mathtt{mdfs}_k$, so u_j is never part of the DFS tree starting from u_i for any two distinct $u_i, u_j \in V[S]$. Hence the adopted terminology of multi-source. Clearly, $\mathtt{mdfs}_k(S)$ is a feasible solution to $\mathtt{dfs}_k(S)$ while the vice versa is not true.

We use the notation $S \sqsubseteq D$ to indicate that $D = \mathtt{mdfs}_k(S)$, and so D is a *certificate* for S: it guarantees that node S in the recursion tree has at least one descending leaf whose corresponding k-tree has not been listed so far. Since the behavior of $\mathtt{mdfs}_k$ is non-ambiguous, relation $\sqsubseteq$ is well defined. We preserve the following invariant on $\mathtt{ListTrees}_{v_i}$, which now has two arguments.

Invariant 1 *For each call to* $\mathtt{ListTrees}_{v_i}(S, D)$, *we have* $S \sqsubseteq D$.

Before showing how to keep the invariant, we detail how to represent the certificate D in a way that it can be efficiently updated. We maintain it as a partition $D = S \cup L \cup F$, where S is the given list of edges, whose endpoints are kept in order as $\hat{V}(S) = \langle u_1, u_2, \dots, u_{k'} \rangle$. Moreover, $L = D \cap C(S)$ are the tree edges of D in the cutset $C(S)$, and F is the forest storing the edges of D whose both endpoints are in $V[D] - V[S]$.

(i) We store the k'-tree S as a sorted doubly-linked list of k' edges $\langle e_1, e_2, \dots, e_{k'} \rangle$, where $e_1 := (\cdot, v_i)$. We also keep the sorted doubly-linked list of vertices $\hat{V}(S) = \langle u_1, u_2, \dots, u_{k'} \rangle$ associated with S, where $u_1 := v_i$. For $1 \leq j \leq k'$, we keep the number of tree edges in the cutset that are incident to u_j, namely $\eta[u_j] = |\{(u_j, x) \in L\}|$.

(ii) We keep $L = D \cap C(S)$ as an ordered doubly-linked list of edges in $\hat{C}(S)$'s order: it can be easily obtained by maintaining the parent edge connecting a root in F to its parent in $\hat{V}(S)$.

(iii) We store the forest F as a sorted doubly-linked list of the roots of the trees in F. The order of this list is that induced by $\hat{C}(S)$: a root r precedes a root t if the (unique) edge in L incident to r appears before the (unique) edge of L incident to t. For each node x of a tree $T \in F$, we also keep its number $\deg(x)$ of children in T, and its predecessor and successor sibling in T.

(iv) We maintain a flag $\mathtt{is\_unary}$ that is true if and only if $|\mathtt{adj}(u_i)| = \eta[u_i] + \sigma(u_i)$ for all $1 \leq i \leq k'$, where $\sigma(u_i) = |\{(u_i, u_j) \in E \mid i \neq j\}|$ is the number of internal edges, namely, having both endpoints in $V[S]$.

Throughout the paper, we identify D with both *(1)* the set of k edges forming it as a k-tree and *(2)* its representation above as a certificate. We also support the following operations on D, under the requirement that $\mathtt{is\_unary}$ is true (i.e. all the edges in the cutset $C(S)$ are tree edges), otherwise they are undefined:

– $\mathtt{treecut}(D)$ returns the last edge in L.
– $\mathtt{promote}(r, D)$, where root r is the last in the doubly-linked list for F: remove r from F and replace r with its children $r_1, r_2, \dots, r_c$ (if any) in that list, so they become the new roots (and so L is updated).

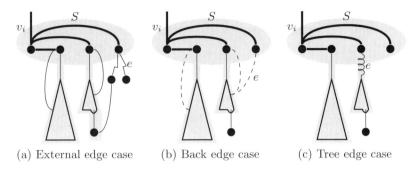

(a) External edge case   (b) Back edge case   (c) Tree edge case

Fig. 2. Choosing edge $e \in C(S)$. The certificate D is shadowed

Lemma 8. *The representation of certificate $D = S \cup L \cup F$ requires $O(|D|) = O(k)$ memory words, and $\mathtt{mdfs}_k(S)$ can build D in $O(k^2)$ time. Moreover, each of the operations $\mathtt{treecut}$ and $\mathtt{promote}$ can be supported in $O(1)$ time.*

4.2   Maintaining the Invariant Using a New choose

We now define $\mathtt{choose}$ in a more refined way to facilitate the task of maintaining the invariant $S \sqsubseteq D$ introduced in Section 4.1. As an intuition, $\mathtt{choose}$ selects an edge $e = (e^-, e^+)$ from the *cutlist* $\hat{C}(S)$ that interferes as least as possible with the certificate D. Recalling that $e^- \in V[S]$ and $e^+ \in V - V[S]$ by definition of cutlist, we consider the following case analysis:

(a) [external edge] Check if there exists an edge $e \in \hat{C}(S)$ such that $e \notin D$ and $e^+ \notin V[D]$. If so, return e, shown as a saw in Figure 2(a).
(b) [back edge] Otherwise, check if there exists an edge $e \in \hat{C}(S)$ such that $e \notin D$ and $e^+ \in V[D]$. If so, return e, shown dashed in Figure 2(b).
(c) [tree edge] As a last resort, every $e \in \hat{C}(S)$ must be also $e \in D$ (i.e. all edges in the cutlist are tree edges). Return $e := \mathtt{treecut}(D)$, the last edge from $\hat{C}(S)$, shown as a coil in Fig. 2(c).

Lemma 9. *For a given k'-tree S, consider its corresponding node in the recursion tree. Then, this node is binary when $\mathtt{choose}$ returns an external or back edge (cases (a)–(b)) and is unary when $\mathtt{choose}$ returns a tree edge (case (c)).*

We now present the new listing approach in Algorithm 3. If the connected component of vertex v_i in the residual graph is smaller than k, we delete its vertices since they cannot provide k-trees, and so we skip them in this way. Otherwise, we launch the new version of $\mathtt{ListTrees}_{v_i}$, shown in Algorithm 4. In comparison with the previous version (Algorithm 2), *we produce the new certificate D' from the current D in $O(1)$ time when we are in a unary node*. On the other hand, we completely rebuild the certificate twice when we are in a binary nodes (since either child could be unary at the next recursion level).

Lemma 10. *Algorithm 4 correctly maintains the invariant $S \sqsubseteq D$.*

Algorithm 3. ListAllTrees($G = (V, E)$, k)

---

    for $v_i \in V$:

1. $S := \langle (\cdot, v_i) \rangle$
2. $D := \mathtt{mdfs}_k(S)$
3. if $|D| < k$ then
    (a) for $u \in V[D]$: $\mathtt{del}(u)$.
4. else
    (a) $\mathtt{ListTrees}_{v_i}(S, D)$
    (b) $\mathtt{del}(v_i)$

Algorithm 4. $\mathtt{ListTrees}_{v_i}(S, D)$ *{Invariant: $S \sqsubseteq D$}*

---

1. if $|S| = k$ then:
    (a) $\mathtt{output}(S)$
    (b) $\mathtt{return}$
2. $e := \mathtt{choose}(S, D)$
3. if $\mathtt{is\_unary}$:
    (a) $D' := \mathtt{promote}(e^+, D)$
    (b) $\mathtt{ListTrees}_{v_i}(S + \langle e \rangle, D')$
4. else:
    (a) $D' := \mathtt{mdfs}_k(S + \langle e \rangle)$
    (b) $\mathtt{ListTrees}_{v_i}(S + \langle e \rangle, D')$
    (c) $\mathtt{del}(e)$
    (d) $D'' := \mathtt{mdfs}_k(S)$
    (e) $\mathtt{ListTrees}_{v_i}(S, D'')$
    (f) $\mathtt{undel}(e)$

4.3   Analysis

We implement $\mathtt{choose}(S, D)$ so that it can now exploit the information in D. At each node S of the recursion tree, when it selects an edge e that belongs to the cutset $C(S)$, it first considers the edges in $C(S)$ that are external or back (cases (a)–(b)) before the edges in D (case (c)).

Lemma 11. *There is an implementation of* $\mathtt{choose}$ *in* $O(1)$ *for unary nodes in the recursion tree and* $O(k^2)$ *for binary nodes.*

Lemma 12. *Algorithm 4 takes* $O(s_i k^2)$ *time and* $O(mk)$ *space, where* s_i *is the number of k-trees reported by* $\mathtt{ListTrees}_{v_i}$.

Theorem 13. *Algorithm 3 solves Problem 1 in* $O(sk^2)$ *time and* $O(mk)$ *space.*

Proof. The vertices belonging to the connected components of size less than k in the residual graph, now contribute with $O(m)$ total time rather than $O(nk^2)$. The rest of the complexity follows from Lemma 12.   □

5   Optimal Approach: Amortization

In this section, we discuss how to adapt Algorithm 4 so that a more careful analysis can show that it takes $O(sk)$ time to list the k-trees. Considering $\mathtt{ListTrees}_{v_i}$, observe that each of the $O(s_i k)$ unary nodes requires a cost of $O(1)$ time and therefore they are not much of problem. On the contrary, each of the $O(s_i)$ binary nodes takes $O(k^2)$ time: our goal is to improve over this situation.

Consider the operations on a binary node S of the recursion tree that take $O(k^2)$ time, namely: (I) $e := \mathtt{choose}(S, D)$; (II) $D' := \mathtt{mdfs}_k(S')$, where $S' \equiv S + \langle e \rangle$; and (III) $D'' := \mathtt{mdfs}_k(S)$ in $G - \{e\}$. In all these operations, while scanning the adjacency lists of vertices in $V[S]$, we visit some edges $e' = (u, v)$, named *internal*, such that $e' \notin S$ with $u, v \in V[S]$. These internal edges of $V[S]$ can be visited even if they were previously visited on an ancestor node to S. In Section 5.1, we show how to amortize the cost induced by the internal edges. In Section 5.2, we show how to amortize the cost induced by the remaining edges and obtain a delay of $t(k) = k^2$ in our optimal output-sensitive algorithm.

5.1   Internal Edges of $V[S]$

To avoid visiting the internal edges of $V[S]$ several times throughout the recursion tree, we remove these edges from the graph G on the fly, and introduce a global data structure, which we call *parking lists*, to store them temporarily. Indeed, out of the possible $O(n)$ incident edges in vertex $u \in V[S]$, less than k are internal: it is simply too costly removing these internal edges by a complete scan of $\mathtt{adj}(u)$. Therefore we remove them as they appear while executing $\mathtt{choose}$ and $\mathtt{mdfs}_k$ operations.

Formally, we define *parking lists* as a global array P of n pointers to lists of edges, where $P[u]$ is the list of internal edges discovered for $u \in V[S]$. When $u \notin V[S]$, $P[u]$ is null. On the implementation level, we introduce a slight modification of the $\mathtt{choose}$ and $\mathtt{mdfs}_k$ algorithms such that, when they meet for the first (and only) time an internal edge $e' = (u, v)$ with $u, v \in V[S]$, they perform $\mathtt{del}(e')$ and add e' at the end of both parking lists $P[u]$ and $P[v]$. We also keep a cross reference to the occurrences of e' in these two lists.

Additionally, we perform a small modification in algorithm $\mathtt{ListTrees}_{v_i}$ by adding a fifth step in Algorithm 4 just before it returns to the caller. Recall that on the recursion node $S + \langle e \rangle$ with $e = (e^-, e^+)$, we added the vertex e^+ to $V[S]$. Therefore, when we return from the call, all the internal edges incident to e^+ are no longer internal edges (and are the only internal edges to change status). On this new fifth step, we scan $P[e^+]$ and for each edge $e' = (e^+, x)$ in it, we remove e' from both $P[e^+]$ and $P[x]$ in $O(1)$ time using the cross reference. Note that when the node is unary there are no internal edges incident to e^+, so $P[e^+]$ is empty and the total cost is $O(1)$. When the node is binary, there are at most $k - 1$ edges in $P[e^+]$, so the cost is $O(k)$.

Lemma 14. *The operations over internal edges done in* $\mathtt{ListTrees}_{v_i}$ *have a total cost of* $O(s_i k)$ *time.*

5.2   Amortization

Let us now focus on the contribution given by the remaining edges, which are not internal for the current $V[S]$. Given the results in Section 5.1, for the rest of this section *we can assume wlog that there are no internal edges* in $V[S]$, namely, $E[S] = S$. We introduce two metrics that help us to parameterize the time complexity of the operations done in binary nodes of the recursion tree.

The first metric we introduce is helpful when analyzing the operation `choose`. For connected edge sets S and X with $S \sqsubseteq X$, define the *cut number* γ_X as the number of edges in the induced (connected) subgraph $G[X] = (V[X], E[X])$ that are in the cutset $C(S)$ (i.e. tree edges plus back edges): $\gamma_X = |E[X] \cap C(S)|$.

Lemma 15. *For a binary node S with certificate D, `choose`(S, D) takes $O(k + \gamma_D)$ time.*

For connected edge sets S and X with $S \sqsubseteq X$, the second metric is the *cyclomatic number* ν_X (also known as circuit rank, nullity, or dimension of cycle space) as the smallest number of edges which must be removed from $G[X]$ so that no cycle remains in it: $\nu_X = |E[X]| - |V[X]| + 1$.

Using the cyclomatic number of a certificate D (ignoring the internal edges of $V[S]$), we obtain a lower bound on the number of k-trees that are output in the leaves descending from a node S in the recursion tree.

Lemma 16. *Considering the cyclomatic number ν_D and the fact that $|V[D]| = k$, we have that $G[D]$ contains at least ν_D k-trees.*

Lemma 17. *For a node S with certificate D, computing $D' = $ `mdfs`$_k(S)$ takes $O(k + \nu_{D'})$ time.*

Recalling that the steps done on a binary node S with certificate D are: (I) $e := $ `choose`(S, D); (II) $D' := $ `mdfs`$_k(S')$, where $S' \equiv S + \langle e \rangle$; and (III) $D'' := $ `mdfs`$_k(S)$ in $G - \{e\}$, they take a total time of $O(k + \gamma_D + \nu_{D'} + \nu_{D''})$. We want to pay $O(k)$ time on the recursion node S and amortize the *rest* of the cost to some suitable nodes descending from its *left child* S' (with certificate D'). To do this we are to relate γ_D with $\nu_{D'}$ and avoid performing step (III) in $G - \{e\}$ by maintaining D'' from D'. We exploit the property that the cost $O(k + \nu_{D'})$ for a node S in the recursion tree can be amortized using the following lemma:

Lemma 18. *Let S' be the left child (with certificate D') of a generic node S in the recursion tree. The sum of $O(\nu_{D'})$ work, over all left children S' in the recursion tree is upper bounded by $\sum_{S'} \nu_{D'} = O(s_i k)$.*

Proof. By Lemma 16, S' has at least $\nu_{D'}$ descending leaves. Charge $O(1)$ to each leaf descending from S' in the recursion tree. Since S' is a left child and we know that the recursion tree is k-left-bounded by Lemma 4, each of the s_i leaves can be charged at most k times, so $\sum_{S'} \nu_{D'} = O(s_i k)$ for all such S'. □

We now show how to amortize the $O(k + \gamma_D)$ cost of step (I). Let us define $comb(S')$ for a left child S' in the recursion tree as its maximal path to the right (its right spine) and the left child of each node in such a path. Then, $|comb(S')|$ is the number of such left children.

Lemma 19. *On a node S in the recursion tree, the cost of* choose(S, D) *is* $O(k + \gamma_D) = O(k + \nu_{D'} + |comb(S')|)$.

Proof. Consider the set E' of γ_D edges in $E[D] \cap C(S)$. Take D', which is obtained from $S' = S + \langle e \rangle$, and classify the edges in E' accordingly. Given $e' \in E'$, one of three possible situations may arise: either e' becomes a tree edge part of D' (and so it contributes to the term k), or e' becomes a back edge in $G[D']$ (and so it contributes to the term $\nu_{D'}$), or e' becomes an external edge for D'. In the latter case, e' will be chosen in one of the subsequent recursive calls, specifically one in $comb(S')$ since e' is still part of $C(S')$ and will surely give rise to another k-tree in a descending leaf of $comb(S')$. □

While the $O(\nu_{D'})$ cost over the leaves of S' can be amortized by Lemma 17, we need to show how to amortize the cost of $|comb(S')|$ using the following:

Lemma 20. $\sum_{S'} |comb(S')| = O(s_i k)$ *over all left children S' in the recursion.*

At this point we are left with the cost of computing the two mdfs$_k$'s. Note that the cost of step (II) is $O(k + \nu_{D'})$, and so is already expressed in terms of the cyclomatic number of its left child, $\nu_{D'}$ (so we use Lemma 16). The cost of step (III) is $O(k + \nu_{D''})$, expressed with the cyclomatic number of the certificate of its *right* child. This cost is not as easy to amortize since, when the edge e returned by choose is a back edge, D' of node $S + \langle e \rangle$ can change *heavily* causing D' to have just S in common with D''. This shows that $\nu_{D''}$ and $\nu_{D'}$ are not easily related.

Nevertheless, note that D and D'' are the same certificate since we only remove from G an edge $e \notin D$. The only thing that can change by removing edge $e = (e^-, e^+)$ is that the right child of node S' is no longer binary (i.e. we removed the last back edge). The question is if we can check quickly whether it is unary in $O(k)$ time: observe that $|\text{adj}(e^-)|$ is no longer the same, invalidating the flag is_unary (item (iv) of Section 4.1). Our idea is the following: instead of recomputing the certificate D'' in $O(k + \nu_{D''})$ time, we update the is_unary flag in just $O(k)$ time. We thus introduce a new operation $D'' = $ unary(D), a valid replacement for $D'' = $ mdfs$_k(D)$ in $G - \{e\}$: it maintains the certificate while recomputing the flag is_unary in $O(k)$ time.

Lemma 21. *Operation* unary(D) *takes $O(k)$ time and correctly computes D''.*

Since there is no modification or impact on unary nodes of the recursion tree, we finalize the analysis.

Lemma 22. *The cost of* ListTrees$_{v_i}(S, D)$ *on a binary node S is $O(k + \nu_{D'} + |comb(S')|)$.*

Lemma 23. *The algorithm* ListTrees$_{v_i}$ *takes $O(s_i k)$ time and $O(mk)$ space.*

Note that our data structures are lists and array, so it is not difficult to replace them with *persistent* arrays and lists, a classical trick in data structures. As a result, we just need $O(1)$ space per pending recursive call, plus the space of the parking lists, which makes a total of $O(m)$ space.

Theorem 24. *Algorithm 3 takes a total of $O(sk)$ time, being therefore optimal, and $O(m)$ space.*

We finally show how to obtain an efficient delay. We exploit the following property on the recursion tree, which allows to associate a unique leaf with an internal node *before* exploring the subtree of that node (recall that we are in a recursion tree). Note that only the rightmost leaf descending from the root is not associated in this way, but we can easily handle this special case.

Lemma 25. *For a binary node S in the recursion tree, $\mathtt{ListTrees}_{v_i}(S, D)$ outputs the k-tree D in the rightmost leaf descending from its left child S'.*

Nakano and Uno [8] have introduced this nice trick. Classify a binary node S in the recursion tree as even (resp., odd) if it has an even (resp., odd) number of ancestor nodes that are binary. Consider the simple modification to $\mathtt{ListTrees}_{v_i}$ when S is binary: if S is even then output D *immediately before* the two recursive calls; otherwise (S is odd), output D *immediately after* the two recursive calls.

Theorem 26. *Algorithm 3 can be implemented with delay $t(k) = k^2$.*

References

1. Alm, E., Arkin, A.P.: Biological networks. Current Opinion in Structural Biology 13(2), 193–202 (2003)
2. Avis, D., Fukuda, K.: Reverse search for enumeration. Discrete Applied Mathematics 65(1-3), 21–46 (1996)
3. Gabow, H.N., Myers, E.W.: Finding all spanning trees of directed and undirected graphs. SIAM Journal on Computing 7(3), 280–287 (1978)
4. Kapoor, S., Ramesh, H.: Algorithms for enumerating all spanning trees of undirected and weighted graphs. SIAM Journal on Computing 24, 247–265 (1995)
5. Milo, R., Shen-Orr, S., Itzkovitz, S., Kashtan, N., Chklovskii, D., Alon, U.: Network motifs: Simple building blocks of complex networks. Science 298, 824–827 (2002)
6. Minty, G.: A simple algorithm for listing all the trees of a graph. IEEE Transactions on Circuit Theory 12(1), 120 (1965)
7. Moon, J.: Counting Labelled Trees, Canadian Mathematical Monographs, No. 1. Canadian Mathematical Congress, Montreal (1970)
8. Nakano, S.I., Uno, T.: Constant time generation of trees with specified diameter. In: Hromkovič, J., Nagl, M., Westfechtel, B. (eds.) Graph -Theoretic Concepts in Computer Science. LNCS, vol. 3353, pp. 33–45. Springer, Heidelberg (2004)
9. Read, R.C., Tarjan, R.E.: Bounds on backtrack algorithms for listing cycles, paths, and spanning trees. Networks (1975)
10. Shioura, A., Tamura, A., Uno, T.: An optimal algorithm for scanning all spanning trees of undirected graphs. SIAM Journal on Computing 26, 678–692 (1994)

Exact Algorithm for the Maximum Induced Planar Subgraph Problem

Fedor V. Fomin[1], Ioan Todinca[2,*], and Yngve Villanger[1]

[1] Department of Informatics, University of Bergen, N-5020 Bergen, Norway
{fomin,yngvev}@ii.uib.no
[2] LIFO, Université d'Orléans, BP 6759 45067 Orléans cedes, France
Ioan.Todinca@univ-orleans.fr

Abstract. We prove that in an n-vertex graph, an induced planar subgraph of maximum size can be found in time $O(1.7347^n)$. This is the first algorithm breaking the trivial $2^n n^{O(1)}$ bound of the brute-force search algorithm for the Maximum Induced Planar Subgraph problem.

1 Introduction

The theory of exact exponential algorithms for hard problems is about the study of algorithms which are better than the trivial exhaustive search, though still exponential [6]. The area is still in a nascent stage and lacks generic tools to identify large classes of NP-complete problems solvable faster than by brute-force search. Maximum Induced Subgraph with Property π, where for a given graph G and hereditary property π one asks for a maximum induced subgraph with property π, defines a large class of problems depending on π. By the result of Lewis and Yannakakis [16], the problem is NP-complete for every non-trivial property π. Different variants of property π like being edgeless, planar, outerplanar, bipartite, complete bipartite, acyclic, degree-constrained, chordal, etc. graph, were studied in the literature, mainly from (in)approximability perspective. As far as property π can be tested in polynomial time, a trivial brute-force search approach trying all possible vertex subsets of G, and selecting a subset of maximum size such that $G[W]$ has property π, solves Maximum Induced Subgraph with Property π in time $2^n n^{O(1)}$ on an n-vertex graph G. The are several examples of problems solvable faster than $2^n n^{O(1)}$: Maximum Independent Set where π means being edgeless graph [18], Maximum Induced Forest where π is being acyclic [19], π being regular [14], being of small treewidth [9], 2 or 3-colorable [1], biclique [10], with a forbidden subgraph [12].

Here, we give the first algorithm breaking the trivial 2^n-barrier for property π being a planar graph. In the Maximum Induced Planar Subgraph we are given a graph G and the task is to find a vertex subset W of maximum size such that the subgraph of G induced by W is planar. This is a classical NP-complete

---

* Partially supported by French National Research Agency (ANR) project AGAPE (ANR-09-BLAN-0159-03).

C. Demetrescu and M.M. Halldórsson (Eds.): ESA 2011, LNCS 6942, pp. 287–298, 2011.
© Springer-Verlag Berlin Heidelberg 2011

problem (GT21 in Garey and Johnson [11]), which has been studied intensively, see the survey of Liebers [17] on related topics. The main result of this paper is the following theorem.

Theorem 1. MAXIMUM INDUCED PLANAR SUBGRAPH *on n-vertex graphs is solvable in time* $O(1.7347^n)$.

Overview of our techniques. To prove Theorem 1, we use the approach built on ideas about minimal triangulations and potential maximal cliques introduced in [3]. We believe that dynamic programing on potential maximal cliques is an interesting tool for exact algorithms. It was shown in [9] that a maximum induced subgraph of treewidth t in an n-vertex graph G can be found in time $O(|\Pi_G| \cdot n^{O(t)})$, where $|\Pi_G|$ is the number of potential maximal cliques in G. By [9], $|\Pi_G| \leq 1.734601^n$. While the treewidth of a planar graph is $O(\sqrt{n})$, unfortunately, the algorithm from [9] cannot be used to obtain a planar graph. Loosely speaking, that algorithm "glues" together, in dynamic programming fashion, large graphs via vertex sets of small size, which are potential maximal cliques and minimal separators. While "essential" potential maximal cliques and separators in planar graphs are still "small", i.e. of size $O(\sqrt{n})$, the result of gluing two planar graphs even via a small vertex set is not necessarily planar. Therefore the approach from [9] cannot be used directly. To overcome this problem, we have to use specific topological properties of potential maximal cliques in planar graphs.

2   Preliminaires

We denote by $G = (V, E)$ a finite, undirected and simple graph with $|V| = n$ vertices and $|E| = m$ edges. Sometimes the vertex set of a graph G is referred to as $V(G)$ and its edge set as $E(G)$. A clique K in G is a set of pairwise adjacent vertices of $V(G)$. For the purposes of this paper it is convenient to assume that the empty set is also a clique. We denote by $\omega(G)$ the maximum clique size in G. The *neighborhood* of a vertex v is $N(v) = \{u \in V : \{u, v\} \in E\}$. For a vertex set $S \subseteq V$ we denote by $N(S)$ the set $\bigcup_{v \in S} N(v) \setminus S$.

Plane graphs, nooses, and θ-structures In this paper we use the expression *plane graph* for any planar graph drawn in a sphere Σ. We do not distinguish between a vertex of a plane graph and the point of Σ used in the drawing to represent the vertex or between an edge and the open line segment representing it. We also consider a plane graph G as the union of the points corresponding to its vertices and edges. We call by *face* of G any connected component of $\Sigma \setminus (E(G) \cup V(G))$. Let G be a plane graph drawn in a sphere Σ. A subset of Σ meeting the drawing only in vertices of G is called *G-normal*. A subset of Σ homeomorphic to the closed interval $[0, 1]$ is called *I-arc*. If the ends of a G-normal I-arc are both vertices of G, then we call it *line* of G. If a simple closed curve $F \subseteq \Sigma$ is G-normal, then we call it *noose*.

Let $x, y \in \Sigma$ be distinct. We call θ-*structure* (L_1, L_2, L_3) of G the union of three internally disjoint lines L_1, L_2, L_3 between vertices x and y of G. Thus for $i, j, 1 \leq i < j \leq 3$, $L_i \cup L_j$ is a noose.

Tree decompositions. The notion of treewidth is due to Robertson and Seymour [20]. A *tree decomposition* of a graph $G = (V, E)$, denoted by $TD(G)$, is a pair (X, T), where T is a tree and $X = \{X_i \mid i \in V(T)\}$ is a family of subsets of V, called *bags*, such that (i) $\bigcup_{i \in V(T)} X_i = V$, (ii) for each edge $e = \{u, v\} \in E(G)$ there exists $i \in V(T)$ such that both u and v are in X_i, and (iii) for all $v \in V$, the set of nodes $\{i \in V(T) \mid v \in X_i\}$ induces a connected subtree of T. The maximum of $|X_i| - 1$, $i \in V(T)$, is called the *width* of the tree decomposition. The *treewidth* of a graph G, denoted by $\mathbf{tw}(G)$, is the minimum width taken over all tree decompositions of G.

Chordal graphs and clique trees. A graph H is *chordal* (or *triangulated*) if every cycle of length at least four has a chord, i.e., an edge between two nonconsecutive vertices of the cycle. By a classical result due to Buneman and Gavril [4,13], every chordal graph G has a tree decomposition $TD(G)$ such that every bag of the decomposition is a maximal clique of G. Such a tree decomposition is referred as a *clique tree* of the chordal graph G.

Minimal triangulations, potential maximal cliques and minimal separators. A *triangulation* of a graph $G = (V, E)$ is a chordal graph $H = (V, E')$ such that $E \subseteq E'$. Graph H is a *minimal triangulation* of G if for every edge set E'' with $E \subseteq E'' \subset E'$, the graph $F = (V, E'')$ is not chordal. It is well known that for any graph G, $\mathbf{tw}(G) \leq k$ if and only if there is a triangulation H of G such that $\omega(H) \leq k + 1$.

Let u and v be two non adjacent vertices of a graph G. A set of vertices $S \subseteq V$ is a u, v-*separator* if u and v are in different connected components of the graph $G[V(G) \setminus S]$. A connected component C of $G[V(G) \setminus S]$ is a *full* component associated to S if $N(C) = S$. Separator S is a *minimal u, v-separator* of G if no proper subset of S is a u, v-separator. Notice that a minimal separator can be strictly included in another one, if they are minimal separators for different pairs of vertices. If G is chordal, then for any minimal separator S and any clique tree T_G of G there is an edge e of T_G such that S is the intersection of the maximal cliques corresponding to endpoints of e [4,13]. We say that S *corresponds* to e in T_G.

A set of vertices $\Omega \subseteq V(G)$ of a graph G is called a *potential maximal clique* if there is a minimal triangulation H of G such that Ω is a maximal clique of H. We denote by Π_G the set of all potential maximal cliques of G. By [3], a subset $S \subseteq \Omega$ is a minimal separator of G if and only if S is the neighborhood of a connected component of $G[V(G) \setminus \Omega]$.

For a minimal separator S and a full connected component C of $G \setminus S$, we say that (S, C) is a *block* associated to S. We sometimes use the notation (S, C) to denote the set of vertices $S \cup C$ of the block. It is easy to see that if $X \subseteq V$ corresponds to the set of vertices of a block, then this block (S, C) is unique:

indeed, $S = N(V \setminus X)$ and $C = X \setminus S$. A block (S, C) is called *full* if C is a full component associated to S. For convenience, the couple $(\emptyset, V)$ is also considered as a full block. For a minimal separator S, a full block (S, C), and a potential maximal clique Ω, we call the triple (S, C, Ω) *good* if $S \subseteq \Omega \subseteq C \cup S$.

Let H_G be a minimal triangulation of a graph G. A clique tree T_G of H_G is called a *connected tree decomposition* of G if it is rooted and, for every node i of T_G, the union of bags of $T_G[i]$ (the subtree of T_G rooted in i) is a full block of G. Such a connected tree decomposition exists for any minimal triangulation of G.

We need the following result about connected tree decompositions. The proof is omitted due to space restrictions.

Lemma 1. *Let F be an induced subgraph of a graph G, let H_F be a minimal triangulation of F. There exists a minimal triangulation H_G of G such that H_F is an induced subgraph of H_G. For any clique tree T_G of H_G, there exists a clique tree T_F of H_F and a surjective map $\mu: V(T_G) \to V(T_F)$ mapping maximal cliques of H_G to maximal cliques of H_F such that:*

1. *For every maximal clique K_F of H_F the vertex set $\mu^{-1}(K_F)$ induces a connected subtree T_{K_F} in T_G;*
2. *For every maximal cliques K_F of H_F and $K_G \in \mu^{-1}(K_F)$, we have that $V(F) \cap K_G \subseteq K_F$;*
3. *Two nodes representing K_F^1 and K_F^2 of T_F are adjacent if and only if the subtrees $T_{K_F^1}$ and $T_{K_F^2}$ of T_G are connected by an edge e in T_G. Moreover, for the minimal separator S_G corresponding to e in T_G, we have that $S_F = V(F) \cap S_G = K_F^1 \cap K_F^2$ is a minimal separator of H_F.*

Lemma 1 holds for *any* clique tree T_G of H_G. In particular, in the proof of the main theorem, we use connected tree decompositions T_G.

We also need the following result from [8].

Proposition 1 (Fomin and Thilikos). *The treewidth of a planar graph on n vertices is at most $c\sqrt{n}$ for some constant $c \le 1.5 \cdot \sqrt{4.5} < 3.182$.*

By Proposition 1, every planar graph F has a minimal triangulation H_F of treewidth at most $3.182\sqrt{|V(F)|}$. Thus every clique in H_F is of size at most $3.182\sqrt{|V(F)|}$. Because of that we can assume that minimal triangulation H_F of F used in Lemma 1 is of treewidth at most $3.182\sqrt{n}$.

3   Minimal Triangulations of Planar Graphs

In this section, we prove a number of structural and topological results on the potential maximal cliques of plane graphs. Since in this paper F is a planar subgraph of graph G, in this section we keep notation F to denote a plane graph which is drawn in a sphere Σ. Eppstein [5] shows that for any minimal separator S of F, there exists a noose N meeting vertices of F exactly in S, i.e. $S = N \cap F$. Moreover, any two vertices adjacent in $F[S]$ appear consecutively in N.

Bouchitté et al. [2] proved that every potential maximal clique in a 2-connected plane graph corresponds to a θ-structure. We show in the full version that this 2-connectivity condition can be omitted.

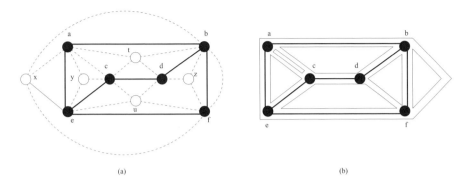

Fig. 1. Part (a) depicts a planar graph (in thin, dashed lines) and a θ-structure (bold lines) corresponding to potential maximal clique $\Omega = \{a, b, c, d, e, f\}$. The cyclic orderings of the θ-structure are $[e, a, b, d, c]$, $[e, c, d, b, f]$ and $[e, f, b, a]$. Part (b) illustrates the nooses corresponding to minimal separators $S_i \subset \Omega$ and to endpoints of edges in $F[\Omega]$.

Lemma 2. *Let H_F be a minimal triangulation of a plane graph F. For every maximal clique Ω of H_F, there is a θ-structure $\theta(\Omega)$ such that $\Omega = \theta(\Omega) \cap F$.*

Let Ω be a potential maximal clique of F. For every connected component C_i of $F \setminus \Omega$, the set $S_i = N(C_i)$ is a minimal separator. Since minimal separators correspond to nooses, it is easy to show that for every C_i there is noose N_i meeting F only in $S_i = N(C_i)$, and for each edge e_j of $F[\Omega]$ a noose meeting F in the endpoints of e_j, such that all these nooses are pairwise non-crossing and do not cross $\theta(\Omega)$. The situation of Lemma 2 is illustrated in Figure 1.

For our purpose, the topological structure of nooses corresponding to minimal separators or θ-structures corresponding to potential maximal cliques is not so crucial. Of much more importance are the orderings induced by the lines of θ-structures. Let a θ-structure corresponding to potential maximal clique Ω be formed by three lines L_1, L_2, L_3 connecting vertices x and y. The three unions $L_i \cup L_j$, $i \neq j$, $1 \leq i, j \leq 3$, form three nooses N_1, N_2, and N_3. Traversing each of the nooses clockwise, we obtain three sequences of vertices, $\mathcal{S}_1 = x, v_1^1, \ldots, v_p^1, y, v_r^2, \ldots, v_1^2$; $\mathcal{S}_2 = x, v_1^2, \ldots, v_r^2, y, v_s^3, \ldots, v_1^3$; $\mathcal{S}_3 = x, v_1^1, \ldots, v_p^1, y,$ $v_s^3, \ldots, v_1^3$, where for each $1 \leq i \leq 3$, vertices v_j^i are the vertices of Ω met by noose N_i. We refer to $(\mathcal{S}_1, \mathcal{S}_2, \mathcal{S}_3)$ as to the *cyclic orderings* of $\theta(\Omega)$. Let us remark that given three sequences of vertices of a plane graph, it is easy to check in polynomial time if these are the cyclic orderings of some θ-structure.

Let $(\mathcal{S}_1, \mathcal{S}_2, \mathcal{S}_3)$ be the cyclic orderings of a θ-structure $\theta(\Omega)$ for some potential maximal clique Ω_F of plane graph F drawn in Σ. The subset of $\Sigma \setminus \theta(\Omega)$ consists of three open regions R_1, R_2, R_3, each region homeomorphic to an open disc. Each of R_i, $1 \leq i \leq 3$, is bordered by noose N_i consisting of two lines of $\theta(\Omega)$. Let $\mathcal{F} = \{F_1, F_2, \ldots, F_r\}$ be the connected components of the graph $F[V(F) \setminus \Omega_F]$. In the drawing of F, each of the connected component of $\mathcal{F}$ is drawn in some R_i, $1 \leq i \leq 3$, and we refer to this component as to a *component assigned* to R_i. If two planar components F_k and F_ℓ are assigned to the same region R_i, then the vertices $N(F_k)$, i.e. the vertices of N_i adjacent to F_k, and the vertices of $N(F_\ell)$ are *non-crossing*. More formally, two vertex subsets X and Y of noose N_i are *non-crossing* if there are two vertices v and w on the noose, splitting it into two subpaths $v, x_1, \ldots, x_q, w$ and $v, y_1, \ldots, y_r, w$ such that X and Y are contained in different subpaths. Note that X and Y may share vertices v and w, but there are no four vertices $a < b < c < d$ in ordering $\mathcal{S}_i$ such that $a, c \in X$ and $b, d \in Y$. The number of non-crossing subsets can be bounded via the number of non-crossing partitions, i.e. sets whose union is $N_i \cap F$. Non-crossing partitions were introduced by Kreweras [15], who showed that the number of non-crossing partitions over n vertices is equal to the n-th Catalan number: $\mathrm{CN}(n) = \frac{1}{n+1}\binom{2n}{n} \sim \frac{4^n}{\sqrt{\pi}n^{\frac{3}{2}}} \approx 4^n$. Thus the number of non-crossing subsets is at most $\sum_{i=1}^{n} \binom{n}{i} \cdot \mathrm{CN}(i) \approx 5^n$.

For each of the three regions of a θ-structure, the neighborhoods of the components of $\mathcal{F}$ should form non-crossing sets. This brings us to the following definition.

Definition 1 (Neighborhood Assignment). *Let $\theta(\Omega_F)$ be a θ-structure on Ω_F and let $(\mathcal{S}_1, \mathcal{S}_2, \mathcal{S}_3)$ be the cyclic orderings corresponding to it. For each $i \in \{1, 2, 3\}$, let $\mathcal{N}_i$ be a set of pairwise non-crossing subsets of $\mathcal{S}_i$, called the neighborhoods assigned to $\mathcal{S}_i$ and such that for each edge $\{u, v\} \in F[\Omega_F]$, there is some $\mathcal{N}_\ell$ containing the pair $\{u, v\}$. The triple $(\mathcal{N}_1, \mathcal{N}_2, \mathcal{N}_3)$ is called a neighborhood assignment for $\theta(\Omega_F)$. If each $\mathcal{N}_i$ is a maximal set of pairwise non-crossing subsets of $\mathcal{S}_i$, then $(\mathcal{N}_1, \mathcal{N}_2, \mathcal{N}_3)$ is called a maximal neighborhood assignment for $\theta(\Omega_F)$ and denoted by $[\theta(\Omega_F)]$.*

Let $W \subseteq \Omega_F$. For each $i \in \{1, 2, 3\}$ let $\mathcal{N}_i'$ be a subset of $\mathcal{N}_i$ such that every element of $\mathcal{N}_i'$ is a subset of W. We denote the triple $(\mathcal{N}_1', \mathcal{N}_2', \mathcal{N}_3')$ by $([\theta(\Omega_F)], [W])$ and refer to it as to a partial neighborhood assignment for $(\theta(\Omega_F), W)$.

The technical condition that for each edge of $F[\Omega_F]$ its endpoints appear as an element of the neighborhood assignment is crucial when we reconstruct F by gluing planar subgraphs that are attached to Ω_F according to its neighborhood assignment. Let us also remark that the number of neighborhood assignments for a given θ-structure $\theta(\Omega_F)$ is $2^{O(|\Omega_F|)}$. Indeed, for each of the three regions, there are at most $5^{|\Omega_F|}$ non-crossing subsets, resulting in at most $5^{3|\Omega_F|} = 2^{O(|\Omega_F|)}$ neighborhood assignments.

For every fixed set W, the number of partial neighborhood assignments for $(\theta(\Omega_F), W)$ is $2^{O(|\Omega_F|)}$ and because the number of subsets of Ω_F is $2^{|\Omega_F|}$, we have that the number of partial neighborhood assignments for Ω_F is $2^{O(|\Omega_F|)}$. Summing up we conclude with the technical lemma which will be used in the analysis of our algorithm.

Lemma 3. *Let F be a plane graph on n vertices. The number of all possible partial neighborhood assignments $([\theta(\Omega_F], [W])$ over all vertex subsets Ω_F of F of size at most $3.182\sqrt{n}$, all θ-structures on Ω_F and all subsets W of Ω_F is $2^{o(n)}$.*

4   Proof of Theorem 1

In this section we prove the main result of the paper, Theorem 1. We first provide the algorithm solving MAXIMUM INDUCED PLANAR SUBGRAPH, then prove its correctness and analyse its running time. As in [9], we proceed by dynamic programming, using all the potential maximal cliques of a graph G. In our dynamic programming we do not only construct the final solution F, but also a minimal triangulation of the final solution, by gluing partial solutions and partial triangulations.

High level description of the algorithm. We start with the following preprocessing step. First we order all full blocks (S, C) of G by their sizes. For each full block we compute the set of its good triples (S, C, Ω), i.e. the set of triples where Ω is a potential maximal clique such that $S \subseteq \Omega \subseteq C \cup S$. For each good triple (S, C, Ω) we associate the corresponding full blocks (S_i, C_i) such that $S_i \subset \Omega$ and $C_i \subseteq C \setminus \Omega$. As it was shown in [7], this preprocessing step can be performed in time $O(n^3 |\Pi_G|)$.

We proceed by dynamic programming over full blocks. Let H_F and H_G be minimal triangulations of F and G as in Lemma 1. Although both triangulations H_F and H_G are not known, we know that maximal cliques of the minimal triangulations define a clique tree, and that each maximal clique of H_G and H_F is a potential maximal clique of the corresponding graph. Because F is planar, by Proposition 1, maximal cliques of H_F are of size at most $3.182\sqrt{n}$. We can enumerate all vertex subsets of G of size at most $3.182\sqrt{n}$ in time $O^*(\binom{n}{3.182\sqrt{n}})$ but we cannot conclude which of these sets are potential maximal cliques in F. To explain our algorithm, let us assume for a moment that we are given a connected tree decomposition T_G of G, as in Lemma 1 corresponding to tree decomposition T_F of maximum induced planar subgraph F. Of course we cannot assume that the decomposition T_G is given. However, it is possible by performing dynamic programming over all full blocks to "browse" through all tree decompositions of G corresponding to minimal triangulations, including T_G.

To compute the maximum induced planar subgraph F of G, we proceed in a dynamic programming fashion which is very similar to techniques used on graphs of bounded treewidth. Recall that the connected tree decomposition T_G is rooted. Let j be a node of the decomposition tree, and let $\Omega = X_j$ be the corresponding bag, which is also a potential maximal clique of G. By definition of a connected tree decomposition, the union of bags $T_G[j]$, corresponding to the subtree of T_G rooted in j, is a full block (S, C) of G. Our dynamic programming algorithm computes for each subset W of S the value $\alpha(S, C, W)$—the size of the largest planar induced subgraph of $G[S \cup C]$ intersecting S exactly in W, and for each subset W of Ω, the value $\beta(S, \Omega, C, W)$, which is the size of the largest planar induced subgraph of $G[S \cup C]$ intersecting Ω exactly in W. In order to compute $\beta(S, \Omega, C, W)$, we need the values $\alpha(S_i, C_i, W_i)$ of smaller blocks (S_i, C_i) corresponding to the subtrees $T_G[i]$, where i runs through the set of children of j in T_G. By [3], the components C_i are exactly the components of $G[C \setminus \Omega]$. The partial solution corresponding to $\beta(S, \Omega, C, W)$ is the union of

W and of partial solutions corresponding to $\alpha(S_i, C_i, W_i)$. The difficulty of such approach is in preserving planarity while gluing partial solutions. The key idea to resolve this difficulty is to use the characterization of potential maximal cliques of Lemma 2. By Lemma 2, W is a subset of a Ω_F of the final solution F, and gluing of partial solutions must proceed along the θ-structure of Ω_F. Therefore α and β also depend on several other parameters: the potential maximal clique Ω_F and its θ-structure, as well as a partial neighborhood assignments of the neighborhoods of connected components of $F[C \setminus \Omega_F]$ in $\theta(\Omega_F)$. The formal definitions of functions α and β are given in the next subsection.

We point out one more difficulty. Recall that the decomposition tree T_G can be partitioned into several subtrees T_G^a, such that for every bag Ω of a fixed subtree T_G^a, the intersection $\Omega \cap F$ is contained in the same clique Ω_F^a of H_F. The dynamic programming technique that we have mentioned above works fine as long as we stay inside a fixed subtree T_G^a. A special treatment is needed for the situation when we "switch" from a subtree T_G^a to a subtree T_G^b.

Eventually, in order to obtain the global solution on G, we simply have to notice that the full block associated to the root of the tree decomposition is $(\emptyset, V)$ and we return the largest value of α.

Formal description of the algorithm. We are now ready to formally define functions α and β. In the following definitions,

- (S, C) is a full block of G and (S, C, Ω) is a good triple;
- Ω_F is a vertex subset of G; $\theta(\Omega_F)$ is a θ-structure of Ω_F and $[\theta(\Omega_F)]$ is a maximal neighborhood assignment for $\theta(\Omega_F)$;
- W' denotes $\Omega_F \cap (S \cup C)$ and $([\theta(\Omega_F)], [W'])$ is a partial neighborhood assignment on $(\theta(\Omega_F), W')$;
- W is a vertex subset corresponding either to $\Omega_F \cap S$ (when we consider the α function) or to $\Omega_F \cap \Omega$ (when we consider the β function).

Definition 2. *Function $\alpha(S, C, \Omega_F, ([\theta(\Omega_F)], [W']), W)$ is the maximum size of a vertex set $C_F \subseteq C \cup W$ such that*

1. *$G[C_F]$ is planar,*
2. *$C_F \cap S = \Omega_F \cap S = W$*
3. *each connected component F_j of $G[C_F \setminus W']$ and each edge $\{x, y\}$ of $G[W']$ is mapped on a neighborhood N_i of $[W']$ where $N_i = N_G(F_j) \cap W'$ or $N_i = \{x, y\}$, such that each N_i of $[W']$ is mapped to at least one component or edge, and to exactly one if $3 \leq |N_i|$.*

Similarily, function $\beta(S, C, \Omega, \Omega_F, ([\theta(\Omega_F)], [W']), W)$ is the maximum size of vertex set $C_F \subseteq (C \setminus \Omega) \cup W$ satisfying condtions 1 and 3, and condition 2 is replaced by

2'. $C_F \cap \Omega = \Omega_F \cap \Omega = W$

If no such induced subgraph exists, the value is set to 0.

Observe that we distinguish neighborhoods of size at least 3 from neighborhood of size at most 2. Indeed on a neighborhood N_i of size at most 2 assigned to some cyclic ordering $\mathcal{S}_\ell$ we can "glue" an arbitrary number of planar graphs inside the face of the θ-structure corresponding to $\mathcal{S}_\ell$. On the contrary, if $3 \leq |N_i|$, we can glue at most one component.

Our algorithm will compute all possible α and β values, for each full block (S, C) by increasing size. For this we need recurrence equations.

Base case. By Lemma 1, for each good triple (S, C, Ω) of H_G where $\Omega = S \cup C$, $V(F) \cap \Omega$ is a subset (possibly empty) of a potential maximal clique Ω_F. As a result we can compute α in the case when $W' = \Omega_F \cap (S \cup C)$ because the neighborhoods assigned to $[W']$ are exactly the ones defined by edges in $G[W']$. Thus

$$\alpha(S, C, \Omega_F, ([\theta(\Omega_F)], [W']), W) = |W'|. \tag{1}$$

Computing $\beta(S, C, \Omega, \Omega_F, \ldots, W)$ *from* $\alpha(S_i, C_i, \Omega_F, \ldots, \Omega_F \cap S_i)$ *on smaller blocks.* Let $C_1, C_2, \ldots, C_r$ be the connected components of $G[C] \setminus \Omega$, $S_i = N_G(C_i)$, and $W_i = \Omega_F \cap S_i$ for $i \in \{1, 2, \ldots, r\}$. Computing β for a good triple (S, C, Ω), a vertex set $\Omega_F \subset V$, and a partial neighborhood assignment $([\theta(\Omega_F)], [W'])$, by only using known values of the α function for blocks (S_i, C_i) (in particular note that $|S_i \cup C_i| < |S \cup C|$) requires intermediate steps. Instead of gluing the r partial solutions in one step, which would be too costly, we glue them one by one. Technically, this is very similar to the classical technique transforming an arbitrary tree decomposition into a tree decomposition with a binary tree. Intuitively, in the decomposition tree of G, bag Ω has r children, with subtrees corresponding to r blocks (S_i, C_i). We need to introduce a new function γ, quite similar to β but also depending on parameter $k \leq r$, where $\gamma(\ldots, k)$ represents the gluing of partial solutions on the first k sons of Ω. We denote $W_k^+ = (S \cup \bigcup_{q=1}^k C_q) \cap \Omega_F$.

Definition 3. *Function* $\gamma(S, C, \Omega, \Omega_F, ([\theta(\Omega_F)], [W_k^+]), W, k)$ *is defined as the size of the maximum vertex set* $C_F \subseteq (W \cup \bigcup_{q=1}^k C_q)$, *such that*

- $G[C_F]$ *is planar,*
- $W = C_F \cap \Omega = \Omega_F \cap \Omega$,
- $W_k^+ = (S \cup \bigcup_{q=1}^k C_q) \cap \Omega_F$,
- *each connected component* F_j *of* $G[C_F \setminus W_k^+]$ *and each edge* $\{x, y\}$ *of* $G[W_k^+]$ *is mapped on a neighborhood* N_i *of* $[W_k^+]$ *where* $N_i = N_G(F_j) \cap W_k^+$ *or* $N_i = \{x, y\}$, *such that each* N_i *of* $[W_k^+]$ *is mapped to at least one component or edge, and to exactly one if* $3 \leq |N_i|$.

If no such induced subgraph exists, the value is set to 0.

Let W_1' be the vertex set $W_1^+ \setminus (W \setminus W_1)$, and let $([\theta(\Omega_F)], [W_1^+])$ be the partial neighborhood assignment containing the neighborhoods of $([\theta(\Omega_F)], [W_1'])$ along with one neighborhood $N_i = \{x, y\}$ contained in $[\theta(\Omega_F)]$ for each edge $\{x, y\}$ of

$G[W]$ where $x \in W \setminus W_1$. For $k = 1$ we can compute the γ function, using the α value for C_1, S_1, that is

$$\gamma(S, C, \Omega, \Omega_F, ([\theta(\Omega_F)], [W_1^+]), W, 1) = |W| - |W_1| + \qquad (2)$$
$$\alpha(S_1, C_1, \Omega_F, ([\theta(\Omega_F)], [W_1']), W_1).$$

Definition 4. *Let $([\theta(\Omega_F)], [W'])$ and $([\theta(\Omega_F)], [W''])$ be two partial neighborhood assignments, obtained as restrictions of the same maximal neighborhood assignment $[\theta(\Omega_F)]$. These assignments are called* disjoint *if there is no neighborhood $N \in [W'] \cap [W'']$ of size at least 3, assigned by both to a same cyclic ordering S_ℓ of $\theta(\Omega_F)$.*

Partial neighborhood assignment $([\theta(\Omega_F)], [W])$ is defined as the union *of $([\theta(\Omega_F)], [W''])$ and $([\theta(\Omega_F)], [W'])$, if $W = W' \cup W''$ and neighborhood N is assigned to cyclic ordering S_ℓ in $[W]$ if and only if N is assigned to S_ℓ in $[W']$ or in $[W'']$.*

Using (2) as the base case, and assuming by induction that values for γ are computed correctly for $k-1$, the equation below can be used to compute function γ for value k. Let $W_k' = W_k^+ \cap (W_k \cup C_k)$. Let $\mathcal{W}$ be the set of disjoint pairs $a = ([\theta(\Omega_F)], [W_{k-1}^+])$ and $b = ([\theta(\Omega_F)], [W_k'])$ such that the union is $([\theta(\Omega_F)], [W_k^+])$. Then

$$\gamma(S, C, \Omega, \Omega_F, ([\theta(\Omega_F)], [W_k^+]), W, k) = \max_{(a,b) \in \mathcal{W}} (\gamma(S, C, \Omega, \Omega_F, a, W, k-1) \quad (3)$$
$$- |W_k| + \alpha(S_k, C_k, \Omega_F, b, W_k)).$$

We emphasis that, by the fact that we "glue" two planar graphs along disjoint partial neighborhood assignments, we guarantee that the resulting graph is also planar. Eventually, function γ is computed for $k = r$, and we are ready to return to β:

$$\beta(S, C, \Omega, \Omega_F, ([\theta(\Omega_F)], [W_r^+]), W) = \gamma(S, C, \Omega, \Omega_F, \qquad (4)$$
$$([\theta(\Omega_F)], [W_r^+]), W, r).$$

Computing $\alpha(S, C, \Omega_F, \ldots, W)$ from $\beta(S, C, \Omega, \Omega_F, \ldots)$. When β is computed for all good triples (S, C, Ω), the value of $\alpha(S, C, \Omega_F, ([\theta(\Omega_F)], [W']), W)$ can be computed using the equation below. Let $\Pi_{S,C}$ be the set of all potential maximal cliques Ω of G, such that (S, C, Ω) is a good triple. Let $W'' = \Omega_F \cap \Omega$.

$$\alpha(S, C, \Omega_F, ([\theta(\Omega_F)], [W']), W) = \max_{\Omega \in \Pi_{S,C}} (\beta(S, C, \Omega, \Omega_F, \qquad (5)$$
$$([\theta(\Omega_F)], [W']), W''))$$

By making use of (5), we recompute $\alpha(S, C, \Omega_F, ([\theta(\Omega_F)], [W']), W)$ each time the value $\beta(S, C, \Omega, \Omega_F, ([\theta(\Omega_F)], [W']), W'')$ is computed.

Switching from one subtree of T_G to another. By making use of (1)–(5), we compute α and β when vertex set Ω_F remains unchanged. By Lemma 1, the clique tree T_G of the minimal triangulation H_G is partitioned into subtrees and these subtrees are in a one-to-one correspondence to the maximal cliques of H_F. It remains to explain how the transition from one subtree to an adjacent subtree can be performed. We only give here the equation used for this case, details will be given in the full version. The "lower" subtree corresponds to bag Ω_F^a of T_F, and the "upper" subtree corresponds to Ω_F^b. Recall that $S_F = \Omega_F^a \cap \Omega_F^b$ is a minimal separator of F.

Let $([\theta(\Omega_F^b], [S_F])$ be a partial neighborhood assignment such that N_i is a neighborhood if and only if $N_i = \{x, y\}$ for some edge in $G[S_F]$ or $N_i = S_F$. Let $\mathcal{W}$ be the set of pairs c, d where $c = \Omega_F^a$ and $d = ([\theta(\Omega_F^a)], [\Omega_F^a])$ is a neighborhood assignments such that: (i) neighborhood $N_i = S_F$ is contained in the maximal neighborhood assignment $[\theta(\Omega_F^a)]$ and not in the partial neighborhood assignment (unless $|S_F| \leq 2$); (ii) there is no neighborhood $N_j = \Omega_F^a$; and (iii) for every pair of non-adjacent vertices $x, y \in G[\Omega_F^a]$, either vertices x, y are contained in S_F or there exists a neighborhood N_l such that $x, y \in N_l$. Then,

$$\alpha(S, C, \Omega_F^b, ([\theta(\Omega_F^b)], [S_F]), S_F) = \max(\ \alpha(S, C, \Omega_F^b, ([\theta(\Omega_F^b)], [S_F]), S_F), \quad (6)$$
$$\max_{(c,d) \in \mathcal{W}} (\alpha(S, C, c, d, S_F)).$$

When all these values are computed we can retrive the solution

$$MIPS(G) = \max(\alpha(\emptyset, V, \Omega_F, ([\theta(\Omega_F)], [\Omega_F]), \emptyset)), \quad (7)$$

where maximum is taken over all Ω_F, θ-structures and all partial neighborhood assignments $([\theta(\Omega_F)], [\Omega_F])$.

Correctness and running time. Due to space restrictions, the proof of correctness is omitted. The algorithm runs through all good triples (S, Ω, C) of G, whose number is $O(1.734601^n)$ by [9]. The most costly equation is (3), which maximizes over triples of partial neighborhood assignments. By Lemma 3 we consider $2^{o(n)}$ partial neighborhood aignments, which concludes the proof of Theorem 1.

5   Open Questions

The natural question is if the results of our work can be generalized to find induced subgraphs from graph families characterized by a finite set of forbidden minors. A "naive" approach of enumerating all possible parts of forbidden minors during dynamic programming does not work directly because the amount of information to keep dominates the running time 2^n. A possible advance here can occur from better understanding of the structure of potential maximal cliques in H-minor-free graphs, the combinatorial question being interesting in its own. Finally, a pure combinatorial question. Let G be an n-vertex graph. Is the number of maximal induced planar subgraphs in $O((2 - \varepsilon)^n)$ for some $\varepsilon > 0$?

References

1. Angelsmark, O., Thapper, J.: Partitioning based algorithms for some colouring problems. In: Hnich, B., Carlsson, M., Fages, F., Rossi, F. (eds.) CSCLP 2005. LNCS (LNAI), vol. 3978, pp. 44–58. Springer, Heidelberg (2006)
2. Bouchitté, V., Mazoit, F., Todinca, I.: Chordal embeddings of planar graphs. Discr. Math. 273(1-3), 85–102 (2003)
3. Bouchitté, V., Todinca, I.: Treewidth and minimum fill-in: Grouping the minimal separators. SIAM J. Comput. 31(1), 212–232 (2001)
4. Buneman, P.: A characterization of rigid circuit graphs. Discr. Math. 9, 205–212 (1974)
5. Eppstein, D.: Subgraph isomorphism in planar graphs and related problems. J. Graph Algorithms Appl. 3(3), 1–27 (1999)
6. Fomin, F.V., Kratsch, D.: Exact Exponential Algorithms. In: An EATCS Series: Texts in Theoretical Computer Science, Springer, Heidelberg (2010)
7. Fomin, F.V., Kratsch, D., Todinca, I., Villanger, Y.: Exact algorithms for treewidth and minimum fill-in. SIAM J. Comput. 38(3), 1058–1079 (2008)
8. Fomin, F.V., Thilikos, D.M.: New upper bounds on the decomposability of planar graphs. Journal of Graph Theory 51(1), 53–81 (2006)
9. Fomin, F.V., Villanger, Y.: Finding induced subgraphs via minimal triangulations. In: Marion, J.-Y., Schwentick, T. (eds.) STACS. LIPIcs, vol. 5, pp. 383–394. Schloss Dagstuhl - Leibniz-Zentrum fuer Informatik (2010)
10. Gapers, S., Kratch, D., Liedloff, M.: On independent sets and bicliques in graphs. WG (2008); to appear. Preliminary version in WG
11. Garey, M.R., Johnson, D.S.: Computers and Intractability: A Guide to the Theory of NP-completness. Freeman, New York (1979)
12. Gaspers, S.: Exponential Time Algorithms: Structures, Measures, and Bounds. Phd thesis, University of Bergen (2008)
13. Gavril, F.: The intersection graphs of a path in a tree are exactly the chordal graphs. Journal of Combinatorial Theory 16, 47–56 (1974)
14. Gupta, S., Raman, V., Saurabh, S.: Fast exponential algorithms for maximum - regular induced subgraph problems. In: Arun-Kumar, S., Garg, N. (eds.) FSTTCS 2006. LNCS, vol. 4337, pp. 139–151. Springer, Heidelberg (2006)
15. Kreweras, G.: Sur les partition non croisées d'un circle. Discr. Math. 1, 333–350 (1972)
16. Lewis, J.M., Yannakakis, M.: The node-deletion problem for hereditary properties is np-complete. J. Comput. Syst. Sci. 20(2), 219–230 (1980)
17. Liebers, A.: Planarizing graphs - a survey and annotated bibliography. Journal of Graph Algorithms and Applications 5 (2001)
18. Moon, J.W., Moser, L.: On cliques in graphs. Israel Journal of Mathematics 3, 23–28 (1965)
19. Razgon, I.: Exact computation of maximum induced forest. In: Arge, L., Freivalds, R. (eds.) SWAT 2006. LNCS, vol. 4059, pp. 160–171. Springer, Heidelberg (2006)
20. Robertson, N., Seymour, P.D.: Graphs minors. II. Algorithmic aspects of treewidth. J. of Algorithms 7, 309–322 (1986)

Scheduling Partially Ordered Jobs Faster Than 2^n

Marek Cygan[1,*], Marcin Pilipczuk[1,*], Michał Pilipczuk[1],
and Jakub Onufry Wojtaszczyk[2]

[1] Institute of Informatics, University of Warsaw, Poland
{cygan,malcin,mp248287}@students.mimuw.edu.pl
[2] Google Inc., Cracow, Poland
onufry@google.com

Abstract. In the SCHED problem we are given a set of n jobs, together with their processing times and precedence constraints. The task is to order the jobs so that their total completion time is minimized. SCHED is a special case of the Traveling Repairman Problem with precedences. A natural dynamic programming algorithm solves both these problems in $2^n n^{O(1)}$ time, and whether there exists an algorithms solving SCHED in $O(c^n)$ time for some constant $c < 2$ was an open problem posted in 2004 by Woeginger. In this paper we answer this question positively.

Keywords: moderately exponential algorithms, jobs scheduling, jobs with precedences.

1 Introduction

It is commonly believed that no NP-hard problem is solvable in polynomial time. However, while all NP-complete problems are equivalent with respect to polynomial time reductions, they appear to be very different with respect to the best exponential time exact solutions. The question asked in the moderately exponential time algorithms area is how small the constant c in the c^n time algorithms can be. Many difficult problems can be solved much faster than by the obvious brute-force algorithm; examples are INDEPENDENT SET [10], DOMINATING SET [10,17], CHROMATIC NUMBER [4] and BANDWIDTH [7]. The race for the fastest exact algorithm inspired several very interesting tools and techniques such as Fast Subset Convolution [3] and Measure&Conquer [10] (for an overview of the field we refer the reader to a recent book by Fomin and Kratsch [9]).

For several problems, including TSP, CHROMATIC NUMBER, PERMANENT, SET COVER, #HAMILTONIAN CYCLES and SAT, the currently best known time complexity is of the form $O(2^n n^{O(1)})$, which is a result of applying dynamic programming over subsets, the inclusion-exclusion principle or a brute force search. The question remains, however, which of those problems are inherently so hard

---

* The first two authors were partially supported by Polish Ministry of Science grant no. N206 355636 and Foundation for Polish Science.

C. Demetrescu and M.M. Halldórsson (Eds.): ESA 2011, LNCS 6942, pp. 299–310, 2011.

that it is not possible to break the 2^n barrier and which are just waiting for new tools and techniques still to be discovered. In particular, the hardness of the k-SAT problem is the starting point for the Strong Exponential Time Hypothesis of Impagliazzo and Paturi [11], which is used as an argument that other problems are hard [6,13,16]. Recently, on the positive side, $O(c^n)$ time algorithms for a constant $c < 2$ have been developed for CAPACITATED DOMINATION [8], IRREDUNDANCE [1] and (a major breakthrough in the field) for the undirected version of the HAMILTONIAN CYCLE problem [2].

In this paper we study the SCHED problem, defined as follows.

SCHED
Input: A partially ordered set $(V, \leq)$, (the elements of which are called *jobs*) together with a nonnegative processing time $t(v)$ for each job $v \in V$.
Task: Compute a bijection $\sigma : V \to \{1, 2, \ldots, |V|\}$ (called an *ordering*) that satisfies the precedence constraints (i.e., if $u < v$, then $\sigma(u) < \sigma(v)$) and minimizes the total completion time of all jobs defined as

$$T(\sigma) = \sum_{v \in V} \sum_{u:\sigma(u) \leq \sigma(v)} t(u) = \sum_{v \in V} (|V| - \sigma(v) + 1)t(v).$$

If $u < v$ for $u, v \in V$ (i.e., $u \leq v$ and $u \neq v$), we say that u *precedes* v, or that u is *required* for v. We denote $|V|$ by n. SCHED is a special case of the precedence constrained Travelling Repairman Problem (prec-TRP), which is a relative of TSP. SCHED was shown to be NP-complete in 1978 by Lenstra and Rinnooy Kan [12], whereas to the best of our knowledge the currently smallest approximation ratio equals 2, due to independently discovered algorithms by Chekuri and Motwani [5] as well as Margot et al. [14].

Woeginger at IWPEC 2004 [18] has posed the question (repeated in 2008 [19]), whether it is possible to construct an $O((2-\varepsilon)^n)$ time algorithm for the SCHED problem. In this paper we present such an algorithm, thus affirmatively answering Woeginger's question. This result is intriguing in particular because the SCHED problem admits arbitrary processing times — and still it is possible to obtain an $O((2 - \varepsilon)^n)$ time algorithm nonetheless. Probably due to arbitrary processing times Woeginger also asked [18,19] whether an $O((2 - \varepsilon)^n)$ time algorithm for one of the problems TRP, TSP, prec-TSP, SCHED implies $O((2 - \varepsilon)^n)$ time algorithms for the other problems. This problem is still open. One should note that Woeginger in his papers asks for an $O(1.99^n)$ algorithm, though the intention is clearly to ask for an $O((2 - \varepsilon)^n)$ algorithm.

The most important ingredient of our algorithm is a combinatorial Lemma (Lemma 4) which allows us to investigate the structure of the SCHED problem. We heavily use the fact that we are solving the SCHED problem and not its more general TSP related version, and for this reason we believe that obtaining $(2 - \varepsilon)^n$ time algorithms for other problems listed by Woeginger is much harder.

Due to space limitations, proofs of lemmata marked with a spade symbol (♠) will appear in the full version of the paper.

2   The Algorithm

2.1   High-Level Overview—Part 1

Let us recall that our task in the SCHED problem is to compute an ordering $\sigma : V \to \{1, 2, \ldots, n\}$ that satisfies the precedence constraints (i.e., if $u < v$ then $\sigma(u) < \sigma(v)$) and minimizes the total completion time of all jobs defined as

$$T(\sigma) = \sum_{v \in V} \sum_{u : \sigma(u) \leq \sigma(v)} t(u) = \sum_{v \in V} (n - \sigma(v) + 1) t(v).$$

We define *the cost of job v at position i* to be $T(v, i) = (n - i + 1) t(v)$. Thus, the total completion time is the total cost of all jobs at their respective positions in the ordering σ.

We begin by describing the algorithm that solves SCHED in $O^\star(2^n)$ time[1], which we call *the DP algorithm* — this will be the basis for our further work. The idea — a standard dynamic programming over subsets — is that if we decide that a particular set $X \subseteq V$ will (in some order) form the prefix of our optimal σ, then the order in which we take the elements of X does not affect the choices we make regarding the ordering of the remaining $V \setminus X$; the only thing that matters are the precedence constraints imposed by X on $V \setminus X$. Thus, for each candidate set $X \subseteq V$ to form a prefix, the algorithm computes a bijection $\sigma[X] : X \to \{1, 2, \ldots, |X|\}$ that minimizes the cost of jobs from X, i.e., it minimizes $T(\sigma[X]) = \sum_{v \in X} T(v, \sigma[X](v))$. The value of $T(\sigma[X])$ is computed using the following easy to check recursive formula:

$$T(\sigma[X]) = \min_{v \in \max(X)} \left[T(\sigma[X \setminus \{v\}]) + T(v, |X|) \right]. \tag{1}$$

Here, by $\max(X)$ we mean the set of maximum elements of X — those which do not precede any element of X. The bijection $\sigma[X]$ is constructed by prolonging $\sigma[X \setminus \{v\}]$ by v, where v is the job at which the minimum is attained. Notice that $\sigma[V]$ is exactly the ordering we are looking for. We calculate $\sigma[V]$ recursively, using formula (1), storing all computed values $\sigma[X]$ in memory to avoid recomputation. Thus, as the computation of a single $\sigma[X]$ value given all the smaller values takes polynomial time, while $\sigma[X]$ for each X is computed at most once the whole algorithm indeed runs in $O^\star(2^n)$ time.

The overall idea of our algorithm is to identify a family of sets $X \subseteq V$ that — for some reason — are not reasonable prefix candidates, and we can skip them in the computations of the DP algorithm; we will call these *unfeasible sets*. If the number of feasible sets will be no larger than c^n for some $c < 2$, we will be done — our recursion will visit only feasible sets, assuming $T(\sigma[X])$ to be ∞ for unfeasible X in formula (1), and the running time will be $O^\star(c^n)$.

---

[1] By $O^\star()$ we denote the standard big O notation where we suppress polynomially bounded terms.

2.2   The Large Matching Case

We begin by noticing that the DP algorithm needs to compute $\sigma[X]$ only for those $X \subseteq V$ that are downward closed, i.e., if $v \in X$ and $u < v$ then $u \in X$. If there are many constraints in our problem, this alone will suffice to limit the number of feasible sets considerably, as follows. Construct an undirected graph G with the vertex set V and edge set $E = \{uv : u < v \vee v < u\}$. Let M be a maximum matching[2] in G, which can be found in polynomial time [15]. If $X \subseteq V$ is downward closed, and $uv \in M$, $u < v$, then it is not possible that $u \notin X$ and $v \in X$. Obviously checking if a subset is downward closed can be performed in polynomial time. This leads to the following lemma:

Lemma 1. *The number of downward closed subsets of V is bounded by $2^{n-2|M|}$ $3^{|M|}$. If $|M| \geq \varepsilon_1 n$, then we can solve the* SCHED *problem in time*

$$T_1(n) = O^\star((3/4)^{\varepsilon_1 n} 2^n).$$

Note that for any small positive constant ε_1 the complexity $T_1(n)$ is of required order, i.e., $T_1(n) = O(c^n)$ for some $c < 2$ that depends on ε_1. Thus, we only have to deal with the case where $|M| < \varepsilon_1 n$.

Let us fix a maximum matching M, let $W_1 \subseteq V$ be the set of endpoints of M, and let $I_1 = V \setminus W_1$. Note that, as M is a maximum matching in G, no two jobs in I_1 are bound by a precedence constraint, while $|W_1| \leq 2\varepsilon_1 n$, $|I_1| \geq (1 - 2\varepsilon_1)n$.

2.3   High-Level Overview—Part 2

We are left in the situation where there is a small number of "special" elements (W_1), and the bulk remainder (I_1), consisting of elements that are tied by precedence constraints only to W_1 and not to each other.

First notice that if W_1 was empty, the problem would be trivial: with no precedence constraints we should simply order the tasks from the shortest to the longest. Now let us consider what would happen if all the constraints between any $u \in I_1$ and $w \in W_1$ would be of the form $u < w$ — that is, if the jobs from I_1 had no prerequisites. For any prefix set candidate $\tilde{X}$ we consider $X = \tilde{X} \cap I_1$. Now for any $x \in X$, $y \in I_1 \setminus X$ we have an alternative prefix candidate: the set $\tilde{X}' = (\tilde{X} \cup \{y\}) \setminus \{x\}$. If $t(y) < t(x)$, there has to be a reason why $\tilde{X}'$ is not a strictly better prefix candidate than $\tilde{X}$ — namely, there has to exist $w \in W_1$ such that $x < w$, but $y \not< w$.

A similar reasoning would hold even if not all of I_1 had no perquisite's, but just some significant fraction J of I — again, the only feasible prefix candidates would be those in which for every $x \in X \cap J$ and $y \in J \setminus X$ there is a reason (either $t(x) < t(y)$ or an element $w \in W_1$ which requires x, but not y) not to exchange them. It turns out that if $|J| > \varepsilon_2 n$, where $\varepsilon_2 > 2\varepsilon_1$, this observation suffices to prove that the number of possible intersections of a feasible set with J is significantly smaller than $2^{|J|}$. This is formalized and proved in Lemma 4, and is the cornerstone of the whole paper.

---

[2] Even an inclusion-maximal matching, which can be found greedily, is enough.

The typical application of this Lemma is as follows: say we have a set $K \subseteq I_1$ of cardinality $|K| > 2j$, while we know for some reason that all the requisites of elements of K appear on positions j and earlier. If K is large (a constant fraction of n), this will be enough to limit the number of feasible sets to $(2 - \varepsilon)^n$. The reasoning is to show there are significantly less than $2^{|K|}$ possible intersections of a feasible set with K. Each such intersection consists of a set of at most j elements (that will be put on positions 1 through j), and then a set in which every element has a reason not to be exchanged with something from outside the set — and there are relatively few of those by Lemma 4 — and when we do the calculations, it turns out the resulting number of possibilities is significantly smaller than $2^{|K|}$.

To apply this reasoning, we need to be able to tell that all the prerequisites of a given element appear at some position or earlier. To achieve this, we need to know the approximate positions of the elements in W_1. We achieve this by branching into $4^{|W_1|}$ cases, for each element $w \in W_1$ choosing to which of the four quarters of the set $\{1, \ldots, n\}$ will $\sigma_{opt}(w)$ belong. This incurs a multiplicative cost of $4^{|W_1|}$, which will be offset by the gains from applying Lemma 4.

We will now repeatedly apply Lemma 4 to obtain information about the positions of various elements of I_1. We will repeatedly say that if "many" elements (by which we always mean more than εn for some ε) do not satisfy something, we can bound the number of feasible sets, and thus finish the algorithm. For instance, look at those elements of I_1 which can appear in the first quarter, i.e., none of their prerequisites appear in quarters two, three and four. If there is significantly over $n/2$ of them, we can apply the above reasoning for $j = n/4$ (Lemma 6). Subsequent lemmata bound the number of feasible sets if there are many elements that cannot appear in any of the two first quarters (Lemma 5), if significantly *less* than $n/2$ elements can appear in the first quarter (Lemma 6) and if a significant number of elements in the second quarter could actually appear in the first quarter (Lemma 7). We also apply similar reasoning to elements that can or cannot appear in the last quarter.

We end up in a situation where we have four groups of elements, each of size roughly $n/4$, split upon whether they can appear in the first quarter and whether they can appear in the last one; moreover, those that can appear in the first quarter will not appear in the second, and those that can appear in the fourth will not appear in the third. This means that there are two pairs of parts which do not interact, as the set of places in which they can appear are disjoint. We use this independence of sorts to construct a different algorithm than the DP we used so far, which solves our problem in this specific case in time $O^\star(2^{3n/4+\varepsilon})$ (Lemma 8).

As can be gathered from this overview, there are many technical details we will have to navigate in the algorithm. This is made more precarious by the need to carefully select all the epsilons. We decided to use symbolic values for them in the main proof, describing their relationship appropriately, using four constants ε_k, $k = 1, 2, 3, 4$. The constants ε_k are very small positive reals, and additionally ε_k is significantly smaller than ε_{k+1} for $k = 1, 2, 3$. At each step, we shortly discuss the existence of such constants. We discuss the choice of optimal values of these constants in Section 3, although the value we perceive in our algorithm lies rather in the existence of an $O^\star((2 - \varepsilon)^n)$ algorithm than in the value of ε (which is admittedly very small).

2.4 Technical Preliminaries

We start with a few simplifications. First, we add a few dummy jobs with no precedence constraints and zero processing times, so that n is divisible by four. Second, by slightly perturbing the jobs' processing times, we can assume that all processing times are pairwise different and, moreover, each ordering has different total completion time. This can be done, for instance, by replacing time $t(v)$ with a pair $(t(v), (n + 1)^{\pi(v)-1})$, where $\pi : V \to \{1, 2, \ldots, n\}$ is an arbitrary numbering of V. The addition of pairs is performed coordinatewise, whereas comparison is performed lexicographically. Note that this in particular implies that the optimal solution is unique, we denote it by σ_{opt}. Third, at the cost of an n^2 multiplicative overhead, we guess the jobs $v_{begin} = \sigma_{opt}^{-1}(1)$ and $v_{end} = \sigma_{opt}^{-1}(n)$ and we add precedence constraints $v_{begin} < v < v_{end}$ for each $v \neq v_{begin}, v_{end}$. If v_{begin} or v_{end} were not in W_1 to begin with, we add them there.

A number of times our algorithm branches into several subcases, in each branch assuming some property of the optimal solution σ_{opt}. Formally speaking, in each branch we seek the optimal ordering among those that satisfy the assumed property. We somewhat abuse the notation and denote by σ_{opt} the optimal solution in the currently considered subcase. Note that σ_{opt} is always unique within any subcase, as each ordering has different total completion time.

For $v \in V$ by $pred(v)$ we denote the set $\{u \in V : u < v\}$ of predecessors of v, and by $succ(v)$ we denote the set $\{u \in V : v < u\}$ of successors of v. We extend this notation to subsets of V: $pred(U) = \bigcup_{v \in U} pred(v)$ and $succ(U) = \bigcup_{v \in U} succ(v)$. Note that for any set $U \subseteq I_1$, both $pred(U)$ and $succ(U)$ are subsets of W_1.

2.5 The Core Lemma

We now formalize the idea of exchanges presented in Section 2.3. In the proof of the first case we exchange u with some v_w whereas in the second case we exchange v with some u_w.

Definition 2. *Consider some set $K \subseteq I_1$, and its subset $L \subseteq K$. If there exists $u \in L$ such that for every $w \in succ(u)$ we can find $v_w \in (K \cap pred(w)) \setminus L$ with $t(v_w) < t(u)$ then we say L is succ-exchangeable with respect to K, otherwise we say L is non-succ-exchangeable with respect to K.*

Similarly, if there exists $v \in (K \setminus L)$ such that for every $w \in pred(v)$ we can find $u_w \in L \cap succ(w)$ with $t(u_w) > t(v)$, we call L pred-exchangeable with respect to K, otherwise we call it non-pred-exchangeable with respect to K.

Whenever it is clear from the context, we omit the set K with respect to which its subset is (non)-exchangeable.

The applicability of this definition is in the following easy observation:

Observation 3 (♠) *Let $K \subseteq I_1$. If for any $v \in K, w \in pred(K)$ we have that $\sigma_{opt}(v) > \sigma_{opt}(w)$, then for any $1 \leq i \leq n$ the set $K \cap \sigma_{opt}(\{1, 2, \ldots, i\})$ is non-succ-exchangeable with respect to K.*

Similarly, if for any $v \in K, w \in succ(K)$ we have $\sigma_{opt}(v) < \sigma_{opt}(w)$, then the sets $K \cap \sigma_{opt}(\{1, 2, \ldots, i\})$ are non-pred-exchangeable with respect to K.

This means that if we manage to identify a set K satisfying the assumptions of the observation, the only sets the DP algorithm has to consider are the non-exchangeable ones. The following core lemma proves that there are few of those, and we can identify them easily.

Lemma 4. *For any set $K \subseteq I_1$ the number of non-pred-exchangeable (non-succ-exchangeable) subsets is at most $\sum_{l \leq |W_1|} \binom{|K|}{l}$. Moreover, there exists an algorithm which checks whether a set is pred-exchangeable (succ-exchangeable) in polynomial time.*

The idea of the proof is to construct a function f that encodes each non-exchangeable set by a subset of K no larger than W_1. To show this encoding is injective, we provide a decoding function g and show that $f \circ g$ is an identity on non-exchangeable sets.

Proof. We prove the Lemma for non-*pred*-exchangeable sets, the proof for non-*succ*-exchangeable sets is analogous (using least expensive predecessors outside of Y instead of most expensive successors in Y) and can be found in the full version of the paper.

For any set $Y \subseteq K$ we define the function $f_Y : W_1 \to K \cup \{nil\}$ as follows: for any element $w \in W_1$ we define $f_Y(w)$ (the *most expensive successor* of w in Y) to be the element of $Y \cap succ(w)$ which has the largest processing time, or nil if $Y \cap succ(w)$ is empty. We now take $f(Y)$ (the set of *most expensive successors in Y*) to be the set $\{f_Y(w) : w \in W_1\} \setminus \{nil\}$. $f(Y)$ is indeed a set of cardinality at most $|W_1|$, we will aim to prove that f is injective on the family of non-*pred*-exchangeable sets.

To this end we define the reverse function g. For a set $Z \subseteq K$ (which we think of as the set of most expensive successors in some Y) let $g(Z)$ be the set of such elements v of K that for any $w \in pred(v)$ there exists a $z_w \in Z \cap succ(w)$ with $t(z_w) \geq t(v)$. Notice, in particular, that $g(Z) \subseteq Z$.

First we prove $Y \subseteq g(f(Y))$ for any $Y \subseteq K$. Indeed — take any $v \in Y$ and consider any $w \in pred(v)$. Then $f_Y(w) \neq$ nil and $t(f_Y(w)) \geq t(v)$, as $v \in Y \cap succ(w)$. Thus $v \in g(f(Y))$, as for any $w \in pred(v)$ we can take $z_w = f_Y(w)$ in the definition of $g(f(Y))$.

In the other direction, let us assume that Y does not satisfy $g(f(Y)) \subseteq Y$. This means we have $v \in g(f(Y)) \setminus Y$. We want to show that Y is *pred*-exchangeable. Consider any $w \in pred(v)$. As $v \in g(f(Y))$ there exists $z_w \in f(Y) \cap succ(w)$ with $t(z_w) \geq t(v)$. But $f(Y) \subseteq Y$, while $v \notin Y$; and as all the values of t are distinct, $t(z_w) > t(v)$ and z_w satisfies the condition for u_w in the definition of *pred*-exchangeability.

Thus if Y is non-exchangeable then $g(f(Y)) = Y$ (in fact it is possible to prove that Y is non-exchangeable iff $g(f(Y)) = Y$). As there are $\sum_{l=0}^{|W_1|} \binom{|K|}{l}$

possible values of $f(Y)$, the first part of the lemma is proven. For the second, it suffices to notice that exchangeability can be checked in time $O(|K|^2|W_1|)$ directly from the definition.

2.6 Important Jobs at $n/2$

As was already mentioned in the overview, the assumptions of Lemma 4 are quite strict; therefore, we need to learn a bit more on how σ_{opt} behaves on W_1 in order to distinguish a suitable place for an application. As $|W_1| \leq 2\varepsilon_1 n$, we can proceed with quite an extensive branching on W_1.

Let $A = \{1, 2, \ldots, n/4\}$, $B = \{n/4 + 1, \ldots, n/2\}$, $C = \{n/2 + 1, \ldots, 3n/4\}$, $D = \{3n/4 + 1, \ldots, n\}$, i.e., we split $\{1, 2, \ldots, n\}$ into quarters. For each $w \in W_1 \setminus \{v_{begin}, v_{end}\}$ we branch into four cases: whether $\sigma_{opt}(w)$ belongs to A, B, C or D. This branching leads to $4^{|W_1|-2} \leq 2^{4\varepsilon_1 n}$ subcases, and thus the same overhead in the time complexity. Of course, we already know that $\sigma_{opt}(v_{begin}) \in A$ and $\sigma_{opt}(v_{end}) \in D$. We terminate all the branches, where the guesses about alignment of jobs from W_1 contradict precedence constraints inside W_1.

In a fixed branch, let W_1^Γ be the set of elements of W_1 to be placed in Γ, for $\Gamma \in \{A, B, C, D\}$. Moreover let $W_1^{AB} = W_1^A \cup W_1^B$ and $W_1^{CD} = W_1^C \cup W_1^D$.

Let us now see what we can learn from the above step about the behaviour of σ_{opt} on I_1. Let $W_2^{AB} = \{v \in I_1 : \exists_w\ w \in W_1^{AB} \wedge v < w\}$ and $W_2^{CD} = \{v \in I_1 : \exists_w\ w \in W_1^{CD} \wedge w < v\}$ — that is W_2^{AB} (resp. W_2^{CD}) are those elements of I_1 which are forced into the first (resp. second) half of σ_{opt} by the choices we made about W_1. If one of the W_2 sets is significantly larger than W_1, we have obtained a gain — by branching into $2^{4\varepsilon_1 n}$ branches we gained additional information about a significant number of other elements (and so we will be able to avoid considering a significant number of sets in the DP algorithm). This is formalized in the following lemma:

Lemma 5 (♠). *If W_2^{AB} or W_2^{CD} has at least $\varepsilon_2 n$ elements, then the DP algorithm can be augmented to solve the remaining instance in time*

$$T_2(n) = \left(\binom{n}{(1/2 - \varepsilon_2/3)n} + 2^{(1-\varepsilon_2)n} \binom{\varepsilon_2 n}{\varepsilon_2/3 \cdot n} \right) n^{O(1)}.$$

Note that we have $2^{4\varepsilon_1 n}$ overhead so far, due to guessing placement of the jobs from W_1. As $\binom{\varepsilon_2 n}{\varepsilon_2/3 n} = O(1.89^{\varepsilon_2 n})$ and $\binom{n}{(1/2-\varepsilon_2/3)n} = O((2 - c(\varepsilon_2))^n)$, for any small fixed ε_2 we can choose ε_1 sufficiently small so that $2^{4\varepsilon_1 n} T_2(n) = O(c^n)$ for some $c < 2$. Note that $2^{4\varepsilon_1 n} T_2(n)$ is an upper bound on the total time spent on processing all the considered subcases.

Let $W_2 = W_2^{AB} \cup W_2^{CD}$ and $I_2 = I_1 \setminus W_2$. From this point we assume that $|W_2^{AB}|, |W_2^{CD}| \leq \varepsilon_2 n$, hence $|W_2| \leq 2\varepsilon_2 n$ and $|I_2| \geq (1 - 2\varepsilon_1 - 2\varepsilon_2)n$. For each $v \in W_2^{AB}$ we branch into two subcases, whether $\sigma_{opt}(v)$ belongs to A or B. Similarly, for each $v \in W_2^{CD}$ we guess whether $\sigma_{opt}(v)$ belongs to C or D. Again, we execute only branches which are not trivially contradicting the constraints. This steps gives us $2^{|W_2|} \leq 2^{2\varepsilon_2 n}$ overhead in the time complexity. We denote the set of elements of W_2 assigned to quarter $\Gamma \in \{A, B, C, D\}$ by W_2^Γ.

2.7   Quarters and Applications of the Core Lemma

In this section we try to apply Lemma 4 as follows: We look which elements of I_2 can be placed in A (the set P^A) and which cannot (the set $P^{\neg A}$). Similarly we define the set P^D (can be placed in D) and $P^{\neg D}$ (cannot be placed in D). For each of these sets, we try to apply Lemma 4 to some subset of it. If we fail, then in the next subsection we infer that the solutions in the quarters are partially independent of each other, and we can solve the problem in time roughly $O(2^{3n/4})$. Let us now proceed with the formal argumentation.

We define the following two partitions of I_2:

$$P^{\neg A} = \{v \in I_2 : \exists_w w \in W_1^B \wedge w < v\},$$
$$P^A = I_2 \setminus P^{\neg A} = \{v \in I_2 : \forall_w w < v \Rightarrow w \in W_1^A\},$$
$$P^{\neg D} = \{v \in I_2 : \exists_w w \in W_1^C \wedge w > v\},$$
$$P^D = I_2 \setminus P^{\neg D} = \{v \in I_2 : \forall_w w > v \Rightarrow w \in W_1^D\}.$$

In other words, the elements of $P^{\neg A}$ cannot be placed in A because some of their requirements are in W_1^B, and the elements of $P^{\neg D}$ cannot be placed in D because they are required by some elements of W_1^C. Note that these definitions are independent of σ_{opt}, so sets P^Δ for $\Delta \in \{A, \neg A, D, \neg D\}$ can be computed in polynomial time. Let

$$p^A = |\sigma_{opt}(P^A) \cap A|,$$
$$p^B = |\sigma_{opt}(P^{\neg A}) \cap B|,$$
$$p^C = |\sigma_{opt}(P^{\neg D}) \cap C|,$$
$$p^D = |\sigma_{opt}(P^D) \cap D|.$$

Note that $p^\Gamma \leq n/4$ for every $\Gamma \in \{A, B, C, D\}$. As $p^A = n/4 - |W_1^A \cup W_2^A|$, $p^D = n/4 - |W_1^D \cup W_2^D|$, these values can be computed by the algorithm. We branch into $(1 + n/4)^2$ subcases, guessing the (still unknown) values p^B and p^C.

Let us focus on the quarter A and assume that p^A is significantly smaller than $|P^A|/2$. We claim that we can apply Lemma 4 as follows. While computing $\sigma[X]$, if $|X| \geq n/4$, we can represent $X \cap P^A$ as a disjoint sum of two subsets $X_A^A, X_{BCD}^A \subseteq P^A$. The first one is of size p^A, and represents the elements of $X \cap P^A$ placed in quarter A, and the second represents the elements of $X \cap P^A$ placed in quarters $B \cup C \cup D$. Note that the elements of X_{BCD}^A have all predecessors in the quarter A, so by Observation 3 the set X_{BCD}^A has to be non-$succ$-exchangeable with respect to $X \cap P^A$; therefore, we can consider only a very narrow choice of X_{BCD}^A. Thus, the whole part $X \cap P^A$ can be represented by its subset of cardinality at most p^A plus some small information about the rest. If p^A is significantly smaller than $|P^A|/2$, this representation is more concise than simply remembering a subset of P^A. Thus we obtain a better bound on the number of feasible sets.

A symmetric situation arises when p^D is significantly smaller than $|P^D|/2$; moreover, we can similarly use Lemma 4 if p^B is significantly smaller than

$|P^{\neg A}|/2$ or p^C than $|P^{\neg D}|/2$. Observe that if any of the sets P^{Δ} for $\Delta \in \{A, \neg A, D, \neg D\}$ is significantly larger than $n/2$, one of these situations indeed occurs, since $p^{\Gamma} \leq n/4$ for $\Gamma \in \{A, B, C, D\}$.

Lemma 6 (♠). *If at least one of the sets P^A, $P^{\neg A}$, P^D and $P^{\neg D}$ is of size at least $(1/2 + \varepsilon_3)n$, then the DP algorithm can be augmented to solve the remaining instance in time bounded by*

$$T_3(n) = 2^{(1/2 - \varepsilon_3)n} \binom{(1/2 + \varepsilon_3)n}{n/4} \binom{(1/2 + \varepsilon_3)n}{2\varepsilon_1 n} n^{O(1)}.$$

Note that we have $2^{(4\varepsilon_1 + 2\varepsilon_2)n} n^{O(1)}$ overhead so far. As $\binom{(1/2 + \varepsilon_3)n}{n/4} = O((2 - c(\varepsilon_3))^{(1/2 + \varepsilon_3)n})$ for some constant $c(\varepsilon_3) > 0$, for any small fixed ε_3 we can choose sufficiently small ε_2 and ε_1 to have $2^{(4\varepsilon_1 + 2\varepsilon_2)n} n^{O(1)} T_3(n) = O(c^n)$ for some $c < 2$.

From this point we assume that $|P^A|, |P^{\neg A}|, |P^D|, |P^{\neg D}| \leq (1/2 + \varepsilon_3)n$. As $P^A \cup P^{\neg A} = I_2 = P^{\neg D} \cup P^D$ and $|I_2| \geq (1 - 2\varepsilon_1 - 2\varepsilon_2)n$, this implies that all these sets are of size at least $(1/2 - 2\varepsilon_1 - 2\varepsilon_2 - \varepsilon_3)n$, i.e., there are of size roughly $n/2$. Having bounded the sizes of the sets P^{Δ} from below, we are able to use the trick from Lemma 6 again: if any of the numbers p^A, p^B, p^C, p^D is significantly smaller than $n/4$, then it is also significantly smaller than half of the cardinality of the corresponding set P^{Δ}.

Lemma 7 (♠). *Let $\varepsilon_{123} = 2\varepsilon_1 + 2\varepsilon_2 + \varepsilon_3$. If at least one of the numbers p^A, p^B, p^C and p^D is smaller than $(1/4 - \varepsilon_4)n$ and $\varepsilon_4 > \varepsilon_{123}/2$, then the DP algorithm can be augmented to solve the remaining instance in time bounded by*

$$T_4(n) = 2^{(1/2 + \varepsilon_{123})n} \binom{(1/2 - \varepsilon_{123})n}{(1/4 - \varepsilon_4)n} \binom{(1/2 - \varepsilon_{123})n}{2\varepsilon_1 n} n^{O(1)}.$$

So far we have $2^{(4\varepsilon_1 + 2\varepsilon_2)n} n^{O(1)}$ overhead. Similarly as before, for any small fixed ε_4 if we choose $\varepsilon_1, \varepsilon_2, \varepsilon_3$ sufficiently small, we have $\binom{(1/2 - \varepsilon_{123})n}{(1/4 - \varepsilon_4)n} = O((2 - c(\varepsilon_4))^{(1/2 - \varepsilon_{123})n})$ and $2^{(4\varepsilon_1 + 2\varepsilon_2)n} n^{O(1)} T_4(n) = O(c^n)$ for some $c < 2$.

Thus we are left with the case when $p^A, p^B, p^C, p^D \geq (1/4 - \varepsilon_4)n$.

2.8   The Remaining Case

In this subsection we infer that in the remaining case the quarters A, B, C and D are somewhat independent, which allows us to develop a faster algorithm. More precisely, note that $p^{\Gamma} \geq (1/4 - \varepsilon_4)n$, $\Gamma \in \{A, B, C, D\}$, means that almost all elements that are placed by σ_{opt} in A belong to P^A, while almost all elements placed in B belong to $P^{\neg A}$. Similarly, almost all elements placed in D belong to P^D and almost all elements placed in C belong to $P^{\neg D}$. As $P^A \cap P^{\neg A} = \emptyset$ and $P^{\neg D} \cap P^D = \emptyset$, this implies that what happens in the quarters A and B, as well as C and D, is (almost) independent. This key observation can be used to develop an algorithm that solves this special case in time roughly $O(2^{3n/4})$.

Let $W_3^B = I_2 \cap (\sigma_{opt}^{-1}(B) \setminus P^{\neg A})$ and $W_3^C = I_2 \cap (\sigma_{opt}^{-1}(C) \setminus P^{\neg D})$. As $p^B, p^C \geq (1/4 - \varepsilon_4)n$ we have that $|W_3^B|, |W_3^C| \leq \varepsilon_4 n$. We branch into at most $n^2 \binom{n}{\varepsilon_4 n}^2$

subcases, guessing the sets W_3^B and W_3^C. Let $W_3 = W_3^B \cup W_3^C$, $I_3 = I_2 \setminus W_3$, $Q^\Delta = P^\Delta \setminus W_3$ for $\Delta \in \{A, \neg A, D, \neg D\}$. Moreover, let $W^\Gamma = W_1^\Gamma \cup W_2^\Gamma \cup W_3^\Gamma$ for $\Gamma \in \{A, B, C, D\}$, using the convention $W_3^A = W_3^D = \emptyset$.

Note that in the current branch any ordering puts into the segment Γ for $\Gamma \in \{A, B, C, D\}$ all the jobs from W^Γ and $q^\Gamma = n/4 - |W^\Gamma|$ jobs from appropriate Q^Δ ($\Delta = A, \neg A, \neg D, D$ for $\Gamma = A, B, C, D$, respectively). Thus, the behaviour of an ordering σ in A influences the behaviour of σ in C by the choice of which elements of $Q^A \cap Q^{\neg D}$ are placed in A, and which in C. Similar dependencies are between A and D, B and C, as well as B and D. Thus, the dependencies form a 4-cycle, and we can compute the optimal arrangement by keeping track of only three out of four dependencies at once, leading us to an algorithm running in time roughly $O(2^{3n/4})$. This is formalized in the following lemma:

Lemma 8 (♠). *If $2\varepsilon_1 + 2\varepsilon_2 + \varepsilon_4 < 1/4$, the remaining case can be solved by an algorithm running in time bounded by*

$$T_5(n) = \binom{n}{\varepsilon_4 n}^2 2^{(3/4+\varepsilon_3)n} n^{O(1)}.$$

So far we have $2^{(4\varepsilon_1 + 2\varepsilon_2)n} n^{O(1)}$ overhead. For sufficiently small ε_4 we have $\binom{n}{\varepsilon_4 n} = O(2^{n/16})$ and then for sufficiently small constants ε_k, $k = 1, 2, 3$ we have $2^{(4\varepsilon_1 + 2\varepsilon_2)n} n^{O(1)} T_5(n) = O(c^n)$ for some $c < 2$.

3   Conclusion

We presented an algorithm that solves SCHED in $O((2 - \varepsilon)^n)$ time for some small ε. This shows that in some sense SCHED appears to be easier than resolving CNF-SAT formulae, which is conjectured to need 2^n time (the so-called Strong Exponential Time Hypothesis). Our algorithm is based on an interesting property of the optimal solution expressed in Lemma 4, which can be of independent interest. However, our best efforts to numerically compute an optimal choice of values of the constants ε_k, $k = 1, 2, 3, 4$ lead us to an ε of the order of 10^{-15}. Although Lemma 4 seems powerful, we lost a lot while applying it. In particular, the worst trade-off seems to happen in Section 2.6, where ε_1 needs to be chosen much smaller than ε_2. The natural question is: can the base of the exponent be significantly improved?

Acknowledgements. We thank Dominik Scheder for very useful discussions on the SCHED problem during his stay in Warsaw.

References

1. Binkele-Raible, D., Brankovic, L., Cygan, M., Fernau, H., Kneis, J., Kratsch, D., Langer, A., Liedloff, M., Pilipczuk, M., Rossmanith, P., Wojtaszczyk, J.O.: Breaking the 2^n-barrier for irredundance: Two lines of attack. Accepted for Publication in Journal of Discrete Algorithms (2011)

2. Björklund, A.: Determinant sums for undirected hamiltonicity. In: 51st Annual IEEE Symposium on Foundations of Computer Science (FOCS), pp. 173–182. IEEE Computer Society, Los Alamitos (2010)
3. Björklund, A., Husfeldt, T., Kaski, P., Koivisto, M.: Fourier meets möbius: fast subset convolution. In: 39th Annual ACM Symposium on Theory of Computing (STOC), pp. 67–74 (2007)
4. Björklund, A., Husfeldt, T., Koivisto, M.: Set partitioning via inclusion-exclusion. SIAM J. Comput. 39(2), 546–563 (2009)
5. Chekuri, C., Motwani, R.: Precedence constrained scheduling to minimize sum of weighted completion times on a single machine. Discrete Applied Mathematics 98(1-2), 29–38 (1999)
6. Cygan, M., Nederlof, J., Pilipczuk, M., Pilipczuk, M., van Rooij, J.M.M., Wojtaszczyk, J.O.: Solving connectivity problems parameterized by treewidth in single exponential time. CoRR abs/1103.0534 (2011)
7. Cygan, M., Pilipczuk, M.: Exact and approximate bandwidth. Theor. Comput. Sci. 411(40-42), 3701–3713 (2010)
8. Cygan, M., Pilipczuk, M., Wojtaszczyk, J.O.: Capacitated domination faster than $o(2^n)$. In: Kaplan, H. (ed.) SWAT 2010. LNCS, vol. 6139, pp. 74–80. Springer, Heidelberg (2010)
9. Fomin, F., Kratsch, D.: Exact Exponential Algorithms. Springer, Heidelberg (2010)
10. Fomin, F.V., Grandoni, F., Kratsch, D.: A measure & conquer approach for the analysis of exact algorithms. J. ACM 56(5), 1–32 (2009)
11. Impagliazzo, R., Paturi, R.: On the complexity of k-SAT. J. Comput. Syst. Sci. 62(2), 367–375 (2001)
12. Lenstra, J.K., Kan, A.R.: Complexity of scheduling under precedence constraints. Operations Research 26, 22–35 (1978)
13. Lokshtanov, D., Marx, D., Saurabh, S.: Known Algorithms on Graphs of Bounded Treewidth are Probably Optimal. In: Proceedings of the Twenty-Second Annual ACM-SIAM Symposium on Discrete Algorithms (SODA), pp. 777–789 (2011)
14. Margot, F., Queyranne, M., Wang, Y.: Decompositions, network flows, and a precedence constrained single-machine scheduling problem. Operations Research 51(6), 981–992 (2003)
15. Mucha, M., Sankowski, P.: Maximum matchings via gaussian elimination. In: 45th Symposium on Foundations of Computer Science (FOCS), pp. 248–255 (2004)
16. Patrascu, M., Williams, R.: On the possibility of faster SAT algorithms. In: Proceedings of the Twenty-First Annual ACM-SIAM Symposium on Discrete Algorithms (SODA), pp. 1065–1075 (2010)
17. van Rooij, J.M.M., Nederlof, J., van Dijk, T.C.: Inclusion/exclusion meets measure and conquer. In: Fiat, A., Sanders, P. (eds.) ESA 2009. LNCS, vol. 5757, pp. 554–565. Springer, Heidelberg (2009)
18. Woeginger, G.J.: Space and time complexity of exact algorithms: Some open problems (Invited talk). In: Downey, R.G., Fellows, M.R., Dehne, F. (eds.) IWPEC 2004. LNCS, vol. 3162, pp. 281–290. Springer, Heidelberg (2004)
19. Woeginger, G.J.: Open problems around exact algorithms. Discrete Applied Mathematics 156(3), 397–405 (2008)

AdCell: Ad Allocation in Cellular Networks

Saeed Alaei[1,*], Mohammad T. Hajiaghayi[1,2,**], Vahid Liaghat[1,***],
Dan Pei[2], and Barna Saha[1,†]

[1] University of Maryland, College Park, MD, 20742
[2] AT&T Labs - Research, 180 Park Avenue, Florham Park, NJ 07932
{saeed,hajiagha,vliaghat,barna}@cs.umd.edu,
peidan@research.att.com

Abstract. With more than four billion usage of cellular phones worldwide, mobile advertising has become an attractive alternative to online advertisements. In this paper, we propose a new targeted advertising policy for Wireless Service Providers (WSPs) via SMS or MMS- namely *AdCell*. In our model, a WSP charges the advertisers for showing their ads. Each advertiser has a valuation for specific types of customers in various times and locations and has a limit on the maximum available budget. Each query is in the form of time and location and is associated with one individual customer. In order to achieve a non-intrusive delivery, only a limited number of ads can be sent to each customer. Recently, new services have been introduced that offer location-based advertising over cellular network that fit in our model (e.g., ShopAlerts by AT&T) .

We consider both online and offline version of the AdCell problem and develop approximation algorithms with constant competitive ratio. For the online version, we assume that the appearances of the queries follow a stochastic distribution and thus consider a Bayesian setting. Furthermore, queries may come from different distributions on different times. This model generalizes several previous advertising models such as online secretary problem [10], online bipartite matching [13,7] and AdWords [18]. Since our problem generalizes the well-known secretary problem, no non-trivial approximation can be guaranteed in the online setting without stochastic assumptions. We propose an online algorithm that is simple, intuitive and easily implementable in practice. It is based on pre-computing a fractional solution for the expected scenario and relies on a novel use of dynamic programming to compute the conditional expectations. We give tight lower bounds on the approximability of some variants of the problem as well. In the offline setting, where full-information is available, we achieve near-optimal bounds, matching the integrality gap of the considered linear program. We believe that our proposed solutions can be used for other advertising settings where personalized advertisement is critical.

Keywords: Mobile Advertisement, AdCell, Online, Matching.

* Supported in part by NSF Grant CCF-0728839.
** Supported in part by NSF CAREER Award, ONR Young Investigator Award, and Google Faculty Research Award.
*** Supported in part by NSF CAREER Award, ONR Young Investigator Award, and Google Faculty Research Award.
† Supported in part by NSF Award CCF-0728839, NSF Award CCF-0937865.

C. Demetrescu and M.M. Halldórsson (Eds.): ESA 2011, LNCS 6942, pp. 311–322, 2011.
© Springer-Verlag Berlin Heidelberg 2011

1 Introduction

In this paper, we propose a new mobile advertising concept called *Adcell*. More than 4 billion cellular phones are in use world-wide, and with the increasing popularity of smart phones, mobile advertising holds the prospect of significant growth in the near future. Some research firms [1] estimate mobile advertisements to reach a business worth over 10 billion US dollars by 2012. Given the built-in advertisement solutions from popular smart phone OSes, such as iAds for Apple's iOS, mobile advertising market is poised with even faster growth.

In the mobile advertising ecosystem, wireless service providers (WSPs) render the physical delivery infrastructure, but so far WSPs have been more or less left out from profiting via mobile advertising because of several challenges. First, unlike web, search, application, and game providers, WSPs typically do not have users' application context, which makes it difficult to provide targeted advertisements. Deep Packet Inspection (DPI) techniques that examine packet traces in order to understand application context, is often not an option because of privacy and legislation issues (i.e., Federal Wiretap Act). Therefore, a targeted advertising solution for WSPs need to utilize *only the information it is allowed to collect by government and by customers via opt-in mechanisms*. Second, without the luxury of application context, targeted ads from WSPs require *non-intrusive delivery methods*. While users are familiar with other ad forms such as banner, search, in-application, and in-game, push ads with no application context (e.g., via SMS) can be intrusive and annoying if not done carefully. The number and frequency of ads both need to be well-controlled. Third, targeted ads from WSPs should be well personalized such that the users have incentive to read the advertisements and take purchasing actions, especially given the requirement that the number of ads that can be shown to a customer is limited.

In this paper, we propose a new mobile targeted advertising strategy, *AdCell*, for WSPs that deals with the above challenges. It takes advantage of the detailed real-time location information of users. Location can be tracked upon users' consent. This is already being done in some services offered by WSPs, such as Sprint's Family Location and AT&T's Family Map, thus there is no associated privacy or legal complications. To locate a cellular phone, it must emit a roaming signal to contact some nearby antenna tower, but the process does not require an active call. GSM localization is then done by multi-lateration[1] based on the signal strength to nearby antenna masts [22]. Location-based advertisement is not completely new. Foursquare mobile application allows users to explicitly "check in" at places such as bars and restaurants, and the shops can advertise accordingly. Similarly there are also automatic proximity-based advertisements using GPS or bluetooth. For example, some GPS models from Garmin display ads for the nearby business based on the GPS locations [23]. ShopAlerts by AT&T[2] is another application along the same line. On the advertiser side, popular stores such as Starbucks are reported to have attracted significant footfalls via mobile coupons.

---

[1] The process of locating an object by accurately computing the time difference of arrival of a signal emitted from that object to three or more receivers.

[2] http://shopalerts.att.com/sho/att/index.html?ref=portal

Most of the existing mobile advertising models are On-Demand, however, AdCell sends the ads via SMS, MMS, or similar methods without any prior notice. Thus to deal with the non-intrusive delivery challenge, we propose user subscription to advertising services that deliver only a *fixed number* of ads per month to its subscribers (as it is the case in AT&T ShopAlerts). The constraint of delivering limited number of ads to each customer adds the main algorithmic challenge in the AdCell model (details in Section 1.1). In order to overcome the incentive challenge, the WSP can "pay" users to read ads and purchase based on them through a reward program in the form of credit for monthly wireless bill. To begin with, both customers and advertisers should sign-up for the AdCell-service provided by the WSP (e.g., currently there are 9 chain-companies participating in ShopAlerts). Customers enrolled for the service should sign an agreement that their *location* information will be tracked; but solely for the advertisement purpose. Advertisers (e.g., stores) provide their advertisements and a maximum chargeable budget to the WSP. The WSP selects proper ads (these, for example, may depend on time and distance of a customer from a store) and sends them (via SMS) to the customers. The WSP charges the advertisers for showing their ads and also for successful ads. An ad is deemed successful if a customer visits the advertised store. Depending on the service plan, customers are entitled to receive different number of advertisements per month. Several logistics need to be employed to improve AdCell experience and enthuse customers into participation. We provide more details about these logistics in the full paper.

1.1 AdCell Model and Problem Formulation

In the AdCell model, advertisers bid for individual customers based on their location and time. The triple (k, ℓ, t) where k is a customer, ℓ is a neighborhood (location) and t is a time forms a *query* and there is a bid amount (possibly zero) associated with each query for each advertiser. This definition of query allows advertisers to customize their bids based on customers, neighborhoods and time. We assume a customer can only be in one neighborhood at any particular time and thus at any time t and for each customer k, the queries (k, ℓ_1, t) and (k, ℓ_2, t) are mutually exclusive, for all distinct l_1, l_2. Neighborhoods are places of interest such as shopping malls, airports, etc. We assume that queries are generated at certain times (e.g., every half hour) and only if a customer stays within a neighborhood for a specified minimum amount of time. The formal problem definition of *AdCell Allocation* is as follows:

AdCell Allocation. *There are m advertisers, n queries and s customers. Advertiser i has a total budget b_i and bids u_{ij} for each query j. Furthermore, for each customer $k \in [s]$, let S_k denote the queries corresponding to customer k and c_k denote the maximum number of ads which can be sent to customer k. The capacity c_k is associated with customer k and is dictated by the AdCell plan the customer has signed up for. Advertiser i pays u_{ij} if his advertisement is shown for query j and if his budget is not exceeded. That is, if x_{ij} is an indicator variable set to 1, when advertisement for advertiser i is shown on query j, then advertiser i pays a total amount of $\min(\sum_j x_{ij} u_{ij}, b_i)$. The goal of AdCell Allocation is to specify an advertisement allocation plan such that the total payment $\sum_i \min(\sum_j x_{ij} u_{ij}, b_i)$ is maximized.*

The AdCell problem is a generalization of the budgeted AdWords allocation problem [4,21] with capacity constraint on each customer and thus is NP-hard. Along with the offline version of the problem, we also consider its online version where queries arrive online and a decision to assign a query to an advertiser has to be done right away. With arbitrary queries/bids and optimizing for the worst case, one cannot obtain any approximation algorithm with ratio better than $\frac{1}{n}$. This follows from the observation that online AdCell problem also generalizes the *secretary problem* for which no deterministic or randomized online algorithm can get approximation ratio better than $\frac{1}{n}$ in the worst case.[3]. Therefore, we consider a stochastic setting.

For the online AdCell problem, we assume that each query j arrives with probability p_j. Upon arrival, each query has to be either allocated or discarded right away. We note that each query encodes a customer id, a location id and a time stamp. Also associated with each query, there is a probability, and a vector consisting of the bids for all advertisers for that query. Furthermore, we assume that all queries with different arrival times or from different customers are independent, however queries from the same customer with the same arrival time are mutually exclusive (i.e., a customer cannot be in multiple locations at the same time).

1.2 Our Results and Techniques

Here we provide a summary of our results and techniques. We consider both the offline and online version of the problem. In the offline version, we assume that we know exactly which queries arrive. In the online version, we only know the arrival probabilities of queries (i.e., $p_1, \cdots, p_m$).

We can write the AdCell problem as the following random integer program in which $\mathbf{I}_j$ is the indicator random variable which is 1 if query j arrives and 0 otherwise:

$$\text{maximize.} \qquad \sum_i \min(\sum_j \mathbf{X}_{ij} u_{ij}, b_i) \qquad (IP_{BC})$$

$$\forall j \in [n]: \qquad \sum_i \mathbf{X}_{ij} \leq \mathbf{I}_j \qquad (F)$$

$$\forall k \in [s]: \qquad \sum_{j \in S_k} \sum_i \mathbf{X}_{ij} \leq c_k \qquad (C)$$

$$\mathbf{X}_{ij} \in \{0, 1\}$$

We will refer to the variant of the problem explained above as IP_{BC}. We also consider variants in which there are either budget constraints or capacity constraints but not both. We refer to these variants as IP_B and IP_C respectively. The above integer program can be relaxed to obtain a linear program LP_{BC}, where we maximize $\sum_i \sum_j \mathbf{X}_{ij} u_{ij}$ with the constraints (F), (C) and additional budget constraint $\sum_j \mathbf{X}_{ij} u_{ij} \leq b_i$ which we refer to by (B). We relax $\mathbf{X}_{ij} \in \{0, 1\}$ to $\mathbf{X}_{ij} \in [0, 1]$. We also refer to the variant of

---

[3] The reduction of the *secretary problem* to AdCell problem is as follows: consider a single advertiser with large enough budget and a single customer with a capacity of 1. The queries correspond to secretaries and the bids correspond to the values of the secretaries. So we can only allocate one query to the advertiser.

this linear program with only either constraints of type (B) or constraints of type (C) as LP_B and LP_C.

In the offline version, for all $i \in [m]$ and $j \in [n]$, the values of $\mathbf{I}_j$ are precisely known. For the online version, we assume to know the $E[\mathbf{I}_j]$ in advance and we learn the actual value of $\mathbf{I}_j$ online. We note a crucial difference between our model and the i.i.d model. In i.i.d model the probability of the arrival of a query is independent of the time, i.e., queries arrive from the same distribution on each time. However, in AdCell model a query encodes time (in addition to location and customer id), hence we may have a different distribution on each time. This implies a prophet inequality setting in which on each time, an onlooker has to decide according to a given value where this value may come from a different distribution on different times (e.g. see [14,11]).

A summary of our results are shown in Table 1. In the online version, we compare the expected revenue of our solution with the expected revenue of the optimal offline algorithm. We should emphasis that we make no assumptions about bid to budget ratios (e.g., bids could be as large as budgets). In the offline version, our result matches the known bounds on the integrality gap.

We now briefly describe our main techniques.

Breaking into smaller sub-problems that can be optimally solved using conditional expectation. Theoretically, ignoring the computational issues, any online stochastic optimization problem can be solved optimally using conditional expectation as follows: At any time a decision needs to be made, compute the total expected objective conditioned on each possible decision, then chose the one with the highest total expectation. These conditional expectations can be computed by backward induction, possibly using a dynamic program. However for most problems, including the AdCell problem, the size of this dynamic program is exponential which makes it impractical. We avoid this issue by using a randomized strategy to break the problem into smaller subproblems such that each subproblem can be solved by a quadratic dynamic program.

Using an LP to analyze the performance of an optimal online algorithm against an optimal offline fractional solution. Note that we compare the expected objective value of our algorithm against the expected objective value of the optimal offline fractional solution. Therefore for each subproblem, even though we use an optimal online algorithm, we still need to compare its expected objective value against the expected objective value of the optimal offline solution for that subproblem. Basically, we need to compare the expected objective of an stochastic online algorithm, which works by maximizing conditional expectation at each step, against the expected objective value

Table 1. Summary of Our Results

Offline Version	Online Version
– A $\frac{3}{4}$-approximation algorithm.	– A $\left(\frac{1}{2} - \frac{1}{e}\right)$-approximation algorithm.
– A $\frac{4-\epsilon}{4}$-approximation algorithm when $\forall_i \max_j u_{ij} \leq \epsilon b_i$.	– A $\left(1 - \frac{1}{e}\right)$-approximation algorithm with only budget constraints.
	– A $\frac{1}{2}$-approximation algorithm with only capacity constraints.

of its optimal offline solution. To do this, we create a minimization linear program that encodes the dynamic program and whose optimal objective is the minimum ratio of the expected objective value of the online algorithm to the expected objective value of the optimal offline solution. We then prove a lower bound of $\frac{1}{2}$ on the objective value of this linear program by constructing a feasible solution for its dual obtaining an objective value of $\frac{1}{2}$.

Rounding method of [20] and handling hard capacities. Handling "hard capacities", those that cannot be violated, is generally tricky in various settings including facility location and many covering problems [5,8,19]. The AdCell problem is a generalization of the budgeted AdWords allocation problem with hard capacities on queries involving each customer. Our essential idea is to iteratively round the fractional LP solution to an integral one based on the current LP structure. The algorithm uses the rounding technique of [20] and is significantly harder than its uncapacitated version.

Due to the interest of the space we differ the omitted proofs to the full paper.

2   Related Work

Online advertising alongside search results is a multi-billion dollar business [15] and is a major source of revenue for search engines like Google, Yahoo and Bing. A related ad allocation problem is the AdWords assignment problem [18] that was motivated by sponsored search auctions. When modeled as an online bipartite assignment problem, each edge has a weight, and there is a budget on each advertiser representing the upper bound on the total weight of edges that might be assigned to it. In the offline setting, this problem is NP-Hard, and several approximations have been proposed [3,2,4,21]. For the online setting, it is typical to assume that edge weights (i.e., bids) are much smaller than the budgets, in which case there exists a $(1 - 1/e)$-competitive online algorithm [18]. Recently, Devanur and Hayes [6] improved the competitive ratio to $(1 - \epsilon)$ in the stochastic case where the sequence of arrivals is a random permutation.

Another related problem is the online bipartite matching problem which is introduced by Karp, Vazirani, and Vazirani [13]. They proved that a simple randomized online algorithm achieves a $(1 - 1/e)$-competitive ratio and this factor is the best possible. Online bipartite matching has been considered under stochastic assumptions in [9,7,17], where improvements over $(1 - 1/e)$ approximation factor have been shown. The most recent of of them is the work of Manshadi et al. [17] that presents an online algorithm with a competitive ratio of 0.702. They also show that no online algorithm can achieve a competitive ratio better than 0.823. More recently, Mahdian et al.[16] and Mehta et al.[12] improved the competitive ratio to 0.696 for unknown distributions.

3   Online Setting

In this section, we present three online algorithms for the three variants of the problem mentioned in the pervious section (i.e., IP_B, IP_C and IP_{BC}).

First, we present the following lemma which provides a means of computing an upper bound on the expected revenue of any algorithm (both online and offline) for the AdCell problem.

Lemma 1 (Expectation Linear Program). *Consider a general random linear program in which* $\mathfrak{b}$ *is a vector of random variables:*

> *(Random LP)*
>
> *maximize.* $\qquad\qquad c^T x$
>
> *s.t.* $\qquad\qquad Ax \leq \mathfrak{b}; \ \ x \geq 0$

Let $OPT(\mathfrak{b})$ *denote the optimal value of this program as a function of the random variables. Now consider the following linear program:*

> *(Expectation LP)*
>
> *maximize.* $\qquad\qquad c^T x$
>
> *s.t.* $\qquad\qquad Ax \leq E[\mathfrak{b}]; \ \ x \geq 0$

We refer to this as the "Expectation Linear Program" *corresponding to the* "Random Linear Program". *Let* $\overline{OPT}$ *denote the optimal value value of this program. Assuming that the original linear program is feasible for all possible draws of the random variables, it always holds that* $E[OPT(\mathfrak{b})] \leq \overline{OPT}$.

Proof. Let $x^*(\mathfrak{b})$ denote the optimal assignment as a function of $\mathfrak{b}$. Since the random LP is feasible for all realizations of $\mathfrak{b}$, we have $Ax^*(\mathfrak{b}) \leq \mathfrak{b}$. Taking the expectation from both sides, we get $AE[x^*(\mathfrak{b})] \leq E[\mathfrak{b}]$. So, by setting $x = E[x^*(\mathfrak{b})]$ we get a feasible solution for the expectation LP. Furthermore, the objective value resulting from this assignment is equal to the expected optimal value of the random LP. The optimal value of the expectation LP might however be higher so its optimal value is an upper bound on the expected optimal value of random LP.

As we will see next, not only does the expectation LP provide an upper bound on the expected revenue, it also leads to a good approximate algorithm for the online allocation as we explain in the following online allocation algorithm. We adopt the notation of using an overline to denote the expectation linear program corresponding to a random linear program (e.g. $\overline{LP}_{BC}$ for LP_{BC}). Next we present an online algorithm for the variant of the problem in which there are only budget constrains but not capacity constraints.

Algorithm 1 (STOCHASTIC ONLINE ALLOCATOR FOR IP_B)

- *Compute an optimal assignment for the corresponding expectation LP (i.e.* $\overline{LP}_B$*). Let* x_{ij}^* *denote this assignment. Note that* x_{ij}^* *might be a fractional assignment.*
- *If query j arrives, for each $i \in [m]$ allocate the query to advertiser i with probability* $\frac{x_{ij}^*}{p_j}$.

Theorem 1. *The expected revenue of 1 is at least $1 - \frac{1}{e}$ of the optimal value of the expectation LP (i.e.,* $\overline{LP}_B$*) which implies that the expected revenue of 1 it is at least $1 - \frac{1}{e}$ of the expected revenue of the optimal offline allocation too. Note that this result holds even if u_{ij}'s are not small compared to b_i. Furthermore, this result holds even if we relax the independence requirement in the original problem and require negative correlation instead.*

Note that allowing negative correlation instead of independence makes the above model much more general than it may seem at first. For example, suppose there is a query that may arrive at several different times but may only arrive at most once or only a limited number of times, we can model this by creating a new query for each possible instance of the original query. These new copies are however negatively correlated.

Remark 1. It is worth mentioning that there is an integrality gap of $1 - \frac{1}{e}$ between the optimal value of the integral allocation and the optimal value of the expectation LP. So the lower bound of Theorem 1 is tight. To see this, consider a single advertiser and n queries. Suppose $p_j = \frac{1}{n}$ and $u_{1j} = 1$ for all j. The optimal value of $\overline{LP}_B$ is 1 but even the expected optimal revenue of the offline optimal allocation is $1 - \frac{1}{e}$ when $n \to \infty$ because with probability $(1 - \frac{1}{n})^n$ no query arrives.

To prove Theorem 1, we use the following theorem:

Theorem 2. *Let C be an arbitrary positive number and let $\mathbf{X}_1, \cdots, \mathbf{X}_n$ be independent random variables (or negatively correlated) such that $\mathbf{X}_i \in [0, C]$. Let $\mu = E[\sum_i \mathbf{X}_i]$. Then:*

$$E[\min(\textstyle\sum_i \mathbf{X}_i, C)] \geq (1 - \tfrac{1}{e^{\mu/C}})C$$

Furthermore, if $\mu \leq C$ then the right hand side is at least $(1 - \frac{1}{e})\mu$.

Proof (Theorem 1). We apply Theorem 2 to each advertiser i separately. From the perspective of advertiser i, each query is allocated to her with probability x_{ij}^* and by constraint (B) we can argue that have $\mu = \sum_j x_{ij}^* u_{ij} \leq b_i = C$ so $\mu \leq C$ and by Theorem 2, the expected revenue from advertiser i is at least $(1 - \frac{1}{e})(\sum_j x_{ij}^* u_{ij})$. Therefore, overall, we achieve at least $1 - \frac{1}{e}$ of the optimal value of the expectation LP and that completes the proof.

Next we present an online algorithm for the variant of the problem in which there are only capacity constrains but not budget constraints.

Algorithm 2 (STOCHASTIC ONLINE ALLOCATOR FOR IP_C)

- *Compute an optimal assignment for the corresponding expectation LP (i.e. $\overline{LP}_C$). Let x_{ij}^* denote this assignment. Note that x_{ij}^* might be a fractional assignment.*
- *Partition the items to sets $T_1, \cdots, T_u$ in increasing order of their arrival time and such that all of the items in the same set have the same arrival time.*
- *For each $k \in [s], t \in [u], r \in [c_k]$, let $E_{k,t}^r$ denote the expected revenue of the algorithm from queries in S_k (i.e., associated with customer k) that arrive at or after T_t and assuming that the remaining capacity of customer k is r. We formally define $E_{k,t}^r$ later.*
- *If query j arrives then choose one of the advertisers at random with advertiser i chosen with a probability of $\frac{x_{ij}^*}{p_j}$. Let k and T_t be respectively the customer and the partition which query j belongs to. Also, let r be the remaining capacity of customer k (i.e. r is c_k minus the number of queries from customer k that have been allocated so far). If $u_{ij} + E_{k,t+1}^{r-1} \geq E_{k,t+1}^r$ then allocate query j to advertiser i otherwise discard query j.*

We can now define $E_{k,t}^r$ recursively as follows:

$$E_{k,t}^r = \sum_{j \in T_t} \sum_{i \in [m]} x_{ij}^* \max(u_{ij} + E_{k,t+1}^{r-1}, E_{k,t+1}^r)$$

$$+ \left(1 - \sum_{j \in T_t} \sum_{i \in [m]} x_{ij}^*\right) E_{k,t+1}^r \qquad (\text{EXP}_k)$$

Also define $E_{k,t}^0 = 0$ and $E_{k,u+1}^r = 0$. Note that we can efficiently compute $E_{k,t}^r$ using dynamic programming.

The main difference between 1 and 2 is that in the former whenever we choose an advertiser at random, we always allocate the query to that advertiser (assuming they have enough budget). However, in the latter, we run a dynamic program for each customer k and once an advertiser is picked at random, the query is allocated to this advertiser only if doing so increases the expected revenue associated with customer k.

Theorem 3. *The expected revenue of 2 is at least $\frac{1}{2}$ of the optimal value of the expectation LP (i.e., $\overline{LP}_C$) which implies that the expected revenue of 2 it is at least $\frac{1}{2}$ of the expected revenue of the optimal offline allocation for IP_C too.*

Remark 2. The approximation ratio of 2 is tight. There is no online algorithm that can achieve in expectation better than $\frac{1}{2}$ of the revenue of the optimal offline allocation without making further assumptions. We show this by providing a simple example. Consider an advertiser with a large enough budget and a single customer with a capacity of 1 and two queries. The queries arrive independently with probabilities $p_1 = 1 - \epsilon$ and $p_2 = \epsilon$ with the first query having an earlier arrival time. The advertiser has submitted the bids $b_{11} = 1$ and $b_{12} = \frac{1-\epsilon}{\epsilon}$. Observe that no online algorithm can get a revenue better than $(1-\epsilon) \times 1 + \epsilon^2 \frac{1-\epsilon}{\epsilon} \approx 1$ in expectation because at the time query 1 arrives, the online algorithm does not know whether or not the second query is going to arrive and the expected revenue from the second query is just $1 - \epsilon$. However, the optimal offline solution would allocate the second query if it arrives and otherwise would allocate the first query so its revenue is $\epsilon \frac{1-\epsilon}{\epsilon} + (1 - \epsilon)^2 \times 1 \approx 2$ in expectation.

Next, we show that an algorithm similar to the previous one can be used when there are both budget constraints and capacity constraints.

Algorithm 3 (STOCHASTIC ONLINE ALLOCATOR FOR IP_{BC})
Run the same algorithm as in 2 except that now x_{ij}^* is a fractional solution of $\overline{LP}_{BC}$ instead of $\overline{LP}_C$.

Theorem 4. *The expected revenue of 3 is at least $\frac{1}{2} - \frac{1}{e}$ of the optimal value of the expectation LP (i.e., $\overline{LP}_{BC}$) which implies that the expected revenue of 3 it is at least $\frac{1}{2} - \frac{1}{e}$ of the expected revenue of the optimal offline allocation too.*

We prove the last two theorems by defining a simple stochastic uniform knapsack problem which will be used as a building block in our analysis. Due to the interest of the space we have moved the proofs to the full paper.

4   Offline Setting

In the offline setting, we explicitly know all the queries, that is all the customers, locations, items triplets on which advertisers put their bids. We want to obtain an allocation of advertisers to queries such that the total payment obtained from all the advertisers is maximized. Each advertiser pays an amount equal to the minimum of his budget and the total bid value on all the queries assigned to him. Since, the problem is NP-Hard, we can only obtain an approximation algorithm achieving revenue close to the optimal. The fractional optimal solution of LP_{BC} (with explicit values for $\mathcal{I}_j, j \in [n]$) acts as an upper bound on the optimal revenue. We round the fractional optimal solution to a nearby integer solution and establish the following bound.

Theorem 5. *Given a fractional optimal solution for LP_{BC}, we can obtain an integral solution for AdCell with budget and capacity constraints that obtains at least a profit of $\frac{4 - \max_i \frac{u_{i,max}}{b_i}}{4}$ of the profit obtained by optimal fractional allocation and maintains all the capacity constraints exactly.*

We note that this approximation ratio is best possible using the considered LP relaxation due to an integrality gap example from [4]. The problem considered in [4] is an uncapacitated version of the AdCell problem, that is there is no capacity constraint (C) on the customers. Capacity constraint restricts how many queries/advertisements can be assigned to each customer. We can represent all the queries associated with each customer as a set; these sets are therefore disjoint and has integer hard capacities associated with them. Our approximation ratio matches the best known bound from [4,21] for the uncapacitated case. For space limitation, most of the details have been moved to the full paper. Here, we give a high-level description of the algorithm. Our algorithm is based on applying the rounding technique of [20] through several iterations. The essential idea of the proposed rounding is to apply a procedure called **Rand-move** to the variables of a suitably chosen subset of constraints from the original linear program. These sub-system must be underdetermined to ensure that the rounding proceeds without violating any constraint and at least one variable becomes integral. The trick lies on choosing a proper sub-system at each step of rounding, which again depends on a detailed case analysis of the LP structure.

Let y^* denote the LP optimal solution. We begin by simplifying the assignment given by y^*. Consider a bipartite graph $G(\mathcal{B}, \mathcal{I}, E^*)$ with advertisers $\mathcal{B}$ on one side, queries $\mathcal{I}$ on the other side and add an edge (i, j) between a advertiser i and query j, if $y^*_{i,j} \in (0, 1)$. That is, define $E^* = \{(i,j)|\ 1 > y^*_{i,j} > 0\}$. Our first claim is that y^* can be modified without affecting the optimal fractional value and the constraints such that $G(\mathcal{B}, \mathcal{I}, E^*)$ is a forest. The proof follows from Claim 2.1 of [4]; we additionally show that such assumption of forest structure maintains the capacity constraints.

Once, we have such a forest structure, several cases arise and depending on the cases, we define a suitable sub-system on which to apply the rounding technique. There are three major cases.

(i) There is a tree with two leaf advertiser nodes: in that case, we show that applying our rounding technique only diminishes the objective function by little and all constraints are maintained.

(ii) No tree contains two leaf advertisers, but there is a tree that contains one leaf advertiser: we start with a leaf advertiser and construct a path spanning several trees such that we either end up with a combined path with advertisers on both side or a query node in one side such that the capacity constraint on the set containing that query is not met with equality (non-tight constraint). This is the most nontrivial case and a detailed discussion is given in the full paper.

(iii) No tree contains any leaf advertiser nodes: in that case we again form a combined path spanning several trees such that the queries on two ends of the combined path come from sets with non-tight capacity constraints.

References

1. Agrawal, M.: Overview of mobile advertising, `http://www.telecomcircle.com/2009/11/overview-of-mobile-advertising`
2. Andelman, N., Mansour, Y.: Auctions with budget constraints. In: Hagerup, T., Katajainen, J. (eds.) SWAT 2004. LNCS, vol. 3111, pp. 26–38. Springer, Heidelberg (2004)
3. Azar, Y., Birnbaum, B., Karlin, A.R., Mathieu, C., Nguyen, C.T.: Improved approximation algorithms for budgeted allocations. In: Aceto, L., Damgård, I., Goldberg, L.A., Halldórsson, M.M., Ingólfsdóttir, A., Walukiewicz, I. (eds.) ICALP 2008, Part I. LNCS, vol. 5125, pp. 186–197. Springer, Heidelberg (2008)
4. Chakrabarty, D., Goel, G.: On the approximability of budgeted allocations and improved lower bounds for submodular welfare maximization and gap. SIAM J. Comp, 2189 (2010)
5. Chuzhoy, J., Naor, J.S.: Covering problems with hard capacities. In: FOCS 2002, p. 481 (2002)
6. Devenur, N.R., Hayes, T.P.: The adwords problem: online keyword matching with budgeted bidders under random permutations. In: EC 2009, pp. 71–78 (2009)
7. Feldman, J., Mehta, A., Mirrokni, V., Muthukrishnan, S.: Online stochastic matching: Beating 1-1/e. In: FOCS 2009, pp. 117–126 (2009)
8. Gandhi, R., Halperin, E., Khuller, S., Kortsarz, G., Srinivasan, A.: An improved approximation algorithm for vertex cover with hard capacities. In: ICALP 2003, pp. 164–175 (2003)
9. Goel, G., Mehta, A.: Online budgeted matching in random input models with applications to adwords. In: SODA 2008, pp. 982–991 (2008)
10. Hajiaghayi, M.T., Kleinberg, R., Parkes, D.C.: Adaptive limited-supply online auctions. In: EC 2004, pp. 71–80 (2004)
11. Hajiaghayi, M.T., Kleinberg, R.D., Sandholm, T.: Automated online mechanism design and prophet inequalities. In: AAAI (2007)
12. Karande, C., Mehta, A., Tripathi, P.: Online bipartite matching with unknown distributions. In: STOC (2011)
13. Karp, R.M., Vazirani, U.V., Vazirani, V.V.: An optimal algorithm for on-line bipartite matching. In: STOC 1990, pp. 352–358 (1990)
14. Krengel, U., Sucheston, L.: Semiamarts and finite values. Bull. Am. Math. Soc (1977)
15. Lahaie, S.: An analysis of alternative slot auction designs for sponsored search. In: EC 2006, pp. 218–227 (2006)
16. Mahdian, M., Yan, Q.: Online bipartite matching with random arrivals: An approach based on strongly factor-revealing lps. In: STOC (2011)
17. Manshadi, V.H., Gharan, S.O., Saberi, A.: Online stochastic matching: Online actions based on offline statistics. In: SODA (2011)
18. Mehta, A., Saberi, A., Vazirani, U., Vazirani, V.: Adwords and generalized online matching. J. ACM 54 (2007)

19. Pál, M., Tardos, E., Wexler, T.: Facility location with nonuniform hard capacities. In: FOCS 2001, pp. 329–338 (2001)
20. Saha, B., Srinivasan, A.: A new approximation technique for resource-allocation problems. In: ICS 2010, pp. 342–357 (2010)
21. Srinivasan, A.: Budgeted allocations in the full-information setting. In: Goel, A., Jansen, K., Rolim, J.D.P., Rubinfeld, R. (eds.) APPROX 2008. LNCS, vol. 5171, pp. 247–253. Springer, Heidelberg (2008)
22. Wang, S., Min, J., Yi, B.K.: Location based services for mobiles: Technologies and standards. In: ICC (2008)
23. Zahradnik, F.: Garmin to offer free, advertising-supported traffic detection and avoidance, http://gps.about.com/b/2008/09/15/garmin-to-offer-free-advertising-supported-traffic-detection-and-avoidance.htm

Submodular Max-SAT

Yossi Azar[1,*], Iftah Gamzu[2], and Ran Roth[1]

[1] Blavatnik School of Computer Science, Tel-Aviv University, Tel-Aviv 69978, Israel
{azar,ranroth}@tau.ac.il
[2] Microsoft R&D Center, Herzliya 46725, Israel
iftah.gamzu@cs.tau.ac.il

Abstract. We introduce the submodular Max-SAT problem. This problem is a natural generalization of the classical Max-SAT problem in which the additive objective function is replaced by a submodular one. We develop a randomized linear-time 2/3-approximation algorithm for the problem. Our algorithm is applicable even for the online variant of the problem. We also establish hardness results for both the online and offline settings. Notably, for the online setting, the hardness result proves that our algorithm is best possible, while for the offline setting, the hardness result establishes a computational separation between the classical Max-SAT and the submodular Max-SAT.

1   Introduction

Max-SAT is one of the most fundamental combinatorial optimization problems in computer science. As input for this problem, we are given a set of boolean variables and a collection weighted CNF clauses. The objective is to find a truth assignment for the variables which maximizes the sum of weights of satisfied clauses. In recent years, there has been a surge of interest in understanding the limits of tractability of optimization problems in which the classic additive objective function was replaced by a submodular one. Submodularity arises naturally in many practical scenarios, most notably in economics due to the property of diminishing returns. In consequence, it seems natural to study the submodular Max-SAT problem. In this variant, the value of a given assignment is the value that a monotone submodular function gives to the set of satisfied clauses. Note that in the special case in which this function is additive then one obtains the classical Max-SAT.

In this paper, we concentrate on developing fast approximation algorithms for submodular Max-SAT. In particular, we are interested in linear time algorithms. Furthermore, we study the online version of submodular Max-SAT. In this variant, the variables arrive one by one. Once a variable arrives, it declares the clauses in which it (or its negation) participates, and then, the algorithm needs to make an irrevocable decision regarding its assignment. Clearly, the algorithm does not have any information about the variables that will arrive in the future and the clauses in which they will appear.

---

* Partially supported by the Israeli Science Foundation grant No. 1404/10.

C. Demetrescu and M.M. Halldórsson (Eds.): ESA 2011, LNCS 6942, pp. 323–334, 2011.

Our contribution. Our results can be briefly described as follows:

- We develop a 2/3-approximation algorithm for the submodular Max-SAT problem. Our algorithm is combinatorial, randomized, simple to implement, and runs in linear time. In fact, our algorithm does a single pass over the input (in some arbitrary order), and therefore, it is also applicable in the online setting.
- We establish that no online algorithm (deterministic or randomized) can attain a competitive ratio better than 2/3 for the online version of the classical Max-SAT problem. This result clearly holds also for the more general online submodular Max-SAT problem. Our result implies that the analysis of our randomized algorithm is tight, and that it is best possible in the online setting.
- We prove an information-theoretic inapproximability result of 3/4 for the offline variant of the submodular Max-SAT problem. This result is based on an observation regarding the equivalence of submodular Max-SAT and the problem of maximizing a monotone submodular function under a so-called binary partition matroid. The hardness result then follows by observing that problem of combinatorial auctions with two submodular bidders is a special instance of our problem in which the monotone submodular function has an additional constraint on its structure. For this problem, Mirrokni, Schapira and Vondrák [20] established a hardness bound of 3/4. We also provide an alternative proof which may be interesting on its own right.

An interesting consequence of our results is an identification of a computational separation between the classical Max-SAT and its submodular counterpart. In particular, this separation is obtained by noticing that the classical Max-SAT can be approximated to within a factor strictly better than 3/4 [2]. In contrast, Online Submodular Max-SAT is shown to be computationally equivalent to Online Classical Max-SAT.

Related work. The classical Max-SAT problem has been given significant attention in the past (see, e.g., [27,13,12,1,17,3]). It is known to be NP-hard to approximate within a factor better than 7/8 = 0.875 [15,21]. The best algorithm for the problem, presented by Asano [2], achieves a provable approximation guarantee of 0.7877 and a conjectured approximation ratio of 0.8353. Specifically, the latter ratio relies on a conjecture of Zwick [28].

The problem of finding combinatorial and possibly online algorithms to Max-SAT has also gained notable attention. In a pioneer work in approximation algorithms, Johnson suggested a greedy algorithm using modified weights [16]. This algorithm was later shown to achieve a 2/3 approximation by Chen, Friesen and Zheng[6] and a simpler proof was given by Engebretsen[9]. Very recently, a randomized combinatorial algorithm achieving a 3/4 approximation to Max-SAT was presented by Poloczek and Schnitger [24]. Their results also hold for online Max-SAT, however, we note that their algorithm does not contradict our hardness result since they assume that the length of the clauses is known in advance to the algorithm. Another online variant of Max-SAT has been studied by Coppersmith et al. [7]. In this variant, clauses rather than variables, arrive in an online fashion and thus it is a different version of the problem.

The submodular Max-SAT problem is equivalent to the problem of maximizing a monotone submodular function under a particular matroid constraint we name a binary partition matroid. This result is established in Subsection 3.1. There has been a long line of research on maximizing monotone submodular functions subject to matroid constraints. Arguably, the simplest scenario is maximizing a submodular function under a uniform matroid. This problem admits a tight approximation of $1 - 1/e \approx 0.632$ [23,22,10]. Recently, Calinescu et al. [4] utilized a continuous relaxation approach to achieve a $(1 - 1/e)$-approximation for monotone submodular maximization under any matroid. In particular, due to the equivalence stated before, this algorithm can be applied to (the offline variant of) our problem. There has also been an ever-growing research on maximizing a monotone submodular function under additional constraints [26,25,14,18,19,5].

2 Preliminaries

Submodular functions. A set function $f : 2^X \to \mathbb{R}$ is called *submodular* if

$$f(S) + f(T) \geq f(S \cup T) + f(S \cap T) ,$$

for all $S, T \subseteq X$, and it is called *monotone* if $f(S) \leq f(T)$ whenever $S \subseteq T$. We remark that without loss of generality we can restrict attention to normalized functions, i.e., $f(\emptyset) = 0$. An alternative definition of submodularity is through the property of decreasing marginal values. Given a function f and a set $S \subseteq X$, the function f_S is defined by $f_S(a) = f(S \cup \{a\}) - f(S)$. The value $f_S(a)$ is called the marginal value of element a to the set S. The *decreasing marginal values* property requires that $f_S(a)$ be a non-increasing function of S for every fixed a. Formally, it requires that $f_S(a) \geq f_T(a)$, for all $S \subseteq T$ and $a \in X \setminus T$. Note that the amount of information necessary to convey an arbitrary submodular function may be exponential. Hence, we assume a value oracle access to the function. A *value oracle* for f allows us to query about the value of $f(S)$ for any set S.

Submodular Max-SAT. An input instance of submodular Max-SAT consists of a set $V = \{v_1, \ldots, v_n\}$ of boolean variables, and a collection $C = \{c_1, \ldots, c_m\}$ of clauses, where each clause is a disjunction of literals over the variables in V. Let $f : C \to \mathbb{R}_+$ be a monotone submodular function over the clauses. Given an assignment $\eta : V \to \{\mathsf{True}, \mathsf{False}\}$, we denote by $C(\eta) \subseteq C$ the subset of clauses satisfied by η. The objective is to find an assignment η that maximizes $f(C(\eta))$ over all possible assignments. We note that the classical Max-SAT problem is obtained as a special case when f is an additive function. An *additive* function can be represented as a sum of weights.

Online Max-SAT. In the online Max-SAT problem, we are given the same input components. However, in this variant, the variables are introduced in an online fashion, namely, one by one. Whenever a variable v_i is introduced, it arrives with two lists of clauses. The first list consists of all the clauses that contain v_i, and the other list consists of all the clauses that contain its negation $\neg v_i$. Any online algorithm must make an irrevocable decision regarding the assignment of v_i without prior knowledge about additional variables that will arrive in the future.

Partition matroids. A *matroid* is a pair $\mathcal{M} = (X, \mathcal{I})$, such that X is a ground set of elements, and $\mathcal{I} \subseteq 2^X$ is a family of independent subsets, containing the empty set and satisfying two additional properties:

1. an *inheritance property*: if $S \in \mathcal{I}$ and $T \subseteq S$ then $T \in \mathcal{I}$, and
2. an *exchange property*: if $S, T \in \mathcal{I}$ and $|T| \leq |S|$ then there is $a \in S$ such that $T \cup \{a\} \in \mathcal{I}$.

A *partition matroid* defines a partition $\bigcup_{t=1}^m P_t = X$ of the ground set elements. Then, the family of independent subsets are defined as $\mathcal{I} = \{S \subseteq X : |S \cap P_t| \leq k_t \ \forall t \in [m]\}$, where each $k_t \in \mathbb{N}$ designates a cardinality bound of the corresponding partition subset. When all the partition subsets have a cardinality of two, we obtain a *binary partition matroid*. Note that without loss of generality we can assume that every independent subset consists of at most one element from each partition subset. More generally, when all the partition subsets have the same cardinality ℓ, and all cardinality bounds are k, we call the matroid a (k, ℓ)-*partition matroid*.

3   A Linear Time Approximation Algorithm

In this section, we describe a fast randomized approximation algorithm for the submodular Max-SAT problem. Our algorithm can be implemented to run in linear time, and it achieves an expected approximation guarantee of 2/3. In fact, our algorithm can be applied when the input is introduced in an online fashion, and thus, it may be employed for the online variant of submodular Max-SAT. Prior to presenting the algorithm, we demonstrate that submodular Max-SAT is equivalent to the problem of maximizing a monotone submodular function subject to a binary partition matroid, formally defined below. We find this observation useful on its own merit as it connects our study with the long line of research on maximizing a monotone submodular function subject to a matroid constraint.

3.1   Submodular Maximization under a Binary Partition Matroid

An instance of the maximizing a monotone submodular function subject to a binary partition matroid consists of a ground set $X = \{a_1, b_1, a_2, b_2, \ldots, a_m, b_m\}$ of $n = 2m$ elements, a monotone submodular function $f : 2^X \to \mathbb{R}_+$, and a binary partition matroid $\mathcal{M} = (X, \mathcal{I})$ where each partition subset $P_t = \{a_t, b_t\}$. In particular, the family of independent subsets of this matroid is $\mathcal{I} = \{S \subseteq X : |S \cap P_t| \leq 1 \ \forall t \in [m]\}$. The objective is to maximize $f(S)$ subject to the constraint that $S \in \mathcal{I}$, that is, S consists of at most one element from each partition subset.

Theorem 1. *Submodular Max-SAT is equivalent to the problem of maximizing a monotone submodular function under a binary partition matroid.*

Proof. In what follows, we prove that any input instance of one problem can be translated to an input instance of the other problem in polynomial time. Note that each of these reductions maintains the solution value, that is, any solution to one problem can be translated to a solution to the other problem with the same value. Assume that we are given an input instance of submodular Max-SAT where we would like to maximize a monotone submodular function $g : 2^C \to \mathbb{R}_+$ for subsets of clauses having no contradiction. Let Lit be the set of literals of V, and let $\alpha : 2^{Lit} \to 2^C$ be the function that given a collection of literals returns the set of clauses that contain at least one of these literals. Formally,

$$\alpha(L) = \{c \in C : \text{there is } \ell \in L \text{ such that } \ell \text{ appears in } c\}$$

Having these definitions in mind, we define an instance of monotone submodular maximization under a binary partition matroid as follows: each partition subset corresponds to the two possible assignments (i.e., two literals) for a variable, and $f : 2^{Lit} \to \mathbb{R}_+$ is defined as $f(L) = g(\alpha(L))$. It is not difficult to verify that maximizing f on this binary partition matroid is equivalent to maximizing g under the original Max-SAT constraints. Hence, we focus on proving that f is indeed monotone and submodular. Notice that

$$
\begin{aligned}
f(S) + f(T) &= g(\alpha(S)) + g(\alpha(T)) \\
&\geq g(\alpha(S) \cup \alpha(T)) + g(\alpha(S) \cap \alpha(T)) \\
&\geq g(\alpha(S \cup T)) + g(\alpha(S \cap T)) \\
&= f(S \cup T) + f(S \cap T) ,
\end{aligned}
$$

where the first inequality is due to the submodularity of g, and the last inequality follows by the monotonicity of g combined with the facts that $\alpha(S \cup T) = \alpha(S) \cup \alpha(T)$ and $\alpha(S \cap T) \subseteq \alpha(S) \cap \alpha(T)$. Consequently, f is submodular. In addition, it is easy to validate that f is monotone since g is monotone and $\alpha(L) \subseteq \alpha(L \cup \{\ell\})$, for any $\ell \in L$.

We now assume that we are given an input instance of monotone submodular maximization under a binary partition matroid. We define an instance of submodular Max-SAT as follows: we create a boolean variable v_t for every partition subset P_t, and arbitrarily associate the literal v_t with a_t and the literal $\neg v_t$ with b_t. Furthermore, we define the set of clauses to be all the singleton clauses on the literals, that is, $C = \{v_1, \neg v_1, v_2, \neg v_2, \ldots, v_n, \neg v_n\}$. Notice that this induces a one-to-one correspondence between the clauses of C and the elements of X. Accordingly, we let $\beta : 2^C \to 2^X$ be the function that takes any subset of clauses and returns its corresponding set of elements under the one-to-one correspondence. Finally, we define $g(C) = f(\beta(C))$. One can easily verify that the function g is monotone and submodular, and that maximizing g under the Max-SAT constraints is equivalent to maximizing f on the original binary partition matroid. ∎

3.2   The Algorithm

In the following, we describe the approximation algorithm for the problem of maximizing a monotone submodular function under a binary partition matroid. As a result of the equivalency described in Subsection 3.1, we obtain an approximation algorithm for submodular Max-SAT. Our proportional select algorithm, formally described below, considers the partition subsets of the matroid in an arbitrary order, and selects one element from each partition subset. Unlike a greedy algorithm, which chooses a partition element with maximal marginal contribution, our algorithm randomly selects one of the two elements in proportion to their marginal contribution. This strategy turns out to be significantly better than using the natural greedy selection rule. In particular, it is straightforward to show the existence of instances to the problem where the greedy algorithm does not yield a better approximation than $1/2$. We note that our algorithm has similarities with the algorithm of Dobzinski and Schapira [8] for the problem of combinatorial auctions with two submodular bidders. Recall that the latter problem is a special instance of our problem in which the monotone submodular function has an additional constraint on its structure. Also recall that $f_S(a)$ is the incremental marginal value of element a to the set S.

Algorithm 1. Proportional Select

Input: A monotone submodular function $f : 2^X \to \mathbb{R}_+$ and a binary matroid $\mathcal{M}$
Output: A set $S \subseteq X$ approximating the maximum of f under $\mathcal{M}$

1: $S_0 \leftarrow \emptyset$
2: **for** $t \leftarrow 1$ to m **do**
3:     $w_t \leftarrow f_{S_{t-1}}(a_t) + f_{S_{t-1}}(b_t)$
4:     $p_{a_t} \leftarrow f_{S_{t-1}}(a_t)/w_t$
5:     $p_{b_t} \leftarrow f_{S_{t-1}}(b_t)/w_t$
6:     Pick s_t at random from $\{a_t, b_t\}$ with respective probabilities (p_{a_t}, p_{b_t})
7:     $S_t \leftarrow S_{t-1} \cup \{s_t\}$
8: **end for**
9: $S \leftarrow S_m$

Analysis. In what follows, we prove that the proportional select algorithm attains $2/3$-approximation. For ease of presentation, we use $+$ and $-$ to denote set union and set difference, respectively. We begin by introducing the definition of an optimal solution O_A constrained by a given (feasible) solution A.

Definition 1. *Let $A \subseteq X$ be an independent set of a matroid $\mathcal{M}$. The set $O_A \subseteq X$ is defined as an optimal solution to the problem of maximizing f under $\mathcal{M}$ that satisfies $A \subseteq O_A$. Formally, $O_A = \operatorname{argmax}_{T \in \mathcal{I}, A \subseteq T} f(T)$. Moreover, we let $\mathrm{OPT}_A = f(O_A)$ be the value that f assigns the set O_A. Note that the value of the optimal solution that maximizes f under $\mathcal{M}$ is $\mathrm{OPT} = \mathrm{OPT}_\emptyset$.*

We turn to bound the loss of O_A when it is constrained to select an element that is not part of it.

Lemma 1. *Let $A \subseteq X$ be an independent set of a matroid $\mathcal{M}$ such that $A \cap P_t = \emptyset$. Moreover, let x_t be the element of P_t that belongs to O_A and y_t the element of P_t that does not appear in O_A. Then,*

$$\mathrm{OPT}_A - \mathrm{OPT}_{A+y_t} \leq f_A(x_t) .$$

Proof. One can easily verify that

$$\begin{aligned}
\mathrm{OPT}_A = f(O_A) &\leq f(O_A + y_t) \\
&= f(O_A + y_t) - f(O_A + y_t - x_t) + f(O_A + y_t - x_t) \\
&= f_{O_A+y_t-x_t}(x_t) + f(O_A + y_t - x_t) ,
\end{aligned}$$

where the inequality results from the monotonicity of f. Notice that $A \subseteq O_A + y_t - x_t$, and therefore, $f_{O_A+y_t-x_t}(x_t) \leq f_A(x_t)$ by the submodularity of f. Furthermore,

$$f(O_A - x_t + y_t) \leq f(O_{A+y_t}) = \mathrm{OPT}_{A+y_t} ,$$

where the inequality follows as O_{A+y_t} is optimal with respect to $A + y_t \subseteq O_A - x_t + y_t$. In consequence, we obtain that $\mathrm{OPT}_A \leq f_A(x_t) + \mathrm{OPT}_{A+y_t}$. ∎

We now analyze the performance of our algorithm. For this purpose, we define the loss of the algorithm at step t of the main loop as $L_t = \mathrm{OPT}_{S_{t-1}} - \mathrm{OPT}_{S_t}$. The following observation makes a connection between the sum of losses along the steps of the algorithm and the difference between the value of the optimal solution, OPT, and the solution of our algorithm, $f(S)$.

Observation 2 $\sum_{t=1}^{m} L_t = \mathrm{OPT} - f(S)$

Proof. Notice that

$$\begin{aligned}
\sum_{t=1}^{m} L_t = \sum_{t=1}^{m} \left(\mathrm{OPT}_{S_{t-1}} - \mathrm{OPT}_{S_t} \right) &= \mathrm{OPT}_{S_0} - \mathrm{OPT}_{S_m} \\
&= \mathrm{OPT}_\emptyset - f(S_m) = \mathrm{OPT} - f(S)
\end{aligned}$$

∎

We are ready to prove the main theorem of this section, stating that the expected value of the solution that our algorithm generates is at least $2/3$ of the value of the optimal solution.

Theorem 3. $\mathbb{E}[f(S)] \geq \frac{2}{3}\mathrm{OPT}$

Proof. Let us focus on step t of the algorithm in which one of the elements of the partition subset P_t is selected. Note that only one of the elements of P_t belongs to $O_{S_{t-1}}$. Let us denote that element by x_t and the element of P_t that is not in $O_{S_{t-1}}$ by y_t.

We begin by bounding the expected loss of the algorithm at step t, given the set of elements selected up to step $t - 1$. Recall that s_t is the element selected at step t of the algorithm.

$$
\begin{aligned}
\mathbb{E}[L_t | S_{t-1}] &= \Pr[s_t = x_t | S_{t-1}] \cdot (\mathrm{OPT}_{S_{t-1}} - \mathrm{OPT}_{S_{t-1}+x_t}) + \\
&\quad \Pr[s_t = y_t | S_{t-1}] \cdot (\mathrm{OPT}_{S_{t-1}} - \mathrm{OPT}_{S_{t-1}+y_t}) \\
&= \Pr[s_t = x_t | S_{t-1}] \cdot 0 + \Pr[s_t = y_t | S_{t-1}] \cdot (\mathrm{OPT}_{S_{t-1}} - \mathrm{OPT}_{S_{t-1}+y_t}) \\
&\leq \Pr[s_t = y_t | S_{t-1}] \cdot f_{S_{t-1}}(x_t) \\
&= \frac{f_{S_{t-1}}(y_t) f_{S_{t-1}}(x_t)}{f_{S_{t-1}}(x_t) + f_{S_{t-1}}(y_t)} ,
\end{aligned}
$$

where the inequality is due to Lemma 1, and the last equality is attained by recalling that the algorithm selects y_t with probability $f_{S_{t-1}}(y_t)/(f_{S_{t-1}}(x_t) + f_{S_{t-1}}(y_t))$. We now turn to calculate the expected gain of the algorithm is step t, given the set of elements selected up to step $t-1$.

$$
\begin{aligned}
\mathbb{E}[f_{S_{t-1}}(s_t) \mid S_{t-1}] &= \Pr[s_t = x_t | S_{t-1}] \cdot f_{S_{t-1}}(x_t) + \Pr[s_t = y_t | S_{t-1}] \cdot f_{S_{t-1}}(y_t) \\
&= \frac{f_{S_{t-1}}(x_t)}{f_{S_{t-1}}(x_t) + f_{S_{t-1}}(y_t)} \cdot f_{S_{t-1}}(x_t) + \frac{f_{S_{t-1}}(y_t)}{f_{S_{t-1}}(x_t) + f_{S_{t-1}}(y_t)} \cdot f_{S_{t-1}}(y_t) \\
&= \frac{f_{S_{t-1}}(x_t)^2 + f_{S_{t-1}}(y_t)^2}{f_{S_{t-1}}(x_t) + f_{S_{t-1}}(y_t)}
\end{aligned}
$$

This implies that the expected loss to gain ratio is

$$
\frac{\mathbb{E}[L_t | S_{t-1}]}{\mathbb{E}[f_{S_{t-1}}(s_t) | S_{t-1}]} \leq \frac{f_{S_{t-1}}(x_t) f_{S_{t-1}}(y_t)}{f_{S_{t-1}}(x_t)^2 + f_{S_{t-1}}(y_t)^2} \leq \frac{1}{2} ,
$$

where the last inequality holds since $2ab \leq a^2 + b^2$, for any $a, b \in \mathbb{R}$. We can now bound the expected loss of the algorithm at step t as follows

$$
\begin{aligned}
\mathbb{E}[L_t] &= \sum_{S_{t-1} \subseteq X} \mathbb{E}[L_t | S_{t-1}] \cdot \Pr\left[\begin{array}{l} \text{the algorithm selects} \\ S_{t-1} \text{ up to step } t-1 \end{array}\right] \\
&\leq \frac{1}{2} \sum_{S_{t-1} \subseteq X} \mathbb{E}[f_{S_{t-1}}(s_t) | S_{t-1}] \cdot \Pr\left[\begin{array}{l} \text{the algorithm selects} \\ S_{t-1} \text{ up to step } t-1 \end{array}\right] \\
&= \frac{1}{2} \mathbb{E}[f_{S_{t-1}}(s_t)]
\end{aligned}
$$

Consequently, we get that

$$
\begin{aligned}
\mathrm{OPT} - \mathbb{E}[f(S)] = \mathbb{E}\left[\sum_{t=1}^{m} L_t\right] &= \sum_{t=1}^{m} \mathbb{E}[L_t] \leq \frac{1}{2} \sum_{t=1}^{m} \mathbb{E}[f_{S_{i-1}}(s_t)] \\
&= \frac{1}{2} \mathbb{E}\left[\sum_{t=1}^{m} f_{S_{i-1}}(s_t)\right] = \frac{1}{2} \mathbb{E}[f(S)] ,
\end{aligned}
$$

where the first equality follows from Observation 2. Thus, $\mathbb{E}[f(S)] \geq 2/3 \cdot \mathrm{OPT}$. ∎

The following corollary summarizes the properties of our algorithm.

Corollary 1. *Algorithm proportional select runs in linear time and achieves an expected approximation ratio of* $2/3$ *for the submodular Max-SAT problem.*

One can easily verify that the order in which our algorithm considers the partition subsets is not really important. This implies that the algorithm may be applied in case the partition subsets are introduced in an online fashion. In this setting, the two elements of a partition subset arrive in each step, and the algorithm needs to make an irrevocable decision which element to select to its solution (without prior knowledge about partition subsets that will arrive in the future). Using the equivalency described in Subsection 3.1, one can observe that the algorithm may be applied to the online Max-SAT problem. Specifically, it is not hard to validate that the two elements of each partition subset can correspond to the two possible assignments for a variable. In consequence, we obtain the following corollary.

Corollary 2. *Algorithm proportional select achieves an expected competitive ratio of* $2/3$ *for the online submodular Max-SAT problem.*

4   A Hardness Bound for the Online Version

In this section, we demonstrate that no *randomized* online algorithm can guarantee a competitive ratio better than $2/3$ for the online version of Max-SAT. We emphasize that this hardness bound also holds for the online variant of the classical Max-SAT. This implies that the analysis of our algorithm from Section 3 is tight, and that this algorithm is optimal for the online setting of both the submodular and classical versions of Max-SAT. We also note that one can easily prove a hardness bound of $1/2$ for any *deterministic* online algorithm. Interestingly, this establishes that randomization provably helps in our online setting.

Theorem 4. *No online algorithm can attain a competitive ratio better than* $2/3$ *for online Max-SAT.*

Proof. In what follows, we present a distribution over inputs such that any deterministic online algorithm, given an input drawn from this distribution, achieves an expected competitive ratio of at most $2/3 + \epsilon$ for any fixed $\epsilon > 0$. By Yao's principle, it then follows that any randomized online algorithm cannot achieve an expected competitive ratio better than $2/3$.

Let $V = \{v_1, \ldots, v_n\}$ be a set of boolean variables in our input. The variables are introduced according to their order, namely, v_1, v_2, and so on. We have $m = 2^n$ clauses with identical weights. Whenever a variable v_i is introduced, the algorithm is given the subsets S_i and T_i. Specifically, S_i consists of all the clauses in which v_i appears, and T_i consists of all the clauses in which $\neg v_i$ appears. We now formally define the sets S_i, T_i that correspond to each variable v_i. At the first step, when v_1 is presented to the algorithm, $S_1 = \{c_1, \ldots, c_{m/2}\}$ and $T_1 = \{c_{m/2+1}, \ldots, c_m\}$, that is, the clauses $c_1, \ldots, c_{m/2}$ contain v_1 and the rest of the clauses contain $\neg v_1$. At the next step, when v_2 is presented to the algorithm, one of the subsets S_1, T_1 is randomly chosen. Let C_1 be the chosen subset, and note that $|C_1| = m/2$. The set S_2 is defined to consist of the first $m/4$ clauses in C_1, while T_2 consists of the remaining $m/4$ clauses in C_1. The same process continues with all the remaining variables. In particular, at step i, when v_i is presented to the algorithm, one of the subsets S_{i-1}, T_{i-1} is randomly chosen as C_{i-1}. Then, the set S_i is defined to consist of the first $|C_{i-1}|/2$ clauses of C_{i-1}, and T_i consists of the remaining $|C_{i-1}|/2$ clauses of C_{i-1}. A concrete construction of clauses is presented in Figure 1.

$$
\begin{aligned}
c_1 &= v_1, & c_5 &= \neg v_1 \vee v_2 \\
c_2 &= v_1, & c_6 &= \neg v_1 \vee v_2 \\
c_3 &= v_1, & c_7 &= \neg v_1 \vee \neg v_2 \vee v_3 \\
c_4 &= v_1, & c_8 &= \neg v_1 \vee \neg v_2 \vee \neg v_3
\end{aligned}
$$

Fig. 1. A construction of clauses for $n = 3$ under the assumption that always $C_i = T_i$.

Notice that each chosen set C_i holds the *active* clauses, namely, clauses that may be affixed with additional variables later in the randomized construction. As a result, by choosing $v_i = \mathsf{False}$ whenever $C_i = S_i$ and $v_i = \mathsf{True}$ whenever $C_i = T_i$, one can obtain an optimal assignment which satisfies all clauses but one. Thus, we have $\mathrm{OPT} = m - 1$. We turn to calculate the expected number of clauses satisfied by any online algorithm. Let us focus on step i. One can validate that choosing $v_i = \mathsf{True}$ if $C_i = S_i$ or $v_i = \mathsf{False}$ if $C_i = T_i$ results in an assignment that cannot satisfy more than $m \cdot (1 - 2^{-i})$ of the clauses. However, since the assignment of v_i is done without any information about the random choice of C_i it follows that any online algorithm makes a wrong choice with probability $1/2$. In consequence, it generates an assignment that cannot satisfy more than $m \cdot (1 - 2^{-i})$ of the clauses. Notice that the probability that the algorithm makes its first wrong choice at step i is 2^{-i}. Hence,

$$
\mathbb{E}\left[\begin{array}{c} \text{number of} \\ \text{satisfied clauses} \end{array}\right] = 2^{-n} \cdot (m-1) + \sum_{i=1}^{n} 2^{-i} \cdot m \cdot (1 - 2^{-i})
$$

$$
\leq m \cdot \left[2^{-n} + \sum_{i=1}^{n} 2^{-i} - \sum_{i=1}^{n} 4^{-i} \right]
$$

$$
= m \cdot \left[1 - \left(\frac{1}{3} - \frac{1}{3 \cdot 4^n} \right) \right]
$$

$$
= \frac{2}{3} m + \frac{m}{3 \cdot 4^n} \leq \frac{2}{3} m + \frac{1}{m} ,
$$

where the first equality holds since an algorithm cannot make a wrong choice in the last step. As m tends to infinity, the expected competitive ratio approaches $2/3$, as required. We note that although the number of clauses in our construction is exponential in n, our hardness result also holds if the number of clauses is required to be polynomial in n. In such case, one can construct an instance as before with $\log m$ variables and then add dummy variables which do not appear in any clause.   ∎

5 A Hardness Bound for the Offline Version

Due to space limitation, this section is deferred to the full version of the paper.

6 Concluding Remarks

In this paper, we introduced the submodular Max-SAT problem. We developed a combinatorial randomized linear-time 2/3-approximation algorithm that is applicable even for the online variant of the problem. We also established hardness results for both the online and offline variants of the problem: for the online setting, the hardness result proves that our algorithm is best possible; for the offline setting, the hardness result establishes a computational separation between the classical Max-SAT and the submodular Max-SAT.

A natural open question is to close the approximation gap for the offline version. In particular, this gap is between 2/3 and 3/4. We have recently learned that independently of our work, Feldman, Naor and Schwartz [11] considered the same problem and attained an improved approximation ratio. However, to the best of our knowledge, their algorithm is not fast and cannot be applied in an online setting.

References

1. Asano, T.: Approximation algorithms for max sat: Yannakakis vs. goemans-williamson. In: Proceedings 5th Israel Symposium on Theory of Computing and Systems, pp. 24–37 (1997)
2. Asano, T.: An improved analysis of goemans and williamson's lp-relaxation for max sat. Theor. Comput. Sci. 354(3), 339–353 (2006)
3. Asano, T., Williamson, D.P.: Improved approximation algorithms for max sat. J. Algorithms 42(1), 173–202 (2002)
4. Calinescu, G., Chekuri, C., Pál, M., Vondrák, J.: Maximizing a monotone submodular function subject to a matroid constraint. In: SICOMP (2010)
5. Chekuri, C., Vondrák, J., Zenklusen, R.: Dependent randomized rounding via exchange properties of combinatorial structures. In: Proceedings 51st FOCS (2010)
6. Chen, J., Friesen, D.K., Zheng, H.: Tight bound on johnson's algorithm for maximum satisfiability. J. Comput. Syst. Sci. 58(3), 622–640 (1999)
7. Coppersmith, D., Gamarnik, D., Hajiaghayi, M.T., Sorkin, G.B.: Random max sat, random max cut, and their phase transitions. Random Struct. Algorithms 24(4), 502–545 (2004)
8. Dobzinski, S., Schapira, M.: An improved approximation algorithm for combinatorial auctions with submodular bidders. In: Proceedings of 17th SODA, pp. 1064–1073 (2006)

9. Engebretsen, L.: Simplified tight analysis of johnson's algorithm. Inf. Process. Lett. 92(4), 207–210 (2004)
10. Feige, U.: A threshold of ln n for approximating set cover. J. ACM 45(4), 634–652 (1998)
11. Feldman, M., Naor, J., Schwartz, R.: Personal Communication (2011)
12. Goemans, M.X., Williamson, D.P.: 878-Approximation algorithms for max cut and max 2sat. In: Proceedings 26th STOC, pp. 422–431 (1994)
13. Goemans, M.X., Williamson, D.P.: New 3/4-approximation algorithms for the maximum satisfiability problem. SIAM J. Discrete Math. 7(4), 656–666 (1994)
14. Goundan, P.R., Schulz, A.S.: Revisiting the greedy approach to submodular set function maximization (2007) (manuscript)
15. Håstad, J.: Some optimal inapproximability results. J. ACM 48(4), 798–859 (2001)
16. Johnson, D.S.: Approximation algorithms for combinatorial problems. J. Comput. Syst. Sci. 9(3), 256–278 (1974)
17. Karloff, H.J., Zwick, U.: A 7/8-approximation algorithm for max 3sat? In: Proceedings 38th FOCS, pp. 406–415 (1997)
18. Kulik, A., Shachnai, H., Tamir, T.: Maximizing submodular set functions subject to multiple linear constraints. In: Proceedings 20th SODA, pp. 545–554 (2009)
19. Lee, J., Sviridenko, M., Vondrák, J.: Submodular maximization over multiple matroids via generalized exchange properties. In: Dinur, I., Jansen, K., Naor, J., Rolim, J. (eds.) APPROX 2009. LNCS, vol. 5687, pp. 244–257. Springer, Heidelberg (2009)
20. Mirrokni, V.S., Schapira, M., Vondrák, J.: Tight information-theoretic lower bounds for welfare maximization in combinatorial auctions. In: Proceedings 9th EC 2008, pp. 70–77 (2008)
21. Moshkovitz, D., Raz, R.: Two-query pcp with subconstant error. J. ACM, 57(5) (2010)
22. Nemhauser, G.L., Wolsey, L.A.: Best algorithms for approximating the maximum of a submodular set function. Math. Operations Research 3(3), 177–188 (1978)
23. Nemhauser, G.L., Wolsey, L.A., Fisher, M.L.: An analysis of approximations for maximizing submodular set functions I. Mathematical Programming 14, 265–294 (1978)
24. Poloczek, M., Schnitger, G.: Randomized variants of johnson's algorithm for max sat. In: SODA, pp. 656–663 (2011)
25. Sviridenko, M.: A note on maximizing a submodular set function subject to a knapsack constraint. Oper. Res. Lett. 32(1), 41–43 (2004)
26. Wolsey, L.A.: Maximising real-valued submodular functions: Primal and dual heuristics for location problems. Math. Operations Research 7(3), 410–425 (1982)
27. Yannakakis, M.: On the approximation of maximum satisfiability. J. Algorithms 17(3), 475–502 (1994)
28. Zwick, U.: Outward rotations: A tool for rounding solutions of semidefinite programming relaxations, with applications to max cut and other problems. In: Proceedings of 31st STOC, pp. 679–687 (1999)

On Variants of the Matroid Secretary Problem

Shayan Oveis Gharan[1,*] and Jan Vondrák[2]

[1] Stanford University, Stanford, CA
shayan@stanford.edu
[2] IBM Almaden Research Center, San Jose, CA
jvondrak@us.ibm.com

Abstract. We present a number of positive and negative results for variants of the matroid secretary problem. Most notably, we design a constant-factor competitive algorithm for the "random assignment" model where the weights are assigned randomly to the elements of a matroid, and then the elements arrive on-line in an adversarial order (extending a result of Soto [20]). This is under the assumption that the matroid is known in advance. If the matroid is unknown in advance, we present an $O(\log r \log n)$-approximation, and prove that a better than $O(\log n / \log \log n)$ approximation is impossible. This resolves an open question posed by Babaioff et al. [3].

As a natural special case, we also consider the classical secretary problem where the number of candidates n is unknown in advance. If n is chosen by an adversary from $\{1, \ldots, N\}$, we provide a nearly tight answer, by providing an algorithm that chooses the best candidate with probability at least $1/(H_{N-1} + 1)$ and prove that a probability better than $1/H_N$ cannot be achieved (where H_N is the N-th harmonic number).

1 Introduction

The secretary problem is a classical problem in probability theory, with obscure origins in the 1950's and early 60's ([11,17,8]; see also [10]). Here, the goal is to select the best candidate out of a sequence revealed one-by-one, where the ranking is uniformly random. A classical solution finds the best candidate with probability at least $1/e$ [10]. Over the years a number of variants have been studied, starting with [12] where multiple choices and various measures of success were considered for the first time.

Recent interest in variants of the secretary problem has been motivated by applications in on-line mechanism design [14,18,3], where items are being sold to agents arriving on-line, and there are certain constraints on which agents can be simultaneously satisfied. Equivalently, one can consider a setting where we want to hire several candidates under certain constraints. Babaioff, Immorlica and Kleinberg [3] formalized this problem and presented constant-factor competitive algorithms for several interesting cases. The general problem formulated in [3] is the following.

---

* This work was done while the author was an intern at IBM Almaden Research Center, San Jose, CA.

C. Demetrescu and M.M. Halldórsson (Eds.): ESA 2011, LNCS 6942, pp. 335–346, 2011.

Matroid secretary problem. Given a matroid $\mathcal{M} = (E, \mathcal{I})$ with non-negative weights assigned to E; the only information known up-front is the number of elements $n := |E|$. The elements of E arrive in a random order, with their weights revealed as they arrive. When an element arrives, it can be selected or rejected. The selected elements must always form an independent set in $\mathcal{M}$, and a rejected element cannot be considered again. The goal is to maximize the expected weight of the selected elements.

Additional variants of the matroid secretary problem have been proposed and studied, depending on how the input ordering is generated, how the weights are assigned and what is known in advance. In all variants, elements with their weights arrive in an on-line fashion and an algorithm must decide irrevocably whether to accept or reject an element once it has arrived. We attempt to bring some order to the multitude of models and we classify the various proposed variants as follows.

Ordering of matroid elements on the input:
- AO = Adversarial Order: the ordering of elements of the matroid on the input is chosen by an adversary.
- RO = Random Order: the elements of the matroid arrive in a random order.

Assignment of weights:
- AA = Adversarial Assignment: weights are assigned to elements of the matroid by an adversary.
- RA = Random Assignment: the weights are assigned to elements by a random permutation of an adversarial set of weights (independent of the input order, if that is also random).

Prior information:
- MK = Matroid Known: the matroid is known beforehand (by means of an independence oracle).
- MN = Matroid - n known: the matroid is unknown but the cardinality of the ground set is known beforehand.
- MU = Matroid - Unknown: nothing about the matroid is known in advance; only subsets of the elements that arrived already can be queried for independence.

For example, the original variant of the matroid secretary problem [3], where the only information known beforehand is the total number of elements, can be described as RO-AA-MN in this classification. We view this as the primary variant of the matroid secretary problem.

We also consider variants of the classical secretary problem; here, only 1 element should be chosen and the goal is to maximize the probability of selecting the best element.

Classical secretary problems:
- CK = Classical - Known n: the classical secretary problem where the number of elements in known in advance.

- CN = Classical - known upper bound N: the classical secretary problem where the number of elements is chosen adversarially from $\{1, \ldots, N\}$, and N is known in advance.
- CU = Classical - Unknown n: the classical secretary problem where no information on the number of elements is known in advance.

Since the independence sets of the underlying matroid in this model are independent of the particular labeling of the ground set, we just use the weight assignment function to characterize different variants of this model. The classical variant of the secretary problem which allows a $1/e$-approximation would be described as RA-CK. The variant where the number of elements n is not known in advance is very natural — and has been considered under different stochastic models where n is drawn from a particular distribution [21,1] — but the worst-case scenario does not seem to have received attention. We denote this model RA-CU, or RA-CN if an upper bound on the number of candidates is given. In the model where the input ordering of weights is adversarial (AA-CK), it is easy to see that no algorithm achieves probability better than $1/n$ [5]. We remark that variants of the secretary problem with other objective functions have been also proposed, such as discounted profits [2], and submodular objective functions [4,13]. We do not discuss these variants here.

1.1   Recent Related Work

The primary variant of matroid secretary problem (RO-AA-MN model) was introduced in [3]. In the following, let n denote the total number of elements and r the rank of the matroid. An $O(\log r)$-approximation for the RO-AA-MN model was given in [3]. It was also conjectured that a constant-factor approximation should exist for this problem and this question is still open. Constant-factor approximations were given in [3] for some special cases such as partition matroids and graphic matroids with a given explicit representation. Further, constant-factor approximations were given for transversal matroids [7,19] and laminar matroids [16]. However, even for graphic matroids in the RO-AA-MK model when the graphic matroid is given by an oracle, no constant factor is known.

Babaioff et al. in [3] also posed as an open problem whether there is a constant-factor approximation algorithm for the following two models: Assume that a set of n numerical values are assigned to the matroid elements using a random one-to-one correspondence but that the elements are presented in an adversarial order (AO-RA in our notation). Or, assume that both the assignment of values and the ordering of the elements in the input are random (RO-RA in our notation). The issue of whether the matroid is known beforehand is left somewhat ambiguous in [3].

In a recent work [20], José Soto partially answered the second question, by designing a constant-factor approximation algorithm in the RO-RA-MK model: An adversary chooses a list of non-negative weights, which are then assigned to the elements using a random permutation, which is independent of the random order at which the elements are revealed. The matroid is known in advance here.

1.2   Our Results

Matroid secretary. We resolve the question from [3] concerning adversarial order and random assignment, by providing a constant-factor approximation algorithm in the AO-RA-MK model, and showing that no constant-factor approximation exists in the AO-RA-MN model. More precisely, we prove that there is
a $40/(1 - 1/e)$-approximation in the AO-RA-MK model, i.e. in the model where
weights are assigned to the elements of a matroid randomly, the elements arrive
in an adversarial order, and the matroid is known in advance. We provide a
simple thresholding algorithm, which gives a constant-factor approximation for
the AO-RA-MK model when the matroid $\mathcal{M}$ is uniformly dense. Then we use
the principal sequence of a matroid to design a constant-factor approximation
for any matroid using the machinery developed by Soto [20].

On the other hand, if the matroid is not known in advance (AO-RA-MN
model), we prove that the problem cannot be approximated better than within
$\Omega(\log n / \log \log n)$. This holds even in the special case of rank 1 matroids; see below. On the positive side, we show an $O(\log r \log n)$-approximation for this model.
We achieve this by providing an $O(\log r)$-approximation thresholding algorithm
for the AO-AA-MU model (when both the input ordering and the assignment of
weights to the elements the matroid are adversarial), when an estimate on the
weight of the largest non-loop element is given. Here, the novel technique is to
employ a dynamic threshold depending on the rank of the elements seen so far.

Classical secretary with unknown n. A very natural question that arises in this
context is the following. Consider the classical secretary problem, where we want
to select 1 candidate out of n. The classical solution relies on the fact that n is
known in advance. However, what if we do not know n in advance, which would
be the case in many practical situations? We show that if an upper bound N on
the possible number of candidates n is given (RA-CN model: i.e., n is chosen by
an adversary from $\{1, \ldots, N\}$), the best candidate can be found with probability
$1/(H_{N-1}+1)$, while there is no algorithm which achieves probability better than
$1/H_N$ (where $H_N = \sum_{i=1}^{N} \frac{1}{i}$ is the N-th harmonic number).

In the model where we maximize the expected value of the selected candidate, and n is chosen adversarially from $\{1, \ldots, N\}$, we prove we cannot achieve
approximation better than $\Omega(\log N / \log \log N)$. On the positive side, even if no
upper bound on n is given, the maximum-weight element can be found with prob-
ability $\epsilon / \log^{1+\epsilon} n$ for any fixed $\epsilon > 0$. We remark that similar results follow from
[15] and [9] where an equivalent problem was considered in the context of online
auctions. More generally, for the matroid secretary problem where no informa-
tion at all is given in advance (RO-AA-MU), we achieve an $O(\frac{1}{\epsilon} \log r \log^{1+\epsilon} n)$
approximation for any $\epsilon > 0$. See Table 1 for an overview of our results.

Organization. In section 2 we provide a $40/(1 - 1/e)$ approximation algorithm
for the AO-RA-MK model. In section 3 we provide an $O(\log n \log r)$ approxima-
tion algorithm for the AO-RA-MN model, and an $O(\frac{1}{\epsilon} \log r \log^{1+\epsilon} n)$ approximation for the RO-AA-MU model. Finally, in section 4 we provide a $(H_{N-1} + 1)$-
approximation and H_N-hardness for the RA-CN model.

Table 1. Summary of results

Problem	New approximation	New hardness
RA-CN	$H_{N-1} + 1$	H_N
RA-CU	$O(\frac{1}{\epsilon} \log^{1+\epsilon} n)$	$\Omega(\log n)$
AO-RA-MK	$40/(1 - 1/e)$	-
AO-RA-MN	$O(\log r \log n)$	$\Omega(\log n / \log \log n)$
AO-RA-MU	$O(\frac{1}{\epsilon} \log r \log^{1+\epsilon} n)$	$\Omega(\log n / \log \log n)$
RO-AA-MU	$O(\frac{1}{\epsilon} \log r \log^{1+\epsilon} n)$	$\Omega(\log n / \log \log n)$

2 Approximation for Adversarial Order and Random Assignment

In this section, we derive a constant-factor approximation algorithm for the AO-RA-MK model, i.e. assuming that the ordering of the elements of the matroid is adversarial but weights are assigned to the elements by a random permutation, and the matroid is known in advance. We build on Soto's algorithm [20], in particular on his use of the *principal sequence of a matroid* which effectively reduces the problem to the case of a uniformly-dense matroid while losing only a constant factor $(1 - 1/e)$. Interestingly, his reduction only requires the randomness in the assignment of weights to the elements but not a random ordering of the matroid on the input. Due to limited space here we do not include the details of the reduction, and we defer it to the full version of the paper.

Recall that the density of a set in a matroid $\mathcal{M} = (E, \mathcal{I})$ is the quantity $\gamma(S) = \frac{|S|}{rank(S)}$. A matroid is uniformly dense, if $\gamma(S) \leq \gamma(E)$ for all $S \subseteq E$. We present a simple thresholding algorithm which works in the AO-RA-MK model (i.e. even for an adversarial ordering of the elements) for any uniformly dense matroid.

Throughout this section we use the following notation. Let $\mathcal{M} = (E, I)$ be a uniformly dense matroid of rank r. This also means that $\mathcal{M}$ contains no loops. Let $|E| = n$ and let $e_1, e_2, \ldots, e_n$ denote the ordering of the elements on the input, which is chosen by an adversary (i.e. we consider the worst case). Furthermore, the adversary also chooses $W = \{w_1 > w_2 > \ldots > w_n\}$, a set of non-negative weights. The weights are assigned to the elements of $\mathcal{M}$ via a random bijection $\omega : E \to W$. For a weight assignment ω, we denote by $w(S) = \sum_{e \in S} \omega(e)$ the weight of a set S, and by $\omega(S) = \{\omega(e) : e \in S\}$ the set of weights assigned to S. We also let $\text{OPT}(\omega)$ be the maximum-weight independent set in $\mathcal{M}$.

Recall that r denotes the rank of the matroid. We show that there is a simple thresholding algorithm which includes each of the topmost $\lfloor r/4 \rfloor$ weights (i.e. $w_1, \ldots, w_{\lfloor r/4 \rfloor}$) with a constant probability. This will give us a constant factor approximation algorithm, as $w(\text{OPT}(\omega)) \leq \sum_{i=1}^{r} w_i$, where $w_1 > w_2 > \ldots > w_r$ are the r largest weights in W. It is actually important that we compare our algorithm to the quantity $\sum_{i=1}^{r} w_i$, because this is needed in the reduction to the uniformly dense case.

The main idea is that the randomization of the weight assignment makes it very likely that the optimum solution contains many of the top weights in W.

Therefore, instead of trying to compute the optimal solution with respect to ω, we can just focus on catching a constant fraction of the top weights in W. Let $A = \{e_1, \ldots, e_{n/2}\}$ denote the first half of the input and $B = \{e_{n/2+1}, \ldots, e_n\}$ the second half of the input. Note that the partition into A and B is determined by the adversary and not random. Our solution is to use the $\lfloor r/4 \rfloor + 1$-st topmost weight in the "sampling stage" A as a threshold and then include every element in B that is above the threshold and independent of the previously selected elements. Details are described in Algorithm 1.

Algorithm 1. Thresholding algorithm for uniformly dense matroids in AO-RA-MK model

---

Input: A uniformly dense matroid $\mathcal{M} = (E, \mathcal{I})$ of rank r.
Output: An independent set $\text{ALG} \subseteq E$.
1: **if** $r < 12$ **then**
2:    run the optimal algorithm for the classical secretary problem, and return the resulting singleton.
3: **end if**
4: $\text{ALG} \leftarrow \emptyset$
5: Observe a half of the input (elements of A) and let w^* be the $(\lfloor r/4 \rfloor + 1)^{st}$ largest weight among them.
6: **for** each element $e \in B$ arriving afterwards **do**
7:    **if** $\omega(e) > w^*$ and $ALG \cup \{e\}$ is independent **then**
8:        $\text{ALG} \leftarrow \text{ALG} \cup \{e\}$
9:    **end if**
10: **end for**
11: **return** ALG

Theorem 2.1. *Let $\mathcal{M}$ be a uniformly dense matroid of rank r, and $\text{ALG}(\omega)$ be the set returned by Algorithm 1 when the weights are defined by a uniformly random bijection $\omega : E \to W$. Then*

$$\mathbf{E}_\omega \left[w(\text{ALG}(\omega)) \right] \geq \frac{1}{40} \sum_{i=1}^{r} w_i$$

where $\{w_1 > w_2 > \ldots > w_r\}$ are the r largest weights in W.

If $r < 12$, the algorithm finds and returns the largest weight w_1 with probability $1/e$ (step 2; the optimal algorithm for the classical secretary problem). Therefore, for $r < 12$, we have $\mathbf{E}_\omega \left[w(\text{ALG}(\omega)) \right] \geq \frac{1}{11e} \sum_{i=1}^{r} w_i > \frac{1}{40} \sum_{i=1}^{r} w_i$.

For $r \geq 12$, we prove that each of the topmost $\lfloor r/4 \rfloor$ weights will be included in $\text{ALG}(\omega)$ with probability at least $1/8$. Hence, we will obtain

$$\mathbf{E}_\omega \left[w(\text{ALG}(\omega)) \right] \geq \frac{1}{8} \sum_{i=1}^{\lfloor r/4 \rfloor} w_i \geq \frac{1}{40} \sum_{i=1}^{r} w_i. \tag{1}$$

Let $t = 2\lfloor r/4 \rfloor + 2$. Define $C'(\omega) = \{e_j : \omega(e_j) \geq w_t\}$ to be the set of elements of $\mathcal{M}$ which get one of the top t weights. Also let $A'(\omega) = C'(\omega) \cap A$ and

$B'(\omega) = C'(\omega) \cap B$. Moreover, for each $1 \leq i \leq t$ we define $C_i'(\omega) = \{e_j : \omega(e_j) \geq w_t \ \& \ \omega(e_j) \neq w_i\}$, $A_i'(\omega) = C_i'(\omega) \cap A$ and $B_i'(\omega) = C_i'(\omega) \cap B$, i.e. the same sets with the element of weight w_i removed.

First, we fix $i \leq \lfloor r/4 \rfloor$ and argue that the size of $B_i'(\omega)$ is smaller than $A_i'(\omega)$ with probability $1/2$. Then we will use the uniformly dense property of $\mathcal{M}$ to show that the span of $B_i'(\omega)$ is also quite small with probability $1/2$ and consequently w_i has a good chance of being included in $\mathrm{ALG}(\omega)$.

Claim 2.2 *Let $\mathcal{M}$ be a uniformly dense matroid of rank r, $t = 2\lfloor r/4 \rfloor + 2$, $1 \leq i \leq \lfloor r/4 \rfloor$, and $B_i'(\omega)$ defined as above. Then we have*

$$\mathbf{P}_\omega \left[|B_i'(\omega)| \leq \lfloor r/4 \rfloor \right] = 1/2. \tag{2}$$

Proof: Consider $C_i'(\omega)$, the set of elements receiving the top t weights except for w_i. This is a uniformly random set of odd size $t - 1 = 2\lfloor r/4 \rfloor + 1$. By symmetry, with probability exactly $1/2$, a majority of these elements are in A, and hence at most $\lfloor r/4 \rfloor$ of these elements are in B, i.e. $|B_i'(\omega)| \leq \lfloor r/4 \rfloor$. □

Now we consider the element receiving weight w_i. We claim that this element will be included in $\mathrm{ALG}(\omega)$ with a constant probability.

Claim 2.3 *Let $\mathcal{M}$ be a uniformly dense matroid of rank r, and $i \leq \lfloor r/4 \rfloor$. Then*

$$\mathbf{P}_\omega \left[\omega^{-1}(w_i) \in \mathrm{ALG}(\omega) \right] \geq 1/8.$$

Proof: Condition on $C_i'(\omega) = S$ for some particular set S of size $t - 1$ such that $|B_i'(\omega)| = |S \cap B| \leq \lfloor r/4 \rfloor$. This fixes the assignment of the top t weights except for w_i. Under this conditioning, weight w_i is still assigned uniformly to one of the remaining $n - t + 1$ elements.

Since we have $|A_i'(\omega)| = |S \cap A| \geq \lfloor r/4 \rfloor + 1$, the threshold w^* in this case is one of the top t weights and the algorithm will never include any weight outside of the top t. Therefore, we have $\mathrm{ALG}(\omega) \subseteq B'(\omega)$. The weight w_i is certainly above w^* because it is one of the top $\lfloor r/4 \rfloor$ weights. It will be added to $\mathrm{ALG}(\omega)$ whenever it appears in B and it is not in the span of previously selected elements. Since all the previously included elements must be in $B_i'(\omega) = S \cap B$, it is sufficient to avoid being in the span of $S \cap B$. To summarize, we have

$$\omega^{-1}(w_i) \in B \setminus span(S \cap B) \Rightarrow \omega^{-1}(w_i) \in \mathrm{ALG}(\omega).$$

What is the probability that this happens? Similar to the proof of [20, Lemma 3.1], since $\mathcal{M}$ is uniformly dense, we have

$$\frac{|span(S \cap B)|}{|S \cap B|} \leq \frac{|span(S \cap B)|}{rank(span(S \cap B))} \leq \frac{n}{r} \Longrightarrow |span(S \cap B)| \leq \frac{n}{r}|S \cap B| \leq \frac{n}{4}$$

using $|S \cap B| \leq \lfloor r/4 \rfloor$. Therefore, there are at least $n/4$ elements in $B \setminus span(S \cap B)$. Given that the weight w_i is assigned uniformly at random among $n - t$ possible elements, we get

$$\mathbf{P}_\omega \left[\omega^{-1}(w_i) \in B \setminus span(S \cap B) \mid C_i'(\omega) = S \right] \geq \frac{n/4}{n - t} \geq \frac{1}{4}.$$

Since this holds for any S such that $|S \cap B| \leq \lfloor r/4 \rfloor$, and $S \cap B = C_i' \cap B = B_i'(\omega)$, it also holds that

$$\mathbf{P}_\omega \left[\omega^{-1}(w_i) \in B \setminus span(B_i'(\omega)) \mid |B_i'(\omega)| \leq \lfloor r/4 \rfloor \right] \geq \frac{1}{4}.$$

Using Claim 2.2, we get $\mathbf{P}_\omega \left[\omega^{-1}(w_i) \in B \setminus span(B_i'(\omega)) \right] \geq 1/8$.   □
This finishes the proof of Theorem 2.1.

Combining our algorithm with Soto's reduction [20, Lemma 4.4], we obtain a constant-factor approximation algorithm for the matroid secretary problem in AO-RA-MK model.

Corollary 2.4. *There exists a $\frac{40}{1-1/e}$-approximation algorithm in the AO-RA-MK model.*

3   Approximation Algorithms for Unknown Matroids

In this section we will be focusing mainly on the AO-RA-MN model. i.e. assuming that the ordering of the elements of the matroid is adversarial, weights are assigned randomly, but the matroid is unknown, and the algorithm only knows n in advance. We present an $O(\log n \log r)$ approximation algorithm for the AO-RA-MN model, where n is the number of elements in the ground set and r is the rank of the matroid. At the end of this section we also give a general framework that can turn any α approximation algorithm for the RO-AA-MN model, (i.e. the primary variant of the matroid secretary problem) into an $O(\alpha \log^{1+\epsilon} n/\epsilon)$ approximation algorithm in the RO-AA-MU model.

It is worth noting that in these models the adversary may set some of the elements of the matroid to be loops, and the algorithm does not know the number of loops in advance. For example it might be the case that after observing the first 10 elements, the rest are all loops and thus the algorithm should select at least one of the first 10 elements with some non-zero probability. This is the idea of the counterexample in section 4 (Corollary 4.4), where we reduce AO-RA-MN, AO-RA-MU models to RA-CN, RA-CU models respectively, and thus we show that there is no constant-factor approximation for either of the models. In fact, no algorithm can do better than $\Omega(\log n/\log \log n)$. Therefore our algorithms are tight within a factor of $O(\log r \log \log n)$ or $O(\log r \log^\epsilon n)$.

We use the same notation as section 2: $\mathcal{M} = (E, I)$ is a matroid of rank r (which is not known to the algorithm), and $e_1, e_2, \ldots, e_n$ is the the adversarial ordering of the elements of $\mathcal{M}$, and $W = \{w_1 > w_2 > \ldots > w_n\}$ is the set of hidden weights chosen by the adversary that are assigned to the elements of $\mathcal{M}$ via a random bijection $\omega : E \to W$.

We start by designing an algorithm for AO-RA-MN model. Our algorithm basically tries to ignore the the loops and only focuses on the non-loop elements. We design our algorithm in two phases. In the first phase we design a randomized algorithm that works even in the AO-AA-MU model assuming that it has a good estimate on the weight of the largest non-loop element. In particular, fix bijection $\omega : W \rightarrow E$, and let e_1^* be the largest non-loop element with respect ω, e_2^* be the second largest one. We assume that the algorithm knows a bound $\omega(e_2^*) < L < \omega(e_1^*)$ on the largest non-loop element in advance. We show there is a thresholding algorithm, with a *non-fixed* threshold, that achieves an $O(\log r)$ fraction of the optimum.

Algorithm 2. for AO-AA-MU model with an estimate of the largest non-loop element

---

Input: The bound L such that $\omega(e_2^*) < L < \omega(e_1^*)$.
Output: An independent set ALG $\subseteq E$.
 1: with probability $1/2$, pick a non-loop element with weight above L and return it.
 2: ALG $\leftarrow \emptyset$ and $r^* \leftarrow 2$; set threshold $w^* \leftarrow L/2$.
 3: **for** each arriving element e_i **do**
 4:    **if** $\omega(e_i) > w^*$ and $ALG \cup \{e_i\}$ is independent **then**
 5:       ALG $\leftarrow$ ALG $\cup \{e_i\}$
 6:    **end if**
 7:    **if** $rank(\{e_1, \ldots, e_i\}) \geq r^*$ **then**
 8:       with probability $\frac{1}{\log 2r^*}$ set $w^* \leftarrow L/2r^*$.
 9:       $r^* \leftarrow 2r^*$.
10:    **end if**
11: **end for**
12: **return** ALG

Theorem 3.1. *For any matroid $\mathcal{M} = (E, \mathcal{I})$ of rank r, and any bijection $\omega : E \rightarrow W$, given the bound $\omega(e_2^*) < L < \omega(e_1^*)$, Algorithm 2 is a $16 \log r$ approximation in the AO-AA-MU model. i.e.*

$$\mathbf{E}\left[w(\mathrm{ALG}(\omega))\right] \geq \frac{1}{16 \log r} w(\mathrm{OPT}(\omega)),$$

where the expectation is over all of the randomization in the algorithm.

In order to solve the original problem, in the second phase we divide the non-loop elements into a set of blocks $B_1, B_2, \ldots, B_{\log n}$, and we use the previous algorithm as a module to get an $O(\log r)$ of optimum within each block.

Theorem 3.2. *For any matroid $\mathcal{M} = (E, \mathcal{I})$ of rank r, there is a polynomial time algorithm with an approximation factor $O(\log r \log n)$ in the AO-RA-MN model.*

Finally, we show how we can use essentially the same technique (decomposing the input into blocks of exponential size) to obtain an algorithm for AO-RA-MU model:

Theorem 3.3. *Let $\mathcal{M}$ be a matroid of rank r on n elements. If there is an α approximation algorithm for the matroid secretary problem on $\mathcal{M}$ in the RO-AA-MN model, then for any fixed $\epsilon > 0$, there is also an $O(\frac{\alpha}{\epsilon} \log^{1+\epsilon} n)$-approximation for the matroid secretary problem on $\mathcal{M}$ with no information given in advance (i.e., the RO-AA-MU model).*

Due to limited space all of the proofs of this section is deferred to the full version of the paper.

4   Classical Secretary with Unknown n

In this section, we consider a variant of the classical secretary problem where we want to select exactly one element (i.e. in matroid language, we consider a uniform matroid of rank 1). However, here we assume that the total number of elements n (which is crucial in the classical $1/e$-competitive algorithm) is not known in advance - it is chosen by an adversary who can effectively terminate the input at any point.

First, let us consider the following scenario: an upper bound N is given such that the actual number of elements on the input is guaranteed to be $n \in \{1, 2, \ldots, N\}$. The adversary can choose any n in this range and we do not learn n until we process the n-th element. (E.g., we are interviewing candidates for a position and we know that the total number of candidates is certainly not going to be more than 1000. However, we might run out of candidates at any point.) The goal is the select the highest-ranking element with a certain probability. Assuming the *comparison model* (i.e., where only the relative ranks of elements are known to the algorithm), we show that there is no algorithm achieving a constant probability of success in this case.

Theorem 4.1. *Given that the number of elements is chosen by an adversary in* $\{1, \ldots, N\}$ *and* N *is given in advance, there is a randomized algorithm which selects the best element out of the first* n *with probability at least* $1/(H_{N-1} + 1)$.

On the other hand, there is no algorithm in this setting which returns the best element with probability more than $1/H_N$. *Here,* $H_N = \sum_{i=1}^{n} \frac{1}{i}$ *is the* N-*th harmonic number.*

Due to limited space, we just sketch the main ideas of the proof. Our proof is based on the method of Buchbinder et al. [6] which bounds the optimal achievable probability by a linear program. In fact the optimum of the linear program is *exactly* the optimal probability that can be achieved.

Lemma 4.2. *Given the classical secretary problem where the number of elements is chosen by an adversary from* $\{1, 2, \ldots, N\}$ *and* N *is known in advance, the best possible probability with which an algorithm can find the optimal element is given by*

$$\max \qquad \alpha :$$
$$\forall n \leq N; \quad \frac{1}{n} \sum_{i=1}^{n} i p_i \geq \alpha, \tag{3}$$
$$\forall i \leq N; \sum_{j=1}^{i-1} p_j + i p_i \leq 1, \tag{4}$$
$$\forall i \leq N; \qquad p_i \geq 0.$$

The only difference between this LP and the one in [6] is that we have multiple constraints (3) instead of what is the objective function in [6]. We use essentially the same proof to argue that this LP captures *exactly* the optimal probability of success α that an algorithm can achieve. For a given N, an algorithm can explicitly solve the LP given by Lemma 4.2 and thus achieve the optimal probability. Theorem 4.1 can be proved by estimating the value of this LP.

A slightly different model arises when elements arrive with (random) weights and we want to maximize the expected weight of the selected element. This model is somewhat easier for an algorithm; any algorithm that selects the best element with probability at least α certainly achieves an α-approximation in this model, but not the other way around. Given an upper bound N on the number of elements (and under a more stringent assumption that weights are chosen i.i.d. from a known distribution), by a careful choice of a probability distribution for the weights, we prove that still no algorithm can achieve an approximation factor better than an $\Omega(\log N/\log\log N)$-approximation.

Theorem 4.3. *For the classical secretary problem with random nonnegative weights drawn i.i.d. from a known distribution and the number of candidates chosen adversarially in the range $\{1, \ldots, N\}$, no algorithm achieves a better than $\frac{\log N}{32 \log\log N}$-approximation in expectation.*

The hard examples are constructed based on a particular exponentially distributed probability distribution. Similar constructions have been used in related contexts [15,9]. The proof is deferred to the full version of the paper. Consequently, we obtain that no algorithm can achieve an approximation factor better than $\Omega(\log N/\log\log N)$ in the AO-RA-MN model.

Corollary 4.4. *For the matroid secretary problem in the AO-RA-MN (and AO-RA-MU, RO-AA-MU) models, no algorithm can achieve a better than $\Omega(\frac{\log N}{\log\log N})$-approximation in expectation.*

References

1. Abdel-Hamid, A.R., Bather, J.A., Trustrum, G.B.: The secretary problem with an unknown number of candidates. J. Appl. Prob. 19, 619–630 (1982)
2. Babaioff, M., Dinitz, M., Gupta, A., Immorlica, N., Talwar, K.: Secretary problems: weights and discounts. In: SODA 2009, pp. 1245–1254 (2009)
3. Babaioff, M., Immorlica, N., Kleinberg, R.: Matroids, secretary problems, and online mechanisms. In: SODA 2007, pp. 434–443 (2007)
4. Bateni, M.H., Hajiaghayi, M.T., Zadimoghaddam, M.: Submodular secretary problem and extensions. In: Serna, M., Shaltiel, R., Jansen, K., Rolim, J. (eds.) APPROX 2010, LNCS, vol. 6302, pp. 39–52. Springer, Heidelberg (2010)
5. Borosan, P., Shabbir, M.: A survey of secretary problem and its extensions (preprint 2009), http://paul.rutgers.edu/~mudassir/Secretary/paper.pdf
6. Buchbinder, N., Jain, K., Singh, M.: Secretary problems via linear programming. In: Eisenbrand, F., Shepherd, F.B. (eds.) IPCO 2010. LNCS, vol. 6080, pp. 163–176. Springer, Heidelberg (2010)
7. Dimitrov, N.B., Plaxton, C.G.: Competitive weighted matching in transversal matroids. In: Aceto, L., Damgård, I., Goldberg, L.A., Halldórsson, M.M., Ingólfsdóttir, A., Walukiewicz, I. (eds.) ICALP 2008, Part I. LNCS, vol. 5125, pp. 397–408. Springer, Heidelberg (2008)
8. Dynkin, E.B.: The optimum choice of the instant for stopping a markov process. Soviet Mathematics, Doklady 4 (1963)

9. Feldman, J., Henzinger, M., Korula, N., Mirrokni, V.S., Stein, C.: Online stochastic packing applied to display ad allocation. In: de Berg, M., Meyer, U. (eds.) ESA 2010. LNCS, vol. 6346, pp. 182–194. Springer, Heidelberg (2010)
10. Ferguson, T.S.: Who solved the secretary problem? Statistical Science 4(3), 282–289 (1989)
11. Gardner, M.: Mathematical Games column. Scientific American (February 1960)
12. Gilbert, J., Mosteller, F.: Recognizing the maximum of a sequence. J. Amer. Statist. Assoc. 61, 35–73 (1966)
13. Gupta, A., Roth, A., Schoenebeck, G., Talwar, K.: Constrained non-monotone submodular maximization: Offline and secretary algorithms. In: Saberi, A. (ed.) WINE 2010. LNCS, vol. 6484, pp. 246–257. Springer, Heidelberg (2010)
14. Hajiaghayi, M.T., Kleinberg, R., Parkes, D.: Adaptive limited-supply online auctions. In: EC 2004, pp. 71–80 (2004)
15. Hajiaghayi, M.T., Kleinberg, R., Sandholm, T.: Automated online mechanism design and prophet inequalities. In: International Conference on Artificial Intelligence 2007, pp. 58–65 (2007)
16. Im, S., Wang, Y.: Secretary problems: Laminar matroid and interval scheduling. In: SODA 2011, pp. 1265–1274 (2011)
17. Lindley, D.V.: Dynamic programming and decision theory. Journal of the Royal Statistical Society. Series C (Applied Statistics) 10(1), 39–51 (1961)
18. Kleinberg, R.: A multiple-choice secretary algorithm with applications to online auctions. In: SODA 2005, pp. 630–631 (2005)
19. Korula, N., Pál, M.: Algorithms for secretary problems on graphs and hypergraphs. In: Albers, S., Marchetti-Spaccamela, A., Matias, Y., Nikoletseas, S., Thomas, W. (eds.) ICALP 2009. LNCS, vol. 5556, pp. 508–520. Springer, Heidelberg (2009)
20. Soto, J.A.: Matroid secretary problem in the random assignment model. In: SODA 2011, pp. 1275–1284 (2011)
21. Stewart, T.J.: The secretary problem with an unknown number of options. Operations Research 1, 130–145 (1981)

Hitting Sets Online and Vertex Ranking

Guy Even and Shakhar Smorodinsky

[1] School of Electrical Engineering, Tel-Aviv Univ.,
Tel-Aviv 69978, Israel
[2] Mathematics Department, Ben-Gurion Univ. of the Negev,
Be'er Sheva 84105, Israel

Abstract. We consider the problem of hitting sets online for several families of hypergraphs.

Given a graph $G = (V, E)$, let $H = (V, R)$ denote the hypergraph whose hyperedges are the subsets $U \subseteq V$ such that the induced subgraph $G[U]$ is connected. We establish a new connection between the best competitive ratio for the online hitting set problem in H and the vertex ranking number of G. This connection states that these two parameters are equal. Moreover, this equivalence is algorithmic in the sense, that given an algorithm to compute a vertex ranking of G with k colors, one can use this algorithm as a black-box in order to design a k-competitive deterministic online hitting set algorithm for H. Also, given a deterministic k-competitive online algorithm for H, we can use it as a black box in order to compute a vertex ranking for G with at most k colors. As a corollary, we obtain optimal online hitting set algorithms for many such hypergraphs including those realized by planar graphs, graphs with bounded tree width, trees, etc. This improves the best previously known and more general bound of Alon et al.

We also consider two geometrically defined hypergraphs. The first one is defined by subsets of a given set of n points in the Euclidean plane that are induced by half-planes. We obtain an $O(\log n)$-competitive ratio. We also prove an $\Omega(\log n)$ lower bound for the competitive ratio in this setting. The second hypergraph is defined by subsets of a given set of n points in the plane induced by unit discs. Since the number of subsets in this setting is $O(n^2)$, the competitive ratio obtained by Alon et al. is $O(\log^2 n)$. We introduce an algorithm with $O(\log n)$-competitive ratio. We also show that any online algorithm for this problem has a competitive ratio of $\Omega(\log n)$, and hence our algorithm is optimal.

Keywords: Online Algorithms, Geometric Hypergraphs, Hitting Set, Set Cover.

1   Introduction

In the minimum hitting set problem, we are given a hypergraph (X, R), where X is the ground set of points and R is a set of hyperedges. The goal is to find a finite set $S \subseteq X$ such that every hyperedge is stabbed by S, namely, every hyperedge has a nonempty intersection with S.

C. Demetrescu and M.M. Halldórsson (Eds.): ESA 2011, LNCS 6942, pp. 347–357, 2011.

The minimum hitting set problem is a classical NP-hard problem [Kar72], and remains hard even for geometrically induced hypergraphs (see [HM85] for several references). A sharp logarithmic threshold for hardness of approximation was proved by Feige [Fei98]. On the other hand, the greedy algorithm achieves a logarithmic approximation ratio [Joh74, Chv79]. Better approximation ratios have been obtained for several geometrically induced hypergraphs using specific properties of the induced hypergraphs [HM85, KR99, BMKM05]. Other improved approximation ratios are obtained using the theory of VC-dimension and ε-nets [BG95, ERS05, CV07]. Much less is known about online versions of the hitting set problem.

In this paper, we consider an online setting in which the set of points X is given in the beginning, and the ranges are introduced one by one. Upon arrival of a new range, the online algorithm may add points (from X) to the hitting set so that the hitting set also stabs the new range. However, the online algorithm may not remove points from the hitting set. We use the competitive ratio, a classical measure for the performance of online algorithms [ST85, BEY98], to analyze the performance of online algorithms.

Alon et al. [AAA$^+$09] considered the online set-cover problem for arbitrary hypergraphs. In their setting, the ranges are known in advance, and the points are introduced one by one. Upon arrival of an uncovered point, the online algorithm must choose a range that covers the point. Hence, by replacing the roles of ranges and points, their setting is equivalent to our setting. The online set cover algorithm presented by Alon et al. [AAA$^+$09] achieves a competitive ratio of $O(\log n \log m)$ where n and m are the number of points and the number of hyperedges respectively. Note that if $m \geq 2^{n/\log n}$, the analysis of the online algorithm only guarantees that the competitive ratio is $O(n)$; a trivial bound if one range is chosen for each point.

Our main focus is hypergraphs H that are realized as the connected induced subgraphs of a given graph G. The input of the algorithm is a graph $G = (V, E)$, and the adversary chooses subsets $V' \subseteq V$ such that the induced subgraph $G[V']$ is connected. Our main result is a proof of an equivalence between the best competitive ratio of a deterministic online hitting set algorithm for such hypergraphs H and the so-called vertex ranking number of G. This equivalence is also algorithmic in the sense that there is a polynomial reduction between the two problems in both directions. We believe that this relation between the online hitting set problem and vertex ranking of graphs is novel and of independent interest.

A vertex ranking is defined as follows: Let $G = (V, E)$ be a simple graph. A *vertex ranking* (also an *ordered coloring*) of G is a coloring of the vertices $c : V \to \{1, \ldots, k\}$ such that whenever two vertices u and v have the same color i then every simple path between u and v contains a vertex with color greater than i. Such a coloring has been studied before and has several applications. It was studied in the context of VLSI design [SDG92] and in the context of parallel Cholesky factorization of matrices [Liu86]. It also has applications in planning efficient assembly of products in manufacturing systems [IRV88]. See also [KMS95, Sch89]

An application for the hitting set problem where the hypergraph H consists of induced connected subgraphs of a given graph G is in the assignment of servers in virtual private networks (VPNs). Each VPN request is a subset of vertices that induce a connected subgraph in the network, and requests for VPNs arrive online. In addition, for each VPN, we need to assign a server (among the nodes in the VPN) that serves the nodes of the VPN. Since setting up a server is expensive, the goal is to select as few servers as possible.

Two more classes of hypergraphs are obtained geometrically as follows. In both settings we are given a set X of n points in the plane. In one hypergraph, the hyperedges are intersections of X with half-planes. In the other hypergraph, the hyperedges are intersections of X with unit discs. Our main result is an online algorithm for the hitting set problem for points in the plane and unit discs (or half-planes) with an optimal competitive ratio of $O(\log n)$. The competitive ratio of this algorithm improves the $O(\log^2 n)$-competitive ratio of Alon et al. by a logarithmic factor.

An application for points and unit discs is the selection of access points or base stations in a wireless network. The points model base stations and the disc centers model clients. The reception range of each client is a disc, and the algorithm has to select a base station that serves a new uncovered client. The goal is to select as few base stations as possible.

Organization. In Section 2 we present the online minimum hitting set problem. In Section 3, we summarize the results of the paper. The proof of our main result is presented in Section 4. An online algorithm for hypergraphs induced by points and half-planes is presented in Section 5. An online algorithm for the case of points and unit discs is presented in Section 6. Lower bounds are presented in Section 7.

2   Preliminaries

Let (X, R) denote a hypergraph, where R is a set of nonempty subsets of the ground set X. Members in X are referred to as *points*, and members in R are referred to as *ranges* (or *hyperedges*). A subset $S \subseteq X$ *stabs* a range r if $S \cap r \neq \emptyset$. A *hitting set* is a subset $S \subseteq X$ that stabs every range in R. In the minimum hitting set problem, the goal is to find a hitting set with the smallest cardinality.

In this paper, we consider the following online setting. The adversary introduces a sequence $\sigma \triangleq \{r_i\}_{i=1}^s$ of ranges. Let σ_i denote the prefix $\{r_1, \ldots, r_i\}$. The online algorithm computes a chain of hitting sets $C_1 \subseteq C_2 \subseteq \cdots$ such that C_i is a hitting set with respect to the ranges in σ_i.

Consider a fixed hypergraph $H = (X, R)$. The competitive ratio of the algorithm is defined as follows. Let $\text{OPT}(\sigma) \subseteq X$ denote a minimum cardinality hitting set for the ranges in σ. Let $\text{ALG}(\sigma) \subseteq X$ denote the hitting set computed by an online algorithm ALG when the input sequence is σ. Note that the sequence of minimum hitting sets $\{\text{OPT}(\sigma_i)\}_i$ is not necessarily a chain of inclusions. The *competitive ratio* of an online hitting set algorithm ALG is defined as the supremum, over all sequences σ of ranges, of the ratio $|\text{ALG}(\sigma)|/|\text{OPT}(\sigma)|$. We denote the competitive ratio of ALG by $\rho(\text{ALG})$. Note that $\rho(\text{ALG})$ refers to the competitive ratio of ALG on the fixed hypergraph H.

For a hypergraph $H = (X, R)$ we define the parameter $\rho(H)$ as the smallest competitive ratio obtainable by any deterministic hitting set online algorithm for H. That is, $\rho(H) \triangleq \min_{\text{ALG}} \rho(\text{ALG})$.

3   Summary of Our Results

3.1   Vertex Ranking and Online Hitting-Set Algorithms

We consider the following setting of a hypergraph induced by connected subgraphs of a given graph. Formally, let $G = (V, E)$ be a graph. Let $H = (V, R)$ denote the hypergraph over the same set of vertices V. A subset $r \subseteq V$ is a hyperedge in R if and only if the subgraph $G[r]$ induced by r is connected.

Our main result is a characterization of the competitive ratio $\rho(H)$ in terms of the vertex ranking number of G [KMS95, Sch89]. A *vertex ranking* of G is a function $c : V \to \mathbb{N}$ that satisfies the following property: For any pair of vertices $u, v \in V$, if $c(u) = c(v)$, then, for any simple path P in G connecting u and v, there is a vertex w in P such that $c(w) > c(u)$. The *vertex ranking number* of G, denoted $\chi_{\text{vr}}(G)$, is the least integer k for which G admits a vertex ranking that uses only k colors.

Notice that, in particular, a vertex ranking of a graph G is also a proper coloring of G since adjacent vertices must get distinct colors. On the other hand, a proper coloring is not necessarily a vertex ranking as is easily seen by taking a path graph P_n. This graph admits a proper coloring with 2 colors but this coloring is not a valid vertex ranking. In fact, in Lemma 7 we prove that $\chi_{\text{vr}}(P_n) = \lfloor \log_2 n \rfloor + 1$.

Let $G = (V, E)$ be a graph and let $H = (V, R)$ be the hypergraph of induced connected subgraphs of G as above. The following theorem establishes the characterization of the competitive ratio for the problem of online hitting set for H in terms of the vertex ranking number of G.

Theorem 1. $\rho(H) = \chi_{\text{vr}}(G)$.

The proof of the theorem uses a constructive reduction in both directions. Namely, given a vertex ranking c for G that uses k colors, we design an online hitting set algorithm for H whose competitive ratio is at most k. Conversely, given an online hitting set algorithm ALG, we construct a vertex ranking c for G that uses at most $\rho(\text{ALG})$ colors. Thus, establishing an equality, as required.

As a corollary, we obtain optimal online hitting set algorithms for a wide class of graphs that admit (hereditary) small balanced separators. For example, consider this problem for planar graphs. Let G be a planar graph on n vertices. It was proved in [KMS95] that $\chi_{\text{vr}}(G) = O(\sqrt{n})$. Therefore, Theorem 1 implies that the competitive ratio of our algorithm for connected subgraphs of planar graphs is $O(\sqrt{n})$. Theorem 1 also implies that this bound is optimal. Indeed, it was proved in [KMS95] that for the $l \times l$ grid graph $G_{l \times l}$ (with l^2 vertices), $\chi_{\text{vr}}(G_{l \times l}) \geq l$. Hence, for $G_{l \times l}$, any deterministic online hitting set algorithm must have a competitive ratio at least l. In Table 1 we list several important classes of such graphs.

Table 1. A list of several graph classes with small separators ($n = |V|$)

graph $G = (V, E)$	competitive ratio	previous result [AAA$^+$09]
path P_n	$\lfloor \log_2 n \rfloor + 1$	$O(\log^2 n)$
tree	$O(\log(\text{diameter}(G)))$	$O(n)$
tree-width d	$O(d \log n)$	$O(n)$
planar graph	$O(\sqrt{n})$	$O(n)$

We note that in the case of a star (i.e., a vertex v with $n - 1$ neighbors), the number of subsets of vertices that induce a connected graph is greater than 2^{n-1}. Hence, the VC-dimension of the hypergraph is linear. However, the star has a vertex ranking that uses just two colors, hence, the competitive ratio of our algorithm in this case is 2. This is an improvement over the analysis of the algorithm of Alon et al. [AAA$^+$09] which only proves a competitive ratio of $O(n)$. Thus, our algorithm is useful even in hypergraphs whose VC-dimension is unbounded.

3.2   Applications to Geometric Range-spaces

Points and Half-Planes. We prove the following results for hypergraphs in which the ground set X is a finite set of n points in $\mathbb{R}^2$ and the ranges are all subsets of X that can be cut off by a half-plane. Namely, each range r is induced by a line L_r such that r is the set of points of X in the half-plane below (respectively, above) the line L_r.

Theorem 2. *The competitive ratio of every deterministic online hitting set algorithm for points and half-planes is $\Omega(\log n)$.*

Theorem 3. *There exists a deterministic online hitting set algorithm for points and half-planes that achieves a competitive ratio of $O(\log n)$.*

Points and Congruent Discs. We prove the following results for hypergraphs in which the ground set X is a finite subset of n points in $\mathbb{R}^2$ and the ranges are intersections of X with unit discs. Namely, a unit disc d induces a range $r = r(d)$ defined by $r = d \cap X$.

Theorem 4. *The competitive ratio of every deterministic online hitting set algorithm for points and unit discs is $\Omega(\log n)$.*

Theorem 5. *There exists a deterministic online hitting set algorithm for points and unit discs that achieves a competitive ratio of $O(\log n)$.*

4   Vertices and Connected Subgraphs

In this section we prove Theorem 1. As mentioned already, the proof is constructive. Let $G = (V, E)$ be a simple graph and let $H = (V, R)$ be the hypergraph with respect to connected induced sub-graphs. We start by proving $\rho(H) \leq \chi_{\text{vr}}(G)$. Assume that we are given a vertex ranking c of G that uses k colors. We introduce an online hitting set algorithm HS for H such that $\rho(\text{HS}) \leq k$.

4.1   An Online Hitting Set Algorithm

A listing of Algorithm HS appears as Algorithm 1. The algorithm is input a graph $G = (V, E)$ and a vertex ranking $c : V \to \{1, \ldots, k\}$. The sequence $\sigma = \{r_i\}_i$ of subsets of vertices that induce connected subgraphs is input online.

For a subset $X' \subseteq V$, let $c_{\max}(X') \triangleq \max\{c(v) \mid v \in X'\}$. It is easy to see that if c is a vertex ranking, then $|\{v \in r \mid c(v) = c_{\max}(r)\}| = 1$, for every subset $r \in R$ (inducing a connected subgraph). Thus, let $v_{\max}(r)$ denote the (unique) vertex v in r such that $c(v) = c_{\max}(r)$.

Algorithm 1. $\text{HS}(G, c)$ - an online hitting set for connected subgraphs, given a vertex ranking c.

Require: $G = (V, E)$ is a graph and $c : V \to \{1, \ldots, k\}$ is a vertex ranking.
1: $C_0 \leftarrow \emptyset$
2: **for** $i = 1$ to ∞ **do** {arrival of a range r_i}
3:      **if** r_i is not stabbed by C_{i-1} **then**
4:          $C_i \leftarrow C_{i-1} \cup \{v_{\max}(r_i)\}$ {add the vertex with the max color in r_i}
5:      **else**
6:          $C_i \leftarrow C_{i-1}$
7:      **end if**
8: **end for**

4.2   Analysis of the Competitive Ratio

Definition 1. *For a color a, let $\sigma(a)$ denote the subsequence of σ that consists of ranges that satisfy: (i) r_i is not stabbed by C_{i-1}, and (ii) $c_{\max}(r_i) = a$.*

The following lemma implies a lower bound on the (offline) minimum hitting set of the ranges in $\sigma(a)$.

Lemma 1. *If $r_i, r_j \in \sigma(a)$, then the subgraph $G[r_i \cup r_j]$ induced by $r_i \cup r_j$ is not connected. Hence, the ranges in $\sigma(a)$ are pairwise disjoint.*

Proof. Clearly, $c_{\max}(r_i \cup r_j) = \max\{c_{\max}(r_i), c_{\max}(r_j)\} = a$. Assume that $r_i \cup r_j$ induces a connected subgraph. Since c is a vertex ranking, we conclude that $r_i \cup r_j$ contains exactly one vertex colored a. This implies that $v_{\max}(r_i) = v_{\max}(r_j)$. If $j > i$, then the range r_j is stabbed by C_{j-1} since it is stabbed by C_i, a contradiction.

Lemma 2. $\rho(\text{HS}) \leq c_{\max}(V) = k$.

Proof. Algorithm HS satisfies $|\text{HS}(\sigma)| = \sum_{a \in \mathbb{N}} |\sigma(a)|$. But $\sum_{a \in \mathbb{N}} |\sigma(a)| \leq c_{\max}(V) \cdot \max_{a \in \mathbb{N}} |\sigma(a)|$. By Lemma 1, each range in $\sigma(a)$ must be stabbed by a distinct vertex, thus $|\text{OPT}(\sigma)| \geq \max_{a \in \mathbb{N}} |\sigma(a)|$, and the theorem follows.

4.3   An Algorithm for Vertex Ranking

Assume that we are given a deterministic online hitting set algorithm ALG for H whose competitive ratio is $\rho(\text{ALG}) = k$. We use the notation $v = \text{ALG}(\sigma, U)$ to denote that ALG uses vertex v to stab the range U after it has been input the sequence of queries σ. We use ALG in order to find a vertex-ranking c for G that uses at most k colors. This proves the inequality $\chi_{\text{vr}}(G) \leq k$.

Algorithm 2. $\text{VR}(U, \text{ALG}, k, \sigma)$ - compute a vertex ranking for G, given an online hitting-set algorithm ALG with $\rho(\text{ALG}) \leq k$ with respect to the hypergraph induced by U.

Require: $G[U]$ is an induced connected subgraph of G and ALG is an online hitting set algorithm for H with $\rho(\text{ALG}) \leq k$. The sequence of ranges input so far to ALG is denoted by σ.
1: If $U = \emptyset$ then return.
2: $v = \text{ALG}(\sigma, U)$ (v is the vertex used by ALG to stab the range U after the sequence of ranges σ)
3: $c(v) \leftarrow k$ (we color v with the integer k)
4: Let σ' denote the sequence obtained by adding U to the end of σ.
5: **for** every connected component $G' = (U', E')$ of $G[U - v]$ **do**
6:     $\text{VR}(U', \text{ALG}, k - 1, \sigma')$ (Recursively color the connected components of $G[V - v]$)
7: **end for**

The algorithm for coloring the nodes of G is called VR and is listed as Algorithm 2. It is a recursive algorithm and is initially invoked by $\text{VR}(V, \text{ALG}, k, \phi)$, where ϕ denotes an empty sequence. Note that in the recursion, we use σ to "remember" the sequence of queries posed to ALG till this point.

Lemma 3. *If* $\rho(\text{ALG}) \leq k$, *then* $\chi_{\text{vr}}(G) \leq k$

In order to prove Lemma 3 we need to prove two lemmas. One states that the number of colors used by VR is at most k colors. The second states that the algorithm VR produces a valid vertex ranking of G.

Lemma 4. *Algorithm VR uses at most k colors.*

Proof. Assume to the contrary that VR uses at least $k+1$ colors. This means that the depth of the recursion tree of the algorithm is at least $k+1$. Thus, there is a sequence of subsets of vertices $V_1, \ldots, V_{k+1}$ such that: (i) $V = V_1$, (ii) $V_{i+1} \subsetneq V_i$ for $i = 1, \ldots, k$, (iii) $V_{k+1} \neq \emptyset$, and (iv) each of the induced subgraphs $G[V_i]$ for $i = 1, \ldots, k+1$ is connected. We now input ALG the sequence $\{V_i\}_{i=1}^{k+1}$. Each subset V_i is stabbed by a vertex $v_i \in V_i \setminus V_{i+1}$. Hence, ALG uses $k+1$ distinct vertices to stab these ranges. However, all these ranges can be stabbed with one vertex (i.e., any vertex in V_{k+1}), and hence, we have $\rho(\text{ALG}) \geq k+1$, a contradiction.

Lemma 5. *Algorithm* VR *computes a valid vertex ranking.*

Proof. We refer to the recursion level in which vertices are colored i as the ith level. For example, in the k'th level only the vertex $v = \text{ALG}(\phi, V)$ is colored with k.

Let x and y be two vertices with the same color i. Let P be a simple path connecting x and y. We need to show that P contains a vertex with color greater than i. Let $\mathcal{U}_i$ be the family of all subsets of vertices used in the execution of VR in the $i'th$ level of the recursion tree. Let V_i be the subset of all vertices colored i. Let X (resp., Y) denote the query in Line 2 of VR that returned the vertex x (resp., y). Note that the sets in $\mathcal{U}_i$ are the connected components with respect to the vertex cut Z_i defined by $Z_i \triangleq \bigcup_{j>i} V_j$. In particular, any path connecting a vertex of X with a vertex of Y must traverse a vertex in Z_i. But, the color of each vertex in Z_i is greater than i. In particular P contains a vertex with a color greater than i. This completes the proof of the lemma.

5 Points and Half-Planes

In this section we consider a special instance of the minimum hitting set problem for a finite set of points in the plane and ranges induced by half-planes. We consider only ranges of points that are below a line; the case of points above a line is dealt with separately. This increases the competitive ratio by at most a factor of two.

Notation. Given a finite planar set of points X, let $V \subseteq X$ denote the subset of extreme points of X. That is, V consists of all points $p \in X$ such that there exists a half-plane h with $h \cap X = \{p\}$. Let $\{p_i\}_{i=1}^{|V|}$ denote an ordering of V in ascending x-coordinate order. Let $P = (V, E_P)$ denote the path graph over V where p_i is a neighbor of p_{i+1} for $i = 1, \ldots, |V|-1$. The intersection of every half-plane with V is a subpath of P. Namely, the intersection of a nonempty range r_i with V is a set of the form $\{p_j \mid j \in [a_i, b_i]\}$. We refer to such an intersection as a discrete interval (or simply an interval, if the context is clear). We often abuse this notation and refer to a point $p_i \in V$ simply by its index i. Thus, the interval of points in the intersection of r_i and V is denoted by $I_i \triangleq [a_i, b_i]$.

Algorithm Description. The algorithm reduces the minimum hitting set problem for points and half-planes to a minimum hitting set of connected subgraphs. The reduction is to the path graph P over the extreme points V of X. To apply Algorithm HS, a vertex ranking c for P is computed, and each half-plane r_i is reduced to the interval I_i. A listing of Algorithm HS_p appears as Algorithm 3. Note that the algorithm HS_p uses only the subset $V \subset X$ of extreme points of X.

Algorithm 3. $\text{HS}_p(X)$ - an online hitting set for half-planes

---

Require: $X \subset \mathbb{R}^2$ is a set of n points.
1: $V \leftarrow$ the extreme points of X.
2: Compute an ordering $\{p_i\}_{i=1}^{|V|}$ of V in ascending x-coordinate order.
3: Let $P = (V, E_P)$ denote the path graph over V, where $E_P \triangleq \{(p_i, p_{i+1})\}_{i=1}^{|V|-1}$.
4: $c \leftarrow$ a vertex ranking of P.
5: Run $\text{HS}(P, c)$ with the sequence of ranges $\{I_i\}_i$, where I_i is the interval $r_i \cap V$.

5.1   Analysis of the Competitive Ratio

Recall that $\sigma(a)$ denotes the subsequence of σ consisting of ranges r_i that are unstabbed upon arrival and stabbed initially by a point colored a. The following lemma is the analog of Lemma 1 for the case of points and half-planes.

Lemma 6. *The ranges in $\sigma(a)$ are pairwise disjoint.*

Proof. Assume for the sake of contradiction that $r_i, r_j \in \sigma(a)$ and $z \in r_i \cap r_j$. Let $[a_i, b_i]$ denote the endpoints of the interval $I_i = r_i \cap V$, and define $[a_j, b_j]$ and I_j similarly. The proof of Lemma 1 proves that $I_i \cup I_j$ is not connected. This implies that $z \notin V$ and that there is a vertex $t \in V$ between I_i and I_j.

Without loss of generality, $b_i < t < a_j$. Let $(z)_x$ denote the x-coordinate of point z. Assume that $(z)_x \leq (t)_x$ (the other case is handled similarly). See Fig. 1 for an illustration. Let L_j denote a line that induces the range r_j, i.e., the set of points below L_j is r_j. Let L_t denote a line that separates t from $X \setminus \{t\}$, i.e., t is the only point below L_t. Then, L_t passes below z, above t, and below a_j. On the other hand, L_j passes above z, below t, and above a_j. Since $(z)_x \leq (t)_x < (a_j)_x$, it follows that the lines L_t and L_j intersect twice, a contradiction, and the lemma follows.

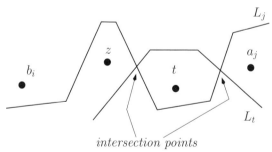

Fig. 1. Proof of Lemma 6. The lines L_j and L_t are depicted as polylines only for the purpose of depicting their above/below relations with the points.

Proposition 1 ([IRV88, KMS95, ELRS03]). *A path of n nodes admits a vertex ranking that uses $\lfloor \log_2(2n) \rfloor$ colors.*

Proof (Proof of Theorem 3). Run Algorithm HS_p with a vertex ranking of P that uses $\lfloor \log_2 2|V| \rfloor$ colors. Follow the proof of Theorem 1 using Lemma 6 instead of Lemma 1.

6   Points and Unit Discs

In this section we consider a special instance of the minimum hitting set problem in which the ground set X is a finite set of n points in $\mathbb{R}^2$. The ranges are subsets of points that are contained in a unit disc. Formally, a unit disc d centered at o is the set $d \triangleq \{x \in \mathbb{R}^2 : ||x-o||_2 \leq 1\}$. The range $r(d)$ induced by a disc d is the set $r(d) \triangleq \{x \in X : x \in d\}$. The circle ∂d is defined by $\partial d \triangleq \{x \in \mathbb{R}^2 : ||x-o||_2 = 1\}$.

Due to lack of space, the proof of Theorem 5 is presented in the Appendix.

7   Lower Bounds

7.1   A Path of n Vertices

Consider the path $P_n = (V, E)$, where $V = \{1, \ldots, n\}$ and $E = \{(i, i+1) : 1 \leq i < n\}$. For the sake of completeness, we present a proof that the vertex ranking of a path P_n is $\Omega(\log n)$. By Theorem 1, it follows that $\rho(H) = \Omega(\log n)$, where H is the hypergraph over V, the hyperedges of which are subpaths in P_n.

Lemma 7. *The vertex ranking number of the path P_n is $\lfloor \log n \rfloor + 1$.*

Proof. Consider a vertex ranking c of P_n that uses $\chi_{\mathrm{vr}}(P_n)$ colors. Let i denote the vertex with the highest color in P_n. The vertex ranking number $\chi_{\mathrm{vr}}(P_n)$ satisfies the recurrence

$$\chi_{\mathrm{vr}}(p_n) \geq 1 + \max\{\chi_{\mathrm{vr}}(P_{i-1}), \chi_{\mathrm{vr}}(P_{n-(i+1)})\}.$$

Clearly, the solution of the recurrence is as required.

7.2   Points and Half-Planes

We now prove Theorem 2.

Proof (Proof of Theorem 2). We reduce the instance of the path P_n and its induced connected subgraphs to an instance of points and half-planes. We place the n points on the parabola $y = x^2$. Namely, point i is mapped to the point (i, i^2). An interval $[i, j]$ of vertices is obtained by points below the line passing through the images of i and j. Hence, the problem of online hitting ranges induced by half-planes is not easier than the problem of online hitting intervals of P_n. The Theorem follows from Lemma 7 and Theorem 1.

The proof of Theorem 4 is similar and uses the fact that if the n points are placed on a line such that the distance between the first and last point is, say 1, then for any interval, there is a unit disc that intersects the set in exactly the same points as the interval. Thus the lower bound for intervals holds also for unit discs.

References

[AAA⁺09]   Alon, N., Awerbuch, B., Azar, Y., Buchbinder, N., Naor, J.S.: The Online
            Set Cover Problem. SIAM Journal on Computing 39, 361 (2009)
[BEY98]   Borodin, A., El-Yaniv, R.: Online computation and competitive analysis,
            vol. 2. Cambridge University Press, Cambridge (1998)
[BG95]   Brönnimann, H., Goodrich, M.T.: Almost optimal set covers in finite VC-
            dimension. Discrete and Computational Geometry 14(1), 463–479 (1995)
[BMKM05]   Ben-Moshe, B., Katz, M.J., Mitchell, J.S.B.: A constant-factor ap-
            proximation algorithm for optimal terrain guarding. In: Proceedings of
            the Sixteenth Annual ACM-SIAM Symposium on Discrete Algorithms,
            pp. 515–524. SIAM, Philadelphia (2005)
[Chv79]   Chvatal, V.: A greedy heuristic for the set-covering problem. Mathematics
            of Operations Research 4(3), 233–235 (1979)
[CV07]   Clarkson, K.L., Varadarajan, K.: Improved approximation algorithms for
            geometric set cover. Discrete and Computational Geometry 37(1), 43–58
            (2007)
[ELRS03]   Even, G., Lotker, Z., Ron, D., Smorodinsky, S.: Conflict-Free Colorings
            of Simple Geometric Regions with Applications to Frequency Assignment
            in Cellular Networks. SIAM Journal on Computing 33, 94–136 (2003)
[ERS05]   Even, G., Rawitz, D., Shahar, S.M.: Hitting sets when the VC-dimension
            is small. Information Processing Letters 95(2), 358–362 (2005)
[Fei98]   Feige, U.: A threshold of ln n for approximating set cover. Journal of the
            ACM (JACM) 45(4), 634–652 (1998)
[HM85]   Hochbaum, D.S., Maass, W.: Approximation schemes for covering and
            packing problems in image processing and VLSI. Journal of the ACM
            (JACM) 32(1), 136 (1985)
[IRV88]   Iyer, A.V., Ratliff, H.D., Vijayan, G.: Optimal node ranking of trees. In-
            formation Processing Letters 28(5), 225–229 (1988)
[Joh74]   Johnson, D.S.: Approximation algorithms for combinatorial problems.
            Journal of Computer and System Sciences 9(3), 256–278 (1974)
[Kar72]   Karp, R.M.: Reducibility among combinatorial problems. In: Proc. Sym-
            pos., IBM Thomas J. Watson Res. Center, Yorktown Heights, NY, pp.
            85–103 (1972)
[KMS95]   Katchalski, M., McCuaig, W., Seager, S.: Ordered colourings. Discrete
            Mathematics 142(1-3), 141–154 (1995)
[KR99]   Kumar, V.S.A., Ramesh, H.: Covering rectilinear polygons with axis-
            parallel rectangles. In: Proceedings of the Thirty-First Annual ACM Sym-
            posium on Theory of Computing, pp. 445–454. ACM, New York (1999)
[Liu86]   Liu, J.W.H.: Computational models and task scheduling for parallel sparse
            cholesky factorization. Parallel Computing 3(4), 327–342 (1986)
[Sch89]   Schäffer, A.A.: Optimal node ranking of trees in linear time. Information
            Processing Letters 33(2), 91–96 (1989)
[SDG92]   Sen, A., Deng, H., Guha, S.: On a graph partition problem with applica-
            tion to VLSI layout. Information Processing Letters 43(2), 87–94 (1992)
[ST85]   Sleator, D.D., Tarjan, R.E.: Amortized efficiency of list update and paging
            rules. Communications of the ACM 28(2), 202–208 (1985)

Fast Sub-exponential Algorithms
and Compactness in Planar Graphs

Dimitrios M. Thilikos

Department of Mathematics, National and Kapodistrian University of Athens,
Panepistimioupolis, 15784 Athens, Greece

Abstract. We provide a new theory, alternative to bidimensionality, of sub-exponential parameterized algorithms on planar graphs, which is based on the notion of *compactness*. Roughly speaking, a parameterized problem is (r, q)-compact when all the faces and vertices of its YES-instances are "r-radially dominated" by some vertex set whose size is at most q times the parameter. We prove that if a parameterized problem can be solved in $c^{\text{branchwidth}(G)} n^{O(1)}$ steps and is (r, q)-compact, then it can be solved by a $c^{r \cdot 2.122 \cdot \sqrt{q \cdot k}} n^{O(1)}$ step algorithm (where k is the parameter). Our framework is general enough to unify the analysis of almost all known sub-exponential parameterized algorithms on planar graphs and improves or matches their running times. Our results are based on an improved combinatorial bound on the branchwidth of planar graphs that bypasses the grid-minor exclusion theorem. That way, our approach encompasses new problems where bidimensionality theory do not directly provide sub-exponential parameterized algorithms.

Keywords: Parameterized Algorithms, Branchwidth, Planar Graphs.

1   Introduction

A parameterized problem can be defined as a language $\Pi \subseteq \Sigma^* \times \mathbb{N}$. Its inputs are pairs $(I, k) \in \Sigma^* \times \mathbb{N}$, where I can be seen as the main part of the problem and k is some parameter of it. A problem $\Pi \subseteq \Sigma^* \times \mathbb{N}$ is *fixed parameter tractable* when it admits an $f(k) \cdot n^{O(1)}$ -time algorithm. In that case, Π is classified in the parameterized complexity class FPT and, when we insist to indicate the parameter dependance (i.e., the function f), we also say that that $\Pi \in f(k)$-FPT.

Sub-exponential parameterized algorithms. A central problem in parameterized algorithm design is to investigate in which cases and under which input restrictions a parameterized problem belongs to FPT and, if so, to find algorithms with the simplest possible parameter dependance. When $f(k) = 2^{o(k)}$, a parameterized problem is said to admit a *sub-exponential parameterized algorithm* (for a survey on this topic, see [15]).

In [5], Cai and Juedes proved that several parameterized problems most probably do not belong to $2^{o(k)}$-FPT. Among them, one can distinguish core problems such as the standard parameterizations of VERTEX COVER, DOMINATING SET,

C. Demetrescu and M.M. Halldórsson (Eds.): ESA 2011, LNCS 6942, pp. 358–369, 2011.
© Springer-Verlag Berlin Heidelberg 2011

and FEEDBACK VERTEX SET. However, it appears that many problems admit sub-exponential parameterized algorithms when their inputs are restricted to planar graphs or other sparse graph classes. Moreover, the results of [5] indicated that this is indeed the best we may expect when the planarity restriction is imposed. The first sub-exponential parameterized algorithm on planar graphs appeared in [1] for DOMINATING SET, INDEPENDENT DOMINATING SET, and FACE COVER. After that, many other problems were classified in $2^{c\sqrt{k}}$-FPT, while there was a considerable effort towards improving the constant c for each one of them [1, 28, 6, 9, 16, 17, 27, 21, 29, 12].

Bidimensionality theory. A major advance towards a theory of sub-exponential parameterized algorithms was made with the introduction of Bidimensionality, in [10]. Bidimensionality theory offered a generic condition for classifying a parameterized problem in $2^{c\sqrt{k}}$-FPT. It also provided a machinery for estimating a (reasonably small) value c for each particular problem. Moreover, it also provided meta-algorithmic results in approximation algorithms [11, 19] and kernelization [20] (for a survey on bidimensionality, see [7]). We stress that alternative approaches for the design of subexponential parameterized algorithms have been appeared in [2, 14, 23, 36].

According to [10], a problem $\Pi \subseteq \mathcal{P} \times \mathbb{N}$ on planar graphs is *minor-bidimensional* with density δ if the following two conditions are satisfied.

(a) the graph G' is a minor of the graph G, then $(G, k) \in \Pi \Rightarrow (G', k) \in \Pi$.
(b) there exists a $\delta > 0$ such that for every $k \in \mathbb{N}$ it holds that $(\Gamma_{\sqrt{k}/\delta}, k) \notin \Pi$.

In the above definition, we denote by $\mathcal{P}$ the class of all planar graphs and we use the term Γ_w for the $(\lceil w \rceil \times \lceil w \rceil)$-grid. Also, we say that G' is a *minor* of G, denoted as $G' \leqslant_m G$, if G' can be obtained by some subgraph of G after a series of edge contractions. We stress that there is an analogous definition for the case where, in **(a)**, we replace minors by contractions only, but, for simplicity, we avoid giving more definitions on this point.

Branchwidth (along with its twin parameter of treewidth) has been a powerful tool in parameterized algorithm design. Roughly speaking, branchwidth is a measure of the topological resemblance of a graph to the structure of a tree. We use the term $\mathbf{bw}(G)$ for the branchwidth of a graph G and we postpone its formal definition until Section 2.

We say that a problem $\Pi \subseteq \mathcal{P} \times \mathbb{N}$ is λ-*single exponentially solvable with respect to branchwidth* if there exists an algorithm that solves it in $2^{\lambda \cdot \mathbf{bw}(G)} n^{O(1)}$ steps. The main idea of [10] was to make use of the grid-minor exclusion theorem in [33] asserting that, for every planar graph G, $\mathbf{bw}(G) \leqslant 4 \cdot \mathbf{gm}(G)$, where $\mathbf{gm}(G) = \max\{w \mid \Gamma_w \leqslant_m G\}$. This result was recently improved by Gu and Tamaki in [25] who proved that $\mathbf{bw}(G) \leqslant 3 \cdot \mathbf{gm}(G)$. This implies that for a bidimensional problem with density δ on planar graphs, all YES-instances have branchwidth at most $\frac{3}{\delta}\sqrt{k}$ and this reduces the problem to its variant where the branchwidth of the inputs are now bounded by $\frac{3}{\delta}\sqrt{k}$. An optimal branch decomposition of a planar graph can be constructed in $O(n^3)$ steps, (see [24, 35]) and it is possible to transform it to a tree decomposition of width at most 1.5 times more [34]. Therefore, the main algorithmic consequence of bidimensionality, as restricted to planar graphs[1], is the following.

Proposition 1. *If* $\Pi \subseteq \mathcal{P} \times \mathbb{N}$ *is minor-bidimensional with density δ and is λ-single exponentially solvable with respect to branchwidth, then* $\Pi \in 2^{(3\lambda/\delta) \cdot \sqrt{k}}$- FPT.

The above result, along with its contraction-bidimensionality counterpart, defined in [10] (see also [18]), reduce the solution of bidimensional problems to the easier task of designing dynamic programming algorithms on graphs with small branchwidth (or treewidth). Dynamic programming is one of the most studied and well developed topics in parameterized algorithms and there is an extensive bibliography on what is the best value of λ that can be achieved for each problem (see e.g., [37]). Especially for planar graphs, there are tools that can make dynamic programming run in single exponential time, even if this is not, so far, possible for general graphs [13]. Lower bounds on the value of λ for problems such as DOMINATING SET appeared recently in [30]. Finally, meta-algorithmic conditions for single exponential solvability with respect to branchwidth have very recently appeared in [32].

A consequence of Proposition 1 and its contraction-counterpart was a massive classification of many parameterized problems in $2^{c \cdot \sqrt{k}}$-FPT and, in many cases, with estimations of c that improved all previously existing bounds. The remaining question was whether it is possible to do even better and when. Indeed, a more refined problem-specific combinatorial analysis improved some of the bounds provided by the bidimensionality framework (see also [12]). For instance, PLANAR DOMINATING SET [21], FACE COVER, PLANAR VERTEX FEEDBACK SET, and CYCLE PACKING [29] where classified in $2^{c \cdot \sqrt{k}}$-FPT where $c = 15.3, 10.1, 15.11$, and 26.3 respectively, improving all previous results on those problems.

Our results. In this paper, we provide an alternative theory for the design of fast sub-exponential parameterized algorithms.

Let us give first some definitions from [4]. Given a plane graph $G = (V, E)$ and a set $S \subseteq V$, we define $\mathbf{R}_G^r(S)$ as the set of all vertices or faces of G whose radial distance from some vertex of S is at most r. The *radial distance* between two vertices x, y is one less than the minimum length of an alternating sequence of vertices and faces starting from x and ending in y, such that every two consecutive elements in this sequence are incident to each other.

The notion of compactness has been introduced in [4] in the context of automated kernel design, while some preliminary concepts had already appeared in [26]. It encompasses a wide number of parameterized problems on graphs; some of them are listed in Table 1.

A parameterized problem $\Pi \subseteq \mathcal{P} \times \mathbb{N}$ is (q, r)-*compact* if for all $(G = (V, E), k) \in \Pi$ and for some planar embedding of a graph H, that is either G or its dual, there

---

[1] The results in [10] apply for more general graph classes and have been further extended in [18, 8].

Table 1. Examples of (q, r)-compact parameterized problems. [a]triconnected instances. [b]biconnected instances. [c]for instances where each vertex of the input graph belongs in some triangle.

	q	r		q	r
DOMINATING SET	1	3	CYCLE DOMINATION	1	4
l-DOMINATING SET	1	$2l+1$	EDGE DOMINATING SET	2	2
l-THRESHOLD DOMINATING SET	1	3	CLIQUE TRANSVERSAL	1	3
PERFECT CODE	1	3	INDEPENDENT DOMINATING SET	1	3
RED BLUE DOMINATING SET	1	3	ODD SET	1	3
INDEPENDENT DIRECTED DOMINATION	1	3	FACE COVER	1	2
VERTEX COVER	1	2	VERTEX TRIANGLE COVERING[c]	1	3
ALMOST OUTERPLANAR[a]	1	3	EDGE TRIANGLE COVERING[c]	2	2
ALMOST SERIES-PARALLEL[a]	1	3	l-CYCLE TRANSVERSAL[b]	1	l
CONNECTED l-DOMINATING SET	1	$2l+1$	l-SCATTERED SET	1	$2l+1$
CONNECTED VERTEX COVER	1	2	CYCLE PACKING	3	3
FEEDBACK VERTEX SET[b]	1	2	INDUCED MATCHING	2	3
FEEDBACK EDGE SET[b]	1	2	MAX INTERNAL SPANNING TREE	1	3
CONNECTED FEEDBACK VERTEX SET	1	3	TRIANGLE PACKING[c]	1	3
MINIMUM-VERTEX FEEDBACK EDGE SET	1	3	MINIMUM LEAF OUT-BRANCHING	1	3
CONNECTED DOMINATING SET	1	3	MAX FULL DEGREE SPANNING TREE	1	3

is a set $S \subseteq V$ such that $|S| \leqslant q \cdot k$ and $\mathbf{R}_H^r(S)$ contains all faces and vertices of H. Intuitively, a parameterized problem is compact if the input graph of every YES-instance can be "covered" by a collection of $O(k)$ balls each of constant radius.

We use the term $\overline{\Pi}$ for the set of NO-instances of the parameterized problem Π. We present below the main algorithmic contribution of our paper.

Theorem 1. *Let $\Pi \subseteq \mathcal{P} \times \mathbb{N}$ be a parameterized problem on planar graphs. If Π is λ-single exponentially solvable with respect to branchwidth and either Π or $\overline{\Pi}$ is (q, r)-compact, then $\Pi \in 2^{\lambda \cdot r \cdot 2.122 \cdot \sqrt{q \cdot k}}$-FPT.*

The advantages of our approach, compared with those of bidimensionality theory, are the following:

- It applies to *many* problems where bidimensionality does not apply directly. This typically happens for problems whose YES-instances are not closed under taking of minors (or contractions) such as INDEPENDENT VERTEX COVER, INDEPENDENT DOMINATING SET, PERFECT CODE, and THRESHOLD DOMINATING SET.
- When applied, it *always* gives better bounds than those provided by bidimensionality theory. A direct comparison of the combinatorial bounds implies that the constants in the exponent provided by compactness are $\sqrt{4.5}/3 \approx 70\%$ of those emerging by the grid-minor exclusion theorem of Gu and Tamaki in [25].
- Matches or improves *all* problem-specific time upper bounds known so far for sub-exponential algorithms in planar graphs (including the results in [9, 21, 29, 28, 12]) and unifies their combinatorial analysis to a single theorem.

Theorem 1 follows from the following theorem that we believe is of independent combinatorial interest.

Theorem 2. *If (G, k) is a* YES-*instance of an (r, q)-compact parameterized problem, then $\mathbf{bw}(G) \leqslant r \cdot \sqrt{4.5 \cdot q \cdot k}$.*

Theorem 2 is our main combinatorial result and the rest of the sections of this paper are devoted to its proof. In Table 1, we give a list of examples of compact parameterized problems together with the corresponding values of q and r. Deriving these estimations of q and r is an easy exercise whose details are omitted.

The paper is organized as follows. In Section 2, we give some necessary definitions and preliminary results. Section 3 is dedicated to the proof of Theorem 2. Finally, some open problems are given in Section 4. The proofs of the Lemmata marked with $(\star)$ are omitted due to space restrictions.

2   Basic Definitions and Preliminaries

All graphs in this paper are simple (i.e., have no multiple edges or loops). For any set $S \subseteq V(G)$, we denote by $G[S]$ the subgraph of G induced by the vertices in S. Given two graphs G_1 and G_2, we define $G_1 \cup G_2 = (V(G_1) \cup V(G_2), E(G_1) \cup E(G_2))$. We also denote by $G \setminus S$ the graph $G[V(G) \setminus S]$. We also denote the distance between two vertices x, y in G by $\mathbf{dist}_G(x, y)$. A *subdivision* of a graph H is any graph that can be obtained from H if we apply a sequence of subdivisions to some (possibly none) of its edges (a subdivision of an edge is the operation of replacing an edge $e = \{x, y\}$ by a (x, y)-path of length two). We say that a graph H is a *topological minor* of a graph G (we denote it by $H \leqslant_t G$) if some subdivision of H is a subgraph of G.

Plane graphs. In this paper, we mainly deal with plane graphs (i.e. graphs embedded in the plane $\mathbb{R}^2$ without crossings). For simplicity, we do not distinguish between a vertex of a plane graph and the point of the plane used in the drawing to represent the vertex or between an edge and the open line segment representing it. Given a plane graph G, we denote its dual by G^*. A *parameter* on plane graphs is any function $\mathbf{p}$ mapping plane graphs to $\mathbb{N}$. Given such a parameter $\mathbf{p}$, we define its *dual* parameter $\mathbf{p}^*$ so that $\mathbf{p}^*(G) = \mathbf{p}(G^*)$.

Given a plane graph G, we denote by $F(G)$ the set of the faces of G (i.e., the connected components of $\mathbb{R}^2 \setminus G$, that are open subsets of the plane). We use the notation $A(G)$ for the set $V(G) \cup F(G)$ and we say that $A(G)$ contains the *elements* of G. If $a_i, i = 1, 2$ is an edge or an element of G, we say that a_1 is *incident* to a_2 if $a_1 \subseteq \bar{a}_2$ or $a_2 \subseteq \bar{a}_1$, where $\bar{x}$ is the closure of the set x. For every face $f \in F(G)$, we denote by $\mathbf{bd}(f)$ the *boundary* of f, i.e., the set $\overline{f} \setminus f$ where $\overline{f}$ is the closure of f.

A *triangulation* H of a plane graph G is a plane graph H where $V(H) = V(G)$, $E(G) \subseteq E(H)$, and where H is triangulated, i.e., every face of H (including the exterior face) has exactly three edges incident upon it.

We use the term *arc* for any subset of the plane homeomorphic to the closed interval $[0, 1]$. Given a plane graph G, an arc I that does not intersect its edges (i.e., $I \cap G \subseteq V(G)$) is called *normal*. The *length* $|I|$ of a normal arc I is equal to the number of elements of $A(G)$ that it intersects minus one. If x and y are the elements of $A(G)$ intersected by the extreme points a normal arc I, then we also call I *normal* (x, y)-*arc*. A *noose* of the plane, where G is embedded, is a Jordan curve that does not intersect the edges of G. We also denote by $V(N)$ the set of vertices of G met by N, i.e., $V(N) = V(G) \cap N$. The *length* $|N|$ of a noose N is $|V(N)|$, i.e., is the number of the vertices it meets.

Let G be a plane graph and let r be a non-negative integer. Given two elements $x, y \in A(G)$, we say that they are *within radial distance at most r* if there is a normal (x, y)-arc of the plane of length at most r and we denote this fact by $\mathbf{rdist}_G(x, y) \leqslant r$.

Observation 1 *Let G be a triangulated plane graph and let $x, y \in V(G)$. Then* $2 \cdot \mathbf{dist}_G(x, y) \leqslant \mathbf{rdist}_G(x, y)$.

Given a vertex set $S \subseteq V(G)$ and a non-negative integer r, we denote by $\mathbf{R}_G^r(S)$ the set of all elements of G that are within radial distance at most r from some vertex in S. We say that a set $S \subseteq V(G)$ is an *r-radial dominating set* of G (or, alternatively we say that S *r-radially dominates* G) if $\mathbf{R}_G^r(S) = A(G)$. We define

$$\mathbf{rds}(G, r) = \min\{k \mid G \text{ contains an } r\text{-radial dominating set of size at most } k\}.$$

The following observation follows easily from the definitions.

Observation 2 *The parameter* $\mathbf{rds}$ *is closed under topological minors. In other words, if $H, G \in \mathcal{P}$, $r \in \mathbb{N}$, and $H \leqslant_t G$, then* $\mathbf{rds}(H, r) \leqslant \mathbf{rds}(G, r)$.

Branchwidth. Let G be a graph on n vertices. A *branch decomposition* (T, μ) of a graph G consists of an unrooted ternary tree T (i.e., all internal vertices are of degree three) and a bijection $\mu : L \to E(G)$ from the set L of leaves of T to the edge set of G. We define for every edge e of T the *middle set* $\omega(e) \subseteq V(G)$, as follows: Let T_1^e and T_2^e be the two connected components of $T \setminus e$. Then, let G_i^e be the graph induced by the edge set $\{\mu(f) : f \in L \cap V(T_i^e)\}$ for $i \in \{1, 2\}$. The *middle set* is the intersection of the vertex sets of G_1^e and G_2^e, i.e., $\omega(e) = V(G_1^e) \cap V(G_2^e)$. The *width* of (T, μ) is the maximum order of the middle sets over all edges of T (in case T has no edges, then the width of (T, μ) is equal to 0). The *branchwidth*, denoted by $\mathbf{bw}(G)$, of G is the minimum width over all branch decompositions of G.

We now state a series of results on branchwidth that are useful for our proofs.

Proposition 2 (See e.g., [34, (4.1)]). *The parameter $\mathbf{bw}$ is closed under topological minors, i.e., if $H \leqslant_t G$, then $\mathbf{bw}(H) \leqslant \mathbf{bw}(G)$.*

Proposition 3 ([31, 35]). *If G is a plane graph with a cycle, then $\mathbf{bw}(G) = \mathbf{bw}^*(G)$.*

Proposition 4 ([22]). *If G is a n-vertex planar graph, then $\mathbf{bw}(G) \leqslant \sqrt{4.5 \cdot n}$.*

Triconnected components. Let G be a graph, let $S \subseteq V(G)$, and let $V_1, \ldots, V_q$ be the vertex sets of the connected components of $G \setminus S$. We define $\mathcal{C}(G, S) = \{G_1, \ldots, G_q\}$ where G_i is the graph obtained from $G[V_i \cup S]$ if we add all edges between vertices in S.

Given a graph G, the set $\mathcal{Q}(G)$ of its *triconnected components* is recursively defined as follows:

- If G is 3-connected or a clique of size $\leqslant 3$, then $\mathcal{Q}(G) = \{G\}$.
- If G contains a separator S where $|S| \leqslant 2$, then $\mathcal{Q}(G) = \bigcup_{H \in \mathcal{C}(G,S)} \mathcal{Q}(H)$.

Observation 3 *Let G be a graph. All graphs in $\mathcal{Q}(G)$ are topological minors of G.*

The following lemma follows easily from Observation 3 and [21, Lemma 3.1].

Lemma 1. *If G is a graph that contains a cycle, then $\mathbf{bw}(G) = \max\{\mathbf{bw}(H) \mid H \in \mathcal{Q}(G)\}$.*

Sphere-cut decompositions. Let G be a plane graph. A branch decomposition (T, μ) of G is called a *sphere-cut decomposition* if for every edge e of T there exists a noose N_e, such that (a) $\omega(e) = V(N_e)$, (b) $G_i^e \subseteq \Delta_i \cup N_e$ for $i = 1, 2$, where Δ_i is the open disc bounded by N_e, and (c) for every face f of G, $N_e \cap f$ is either empty or connected (i.e., if the noose traverses a face then it traverses it once).

The following theorem is a useful tool when dealing with branch decompositions of planar graphs.

Proposition 5 ([35, Theorem (5.1)]). *Let G be a planar graph without vertices of degree one and with branchwidth at most k embedded on a sphere. Then there exists a sphere-cut decomposition of G of width at most k.*

3   Radially Extremal Sets and Branchwidth

Before we proceed with the proof of Theorem 2, we prove first some auxiliary results.

Let G be a plane graph, $y \in \mathbb{N}$, and $S \subseteq V(G)$. We say that S is *y-radially scattered* if for any $a_1, a_2 \in S$, $\mathbf{rdist}_G(a_1, a_2) \geqslant y$. We say that S is *r-radially extremal in G* if S is an r-radial dominating set of G and S is $2r$-radially scattered in G.

The proof of Theorem 2 is based on a suitable "padding" of a graph that is r-radially dominated so that it becomes r-radially extremal. This is done by the following lemma.

Lemma 2. *Let G be a 3-connected plane graph and S be an r-radial dominating set of G. Then G is the topological minor of a triangulated 3-connected plane graph H where S is r-radially extremal in H.*

We postpone the proof of Lemma 2 until after presentation of the following two lemmata.

Lemma 3 (⋆). *Let G be a 3-connected plane graph and S an r-radial dominating set of G. Then G has a planar triangulation H where S is an r-radial dominating set of H.*

Given a number $k \in \mathbb{N}$, we denote by $\mathcal{A}^{\langle k \rangle}$ the set of all sequences of numbers in $\mathbb{N}$ with length k. We denote $\mathbf{0}_k$, the sequence containing k zero's. Given $\alpha^i = (a_1^i, \ldots, a_k^i)$, $i = 1, 2$, we say that $\alpha^1 \prec \alpha^2$ if, there is some integer $j \in 1, \ldots, k$, such that $a_h^1 = a_h^2$ for all $h \leqslant j$ and $a_j^1 < a_j^2$. For example $(1, 1, 2, 4, 15, 3, 82, 2) \prec (1, 1, 3, 1, 6, 29, 1, 3)$. A sequence $A = (\alpha^i \mid i \in \mathbb{N})$ of sequences in $\mathcal{A}^{\langle k \rangle}$ is *properly decreasing* if for any two consecutive elements α^j, α^{j+1} of A it holds that $\alpha^j \prec \alpha^{j+1}$. We will use the following known lemma.

Observation 4 *For every $k \in \mathbb{N}$, every properly decreasing sequence of sequences in $\mathcal{A}^{\langle k \rangle}$ is finite.*

Given a graph G and a subset $S \subseteq V(G)$, we call a path an S-*path* if its extremes are vertices of S. The proof of the following lemma is based on Observations 1 and 4.

Lemma 4 (⋆). *Let G be a triangulated plane graph and let S be an r-radial dominating set of G. Then G is the topological minor of a graph H that is $2r$-radially extremal.*

Proof (of Lemma 2). As G is triangulated and thus 3-connected we can apply Lemma 3, we obtain a planar triangulation H' of G where the set S is an r-radial dominating. Then, applying Lemma 4 on H', we obtain a planar triangulated graph H that contains of H' as a topological minor and such that S is $2r$-radially scattered in H. The lemma follows as G is a topological minor of H.

We are now ready to prove our main combinatorial result.

Theorem 3. *Let r be a positive integer and let G be a plane graph. Then $\mathbf{bw}(G) \leqslant r \cdot \sqrt{4.5 \cdot \mathbf{rds}(G, r)}$.*

Proof. We use induction on r. If $r = 1$ then $|V(G)| = \mathbf{rds}(G, 1)$ and the result follows from Proposition 4. Assume now that the lemma holds for values smaller than r and we will prove that it also holds for r, where $r \geqslant 2$. Let G be a plane graph where $\mathbf{rds}(G, r) \leqslant k$. Using Lemma 1, we choose $H \in \mathcal{Q}(G)$ such that $\mathbf{bw}(H) = \mathbf{bw}(G)$ (we may assume that G contains a cycle, otherwise the result follows trivially). By Observations 2 and 3, $\mathbf{rds}(H, r) \leqslant \mathbf{rds}(G, r) \leqslant k$. Let S be an r-radial dominating set of H where $|S| \leqslant k$. From Lemma 2, H is the topological minor of a 3-connected plane graph H_1 where S is r-radially extremal.

Let H_2 be the graph obtained if we remove from H_1 all vertices of S. By the 3-connectivity of H_1, it follows that, for any $v \in S$, the graph $H_1[N_{H_1}(v)]$ is a cycle and each such cycle is the boundary of some face of H_2. We denote by F the set of these faces and observe that F^* is a $(r-1)$-radial dominating set of H_2^* (we denote by F^* the vertices of H_2^* that are duals of the faces of F in H_2).

Moreover, the fact that S is a $2r$-scattered dominating set in H_1, implies that F^* is a $2(r-1)$-scattered dominating set in H_2^*.

From the induction hypothesis and the fact that $|F^*| = |S|$, we obtain that $\mathbf{bw}(H_2^*) \leqslant (r-1) \cdot \sqrt{4.5 \cdot k}$. This fact, along with Proposition 3, implies that $\mathbf{bw}(H_2) \leqslant (r-1) \cdot \sqrt{4.5 \cdot k}$.

In graph H_2, for any face $f_i \in F$, let $(x_0^i, \ldots, x_{m_i-1}^i)$ be the cyclic order of the vertices in its boundary cycle (as H_1 is 3-connected we have that $m_i \geqslant 3$). We also denote by x^i the vertex in H_1 that was removed in order f_i to appear in H_2. Let (T, τ) be a sphere cut decomposition of H_2 of width $\leqslant (r-1) \cdot \sqrt{4.5 \cdot k}$. By Proposition 5, we may assume that (T, τ) is a sphere-cut decomposition. We use (T, τ) in order to construct a branch decomposition of H_1, by adding new leaves in T and mapping them to the edges of $E(H_1) \setminus E(H_2) = \bigcup_{i=1,\ldots,|F|} \{\{x^i, x_h^i\} \mid h = 0, \ldots, m_i - 1\}$ in the following way: For every $i = 1, \ldots, |F|$ and every $h = 0, \ldots, m_i - 1$, we set $t_h^i = \tau^{-1}(\{x_h^i, x_{h+1 \mod m_i}^i\})$ and let $e_h^i = \{y_h^i, t_h^i\}$ be the unique edge of T that is incident to t_h^i. We subdivide e_h^i and we call the subdivision vertex s_h^i. We also add a new vertex z_h^i and make it adjacent to s_h^i. Finally, we extend the mapping of τ by mapping the vertex z_h^i to the edge $\{x^i, x_h^i\}$ and we use the notation (T', τ') for the resulting branch decomposition of H_1.

Claim. The width of (T', τ') is at most $r \cdot \sqrt{4.5 \cdot k}$.

Proof. We use the functions ω and ω' to denote the middle sets of (T, τ) and (T', τ') respectively. Let e be an edge of T'. If e is not an edge of T (i.e., is an edge of the form $\{z_h^i, s_h^i\}$ or $\{t_h^i, s_h^i\}$ or $\{y_h^i, s_h^i\}$), then $|\omega'(e)| \leqslant 3$, therefore we may fix our attention to the case where e is also an edge of T. Let N_e be the noose of H_2 meeting the vertices of $\omega(e)$. We distinguish the following cases.

Case 1. N_e does not meet any face of F, then clearly $\omega'(e) = \omega(e)$. Thus $|\omega'(e)| \leqslant (r-1) \cdot \sqrt{4.5 \cdot k}$.

Case 2. If N_e meets only one, say f_i, of the faces of F, then the vertices in $\omega'(e)$ are the vertices of a noose N_e' of H_1 meeting all vertices of $\omega(e)$ plus x^i. Therefore, $\omega'(e) = \omega(e) \cup \{x^i\}$ and thus $|\omega'(e)| \leqslant (r-1) \cdot \sqrt{4.5 \cdot k} + 1$.

Case 3. N_e meets $p \geqslant 2$ faces of F. We denote by $\{f_0', \ldots, f_{p-1}'\}$ the set of these faces and let $J_0, \ldots, J_{p-1}$ be the normal arcs corresponding to the connected components of $N_i - \bigcup_{i=0,\ldots,p-1} f_i'$. Let also $I_0, \ldots, I_{p-1}$ be the normal arcs corresponding to the closures of the connected components of $N_i \cap (\bigcup_{i=0,\ldots,p-1} f_i')$, assuming that $I_i \subseteq \overline{f_i'}$, for $i = 0, \ldots, p-1$. Recall that F^* is a $(r-1)$-scattered dominating set of H_2^*. This implies that each J_i meets at least $r-1$ vertices of H_2 and therefore $p \cdot (r-1) \leqslant \omega(e) \leqslant (r-1) \cdot \sqrt{4.5 \cdot k}$. We conclude that $p \leqslant \sqrt{4.5 \cdot k}$. Observe now that the vertices of $\omega'(e)$ are the vertices of a noose N_e' of H_1 where $N_e' = (\bigcup_{i=0,\ldots,p-1} J_i) \cup (\bigcup_{i=0,\ldots,p-1} I_i')$ and such that, for each $i = 0, \ldots, p-1$, I_i' is a replacement of I_i so that it is still a subset of $\overline{f_i'}$, has the same extremes as I_i, and also meets the unique vertex in $S \cap f_i'$. As N_e' meets in H_1 all vertices of $N_e \cap V(H_2)$ plus p more, we obtain that $|\omega'(e)| \leqslant (r-1) \cdot \sqrt{4.5 \cdot k} + p \leqslant r \cdot \sqrt{4.5 \cdot k}$.

According to the above case analysis, $|\omega'(e)| \leqslant \max\{3, \sqrt{4.5 \cdot k} + 1, r \cdot \sqrt{4.5 \cdot k}\} = r \cdot \sqrt{4.5 \cdot k}$ and the claim follows.

We just proved that $\mathbf{bw}(H_1) \leqslant r \cdot \sqrt{4.5 \cdot k}$. As H is a topological minor of H_1, from Proposition 2, we also have that $\mathbf{bw}(H) \leqslant r \cdot \sqrt{4.5 \cdot k}$. The lemma follows as $\mathbf{bw}(G) = \mathbf{bw}(H)$. □

Proof (of Theorem 2). By the definition of compactness, G has an embedding such that either G or G^* contains an r-radial dominating set of size at most $q \cdot k$. Without loss of generality, assume that this is the case for G (here we use Proposition 3). Then $\mathbf{rds}(G, r) \leqslant q \cdot k$ and, from Theorem 3, $\mathbf{bw}(G) \leqslant r \cdot \sqrt{4.5 \cdot q \cdot \mathbf{p}(G)}$. □

4   Conclusions and Open Problems

The concept of compactness for parameterized problems appeared for the first time in [4] in the context of kernelization. In this paper, we show that it can also be used to improve the running time analysis of a wide family of sub-exponential parameterized algorithms. Essentially, we show that such an analysis can be done without the grid-minor exclusion theorem. Instead, our better combinatorial bounds emerge from the result in [22] that, in turn, is based on the "planar separators theorem" of Alon, Seymour, and Thomas in [3]. This implies that any improvement of the constant $\sqrt{4.5}$ in [3] would improve all the running times emerged from the framework of our paper.

It follows that there are bidimensional parameterized problems that are not compact and vice versa. For instance, INDEPENDENT DOMINATING SET is compact but not bidimensional while LONGEST PATH is bidimensional but not compact. Is it possible to extend both frameworks to a more powerful theory, at least in the context of sub-exponential parameterized algorithms?

References

1. Alber, J., Bodlaender, H.L., Fernau, H., Kloks, T., Niedermeier, R.: Fixed parameter algorithms for dominating set and related problems on planar graphs. Algorithmica 33, 461–493 (2002)
2. Alon, N., Lokshtanov, D., Saurabh, S.: Fast FAST. In: Albers, S., Marchetti-Spaccamela, A., Matias, Y., Nikoletseas, S., Thomas, W. (eds.) ICALP 2009. LNCS, vol. 5555, pp. 49–58. Springer, Heidelberg (2009)
3. Alon, N., Seymour, P., Thomas, R.: Planar separators. SIAM J. Discrete Math. 7, 184–193 (1994)
4. Bodlaender, H.L., Fomin, F.V., Lokshtanov, D., Penninkx, E., Saurabh, S., Thilikos, D.M.: (Meta) Kernelization. In: 50th Annual IEEE Symposium on Foundations of Computer Science (FOCS 2009), pp. 629–638. IEEE, Los Alamitos (2009)
5. Cai, L., Juedes, D.: On the existence of subexponential parameterized algorithms. Journal of Computer and System Sciences 67, 789–807 (2003)
6. Chen, J., Kanj, I., Perkovic, L., Sedgwick, E., Xia, G.: Genus characterizes the complexity of graph problems: some tight results. In: Baeten, J.C.M., Lenstra, J.K., Parrow, J., Woeginger, G.J. (eds.) ICALP 2003. LNCS, vol. 2719, pp. 845–856. Springer, Heidelberg (2003)

7. Demaine, E., Hajiaghayi, M.: The bidimensionality theory and its algorithmic applications. The Computer Journal 51, 292–302 (2007)
8. Demaine, E.D., Fomin, F.V., Hajiaghayi, M., Thilikos, D.M.: Bidimensional parameters and local treewidth. SIAM J. Discrete Math. 18, 501–511 (2005)
9. Demaine, E.D., Fomin, F.V., Hajiaghayi, M., Thilikos, D.M.: Fixed-parameter algorithms for (k, r)-center in planar graphs and map graphs. ACM Trans. Algorithms 1, 33–47 (2005)
10. Demaine, E.D., Fomin, F.V., Hajiaghayi, M., Thilikos, D.M.: Subexponential parameterized algorithms on bounded-genus graphs and H-minor-free graphs. J. Assoc. Comput. Mach. 52, 866–893 (2005)
11. Demaine, E.D., Hajiaghayi, M.: Bidimensionality: new connections between FPT algorithms and PTASs. In: Sixteenth Annual ACM-SIAM Symposium on Discrete Algorithms, pp. 590–601. ACM, New York (2005) (electronic)
12. Demaine, E.D., Hajiaghayi, M., Thilikos, D.M.: Exponential speedup of fixed-parameter algorithms for classes of graphs excluding single-crossing graphs as minors. Algorithmica 41, 245–267 (2005)
13. Dorn, F.: Dynamic Programming and Fast Matrix Multiplication. In: Azar, Y., Erlebach, T. (eds.) ESA 2006. LNCS, vol. 4168, pp. 280–291. Springer, Heidelberg (2006)
14. Dorn, F., Fomin, F.V., Lokshtanov, D., Raman, V., Saurabh, S.: Beyond bidimensionality: Parameterized subexponential algorithms on directed graphs. In: 27th International Symposium on Theoretical Aspects of Computer Science (STACS 2010). LIPIcs, vol. 5, pp. 251–262. Schloss Dagstuhl - Leibniz-Zentrum fuer Informatik (2010)
15. Dorn, F., Fomin, F.V., Thilikos, D.M.: Subexponential parameterized algorithms. Computer Science Review 2, 29–39 (2008)
16. Fernau, H.: Graph Separator Algorithms: A Refined Analysis. In: Kučera, L. (ed.) WG 2002. LNCS, vol. 2573, pp. 186–197. Springer, Heidelberg (2002)
17. Fernau, H., Juedes, D.: A Geometric Approach to Parameterized Algorithms for Domination Problems on Planar Graphs. In: Fiala, J., Koubek, V., Kratochvíl, J. (eds.) MFCS 2004. LNCS, vol. 3153, pp. 488–499. Springer, Heidelberg (2004)
18. Fomin, F.V., Golovach, P., Thilikos, D.M.: Contraction bidimensionality: The accurate picture. In: Fiat, A., Sanders, P. (eds.) ESA 2009. LNCS, vol. 5757, pp. 706–717. Springer, Heidelberg (2009)
19. Fomin, F.V., Lokshtanov, D., Raman, V., Saurabh, S.: Bidimensionality and EPTAS. In: 22st ACM–SIAM Symposium on Discrete Algorithms (SODA 2011), pp. 748–759. ACM-SIAM, San Francisco (2011)
20. Fomin, F.V., Lokshtanov, D., Saurabh, S., Thilikos, D.M.: Bidimensionality and kernels. In: 21st Annual ACM-SIAM Symposium on Discrete Algorithms (SODA 2010), Austin, Texas, pp. 503–510. ACM-SIAM (2010)
21. Fomin, F.V., Thilikos, D.M.: ominating sets in planar graphs: branch-width and exponential speed-up. SIAM J. Comput. 36, 281–309 (2006) (electronic)
22. Fomin, F.V., Thilikos, D.M.: New upper bounds on the decomposability of planar graphs. Journal of Graph Theory 51, 53–81 (2006)
23. Fomin, F.V., Villanger, Y.: Subexponential parameterized algorithm for minimum fill-in. CoRR, abs/1104.2230 (2011)
24. Gu, Q.-P., Tamaki, H.: Optimal branch decomposition of planar graphs in $O(n^3)$ time. ACM Trans. Algorithms 4, 1–13 (2008)
25. Gu, Q.-P., Tamaki, H.: Improved Bounds on the Planar Branchwidth with Respect to the Largest Grid Minor Size. In: Cheong, O., Chwa, K.-Y., Park, K. (eds.) ISAAC 2010, Part II. LNCS, vol. 6507, pp. 85–96. Springer, Heidelberg (2010)

26. Guo, J., Niedermeier, R.: Linear Problem Kernels for NP-Hard Problems on Planar Graphs. In: Arge, L., Cachin, C., Jurdziński, T., Tarlecki, A. (eds.) ICALP 2007. LNCS, vol. 4596, pp. 375–386. Springer, Heidelberg (2007)
27. Kanj, I., Perković, L.: Improved Parameterized Algorithms for Planar Dominating Set. In: Diks, K., Rytter, W. (eds.) MFCS 2002. LNCS, vol. 2420, pp. 399–410. Springer, Heidelberg (2002)
28. Kloks, T., Lee, C.M., Liu, J.: New Algorithms for k-Face Cover, k-Feedback Vertex Set, and k-Disjoint Cycles on Plane and Planar Graphs. In: Kučera, L. (ed.) WG 2002. LNCS, vol. 2573, pp. 282–295. Springer, Heidelberg (2002)
29. Koutsonas, A., Thilikos, D.: Planar feedback vertex set and face cover: Combinatorial bounds and subexponential algorithms. Algorithmica, 1–17 (2010)
30. Lokshtanov, D., Marx, D., Saurabh, S.: Known algorithms on graphs of bounded treewidth are probably optimal. In: 22st ACM–SIAM Symposium on Discrete Algorithms (SODA 2011), pp. 777–789. ACM-SIAM, San Francisco, California (2011)
31. Mazoit, F., Thomassé, S.: Branchwidth of graphic matroids. In: Surveys in combinatorics 2007. London Math. Soc. Lecture Note Ser., vol. 346, pp. 275–286. Cambridge Univ. Press, Cambridge (2007)
32. Pilipczuk, M.: Problems parameterized by treewidth tractable in single exponential time: a logical approach. Tech. Rep. arXiv:1104.3057, Cornell University (April 2011)
33. Robertson, N., Seymour, P., Thomas, R.: Quickly excluding a planar graph. J. Combin. Theory Ser. B 62, 323–348 (1994)
34. Robertson, N., Seymour, P.D.: Graph minors. X. Obstructions to tree-decomposition. J. Combin. Theory Ser. B 52, 153–190 (1991)
35. Seymour, P.D., Thomas, R.: Call routing and the ratcatcher. Combinatorica 14, 217–241 (1994)
36. Tazari, S.: Faster approximation schemes and parameterized algorithms on H-minor-free and odd-minor-free graphs. In: Hliněný, P., Kučera, A. (eds.) MFCS 2010. LNCS, vol. 6281, pp. 641–652. Springer, Heidelberg (2010)
37. van Rooij, J.M.M., Bodlaender, H.L., Rossmanith, P.: Dynamic Programming on Tree Decompositions Using Generalised Fast Subset Convolution. In: Fiat, A., Sanders, P. (eds.) ESA 2009. LNCS, vol. 5757, pp. 566–577. Springer, Heidelberg (2009)

Isomorphism of (mis)Labeled Graphs

Pascal Schweitzer*

The Australian National University, Canberra, Australia
pascal.schweitzer@anu.edu.au

Abstract. For similarity measures of labeled and unlabeled graphs, we study the complexity of the graph isomorphism problem for pairs of input graphs which are close with respect to the measure. More precisely, we show that for every fixed integer k we can decide in quadratic time whether a labeled graph G can be obtained from another labeled graph H by relabeling at most k vertices. We extend the algorithm solving this problem to an algorithm determining the number ℓ of vertices that must be deleted and the number k of vertices that must be relabeled in order to make the graphs equivalent. The algorithm is fixed-parameter tractable in $k + \ell$.

Contrasting these tractability results, we also show that for those similarity measures that change only by finite amount d whenever one edge is relocated, the problem of deciding isomorphism of input pairs of bounded distance d is equivalent to solving graph isomorphism in general.

1 Introduction

Given two graphs G and H, the graph isomorphism problem asks whether there exists an isomorphism from G to H. That is, the problem asks whether there exists an adjacency and non-adjacency preserving bijection from the vertices of G to the vertices of H. Graph isomorphism is a computational decision problem contained in NP, since the isomorphism represents a witness checkable in quadratic time. However, it is not known whether the problem is NP-complete and not known whether the problem is polynomial-time solvable (see [14] or [19] for an introduction to the problem).

Since graph isomorphism still has unknown complexity, researchers have considered the complexity of the isomorphism problem on subclasses of graphs. The isomorphism problem is for example isomorphism complete (i.e., polynomially equivalent to graph isomorphism) on bipartite graphs and regular graphs (see [24]). Among many algorithms that have been developed for specific graph classes, there is Luks' [16] polynomial-time algorithm for graphs of bounded degree, and polynomial-time algorithms for graphs of bounded genus developed by Filotti and Mayer [10] and by Miller [18]. For the known bounded degree algorithm and the known bounded genus algorithms, the degree of the polynomial bounding the running time increases with increasing parameter (i.e., they have a

---

* Supported by the National Research Fund, Luxembourg, and cofunded under the Marie Curie Actions of the European Commission (FP7-COFUND).

C. Demetrescu and M.M. Halldórsson (Eds.): ESA 2011, LNCS 6942, pp. 370–381, 2011.

running time of $\mathcal{O}(n^{f(k)})$). Algorithms with uniformly polynomial running time (i.e., having a running time of $\mathcal{O}(f(k) \cdot n^d)$ with d fixed) have only been devised for the parameters eigenvalue multiplicity [9], color multiplicity [12], feedback vertex set number [15], and rooted tree distance width [21]. In parametrized complexity theory such algorithms are called *fixed-parameter tractable*. See [8] or [11] for general parameterized complexity theory, and the introduction of [15] for a graph isomorphism specific overview.

In the context of the isomorphism problem, the subproblems that have traditionally been investigated, including all of the ones just mentioned, impose restrictions on both input graphs. In this paper we investigate the effect on the complexity when the input graphs are related to each other, i.e., when a certain similarity between the input graphs is required. This allows each input to be an arbitrary graph, but of course the input graphs are not independent. For any given similarity (or distance) measure we thus investigate:

What is the complexity of graph isomorphism when the input is restricted to be a pair of similar graphs with respect to a given measure?

The permutation distance. Throughout the paper, we always assume the input graphs are defined over the same vertex set. With regard to our question, we consider the following measure for labeled graphs: Two isomorphic graphs have a distance of at most k, if there exists a permutation of the vertices that is an isomorphism and permutes at most k vertices. For a pair of non-isomorphic graphs the distance is infinite.

The motivation for this definition is the following question: Suppose we are to decide whether a labeled graph G is isomorphic to a labeled blueprint graph H. A one by one comparison of the edges trivially determines whether they are isomorphic as labeled graphs. If however a small number of vertices (say k) in G have been mislabeled, how do we determine whether G matches the blueprint H, and how do we determine the vertices that have been mislabeled?

We show in Section 2 that there is a fixed-parameter tractable algorithm with a running time of $\mathcal{O}(f(k)n^2)$ for this problem. In other words, if the number of mislabeled vertices is bounded, an isomorphism can be found in time quadratic in the total number of vertices.

As a generalization, we show in Section 3 that the computation of the maximum common subgraph distance of a graph H to all graphs obtained from G by permuting at most k vertices is fixed-parameter tractable. That is, in the sense of the previous motivation, this extended problem asks: Given a labeled blueprint graph H and a labeled graph G in which not only k vertices have been mislabeled, but additionally for some vertices the adjacencies to other vertices have been altered, how do we detect the mislabeled vertices and the vertices for which the adjacencies have been altered?

Other similarity measures. In Section 4 we consider alternative similarity measures that have been defined for graphs in the literature. Formally, for a similarity measure d we denote by $d(G, H)$ the distance of G and H. Given a bound $k \in \mathbb{N}$ on the similarity, the question highlighted above asks for the complexity of graph isomorphism on input graphs G and H with $d(G, H) \leq k$.

There are two substantially different kinds of graph similarity measures, namely similarity measures that apply to labeled graphs and similarity measures that apply to unlabeled graphs. From a labeled distance measure one can obtain an unlabeled distance measure in the following general way: The (unlabeled) distance from G to H is the shortest labeled distance of G to some permutation of H. (Recall that we always assume that the input graphs are defined on the same vertex set and they thus in particular have equal size.)

Among the labeled (and thus also among the unlabeled) distance measures that have been considered are the maximum common subgraph and the maximum common contract distance, both first defined by Zelinka [22,23]. These distances measure the difference of the size of the input graphs to the size of the largest common subgraph and to the size of the largest common contract, respectively. Recall that a contract is a graph obtained by repeatedly contracting an edge, i.e., identifying two adjacent vertices. Further measures applicable to both labeled and unlabeled graphs are the edge slide distance (Johnsen [13]), the edge rotation distance (Chartrand, Saba, and Zou [4]) and with them also the edit distance (see [2]). These measures are based on local edge replacements according to certain allowed moves. Finally, the more recent cut distance [1] considers all bipartitions of the vertices and among those measures the maximum number of crossing edges contained in exactly one of the input graphs.

An intrinsically unlabeled distance measure is the spectral distance [20], which is defined as the square root of the sum of the squared differences of the eigenvalues of two graphs.

Using simple reductions, we show (Section 4) that for a large class of measures, including all of the above, the isomorphism problem of similar graphs is isomorphism complete, both in the labeled and the unlabeled case, except of course for the permutation distance treated in Section 2.

2   The Permutation Distance as Similarity Measure

We now consider the similarity measure which, for two labeled graphs over the same vertex set, measures how many vertices have to be permuted in order to obtain identical graphs. The distance of two non-isomorphic graphs is infinite.

For every fixed integer k, there is a trivial algorithm running in polynomial time $\mathcal{O}(k!\binom{n}{k}n^2) \subseteq \mathcal{O}(n^{k+2})$ that checks whether two graphs have distance at most k: Indeed, by testing all permutations of all k-tuples of vertices and checking whether the obtained graphs are identical, every isomorphism with the required properties will be found. Before we develop an $\mathcal{O}(n^2)$ algorithm we review basic definitions and notation for permutation groups.

Basic definitions and notation for permutation groups: Let π be a permutation of some finite set V. The *orbit* of an element $u \in V$ is the set $\{u, \pi(u), \pi^2(u), \ldots\}$. The *support* of π, denoted by $\mathrm{supp}(\pi)$, is the set $\{u \in V \mid \pi(u) \neq u\}$ of elements that are not fixed. The composition of two permutations $\pi \cdot \pi'$ is the permutation obtained by first applying π and then π'. The permutation π gives rise to a permutation on the set of all pairs of elements of V. For a pair of elements (u, v)

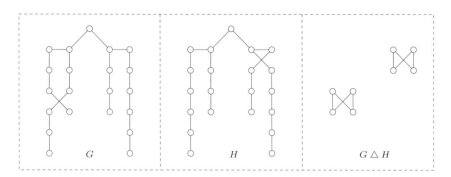

Fig. 1. Only one isomorphism exists between the depicted graphs G and H. This isomorphism fixes the vertex at the top and swaps the left part with the right part hanging from that vertex. The example shows that there does not necessarily exist an isomorphism from G to H that maps a vertex in $G \triangle H$ to a vertex in $G \triangle H$.

we define $\pi((u, v))$ as $(\pi(u), \pi(v))$. Abusing notation we denote this as $\pi(u, v)$. For the purpose of this paper we define the complexity of a permutation.

Definition 1. *For a permutation π its* complexity $\operatorname{compl}(\pi)$ *is the size of the support of π minus the number of orbits of π which are of size at least 2.*

Observe that for any non-trivial permutation π it holds that $0 \leq \operatorname{compl}(\pi) < |\operatorname{supp}(\pi)|$. Furthermore, if $\operatorname{compl}(\pi) = 0$ then π is the identity. The complexity of a permutation is the minimum number of transpositions whose multiplication gives the permutation [7], but we will not need this fact.

To develop our algorithm, we use the symmetric difference as a concise way of addressing vertices and edges causing two labeled graphs to be non-identical.

Definition 2. *For two graphs $G = (V, E_G)$ and $H = (V, E_H)$ we define their* symmetric difference *to be the graph $G \triangle H := (V', E_G \triangle E_H)$, where V' is the set of vertices that are endpoints of edges in $E_G \triangle E_H$.*

Intuitively, $G \triangle H$ is the unique smallest graph (w.r.t. inclusion) that has an edge for all vertex pairs which are adjacent in exactly one of the graphs G and H.

Since we assume that the input graphs G and H are defined over the same vertex set V, every isomorphism from G to H is represented by a permutation π of V. (Abusing terminology, we use the terms interchangeably.) To design an algorithm with a polynomial running time whose degree is independent of k, we require a small candidate subset of vertices that should be permuted. Intuitively, such a candidate subset could be $V(G \triangle H)$, since for every edge in $G \triangle H$ one of its endpoints has to be permuted. However, $V(G \triangle H)$ may be of linear size, even for bounded k, and more importantly there are examples of isomorphic labeled graphs for which no isomorphism maps a vertex in $V(G \triangle H)$ to a vertex in $V(G \triangle H)$. An example of such a pair of graphs is shown in Figure 1.

Algorithm 1. Labeled isomorphism algorithm $\mathrm{ISO}_k(G, H, c, \psi)$

Input: Two labeled graphs G and H over a vertex set V, a recursion depth c for the
algorithm and a permutation ψ of V that permutes at most k vertices.

Output: An isomorphism $\phi \colon G \to H$ that has complexity at most c so that the composition $\phi \cdot \psi$ permutes at most k vertices, or report **false**, if no such isomorphism
exists. The algorithm may also report **false** if all isomorphisms ϕ with the described
properties do not satisfy $|\operatorname{supp}(\psi) \cup \operatorname{supp}(\phi)| \leq k$.

```
 1: compute G △ H
 2: if G △ H is empty then
 3:    return  φ = id                                      // where id is the identity
 4: end if
 5: if c > 0 then                                          // compute set of candidates C
 6:    if |V(G △ H)| ≤ 2k then
 7:       C ← V(G △ H)
 8:    else
 9:       if G △ H has vertex cover number greater than k then
10:          return false
11:       else
12:          compute a vertex cover C of G △ H of size at most k
13:       end if
14:    end if
15:    for all v₁, v₂ ∈ C do                               // try swapping all pairs from C
16:       ψ' ← (v₁, v₂) · ψ
17:       form graph H' from H by swapping the neighbors of the vertices v₁ and v₂
18:       if ψ' permutes at most k vertices then
19:          call ISOₖ(G, H', c − 1, ψ') and record the result in φ'
20:          if φ' ≠ false then
21:             return φ = φ' · (v₁, v₂)
22:          end if
23:       end if
24:    end for
25: end if
26: return  false
```

Description of the algorithm: For a fixed integer k (the number of vertices that
may be permuted) algorithm ISO_k takes as input two labeled graphs G and H,
a recursion depth c and a permutation ψ that keeps track of the alterations that
have been applied to H. The integer c is an upper bound on the complexity
of the permutation representing the isomorphism. To determine whether there
exists an isomorphism permuting at most k vertices, the algorithm is called
as $\mathrm{ISO}_k(G, H, k, \mathrm{id})$ (i.e., c is set to k, and ψ is the identity). The algorithm
either outputs an isomorphism that permutes at most k vertices, or reports
with **false** that no such isomorphism exists. (See Algorithm 1.)

The algorithm first determines whether the two labeled input graphs are identical. If not, it computes the symmetric difference of the two input graphs, and
then, depending on the number of vertices in the symmetric difference, performs
one of two operations: If the number of vertices in the symmetric difference is at

most $2k$, for all pairs of vertices in the symmetric difference, it calls itself after having transposed the two vertices in the graph H. Otherwise it tries to compute a vertex cover of the symmetric difference of size at most k and, if successful, performs the transposition with every pair of vertices in the vertex cover, before performing the recursive calls. Returning from a successful recursive call, the obtained isomorphism is altered by the transposition of the two vertices that have been swapped, and subsequently returned. If a call is unsuccessful, algorithm proceeds with the next pair. At any point in time the algorithm keeps track of the permutation that has been performed on graph H with the help of input parameter ψ. If this permutation moves more than k vertices, no recursive call is performed.

The intention not to return any isomorphism that moves more than k vertices complicates the intermediate calls. More precisely, it may be the case that the permutation ψ that has been performed, together with some permutation ϕ of complexity c yields an isomorphism $\psi \cdot \phi$ that permutes at most k vertices, but that the intermediate steps require to consider permutations that permute more than k vertices. If no isomorphism avoids this issue, the algorithm may output **false** independent of these existing isomorphisms. Intuitively, this situation only happens, when on a previous recursion level the algorithm moved away from the sought isomorphism by swapping a wrong pair of candidate vertices.

2.1   Correctness:

Since the algorithm keeps track of the permutation performed on H, any isomorphism permutes at most k vertices. By definition of the algorithm, it is immediate that it never returns a map that is not an isomorphism (i.e., it has no false positives).

To establish the correctness, we show that if there exists an isomorphism that permutes k vertices or less, it will be found within a recursion depth of k. By induction, a product of c transpositions has complexity at most c, thus the algorithm will not output an isomorphism of complexity larger than c.

We show that if an isomorphism exists that permutes k vertices or less with a complexity of c, then the algorithm will call itself on an instance for which an isomorphism of complexity at most $c-1$ exists that also permutes k vertices or less. This is ensured by identifying two vertices that lie in the same orbit, and by using a basic fact on permutations:

Lemma 1. *If π is a permutation and v, v' are distinct elements in an orbit of π, then for the permutation $\pi' = \pi \cdot (v, v')$ (i.e., π followed by the transposition of v and v') it holds that* $\mathrm{supp}(\pi') \subseteq \mathrm{supp}(\pi)$ *and* $\mathrm{compl}(\pi') < \mathrm{compl}(\pi)$.

Proof. Transposing v and v' either splits the orbit that contains both v and v' into two orbits, of which at least one has size larger than 1, or v and v' form a complete orbit of π, in which case the support of the permutation decreases by two, while the number of orbits of size larger than 1 is only reduced by one. $\square$

To identify two vertices lying in the same orbit we consider the symmetric difference $G \triangle H$ of the input graphs.

Lemma 2. *Let G and H be non-identical graphs on the same vertex set. If there is an isomorphism from G to H represented by a permutation π, then there exist two distinct vertices $v, v' \in G \bigtriangleup H$ contained in the same orbit under π.*

Proof. Suppose otherwise, i.e., that for every vertex $v \in G \bigtriangleup H$ no other vertex in the orbit of v is contained in $G \bigtriangleup H$. Let (u, v) be an edge in $G \bigtriangleup H$. This implies that u or v is not fixed. By assumption v and u are the only vertices in their respective orbits which are contained in $G \bigtriangleup H$.

W.l.o.g. suppose $(u, v) \in E(G)$ and therefore $(u, v) \notin E(H)$. Let i be the least positive integer for which $(\pi^i(u), \pi^i(v)) = (u, v)$. We now show by induction that for all $j \in \{0, \ldots, i - 1\}$ the pair $\pi^j(u, v) = (\pi^j(u), \pi^j(v))$ is an edge in G: By assumption $\pi^0(u, v) = (u, v) \in E(G)$ and for $0 < j < i$ if $\pi^{j-1}(u, v)$ is an edge in G, then $\pi^j(u, v)$ is an edge in H, since π is an isomorphism from G to H. By the definition of i we know that $\pi^j(u)$ is different from u, and therefore not contained in $V(G \bigtriangleup H)$ or $\pi^j(v)$ is different from v, and therefore not contained in $V(G \bigtriangleup H)$. Either way $\pi^j(u, v)$ is not in $G \bigtriangleup H$. Thus the fact that $\pi^j(u, v)$ is an edge in H implies that it is also an edge in G.

Finally, since $\pi^{i-1}(u, v)$ is an edge in G and π is an isomorphism, $\pi^i(u, v) = (u, v)$ is an edge in H. This yields a contradiction. $\qquad\square$

To address the issue that the set $G \bigtriangleup H$ may be of linear size, and thus too large in order to be a candidate set for vertices to be permuted, we consider a vertex cover of $G \bigtriangleup H$.

Lemma 3. *Suppose G and H are graphs on the same vertex set. If π is an isomorphism from G to H, then the support of π is a vertex cover of $G \bigtriangleup H$.*

Proof. Since π is an isomorphism, no edge $(u, v) \in E(G \bigtriangleup H)$ can be fixed by π. Thus for every edge (u, v), one of the vertices u, v is not fixed by π. $\qquad\square$

The lemma implies that $G \bigtriangleup H$ has vertex cover number at most k if there is an isomorphism that maps G to H and leaves at most k vertices unfixed.

Lemma 2 shows that there are vertices in $G \bigtriangleup H$ that lie in the same orbit. However when the candidate set C is restricted to a vertex cover of $G \bigtriangleup H$, to apply Lemma 1 we require that two vertices from C lie in the same orbit. Such two vertices do not exist in general, as shown by the example in Figure 2. The next lemma helps us to circumvent this problem, by performing a case distinction depending on the size of $G \bigtriangleup H$ relative to $2k$.

Lemma 4. *Let G and H be two graphs on the same vertex set V and let C be a vertex cover of $G \bigtriangleup H$ of size at most k. Suppose there is an isomorphism from G to H represented by a permutation π that leaves at most k vertices unfixed. If every orbit of π contains at most one vertex from C, then $|V(G \bigtriangleup H)| \leq 2k$.*

Proof. Since $|C| \leq k$ and π permutes at most k vertices, it suffices to show that π has no fix-point in $V(G \bigtriangleup H) \setminus C$. For contradiction assume $u \notin C$ is a vertex fixed under π and there is an edge $(u, v) \in G \bigtriangleup H$. W.l.o.g. assume $(u, v) \in E(G)$, and therefore $(u, v) \notin E(H)$. Since (u, v) cannot be fixed under π, the vertex v

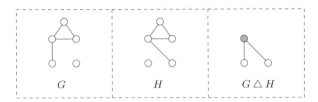

Fig. 2. The figure depicts isomorphic graphs G and H and their symmetric difference $G \triangle H$. In $G \triangle H$ the only minimum vertex cover C is highlighted by the gray shaded vertex. No isomorphism from G to H has an orbit that contains two vertices of C.

is in $C \cap \operatorname{supp}(\pi)$. Consider the orbit of (u, v) under π. Let i be the least positive integer for which $\pi^i(v) = v$. By assumption for no positive $j < i$ is $\pi^j(v)$ in C. Therefore, for $0 < j < i$, the pair $\pi^j(u, v) = (u, \pi^j(v))$ is not an edge in $G \triangle H$. By induction we show that for all $j \in \{0, \ldots, i-1\}$ the vertex pair $\pi^j(u, v)$ is an edge in G: This is true for $j = 0$, and if $\pi^{j-1}(u, v)$ is an edge in G, then since π is an isomorphism from G to H, $\pi^j(u, v)$ is an edge in H. Since $\pi^j(u, v)$ is not in $G \triangle H$ we conclude that $\pi^j(u, v)$ is an edge in G.

Finally since $\pi^{i-1}(u, v)$ is an edge in G and π is an isomorphism, $\pi^i(u, v)$ is an edge in H, which yields a contradiction.   □

Assembling Lemmas 1–4 we now conclude that Algorithm 1 finds two vertices whose transposition reduces the complexity of the sought isomorphism.

Lemma 5. *Let ψ be a permutation. If two labeled graphs G and H differ by an isomorphism ϕ of complexity c such that $|\operatorname{supp}(\psi) \cup \operatorname{supp}(\phi)| \leq k$ then $\mathrm{ISO}_k(G, H, c, \psi)$ returns an isomorphism from G to H.*

Proof. We show the statement by induction on c. If $c = 0$ then G and H are identical and $\mathrm{ISO}_k(G, H, 0)$ returns the identity. Suppose $c > 0$. Let ϕ be the isomorphism that maps G to H and that fulfills the assumptions. If $|V(G \triangle H)| \leq 2k$ the algorithm simulates the permutation of all pairs of vertices in $V(G \triangle H)$ by swapping vertices in H. By Lemma 2 there exist two distinct vertices v_1, v_2 in $V(G \triangle H)$ that lie in the same orbit under ϕ. By Lemma 1 their transposition reduces the complexity of ϕ and does not increase the number of vertices that have to be permuted.

If on the other hand $|V(G \triangle H)| > 2k$, then by Lemma 3 the graph $G \triangle H$ has vertex cover number at most k. By Lemma 4, for any vertex cover C of $G \triangle H$ of size at most k there exist two vertices v_1, v_2 in C that lie in the same orbit. Again by Lemma 1 their transposition reduces the complexity.

Note that in both cases $v_1, v_2 \in \operatorname{supp}(\phi)$. Therefore $|\operatorname{supp}((v_1, v_2) \cdot \psi) \cup \operatorname{supp}(\phi \cdot (v_1, v_2))| \leq |\operatorname{supp}(\psi) \cup \operatorname{supp}(\phi)| \leq k$. Thus, by induction, the call to $\mathrm{ISO}_k(G, H', c-1, \phi \cdot (v_1, v_2))$ with a permuted H returns an isomorphism. Since π is an isomorphism for the input pair (G, H) if and only if $\pi \cdot (v_1, v_2)$ is an isomorphism for the modified input pair (G, H'), the returned isomorphism is altered to an isomorphism from G to H and subsequently returned.   □

The lemma establishes that the call $\mathrm{ISO}_k(G, H, k, \mathrm{id})$ of Algorithm 1 finds an isomorphism if there exists an isomorphism that permutes at most k vertices.

2.2 Running Time:

Having established the correctness of the algorithm, we now analyze its running time. The following theorem shows that our problem is fixed-parameter tractable.

Theorem 1. *For any integer k Algorithm 1 solves graph isomorphism for labeled graphs that differ by a permutation of at most k vertices in $\mathcal{O}(f(k)n^2)$ time.*

Proof. The time spent within each call of the algorithm is dominated by the computation of $G \triangle H$ and the computation of a vertex cover of $G \triangle H$ of size up to k. The computation of $G \triangle H$ can be performed in $\mathcal{O}(n^2)$ time, as it only requires a simple one-by-one comparison of the edges in G and H. Having a representation of the graph $G \triangle H$, with the classical fpt-algorithms (see [11]), the vertex cover problem can then be solved in $\mathcal{O}(f(k)n^2)$ time.

It remains to bound the number of selfcalls of the algorithm. For this we observe that each iteration calls its own algorithm at most $(2k)^2$ times: Indeed this bound holds for both cases of the size of $V(G \triangle H)$. The recursion depth of the algorithm is at most k, thus there are at most $((2k)^2)^k$ calls of the algorithm. The overall running time is thus in $\mathcal{O}((2k)^{2k} \cdot f(k) \cdot n^2)$. $\square$

Note that currently, for the parameter vertex cover number, the best known fpt-algorithm for the vertex cover problem, by Chen Kanj and Xia [5], runs in time $\mathcal{O}(1.2738^k + kn) \subseteq \mathcal{O}(1.2738^k \cdot n^2)$. This gives an overall running time of $\mathcal{O}((2k)^{2k} \cdot 1.2738^k \cdot n^2)$ for our algorithm.

3   Extension to Maximum Common Subgraph Distance

For two labeled graphs G and H defined on a vertex set of n vertices, let $d_0(G, H)$ be n minus the size of the maximum common (labeled) subgraph of G and H. That is, $d_0(G, H)$ is the number of vertices that have to be removed from the vertex set of the graphs G and H to obtain identical labeled graphs. It is easy to show that $d_0(G, H)$ is equal to the vertex cover number of $G \triangle H$. This implies that the computation of $d_0(G, H)$ is fixed-parameter tractable.

We define $d_k(G, H)$ to be the minimum of all $d_0(G', H)$ where G' ranges over all graphs obtained from G by permuting at most k vertices. For the complete graph K_n, the distance $d_0(G, K_n)$ is equal to n minus the size of the largest clique in G. Thus, with k as parameter, computation of $d_k(G, H)$ is W[1]-hard. (In fact, it is even NP-complete for $k = 0$.)

However, if we take both the parameters k and d_k for the input graphs into account, then the computation is fixed-parameter tractable.

Theorem 2. *There is an algorithm that computes for two graphs G and H the distance $d_k(G, H)$ (i.e., the maximum common subgraph distance between H and all graphs obtained from G by a permutation of at most k vertices) in time $\mathcal{O}(f(k, d_k(G, H)) \cdot n^2)$. Here $f(k, d) = (k + d)^{2k} \cdot 2^{k \cdot (k+d)}$.*

We omit the description of the algorithm, which is an adaptation of Algorithm 1, and the proof of the theorem, which generalizes the proof of Theorem 1.

4   Similarity Measures and Intractability

Contrasting the tractability of the isomorphism problem for graphs of low permutation distance, we now show intractability of the graph isomorphism problem for a wide range of other measures. The following folklore observation relates the evaluation of similarity measures to the complexity of similar graphs.

Theorem 3. *Let $d(\cdot, \cdot)$ be an arbitrary real-valued function that takes two labeled graphs as input.*

1. *If $d(G, H) \neq 0$ implies $G \not\cong H$ and $d(G, H) = 0$ can be decided in polynomial-time, then the problem of deciding isomorphism for graphs of bounded distance (i.e., $d(G, H) \leq k$ for a fixed constant k) is isomorphism complete.*
2. *If $d(G, H) = 0$ is equivalent to $G \cong H$ then evaluation of $d(\cdot, \cdot)$ is at least as hard as deciding graph isomorphism.*

Proof. For the first part of the theorem observe that graph isomorphism reduces to the problem of deciding isomorphism of graphs of bounded distance in the following way: By assumption, it can be decided in polynomial time whether $d(G, H) = 0$. If $d(G, H) > 0$ then the graphs are non-isomorphic. Otherwise any algorithm that solves the isomorphism problem for graphs of bounded distance decides whether G and H are isomorphic.

The second claim follows from the fact that under the assumption, deciding $d(G, H) = 0$ is equivalent to deciding whether G and H are isomorphic. □

The theorem applies to labeled similarity measures in general, but the assumptions are typically fulfilled by unlabeled similarity measures (i.e., measures invariant when replacing the inputs with isomorphic copies). For example, Part 1 applies to the spectral distance: Indeed the characteristic polynomial of both graphs can be computed in polynomial time, and their comparison determines whether the graphs have the same eigenvalues. This is of course well known (see [6, Section 1.2]). Part 2 applies to all other unlabeled measures from the introduction. The evaluation of many of them (e.g., rotation distance [17] and edit distance [3]) is even known to be NP-hard.

Independent of the complexity of the similarity measures' evaluation, we still obtain hardness results for similarity measures which govern graph as close, whenever the graphs differ by the repositioning of one edge: We say that two graphs G and H over the same vertex set *differ by the repositioning of one edge*, if G and H have the same number of edges and $G \bigtriangleup H$ has exactly two edges.

Theorem 4. *Deciding whether two labeled graphs G and H that differ by the repositioning of one edge are isomorphic is graph isomorphism complete. Moreover, the problem remains graph isomorphism complete for input graphs that have a universal vertex (i.e., a vertex adjacent to every other vertex).*

Proof. We reduce the graph isomorphism problem to the problem of deciding whether two labeled graphs G and H that differ by the repositioning of an edge are isomorphic. Let G and H be two input graphs. W.l.o.g. we assume that G

and H are connected, have the same number of edges, neither graph is complete, and their vertex-sets are disjoint. Let v and v' be two non-adjacent vertices of G. We construct a set of pairs of new input graphs as follows:

The first graph is always $U_1 = (G + (v, v')) \,\dot\cup\, H$, i.e., the disjoint union of the graph H with the graph obtained from G by adding the edge (v, v').

For the second graph we take all choices of non-adjacent vertices $u, u' \in H$ and form the graph $U_2(u, u') = G \,\dot\cup\, (H + (u, u'))$. Note that there are at most $|V(H)|^2$ such graphs. For any choice of non-adjacent vertices u, u' the graphs U_1 and $U_2(u, u')$ differ by at most two edges.

It suffices now to show that G and H are isomorphic if and only if there exist u, u' such that U_1 is isomorphic to $U_2(u, u')$. Suppose ϕ is an isomorphism from G to H, then by construction U_1 is isomorphic to $U_2(\phi(v), \phi(v'))$. Suppose now that G and H are non-isomorphic. Since G and H have the same number of edges, for any choice of u, u', any isomorphism must map the components of U_1 to components of $U_2(u, u')$. Moreover, due to the number of edges, the component of U_1 that is an exact copy of G must be mapped to the component of U_2 that is an exact copy of H, which yields a contradiction.

To see that the input graphs can also be required to have a universal vertex, note that the addition to both input graphs of a vertex that is adjacent to every other vertex preserves isomorphism, non-isomorphism and the symmetric difference. This operation thus reduces the problem to the special case of input graphs having a universal vertex. □

The theorem has intractability implications for all graph similarity measures which change only by a bounded amount whenever an edge is added or removed.

Corollary 1. *Let $d(\cdot, \cdot)$ be the labeled or unlabeled version of the maximum common subgraph distance, the maximum common contract distance, the edge rotation distance, the edge slide distance or the cut distance. There is a $k \in \mathbb{N}$, such that the graph isomorphism problem is graph isomorphism complete for the class of input pairs G and H with $d(G, H) \le k$.*

Proof. By definition, for of each of the distances, graphs that differ by the repositioning of an edge and have a universal vertex are of bounded distance. □

Acknowledgements. I thank Danny Hermelin for posing to me the core question of the paper answered by Theorem 1. I also thank Reto Spöhel for helpful comments and suggestions.

References

1. Borgs, C., Chayes, J.T., Lovász, L., Sós, V.T., Szegedy, B., Vesztergombi, K.: Graph limits and parameter testing. In: STOC 2006, New York, pp. 261–270 (2006)
2. Bunke, H.: On a relation between graph edit distance and maximum common subgraph. Pattern Recognition Letters 18(8), 689–694 (1997)
3. Bunke, H.: Error correcting graph matching: on the influence of the underlying cost function. Pattern Analysis and Machine Intelligence 21(9), 917–922 (1999)

4. Chartrand, G., Saba, F., Zou, H.-B.: Edge rotations and distance between graphs. Časopis Pěst. Mat. 110(1), 87–91 (1985)
5. Chen, J., Kanj, I.A., Xia, G.: Improved parameterized upper bounds for vertex cover. In: Královič, R., Urzyczyn, P. (eds.) MFCS 2006. LNCS, vol. 4162, pp. 238–249. Springer, Heidelberg (2006)
6. Cvetković, D.M., Rowlinson, P., Simić, S.: Eigenspaces of Graphs. Cambridge University Press, Cambridge (1997)
7. Dénes, J.: The representation of a permutation as the product of a minimal number of transpositions, and its connection with the theory of graphs. Magyar Tud. Akad. Mat. Kutató Int. Közl. 4, 63–71 (1959)
8. Downey, R.G., Fellows, M.R.: Parameterized Complexity. Monographs in Computer Science. Springer, London (1998)
9. Evdokimov, S., Ponomarenko, I.N.: Isomorphism of coloured graphs with slowly increasing multiplicity of Jordan blocks. Combinatorica 19(3), 321–333 (1999)
10. Filotti, I.S., Mayer, J.N.: A polynomial-time algorithm for determining the isomorphism of graphs of fixed genus. In: STOC 1980, pp. 236–243 (1980)
11. Flum, J., Grohe, M.: Parameterized Complexity Theory. Texts in Theoretical Computer Science. An EATCS Series. Springer, London (2006)
12. Furst, M.L., Hopcroft, J.E., Luks, E.M.: Polynomial-time algorithms for permutation groups. In: FOCS 1980, Washington, USA, pp. 36–41 (1980)
13. Johnson, M.: An ordering of some metrics defined on the space of graphs. Czechoslovak Math. J. 37(112), 75–85 (1987)
14. Köbler, J., Schöning, U., Torán, J.: The graph isomorphism problem: its structural complexity. Birkhäuser Verlag, Basel (1993)
15. Kratsch, S., Schweitzer, P.: Isomorphism for graphs of bounded feedback vertex set number. In: Kaplan, H. (ed.) SWAT 2010. LNCS, vol. 6139, pp. 81–92. Springer, Heidelberg (2010)
16. Luks, E.M.: Isomorphism of graphs of bounded valence can be tested in polynomial time. Journal of Computer and System Sciences 25(1), 42–65 (1982)
17. Marcu, D.: Note on the edge rotation distance between trees. International Journal of Computer Mathematics 30(1), 13–15 (1989)
18. Miller, G.L.: Isomorphism testing for graphs of bounded genus. In: STOC 1980, New York, pp. 225–235 (1980)
19. Schweitzer, P.: Problems of unknown complexity: graph isomorphism and Ramsey theoretic numbers. Phd thesis, Universität des Saarlandes, Saarbrücken (2009)
20. Wilson, R.C., Zhu, P.: A study of graph spectra for comparing graphs and trees. Pattern Recognition 41(9), 2833–2841 (2008)
21. Yamazaki, K., Bodlaender, H.L., de Fluiter, B., Thilikos, D.M.: Isomorphism for graphs of bounded distance width. Algorithmica 24(2), 105–127 (1999)
22. Zelinka, B.: On a certain distance between isomorphism classes of graphs. Časopis Pěst. Mat. 100(4), 371–373 (1975)
23. Zelinka, B.: Contraction distance between isomorphism classes of graphs. Časopis Pěst. Mat. 115(2), 211–216 (1990)
24. Zemlyachenko, V.N., Korneenko, N.M., Tyshkevich, R.I.: Graph isomorphism problem. Journal of Mathematical Sciences 29(4), 1426–1481 (1985)

Paths, Flowers and Vertex Cover

Venkatesh Raman, M.S. Ramanujan, and Saket Saurabh

The Institute of Mathematical Sciences, Chennai, India
{vraman,msramanujan,saket}@imsc.res.in

Abstract. It is well known that in a bipartite (and more generally in a König) graph, the size of the minimum vertex cover is equal to the size of the maximum matching. We first address the question whether (and if not when) this property still holds in a König graph if we insist on forcing one of the two vertices of some of the matching edges in the vertex cover solution. We characterize such graphs using the classical notions of augmenting paths and flowers used in Edmonds' matching algorithm. Using this characterization, we develop an $O^*(9^k)$[1] algorithm for the question of whether a general graph has a vertex cover of size at most $m + k$ where m is the size of the maximum matching. Our algorithm for this well studied ABOVE GUARANTEE VERTEX COVER problem uses the technique of iterative compression and the notion of important separators, and improves the runtime of the previous best algorithm that took $O^*(15^k)$ time. As a consequence of this result we get that well known problems like ALMOST 2 SAT (deleting at most k clauses to get a satisfying 2 SAT formula) and KÖNIG VERTEX DELETION (deleting at most k vertices to get a König graph) also have an algorithm with $O^*(9^k)$ running time, improving on the previous bound of $O^*(15^k)$.

1 Introduction

The classical notions of *matchings* and *vertex covers* have been at the center of serious study for several decades in the area of Combinatorial Optimization [9]. In 1931, König and Egerváry independently proved a result of fundamental importance: in a bipartite graph the size of a maximum matching equals that of a minimum vertex cover [9]. This led to a polynomial-time algorithm for finding a minimum vertex cover in bipartite graphs. Interestingly, this min-max relationship holds for a larger class of graphs known as König-Egerváry graphs and it includes bipartite graphs as a proper subclass. König-Egerváry graphs will henceforth be called König graphs. Our first result in this paper is an extension of this classical result. That is, we address the following question:

When does a König graph have a minimum vertex cover equal to the size of a maximum matching when we insist on forcing one of the two vertices of some of the matching edges in the vertex cover solution?

---

[1] O^* notation suppresses polynomial factors.

C. Demetrescu and M.M. Halldórsson (Eds.): ESA 2011, LNCS 6942, pp. 382–393, 2011.
© Springer-Verlag Berlin Heidelberg 2011

We resolve this problem by obtaining a excluded-subgraph characterization for König graphs G satisfying this property. More precisely, let G be a König graph, M be a maximum matching of G and let S be a set of vertices containing exactly one vertex from some of the edges of M. Then G has a minimum vertex cover of size $|M|$ containing S if and only if it does not contain "certain kind of M-augmenting paths and M-flowers." These notions of augmenting paths and flowers are the same as the one used in the classical maximum matching algorithm of Edmonds [6] on general graphs.

Our main motivation for this excluded-subgraph characterization stems from obtaining a faster algorithm for a version of the VERTEX COVER problem studied in parameterized complexity. For decision problems with input size n, and a parameter k, the goal in parameterized complexity is to design an algorithm with runtime $f(k)n^{O(1)}$ where f is a function of k alone, as contrasted with a trivial $n^{k+O(1)}$ algorithm. Such algorithms are said to be fixed parameter tractable (FPT). We also call an algorithm with a runtime of $f(k)n^{O(1)}$, as an FPT algorithm, and such a runtime as FPT runtime. The theory of parameterized complexity was developed by Downey and Fellows [5]. For recent developments, see the book by Flum and Grohe [7]. The version of classical VERTEX COVER that we are interested in is following.

ABOVE GUARANTEE VERTEX COVER (AGVC)
Input: $(G = (V, E), M, k)$, where G is an undirected graph, M is a maximum matching for G, k a positive integer
Parameter: k
Question: Does G have a subset S of size at most $|M| + k$ that covers all the edges?

The only known parameterized algorithm for AGVC is using a parameter preserving reduction to ALMOST 2-SAT. In ALMOST 2-SAT, we are given a 2-SAT formula ϕ, a positive integer k and the objective is to check whether there exists at most k clauses whose deletion from ϕ can make the resulting formula satisfiable. The ALMOST 2-SAT problem was introduced in [10] and a decade later it was shown by Razgon and Barry O'Sullivan [15] to have an $O^*(15^k)$ time algorithm, thereby proving fixed-parameter tractability of the problem when k is the parameter. The ALMOST 2-SAT problem is turning out to be a fundamental problem in the context of designing parameterized algorithms. This is evident from the fact that there is a polynomial time parameter preserving reduction from problems like ODD CYCLE TRANSVERSAL [8] and AGVC [14] to it. An FPT algorithm for ALMOST 2-SAT led to FPT algorithms for several problems, including AGVC and KÖNIG VERTEX DELETION [14]. In recent times this has been used as a subroutine in obtaining a parameterized approximation as well as an FPT algorithm for MULTI-CUT [12,13]. Our second motivation for studying AGVC is that it also implies a faster FPT algorithm for ALMOST 2-SAT. This is obtained through a parameter preserving reduction from ALMOST 2-SAT to AGVC and hence this also shows that these two problems are equivalent.

The standard version of VERTEX COVER, where we are interested in finding a vertex cover of size at most k for the given parameter k was one of the earliest problems that was shown to be FPT [5]. After a long race, the current best algorithm for VERTEX COVER runs in time $O(1.2738^k + kn)$ [1]. However, when $k < m$, the size of the maximum matching, the standard version of VERTEX COVER is not interesting, as the answer is trivially NO. And if m is large (suppose, for example, the graph has a perfect matching), then for the cases the problem is interesting, the running time of the standard version is not practical, as k, in this case, is quite large. This also motivates the study of AGVC.

Our results and methodology. Many of the recent FPT algorithms, including the ones for ALMOST 2-SAT [15], DIRECTED FEEDBACK VERTEX SET [3], MULTIWAY CUT [2], MULTICUT [13] are based on a combination of the method of iterative compression introduced in [16] and graph separation. In the iterative compression method, we assume that a solution of size $k+1$ is part of the input, and attempt to compress it to a solution of size k. The method adopted usually is to begin with a subgraph that trivially admits a $(k + 1)$-sized solution and then expand it iteratively. The main ingredient of the graph separation part of these algorithms is to find the right structures to eliminate, which in most of these cases are certain kind of paths, and to be able to eliminate them in FPT time. Notions of "important sets" and "important separators" have turned out to be very useful in these cases [2,11,13]. We follow the same paradigm here and using our excluded subgraph characterization find a set of structures that we need to eliminate to solve AGVC faster. More precisely, using our graph theoretic results together with algorithmic technique of iterative compression and the notion of important separators, we develop an $O^*(9^k)$ algorithm for AGVC. This improves the runtime of the previous best algorithm that took $O^*(15^k)$ time. This in turn together with known parameterized reductions implies faster algorithms $(O^*(9^k))$ for a number of problems including ALMOST 2-SAT (both variable and clause variants) and KÖNIG VERTEX DELETION (the definitions of these problems are available in the full version of the paper).

Organization of the paper. In Section 3 we give a general outline of our algorithm and describe the method of iterative compression applied to the AGVC problem. In Section 4 we show that the structures we need to eliminate in the case of the AGVC problem are precisely the augmenting paths and flowers seen in the classical maximum matching algorithm of Edmonds [6] on general graphs. In Section 5, we then exploit the structure given by the combinatorics of vertex covers and maximum matchings to obtain an FPT algorithm for AGVC that runs in time $O^*(9^k)$. Finally, in Section 5 we prove that ALMOST 2 SAT has an $O^*(9^k)$ algorithm by giving a polynomial time parameter preserving reduction from AGVC to ALMOST 2 SAT.

2   Preliminaries

Let $G = (V, E)$ be a graph and $D = (V, A)$ be a directed graph. We call the ordered pair $(u, v) \in A$ arc and the unordered pair $(u, v) \in E$ edge. For a subset S of V, the *subgraph of G induced by S* is denoted by $G[S]$. By

$N_G(u)$ we denote (open) neighborhood of u that is set of all vertices adjacent to u. Similarly, for a subset $T \subseteq V$, we define $N_G(T) = (\cup_{v \in T} N_G(v)) \setminus T$. Given a graph $G = (V, E)$ and two disjoint vertex subsets V_1, V_2 of V, we let (V_1, V_2) denote the bipartite graph with vertex set $V_1 \cup V_2$ and edge set $\{(u, v) : (u, v) \in E \text{ and } u \in V_1, v \in V_2\}$. Given a graph G, we use $\mu(G)$ and $\beta(G)$ to denote, respectively, the size of a maximum matching and a minimum vertex cover. A graph $G = (V, E)$ is said to be *König* if $\beta(G) = \mu(G)$. If M is a matching and $(u, v) \in M$ then we say that u *is the partner of v in M*. If the matching being referred to is clear from the context we simply say u *is a partner of v*. The vertices of G that are the endpoints of edges in the matching M are said to be *saturated by M* and we denote the set of these vertices by $V(M)$; all other vertices are *unsaturated by M*. Given graph $G = (V, E)$, and a (not necessarily simple) path $P = v_1, \ldots, v_t$, we define by $Rev(P)$ the path $v_t, v_{t-1}, \ldots, v_1$. Even though it may not make sense to talk about the direction of a path in an undirected graph, we will use this notation to simplify our presentation. We will call the number of edges in P the length of P and represent it by $|P|$. Let $P_1 = v_1, \ldots, v_t$ and $P_2 = v_t, \ldots, v_x$ be two paths which have only the vertex v_t in common. We represent by $P_1 + P_2$ the concatenated path $v_1, \ldots, v_t, v_{t+1}, \ldots, v_x$. We also need the following characterization of König graphs.

Lemma 1 (see for example [14]). *A graph $G = (V, E)$ is König if and only if there exists a bipartition of V into $V_1 \uplus V_2$, with V_1 a vertex cover of G such that there exists a matching across the cut (V_1, V_2) saturating every vertex of V_1.*

Definition 1. *Let Z be a finite set. A function $f : 2^Z \to \mathbb{N}$ is submodular if for all subsets A and B of Z, $f(A \cup B) + f(A \cap B) \leq f(A) + f(B)$*

3   Outline of the Algorithm

We first make use of a known reduction that allows us to assume that the input graph has a perfect matching. Given an instance $(G = (V, E), M, k)$ of AGVC, in polynomial time we obtain an equivalent instance with a perfect matching using [14, Lemma 5]. That is, if (G, M, k) is an instance of AGVC and G is a graph without a perfect matching, then in polynomial time we can obtain an instance (G', M', k) such that G' has a perfect matching M' and (G, M, k) is a YES instance of AGVC if and only if (G', M', k) is a YES instance of AGVC. Because of this reduction, throughout this paper we assume that in our input instance (G, M, k), the matching M is a perfect matching of G. We now describe the iterative compression step, which is central to our algorithm, in detail.

Iterative Compression for AGVC. Given an instance $(G = (V, E), M, k)$ of AGVC let $M = \{m_1, \ldots, m_{\frac{n}{2}}\}$ be a perfect matching for G, where $n = |V|$. Define $M_i = \{m_1, \ldots, m_i\}$ and $G_i = G[V(M_i)]$, $1 \leq i \leq \frac{n}{2}$. We iterate through the instances (G_i, M_i, k) starting from $i = k+1$ and for the ith instance, with the help of a *known* solution S_i of size at most $|M_i| + k + 1$ we try to find a solution $\hat{S}_i$ of size at most $|M_i| + k$. Formally, the compression problem we address is as follows.

ABOVE GUARANTEE VERTEX COVER COMPRESSION (AGVCC)

Input: $(G = (V, E), S, M, k)$, where G is an undirected graph, M is a perfect matching for G, S is a vertex cover of G of size at most $|M| + k + 1$, k a positive integer

Parameter: k

Question: Does G have a vertex cover $\hat{S}$ of size at most $|M| + k$?

We will reduce the AGVC problem to $\frac{n}{2} - k$ instances of the AGVCC problem as follows. Let $I_i = (G_i, M_i, S_i, k)$ be the ith instance. Clearly, the set $V(M_{k+1})$ is a vertex cover of size at most $|M_{k+1}| + k + 1$ for the instance I_{k+1}. It is also easy to see that if $\hat{S}_{i-1}$ is a vertex cover of size at most $|M_{i-1}| + k$ for instance I_{i-1}, then the set $\hat{S}_{i-1} \cup V(m_i)$ is a vertex cover of size at most $|M_i| + k + 1$ for the instance I_i. We use these two observations to start off the iteration with the instance $(G_{k+1}, M_{k+1}, S_{k+1} = V(M_{k+1}), k)$ and look for a vertex cover of size at most $|M_{k+1}| + k$ for this instance. If there is such a vertex cover $\hat{S}_{k+1}$, we set $S_{k+2} = \hat{S}_{k+1} \cup V(m_{k+2})$ and ask for a vertex cover of size at most $|M_{k+2}| + k$ for the instance I_{k+2} and so on. If, during any iteration, the corresponding instance does not have a vertex cover of the required size, it implies that the original instance is also a NO instance. Finally the solution for the original input instance will be $\hat{S}_{\frac{n}{2}}$. Since there can be at most $\frac{n}{2}$ iterations, the total time taken is bounded by $\frac{n}{2}$ times the time required to solve the AGVCC problem.

Our algorithm for AGVCC is as follows. Let the input instance be $I = (G = (V, E), S, M, k)$. Let M' be the edges in M which have both vertices in S. Note that $|M'| \leq k + 1$. Then, $G \setminus V(M')$ is a König graph and by Lemma 1 has a partition (A, B) such that A is a minimum vertex cover and there is a matching saturating A across the cut (A, B), which in this case is $M \setminus M'$. We guess a subset $Y \subseteq M'$ with the intention of picking both vertices of these edges in our solution. For the remaining edges of M', exactly one vertex from each edge will be part of our eventual solution. For each edge of $M' \setminus Y$, we guess the vertex which is not going to be part of our eventual solution. Let the set of vertices guessed this way as not part of the solution be T. Define $L = A \cap N_G(T)$ and $R = B \cap N_G(T)$. Clearly our guess forces $L \cup R$ to be part of the solution. We have thus reduced this problem to checking if the instance $(G[V(M \setminus M')], A, M \setminus M', k - |M'|)$ has a vertex cover of size at most $|M \setminus M'| + k - |M'|$ which contains L and R. We formally define this annotated variant as follows.

ANNOTATED ABOVE GUARANTEE VERTEX COVER (A-AGVC)

Input: $(G = (A, B, E), M, L, R, k)$, where G is an undirected König graph, (A, B) is a partition of the vertex set of G, A is a minimum vertex cover for G, M is a perfect matching for G saturating A and B, $L \subseteq A$ and $R \subseteq B$, k a positive integer

Parameter: k

Question: Does G have a vertex cover of size at most $|M| + k$ such that it contains $L \cup R$?

Our main result is the following Lemma.

Lemma 2. A-AGVC *can be solved in* $O^*(4^k)$ *time.*

Given Lemma 2 the running time of our algorithm for AGVCC is bounded as follows. For every $0 \leq i \leq k$, for every i sized subset Y, for every guess of T, we run the algorithm for A-AGVC with parameter $k - i$. For each i, there are $\binom{k+1}{i}$ subsets of M' of size i, and for every choice of Y of size i, there are 2^{k+1-i} choices for T and for every choice of T, running the algorithm for A-AGVC given by Lemma 2 takes time $O^*(4^{k-i})$. Hence, the running time of our algorithm for AGVCC is bounded by $O^*(\Sigma_{i=0}^{k}\binom{k+1}{i}2^{k+1-i}4^{k-i}) = O^*(9^k)$ and hence our algorithm for AGVC runs in time $O^*(9^k)$. Thus we have the following Theorem.

Theorem 1. AGVC *can be solved in* $O^*(9^k)$ *time.*

Sections 4 and 5 are devoted to proving Lemma 2.

4   König Graphs with Extendable Vertex Covers

In this section we obtain a characterization of those König graphs, in which, a given subset of vertices can be extended to a minimum vertex cover. Recall that whenever we say a *minimum vertex cover* of a König graph, we mean a vertex cover that has size equal to the size of a maximum matching. We start off with a couple of definitions.

Definition 2. *Given a graph* $G = (V, E)$ *and a matching* M, *we call a path* $P = v_1, \ldots, v_t$ *in the graph, an* M−*alternating path if the edge* $(v_1, v_2) \in M$, *every subsequent alternate edge is in* M *and no other edge of* P *is in* M. *An odd length* M−*alternating path is called an odd* M−*path and an even length* M−*alternating path is called an even* M−*path.*

A simple and useful observation to the above definition is the following.

Observation 1. *In odd* M−*paths, the last edge of the path is a matched edge, while in even* M−*paths, the last edge is not a matching edge.*

Note that, by our definition, a single matching edge is an odd M− path. In addition, we consider a path consisting of a single vertex to be an even M−path by convention. Let $P = v_1, \ldots, v_t$ be an odd (similarly even) M−path and let $Q_1, Q_2 \subseteq V(G)$ such that $v_1 \in Q_1$ and $v_t \in Q_2$. Then, we say that P is an odd (similarly even) M−path from Q_1 to Q_2.

Definition 3. *Given a graph* G *and a matching* M, *we define an* M−*flower as a walk* $W = v_1, \ldots, v_b, v_{b+1} \ldots v_{t-1}, v_t$ *with the following properties.*

 - *The vertex* $v_t = v_b$ *and all other vertices of* W *are distinct.*
 - *The subpaths* $P_1 = v_1, \ldots, v_b$ *and* $P_2 = v_1, \ldots, v_{t-1}$ *are odd* M−*paths from* v_1 *to* v_b *and* v_1 *to* v_{t-1} *respectively.*
 - *The odd cycle* $C = v_b, v_b + 1, \ldots, v_t$, *which has length* $t - b$ *and contains exactly* $\lfloor \frac{t-b}{2} \rfloor$ *edges from* M *is called the blossom.*
 - *The odd* M−*path* P_1 *is called the stem of the flower, the vertex* v_b *is called the base and* v_1 *is called the root (see Fig. 1).*

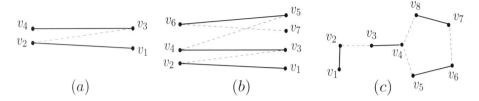

Fig. 1. Illustrations of the two types of $M-$alternating paths and an $M-$flower. The non matching edges are represented by the dashed lines. (a) An odd $M-$path v_1, v_2, v_3, v_4. (b) An even $M-$path $v_1, v_2, v_3, v_4, v_5, v_6, v_7$. (c) An $M-$flower with root v_1, base v_4, stem v_1, v_2, v_3, v_4, blossom $v_4, v_5, v_6, v_7, v_8, v_4$, and a blossom path v_5, v_6, v_7, v_8.

Given a set $X \subseteq V(G)$, we say that G has an X $M-$flower if there is a $M-$flower in G such that the root is in X. The odd $M-$path $v_{b+1}, v_{b+2}, \ldots, v_{t-1}$ is called a blossom $M-$path from v_{b+1} to v_{t-1}. Blossom $M-$ paths are defined only between the two neighbors of the base which lie in the blossom.

The following consequences follow from the above definitions.

Lemma 3. [⋆][2] Let $(G = (A, B, E), M, L, R, k)$ be an instance of A-AGVC.

(a) There cannot be an odd $M-$path from A to A.

(b) Any odd $M-$path from B to B has to contain exactly one edge between two vertices in A.

(c) There cannot be an R $M-$flower with its base in B.

(d) Let $\mathcal{P}$ be an R $M-$flower with base v and let the neighbors of v in the blossom be u_1 and u_2. Then, $u_1 \in B$ or $u_2 \in B$.

(e) Let $P = v_1, \ldots, v_t$ be an odd $M-$path in G and suppose S is a minimum vertex cover of G. If $v_1 \in S$, then $v_t \notin S$.

(f) Let $P = v_1, \ldots, v_t$ be an odd $M-$path from B to B. Then there is an edge (u, v) such that $u, v \in A$ and there is an odd $M-$path P_1 from v_1 to u and an odd $M-$path P_2 from v_t to v and P_1 and P_2 are subpaths of P.

(g) Consider a vertex v and an even $M-$path P from some vertex u to v. Then P does not contain the matching partner of v.

Using all these observations, we prove the following characterization.

Lemma 4. [⋆] Given an instance $(G = (A, B, E), M, L, R, k)$ of A-AGVC, G has a minimum vertex cover S such that $L \cup R \subseteq S$ if and only if there is neither an odd $M-$path from $L \cup R$ to $L \cup R$, nor an R $M-$flower.

We also have the following simple Lemma.

Lemma 5. [⋆] Given an instance $(G = (A, B, E), M, L, R, k)$ of A-AGVC, if G has a minimum vertex cover containing $L \cup R$, then one such minimum vertex cover can be computed in polynomial time.

---

[2] Proofs of results labeled with $\star$ are available in the full version of the paper.

5   Important Separators and the Algorithm

In this Section we use Lemma 4 to model the A-AGVC problem as a problem of eliminating certain structures in the graph and develop an efficient algorithm for the problem. The overall idea of our algorithm is that we use important separators (defined below) to eliminate odd $M-$paths from $L \cup R$ to $L \cup R$ and when no such path exists, we find an edge and recursively try to solve the problem by including one of the end-points in our potential vertex cover. A well chosen measure allows us to capture this progress and finally lead to the faster algorithm.

Important Separators in Directed Graphs. The notion of important separators was formally introduced in [11]. Here we extend these definitions to directed graphs in a very natural way. Given a directed graph $D = (V, A)$, consider a set $X \subseteq V$. We denote by $\delta_G^+(X)$, the set of arcs going out of X in D and we define a function $f : 2^V \to \mathbb{N}$ where $f(X) = |\delta_G^+(X)|$. It is easy to verify that the function f is submodular.

Let $X, Y \subset V$ be two disjoint vertex sets. A set $S \subseteq A$ is called an $X - Y$ *separator* or an *arc separator* if no vertex in Y is reachable from any vertex in X in $D \setminus S$. We call S a minimal separator if no proper subset of S is an $X - Y$ separator and denote by $K_{X,S}$ the set of vertices reachable from X in the graph $D \setminus S$. We drop the explicit reference to X if it is clear from the context and just call this set K_S. We let $\lambda_D(X,Y)$ denote the size of the smallest $X - Y$ separator in D. We drop the subscript D when it is clear from the context.

Definition 4. *Let $X, Y \subset V$ be two disjoint vertex sets of D and let $S, S_1 \subseteq V$ be two $X - Y$ separators. We say that S_1 dominates S with respect to X if $|S_1| \leq |S|$ and $K_{X,S} \subset K_{X,S_1}$. We drop the explicit reference to X if it is clear from the context. We call S an important $X - Y$ separator if it is minimal and there is no $X - Y$ separator that dominates S with respect to X.*

Note that, if Y is not reachable from X in D, then the empty set is a trivial important $X-Y$ separator. We make the following observations about important separators, which we will use later in the algorithm.

Lemma 6. [$\star$] *Let $D = (V, A)$ be a directed graph where $|V| = n$, $X, Y \subset V$ be two disjoint vertex sets and S be an important $X - Y$ separator.*

1. *For every $e = (u, v) \in S$, $S \setminus \{e\}$ is an important $X - Y$ separator in the graph $D \setminus \{e\}$.*
2. *If S is an $X' - Y$ arc separator for some $X' \supset X$ such that X' is reachable from X in $D[X']$ where $D[X']$ is the subgraph of D induced on the vertices of X', then S is also an important $X' - Y$ separator.*
3. *There is a unique important $X - Y$ separator S^* of size $\lambda(X,Y)$ and it can be found in polynomial time. Furthermore, $K_{S^*} \subseteq K_S$.*

Note that, given an instance $(G = (A, B, E), M, L, R, k)$ of A-AGVC, in order to find a minimum vertex cover containing $L \cup R$, it is *sufficient* to find the set M' of matched edges which have both end points in this minimum vertex cover. This follows from the fact that the graph $G \setminus V(M')$ is König and has a minimum vertex cover that contains $(L \cup R) \setminus V(M')$. Thus, by Lemma 5, a minimum vertex cover of $G \setminus V(M')$ containing $(L \cup R) \setminus V(M')$ can be computed in polynomial time. Hence, in the rest of the paper whenever we talk about *a solution S for an instance of* A-AGVC, *we will mean the set of edges of M which have both endpoints in the vertex cover.* Given an instance of A-AGVC we define a directed graph $D(G)$ corresponding to this instance as follows. Remove all the edges in $G[A]$, orient all the edges of M from A to B and all the other edges from B to A. An immediate observation to this is the following.

Observation 2. *There is a path from L to R in $D(G)$ if and only if there is an odd $M-$path from L to R in G*

Even though the edges of $D(G)$ are directed (and henceforth will be called arcs), they come from G and have a fixed direction. Hence we will occasionally use the same set of arcs/edges in both the undirected and directed sense. For example we may say that a set S of edges of G is both a solution for the corresponding instance (undirected) and an arc separator in the graph $D(G)$ (directed). The next Lemma characterizes the $L-R$ separators in $D(G)$ and the lemma following it gives an upper bound on the number of such separators.

Lemma 7. [⋆] *Given an instance $(G = (A, B, E), M, L, R, k)$, any important $L-R$ separator in $D(G)$ comprises precisely arcs corresponding to some subset of M.*

Lemma 8. [⋆] *Let $(G = (A, B, E), M, L, R, k)$ be an instance of A-AGVC and let $D = D(G) = (V, A)$ be defined as above. Then the number of important $L-R$ separators of size at most k is bounded by 4^k and these can be enumerated in time $O^*(4^k)$.*

Lemma 9. [⋆] *Let $(G = (A, B, E), M, L, R, k)$ be an instance of A-AGVC. If $(G = (A, B, E), M, L, R, k)$ is a YES instance, then it has a solution $\hat{S}$ which contains an important $L-R$ separator in $D(G)$.*

The next lemma is used to handle the case when the instance does not have odd $M-$paths from L to R.

Lemma 10. [⋆] *Let $(G = (A, B, E), M, L, R, k)$ be an instance of A-AGVC such that there are no odd $M-$paths from L to R in G. If there is either an odd $M-$path P from R to R or an R $M-$flower $\mathcal{P}$, then there is an edge (u, v) such that $u, v \in A \setminus L$ and there is an odd $M-$path from u to R and an odd $M-$path from v to R. Moreover, this edge can be found in polynomial time.*

We are now ready to prove Lemma 2 by describing an algorithm (Algorithm. 5.1) for A-AGVC. The idea of the algorithm is as follows. If there is an odd $M-$path from L to R in G, then by Lemma 9 we know that if there is a solution, there

Input : An instance (G, M, L, R, k) of A-AGVC
Output: A solution of size at most k for the instance (G, M, L, R, k) if it exists
and No otherwise

1 **if** $k < 0$ **then return** No
2 *Compute a minimum size $L - R$ arc separator S in the directed graph $D(G)$*
3 **if** $|S| = 0$ **then**
4   there is no odd L to R $M-$ path
5   **if** *there an odd $M-$path from R to R or an R $M-$flower* **then**
6   *compute the edge $e = (u, v)$ given by Lemma 10*
7          $S_1 \leftarrow Solve - AAGVC(G, M, L \cup \{u\}, R, k)$
8   **if** S_1 *is not* No **then return** S_1
9          $S_2 \leftarrow Solve - AAGVC(G, L \cup \{v\}, R, k)$
10   **return** S_2
11   **end**
12   **else return** ϕ
13 **end**
14 **if** $|S| > k$ **then return** No
15 **else** *Compute the unique minimum size important $L - R$ separator S^* in $D(G)$*
(Lemma 6(c)) and select an arc $e = (w, z) \in S^*$
16 $S_3 \leftarrow Solve - AAGVC(G \setminus \{e\}, M \setminus \{e\}, L, R, k - 1)$
17 **if** S_3 *is not* No **then return** $S_3 \cup \{e\}$
18 $S_4 \leftarrow Solve - AAGVC(G, M, A \cap (\delta^+_{D(G)}(z) \cup K_{S^*}), R, k)$
19 **return** S_4

Algorithm 5.1. Algorithm $Solve - AAGVC$ for A-AGVC

is one which contains an important $L - R$ separator in $D(G)$. Hence we branch
on a choice of an important $L - R$ separator. If there are no odd $M-$paths from
L to R, but there is either an odd $M-$path from R to R or an R $M-$flower,
we use Lemma 10 to the get an edge (u, v) between two vertices in A and guess
the vertex which covers this edge and continue. If neither of these two cases
occur, then by Lemma 4 the graph has a minimum vertex cover containing
$L \cup R$. Such a minimum vertex cover will not contain both end points of any
edge of the perfect matching and hence the algorithm returns the empty set. In
order to make the analysis of our algorithm simpler, we embed the algorithm for
enumerating important separators (Lemma 8) into our algorithm for A-AGVC.

Correctness. The Correctness of Step 1 is obvious. In Steps 6 and 8 we are
merely guessing the vertex which covers the edge (u, v), while Step 11 is correct
due to Lemma 4. Step 13 is correct because the size of the minimum $L - R$
separator in $D(G)$ is a lower bound on the solution size. Steps 15 and 17 are
part of enumerating the important $L - R$ separators as seen in Lemma 8. Since
we have shown in Lemma 9 that if there is a solution, there is one which contains
an important $L - R$ separator in $D(G)$, these steps are also correct.

Running Time. In order to analyze the algorithm, we define the search tree
$\mathbb{T}(G, M, L, R, k)$ resulting from a call to $Solve - AAGVC(G, M, L, R, k)$ inductively as follows. The tree $\mathbb{T}(G, M, L, R, k)$ is a rooted tree whose root node
corresponds to the instance (G, M, L, R, k). If $Solve - AAGVC(G, M, L, R, k)$

does not make a recursive call, then (G, M, L, R, k) is said to be the only node of this tree. If it does make recursive calls, then the children of (G, M, L, R, k) correspond to the instances given as input to the recursive calls made inside the current procedure call. The subtree rooted at a child node (G', M', L', R', k') is the search tree $\mathbb{T}(G', M', L', R', k')$.

Given an instance $I = (G, M, L, R, k)$, we prove by induction on $\mu(I) = 2k - \lambda_{D(G)}(L, R)$ that the number of leaves of the tree $\mathbb{T}(I)$ is bounded by $max\{2^{2\mu(I)}, 1\}$. In the base case, if $\mu(I) < k$, then $\lambda(L, R) > k$ in which case the number of leaves is 1. Assume that $\mu(I) \geq k$ and our claim holds for all instances I' such that $\mu(I') < \mu(I)$.

Suppose $\lambda(L, R) = 0$. In this case, the children I_1 and I_2 of this node correspond to the recursive calls made in Steps 6 and 8. By Lemma 10 there are odd $M-$paths from u to R and from v to R. Hence, $\lambda(L \cup \{u\}, R) > 0$ and $\lambda(L \cup \{v\}, R) > 0$. This implies that $\mu(I_1), \mu(I_2) < \mu(I)$. By the induction hypothesis, the number of leaves in the search trees rooted at I_1 and I_2 are at most $2^{\mu(I_1)}$ and $2^{\mu(I_2)}$ respectively. Hence the number of leaves in the search tree rooted at I is at most $2.2^{\mu(I)-1} = 2^{\mu(I)}$.

Suppose $\lambda(L, R) > 0$. In this case, the children I_1 and I_2 of this node correspond to the recursive calls made in Steps 15 and 17. But in these two cases, as seen in the proof of Lemma 8, $\mu(I_1), \mu(I_2) < \mu(I)$ and hence applying induction hypothesis on the two child nodes and summing up the number of leaves in the sub trees rootes at each, we can bound the number of leaves in the sub tree of I by $2^{\mu(I)}$.

Hence the number of leaves of the search tree $\mathbb{T}$ rooted at the input instance $I = (G, M, L, R, k)$ is $2^{\mu(I)} \leq 2^{2k}$. The time spent at a node I' is bounded by the time required to compute the unique smallest $L - R$ separator in $D(G)$ which is polynomial(Lemma 6). Along any path from the root to a leaf, at any internal node, the size of the set L increases or an edge is removed from the graph. Hence the length of any root to leaf path is at most n^2. Therefore the running time of the algorithm is $O^*(4^k)$. This completes the proof of Lemma 2.

Consequences. Theorem 1 has some immediate consequences. By [14, Theorem 4], the following corollary follows.

Corollary 1. KÖNIG VERTEX DELETION *can be solved in time* $O^*(9^k)$ *time.*

It has been mentioned without proof in [4, Open Problem Posed by M. Fellows] that AGVC and ALMOST 2 SAT are equivalent. However, for the sake of completeness, we give a polynomial time parameter preserving reduction from ALMOST 2 SAT to AGVC and hence prove the following Lemma.

Lemma 11. [⋆] ALMOST 2 SAT *can be solved in* $O^*(9^k)$ *time.*

By [13, Theorem 3.1], we have the following Corollary.

Corollary 2. ALMOST 2 SAT(VARIABLE) *can be solved in time* $O^*(9^k)$ *time.*

6 Conclusion

In this paper we obtained a faster FPT algorithm for AGVC through a structural characterization of König graphs in which a minimum vertex cover is forced to contain some vertices. This also led to faster FPT algorithms for ALMOST 2 SAT, ALMOST 2 SAT(VARIABLE) and KÖNIG VERTEX DELETION. It will be interesting to improve the running time of these algorithms. One fundamental problem that remains elusive about these problems is their kernelization complexity. We leave this as one of the main open problems.

References

1. Chen, J., Kanj, I.A., Xia, G.: Improved upper bounds for vertex cover. Theor. Comput. Sci. 411(40-42), 3736–3756 (2010)
2. Chen, J., Liu, Y., Lu, S.: An improved parameterized algorithm for the minimum node multiway cut problem. Algorithmica 55(1), 1–13 (2009)
3. Chen, J., Liu, Y., Lu, S., O'Sullivan, B., Razgon, I.: A fixed-parameter algorithm for the directed feedback vertex set problem. J. ACM 55(5) (2008)
4. Demaine, E., Gutin, G., Marx, D., Stege, U.: Open problems from dagstuhl seminar 07281 (2007)
5. Downey, R.G., Fellows, M.R.: Parameterized Complexity. Springer, New York (1999)
6. Edmonds, J.: Paths, trees, and flowers. Canad. J. Math. 17, 449–467 (1965)
7. Flum, J., Grohe, M.: Parameterized Complexity Theory. In: Texts in Theoretical Computer Science. Springer, Berlin (2006)
8. Khot, S., Raman, V.: Parameterized complexity of finding subgraphs with hereditary properties. Theor. Comput. Sci. 289(2), 997–1008 (2002)
9. Lovász, L., Plummer, M.D.: Matching Theory. North-Holland, Amsterdam (1986)
10. Mahajan, M., Raman, V.: Parameterizing above guaranteed values: Maxsat and maxcut. J. Algorithms 31(2), 335–354 (1999)
11. Marx, D.: Parameterized graph separation problems. Theoret. Comput. Sci. 351(3), 394–406 (2006)
12. Marx, D., Razgon, I.: Constant ratio fixed-parameter approximation of the edge multicut problem. Inf. Process. Lett. 109(20), 1161–1166 (2009)
13. Marx, D., Razgon, I.: Fixed-parameter tractability of multicut parameterized by the size of the cutset. In: STOC, pp. 469–478 (2011)
14. Mishra, S., Raman, V., Saurabh, S., Sikdar, S., Subramanian, C.R.: The complexity of könig subgraph problems and above-guarantee vertex cover. Algorithmica 58 (2010)
15. Razgon, I., O'Sullivan, B.: Almost 2-sat is fixed-parameter tractable. J. Comput. Syst. Sci. 75(8), 435–450 (2009)
16. Reed, B.A., Smith, K., Vetta, A.: Finding odd cycle transversals. Oper. Res. Lett. 32(4), 299–301 (2004)

Hitting and Harvesting Pumpkins

Gwenaël Joret[1], Christophe Paul[2], Ignasi Sau[2],
Saket Saurabh[3], and Stéphan Thomassé[4]

[1] Département d'Informatique, Université Libre de Bruxelles, Brussels, Belgium
gjoret@ulb.ac.be
[2] AlGCo project-team, CNRS, LIRMM, Montpellier, France
{paul,sau}@lirmm.fr
[3] The Institute of Mathematical Sciences, Chennai, India
saket@imsc.res.in
[4] Laboratoire LIP (U. Lyon, CNRS, ENS Lyon, INRIA, UCBL), Lyon, France
stephan.thomasse@ens-lyon.fr

Abstract. The *c-pumpkin* is the graph with two vertices linked by $c \geq 1$ parallel edges. A *c-pumpkin-model* in a graph G is a pair $\{A, B\}$ of disjoint subsets of vertices of G, each inducing a connected subgraph of G, such that there are at least c edges in G between A and B. We focus on hitting and packing c-pumpkin-models in a given graph: On the one hand, we provide an FPT algorithm running in time $2^{\mathcal{O}(k)} n^{\mathcal{O}(1)}$ deciding, for any fixed $c \geq 1$, whether all c-pumpkin-models can be hit by at most k vertices. This generalizes the *single-exponential* FPT algorithms for VERTEX COVER and FEEDBACK VERTEX SET, which correspond to the cases $c = 1, 2$ respectively. For this, we use a combination of iterative compression and a kernelization-like technique. On the other hand, we present an $\mathcal{O}(\log n)$-approximation algorithm for both the problems of hitting all c-pumpkin-models with a smallest number of vertices, and packing a maximum number of vertex-disjoint c-pumpkin-models. Our main ingredient here is a combinatorial lemma saying that any *properly reduced* n-vertex graph has a c-pumpkin-model of size at most $f(c) \log n$, for a function f depending only on c.

Keywords: Hitting and packing, parameterized complexity, approximation algorithm, single-exponential algorithm, iterative compression, graph minors.

1 Introduction

The *c-pumpkin* is the graph with two vertices linked by $c \geq 1$ parallel edges. A *c-pumpkin-model* in a graph G is a pair $\{A, B\}$ of disjoint subsets of vertices of G, each inducing a connected subgraph of G, such that there are at least c edges in G between A and B. In this article we study the problems of hitting all c-pumpkin-models of a given graph G with few vertices, and packing as many disjoint c-pumpkin-models in G as possible. As discussed below, these problems generalize several well-studied problems in algorithmic graph theory. We focus

C. Demetrescu and M.M. Halldórsson (Eds.): ESA 2011, LNCS 6942, pp. 394–407, 2011.

on FPT algorithms for the parameterized version of the hitting problem, as well as on poly-time approximation algorithms for the optimization version of both the packing and hitting problems.

FPT algorithms. From the parameterized complexity perspective, we study the following problem for every fixed integer $c \geq 1$.

---

p-c-PUMPKIN-HITTING (p-c-HIT for short)

  Instance: A graph G and a non-negative integer k.
  Parameter: k
  Question: Does there exist $S \subseteq V(G)$, $|S| \leq k$, such that
        $G \setminus S$ does not contain the c-pumpkin as a minor?

When $c = 1$, the p-c-HIT problem is the p-VERTEX COVER problem [2,8]. For $c = 2$, it is the p-FEEDBACK VERTEX SET problem [16,9]. When $c = 3$, this corresponds to the recently introduced p-DIAMOND HITTING SET problem [12].

The p-c-HIT problem can also be seen as a particular case of the following problem, recently introduced by Fomin *et al.* [14] and studied from the kernelization perspective: Let $\mathcal{F}$ be a finite set of graphs. In the p-$\mathcal{F}$-HIT problem, we are given an n-vertex graph G and an integer k as input, and asked whether at most k vertices can be deleted from G such that the resulting graph does not contain any graph from $\mathcal{F}$ as a minor. Among other results, it is proved in [14] that if $\mathcal{F}$ contains a c-pumpkin for some $c \geq 1$, then p-$\mathcal{F}$-HIT admits a kernel of size $\mathcal{O}(k^2 \log^{3/2} k)$. As discussed in Section 3, this kernel leads to a simple FPT algorithm for p-$\mathcal{F}$-HIT in this case, and in particular for p-c-HIT, with running time $2^{\mathcal{O}(k \log k)} \cdot n^{\mathcal{O}(1)}$. A natural question is whether there exists an algorithm for p-c-HIT with running time $2^{\mathcal{O}(k)} \cdot n^{\mathcal{O}(1)}$ for every fixed $c \geq 1$. Such algorithms are called *single-exponential*. For the p-VERTEX COVER problem the existence of single-exponential algorithms is well-known since almost the beginnings of the field of Parameterized Complexity [2], the best current algorithm being by Chen *et al.* [8]. On the other hand, the question about the existence of single-exponential algorithms for p-FEEDBACK VERTEX SET was open for a while and was finally settled independently by Guo *et al.* [16] (using iterative compression) and by Dehne *et al.* [9].

We present in Section 3 a single-exponential algorithm for p-c-HIT for every fixed $c \geq 1$, using a combination of a kernelization-like technique and iterative compression. Notice that this generalizes the above results for p-VERTEX COVER and p-FEEDBACK VERTEX SET. We remark that asymptotically these algorithms are optimal, that is, it is known that unless ETH fails neither p-VERTEX COVER nor p-FEEDBACK VERTEX SET admit an algorithm with running time $2^{o(k)} \cdot n^{\mathcal{O}(1)}$ [7,17]. It is worth mentioning here that a similar quantitative approach was taken by Lampis [20] for graph problems expressible in MSOL parameterized by the sum of the formula size and the size of a minimum vertex cover of the input graph.

Approximation algorithms. For a fixed integer $c \geq 1$, we define the following two optimization problems.

MINIMUM c-PUMPKIN-HITTING (MIN c-HIT for short)
> *Input:* A graph G.
> *Output:* A subset $S \subseteq V(G)$ such that $G \setminus S$
>   does not contain the c-pumpkin as a minor.
> *Objective:* Minimize $|S|$.

MAXIMUM c-PUMPKIN-PACKING (MAX c-PACK for short)
> *Input:* A graph G.
> *Output:* A collection $\mathcal{M}$ of vertex-disjoint subgraphs of G,
>   each containing the c-pumpkin as a minor.
> *Objective:* Maximize $|\mathcal{M}|$.

Let us now discuss how the above problems encompass several well-known problems. For $c = 1$, MIN 1-HIT is the MINIMUM VERTEX COVER problem, which can be easily 2-approximated by finding any maximal matching, whereas MAX 1-PACK corresponds to finding a MAXIMUM MATCHING, which can be done in polynomial time. For $c = 2$, MIN 2-HIT is the MINIMUM FEEDBACK VERTEX SET problem, which can be also 2-approximated [1, 3], whereas MAX 2-PACK corresponds to MAXIMUM VERTEX-DISJOINT CYCLE PACKING, which can be approximated to within an $\mathcal{O}(\log n)$ factor [19]. For $c = 3$, MIN 3-HIT is the DIAMOND HITTING SET problem studied by Fiorini *et al.* in [12], where a 9-approximation algorithm is given.

We provide in Section 4 an algorithm that approximates both the MIN c-HIT and the MAX c-PACK problems to within a factor $\mathcal{O}(\log n)$ for every fixed $c \geq 1$. Note that this algorithm matches the best existing algorithms for MAX c-PACK for $c = 2$ [19]. For the MIN c-HIT problem, our result is only a slight improvement on the $\mathcal{O}(\log^{3/2} n)$-approximation algorithm given in [14]. However, for the MAX c-PACK problem, there was no approximation algorithm known before except for the case $c = 2$. Also, let us remark that, for $c \geq 2$ and every fixed $\varepsilon > 0$, MAX c-PACK is quasi-NP-hard to approximate to within an $\mathcal{O}(\log^{1/2-\varepsilon} n)$ factor: For $c = 2$ this was shown by Friggstad and Salavatipour [15], and their result can be extended to the case $c > 2$ in the following straightforward way: given an instance G of MAX 2-HIT, we build an instance of MAX c-HIT by replacing each edge of G with $c - 1$ parallel edges.

The main ingredient of our approximation algorithm is the following combinatorial result: We show that every n-vertex graph G either contains a small c-pumpkin-model or has a structure that can be reduced in polynomial time, giving a smaller equivalent instance for both the MIN c-HIT and the MAX c-PACK problems. Here by a "small" c-pumpkin-model, we mean a model of size at most $f(c) \cdot \log n$ for some function f independent of n. This result extends one of Fiorini *et al.* [12], who dealt with the $c = 3$ case.

2   Preliminaries

Graphs. We use standard graph terminology, see for instance [10]. All graphs in this article are finite and undirected, and may have parallel edges but no loops. We will sometimes restrict our attention to simple graphs, that is, graphs without parallel edges.

Given a graph G, we denote the vertex set of G by $V(G)$ and the edge set of G by $E(G)$. We use the shorthand $|G|$ for the number of vertices in G. For a subset $X \subseteq V(G)$, we use $G[X]$ to denote the subgraph of G induced by X. For a subset $Y \subseteq E(G)$ we let $G[Y]$ be the graph with $E(G[Y]) := Y$ and with $V(G[Y])$ being the set of vertices of G incident with some edge in Y. For a subset $X \subseteq V(G)$, we may use the notation $G \setminus X$ to denote the graph $G[V(G) \setminus X]$.

The set of neighbors of a vertex v of a graph G is denoted by $N_G(v)$. The *degree* $\deg_G(v)$ of a vertex $v \in V(G)$ is defined as the number of edges incident with v (thus parallel edges are counted). We write $\deg_G^*(v)$ for the number of neighbors of v, that is, $\deg_G^*(v) := |N_G(v)|$. Similarly, given a subgraph $H \subseteq G$ with $v \in V(H)$, we can define in the natural way $N_H(v)$, $\deg_H(v)$, and $\deg_H^*(v)$, that is, $N_H(v) = N_G(v) \cap V(H)$, $\deg_H(v)$ is the number of edges incident with v with both endpoints in H, and $\deg_H^*(v) = |N_H(v)|$. In these notations, we may drop the subscript if the graph is clear from the context. The minimum degree of a vertex in a graph G is denoted $\delta(G)$, and the maximum degree of a vertex in G is denoted $\Delta(G)$. We use the notation $\mathbf{cc}(G)$ to denote the number of connected components of G. Also, we let $\mu(G)$ denote the maximum multiplicity of an edge in G.

Minors and models. Given a graph G and an edge $e \in E(G)$, let $G \setminus e$ be the graph obtained from G by removing the edge e, and let G/e be the graph obtained from G by contracting e (we note that parallel edges resulting from the contraction are kept but loops are deleted). If H can be obtained from a subgraph of G by a (possibly empty) sequence of edge contractions, we say that H is a *minor* of G, and we denote it by $H \preceq_m G$. A graph G is H-*minor-free*, or simply H-*free*, if G does not contain H as a minor. A *model* of a graph H, or simply H-*model*, in a graph G is a collection $\{S_v \subseteq V(G) \mid v \in V(H)\}$ such that

(i) $G[S_v]$ is connected for every $v \in V(H)$;
(ii) S_v and S_w are disjoint for every two distinct vertices v, w of H, and
(ii) there are at least as many edges between S_v and S_w in G as between v and w in H, for every $vw \in E(H)$.

The *size* of the model is defined as $\sum_{v \in V(H)} |S_v|$. Clearly, H is a minor of G if and only if there exists a model of H in G. In this paper, we will almost exclusively consider H-models with H being isomorphic to a c-pumpkin for some $c \geq 1$. Thus a c-pumpkin-model in a graph G is specified by an unordered pair $\{A, B\}$ of disjoint subsets of vertices of G, each inducing a connected subgraph of G, such that there are at least c edges in G between A and B. A c-pumpkin-model $\{A, B\}$ of G is said to be *minimal* if there is no c-pumpkin-model $\{A', B'\}$ of G with $A' \subseteq A$, $B' \subseteq B$, and $|A'| + |B'| < |A| + |B|$.

A subset X of vertices of a graph G such that $G \setminus X$ has no c-pumpkin-minor is called a c-*pumpkin-hitting set*, or simply c-*hitting set*. We denote by $\tau_c(G)$ the minimum size of a c-pumpkin-hitting set in G. A collection $\mathcal{M}$ of vertex-disjoint subgraphs of a graph G, each containing a c-pumpkin-model, is called a *c-pumpkin-packing*, or simply *c-packing*. We denote by $\nu_c(G)$ the maximum size of a *c*-pumpkin-packing in G. Obviously, for any graph G it holds that $\nu_c(G) \leq \tau_c(G)$, but the converse is not necessarily true.

The following lemma on models will be useful in our algorithms. The proof is straightforward and hence is omitted.

Lemma 1. *Suppose G' is obtained from a graph G by contracting some vertex-disjoint subgraphs of G, each of diameter at most k. Then, given an H-model in G' of size s, one can compute in polynomial time an H-model in G of size at most $k \cdot \Delta(H) \cdot s$.*

Tree-width. We refer the reader to Diestel's book [10] for the definition of tree-width. It is an easy exercise to check that the tree-width of a simple graph is an upper bound on its minimum degree. This implies the following lemma.

Lemma 2. *Every n-vertex simple graph with tree-width k has at most $k \cdot n$ edges.*

We will need the following result of Bodlaender *et al.*

Theorem 1 (Bodlaender *et al.* [6]). *Every graph not containing a c-pumpkin as a minor has tree-width at most $2c - 1$.*

The following corollary is an immediate consequence of the above theorem.

Corollary 1. *Every n-vertex graph with no minor isomorphic to a c-pumpkin has at most $(c - 1) \cdot (2c - 1) \cdot n$ edges.*

3   A Single-Exponential FPT Algorithm

As mentioned in the introduction, it is proved in [14] that given an instance (G, k) of p-$\mathcal{F}$-HIT such that $\mathcal{F}$ contains a *c*-pumpkin for some $c \geq 1$, one can obtain in polynomial time an equivalent instance with $\mathcal{O}(k^2 \log^{3/2} k)$ vertices. (We assume that the reader is familiar with the basic concepts of Parameterized Complexity; c.f. for instance [11].) This kernel leads to the following simple FPT algorithm for p-$\mathcal{F}$-HIT: First compute the kernel K in polynomial time, and then for every subset $S \subseteq V(K)$ of size k, test whether $K[V(K) \setminus S]$ contains any of the graphs in $\mathcal{F}$ as a minor, using the poly-time algorithm of Robertson and Seymour [22] for every graph in $\mathcal{F}$. If for some S we have that $K[V(K) \setminus S]$ contains no graph from $\mathcal{F}$ as a minor, we answer YES; otherwise the answer is NO. The running time of this algorithm is clearly bounded by $\binom{k^2 \log^{3/2} k}{k} \cdot n^{\mathcal{O}(1)} = 2^{\mathcal{O}(k \log k)} \cdot n^{\mathcal{O}(1)}$.

In this section we give an algorithm for p-c-PUMPKIN-HITTING that runs in time $d^k \cdot n^{\mathcal{O}(1)}$ for any fixed $c \geq 1$, where d only depends on the fixed constant c. Towards this, we first introduce a variant of p-c-PUMPKIN-HITTING, namely p-c-PUMPKIN-DISJOINT HITTING, or p-c-DISJOINT HIT for short. In p-c-DISJOINT HIT, apart from a graph G and a positive integer k, we are also given a set S of size at most $k + 1$ such that $G \setminus S$ does not contain the *c*-pumpkin as a minor. Here the objective is to find a set $S' \subseteq V(G) \setminus S$ such that $|S'| \leq k$ and $G \setminus S'$ does not contain the *c*-pumpkin as a minor. Next we show a lemma that allows us to relate the two problems mentioned above.

Lemma 3 ($\star^1$). *If p-c-DISJOINT HIT can be solved in time $\eta(c)^k \cdot n^{\mathcal{O}(1)}$, then p-c-HIT can be solved in time $(\eta(c) + 1)^k \cdot n^{\mathcal{O}(1)}$.*

Lemma 3 allows us to focus on p-c-DISJOINT HIT. In what follows we give an algorithm for p-c-DISJOINT HIT that runs in single-exponential time.

Overview of the algorithm. The algorithm for p-c-DISJOINT HIT is based on a combination of branching and poly-time preprocessing. Let (G, S, k) be the given instance of p-c-DISJOINT HIT and let $V := V(G)$. Our main objective is to eventually show that $A := \{v \in V \setminus S : N_G(v) \cap S \neq \emptyset\}$ has cardinality $\mathcal{O}(k)$. As far as we cannot guarantee this upper bound, we will see that we can branch suitably to obtain a few subproblems. Once we have the desired upper bound, we use a protrusion-based reduction rule from [14] to give a poly-time procedure that given an instance (G, S, k) of p-c-DISJOINT HIT, returns an equivalent instance (G', S, k') such that G' has $\mathcal{O}(k)$ vertices. That is, we obtain a linear vertex kernel for p-c-DISJOINT HIT in this particular case. Notice that once we have a linear vertex kernel of size αk for p-c-DISJOINT HIT, we can solve the problem in $\binom{\alpha k}{k} \cdot k^{\mathcal{O}(1)}$. So the overall algorithm consists of a recursion tree where in leaf nodes we obtain a linear kernel and then solve the problem using brute force enumeration.

We can now proceed to the formal description of the algorithm. Let

$$V_1 := \{v \in V \setminus S : |N_G(v) \cap S| = 1\}$$
$$V_{\geq 2} := \{v \in V \setminus S : |N_G(v) \cap S| \geq 2\}.$$

Linear kernel. We start off with a procedure, called *protrusion rule*, that bounds the size of our graph when $|V_1 \cup V_{\geq 2}| = \mathcal{O}(k)$. Given $R \subseteq V(G)$, we define $\partial_G(R)$ as the set of vertices in R that have a neighbor in $V(G) \setminus R$. Thus the neighborhood of R is $N_G(R) = \partial_G(V(G) \setminus R)$. We say that a set $X \subseteq V(G)$ is an r-*protrusion* of G if $\mathbf{tw}(G[X]) \leq r$ and $|\partial_G(X)| \leq r$. We now state the protrusion rule.

P Let (G, S, k) be an instance and let $\gamma : \mathbb{N} \to \mathbb{N}$ be the function defined in [14, Lemma 5] (originally shown in [4]). If G contains a $4c$-protrusion Y of size at least $\gamma(4c)$, then replace Y to obtain an equivalent instance (G^*, S, k^*) such that $|V(G^*)| < |V(G)|$.

The proof of the next lemma is identical to the proof of [14, Theorem 1].

Lemma 4 ($\star$). *If $|V_1 \cup V_{\geq 2}| = \mathcal{O}(k)$ then p-c-DISJOINT HIT has a kernel with $\mathcal{O}(k)$ vertices.*

Branching procedure. For our branching algorithm we take the following measure:

$$\mu = \mathbf{cc}(G[S]) + k. \tag{1}$$

---

[1] The proofs of the results marked with "$\star$" can be found in [18].

For simplicity we call the vertices in V_1 *white*. We start with some simple reduction rules (depending on c) which will be applied in the compression routine. We also prove, together with the description of each rule, that they are valid for our problem.

R1. Let C be a connected component of $G[V \setminus S]$. If C does not have any neighbor in S, we can safely delete C, as its vertices will never participate in a c-pumpkin-model.

R2. Let C be a connected component of $G[V \setminus (S \cup \{v\})]$ for some vertex $v \in V \setminus S$. If C does not have any neighbor in S, we can safely delete C, as its vertices will never participate in a minimal c-pumpkin-model.

R3. Let C be a connected component of $G[V \setminus S]$. If C has exactly one neighbor v in S and $G[V(C) \cup \{v\}]$ is c-pumpkin-free, we can safely delete it, as its vertices will never participate in a minimal c-pumpkin-model.

R4. Let B be a connected induced subgraph (not necessarily a connected component) of $G[V \setminus S]$, let $v \in V(B)$, and let $P_1, \ldots, P_\ell$ be the connected components of $G[V(B) \setminus \{v\}]$. Assume that $\ell \geq c$ and that the following conditions hold:

  (i) For $1 \leq i \leq \ell$, all neighbors of vertices in P_i are in $S \cup V(P_i) \cup \{v\}$.
  (ii) For $1 \leq i \leq \ell$, there exists a vertex $u_i \in S$ such that $\bigcup_{w \in V(P_i)} N_G(w) \cap S = \{u_i\}$.
  (iii) For $1 \leq i \leq \ell$, $G[V(P_i) \cup \{u_i\}]$ is c-pumpkin-free.
  (iv) The vertices $u_1, \ldots, u_\ell$ belong to the same connected component D of $G[S]$.

  Then we include vertex v in the solution and we decrease the parameter by one. Indeed, note that as $\ell \geq c$ and by conditions (ii) and (iv), $G[V(B) \cup V(D)]$ contains a c-pumpkin, so any solution needs to contain at least one vertex of B, since we are looking for a solution that does not include vertices from S. On the other hand, by condition (iii) every minimal c-pumpkin-model intersecting B necessarily contains vertex v, and condition (i) guarantees that no minimal c-pumpkin-model of $G \setminus \{v\}$ contains a vertex of B. Therefore, we can safely include vertex v in the solution and decrease the parameter accordingly. See Fig. 1 for an illustration for $c = 4$.

We say that the instance (G, S, k) is (S, c)-*reduced* if rules **R1**, **R2**, **R3**, or **R4** cannot be applied anymore. Note that these reduction rules can easily be applied in polynomial time.

We now describe a branching rule which will be used in our compression routine.

B Let P be a simple path in $G[V \setminus S]$ with $|V(P)| = \ell$ and let v_1 and v_2 be the endpoints of P. Suppose that v_1 (resp. v_2) has a neighbor in a connected component C_1 (resp. C_2) of $G[S]$, with $C_1 \neq C_2$. Then we branch for all 2^ℓ subsets of vertices in P. Note that in every case we decrease the measure of the iterative compression, as either we include at least one vertex in the solution or we decrease the number of connected components of $G[S]$.

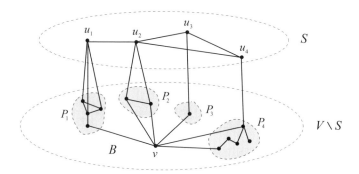

Fig. 1. Illustration of reduction rule **R4** for $c = 4$

The branching rule above will be used on paths P that are "small enough". The following Lemmas 5 and 8 are key to our algorithm. We also need two intermediate technical results, which will be used in the proof of Lemma 8.

Lemma 5 ($\star$). *There is a function $f(c)$ such that if (G, S, k) is an (S, c)-reduced* YES-*instance to the p-c-*DISJOINT HIT *problem, then $|V_{\geq 2}| \leq f(c) \cdot k$.*

Lemma 6 ($\star$). *There is a function $g(c)$ such that if (G, S, k) is an (S, c)-reduced* YES-*instance to the p-c-*DISJOINT HIT *problem and $\mathcal{C}$ is a collection of disjoint connected subgraphs of $G[V \setminus S]$ such that each subgraph has at least two distinct neighbors in S, then $|\mathcal{C}| \leq g(c) \cdot k$.*

Lemma 7 ($\star$). *In an (S, c)-reduced* YES-*instance (G, S, k), the number of connected components of $G[V \setminus S]$ is $\mathcal{O}(k)$.*

Now we prove our key structural lemma.

Lemma 8. *There is a function $h(c)$ such that if (G, S, k) is an (S, c)-reduced* YES-*instance to the p-c-*DISJOINT HIT *problem, then either $|V_1| \leq h(c) \cdot k$, or we can apply protrusion rule $\mathbf{P}$, or we can branch according to branching rule $\mathbf{B}$.*

Proof: We proceed to find a packing of disjoint connected subgraphs $\mathcal{P} = \{B_1, \ldots, B_\ell\}$ of $G[V \setminus S]$ containing all white vertices except for $\mathcal{O}(k)$ of them. This will help us in bounding $|V_1|$. We call the subgraphs in $\mathcal{P}$ *blocks*. For a graph $H \subseteq G[V \setminus S]$, let $w(H)$ be the number of white vertices in H. The idea is to obtain, as far as possible, blocks B with $c \leq w(B) \leq c^3$; we call these blocks *suitable*, and the other blocks are called *unsuitable*. If at some point we cannot refine the packing anymore in order to obtain suitable blocks, we will argue about its structural properties, which will allow us to either bound the number of white vertices, or to apply protrusion rule $\mathbf{P}$, or to branch according to branching rule $\mathbf{B}$.

We start with $\mathcal{P}$ containing all the connected components C of $G[V \setminus S]$ such that $w(C) > c^3$, and we recursively try to refine the current packing. By Lemma 7, we know that the number of connected components is $\mathcal{O}(k)$, and hence the number of white vertices, that are not included in $\mathcal{P}$ is $\mathcal{O}(c^3 k) = \mathcal{O}(k)$.

More precisely, for each block B with $w(B) > c^3$, we build a spanning tree T of B, and we orient each edge $e \in E(T)$ towards the components of $T \setminus \{e\}$ containing at least c white vertices. (Note that, as $w(B) > c^3$, each edge gets at least one orientation, and that edges may be oriented in both directions.) If some edge $e \in E(T)$ is oriented in both directions, we replace in $\mathcal{P}$ block B with the subgraphs induced by the vertices in each of the two subtrees. We stop this recursive procedure whenever we cannot find more suitable blocks using this orientation trick. Let $\mathcal{P}$ be the current packing.

Now let B be an unsuitable block in $\mathcal{P}$, that is, $w(B) > c^3$ and no edge of its spanning tree T is oriented in both directions. This implies that there exists a vertex $v \in V(T)$ with all its incident edges pointing towards it. We call such a vertex v a *sink*. Let $T_1, \dots, T_p$ be the connected components of $T \setminus \{v\}$. Note that as v is a sink, $w(T_i) < c$ for $1 \le i \le p$, using the fact that $w(B) > c^3$ we conclude that $p \ge c^2$. Now let $P_1, \dots, P_\ell$ be the connected components of $G[V(T_1) \cup \dots \cup V(T_p)] = G[V(B) \setminus \{v\}]$, and note that $\ell \le p$. We call these subgraphs P_i the *pieces* of the unsuitable block B. For each unsuitable block, we delete the pieces with no white vertex. This completes the construction of $\mathcal{P}$. The next claim bounds the number of white vertices in each piece of an unsuitable block in $\mathcal{P}$.

Claim 1. *Each of the pieces of an unsuitable block contains less than c^2 white vertices.*

Proof: Assume for contradiction that there exists a piece P of an unsuitable block with $w(P) \ge c^2$, and let v be the sink of the unsuitable block obtained from tree T. By construction $V(P)$ is the union of the vertices in some of the trees in $T \setminus \{v\}$; let without loss of generality these trees be $T_1, \dots, T_q$. As $w(P) \ge c^2$ and $w(T_i) < c$ for $1 \le i \le q$, it follows that $q \ge c$. As v has at least one neighbor in each of the trees T_i, $1 \le i \le q$, and P is a connected subgraph of G, we can obtain a c-pumpkin-model $\{A, B\}$ in $G[V \setminus S]$ by setting $A := \{v\}$ and $B := V(P)$, contradicting the fact that $G[V \setminus S]$ is c-pumpkin-free. $\square$

Hence in the packing $\mathcal{P}$ we are left with a set of suitable blocks with at most c^3 white vertices each, and a set of unsuitable blocks, each one broken up into pieces linked by a sink in a star-like structure. By Claim 1, each piece of the remaining unsuitable blocks contains at most c^2 white vertices.

Now we need two claims about the properties of the constructed packing.

Claim 2 ($\star$). *In the packing $\mathcal{P}$ constructed above, the number of suitable or unsuitable blocks is $\mathcal{O}(k)$.*

Claim 3 ($\star$). *In an (S, c)-reduced instance, either the total number of pieces in the packing $\mathcal{P}$ constructed above is $\mathcal{O}(k)$ or we can branch according to branching rule **B**.*

More precisely, the proof of Claim 3 shows that if the number of pieces is not $\mathcal{O}(k)$, then we can apply protrusion rule **P** (and possibly reduction rules **R2** and **R3**) to eventually branch according to branching rule **B**.

To conclude, recall that the constructed packing $\mathcal{P}$ contains all but $\mathcal{O}(k)$ white vertices, either in suitable blocks or in pieces of unsuitable blocks. As by construction each suitable block has at most c^3 white vertices and by Claim 2 the number of such blocks is $\mathcal{O}(k)$, it follows that the number of white vertices contained in suitable blocks is $\mathcal{O}(k)$. Similarly, by Claim 1 each piece contains at most c^2 white vertices, and the total number of pieces is $\mathcal{O}(k)$ by Claim 3 unless we can branch according to branching rule **B**, so if we cannot branch the number of white vertices contained in pieces of unsuitable blocks is also $\mathcal{O}(k)$.□

Final algorithm. Finally we combine everything to obtain the following result.

Theorem 2. *For any fixed $c \geq 1$, the p-c-PUMPKIN-HITTING problem can be solved in time $2^{\mathcal{O}(k)} \cdot n^{\mathcal{O}(1)}$.*

Proof: To obtain the desired result, by Lemma 3 it is sufficient to obtain an algorithm for p-c-DISJOINT HIT. Recall the measure $\mu = \mathbf{cc}(G[S]) + k$. Now by Lemma 8 either we branch according to branching rule **B**, or we can apply protrusion rule **P** and get a smaller instance, or we have that $|V_1| \leq h(c) \cdot k$. In the first case we branch into $\alpha(c)$ ways and in each branch the measure decreases by one as either we include a vertex in our potential solution or we decrease the number of connected components of $G[S]$. Here $\alpha(c)$ is a constant that only depends on c. This implies that the number of nodes in the branching tree is upper-bounded by $\alpha(c)^\mu \leq \alpha(c)^{2k+1} = 2^{\mathcal{O}(k)}$. In the second case, we get a smaller instance in polynomial time. Finally, in the third case, using Lemma 5 we get that $|V_1 \cup V_{\geq 2}| = \mathcal{O}(k)$. Thus using Lemma 4 we can get an equivalent instance (G^*, S, k^*) with $\mathcal{O}(k)$ vertices and hence the problem can be solved by enumerating all subsets of size at most k^* of $G^* \setminus S$. This concludes the proof. □

4 An Approximation Algorithm for Hitting and Packing Pumpkins

In this section we show that every n-vertex graph G either contains a small c-pumpkin-model or has a structure that can be reduced, giving a smaller equivalent instance for both the MINIMUM c-PUMPKIN-HITTING and the MAXIMUM c-PUMPKIN-PACKING problems. Here by a "small" c-pumpkin-model, we mean a model of size at most $f(c) \cdot \log n$ for some function f independent of n. We finally use this result to derive an $\mathcal{O}(\log n)$-approximation algorithm for both problems.

Reduction rules. We describe two reduction rules for hitting/packing c-pumpkin-models, which given an input graph G, produce a graph H with less vertices than G and satisfying $\tau_c(G) = \tau_c(H)$ and $\nu_c(G) = \nu_c(H)$. Moreover, these operations are defined in such a way that, for both problems, an optimal (resp. approximate) solution for G can be retrieved in polynomial time from an optimal (resp. approximate) solution for H.

An *outgrowth* of a graph G is a triple (C, u, v) such that (i) u, v are two distinct vertices of G; (ii) C is a connected component of $G \setminus \{u, v\}$ with $|V(C)| \geq 2$; and (iii) u and v both have at least one neighbor in C in the graph G.

Given an outgrowth (C, u, v) of a graph G, we let $\Gamma(C, u, v)$ denote the subgraph of G induced by $V(C) \cup \{u, v\}$ where all the edges between u and v are removed. We let $\Lambda(C, u, v)$ be the graph $\Gamma(C, u, v)$ where an edge between u and v has been added. Also, we define $\gamma(C, u, v)$ as the largest integer k such that $\Gamma(C, u, v)$ has a k-pumpkin-model $\{A, B\}$ with $u \in A$ and $v \in B$, and $\lambda(C, u, v)$ as the largest integer k such that $\Lambda(C, u, v)$ has a k-pumpkin-model $\{A, B\}$ with $u, v \in A$.[2] See Fig. 2 for an illustration.

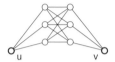

Fig. 2. The graph $\Gamma(C, u, v)$ of two outgrowths (C, u, v). We have $\lambda(C, u, v) = 8 < 9 = \gamma(C, u, v)$ in the left one, while $\lambda(C, u, v) = 7 > 5 = \gamma(C, u, v)$ holds for the right one.

Now that we are equipped with these definitions and notations, we may describe the two reduction rules, which depend on the value of c.

Z1 Suppose v is a vertex of G such that no block (here, a *block* is a maximal 2-connected component) of G containing v has a c-pumpkin-minor. Then define H as the graph obtained from G by removing v.

Z2 Suppose (C, u, v) is an outgrowth of G such that $\Gamma(C, u, v)$ has no c-pumpkin-minor. Assume further that **Z1** cannot be applied on G. Let $\gamma := \gamma(C, u, v)$ and $\lambda := \lambda(C, u, v)$. If $\lambda \leq \gamma$, define H as the graph obtained from $G \setminus V(C)$ by adding γ parallel edges between u and v. Otherwise, define H as the graph obtained from $G \setminus V(C)$ by adding a new vertex v_C and linking v_C to u with γ parallel edges, and to v with $\lambda - \gamma$ parallel edges.

See Fig. 3 for an illustration of **Z2**. A graph G is said to be *c-reduced* if neither **Z1** nor **Z2** can be applied to G. The next lemma shows the validity of these reduction rules.

Lemma 9 ($\star$). *Let c be a fixed positive integer. Suppose that H results from the application of **Z1** or **Z2** on a graph G. Then*

(a) $\tau_c(G) = \tau_c(H)$ and moreover, given a c-hitting set X' of H, one can compute in polynomial time a c-hitting set X of G with $|X| \leq |X'|$.

(b) $\nu_c(G) = \nu_c(H)$ and moreover, given a c-packing $\mathcal{M}'$ of H, one can compute in polynomial time a c-packing $\mathcal{M}$ of G with $|\mathcal{M}| = |\mathcal{M}'|$.

Small pumpkins in reduced graphs. Our goal is to prove that every n-vertex c-reduced graph G has a c-pumpkin-model of size $\mathcal{O}_c(\log n)$. We will use the following recent result by Fiorini *et al.* [13] about the existence of small minors in *simple* graphs with large average degree.

---

[2] We note that $\lambda(C, u, v)$ can equivalently be defined as the largest integer k such that $\Lambda(C, u, v)/uv$ has a k-pumpkin-model $\{A, B\}$ with $u \in A$. Thus $\lambda(C, u, v)$ can be computed in polynomial time.

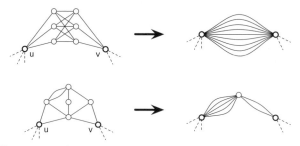

Fig. 3. Illustration of reduction rule **Z2** on the two outgrowths from Fig. 2

Theorem 3 (Fiorini *et al.* **[13]).** *There is a function h such that every n-vertex simple graph G with average degree at least 2^t contains a K_t-model with at most $h(t) \cdot \log n$ vertices.*

Even though the authors of [13] do not mention it explicitly, we note that the proof given in that paper can easily be turned into a poly-time algorithm finding such a K_t-model. Since a K_t-model in a graph directly gives a c-pumpkin-model of the same size for $c = (\lfloor t/2 \rfloor)^2$, we have the following corollary from Theorem 3, which is central in the proof of Lemma 10.

Corollary 2. *There is a function h such that every n-vertex simple graph G with average degree at least $2^{2\sqrt{c}+1}$ contains a c-pumpkin-model with at most $h(c) \cdot \log n$ vertices. Moreover, such a model can be computed in polynomial time.*

The next lemma states the existence of small c-pumpkin-models in c-reduced graphs; its proof strongly relies on a graph structure that we call *hedgehog* (see [18] for more details). The idea is that a hedgehog witnesses the existence of either a small c-pumpkin-model or an outgrowth, which is impossible in a c-reduced graph.

Lemma 10 ($\star$). *There is a function f such that every n-vertex c-reduced graph G contains a c-pumpkin-model of size at most $f(c) \cdot \log n$. Moreover, such a model can be computed in polynomial time.*

Algorithmic consequences. Lemma 10 can be used to obtain $\mathcal{O}(\log n)$-approximation algorithms for both the MINIMUM c-PUMPKIN-HITTING and the MAXIMUM c-PUMPKIN-PACKING problems for every fixed $c \geq 1$, as we now show.

Theorem 4 ($\star$). *Given an n-vertex graph G, an $\mathcal{O}(\log n)$-approximation for both the MINIMUM c-PUMPKIN-HITTING and the MAXIMUM c-PUMPKIN-PACKING problems on G can be computed in polynomial time using Algorithm 1, for any fixed integer $c \geq 1$.*

5   Concluding Remarks

On the one hand, we provided an FPT algorithm running in time $2^{\mathcal{O}(k)} \cdot n^{\mathcal{O}(1)}$ deciding, for any fixed $c \geq 1$, whether all c-pumpkin-models of a given graph

---

Algorithm 1. An $\mathcal{O}(\log n)$-approximation algorithm.

INPUT: An arbitrary graph G

OUTPUT: A c-packing $\mathcal{M}$ of G and a c-hitting set X of G s.t. $|X| \leq (f(c) \log |G|) \cdot |\mathcal{M}|$

$\mathcal{M} \leftarrow \emptyset;\ X \leftarrow \emptyset$

If $|G| \leq 1$: Return $\mathcal{M},\ X$ /* G cannot have a c-pumpkin-minor */

Else:

    If G is not c-reduced:

        Apply a reduction rule on G, giving a graph H

        Call the algorithm on H, giving a packing $\mathcal{M}'$ and a c-hitting set X' of H

        Compute using Lemma 9(b) a c-packing $\mathcal{M}$ of G with $|\mathcal{M}| = |\mathcal{M}'|$

        Compute using Lemma 9(a) a c-hitting set X of G with $|X| \leq |X'|$

        Return $\mathcal{M},\ X$

    Else:

        Compute using Lemma 10 a c-pumpkin-model $M = \{A, B\}$ of G with $|A \cup B| \leq f(c) \log |G|$

        $H \leftarrow G \setminus (A \cup B)$

        Call the algorithm on H, giving a packing $\mathcal{M}'$ and a c-hitting set X' of H

        $\mathcal{M} \leftarrow \mathcal{M}' \cup \{M\}$

        $X \leftarrow X' \cup A \cup B$

        Return $\mathcal{M},\ X$

---

can be hit by at most k vertices. In our algorithms we used protrusions, but it may be possible to avoid it by further exploiting the structure of the graphs during the iterative compression routine (for example, a graph excluding the 3-pumpkin is a forest of cacti). It is natural to ask whether there exist faster algorithms for sparse graphs. Also, it would be interesting to have lower bounds for the running time of parameterized algorithms for this problem, in the spirit of those recently provided in [21]. A more difficult problem seems to find single-exponential algorithms for the problem of deleting at most k vertices from a given graph so that the resulting graph has tree-width bounded by some constant. One could also consider the parameterized version of packing disjoint c-pumpkin-models, as it has been done for $c = 2$ in [5].

On the other hand, we provided an $\mathcal{O}(\log n)$-approximation for the problems of packing the maximum number of vertex-disjoint c-pumpkin-models, and hitting all c-pumpkin-models with the smallest number of vertices. It may be possible that the hitting version admits a constant-factor approximation; so far, such an algorithm is only known for $c \leq 3$.

References

1. Bafna, V., Berman, P., Fujito, T.: A 2-approximation algorithm for the undirected feedback vertex set problem. SIAM Journal on Discrete Mathematics 12(3), 289–297 (1999)
2. Balasubramanian, R., Fellows, M., Raman, V.: An improved fixed parameter algorithm for Vertex Cover. Information Processing Letters 65, 163–168 (1998)

3. Becker, A., Geiger, D.: Optimization of pearl's method of conditioning and greedy-like approximation algorithms for the vertex feedback set problem. Artificial Intelligence 83, 167–188 (1996)

4. Bodlaender, H.L., Fomin, F.V., Lokshtanov, D., Penninkx, E., Saurabh, S., Thilikos, D.M.: (Meta) kernelization. In: Proc. of the 50th Annual IEEE Symposium on Foundations of Computer Science (FOCS), pp. 629–638 (2009)

5. Bodlaender, H.L., Thomassé, S., Yeo, A.: Kernel bounds for disjoint cycles and disjoint paths. In: Fiat, A., Sanders, P. (eds.) ESA 2009. LNCS, vol. 5757, pp. 635–646. Springer, Heidelberg (2009)

6. Bodlaender, H.L., van Leeuwen, J., Tan, R.B., Thilikos, D.M.: On interval routing schemes and treewidth. Information and Computation 139(1), 92–109 (1997)

7. Chen, J., Chor, B., Fellows, M., Huang, X., Juedes, D.W., Kanj, I.A., Xia, G.: Tight lower bounds for certain parameterized NP-hard problems. Information and Computation 201(2), 216–231 (2005)

8. Chen, J., Kanj, I.A., Xia, G.: Improved upper bounds for vertex cover. Theoretical Computer Science 411(40-42), 3736–3756 (2010)

9. Dehne, F.K.H.A., Fellows, M.R., Langston, M.A., Rosamond, F.A., Stevens, K.: An $O(2^{O(k)}n^3)$ FPT Algorithm for the Undirected Feedback Vertex Set Problem. Theory of Computing Systems 41(3), 479–492 (2007)

10. Diestel, R.: Graph Theory, vol. 173. Springer, Heidelberg (2005)

11. Downey, R.G., Fellows, M.R.: Parameterized complexity. Springer, Heidelberg (1999)

12. Fiorini, S., Joret, G., Pietropaoli, U.: Hitting Diamonds and Growing Cacti. In: Eisenbrand, F., Shepherd, F.B. (eds.) IPCO 2010. LNCS, vol. 6080, pp. 191–204. Springer, Heidelberg (2010)

13. Fiorini, S., Joret, G., Theis, D.O., Wood, D.R.: Small Minors in Dense Graphs (2010) (manuscript), http://arxiv.org/abs/1005.0895

14. Fomin, F.V., Lokshtanov, D., Misra, N., Philip, G., Saurabh, S.: Hitting forbidden minors: Approximation and Kernelization. In: Proc. of the 28th Symposium on Theoretical Aspects of Computer Science (STACS). LIPIcs, vol. 9, pp. 189–200 (2011)

15. Friggstad, Z., Salavatipour, M.: Approximability of packing disjoint cycles. Algorithmica 60, 395–400 (2011)

16. Guo, J., Gramm, J., Hüffner, F., Niedermeier, R., Wernicke, S.: Compression-based fixed-parameter algorithms for feedback vertex set and edge bipartization. Journal of Computer and System Sciences 72(8), 1386–1396 (2006)

17. Impagliazzo, R., Paturi, R., Zane, F.: Which problems have strongly exponential complexity? Journal of Computer and System Sciences 63(4), 512–530 (2001)

18. Joret, G., Paul, C., Sau, I., Saurabh, S., Thomassé, S.: Hitting and Harvesting Pumpkins (2011) (manuscript), http://arxiv.org/abs/1105.2704

19. Krivelevich, M., Nutov, Z., Salavatipour, M.R., Yuster, J., Yuster, R.: Approximation algorithms and hardness results for cycle packing problems. ACM Transactions on Algorithms 3(4) (2007)

20. Lampis, M.: Algorithmic meta-theorems for restrictions of treewidth. In: de Berg, M., Meyer, U. (eds.) ESA 2010. LNCS, vol. 6346, pp. 549–560. Springer, Heidelberg (2010)

21. Lokshtanov, D., Marx, D., Saurabh, S.: Known Algorithms on Graphs of Bounded Treewidth are Probably Optimal. In: Proc. of the 22nd annual ACM-SIAM Symposium on Discrete algorithms (SODA), pp. 777–789 (2011)

22. Robertson, N., Seymour, P.: Graph Minors. XIII. The Disjoint Paths Problem. Journal of Combinatorial Theory, Series B 63(1), 65–110 (1995)

Deterministic Discrepancy Minimization

Nikhil Bansal[1] and Joel Spencer[2]

[1] IBM T.J. Watson, Yorktown Hts., NY 10598
`nikhil@us.ibm.com`
[2] New York University, New York
`spencer@cims.nyu.edu`

Abstract. We derandomize a recent algorithmic approach due to Bansal [2] to efficiently compute low discrepancy colorings for several problems. In particular, we give an efficient deterministic algorithm for Spencer's six standard deviations result [13], and to a find low discrepancy coloring for a set system with low hereditary discrepancy.

1   Introduction

Let $(V, \mathcal{S})$ be a set-system, where $V = \{1, \ldots, n\}$ are the elements and $\mathcal{S} = \{S_1, \ldots, S_m\}$ is a collection of subsets of V. Given a $\{-1, +1\}$ coloring $\mathcal{X}$ of elements in V, let $\mathcal{X}(S_j) = \sum_{i \in S_j} \mathcal{X}(i)$ denote the discrepancy of $\mathcal{X}$ for set S_j. The discrepancy of the collection $\mathcal{S}$ is defined as $\mathrm{disc}(\mathcal{S}) = \min_{\mathcal{X}} \max_{j \in [m]} |\mathcal{X}(S_j)|$. Perhaps surprisingly, questions about the discrepancy of various set systems are intimately related to fundamental questions in several diverse areas such as probabilistic and approximation algorithms, computational geometry, numerical integration, derandomization, communication complexity, machine learning, optimization and others(see for example [4,5,8]).

Until recently, most of the best known bounds for discrepancy problems were based on non-constructive techniques, particularly on the so-called entropy method or the partial coloring lemma (cf. section 2). Recently, Bansal [2] gave a technique for converting many of these entropy method based results into efficient randomized algorithms. His result also has other new applications, such as finding a low discrepancy coloring if the hereditary discrepancy of the underlying system is small [2]. The main idea of [2] is to start from the coloring $(0, \ldots, 0)$ and let the colors of the elements perform (tiny) correlated random walks over time until they reach $\{-1, +1\}$. At each time step, the increments for these random walks are determined by solving and rounding certain semidefinite programs (SDPs), and the choice of these SDPs themselves is based on the non-constructive entropy method.

In this paper we give a technique to derandomize the algorithm of Bansal [2]. In addition to using standard derandomization techniques such as k-wise independence and potential functions, one of our key and novel ideas is to use the entropy method itself to assist with the derandomization. Roughly speaking (we give more details later), randomization is used crucially in [2] to argue that at each time step, the algorithm makes progress on different (and opposing)

C. Demetrescu and M.M. Halldórsson (Eds.): ESA 2011, LNCS 6942, pp. 408–420, 2011.

objectives, in *expectation*. Now, to derandomize this, at the very least one needs a way to find at each time step, some (good) deterministic move that makes appropriate progress for each of the objectives. But, a priori it is unclear why such good moves should even exist: it could be that every move that makes positive progress for one objective makes negative progress for the others. To get around this problem, our idea is to "force" certain good moves to lie in the space of available moves, by adding certain new constraints to the underlying SDP. These constraints are chosen such that the entropy method can ensure that the new (more constrained) SDP still remains feasible.

We illustrate our techniques with two results.

Theorem 1. *Given any arbitrary set system on n elements and $m = O(n)$ sets[1], there is a deterministic algorithm to find a coloring with discrepancy $O(\sqrt{n})$.*

This gives a deterministic algorithm to find a coloring that matches the guarantees of the celebrated "six standard deviations suffice" result of Spencer [13] up to constant factors.

Let us compare theorem 1 with results for derandomizing Chernoff bounds. Via the connection to linear discrepancy ([1], chapter 12.3), theorem 1 implies that given any fractional solution $x_i \in [0,1]$ for $i \in [n]$ and $m = O(n)$ linear relations $\sum_i a_{ij} x_i$ for $j \in [m]$ with $|a_{ij}| \leq 1$, there exists a rounded solution $\tilde{x} \in \{0,1\}^n$ such that the roundoff error $|\sum_i a_{ij}(\tilde{x}_i - x_i)|$ is $O(\sqrt{n})$ for every j. In contrast, Chernoff bounds only give a weaker error bound of $O(\sqrt{n \log n})$.

In general, several potential based techniques such as pessimistic estimators [11] or the hyperbolic cosine method [12] have been developed to derandomize probabilistic tail bounds. However, it is known that a standard application of these techniques does not suffice to show theorem 1. In particular, they lose an additional factor of $O(\sqrt{\log n})$ (see [1], Chapter 14.5 for the precise result). So, our theorem 1 can be viewed as an extension and refinement of these techniques.

Theorem 2. *Given a set system $(V, \mathcal{S})$, $V = [n]$ and $\mathcal{S} = \{S_1, \ldots, S_m\}$ with hereditary discrepancy λ. There is a deterministic polynomial time algorithm to find an $O(\lambda \log mn)$ discrepancy coloring.*

This matches the guarantee due to [2] up to constant factors. Recall that the hereditary discrepancy of a set system $(V, \mathcal{S})$ is defined as the maximum value of discrepancy over all restrictions $W \subseteq V$. Specifically, given $W \subseteq V$, let $\mathcal{S}_{|W}$ denote the collection $\{S \cap W : S \in \mathcal{S}\}$. Then, the hereditary discrepancy of $(V, \mathcal{S})$ is defined as $\mathrm{herdisc}(\mathcal{S}) = \max_{W \subseteq V} \mathrm{disc}(\mathcal{S}_{|W})$.

We begin by discussing the algorithm of Bansal [2] and the challenges in derandomizing it, and then give an overview of our ideas.

The Non-constructive Entropy Method: This method is based on applying a pigeonhole principle argument to the set of all 2^n colorings, and gives the best known bounds for most discrepancy problems. Roughly, one shows that since

---

[1] The result can be generalized to arbitrary m, but we focus here on the most interesting case of $m = O(n)$.

there are (too many) 2^n colorings, there must exist two of them, say χ_1 and χ_2 that are far apart from each other in hamming distance, and yet have similar discrepancies for all sets. Then $\chi = (\chi_1 - \chi_2)/2$ defines a partial $\{-1, 0, +1\}$ coloring, that is non-zero on a large fraction of coordinates, and has low discrepancy for each set (c.f. lemma 1 below). This partial coloring procedure is then applied iteratively for $O(\log n)$ phases until all the elements are colored, and the overall discrepancy is at most the total discrepancy incurred in the various phases. We note that this method is completely existential and does not give any efficient algorithm to find a good partial coloring.

The algorithm of [2]: The algorithm works with fractional colorings, where a coloring is a vector in $[-1, +1]^n$ instead of $\{-1, +1\}^n$. Starting with the coloring $x_0 = (0, 0, \ldots, 0)$ the algorithm constructs a $\{-1, +1\}^n$ coloring gradually in several steps. At each time step t, the coloring $x_t = x_{t-1} + \gamma_t$ is obtained by adding a suitably chosen tiny update vector $\gamma_t \in \mathbb{R}^n$ to the coloring at $t-1$. Thus the color $x_t(i)$ of each element $i \in [n]$ evolves over time. When it reaches -1 or $+1$, it is considered *fixed* and is never updated again. The procedure continues until all the elements are colored either -1 or $+1$.

For this to work, the updates γ_t need to satisfy two crucial properties: First, they should make progress towards eventually producing a $\{-1, +1\}$ coloring (for example, the coloring should not oscillate forever around $(0, \ldots, 0)$). Let us call this the *Progress Property*. Second, for every set S_j, the accumulated discrepancy over time must be small. Let us call this the *Low Discrepancy Property*.

The algorithm of [2] ensures that both these properties hold in expectation at each step. In particular, the update γ_t is obtained by rounding the solution to a certain SDP relaxation for a partial coloring problem (defined suitably based on the current state of the algorithm). The entries $\gamma_t(i)$ of γ_t are (Gaussian) random variables with mean 0, but with large variance on average, which ensures that the progress property holds. Next, the entries $\gamma_t(i)$ are correlated so that the discrepancy increment $\sum_{i \in S_j} \gamma_t(i)$ for every set S_j has small variance. Note that these requirements on γ_t are conflicting. The progress property needs high variance to argue that each color reaches -1 or $+1$ quickly. But the low discrepancy property needs low variance to keep the discrepancy low. As we shall see, these conflicting requirements make the derandomization difficult.

Challenges to Derandomization: To derandomize the above algorithm, one first needs an efficient deterministic procedure to obtain the update vector γ_t from the SDP solution at each time t step. This by itself is simple, and there are several known deterministic SDP rounding techniques [10,6,14]. Second, as our algorithm proceeds over time, the updates γ_t at different time steps must be related (otherwise, for example, the coloring can oscillate forever around $(0, \ldots, 0)$). To handle this, a natural idea is to use some potential function to track the progress. So, we can define two potentials $\Phi_1(t)$ and $\Phi_2(t)$, one for the low discrepancy property and the other for the progress property, and try to argue that there some exists choice of γ_t that makes progress wrt both Φ_1 and Φ_2.

It is here that the problem arises. There is no a priori reason why some γ_t should exist that increases both Φ_1 and Φ_2. In particular, even though a random

γ_t increases both Φ_1 and Φ_2 in expectation, it could be that half the choices increase Φ_1 but decrease Φ_2, while the other half increase Φ_2 but decrease Φ_1.

Our Approach: To fix this problem, our idea is to introduce extra constraints in the SDP which help ensure the existence of γ_t's that will increase both the potentials sufficiently (we will also show how to find such γ_t's efficiently) . Let us look at the underlying issue in more detail. Consider the second potential $\Phi_2(t)$, which we set to be the energy $\sum_i x_t(i)^2$, as in [2]. The increase in $\Phi_2(t)$ at time t is $\Delta = (x_{t-1}(i) + \gamma_t(i))^2 - x_{t-1}^2 = \sum_i 2x_{t-1}(i)\gamma_t(i) + \gamma_t(i)^2$. In [2], roughly each $\gamma_t(i) \sim N(0, \delta_i^2)$, where $\delta_i \ll 1$ and hence $\mathbb{E}[\Delta] = \sum_i \delta_i^2$. But as δ_i is much smaller than the typical values $x_t(i)$, the term $\sum_i 2x_{t-1}(i)\gamma_t(i)$ in Δ completely swamps $\mathbb{E}[\Delta]$, and hence Δ will be both positive and negative roughly with half probability. Thus it could be the case that whenever $\Delta > 0$, the change in Φ_1 (by an analogous argument) is negative.

Our fix will be to essentially impose an additional "orthogonality" constraint of the type $\sum_i x_{t-1}(i)\gamma_t(i) = 0$ on the increment γ_t. This ensures that the Φ_2 increment $\Delta = \sum_i 2x_{t-1}(i)\gamma_t(i) + \sum_i \gamma_t(i)^2 = \sum_i \gamma_t(i)^2$ which is always non-negative. While we cannot add such a constraint directly, as $\gamma_t(i)$ is obtained by rounding SDP, we could instead impose the SDP constraint $\|\sum_i x_{t-1}(i)v_{t,i}\|_2^2 = 0$, where v_t's are the SDP variables from which γ_t is obtained. But the problem could be that the SDP may become infeasible up on adding this constraint. However, it turns out that the entropy method can be used to show that the SDP remains feasible if we impose $\|x_t(i)v_{t,i}\|_2^2$ be $1/\text{poly}(n)$ (instead of 0). Adding this constrains helps us ensure that there are a large fraction of choices for rounding the SDP for which Φ_2 increases sufficiently. We also impose an analogous constraint for the low discrepancy potential Φ_1, which helps us ensure that there are a large fraction of rounding choices for the SDP that do not increase Φ_1 by much.

2   Preliminaries

The Entropy Method: We recall here the partial coloring lemma of Beck [3], based on the Entropy Method. We also describe how it is used to obtain Spencer's result [13]. The form we use is from [9], but modified slightly for our purposes.

Lemma 1 (Entropy Method). *Let $\mathcal{S}$ be a set system on n elements, and let a number $\Delta_S > 0$ be given for each set $S \in \mathcal{S}$. Then there exists a partial coloring $\mathcal{X}$ that assigns -1 or $+1$ to at least $n/2$ variables (and 0 to other variables), and satisfies $|\mathcal{X}(S)| \le \Delta_S$ for each $S \in \mathcal{S}$ if the Δ_S satisfy the condition*

$$\sum_{S \in \mathcal{S}} g\left(\Delta_S / \sqrt{|S|}\right) \le n/5 \tag{1}$$

where
$$g(\lambda) = \begin{cases} Ke^{-\lambda^2/9} & \text{if } \lambda > 0.1 \\ K\ln(\lambda^{-1}) & \text{if } \lambda \le 0.1 \end{cases}$$

for some absolute constant K. Additionally, $\mathcal{X}$ can be assumed to simultaneously satisfy extra linear constraints $|\sum a_{ij}\mathcal{X}(i)| \leq \delta_j$, for $j = 1, \ldots, k$ provided

$$\sum_{j=1}^{k} K \cdot \ln\left(\frac{|a_j|_1}{\delta_j} + 1\right) + \sum_{S \in \mathcal{S}} g\left(\frac{\Delta_S}{\sqrt{|S|}}\right) \leq \frac{n}{5} \qquad (2)$$

where $|a_j|_1 = \sum_i |a_{ij}|$.

Spencer's Result [13]: The coloring is constructed in phases. In phase i, for $i = 0, \ldots, \log n$, the number of uncolored elements is at most $n_i \leq n/2^i$. In phase i, apply lemma 1 to these n_i elements with $\Delta_S^i = c(n_i \log(2m/n_i))^{1/2}$. It is easily verified that (1) holds for a large enough constant c. This gives a partial coloring on at least $n_i/2$ elements, with discrepancy for any set S at most Δ_S^i. Summing up over the phases, the overall discrepancy for any set is at most

$$\Delta_S \leq \sum_i \Delta_S^i = \sum_i c\left(n2^{-i}\log\left(\frac{2m}{n2^{-i}}\right)\right)^{1/2} = O((n\log(2m/n))^{1/2}).$$

k-wise Independent Distributions: A multiset $I \subset \{-1, 1\}^n$ is called a k-wise independent sample space, if for every $j \leq k$ indices $i_1, \ldots, i_j \in [n]$ and for every choice of $z_1, \ldots, z_j \in \{-1, 1\}$, it holds that $\{x \in I : (x_{i_1}, \ldots, x_{i_j}) = (z_1, \ldots, z_j)\} = \frac{|I|}{2^j}$.

It is known for that for every $k \geq 1$, there exists an explicit construction of a k-wise independent sample space of size $O(n^{k/2})$. See [1] for more details.

A Probabilistic Inequality: We will repeatedly use the following simple lemma, the proof of which is omitted due to lack of space.

Lemma 2. *Let X be a random variable with $0 \leq \mu \leq \mathbb{E}[X] \leq \mu'$, and $X \geq -\epsilon\mu'$. Then, for $t \geq 1$, $\Pr[X \geq t\mu'] \leq (1 + \epsilon)/t$. Moreover, if $\mathbb{E}[X^2] \leq c\mu^2$, then for $\beta \leq 2/3$, $\Pr[X \geq \beta\mu] \geq ((1 - 3\beta/2)\beta)/2c$.*

3   Spencer's Bound

In this section we prove theorem 1. By adding dummy elements or sets as needed, we will assume henceforth that $m = n$. Our algorithm proceeds in phases, similar to the original proof of [13] (sketched in section 2) and as in the randomized algorithm of [2]. In each phase, at least half the remaining elements are colored, and the discrepancy incurred falls geometrically. For simplicity we only describe the first phase. This phase captures all the hardness of the problem.

In particular, our algorithm will start with the coloring $x_0 = (0, \ldots, 0)$. At the end, we produce a coloring x where at least $n/2$ variables are set to -1 or $+1$ (the other variables are possibly set to a value in the range $[-1, +1]$), and each set S_j has discrepancy $\sum_{i \in S_j} x(i)$ in the range $[-c\sqrt{n}, c\sqrt{n}]$. Here c is a constant that can be computed explicitly from our analysis.

Consider the function

$$f(z) = \left(\frac{c\sqrt{n}}{c\sqrt{n} - |z|}\right)^4 - 1$$

with domain $(-c\sqrt{n}, c\sqrt{n})$. Note that $f(0) = 0$, and it rises to ∞ as z approaches $\pm c\sqrt{n}$. We set $f(z) = \infty$ if $|z| \geq c\sqrt{n}$. We will use f as a "barrier" function that will prevent the discrepancy for any set from exceeding $c\sqrt{n}$. We also note that $|f''(z)| = 20(1/c^2 n)(1 - |z|/c\sqrt{n})^{-6}$.

Let $x_t(i)$ denote the color of element i at the end of time t. For a set S, let us also denote $x_t(S) = \sum_{i \in S} x_t(i)$, the discrepancy of set S at the end of time t. Let $A(t)$ denote the set of elements that are still alive at the beginning of time t, i.e. they did not reach $+1$ or -1 at the end of time $t - 1$.

3.1 Algorithm (Phase 1)

Let d be a constant that we make explicit later. Let $\gamma = 1/n^5$ and $\delta = 1/n^{15}$. For each time $t = 1, 2, \ldots$ do the following, until at least $n/2$ elements are colored -1 or $+1$.

1. Find a feasible solution to the following semidefinite program:

$$\sum_{i \in [n]} ||v_i||_2^2 \geq |A(t)|/2 \tag{3}$$

$$||\sum_{i \in S_j} v_i||_2^2 \leq 20dn/(c^2 n f''(x_{t-1}(S_j))) \qquad \forall j \tag{4}$$

$$||\sum_i x_{t-1}(i) v_i||_2^2 \leq \delta^2 \tag{5}$$

$$||\sum_j f'(x_{t-1}(S_j)) \sum_{i \in S_j} v_i||_2^2 \leq \delta^2 \tag{6}$$

$$||v_i||_2^2 \leq 1 \qquad \forall i \in A(t) \tag{7}$$

$$||v_i||_2^2 = 0 \qquad \forall i \notin A(t) \tag{8}$$

If the SDP is infeasible, abort the algorithm and return fail.

Otherwise, let $v_i \in \mathbb{R}^n$, $i = 1, \ldots, n$ be the solution returned by the SDP.

2. Consider a 4-wise independent sample space I of $\{-1, 1\}^n$ vectors. Pick an $r \in I$ that satisfies the following conditions.

$$\sum_j f(x_{t-1}(S_j) + \gamma\langle r, \sum_{i \in S_j} v_i\rangle) - \sum_j f(x_{t-1}(S_j)) \leq 100\gamma^2 dn/c^2 \tag{9}$$

$$\sum_i ((x_{t-1}(i) + \gamma\langle r, v_i\rangle)^2 - x_{t-1}^2(i)) \geq \gamma^2 n/8. \tag{10}$$

If no such r exists, then abort the algorithm.

Update $x_t(i) = x_{t-1}(i) + \gamma\langle r, v_i\rangle$.

3. For each $i \in A(t)$, if $x_t(i) \geq 1$ (resp. $x_t(i) \leq -1$), then set $x_t(i) = 1$ (resp. $x_t(i) = -1$) and update $A(t+1)$ accordingly.

We note that SDP above in addition to being a relaxation of finding a partial coloring has two additional constraints (5) and (6). In particular, constraint (3) requires that at least half the elements be colored. Constraint (4) essentially says that a set that has discrepancy of about $(1 - 1/k)c\sqrt{n}$ thus far (based on the coloring x_{t-1}) should have a discrepancy of at most $d\sqrt{n}/k^6$ in the partial coloring. The constraints (5) and (6) are new and non-obvious, and as we shall see later, are crucial for arguing the existence of r satisfying (9) and (10) in step 2 of the algorithm at time t.

3.2   Analysis

To show the correctness of the algorithm, we need to prove two things. First, that the algorithm never aborts, either due to the SDP being infeasible, or because it cannot find an r satisfying (9) and (10). Second, when the number of alive variables falls below $n/2$ and the algorithm terminates, the discrepancy of each set is at most $c\sqrt{n}$. To this end, we define two potential functions

$$\Phi_1(t) = \sum_j f(x_t(S_j)) \qquad \text{and} \qquad \Phi_2(t) = \sum_i x_t(i)^2.$$

The function Φ_1 is designed to capture the discrepancies of various sets and Φ_2 is designed to track the progress property. Note that both $\Phi_1(t)$ and $\Phi_2(t)$ are completely determined by the coloring x_t and that $\Phi_1(0) = \Phi_2(0) = 0$. We will show the following result, and then show how theorem 1 follows directly from it.

Theorem 3. *Let $t \geq 1$ be any time such that at least $n/2$ variables are alive. If the coloring x_{t-1} satisfies $\Phi_1(t - 1) \leq c'n$, where $c' = 800d/c^2$, then*

1. *The SDP defined at time t is feasible.*
2. *Moreover, given any feasible solution to the SDP at time t, there exists a r satisfying (9) and (10). That is, there exists a r satisfying*

$$\Phi_1(t) - \Phi_1(t-1) \leq 100\gamma^2 dn/c^2 = c'\gamma^2 n/8 \quad \text{and} \quad \Phi_2(t) - \Phi_2(t-1) \geq \gamma^2 n/8.$$

Let us see how theorem 3 implies theorem 1. The argument is by induction over time. Initially, the coloring $x_0 = (0, \ldots, 0)$ and hence $\Phi_1(0) = 0 \leq c'n$. So by theorem 3, the SDP at time 1 is feasible. Suppose that the SDP is feasible until some time $t \geq 1$. Then, again by theorem 3 $\Phi_1(t') - \Phi_1(t' - 1) \leq c'\gamma^2 n/8$ for each $t' \leq t$ and hence $\Phi_1(t) \leq tc'\gamma^2 n/8$. Also $\Phi_2(t') - \Phi_2(t' - 1) \geq \gamma^2 n/8$ for each $t' \leq t$ and hence $\Phi_2(t) \geq \gamma^2 nt/8$. Thus, the algorithm can last no more than $T = 8/\gamma^2$ steps (as Φ_2 can never exceed n, as no entry $x_t(i)$ can exceed 1). But this implies that $\Phi_1(t + 1) \leq Tc'\gamma^2 n/8 = c'n$ for any time $t + 1 \leq T$, and hence condition of the theorem will also be satisfied at $t + 1$. Thus, the algorithm will not abort as long as there are at least $n/2$ alive variables.

To bound the discrepancy of each set when the algorithm terminates, we simply use the fact that $\Phi_1 \leq c'n$ when the algorithm terminates. This implies that $f(x(S_j)) \leq c'n$ for each set S_j and hence the discrepancy $|x(S_j)|$ is at most $c\sqrt{n}$ (recall that $f(x) = \infty$ if $|x| \geq c\sqrt{n}$). This implies theorem 1.

We now focus on proving theorem 3.

Lemma 3. *If x_{t-1} is a coloring such that $\Phi_1(t-1) \leq c'n$, then the SDP at time t is feasible.*

Proof. It suffices to show that there is some partial $\{-1, +1\}$ coloring on at least half the alive elements, that satisfies the constraints (4), (5) and (6) of the SDP. To do this, we apply lemma 1 suitably.

We first bound the contribution in (2) due to SDP constraints (4). Let $k \geq 1$ be an integer. We say a set S is k-dangerous if $|x_{t-1}(S)|/(c\sqrt{n}) \in [1 - 1/k, 1 - 1/(k+1)]$. For any k-dangerous set, $f(x_{t-1}(S)) \in [k^4, (k+1)^4]$ and hence our hypothesis $\Phi_1(t-1) \leq c'n$ implies that there are at most $c'n/k^4$ k-dangerous sets. In the SDP constraint (4), the discrepancy bound Δ_S in the partial coloring that we impose for any such set is no less than $\beta(k) = \sqrt{dn/(k+1)^6}$.

Consider $k = 1$. Let us choose $d > 1$ large enough (this completely determines d) such that the entropy penalty $g(\beta(1)) \leq 1/20$ for any 1-dangerous set. Thus, the entropy contribution of 1-dangerous sets is at most $n/20$. For $k \geq 2$, the entropy contribution of a k-dangerous set is $g(\beta(k)) \leq g(n/(k+1)^6) \leq 6K \ln(k+1)$. By our upper bound on the number of k-dangerous sets, the total entropy contribution is at most $\sum_{k \geq 2}(c'n/k^4) \cdot 6K \ln(k+1)$. As $c' = 800d/c^2$ and having fixed d above, we choose c large enough such that this contribution is at most $n/20$. Thus the total entropy contribution due to constraints (4) is at most $n/10$.

Finally we note that both constraints (5) and (6) contribute only $O(\log n)$ to (2). This is because $1/\delta$ is polynomially bounded in n, and the sum of coefficients of these constraints are $\sum_S |f(x_{t-1}(S))| = \Phi_1 = O(n)$ and $\sum_i |x_{t-1}(i)| \leq n$. Thus the total contribution of all the constraints to (2) is at most $n/5$ and hence the claimed partial coloring exists by lemma 1. $\qquad\square$

We now show the second statement of the theorem. That is, Given any feasible solution to the SDP at time t, there exists an r from a 4-wise independent sample space satisfying conditions (9) and (10). It suffices to that at least 0.02 fraction of the r satisfy (10), and at least 0.99 fraction of r satisfy (9). We prove these below.

Lemma 4. *Given any feasible solution to the SDP at time t, at least 0.02 fraction of r from a 4-wise independent sample space satisfy (10).*

Proof. Let $x = (x(1), \ldots, x(n))$ denote the coloring at time $t - 1$. For a vector r chosen randomly from the 4-wise independent sample space, consider the random coloring $y = (y(1), \ldots, y(n))$ obtained as $y(i) = x(i) + \gamma\langle r, v_i \rangle$. The change in potential Φ_2 is the random variable $\Delta(r) = \sum_i y(i)^2 - x(i)^2 = 2\gamma\langle\sum_i x(i)v_i, r\rangle + \gamma^2 \sum_i \langle v_i, r\rangle^2$.

Our goal is to apply lemma 2 to $\Delta(r)$. We first show that the condition of lemma 2 holds with $\epsilon = o(1)$. By Cauchy-Schwarz, we have that $|\langle\sum_i x_i v_i, r\rangle| \leq$

$|| \sum_i x_i v_i ||_2 ||r||_2 \leq || \sum_i x_i v_i ||_2 \sqrt{n}$ and hence by the SDP constraint (5), it follows that $|\langle \sum_i x_i v_i, r \rangle| \leq \delta \sqrt{n}$. Since $\sum_i \langle v_i, r \rangle^2 \geq 0$, it follows that $\Delta(r) \geq -2\gamma\delta\sqrt{n}$ for all r. Moreover,

$$\mathbb{E}_r[\Delta(r)] = \gamma^2 \mathbb{E}_r \left[\sum_i \langle v_i, r \rangle^2 \right] = \gamma^2 \mathbb{E}_r \left[\sum_i \left(\sum_{j,j} v_i(j) v_i(j') r(j) r(j') \right) \right] \tag{11}$$

$$= \gamma^2 \sum_i \sum_j v_i(j)^2 = \gamma^2 \sum_i ||v_i||_2^2 \geq \gamma^2 |A(t)|/2 \geq \gamma^2 n/4 \quad \text{(By SDP constraint 3)}.$$

The second step follows since r is 4-wise independent, and hence for $j \neq j'$ we have $E_r[r(j) \cdot r(j')] = 0$.

Note that $|2\gamma\delta\sqrt{n}| \ll \mathbb{E}_r[\Delta(r)]$ and hence the setting of lemma 2 holds with $\epsilon = o(1)$. Similarly as $\delta n^\beta \ll \gamma$ for $\beta \leq 10$, it follows that $\mathbb{E}_r[\Delta(r)^2] \leq (1 + o(1))\gamma^4 \mathbb{E}_r \left[\left(\sum_i \langle v_i, r \rangle^2 \right)^2 \right]$ and hence to upper bound $\mathbb{E}_r\left[\Delta(r)^2\right]$ we focus on $\mathbb{E}_r \left[\left(\sum_i \langle v_i, r \rangle^2 \right)^2 \right] = \mathbb{E}_r \left[\sum_{i,i'} \langle v_i, r \rangle^2 \langle v_{i'}, r \rangle^2 \right]$ which we write as

$$\mathbb{E}_r \left[\sum_{i,i'} \sum_{j_1, j_2, j_3, j_4} v_i(j_1) r(j_1) v_i(j_2) r(j_2) v_{i'}(j_3) r(j_3) v_{i'} r(j_4) \right] \tag{12}$$

By the 4-wise independence of r, the terms $\mathbb{E}_r[r(j_1) r(j_2) r(j_3) r(j_4)]$ vanish for every choice of indices $j_1, \ldots, j_4$, expect in the following cases.

1. Case 1: $j_1 = j_2$ and $j_3 = j_4$. In this case, the contribution of (12) is

$$\sum_{i,i',j,j'} v_i(j)^2 v_{i'}(j')^2 = (\sum_i ||v_i||_2^2)^2.$$

2. Case 2: $j_1 = j_3$ and $j_2 = j_4$ (and the symmetric case of $j_1 = j_4$ and $j_2 = j_3$). Here the contribution of (12) is

$$\sum_{i,i',j,j'} v_i(j) v_{i'}(j) v_i(j') v_{i'}(j') = \sum_{i,i'} (v_i \cdot v_{i'})^2 \leq \sum_{i,i'} (||v_i||_2^2 \cdot ||v_{i'}||_2^2) = (\sum_i ||v_{i'}||_2^2)^2.$$

   where the inequality follows from Cauchy-Schwarz.

So $\mathbb{E}_r[(\sum_i \langle v_i, r \rangle^2)^2] \leq 3(\mathbb{E}_r[\sum_i \langle v_i, r \rangle^2])^2$, and hence $\mathbb{E}_r[\Delta(r)^2] \leq (3 + o(1)) \mathbb{E}_r[\Delta(r)]^2$. By second part of lemma 2, with $\epsilon = o(1)$ and $c = 3 + o(1)$ and $\beta = 1/2 + o(1)$, it follows that $\Pr[\Delta(r) \geq \gamma^2 n/8] \geq 0.02$ as desired. □

Lemma 5. *Let x_{t-1} be any coloring with $\Phi_1(x_{t-1}) \leq c'n$, then given any feasible solution to the SDP at time t, at least 0.99 fraction of r from a 4-wise independent sample space satisfy (9).*

Proof. Given the coloring x_{t-1}, and a feasible SDP solution, consider the random coloring x_t obtained by choosing r from the 4-wise independent family. Let us define $w_j = \sum_{i \in S_j} v_i$. We wish to bound

$$\Delta(r) = \sum_j (f(x_t(S_j)) - f(x_{t-1}(S_j))) = \sum_j (f(x_{t-1}(S_j) + \gamma\langle w_j, r \rangle) - f(x_{t-1}(S_j)))$$

By Taylor expansion for any x and y in the domain of f, we have

$$f(y) \geq f(x) + f'(x)(y-x) + \frac{1}{2!}f''(x)(y-x)^2 - \frac{1}{3!}|\text{supp}_{z \in [x,y]}f'''(z)(y-x)^3|.$$

Setting $x_j = x_{t-1}(S_j)$ and $y_j = x_j + \gamma \langle w_j, r \rangle$, and summing over $j = 1, \ldots, n$, we obtain that $\Delta(r)$ is at least

$$\gamma \langle \sum_j f'(x_j)w_j, r \rangle + \sum_j \frac{1}{2!}f''(x_j)\gamma^2\langle w_j, r \rangle^2 - \frac{1}{3!}\sum_j |\text{supp}_{z_j \in [x_j, y_j]}f'''(z)\gamma^3\langle w_j, r \rangle^3|.$$

$$(13)$$

We now show that $\Delta(r)$ satisfies the conditions of lemma 2 with $\epsilon = o(1)$. As the second term in (13) is always non-negative, we lower bound the other terms.

To bound the first term, we note by Cauchy-Schwarz and (6) that

$$\gamma \langle \sum_j f'(x_j)w_j, r \rangle \leq \gamma || \sum_j f'(x_j)w_j ||_2 ||r||_2 \leq \gamma \delta \sqrt{n}.$$

The second term is always non-negative with expectation $\mathbb{E}_r[f''(x_j)\gamma^2\langle w_j, r \rangle^2] = f''(x_j)\gamma^2||w_j||_2^2 \leq \gamma^2 d/c^2$ where the inequality above follows from the SDP constraint (4). Summing up j over all the n constraints, we get that

$$\mathbb{E}_r[\sum_j f''(x_j)\gamma^2\langle w_j, r \rangle^2] \leq \gamma^2 dn/c^2.$$

For the third term we bound $|\text{supp}_{z_j \in [x_j, y_j]}f'''(z_j)\gamma^3\langle w_j, r \rangle^3|$. As $\sum_j f(x_j) = \Phi_1(x_{t-1}) \leq c'n$, we have that $f(x_j) \leq c'(n)$ and hence that $|x_j| \leq c\sqrt{n} - c''n^{1/4}$ for every j, for some constant c''. As $z_j = x_j + \gamma\langle w_j, r \rangle$ and $\gamma = n^{-5}$ we get,

$$|z_j| \leq |x_j| + \gamma||w_j||_2||r||_2 \leq |x_j| + \gamma n^{3/2} \leq c\sqrt{n} - (c''/2)n^{1/4}.$$

Thus for any j, $f'''(z_j) \leq \frac{140}{(c\sqrt{n})^3}(1 - |z_j|/c\sqrt{n})^{-7} = O(n^{7/4} \cdot n^{-3/2}) = O(n^{1/4})$. So, the third term is $O(n \cdot n^{1/4} \cdot \gamma^3 \cdot (||w_j||_2||r||_2)^3) = O(n^{17/4})\gamma^3 = o(\gamma^2)$.

The result now follows by the first part of lemma 2, with $\mu' = \gamma^2 dn/2c^2$, $t = 200$ and $\epsilon = o(1) \leq 1$. □

4   Pseudo-approximation for Discrepancy

In this section we show theorem 2. As previously, our algorithm constructs the desired coloring in several steps. Let x_t denote the fractional coloring $\in [-1, +1]^n$ at time t. We start with the $x_0 = (0, \ldots, 0)$ and update it over time. During the algorithm, if the color of element i reaches $+1$ or -1 at time t, it is fixed and never updated again. Otherwise, the variable is alive. Let $A(t)$ denote the set of alive variables at beginning of time t. Let us also assume that the algorithm knows λ (it can try out all possible values for λ). Unlike the previous section where we only gave an algorithm for phase 1, here we describe the algorithm to obtain the complete coloring.

4.1 Algorithm

Initialize, $x_0(i) = 0$ for all $i \in [n]$. Let $\gamma = (mn)^{-c}$ and $\delta = (mn)^{-3c}$. Here c is an explicit constant that can be determined by our analysis below. Let $f(x)$ denote the function $e^{\alpha x} + e^{-\alpha x}$, where $\alpha = 1/(2\lambda)$.

For each time step $t = 1, 2, \ldots, \ell$ repeat the following, until the number of alive variables falls below $c'\lambda \log mn$. Here c' is a constant depending up on c.

1. Find a feasible solution to the following semidefinite program:

$$|| \sum_{i \in S_j} v_i ||_2^2 \le 4\lambda^2 \qquad \text{for each set } S_j \qquad (14)$$

$$\sum_i ||v_i||_2^2 \ge |A(t)|/10 \qquad \forall i \in A(t) \qquad (15)$$

$$|| \sum_j f'(x_{t-1}(S_j)) \sum_{i \in S_j} v_i ||_2^2 \le \delta^2 \qquad (16)$$

$$|| \sum_i x_{t-1}(i)v_i ||_2^2 \le \delta^2 \qquad (17)$$

$$||v_i||_2^2 \le 1 \qquad \forall i \in A(t) \qquad (18)$$

$$||v_i||_2^2 = 0 \qquad \forall i \notin A(t) \qquad (19)$$

Let $v_i \in \mathbb{R}^n$, $i \in [n]$ denote some arbitrary feasible solution.
2. Choose an $r \in \{-1, +1\}^n$ from a 4-wise independent family such that the following two conditions are satisfied.

$$\sum_i (f(x_{t-1}(i) + \gamma\langle v_i, r\rangle)) \le (\sum_i f(x_{t-1}(i))) \cdot (1 + 100\gamma^2) \qquad (20)$$

$$\sum_i (x_{t-1}(i) + \gamma\langle v_i, r\rangle)^2 - \sum_i x_{t-1}(i)^2 \ge \gamma^2 |A(t)|/40. \qquad (21)$$

If no such choice of r exists then abort the algorithm.
  For each i, update $x_t(i) = x_{t-1}(i) + \gamma\langle r, v_i\rangle$.
3. For each i, set $x_t(i) = 1$ if $x_t(i) \ge 1$ or set $x_t(i) = -1$ if $x_t(i) < -1$. Update $A(t)$ accordingly.

When the above procedure terminates we arbitrarily color the $c\lambda \log mn$ remaining alive variables.

4.2 Analysis

We wish to show that the discrepancy at the end of the algorithm is $O(\lambda \log mn)$, and that the algorithm never aborts, i.e it always finds a feasible SDP solution and an r satisfying (20) and (21). As previously, we define two potential functions, the first controls the discrepancy and the second controls the progress.

$$\Phi_1(t) = \sum_j f(x_t(S_j)) \qquad \text{and} \qquad \Phi_2(t) = \sum_i x_t(i)^2.$$

As $x_0 = (0, \ldots, 0)$, $\Phi_1(0) = 2m$ and $\Phi_2(0) = 0$. Also, (20) and (21) can be written as

$$\Phi_1(t) \leq \Phi_1(t-1)(1 + 100\gamma^2) \qquad \text{and} \qquad \Phi_2(t) - \Phi_2(t-1) \geq \gamma^2 |A(t)|/40.$$

If the claimed r exists at each step, then clearly the algorithm terminates in $O(\log n/\gamma^2)$ steps. This is because if $\Phi_2(t) \leq n - k$, then this implies that least k variables are alive at the end of t. Thus, by time $t + 40/\gamma^2$, condition (21) would ensure that $\Phi_2(t)$ reaches $n - k/2$. So, starting from $\Phi_2(0) = 0$, Φ_2 will certainly reach n in $O((\log n)/\gamma^2)$ time steps.

We show the following result and describe how theorem 2 follows from it.

Theorem 4. *Let $t \geq 1$ be any time such that the number of alive variables is more than $c\lambda \log mn$. If the coloring x_{t-1} satisfies $\Phi_1(t-1) \leq (mn)^{O(1)}$, then*

1. *The SDP defined at time t is feasible.*
2. *Given any feasible solution to the SDP at time t, there exists a r satisfying (20) and (21).*

Theorem 4 implies theorem 2 by induction over time. As $\Phi_1(0) = 2m$, the condition $\Phi_1(0) \leq (mn)^{O(1)}$ holds, and hence the SDP at time 1 is feasible. Suppose the SDP is feasible until time t, for some $t \geq 1$. Then as, $\Phi_1(t') \leq \Phi_1(t'-1)(1 + 100\gamma^2)$ for all $t' \leq t$. It follows $\Phi_1(t) \leq 2m \exp(100t\gamma^2)$. Similarly, as $\Phi_2(t') - \Phi_2(t'-1) \geq \gamma^2 |A(t')|/40$ at each time $t' \leq t$, by the argument above, the algorithm cannot last more than $T = O((\log n)/\gamma^2)$ steps. As $2m \exp(100T\gamma^2) = (mn)^{O(1)}$, the condition $\Phi_1(t+1) = (mn)^{O(1)}$ also holds and hence the algorithm will not abort before it terminates.

To bound the discrepancy, we note that $\Phi_1 \leq (mn)^{O(1)}$ when the algorithm terminates. Hence $f(x(S_j)) \leq (mn)^{O(1)}$ for each set S_j and hence the discrepancy $|x(S_j)| = O(\log(mn))/\alpha = O(\lambda \log mn)$, implying theorem 2. The proof of theorem 4 is similar to that of theorem 3 and is omitted due to lack of space.

References

1. Alon, N., Spencer, J.: The Probabilistic Method. John Wiley, Chichester (2000)
2. Bansal, N.: Constructive algorithm for discrepancy minimization. In: FOCS 2010 (2010)
3. Beck, J.: Roth's estimate on the discrepancy of integer sequences is nearly sharp. Combinatorica 1, 319–325 (1981)
4. Beck, J., Sos, V.: Discrepancy theory. In: Graham, R.L., Grotschel, M., Lovasz, L. (eds.) Handbook of Combinatorics, pp. 1405–1446. North-Holland, Amsterdam (1995)
5. Chazelle, B.: The discrepancy method: randomness and complexity. Cambridge University Press, Cambridge (2000)
6. Engebretsen, L., Indyk, P., O'Donnell, R.: Derandomized dimensionality reduction with applications. In: SODA 2002, pp. 705–712 (2002)
7. Kim, J.H., Matousek, J., Vu, V.H.: Discrepancy After Adding A Single Set. Combinatorica 25(4), 499–501 (2005)

8. Matousek, J.: Geometric Discrepancy: An Illustrated Guide, Algorithms and Combinatorics, vol. 18. Springer, Heidelberg (1999)
9. Matousek, J.: An Lp version of the Beck-Fiala conjecture. European Journal of Combinatorics 19(2), 175–182 (1998)
10. Mahajan, S., Ramesh, H.: Derandomizing Approximation Algorithms Based on Semidefinite Programming. SIAM J. Comput. 28(5), 1641–1663 (1999)
11. Raghavan, P.: Probabilistic construction of deterministic algorithms: approximating packing integer programs. J. of Computer and Systems Sciences 37, 130–143
12. Spencer, J.: Balancing Games. J. Comb. Theory, Ser. B 23(1), 68–74 (1977)
13. Spencer, J.: Six standard deviations suffice. Trans. Amer. Math. Soc. 289, 679–706 (1985)
14. Sivakumar, D.: Algorithmic Derandomization via Complexity Theory. In: IEEE Conference on Computational Complexity (2002)

Pattern Matching in Lempel-Ziv Compressed Strings: Fast, Simple, and Deterministic

Paweł Gawrychowski[⋆]

Institute of Computer Science,
University of Wrocław,
ul. Joliot-Curie 15, 50–383 Wroclaw, Poland
`gawry@cs.uni.wroc.pl`

Abstract. Countless variants of the Lempel-Ziv compression are widely used in many real-life applications. This paper is concerned with a natural modification of the classical pattern matching problem inspired by the popularity of such compression methods: given an uncompressed pattern $p[1 .. m]$ and a Lempel-Ziv representation of a string $t[1 .. N]$, does p occur in t? Farach and Thorup [5] gave a randomized $\mathcal{O}(n \log^2 \frac{N}{n} + m)$ time solution for this problem, where n is the size of the compressed representation of t. Building on the methods of [3] and [6], we improve their result by developing a faster and fully deterministic $\mathcal{O}(n \log \frac{N}{n} + m)$ time algorithm with the same space complexity. Note that for highly compressible texts, $\log \frac{N}{n}$ might be of order n, so for such inputs the improvement is very significant. A small fragment of our method can be used to give an asymptotically optimal solution for the substring hashing problem considered by Farach and Muthukrishnan [4].

Keywords: pattern matching, compression, Lempel-Ziv.

1 Introduction

Effective compression methods allow us to decrease the space requirements which is clearly worth pursuing on its own. On the other hand, we do not want to store the data just for the sake of having it: we want to process it efficiently on demand. This suggest an interesting direction: can we process the data without actually decompressing it? Or, in other words, can we speed up the processing if the compression ratio is high? Answer to such questions clearly depends on the particular compression and processing method chosen. In this paper we focus on Lempel-Ziv (also known as LZ77, or simply LZ for the sake of brevity), one of the most commonly used compression methods being the basis of the widely popular `zip` and `gz` archive file formats, and on pattern matching, one of the most natural text processing problem we might encounter. More specifically, we deal with the compressed pattern matching problem: given an uncompressed pattern

---

[⋆] Supported by MNiSW grant number N N206 492638, 2010–2012 and START scholarship from FNP.

C. Demetrescu and M.M. Halldórsson (Eds.): ESA 2011, LNCS 6942, pp. 421–432, 2011.

$p[1 .. m]$ and a LZ representation of a string $t[1 .. N]$, does p occur in t? This line of research has been addressed before quite a few times already. Amir, Benson, and Farach [1] considered the problem with LZ replaced by Lempel-Ziv-Welch (a simpler and easier to implement specialization of LZ), giving two solutions with complexities $\mathcal{O}(n \log m + m)$ and $\mathcal{O}(n + m^2)$, where n is the size of the compressed representation. The latter was soon improved [11] to $\mathcal{O}(n + m^{1+\epsilon})$. Then Farach and Thorup [5] considered the problem in its full generality and gave a (randomized) $\mathcal{O}(n \log^2 \frac{N}{n} + m)$ time algorithm for the LZ case. Their solution consists of two phases, called *winding* and *unwinding*, the first one uses a cleverly chosen potential function, and the second one adds fingerprinting in the spirit of string hashing of Karp and Rabin [9]. While a recent result of [7] shows that the winding can be performed in just $\mathcal{O}(n \log \frac{N}{n})$, it is not clear how to use it to improve the whole running time (or remove randomization). In this paper we take a completely different approach, and manage to develop a $\mathcal{O}(n \log \frac{N}{n} + m)$ time algorithm. This complements our recent result from SODA'11 [6] showing that in case of Lempel-Ziv-Welch, the compressed pattern matching can be solved in optimal linear time. The space usage of the improved algorithm is the same as in the solution of Farach and Thorup, $\mathcal{O}(n \log \frac{N}{n} + m)$.

Besides the algorithm of Farach and Throup, the only other result that can be applied to the LZ case we are aware of is the work of Kida *et al.* [10]. They considered the so-called *collage systems* allowing to capture many existing compression schemes, and developed an efficient pattern matching algorithm for them. While it does not apply directly to the LZ compression, we can transform a LZ parse into a non-truncating collage system with a slight increase in the size, see Section 5. The running time (and space usage) of the resulting algorithm is $\mathcal{O}(n \log \frac{N}{n} + m^2)$. While m^2 might be acceptable from a practical point of view, removing the quadratic dependency on the pattern length seems to be a non-trivial and fascinating challenge from a more theoretical angle, especially given that for some highly compressible texts n might be much smaller than m. Citing [10], even decreasing the dependency to $m^{1.5} \log m$ (the best preprocessing complexity known for the LZW case [11] at the time) "is a challenging problem".

While we were not able to achieve linear time for the general LZ case, the algorithm developed in this paper not only significantly improves the previously known time bounds, but also is fully deterministic and (relatively) simple. Moreover, LZ compression allows for an exponential decrease in the size of the compressed text, while in LZW n is at least $\sqrt{N}$. In order to deal with such highly compressible texts efficiently we need to combine quite a few different ideas, and the nonlinear time of our (and the previously known) solution might be viewed as an evidence that LZ is substantially more difficult to deal with than LZW. While most of those ideas are simple, they are very carefully chosen and composed in order to guarantee the $\mathcal{O}(n \log \frac{N}{n} + m)$ running time. We believe the simplicity of those basic building blocks should not be viewed as a drawback. On the contrary, it seems to us that improving a previously known result (which used fairly complicated techniques) by a careful combination of simple tools should be seen as an advantage. We also argue that in a certain sense, our result is the best possible: if integer division is not allowed, our algorithm can be implemented in $\mathcal{O}(n \log N + m)$ time, and this is the best time possible.

2   Overview of the Algorithm

Our goal is to detect an occurrence of p in a given Lempel-Ziv compressed text $t[1 .. N]$. The Lempel-Ziv representation is quite difficult to work with efficiently, even for such a simple task as extracting a single letter. The starting point of our algorithm is thus transforming the input into a *straight-line program*, which is a context-free grammar with each nonterminal generating exactly one string. For that we use the method of Charikar *et al.* [3] to construct a SLP of size $\mathcal{O}(n \log \frac{N}{n})$ with additional property that all productions are *balanced*, meaning that the right sides are of the form XY with $\frac{\alpha}{1-\alpha} \leq \frac{|X|}{|Y|} \leq \frac{1-\alpha}{\alpha}$ for some constant α, where $|X|$ is the length of the (unique) string generated by X. Note that Rytter gave a much simpler algorithm [12] with the same size guarantee, using the so-called AVL grammars but we need the grammar to be balanced. We also need to add a small modification to allow self-referential LZ.

After transforming the text into a balanced SLP, for each nonterminal we try to check if the string it represents occurs inside p, and if so, compute the position of (any) its occurrence. Otherwise we would like to compute the longest prefix (suffix) of this string which is a suffix (prefix) of p. At first glance this might seem like a different problem that the one we promised to solve: instead of locating an occurrence of the pattern in the text, we retrieve the positions of fragments of the text in the pattern. Nevertheless, solving it efficiently gives us enough information to answer the original question due to a constant time procedure which detects an occurrence of p in a concatenation of two its substrings.

The first (simple) algorithm for processing a balanced SLP we develop requires as much as $\mathcal{O}(\log m)$ time per nonterminal, which results in $\mathcal{O}(n \log \frac{N}{n} \log m + m)$ total complexity. This is clearly not enough to beat [5] on all possible inputs. Hence instead of performing the computation for each nonterminal separately, we try to process them in $\mathcal{O}(\log N)$ groups corresponding to the (truncated) logarithm of their length. Using the fact that the grammar is balanced, we are then able to achieve $\mathcal{O}(n \log \frac{N}{n} + m \log m)$ time. Because of some technical difficulties, in order to decrease this complexity we cannot really afford to check if the represented string occurs in p for each nonterminal exactly, though. Nevertheless, we can compute some approximation of this information, and by using a tailored variant of binary search applied to all nonterminals in a single group at once, we manage to process the whole grammar in time proportional to its size while adding just $\mathcal{O}(m)$ to the running time.

Some proofs are omitted because of the space constraints.

3   Preliminaries

The computational model we are going to use is the standard RAM allowing direct and indirect addressing, addition, subtraction, integer division and conditional jump with word size $w \geq \max\{\log n, \log N\}$. One usually allows multiplication as well in this model but we do not need it, and the only place where we use integer division (which in some cases is known to significantly increase the computational power), is the proof of Lemma 8.

We do not assume that any other operation (like, for example, taking logarithms) can be performed in constant time on arbitrary words of size w. Nevertheless, because of the n addend in the final running time, we can afford to preprocess the results on words of size $\log n$ and hence assume that some additional (reasonable) operations can be performed in constant time on such inputs.

As usually, $|w|$ stands for the length of w, $w[i \mathinner{.\,.} j]$ refers to its fragment of length $j - i + 1$ beginning at the i-th character, where characters are numbered starting from 1. All strings are over an alphabet of polynomial cardinality, namely $\Sigma = \{1, 2, \ldots, (n+m)^c\}$. A border of $w[1 \mathinner{.\,.} |w|]$ is a fragment which is both a proper prefix and a suffix of w, i.e., $w[1 \mathinner{.\,.} i] = w[|w| - i + 1 \mathinner{.\,.} |w|]$ with $i < |w|$. We identify such fragment with its length and say that $\mathrm{border}(t) = \{i_1, \ldots, i_k\}$ is the set of all borders of t. A period of a string $w[1 \mathinner{.\,.} |w|]$ is an integer d such that $w[i] = w[i + d]$ for all $1 \leq i \leq |w| - d$. Note that d is a period of w iff $|w| - d$ is a border. The following lemma is a well-known property of periods.

Lemma 1 (Periodicity lemma). *If p and q are both periods of w, and $p + q \leq |w| + \gcd(p, q)$, then $\gcd(p, q)$ is a period as well.*

The Lempel-Ziv representation of a string $t[1 \mathinner{.\,.} N]$ is a sequence of triples ($start_i$, len_i, $next_i$) for $i = 1, 2, \ldots, n$, where n is the size of the representation. $start_i$ and len_i are nonnegative integers, and $next_i \in \Sigma$. Such triple refers to a fragment of the text $t[start_i \mathinner{.\,.} start_i + len_i - 1]$ and defines $t[1 + \sum_{j<i} len_j \mathinner{.\,.} \sum_{j \leq i} len_j] = t[start_i \mathinner{.\,.} start_i + len_i - 1]next_i$. We require that $start_i \leq \sum_{j<i} len_j$ if $len_i > 0$. The representation is not self-referential if all fragments we are referring to are already defined, i.e., $start_i + len_i - 1 \leq \sum_{j<i} len_j$ for all i. The sequence of triples is often called the *LZ parse* of text.

Straight-line program is a context-free grammar in the Chomsky normal form such that the nonterminals $X_1, X_2, \ldots, X_s$ can be ordered in such a way that each X_i occurs exactly once as a left side, and whenever $X_i \to X_j X_k$ it holds that $j, k < i$. We identify each nonterminal with the unique string it derives, so $|X|$ stands for the length of the string derived from X. We call a straight-line program (SLP) *balanced* if for each production $X \to YZ$ both $|Y|$ and $|Z|$ are bounded by a constant fraction of $|X|$.

We preprocess the pattern p using standard tools (suffix trees [13] built for p and reversed p, and LCA queries [2]) to get the following primitives.

Lemma 2. *Pattern p can be preprocessed in linear time so that given i, j, k representing any two fragments $p[i \mathinner{.\,.} i + k]$ and $p[j \mathinner{.\,.} j + k]$ we can find their longest common prefix (suffix) in constant time.*

Lemma 3. *Pattern p can be preprocessed in linear time so that given any fragment $p[i \mathinner{.\,.} j]$ we can find its longest prefix (suffix) which is a suffix (prefix) of the whole pattern in constant time, assuming we know the (explicit or implicit) vertex corresponding to $p[i \mathinner{.\,.} j]$ in the suffix tree built for p (reversed p).*

We will also use the suffix array SA built for p [8]. For each suffix of p we store its position inside SA, and treat the array as a sequence of strings rather than a permutation of $\{1, 2, \ldots, |p|\}$. Given any word w, we will say that it occurs at position i in the SA if w begins $p[SA[i] \mathinner{\ldotp\ldotp} |p|]$. Similarly, the fragment of SA corresponding to w is the (maximal) range of entries at which w occurs.

4   Snippets Toolbox

In this section we develop a few efficient procedures operating on fragments of the pattern, which we call *snippets*:

Definition 1. *A snippet is a substring $p[i \mathinner{\ldotp\ldotp} j]$ of the pattern. If $i = 1$ we call it a prefix snippet, if $j = m$ a suffix snippet.*

We identify snippets with the substrings they represent, and use $|s|$ to denote the length of the string represented by s. A snippet is stored as a pair (i, j).

The two results of this section that we are going to use later build heavily on the contents of [6]. Specifically, Lemma 6 appears there as Lemma 5. To prove it, we first need the following simple and relatively well known property of borders.

Lemma 4. *If the longest border of t is of length $b \geq \frac{|t|}{2}$ then all borders of length at least $\frac{|t|}{2}$ create one arithmetic progression. More specifically, $\mathrm{border}(t) \cap \left\{ \frac{|t|}{2}, \ldots, |t| \right\} = \left\{ |t| - \alpha p : 1 \leq \alpha \leq \frac{|t|}{2d} \right\}$, where $d = |t| - b$ is the period of t. We call this set the long borders of t.*

By applying the preprocessing from the Knuth-Morris-Pratt algorithm to p and p^r we can extract borders of prefix and suffix snippets efficiently.

Lemma 5. *Pattern p can be preprocessed in linear time so that we can find the longest border of each its prefix (suffix) in constant time.*

The first result shows how to detect an occurrence in a concatenation of two snippets. We will perform a lot of such operations.

Lemma 6 (Lemma 5 of [6]). *Given a prefix snippet and a suffix snippet we can detect an occurrence of the pattern in their concatenation in constant time.*

The second result can be deduced from Lemma 6 and Lemma 8 of [6], but we prefer to give an explicit proof for the sake of completeness. Its running time is constant as long as $|s_1|$ is bounded from above by a constant fraction of $|s_2|$.

Lemma 7. *Given a prefix snippet s_1 and a snippet s_2 for which we know the corresponding (explicit or implicit) node in the suffix tree, we can compute the longest prefix of p which is a suffix of $s_1 s_2$ in time $\mathcal{O}\left(\max\left(1, \log \frac{|s_1|}{|s_2|}\right)\right)$.*

5   Constructing Balanced Grammar

Recall that a LZ parse is a sequence of triples $(start_i, len_i, next_i)$ for $i = 1, 2, \ldots, n$. In the not self-referential variant considered in [3], we require that $start_i + len_i - 1 \leq \sum_{j<i} len_j$ so that each triple refers only to the prefix generated so far. Although such assumption is made by some LZ-based compressors, [5] deals with the compressed pattern matching problem in its full generality, allowing self-references. Thus for the sake of completeness we need to construct a balanced grammar from a potentially self-referential LZ parse. It turns out that a small modification of a known method is enough for this task.

Lemma 8 (see Theorem 1 of [3]). *Given a (potentially self-referential) LZ parse of size n, we can build an α-balanced SLP of size $\mathcal{O}(n \log \frac{N}{n})$ describing the same string of length N, for any constant $0 < \alpha \leq 1 - \frac{\sqrt{2}}{2}$. Running time of the construction is proportional to the size of the output.*

As a result we get a context-free grammar in which all nonterminals derive exactly one string, and right sides of all productions are of the form XY with $\frac{\alpha}{1-\alpha} \leq \frac{|X|}{|Y|} \leq \frac{1-\alpha}{\alpha}$. The exact value of α is not important, we only need the fact that both $\frac{|X|}{|Y|}$ and $\frac{|Y|}{|X|}$ are bounded from above. For the sake of concreteness we assume $\alpha = 0.25$. We also need to compute $|X|$ for each nonterminal X, and to group the nonterminals according to the (rounded down) logarithm of their length, with the base of the logarithm to be chosen later. Note that taking logarithms of large numbers (i.e., substantially longer than $\log n$ bits) is not necessarily a constant time operations in our model. We can use the fact that the grammar is balanced here: if $X \to YZ$, then $\log_b |X| \leq \beta + \max(\log_b |Y|, \log_b |Z|)$ for some constant β depending only on α and b, and the logarithms can be computed for all nonterminals in a bottom-up fashion using just linear time.

6   Processing Balanced Grammar

While the final goal of this section is a $\mathcal{O}(n \log \frac{N}{n} + m)$ time algorithm, we start with a simple $\mathcal{O}(n \log \frac{N}{n} \log m + m)$ time solution, which then is modified to take just $\mathcal{O}(n \log \frac{N}{n} + m \log m)$, and finally $\mathcal{O}(n \log \frac{N}{n} + m)$ time.

For each nonterminal X we would like to check if the string it represents occurs inside p. If it does not, we would like to compute $\text{prefix}(X)$ and $\text{suffix}(X)$, the longest prefix (suffix) which is a suffix (prefix) of the whole p. Given such information for all possible nonterminals, we can easily detect an occurrence.

Lemma 9. *If p occurs in a string represented by a SLP then there exists a production $X \to YZ$ such that p occurs in $\text{suffix}(Y)\,\text{prefix}(Z)$.*

Lemma 10. *Given a (potentially self-referential) Lempel-Ziv parse of size n describing a text $t[1 \ldots N]$ and a pattern $p[1 \ldots m]$, we can detect an occurrence of p inside t deterministically in time $\mathcal{O}(n \log \frac{N}{n} \log m + m)$.*

Proof. By Lemma 8 and Lemma 9, we have to compute for each nonterminal X its corresponding snippet (if any) and both prefix(X) and suffix(X). We process the productions in a bottom-up order. Assume that we have the information concerning Y and Z available and would like to process $X \rightarrow YZ$. If both Y and Z correspond to substrings of p, we can apply binary search in the suffix array to find their concatenation in $\mathcal{O}(\log m)$ steps. To compute prefix(X) and suffix(X) we use Lemma 7 with some additional preprocessing.   □

We would like to remove the $\log m$ factor from the above complexity. It seems that the main difficulty here is that we need to implement a procedure for detecting if a concatenation of two substrings of p occurs in p as well, and in order to get the claimed running time we would need to answer such queries in constant time after a linear (or close to linear) preprocessing. We overcome the obstacle by choosing to work with an approximation of this information instead and using the fact that the grammar we are working with is balanced.

Definition 2. *A cover of a nonterminal X is pair of snippets $p[i .. i + 2^k - 1]$ and $p[j .. j + 2^k - 1]$ such that $2^k < |X| \leq 2^{k+1}$, $p[i .. i + 2^k - 1]$ is a prefix of the string represented by X, and $p[j .. j + 2^k - 1]$ is a suffix of the string represented by X. We call k the order of X's cover.*

We try to find the cover of each nonterminal X. If there is none, we know that the string it represents does not occur inside p. In such case we compute prefix(X) and suffix(X). More precisely, we either:

1. compute the cover, in such case the string represented by X might or might no occur in p,
2. do not compute the cover, in such case the string represented by X does not occur in p.

Having prefix(X) and suffix(X) for each nonterminal X is enough to detect an occurrence, and as the lemma below shows, it is possible to extract prefix(X) and suffix(X) from the cover of X using Lemma 7 in constant time.

Lemma 11. *Given the covers of Y and Z, we can compute prefix(X) and suffix(X) in constant time as long as $\frac{|Y|}{|Z|}$ and $\frac{|Z|}{|Y|}$ are bounded from above by a constant and for any snippet $p[i .. i + 2^k - 1]$ we can retrieve the corresponding node in the suffix tree in constant time. To compute prefix(X) we can use prefix(Z) instead of the cover of Z, and to compute suffix(X) we can use suffix(Y) instead of the cover of Y.*

To find the covers we process the nonterminals in groups. Nonterminals in the k-th group $\mathcal{G}_\ell = \{X_1, X_2, \ldots X_s\}$ are chosen so that $(\frac{4}{3})^\ell < |X_i| \leq (\frac{4}{3})^{\ell+1}$. The groups are disjoint so $\sum_\ell |\mathcal{G}_\ell| = \mathcal{O}(n \log \frac{N}{n})$. Furthermore, the partition can be constructed in linear time. We start with computing the covers of nonterminals in $\mathcal{G}_1$ naively. Then we assume that all nonterminal in $\mathcal{G}_{\ell-1}$ are already processed, and we consider $\mathcal{G}_\ell$. Because the grammar is 0.25-balanced, if $X_i \rightarrow Y_i Z_i$ then $|Y_i|, |Z_i| \leq \frac{3}{4}|X_i|$, and Y_i, Z_i belong to already processed $\mathcal{G}_{\ell'}$ with $\ell - 5 \leq \ell' < \ell$.

If for some Y_i or Z_i we do not have the corresponding cover, neither must have X_i, so we use Lemma 11 to calculate prefix(X_i), suffix(X_i), and remove X_i from $\mathcal{G}_\ell$.

For all remaining X_i we are left with the following task: given the covers of Y_i and Z_i, compute the cover of X_i, or detect that the represented string does not occur in p and so we do not need to compute the cover. Note that the known covers are of order k with $k_{min} = \lfloor \ell \log \frac{4}{3} \rfloor - 3 \le k \le \lceil \ell \log \frac{4}{3} \rceil = k_{max}$.

We reduce computing the covers to a sequence of batched queries of the form: given a sequence of pairs of snippets $p[i \mathinner{..} i + 2^{k_1} - 1]$, $p[j \mathinner{..} j + 2^{k_2} - 1]$ does their concatenation occur in p, and if so, what is the corresponding snippet? We call this merging the pair. For each ℓ we will require solving a constant number of such problems with $k_{min} \le k_1, k_2 \le k_{max}$, each containing $\mathcal{O}(|\mathcal{G}_\ell|)$ queries. We call this problem BATCHED-POWERS-MERGE. Before we develop an efficient solution for such question, lets see how it can be used to compute covers.

Lemma 12. *Computing covers of the nonterminals in any $\mathcal{G}_\ell$ can be reduced in linear time to a constant number of calls to* BATCHED-POWERS-MERGE, *with the number of pairs in each call bounded by $|\mathcal{G}_\ell|$.*

Now we only have to develop an algorithm for BATCHED-POWERS-MERGE. A simple solution would be to do a binary search in the suffix array built for p for each pair separately: we can compare $p[i \mathinner{..} i + 2^{k_1} - 1]p[j \mathinner{..} j + 2^{k_2} - 1]$ with any suffix of p in constant time using at most two longest common prefix queries so a single search takes $\mathcal{O}(\log m)$ time, which gets us back to the bounds from Theorem 10. In order to get a better running time we aim to exploit the fact that we are given many pairs at once. First observe that we can order all concatenations from a single problem efficiently.

Lemma 13. *Given $\mathcal{O}(|\mathcal{G}_\ell|)$ pairs of words of the form $p[i \mathinner{..} i + 2^{k_1} - 1]$, $p[j \mathinner{..} j + 2^{k_2} - 1]$ with $k_{min} \le k_1, k_2 \le k_{max}$ we can lexicographically sort their concatenations in time $\mathcal{O}(|\mathcal{G}_\ell| + m^\epsilon)$ if $|k_{max} - k_{min}| \in \mathcal{O}(1)$.*

We apply the above lemma to all calls to BATCHED-POWERS-MERGE corresponding to nonempty $\mathcal{G}_\ell$. If $(\frac{4}{3})^\ell > m$ then clearly the corresponding $\mathcal{G}_\ell$ is empty, so the total running time of this part is just $\mathcal{O}(m^\epsilon \log m + \sum_\ell |\mathcal{G}_\ell|) = \mathcal{O}(m + n \log \frac{N}{n})$. Now that the queries in a single call to BATCHED-POWERS-MERGE are sorted, instead of performing a separate binary search for each of them we can scan the queries and the suffix array at once, resulting in a $\mathcal{O}(|\mathcal{G}_\ell| + m)$ running time for each different ℓ. As after a $\mathcal{O}(m \log m)$ preprocessing we can compute the corresponding node in the suffix tree for all snippets $p[i \mathinner{..} i + 2^k - 1]$ used in Lemma 11, this gives us the following total running time.

Theorem 1. *Given a (potentially self-referential) Lempel-Ziv parse of size n describing a text $t[1 \mathinner{..} N]$ and a pattern $p[1 \mathinner{..} m]$, we can detect an occurrence of p inside t deterministically in time $\mathcal{O}(n \log \frac{N}{n} + m \log m)$.*

This is still not enough to improve [5] on all possible inputs. We would like to replace $m \log m$ by m in the above complexity by focusing on improving the running time of BATCHED-POWERS-MERGE. At a high level the idea is to consider the queries in a single call in sorted order, and start the binary search where the lexicographically previous pair was found. This might be still too slow though. To accelerate the search we develop a constant time procedure for locating the fragment of the suffix array corresponding to all occurrences of a given $p[i \mathrel{..} i + 2^k - 1]$. This procedure will be also used in Lemma 11.

Lemma 14. *The pattern p can be processed in linear time so that given any $p[i \mathrel{..} i + 2^k - 1]$ we can compute its first and last occurrence in the suffix array of p in constant time.*

Observe that the above lemma can be used to give an optimal solution for a slight relaxation of the *substring fingerprints* problem considered in [4]. This problem is defined as follows: given a string s, preprocess it to compute any *substring hash* $h_s(s[i \mathrel{..} j])$ efficiently. We require that:

1. $h_s(s[i \mathrel{..} j]) \in [1, \mathcal{O}(|s|^2)]$ so that the values can be operated on efficiently,
2. $h_s(s[i \mathrel{..} j]) = h_s(s[k \mathrel{..} l])$ iff $s[i \mathrel{..} j] = s[k \mathrel{..} l]$.

If we allow the range of h_s to be slightly larger, say $\mathcal{O}(|s|^3)$, a direct application of the above lemma allows us to evaluate the fingerprints in constant time.

Lemma 15. *Substring fingerprints of size $\mathcal{O}(|s|^3)$ can be computed in constant time after a linear time preprocessing.*

Now getting back to the original question, the input to BATCHED-POWER-MERGE is a sequence of pairs of snippets $w_1, w_2, \ldots, w_{|\mathcal{G}_\ell|}$. By Lemma 13 we can consider them in a sorted order. For each such pair $w = p[i \mathrel{..} i + 2^{k_1} - 1]p[j \mathrel{..} j + 2^{k_2} - 1]$, we first look up the fragment of the suffix array corresponding to its prefix $p[i \mathrel{..} i + 2^{k_{min}} - 1]$ using Lemma 14. Then we apply binary search in this fragment, with the exception that if the previous binary search was in this fragment as well, we start from the position it finished, not the beginning of the fragment. Additionally, the binary search is performed from the beginning and the end of the interval at the same time. If the initial interval is $[a, b]$ and the position we are after is r, such modified search uses just $\mathcal{O}(\log \min(r - a + 1, b - r + 1))$ applications of Lemma 2 instead of $\mathcal{O}(\log(b - a + 1))$ time, which is important.

While a single binary search might require a non-constant time, we will show that their amortized complexity is constant. To analyze the whole sequence of those searches, we keep a partition of the whole $[1, |p|]$ into a number of disjoint intervals. Doing a single search splits at most one interval into two parts at the position of the first occurrence. If the first occurrence is exactly at an already existing boundary, there is no split, otherwise we say that those two smaller intervals have been created in phase k_{min} (recall that k_{min} linearly depends on ℓ), and intervals created in phase k_{min} are kept in a list $I_{k_{min}}$. We do not want to split an interval more than once and hence each call to BATCHED-POWERS-MERGE starts with finding for each w_i its corresponding interval in $I_{k_{min}}$. After processing all concatenations, we add the new intervals to $I_{k_{min}}$ and prune it to contain the intervals which are minimal under inclusion. Scanning and pruning $I_{k_{min}}$ takes linear time in its size, and we will show that this size is small.

Algorithm 1. BATCHED-POWERS-MERGE$(w_1, w_2, \ldots, w_{|\mathcal{G}_\ell|})$

1: sort all w_i ▷ **Lemma 13**
2: scan $I_{k_{min}}$ to find the intervals containing w_i
3: $L \leftarrow \emptyset$
4: $r_0 \leftarrow 1$
5: **for** $i \leftarrow 1$ to $|\mathcal{G}_\ell|$ **do**
6:     $[a,b] \leftarrow$ the interval corresponding to $w_i[1 .. 2^{k_{min}}]$ in SA ▷ **Lemma 14**
7:     choose $[c,d] \in I_{k_{min}}$ containing the first occurrence of w_i in SA
8:     **if** $[c,d]$ is defined **then**
9:         $a \leftarrow \max(a,c)$
10:       $b \leftarrow \min(b,d)$
11:     **end if**
12:     $a \leftarrow \max(r_{i-1}, a)$
13:     $r_i \leftarrow$ TWO-WAY-BINARY-SEARCH(a, b, w_i)
14:     add $[a, r_i]$ and $[r_i, b]$ to L
15: **end for**
16: sort L and merge it with $I_{k_{min}}$, removing non-minimal intervals
17: **return** all answers r_i

Lemma 16. *All $\mathcal{O}(\log m)$ calls to* BATCHED-POWERS-MERGE *run in total time* $\mathcal{O}(m + \sum_\ell |\mathcal{G}_\ell|)$.

Proof. First note that the sorting in line 16 can be performed in time $\mathcal{O}(m^\epsilon + |I_{k_{min}}| + |\mathcal{G}_\ell|)$ using radix sort. Line 1 takes time $\mathcal{O}(m^\epsilon + |\mathcal{G}_\ell|)$ due to Lemma 13, and line 2 requires $\mathcal{O}(|I_{k_{min}}| + |\mathcal{G}_\ell|)$. All executions of line 7 take time $\mathcal{O}(|I_{k_{min}}|)$ because the words w_i are already sorted. For the time being assume that the binary search in line 13 is for free. Then the total complexity becomes $\mathcal{O}(\sum_i m^\epsilon + \left|I_{k_{min}}^{(i)}\right| + |\mathcal{G}_\ell|)$ where $\left|I_{k_{min}}^{(i)}\right|$ is the size of $I_{k_{min}}$ just before the i-th call to BATCHED-POWERS-MERGE. There is a constant number of those calls for each value of $1 \leq \ell \leq m$, and each k_{min} corresponds to at most constant number of different continuous values of ℓ, thus the sum is in fact $\mathcal{O}(m + \sum_\ell |\mathcal{G}_\ell|)$.

To finish the proof we have to bound the time taken by all binary searches. For that to happen we will view the intervals as vertices of a tree. Whenever performing a binary search splits an interval into two, we add a left and right child to the corresponding leaf v, see Figure 1. The *rank* rank(v) of a vertex v is the rounded logarithm of its *weight*, which is the length of the corresponding interval. Then the cost of line 13 is simply $\mathcal{O}(1 + \min(\text{rank}(\text{left}(v)), \text{rank}(\text{right}(v))))$ where left(v) and right(v) are the left and right child of v, respectively. Hence we should bound the sum $\sum_v \min(\text{rank}(\text{left}(v)), \text{rank}(\text{right}(v)))$, where v is a non-leaf. We say that a vertex is *charged* when its weight does not exceed the weight of its brother. Now we claim that there are at most $\frac{m}{2^k}$ charged vertices of rank k: assume that there are u and v such that u is an ancestor of v, both are charged and of rank k, then weight of v plus weight of its brother is at least twice as large as the weight of v alone, thus the rank of their parent is larger than the rank of v, contradiction. So all charged vertices of the same rank correspond to disjoint intervals, and there cannot be more than $\frac{m}{2^k}$ disjoint intervals of length at least 2^k on a segment of length m. Bounding the sum gives the claim:

$$\sum_{v} \min(\mathrm{rank}(\mathrm{left}(v)), \mathrm{rank}(\mathrm{right}(v))) \leq \sum_{k \geq 0}^{\log m} k\frac{m}{2^k} \leq m \sum_{k \geq 0}^{\infty} \frac{k}{2^k} = 2m \qquad \square$$

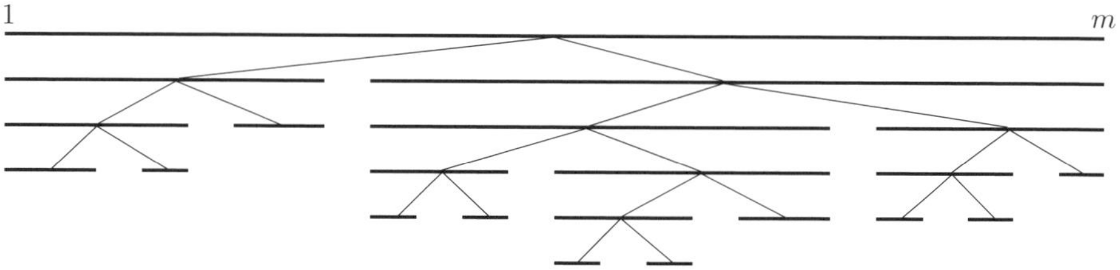

Fig. 1. Interpreting the intervals as a tree

Theorem 2. *Given a 0.25-balanced SLP of size $\mathcal{O}(n \log \frac{N}{n})$ and a pattern $p[1 .. m]$, we can detect an occurrence of p in the represented text in time $\mathcal{O}(n \log \frac{N}{n} + m)$.*

Proof. By Lemma 12 and Lemma 16 we compute the covers of all nonterminals which represent subwords of p in time $\mathcal{O}(n \log \frac{N}{n} + m)$. For the remaining nonterminals X we use Lemma 11 to compute $\mathrm{prefix}(X)$ and $\mathrm{suffix}(X)$ in linear time considering the nonterminals in bottom-up order. Then due to Lemma 9 if there is an occurrence of p, there is an occurrence in $\mathrm{prefix}(Y) \mathrm{suffix}(Z)$ for some production $X \to YZ$. We consider every nonterminal X, either lookup the already computed $\mathrm{prefix}(Y)$ and $\mathrm{suffix}(Z)$ or compute them using the known covers and Lemma 11, and use Lemma 6 to detect a possible occurrence. $\square$

Theorem 3. *Given a (potentially self-referential) Lempel-Ziv parse of size n describing a text $t[1 .. N]$ and a pattern $p[1 .. m]$, we can detect an occurrence of p inside t deterministically in time $\mathcal{O}(n \log \frac{N}{n} + m)$.*

7   Conclusions

Recall that in order to guarantee a $\mathcal{O}(n \log \frac{N}{n} + m)$ running time, it was necessary to use integer division in the proof of Lemma 8. This was the only such place, though. If we assume that integer division is not allowed, and the only operations on the integers $start_i, len_i$ appearing in the input triples are addition, subtraction, multiplication and comparing with 0 (which are the only operations used by the $\mathcal{O}(n \log N + m)$ version of our algorithm), we can prove a matching lower bound by looking at the corresponding algebraic computation trees. More precisely, using standard tools [14] one can show that the depth of such tree which recognizes the set of integers $t, x_1, x_2, \ldots, x_n$ such that for all i it holds that $x_i = (2\alpha_i + 1)t + \beta_i$ with $0 \leq \beta_i < t$ and $0 \leq \alpha_i < N$ is $\Omega(n \log N)$. On the other hand, one can construct a self-referential LZ of constant size deriving $(1^t 0^t)^N$. Hence one can construct a LZ of size $\mathcal{O}(n)$ deriving $(1^t 0^t)^N b_1 1 \ldots b_n 1$ where $b_i = \lfloor \frac{x_i}{t} \rfloor \bmod 2$ which does not contain 11 as a substring iff all x_i are of the form $x_i = (2\alpha_i + 1)t + \beta_i$ and the lower bound follows.

References

1. Amir, A., Benson, G., Farach, M.: Let sleeping files lie: pattern matching in z-compressed files. In: SODA 1994: Proceedings of the Fifth Annual ACM-SIAM Symposium on Discrete Algorithms, pp. 705–714. SIAM, Philadelphia (1994)
2. Bender, M.A., Farach-Colton, M.: The lca problem revisited. In: Gonnet, G.H., Viola, A. (eds.) LATIN 2000. LNCS, vol. 1776, pp. 88–94. Springer, Heidelberg (2000)
3. Charikar, M., Lehman, E., Liu, D., Panigrahy, R., Prabhakaran, M., Rasala, A., Sahai, A., Shelat, A.: Approximating the smallest grammar: Kolmogorov complexity in natural models. In: STOC 2002: Proceedings of the Thiry-Fourth Annual ACM Symposium on Theory of Computing, pp. 792–801. ACM, New York (2002)
4. Farach, M., Muthukrishnn, S.: Perfect hashing for strings: Formalization and algorithms. In: Hirschberg, D.S., Meyers, G. (eds.) CPM 1996. LNCS, vol. 1075, pp. 130–140. Springer, Heidelberg (1996)
5. Farach, M., Thorup, M.: String matching in Lempel-Ziv compressed strings. In: STOC 1995: Proceedings of the Twenty-Seventh Annual ACM Symposium on Theory of Computing, pp. 703–712. ACM, New York (1995)
6. Gawrychowski, P.: Optimal pattern matching in LZW compressed strings. In: SODA 2011: Proceedings of the Twenty-Second Annual ACM-SIAM Symposium on Discrete Algorithms (2011)
7. Iacono, J., Özkan, Ö.: Mergeable dictionaries. In: Abramsky, S., Gavoille, C., Kirchner, C., Meyer auf der Heide, F., Spirakis, P.G. (eds.) ICALP 2010. LNCS, vol. 6198, pp. 164–175. Springer, Heidelberg (2010)
8. Kärkkäinen, J., Sanders, P., Burkhardt, S.: Linear work suffix array construction. J. ACM 53(6), 918–936 (2006)
9. Karp, R.M., Rabin, M.O.: Efficient randomized pattern-matching algorithms. IBM J. Res. Dev. 31(2), 249–260 (1987)
10. Kida, T., Matsumoto, T., Shibata, Y., Takeda, M., Shinohara, A., Arikawa, S.: Collage system: a unifying framework for compressed pattern matching. Theor. Comput. Sci. 298, 253–272 (2003)
11. Kosaraju, S.R.: Pattern matching in compressed texts. In: Thiagarajan, P.S. (ed.) FSTTCS 1995. LNCS, vol. 1026, pp. 349–362. Springer, Heidelberg (1995)
12. Rytter, W.: Application of Lempel-Ziv factorization to the approximation of grammar-based compression. Theor. Comput. Sci. 302(1-3), 211–222 (2003)
13. Ukkonen, E.: On-line construction of suffix trees. Algorithmica 14(3), 249–260 (1995)
14. Yao, A.C.C.: Lower bounds for algebraic computation trees with integer inputs. In: Proceedings of the 30th Annual Symposium on Foundations of Computer Science, pp. 308–313. IEEE Computer Society, Washington, DC, USA (1989)

An Experimental Study on Approximating K Shortest Simple Paths

Asaf Frieder and Liam Roditty

Department of Computer Science, Bar-Ilan University, Ramat-Gan, Israel
{asaffrr,liam.roditty}@gmail.com

Abstract. We have conducted an extensive experimental study on approximation algorithms for computing k shortest simple paths in weighted directed graphs. Very recently, Bernstein [2] presented an algorithm that computes a $1 + \varepsilon$ approximated k shortest simple paths in $O(\varepsilon^{-1}k(m + n\log n)\log^2 n)$ time. We have implemented Bernstein's algorithm and tested it on synthetic inputs and real world graphs (road maps). Our results reveal that Bernstein's algorithm has a practical value in many scenarios. Moreover, it produces in most of the cases exact paths rather than approximated. We also present a new variant for Bernstein's algorithm. We prove that our new variant has the same upper bounds for the running time and approximation as Bernstein's original algorithm. We have implemented and tested this variant as well. Our testing show that this variant, which is based on a simple theoretical observation, is better than Bernstein's algorithm in practice.

1 Introduction

Computing a shortest path connecting between a pair of vertices in a graph is one of the most fundamental algorithmic problems in graph theory. Its natural generalization is to compute a set of k shortest paths that connect between a given pair of vertices, for some integer $k > 1$. Formally, let $G(V, E)$ be a graph with non-negative edge weights, where $|V| = n$ and $|E| = m$. Let $k > 1$ be a positive integer, and let $s, t \in V$ be two arbitrary vertices. The algorithm has to compute the k shortest paths from s to t, where any two paths have to differ by at least one edge.

This problem was studied extensively in the last 4 decades. We outline here the main results. In the beginning of the seventies Yen [19] and Lawler [12] showed how to compute the k shortest **simple** paths in weighted directed graphs. The running time of their algorithms, when modern data structures are used, is $O(k(mn + n^2 \log n))$. For the restricted case of undirected graphs much faster algorithms are available. Katoh, Ibaraki and Mine [11] and Hershberger and Suri [10] presented an algorithm with a running time of $O(k(m + n\log n))$. For the case that the paths are not required to be simple Eppstein [6] presented an algorithm, for weighted directed graphs, that finds the k shortest paths in $O(m+n\log n+k)$. Roditty and Zwick [16] presented a randomized algorithm with a running time of $O(km\sqrt{n}\log^2 n)$ for unweighted directed graphs. Throughout

C. Demetrescu and M.M. Halldórsson (Eds.): ESA 2011, LNCS 6942, pp. 433–444, 2011.
© Springer-Verlag Berlin Heidelberg 2011

the last three decades there were also many improved implementations of Yen's and Lawler's algorithms [3,8,13,14]; however, the worst case bound of $O(k(mn + n^2 \log n))$ for weighted directed graphs, remains unbeaten.

Closely related to the k shortest simple paths problem is the replacement paths problem. In the *replacement paths* problem we are required to find, for every edge e on the shortest path from s to t, a shortest path from s to t that avoids e (see Demetrecu et. al. [4] and Gotthilf and Lewenstein [7] for more details). Recently, Vassilevska W. [18] showed that replacement paths can be computed in $\tilde{O}(n^\omega)$ time using fast matrix multiplications.

Hershberger at al [9] used their algorithm [10] for replacement paths in undirected graphs that has an $O(m + n \log n)$ running time. However, running the original algorithm on directed graphs risks that it will generate non-simple paths in some cases. In such cases the algorithm detects the failure and switch to a slower algorithm that computes a simple path in $O(mn + n^2 \log n)$ time. They have tested the algorithm on various types of data and showed that in most cases there is no need to switch to the slower algorithm.

The lack of theoretical improvement in directed graphs can be explained by a recent result of Vassilevska W. and Williams [17] in which they showed that the second shortest simple path problem in weighted directed graphs is equivalent to many other graph and matrix problems for which no truly subcubic ($O(n^{3-\varepsilon})$-time for constant $\varepsilon > 0$) algorithms are known. In particular, they showed that if there is a truly subcubic algorithm for the second shortest simple path, then the all pairs shortest paths (APSP) problem also has a truly subcubic algorithm.

It is very common to relax the requirements to gain efficiency for problems with seemingly no subcubic time exact algorithms. Many such algorithms have been developed in the context of all pair of shortest paths computation (see for example [20,1,5]). In the context of k shortest simple paths Roditty [15] presented an approximation for weighted directed graphs. The $i \leq k$ path returned by the algorithm is at most $3/2$ times more than the exact ith shortest simple path. The running time of the algorithm is $O(k(m\sqrt{n}+n^{3/2} \log n))$. Roditty [15] based his algorithm on an earlier observation made by Roditty and Zwick [16]. They observed that the k shortest simple paths can be computed by $O(k)$ applications of an algorithm that computes the second shortest simple path. Roditty [15] presented an algorithm that computes a $3/2$ multiplicative approximation of the second shortest simple path and used it to obtain an approximation for the k shortest simple paths.

Very recently, in a major breakthrough, Bernstein [2] significantly improved both the running time and the approximation factor of Roditty's algorithm. In particular, Bernstein [2] presented an algorithm that computes a $1 + \varepsilon$ multiplicative approximation of the second shortest simple path in $O(\varepsilon^{-1}(m + n \log n) \log^2 n)$ time. Using his algorithm with the observation of Roditty and Zwick [16] he obtained an $O(\varepsilon^{-1} k(m + n \log n) \log^2 n)$ time algorithm for the k shortest simple paths problem.

In this paper we report on an extensive experimental study that we performed on Bernstein's algorithm. The first goal of our study was to find out whether his algorithm is practical. In many cases algorithms with an efficient worst case upper bound hide large constants and poly-logarithmic factors that make them impractical for inputs of reasonable size. In such cases an algorithm with a slower worst case upper bound can turn out to be more efficient in practice. Thus, we have compared Bernstein's algorithm to an algorithm that computes exact k shortest simple paths using a naive second shortest path algorithm. This baseline algorithm can be viewed as a variant of Yen's algorithm. Our experiments show that Bernstein's algorithm is significantly faster then the naive algorithm in most of the cases.

One of the reasons Bernstein's algorithm is very practical is that it computes approximated paths rather than exact paths. This leads us to another goal of our study. We analyzed the approximation factor of the algorithm. Our testing show that the actual error is less than 0.1% of the theoretical upper bound. Moreover, in many cases the algorithm finds the exact paths.

While working on the implementation of Bernstein's algorithm we became aware to a simple improvement that does not change the theoretical upper bound but might improve the running time in practice. We have implemented this variant as well and compared it both to Bernstein's algorithm and to the naive algorithm. This variant improves significantly the results of Bernstein's algorithm. It produced only exact paths and it was faster than Bernstein's algorithm in all cases.

The rest of this paper is organized as follow. In the next Section we describe the algorithm of Bernstein. In Section 3 we describe our improvement to Bernstein's algorithm and prove that its running time and approximation factor are the same. In Section 4 we discuss a couple of implementation issues that were not addressed explicitly by Bernstein and required a careful care. Finally, in Section 5 we present our experimental results. Due to the strict space restriction we differ all Figures and Tables to the Appendix.

2   Bernstein's Approach

Let $G = (V, E, w)$ be a directed graph with nonnegative edge weights and let $s, t \in V$. Let $\delta(s, t)$ be the length of the shortest paths from s to t. Let $P(s, t)$ be a set of vertices on a path from s to t, that is, $P(s, t) = \{s = u_1, u_2, \ldots, u_\ell = t\}$ and $(u_i, u_{i+1}) \in E$, for every $1 \leq i \leq \ell - 1$. Let $w(P(s, t)) = \sum_{i=1}^{\ell-1} w(u_i, u_{i+1})$ be the weight of the path. Let $|P(s, t)|$ be the hop length of the path. Let $P_1(s, t)$ be a shortest path from s to t. We define the i-th shortest path $P_i(s, t)$ as the shortest simple path that is also different from all paths $P_j(s, t)$ for every $j < i$. Let $\delta_i(s, t) = w(P_i(s, t))$. The objective of a k-shortest path algorithm is to output k shortest simple paths. Roditty and Zwick [16] showed that k-shortest simple paths can be computed by $O(k)$ computations of the second shortest path. Hence, it suffices to compute efficiently the second shortest path. Bernstein [2] showed how to approximate the second shortest path in almost linear time. His algorithm outputs a path of length at most $(1 + \varepsilon) \cdot \delta_2(s, t)$. Below we describe the details of his algorithm.

We start with the definition of a detour and its span:

Definition 1 (Detour and detour span). *Let* $P(s,t) = \{s = u_1, u_2, \ldots, u_\ell = t\}$ *be a shortest simple path from* s *to* t. *A simple path* $D(u_i, u_j)$ *from* u_i *to* u_j, *where* $j > i$, *is a detour of* $P(s,t)$ *if* $D(u_i, u_j) \cap P(s,t) = \{u_i, u_j\}$. *The span of the detour* $D(u_i, u_j)$ *is* $j - i$. *The detour span of a path* $P(s, u_i) \cdot D(u_i, u_j) \cdot P(u_j, t)$[1] *is the span of* $D(u_i, u_j)$.

Bernstein [2] splits the search for the second shortest path into $\log q$ phases, where $q = |P_1(s,t)|$. In the i-th phase the algorithm tries to compute a shortest path from s to t with detour span in the range $[q/2^i, q/2^{i-1}]$. Let U_i be the length of a shortest s-t path with a detour span at least $q/2^i$

In phase i the vertices of $P_1(s,t)$ are divided into 2^i intervals each of size $q/2^i$ (assuming that q is a power of 2). Let $I_1, I_2 \ldots, I_{2^i}$ be the resulting intervals. Each phase is divided into 4 sub-phases. In sub-phase $j \in [1,4]$ the vertices of the intervals $I_j, I_{j+4}, I_{j+8} \ldots$ are labeled as start vertices and all other vertices are labeled as finish vertices. Let $L_{i,j}$ be the labeling of vertices into intervals in the j-th sub-phase of the i-th phase. For the special case of $q/2$ we only have one start interval and one finish interval so we denote this labeling with L_1. For a labeling L of the vertices let $\delta(L)$ be the length of the shortest path of the form $P(s, v_x) \cdot D(v_x, v_y) \cdot P(v_y, t)$, where v_x is labeled with start, v_y is labeled with finish, and $y > x$.

It is straightforward to see that for any pair of vertices $u_x, u_y \in P_1(s,t)$ such that $y - x \in [q/2^i, q/2^{i-1}]$ there is a sub-phase in which u_x is labeled as a start vertex and u_y is labeled as a finish vertex. Thus, any path with detour span from the range $[q/2^i, q/2^{i-1}]$ is represented in one of the four sub-phases. For each start interval $I_h \in I_j, I_{j+4}, I_{j+8} \ldots$ in sub-phase j of phase i we use the graph $G_{i,j,I_h}(V', E')$. We initialize V' with V and remove every vertex that appears in interval prior to I_h to prevent non-simple paths. We initialize E' with E. We then remove the incoming edges of every vertex that is labeled as a start vertex and the outgoing edges of every vertex that is labeled as a finish vertex. We also remove the edges of $P_1(s,t)$. We add to E' the following edges. For each finish vertex u we add an edge (u,t) whose weight is $\delta(u,t)$. We add a new vertex s' with an edge to every vertex u in the start interval I_h whose weight is $\delta(s,u)$. We will soon explain our reason to do so.

Ideally, we would like to compute a shortest path from s to t for every start interval. However, this requires running Dijkstra's algorithm for each start interval and there are $\Omega(n)$ start intervals. Roughly speaking, Bernstein's main idea is to run Dijkstra's algorithm in a progressive manner for all start intervals of a given sub-phase. The algorithm relaxes an edge only if it significantly improves the current shortest path estimation.

Assume that we are in phase i and sub-phase $j = 1$. We have the start intervals $I_1, I_5, I_9 \ldots$. We start by running Dijkstra from s' on the graph $G_{i,1,I_1}(V', E')$. We maintain an estimated distance $c(u)$ for every vertex as in the regular Dijkstra. We initialize $c(s') = 0$ and $c(u) = \infty$ for every $u \neq s'$. When we finish

---

[1] We use $\cdot$ for concatenating two paths.

running Dijkstra on the graph $G_{i,1,I_1}(V', E')$ we proceed and run Dijkstra this time for the graph $G_{i,1,I_5}(V', E')$ but with the values of $c(u)$ that were updated during the previous run. Moreover, we do not build the graph $G_{i,1,I_5}(V', E')$ from scratch as it is too time consuming, instead we use $G_{i,1,I_1}(V', E')$ and make only the required changes to obtain $G_{i,1,I_5}(V', E')$. In each run we start the progressive Dijkstra from s' that only has edges to the current start interval. We cannot use s directly as it might have many edges and we cannot afford scanning these edges in each step. Now in the second run (and the ones that follow) we do not use the regular relax procedure. Instead when we relax an edge (u', u) we only update the value $c(u)$ to $c(u') + w(u', u)$ if $c(u') + w(u', u) \leq c(u) - \alpha$ (where $\alpha = \varepsilon \cdot UB/8\log q$ and UB is the cost of the best path found so far). Once $c(u)$ is changed during the run of Dijkstra for a start interval we only require that $c(u') + w(u', u) \leq c(u)$ (as it is already in the heap). Moreover, when a vertex v is extracted from the heap its edges are scanned only if $c(v) < UB$. At the end of each sub-phase j the algorithm set $R_{i,j}$ to $c(t)$ and reset $c(v)$ for every $v \in V$.

Using these relaxation rules a vertex can only be explored $O(\log n \cdot 1/\varepsilon)$ times during a sub-phase and the running time is bounded by $O(\varepsilon^{-1}(m + n\log n)\log^2 n)$. We refer to this relaxation rule as *additive* relaxation and denote the algorithm that uses it with $BR(\varepsilon)$.

3   A Practical Improvement

In this section we present a new idea that reduces even further the number of times we explore each vertex. This improvement does not reduce the worst-case upper bound on the running time but our experiments indicate the running time improves in practice for most cases.

Our idea is based on the following observation. In phase i, the Bernstein's algorithm only needs to compute paths with detour span in the range $[q/2^i, q/2^{i-1}]$. In particular, in sub-phase j it computes paths with detours that start in a start interval I_h, where $I_h \in \{I_j, I_{j+4}, I_{j+8} \ldots\}$ and end in one of the finish intervals $I_{h+1}, I_{h+2}, I_{h+3}$. Let $v(x)$ be the last vertex of interval I_x. Let UB be the best path that we have obtained in previous phases. If we currently perform a progressive Dijkstra for a start interval I_h then there is no point in relaxing an edge (u', u) in case that $c(u') + w(u', u) \geq UB - \delta(v(h+3), t)$, as in the most optimistic case the path from u has to traverse at least $\delta(v(h+3), t)$ to reach t from a vertex in $I_{h+1}, I_{h+2}, I_{h+3}$. Although the amount that we subtract from UB is different for each start interval it is easy to obtain it as it is part of the shortest path computed from s to t. We refer to this amount as the *detour limit* and for start interval I_x we denote it with DL_x. In general we will relax an edge (u', u) for start interval I_x only if $c(u') + w(u', u) < UB - DL_x$ or $u = t$. We refer to this relaxation rule as *limited* relaxation.

The second algorithm that we obtain combines between the additive relaxation and the limited relaxation rules, that is, an edge must satisfy both rules in order to be relaxed. We denote this hybrid approach with $BN(\varepsilon)$.

The running time of the algorithm is still $O((m+n\log n)\log^2 n \cdot 1/\varepsilon)$. Bernstein proved that the additive factor limits the number of times in which a vertex

participates in a relaxation in each a sub-phase. Our algorithm still demands
an improvement by at least $\alpha = \varepsilon \cdot UB/8 \log q$ every time a vertex is added to
the heap. Therefore, a vertex is still explored only $O(\log n \cdot 1/\varepsilon)$ times during a
sub-phase.

Recall that $\delta(L_{i,j})$ is the length of the shortest path satisfies the labeling $L_{i,j}$.
Bernstein's algorithms compute a sequence of values R_1 and $R_{i,j}$ for every $i > 1$
and $j \in [1, 4]$ that satisfy the following requirements:

1. $R_1 = \delta(L_1)$
2. If $\delta(L_{i,j}) < U_{i-1}/(1 + \varepsilon')$ then $R_{i,j} = \delta(L_{i,j})$
3. If $\delta(L_{i,j}) \geq U_{i-1}/(1 + \varepsilon')$ then $R_{i,j} \geq \delta(L_{i,j})$

He then showed that $R = min_{i,j}(R_{i,j})$ is a $1 + \varepsilon$ approximation of the second
shortest path. We will now show that even when we combine the additive relaxation with the limited relaxation a sequence of values that satisfy the above
requirements is computed.

The first requirement, $R_1 = \delta(L_1)$, holds because in the first phase we perform
a full exploration. Moreover, $R_{i,j} \geq \delta(L_{i,j})$ because the combined algorithm
considered only paths that satisfies the labeling $L_{i,j}$ and hence their length is at
least $\delta(L_{i,j})$.

Let $P = P(s, v_x) \cdot D(v_x, v_y) \cdot P(v_y, t)$ be a shortest path in sub-phase j of
phase i such that $v_x \in I_h$. Assume that algorithm $BN(\varepsilon)$ did not find this path.
There must be an edge (u', u) of P that was not relaxed. Moreover, as we relax
all edges going out from s' and all edges getting into t this edge must be part
of the detour $D(v_x, v_y)$. Hence, the edge (u', u) does not satisfy at least one of
the relaxation rules. Bernstein showed that if the additive relaxation rule does
not hold then $\delta(L_{i,j}) \geq U_{i-1}/(1 + \varepsilon')$. When this is the case we only need to
satisfy the third requirement and as we mentioned above $R_{i,j} \geq \delta(L_{i,j})$. Thus,
we need to deal with the case that the edge (u', u) does not satisfy the limited
relaxation rule. We show that in such a case $\delta(L_{i,j}) \geq U_{i-1}/(1 + \varepsilon')$ as well. If
we did not relax (u', u) then the length of $P(s, u)$ is more than $UB - \delta(v(h), t)$.
If $y - x \geq q/2^{i-1}$ then P is a valid path also in phase $i - 1$ thus $U_{i-1} \leq \delta(L_{i,j})$.
For the case that $y - x < q/2^{i-1}$ we get:

$$\delta(L_{i,j}) = w(P(s, u)) + w(P(u, t)) \geq UB - \delta(v(h), t) + w(P(u, t)) \geq$$

$$\geq UB - \delta(v(h), t) + \delta(v(h), t) = UB \geq U_{i-1}$$

We conclude that the $BN(\varepsilon)$ algorithm computes values R_1 and $R_{i,j}$ for every
$i > 1$ and $j \in [1, 4]$ that satisfy the desired requirements and hence obtains a
$1 + \varepsilon$ approximation.

4   Implementation Issues

While implementing Bernstein's algorithm we became aware of several issues
that were not explicitly explained in his paper and required careful treatment.

In his paper Bernstein assumed that the number of vertices of the shortest path is a power of 2. For a path of arbitrary size, we rounded up the number of vertices in each interval and the number of phases.

Bernstein also does not provide the details of how to produce an actual path by its algorithm. If we reconstruct a path, from t backwards using the last edges that updated the cost of each vertex, the path we will produce may not be simple. The correct path can be reconstructed using only the edges relaxed in the interval where that path was originally found.

Due to the strict space restriction we differ the details of these problems and our solutions to the full version of this paper.

5   Experimental Results

There were several objectives to our experiments. The first and the most obvious objective was to see if Bernstein's algorithm is practical and does not hide large constant factors that make it impossible to run on real world inputs. As our results indicate the algorithm is very practical and performs very well both on real world graphs that represent road maps and synthetic generated data.

A more specific question was to discover the practical approximation factor by comparing the algorithm results to the exact paths that can be found using a simple base-line algorithm. Moreover, as Bernstein's algorithm works for different values of ε and its running time depends on ε it is important to understand the affect of different values of ε on the running time. For example, a natural question is whether it worth to have $\varepsilon = 1/16$ and large running time or it is ok to settle for $\varepsilon = 1$ for a better running time.

Another byproduct of our research is the new version of the algorithm that limits the search. We have compared this approach as well. For the naive base line algorithm we used the following approach. Given a shortest path compute the second shortest path by removing each edge of the shortest path in its turn and computing the shortest path in the resulted graph. The cost of such an algorithm is obviously $\tilde{O}(mn)$ but it produces exact paths. We have implemented and tested the three variants of Bernstein's algorithm mentioned before: $BM(\varepsilon)$, $BR(\varepsilon)$ and $BN(\varepsilon)$. We used in our testing both road networks and synthetic data. Since $BM(\varepsilon)$ and $BR(\varepsilon)$ were both suggested by Bernstein but $BR(\varepsilon)$ has a better running time we mainly used $BR(\varepsilon)$. We have conducted also a comparison between the two for special graphs. Through this section we refer to $BR(\varepsilon)$ as Bernstein's algorithm and to $BN(\varepsilon)$ as our variant of the algorithm. We have implemented the algorithms using C++ Visual Studio 2008 compiler. We ran our tests on a Dell Studio XPS1340 with 2 duo CPU 2.4 GHz.

5.1   Results for Road Maps

We used our algorithms to compute approximation of the 10 shortest simple paths between 50 vertex pairs that were chosen at random in four different road maps. The road maps are taken from the 9th DIMACS Implementation Challenge and represent different road systems in the United States. The cost of an edge in these graphs is the expected time that it takes to traverse the road that connects between the two endpoints of that edge. We used the following maps:

(i)   New York City (NYC) map with 264,346 vertices and 733,846 edges
(ii) San Francisco Bay area (SFB) map with 321,270 vertices and 800,172 edges.
(iii) Colorado state (Col) map with 435,666 vertices and 1,057,066 edges.
(iv) Florida state (Flo) map with 1,070,376 vertices and 2,712,798 edges.

In our first experiment we tested how well the algorithm approximates paths. We computed the exact 10-shortest simple paths using the naive algorithm and then compare it to the paths produced by $BR\left(\frac{1}{16}\right)$ and $BR(1)$. The percentage of exact paths that were produced using $BR\left(\frac{1}{16}\right)$ was 99.6% (only 8 of the 2000 were replaced by approximations). The percentage of exact paths that were produced using $BR(1)$ was 95.8%.

Another approach to evaluate the quality of the approximation of Bernstein's algorithm is to check the error of each non-optimal path that it produces. Let $\hat{\delta}_i(s,t)$ be the length of a path produced by the approximation algorithm. We define $err = (\hat{\delta}_i(s,t) - \delta_i(s,t))/\delta_i(s,t)$. Obviously, $err \leq \varepsilon$ for any path. Our results show that in most of the cases err is significantly smaller. For 2000 paths the maximal error for $BR(1)$ was only $4.46 * 10^{-4}$. This means that an approximated path of an exact path that should take an hour will be at most 1.6 seconds longer, rather than an hour longer as the upper bound suggest. Moreover, as the average error is $6.29 * 10^{-6}$, the expected error is only 23 milliseconds. Although these errors are very small the results of our variant of the algorithm $BN(\varepsilon)$ are even more impressive. In all the tests both $BN\left(\frac{1}{16}\right)$ and $BN(1)$ found the exact paths. We summarize the results for the precision test in Table 1.

We then turn to analyze the running time of Bernstein's algorithm. In particular, we checked the effect of different values of ε on the running time. We also compared the running time to the running time of the naive algorithm. Our finding indicates that the speedup obtained by Bernstein's algorithm is quite impressive. For the the NYC map the total running time of $BR\left(\frac{1}{16}\right)$ was 4.4 times faster than the naive algorithm. For the other maps the speedup of $BR\left(\frac{1}{16}\right)$ was 5.7, 8.7, and 7.4, respectively.

Table 1. Precision test

Input	$BR\left(\frac{1}{16}\right)$	$BR(1)$	$BN\left(\frac{1}{16}\right)$	$BN(1)$
NYC	98.8%	95.4%	100%	100%
SFB	100%	97.6%	100%	100%
Col	99.8%	94.2%	100%	100%
Flo	99.8%	96%	100%	100%
200×200	100%	91.4%	100%	100%
10×4000	100%	49.5%	100%	100%

Table 2. Running time in hours

Input	Naive	$BR\left(\frac{1}{16}\right)$	$BR(1)$	$BN\left(\frac{1}{16}\right)$	tests
NYC	9.61	2.17	2.13	0.71	50
SFB	15.66	2.76	2.49	0.84	50
Col	40.63	4.64	4.11	1.27	50
Flo	85.67	11.58	8.57	2.59	50
200×200	2.24	0.74	0.56	0.246	50
10×4000	8.84	4.84	0.78	0.17	20

As it can be seen from the table, while $BR\left(\frac{1}{16}\right)$ is significantly faster than the naive algorithm, choosing a larger ε does not significantly improve the running time for road maps. For $BR(1)$ the running time was faster by roughly 11%, although the $1/\varepsilon$ factor in the running time suggests a larger improvement. In all pairs $BR(1)$ was no more than 40% faster than $BR\left(\frac{1}{16}\right)$. In Figure1 we present the ratio between $BR\left(\frac{1}{16}\right)$ to the naive algorithm and the ratio between $BR(1)$ to the naive algorithm for the Colorado map with respect to the length (in hops) of the path of each individual pair of vertices that was tested. We summarize the running time of the algorithms in the Table 2.

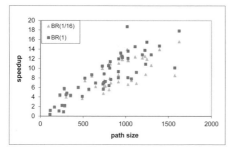

Fig. 1. The naive to BR ratio in Col different path hop lengths

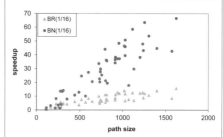

Fig. 2. The naive to BR and BN ratio in Col different path hop lengths

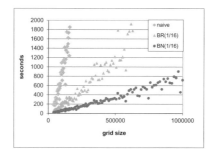

Fig. 3. Duration in seconds for the naive algorithm, $BR\left(\frac{1}{16}\right)$ and $BN\left(\frac{1}{16}\right)$ over square grids of different sizes

We established that $BR\left(\frac{1}{16}\right)$ and $BR(1)$ are significantly better than the naive algorithm. We also tested our version of the algorithm and got even more impressive results. In the NYC map, $BN\left(\frac{1}{16}\right)$ was 13.7 times faster than the naive algorithm. In the other three maps $BN\left(\frac{1}{16}\right)$ was 18.8, 32 and 33 times faster than the naive. These results are 3.1, 3.3, 3.7 and 4.6 times faster than $BR\left(\frac{1}{16}\right)$ respectively. Surprisingly, these results are also faster than $BR(1)$ (2.8, 3, 3.2 and 3.3 times faster for the four maps, respectively).

While the BR algorithm was not influenced by the value of ε as much as we expected, the BN is hardly influenced by the value of ε at all. Not only did $BN\left(\frac{1}{16}\right)$ and $BN(1)$ find the same paths, $BN(1)$ was no more than 0.5% faster than $BN\left(\frac{1}{16}\right)$ in any of the graphs. In Figure2 we present the ratio between $BR\left(\frac{1}{16}\right)$ to the naive algorithm and the ratio between $BN\left(\frac{1}{16}\right)$ to the naive algorithm for the Colorado map with respect to the length (in hops) of the path of each individual pair of vertices that was tested.

As one can expect, BR and BN algorithms yield greater speedups for larger graphs. Although there are many other factors that may influence on the running time of 10 shortest simple path search, such as the positions of the detours, we can see in Figure2 that the speedup of the algorithm is highly correlated to the number of hops in the first shortest path. The ratio between the different algorithms is also highly correlated to the number of hops in the paths. In all four maps the $BR\left(\frac{1}{16}\right)$ and $BR(1)$ are almost the same for small values of $\mathrm{O}(|P(s,t)|)$, while they are set apart for large values of $\mathrm{O}(|P(s,t)|)$ (see Figure 2).

5.2   Results for Square Grids

We have tested the algorithms on synthetic generated random grids. We followed the scheme described in [9]. We created grids of different sizes where a vertex is connected to its 8 neighbors with an edge whose weight is chosen uniformly at a random from the range $[0, 1000]$. The source and the destination are selected to be the opposite corners of the grid.

We used Bernstein's algorithm to find 10-shortest simple paths on 50 random 200×200 neighborhood grids. All paths that were found by Algorithm $BR\left(\frac{1}{16}\right)$ were optimal paths. Its running time was 3 times faster then the naive algorithm. This speedup is very close to the results in road maps for pairs connected by paths with similar hop-length. It seems that changing the value of ε has a greater effect in square grids than in road maps. Algorithm $BR(1)$ only found 91.4% of the exact paths, and the average error was $1.76 * 10^{-5}$. The running time of $BR(1)$ was also 32% faster than the running time of $BR\left(\frac{1}{16}\right)$ (as opposed to roughly 10% in most road maps).

Similar to the case with road maps our version of the algorithm was more accurate and faster. In particular, $BN\left(\frac{1}{16}\right)$ found the exact paths, while being 9.26 times faster than the naive algorithm and 2.3 times faster than $BR(1)$.

Finally, we tested the effect of grid size on the runtime of the algorithms We create a sequence of square grids starting with a 200×200 grid and adding 4 row and 4 columns in each step. In Figure 3 we present the connection between the size of the grid and the running time. The chart clearly demonstrates how the running time of the naive algorithm is highly affected by the size of the grid. It takes around 800 seconds for the naive algorithm to find 10 paths in square grids with $100,000$ nodes. Algorithm $BR\left(\frac{1}{16}\right)$ can accomplish this task in the same time for grids with $400,000$ nodes and $BN\left(\frac{1}{16}\right)$ only crosses the 800 second barrier for around $1,000,000$ nodes.

5.3   Results for Rectangle Grids

In this section we discuss the emperic results for rectangular grids with a large difference between the number of rows and the number of columns. These grids are interesting because in such grids $|P_1(s,t)| = O(n)$ rather than $O(\sqrt{n})$ as in square grids. The naive second shortest path algorithm runs Dijkstra for each edge in the path, therefore, its running time is dependent on the path hop length. The number of iterations in the last phases of Bernstein's algorithm are also directly dependent on the number of hops in the path.

We searched for 10-shortest simple paths in 20 random 10×4000 grids. Algorithm $BR\left(\frac{1}{16}\right)$ was only 1.85 times faster on average than the naive algorithm. This is very discouraging considering that in square grids of the same size (i.e., 200×200) the algorithm was 3 times faster. On the other hand $BR(1)$ is about 11.47 times faster. Its precision changes drastically as well: from 100% to 49.5%. The maximum error was only $3.24 * 10^{-6}$. This can be partially explained by the large number of efficient detours that can be found in large paths.

Although Bernstein's approximation proof only covers $\varepsilon \in (0, 1]$, we decided to test the effect of high ε values in these grids. Algorithm $BR(64)$ was 44.7 times faster than the naive algorithm 24.1 times faster than $BR\left(\frac{1}{16}\right)$. It was very accurate as well. The maximal error found for all 200 paths was only $3.59 * 10^{-5}$ despite the fact that only 16% of the paths were optimal (10% of which being the first paths).

While Bernstein's algorithm was drastically affected by the value of ε, in our version the difference in the running time was not significant. Both $BN\left(\frac{1}{16}\right)$ and $BN(64)$ outperformed the original algorithm. They were only 18% faster than $BR(64)$ (49 times faster than the naive algorithm), but they found all the optimal paths, while $BR(64)$ found only 16%. This indicates that our version of Bernstein's algorithm is faster than the original for any value of ε without compromising the precision, thus eliminating the user need to select an ε that fits her requirements.

References

1. Aingworth, D., Chekuri, C., Indyk, P., Motwani, R.: Fast estimation of diameter and shortest paths (without matrix multiplication). SIAM J. Comput. 28(4), 1167–1181 (1999)
2. Bernstein, A.: A nearly optimal algorithm for approximating replacement paths and k shortest simple paths in general graphs. In: Proc. of 21st SODA, pp. 742–755 (2010)
3. Brander, A.W., Sinclair, M.C.: A comparative study of k-shortest path algorithms. In: Proc. 11th UK Performance Engineering Worksh. for Computer and Telecommunications Systems (September 1995)
4. Demetrescu, C., Thorup, M., Chowdhury, R., Ramachandran, V.: Oracles for distances avoiding a failed node or link. SIAM J. Comput. 37(5), 1299–1318 (2008)
5. Dor, D., Halperin, S., Zwick, U.: All-pairs almost shortest paths. SIAM J. Comput. 29(5), 1740–1759 (2000)
6. Eppstein, D.: Finding the k shortest paths. SIAM Journal on Computing 28(2), 652–673 (1998)

7. Gotthilf, Z., Lewenstein, M.: Improved algorithms for the k simple shortest paths and the replacement paths problems. Inf. Process. Lett. 109(7), 352–355 (2009)
8. Hadjiconstantinou, E., Christofides, N.: An efficient implementation of an algorithm for finding K shortest simple paths. Networks 34(2), 88–101 (1999)
9. Hershberger, J., Maxel, M., Suri, S.: Finding the k shortest simple paths: A new algorithm and its implementation. ACM Transactions on Algorithms 3, 45 (2007)
10. Hershberger, J., Suri, S.: Vickrey prices and shortest paths: what is an edge worth? In: Proc. of 42nd FOCS, pp. 252–259 (2001)
11. Katoh, N., Ibaraki, T., Mine, H.: An efficient algorithm for K shortest simple paths. Networks 12(4), 411–427 (1982)
12. Lawler, E.L.: A procedure for computing the K best solutions to discrete optimization problems and its application to the shortest path problem. Management Science 18, 401–405 (1971/1972)
13. Martins, E., Pascoal, M., Esteves dos Santos, J.: Deviation algorithms for ranking shortest paths. Int. J. Found. Comp. Sci. 10(3), 247–263 (1999)
14. Perko, A.: Implementation of algorithms for K shortest loopless paths. Networks 16, 149–160 (1986)
15. Roditty, L.: On the k shortest simple paths problem in weighted directed graphs. SIAM Journal on Computing, 2363–2376 (2010)
16. Roditty, L., Zwick, U.: Replacement paths and k simple shortest paths in unweighted directed graphs. In: Caires, L., Italiano, G.F., Monteiro, L., Palamidessi, C., Yung, M. (eds.) ICALP 2005. LNCS, vol. 3580, pp. 249–260. Springer, Heidelberg (2005)
17. Vassilevska Williams, V., Williams, R.: Subcubic equivalences between path, matrix and triangle problems. In: Proc. FOCS (2010)
18. Vassilevska Williams, V.: Faster replacement paths. In: Proc. of 22th SODA (2011)
19. Yen, J.Y.: Finding the K shortest loopless paths in a network. Management Science 17, 712–716 (1970/1971)
20. Zwick, U.: All pairs shortest paths using bridging sets and rectangular matrix multiplication. JACM 49(3), 289–317 (2002)

Scope-Based Route Planning*

Petr Hliněný and Ondrej Moriš

Faculty of Informatics, Masaryk University
Botanická 68a, 602 00 Brno, Czech Republic
{hlineny,xmoris}@fi.muni.cz

Abstract. A new approach to the static route planning problem, based on a multi-staging concept and a *scope* notion, is presented. The main goal (besides implied efficiency of planning) of our approach is to address—with a solid theoretical foundation—the following two practically motivated aspects: a *route comfort* and a very *limited storage* space of a small navigation device, which both do not seem to be among the chief objectives of many other studies. We show how our novel idea can tackle both these seemingly unrelated aspects at once, and may also contribute to other established route planning approaches with which ours can be naturally combined. We provide a theoretical proof that our approach efficiently computes exact optimal routes within this concept, as well as we demonstrate with experimental results on publicly available road networks of the US the good practical performance of the solution.

1 Introduction

The single pair shortest path SPSP problem in world road networks, also known as the *route planning problem*, has received considerable attention in the past decades. However classical algorithms like Dijkstra's or A* are in fact "optimal" in a theoretical sense, they are not well suitable for huge graphs representing real-world road networks having up to tens millions of edges. In such situation even an algorithm with a linear time complexity cannot be feasibly run on a mobile device lacking computational power and working memory.

What can be done better? We focus on the *static route planning problem* where the road network is static in the sense that the underlying graph and its attributes do not change in time. Thus, a feasible solution lies in suitable *preprocessing* of the road network in order to improve both time and space complexity of subsequent *queries* (to find an optimal route from one position to another). However, to what extent such a preprocessing is *limited in the size* of the precomputed auxiliary data? It is not hard to see that there is always some trade-off between this storage space requirement and the efficiency of queries—obviously, one cannot precompute and store all the optimal routes in advance. See also a closer discussion below.

* This research has been supported by the Czech Science Foundation, grants P202/11/0196 and Eurocores GIG/11/E023.

Related Work. Classical techniques of the static route planning are represented by Dijkstra's algorithm [9], A* algorithm [16] and their bidirectional variations [21]. In the last decade, two sorts of more advanced techniques have emerged and become popular. The first one prunes the search of Dijkstra's or A* algorithms using preprocessed information This includes, in particular, reach-based [15,12], landmarks [11,14], combinations of those [13], and recent hub-based labeling [1].

The second sort of techniques (where our approach conceptually fits, too) exploits a road network structure with levels of hierarchy to which a route can be decomposed into. For instance, highway and contraction hierarchies [22,25,10], transit nodes [3], PCD [19] and SHARC routing [4,5] represent this sort. Still, there are also many other techniques and combinations, but—due to lack of space—we just refer to Cherkassky et al. [6], Delling et al. [8,7], and Schultes [24]. Finally, we would like to mention the interesting notion of highway dimension [2], and the ideas of customizable [1] and mobile [23] route planning.

Our Contribution. We summarize the essence of all our contribution already here, while we implicitly refer to the subsequent sections (and to the full technical report [18]) for precise definitions, algorithms, proofs, and further details.

First of all, we mention yet another integral point of practical route planning implementations—human-mind intuitiveness and comfortability of the computed route. This is a rather subjective requirement which is not easy to formalize via mathematical language, and hence perhaps not often studied together with the simple precise "shortest/fastest path" utility function in the papers.

> *Intuitiveness and comfort of a route*: Likely everyone using car navigation systems has already experienced a situation in which the computed route contained unsuitable segments, e.g. tricky shortcuts via low-category roads in an unfamiliar area. Though such a shortcut might save a few seconds on the route in theory, regular drivers would certainly not appreciate it and the real practical saving would be questionable. This should better be avoided.

Nowadays, the full (usually commercially) available road network data contain plenty of additional metadata which allow it to detect such unreasonable routes. Hence many practical routing implementations likely contain some kinds of rather *heuristic penalization* schemes dealing with this comfortability issue. We offer here a mathematically sound and precise formal solution to the route comfort issue which builds on a new theoretical concept of *scope* (Sec. 3).

The core idea of a scope and of scope-admissible routes can be informally outlined as follows: The elements (edges) of a road network are spread into several *scope levels*, each associated with a *scope value*, such that an edge e assigned scope value s_e is *admissible* on a route R if, before or after reaching e, such R travels distance less than s_e on edges of higher scope level than e. Intuitively, the scope levels and values describe suitability and/or importance of particular edges for long-range routing. This is in some sense similar to the better known concept of *reach* [15]; but in our case the importance of an edge is to be decided from available network metadata and hence its comfort and intuitive suitability is reflected, making a fundamental difference from the reach concept.

The effect seen on *scope admissible* routes (Def. 3.1, 3.3,3.4) is that low-level roads are fine "near" the start or target positions (roughly until higher-level roads become available), while only roads of the highest scope levels are admissible in the long middle sections of distant routing queries. This nicely corresponds with human thinking of intuitive routes, where the driver is presumably familiar with neighborhoods of the start and the target of his route (or, such a place is locally accessible only via low-level roads anyway). On contrary, on a long drive the mentally demanding navigation through unknown rural roads or complicated city streets usually brings no overall benefit, even if it were faster in theory.

To achieve good practical usability, too, road network segments are assigned scope levels (cf. Table 1) according to expectantly available metadata of the network (such as road categories, but also road quality and other, e.g. statistical information). It is important that already experiments with publicly available TIGER/Line 2009 [20] US road data, which have metadata of questionable quality, show that the restriction of admissible routes via scope has only a marginal statistical effect on shortest distances in the network.

Furthermore, a welcome added benefit of our categorization of roads into scope levels and subsequent scope admissibility restriction is the following.

Storage space efficiency: We suggest that simply allowing to store "linearly sized" precomputed auxiliary data (which is the case of many studies) may be too much in practice. Imagine that setting just a few attributes of a utility function measuring route optimality results in an exponential number of possibilities to which the device has to keep separate sets of preprocessed auxiliary data. In such a case a stricter storage limits should be imposed.

In our approach, preprocessed data for routing queries have to deal only with the elements of the highest scope level(s). This allows us to greatly reduce the amount of auxiliary precomputed information needed to answer queries quickly. We use (Sec. 4.1) a suitably adjusted fast vertex-separator approach which stores only those precomputed boundary-pair distances (in the cells) that are admissible on the highest scope level. This way we can shrink the *auxiliary data size to less than* 1% of the road network size (Table 2) which is a huge improvement.

Not to forget, our subsequent routing query algorithm (Sec. 4.2) then answers quickly and exactly (not heuristically) an optimal route among all scope admissible ones between the given positions. It is briefly summarized as follows.

Multi-staging approach: The computation of an optimal route is split into two different stages. In the local – *cellular*, stage, a modification of plain Dijkstra's algorithm is used to reach the cell boundaries in such a way that lower scope levels are no longer admissible. Then in the global – *boundary*, stage, an optimal connection between the previously reached boundary vertices is found on the (much smaller) boundary graph given by auxiliary data.

After all, the domain of a cellular stage is a small local neighbourhood, and Dijkstra's search can thus be very fast on it with additional help of a reach-like parametrization (Def. 3.8). Then, handling the precomputed boundary graph in the global stage is very flexible—since no side restrictions exist there—and can be combined with virtually any other established route planning algorithm (see Sec. 4.2). The important advantage is that the boundary graph is now much smaller (recall, $< 1\%$ in experiments) than the original network size, and hence computing on it is not only faster, but also more working-memory efficient.

Paper organization. After the informal outline of new contributions, this paper continues with the relevant formal definitions—Section 2 for route planning basics, and Section 3 for thorough description of the new scope admissibility concept. An adaptation of Dijkstra for scope is sketched in Sec. 3.1. Then Section 4 shows further details of the road network preprocessing (4.1) and query (4.2) algorithms. A summary of their experimental results can be found in Tables 2 and 3. Unfortunately, due to imposed space restrictions on the main paper, most of the details are to be left for the full technical report [18].

2   Preliminaries

A *directed graph* G is a pair of a finite set $V(G)$ of vertices and a finite multiset $E(G) \subseteq V(G) \times V(G)$ of edges. A *walk* $P \subseteq G$ is an alternating sequence $(u_0, e_1, u_1, \ldots, e_k, u_k)$ of vertices and edges of G such that $e_i = (u_{i-1}, u_i)$ for $i = 1, \ldots, k$. The *weight* of a walk $P \subseteq G$ wrt. a weighting $w : E(G) \mapsto \mathbb{R}$ of G is defined as $|P|_w = w(e_1) + w(e_2) + \cdots + w(e_k)$ where $P = (u_0, e_1, \ldots, e_k, u_k)$. The *distance* $\delta_w(u, v)$ from u to v in G is the minimum weight over all walks from u to v (achieved by an *optimal walk*), or ∞ if there is no such walk. A *subwalk* is a subsequence of a walk, and a *path* is a walk not repeating vertices.

A road network can be naturally represented by a graph G such that the junctions are represented by $V(G)$ and the roads (or road segments) by $E(G)$. Chosen cost function (e.g. travel time, distance, expenses, etc.) is represented by a *non-negative* edge weighting $w : E(G) \mapsto \mathbb{R}_0^+$ of G.

Definition 2.1 (Road Network). *Let G be a graph with a non-negative edge weighting w. A* road network *is the pair (G, w).*

We assume familiarity with classical Dijkstra's and A* algorithms and their bidirectional variants for shortest paths. In this context we also recall the useful notion of a reach given by Gutman [15].

Definition 2.2 (Reach [15]). *Consider a walk P in a road network (G, w) from s to t where $s, t \in V(G)$. The* reach *of a vertex $v \in V(P)$ on P is $r_P(v) = \min\{|P^{sv}|_w, |P^{vt}|_w\}$ where P^{sv} and P^{vt} is a subwalk of P from s to v and from v to t, respectively. The reach of v in G, $r(v)$, is the maximum value of $r_Q(v)$ over all optimal walks Q between pairs of vertices in G such that $v \in V(Q)$.*

Dijkstra's and A* algorithms can be accelerated by reach as follows: when discovering a vertex v from u, the algorithm first tests whether $r(v) \geq d[u] + w(u, v)$ (the current distance estimate from the start) or $r(v) \geq lower(v)$ (an auxiliary lower bound on the distance from v to target), and only in case of success it inserts v into the queue of vertices for processing.

Table 1. A very simple demonstration of a scope mapping which is based just on the road categories. Highways and other important roads have unbounded (∞) scope, while local, private or restricted roads have smaller scope. The zero scope is reserved for roads that physically cannot be driven through (including, for instance, long-term road obstructions). The weight function of this example is the travel distance in meters.

Scope level i	Value ν_i^S	Handicap κ_i^S	Road category
0	0	1	Alley, Walkway, Bike Path, Bridle Path
1	250	50	Parking Lot Road, Restricted Road
2	2000	250	Local Neighborhood Road, Urban Roads
3	5000	600	Rural Area Roads, Side Roads
∞	∞	(∞)	Highway, Primary (Secondary) Road

3   The New Concept – Scope

The main purpose of this section is to provide a theoretical foundation for the aforementioned vague objective of "comfort of a route". Recall that the scope levels referred in Definition 3.1 are generally assigned according to auxiliary metadata of the road network, e.g. the road categories and additional available information which is presumably included with it; see Table 1. Such a scope level assignment procedure is *not* the subject of the theoretical foundation.

Definition 3.1 (Scope). *Let (G, w) be a road network. A scope mapping is defined as $S : E(G) \mapsto \mathbb{N}_0 \cup \{\infty\}$ such that $0, \infty \in Im(S)$. Elements of the image $Im(S)$ are called* scope levels. *Each scope level $i \in Im(S)$ is assigned a constant value of scope $\nu_i^S \in \mathbb{R}_0 \cup \{\infty\}$ such that $0 = \nu_0^S < \nu_1^S < \cdots < \nu_\infty^S = \infty$.*

Actually, there is one more formal ingredient missing to make the scope concept a perfect fit: imagine that one prefers to drive a major highway, then she should better not miss available slip-roads to it. This is expressed with a "handicap" assigned to the situations in which a turn to a next road of higher scope level is possible, as follows:

Definition 3.2 (Turn-Scope Handicap). *Let S be a scope mapping in (G, w). The turn-scope handicap $h_S(e) \in \mathbb{N}_0 \cup \{\infty\}$ is defined, for every $e = (u, v) \in E(G)$, as the maximum among $S(e)$ and all $S(f)$ over $f = (v, w) \in E(G)$. Each handicap level i is assigned a constant κ_i^S such that $0 < \kappa_0^S < \cdots < \kappa_\infty^S$.*

The desired effect of admitting low-level roads only "near" the start or target positions—until higher level roads become widely available—is formalized in next Def. 3.3, 3.4. We remark beforehand that the seemingly complicated formulation is actually the right simple one for a smooth integration into Dijkstra.

Definition 3.3 (Scope Admissibility). *Let (G, w) be a road network and let $x \in V(G)$. An edge $e = (u, v) \in E(G)$ is x-admissible in G for a scope mapping S if, and only if, there exists a walk $P \subseteq G - e$ from x to u such that*

1. each edge of P is x-admissible in $G - e$ for $\mathcal{S}$,
2. P is optimal subject to (1), and
3. for $\ell = \mathcal{S}(e)$, $\sum_{f \in E(P), \mathcal{S}(f) > \ell} w(f) + \sum_{f \in E(P), h_{\mathcal{S}}(f) > \ell \geq \mathcal{S}(f)} \kappa_\ell^{\mathcal{S}} \leq \nu_\ell^{\mathcal{S}}$.

Note; every edge e such that $\mathcal{S}(e) = \infty$ (*unbounded scope* level) is always admissible, and with the values of $\nu_i^{\mathcal{S}}$ growing to infinity, Def. 3.3 tends to admit more and more edges (of smaller scope).

Definition 3.4 (Admissible Walks). *Let (G, w) be a road network and $\mathcal{S}$ a scope mapping. For a walk $P = (s = u_0, e_1, \ldots e_k, u_k = t) \subseteq G$ from s to t;*

- *P is s-admissible in G for $\mathcal{S}$ if every $e_i \in E(P)$ is s-admissible in G for $\mathcal{S}$,*
- *and P is st-admissible in G for $\mathcal{S}$ if there exists $0 \leq j \leq k$ such that every $e_i \in E(P)$, $i \leq j$ is s-admissible in G, and the reverse of every $e_i \in E(P)$, $i > j$ is t-admissible in G^R – the graph obtained by reversing all edges.*

In a standard connectivity setting, a graph (road network) G is *routing-connected* if, for every pair of edges $e, f \in E(G)$, there exists a walk in G starting with e and ending with f. This obviously important property can naturally be extended to our scope concept as follows.

Definition 3.5 (Proper Scope). *A scope mapping $\mathcal{S}$ of a routing-connected graph G is proper if, for all $i \in Im(\mathcal{S})$, the subgraph $G^{[i]}$ induced by those edges $e \in E(G)$ such that $\mathcal{S}(e) \geq i$ is routing-connected, too.*

Theorem 3.6. *Let (G, w) be a routing-connected road network and let $\mathcal{S}$ be a proper scope mapping of it. Then, for every two edges $e = (s, x), f = (y, t) \in E(G)$, there exists an st-admissible walk $P \subseteq G$ for scope $\mathcal{S}$ such that P starts with the edge e and ends with f.*

Note that validity of Definition 3.5 should be enforced in the scope-assignment phase of preprocessing (e.g., the assignment should reflect known long-term detours on a highway[1] accordingly). This and further practical-oriented aspect of a scope assignment are left for discussion in [18].

3.1 $\mathcal{S}$-Dijkstra's Algorithm and $\mathcal{S}$-Reach

As noted beforehand, Def. 3.4 smoothly integrates into bidirectional Dijkstra's or A* algorithm, simply by keeping track of the admissibility condition (3.):

$\mathcal{S}$-Dijkstra's Algorithm (one direction of the search).

- For every accessed vertex v and each scope level $\ell \in Im(\mathcal{S})$, the algorithm keeps, as $\sigma_\ell[v]$, the best achieved value of the sum. formula in Def. 3.3 (3.).
- The s-admissibility of edges e starting in v depends then simply on $\sigma_{\mathcal{S}(e)}[v] \leq \nu_{\mathcal{S}(e)}^{\mathcal{S}}$, and only s-admissible edges are relaxed further.

---

[1] Note, regarding real-world navigation with unexpected road closures, that the proper-scope issue is not at all a problem—a detour route could be computed from the spot with "refreshed" scope admissibility constrains. Here we solve the static case.

Theorem 3.7. $\mathcal{S}$-*Dijkstra's algorithm, for a road network* (G, w)*, a scope mapping* $\mathcal{S}$*, and a start vertex* $s \in V(G)$*, computes an optimal s-admissible walk from s to every* $v \in V(G)$ *in time* $\mathcal{O}\big(|E(G)| \cdot |Im(\mathcal{S})| + |V(G)| \cdot \log |V(G)|\big)$.

Furthermore, practical complexity of this algorithm can be largely decreased by a suitable adaptation of the *reach* concept (Def. 2.2), given in Def. 3.8. For $x \in V(G)$ in a road network with scope $\mathcal{S}$, we say that a vertex $u \in V(G)$ is *x-saturated* if no edge $f = (u, v)$ of G from u of bounded scope (i.e., $\mathcal{S}(f) < \infty$) is x-admissible for $\mathcal{S}$. A walk P with ends s, t is *saturated* for $\mathcal{S}$ if some vertex of P is both s-saturated in G and t-saturated in the reverse network G^R.

Definition 3.8 ($\mathcal{S}$-reach). *Let* (G, w) *be a road network and* $\mathcal{S}$ *its scope mapping. The* $\mathcal{S}$*-reach of* $v \in V(G)$ *in* G*, den.* $r^{\mathcal{S}}(v)$*, is the maximum value among* $|P^{xv}|_w$ *over all* $x, y \in V(G)$ *such that* y *is x-saturated while* v *is not x-saturated, and there exists an optimal x-admissible walk* $P \subseteq G$ *from* x *to* y *such that* P^{xv} *is a subwalk of* P *from* x *to* v*.* $r^{\mathcal{S}}(v)$ *is undefined* (∞) *if there is no such walk.*

There is no general easy relation between classical reach and $\mathcal{S}$-reach; they both just share the same conceptual idea. Moreover, $\mathcal{S}$-reach can be computed more efficiently (unlike reach) since the set of non-x-saturated vertices is rather small and local in practice, and only its x-saturated neighbors are to be considered among the values of y in Def. 3.8.

The way $\mathcal{S}$-reach of Def. 3.8 is used to amend $\mathcal{S}$-Dijkstra's algorithm is again rather intuitive; an edge $f = (u, v) \in E(G)$ is relaxed from u only if $\mathcal{S}(f) = \infty$, or $r^{\mathcal{S}}(v) \geq d[u] + w(f)$. Notice however, that this $\mathcal{S}$-reach amending scheme has one inevitable limit of usability—it becomes valid only if the both directions of Dijkstra's search get to the "saturated" state. (In the opposite case, the start and target are close to each other in a local neighborhood, and the shortest route is quickly found without use of $\mathcal{S}$-reach, anyway.) Hence we conclude:

Theorem 3.9. *Let* $s, t \in V(G)$ *be vertices in a road network* (G, w) *with a scope mapping* $\mathcal{S}$*. Bidirectional* $\mathcal{S}$*-reach* $\mathcal{S}$*-Dijkstra's algorithm computes an optimal one among all st-admissible walks in* G *from s to t which are saturated for* $\mathcal{S}$*.*

4   The Route-Planning Algorithm

Following the informal outline from the introduction, we now present the second major ingredient for our approach; a separator based partitioning of the road network graph with respect to a given scope mapping.

4.1   Preprocessing into a Boundary Graph

At first, a road network is partitioned into a set of pairwise edge-disjoint subgraphs called *cells* such that their *boundaries* (i.e., the vertex-separators shared between a cell and the rest) contain as few as possible vertices incident with edges of unbounded scope. The associated formal definition follows.

Definition 4.1 (Partitioning and Cells). *Let* $\mathcal{E} = \{E_1, \ldots, E_\ell\}$ *be a partition of the edge set* $E(G)$ *of a graph* G. *We call* cells *of* $(G, \mathcal{E})$ *the subgraphs* $C_i = G[E_i] \subseteq G$, *for* $i = 1, 2, \ldots, \ell$. *The* cell boundary $\Gamma(C_i)$ *of* C_i *is the set of all vertices that are incident both with some edge in* E_i *and some in* $E(G) \setminus E_i$, *and the* boundary *of* $\mathcal{E}$ *is* $\Gamma(G, \mathcal{E}) = \bigcup_{1 \leq i \leq \ell} \Gamma(C_i)$.

Practically, we use a graph partitioning algorithm hierarchically computing a so-called *partitioned branch-decomposition* of the road network. The algorithm employs an approach based on max-flow min-cut which, though being heuristic, performs incredibly well—being fast in finding really good small vertex separators.[2] More details can be found in Table 2 and [18].

Secondly, the in-cell distances between pairs of boundary vertices are precomputed such that only the edges of unbounded scope are used. This simplification is, on one hand, good enough for computing optimal routes on a "global level" (i.e., as saturated for scope in the sense of Sec. 3.1). On the other hand, such a simplified precomputed distance graph (cf. $B_{\mathcal{E}}$) is way much smaller than if all boundary-pair distance were stored for each cell. See in Table 2, the last column.

We again give the associated formal definition and a basic statement whose proof is trivial from the definition.

Definition 4.2 (Boundary Graph; $\mathcal{S}$-restricted). *Assume a road network* (G, w) *together with a partition* $\mathcal{E} = \{E_1, \ldots, E_\ell\}$ *and the notation of Def. 4.1. For a scope mapping* $\mathcal{S}$ *of* G, *let* $G^{[\infty]}$ *denote the subgraph of* G *induced by the edges of unbounded scope level* ∞, *and let* $C_i^{\infty} = G[E_i] \cap G^{[\infty]}$.

The ($\mathcal{S}$-restricted) in-cell distance graph D_i *of the cell* C_i *is defined on the vertex set* $V(D_i) = \Gamma(C_i)$ *with edges and weighting* p_i *as follows. For* $u, v \in \Gamma(C_i) \cap V(C_i^{\infty})$ *only, let* $\delta_w^{\infty}(u, v)$ *be the distance in* C_i^{∞} *from* u *to* v, *and let* $f = (u, v) \in E(D_i)$ *iff* $p_i(f) := \delta_w^{\infty}(u, v) < \infty$.

The weighted ($\mathcal{S}$-restricted) boundary graph $B_{\mathcal{E}}$ *of a road network* (G, w) *wrt. scope mapping* $\mathcal{S}$ *and partition* $\mathcal{E}$ *is then obtained as the union of all the cell-distance graphs* D_i *for* $i = 1, 2, \ldots, \ell$, *simplified such that for each bunch of parallel edges only one of the smallest weight is kept in* $B_{\mathcal{E}}$.

Proposition 4.3. *Let* (G, w) *be a road network,* $\mathcal{S}$ *a scope mapping of it, and* $\mathcal{E}$ *a partition of* $E(G)$. *For any* $s, t \in \Gamma(G, \mathcal{E})$, *the minimum weight of a walk from* s *to* t *in* $G^{[\infty]}$ *equals the distance from* s *to* t *in* $B_{\mathcal{E}}$.

Experimental evaluation. We have run our partitioning algorithm (Def. 4.1), in-cell distance computations (Def. 4.2), and $\mathcal{S}$-reach computation (Def. 3.8) in parallel on a quad-core XEON machine with 16 GB in 32 threads. A decomposition of the complete *TIGER/Line 2009* [20] US road network into the boundary graph $B_{\mathcal{E}}$, together with computations of $\mathcal{S}$-reach in G (and $\mathcal{S}$-reach in G^R), distances in D_is, and standard reach estimate for $B_{\mathcal{E}}$, took only *192 minutes altogether*. This, and the tiny size of $B_{\mathcal{E}}$, are both very promising for potential

---

[2] It is worth to mention that max-flow based heuristics for a branch-decomposition have been used also in other combinatorial areas recently, e.g. in the works of Hicks.

Table 2. Partitioning results for the TIGER/Line 2009 [20] US road network. The left section identifies the (sub)network and its size, the middle one the numbers and average size of partitioned cells, and the right section summarizes the results—the boundary graph size data, with percentage of the original network size. This boundary graph $B_{\mathcal{E}}$ is wrt. the simple scope assignment of Table 3.1, and $V(B_{\mathcal{E}})$ includes also isolated boundary vertices (i.e. those with no incident edge of unbounded scope). Notice the tiny size ($<1\%$) of $B_{\mathcal{E}}$ compared to the original road network, and the statistics regarding *cell boundary sizes*: maximum is 74, average 19, median 18, and 9-decil is 31 vertices.

Input G		Partitioning		Boundary Graph $B_{\mathcal{E}}$	
Road network	#Edges	#Cells	Cell sz.	#Vertices	#Edges / % size
USA-all	88 742 994	15 862	5 594	253 641	524 872 / 0.59%
USA-east	24 130 457	4 538	5 317	62 692	107 986 / 0.45%
USA-west	12 277 232	2 205	5 567	23 449	42 204 / 0.34%
Texas	7 270 602	1 346	5 366	17 632	36 086 / 0.50%
California	5 503 968	1 011	5 444	11 408	16 978 / 0.31%
Florida	3 557 336	662	5 373	5 599	25 898 / 0.73%

practical applications in which the preprocessing may have to be run and the small boundary graphs separately stored for multiple utility weight functions (while the $\mathcal{S}$-reach values could still be kept in one maximizing instance).

4.2   Route Planning Query

Having already computed the boundary graph and $\mathcal{S}$-reach in the preprocessing phase, we now describe a natural simplified two-stage query algorithm based on the former. In its *cellular* stage, as outlined in the introduction, the algorithm runs $\mathcal{S}$-Dijkstra's search until all its branches get saturated at cell boundaries (typically, only one or two adjacent cells are searched). Then, in the *boundary* stage, virtually any established route planning algorithm may be used to finish the search (cf. Prop. 4.3) since $B_{\mathcal{E}}$ is a relatively small graph (Table 2) and is free from scope consideration. E.g., we use the standard reach-based A* [12].

Two-stage Query Algorithm (simplified). Let a road network (G, w), a proper scope mapping $\mathcal{S}$, an $\mathcal{S}$-reach $r^{\mathcal{S}}$ on G, and the boundary graph $B_{\mathcal{E}}$ associated with an edge partition $\mathcal{E}$ of G, be given. Assume start and target positions $s, t \in V(G)$; then the following algorithm computes, from s to t, an optimal st-admissible and saturated walk in G for the scope $\mathcal{S}$.

1. *Opening cellular stage.* Let $I_s \subseteq G$ initially be the subgraph formed by the cell (or a union of such) containing the start s. Let $\Gamma(I_s) \subseteq V(I_s)$ denote the actual boundary of I_s, i.e. those vertices incident both with edges of I_s and of the complement (not the same as the union of cell boundaries).

   (a) Run $\mathcal{S}$-*reach* $\mathcal{S}$-*Dijkstra's* algorithm (unidir.) on (I_s, w) starting from s.

   (b) Let U be the set of non-s-saturated vertices in $\Gamma(I_s)$ accessed in (1a). As long as $U \neq \emptyset$, let $I_s \leftarrow I_s \cup J_U$ where J_U is the union of all cells containing some vertex of U, and continue with (1a).

P. Hliněný and O. Moriš

Table 3. Experimental queries for the preprocessed continental US road network (USA-all) from Table 2. Each row carries statistical results for 1000 uniformly chosen start–target pairs with saturated optimal walks. The table contains average values for the numbers of cells hit by the resulting optimal walk, the overall numbers of vertices scanned in G during the cellular and in $B_{\mathcal{E}}$ during the boundary stage, the maximal numbers of elements in the processing queue during the cellular and the boundary stages, and the average time spent in the cellular and the boundary stages. We remark that this statistics skips the closing cellular stage since the computed walk can be "unrolled" inside each cell on-the-fly while displaying details of the route.

Query distance (km)	Hit cells	Scanned vertices cellular / boundary	Max. queue size cellular / boundary	Query time (ms)
3000 - ∞	277	1 392 / 3 490	60 / 58	8.2 + 29.8
2000 - 3000	139	1 411 / 1 814	64 / 52	7.9 + 26.9
1000 - 2000	57	1 343 / 1 511	57 / 49	7.7 + 22.8
500 - 1000	25	1 113 / 1 192	53 / 38	8.1 + 19.0
0 - 500	10	998 / 716	41 / 34	6.9 + 16.1

An analogical procedure is run concurrently in the reverse network (G^R, w) on the target t and I_t. If it happens that I_s, I_t intersect and a termination condition of bidirectional Dijkstra is met, then the algorithm stops here.

2. *Boundary stage.* Let $B_{s,t}$ be the graph created from $B_{\mathcal{E}} \cup \{s, t\}$ by adding the edges from s to each vertex of $\Gamma(I_s)$ and from each one of $\Gamma(I_t)$ to t. The weights of these new edges equal the distance estimates computed in (1). (Notice that many of the weights are actually ∞—can be ignored—since the vertices are inaccessible, e.g., due to scope admissibility or $\mathcal{S}$-reach.)

   Run the *standard reach-based A^** algorithm on the weighted graph $B_{s,t}$ (while the reach refers back to $B_{\mathcal{E}}$), to find an optimal path Q from s to t.

3. *Closing cellular stage.* The path Q computed in (2) is easily "unrolled" into an optimal st-admissible saturated walk P from s to t in the network (G, w).

A simplification in the above algorithm lies in neglecting possible non-saturated walks between s, t (cf. Theorem 3.9), which may not be found by an $\mathcal{S}$-reach-based search in (1). This happens only if s, t are very close in a local neighborhood wrt. $\mathcal{S}$.

Experimental evaluation. Practical performance of the query algorithm has been evaluated by multiple simulation runs on Intel Core 2 Duo mobile processor T6570 (2.1 GHz) with 4GB RAM. Our algorithms are implemented in C and compiled with gcc-4.4.1 (with -O2). We keep track of parameters such as the number of scanned vertices and the queue size that influence the amount of working memory needed, and are good indicators of suitability of our algorithm for the mobile platforms. The collected statistical data in Table 3—namely the total numbers of scanned vertices and the maximal queue size of the search—though, reasonably well estimate also the expected runtime and mainly low memory demands of the same algorithm on a mobile navigation device.

5   Conclusions

We have introduced a new concept of *scope* in the static route planning problem, aiming at a proper formalization of vague "route comfort" based on anticipated additional metadata of the road network. At the same time we have shown how the scope concept *nicely interoperates* with other established tools in route planning; such as with vertex-separator partitioning and with the reach concept. Moreover, our approach allows also a smooth incorporation of local route restrictions and traffic regulations modeled by so-called *maneuvers* [17].

On the top of formalizing desired "comfortable routes", the proper mixture of the aforementioned classical concepts with scope brings more added values; very small size of auxiliary metadata from preprocessing (Table 2) and practically very efficient optimal routing query algorithm (Table 3). The *small price to be paid* for this route comfort, fast planning, and small size of auxiliary data altogether, is a marginal increase in the weight of an optimal scope admissible walk as compared to the overall optimal one (scope admissible walks form a proper subset of all walks). Simulations with the very basic scope mapping from Table 1 reveal an average increase of less than 3% for short queries up to 500km, and 1.5% for queries above 3000km. With better quality road network metadata and a more realistic utility weight function (such as travel time) these would presumably be even smaller numbers.

At last we very briefly outline two directions for further research on the topic.

 i. With finer-resolution road metadata, it could be useful to add a few more scope levels and introduce another query stage(s) "in the middle".
 ii. The next natural step of our research is to incorporate dynamic road network changes (such as live traffic info) into our approach—more specifically into the definition of scope-admissible walks; e.g., by locally re-allowing roads of low scope level nearby such disturbances.

References

1. Abraham, I., Delling, D., Goldberg, A.V., Werneck, R.F.: A hub-based labeling algorithm for shortest paths in road networks. In: Pardalos, P.M., Rebennack, S. (eds.) SEA 2011. LNCS, vol. 6630, pp. 230–241. Springer, Heidelberg (2011)
2. Abraham, I., Fiat, A., Goldberg, A.V., Werneck, R.F.: Highway dimension, shortest paths, and provably efficient algorithms. In: SODA 2010: Proceedings of the 21st Annual ACM-SIAM Symposium on Discrete Algorithms, pp. 782–793 (2010)
3. Bast, H., Funke, S., Matijevic, D., Sanders, P., Schultes, D.: In transit to constant shortest-path queries in road networks. In: ALENEX 2007: Proceedings of the 9th Workshop on Algorithm Engineering and Experiments, pp. 46–59 (2007)
4. Bauer, R., Delling, D.: SHARC: Fast and robust unidirectional routing. J. Exp. Algorithmics 14(4), 2.4–4:2.29 (2010)
5. Brunel, E., Delling, D., Gemsa, A., Wagner, D.: Space-efficient SHARC-routing. In: SEA 2010: Proceedings of the 9th International Symposium on Experimental Algorithms, pp. 47–58 (2010)

456   P. Hliněný and O. Moriš

6. Cherkassky, B., Goldberg, A.V., Radzik, T.: Shortest paths algorithms: Theory and experimental evaluation. Mathematical Programming 73(2), 129–174 (1996)
7. Delling, D., Sanders, P., Schultes, D., Wagner, D.: Engineering route planning algorithms. In: Lerner, J., Wagner, D., Zweig, K.A. (eds.) Algorithmics of Large and Complex Networks. LNCS, vol. 5515, pp. 117–139. Springer, Heidelberg (2009)
8. Delling, D., Wagner, D.: Time-dependent route planning. In: Ahuja, R.K., Möhring, R.H., Zaroliagis, C.D. (eds.) Robust and Online Large-Scale Optimization. LNCS, vol. 5868, pp. 207–230. Springer, Heidelberg (2009)
9. Dijkstra, E.: A note on two problems in connexion with graphs. Numerische Mathematik 1, 269–271 (1959)
10. Geisberger, R., Sanders, P., Schultes, D., Delling, D.: Contraction hierarchies: Faster and simpler hierarchical routing in road networks. In: McGeoch, C.C. (ed.) WEA 2008. LNCS, vol. 5038, pp. 319–333. Springer, Heidelberg (2008)
11. Goldberg, A.V., Harrelson, C.: Computing the shortest path: A* search meets graph theory. In: SODA 2005: Proceedings of the 16th Annual ACM-SIAM Symposium on Discrete Algorithms, pp. 156–165 (2005)
12. Goldberg, A.V., Kaplan, H., Werneck, R.F.: Reach for A*: Efficient point-to-point shortest path algorithms. Technical report, Microsoft Research (2005)
13. Goldberg, A.V., Kaplan, H., Werneck, R.F.: Better landmarks within reach. In: Demetrescu, C. (ed.) WEA 2007. LNCS, vol. 4525, pp. 38–51. Springer, Heidelberg (2007)
14. Goldberg, A.V., Werneck, R.F.: Computing point-to-point shortest paths from external memory. In: ALENEX/ANALCO 2005: Proceedings of the 7th Workshop on Algorithm Engineering and Experiments and the 2nd Workshop on Analytic Algorithmics and Combinatorics, pp. 26–40 (2005)
15. Gutman, R.: Reach-based routing: A new approach to shortest path algorithms optimized for road networks. In: ALENEX 2004: Proceedings of the 6th Workshop on Algorithm Engineering and Experiments, pp. 100–111 (2004)
16. Hart, P.E., Nilsson, N.J., Raphael, B.: A formal basis for the heuristic determination of minimum cost paths. IEEE Transactions on Systems Science and Cybernetics SSC4 4(2), 100–107 (1968)
17. Hliněný, P., Moriš, O.: Generalized maneuvers in route planning. ArXiv e-prints, arXiv:1107.0798 (July 2011)
18. Hliněný, P., Moriš, O.: Multi-Stage Improved Route Planning Approach: theoretical foundations. ArXiv e-prints, arXiv:1101.3182 (January 2011)
19. Maue, J., Sanders, P., Matijevic, D.: Goal-directed shortest-path queries using precomputed cluster distances. J. Exp. Algorithmics 14, 3.2–3.27 (2009)
20. Murdock, S.H.: 2009 TIGER/Line Shapefiles. Technical Documentation published by U.S. Census Bureau (2009)
21. Pohl, I.S.: Bi-directional and heuristic search in path problems. PhD thesis, Stanford University, Stanford, CA, USA (1969)
22. Sanders, P., Schultes, D.: Engineering highway hierarchies. In: Azar, Y., Erlebach, T. (eds.) ESA 2006. LNCS, vol. 4168, pp. 804–816. Springer, Heidelberg (2006)
23. Sanders, P., Schultes, D., Vetter, C.: Mobile route planning. In: Halperin, D., Mehlhorn, K. (eds.) ESA 2008. LNCS, vol. 5193, pp. 732–743. Springer, Heidelberg (2008)
24. Schultes, D.: Route Planning in Road Networks. PhD thesis, Karlsruhe University, Karlsruhe, Germany (2008)
25. Schultes, D., Sanders, P.: Dynamic highway-node routing. In: Demetrescu, C. (ed.) WEA 2007. LNCS, vol. 4525, pp. 66–79. Springer, Heidelberg (2007)

Maximum Flows by
Incremental Breadth-First Search

Andrew V. Goldberg[1], Sagi Hed[2], Haim Kaplan[2],
Robert E. Tarjan[3], and Renato F. Werneck[1]

[1] Microsoft Research Silicon Valley
{goldberg,renatow}@microsoft.com
[2] Tel Aviv University
{sagihed,haimk}@tau.ac.il
[3] Princeton University and HP Labs
ret@cs.princeton.edu

Abstract. Maximum flow and minimum s-t cut algorithms are used to solve several fundamental problems in computer vision. These problems have special structure, and standard techniques perform worse than the special-purpose *Boykov-Kolmogorov* (BK) algorithm. We introduce the *incremental breadth-first search* (IBFS) method, which uses ideas from BK but augments on shortest paths. IBFS is theoretically justified (runs in polynomial time) and usually outperforms BK on vision problems.

1 Introduction

Computing maximum flows is a classical optimization problem that often finds applications in new areas. In particular, the minimum cut problem (the dual of maximum flows) is now an important tool in the field of computer vision, where it has been used for segmentation, stereo images, and multiview reconstruction. Input graphs in these applications typically correspond to images and have special structure, with most vertices (representing pixels or voxels) arranged in a regular 2D or 3D grid. The source and sink are special vertices connected to all the others with varying capacities. See [2,3] for surveys of these applications.

Boykov and Kolmogorov [2] developed a new algorithm that is superior in practice to general-purpose methods on many vision instances. Although it has been extensively used by the vision community, it has no known polynomial-time bound. No exponential-time examples are known either, but the algorithm performs poorly in practice on some non-vision problems.

The lack of a polynomial time bound is disappointing because the maximum flow problem has been extensively studied from the theoretical point of view and is one of the better understood combinatorial optimization problems. Known solutions to this problem include the augmenting path [9], network simplex [7], blocking flow [8,15] and push-relabel [12] methods. A sequence of increasingly better time bounds has been obtained, with the best bounds given in [16,11].

Experimental work on the maximum flow problem has a long history and includes implementations of algorithms based on blocking flows (e.g., [5,13]) and on the push-relabel method (e.g., [6,10,4]), which is the best general-purpose

C. Demetrescu and M.M. Halldórsson (Eds.): ESA 2011, LNCS 6942, pp. 457–468, 2011.
© Springer-Verlag Berlin Heidelberg 2011

approach in practice. With the extensive research in the area and its use in computer vision, the *Boykov-Kolmogorov* (BK) algorithm is an interesting development from a practical point of view.

In this paper we develop an algorithm that combines ideas from BK with those from the shortest augmenting path algorithms. In fact, our algorithm is closely related to the blocking flow method. However, we build the auxiliary network for computing augmenting paths in an incremental manner, by updating the existing network after each augmentation while doing as little work as we can. Since for the blocking flow method network construction is the bottleneck in practice, this leads to better performance. Like BK, and unlike most other current algorithms, we build the network in a bidirectional manner, which also improves practical performance.

We call the resulting algorithm *Incremental Breadth First Search* (IBFS). It is theoretically justified in the sense that it gets good (although not the best) theoretical time bounds. Our experiments show that IBFS is faster than BK on most vision instances. Like BK, the algorithm does not perform as well as state-of-the-art codes on some non-vision instances. Even is such cases, however, IBFS appears to be more robust than BK. BK is heavily used to solve vision problems in practice. IBFS offers a faster and theoretically justified alternative.

2   Definitions and Notation

The input to the maximum flow problem is (G, s, t, u), where $G = (V, A)$ is a directed graph, $s \in V$ is the *source*, $t \in V$ is the *sink* (with $s \neq t$), and $u : A \Rightarrow [1, \ldots, U]$ is the *capacity function*. Let $n = |V|$ and $m = |A|$.

Let a^R denote the *reverse* of an arc a, let A^R be the set of all reverse arcs, and let $A' = A \cup A^R$. A function g on A' is *anti-symmetric* if $g(a) = -g(a^R)$. Extend u to be an anti-symmetric function on A', i.e., $u(a^R) = -u(a)$.

A flow f is an anti-symmetric function on A' that satisfies *capacity constraints* on all arcs and *conservation constraints* at all vertices except s and t. The capacity constraint for $a \in A$ is $0 \le f(a) \le u(a)$ and for $a \in A^R$ it is $-u(a^R) \le f(a) \le 0$. The conservation constraint for v is $\sum_{(u,v) \in A} f(u, v) = \sum_{(v,w) \in A} f(v, w)$. The *flow value* is the total flow into the sink: $|f| = \sum_{(v,t) \in A} f(v, t)$. A *cut* is a partitioning of vertices $S \cup T = V$ with $s \in S, t \in T$. The capacity of a cut is defined by $u(S, T) = \sum_{v \in S, w \in T, (v,w) \in A} u(v, w)$. The max-flow/min-cut theorem [9] says that the maximum flow value is equal to the minimum cut capacity.

The *residual capacity* of an arc $a \in A'$ is defined by $u_f(a) = u(a) - f(a)$. Note that if f satisfies capacity constraints, then u_f is nonnegative. The *residual graph* $G_f = (V, A_f)$ is the graph induced by the arcs in A' with strictly positive residual capacity. An *augmenting path* is an s–t path in G_f.

When we talk about distances (and shortest paths), we mean the distance in the residual graph for the unit length function. A *distance labeling from s* is an integral function d_s on V that satisfies $d_s(s) = 0$. Given a flow f, we say that d_s is valid if for all $(v, w) \in A_f$ we have $d_s(w) \le d_s(v) + 1$. A (valid) distance labeling to t, d_t, is defined symmetrically. We say that an arc (v, w) is *admissible w.r.t. d_s* if $(v, w) \in A_f$ and $d_s(v) = d_s(w) - 1$, and *admissible w.r.t. d_t* if $(v, w) \in A_f$ and $d_t(w) = d_t(v) - 1$.

3   BK Algorithm

In this section we briefly review the BK algorithm [2]. It is based on augmenting paths. It maintains two trees of residual arcs, S rooted from s and T rooted into t. Initially S contains only s and T contains only t. At each step, a vertex is in S, in T, or *free*. Each tree has *active* and *internal* vertices. The outer loop of the algorithm consists of three stages: *growth*, *augmentation*, and *adoption*.

The growth stage expands the trees by scanning their active vertices and adding newly-discovered vertices to the tree from which they have been discovered. The newly-added vertices become active. Vertices become internal after being scanned. If no active vertices remain, the algorithm terminates. If a residual arc from S to T is discovered, then the augmentation stage starts.

The augmentation stage takes the path found by the growth stage and augments the flow on it by its bottleneck residual capacity. Some tree arcs become saturated, and their endpoints farthest from the corresponding root become *orphans*. If an arc (v, w) becomes saturated and both v and w are in S, then w becomes an S-orphan. If both v and w are in T, v becomes a T-orphan. If v is in S and w is in T, then a saturation of (v, w) does not create orphans. Orphans are placed on a list and processed in the adoption stage.

The adoption stage processes orphans until there are none left. Consider an S-orphan v (T-orphans are processed similarly). We examine residual arcs (u, v) in an attempt to find a vertex u in S. If we find such u, we check whether the tree path from u to s is valid (it may not be if it contains an orphan, including u). If a vertex u with a valid path is found, we make u the parent of v.

If we fail to find a new parent for v, we make v a free vertex and make all children of v orphans. Then we examine all residual arcs (u, v) and for each u in S, we make u active. Note that for each such u, the tree path from s to u contains an orphan (otherwise u would have been picked as v's parent) and this orphan may find a new parent. Making u active ensures that we find v again.

The only known way to analyze BK is as a generic augmenting path algorithm, which does not give polynomial bounds.

4   Incremental Breadth-First Search

The main idea IBFS is to modify BK to maintain breadth-first search trees, which leads to a polynomial time bound ($O(n^2 m)$). Existing techniques can improve this further, matching the best known bounds for blocking flow algorithms.

The algorithm maintains distance labels d_s and d_t for every vertex. The two trees, S and T, satisfy the *tree invariants*: for some values D_s and D_t, the trees contain all vertices at distances up to D_s from s and up to D_t to t, respectively. We also maintain the invariant that $L = D_s + D_t + 1$ is a lower bound on the augmenting path length, so the trees are disjoint.

A vertex can be an S-vertex, T-vertex, S-orphan, T-orphan, or N-vertex (not in any tree). Each vertex maintains a parent pointer p, which is *null* for N-vertices and orphans. We maintain the invariant that tree arcs are admissible.

During the adoption step, the trees are rebuilt and are not well-defined. Some invariants are violated and some orphans may leave the trees. We say that a vertex is in S if it is an S-vertex or an S-orphan. In a growth step, there are no orphans, so S is the set of S-vertices. Similarly for T.

If a vertex v is in S, $d_s(v)$ is the meaningful label value and $d_t(v)$ is unused. The situation is symmetric for vertices in T. Labels of N-vertices are irrelevant. Since at most one of $d_s(v)$ and $d_t(v)$ is used at any given time, one can use a single variable to represent both labels.

Initially, S contains only s, T contains only t, $d_s(s) = d_t(t) = 0$, and all parent pointers are *null*. The algorithm proceeds in passes. At the beginning of a pass, all vertices in S are S-vertices, all vertices in T are T-vertices, and other vertices are N-vertices. The algorithm chooses a tree to grow in the pass, either S (*forward*) or T (*reverse*). Assume we have a forward pass; the other case is symmetric. The goal of a pass is to grow S by one level and to increase D_s (and L) by one. We make all vertices v of S with $d_s(v) = D_s$ *active*. The pass executes growth steps, which may be interrupted by augmentation steps (when an augmenting path is found) followed by adoption steps (to fix the invariants violated when some arcs get saturated). At the end of the pass, if S has any vertices at level $D_s + 1$, we increment D_s; otherwise we terminate.

For efficiency, we use the *current arc* data structure, which ensures that each arc into a vertex is scanned at most once between its distance label increases during the adoption step. When an N-vertex is added to the tree or when the distance label of a vertex changes, we set the current arc to the first arc in its adjacency list. We maintain the invariant that the arcs preceding the current arc on the adjacency list are not admissible.

The growth step picks an active vertex v and scans v by examining residual arcs (v, w). If w is an S-vertex, we do nothing. If w is an N-vertex, we make w an S-vertex, set $p(w) = v$, and set $d_s(w) = D_s + 1$. If w is in T, we perform an augmentation step as described below. Once all arcs out of v are scanned, v becomes inactive. If a scan of v is interrupted by an augmentation step, we remember the outgoing arc that triggered it. If v is still active after the augmentation, we resume the scan of v from that arc.

The augmentation step applies when we find a residual arc (v, w) with v in S and w in T. The path P obtained by concatenating the s–v path in S, the arc (v, w), and the w–t path in T is an augmenting path. We augment on P, saturating some of its arcs. Saturating an arc $(x, y) \neq (v, w)$ creates orphans. Note that x and y are in the same tree. If they are in S, we make y an S-orphan, otherwise we make x a T-orphan. At the end of the augmentation step, we have (possibly empty) sets O_s and O_t of S- and T-orphans, respectively. These sets are processed during the adoption step.

We describe the adoption step assuming we grow S (the case for T is symmetric). S has a partially completed level $D_s + 1$. To avoid rescanning vertices at level D_s, we allow adding vertices to this level during orphan processing.

Our implementation of the adoption step is based on the *relabel* operation of the push-relabel algorithm. To process an S-orphan v, we first scan the arc list starting from the current arc and stop as soon as we find a residual arc (u, v) with $d_s(u) = d_s(v) - 1$. If such a vertex u is found, we make v an S-vertex, set the current arc of v to (v, u), and set $p(v) = u$. If no such u is found, we apply the *orphan relabel* operation to v. The operation scans the whole list to find the vertex u for which $d_s(u)$ is minimum and (u, v) is residual. If no such u exists, or if $d_s(u) > D_s$, we make v an N-vertex and make vertices w such that $p(w) = v$ S-orphans. Otherwise we choose u to be the first such vertex and set the current arc of v to be (v, u), set $p(v) = u$, set $d_s(v) = d_s(u) + 1$, make v an S-vertex, and make vertices w such that $p(w) = v$ S-orphans. If v was active and now $d_s(v) = D_s + 1$, we make v inactive.

The adoption step for T-vertices is symmetric except we make v an N-vertex if $d_t(u) \geq D_t$ (not just $d_t(u) > D_t$) because we are in the forward pass. Once both adoption steps finish, we continue the growth step.

4.1   Correctness and Running Time

We now prove that IBFS is correct and bound its running time. When analyzing individual passes, we assume we are in a forward pass; the reverse pass is similar.

We start the analysis by considering what happens on tree boundaries.

Lemma 1. *If (u, v) is residual:*

1. *If $u \in S$, $d_s(u) \leq D_s$, and $v \notin S$, then u is an active S-vertex.*
2. *If $v \in T$ and $u \notin T$, then $d_t(v) = D_t$.*
3. *After the increase of D_s, if $u \in S$ and $v \notin S$, then $d_s(u) = D_s$.*

Proof. The proof is by induction on the number of growth, augmentation, and adoption steps and passes.

At the beginning of a pass, all S-vertices u with $d_s(u) = D_s$ are active. Moreover, (2) and (3) hold at the end of the previous pass (or after the initialization for the first pass). This implies (1) and (2).

A growth step on u without an augmentation makes u inactive, but only after completing a scan of arcs (u, v) and adding all vertices $v \neq t$ with a residual arc (u, v) to S, so (1) is maintained. A growth step does not change T, so it cannot affect the validity of (2).

An augmentation can make an arc (u, v) non-residual, which cannot cause any claim to be violated. An augmentation can create a new residual arc (u, v) with $u \in S$, if flow is pushed along (v, u). In this case $v = p(u)$, so v must also be in S and (1) does not apply for (u, v). The symmetric argument shows that (2) does not apply for a new residual arc either.

An orphan relabel step can remove a vertex v from S. However, if a residual arc (u, v) exists with $u \in S$ and $d_s(u) \leq D_s$, then by definition of the orphan relabel step, v remains an S-vertex. So (1) is maintained after an orphan relabel step. The symmetric argument shows that (2) is maintained as well.

Finally, if there are no active vertices, then (u, v) can be a residual arc with $u \in S$ and $v \notin S$ only if $d_s(u) > D_s$. Since we grow the tree by one level, $d_s(u) = D_s + 1$. This implies that (3) holds after the increase of D_s.   □

We now consider the invariants maintained by the algorithm.

Lemma 2. *The following invariants hold:*

1. *Vertices in S and T have valid labelings, d_s and d_t.*
2. *For every vertex u in S, u's current arc either precedes or is equal to the first admissible arc to u. For every vertex u in T, u's current arc either precedes or is equal to the first admissible arc from u.*
3. *If u is an S-vertex, then $(p(u), u)$ is admissible. If u is a T-vertex, then $(u, p(u))$ is admissible.*
4. *For every vertex v, $d_s(v)$ and $d_t(v)$ never decrease.*

Proof. The proof is by induction on the growth, augmentation and adoption steps. We prove the claim for S; the proof for T is symmetric.

Augmentations do not change labels and therefore (4) does not apply. An augmentation can create a new residual arc $(u, p(u))$ by pushing flow on $(p(u), u)$. Using the induction assumption of (3), however, $(p(u), u)$ is admissible, so $(u, p(u))$ cannot be admissible and thus (2) still applies. In addition, $d_s(p(u)) = d_s(u) - 1$, so (1) is maintained. An augmentation can make an arc $(p(u), u)$ non-admissible by saturating it. However, this cannot violate claims (1) or (2) and vertex u becomes an orphan, so (3) is not applicable.

Consider a growth step on u that adds a new vertex v to S. We set $d_s(v) = d_s(u) + 1 = D_s + 1$, so (3) holds. For every residual arc (w, v) with $w \in S$, w must be active by Lemma 1. Since the d_s value of every active vertex is D_s, we get $d_s(w) = D_s = d_s(v) - 1$, so (1) holds. The current arc of v is v's first arc, so (2) holds. Since v is added at the highest possible label, it is clear that the label of v did not decrease and (4) is maintained.

Consider an adoption step on v. The initial scan of the orphan's arc list does not change labels and therefore cannot break (1) or (4). An orphan scan starts from the current arc, which precedes the first admissible arc by the induction assumption of (2), therefore it will find the first admissible arc to v. So if v finds a new parent, the new current arc is the first admissible arc to v, as required by (2) and (3). An orphan relabel finds the first lowest label $d_s(u)$ such that (u, v) is residual. So the labeling remains valid and the current arc is the first admissible arc, as required by (1), (2) and (3). Using the induction assumption of (1), labeling validity ensures that an orphan relabel cannot decrease the label of a vertex, by definition, so (4) is maintained.   □

At the end of the forward step there are no active vertices, so if the level $D_s + 1$ of S is empty, then by Lemma 1 there are no residual arcs from a vertex in S to a vertex not in S, and therefore the current flow is a maximum flow.

The following two lemmas are useful to provide some intuition on the algorithm. They are not needed for the analysis, so we state them without proofs.

Lemma 3. *During a growth phase, for every vertex $v \in S$, the path in S from s to v is a shortest path, and for every vertex $v \in T$, the path in T from v to t is a shortest path.*

Lemma 4. *The algorithm maintains the invariant that $L = D_s + D_t + 1$ is a lower bound on the augmenting path length, and always augments along the shortest augmenting path.*

The next lemma allows us to charge the time spent on orphan arc scans.

Lemma 5. *After an orphan relabel on v in S, $d_s(v)$ increases. After an orphan relabel on v in T, $d_t(v)$ increases.*

Proof. Consider an orphan relabel on an orphan $v \in S$. The analysis for an orphan $v \in T$ is symmetric.

Let U be the set of vertices u such that $u \in S$ and (u, v) is residual during the orphan relabel. By Lemma 2, v's current arc precedes the first admissible arc to v. Since during the orphan scan we did not find any admissible arc after v's current arc, there are no admissible arcs to v. By Lemma 2, the labeling is valid, so $d_s(u) \geq d_s(v) - 1$ for every $u \in U$. Since no admissible arc to v exists, we have that $d_s(u) \geq d_s(v)$ for every $u \in U$. So if the relabel operation does not remove v from S, it will increase $d_s(v)$.

Assume the relabel operation removes v from S. Let $d'_s(v)$ be the value of $d_s(v)$ when v was removed from S. Vertex v might be added to S later, during a growth step on some vertex $w \in S$. If $w \in U$, then $d_s(w)$ did not decrease since the relabel on v (by Lemma 2), so v will be added to S with a higher label. If $w \notin U$ then (w, v) became residual after v was removed from S. This means flow was pushed along (v, w) with $v \notin S$. This is only possible with $w \notin S$. So w was at some point removed from S and then added back to S at label $D_s + 1 \geq d'_s(v)$. Using Lemma 2, $d_s(w)$ did not decrease since that time, so when v is added to S, we get $d_s(v) = d_s(w) + 1 \geq d'_s(v) + 1$. $\square$

We are now ready to bound the running time of the algorithm.

Lemma 6. *IBFS runs in $O(n^2 m)$ time.*

Proof. There are three types of operations we must account for: adoption steps, growth steps with augmentations, and growth steps without augmentations.

Consider a growth step on v without an augmentation. We charge a scan of a single arc during the step to the label of v. Since we do not perform augmentations, v becomes inactive once the scan of its arcs is done. Vertex v can become active again only when its label increases. Thus every arc (v, u) scanned during such a growth step charges the distance label at most once. There are at most $n - 1$ different label values for each side (S or T), so the total time spent scanning arcs in growth steps without augmentations is $O(\Sigma_v \text{degree}(v) \cdot (n-1)) = O(nm)$.

We charge a scan of a single arc during an adoption step on v to the label of v. By Lemma 5 and Claim (4) of Lemma 2, after every orphan relabel $d_s(v)$ or $d_t(v)$ increases and cannot decrease afterwards. So every arc charges each label at most twice, once in an orphan scan and once in an orphan relabel. Since there are $O(n)$ labels, the time spent scanning arcs in adoption steps is also $O(nm)$.

We divide the work of a growth step with an augmentation on v into scanning arcs of v to find the arc to T and performing the augmentation. For the

former, since we remember the arc used in the last augmentation, an arc of v
not participating in an augmentation is scanned only once per activation of v.
An analysis similar to that for the growth steps without augmentation gives an
$O(nm)$ bound on such work for the whole algorithm. For the latter, the work
per augmentation is $O(n)$. If the saturated arc (u, v) is in S or T, the work can
be charged to the previous scan of the arc after which it was added to the tree.
It remains to account for augmentations that saturate an arc (u, v) with $u \in S$
and $v \in T$. We charge every such saturation to the label $d_s(u)$. While u remains
active, (u, v) cannot be saturated again. As with growth steps without augmentations, u can only become active again when its label increases. So a saturation
of (u, v) charges the label $d_s(u)$ at most once. There are at most $n - 1$ distinct
label values, so the total number of such charges is $O(nm)$. An augmentation
during a growth of v, including the scan of v's arcs until the augmentation,
takes $O(n)$ time. So the total time spent on growth steps with augmentations is
$O(n^2m)$.   □

This bound can be improved to $O(nm \log n)$ using the dynamic trees data structure [17], but in practice the simple $O(n^2m)$ version is faster on vision instances.

5   Variants of IBFS

We briefly discuss two variants of IBFS, incorporating blocking flows and delays.
According to our preliminary experiments, these variants have higher constant
factors and are somewhat slower than the standard algorithm on vision instances,
which are relatively simple. These algorithms are interesting from a theoretical
viewpoint, however, and are worth further experimental evaluation as well.

A blocking flow version. Note that at the beginning of a growth pass, we have
an auxiliary network on which we can compute a blocking flow (see e.g. [15]).
The network is induced by the arcs (v, w) such that either both v and w are in
the same tree and the arc is admissible, or v is in S and w is in T and (v, w)
is residual. We get a blocking flow algorithm by delaying vertex relabelings: a
vertex whose parent arc becomes saturated, or whose parent becomes an orphan,
tries to reconnect at the same level of the same tree and becomes an orphan if it
fails. In this case its distance from s (if it is an S-orphan) or from t (T-orphan)
increased. We process orphans at the end of the growth/augment pass.

It may be possible to match the bound on the number of iterations of the
binary blocking flow algorithm bound [11].

A delayed version. The standard version of IBFS ignores some potentially useful
information. For example, suppose that $D_s = D_t = 10$, $L = 21$, and for an S-
vertex v, $d_s(v) = 2$. Then a lower bound on the distance from v to t is $21 - 2 = 19$.
Suppose that, after an augmentation and an adoption step, v remains an S-vertex
but $d_s(v) = 5$. Because distances to t are monotone, 19 is still a valid lower
bound, and we can delay the processing of v until L increases to $5 + 19 = 24$.

The *delayed IBFS* algorithm takes advantage of such lower bounds to delay
processing vertices known not to be on shortest paths of length L. Furthermore,

the algorithm is lazy: it does not scan delayed vertices. As a result, vertices reachable only through delayed vertices (not "touching" tree vertices) are implicitly delayed as well. Compared to standard IBFS, the delayed variant is more complicated, and so is its analysis: it maintains a lot of information implicitly, and more state transitions can occur.

6 Experimental Results

6.1 Implementation Details

We now give details of our implementation of IBFS, which we call *IB*. Instead of performing a forward or reverse pass independently, we grow both trees by one level simultaneously. This may result in augmenting paths one arc longer than shortest augmenting paths: for example, during the growth step of an S-vertex v with label D_s we may find a T-vertex w with label $D_t + 1$. Since the s–v path in S and the w–t path in T are shortest paths, one can still show that the distances are monotone and the analysis remains valid. Note that BK runs in the same manner, growing both trees simultaneously.

We process orphans in FIFO order. If an augmentation saturates a single arc (which is quite common), FIFO order means that all subsequent orphans (in the original orphan's subtree) will be processed in ascending order of labels.

We maintain *current arcs* implicitly. The invariants of IBFS ensure the current arc of v is either its parent arc or the first arc in its adjacency list. A single bit suffices to distinguish between these cases.

For each vertex v in a tree, we keep its children in a linked list, allowing them to be easily added to the list of orphans when v is relabeled.

During an orphan relabel step on a vertex v in S, if a potential parent u is found with $d_s(u) = d_s(v)$, then the scan halts and u is taken as the parent. It is easy to see that such a vertex u must have the minimum possible label. A similar rule is applied to vertices in T.

On vision instances, orphan relabels often result in increasing the label of the orphan by one. To make this case more efficient, we use the following heuristic. When an orphan v is relabeled, its children become orphans. For every child u of v, we make (v, u) the first arc in u's adjacency list. If v's label does increase by one, a subsequent orphan relabel step on u will find (u, v) immediately and halt (due to the previous heuristic), saving a complete scan of u's arc list.

We also make some low-level optimizations for improved cache efficiency. Every arc (u, v) maintains a bit stating whether the residual capacity of (v, u) is zero. This saves an extra memory access to the reverse arc during growth steps in T and during orphan steps in S. The bit is updated during augmentations, when residual capacities change. Moreover, we maintain the adjacency list of a vertex v in an array. To make the comparison fair, we make these low-level optimizations to BK as well. We compared our improved code, *UBK*, to BK version 3.0.1 from *http://www.cs.ucl.ac.uk/staff/V.Kolmogorov/software.html*. Overall, UBK is about 20% faster than the original BK implementation, although the speedup is not uniform and BK is slightly faster on some instances.

Table 1. Performance of IBFS and BK on various instances

INSTANCE			TIME [s]			PU		GS		OS		OT
NAME	n	$\frac{m}{n}$	IB	UBK	SPD	IB	UBK	IB	UBK	IB	UBK	UBK
diggedshweng	301035	5.0	0.42	1.26	3.00	16.9	160.0	6.7	7.7	87.8	7.7	38.4
hessi1a	494402	5.0	5.81	6.43	1.11	108.4	353.2	7.3	25.4	601.7	43.9	126.5
monalisa	789419	5.0	2.92	4.33	1.48	30.9	181.9	8.1	11.7	239.4	17.1	59.2
house	967874	5.0	2.54	3.16	1.24	33.0	122.2	6.3	10.2	129.6	13.3	43.7
anthra	1061920	5.0	6.28	6.73	1.07	53.5	153.0	6.8	17.3	348.3	27.3	83.3
bone_subx10	3899394	7.0	2.73	3.20	1.17	0.6	1.3	6.6	8.2	25.0	5.5	11.7
bone_subx100	3899394	7.0	3.30	5.32	1.61	2.8	10.9	6.8	8.8	30.1	6.8	23.0
liver10	4161602	7.0	4.91	5.98	1.22	1.0	2.1	6.5	9.6	45.6	8.7	22.2
liver100	4161602	7.0	6.62	14.21	2.15	7.5	23.2	6.9	12.3	56.0	13.6	66.5
babyface10	5062502	7.0	4.98	5.72	1.15	0.5	1.0	6.4	9.3	38.6	7.0	15.4
babyface100	5062502	7.0	6.44	11.33	1.76	4.5	12.7	6.6	10.7	46.3	9.5	39.5
bone10	7798786	7.0	6.24	4.21	0.67	0.1	0.1	6.9	7.5	30.7	3.6	3.6
bone100	7798786	7.0	7.01	5.56	0.79	0.5	2.0	6.9	8.1	35.6	5.1	7.0
bunny-med	6311088	7.0	1.04	1.28	1.23	0.3	0.5	6.2	6.2	0.6	0.4	0.6
gargoyle-sml	1105922	5.0	0.89	8.56	9.57	7.8	212.8	7.5	6.8	33.5	10.7	143.2
gargoyle-med	8847362	5.0	22.58	139.06	6.16	22.7	337.2	8.7	12.1	121.6	20.7	250.5
camel-sml	1209602	5.0	0.84	1.31	1.56	5.3	27.6	6.6	6.8	27.5	8.0	23.1
camel-med	9676802	5.0	21.00	32.33	1.54	20.4	74.0	6.8	9.4	92.4	13.0	61.2
BVZ-tsukuba	—	—	0.42	0.45	1.09	1.2	1.7	5.1	5.5	10.8	3.9	2.8
BVZ-sawtooth	—	—	0.70	0.84	1.20	1.6	2.5	5.1	5.5	6.1	3.7	2.7
BVZ-venus	—	—	1.06	1.19	1.11	2.3	4.1	5.7	6.2	13.5	6.0	5.1
KZ2-sawtooth	—	—	1.68	2.49	1.48	2.6	4.3	8.1	9.3	7.5	8.8	4.0
KZ2-venus	—	—	2.98	4.14	1.39	3.3	6.2	8.8	11.2	18.0	13.5	8.1
rmf-wide-14	16807	6.6	0.17	0.57	3.35	99.6	385.5	57.1	113.7	492.1	339.9	1659.5
rmf-wide-16	65025	6.7	2.06	13.22	6.43	184.6	1339.2	97.7	413.0	1161.0	982.6	8835.8
rmf-wide-18	259308	6.8	25.37	641.83	25.30	334.3	5923.9	150.4	3417.3	2626.8	6635.4	85807.3

All implementations (BK, UBK, and IB) actually eliminate the source and target vertices (and their incident arcs) during a preprocessing step. For each vertex v, they perform a trivial augmentation along the path $(s, v) \cdot (v, t)$ and assign either a demand or an excess to v, depending on whether (s, v) or (v, t) is saturated. The running times we report to not include preprocessing.

6.2   Experiments

We ran our experiments on a 32-bit Windows 7 machine with 4 GB of RAM and a 2.13 GHz Intel Core i3-330M processor (64 KB L1, 256 KB L2, and 3 MB L3 cache). We used the Microsoft Visual C++ 6.0 compiler with default "Release" optimization settings. We report system times (obtained with the *ftime* function) of the maximum flow computation, which excludes the time to read and initialize the graph. For all problems, capacities are integral.

Table 1 has the results. For each instance, we give the number of vertices, n, and density, m/n. We then report the running times (in seconds) of IB and UBK, together with the relative speedup (SPD), i.e., the ratio between them. Values greater than 1.0 favor IB. The remaining columns contain some useful operation counts. PU is the combined length of all augmenting paths. GS is the number of arc scans during growth steps. OS is the number of arc scans during orphan steps. Finally, OT is the number of arcs scanned by UBK when traversing the paths from a potential parent to the root of its tree (these are not included in OS). Note that all counts are *per vertex* (i.e, they are normalized by n).

The instances in the table are split into six blocks. Each represents a different family: image segmentation using scribbles, image segmentation, surface fitting, multiview reconstruction, stereo images, and a hard DIMACS family.

The first five families are vision instances. The scribble instances were created by the authors, and are available upon request. The four remaining vision families are available at *http://vision.csd.uwo.ca/maxflow-data/*, together with detailed descriptions. (Other instances from these families are available as well; we took a representative sample due to space constraints.) Note that each image segmentation instance has two versions, with maximum capacity 10 or 100. For the vision problems, the running times are the average of three runs for every instance. Because stereo image instances are solved extremely fast, we present the total time for solving all instances of each subfamily.

Note that IB is faster than BK on all vision instances, except `bone10` and `bone100`. The speedup achieved by IB is usually modest, but can be close to an order of magnitude in some cases (such as `gargoyle`). IB is also more robust. It has similar performance on `gargoyle` and `camel`, which are problems from the same application and of similar size; in contrast, UBK is much slower on `gargoyle` than on `camel`.

Operation counts show that augmenting on shortest paths leads to fewer arc flow changes and growth steps, but to more orphan processing. This is because IB has more restrictions on how a disconnected vertex can be reconnected. UBK also performs OT operations, which are numerous on some instances (e.g., `gargoyle`).

Most vision instances are easy, with few operations per vertex. To see what happens on harder problems, and to observe asymptotic trends, we use the DIMACS [14] family that is hardest for modern algorithms, `rmf-wide`. In this case, each entry in the table is the average of five instances with the same parameters and different seeds. On this family, IB is asymptotically faster than UBK, but not competitive with good general-purpose codes [10]. For larger instances, UBK performs more operations of every kind, including orphan processing. In addition, it performs a large number of OT operations.

We also experimented with other DIMACS problem families. On all of them IBFS outperformed UBK, in some cases by a very large margin.

7   Concluding Remarks

We presented a theoretically justified analog to the BK algorithm and showed that it is more robust in practice. We hope that the algorithm will be adopted by the vision community. Recently, Arora et al. [1] presented a new push-relabel algorithm that runs in polynomial time and outperforms BK on vision instances. It may outperform ours on some instances as well, but unfortunately we were unable to perform a direct comparison. Unfortunately we were unable to compare this algorithm to ours directly.

Note that our algorithm also applies in the semi-dynamic setting where we want to maintain shortest path trees when arbitrary arcs can be deleted from the graph, and arcs not on shortest paths can be added. We believe that the variants of the IB algorithm introduced in Section 5 are interesting and deserve further investigation.

Motivated by the BK algorithm, we give its theoretically justified analog with appears to be more robust in practice.

References

1. Arora, C., Banerjee, S., Kalra, P., Maheshwari, S.: An Efficient Graph Cut Algorithm for Computer Vision Problems. In: Daniilidis, K. (ed.) ECCV 2010, Part III. LNCS, vol. 6313, pp. 552–565. Springer, Heidelberg (2010)
2. Boykov, Y., Kolmogorov, V.: An Experimental Comparison of Min-Cut/Max-Flow Algorithms for Energy Minimization in Vision. IEEE transactions on Pattern Analysis and Machine Intelligence 26(9), 1124–1137 (2004)
3. Boykov, Y., Veksler, O.: Graph Cuts in Vision and Graphics: Theories and Applications. In: Paragios, N., Chen, Y., Faugeras, O. (eds.) Handbook of Mathematical Models in Computer Vision, pp. 109–131. Springer, Heidelberg (2006)
4. Chandran, B., Hochbaum, D.: A Computational Study of the Pseudoflow and Push-Relabel Algorithms for the Maximum Flow Problem. Operations Research 57, 358–376 (2009)
5. Cherkassky, B.V.: A Fast Algorithm for Computing Maximum Flow in a Network. In: Karzanov, A.V. (ed.) Collected Papers. Combinatorial Methods for Flow Problems, vol. 3, pp. 90–96. The Institute for Systems Studies, Moscow (1979) (in Russian); English translation appears in AMS Trans. 158, 23–30 (1994)
6. Cherkassky, B.V., Goldberg, A.V.: On Implementing Push-Relabel Method for the Maximum Flow Problem. Algorithmica 19, 390–410 (1997)
7. Dantzig, G.B.: Application of the Simplex Method to a Transportation Problem. In: Koopmans, T.C. (ed.) Activity Analysis and Production and Allocation, pp. 359–373. Wiley, New York (1951)
8. Dinic, E.A.: Algorithm for Solution of a Problem of Maximum Flow in Networks with Power Estimation. Soviet Math. Dokl. 11, 1277–1280 (1970)
9. Ford Jr., L.R., Fulkerson, D.R.: Maximal Flow Through a Network. Canadian Journal of Math. 8, 399–404 (1956)
10. Goldberg, A.V.: Two-Level Push-Relabel Algorithm for the Maximum Flow Problem. In: Goldberg, A.V., Zhou, Y. (eds.) AAIM 2009. LNCS, vol. 5564, pp. 212–225. Springer, Heidelberg (2009)
11. Goldberg, A.V., Rao, S.: Beyond the Flow Decomposition Barrier. J. Assoc. Comput. Mach. 45, 753–782 (1998)
12. Goldberg, A.V., Tarjan, R.E.: A New Approach to the Maximum Flow Problem. J. Assoc. Comput. Mach. 35, 921–940 (1988)
13. Goldfarb, D., Grigoriadis, M.D.: A Computational Comparison of the Dinic and Network Simplex Methods for Maximum Flow. Annals of Oper. Res. 13, 83–123 (1988)
14. Johnson, D.S., McGeoch, C.C.: Network Flows and Matching: First DIMACS Implementation Challenge. AMS, Providence (1993)
15. Karzanov, A.V.: Determining the Maximal Flow in a Network by the Method of Preflows. Soviet Math. Dok. 15, 434–437 (1974)
16. King, V., Rao, S., Tarjan, R.: A Faster Deterministic Maximum Flow Algorithm. J. Algorithms 17, 447–474 (1994)
17. Sleator, D.D., Tarjan, R.E.: A Data Structure for Dynamic Trees. J. Comput. System Sci. 26, 362–391 (1983)

Engineering Multilevel Graph Partitioning Algorithms*

Peter Sanders and Christian Schulz

Karlsruhe Institute of Technology (KIT), 76128 Karlsruhe, Germany
{sanders,christian.schulz}@kit.edu

Abstract. We present a multi-level graph partitioning algorithm using novel local improvement algorithms and global search strategies transferred from multigrid linear solvers. Local improvement algorithms are based on max-flow min-cut computations and more localized FM searches. By combining these techniques, we obtain an algorithm that is fast on the one hand and on the other hand is able to improve the best known partitioning results for many inputs. For example, in Walshaw's well known benchmark tables we achieve 317 improvements for the tables at 1%, 3% and 5% imbalance. Moreover, in 118 out of the 295 remaining cases we have been able to reproduce the best cut in this benchmark.

1   Introduction

Graph partitioning is a common technique in computer science, engineering, and related fields. For example, good partitionings of unstructured graphs are very valuable for parallel computing. In this area, graph partitioning is mostly used to partition the underlying graph model of computation and communication. Roughly speaking, vertices in this graph represent computation units and edges denote communication. This graph needs to be partitioned such that there are few edges between the blocks (pieces). In particular, if we want to use k processors we want to partition the graph into k blocks of about equal size. In this paper we focus on a version of the problem that constrains the maximum block size to $(1 + \epsilon)$ times the average block size and tries to minimize the total cut size, i.e., the number of edges that run between blocks.

A successful heuristic for partitioning large graphs is the *multilevel graph partitioning* (MGP) approach depicted in Figure 1 where the graph is recursively *contracted* to achieve smaller graphs which should reflect the same basic structure as the input graph. After applying an *initial partitioning* algorithm to the smallest graph, the contraction is undone and, at each level, a *local refinement* method is used to improve the partitioning induced by the coarser level.

Although several successful multilevel partitioners have been developed in the last 13 years, we had the impression that certain aspects of the method are not well understood. We therefore have built our own graph partitioner KaPPa [4] (Karlsruhe Parallel Partitioner) with focus on scalable parallelization. Somewhat astonishingly, we also obtained improved partitioning quality through rather simple methods. This motivated us to make a fresh start putting all aspects of MGP on trial. Our focus is on solution quality and sequential speed for large graphs. This paper reports the first results we have obtained which relate to the local improvement methods and overall

---

* This paper is a short version of the technical report [10].

C. Demetrescu and M.M. Halldórsson (Eds.): ESA 2011, LNCS 6942, pp. 469–480, 2011.
© Springer-Verlag Berlin Heidelberg 2011

search strategies. We obtain a system that can be configured to either achieve the best known partitions for many standard benchmark instances or to be the fastest available

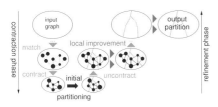

system for large graphs while still improving partitioning quality compared to the previous fastest system. We begin in Section 2 by introducing basic concepts. After shortly presenting Related Work in Section 3 we continue describing novel local improvement methods in Section 4. This is followed by Section 5 where we present new global search methods. Section 6 is a summary of extensive

Fig. 1. Multilevel graph partitioning

experiments done to tune the algorithm and evaluate its performance. We have implemented these techniques in the graph partitioner KaFFPa (Karlsruhe Fast Flow Partitioner) which is written in C++. Experiments reported in Section 6 indicate that KaFFPa scales well to large networks and is able to compute partitions of very high quality.

2   Preliminaries

2.1   Basic Concepts

Consider an undirected graph $G = (V, E, c, \omega)$ with edge weights $\omega : E \to \mathbb{R}_{>0}$, node weights $c : V \to \mathbb{R}_{\geq 0}$, $n = |V|$, and $m = |E|$. We extend c and ω to sets, i.e., $c(V') := \sum_{v \in V'} c(v)$ and $\omega(E') := \sum_{e \in E'} \omega(e)$. $\Gamma(v) := \{u : \{v, u\} \in E\}$ denotes the neighbors of v. We are looking for *blocks* of nodes $V_1, \ldots, V_k$ that partition V, i.e., $V_1 \cup \cdots \cup V_k = V$ and $V_i \cap V_j = \emptyset$ for $i \neq j$. The *balancing constraint* demands that $\forall i \in 1..k : c(V_i) \leq L_{\max} := (1 + \epsilon)c(V)/k + \max_{v \in V} c(v)$ for some parameter ϵ. The last term in this equation arises because each node is atomic and therefore a deviation of the heaviest node has to be allowed. The objective is to minimize the total *cut* $\sum_{i<j} w(E_{ij})$ where $E_{ij} := \{\{u, v\} \in E : u \in V_i, v \in V_j\}$. A vertice $v \in V_i$ that has a neighbor $w \in V_j, i \neq j$, is a boundary vertice. An abstract view of the partitioned graph is the so called *quotient graph*, where vertices represent blocks and edges are induced by connectivity between blocks. An example can be found in Figure 2. By default, our initial inputs will have unit edge and node weights. However, even those will be translated into weighted problems in the course of the algorithm.

A matching $M \subseteq E$ is a set of edges that do not share any common nodes, i.e., the graph (V, M) has maximum degree one. *Contracting* an edge $\{u, v\}$ means to replace the nodes u and v by a new node x connected to the former neighbors of u and v. We set $c(x) = c(u) + c(v)$ so the weight of a node at each level is the number of nodes it is representing in the original graph. If replacing edges of the form $\{u, w\}, \{v, w\}$ would generate two parallel edges $\{x, w\}$, we insert a single edge with $\omega(\{x, w\}) = \omega(\{u, w\}) + \omega(\{v, w\})$. *Uncontracting* an edge e undoes its contraction. In order to avoid tedious notation, G will denote the current state of the graph before and after a (un)contraction unless we explicitly want to refer to different states of the graph.

The multilevel approach to graph partitioning consists of three main phases. In the *contraction* (coarsening) phase, we iteratively identify matchings $M \subseteq E$ and contract the edges in M. Contraction should quickly reduce the size of the input and each computed level should reflect the global structure of the input network. A rating function indicates how much sense it makes to contract an edge based on local information. A matching algorithm tries to maximize the sum of the ratings of the contracted edges looking at the global structure of the graph. In KaPPa [4] we have shown that the rating function expansion$^{*2}(\{u,v\}) := \omega(\{u,v\})^2/c(u)c(v)$ works best among other edge rating functions. Contraction is stopped when the graph is small enough to be directly partitioned using some expensive other algorithm. In the *refinement* (or uncoarsening) phase, the matchings are iteratively uncontracted. After uncontracting a matching, the refinement algorithm moves nodes between blocks in order to improve the cut size or balance. The succession of movements is based on priorities called *gain*,

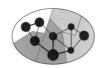

Fig. 2. A graph partitioned into five blocks and its quotient graph $\mathcal{Q}$. Two pairs of blocks are highlighted in red and green.

i.e., the decrease in edge cut when the node is moved to the other side. There are two main types of local search heuristics: *k-way* and *two-way* local search. *k*-way local search is allowed to move a node to an arbitrary block whereas two-way local search is restriced to move nodes only between a pair of blocks. The latter is usually applied to all pairs of blocks sharing a non-empty boundary. The intuition behind this approach is that a good partition at one level of the hierarchy will also be a good partition on the next finer level so that refinement will quickly find a good solution. KaFFPa *makes use of techniques proposed* in KaPPa [4] and KaSPar [7]. These techniques concern coarsening (edge ratings, global paths algorithm (GPA) as matching algorithm), initial partititioning (using Scotch) and a flexible stopping criterion for local search. They are described in the TR [10].

3   Related Work

There has been a huge amount of research on graph partitioning so that we refer the reader to [3,15] for more material. All general purpose methods that are able to obtain good partitions for large real world graphs are based on the multilevel principle outlined in Section 2. The basic idea can be traced back to multigrid solvers for solving systems of linear equations [12] but more recent practical methods are based on mostly graph theoretic aspects in particular edge contraction and local search. Well known software packages based on this approach include, Jostle [15], Metis [11], and Scotch [8]. KaSPar [7] is a graph partitioner based on the central idea to (un)contract only a single edge between two levels. KaPPa [4] is a "classical" matching based MGP algorithm designed for scalable parallel execution. DiBaP [6] is a multi-level graph partitioning package where local improvement is based on diffusion. MQI [5] and Improve [1] are flow-based methods for improving graph cuts when cut quality is measured by quotient-style metrics such as *expansion* or *conductance*. This approach is only feasible for $k = 2$. Improve uses several minimum cut computations to improve the *quotient cut* score of a proposed partition.

The concept of *iterated multilevel algorithms* was introduced by [13]. The main idea is to iterate the coarsening and uncoarsening phase. Once the graph is partitioned, edges that are between two blocks are not contracted. This ensures increased quality of the partition if the refinement algorithms guarantees no worsening.

4   Local Improvement

Recall that once a matching is uncontracted a local improvement method tries to reduce the cut size of the projected partition. We now present two novel local improvement methods. The first method is based on max-flow min-cut computations between pairs of blocks, i.e., improving a given 2-partition. Roughly speaking, this improvement method is then applied between all pairs of blocks that share a non-empty boundary. In contrast to previous flow-based methods we improve the edge cut whereas previous systems improve conductance or expansion. The second method which is described in Section 4.2 is called multi-try FM. Roughly speaking, a k-way local search initialized with a *single* boundary node is repeatedly started. Previous methods are initialized with *all* boundary nodes. At the end of the section we show how pairwise refinements can be scheduled and how multi-try FM local search can be incorporated with this scheduling.

4.1   Max-Flow Min-Cut Computations for Local Improvement

We now explain how flows can be employed to improve a partition of *two blocks V_1, V_2* without violating the balance constraint. That yields a local improvement algorithm.

First we introduce a few notations. Given a set of nodes $B \subset V$ we define its *border* $\partial B := \{u \in B \mid \exists (u, v) \in E : v \notin B\}$. The set $\partial_1 B := \partial B \cap V_1$ is called *left border*

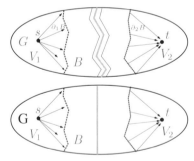

of B and the set $\partial_2 B := \partial B \cap V_2$ is called *right border* of B. A B *induced subgraph* G' is the node induced subgraph $G[B]$ plus two nodes s, t that are connected to the border of B. More precisely s is connected to all left border nodes $\partial_1 B$ and all right border nodes $\partial_2 B$ are connected to t. All of these new edges get the edge weight ∞. Note that the additional edges are directed. G' has the *cut property* if each (s,t)-min-cut induces a cut within the balance constraint in G.

Fig. 3. The construction of a feasible flow problem G' is shown on the top and an improved cut within the balance constraint in G is shown on the bottom

The basic idea is to construct a B induced subgraph G' having the cut property. Each mincut will then yield a feasible improved cut within the balance constraint in G. By performing two Breadth First Searches (BFS) we can find such a set B. Each node touched during these searches belongs to B. The first BFS is done in the subgraph of G induced by V_1. It is initialized with the boundary nodes of V_1. As soon as the weight of the area found by this BFS would exceed $(1 + \epsilon)c(V)/2 - w(V_1)$, we stop the BFS. The second BFS is done for V_2 in an analogous fashion. The constructed subgraph G' has the cut property since the worst case new weight of V_2 is lower or equal to $w(V_2) + (1 + \epsilon)c(V)/2 - w(V_2) = (1 + \epsilon)c(V)/2$. Indeed the same holds for the worst case new weight of V_1.

There are multiple ways to improve this method. First, if we found an improved cut, we can apply this method again since the initial boundary has changed, i.e., the set B will also change. Second, we can adaptively control the size of the set B found by the BFS. This enables us to search for cuts that fulfill our balance constraint in a larger subgraph (say $\epsilon' = \alpha\epsilon$ for some parameter α). To be more precise if the found min-cut in G' for ϵ' fulfills the balance constraint in G, we accept it and increase α to $\min(2\alpha, \alpha')$ where α' is an upper bound for α. Otherwise the cut is not accepted and we decrease α to $\max(\frac{\alpha}{2}, 1)$. This method is iterated until a maximal number of iterations is reached or if the computed cut yields a feasible partition without a decreased cut. We call this method *adaptive flow iterations*.

Most Balanced Minimum Cuts. Picard and Queyranne have been able to show that *one* (s, t)-max-flow contains information about *all* minimum (s,t)-cuts in the graph. Thus the idea to search for feasable cuts in larger subgraphs becomes even more attractive. Roughly speaking, we present a heuristic that, given a max-flow, selects min-cuts with better balance in G. First we need a few notations. For a graph $G = (V, E)$ a set $C \subseteq V$ is a *closed vertex set* iff for all vertices $u, v \in V$, the conditions $u \in C$ and $(u, v) \in E$ imply $v \in C$. An example can be found in Figure 4.

Lemma 1 (Picard and Queyranne [9]). *There is a 1-1 correspondence between the minimum (s, t)-cuts of a graph and the closed vertex sets containing s in the residual graph of a maximum (s, t)-flow.*

For a given closed vertex set C of the residual graph containing s, the corresponding min-cut is $(C, V \setminus C)$. Note that distinct maximum flows may produce different residual graphs but the closed vertex sets remain the same. To enumerate all minimum cuts of a graph [9] a further reduced graph is computed which is described below. However, the problem of finding the most balanced minimum cut is NP-hard [9].

We now define how the representation of the residual graph can be made more compact [9] and then explain our heuristic to obtain closed vertex sets on this graph in order to select min-cuts with better balance. First we take a maximum (s, t)-flow and compute the strongly connected components of its residual graph. We make the representation more compact by contracting the components and refer to it as *minimum cut representation*. The reduction is possible since two vertices that lie on a cycle have to be in the same closed vertex set of the residual graph. The result is a weighted, directed and acyclic graph (DAG). Note that each closed vertex set of the minimum cut representation induces a minimum cut as well. On this graph we search for closed vertex sets

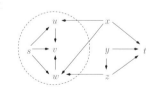

Fig. 4. The set $\{s, u, v, w\}$ is a closed vertex set

(containing the component S that contains the source) since they still induce (s, t)-min-cuts in the original graph. This is done by using the following heuristic which is repeated a few times. The main idea is that a topological order yields complements of closed vertex sets quite easily. Therefore, we first compute a random topological order using a randomized DFS[1].

We sweep through this topological order and sequentially add the components to the complement of the closed vertex set. By sweeping through the topological order we compute closed vertex sets each inducing a min-cut having a different balance. We stop when we have reached the best balanced minimum cut induced through this particular topological order. The closed vertex set with the best balance occurred for different topological orders is returned. Note that this procedure may still finds cuts that are not feasible in oversized subgraphs, e.g. if there is no feasible minimum cut. Therefore the algorithm is combined with the adaptive strategy from above. We call this method *balanced adaptive flow iterations*.

4.2   Multi-try FM

This local improvement method moves nodes between blocks in order to decrease the cut. Previous k-way methods were initialized with *all* boundary nodes, i.e., all boundary nodes are eligible for movement at the beginning. Our method is repeatedly initialized with a *single* boundary node. More details about k-way methods can be found in the TR [10].

Multi-try FM is organized in rounds. In each round we put *all* boundary nodes of the current block pair into a todo list T. Subsequently, we begin a k-way local search starting with a *single* random node v of T if it is still a boundary node. Note that the difference to the global k-way search is in the initialisation of the search. The local search is only started from v if it was not touched by a previous localized k-way search in this round. Either way, the node is removed from the todo list. A localized k-way search is not allowed to move a node that has been touched in a previous run. This assures that at most n nodes are touched during a round of the algorithm. The algorithm uses the adaptive stopping criterion from KaSPar [7].

4.3   Scheduling Quotient Graph Refinement

Our algorithm to schedule two-way local searches on pairs of blocks is called *active block scheduling*. The main idea is that local search should be done in areas in which change still happens. The algorithm begins by setting every block of a partition *active*. The scheduling then is organized in rounds. In each round, the algorithm refines adjacent pairs of blocks that have *at least* one active block in a random order. If changes occur during this search both blocks are marked active for the next round of the algorithm. In this case a refinement of adjacent pairs of blocks can be both, two-way local search and local improvement by using flow, depending on the configuration. After each pair-wise refinement a multi-try FM search (k-way) is started. The todo list T is initialized with all boundaries of the current pair of blocks. Each block that changed during this search is also marked active for the next round. The algorithm stops if no active block is left.

---

[1] We also tried an algorithm that iteratively removes vertices having outdegree zero to compute a topological order. Improvements obtained by using this algorithm were negligible.

5   Global Search

Iterated Multilevel Algorithms (V-cycles) were introduced by [13]. The main idea is to iterate coarsening and refinement several times using different seeds for random tiebreaking. Edges between blocks are not contracted as soon as the graph is partitioned. Thus a given partition can be used as initial partition of the coarsest graph. This ensures increased quality if the refinement algorithm guarantees no worsening. In multigrid linear solvers Full-Multigrid methods are preferable to simple V-cycles [2]. Therefore, we now introduce two novel global search strategies namely *W-cycles* and *F-cycles* for graph partitioning. A W-cycle works as follows: on *each* level we perform *two recursive calls* using different random seeds during contraction and local search. As soon as the graph is partitioned, edges that are between blocks are not contracted. An F-cycle works similar to a W-cycle with the difference that further recursive calls are only made the second time that the algorithm reaches a particular level. In most cases the initial partitioner is not able to improve a given partition from scratch or even to find this partition. Therefore no further initial partitioning is used as soon as the graph is partitioned. Experiments in Section 6 show that all cycle variants are more efficient than simple restarts of the algorithm. In order to bound the execution time we introduce a level split parameter d such that further recursive calls are only performed every d'th level. We go into more detail after we have analysed the run time of the global search strategies.

Analysis. We now roughly analyse the run time of the different global search strategies under a few assumptions. In the following the shrink factor a names the factor that the graph shrinks (nodes and edges uniformly) during one coarsening step.

Theorem 1. *If the time for coarsening and refinement is $T_{cr}(n) := bn$ and a constant shrink factor $a \in [1/2, 1)$ is given. Then:*

$$
T_{W,d}(n) \begin{cases} \lesssim \frac{1-a^d}{1-2a^d} T_V(n) & \text{if } 2a^d < 1 \\ \in \Theta(n \log n) & \text{if } 2a^d = 1 \\ \in \Theta(n^{\log_d \log_{1/a} 2}) & \text{if } 2a^d > 1 \end{cases} \tag{1}
$$

$$
T_{F,d}(n) \le \frac{1}{1-a^d} T_V(n) \tag{2}
$$

where T_V is the time for a single V-cycle and $T_{W,d}, T_{F,d}$ are the time for a W-cycle and F-cycle with level split parameter d.

The proof can be found in the TR [10]. For the optimistic assumption that $a = 1/2$ and $d = 1$, a F-cycle is only twice as expensive as a single V-cycle. If we use the same parameters for a W-cycle we get a factor $\log n$ asymptotic larger execution times. However in practice the shrink factor is usually worse than $1/2$. That yields an even larger asymptotic run time for the W-cycle (since for $d = 1$ we have $2a > 1$). Therefore, in order to bound the run time of the W-cycle the choice of the level split parameter d is crucial. Our default value for d for W- and F-cycles is 2.

6   Experiments

Implementation / Instances / System. We have implemented the algorithm decribed above using C++. We report experiments on two suites of instances (medium sized graphs used in Subsection 6.1-6.3 and large sized graphs used in Subsection 6.4). The medium sized testset contains 20 graphs having between thirteen thousand and five houndred thousand vertices. The large sized testset contains 12 graphs having between seventy thousand and eighteen millon vertices. They are the same as in [7] and can be found the TR [10]. All implementation details, system information and more information about the graphs can be found in the TR [10].

Configuring the Algorithm. We currently define three configurations of our algorithm: Strong, Eco and Fast. The strong configuration is able to achieve the best known partitions for many standard benchmark instances, the eco version is a good tradeoff between quality and speed and the fast version of KaFFPa is the fastest available system for large graphs while still improving partitioning quality to the previous fastest system. All configurations use the FM algorithm. The strong configuration further employs Flows, Multi-Try FM and Global Search. The eco configuration also employs Flows. For a full description of the configurations we refer the reader to the TR [10].

Experiment Description. We performed two types of experiments namely normal tests and tests for effectiveness. Both are described below.

Normal Tests: Here we perform 10 repetitions for the small networks and 5 repetitions for the other. We report the arithmetic average of computed cut size, running time and the best cut found. When further averaging over multiple instances, we use the geometric mean in order to give every instance the same influence on the *final score*.[2]

Effectiveness Tests: Here each algorithm configuration has the same time for computing a partition. Therefore, for each graph and k each configuration is executed once and we remember the largest execution time t that occurred. Now each algorithm gets time $3t$ to compute a good partition, i.e., taking the best partition out of repeated runs. If a variant can perform a next run depends on the remaining time, i.e., we flip a coin with corresponding probabilities such that the expected time over multiple runs is $3t$. This is repeated 5 times. The final score is computed as above using these values. Note that on the middlesized testset the final eff. score of an algorithm configuration is the result of at least 1 800 algorithm runs.

6.1   Insights about Flows

We now evaluate max-flow min-cut based improvement algorithms. First we define a basic two-way FM configuration to compare with. It uses the GPA algorithm as a matching algorithm and performs five initial partitioning attempts using Scotch as initial partitioner. It further employs the active block scheduling algorithm equipped with the two-way FM algorithm. The FM algorithm stops as soon as 5% of the number of nodes

---

[2] Because we have multiple repetitions for each instance (graph, k), we compute the geometric mean of the average (**Avg.**) edge cut values for each instance or the geometric mean of the best (**Best.**) edge cut value occurred. The same is done for the run time **t**.

in the current block pair have been moved without yielding an improvement. Edge rating functions are used as in KaFFPa Strong. Note that during this test our main focus is the evaluation of flows and therefore we *don't* use k-way refinement or multi-try FM search.

To evaluate the performance of specific algorithmic components the basic configuration is extended by specific algorithms. A configuration that uses Flow, FM and the most balanced cut heuristics (MB) will be indicated by (+F, +MB, +FM). In Table 1 we see that by Flow on its own we obtain cuts and run times which are worse than those of the basic two-way FM configuration. The results improve in terms of quality and run time if we enable the most balanced minimum cut heuristic.

Table 1. The final score of different algorithm configurations. α' is the flow region upper bound factor. All average and best cut values are improvements relative to the basic configuration in %.

Variant	(+F, -MB, -FM)			(+F, +MB, -FM)			(+F, -MB, +FM)			(+F, +MB, +FM)		
α'	Avg.	Best.	$t[s]$	Avg.	Best.	$t[s]$	Avg.	Best.	$t[s]$	Avg.	Best.	$t[s]$
16	−1.88	−1.28	4.17	0.81	0.35	3.92	6.14	5.44	4.30	7.21	6.06	5.01
8	−2.30	−1.86	2.11	0.41	−0.14	2.07	5.99	5.40	2.41	7.06	5.87	2.72
4	−4.86	−3.78	1.24	−2.20	−2.80	1.29	5.27	4.70	1.62	6.21	5.36	1.76
2	−11.86	−10.35	0.90	−9.16	−8.24	0.96	3.66	3.37	1.31	4.17	3.82	1.39
1	−19.58	−18.26	0.76	−17.09	−16.39	0.80	1.64	1.68	1.19	1.74	1.75	1.22
Ref.	(-F, -MB, +FM)			2 974	2 851	1.13						

In some cases, flows and flows with the MB heuristic are not able to produce results that are comparable to the basic two-way FM configuration. Perhaps, this is due to the lack of the method to accept suboptimal cuts which yields small flow problems and therefore bad cuts. Consequently, we also combined both methods to fix this problem. In Table 1 we can see that the combination of flows with local search produces up to 6.14% lower cuts on average than the basic configuration. If we enable the most balancing cut heuristic we get on average 7.21% lower cuts than the basic configuration. Experiments in the TR [10] show that these combinations are more effective than the repeated executions of the basic two-way FM configuration. The most effective configuration is the basic two-way FM configuration using flows with $\alpha' = 8$ combined with the most balanced cut heuristic. It yields 4.73% lower cuts than the basic configuration in the effectiveness test.

6.2   Insights about Global Search Strategies

In Table 2 we compared different global search strategies against a single V-cycle. This time we choose a relatively fast configuration of the algorithm as basic configuration since the global search strategies are at focus. The coarsening phase is the same as in KaFFPa Strong. We perform one initial partitioning attempt using Scotch. The refinement employs k-way local search followed by quotient graph style refinements. Flow algorithms are not enabled for this test. The only parameter varied during this test is the global search strategy. Clearly, more sophisticated global search strategies decrease the cut but also increase the run time of the algorithm. However, the effectiveness results in Table 2 indicate that repeated executions of more sophisticated global search strategies are always superior to repeated executions of one single V-cycle. The increased effectiveness of more sophisticated global search strategies is due to different reasons. First of all by using a given partition in later cycles we obtain a very good initial partitioning for the coarsest graph which yields good starting points for local improvement on each level of refinement. Furthermore, the increased effectiveness is due to time saved using the active block strategy which converges very quickly in later cycles. On the other hand we save time for initial partitioning since it is only performed the first time the algorithm arrives in the initial partitioning phase. It is interesting

Table 2. Test results for normal and effectiveness tests for different global search strategies. Shown are improvements in % relative to the basic configuration

Algorithm	Avg.	$t[s]$	Eff. Avg.
2 F-cycle	2.69	2.31	2 806
3 V-cycle	2.69	2.49	2 810
2 W-cycle	2.91	2.77	2 810
1 W-cycle	1.33	1.38	2 815
1 F-cycle	1.09	1.18	2 816
2 V-cycle	1.88	1.67	2 817
1 V-cycle	2 973	0.85	2 834

to see that although the analysis in Section 5 makes some simplified assumptions the measured run times in Table 2 are very close to the values obtained by the analysis.

6.3   Removal / Knockout Tests

We now turn into two kinds of experiments to evaluate interactions and relative importance of our algorithmic improvements. In the component *removal tests* we take KaFFPa Strong and remove components step by step yielding weaker and weaker variants of the algorithm. For the *knockout tests* only one component is removed at a time, i.e., each variant is exactly the same as KaFFPa Strong minus the specified component. Table 3 summarizes the knockout test results. More detailed results of the tests can be found in the TR [10]. We shortly summarize the main results. First, in order to achieve high quality partitions we don't need to perform classical global k-way refinement. The changes in solution quality are negligible and both configurations (Strong without global k-way and Strong with global k-way) are equally effective. However, the global k-way refinement algorithm speeds up overall run time of the algorithm; hence we included it into

Table 3. Removal tests: each configuration is same as its predecessor minus the component shown in the first column. All average cuts and best cuts are shown as increases in cut (%) relative to the values obtained by KaFFPa Strong.

Variant	Avg.	Best.	$t[s]$	Eff. Avg.	Eff. Best.
Strong	2 683	2 617	8.93	2 636	2 616
-KWay	−0.04	−0.11	9.23	0.00	0.08
-Multitry	1.71	1.49	5.55	1.21	1.30
-Cyc	2.42	1.95	3.27	1.25	1.41
-MB	3.35	2.64	2.92	1.82	1.91
-Flow	9.36	7.87	1.66	6.18	6.08

KaFFPa Strong. In contrast, removing the multi-try FM algorithm increases average cuts by almost two percent and decreases the effectiveness of the algorithm. It is also interesting to see that as soon as a component is removed from KaFFPa Strong (except for the global k-way search) the algorithm gets less effective.

6.4   Comparison with Other Partitioners

We now switch to our suite of larger graphs since that's what KaFFPa was designed for. We compare ourselves with KaSPar Strong, KaPPa Strong, DiBaP Strong[3], Scotch and Metis. Table 4 summarizes the results. Detailed per instance results can be found in the TR [10]. kMetis produces about 33% larger cuts than KaFFPa Strong. Scotch, DiBaP, KaPPa, and KaSPar produce 20%,11%, 12% and 3% larger cuts than KaFFPa Strong respectively. In 57 out of 66 cases KaFFPa produces a better best cut than the best cut obtained by KaSPar. KaFFPa Eco now outper-

Table 4. Averaged quality of the different partitioning algorithms

Algorithm	large graphs		
	Best	Avg.	$t[s]$
KaFFPa Strong	12 054	12 182	121.50
KaSPar Strong	+3%	+3%	87.12
KaFFPa Eco	+6%	+6%	3.82
KaPPa Strong	+10%	+12%	28.16
Scotch	+18%	+20%	3.55
KaFFPa Fast	+25%	+24%	0.98
kMetis	+26%	+33%	0.83

forms Scotch and DiBaP producing 4.7 % and 12% smaller cuts than DiBaP and Scotch respectively. Note that DiBaP has a factor 3 larger run times than KaFFPa Eco on average. In the TR [10] we take two graph families and study the behaviour of our algorithms when the graph size increases. As soon as the graphs have more than 2^{19} nodes, KaFFPa Fast outperforms kMetis in terms of speed and quality. In general the speed up of KaFFPa Fast relative to kMetis increases with increasing graph size. The largest difference is obtained on the largest graphs where kMetis has up to 70% larger run times than our fast configuration which still produces 2.5% smaller cuts.

6.5   The Walshaw Benchmark

We now apply KaFFPa Strong to Walshaw's benchmark archive [14] using the rules used there, i.e., running time is no issue but we want to achieve minimal cut values for $k \in \{2, 4, 8, 16, 32, 64\}$ and balance parameters $\epsilon \in \{0, 0.01, 0.03, 0.05\}$.

We ran KaFFPa Strong with a time limit of two hours per graph and k (we excluded $\epsilon = 0$ since flows are not made for this case). KaFFPa computed 317 partitions which are better that previous best partitions reported there: 99 for 1%, 108 for 3% and 110 for 5%. Moreover, it reproduced equally sized cuts in 118 of the 295 remaining cases. After the partitions were accepted, we ran KaFFPa Strong as before and took the previous entry as input. Now overall in 560 out of 612 cases we where able to improve a given entry or have been able to reproduce the current result (in the first run). The complete list of improvements is available at [10].

---

[3] We exluded the European and German road network as well as the Random Geometric Graphs for the comparison with DiBaP since DiBaP can't handle singletons. In general, we excluded the case $k = 2$ for the European road network since KaPPa runs out of memory for this case.

7 Conclusions and Future Work

KaFFPa is an approach to graph partitioning that can be configured to either achieve the best known partitions for many standard benchmark instances or to be the fastest available system for large graphs while still improving partitioning quality compared to the previous fastest system. This success is due to new local improvement methods and global search strategies which were transferred from multigrid linear solvers. Regarding future work, we want to try other initial partitioning algorithms and ways to integrate KaFFPa into metaheuristics like evolutionary search.

Acknowledgements. We would like to thank Vitaly Osipov for supplying data for KaSPar and Henning Meyerhenke for providing a DiBaP-full executable. We also thank Tanja Hartmann, Robert Görke and Bastian Katz for valuable advice regarding balanced min cuts.

References

1. Andersen, R., Lang, K.J.: An algorithm for improving graph partitions. In: Proceedings of the Nineteenth Annual ACM-SIAM Symposium on Discrete Algorithms, pp. 651–660. SIAM, Philadelphia (2008)
2. Briggs, W.L., McCormick, S.F.: A multigrid tutorial. Soc. for Ind. Mathe (2000)
3. Fjallstrom, P.O.: Algorithms for graph partitioning: A survey. Linkoping Electronic Articles in Computer and Information Science 3(10) (1998)
4. Holtgrewe, M., Sanders, P., Schulz, C.: Engineering a Scalable High Quality Graph Partitioner. In: 24th IEEE International Parallal and Distributed Processing Symposium (2010)
5. Lang, K., Rao, S.: A flow-based method for improving the expansion or conductance of graph cuts. In: Bienstock, D., Nemhauser, G.L. (eds.) IPCO 2004. LNCS, vol. 3064, pp. 325–337. Springer, Heidelberg (2004)
6. Meyerhenke, H., Monien, B., Sauerwald, T.: A new diffusion-based multilevel algorithm for computing graph partitions of very high quality. In: IEEE International Symposium on Parallel and Distributed Processing, IPDPS 2008, pp. 1–13 (2008)
7. Osipov, V., Sanders, P.: n-Level Graph Partitioning. In: 18th European Symposium on Algorithms (2010); see also arxiv preprint arXiv:1004.4024
8. Pellegrini, F.: Scotch home page, http://www.labri.fr/pelegrin/scotch
9. Picard, J.C., Queyranne, M.: On the structure of all minimum cuts in a network and applications. Mathematical Programming Studies 13, 8–16 (1980)
10. Sanders, P., Schulz, C.: Engineering Multilevel Graph Partitioning Algorithms. Technical report, Karlsruhe Institute of Technology (2010); see ArXiv preprint arXiv:1012.0006v3
11. Schloegel, K., Karypis, G., Kumar, V.: Graph partitioning for high performance scientific simulations. In: Dongarra, J., et al. (eds.) CRPC Par. Comp. Handbook. Morgan Kaufmann, San Francisco (2000)
12. Southwell, R.V.: Stress-calculation in frameworks by the method of "Systematic relaxation of constraints". Proc. Roy. Soc. Edinburgh Sect. A, 57–91 (1935)
13. Walshaw, C.: Multilevel refinement for combinatorial optimisation problems. Annals of Operations Research 131(1), 325–372 (2004)
14. Walshaw, C., Cross, M.: Mesh Partitioning: A Multilevel Balancing and Refinement Algorithm. SIAM Journal on Scientific Computing 22(1), 63–80 (2000)
15. Walshaw, C., Cross, M.: JOSTLE: Parallel Multilevel Graph-Partitioning Software – An Overview. In: Magoules, F. (ed.) Mesh Partitioning Techniques and Domain Decomposition Techniques, pp. 27–58. Civil-Comp Ltd (2007) (invited chapter)

A Nearly Optimal Algorithm for Finding L_1 Shortest Paths among Polygonal Obstacles in the Plane[*]

Danny Z. Chen and Haitao Wang[**]

Department of Computer Science and Engineering
University of Notre Dame, Notre Dame, IN 46556, USA
{dchen,hwang6}@nd.edu

Abstract. Given a set of h pairwise disjoint polygonal obstacles of totally n vertices in the plane, we study the problem of computing an L_1 (or rectilinear) shortest path between two points avoiding the obstacles. Previously, this problem has been solved in $O(n \log n)$ time and $O(n)$ space, or alternatively in $O(n + h \log^{1.5} n)$ time and $O(n + h \log^{1.5} h)$ space. A lower bound of $\Omega(n + h \log h)$ time and $\Omega(n)$ space can be established for this problem. In this paper, we present a nearly optimal algorithm of $O(n + h \log^{1+\epsilon} h)$ time and $O(n)$ space for the problem, where $\epsilon > 0$ is an arbitrarily small constant. Specifically, after the free space is triangulated in $O(n + h \log^{1+\epsilon} h)$ time, our algorithm finds a shortest path in $O(n + h \log h)$ time and $O(n)$ space. Our algorithm can also be extended to obtain improved results for other related problems, e.g., finding shortest paths with fixed orientations, finding approximate Euclidean shortest paths, etc. In addition, our techniques yield improved results on some shortest path query problems.

1 Introduction

Computing shortest obstacle-avoiding paths in the plane is a fundamental problem in computational geometry. The Euclidean version in which the path length is measured by the Euclidean distance has been well studied (e.g., see [4,9,11,15,20,22]). In this paper, we consider the L_1 version, defined as follows. Given a set of h pairwise disjoint polygonal obstacles, $\mathcal{P} = \{P_1, P_2, \ldots, P_h\}$, of totally n vertices and two points s and t in the plane, the plane minus the interior of the obstacles is called the *free space*. Two objects are *disjoint* if they do not intersect in their interior. The L_1 *shortest path problem*, denoted by L_1-SPP, seeks a polygonal path in the free space from s to t with the minimum L_1 distance.

Note that the rectilinear version problem is also solved by our problem. As shown in [6,17,18,19], it is easy to convert an arbitrary polygonal path to a rectilinear path with the same L_1 length. Thus, in this paper, we focus on computing polygonal paths measured by the L_1 distance.

---

[*] This research was supported in part by NSF under Grant CCF-0916606.
[**] Corresponding author.

C. Demetrescu and M.M. Halldórsson (Eds.): ESA 2011, LNCS 6942, pp. 481–492, 2011.

The L_1-SPP problem has been studied extensively (e.g., see [3,6,7,17,18,19,23]). In general, there are two approaches for solving this problem: Constructing a sparse "path preserving" graph (analogous to a visibility graph), or applying the continuous Dijkstra paradigm. Clarkson, Kapoor, and Vaidya [6] constructed a graph of $O(n \log n)$ nodes and $O(n \log n)$ edges such that a shortest L_1 path can be found in the graph in $O(n \log^2 n)$ time; subsequently, they gave an algorithm of $O(n \log^{1.5} n)$ time and $O(n \log^{1.5} n)$ space [7]. Based on some observations, Chen, Klenk, and Tu [3] showed that the problem was solvable in $O(n \log^{1.5} n)$ time and $O(n \log n)$ space. By using the continuous Dijkstra paradigm, Mitchell [18,19] solved the problem in $O(n \log n)$ time and $O(n)$ space. An $\Omega(n + h \log h)$ time lower bound can be established for solving L_1-SPP (e.g., based on the results in [21]). Hence, Mitchell's algorithm is worst-case optimal. Recently, by using a corridor structure and building a smaller size path preserving graph, Inkulu and Kapoor [13] solved the problem in $O(n + h \log^{1.5} n)$ time and $O(n + h \log^{1.5} h)$ space. Note that when all polygonal obstacles in $\mathcal{P}$ are convex, to our best knowledge, there is previously no better result than those mentioned above.

Our Results

We propose a new algorithm for L_1-SPP. After a triangulation of the free space is computed (say, in $O(n + h \log^{1+\epsilon} h)$ time [1] for an arbitrarily small constant $\epsilon > 0$), our algorithm finds a shortest s-t path in $O(n + h \log h)$ time and $O(n)$ space. Thus, if the triangulation can be done optimally (i.e., in $O(n + h \log h)$ time), our algorithm is also optimal. For the convex case of L_1-SPP in which all obstacles in $\mathcal{P}$ are convex, our algorithm is optimal since the triangulation can be done in $O(n + h \log h)$ time (e.g., by the approaches in [1,12]).

As in [18,19], we can also extend our approach to "fixed orientation metrics" [18,19,24], in which case a sought path is allowed to follow only a given set of orientations. For a number c of given orientations, Mitchell's algorithm finds such a shortest path in $O(cn \log n)$ time and $O(cn)$ space, and our algorithm takes $O(n + h \log^{1+\epsilon} h + c^2 h \log h)$ time and $O(n + c^2 h)$ space. Our approach also leads to an $O(n + h \log^{1+\epsilon} h + (1/\delta)h \log h)$ time algorithm for computing a δ-optimal Euclidean shortest path among polygonal obstacles for any constant $\delta > 0$. For this problem, Mitchell's algorithm [18,19] takes $O((\sqrt{1/\delta})n \log n)$ time, and Clarkson's algorithm [5] runs in $O((1/\delta)n \log n)$ time. The best announced exact algorithms for the Euclidean shortest path problem take $O(n \log n)$ time [11], $O(n + h \log h \log n)$ time [14], and $O(n + h \log^{1+\epsilon} h + k)$ time [4], where k is sensitive to the input and is bounded by $O(h^2)$.

Mitchell's algorithm [18,19] further builds a shortest path map (with respect to a given source point) of $O(n)$ space in $O(n \log n)$ time such that for any query point p, the L_1 shortest path length from s to p can be reported in $O(\log n)$ time. While Mitchell's algorithm works for general polygons, this result was also the best known for the convex case of L_1-SPP. In addition, the $O(n)$ space and $O(\log n)$ query time seems optimal. As a by-product of our techniques, we can build a shortest path map whose space size is $O(h)$ (instead of $O(n)$) for the convex case of L_1-SPP in $O(n + h \log h)$ time (instead of $O(n \log n)$ time) such

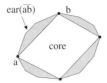

Fig. 1. Illustrating the core and ears of a convex obstacle; $ear(\overline{ab})$ is indicated

Fig. 2. The line segment $\overline{cd}$ penetrates $ear(\overline{ab})$; $\overline{cd}$ intersects the obstacle path of $ear(\overline{ab})$ at e and f

that any shortest path length query is answered in $O(\log h)$ time (instead of $O(\log n)$ time). This result may be of independent interest.

All our results also hold for computing L_∞ shortest paths (by rotating the plane by 45°). As in [18,19], for simplicity of discussion, we assume the free space is connected (thus, a feasible s-t path always exists), and no two obstacle vertices lie on the same horizontal or vertical line. These assumptions can be removed without deteriorating the performances of the algorithms asymptotically. In the rest of this paper, unless otherwise stated, a shortest path always refers to an L_1 shortest path and a length is always in the L_1 metric. Due to the space limit, some details are omitted and can be found in the full paper.

Below, we first give our algorithm for the convex case of L_1-SPP, which is used as a key procedure in our algorithm for the general L_1-SPP.

2   Shortest Paths among Convex Obstacles

Let $\mathcal{P}' = \{P_1', P_2', \ldots, P_h'\}$ be a set of h pairwise disjoint convex polygonal obstacles of totally n vertices. Given two points s and t in the free space, our algorithm finds a shortest s-t path in $O(n + h \log h)$ time and $O(n)$ space. A shortest path map (SPM for short) for handling shortest path queries can also be computed.

2.1   Notation and Observations

For each polygon $P_i' \in \mathcal{P}'$, we define its *core*, $core(P_i')$, as the simple polygon by connecting the leftmost, topmost, rightmost, and bottommost vertices of P_i' with line segments (see Fig. 1). Note that $core(P_i')$ is contained in P_i' and has at most four edges. Let $core(\mathcal{P}')$ be the set of the cores of all obstacles in $\mathcal{P}'$. The cores of the obstacles play a critical role in our algorithm. A key observation is that a shortest s-t path avoiding the cores in $core(\mathcal{P}')$ corresponds to a shortest s-t path avoiding the obstacles in $\mathcal{P}'$ with the same L_1 length.

To prove the above key observation, we first define some concepts. Consider an obstacle P_i' and $core(P_i')$. For each edge $\overline{ab}$ of $core(P_i')$ with vertices a and b, if $\overline{ab}$ is not an edge of P_i', then it divides P_i' into two polygons, one of them containing $core(P_i')$; we call the one that does not contain $core(P_i')$ an *ear* of P_i' *based on* $\overline{ab}$, denoted by $ear(\overline{ab})$ (see Fig. 1). If $\overline{ab}$ is also an edge of P_i, then $ear(\overline{ab})$ is not defined. Note that $ear(\overline{ab})$ has only one edge bounding $core(P_i')$, i.e., $\overline{ab}$, which we call its *core edge*. The other edges of $ear(\overline{ab})$ are on the boundary of P'_i, which we call *obstacle edges*. There are two paths between a and b along the boundary of $ear(\overline{ab})$: One path is the core edge $\overline{ab}$ and the other consists of all its obstacle edges. We call the latter path the *obstacle path* of the ear. A line segment is *positive-sloped* (resp., *negative-sloped*) if its slope is positive (resp., negative). An ear is *positive-sloped* (resp., *negative-sloped*) if its core edge is positive-sloped (resp., negative-sloped). A point p is *higher* (resp., *lower*) than another point q if the y-coordinate of p is no smaller (resp., no larger) than that of q. The next observation is self-evident.

Observation 1. *For any ear, its obstacle path is monotone in both the x- and y-coordinates. Specifically, consider an ear $ear(\overline{ab})$ and suppose the vertex a is lower than the vertex b. If $ear(\overline{ab})$ is positive-sloped, then the obstacle path from a to b is monotonically increasing in both the x- and y-coordinates; if it is negative-sloped, then the obstacle path from a to b is monotonically decreasing in the x-coordinates and monotonically increasing in the y-coordinates.*

For an ear $ear(\overline{ab})$ and a line segment $\overline{cd}$, we say that $\overline{cd}$ *penetrates* $ear(\overline{ab})$ if the following hold (see Fig. 2): (1) $\overline{cd}$ intersects the interior of $ear(\overline{ab})$, (2) neither c nor d is in the interior of $ear(\overline{ab})$, and (3) $\overline{cd}$ does not intersect the core edge $\overline{ab}$ at its interior. The next two lemmas are due to the definition of an ear and their proofs are omitted.

Lemma 1. *Suppose a line segment $\overline{cd}$ penetrates $ear(\overline{ab})$. If $\overline{cd}$ is positive-sloped (resp., negative-sloped), then $ear(\overline{ab})$ is also positive-sloped (resp., negative-sloped).*

Clearly, if $\overline{cd}$ penetrates the ear $ear(\overline{ab})$, $\overline{cd}$ intersects the boundary of $ear(\overline{ab})$ at two points and both points lie on the obstacle path of $ear(\overline{ab})$ (see Fig. 2).

Lemma 2. *Suppose a line segment $\overline{cd}$ penetrates an ear $ear(\overline{ab})$. Let e and f be the two points on the obstacle path of $ear(\overline{ab})$ that $\overline{cd}$ intersects. Then the L_1 length of the line segment $\overline{ef}$ is equal to that of the portion of the obstacle path of $ear(\overline{ab})$ between e and f (see Fig. 2).*

If $\overline{cd}$ penetrates $ear(\overline{ab})$, by Lemma 2, we can obtain another path from c to d by replacing $\overline{ef}$ with the portion of the obstacle path of $ear(\overline{ab})$ between e and f such that the new path has the same L_1 length as $\overline{cd}$ and the new path does not intersect the interior of $ear(\overline{ab})$.

The results in the following lemma have been proved in [18,19].

Lemma 3. *[18,19] There exists a shortest s-t path in the free space such that if the path makes a turn at a point p, then p must be an obstacle vertex.*

We call a shortest path that satisfies the property in Lemma 3 a *vertex-preferred shortest path*. Mitchell's algorithm [18,19] can find a vertex-preferred shortest s-t path. Denote by $Tri(\mathcal{P}')$ a triangulation of the free space and the space inside all obstacles. Note that the free space can be triangulated in $O(n + h \log h)$ time [1,12] and the space inside all obstacles can be triangulated in totally $O(n)$ time [2]. Hence, $Tri(\mathcal{P}')$ can be computed in $O(n + h \log h)$ time. The next lemma gives our key observation.

Lemma 4. *Given a vertex-preferred shortest s-t path that avoids the polygons in core($\mathcal{P}'$), we can find in $O(n)$ time a shortest s-t path with the same L_1 length that avoids the obstacles in $\mathcal{P}'$.*

Proof. Consider a vertex-preferred shortest s-t path for $core(\mathcal{P}')$, denoted by $\pi_{core}(s,t)$. Suppose it makes turns at $p_1, p_2, \ldots, p_k$, ordered from s to t along the path, and each p_i is a vertex of a core in $core(\mathcal{P}')$. Let $p_0 = s$ and $p_{k+1} = t$. Then for each $i = 0, 1, \ldots, k$, the portion of $\pi_{core}(s,t)$ from p_i to p_{i+1} is the line segment $\overline{p_i p_{i+1}}$. Below, we first show that we can find a path from p_i to p_{i+1} such that it avoids the obstacles in $\mathcal{P}'$ and has the same L_1 length as $\overline{p_i p_{i+1}}$.

If $\overline{p_i p_{i+1}}$ does not intersect the interior of any obstacle in $\mathcal{P}'$, then we are done with $\overline{p_i p_{i+1}}$. Otherwise, because $\overline{p_i p_{i+1}}$ avoids $core(\mathcal{P}')$, it can intersects only the interior of some ears. Consider any such ear $ear(\widehat{ab})$. Below, we prove that $\overline{p_i p_{i+1}}$ penetrates $ear(\widehat{ab})$.

First, we already know that $\overline{p_i p_{i+1}}$ intersects the interior of $ear(\widehat{ab})$. Second, it is obvious that neither p_i nor p_{i+1} is in the interior of $ear(\widehat{ab})$. It remains to show that $\overline{p_i p_{i+1}}$ cannot intersect the core edge $\overline{ab}$ of $ear(\widehat{ab})$ at the interior of $\overline{ab}$. Denote by $A' \in \mathcal{P}'$ the obstacle that contains $ear(\widehat{ab})$. The interior of $\overline{ab}$ is in the interior of A'. Since $\overline{p_i p_{i+1}}$ does not intersect the interior of A', $\overline{p_i p_{i+1}}$ cannot intersect $\overline{ab}$ at its interior. Therefore, $\overline{p_i p_{i+1}}$ penetrates $ear(\widehat{ab})$.

Recall that we have assumed that no two obstacle vertices lie on the same horizontal or vertical line. Since both p_i and p_{i+1} are obstacle vertices, the segment $\overline{p_i p_{i+1}}$ is either positive-sloped or negative-sloped. Without loss of generality, assume $\overline{p_i p_{i+1}}$ is positive-sloped. By Lemma 1, $ear(\widehat{ab})$ is also positive-sloped. Let e and f denote the two intersection points between $\overline{p_i p_{i+1}}$ and the obstacle path of $ear(\widehat{ab})$, and $\widehat{ef}$ denote the portion of the obstacle path of $ear(\widehat{ab})$ between e and f. By Lemma 2, we can replace the line segment $\overline{ef}$ ($\subseteq \overline{p_i p_{i+1}}$) by $\widehat{ef}$ to obtain a new path from p_i to p_{i+1} such that the new path has the same L_1 length as $\overline{p_i p_{i+1}}$. Further, as a portion of the obstacle path of $ear(\widehat{ab})$, $\widehat{ef}$ is a boundary portion of the obstacle A' that contains $ear(\widehat{ab})$, and thus $\widehat{ef}$ does not intersect the interior of any obstacle in $\mathcal{P}'$.

By processing each ear whose interior is intersected by $\overline{p_i p_{i+1}}$ as above, we find a new path from p_i to p_{i+1} such that the path has the same L_1 length as $\overline{p_i p_{i+1}}$ and the path does not intersect the interior of any obstacle in $\mathcal{P}'$.

By processing each segment $\overline{p_i p_{i+1}}$ in $\pi_{core}(s,t)$ as above for $i = 0, 1, \ldots, k$, we obtain another s-t path $\pi(s,t)$ such that the L_1 length of $\pi(s,t)$ is equal to that of $\pi_{core}(s,t)$ and $\pi(s,t)$ avoids all obstacles in $\mathcal{P}'$. We claim that $\pi(s,t)$ is a shortest s-t path avoiding the obstacles in $\mathcal{P}'$. Indeed, since each core in $core(\mathcal{P}')$ is contained in an obstacle in $\mathcal{P}'$, the length of a shortest s-t path avoiding $core(\mathcal{P}')$ cannot be longer than that of a shortest s-t path avoiding $\mathcal{P}'$. Since the length of $\pi(s,t)$ is equal to that of $\pi_{core}(s,t)$ and $\pi_{core}(s,t)$ is a shortest s-t path avoiding $core(\mathcal{P}')$, $\pi(s,t)$ is a shortest s-t path avoiding $\mathcal{P}'$.

The above discussion also provides a way to construct $\pi(s,t)$, which can be easily done in $O(n)$ time with the help of $Tri(\mathcal{P}')$. The lemma thus follows.

2.2 The Shortest Path Algorithm

Our algorithm works as follows: (1) Apply Mitchell's algorithm [18,19] on $core(\mathcal{P}')$ to find a vertex-preferred shortest s-t path avoiding the cores in $core(\mathcal{P}')$; (2) by Lemma 4, find a shortest s-t path that avoids the obstacles in $\mathcal{P}'$. The first step takes $O(h \log h)$ time and $O(h)$ space since the cores in $core(\mathcal{P}')$ have totally $O(h)$ vertices. The second step takes $O(n)$ time and $O(n)$ space.

Theorem 1. *A shortest path for the convex case of L_1-SPP can be found in $O(n + h \log h)$ time and $O(n)$ space.*

2.3 Computing a Shortest Path Map

Mitchell's algorithm [18,19] can also compute an SPM of size $O(n)$ for the general L_1-SPP in $O(n \log n)$ time such that a shortest path length query is answered in $O(\log n)$ time and an actual path is reported in additional time proportional to the number of turns of the path.

By applying Mitchell's algorithm [18,19] on the core set $core(\mathcal{P}')$ and a source point s, we can compute an SPM of size $O(h)$ in $O(h \log h)$ time, denoted by $SPM(core(\mathcal{P}'), s)$. With a planar point location data structure [8,16], for any query point p in the free space among $\mathcal{P}'$, the length of a shortest s-p path avoiding $core(\mathcal{P}')$ can be reported in $O(\log h)$ time, which is also the length of a shortest s-p path avoiding $\mathcal{P}'$, by Lemma 4. We thus have the following result.

Theorem 2. *For the convex L_1-SPP, in $O(n + h \log h)$ time, we can build an SPM of size $O(h)$ with respect to s, such that the length of an L_1 shortest path between s and any query point in the free space can be reported in $O(\log h)$ time.*

However, with the SPM for Theorem 2, an actual shortest path avoiding $\mathcal{P}'$ between s and a query point p cannot be reported in additional time proportional to the number of turns of the path. To process queries on actual shortest paths, in Lemma 5 below, using $SPM(core(\mathcal{P}'), s)$, we compute an SPM for $\mathcal{P}'$, denoted by $SPM(\mathcal{P}', s)$, of size $O(n)$, which can answer a shortest path length query in $O(\log h)$ time and report an actual path in time proportional to $O(\log n)$ plus the number of turns of the path.

Lemma 5. *Given $SPM(core(\mathcal{P}'), s)$ for the core set $core(\mathcal{P}')$, we can compute a shortest path map $SPM(\mathcal{P}', s)$ for the obstacle set $\mathcal{P}'$ in $O(n)$ time.*

Proof. We only sketch the main idea. Consider a cell $C_{core}(r)$ with the root r in $SPM(core(\mathcal{P}'), s)$. Recall that r is a vertex of a core in $core(\mathcal{P}')$ and $C_{core}(r)$ is star-shaped. Denote by $\mathcal{F}(\mathcal{P}')$ (resp., $\mathcal{F}(core(\mathcal{P}'))$) the free space with respect to $\mathcal{P}'$ (resp., $core(\mathcal{P}')$). The cell $C_{core}(r)$ may intersect some ears. Let $C(r)$ be the subregion of $C_{core}(r)$ by removing from $C_{core}(r)$ the space occupied by all ears except their obstacle paths. Thus $C(r)$ lies in $\mathcal{F}(\mathcal{P}')$. However, for each point $p \in C(r)$, p may not be visible to r with respect to $\mathcal{P}'$. Our task here is to further decompose $C(r)$ into a set of *SPM regions* such that each such region has a root visible to all points in the region with respect to $\mathcal{P}'$; further, we need to make sure that each point q in an SPM region has a shortest path in $\mathcal{F}(\mathcal{P}')$ from s that contains the line segment connecting q and the root of the region.

Since $C_{core}(r)$ is a simple polygon, we can show $C(r)$ is also a (possibly degenerate) simple polygon. Thus, $SPM(C(r))$, which is an SPM of $C(r)$ with respect to the point r, can be easily computed in linear time in terms of the number of edges of $C(r)$. Further, we can show that for any point $p \in C(r)$ there is a shortest path in $\mathcal{F}(\mathcal{P}')$ from s to p that contains r. The lemma thus follows.

Theorem 3. *For the convex L_1-SPP, in $O(n + h \log h)$ time, we can construct an SPM of size $O(n)$ with respect to s, such that given any query point p, the length of a shortest s-p path can be reported in $O(\log h)$ time and an actual path can be found in $O(\log n + k)$ time, where k is the number of turns of the path.*

3 Shortest Paths among General Polygonal Obstacles

In this section, we consider the general L_1-SPP problem. Let $\mathcal{P} = \{P_1, P_2, \ldots, P_h\}$ be a set of h pairwise disjoint arbitrary polygonal obstacles of totally n vertices in the plane. We seek to find a shortest s-t path in the free space.

3.1 Preprocessing

For simplicity of discussion, we assume that all obstacles are contained in a large rectangle $\mathcal{R}$ (see Fig. 3). Denote by $\mathcal{F}$ the free space inside $\mathcal{R}$. We also view s and t as two special obstacles in $\mathcal{P}$. Denote by $Tri(\mathcal{F})$ a triangulation of $\mathcal{F}$.

Based on $Tri(\mathcal{F})$, we compute a corridor structure, which was also used in [13,15]. This corridor structure can help reduce our problem to the convex case of L_1-SPP to some extent, as discussed below. Let $G(\mathcal{F})$ denote the (planar) dual graph of $Tri(\mathcal{F})$. The degree of each node in $G(\mathcal{F})$ is at most three. Suppose there is a feasible path from s to t. As in [15], at least one node dual to a triangle incident to each of s and t is of degree three. Based on $G(\mathcal{F})$, we compute a planar 3-regular graph, denoted by G^3 (the degree of each node in G^3 is three), possibly with loops and multi-edges, as follows. First, we remove every degree-one node from $G(\mathcal{F})$ along with its incident edge; repeat this process until no degree-one node exists. Second, remove every degree-two node from $G(\mathcal{F})$ and replace its two incident edges by a single edge; repeat this process until no degree-two node exists. The resulting graph is G^3 (e.g., see Fig. 3). By Euler's formula, the resulting graph G^3 has $h + 1$ faces, $2h - 2$ nodes, and $3h - 3$ edges [15]. Each node of G^3 corresponds to a triangle in $Tri(\mathcal{F})$, which is called a *junction triangle* (e.g., see Fig. 3). The removal of all junction triangles from G^3 results in $O(h)$ *corridors*, each of which corresponds to one edge of G^3 [15].

The boundary of a corridor C consists of four parts (see Fig. 4): (1) A boundary portion of an obstacle $P_i \in \mathcal{P}$, from a point a to a point b; (2) a diagonal of a junction triangle from b to a boundary point e on an obstacle $P_j \in \mathcal{P}$ ($P_i = P_j$ is possible); (3) a boundary portion of the obstacle P_j from e to a point f; (4) a diagonal of a junction triangle from f to a. The two diagonals $\overline{be}$ and $\overline{af}$ are called the *doors* of C. The corridor C is a simple polygon. Let $|C|$ denote the

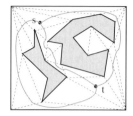
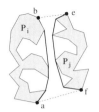
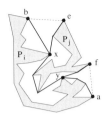

Fig. 3. Illustrating a triangulation of the free space among two obstacles and the corridors (with red solid curves). There are two junction triangles indicated by the large dots inside them, connected by three solid (red) curves. Removing the two junction triangles results in three corridors.

Fig. 4. Illustrating an open hourglass (left) and a closed hourglass (right) with a corridor path linking the apices x and y of the two funnels. The dashed segments are diagonals. The paths $\pi(a, b)$ and $\pi(e, f)$ are shown with thick solid curves.

number of obstacle vertices on the boundary of C. In $O(|C|)$ time, we can compute the shortest path $\pi(a, b)$ (resp., $\pi(e, f)$) from a to b (resp., e to f) inside C. The region H_C bounded by $\pi(a, b), \pi(e, f)$, and the two diagonals $\overline{be}$ and $\overline{fa}$ is called an *hourglass*, which is *open* if $\pi(a, b) \cap \pi(e, f) = \emptyset$ and *closed* otherwise (see Fig. 4). If H_C is open, then both $\pi(a, b)$ and $\pi(e, f)$ are convex chains and are called the *sides* of H_C; otherwise, H_C consists of two "funnels" and a path $\pi_C = \pi(a, b) \cap \pi(e, f)$ joining the two apices of the two funnels, called the *corridor path* of C. The two funnel apices connected by the corridor path are called the *corridor path terminals*. Each funnel side is also convex. We process each corridor as above. The total time for processing all corridors is $O(n)$.

Let Q be the union of all junction triangles and hourglasses. Then Q consists of $O(h)$ junction triangles, open hourglasses, funnels, and corridor paths. As shown in [13], there exists a shortest s-t path $\pi(s, t)$ avoiding the obstacles in $\mathcal{P}$ which is contained in Q. Consider a corridor C. If $\pi(s, t)$ contains an interior point of C, then the path $\pi(s, t)$ must intersect both doors of C; further, if the hourglass H_C of C is closed, then we claim that we can make the corridor path of C entirely contained in $\pi(s, t)$. Suppose $\pi(s, t)$ intersects the two doors of C, say, at two points p and q respectively. Then since C is a simple polygon, a Euclidean shortest path between p and q inside C, denoted by $\pi_E(p, q)$, is also an L_1 shortest path in C [10]. Note that $\pi_E(p, q)$ must contain the corridor path of C. If we replace the portion of $\pi(s, t)$ between p and q by $\pi_E(p, q)$, then we obtain a new L_1 shortest s-t path that contains the corridor path π_C. For simplicity, we still use $\pi(s, t)$ to denote the new path. In other words, $\pi(s, t)$ has the property that if $\pi(s, t)$ intersects both doors of C and the hourglass H_C is closed, then the corridor path of C is contained in $\pi(s, t)$.

Let Q' be Q minus the corridor paths. Then the boundary of Q' consists of $O(h)$ reflex vertices and $O(h)$ convex chains, implying that the complementary region $\mathcal{R} \setminus Q'$ consists of a set of polygons of totally $O(h)$ reflex vertices and $O(h)$ convex chains. As shown in [15], the region $\mathcal{R} \setminus Q'$ can be partitioned into a set $\mathcal{P}'$ of $O(h)$ convex polygons of totally $O(n)$ vertices (e.g., by extending an angle-bisecting segment inward from each reflex vertex). In addition, for each corridor path, no portion of it lies in the free space with respect to $\mathcal{P}'$. Further, the shortest path $\pi(s,t)$ is a shortest s-t path avoiding all convex polygons in $\mathcal{P}'$ and possibly utilizing some corridor paths. The set $\mathcal{P}'$ can be easily obtained in $O(n + h \log h)$ time. Therefore, other than the corridor paths, we reduce our original L_1-SPP problem to the convex case.

3.2   The Main Algorithm

With the convex polygon set $\mathcal{P}'$, to find a shortest s-t path in $\mathcal{F}$ (i.e., the free space in $\mathcal{R}$ with respect to $\mathcal{P}$), if there is no corridor path, then we can simply apply our algorithm in Section 2. Otherwise, the situation is more complicated because the corridor paths can give possible "shortcuts" for the sought s-t path, and we must take these possible "shortcuts" into consideration while running the continuous Dijkstra paradigm. The details are given below.

First, we compute the core set $core(\mathcal{P}')$ of $\mathcal{P}'$. However, the way we construct $core(\mathcal{P}')$ here is slightly different from that in Section 2. For each convex polygon $A' \in \mathcal{P}'$, in addition to its leftmost, topmost, rightmost, and bottommost vertices, if a vertex v of A' is a corridor path terminal, then v is also kept as a vertex of the core $core(A')$. In other words, $core(A')$ is a simple (convex) polygon whose vertex set consists of the leftmost, topmost, rightmost, and bottommost vertices of A' and all corridor path terminals on A'. Since there are $O(h)$ terminal vertices, the cores in $core(\mathcal{P}')$ still have totally $O(h)$ vertices and edges. Further, the core set thus defined still has the properties discussed in Section 2 for computing shortest L_1 paths, e.g., Observation 1 and Lemmas 1, 2, and 4. Hence, by using our scheme in Section 2, we can first find a shortest s-t path avoiding the cores in $core(\mathcal{P}')$ in $O(h \log h)$ time by applying Mitchell's algorithm [18,19], and then obtain a shortest s-t path avoiding $\mathcal{P}'$ in $O(n)$ time by Lemma 4. But, the path thus computed may not be a true shortest path in $\mathcal{F}$ since the corridor paths are not utilized. To find a true shortest path in $\mathcal{F}$, we need to modify the continuous Dijkstra paradigm when applying it on the core set $core(\mathcal{P}')$, as follows.

In Mitchell's algorithm [18,19], when an obstacle vertex v is hit by the wavefront for the first time, it will be "permanently labeled" with a value $d(v)$, which is the length of a shortest path from s to v in the free space. The wavefront consists of many "wavelets" (each wavelet is a line segment of slope 1 or -1). The algorithm maintains a priority queue (called "event queue"), and each element in the queue is a wavelet associated with an "event point" and an "event distance", which means that the wavelet will hit the event point with the event distance. The algorithm repeatedly takes (and removes) an element from the event queue with the smallest event distance, and processes the event. After an event is processed, some new events may be added to the event queue. The algorithm stops when the point t is hit by the wavefront for the first time.

To handle the corridor paths in our problem, consider a corridor path π_C with x and y as its terminals and let l be the length of π_C. Recall that x and y are vertices of a core in $core(\mathcal{P}')$. Consider the moment when the vertex x

is permanently labeled with the distance $d(x)$. Suppose the wavefront that first hits x is from the funnel whose apex is x. Then according to our discussions above, the only way that the wavefront at x can affect a shortest s-t path is through the corridor path π_C. If y is not yet permanently labeled, then y has not been hit by the wavefront. We initiate a "pseudo-wavelet" that originates from x with the event point y and event distance $d(x)+l$, meaning that y will be hit by this pseudo-wavelet at the distance $d(x) + l$. We add the pseudo-wavelet to the event queue. If y has been permanently labeled, then the wavefront has already hit y and is currently moving along the corridor path π_C from y to x. Thus, the wavefront through x will meet the wavefront through y somewhere on the path π_C, and these two wavefronts will "die" there and never affect the free space outside the corridor. Thus, if y has been permanently labeled, then we do not need to do anything on y. In addition, at the moment when the vertex x is permanently labeled, if the wavefront that hits x is from the corridor path π_C (i.e., through y), then the wavefront will keep going to the funnel of x through x; therefore, we process this event on x as usual (i.e., as in [18,19]), by initiating wavelets that originate from x.

Intuitively, the above treatment of corridor path terminals makes corridor paths act as possible "shortcuts" when we propagate the wavefront. The rest of the algorithm proceeds in the same way as in [18,19] (e.g., processing the segment dragging queries). The algorithm stops when the wavefront first hits the point t, at which moment a shortest s-t path in $\mathcal{F}$ has been found.

Since there are $O(h)$ corridor paths, with the above modifications to Mitchell's algorithm as applied to $core(\mathcal{P}')$, its running time is still $O(h \log h)$. Indeed, comparing with the original continuous Dijkstra scheme [18,19] (as applied to $core(\mathcal{P}')$), there are $O(h)$ additional events on the corridor path terminals, i.e., events corresponding to those pseudo-wavelets. To handle these additional events, we may, for example, as preprocessing, for each corridor path, associate with each its corridor path terminal x the other terminal y as well as the corridor path length l. Thus, during the algorithm, when we process the event point at x, we can find y and l immediately. In this way, each additional event is handled in $O(1)$ time in addition to adding a new event for it to the event queue. Hence, processing all events still takes $O(h \log h)$ time. Note that the shortest s-t path thus computed may penetrate some ears of $\mathcal{P}'$. As in Lemma 4, we can obtain a shortest s-t path in the free space $\mathcal{F}$ in additional $O(n)$ time. Since applying Mitchell's algorithm on $core(\mathcal{P}')$ takes $O(h)$ space, the space used in our entire algorithm is $O(n)$. In summary, we have the following result.

Theorem 4. *A shortest path for L_1-SPP can be found in $O(n + h \log^{1+\epsilon} h)$ time (or $O(n + h \log h)$ time if a free space triangulation is given) and $O(n)$ space.*

4   Applications

This section gives some applications of our shortest path algorithm for L_1-SPP.

A *C-oriented path* is a polygonal path with each edge parallel to one of a given set C of fixed orientations [24]. A shortest C-oriented path between two points is a C-oriented path with the minimum Euclidean distance. Rectilinear paths are a special case with two fixed orientations of 0 and $\pi/2$. Let $c = |C|$. Mitchell's algorithm [18,19] can compute a shortest C-oriented path in $O(cn \log n)$ time and $O(cn)$ space among h pairwise disjoint polygons of totally n vertices in the plane. Similarly, our algorithm also works for this problem, as follows.

We first consider the convex case (i.e., all polygons are convex). We compute a core for each convex polygon based on the orientations in C. Note that in this case, a core has $O(c)$ vertices. Thus, we obtain a core set of totally $O(ch)$ vertices. We then apply Mitchell's algorithm for the fixed orientations of C on the core set to compute a shortest path avoiding the cores in $O(c^2 h \log h)$ time and $O(c^2 h)$ space, after which we find a shortest path avoiding the input polygons in additional $O(n)$ time as in Lemma 4. Thus, a shortest path can be found in totally $O(n + c^2 h \log h)$ time and $O(n + c^2 h)$ space. For the general case when the polygons need not be convex, the algorithm scheme is similar to our L_1 algorithm in Section 3. In summary, we have the following result.

Theorem 5. *Given a set C of orientations and a set of h pairwise disjoint polygonal obstacles of totally n vertices in the plane, we can compute a C-oriented shortest s-t path in the free space in $O(n + h \log^{1+\epsilon} h + c^2 h \log h)$ time (or $O(n + c^2 h \log h)$ time if a triangulation is given) and $O(n + c^2 h)$ space, where $c = |C|$.*

This also yields an approximation algorithm for computing a Euclidean shortest path between two points among polygonal obstacles. Since the Euclidean metric can be approximated within an accuracy of $O(1/c^2)$ if we use c equally spaced orientations, as in [18,19], Theorem 5 leads to an algorithm that computes a path guaranteed to have a length within a factor $(1 + \delta)$ of the Euclidean shortest path length, where c is chosen such that $\delta = O(1/c^2)$.

Corollary 1. *A δ-optimal Euclidean shortest path between two points among h pairwise disjoint polygons of totally n vertices in the plane can be computed in $O(n + h \log^{1+\epsilon} h + (1/\delta) h \log h)$ time (or $O(n + (1/\delta) h \log h)$ time if a triangulation is given) and $O(n + (1/\delta) h)$ space.*

References

1. Bar-Yehuda, R., Chazelle, B.: Triangulating disjoint Jordan chains. International Journal of Computational Geometry and Applications 4(4), 475–481 (1994)
2. Chazelle, B.: Triangulating a simple polygon in linear time. Discrete and Computational Geometry 6, 485–524 (1991)
3. Chen, D., Klenk, K., Tu, H.Y.: Shortest path queries among weighted obstacles in the rectilinear plane. SIAM Journal on Computing 29(4), 1223–1246 (2000)
4. Chen, D., Wang, H.: Computing shortest paths among curved obstacles in the plane. In: arXiv:1103.3911 (2011)
5. Clarkson, K.: Approximation algorithms for shortest path motion planning. In: Proc. of the 19th Annual ACM Symposium on Theory of Computing, pp. 56–65 (1987)
6. Clarkson, K., Kapoor, S., Vaidya, P.: Rectilinear shortest paths through polygonal obstacles in $O(n \log^2 n)$ time. In: Proc. of the 3rd Annual Symposium on Computational Geometry, pp. 251–257 (1987)

7. Clarkson, K., Kapoor, S., Vaidya, P.: Rectilinear shortest paths through polygonal obstacles in $O(n \log^{2/3} n)$ time (1988) (manuscript)
8. Edelsbrunner, H., Guibas, L., Stolfi, J.: Optimal point location in a monotone subdivision. SIAM Journal on Computing 15(2), 317–340 (1986)
9. Ghosh, S., Mount, D.: An output-sensitive algorithm for computing visibility. SIAM Journal on Computing 20(5), 888–910 (1991)
10. Hershberger, J., Snoeyink, J.: Computing minimum length paths of a given homotopy class. Computational Geometry: Theory and Applications 4(2), 63–97 (1994)
11. Hershberger, J., Suri, S.: An optimal algorithm for Euclidean shortest paths in the plane. SIAM Journal on Computing 28(6), 2215–2256 (1999)
12. Hertel, S., Mehlhorn, K.: Fast triangulation of the plane with respect to simple polygons. Information and Control 64, 52–76 (1985)
13. Inkulu, R., Kapoor, S.: Planar rectilinear shortest path computation using corridors. Computational Geometry: Theory and Applications 42(9), 873–884 (2009)
14. Inkulu, R., Kapoor, S., Maheshwari, S.: A near optimal algorithm for finding Euclidean shortest path in polygonal domain. In: arXiv:1011.6481v1 (2010)
15. Kapoor, S., Maheshwari, S., Mitchell, J.: An efficient algorithm for Euclidean shortest paths among polygonal obstacles in the plane. Discrete and Computational Geometry 18(4), 377–383 (1997)
16. Kirkpatrick, D.: Optimal search in planar subdivisions. SIAM Journal on Computing 12(1), 28–35 (1983)
17. Larsona, R., Li, V.: Finding minimum rectilinear distance paths in the presence of barriers. Networks 11, 285–304 (1981)
18. Mitchell, J.: An optimal algorithm for shortest rectilinear paths among obstacles. In: Abstracts of the 1st Canadian Conference on Computational Geometry (1989)
19. Mitchell, J.: L_1 shortest paths among polygonal obstacles in the plane. Algorithmica 8(1), 55–88 (1992)
20. Mitchell, J.: Shortest paths among obstacles in the plane. International Journal of Computational Geometry and Applications 6(3), 309–332 (1996)
21. de Rezende, P., Lee, D., Wu, Y.: Rectilinear shortest paths with rectangular barriers. In: Proc. of the 1st Annual Symposium on Computational Geometry, pp. 204–213 (1985)
22. Storer, J., Reif, J.: Shortest paths in the plane with polygonal obstacles. Journal of the ACM 41(5), 982–1012 (1994)
23. Widmayer, P.: On graphs preserving rectilinear shortest paths in the presence of obstacles. Annals of Operations Research 33(7), 557–575 (1991)
24. Widmayer, P., Wu, Y., Wong, C.: On some distance problems in fixed orientations. SIAM Journal on Computing 16(4), 728–746 (1987)

Motion Planning via Manifold Samples[*]

Oren Salzman[1], Michael Hemmer[1], Barak Raveh[1,2], and Dan Halperin[1]

[1] Tel-Aviv University, Israel
[2] Hebrew University, Israel

Abstract. We present a general and modular algorithmic framework for path planning of robots. Our framework combines geometric methods for exact and complete analysis of low-dimensional configuration spaces, together with sampling-based approaches that are appropriate for higher dimensions. We suggest taking samples that are *entire low-dimensional manifolds of the configuration space*. These samples capture the connectivity of the configuration space much better than isolated point samples. Geometric algorithms then provide powerful primitive operations for complete analysis of the low-dimensional manifolds. We have implemented our framework for the concrete case of a polygonal robot translating and rotating amidst polygonal obstacles. To this end, we have developed a primitive operation for the analysis of an appropriate set of manifolds using arrangements of curves of rational functions. This modular integration of several carefully engineered components has lead to a significant speedup over the PRM sampling-based algorithm, which represents an approach that is prevalent in practice.

1 Introduction

Motion planning is a fundamental research topic in robotics with applications in diverse domains such as graphical animation, surgical planning, computational biology and computer games. For a general overview of the subject and its applications see [1], [2], [3]. In its basic form, the motion-planning problem is to find a collision-free path for a robot or a moving object R in a *workspace* cluttered with static obstacles. The spatial pose of R, or the *configuration* of R, is uniquely defined by some set of parameters, the degrees of freedom (*dof*s) of R. The set of all robot configurations $\mathcal{C}$ is termed the *configuration space* of the robot, and decomposes into the disjoint sets of free and forbidden configurations, which we denote by $\mathcal{C}_{\text{free}}$ and $\mathcal{C}_{\text{forb}}$, respectively. Thus, it is common to rephrase the motion-planning problem as the problem of moving R from a start configuration q_s to a target configuration q_t in a path that is fully contained within $\mathcal{C}_{\text{free}}$.

Analytic solutions to the general motion planning problem: The motion-planning problem is computationally hard with respect to the number of *dof*s [4],

[*] This work has been supported in part by the 7th Framework Programme for Research of the European Commission, under FET-Open grant number 255827 (CGL—Computational Geometry Learning), by the German-Israeli Foundation (grant no. 969/07), and by the Hermann Minkowski–Minerva Center for Geometry at Tel Aviv University.

C. Demetrescu and M.M. Halldórsson (Eds.): ESA 2011, LNCS 6942, pp. 493–505, 2011.

yet much research has been devoted to solving the general problem and its various instances using geometric, algebraic and combinatorial tools. The configuration-space formalism was introduced by Lozano-Perez [5]. Schwartz and Sharir proposed the first general algorithm for solving the motion planning problem, with running time that is doubly-exponential in the number of *dof*s [6]. Singly exponential-time algorithms have followed [7], [8], [9], but are generally considered too complicated to be implemented in practice.

Solutions to low-dimensional instances of the problem: Although the general motion-planning problem cannot be efficiently solved analytically, more efficient algorithms have been proposed for various low-dimensional instances [2], such as translating a polygonal or polyhedral robot [10], [5], and translation with rotation of a polygonal robot in the plane [11], [12], [13]. For a survey of related approaches see [14]. Moreover, considerable advances in robust implementation of computational geometry algorithms in recent years have led to a set of implemented tools that are of interest in this context. Minkowski sums, which enable the representation of the configuration space of a translating robot, have robust and exact planar and 3-dimensional implementations [15], [16], [17].

Sampling-based approaches to motion planning: The sampling-based approach to motion-planning has extended the applicability of motion planning algorithms beyond the restricted subset of problems that can be solved efficiently by exact algorithms [1], [3]. Sampling-based motion planning algorithms, such as Probabilistic Roadmaps (PRM), Expansive Space Trees (EST) and Rapidly-exploring Random Trees (RRT) (see, e.g. [1, C.7], [3]) as well as their many variants, aim to capture the connectivity of $\mathcal{C}_{\text{free}}$ in a graph data structure, via random sampling of robot configurations. For a general survey on the field see [1]. Importantly, the PRM and RRT algorithms were both shown to be probabilistically complete [18], [19], [20], that is, they are guaranteed to find a valid solution, if one exists. However, the required running time for finding such a solution cannot be computed for new queries at run-time, and the proper usage of sampling-based approaches may still be considered somewhat of an art. Moreover, sampling-based methods are also considered sensitive to tight passages in the configuration space, due to the high-probability of missing such passages.

Hybrid methods for motion-planning: Few hybrid methods attempt to combine both deterministic and probabilistic planning strategies. Hirsch and Halperin [21] studied two-disc motion planning by exactly decomposing the configuration space of each robot, then combining the two solutions to a set of free, forbidden and mixed cells, and using PRM to construct the final connectivity graph. Zhang et al. [22] used PRM in conjunction with approximate cell decomposition, which also divides space into free, forbidden and mixed cells. Other studies have suggested to connect a dense set of near-by configuration space "slices". Each slice is decomposed to free and forbidden cells, but adjacent slices are connected in an inexact manner, by e.g., identifying overlaps between adjacent slices [23, pp. 283-287], or heuristic interpolation and local-planning [24].

In [25] a 6 *dof* RRT planner is presented with a 3 *dof* local planner hybridizing probabilistic, heuristic and deterministic methods.

1.1 Contribution

In this study, we present a novel general scheme for motion planing via manifold samples (MMS), which extends sampling-based techniques like PRM as follows: Instead of sampling isolated robot configurations, we sample *entire low-dimensional manifolds*, which can be analyzed by complete and exact methods for decomposing space. This yields an explicit representation of maximal connected patches of free configurations on each manifold, and provides a much better coverage of the configuration space compared to isolated point samples. At the same time, the manifold samples are deliberately chosen such that they are likely to intersect each other, which allows to establish connections among different manifolds. The general scheme of MMS is illustrated in Figure 1. A detailed discussion of the scheme is presented in Section 2.

In Section 3, we discuss the application of MMS to the concrete case of a polygonal robot translating and rotating in the plane amidst polygonal obstacles. We present in detail appropriate families of manifolds as well as filtering schemes that should also be of interest for other scenarios. Although our software is prototypical, we emphasize that the achieved results are due to careful design and implementation on all levels. In particular, in Section 4 we present an exact analytic solution and efficient implementation to a motion planning problem instance: moving a polygonal robot in the plane with rotation and translation *along an arbitrary axis*. To the best of our knowledge the problem has not been analytically studied before. The implementation involves advanced algebraic and extension of state-of-the-art applied geometry tools. In Section 5 we present experimental results, which show our method's superior behavior for several test cases vis-à-vis a common implementation of the sampling-based PRM algorithm. For example, in a tight passage scenario we demonstrate a 27-fold speedup in running time. We conclude with a discussion of extensions of our scheme, which we anticipate could greatly widen the scope of applicability of sampling-based methods for motion planning by combining them with strong analytic tools in a natural and straightforward manner.

2 General Scheme for Planning with Manifold Samples

Preprocessing—Constructing the connectivity graph: We propose a multi-query planner for motion planning problems in a possibly high-dimensional configuration space. The preprocessing stage constructs the *connectivity graph* of $\mathcal{C}$, a data structure that captures the connectivity of $\mathcal{C}$ using manifolds as samples. The manifolds are decomposed into cells in $\mathcal{C}_{\text{free}}$ and $\mathcal{C}_{\text{forb}}$ in a complete and exact manner; we call a cell of the decomposed manifold that lies in $\mathcal{C}_{\text{free}}$ a *free space cell* (*FSC*) and refer to the connectivity graph as $\mathcal{G}$. The *FSC*s serve as nodes in $\mathcal{G}$ while two nodes in $\mathcal{G}$ are connected by an edge if their corresponding *FSC*s intersect. See Figure 1 for an illustration.

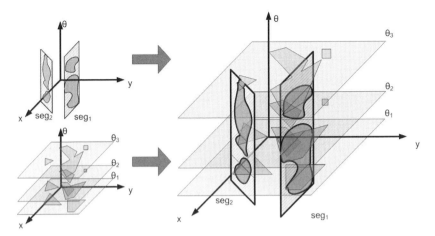

Fig. 1. Three-dimensional configuration space: The left side illustrates two families of manifolds where the decomposed cells are darkly shaded. The right side illustrates their intersection that induces the graph $\mathcal{G}$.

We formalize the preprocessing stage by considering manifolds induced by a family of constraints Ψ, such that $\psi \in \Psi$ defines a manifold m_ψ of the configuration space and FSC_{m_ψ} is the set of FSCs of m_ψ. The construction of a manifold m_ψ and its decomposition into FSCs are carried out via a Ψ-primitive, denoted P_Ψ, applied to an element $\psi \in \Psi$. By a slight abuse of notations we refer to an FSC both as a cell and as a node in the graph. Using this notation, Algorithm 1 summarizes the construction of $\mathcal{G}$. In lines 3-4, a new manifold constraint is generated and added to the collection X of manifold constraints. In lines 5-7, the manifold induced by the new constraint is decomposed by the appropriate primitive and its FSCs are added to $\mathcal{G}$.

Query: Once the connectivity graph $\mathcal{G}$ has been constructed it can be queried for paths between two free configurations q_s and q_t in the following manner: A manifold that contains q_s (respectively q_t) in one of its FSCs is generated and

---

Algorithm 1. Construct Connectivity Graph

1: $V \leftarrow \emptyset$, $E \leftarrow \emptyset$, $X \leftarrow \emptyset$
2: **repeat**
3:     $\psi \leftarrow$ generate_constraint(V,E,X)
4:     $X \leftarrow X \cup \{\psi\}$
5:     $FSC_{m_\psi} \leftarrow P_\Psi(m_\psi)$
6:     $V \leftarrow V \cup \{\text{fsc}|\ \text{fsc} \in FSC_{m_\psi}\}$
7:     $E \leftarrow E \cup \{(\text{fsc}_1, \text{fsc}_2) \mid\ \text{fsc}_1 \in V,\ \text{fsc}_2 \in FSC_{m_\psi}$,
                        $\text{fsc}_1 \cap \text{fsc}_2 \neq \emptyset\ \&\ \text{fsc}_1 \neq \text{fsc}_2\}$
8: **until** stopping_condition
9: **return** $G(V,E)$

---

decomposed. Its *FSCs* and their appropriate edges are added to $\mathcal{G}$. We compute a path γ in $\mathcal{G}$ between the *FSCs* that contain q_s and q_t. A path in $\mathcal{C}_{\text{free}}$ may then be computed by planning a path within each *FSC* in γ.

2.1 Desirable Properties of Manifold Families

Choosing the specific set of manifold families may depend on the concrete problem at hand, as detailed in the next section. However, it seems desirable to retain some general properties. First, each manifold should be simple enough such that it is possible to decompose it into free and forbidden cells in a computationally effcient manner. The choice of manifold families should also *cover* the configuration space, such that each configuration intersects at least a single manifold m_ψ. In addition, local transitions between close-by configurations should be made possible via cross-connections of several intersecting manifolds, which we term the *spanning* property. We anticipate that these simple and intuitive properties (perhaps subject to some fine tuning) may lead to a proof of probabilistic completeness of the approach.

2.2 Exploration and Connection Strategies

A naïve way to generate constraints that induce manifolds is by random sampling. Primitives may be computationally complex and should thus be applied sparingly. We suggest a general exploration/connection scheme and additional optimization heuristics that may be used in concrete implementations of the proposed general scheme. We describe strategies in general terms, providing conceptual guidelines for concrete implementations, as demonstrated in Section 3.

Exploration and connection phases: Generation of constraints is done in two phases: *exploration* and *connection*. In the exploration phase constraints are generated such that primitives will produce *FSCs* that introduce new connected components in $\mathcal{C}_{\text{free}}$. The aim of the exploration phase is to increase the coverage of the configuration space as efficiently as possible. In contrast, in the connection phase constraints are generated such that primitives will produce *FSCs* that connect existing connected components in $\mathcal{G}$. Once a constraint is generated, $\mathcal{G}$ is updated as described above. Finally, we note that we can alternate between exploration and connection, namely we can decide to further explore after some connection work has been performed.

Region of interest (RoI): Decomposing an entire manifold m_ψ by a primitive P_Ψ may be unnecessary. Patches of m_ψ may intersect $\mathcal{C}_{\text{free}}$ in highly explored parts or connect already well-connected parts of $\mathcal{G}$ while others may intersect $\mathcal{C}_{\text{free}}$ in sparsely explored areas or less well connected parts of $\mathcal{G}$. Identifying the regions where the manifold is of good use (depending on the phase) and constructing m_ψ only in those regions increases the effectiveness of P_Ψ while desirably biasing the samples. We refer to a manifold patch that is relevant in a specific phase as the *Region of Interest* - RoI of the manifold.

Constraint filtering: Let $\psi \in \Psi$ be a constraint such that applying P_Ψ to ψ yields the set of *FSCs* on m_ψ. If we are in the *connection* phase, inserting the associated nodes into $\mathcal{G}$ and intersecting them with the existing *FSCs* should connect existing connected components of $\mathcal{G}$. Otherwise, the primitive's contribution is poor. We suggest applying a filtering predicate immediately after generating a constraint ψ to check if $P_\psi(\psi)$ *may* connect existing connected components of $\mathcal{G}$. If not, m_ψ should not be constructed and ψ should be discarded.

3   The Case of Rigid Polygonal Motion

We demonstrate the scheme suggested in Section 2 by considering a (not necessarily convex) polygonal robot R translating and rotating in the plane amidst polygonal obstacles. A configuration of R describes the position of the reference point (chosen arbitrarily) of R and the orientation of R. As we consider full rotations, the configuration space $\mathcal{C}$ is the three dimensional space $\mathbb{R}^2 \times S^1$.

3.1   Manifold Families

As defined in Section 2, we consider manifolds defined by *constraints* and construct and decompose them using *primitives*. We suggest the following constraints restricting motions of the robot R and describe their associated primitives: (i) The **Angle Constraint** fixes the orientation of R while it is still free to translate anywhere within the workspace, and (ii) the **Segment Constraint** restricts the position of the reference point to a segment in the workspace while R is free to rotate.

The left part of Figure 1 demonstrates decomposed manifolds associated to the angle (left bottom) and segment (left top) constraints. The angle constraint induces a two-dimensional horizontal plane where the cells are polygons. The segment constraint induces a two-dimensional vertical slab where the cells are defined by the intersection of rational curves (as explained in Section 4).

We delay the discussion of creating and decomposing manifolds to Section 4. For now, notice that the Segment-Primitive is computationally far more time-consuming than the Angle-Primitive, since it involves arrangements[1] of curves of higher algebraic degree.

3.2   Exploration and Connection Strategies

We use manifolds constructed by the Angle-Primitive for the exploration phase and manifolds constructed by the Segment-Primitive for the connection phase. Since the Segment-Primitive is far more costly than the Angle-Primitive, we focused our efforts on optimizing the former.

Region of interest - RoI: As suggested in Section 2.2 we may consider the Segment-Primitive in a subset of the range of angles. This results in a somewhat "weaker" yet more efficient primitive than considering the whole range. If the connectivity of a local area of the configuration space is desired, then using this optimization may suffice while considerably speeding up the algorithm.

Generating segments: Let fsc be a random *FSC*. Consecutive layers (manifolds of the Angle Constraint) have a similar structure unless topological criticalities

---

[1] A subdivision of the plane into zero-dimensional, one-dimensional and two-dimensional cells, called vertices, edges and faces, respectively induced by the curves.

occur in $\mathcal{C}$. Once such a criticality occurs, an *FSC* either appears and grows or shrinks and disappears. We thus suggest a heuristic for generating a segment in the workspace for the Segment-Primitive using the size of fsc as a parameter where we refer to small and large cells according to pre-defined constants. The RoI used will be proportional to the size of fsc. The segment generated will be chosen with one of the following procedures used in Algorithm 2:

Algorithm 2. Generate Segment Constraint (V,E)

1: $r \leftarrow$ random_num $([0,1])$
2: **if** $r \geq$ random_threshold **then**
3:    **return** random_segment_procedure()
4: **else**
5:    fsc $\leftarrow$ random_fsc(V)
6:    $\alpha \leftarrow$ [size(fsc) - small_cell_size] / [large_cell_size - small_cell_size]
7:    **if** $r \geq \alpha$ **then**
8:        **return** small_cell_procedure(fsc,V)
9:    **else**
10:        **return** large_cell_procedure(fsc,V)
11:    **end if**
12: **end if**

Algorithm 3. Filter Segment (s,RoI,V,E)

1: $cc_{ids} \leftarrow \emptyset$
2: **for all** $v \in V$ **do**
3:    fsc $\leftarrow$ free_space_cell(v)
4:    **if** constraining_angle(fsc)$\in$ *RoI* **then**
5:        $cc_{ids} \leftarrow cc_{ids}\cup$ connected_component_id(v)
6:    **end if**
7: **end for**
8: **if** $|cc_{ids}| \leq 1$ **then**
9:    **return** filter_out
10: **end if**

(i) **Random procedure:** Return a random segment from the workspace. (ii) **Large cell procedure:** Return a random segment from the projection of fsc onto the xy-plane. (iii) **Small cell procedure:** Return a segment by considering fsc and an adjacent layer. For full details see [26].

Constraint Filtering: As suggested in Section 2.2, we avoid computing unnecessary primitives. All *FSCs* that will intersect a "candidate" constraint s, namely all *FSCs* of layers in its RoI, are tested. If they are all in the same connected component in $\mathcal{G}$, s can be discarded as demonstrated in Algorithm 3.

3.3 Path Planning Query

A configuration q is marked by (x,y,θ) where x,y is the location of the reference point and θ is the amount of counterclockwise rotation of R relative to its original placement. For a query $q = (q_1,q_2)$, where $q_i = (x_i,y_i,\theta_i)$, $i \in 1,2$, m_{θ_1} and m_{θ_2}

are constructed and $FSC_{m_{\theta_1}}$, $FSC_{m_{\theta_2}}$ are added to $\mathcal{G}$. A path of $FSCs$ between the $FSCs$ containing q_1 and q_2 is searched for. A local path in an Angle-Primitive's FSC (which is a polygon) is constructed by computing the shortest path on the visibility graph defined inside the polygon by the vertices of the polygon. A local path in an FSC of a Segment-Primitive (which is a two-dimensional region bounded by rational arcs) is constructed by applying cell decomposition and computing the shortest path on the graph induced by the decomposed cells.

4   Efficient Implementation of Manifold Decomposition

The algorithm discussed in Section 3 is implemented in C++. It is based on CGAL's arrangement package, which is used for the geometric primitives, and the BOOST graph library [27], which is used to represent the connectivity graph $\mathcal{G}$. We next discuss the manifold decomposition methods in more detail.

Angle-Primitive: The Angle-Primitive for a constraining angle θ (denoted $P_\Theta(\theta)$) is constructed by computing the Minkowski sum of $-R_\theta$ with the obstacles[2]. The implementation is an application of CGAL's Minkowski sums package [28, C.24]. We remark that we ensure (using the method of Canny et al. [29]) that the angle θ is chosen such that $\sin\theta$ and $\cos\theta$ are rational. This allows for an exact rotation of the robot and an exact computation of the Minkowski Sum.

Segment-Primitive: Limiting the possible positions of the robot's reference point r to a given segment s, results in a two-dimensional configuration space. Each vertex (or edge) of the robot in combination with each edge (or vertex) of an obstacle gives rise to a critical curve in this configuration space. Namely the set of all configurations that put the two features into contact, and thus mark a potential transition between $\mathcal{C}_{\mathrm{forb}}$ and $\mathcal{C}_{\mathrm{free}}$. Our analysis [26] shows that these critical curves can be expressed by rational functions only. Thus, the implementation of the Segment-Primitive is first of all a computation of an arrangement of rational functions.

CGAL follows the *generic programming paradigm* [30], that is, algorithms are formulated and implemented such that they abstract away from the actual types, constructions and predicates. Using the C++ programming language this is realized by means of class and function templates. In particular, the arrangement package is written such that it takes a traits class as a template argument. This traits class defines the supported curve type and provides the operations that are required for this type. Since the old specialized traits class was too slow for MMS to be effective (even slower than the solution for general algebraic curves presented in [31]), we devised a new efficient traits class for rational functions.

The new traits class [26] is written such that it takes maximal advantage of the fact that the supported curves are functions. As opposed to the general traits in [31], we never have to shear the coordinate system and we only require tools provided by the univariate algebraic kernel of CGAL [28, C.8]. A comparison using the benchmark instances that were also used in [31] shows that the new

---

[2] We use $-R_\theta$ to denote R rotated by θ and reflected about the origin.

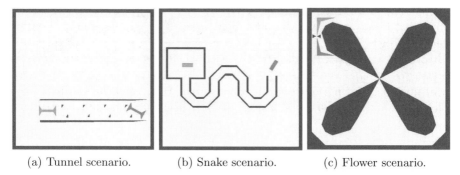

(a) Tunnel scenario.   (b) Snake scenario.   (c) Flower scenario.

Fig. 2. Experimental scenarios

traits class is about 3-4 times faster then the general traits class; this is a total speed up of about 10 when compared to the old dedicated traits class.

The development of this new traits class represents the low tier of our efforts to produce an effective motion planner and relies on a more intimate acquaintance with CGAL in general and the arrangement traits for algebraic curves in particular; for further details see [26]. We note that the new traits class has been integrated into CGAL and will be available in the upcoming CGAL release 3.9.

5   Experimental Results

We demonstrate the performance of our planner using three different scenarios. All scenarios consist of a workspace, a robot with obstacles and one query (source and target configurations). Figure 2 illustrates the scenarios where the obstacles are drawn in blue and the source and target configurations are drawn in green and red, respectively. All reported tests were measured on a Dell 1440 with one 2.4GHz P8600 Intel Core 2 Duo CPU processor and 3GB of memory running with a Windows 7 32-bit OS. Preprocessing times presented (in seconds) are times that yielded at least 80% (minimum of 5 runs) success rate in solving queries.

5.1   Algorithm Properties

Our planner has two parameters: the number n_θ of layers to be generated and the number n_s of segment constraints to be generated. We chose the following values for these parameters: $n_\theta \in \{10, 20, 40, 80\}$ and $n_s \in \{2^i | i \in \mathbb{N}, i \leq 14\}$. For a pair of parameters (n_θ, n_s) we report the preprocessing time t and whether a path was found (marked $\checkmark$) or not (marked $\times$) once the query was issued. The results for the flower scenario are reported in Table 1. We show that a considerable increase in parameters has only a limited effect on the preprocessing time.

In order to test the effectiveness of our optimizations, we ran the planner with or without any heuristic for choosing segments and with or without segment filtering. We also added a test with all optimizations using the old traits class. The results for the flower scenario can be viewed in Table 2. We remark that the

Table 1. Parameter sensitivity

		n_θ			
		10	20	40	80
n_s	256	(6,×)	(11,×)	(12,×)	(16,×)
	512	(7,×)	(13,×)	(14,×)	(25,×)
	1024	(16,×)	(20,✓)	(23,✓)	(35,✓)
	2048	(30,×)	(35,✓)	(38,✓)	(51,✓)
	4096	(46,✓)	(53,✓)	(60,✓)	(82,✓)

Table 2. Optimization results

Traits	Segment Generation	Filtering	n_θ	n_s	t
New	random	not used	20	8192	1418
		used	20	8192	112
	heuristic	not used	40	512	103
		used	20	1024	20
Old	heuristic	used	20	1024	138

engineering work invested in optimizing MMS yielded an algorithm comparable and even surpassing a motion planner that is in prevalent use as shown next.

5.2 Comparison with PRM

We used an implementation of the PRM algorithm as provided by the OOPSMP package [32]. For fair comparison, we did not use cycles in the roadmap as cycles increase the preprocessing time significantly. We manually optimized the parameters of each planner over a concrete set. As with previous tests, the parameters for MMS are n_θ and n_s. The parameters used for the PRM are the number of neighbors (denoted k) to which each milestone should be connected and the percentage of time used to sample new milestones (denoted % st in Table 3).

Furthermore, we ran the flower scenario several times, progressively increasing the robot size. This caused a "tightening" of the passages containing the desired path. Figure 3 demonstrates the preprocessing time as a function of the tightness of the problem for both planners. A tightness of zero denotes the base scenario (Figure 2c) while a tightness of one denotes the tightest problem solved.

The results show a speedup for all scenarios when compared to the PRM implementation. Moreover, our algorithm has little sensitivity to the tightness of the problem as opposed to the PRM algorithm. In the tightest experiment, MMS runs 27 times faster than the PRM implementation.

6 Further Directions

To conclude, we outline directions for extending and enhancing the current work. Our primary goal is to use the MMS framework to solve progressively more com-

Table 3. Comparison with PRM

Scenario	MMS			PRM			Speedup
	n_θ	n_s	t	k	% st	t	
Tunnel	20	512	100	20	0.0125	180	1.8
Snake	40	256	22	20	0.025	140	6.3
Flower	20	1024	20	24	0.0125	40	2

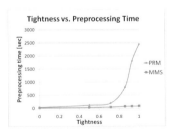

Fig. 3. Tightness results

plicated motion-planning problems. As suggested earlier, we see the framework as a platform for convenient transfer of strong geometric primitives into motion planning algorithms. For example, among the recently developed tools are efficient and exact solutions for computing the Minkowski sums of polytopes in $\mathbb{R}^3$ (see Introduction) as well as for exact update of the sum when the polytopes rotate [33]. These could be combined into an MMS for planning full rigid motion of a polytope among polytopes, which, extrapolating from the current experiments could outperform more simplistic solutions in existence.

Looking at more intricate problems, we anticipate some difficulty in turning constraints into manifolds that can be exactly decomposed. We propose to have manifolds where the decomposition yields some *approximation* of the *FSCs*, using recent advanced meshing tools for example. We can endow the connectivity-graph nodes with an attribute describing their approximation quality. One can then decide to only look for paths all whose nodes are above a certain approximation quality. Alternatively, one can extract any solution path and then refine only those portions of the path that are below a certain quality.

Beyond motion planning: We foresee an extension of the framework to other problems that involve high-dimensional arrangements of critical hypersurfaces. It is difficult to describe the entire arrangement analytically, but there are often situations where constraint manifolds could be computed analytically. Hence, it is possible to shed light on problems such as loop closure and assembly planning where we can use manifold samples to analytically capture pertinent information of high-dimensional arrangements of hypersurfaces. Notice that although in Section 3 we used only planar manifolds, there are recently developed tools to construct two-dimensional arrangement of curves on curved surfaces [34] which gives further flexibility in choosing the manifold families.

For supplementary material and updates the reader is referred to our webpage http://acg.cs.tau.ac.il/projects/mms.

References

1. Choset, H., Burgard, W., Hutchinson, S., Kantor, G., Kavraki, L.E., Lynch, K., Thrun, S.: Principles of Robot Motion: Theory, Algorithms, and Implementation. MIT Press, Cambridge (2005)
2. Latombe, J.C.: Robot Motion Planning. Kluwer Academic Publishers, Norwell (1991)
3. LaValle, S.M.: Planning Algorithms. Cambridge University Press, Cambridge (2006)
4. Reif, J.H.: Complexity of the mover's problem and generalizations. In: FOCS, pp. 421–427. IEEE Computer Society, Washington, DC, USA (1979)
5. Lozano-Perez, T.: Spatial planning: A configuration space approach. MIT AI Memo 605 (1980)
6. Schwartz, J.T., Sharir, M.: On the "piano movers" problem: II. General techniques for computing topological properties of real algebraic manifolds. Advances in Applied Mathematics 4(3), 298–351 (1983)

7. Basu, S., Pollack, R., Roy, M.F.: Algorithms in Real Algebraic Geometry. Algorithms and Computation in Mathematics. Springer, Heidelberg (2003)
8. Canny, J.F.: Complexity of Robot Motion Planning (ACM Doctoral Dissertation Award). MIT Press, Cambridge (1988)
9. Chazelle, B., Edelsbrunner, H., Guibas, L.J., Sharir, M.: A singly exponential stratification scheme for real semi-algebraic varieties and its applications. Theoretical Computer Science 84(1), 77–105 (1991)
10. Aronov, B., Sharir, M.: On translational motion planning of a convex polyhedron in 3-space. SIAM J. Comput. 26(6), 1785–1803 (1997)
11. Avnaim, F., Boissonnat, J., Faverjon, B.: A practical exact motion planning algorithm for polygonal object amidst polygonal obstacles. In: Boissonnat, J.-D., Laumond, J.-P. (eds.) Geometry and Robotics. LNCS, vol. 391, pp. 67–86. Springer, Heidelberg (1989)
12. Halperin, D., Sharir, M.: A near-quadratic algorithm for planning the motion of a polygon in a polygonal environment. Disc. Comput. Geom. 16(2), 121–134 (1996)
13. Schwartz, J.T., Sharir, M.: On the "piano movers" problem: I. The case of a two-dimensional rigid polygonal body moving amidst polygonal barriers. Commun. Pure appl. Math. 35, 345–398 (1983)
14. Sharir, M.: Algorithmic Motion Planning. In: Handbook of Discrete and Computational Geometry, 2nd edn., CRC Press, Inc., Boca Raton (2004)
15. Fogel, E., Halperin, D.: Exact and efficient construction of Minkowski sums of convex polyhedra with applications. CAD 39(11), 929–940 (2007)
16. Hachenberger, P.: Exact Minkowksi sums of polyhedra and exact and efficient decomposition of polyhedra into convex pieces. Algorithmica 55(2), 329–345 (2009)
17. Wein, R.: Exact and efficient construction of planar minkowski sums using the convolution method. In: Azar, Y., Erlebach, T. (eds.) ESA 2006. LNCS, vol. 4168, pp. 829–840. Springer, Heidelberg (2006)
18. Kavraki, L.E., Kolountzakis, M.N., Latombe, J.C.: Analysis of probabilistic roadmaps for path planning. IEEE Trans. Robot. Automat. 14(1), 166–171 (1998)
19. Kuffner, J.J., Lavalle, S.M.: RRT-Connect: An efficient approach to single-query path planning. In: ICRA, pp. 995–1001. IEEE, Los Alamitos (2000)
20. Ladd, A.M., Kavraki, L.E.: Generalizing the analysis of PRM. In: ICRA, pp. 2120–2125. IEEE, Los Alamitos (2002)
21. Hirsch, S., Halperin, D.: Hybrid motion planning: Coordinating two discs moving among polygonal obstacles in the plane. In: WAFR 2002, pp. 225–241 (2002)
22. Zhang, L., Kim, Y.J., Manocha, D.: A hybrid approach for complete motion planning. In: IROS, pp. 7–14 (2007)
23. De Berg, M., Cheong, O., van Kreveld, M., Overmars, M.: Computational Geometry: Algorithms and Applications. Springer, Heidelberg (2008)
24. Lien, J.M.: Hybrid motion planning using Minkowski sums. In: RSS 2008 (2008)
25. Yang, J., Sacks, E.: RRT path planner with 3 DOF local planner. In: ICRA, pp. 145–149. IEEE, Los Alamitos (2006)
26. Salzman, O., Hemmer, M., Raveh, B., Halperin, D.: Motion planning via manifold samples. In: arXiv:1107.0803 (2011)
27. Siek, J.G., Lee, L.-Q., Lumsdaine, A.: The Boost Graph Library: User Guide and Reference Manual. Addison-Wesley Professional, Reading (2001)
28. The CGAL Project: CGAL User and Reference Manual. 3.7 edn. CGAL Editorial Board (2010), http://www.cgal.org/

29. Canny, J., Donald, B., Ressler, E.K.: A rational rotation method for robust geo-metric algorithms. In: SoCG 1992, pp. 251–260. ACM, New York (1992)
30. Austern, M.H.: Generic Programming and the STL. Addison-Wesley, Reading (1998)
31. Berberich, E., Hemmer, M., Kerber, M.: A generic algebraic kernel for non-linear geometric applications. In: SoCG 2011 (2011)
32. Plaku, E., Bekris, K.E., Kavraki, L.E.: OOPS for motion planning: An online open-source programming system. In: ICRA, pp. 3711–3716. IEEE, Los Alamitos (April 2007)
33. Mayer, N., Fogel, E., Halperin, D.: Fast and robust retrieval of Minkowski sums of rotating convex polyhedra in 3-space. In: SPM, pp. 1–10 (2010)
34. Berberich, E., Fogel, E., Halperin, D., Mehlhorn, K., Wein, R.: Arrangements on parametric surfaces I: General framework and infrastructure. Mathematics in Computer Science 4(1), 45–66 (2010)

Ray-Shooting Depth: Computing Statistical Data Depth of Point Sets in the Plane

Nabil H. Mustafa[1] *, Saurabh Ray[2] **, and Mudassir Shabbir[3]

[1] Dept. of Computer Science, LUMS, Pakistan
nabil@lums.edu.pk
[2] Max-Plank-Institut für Informatik, Saarbrücken, Germany
saurabh@mpi-inf.mpg.de
[3] Dept. of Computer Science, Rutgers, USA
mudassir@cs.rutgers.edu

Abstract. Over the past several decades, many combinatorial measures have been devised for capturing the statistical data depth of a set of n points in $\mathbb{R}^2$. These include Tukey depth [15], Oja depth [12], Simplicial depth [10] and several others. Recently Fox *et al.* [7] have defined the Ray-Shooting depth of a point set, and given a topological proof for the existence of points with high Ray-Shooting depth in $\mathbb{R}^2$. In this paper, we present an $O(n^2 log^2 n)$-time algorithm for computing a point of high Ray-Shooting depth. We also present a linear time approximation algorithm.

1 Introduction

The area of statistical data analysis deals with the following kind of question: given some multivariate data in $\mathbb{R}^d$, possibly with noise, what one point in $\mathbb{R}^d$ best describes that data. There are several considerations to take into account when quantitatively formulating the measure that accurately captures the notion of 'best': robustness to noise, computational ease, invariance under translation and rotation, invariance to change-of-axis and so on. (see e.g. [13,3,1,2,10,11,7,16,12]).

A robust measure can be constructed by generalizing the concept of the *median* of numbers to points in $\mathbb{R}^d$. Given a set P of n points in $\mathbb{R}^d$, define the Tukey depth [15] of a point $q \in \mathbb{R}^d$ as the smallest-number of points contained in any halfspace containing q. The classical Centerpoint theorem states that given any set P of n points in $\mathbb{R}^d$, there exists a point with Tukey depth at least $n/(d+1)$; furthermore, this is the best possible. Finally, the Tukey depth of a point set P is defined as the maximum Tukey depth of any point in $\mathbb{R}^d$. There have been a sequence of papers related to finding efficient algorithms for computing a point

---

* This work was done on a one-year visit to the wonderful DCG group at EPFL, Switzerland.
** The first and second authors gratefully acknowledge support from the Bernoulli Center at EPFL and from the Swiss National Science Foundation, Grant No. 200021-125287/1.

C. Demetrescu and M.M. Halldórsson (Eds.): ESA 2011, LNCS 6942, pp. 506–517, 2011.

with high Tukey depth, and a point realizing the Tukey depth of a given point set. In $\mathbb{R}^2$, the naive algorithm to find the point with the highest Tukey depth takes $O(n^6)$ time, which was improved by Rousseaw and Ruts [14] to $O(n^2 \log n)$. This was further improved by Matousek to $O(n \log^5 n)$. Finally Chan [4] gave an $O(n \log n)$ time randomized algorithm. A deterministic $O(n \log^3 n)$ time algorithm was given by Langerman and Steiger [9]. A point with Tukey depth at least $n/3$ can be computed in $O(n)$ time [8]. In $\mathbb{R}^d$, the current-best algorithms for both finding a point of depth $n/(d+1)$ and the highest depth point take $O(n^{d-1})$ time [4].

Recently an elegant new measure, which we call Ray-Shooting depth, has been proposed in [7]. Given a set P of n points, let E be the set of all $\binom{n}{d}$ $(d-1)$-simplices spanned by P. Then define the Ray-Shooting depth (called RS depth from now on) of a point q as the minimum number of simplices in E that any ray from q must intersect. While the problem of existence of a point with high RS depth is open in $\mathbb{R}^d$, $d \geq 3$, it is shown in [7] that given any P in $\mathbb{R}^2$, there exists a point with RS depth at least $n^2/9$. An easy construction shows that this is optimal. As the status of the problem is not known even for $\mathbb{R}^3$, from now onwards, we only consider the $d = 2$ case in this paper.

Unfortunately the proof given in [7] is completely existential, as it follows from a variant of Brouwer's fixed point theorem. A straightforward algorithm can be derived from it with running time $O(n^5 \log^5 n)$ by an exhaustive search. The main focus of this paper will be to present faster algorithms, both exact and approximate, for computing RS depth of a planar point set. Admittedly, one would like an algorithm for computing points of high RS depth in higher dimensions as well. However, it is not clear how to extend the current topological machinery that we use for the $d = 2$ case to higher dimensions.

Our Contributions. We show that one can compute a point of RS depth $n^2/9$ in the plane in $O(n^2 \log^2 n)$ time. One can also compute an approximation to such a point in linear-time. Specifically, one can compute a point with RS depth at least $n^2/9 - o(n^2)$ in linear-time. These two results are presented in Section 2. We have also implemented the approximation algorithm, and made it available as a package in the statistical computing software R; this is explained in Section 4.

2   Computing a Point of Ray-Shooting Depth at Least $n^2/9$

In this section, we present an algorithm that takes as input a finite set P of n points in the plane and returns a point p in the plane whose Ray-Shooting depth is at least $n^2/9$. Given a point $q \in \mathbb{R}^2$ and a vector $u \in \mathbb{S}^1$, denote by $\rho_{q,u}$ the closed ray emanating from q in the direction u. For any $p \in \mathbb{R}^2$, $p \neq q$, denote the closed ray emanating from q and passing through p as $\rho_{q,p}$. We will denote the unit vector in the direction of $\rho_{q,p}$ as $\delta_{q,p}$. A ray r is *bad* if it intersects fewer than $n^2/9$ edges (segments) spanned by the input points, and *good* otherwise. The following lemma is from [7].

Lemma 1. *For any $q \in \mathbb{R}^2$ and for any two bad rays ρ_1 and ρ_2 emanating from q, one of the two closed cones defined by ρ_1 and ρ_2 contains fewer than $n/3$ input points.*

Proof. If both cones defined by ρ_1 and ρ_2 have at least $n/3$ points, the rays ρ_1 and ρ_2 together intersect at least $2n^2/9$ edges. This is a contradiction to the assumption that both of them are bad rays. □

Using the above lemma, we prove the following.

Lemma 2. *For any $q \in \mathbb{R}^2$, there exists a cone $\mathcal{C}_q$ with apex q containing more than $2n/3$ input points, and where all the rays emanating from q and contained in $\mathcal{C}_q$ are good.*

Proof. If there are no bad rays emanating from q then we take $\mathcal{C}_q$ as the entire plane. Otherwise, let ρ be a bad ray emanating from q. Let us pick a ray ρ' emanating from q such that both the closed cones defined by ρ and ρ' contain at least $n/2$ input points. Note that ρ' is a good ray (Lemma 1).

Imagine starting with the ray ρ', and rotating it in either direction about q as long as it is good. See Figure 2. Let ρ_1 (respectively ρ_2) be the last good ray before we hit a bad ray, in the counter-clockwise (resp. clockwise) direction. Clearly ρ_1 and ρ_2 pass through input points. Let ρ_1' be a bad ray close-enough to ρ_1 so that there are no points of P between these rays. Similarly let ρ_2' be a close-enough bad ray to ρ_2.

The cone C_1 defined by ρ and ρ_1' (Figure 2) contains fewer than $n/3$ input points (by Lemma 1 applied to ρ and ρ_1', as the cone complement to C_1 contains at least $n/2$ points). Similarly, the cone C_2 contains fewer than $n/3$ points. Consider now the bad rays ρ_1' and ρ_2' . Let C_3 be the cone defined by these rays that contains the ray ρ'. C_3 cannot contain fewer than $n/3$ points since $C_1 \cup C_2 \cup C_3$ covers $\mathbb{R}^2$ and C_1 and C_2 contain fewer than $n/3$ points each. Hence the other cone defined by ρ_1' and ρ_2' contains fewer than $n/3$. Therefore, the cone C_q defined by ρ_1 and ρ_2 that contains ρ' contains more than $2n/3$ points. By construction, all rays in C_q are good. □

For any point $q \in \mathbb{R}^2$, define the good cone at q to be the closed cone C with apex q containing the maximum number of input points such that all rays emanating from q and lying in C are good. Clearly, the cone $\mathcal{C}_q$ obtained in the proof of the previous lemma is the unique good cone at the point q since it contains more than $2n/3 > n/2$ points and is maximal. For any $u \in \mathbb{S}^1$, if $\rho_{q,u} \in \mathcal{C}_q$, we say that u is a *good direction* at q. Then the cone $\mathcal{C}_q$ defines the set D_q of good directions at q. We will call the set of input points lying in $\mathcal{C}_q$ the *good set* of q and denote it by $\mathcal{G}_q$.

For computational purposes, we will need a data structure which we now describe. Let X be a set of n distinct fixed points on a circle S which we will refer to as *locations*. Let $x_1, x_2, \cdots, x_n$ be the circular order of these locations (the precise coordinates of these points does not matter). Let $T = \{t_1, \cdots, t_n\}$ be the set of n open intervals on S defined by consective locations in X. Let

$Y = \{Y_1, \cdots, Y_n\}$ be a set of n points, each placed at one of the locations in X. For any points $u, v \in S$, we will denote by $A_{u,v}$ the arc going from u to v counter-clockwise along S (so $A_{u,v}$ is different from $A_{v,u}$). Our data-structure, denoted by Ψ, will store Y at the locations in X and a number of arcs defined by points of Y.

Define the *depth* of any point $q \in S$ as the number of arcs in Ψ containing q. Since each point in any of the open intervals t_i has the same depth, we can define the depth of t_i as the depth of any point in it. We need Ψ to support the following operations:

1. Insert an arc $A_{u,v}$, $u, v \in Y$ into Ψ
2. Delete an arc $A_{u,v}$, $u, v \in Y$ from Ψ
3. Move a point $y \in Y$ to a neighboring location.
4. Given a query arc $A_{u,v}$, $u, v \in S$ and an integer k, report the first and last intervals, if any, on $A_{u,v}$ with depth smaller than k.
5. Given an integer k report an interval, if any, with depth smaller than k.

Notice that the endpoints of the arcs we add or delete are in Y. When we move a point $y \in Y$ to a neighboring location, the endpoints of the arcs incident to y move with it. However, in the fourth operation listed above, the endpoints of the query arc are arbitrary points in S. Using standard data-structuring techniques in computational geometry (augmented interval trees), it is possible to build the data structure in $O(n \log n)$ time such that each of the operations take $O(\log n)$ time. We skip the easy details. Let us see how we can use this data structure to compute the good cone C_z at any point $z \in \mathbb{R}^2$.

Lemma 3. *The good cone of any point $z \in \mathbb{R}^2$ can be computed in $O(n^2 \log n)$ time.*

Proof. We take a unit circle and fix X to be any n points $x_1, \cdots, x_n$, in that order around S. We put a point y_i at the location x_i for $i \in [1, n]$ as follows: angularly sort the input points around the point z and for every input point p_i put a representative point y_i in the location x_r where r is the rank of p_i in the sorted order.

For each pair of input points p_i and p_j, if the line $l_{i,j}$ through p_i and p_j does not pass through z, we insert either the arc A_{y_i, y_j} or A_{y_j, y_i} depending on whether z is to the left or right of the line $l_{i,j}$ oriented in the direction from p_i to p_j. If $l_{i,j}$ passes through z we either add both or none of the arcs depending on whether or not z lies on the edge $p_i p_j$. We then query the data structure to see if there are any intervals of depth smaller than $n^2/9$. If there are no such intervals, all rays emanating from z are good. If not, the data structure returns an interval I with depth less than $n^2/9$.

From any point on this interval, we obtain a corresponding bad ray r_b emanating from z. As in the proof of Lemma 2, we pick a good ray r_g such that each of the closed cones defined by r_b and r_g contain at least $n/2$ points. As before, one can find the good cone at z by rotating r_g to either side and stopping at the last good ray before we hit bad rays. We can find the first bad interval (interval of depth smaller than $n^2/9$ in either direction by using queries of type 4. Since we insert $O(n^2)$ arcs and each insertion takes $O(\log n)$ time, the time taken for inserting the arcs is $O(n^2 \log n)$. Once we have inserted all the arcs, it takes only a constant number of queries, each taking $O(\log n)$ time, to determine the good cone. The overall running time is therefore $O(n^2 \log n)$.   □

This is not the best algorithm for computing the good cone at a point. The good cone can be computed in $O(n \log n)$ time but we will not need it for this paper since this is not the bottleneck for the main algorithm.

Let $\mathcal{L}$ be the set of the $\binom{n}{2}$ lines passing through each pair of input points. Let $\mathcal{A}$ be the arrangement of the lines in $\mathcal{L}$.

Lemma 4. *If* $q, r \in \mathbb{R}^2$ *lie in the interior of the same cell of* $\mathcal{A}$ *then* $\mathcal{G}_q = \mathcal{G}_r$. *If* q *lies in the interior of a cell and* r *lies on the boundary of the same cell,* $\mathcal{G}_q \subseteq \mathcal{G}_r$.

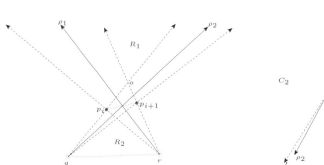

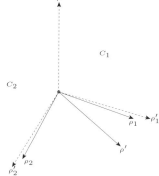

Fig. 1. The good set is the same for all points in a cell

Fig. 2. The good rays are solid, bad rays are dashed

Proof. Let q and r be points in the interior of the same cell of $\mathcal{A}$. The angular order of the input points is the same around both points. Let $p_1, \cdots, p_k$ be the points in the good set of q in the angular order. Any ray in the good cone of q intersects some edge $p_i p_{i+1}$ formed by two consecutive points in the good set of q. We will show that for each such edge, any ray emanating from r and intersecting that edge is also good. This will show that all points in the good set of q are also in the good set of r i.e. $\mathcal{G}_q \subseteq \mathcal{G}_r$. The same argument with the roles of q and r exchanged shows that $\mathcal{G}_r \subseteq \mathcal{G}_q$, implying that $\mathcal{G}_q = \mathcal{G}_r$.

Let ρ_1 be a ray emanating from r and intersecting the edge $p_i p_{i+1}$. Let ρ_2 be a ray emanating from q and intersecting the edge $p_i p_{i+1}$ at the same point as ρ_1. Since p_i and p_{i+1} are in the good set of q, we know that the ρ_2 is good. We will show that ρ_1 intersects all the edges that ρ_2 intersects. Let C_1 be the cone defined by ρ_1 and ρ_2 containing the region R_1 and let C_2 be the cone defined by them containing the region R_2 (see Figure 1). If there is an edge that intersects, ρ_2 but not ρ_1 there must be an input point in at least one of these cones. Let

A_q (respectively A_r) be the cone with apex q (resp. r), containing $p_i p_{i+1}$ and bounded by rays through p_i and p_{i+1}. Since p_i and p_{i+1} are consecutive points in the angular order around the points q and r, there are no input points in the cones A_q and A_r. Therefore, if there are any input points in the cones C_1 or C_2, they must lie in the regions R_1 or R_2. However if there is any input point p in one of these regions then pp_i intersects the segment qr contradicting the assumption that q and r lie in the interior of the same cell. The same argument goes through if q lies in the interior of a cell and r lies on the boundary of a cell, except if r lies on the line through p_i and p_j. In this case, one can still show that $\mathcal{G}_q \subseteq \mathcal{G}_r$ but ρ_1 may intersect edges incident to p_i or p_{i+1}.   □

In the following, to any continuous non-self-intersecting curve ω with distinct endpoints a and b, we associate a continuous bijective function $\hat{\omega} : [0,1] \mapsto \omega$ so that $\hat{\omega}(0) = a$ and $\hat{\omega}(1) = b$. Similarly to any continuous non-self-intersecting loop ω, we associate a continuous bijective function $\hat{\omega} : \mathbb{S}^1 \mapsto \omega$. We orient $[0,1]$ from 0 to 1 and $\mathbb{S}^1$ in the clockwise direction. This gives an orientation to ω.

Let $J \subseteq \mathbb{R}^2 \times \mathbb{S}^1$ be the set $\{(w,u) : w \in \mathbb{R}^2, u \in D_w\}$. Let π_1 and π_2 be projection functions that map a point (w,u) in J to w and u respectively. Let us also denote by ω_i, $i \in \{1,2\}$, the curve defined by the function $\hat{\omega}_i(t) = \pi_i(\hat{\omega}(t))$. The domain of $\hat{\omega}_i$ is the same as the domain of $\hat{\omega}$ (either $\mathbb{S}^1$ or $[0,1]$ depending on whether ω is a closed loop). The orientation of ω gives the orientation of ω_i. When ω is a closed loop, we define the winding number of ω as the winding number of $\hat{\omega}_2$.

Let γ be a non-self-intersecting continuous curve in the plane with distinct endpoints and let ω be a continuous curve in J so that $\gamma = \omega_1$. We call ω a *walk* along γ. The next lemma shows that there exists a walk along any line segment in the plane.

Lemma 5. *Let $s = (w_1, u_1)$ and $t = (w_2, u_2)$ be two points in J. Let σ be the segment joining w_1 and w_2. There exists a curve ω in J joining s and t so that $\omega_1 = \sigma$, i.e., ω is a walk over σ with endpoints s and t. Furthermore, ω can be computed in $O(n^2 \log n)$ time.*

Intuitive meaning: The statement of the lemma uses a lot of notation in order to be precise. However, since this may make it seem more complicated, here is the intuitive meaning. We want to move from w_1 to w_2 along the segment σ always maintaining a ray in the good cone of the current point. We start with the ray in the direction u_1 when we are at w_1 and the direction of the ray changes continuously as we move to w_2 and we finish with the direction u_2 at w_2. Along the motion from w_1 to w_2, we are allowed to *stand* at a point p on σ and move the ray continuously within the good cone of p.

Proof. We first compute the intersection of σ with each line in $\mathcal{L}$. Let $p_1, p_2, \cdots$, p_{k-1} be these points in the order of intersection. Let $p_0 = w_1$ and $p_k = w_2$. Let I_i be the interval $[p_i, p_{i+1}]$. We will traverse σ from w_1 to w_2, and construct ω as we go along. The events in this traversal will be the intervals I_i and the points p_i in the order that they appear in the segment from w_1 to w_2. As we sweep, we will maintain a data structure that gives us information about the good set of the interval or point that we are currently in. Using this information, we will compute a walk along each of the points p_i and each of the intervals I_i which when put together gives us a walk along σ with endpoints w_1 and w_2. We construct $2k$ points in J, $(p_1 = w_1, s_1), (p_1, t_1), \ldots, (p_i, s_i), (p_i, t_i), \ldots, (p_k = w_2, t_k)$; the curve they define gives the required walk.

Each of the intervals I_i lies entirely within a single cell of $\mathcal{A}$. Therefore, there is some $g_i \in P$ so that $g_i \in \mathcal{G}_x$ for all $x \in I_i$. For point $i \in [1, \cdots, k-1]$, both g_{i-1} and g_i are in $\mathcal{G}_{p_i}$. Set $s_i = \delta_{p_i, g_{i-1}}$ and $t_i = \delta_{p_i, g_i}$; note that both these directions are in $\mathcal{D}_{p_i}$. Therefore, there is an interval $K_i \subseteq \mathcal{D}_{p_i}$ with endpoints s_i and t_i. We define K_0 as the interval contained in $\mathcal{D}_{p_0}$ with endpoints u_1 and δ_{p_0, g_0} and we define K_k as the interval contained in $\mathcal{D}_{p_k}$ with endpoints $\delta_{p_k, g_{k-1}}$ and u_2. For the interval I_i, we define the walk $\omega_{I_i} = \{(x, \delta_{x, g_i}) : x \in I_i\}$ and for each point p_i we define the walk $\omega_{p_i} = \{(p_i, x) : x \in K_i\}$. The walk ω_{p_0} starts at the point (p_0, u_1) and ends at the point (p_0, δ_{p_0, g_0}) which is where ω_{I_0} starts. ω_{I_0} ends at (p_1, δ_{p_1, g_0}) which is where ω_{p_1} starts and so on. Putting together these walks we get the required walk ω from (w_1, u_1) to (w_2, u_2).

In order to compute these walks, we will need to know the good cones in each of the intervals I_i and at each of the points p_i. We start by computing the good cone of p_0 and the good cone of some point z in I_0. The good cone at z gives us g_0 with which we compute ω_{p_0} and ω_{I_0}. We will update the data structure used to compute the good cone of z and obtain the good cone of p_1. In the data structure we have a representative y_i for each input point in p_i whose locations reflect their angular ordering around z. For convenience we will refer to the representatives by the points themselves. So, when we write "move p_i to a neighboring location", we mean "move y_i to a neighboring location". The point p_1 is the intersection of σ with some line in $\mathcal{L}$ passing through two input points a and b. There are two cases to consider depending on whether p_1 lies on the edge ab or not. Assume that p_1 lies on the edge ab. In this case we add the arc joining a and b which is not already in the data structure so that both arcs are present when we are at p_1. This reflects the fact that any ray emanating from p_1 intersects the edge ab. When we move from p_1 to I_1, we will remove the arc that was added first and keep the second one. Effectively as we move across p_1, we switch from one arc formed by a and b to the other arc formed by a and b. Assume now that p_1 does not lie on the edge ab. In this case, we move the point a to the location of b as we move from I_0 to p_1. When we move from p_1 to I_1, we move b to the previous location of a. This reflects the fact that as we move across p_1 from I_0 to I_1, the points a and b switch their positions in the angular order. When we are at p_1, they are at the same position. The rest of the sweep is done in a similar fashion. Each update takes $O(\log n)$ time. Hence the total time required for the sweep is $O(n^2 \log n)$.   □

Let ω be a walk along a rectangle R in the plane and $p_1 = (w_1, u_1)$ and $p_2 = (w_2, u_2)$ be two points on ω s.t. w_1 and w_2 lie on different edges of R. Let σ be a chord of R joining w_1 and w_2. From Lemma 5, we obtain walk $\tilde{\sigma}$ joining p_1 and p_2 in J such that $\tilde{\sigma}_1 = \sigma$. The points p_1 and p_2 split ω into two arcs, one oriented from p_1 to p_2 and the other oriented from p_2 to p_1. The curve $\tilde{\sigma}$ splits the loop ω into two loops α and β. The loop α traverses the arc of ω from p_1 to p_2 followed by the curve $\tilde{\sigma}$ from p_2 to p_1. The loop β traverses curve σ from p_1 to p_2 followed by the the arc of ω from p_2 to p_1. Observe that the winding numbers of the loops α and β add up to the winding number of the loop ω because if we traverse α followed by β, we traverse $\tilde{\sigma}$ consecutively in opposite direction cancelling its effect with respect to the winding number. Therefore, if the winding number of ω is non-zero, the winding number of one of the loops α or β is non-zero. This gives us a way to find, from any loop of non-zero winding number, a *smaller* loop of non-zero winding number.

Lemma 6. *Let R be a rectangle with a walk ω of non-zero winding number over it. There is point p inside R with Ray-Shooting depth at least $n^2/9$.*

Proof. Let σ be a vertical or horizontal chord of R that splits R along its longer side, into two rectangles R_1 and R_2 of equal area. From the above discussion, it follows that there is a walk of non-zero winding number along one of the rectangles R_1 or R_2. We repeat this process with that rectangle. In this process, we get nested rectangles with smaller and smaller longer side and hence converge to a point p which has a non-zero winding number over it. This means that any ray emanating from p is good and hence p has Ray-Shooting depth at least $n^2/9$. $\square$

The above lemma remains true even if R is any closed curve instead of a rectangle. We state it in terms of a rectangle because that is what we use for computational purposes.

Let R be a rectangle containing P. We will show that there is a walk ω along R with a non-zero winding number. We will then split R into two rectangles R_1 and R_2 using a vertical chord σ of R which bisects the set of vertices of $\mathcal{A}$ within R. We will find a walk $\tilde{\sigma}$ along σ that splits ω into two walks α and β along R_1 and R_2 respectively. One of these will have a non-zero winding number. We will replace R by the rectangle that has a walk of non-zero winding number along it and repeat the process. In each iteration, we reduce the number of vertices of $\mathcal{A}$ in R by a factor of two. Since there are at most $O(n^4)$ vertices to begin with, in $O(\log n)$ iterations, we will find a rectangle R so that there is walk ω along R with non-zero winding number and R has at most one vertex of $\mathcal{A}$. From Lemma 6, we can conclude that there is a point p of Ray-Shooting depth at least $n^2/9$ in the region bounded by R. Since all points in the region bounded by R belong to a cell intersecting R, we can just check the $O(n^2)$ cells intersecting R to find the required point. We will finally show that each iteration can be implemented in $O(n^2 \log n)$ time. The overall running time of the algorithm will therefore be $O(n^2 \log^2 n)$.

We now show that there is a walk of non-zero winding number along any non-self-intersecting continuous loop γ enclosing the input point set. Let $\mu = (3 - \sqrt{5})/6$ so that $\mu(1 - \mu) = 1/9$. It is easy to see that for any point p on γ, a ray r emanating from p is good if and only if it has at least μn input points on either side of the line containing r. This means that any point $q \in \mathbb{R}^2$ with Tukey depth more than μn (w.r.t. P) is in the good cone of every point p on γ.

Since $\mu < 1/3$, the Centerpoint theorem guarantees that there is such a point o. For any point p on γ, let $u_p \in \mathbb{S}^1$ be the direction $\delta_{p,o}$. Consider the walk ω along γ such that $\hat{\omega}(t) = (\hat{\gamma}(t), u_{\hat{\gamma}(t)})$. Clearly the winding number of ω is either $+1$ or -1 depending on whether γ itself has a winding number $+1$ or -1 around o.

For the purpose of our algorithm, we will start with a rectangle R which encloses the input points and none of the lines in $\mathcal{L}$ intersect the horizontal sides of $\mathcal{A}$. We can assume without loss of generality that no two of the input points lie on the same vertical line. Therefore, such a rectangle always exists and can be computed in $O(n^2)$ time. We then compute a point of Tukey depth at least μn. This can be done in $O(n)$ time [8]. This gives us a walk ω over R.

We can compute the number of vertices of $\mathcal{A}$ between any two vertical lines l_1 and l_2 by comparing the top to bottom order of the intersection of the lines in $\mathcal{L}$ with these two lines. Finding a vertical chord σ of R that splits the number of vertices evenly is therefore a simple slope selection problem and can be done in $O(n \log n)$ time [6]. Let a and b be the endpoints on σ and let R_1 and R_2 be the rectangles that σ splits R into.

Using Lemma 5, we compute a walk $\tilde{\sigma}$ over σ joining some points (a, u_a) and (b, u_b) on ω. $\tilde{\sigma}$ splits ω into two walks α and β along R_1 and R_2 respectively. Each of the walks consists of $O(n^2)$ pieces which form $\tilde{\sigma}$ and one piece along ω. The curve α_2 (and similarly β_2) is monotonic on each of these pieces. Hence the winding number of the walks α and β can easily be computed from these pieces. We pick one of the rectangles R_1 or R_2 which has a walk of non-zero winding number along it and recurse. We do this until we have a rectangle that contains at most one vertex of $\mathcal{A}$. From Lemma 6, we know that there is a point of Ray-Shooting depth at least $n^2/9$ inside the rectangle. Since the rectangle contains at most one vertex there is no cell contained completely in the interior of R. In our divide and conquer process, we have already checked all the cells crossed by R and computed a good cone for a point in it. One of those points must have Ray-Shooting depth at least $n^2/9$. We have therefore shown that:

Theorem 1. *Given a set P of n points in the plane, a point of Ray Shooting depth at least $n^2/9$ with respect to P can be computed in $O(n^2 \log^2 n)$ time.*

3   Computing a Point of Ray-Shooting Depth Approximately $n^2/9$

Suppose that we want to compute a point with Ray Shooting depth at least $(1-\epsilon)n^2/9$. We can do this by just taking a small random sample Q of the input point set P and computing a point of Ray-Shooting depth $|Q|^2/9$ with respect to Q using Theorem 1. The same point has a Ray Shooting depth at least $(1-\epsilon)n^2/9$ with respect to P with high probability. The sample Q is obtained by picking each point in P with probability $p = O(\frac{1}{\epsilon^2} \log \frac{1}{\epsilon} \cdot \frac{1}{n})$. The expected size of Q is $pn = O(\frac{1}{\epsilon^2} \log \frac{1}{\epsilon})$ which is independent of n and depends only on the desired relative error ϵ.

To prove that the above simple algorithm works, we will need to use the notion of ϵ-approximations. Given a range space $(V, \mathcal{R})$, where V is the ground set and $\mathcal{R}$ is a set of subsets of V, and a parameter $0 < \epsilon < 1$, an ϵ-approximation for $(V, \mathcal{R})$ is a subset $X \subseteq V$ such that for each range $R \in \mathcal{R}$,

$$\left| \frac{|R \cap X|}{|X|} - \frac{|R|}{|V|} \right| \leq \epsilon.$$

In other words, the relative size of R with respect to X does not differ from its relative size with respect to V by more than ϵ. It is known that if the VC dimension of the range space is d then ϵ-approximations of size $O(d/\epsilon^2 \log 1/\epsilon)$ exist. In fact any random sample of this size is an ϵ-approximation with high probability.

One way to exploit the existence of small ϵ-approximations in our case would be to set V to be the set of all $\binom{n}{2}$ edges defined by pairs of input points, and set $\mathcal{R}$ to be precisely those subsets of the edges that could be intersected by a ray. Such a subset has a finite VC dimension and hence we could pick a random subset of the edges and compute a point of high Ray-Shooting depth with respect to these edges. Unfortunately, since in the random sample we have not picked all edges spanned by a subset of the points, it is not clear whether such a point exists and how it can be computed. Fortunately, it can be shown that instead of picking a random subset of the edges, it suffices to pick a random subset of the points and consider all edges spanned by them. This can be done by using the notion of product range spaces.

The following definitions and results are from [5]. Given two range spaces $\Sigma_1 = (\mathcal{X}_1, \mathcal{R}_1)$ and $\Sigma_2 = (\mathcal{X}_2, \mathcal{R}_2)$, the *product range space* $\Sigma_1 \otimes \Sigma_2$ is defined as $(\mathcal{X}_1 \times \mathcal{X}_2, \mathcal{T})$ where $\mathcal{T}$ consists of all subsets $T \subseteq \mathcal{X}_1 \times \mathcal{X}_2$ such that each *cross-section* $T_{x_2}^1 = \{x \in \mathcal{X}_1 : (x, x_2) \in T\}$ is a set of $\mathcal{R}_1$ and each *cross-section* $T_{x_1}^2 = \{x \in \mathcal{X}_2 : (x_1, x) \in T\}$ is a set of $\mathcal{R}_2$. It is shown in [5] that if A_1 is an ϵ_1-approximation for Σ_1 and A_2 is an ϵ_2-approximation for Σ_2, then $A_1 \times A_2$ is an $\epsilon_1 + \epsilon_2$ approximation for $\Sigma_1 \otimes \Sigma_2$.

For any input point $p \in P$ and any ray ρ emanating from a point $q \in \mathbb{R}^2$, the ray ρ and the line l through p and q define two closed cones (i.e., those contained in the halfspace defined by l that contains ρ). Denote by $C_{p,\rho}$ the cone that does not contain p. Let $\Sigma = (P, \{C_{p,\rho} \cap P : p \in P, \rho \in \mathbb{R}^2 \times \mathbb{S}^1\})$ be the range space obtained by intersecting P with such cones.

For any ray ρ emanating from some point q, let E_ρ be the set of edges spanned by P that ρ intersects. Let $\mathcal{T} = (P \times P, \{E_\rho : \rho \in \mathbb{R}^2 \times \mathbb{S}^1\})$ be the range space in which the ground set is the set of edges spanned by P and each range consists of the set of edges intersected by a ray.

It can be checked that $\mathcal{T}$ is the product range space $\Sigma \otimes \Sigma$. Since Σ has a finite VC-dimension, picking a sample Q in which every point in P is chosen with probability $p = O(\frac{1}{\epsilon^2} \log \frac{1}{\epsilon} \cdot \frac{1}{n})$ gives an $\epsilon/2$ approximation for Σ. Therefore, $Q \times Q$ gives an ϵ approximation for $\mathcal{T}$. This justifies the algorithm described before. The running time of that algorithm is $O(n + k^2 \log^2 k)$, where $k = pn = |Q|$. It takes $O(n)$ time to pick the sample Q. Once we have picked the sample, we run our previous algorithm on the sample Q. Thus we have the following theorem:

Theorem 2. *Given a set P of n points in the plane, a point z of Ray-Shooting depth at least $(1-\epsilon)n^2/9$ with respect to P can be computed in $O((1/\epsilon \log 1/\epsilon)^4 + n)$.*

Setting $\epsilon = \log n/n^{1/4}$ we see that a point of Ray Shooting depth at least $n^2/9 - n^{7/4} \log n$ can be computed in linear time.

Once we have chosen a sample Q as above so that $Q \times Q$ is an ϵ-approximation for our product range space, we can also use it to compute a point of approximately the highest Ray-Shooting depth. This can be done by considering the set of lines $\mathcal{L}$ defined by pairs of points in Q and computing the cell of highest depth in their arrangement $\mathcal{A}$. A brute force way is to separately compute the depth of each of the cells in $\mathcal{A}$. However, as we have seen in the proof of Lemma 5, the depth of all cells crossed by a line can be computed incrementally in $O(k^2 \log k)$ time where $k = |Q|$. For each line $l \in \mathcal{L}$, we take two lines which are very close and parallel to l, one on each side of l. These lines together cross all cells of $\mathcal{A}$. We can therefore compute the cell of highest depth along each of these lines and then take the overall highest. This takes $O(k^4 \log k)$ time. Since picking Q takes $O(n)$ time, and $k = O(1/\epsilon^2 \log 1/\epsilon)$, we have:

Theorem 3. *Given a set P of n points in the plane, a point z of Ray-Shooting depth at least $(1-\epsilon)d$, where d is the maximum Ray-Shooting depth of any point with respect to P, can be computed in $O(1/\epsilon^8 \log^5 1/\epsilon + n)$ time.*

4   Implementation

We have implemented the sampling-based algorithm from Theorem 3 to approximately find the point of highest RS depth. This has been integrated as a package in the popular statistical computing software R, and is available here:

http://cran.r-project.org/web/packages/rsdepth/.

Acknowledgements. The first author would like to thank Janos Pach for several enlightening discussions. The second author is thankful to Hans Raj Tiwary and Ujjyini Singh Ray for important ideas in the paper.

References

1. Basit, A., Mustafa, N.H., Ray, S., Raza, S.: Centerpoints and tverberg's technique. Computational Geometry: Theory and Applications 43(7), 593–600 (2010)
2. Basit, A., Mustafa, N.H., Ray, S., Raza, S.: Hitting simplices with points in $\mathbb{R}^3$. Discrete & Computational Geometry 44(3), 637–644 (2010)
3. Boros, E., Füredi, Z.: The number of triangles covering the center of an n-set. Geometriae Dedicata 17(1), 69–77 (1984)
4. Chan, T.M.: An optimal randomized algorithm for maximum tukey depth. In: SODA, pp. 430–436 (2004)
5. Chazelle, B.: The Discrepancy Method. Cambridge University Press, Cambridge (2000)

6. Cole, R., Salowe, J.S., Steiger, W.L., Szemerédi, E.: An optimal-time algorithm for slope selection. SIAM J. Comput. 18(4), 792–810 (1989)

7. Fox, J., Gromov, M., Lafforgue, V., Naor, A., Pach, J.: Overlap properties of geometric expanders. In: SODA (2011)

8. Jadhav, S., Mukhopadhyay, A.: Computing a centerpoint of a finite planar set of points in linear time. In: Symposium on Computational Geometry, pp. 83–90 (1993)

9. Langerman, S., Steiger, W.L.: Optimization in arrangements. In: Alt, H., Habib, M. (eds.) STACS 2003. LNCS, vol. 2607, pp. 50–61. Springer, Heidelberg (2003)

10. Liu, R.: A notion of data depth based upon random simplices. The Annals of Statistics 18, 405–414 (1990)

11. Mustafa, N.H., Ray, S.: An optimal extension of the centerpoint theorem. Computational Geometry: Theory and Applications 42(7), 505–510 (2009)

12. Oja, H.: Descriptive statistics for multivariate distributions. Statistics and Probability Letters 1, 327–332 (1983)

13. Rado, R.: A theorem on general measure. J. London. Math. Soc. 21, 291–300 (1947)

14. Rousseeuw, P.J., Ruts, I.: Constructing the bivariate tukey median. Statistica Sinica 8, 827–839 (1998)

15. Tukey, J.: Mathematics and the picturing of data. In: Proc. of the International Congress of Mathematicians, pp. 523–531 (1975)

16. Vardi, Y., Zhang, C.: The multivariate L1-median and associated data depth. Proceedings of the National Academy of Sciences of the United States of America 97(4), 14–23 (2000)

Improved Algorithms for Partial Curve Matching*

Anil Maheshwari[1], Jörg-Rüdiger Sack[1], Kaveh Shahbaz[1],
and Hamid Zarrabi-Zadeh[1,2]

[1] School of Computer Science, Carleton University, Ottawa,
Ontario K1S 5B6, Canada
{anil,sack,kshahbaz,zarrabi}@scs.carleton.ca
[2] Department of Computer Engineering, Sharif University of Technology,
Tehran, Iran
zarrabi@ce.sharif.edu

Abstract. Back in 1995, Alt and Godau gave an efficient algorithm for
deciding whether a given curve resembles some *part* of a larger curve under a fixed Fréchet distance, achieving a running time of $O(nm \log(nm))$,
for n and m being the number of segments in the two curves, respectively.
We improve this long-standing result by presenting an algorithm that
solves this decision problem in $O(nm)$ time. Our solution is based on
constructing a simple data structure which we call *free-space map*. Using
this data structure, we obtain improved algorithms for several variants
of the Fréchet distance problem, including the Fréchet distance between
two closed curves, and the so-called minimum/maximum walk problems.
We also improve the map matching algorithm of Alt *et al.* for the case
when the map is a directed acyclic graph.

1 Introduction

The Fréchet distance is a widely-used metric for measuring the similarity of
the curves. It finds applications in morphing [8], handwriting recognition [12],
protein structure alignment [9], etc. This measure is often illustrated as the
minimum-length leash needed for a person to walk a dog, while each of them is
traversing a pre-specified polygonal curve without backtracking.

Alt and Godau [3] showed how the Fréchet distance between two polygonal
curves with n and m vertices can be computed in $O(nm \log(nm))$ time. For their
solution, they introduced a data structure, called *free-space diagram*. The free-
space diagram and its variants have been proved to be useful in other applications
involving the Fréchet distance (see e.g. [2,4,7]).

In their seminal work, Alt and Godau [3] also studied a *partial curve matching*
problem in which one wants to see if a given curve resembles some "part" of a
larger curve. Given two polygonal curves P and Q of size n and m, respectively,
they presented an algorithm that decides in $O(nm \log(nm))$ time whether there

---

* Research supported by NSERC, HPCVL, and SUN Microsystems.

C. Demetrescu and M.M. Halldórsson (Eds.): ESA 2011, LNCS 6942, pp. 518–529, 2011.

is a subcurve R of P whose Fréchet distance to Q is at most ε, for a given $\varepsilon \geqslant 0$. Using parametric search, they solved the optimization problem of finding the minimum such ε in $O(nm \log^2(nm))$ time.

Later, Alt *et al.* [2] proposed a generalization of the partial curve matching problem to measure the similarity of a curve to some part of a graph. Given a polygonal curve P and a graph G, they presented an $O(nm \log m)$-time algorithm to decide whether there is a path π in G whose Fréchet distance to P is at most ε, where n is the size of P and m is the complexity of G. A variant of the partial curve matching in the presence of outliers is studied by Buchin *et al.* [6], leading to an algorithm with $O(nm(n + m) \log(nm))$ running time.

Our results. In this paper, we present a simple data structure, which we call *free-space map*, that enables us to solve several variants of the Fréchet distance problem efficiently. The results we obtain using this data structure are summarized below. In the following, n and m represent the size of the two given polygonal curves P and Q, respectively, and $\varepsilon \geqslant 0$ is a fixed input parameter.

- *Partial curve matching.* Given two polygonal curves P and Q, we present an algorithm to decide in $O(nm)$ time whether there is a subcurve $R \subseteq P$ whose Fréchet distance to Q is at most ε. This improves the best previous algorithm for this decision problem due to Alt and Godau [3], that requires $O(nm \log(nm))$ time. This also leads to an $O(\log(nm))$ faster algorithm for solving the optimization version of the problem, using parametric search.

- *Closed Fréchet metric.* Alt and Godau [3] showed that for two closed curves P and Q, the decision problem of whether the closed Fréchet distance between P and Q is at most ε can be solved in $O(nm \log(nm))$ time. We improve this long-standing result by giving an algorithm that runs in $O(nm)$ time. As a result, we also improve by a $\log(nm)$-factor the running time of the optimization algorithm for computing the minimum such ε.

- *Maximum walk.* Given two curves P and Q and a fixed $\varepsilon \geqslant 0$, the *maximum walk* problem asks for the maximum-length subcurve of Q whose Fréchet distance to P is at most ε. We show that this optimization problem can be solved efficiently in $O(nm)$ time, without additional $\log(nm)$ factors. The *minimum walk* problem is analogously defined, and can be solved by an extension of the free-space map, as described in the full version of this paper.

- *Graph matching.* Given a directed acyclic graph G, we present an algorithm to decide in $O(nm)$ time whether a curve P matches some part of G under a Fréchet distance of ε, for n and m being the size of P and the complexity of G, respectively. This improves the map matching algorithm of Alt *et al.* [2] for the case of DAGs. Note that Alt *et al.*'s algorithm has a running time of $\Theta(nm \log m)$ in the worst case even if the input graph is a DAG or a simple path.

The above improved results are obtained using a novel simple approach for propagating the reachability information "sequentially" from bottom side to the top side of the free-space diagram. Our approach is different from and simpler than the divide-and-conquer approach used by Alt and Godau [3], and also, the approach taken by Alt *et al.* [2] which is a mixture of line sweep, dynamic programming, and Dijsktra's algorithm.

The free-space map introduced in this paper encapsulates all the information available in the standard free-space diagram, yet it is capable of answering a more general type of queries efficiently. Namely, for any query point on the bottom side of the free-space diagram, our data structure can efficiently report all points on the top side of the diagram which are reachable from that query point. Given that our data structure has the same size and construction time as the standard free-space diagram, it can be viewed as a powerful alternative.

The current lower bound for deciding whether the Fréchet distance between two polygonal curves with total n vertices is at most a given value ε, is $\Omega(n \log(n))$ [5]. However, no subquadratic algorithm is known for this decision problem, and hence, one might conjecture that the problem may be 3SUM-hard (see [1]). If this conjecture holds, then the results obtained in this paper do not only represent improvements, but are also optimal.

The remainder of the paper is organized as follows. In Section 3, we define the free-space map and show how it can be efficiently constructed. In Section 4, we present some applications of the free-space map to problems such as partial curve matching, maximum walk, and closed Fréchet metric. In Section 5, we provide an improved algorithm for matching a curve in a DAG. We conclude in Section 6 with some open problems.

2 Preliminaries

A *polygonal curve* in $\mathbb{R}^d$ is a continuous function $P : [0, n] \to \mathbb{R}^d$ such that for each $i \in \{1, \ldots, n\}$, the restriction of P to the interval $[i-1, i]$ is affine (i.e., forms a line segment). The integer n is called the *size* of P. For each $i \in \{1, \ldots, n\}$, we denote the line segment $P|_{[i-1,i]}$ by P_i.

A *monotone reparametrization* of $[0, n]$ is a continuous non-decreasing function $\alpha : [0, 1] \to [0, n]$ with $\alpha(0) = 0$ and $\alpha(1) = n$. Given two polygonal curves P and Q of size n and m, respectively, the *Fréchet distance* between P and Q is defined as

$$\delta_F(P, Q) = \inf_{\alpha, \beta} \max_{t \in [0,1]} \|P(\alpha(t)), Q(\beta(t))\|,$$

where $\|\cdot\|$ denotes the Euclidean metric, and α and β range over all monotone reparameterizations of $[0, n]$ and $[0, m]$, respectively. Given a parameter $\varepsilon \geqslant 0$, the *free space* of the two curves P and Q is defined as

$$\mathcal{F}_\varepsilon(P, Q) = \{(s, t) \in [0, n] \times [0, m] \mid \|P(s), Q(t)\| \leqslant \varepsilon\}.$$

We call points in $\mathcal{F}_\varepsilon(P, Q)$ *feasible*. The partition of the rectangle $[0, n] \times [0, m]$ into regions formed by feasible and infeasible points is called the *free-space diagram* of P and Q, denoted by $\mathcal{FD}_\varepsilon(P, Q)$ (see Figure 1.a). For $0 \leqslant j \leqslant m$, we denote by $\mathcal{FD}_j$ the one-dimensional free-space diagram $\mathcal{FD}_\varepsilon(P, Q) \cap ([0, n] \times \{j\})$, corresponding to the curve P and the point $Q(j)$. For each $(i, j) \in \{1 \cdots n\} \times$

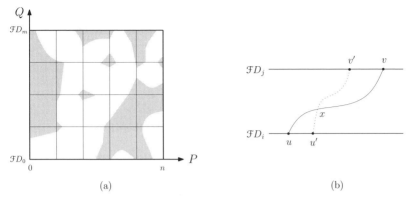

Fig. 1. (a) An example of a free-space diagram. (b) Proof of the crossing lemma.

$\{1\cdots m\}$, the intersection of the free space diagram with the square $[i-1,i]\times$ $[j-1,j]$ is called a *cell* of the diagram. Likewise, we call the intersection of $\mathcal{FD}_j$ with each interval $[i-1,i]$ a cell (or more precisely, the i-th cell) of $\mathcal{FD}_j$.

For an interval I on the line, we denote by left(I) and right(I) the left and the right endpoint of I, respectively. Given two points a and b in the plane, we write $a < b$ if $a_x < b_x$. The following simple lemma serves as a building block in our algorithm.

Lemma 1. *Given two sequences A and B of points on a line sorted from left to right, we can preprocess the two sequences into a data structure of size $O(|A|)$ in $O(|A|+|B|)$ time, such that for any query point $a \in A$, the leftmost point $b \in B$ with $a < b$ can be reported in $O(1)$ time.*

3   The Data Structure

Throughout this section, let P and Q be two polygonal curves of size n and m, respectively, and $\varepsilon \geqslant 0$ be a fixed parameter. We call a curve *feasible* if it lies completely within $\mathcal{F}_\varepsilon(P,Q)$. We call the curve *monotone* if it is monotone in both x- and y-coordinates. Alt and Godau [3] showed that $\delta_F(P,Q) \leqslant \varepsilon$ if and only if there is a monotone feasible curve in $\mathcal{F}_\varepsilon(P,Q)$ from $(0,0)$ to (n,m).

For $0 \leqslant j \leqslant m$, let $\mathcal{F}_j$ be the set of feasible points in $\mathcal{FD}_j$. $\mathcal{F}_j$ consists of $O(n)$ feasible intervals, where each feasible interval is a maximal continuous set of feasible points, restricted to be within a cell.

Given two points u and v in the free space, we say that v is *reachable* from u, denoted by $u \leadsto v$, if there is a monotone feasible curve in $\mathcal{F}_\varepsilon(P,Q)$ from u to v. Clearly, reachability is "transitive": if $u \leadsto v$ and $v \leadsto w$, then $u \leadsto w$.

Lemma 2 (Crossing Lemma). *Let $u, u' \in \mathcal{F}_i$ and $v, v' \in \mathcal{F}_j$ $(i < j)$ such that $u \leqslant u'$ and $v' \leqslant v$. If $u \leadsto v$ and $u' \leadsto v'$, then $u \leadsto v'$ and $u' \leadsto v$.*

Proof. Let π be a monotone feasible curve that connects u to v. Since u' and v' are in different sides of π, any monotone curve that connects u' to v' in $\mathcal{F}_\varepsilon(P,Q)$

intersects π at some point x (see Figure 1.b). The concatenation of the subcurve from u to x and the one from x to v' gives a monotone feasible curve from u to v'. Similarly, v is connected to u' by a monotone feasible curve through x.   □

Let S be a set of points in $\mathcal{F}_\varepsilon(P, Q)$. For $0 \leqslant j \leqslant m$, we define

$$\mathcal{R}_j(S) := \{v \in \mathcal{F}_j \mid \exists u \in S \text{ s.t. } u \rightsquigarrow v\}.$$

For an interval I on $\mathcal{F}_i$ with $i \leqslant j$, we define the *left pointer* of I on $\mathcal{F}_j$, denoted by $\ell_j(I)$, to be the leftmost point in $\mathcal{R}_j(I)$. Similarly, the *right pointer* of I on $\mathcal{F}_j$, denoted by $r_j(I)$, is defined to be the rightmost point in $\mathcal{R}_j(I)$. If $\mathcal{R}_j(I)$ is empty, both pointers $\ell_j(I)$ and $r_j(I)$ are set to null. Such pointers have been previously used in [2,3]. For a single point u, we use $\mathcal{R}_j(u)$, $\ell_j(u)$, and $r_j(u)$ instead of $\mathcal{R}_j(\{u\})$, $\ell_j(\{u\})$, and $r_j(\{u\})$, respectively.

For $0 \leqslant j \leqslant m$, we define the *reachable set* $\mathcal{R}(j) := \mathcal{R}_j(\mathcal{F}_0)$ to be the set of all points in $\mathcal{F}_j$ reachable from $\mathcal{F}_0$. We call each interval of $\mathcal{R}(j)$, contained in a feasible interval of $\mathcal{F}_j$, a *reachable interval*. It is clear by our definition that $\mathcal{R}(0) = \mathcal{F}_0$.

Observation 1. *For $0 \leqslant i \leqslant j \leqslant m$, we have $\mathcal{R}(j) = \mathcal{R}_j(\mathcal{R}(i))$.*

The following lemma describes an important property of reachable sets.

Lemma 3. *For any two indices i, j ($0 \leqslant i \leqslant j \leqslant m$) and any point $u \in \mathcal{R}(i)$, $\mathcal{R}_j(u) = \mathcal{R}(j) \cap [\ell_j(u), r_j(u)]$.*

Proof. Let $S = [\ell_j(u), r_j(u)]$. By Observation 1, $\mathcal{R}(j) = \mathcal{R}_j(\mathcal{R}(i))$. Thus, it is clear by the definition of pointers that $\mathcal{R}_j(u) \subseteq \mathcal{R}(j) \cap S$. Therefore, it remains to show that $\mathcal{R}(j) \cap S \subseteq \mathcal{R}_j(u)$. Suppose, by way of contradiction, that there is a point $v \in \mathcal{R}(j) \cap S$ such that $v \notin \mathcal{R}_j(u)$. Since $v \in \mathcal{R}(j)$, there exists some point $u' \in \mathcal{R}(i)$ such that $u' \rightsquigarrow v$. If u' is to the left (resp., to the right) of u, then the points u, u', v, and $\ell_j(u)$ (resp., $r_j(u)$) satisfy the conditions of Lemma 2. Therefore, by Lemma 2, $u \rightsquigarrow v$, which implies that $v \in \mathcal{R}_j(u)$; a contradiction.   □

Lemma 3 provides an efficient way for storing the reachable sets $\mathcal{R}_j(I)$, for all feasible intervals I on $\mathcal{F}_0$: instead of storing each reachable set $\mathcal{R}_j(I)$ separately, one set per feasible interval I, which takes up to $\Theta(n^2)$ space, we only need to store a single set $\mathcal{R}(j)$, along with the pointers $\ell_j(I)$ and $r_j(I)$ which takes only $O(n)$ space in total. The reachable set $\mathcal{R}_j(I)$, for each interval I on $\mathcal{F}_0$, can be then obtained by $\mathcal{R}(j) \cap [\ell_j(I), r_j(I)]$. For each interval I on $\mathcal{F}_0$, we call the set $\{\ell_j(I), r_j(I)\}$ a *compact representation* of $\mathcal{R}_j(I)$.

Lemma 4. *For $0 < j \leqslant m$, if $\mathcal{R}(j-1)$ is given, then $\mathcal{R}(j)$ can be computed in $O(n)$ time.*

Proof. Let $\mathcal{D}$ be the restriction of the free space diagram to the rectangle $[1, n] \times [j-1, j]$. Alt *et al.* [2] showed that for all feasible intervals I on the bottom side of $\mathcal{D}$ (i.e., on $\mathcal{F}_{j-1}$), the left and the right pointers of I on the top side of $\mathcal{D}$ (i.e., on $\mathcal{F}_j$) can be computed using a series of linear scans in $O(n)$ time. Let $\mathcal{D}'$ be a copy of $\mathcal{D}$ in which all points in $\mathcal{F}_{j-1} \setminus \mathcal{R}(j-1)$ are marked infeasible. $\mathcal{D}'$ can be computed from $\mathcal{D}$ in $O(n)$ time. By running the algorithm of [2] on $\mathcal{D}'$, we obtain all pointers $\ell_j(I)$ and $r_j(I)$, for all reachable intervals I on $\mathcal{R}(j-1)$, in $O(n)$ total time. Now, we can produce $\mathcal{R}_j(\mathcal{R}(j-1))$ easily by identifying those (portions of) intervals on $\mathcal{F}_j$ that lie in at least one interval $[\ell_j(I), r_j(I)]$. Since for all intervals I on $\mathcal{R}(j-1)$ sorted from left to right, $\ell_j(I)$'s and $r_j(I)$'s are in sorted order (this is an easy corollary of the Lemma 2), we can accomplish this step by a linear scan over the left and right pointers in $O(n)$ time. The proof is complete, as $\mathcal{R}(j) = \mathcal{R}_j(\mathcal{R}(j-1))$ by Observation 1.   □

We now describe our main data structure, which we call *free-space map*. The data structure maintains reachability information on each row of the free-space diagram, using some additional pointers that help answering reachability queries efficiently. The free-space map of two curves P and Q consists of the following:

(i) the reachable sets $\mathcal{R}(j)$, for $0 \leqslant j \leqslant m$,
(ii) the right pointer $r_j(I)$ for each reachable interval I on $\mathcal{R}(j-1)$, $0 < j \leqslant m$,
(iii) the next reachable point for each cell in $\mathcal{FD}_j$, for $0 < j \leqslant m$, and
(iv) the previous take-off point for each cell in $\mathcal{FD}_j$, for $0 \leqslant j < m$,

where a *take-off* point on $\mathcal{FD}_j$ is a reachable point in $\mathcal{R}(j)$ from which a point on $\mathcal{FD}_{j+1}$ is reachable. For example, in Figure 2, ℓ_j is the next reachable point of ℓ', and r' is the previous take-off point of r_{j-1}.

Theorem 1. *Given two polygonal curves P and Q of sizes n and m, respectively, we can build the free-space map of P and Q in $O(nm)$ time.*

Proof. We start from $\mathcal{R}(0) = \mathcal{F}_0$, and construct each $\mathcal{R}(j)$ iteratively from $\mathcal{R}(j-1)$, for j from 1 to m, using Lemma 4. The total time needed for this step is $O(nm)$. The construction of $\mathcal{R}(j)$, as seen in the proof of Lemma 4, involves computing all right (and left) pointers, for all reachable intervals on $\mathcal{R}(j-1)$. Therefore, item (ii) of the data structure can be obtained at no additional cost. Item (iii) is computed as follows. Let S be the set of all left pointers obtained upon constructing $\mathcal{R}(j)$. For each cell c in $\mathcal{FD}_j$, the next reachable point of c, if any, is a member of S. We can therefore compute item (iii) for each row $\mathcal{FD}_j$ by a linear scan over the cells and the set S using Lemma 1 in $O(n)$ time. For each row, item (iv) can be computed analogous to item (iii), but in a reverse order. Namely, given the set $\mathcal{R}(j)$, we compute the set of points on $\mathcal{FD}_{j-1}$ reachable from $\mathcal{R}(j)$ in the free-space diagram rotated by 180 degrees. Let S be the set of all left pointers obtained in this reverse computation. For each cell c in $\mathcal{FD}_{j-1}$, the previous take-off point of c, if there is any, is a member of S. We can therefore compute item (iv) for each row by a linear scan over the cells and the set S using Lemma 1 in $O(n)$ time. The total time for constructing the free-space map is therefore $O(nm)$.   □

In the following, we show how the reachability queries can be efficiently answered, using the free-space map. For the sake of describing the query algorithm, we introduce two functions as follows. Given a point $u \in \mathcal{FD}_j$, we define

---

Algorithm 1. QUERY(u), where $u \in \mathcal{F}_0$

1: let $\ell_0 = r_0 = u$

2: **for** $j = 1$ to m **do**

3:     let ℓ' be the orthogonal projection of ℓ_{j-1} onto $\mathcal{FD}_j$

4:     $\ell_j \leftarrow$ LEFTMOST-REACHABLE(ℓ')

5:     let $r' =$ RIGHTMOST-TAKE-OFF(r_{j-1})

6:     **if** $r' < \ell_{j-1}$ or $r' =$ null **then**

7:         $r_j \leftarrow$ null

8:     **else**

9:         $r_j \leftarrow r_j(I)$, for I being the reachable interval containing r'

10:     **if** ℓ_j or r_j is null **then**

11:         return null

12: return ℓ_m, r_m

LEFTMOST-REACHABLE(u) to be the leftmost point on or after u on $\mathcal{FD}_j$. Analogously, we define RIGHTMOST-TAKE-OFF(u) to be the rightmost take-off point on or before u on $\mathcal{FD}_j$. Note that both these functions can be computed in $O(1)$ time using the pointers stored in the free-space map.

Theorem 2. *Let the free-space map of P and Q be given. Then, for any query point $u \in \mathcal{F}_0$, $\ell_m(u)$ and $r_m(u)$ can be computed in $O(m)$ time.*

Proof. The procedure for computing $\ell_m(u)$ and $r_m(u)$ for a query point $u \in \mathcal{F}_0$ is described in Algorithm 1. The following invariant holds during the execution of the algorithm: After the j-th iteration, $\ell_j = \ell_j(u)$ and $r_j = r_j(u)$. We prove this by induction on j. The base case, $\ell_0 = r_0 = u$, trivially holds. Now, suppose inductively that $\ell_{j-1} = \ell_{j-1}(u)$ and $r_{j-1} = r_{j-1}(u)$. We show that after the j-th iteration, the invariant holds for j. We assume, w.l.o.g., that $\mathcal{R}_j(u)$ is non-empty, i.e., $\ell_j(u) \leqslant r_j(u)$. Otherwise, the last take-off point from $\mathcal{R}(j-1)$ will be either null, or smaller than r_{j-1}, which is then detected and handled by lines 6–7.

We first show that $\ell_j = \ell_j(u)$. Suppose by contradiction that $\ell_j \neq \ell_j(u)$. If $\ell_j < \ell_j(u)$, then we draw a vertical line from ℓ_j to $\mathcal{FD}_{j-1}$ (see Figure 2). This line crosses any monotone path from $\ell_{j-1} = \ell_{j-1}(u)$ to $\ell_j(u)$ at a point x. The line segment $x\ell_j$ is completely in the free space, because otherwise, it must be cut by an obstacle, which contradicts the fact that the free space inside a cell

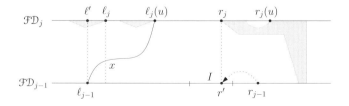

Fig. 2. Proof of Theorem 2

is convex. But then, ℓ_j becomes reachable from ℓ_{j-1} through x, contradicting the fact that $\ell_j(u)$ is the leftmost reachable point in $\mathcal{R}(j)$. The case, $\ell_j > \ell_j(u)$, cannot arise, because then, $\ell_j(u)$ is a reachable point after ℓ' and before ℓ_j, which contradicts our selection of ℓ_j as the leftmost reachable point of ℓ' in line 4.

We can similarly show that $r_j = r_j(u)$. Suppose by contradiction that $r_j \neq r_j(u)$. The case $r_j > r_j(u)$ is impossible, because then, r_j is a point on $\mathcal{R}(j)$ reachable from $\mathcal{R}(j-1)$ which is after $r_j(u)$. This contradicts the fact that $r_j(u)$ is the rightmost point on $\mathcal{R}(j)$. If $r_j < r_j(u)$ (see Figure 2), then $r_j(u)$ is reachable from a point $x \in \mathcal{R}(j-1)$ with $x < r'$, because r' is the rightmost take-off point on or before r_{j-1}. But then, by Lemma 2, $r_j(u)$ is reachable from r', which contradicts the fact that r_j is the left pointer of the reachable interval I containing r'. $\square$

Theorem 3. *Given two polygonal curves P and Q of size n and m, respectively, we can build in $O(nm)$ time a data structure of size $O(nm)$, such that for any query point $u \in \mathcal{F}_0$, a compact representation of $\mathcal{R}_m(u)$ can be reported in $O(m)$ time.*

Remark. The query time in Theorem 3 can be improved to $O(\log m)$, as shown in the full version of this paper. However, we only use the $O(m)$ query time for the applications provided in the next section.

4   Applications

In this section, we provide some of the applications of the free-space map.

Partial Curve Matching.   Given two polygonal curves P and Q, and an $\varepsilon \geqslant 0$, the partial curve matching problem involves deciding whether there exists a subcurve $R \subseteq P$ such that $\delta_F(R, Q) \leqslant \varepsilon$. As noted in [3], this is equivalent to deciding whether there exists a monotone path in the free space from $\mathcal{F}\mathcal{D}_0$ to $\mathcal{F}\mathcal{D}_m$. This decision problem can be efficiently solved using the free-space map. For each feasible intervals I on $\mathcal{F}\mathcal{D}_0$, we compute a compact representation of $\mathcal{R}_m(\text{left}(I))$ using Theorem 3 in $O(m)$ time. Observe that $\mathcal{R}_m(I) = \emptyset$ if and only if $\mathcal{R}_m(\text{left}(I)) = \emptyset$. Therefore, we can decide in $O(nm)$ time whether there exists a point on $\mathcal{F}\mathcal{D}_m$ reachable from $\mathcal{F}\mathcal{D}_0$. Furthermore, we can use parametric search as in [3] to find the smallest ε for which the answer to the above decision problem is "yes" in $O(nm \log(nm))$ time. Therefore, we obtain:

Theorem 4. *Given two polygonal curves P and Q of size n and m, respectively, we can decide in $O(nm)$ time whether there exists a subcurve $R \subseteq P$ such that $\delta_F(R, Q) \leqslant \varepsilon$, for a given $\varepsilon \geqslant 0$. A subcurve $R \subseteq P$ minimizing $\delta_F(R, Q)$ can be computed in $O(nm \log(nm))$ time.*

Closed Curves.   An important variant of the Fréchet metric considered by Alt and Godau [3] is the following. Given two closed curves P and Q, define

$$\delta_C(P, Q) = \inf_{s_1, s_2 \in \mathbb{R}} \delta_F(R \text{ shifted by } s_1, Q \text{ shifted by } s_2)$$

to be the *closed Fréchet metric* between P and Q. This metric is of significant importance for comparing shapes.

Consider a diagram $\mathcal{D}$ of size $2n \times m$ obtained from concatenating two copies of the standard free-space diagram of P and Q. Alt and Godau showed that $\delta_C(P, Q) \leqslant \varepsilon$ if and only if there exists a monotone feasible path in $\mathcal{D}$ from $(t, 0)$ to $(n + t, m)$, for a value $t \in [0, n]$. We show how such a value t, if there is any, can be found efficiently using a free-space map built on top of $\mathcal{D}$.

Observation 2. *Let i be a fixed integer $(0 < i \leqslant n)$, $I = [a, b]$ be the feasible interval on the i-th cell of $\mathcal{FD}_0$, and $J = [c, d]$ be the feasible interval on the $(i + n)$-th cell of $\mathcal{FD}_m$. Then there is a value $t \in [i - 1, i]$ with $(t, 0) \rightsquigarrow (n + t, m)$ if and only if $\max((\ell_m(I))_x, c) \leqslant b + n$ and $\min((r_m(I))_x, d) \geqslant a + n$.*

We iterate on i from 1 to n, and check for each i if a desired value $t \in [i - 1, i]$ exists using Observation 2. Each iteration involves the computation of $\ell_m(I)$ and $r_m(I)$ that can be done in $O(m)$ time using the free-space map. The total time is therefore $O(nm)$.

Theorem 5. *Given two closed polygonal curves P and Q of size n and m, respectively, we can decide in $O(nm)$ time whether $\delta_C(P, Q) \leqslant \varepsilon$, for a given $\varepsilon \geqslant 0$. Furthermore, $\delta_C(P, Q)$ can be computed in $O(nm \log(nm))$ time.*

Maximum Walk. An interesting variant of the Fréchet distance problem is the following: Given two curves P and Q and a fixed $\varepsilon \geqslant 0$, find a maximum-length subcurve of Q whose Fréchet distance to P does not exceed ε. In the dog-person illustration, this problem corresponds to finding the best starting point on P such that when the person walks the whole curve Q, his or her dog can walk the maximum length on P, without exceeding a leash of length ε. We show that this optimization problem can be solved efficiently in $O(nm)$ time using the free space map. The following observation is the main ingredient.

Observation 3. *Let R be a maximum-length subcurve of P such that $\delta_F(R, Q) \leqslant \varepsilon$. The starting point of R corresponds to the left endpoint of a feasible interval I on $\mathcal{FD}_0$, and its ending point corresponds to $r_m(I)$.*

By Observation 3, we only need to test n feasible intervals on $\mathcal{FD}_0$, and their right pointer on $\mathcal{FD}_m$ to find the best subcurve R. Note that, given the two endpoints of a subcurve R on the free-space map, finding the actual length of R on the original curve P can take up to $O(n)$ time. To speed up the length computation step, we can use a table lookup method used by the authors in [10] to answer each length query in $O(1)$ time, after $O(n)$ preprocessing. The total time for computing the maximum-length subcurve R will be therefore $O(nm)$.

Theorem 6. *Given two polygonal curves P and Q of size n and m, respectively, and a parameter $\varepsilon \geqslant 0$, we can find in $O(nm)$ time a maximum-length subcurve $R \subseteq P$ such that $\delta_F(R, Q) \leqslant \varepsilon$.*

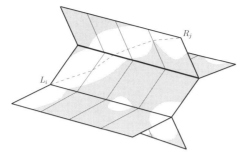

Fig. 3. An example of a free-space surface

5   Matching a Curve in a DAG

Let P be a polygonal curve of size n, and G be a connected graph with m edges. Alt *et al.* [2] presented an $O(nm \log m)$-time algorithm to decide whether there is a path π in G with Fréchet distance at most ε to P, for a given $\varepsilon \geqslant 0$. In this section, we improve this result for the case when G is a directed acyclic graph (DAG), by giving an algorithm that runs in only $O(nm)$ time. The idea is to use a sequential reachability propagation approach similar to the one used in Section 3. Our approach is structurally different from the one used by Alt *et al.* [2]. In particular, their algorithm has $\Theta(nm \log m)$ running time in the worst case even if the input graph is a DAG or a simple path.

We first borrow some notation from [2]. Let $G = (V, E)$ be a connected DAG with m edges, such that $V = \{1, \ldots, \nu\}$ corresponds to points $\{v_1, \ldots, v_\nu\} \subseteq \mathbb{R}^2$, for $\nu \leqslant m+1$. We assume, w.l.o.g., that the elements of V are numbered according to a topological ordering of the vertices of G. Such a topological ordering can be computed in $O(m)$ time. We embed each edge $(i, j) \in E$ as an oriented line segment s_{ij} from v_i to v_j. Each s_{ij} is continuously parametrized by values in $[0, 1]$ according to its natural parametrization, namely, $s_{ij} : [0, 1] \to \mathbb{R}^2$.

For each vertex $j \in V$, let $\mathcal{FD}_j := \mathcal{FD}_\varepsilon(P, v_j)$ be the one-dimensional free-space diagram corresponding to the path P and the vertex j. We denote by L_j and R_j the left endpoint and the right endpoint of $\mathcal{FD}_j$, respectively. Moreover, we denote by $\mathcal{F}_j$ the set of feasible points on $\mathcal{FD}_j$. For each $(i, j) \in E$, let $\mathcal{FD}_{ij} := \mathcal{FD}_\varepsilon(P, s_{ij})$ be a two-dimensional free-space diagram, which consists of a row of n cells. We glue together the two-dimensional free-space diagrams according to the adjacency information of G, as shown in Figure 3. The resulting structure is called the *free-space surface* of P and G, denoted by $\mathcal{FS}_\varepsilon(P, G)$. We denote the set of feasible points in $\mathcal{FS}_\varepsilon(P, G)$ by $\mathcal{F}_\varepsilon(P, G)$.

Given two points $u, v \in \mathcal{F}_\varepsilon(P, G)$, we say that v is *reachable* from u, denoted by $u \leadsto v$, if there is a monotone feasible curve from u to v in $\mathcal{F}_\varepsilon(P, G)$, where monotonicity in each cell of the surface is with respect to the orientation of the edges of P and G defining that cell. Given a set of points $S \subseteq \mathcal{F}_\varepsilon(P, G)$, we define $\mathcal{R}_j(S) := \{v \in \mathcal{F}_j \mid \exists u \in S \text{ s.t. } u \leadsto v\}$. Let $\mathbb{L} = \cup_{j \in V}(L_j \cap F_j)$. For each $j \in V$, we define the *reachable set* $\mathcal{R}(j) := \mathcal{R}_j(\mathbb{L})$. Observe that there is a path π in G with $\delta_F(P, \pi) \leqslant \varepsilon$ if and only if there is a vertex $j \in V$ with $R_j \in \mathcal{R}(j)$.

Algorithm 2. DECISION-DAG-MATCHING(P, G, ε)

1: **for each** $j \in V$ in a topological order **do**
2:      $\mathcal{R}(j) \leftarrow \mathcal{R}_j(L_j \cap \mathcal{F}_j) \cup (\cup_{(i,j) \in E} \mathcal{R}_j(\mathcal{R}(i)))$
3: let $S = \cup_{j \in V} (R_j \cap \mathcal{R}(j))$
4: return TRUE if $S \neq \emptyset$, otherwise return FALSE

Theorem 7. *Given a polygonal curve P of size n and a directed acyclic graph G of size m, we can decide in $O(nm)$ time whether there is a path π in G with $\delta_F(P, \pi) \leqslant \varepsilon$, for a given $\varepsilon \geqslant 0$. A path π in G minimizing $\delta_F(P, \pi)$ can be computed in $O(nm \log(nm))$ time.*

Proof. Algorithm 2 computes, for each vertex $j \in V$, the reachable set $\mathcal{R}(j)$ in a topological order. It then returns true only if there is a vertex $j \in V$ such that R_j is reachable which indicates the existence of a path π in G with $\delta_F(P, \pi) \leqslant \varepsilon$. To prove the correctness, we only need to show that for every vertex $j \in V$, the algorithm computes $\mathcal{R}(j)$ correctly. We prove this by induction on j. Suppose by induction that the set $\mathcal{R}(i)$ for all $i < j$ is computed correctly. Now consider a point $u \in F_j$. If $u \in \mathcal{R}(j)$, then there exists a vertex $k < j$ such that L_k is connected to u by a monotone feasible curve $\mathcal{C}$ in $\mathcal{FS}_\varepsilon(P, G)$. If $k = j$, then $u \in \mathcal{R}(j)$ because $\mathcal{R}_j(L_j \cap \mathcal{F}_j)$ is added to $\mathcal{R}(j)$ in line 2. If $k < j$, then the curve $\mathcal{C}$ must pass through a vertex i with $(i, j) \in E$. Since the vertices of V are sorted in a topological order, we have $i < j$, and hence, $\mathcal{R}(i)$ is computed correctly by the induction hypothesis. Hence, letting $x = \mathcal{C} \cap \mathcal{F}_i$, we have $x \in \mathcal{R}(i)$. Furthermore, we know that x is connected to u using the curve $\mathcal{C}$. Therefore, the point u is in $\mathcal{R}_j(\mathcal{R}(i))$, and hence, is added to $\mathcal{R}(j)$ in line 2. Similarly, we can show that if $u \notin \mathcal{R}(j)$, then u is not added to $\mathcal{R}(j)$ by the algorithm. Suppose by contradiction that u is added to $\mathcal{R}(j)$ in line 2. Then either $u \in \mathcal{R}_j(L_j \cap \mathcal{F}_j)$ or $u \in \mathcal{R}_j(\mathcal{R}(i))$, for some $i < j$. But by the definition of reachability, both cases imply that u is reachable from a point in $\mathbb{L}$, which is a contradiction.

For the time complexity, note that each $\mathcal{R}_j(\mathcal{R}(i))$ in line 2 can be computed in $O(n)$ time using Lemma 4. Moreover, $\mathcal{R}_j(L_j \cap \mathcal{F}_j)$, for each $j \in V$, can be computed by finding the largest feasible interval on F_j containing L_j in $O(n)$ time. Therefore, processing each edge (i, j) takes $O(n)$ time, and hence, the whole computation takes $O(nm)$ time. Once the algorithm finds a reachable left endpoint v, we can construct a feasible monotone path connecting a right endpoint $u \in \mathbb{L}$ to v by keeping, for each reachable interval I on $R(j)$, a back pointer to a reachable interval J on $R(i)$, $(i, j) \in E$, from which I is reachable. The path $u \rightsquigarrow v$ can be constructed by following the back pointers from v to u, in $O(m)$ time. For the optimization problem, we use parametric search as in [2,3], to find the value of $\delta_F(P, \pi)$ by an extra $\log(nm)$-factor, namely, in $O(nm \log(nm))$ time. $\square$

Remark. As noted in [2], it is straight-forward to modify the algorithm to allow paths in G to start and end anywhere inside edges of the graph, not necessarily at the vertices. This can be easily done by allowing the feasible path found by our algorithm to start and end at any feasible point on the left and right boundary of $\mathcal{FD}_{ij}$, for each edge $(i,j) \in E$.

6   Conclusions

In this paper, we presented improved algorithms for some variants of the Fréchet distance problem, including partial curve matching, closed Fréchet distance, maximum walk, and matching a curve in a DAG. Our improved results are based on a new data structure, called free-space map, that might be applicable to other problems involving the Fréchet metric. It remains open whether the same improvements obtained here can be achieved for matching curves inside general graphs (see [11] for a recent improvement on complete graphs). Proving a lower bound better than $\Omega(n \log n)$ is another major problem left open.

References

1. Alt, H.: The computational geometry of comparing shapes. In: Efficient Algorithms: Essays Dedicated to Kurt Mehlhorn on the Occasion of His 60th Birthday, pp. 235–248. Springer, Heidelberg (2009)
2. Alt, H., Efrat, A., Rote, G., Wenk, C.: Matching planar maps. J. Algorithms 49(2), 262–283 (2003)
3. Alt, H., Godau, M.: Computing the Fréchet distance between two polygonal curves. Int. J. of Comput. Geom. Appl. 5, 75–91 (1995)
4. Buchin, K., Buchin, M., Gudmundsson, J.: Constrained free space diagrams: a tool for trajectory analysis. Int. J. of Geogr. Inform. Sci. 24(7), 1101–1125 (2010)
5. Buchin, K., Buchin, M., Knauer, C., Rote, G., Wenk, C.: How difficult is it to walk the dog? In: Proc. 23rd EWCG, pp. 170–173 (2007)
6. Buchin, K., Buchin, M., Wang, Y.: Exact algorithms for partial curve matching via the Fréchet distance. In: Proc. 20th ACM-SIAM Sympos. Discrete Algorithms, pp. 645–654 (2009)
7. Cook, A.F., Wenk, C.: Geodesic Fréchet distance inside a simple polygon. In: Proc. 25th Sympos. Theoret. Aspects Comput. Sci. LNCS, vol. 5664, pp. 193–204 (2008)
8. Efrat, A., Guibas, L.J., Har-Peled, S., Mitchell, J.S.B., Murali, T.M.: New similarity measures between polylines with applications to morphing and polygon sweeping. Discrete Comput. Geom. 28(4), 535–569 (2002)
9. Jiang, M., Xu, Y., Zhu, B.: Protein structure-structure alignment with discrete Fréchet distance. J. Bioinform. Comput. Biol. 6(1), 51–64 (2008)
10. Maheshwari, A., Sack, J.-R., Shahbaz, K., Zarrabi-Zadeh, H.: Fréchet distance with speed limits. Comput. Geom. Theory Appl. 44(2), 110–120 (2011)
11. Maheshwari, A., Sack, J.-R., Shahbaz, K., Zarrabi-Zadeh, H.: Staying close to a curve. In: Proc. 23rd Canad. Conf. Computat. Geom. (to appear, 2011)
12. Sriraghavendra, E., Karthik, K., Bhattacharyya, C.: Fréchet distance based approach for searching online handwritten documents. In: Proc. 9th Internat. Conf. Document Anal. Recognition, pp. 461–465 (2007)

On the Configuration-LP for Scheduling on Unrelated Machines[⋆]

José Verschae and Andreas Wiese

Technische Universität Berlin, Straße des 17. Juni 136, 10623 Berlin, Germany
{verschae,wiese}@math.tu-berlin.de

Abstract. Closing the approximability gap between 3/2 and 2 for the minimum makespan problem on unrelated machines is one of the most important open questions in scheduling. Almost all known approximation algorithms for the problem are based on linear programs (LPs). In this paper, we identify a surprisingly simple class of instances which constitute the core difficulty for LPs: the so far hardly studied *unrelated graph balancing* case in which each job can be assigned to at most two machines. We prove that already for this basic setting the strongest known LP-formulation – the *configuration-LP* – has an integrality gap of 2, matching the best known approximation factor for the general case. This points towards an interesting direction of future research. The result is shown by a sophisticated construction of instances, based on deep insights on two key weaknesses of the configuration-LP.

For the objective of maximizing the minimum machine load in the unrelated graph balancing setting we present an elegant purely combinatorial 2-approximation algorithm with only quadratic running time. Our algorithm uses a novel preprocessing routine that estimates the optimal value as good as the configuration-LP. This improves on the computationally costly LP-based $(2 + \varepsilon)$-approximation algorithm by Chakrabarty et al. [6].

1 Introduction

The problem of minimizing the makespan on unrelated machines, usually denoted $R||C_{\max}$, is one of the most prominent and important problems in the area of machine scheduling. In this setting we are given a set of n jobs and a set of m unrelated machines to process the jobs. Each job j requires $p_{i,j} \in \mathbb{N}^+ \cup \{\infty\}$ time units of processing if it is assigned to machine i. The scheduler must find an assignment of jobs to machines with the objective of minimizing the makespan, i.e., the largest completion time of a job.

In a seminal work, Lenstra, Shmoys, and Tardos [16] give a 2-approximation algorithm based on a natural LP-relaxation. On the other hand, they show that the problem is NP-hard to approximate within a better factor than 3/2, unless

⋆ This work was partially supported by Berlin Mathematical School (BMS) and by the DFG Focus Program 1307 within the project "Algorithm Engineering for Real-time Scheduling and Routing".

C. Demetrescu and M.M. Halldórsson (Eds.): ESA 2011, LNCS 6942, pp. 530–542, 2011.

$P = NP$. Reducing this gap is considered to be one of the most important open questions in the area of machine scheduling [20] and it has been opened for more than 20 years.

The best known approximation algorithms for this problem and its special cases are derived by linear programming techniques [8,16,23]. A special role plays the *configuration-LP* (which has been successfully used for Bin-Packing [13] and was first used for a scheduling problem by Bansal and Sviridenko [4]). It is the strongest linear program for the problem considered in the literature and it implicitly contains a vast class of inequalities. In fact, for the most relevant cases of $R||C_{\max}$ the best known approximation factors match the best known upper bounds on the integrality gap of the configuration-LP.

Given the apparent difficulty of this problem, researchers have turned to consider simpler cases. One special case that has drawn a lot of attention is the *restricted assignment* problem. In this setting each job can only be assigned to a subset of machines, and it has the same processing time on all its available machines. That is, the processing times $p_{i,j}$ of a job j equal either a machine-independent processing time $p_j \in \mathbb{N}^+$ or infinity. Surprisingly, the best known approximation algorithm for this problem continues to be the 2-approximation algorithm by Lenstra et al. [16]. However, Svensson [23] shows that the configuration-LP has an integrality gap of $33/17 \approx 1.94$. Thus, it is possible to compute in polynomial time a lower bound that is within a factor $33/17 + \varepsilon$ to the optimum. However, no polynomial time algorithm is known to construct an α-approximate solution for $\alpha < 2$.

The restricted assignment case seems to capture the complexity of the general case to a major extend. However, we show that the core complexity of the problem, in terms of the configuration-LP, lies in the *unrelated graph balancing case*, where each job can be assigned to at most two machines, but with possibly different processing times on each of them.

In the second part of this paper we study a different objective function which has been actively studied by the scheduling community in recent years, see e. g. [1,3,4,6]. In the MaxMin-allocation problem we are also given a set of jobs, a set of unrelated machines, and processing times $p_{i,j}$ as before. The load of a machine i, denoted by ℓ_i, is the sum of the processing times assigned to machine i. The objective is to maximize the minimum load of the machines, i. e., to maximize $\min_i \ell_i$. The idea behind this objective function is a fairness property: Consider that jobs represent resources that must be assigned to machines. Each machine i has a personal valuation of job (resource) j, namely $p_{i,j}$. The objective of maximizing the minimum machine load is equivalent to maximizing the total valuation of the machine that receives the least total valuation.

1.1 Related Work

Minimum Makespan Problem. Besides the paper by Lenstra et al. [16] that we have already mentioned, there has not been much progress on how to diminish the approximability gap for $R||C_{\max}$. Shchepin and Vakhania [21] give a more sophisticated rounding procedure for the LP by Lenstra et al. and improve the approximation guarantee to $2 - 1/m$, which is best possible among all rounding algorithms for this LP. On the other hand, Gairing, Monien, and Woclaw [11] propose a more efficient combinatorial 2-approximation algorithm based on unsplittable flow techniques. Also, the preemptive version of this problem has been studied [7,14].

An important special case is the restricted assignment case. Besides the result by Svensson [23] mentioned above, some special cases are studied depending on the structure of the machines and the processing times of the jobs [17,18].

For the graph balancing case—where each job can be processed on at most two machines with the same processing time on both machines—Ebenlendr, Krčál, and Sgall [8] give a 1.75-approximation algorithm based on an tighter version of the LP-relaxation by Lenstra et al. [16]. They also show that the problem is NP-hard to approximate with a better factor than $3/2$, which matches the best known NP-hardness result for the general case. If the underlying graph is a tree, there is an FPTAS [15]. In this paper we consider a slight generalization of the graph balancing problem, where each job can have different processing times on its two available machines. We call this setting the *unrelated graph balancing* problem. To the best of our knowledge, this problem has not been studied in its own right before and, hence, all results for this case follow from the general case.

Generalized Assignment. A well-known generalization of the minimum makespan problem is the *generalized assignment* problem. In this setting we are given a budget B, and jobs have machine-dependent processing times and costs. The objective is to find a schedule minimizing the makespan among all schedules with cost at most B. Shmoys and Tardos [22] extend the rounding procedure given by Lenstra et al. [16] to this setting, obtaining a 2-approximation algorithm for this problem.

MaxMin-Allocation Problem. The MaxMin-allocation problem has drawn a lot of attention recently. In contrast to $R||C_{\max}$, for the general case Bansal and Sviridenko [4] show that the configuration-LP has a super-constant integrality gap of $\Omega(\sqrt{m})$. On the other hand, Asadpour and Saberi [3] show constructively that this is tight up to logarithmic factors and provide an algorithm with approximation ratio $O(\sqrt{m} \log^3 m)$. Relaxing the bound on the running time, Chakrabarty, Chuzhoy, and Khanna [6] present a poly-logarithmic approximation algorithm that runs in quasi-polynomial time.

For the restricted assignment case of MaxMin-allocation, the integrality gap of the configuration-LP is much better. In a series of papers, it is shown to be at most $O(\log \log m / \log \log \log m)$ by Bansal et al. [4], in $O(1)$ by Feige [10], and bounded by 5 and subsequently by 4 by Asadpour, Feige, and Saberi [1,2]. Only the first bound is given by a constructive proof. However, Haeupler et al. [12] make the proof by Feige [10] constructive, yielding a polynomial time constant factor approximation algorithm.

For the unrelated graph balancing case of MaxMin-allocation (each job can be assigned to at most two machines with possibly different execution times on them) Bateni et al. [5] give a 4-approximation algorithm. Chakrabarty et al. [6] improve this result by showing that the configuration-LP has an integrality gap of 2, yielding a $(2 + \varepsilon)$-approximation algorithm. Moreover, it is NP-hard to approximate even this special case with a better ratio than 2 [5,6]. This bound matches the best known complexity result for the general case. Interestingly, the case that every job can be assigned to at most three machines is essentially equivalent to the general case [5]. *Bin-packing.* In the *bin-packing* problem we are given a list of n items that need to be assigned to the least amount of unit size bins, without overloading them. For this problem Karamarkar, and Karp [13] show that an analogue to the Configuration-LP approximates the optimal value up to an additive error of $O(\log^2(n))$. It has been further conjectured that this additive error is at most one [9,19].

1.2 Our Contribution

Almost all known approximation algorithms for $R||C_{\max}$ and its special cases are based on linear programs [8,16,23]. The strongest known LP is the configuration-LP which implicitly contains a vast class of inequalities. In this paper, we identify a surprisingly basic class of instances which captures the core complexity of the problem for LPs: the unrelated graph balancing setting. We show that even the configuration-LP has an integrality gap of 2 in the unrelated graph balancing setting and hence cannot help to improve the best known approximation factor. Interestingly, if one additionally requires that each job has the same processing time on its two machines, the integrality gap of the configuration-LP is at most 1.75 (implicitly in [8]). We prove our result by presenting a sophisticated family of instances for which the configuration-LP has an integrality gap of 2. The instances have two novel technical properties which together lead to this large integrality gap. The first property is the usage of gadgets that we call *high-low-gadgets*. These gadgets form the seed of the inaccuracy of the configuration-LP. Secondly, the machines of our instances are organized in a large number of *layers*. Through the layers, the introduced inaccuracy is amplified such that the integrality gap reaches 2. To the best of our knowledge, the unrelated graph balancing case has not been considered in its own right before. Therefore, our result points to an interesting direction of future research to eventually improve the approximation factor of 2 for the general case. We note that for the restricted assignment case the configuration-LP has an integrality gap of $33/17 < 2$ [23]. We conclude that—at least for the configuration-LP—the restricted assignment case is easier than the unrelated graph balancing case. Table 1 shows an overview of the integrality gap of the configuration-LP for the respective cases.

Only few approximation algorithms are known for scheduling unrelated machines which do not rely on solving a linear program. As seen above, LP-based algorithms have certain limitations and can be costly to solve. It is then preferable to have combinatorial algorithm with lower running times. For the unrelated graph balancing case of the MaxMin-allocation problem, we present an elegant combinatorial approximation algorithm with only quadratic running time and an approximation guarantee of 2. In the algorithm we use a new method of preprocessing that simplifies the complexity of a given instance and also yields a lower bound on the optimal makespan. This lower bound is as strong as the

Table 1. The integrality gap of the configuration-LP for $R||C_{\max}$ in the respective scenarios. *Job degree*: maximum number of machines on which a job can have a finite processing time. *Proc. times*: whether each job has the same processing time on all its available machines.

Processing Times / Job Degree	Unbounded	≤ 2
Arbitrary $p_{i,j}$	2	2
Restricted Assignment	$\left[\frac{3}{2}, \frac{33}{17}\right]$ [8,16,23]	$\left[\frac{3}{2}, \frac{7}{4}\right]$ [8]

worst case bound given by the configuration-LP. Then, we introduce a novel construction of a bipartite graph to link pairs of jobs which can be assigned to the same machine. A coloring for the graph then implies an assignment of the jobs to machines which ensures the claimed approximation factor. This improves on the LP-based $(2 + \varepsilon)$-approximation algorithm by Chakrabarty et al. [6]. Their algorithm resorts to the ellipsoid method to approximately solve the configuration-LP with its exponentially many variables. Our new algorithm is fast, very simply to implement, and moreover best possible, unless $P = NP$.

2   LP-Based Approaches

In this section we revise the classical LP-formulation by Lenstra et al. [16] as well the configuration-LP. Moreover we study the relationship of these two LPs and give an explicit characterization of which kind of inequalities are implicitly implied by the configuration-LP. In the sequel, we denote by J the set of jobs and M the set of machines of a given instance.

The Natural LP-Relaxation. The natural IP-formulation used by Lenstra et al. [16] uses assignment variables $x_{i,j} \in \{0,1\}$ that denote whether job j is assigned to machine i. This formulation, which we denote by LST-IP, takes a target value for the makespan T (which will be determined later by a binary search) and does not use any objective function: $\sum_{i \in M} x_{i,j} = 1$ for all $j \in J$; $\sum_{j \in J} p_{i,j} \cdot x_{i,j} \leq T$ for all $i \in M$; $x_{i,j} = 0$ for all i, j such that $p_{i,j} > T$. Here the variables x_{ij} are binary for all $i \in M, j \in J$. The corresponding LP-relaxation of this IP, which we denote by LST-LP, can be obtained by replacing the integrality condition by $x_{i,j} \geq 0$. Let C_{LP} be the smallest integer value of T so that LST-LP is feasible, and let C^* be the optimal makespan of our instance. Thus, since the LP is feasible for $T = C^*$ we have that C_{LP} is a lower bound on C^*. Moreover, we can easily find C_{LP} in polynomial time with a binary search procedure.

Lenstra et al. [16] give a rounding procedure that takes a feasible solution of LST-LP with target makespan T and returns an integral solution with makespan at most $2T$. By taking $T = C_{LP} \leq C^*$ this yields a 2-approximation algorithm.

Integrality gaps and the configuration-LP. Shmoys and Tardos [22] implicitly show that the rounding just mentioned is best possible by means of the *integrality gap* of LST-LP. For an instance I of $R||C_{\max}$, let $C_{LP}(I)$ be the smallest integer value of T so that LST-LP is feasible, and let $C^*(I)$ the minimum makespan of this instance. Then the integrality gap of this LP is defined as $\sup_I C^*(I)/C_{LP}(I)$. It is easy to see that the integrality gap is the best possible approximation guarantee that can be shown by using C_{LP} as a lower bound. Shmoys and Tardos [22] give an example showing that the the integrality gap of LST-LP is arbitrarily close to 2, and thus the rounding procedure by Lenstra et al. [16] is best possible. Equivalently, the integrality gap of LST-LP equals 2.

It is natural to ask whether adding a family of cuts can help to obtain a formulation with smaller integrality gap. Indeed, for special cases of our problem it has been shown that adding certain inequalities reduces the integrality gap. In particular, Ebenlendr et al. [8] consider the *star-cuts*, $\sum_{j \in J: p_{i,j} > T/2} x_{i,j} \leq 1$, for all $i \in M$. They show that adding these inequalities to LST-LP yields an integrality gap of at most 1.75 in the graph balancing setting.

The results of this paper have important implications on whether it is possible to add similar cuts to strengthen the LP for the unrelated graph balancing problem or for the general case of $R||C_{\max}$. We study the *configuration-LP* which we define below. Let T be a target makespan, and define $\mathcal{C}_i(T)$ as the collection of all subsets of jobs with total processing time at most T, i.e., $\mathcal{C}_i(T) := \left\{ C \subseteq J : \sum_{j \in C} p_{i,j} \leq T \right\}$. We introduce a binary variable $y_{i,C}$ for all $i \in M$ and $C \in \mathcal{C}_i(T)$, representing whether the set of jobs assigned to machine i equals C. The configuration-LP is defined as follows:

$$\sum_{C \in \mathcal{C}_i(T)} y_{i,C} = 1 \qquad \text{for all } i \in M,$$

$$\sum_{i \in M} \sum_{C \in \mathcal{C}_i(T): C \ni j} y_{i,C} = 1 \qquad \text{for all } j \in J,$$

$$y_{i,C} \geq 0 \qquad \text{for all } i \in M, C \in \mathcal{C}_i(T).$$

It is not hard to see that an integral version of this LP is a formulation for $R||C_{\max}$. Also notice that the configuration-LP suffers from an exponential number of variables, and thus it is not possible to solve it directly in polynomial time. However, it is easy to show that the separation problem of the dual corresponds to an instance of KNAPSACK and thus we can solve the LP approximately in polynomial time. More precisely, given a target makespan T there is a polynomial time algorithm that either asserts that the configuration-LP is infeasible or computes a solution which uses only configurations whose makespan is at most $(1+\varepsilon)T$, for any constant $\varepsilon > 0$ [23]. The following section, we will show that the integrality gap of this formulation is as large as the integrality gap of LST-LP even for the unrelated graph balancing case.

Notice that a solution of the configuration-LP yields a feasible solution to LST-LP with the same target makespan. On the other hand, there are solutions to LST-LP that do not have corresponding feasible solutions to the configuration-LP.

Now we elaborate on the relation of the two LPs, by giving a formulation in the space of the $x_{i,j}$ variables that is equivalent to the configuration-LP. Intuitively, the configuration-LP contains all possible (local) information for any single machine. Indeed, we show that any cut in the $x_{i,j}$ variables that involves only one machine is implied by the configuration-LP. Indeed, let $\alpha \in \mathbb{Q}^J$ be an arbitrary row column. The configuration-LP will imply any cut of the form $\sum_{j \in J} \alpha_j x_{i,j} \leq \delta_{\alpha,i}$, where the right-hand-side is properly chosen so that no single machine schedule for machine i is removed by the cut. The proof of the following proposition can be found in [24].

Proposition 1. *Fix a target makespan T. For each $\alpha \in \mathbb{Q}^J$ we define $\delta_{\alpha,i} := \max\{\sum_{j \in S} \alpha_j : S \in \mathcal{C}_i(T)\}$. The feasibility of the configuration-LP is equivalent to the feasibility of the linear program: $\sum_{i \in M} x_{ij} = 1$ for all $j \in J$ and $\sum_{j \in J} \alpha_j x_{i,j} \leq \delta_{\alpha,i}$ for all $\alpha \in \mathbb{Z}^J, i \in M$.*

As an example of the implications of this proposition, we note that adding the star-cuts does not help diminishing the integrality gap of LST-LP for unrelated graph balancing. This follows by taking α as the characteristic vector of the set $\{j \in J : p_{i,j} > T/2\}$ for each $i \in M$.

3   Integrality Gap of the Configuration-LP

In this section we prove that the configuration-LP has an integrality gap of 2, already for the special case of unrelated graph balancing. This shows that the core complexity of the configuration-LP is already captured by the unrelated graph balancing case.

We construct a family of instances I_k of the unrelated graph balancing case such that $p_{i,j} \in \{\frac{1}{k}, 1, \infty\}$ for each machine i and each job j for some integer k. We will show that for I_k there is a solution of the configuration-LP which uses only configurations with makespan $1 + \frac{1}{k}$. However, every integral solution for I_k has a makespan of at least $2 - \frac{1}{k}$.

Let $k \in \mathbb{N}$ and let N be the smallest integer satisfying $k^N/(k-1)^{N+1} \geq \frac{1}{2}$. Consider two k-ary trees of height $N - 1$, i.e., two trees of height $N - 1$ in which apart from the leaves every vertex has k children. We say that all vertices with the same distance to the root are in the same *layer*. For every leaf v, we introduce another vertex v' and k edges between v and v'. (Hence, v is no longer a leaf.) We call such a pair of vertices v, v' a *high-low-gadget*. Observe that the resulting "tree" has height N.

Based on this, we describe our instance of unrelated graph balancing. For each vertex v we introduce a machine m_v. For each edge $e = \{u, v\}$ we introduce a job j_e. Assume that u is closer to the root than v. We define that j_e has processing time $\frac{1}{k}$ on machine m_u, processing time 1 on machine m_v, and infinite processing time on any other machine. This motivates the term "high-low-gadget": each job inside such a gadget – given by two vertices v and v' as above – has a high processing time on $m_{v'}$ and a low processing time on m_v. Finally, let $m_r^{(1)}$ and $m_r^{(2)}$ denote the two machines corresponding to the two root vertices. We introduce a job j_{big} which has processing time 1 on $m_r^{(1)}$ and $m_r^{(2)}$. Denote by I_k the resulting instance.

To gain some intuition for the construction, consider a high-low-gadget consisting of two machines m_v, $m_{v'}$ where v' is a leaf. In any solution with a makespan of at most $1 + \frac{1}{k}$ it is clear that m_v can schedule only the jobs whose respective edges connect v and v'. However, we will see in the sequel that there are solutions for the configuration-LP with makespan $1 + \frac{1}{k}$ in which also a fraction of the job with processing time 1 is scheduled on m_v. Since we chose a large number of layers, this fraction will be amplified through the layers to the root until we obtain a feasible solution to the configuration-LP using only configurations with makespan at most $1 + \frac{1}{k}$. However, any integral solution has a makespan of at least $2 - \frac{1}{k}$ as we will prove in the following lemma. This implies that the configuration-LP has an integrality gap of 2.

Lemma 1. *Any integral solution for I_k has a makespan of at least $2 - \frac{1}{k}$.*

Proof. We can assume w.l.o.g. that job j_{big} is assigned to machine $m_r^{(1)}$. If the makespan of the whole schedule is less than 2 then there must be at least one job which has processing time $\frac{1}{k}$ on $m_r^{(1)}$ which is *not* assigned to $m_r^{(1)}$ but to some other machine m. We can apply the same argumentation to machine m. Iterating the argument shows that there must be a leaf v such that machine m_v has a job with processing time 1 assigned to it. Hence, either m_v has a load of at least $2 - \frac{1}{k}$ or machine $m_{v'}$ has a load of at least 2.   □

Now we want to show that there is a feasible solution of the configuration-LP for I_k which uses only configurations with makespan $1 + \frac{1}{k}$. To this end, we introduce the concept of *j-α-solutions* for the configuration-LP. A j-α-solution is a solution for the configuration-LP whose right hand side is modified as follows: job j does not need to be fully assigned but only to an extent of $\alpha \leq 1$. This value α corresponds to the fraction of the big job assigned to a machine like m_v as described above.

For any $h \in \mathbb{N}$ denote by $I_k^{(h)}$ a subinstance of I_k defined as follows: Take a vertex v of height h and consider the subtree $T(v)$ rooted at v. For the subinstance $I_k^{(h)}$ we take all machines and jobs which correspond to vertices and edges in $T(v)$. (Note that since our construction is symmetric it does not matter which vertex of height h we take.) Additionally, we take the job which has processing time 1 on m_v. We denote the latter by $j^{(h)}$. We prove inductively that there are $j^{(h)}$-$\alpha^{(h)}$-solutions for the subinstances $I_k^{(h)}$ for values $\alpha^{(h)}$ which depend only on h. These values $\alpha^{(h)}$ increase for increasing h. The important point is that $\alpha^{(N)} \geq \frac{1}{2}$. Hence, there are solutions for the configuration-LP which distribute j_{big} on the two machines $m_r^{(1)}$ and $m_r^{(2)}$ (which correspond to the two root vertices).

The following lemma gives the base case of the induction. It explains the inaccuracy of the configuration-LP in the high-low-gadgets.

Lemma 2. *There is a $j^{(1)}$-$\frac{1}{k-1}$-solution for the configuration-LP for $I_k^{(1)}$ which uses only configurations with makespan at most $1 + \frac{1}{k}$.*

Proof. Let m_v be the machine in $I_k^{(1)}$ which corresponds to the root of $I_k^{(1)}$. Similarly, let $m_{v'}$ denote the machine which corresponds to the leaf v'. For $\ell \in \{1, ..., k\}$ let $j_\ell^{(0)}$ be the jobs which have processing time 1 on $m_{v'}$ and processing time $\frac{1}{k}$ on m_v.

For $m_{v'}$ the configurations with makespan at most $1 + \frac{1}{k}$ are $C_\ell := \left\{ j_\ell^{(0)} \right\}$ for each $\ell \in \{1, ..., k\}$. We define $y_{m_{v'}, C_\ell} := \frac{1}{k}$ for each ℓ. Hence, for each job $j_\ell^{(0)}$ a fraction of $\frac{k-1}{k}$ remains unassigned. For machine m_v there are the following (maximal) configurations: $C_{small} := \left\{ j_1^{(0)}, ..., j_k^{(0)} \right\}$ and $C_{big}^\ell := \left\{ j^{(1)}, j_\ell^{(0)} \right\}$ for each $\ell \in \{1, ..., k\}$. We define $y_{m_v, C_{big}^\ell} := \frac{1}{k(k-1)}$ for each ℓ and $y_{m_v, C_{small}} := 1 - \frac{1}{k-1}$. This assigns each job $j_\ell^{(0)}$ completely and job $j^{(1)}$ to an extent of $k \cdot \frac{1}{k(k-1)} = \frac{1}{k-1}$. □

After having proven the base case, the following lemma yields the inductive step. It shows how the value α of our j-α-solutions is increased by the layers of our construction, and thus the effect of the high-low-gadgets is amplified. Note that the intermediate nodes have similarities to the nodes which constitute the high-low-gadgets.

Lemma 3. *Assume that we are given a $j^{(n)}$-$\left(k^n / (k-1)^{n+1} \right)$-solution for the configuration-LP for $I_k^{(n)}$ which uses only configurations with makespan at most $1 + \frac{1}{k}$. Then, there is a $j^{(n+1)}$-$\left(k^{n+1} / (k-1)^{n+2} \right)$-solution for the configuration-LP for $I_k^{(n+1)}$ which uses only configurations with makespan at most $1 + \frac{1}{k}$.*

Proof. Note that $I_k^{(n+1)}$ consists of k copies of $I_k^{(n)}$, one additional machine and one additional job. Denote by m_v the additional machine (which forms the "root" of $I_k^{(n+1)}$). Recall that $j^{(n+1)}$ is the (additional) job that can be assigned to m_v but to no other machine in $I_k^{(n+1)}$. For $\ell \in \{1, ..., k\}$ let $j_\ell^{(n)}$ be the jobs which have processing time $\frac{1}{k}$ on m_v.

Inside of the copies of $I_k^{(n)}$ we use the solution defined in the induction hypothesis. Hence, each job $j_\ell^{(n)}$ is already assigned to an extent of $\left(k^n / (k-1)^{n+1} \right)$. Like in Lemma 2 the (maximal) configurations for m_v are given by $C_{small} := \left\{ j_1^{(n)}, ..., j_k^{(n)} \right\}$ and $C_{big}^\ell := \left\{ j^{(n+1)}, j_\ell^{(n)} \right\}$ for each $\ell \in \{1, ..., k\}$. We define the value $y_{m_v, C_{big}^\ell} := k^n / (k-1)^{n+2}$ for each ℓ and $y_{m_v, C_{small}} := 1 - k^{n+1} / (k-1)^{n+2}$. This assigns each job $j_\ell^{(n)}$ completely and the job $j^{(n+1)}$ is assigned to an extent of $k \cdot k^n / (k-1)^{n+2} = k^{n+1} / (k-1)^{n+2}$. □

Theorem 1. *The integrality gap of the configuration-LP is 2 for the unrelated graph balancing problem.*

Proof. Due to the above reasoning and the choice of N for each of the two subinstances $I_k^{(N)}$ there are j_{big}-$\frac{1}{2}$-solutions. Hence, there is a solution for the configuration-LP using only configurations with makespan at most $1 + \frac{1}{k}$. Since by Lemma 1 any integral solution has a makespan of at least $2 - \frac{1}{k}$ the claim follows (as k can be chosen arbitrarily large). $\qquad\square$

4   Unrelated Graph Balancing Case of MaxMin-Allocation

In this section we study the MaxMin-allocation problem in the unrelated graph balancing setting. We present an elegant purely combinatorial 2-approximation algorithm with quadratic running time which is quite easy to implement. To this end we introduce a new preprocessing procedure that already gives an estimate to the optimal makespan that is within a factor of 2, and thus it is as strong as the worst-case bound of the configuration-LP.

Let I be an instance of the problem and let T be a positive integer. Our algorithm either finds a solution with value $T/2$ or asserts that there is no solution with value T or larger. With an additional binary search this yields a 2-approximation algorithm. For each machine i denote by $J_i = \{j_{i,1}, j_{i,2}, ...\}$ the list of all jobs which can be assigned to i. We partition this set into the sets $A_i \dot\cup B_i$ where $A_i = \{a_{i,1}, a_{i,2}, ...\}$ denotes the jobs in J_i which can be assigned to two machines (machine i and some other machine) and B_i denotes the jobs in J_i which can only be assigned to i. We define A_i' to be the set A_i without the job with largest processing time (or one of those jobs in case there is a tie). For any machine i and any set of jobs J_i' we define $p(J_i') := \sum_{j \in J'} p_{i,j}$.

Denote by $p_{i,\ell}$ the processing time of job $a_{i,\ell}$ on machine i. We assume that the elements of A_i are ordered non-increasingly by processing time, i.e., $p_{i,\ell} \geq p_{i,\ell+1}$ for all respective values of ℓ. The preprocessing phase of our algorithm works as follows. If there is a machine i such that $p(A_i) + p(B_i) < T$ we output that there is no solution with value T or larger. So now assume that $p(A_i) + p(B_i) \geq T$ for all machines i. If there is a machine i such that $p(A_i') + p(B_i) < T$ (but $p(A_i) + p(B_i) \geq T$) then any solution with value at least T has to assign $a_{i,1}$ to i. Hence, we assign $a_{i,1}$ to i. This can be understood as moving $a_{i,1}$ from A_i to B_i. We rename the remaining jobs in A_i accordingly and update the values $p(A_i)$, $p(A_i')$, and $p(B_i)$. We do this procedure until either one of the following two conditions holds: (1) There is one machine i such that $p(A_i) + p(B_i) < T$; or (2) for all machines i we have that $p(A_i') + p(B_i) \geq T$. In case (1) holds we output that there is no solution with value T or larger.

If during the preprocessing phase the algorithm outputs that no solution with value T or larger exists, then clearly there can be no such solution. As we will see below, this preprocessing phase together with the binary search procedure already gives a lower bound that is within a factor 2 to the optimal solution. This matches the worst-case bound given by the configuration-LP.

Now we construct a graph G as follows: For each machine i and each job $a_{i,\ell} \in A_i$ we introduce a vertex $\langle a_{i,\ell} \rangle$. We connect two vertices $\langle a_{i,\ell} \rangle, \langle a_{i',\ell'} \rangle$ if $a_{i,\ell}$ and $a_{i',\ell'}$ represent the same job (but on different machines). Also, for each machine i we introduce an edge between the vertices $\langle a_{i,2k+1} \rangle$ and $\langle a_{i,2k+2} \rangle$ for each respective value $k \geq 0$. It can be shown that G is bipartite (see [24] for details). It follows that we can color G with two colors, black and white, such that two adjacent vertices have different colors. Let i be a machine. We assign each job $a_{i,\ell}$ to i if and only if $\langle a_{i,\ell} \rangle$ is black. Also, we assign each job in B_i to i. We will prove in Theorem 2 that this procedure assigns each machine at least a load of $T/2$. In order to transform the previously described procedure into a 2-approximation algorithm, it is necessary to find the largest value T for which the preprocessing phase does not assert that T is too large. This can be easily achieved by embedding the preprocessing algorithm into a binary search framework. With appropriate data structures and a careful implementation the whole algorithm has a running time of $O(|I|^2)$ where $|I|$ denotes the overall input length in binary encoding, see [24] for details.

Theorem 2. *There is a 2-approximation algorithm with running time $O(|I|^2)$ for the unrelated graph balancing case of the MaxMin-allocation problem.*

Proof. It remains to prove the approximation ratio. Let i be a machine. We show that the total weight of jobs assigned to i is at least $p(A'_i)/2 + p(B_i)$. For each connected pair of vertices $\langle a_{i,2k+1} \rangle, \langle a_{i,2k+2} \rangle$ we have that either $a_{i,2k+1}$ or $a_{i,2k+2}$ is assigned to i. We calculate that $\sum_{k \in \mathbb{N}} p_{i,2k+2} \geq p(A'_i)/2$. Since $p_{i,2k+1} \geq p_{i,2k+2}$ (for each respective value of k) we conclude that the total weight of the jobs assigned to i is at least $p(A'_i)/2 + p(B_i)$. Since $p(A'_i) + p(B_i) \geq T$ the claim follows. □

Further Results. The best known NP-hardness reductions for $R||C_{\max}$ create instances with the property that $p_{i,j} \in \{1, 2, 3, \infty\}$ for all jobs j and all machines i (up to scaling), see [8,16]. In fact, we can show that if the (finite) execution times of the jobs in an instance of $R||C_{\max}$ differ by at most a factor of $10/3$ then there is a $(1 + \frac{5}{6})$-approximation algorithm, see [24]. Also, if the greatest common divisor of the processing times arising in an instance is bounded, we also obtain a better approximation factor than 2, see [24]. These results imply key properties for reductions that rule out an approximation factor of $(2 - \varepsilon)$ for $R||C_{\max}$.

Acknowledgments. We would like to thank Annabell Berger, Matthias Müller-Hannemann, Thomas Rothvoß, and Laura Sanità for helpful discussions.

References

1. Asadpour, A., Feige, U., Saberi, A.: Santa claus meets hypergraph matchings. In: Goel, A., Jansen, K., Rolim, J.D.P., Rubinfeld, R. (eds.) APPROX and RANDOM 2008. LNCS, vol. 5171, pp. 10–20. Springer, Heidelberg (2008)
2. Asadpour, A., Feige, U., Saberi, A.: Santa claus meets hypergraph matchings. Technical report, Standford University (2009), Available for download at http://www.stanford.edu/~asadpour/publication.htm

3. Asadpour, A., Saberi, A.: An approximation algorithm for max-min fair allocation of indivisible goods. In: Proceedings of the 39th annual ACM symposium on Theory of computing (STOC 2007), pp. 114–121 (2007)

4. Bansal, N., Sviridenko, M.: The santa claus problem. In: Proceedings of the 38th Annual ACM Symposium on Theory of Computing (STOC 2006), pp. 31–40 (2006)

5. Bateni, M., Charikar, M., Guruswami, V.: Maxmin allocation via degree lower-bounded arborescences. In: Proceedings of the 41st Annual ACM Symposium on Theory of Computing (STOC 2009), pp. 543–552 (2009)

6. Chakrabarty, D., Chuzhoy, J., Khanna, S.: On allocating goods to maximize fairness. In: Proceedings of the 50th Annual Symposium on Foundations of Computer Science (FOCS 2009), pp. 107–116 (2009)

7. Correa, J.R., Skutella, M., Verschae, J.: The power of preemption on unrelated machines and applications to scheduling orders. In: Dinur, I., Jansen, K., Naor, J., Rolim, J. (eds.) APPROX 2009. LNCS, vol. 5687, pp. 84–97. Springer, Heidelberg (2009)

8. Ebenlendr, T., Krčál, M., Sgall, J.: Graph balancing: a special case of scheduling unrelated parallel machines. In: Proceedings of the 19th Annual ACM-SIAM Symposium on Discrete Algorithms (SODA 2008), pp. 483–490 (2008)

9. Eisenbrand, F., Palvoelgyi, D., Rothvoss, T.: Bin packing via discrepancy of permutations. In: Proceedings of the 22th Annual ACM-SIAM Symposium on Discrete Algorithms (SODA 2011), pp. 476–481 (2011)

10. Feige, U.: On allocations that maximize fairness. In: Proceedings of the 19th Annual ACM-SIAM Symposium on Discrete Algorithms (SODA 2008), pp. 287–293 (2008)

11. Gairing, M., Monien, B., Woclaw, A.: A faster combinatorial approximation algorithm for scheduling unrelated parallel machines. Theoretical Computer Science 380, 87–99 (2007)

12. Haeupler, B., Saha, B., Srinivasan, A.: New constructive aspects of the lovasz local lemma. In: Proceedings of the 51st Annual IEEE Symposium on Foundations of Computer Science (FOCS 2010), pp. 397–406 (2010)

13. Karmarkar, N., Karp, R.M.: An efficient approximation scheme for the one-dimensional bin-packing problem. In: Proceedings of the 23rd Annual IEEE Symposium on Foundations of Computer Science (FOCS 1982), pp. 312–320 (1982)

14. Lawler, E.L., Labetoulle, J.: On preemptive scheduling of unrelated parallel processors by linear programming. Journal of the ACM 25, 612–619 (1978)

15. Lee, K., Leung, J.Y., Pinedo, M.L.: A note on graph balancing problems with restrictions. Information Processing Letters 110, 24–29 (2009)

16. Lenstra, J.K., Shmoys, D.B., Tardos, E.: Approximation algorithms for scheduling unrelated parallel machines. Mathematical Programming 46, 259–271 (1990)

17. Leung, J.Y., Li, C.: Scheduling with processing set restrictions: A survey. International Journal of Production Economics 116, 251–262 (2008)

18. Lin, Y., Li, W.: Parallel machine scheduling of machine-dependent jobs with unit-length. European Journal of Operational Research 156, 261–266 (2004)

19. Scheithauer, G., Terno, J.: Theoretical investigations on the modified integer round-up property for the one-dimensional cutting stock problem. Operations Research Letters 20, 93–100 (1997)

20. Schuurman, P., Woeginger, G.J.: Polynomial time approximation algorithms for machine scheduling: Ten open problems. Journal of Scheduling 2, 203–213 (1999)

21. Shchepin, E.V., Vakhania, N.: An optimal rounding gives a better approximation for scheduling unrelated machines. Operations Research Letters 33, 127–133 (2005)

22. Shmoys, D.B., Tardos, E.: An approximation algorithm for the generalized assignment problem. Mathematical Programming 62, 461–474 (1993)
23. Svensson, O.: Santa claus schedules jobs on unrelated machines. In: Proceedings of the 43th Annual ACM Symposium on Theory of Computing, STOC 2011 (2011) (to appear)
24. Verschae, J., Wiese, A.: On the configuration-LP for scheduling on unrelated machines. Technical Report 025-2010, Technische Universität Berlin (November 2010)

Resource Allocation for Covering Time Varying Demands*

Venkatesan T. Chakaravarthy[1], and Amit Kumar[2] and Sambuddha Roy[1], and Yogish Sabharwal[1]

[1] IBM Research - India, New Delhi
[2] Indian Institute of Technology, New Delhi
{vechakra,sambuddha,ysabharwal}@in.ibm.com; amitk@cse.iitd.ac.in

Abstract. We consider the problem of allocating resources to satisfy demand requirements varying over time. The input specifies a demand for each timeslot. Each resource is specified by a start-time, end-time, an associated cost and a capacity. A feasible solution is a multiset of resources such that at any point of time, the sum of the capacities offered by the resources is at least the demand requirement at that point of time. The goal is to minimize the total cost of the resources included in the solution. This problem arises naturally in many scenarios such as workforce management, sensor networks, cloud computing, energy management and distributed computing. We study this problem under the partial cover setting and the zero-one setting. In the former scenario, the input also includes a number k and the goal is to choose a minimum cost solution that satisfies the demand requirements of at least k timeslots. For this problem, we present a 16-approximation algorithm; we show that there exist "well-structured" near-optimal solutions and that such a solution can be found in polynomial time via dynamic programming. In the zero-one setting, a feasible solution is allowed to pick at most one copy of any resource. For this case, we present a 4-approximation algorithm; our algorithm uses a novel LP relaxation involving flow-cover inequalities.

1 Introduction

We consider the problem of allocating resources to satisfy demand requirements varying over time. We assume that time is uniformly divided into discrete timeslots. The input specifies a demand for each timeslot. Each resource is specified by its start-time, end-time, the capacity of the resource that it offers during this interval and its associated cost. A feasible solution is a set of resources satisfying the constraint that at any timeslot, the total sum of the capacities offered by the resources is at least the demand required at that timeslot, i.e. the selected resources must cover the demands.

The above problem is motivated by applications in workforce management. For instance, consider the problem of scheduling employees in call centers. Typically, in these settings, we have a reasonably accurate forecast of demands which

---

* A full version of this paper is available at http://www.cs.wisc.edu/~venkat

C. Demetrescu and M.M. Halldórsson (Eds.): ESA 2011, LNCS 6942, pp. 543–554, 2011.
© Springer-Verlag Berlin Heidelberg 2011

will arrive at any timeslot and for each employee, we know the time interval in which they can work. Employees have proficiencies (or capacities) determining the number of calls they can attend on an average in a timeslot. They also have cost or wages associated with them. The goal is to choose a minimum cost set of employees whose schedule meets the demand at all the timeslots. The problem framework is quite general and captures many other situations arising in sensor networks, cloud computing, energy management and distributed computing (see [12,7,4]).

Motivated by such applications, Chakaravarthy et al. [4] studied the above problem. They considered the version (called MULTIRESALL) wherein the solution is allowed to pick multiple copies of a resource by paying as many units of the associated cost. In the context of workforce management, employees can be classified based on their proficiency and the shifts they work in. Choosing multiple units of a single resource corresponds to selecting multiple employees of the same class. They presented a 4-approximation algorithm for the MULTIRESALL problem. In this paper, we consider two generalizations of the MULTIRESALL problem.

The first variant (called partial MULTIRESALL) considers the partial covering scenario, wherein the input also specifies a number k and a feasible solution is only required to satisfy the demand for at least k timeslots. In the workforce management setting, this corresponds to the concept of service level agreements (SLA's), which stipulates that the requirements of a large fraction of timeslots are satisfied.

The second variant considers a more natural bounded availability setting, wherein for each resource, the input specifies a bound on the maximum number of copies that can be selected. In this paper, we shall study the version (called (0,1)-RESALL) in which a solution can pick a resource at most once. Note that the bounded setting can be reduced to the zero-one setting, by duplicating the resources (albeit at an increased running time). We now formally define these problems.

Problem Definitions: We first define the MULTIRESALL problem. We consider time to be uniformly divided into discrete units ranging from 1 to T. We refer to each integer t in the range $[1, T]$ as a *timeslot*. The input specifies a *demand profile* $d : [1, T] \to \mathbb{Z}$; here $d(t)$ (also denoted by d_t) is the *demand* at timeslot t for $t \in [1, T]$. The input further consists of a set of *resources* (or *intervals*) $\mathcal{I}$. Each resource $i \in \mathcal{I}$ is specified by an interval $I_i = [s_i, e_i]$, where s_i and e_i are the *start-time* and the *end-time* of the resource i; we assume that s_i and e_i are integers in the range $[1, T]$. The resource i is also associated with a *capacity* (or *height*) h_i and a cost c_i. We say that the resource i is *active* at a timeslot t, if $t \in I_i$. For a timeslot t, let $A(t)$ denote the set of all resources active at timeslot t.

Let S be a multiset of resources. For a resource $i \in S$, let $f_S(i)$ denote the number of times i appears in S. The multiset S is said to *cover* a timeslot t, if the sum of the capacities of the resources from S active at the timeslot t is at least d_t, i.e., $\sum_{i \in A(t)} f_S(i) \cdot h_i \geq d_t$. The multiset S is said to be a *full cover*, if

S covers all the timeslots $t \in [1, T]$. The cost of the multiset S is defined to be $\mathsf{cost}(S) = \sum_{i \in S} f_S(i) \cdot c_i$ (where the summation is over distinct intervals in S).

- MULTIRESALL Problem: In this problem, the goal is to find a full cover having the minimum cost.
- Partial MULTIRESALL Problem: In this problem, the input also includes an integer k. A multiset S is said to be a k-*partial cover*, if S covers at least k timeslots. The goal is to find a minimum cost k-partial cover.
- (0,1)-RESALL Problem: In this problem, a feasible solution is allowed to pick each resource at most once. The goal is to find a minimum cost full cover.

Prior Work: As mentioned earlier, a 4-approximation algorithm for the MULTIRESALL problem was presented in [4]. The algorithm was based on the primal dual approach.

The special case of the (0,1)-RESALL problem wherein there is only a single timeslot $(T = 1)$ corresponds to the classical minimum knapsack cover problem (MKP). It is well known that the problem is NP-hard and that it admits a FPTAS [11]. En route to proving more general results, Carnes and Shmoys [3] gave a 2-approximation algorithm based on the primal dual approach.

The (0,1)-RESALL problem is related to the well-studied unsplittable flow problem on line (UFP). In the (0,1)-RESALL problem, we need to select a set of intervals covering a given demand profile. In the UFP problem, the goal is to select a set of intervals that can be packed within a given bandwidth profile. Thus, while UFP is a packing problem, (0,1)-RESALL is its analogous covering version. Bansal et al.[1] presented a quasi-PTAS for the restricted case of the UFP problem, when all the capacities and demands are quasi-polynomial. In a recent breakthrough, Bonsma et al. [2] designed a polynomial time $(7 + \epsilon)$-approximation algorithm. Under the so called "no-bottleneck assumption", Chekuri et al. [6] presented a $(2 + \epsilon)$-approximation algorithm. Prior work have addressed partial versions of many other covering problems, such as vertex cover [8], multicut [10] and spanning trees [9].

Chakrabarty et al.[5] studied the more general version of the (0,1)-RESALL problem called column restricted covering integer programs. Their framework yields a 40-approximation algorithm for the (0,1)-RESALL problem. Independent of our work, Korula obtained a 4-approximation algorithm for the (0,1)-RESALL problem (see [5]). To the best of our knowledge, partial versions of the MULTIRESALL and (0,1)-RESALL problems have not been studied.

Our Results and Techniques: The main results of this paper are as follows:

- We present a 16-approximation algorithm for the partial MULTIRESALL problem.
- For the special case, of the partial MULTIRESALL problem, where the input demands are uniform, our approach can be fine tuned to yield an improved 2-approximation algorithm. The details of this result will be presented in the full version of the paper.
- Our next result is a 4-approximation algorithm for the (0,1)-RESALL problem. This result generalizes the known 4-approximation algorithm for the MULTIRESALL problem [4].

Our techniques do not extend to the partial version of the (0,1)-RESALL problem. Obtaining a constant factor approximation for this problem remains open.

The 4-approximation algorithm for the MULTIRESALL problem [4] is based on the primal-dual approach applied to a natural LP formulation. In the case of (0,1)-RESALL, the corresponding natural LP has an unbounded integrality gap. We circumvent the issue by strengthening the LP with "flow-cover inequalities", à la the approach that Carnes and Shmoys [3] adopt for the MKP problem. An additional advantage of the new LP is that it admits a simpler primal-dual analysis, while matching the approximation factor of the MULTIRESALL problem.

We now discuss two approaches that have been widely successful for solving partial covering problems (such as vertex cover) and briefly outline the difficulties in applying either of them to our problem. One of the approaches typically starts with the natural LP relaxation, which may have unbounded integrality gap. It turns out that that by augmenting it with some extra information, the integrality gap can be brought down to a constant (see for instance: partial vertex cover [8], k-MST [9]). In the MULTIRESALL problem, it is not clear how to construct such a strengthened LP.

A second approach goes via Langrangian relaxations [13] and has been successful applied to problems such as such as multicut [10] and k-MST [9]. Under this paradigm, one first designs a primal-dual approximation algorithm for the prize collecting version with certain additional properties, which is then used to solve the partial covering problem. We do not know how to design such an algorithm for the prize collecting version of the MULTIRESALL problem; however, we note that our techniques discussed below yield a constant factor approximation for prize collecting version.

Our algorithm for the partial MULTIRESALL problem builds on two main insights regarding the MULTIRESALL problem. The first insight is that there always exists a near-optimal solution satisfying the following simple structural property: for every timeslot, there exists an interval in the solution whose copies are enough to cover the demand at the timeslot (without the aid of other intervals). The second main insight is that among all the solutions satisfying the above property, the optimal one can be found in polynomial time using dynamic programming. Combining the two ideas, we get an alternative constant factor approximation algorithm for the MULTIRESALL problem. We show that both the steps extend to the partial cover setting and thereby we get a constant factor approximation algorithm for the partial MULTIRESALL problem. It is interesting to note that the recent $(7 + \epsilon)$-approximation algorithm for the UFP problem [2] also uses a similar strategy of first showing the existence of solutions with special properties and then invoking dynamic programming.

2   Approximating the Partial MULTIRESALL Problem

In this section, we present a 16-approximation algorithm for the partial MULTIRESALL problem. The notion of *single resource assignment* (SRA) covers, defined next, is useful for this purpose.

Single Resource Assignment (SRA) Solution: Let S be a multiset of resources. The multiset S is said to cover a timeslot $t \in [1, T]$ in an *SRA fashion*, if we can find a single resource $i \in S$ such that the demand at t can be covered by the copies of i alone i.e., there exists $i \in S$ such that $f_S(i) \cdot h_i \geq d_t$. The multiset S is said to be an *SRA full cover*, if S covers all the timeslots in an SRA fashion. A k-partial cover S is said to be an *SRA k-partial cover*, if S covers at least k timeslots in an SRA fashion.

Let Opt and pOpt denote the optimal full cover and k-partial cover, respectively. Similarly, let $\widehat{\text{Opt}}$ and $\widehat{\text{pOpt}}$ denote the optimal SRA full cover and SRA k-partial cover, respectively. Our constant factor approximation algorithm is based on the following two theorems and the subsequent corollaries.

Theorem 1. *There exists an SRA full cover $\widehat{S}$ such that* $\text{cost}(\widehat{S}) \leq 16 \cdot \text{cost}(\text{Opt})$.

Theorem 2. *Given an instance of the* MultiResAll *problem, we can find the optimal SRA full cover in polynomial time.*

Corollary 1. *There exists an SRA k-partial cover $\widehat{S}$ such that* $\text{cost}(\widehat{S}) \leq 16 \cdot \text{cost}(\text{pOpt})$.

Corollary 2. *Given an instance of the partial* MultiResAll *problem, we can find the optimal SRA k-partial cover $\widehat{\text{pOpt}}$ in polynomial time.*

We note that an alternative proof of Theorem 2 is given in [5]. However, our proof is significantly simpler. By combining the two corollaries, we obtain a constant factor approximation algorithm for the partial MultiResAll problem, as follows. Our overall algorithm simply runs the algorithm claimed in Corollary 2 and outputs the SRA k-partial cover $\widehat{\text{pOpt}}$. Corollary 1 implies that $\text{cost}(\widehat{\text{pOpt}}) \leq \text{cost}(\widehat{S}) \leq 16 \cdot \text{cost}(\text{pOpt})$. We have established the main result of the section: the partial MultiResAll problem can be approximated within a factor of 16.

2.1   Existence of Good SRA Covers

In this section, we first prove Theorem 1 and then derive Corollary 1. In proving Theorem 1, even though it suffices to show only the existence of $\widehat{S}$, we shall in fact present a polynomial time algorithm for producing the claimed SRA full cover. The algorithm goes via the primal-dual approach.

Let us first consider a naive LP formulation. We associate a variable x_i with each $i \in \mathcal{I}$, which specifies the number of copies of the resource i in the solution. For each timeslot t, we will have a constraint that enforces the coverage requirement for the timeslot ($\sum_{i \in A(t)} h_i \cdot x_i \geq d_t$). However, it is not hard to show that this LP has an unbounded integrality gap. For instance, consider a single timeslot t (i.e., $T = 1$) with $d_t = 1$ and let there be a single resource i of cost $c_i = 1$, with height $h_i = B$ (for some B). Then, the optimal integral solution will have cost 1. On the other hand, the LP can set $x_i = 1/B$ and get a cost of $1/B$. One can easily handle the issue via adjusting heights as follows. For a resource $i \in I$ and a timeslot t where i is active, denote $\widetilde{h}_i(t) = \min\{h_i, d_t\}$. We obtain a strengthened LP formulation by utilizing the adjusted heights $\widetilde{h}$.

$$\min \qquad \sum_{i \in \mathcal{I}} x_i \cdot c_i$$

$$\sum_{i \in A(t)} \widetilde{h}(i,t) \cdot x_i \geq d_t \qquad \text{for all time-slots } t$$

$$x_i \geq 0 \qquad \text{for all } i \in \mathcal{I}$$

Each integral primal feasible solution $\mathbf{x}$ corresponds to a full cover multiset S; similarly, every full cover S corresponds to a primal integral feasible solution $\mathbf{x}$, given by $\mathbf{x}_i = f_S(i)$. In our discussion, we shall use the viewpoint of multisets.

The dual of the LP is shown next. It has a variable y_t for each timeslot t.

$$\max \qquad \sum_{t \in [1,T]} y_t \cdot d_t$$

$$\sum_{i \in A(t)} y_t \cdot \widetilde{h}(i,t) \leq c_i \qquad \text{for all } i \in \mathcal{I}$$

$$y_t \geq 0 \qquad \text{for all } t$$

We will produce a primal integral solution $\widehat{S}$ and a dual feasible solution $\mathbf{y}$ such that the complementary slackness conditions are satisfied approximately: (i) Primal slackness conditions: for any $i \in \mathcal{I}$, if $f_{\widehat{S}}(i) > 0$ then the corresponding dual constraint is tight:

$$\sum_{t:i \in A(t)} y_t \cdot \widetilde{h}(i,t) = c_i. \tag{1}$$

(ii) Approximate dual slackness conditions: for any timeslot t, if $y_t > 0$ then the corresponding primal constraint is tight within a factor of 16:

$$\sum_{i:i \in A(t)} f_{\widehat{S}}(i) \cdot \widetilde{h}(i,t) \leq 16 \cdot d_t. \tag{2}$$

Using well-known arguments via complementary slackness and weak duality, we can show that the above conditions imply $\mathtt{cost}(\widehat{S}) \leq 16 \cdot \mathtt{cost}(\mathtt{Opt})$. This would prove Theorem 1. We now present the algorithm meeting the above requirements.

The algorithm runs in three phases, a construction phase, a deletion phase and a doubling phase. In the construction phase, we shall construct a dual feasible solution $\mathbf{y}$ and also obtain an SRA full cover $\mathbf{A}$. In the deletion phase, we delete certain carefully chosen intervals from $\mathbf{A}$ and obtain a multiset S'. The set S' may not be a full cover, but it will satisfy at least half the demand at each timeslot in an SRA fashion. In the final doubling phase, we simply double the number of copies each interval in S' and obtain an SRA full cover $\mathbf{B}$. We shall argue that the primal feasible solution $\widehat{S} = \mathbf{B}$ and the dual feasible solution $\mathbf{y}$ satisfy the slackness conditions (1) and (2).

Construction Phase: We employ an greedy procedure for constructing a dual feasible solution $\mathbf{y}$ having high objective value (i.e., $\sum_t \mathbf{y}_t \cdot d_t$). The algorithm runs in multiple iterations, wherein each iteration we greedily choose the dual variable y_t that potentially gives us the maximum benefit; thus, we choose the timeslot t having the maximum demand d_t. The algorithm is formally described next. To start with all the timeslots are marked as *alive*. Consider any iteration $j \geq 0$. Let t_j be the timeslot having the highest demand d_{t_j} among all the currently alive timeslots. We raise the dual variable y_{t_j} to the maximum possible value until some dual constraint, say corresponding to an interval i_j, becomes tight. We add $\lceil d_{t_j}/\widetilde{h}(i_j, t_j) \rceil$ copies of i_j in our solution $\mathbf{A}$. We call t_j the *main timeslot* of i_j. The interval i_j then marks as *dead* all the timeslots in its span that are currently alive. We say that the main timeslot t_j along with all the other timeslots marked dead by i_j are *associated* with i_j. We proceed in this manner until our solution $\mathbf{A}$ covers all the timeslots. This completes the construction of $\mathbf{y}$ and $\mathbf{A}$.

Each interval $i \in \mathbf{A}$ covers all the timeslots associated with it in an SRA fashion and hence, $\mathbf{A}$ is an SRA full cover. Let ℓ be the number of main timeslots, given as $t_1, t_2, \ldots, t_\ell$. For each interval i chosen by $\mathbf{A}$, let the collection of copies of i be called the *bundle* of i. For an interval i and a timeslot t, let $\widetilde{H}(i, t)$ denote the *adjusted bundle height at t*, given by $\widetilde{H}(i, t) = f_{\mathbf{A}}(i) \cdot \widetilde{h}(i, t)$. The solution $\mathbf{A}$ satisfies the following property.

Lemma 1. *Let i be any interval chosen in $\mathbf{A}$ and t^* be any main timeslot within the span of i. Then, $\widetilde{H}(i, t^*) \leq 2d_{t^*}$.*

Proof. Let t be the main timeslot associated with i. By definition, $\widetilde{H}(i, t^*) = \lceil d_t/h_i \rceil \cdot \min\{h_i, d_t^*\}$. We see that $d_t \leq d_{t^*}$ (for otherwise t^* cannot a main timeslot). A simple calculation yields the lemma. □

Deletion Phase: In the deletion phase, we delete certain redundant intervals from $\mathbf{A}$ and obtain a multiset S'. Consider the main timeslots in an arbitrary order, say the original order $t_1, t_2, \ldots, t_\ell$. Let t^* be any main timeslot in this ordering. Lemma 1 shows that for any $i \in A(t^*)$, $\widetilde{H}(i, t^*) \leq 2d_{t^*}$. We partition the set $A(t^*)$ of intervals active at t^* in a geometric fashion into a set of *bands* as follows. Let $m = \lceil \log(2d_{t^*}) \rceil$. For $1 \leq j \leq m$, define the *band* B_j to be

$$B_j = \{ i \in \mathbf{A} \cap A(t^*) \ : \ 2d_{t^*}/2^j < \widetilde{H}(i, t^*) \leq 2d_{t^*}/2^{j-1} \}.$$

For each band B_j, let i_1 be the interval in B_j extending farthest to the left (i.e., having the minimum start-time); similarly, let i_2 be the interval active at t^* extending farthest to the right (i.e., having the maximum end-time). We shall retain these two interval (bundles) and delete all the other interval (bundles) of B_j from $\mathbf{A}$. Let the constructed multiset be S'.

We note that S' need not be an SRA full cover (it may not even be a full cover). However, we can make the following claim about S' that at every timeslot t, the solution S' covers at least half the demand at t in an SRA fashion. The lemma is proved as follows. If i, the interval covering t in $\mathbf{A}$, got deleted while considering a main timeslot t^*, then either the leftmost or the rightmost interval appearing in the same band as i cover at least half the demand at t.

Lemma 2. *For any timeslot t, there exists an interval $i' \in S'$ such that i' is active at t and $f_{S'}(i')h_{i'} \geq d_t/2$.*

Proof. Let i be the interval associated with t. Clearly, if i is included in S', then $f_{S'}(i)h_i \geq d_t$ and we can take $i' = i$. Now suppose i got deleted while considering a main timeslot t^*. With respect to t^*, let B_j be the band to which i belongs and let i_1 and i_2 be the leftmost and rightmost intervals that were retained for this band. Then, at least one of these two intervals (say i_1) is also active at t. Since i_1 is in the same band as i, $\widetilde{H}(i_1, t^*) \geq \widetilde{H}(i, t^*)/2$. First consider the case where $h_i \leq d_{t^*}$. In this case, $\widetilde{H}(i, t^*) = f_{S'}(i) \cdot h_i \geq d_t$. It follows that $f_{S'}(i_1) \cdot h_{i_1} \geq \widetilde{H}(i_1, t^*) \geq d_t/2$. Now suppose $h_i \geq d_{t^*}$. This implies that $\widetilde{H}(i, t^*) \geq d_{t^*}$ and so, i belongs to the band B_1 (corresponding to the range $[d_{t^*}, 2d_{t^*}]$). Since i_1 also belongs to the same band as i, $\widetilde{H}(i_1, t^*) \geq d_{t^*}$, which implies that $f_{S'}(i_1) \cdot h_{i_1} \geq d_{t^*}$. Notice that $d_{t^*} \geq d_{t'}$, where t' is main timeslot of i; otherwise, t' would have been considered earlier than t^* and so, i would have marked t^* as dead and so, t^* could not be a main timeslot. Notice that $d_{t'} \geq d_t$; otherwise t' cannot be the main timeslot of i. It follows that $d_{t^*} \geq d_t$. This implies that that $f_{S'}(i_1) \cdot h_{i_1} \geq d_t$. Thus, we can take $i' = i_1$. $\qquad\square$

Doubling Phase: In this phase, we transform S' into an SRA full cover $\mathbf{B}$, by simply doubling the number of copies of every interval in S'. Lemma 2 implies that the solution $\mathbf{B}$ is an SRA full cover.

Complementary Slackness Conditions: We now argue that the complementary slackness conditions (1) and (2) are satisfied by the solutions $\mathbf{y}$ and $\mathbf{B}$. First notice that the primal slackness conditions (1) are automatically satisfied after the construction phase; in the latter two phases, we do not change the dual solution $\mathbf{y}$. Let us now focus on the dual slackness condition (2). The dual variable $y_t > 0$ only for main timeslots t. The lemma below proves the slackness condition for all these main timeslots.

Lemma 3. *For any main timeslot t^*,*

$$\sum_{i \in A(t^*)} f_{\mathbf{B}}(i)\widetilde{h}(i, t^*) \leq 16 \cdot d_{t^*}.$$

Proof. We first derive a bound on the quantity $W = \sum_{i \in A(t^*)} f_{S'}(i)\widetilde{h}(i, t^*)$. Consider the bands $B_1, B_2, \ldots, B_m$ corresponding to the timeslot t^*. At the end of the deletion phase, only two intervals were retained in S' for each band B_j: they contribute at most $2 \cdot (2d_{t^*}/2^{j-1})$ to the quantity W. So, the total contribution across all the bands can be computed as:

$$W \leq \sum_{j=1}^{m} 2 \cdot (2d_{t^*}/2^{j-1}) \leq 8 \cdot d_{t^*}.$$

The doubling procedure implies that for any interval $i \in S'$, $f_{\mathbf{B}}(i) = 2 \cdot f_{S'}(i)$. This means the LHS of the inequality in the lemma is at most $2W \le 16 \cdot d_{t^*}$.   □

This completes the proof of Theorem 1.

Proof of Corollary 1:   Note that the proof of Theorem 1 generalizes to the following partial covering scenario. Given a subset of timeslots $X \subseteq [1, T]$, the primal-dual algorithm can produce a solution $\widehat{S}$ such that $\widehat{S}$ covers all the timeslots in X in an SRA fashion and $\mathtt{cost}(\widehat{S}) \le 16 \cdot \mathtt{Opt}(\mathtt{X})$, where $\mathtt{Opt}(\mathtt{X})$ is the optimum solution covering all the timeslots in X. The optimum partial k-cover pOpt covers a subset of timeslots Z with $|Z| \ge k$. We focus on only the timeslots in Z and ignore the rest. By invoking the generalization (with $X = Z$), we can get an SRA k-partial cover $\widehat{S}$ such that $\mathtt{cost}(\widehat{S}) \le 16 \cdot \mathtt{cost}(\mathtt{pOpt})$ (since $\mathtt{Opt}(\mathtt{Z}) = \mathtt{pOpt}$).

2.2   Computing Optimal SRA Covers

Here, we first prove Theorem 2 and then derive Corollary 2. The algorithm goes via a reduction to a problem that we call the *layered interval covering* problem (LIC), defined next.

Layered Interval Covering Problem (LIC):   The input consists of a set of intervals $\mathcal{I}$ over timeslots $[1, T]$. Each interval is specified by its range $[s_i, e_i]$, where s_i is the start-time and e_i is the end-time) and a cost c_i. The input includes a set of *colors* $\mathcal{L} = \{1, 2, \ldots, L\}$ and specifies a color $\chi(i)$, for each interval $i \in \mathcal{I}$ and each timeslot $t \in [1, T]$. An interval i is said to *cover* a timeslot t, if i is active at t and $\chi(i) \ge \chi(t)$. A feasible solution S is a set of intervals such that for each timeslot $t \in [1, T]$, at least one interval i covers t. The goal is to compute a feasible solution of minimum cost.

Reduction: It is not hard to reduce the problem of finding the optimal SRA full cover to the LIC problem. Let $\mathcal{A}$ be the input SRA full cover problem instance and we will produce a LIC problem instance $\mathcal{B}$. The number of timeslots for $\mathcal{B}$ is declared to be the same as that of $\mathcal{A}$ (i.e., T is retained as such). Let $\mathcal{D}$ be the set of all distinct demand values (i.e., $\mathcal{D} = \{d_t : t \in [1, T]\}$) and let $L = |\mathcal{D}|$. Notice that $L \le |T|$. Let $a_1, a_2, \ldots, a_L$ be the values in $\mathcal{D}$ sorted in the increasing order. We declare the number of colors in $\mathcal{B}$ to be L, so that each a_j corresponds to a color j. For each interval $i \in \mathcal{A}$ we introduce at most i in $\mathcal{B}$ as follows. For each $1 \le j \le L$, let r be the smallest integer such that $a_j \le rh_i < a_{j+1}$; if such an r does not exist, we ignore this j; otherwise, we introduce a copy i' of i in $\mathcal{B}$ with color $\chi(i') = j$ and cost $c_{i'} = rc_i$ (where c_i is the cost of i in $\mathcal{A}$). This completes the reduction. It is easy to see that any feasible solution S of $\mathcal{B}$ can be transformed into an SRA full cover of $\mathcal{A}$ with the same cost and vice versa.

Below we show that the LIC problem can be solved optimally in polynomial time. It follows that optimum SRA full covers can be found in polynomial time, establishing Theorem 2.

Solving the LIC Problem Optimally: Our polynomial time algorithm is based on dynamic programming and it builds on a decomposition lemma for feasible solutions. The lemma needs the notion of *time-cuts*, defined next.

Let S be a set of intervals that cover all the timeslots in some range $[a, b] \subseteq [1, T]$. Let S_1 and S_2 be a partition of the set S and t be a timeslot satisfying $a \leq t \leq b - 1$. Then the triplet $\langle S_1, S_2, t \rangle$ is said to be a *time-cut* for S, if S_1 covers all the timeslots in the range $[a, t]$ and S_2 covers all the timeslots in the range $[(t + 1), b]$. Intuitively, some intervals in S_1 may span (and even cover) some timeslots in $[(t + 1), b]$, but S_2 is responsible for covering all the timeslots in this range (and vice versa). The decomposition lemma says that we can always find a time-cut, but for an exceptional scenario. The restriction that $t \leq b - 1$ ensures that the time-cut is non-trivial.

Lemma 4. *Let S be a set of intervals covering all the timeslots in some range $[a, b]$. Then, at least one of the following conditions holds: (i) there exists a time-cut for S; (ii) there exists an interval $i \in S$ spanning the entire range $[a, b]$.*

Proof. Consider any timeslot $t \in [a, b]$. Among all the intervals in S active at t, let i be the one having the maximum color; we say that i is *responsible* for t. Let i^* be any interval active at the initial timeslot a. If i^* spans the range $[a, b]$ then the (ii) is satisfied and we are done. So, assume that $e_{i^*} \leq b - 1$. Let t^* be the maximum timeslot for which i^* is responsible. Define S_1 to be the set consisting of i^* and all intervals $i \in S$ having end-time $e_i \leq t^* - 1$ and let $S_2 = S - S_1$. We claim that the triple $\langle S_1, S_2, t^* \rangle$ is a time-cut. To see this first consider any timeslot $t \in [(t^* + 1), b]$ and we will argue that some interval in S_2 covers t. Since i^* is not responsible for any timeslot in $[(t^* + 1), b]$, there must exist an interval $i \neq i^*$ covering t. Therefore, such an interval i belongs to S_2. Now consider any timeslot $t \in [a, t^*]$ and let us argue that some interval in S_1 covers t. Let i be any interval $i \in S$ covering t; so, $\chi(i) \geq \chi(t)$. If i belongs to S_1, we are done. Hence, consider the case where $i \in S_2$. We see that i is active at t^*, because i is active at t and has end-time $e_i \geq t^*$. Since i^* is responsible for t^*, we have that $\chi(i^*) \geq \chi(i)$. Hence, i^* also covers t. Finally, notice that $t^* \leq b - 1$.   □

Lemma 4 can easily be generalized to the partial setting as follows.

Corollary 3. *Let $[a, b]$ be a range of timeslots and let S a set of intervals covering all the timeslots in a subset $X \subseteq [a, b]$. Then, at least one of the following is true: (i) there exists a partition of S into S_1 and S_2, and a timeslot $t \in [a, b - 1]$ such that S_1 covers all the timeslots in $X \cap [a, t]$ and S_2 covers all those in $X \cap [t + 1, b]$; (ii) there exists an interval $i \in S$ spanning all the timeslots in X.*

For a range of timeslots $[a, b] \subseteq [1, T]$ and a color p, let $U([a, b], p)$ denote all the timeslots $t \in [a, b]$ such that $\chi(t) \geq p$. Let $\mathrm{DP}([a, b], p)$ be the cost of the optimum set of intervals covering all the timeslots in $U([a, b], p)$. Corollary 3 establishes the following recurrence relation: $\mathrm{DP}([a, b], p) = \min\{Q_1, Q_2\}$, where

$$Q_1 = \min_{a \leq t \leq b - 1} \{\mathrm{DP}([a, t], p) + \mathrm{DP}([t + 1, b], p)\}$$

$$Q_2 = \min_{i \text{ spans } U([a,b],p)} \{c_i + \mathrm{DP}([a,b], \chi(i) + 1)\}.$$

The quantity Q_1 corresponds to the situation where (i) in Corollary 3 applies. In Q_2, we try all the possible intervals i spanning $U([a,b],p)$. Any such i can cover all the timeslots t with $p \leq \chi(t) \leq \chi(i)$ and so the recursive component focuses on $U([a,b], \chi(i) + 1)$.

Proof of Corollary 2: We now focus on the partial version of the LIC problem; here the input also includes an integer k and the goal is to find an optimum solution that covers at least k timeslots. We can extend the recurrence relation used in solving LIC to the partial cover setting. For a range of timeslots $[a,b]$, a color p and a number r, let $\mathrm{pDP}([a,b],p,r)$ be the cost of the optimum set of intervals covering at least r of the timeslots in $U([a,b],p)$. Using Corollary 3, we can write a recurrence relation for $\mathrm{pDP}([a,b],p,r)$. For a color $q \geq p$, let $\lambda([a,b],[p,q])$ be the number of timeslots t such that $p \in [a,b]$ and $\chi(t) \in [p,q]$. Corollary 3 establishes the following recurrence relation: $\mathrm{pDP}([a,b],p,r) = \min\{Q_1, Q_2\}$, where

$$Q_1 = \min_{\substack{a \leq t \leq b-1 \\ r_1 \leq r}} \{\mathrm{pDP}([a,t],p,r_1) + \mathrm{pDP}([t+1,b],p,r-r_1)\}$$

$$Q_2 = \min_{i \text{ spans } U([a,b],p)} \{c_i + \mathrm{pDP}(\ [a,b],\ \chi(i)+1,\ r - \lambda([a,b],[p,\chi(i)])\ )\}.$$

In Q_1, in addition iterating over all possible cuts t, we also iterate over r_1, which symbolizes the number of timeslots that would be covered in the first range. In Q_2, we try all the possible intervals i spanning $U([a,b],p)$, while keeping count of the timeslots that got covered by the choice of i (i.e., $\lambda([a,b],[p,\chi(i)])$). Based on the above recurrence relation, we can write an dynamic program based algorithm; the final solution corresponds to the entry $\mathrm{pDP}([1,T],1,k))$. The running time of the algorithm is $O(km^4)$; here $m = \max\{n,T\}$, where k is the number timeslots required to be covered, n is the total number of intervals and T is the total number of timeslots. The problem of finding the optimum SRA k-partial cover can be reduced to the partial LIC problem, thereby establishing Corollary 2.

3   Approximation Algorithm for the (0,1)-RESALL Problem

In this Section, we briefly discuss the (0,1)-RESALL problem. As indicated in the introduction, we consider the strengthened LP augmented with so called flow-cover inequalities. We associate a variable x_i with each $i \in \mathcal{I}$; this variable is an indicator variable for whether the resource i is selected in the solution. There will be a constraint for every set $S \subseteq \mathcal{I}$ of intervals: given that the set S is already chosen, the solution must pick enough intervals from the remaining intervals in $\mathcal{I}$, such that the *residual* demand is covered. With this goal in mind, given a set S, let $d_t(S)$ denote the *residual* demand, i.e. $d_t(S) = d_t - \sum_{i \in S \cap A(t)} h_i$. We define $\widetilde{h}(i,t,S) = \min\{h_i, d_t(S)\}$. Then, the *IP* is as follows.

$$\min \qquad \sum_{i \in \mathcal{I}} x_i \cdot c_i$$

$$\sum_{i \in A(t), i \notin S} \widetilde{h}(i, t, S) \cdot x_i \geq d_t(S) \qquad \text{for all time-slots } t \text{ and all subsets } S \subseteq \mathcal{I}$$

$$x_i \in \{0, 1\} \qquad \text{for all } i \in \mathcal{I}$$

One could have written a simpler IP with h_i replacing $\widetilde{h}(i, t, S)$ in the constraints above. However, it can be shown that the corresponding LP also has an unbounded integrality gap. Our 4-approximation algorithm goes via the primal-dual approach applied to the LP relaxation of the above IP. The details are in the full version of the paper.

References

1. Bansal, N., Chakrabarti, A., Epstein, A., Schieber, B.: A quasi-PTAS for unsplittable flow on line graphs. In: STOC (2006)
2. Bonsma, P., Schulz, J., Wiese, A.: A constant factor approximation algorithm for unsplittable flow on paths. In: FOCS (2011)
3. Carnes, T., Shmoys, D.: Primal-dual schema for capacitated covering problems. In: Lodi, A., Panconesi, A., Rinaldi, G. (eds.) IPCO 2008. LNCS, vol. 5035, pp. 288–302. Springer, Heidelberg (2008)
4. Chakaravarthy, V., Kumar, A., Parija, G., Roy, S., Sabharwal, Y.: Minimum cost resource allocation for meeting job requirements. In: IPDPS (2011)
5. Chakrabarty, D., Grant, E., Könemann, J.: On column-restricted and priority covering integer programs. In: IPCO (2010)
6. Chekuri, C., Mydlarz, M., Shepherd, F.: Multicommodity demand flow in a tree and packing integer programs. ACM Transactions on Algorithms 3(3) (2007)
7. Dhesi, A., Gupta, P., Kumar, A., Parija, G., Roy, S.: Contact center scheduling with strict resource requirements. In: IPCO (2011)
8. Gandhi, R., Khuller, S., Srinivasan, A.: Approximation algorithms for partial covering problems. J. Algorithms 53(1), 55–84 (2004)
9. Garg, N.: Saving an ϵ: a 2-approximation for the k-MST problem in graphs. In: STOC, pp. 396–402 (2005)
10. Golovin, D., Nagarajan, V., Singh, M.: Approximating the k-multicut problem. In: SODA (2006)
11. Ibarra, O., Kim, C.: Fast approximation algorithms for the knapsack and sum of subset problems. J. ACM 22, 463–468 (1975)
12. Ingolfsson, A., Campello, F., Wu, X., Cabral, E.: Combining Integer Programming and the Randomization Method to Schedule Employees. European J. Operations Research 202(1), 153–163 (2010)
13. Jain, K., Vazirani, V.: Approximation algorithms for metric facility location and k-median problems using the primal-dual schema and Lagrangian relaxation. J. ACM 48(2), 274–296 (2001)

Mixed-Criticality Scheduling
of Sporadic Task Systems

Sanjoy K. Baruah[1], Vincenzo Bonifaci[2], Gianlorenzo D'Angelo[3],
Alberto Marchetti-Spaccamela[4], Suzanne van der Ster[5], and Leen Stougie[5,6]

[1] University of North Carolina at Chapel Hill, USA
baruah@cs.unc.edu
[2] Max-Planck Institut für Informatik, Saarbrücken, Germany
bonifaci@mpi-inf.mpg.de
[3] University of L'Aquila, Italy
gianlorenzo.dangelo@univaq.it
[4] Sapienza Università di Roma, Rome, Italy
alberto@dis.uniroma1.it
[5] Vrije Universiteit Amsterdam, The Netherlands
suzanne.vander.ster@vu.nl, l.stougie@vu.nl
[6] CWI, Amsterdam, The Netherlands
stougie@cwi.nl

Abstract. We consider the scheduling of mixed-criticality task systems,
that is, systems where each task to be scheduled has multiple levels of
worst-case execution time estimates. We design a scheduling algorithm,
EDF-VD, whose effectiveness we analyze using the processor speedup
metric: we show that any 2-level task system that is schedulable on a
unit-speed processor is correctly scheduled by EDF-VD using speed ϕ;
here $\phi < 1.619$ is the golden ratio. We also show how to generalize the
algorithm to $K > 2$ criticality levels. We finally consider 2-level instances
on m identical machines. We prove speedup bounds for scheduling an
independent collection of jobs and for the partitioned scheduling of a
2-level task system.

1 Introduction

We study a scheduling problem occurring in safety-critical systems that are subject to certification requirements. Our work is motivated by the increasing trend
in embedded systems towards integrating multiple functionalities on a common
platform – consider, for example, the IMA (Integrated Modular Avionics) ini-
tiative for aerospace and AUTOSAR (AUTomotive Open System ARchitecture)
for the automotive industry. Such platforms may support functionalities of different degrees of importance or *criticalities*. Some of the more safety-critical
functionalities may be subject to mandatory certification by statutory certification authorities (CAs). The increasing prevalence of platform integration means
that even in highly safety-critical systems, typically only a relatively small fraction of the overall system is actually of critical functionality and needs to be

C. Demetrescu and M.M. Halldórsson (Eds.): ESA 2011, LNCS 6942, pp. 555–566, 2011.
© Springer-Verlag Berlin Heidelberg 2011

certified. The definition of procedures that will allow for the cost-effective certification of such mixed-criticality systems has been identified as a challenging collection of problems [1]. As a recognition of the importance of these challenges we mention the Mixed Criticality Architecture Requirements (MCAR) program led by several federal agencies and by industry in the US, aimed at streamlining the certification process for safety-critical embedded systems with the long term goal of devising efficient and cost-effective certification processes.

During the certification process, the CA makes certain assumptions about the worst-case run-time behavior of the system. We focus on one particular aspect of run-time behavior: the worst case execution time (WCET) of pieces of code. CAs tend to be very conservative, and hence it is typically the case that the WCET-analysis tools, techniques, and methodologies used by the CA will yield WCET estimates that are far more pessimistic (i.e., larger) than those the system designer would use during the design process. On the other hand, while the CA is only concerned with the correctness of the part of the system that is subject to certification the system designer wishes to ensure that the entire system is correct, including the non-critical parts. We illustrate this by an example from the domain of unmanned aerial vehicles. The functionalities on board such vehicles may be classified into two levels of criticality:

- Level 1: the *mission-critical* functionalities, concerning reconnaissance and surveillance objectives, like capturing images from the ground, transmitting these images to the base station, etc.
- Level 2: the *flight-critical* functionalities: to be performed by the aircraft to ensure its safe operation.

For permission to operate such vehicles over civilian airspace, it is mandatory that its flight-critical (i.e., level 2) functionalities be certified by statutory CAs such as the Federal Aviation Authority (FAA) in the US, or the European Aviation Safety Agency (EASA) in Europe. These CAs are not concerned with the mission-critical functionalities, which must be validated separately by the clients and the vendor-manufacturer. The latter are also interested in the level 2 functionalities, but typically to standards that are less rigorous than the ones used by the civilian CAs.

In Section 2 we will formally define the scheduling problem that we study in this work but as an example of the previous scenario let us consider a simple instance of two jobs, J_1 of criticality level 1 and J_2 of criticality level 2. Job J_1 is characterized by a release time, a deadline and a WCET $c_1(1)$, while job J_2 is characterized by a release time, a deadline and a pair $(c_2(1), c_2(2))$ giving the WCET for level 1 and level 2, respectively – $c_2(1)$ is the WCET estimate used by the system designer, whereas $c_2(2)$ is the (typically much larger) WCET estimate used by the CA. The scheduling goal is to find a schedule such that the following conditions are both verified

1. *(Validation by client/ manufacturer)*. If the execution time of J_1 is no larger than $c_1(1)$ and that of J_2 no larger than $c_2(1)$ then we require that both J_1 and J_2 complete by their deadlines.

2. *(Certification by CA)*. If the execution time of J_2 is no larger than $c_2(2)$ then we require that J_2 must complete by its deadline. (Since the CA is not concerned with the level 1 job J_1, it follows that if J_2 executes for more than $c_2(1)$ but no more than $c_2(2)$, then it is not mandatory anymore to complete J_1 by its deadline.)

The difficulty in finding such a schedule is that scheduling decisions must be made *before* knowing whether we are in case 1 or 2 above: the scheduling algorithm becomes aware of J_2's actual execution time only by executing it (i.e. if the job does not finish after $c_2(1)$ execution time).

If there are more than two jobs then we require that if all jobs of level 2 meet their level 1 WCET estimates then all jobs must be completed by their deadline; otherwise we require that only level 2 jobs should meet their deadlines. This model is easily generalized to more than two criticality levels (see Sect. 2 for a formal definition);

Many real-time systems are better modeled as collections of recurrent processes (known as *sporadic tasks*) rather than as a finite set of jobs. We refer to [3] for an introduction to sporadic task systems; mixed-criticality sporadic task systems are defined in Section 2 below. We observe that schedulability analysis of such systems is typically far more difficult than the analysis of systems modeled as collections of independent jobs, since (i) a sporadic task system can generate infinitely many jobs during any one run; and (ii) the collection of jobs generated during different runs of the system may be different: in general, a single system may legally give rise to infinitely many different collections of jobs.

One focus of this paper is to study this more difficult problem of scheduling mixed-criticality systems modeled as collections of sporadic tasks, upon a single shared preemptive processor.

Related work. The mixed-criticality model that we follow was studied, for independent collections of jobs, in [4]. The authors considered a finite collection of jobs of two criticality levels to be scheduled on one machine. They analysed, using the processor speedup factor (cf. resource augmentation in performance analysis of approximation algorithms, as initiated in [6]), the effectiveness of two techniques that are widely used in practice in real time scheduling. They proposed an algorithm called OCBP and showed that OCBP has a processor speedup factor equal to the golden ratio ϕ. These results were later extended [2] to any number of levels showing also that OCBP is tight for two levels.

The mixed-criticality model has been extended to recurrent task systems in [7]; however the proposed algorithm has a pseudopolynomial running time per scheduling decision and, therefore, cannot be applied in practice.

To the best of our knowledge no results are known for multiple machines.

Our results. Our main result concerns the scheduling of a mixed-criticality task system on a single machine. We design a sufficient schedulability condition and a scheduling algorithm (EDF-VD) whose effectiveness we analyze using the processor speedup metric. We show that any 2-level task system that is schedulable by any algorithm on a single processor of unit speed is correctly scheduled by

EDF-VD on a speed ϕ processor; here $\phi < 1.619$ is the golden ratio. We then show how to generalize the schedulability condition and the algorithm to $K > 2$ criticality levels.

The other main contribution is the study of 2-level instances on multiple identical machines. We generalize the OCBP algorithm for an independent collection of jobs to the case of m machines, proving a speedup bound of $\phi + 1 - 1/m$. Finally we consider the *partitioned* scheduling of a 2-level task system, in which the tasks are assigned to the machines and then scheduled on each machine independently, without migration; we extend the results for the single machine case to derive a $\phi + \epsilon$ speedup algorithm for any $\epsilon > 0$.

Organization of the paper. We give a formal description of the mixed-criticality model along with basic concepts and notation in Section 2. In Section 3 we treat the case of a single machine, for two (Section 3.1) and more than two (Section 3.2) criticality levels. Section 4 is devoted to the case of multiple machines; collection of independent jobs are treated in Section 4.1, while the partitioned scheduling of a task system is considered in Section 4.2.

2   Model

MC jobs. A job in a K-level MC system is characterized by a 4-tuple of parameters: $J_j = (r_j, d_j, \chi_j, c_j)$, where r_j is the release time, d_j is the deadline $(d_j \geq r_j)$, $\chi_j \in \{1, 2, \ldots, K\}$ is the *criticality level* of the job and c_j is a vector $(c_j(1), c_j(2), \ldots, c_j(K))$ representing the *worst-case execution times* (WCET) of job J_j at each level; it is assumed that $c_j(1) \leq c_j(2) \leq \cdots \leq c_j(K)$ and $c_j(i) = c_j(\chi_j)$, for each $i > \chi_j$. Each job J_j in a collection $J_1, \ldots, J_n$ should receive execution time C_j within time window $[r_j, d_j]$. The value of C_j is not known but is discovered by executing job J_j until it signals completion. A collection of realized values $(C_1, C_2, \ldots, C_n)$ is called a scenario. The criticality level of a scenario $(C_1, \ldots, C_n)$ is defined as the smallest integer ℓ such that $C_j \leq c_j(\ell)$ for each job J_j. (We only consider scenarios where such an ℓ exists; i.e. $C_j \leq c_j(K)$, $\forall j$.) The crucial aspect of the model is that, in a scenario of level ℓ, it is necessary to guarantee only that jobs of criticality at least ℓ are completed before their deadlines. In other words, once a scenario is known to be of level ℓ, the jobs of criticality $\ell - 1$ or less can be safely dropped.

MC task systems. Let $T = (\tau_1, \ldots, \tau_n)$ be a system of n tasks, each task $\tau_j = (c_j, p_j, \chi_j)$ having a worst-case *execution time vector* $c_j = (c_j(1), c_j(2), \ldots, c_j(K))$, a *period* p_j, and a *criticality level* $\chi_j \in \{1, 2, \ldots, K\}$. Again we assume that $c_j(1) \leq \cdots \leq c_j(K)$. Task τ_j generates a potentially infinite sequence of jobs, with successive jobs being released at least p_j units apart. Each such job has a deadline that is p_j time units after its release (*implicit deadlines*). The criticality of such job is χ_j, and its WCET vector is given by c_j.

MC-schedulability. An (online) algorithm schedules a sporadic task system T correctly if it is able to schedule *every* job sequence generated by T such that if the criticality level of the corresponding scenario is ℓ, then all jobs of level at least ℓ are completed between their release time and deadline. A system is called *MC-schedulable* if it admits some correct scheduling algorithm.

Utilization in task systems. Let $L_k = \{j \in \{1,\ldots,n\} : \chi_j = k\}$. Define the *utilization* of task j at level k as

$$u_j(k) = \frac{c_j(k)}{p_j}, \qquad j = 1,\ldots,n, \ k = 1,\ldots,\chi_j;$$

Define the total utilization at level k of jobs with criticality level i as

$$U_i(k) = \sum_{j \in L_i} u_j(k), \qquad i = 1,\ldots,K, \ k = 1,\ldots,i.$$

It is well-known that in the case of a *single* criticality level, an (implicit-deadline) task system is feasible on m processors of speed σ if and only if $U_1(1) \leq \sigma m$ and $u_j(1) \leq \sigma$ for all j. The following necessary condition for MC-schedulability is an easy consequence of that fact.

Proposition 1. *If T is MC-schedulable on a unit speed processor, then for each $k = 1,\ldots,K$,*

$$\sum_{i=k}^{K} U_i(k) \leq 1.$$

In particular, when $K = 2$,

$$U_1(1) + U_2(1) \leq 1, \ and \tag{1}$$
$$U_2(2) \leq 1. \tag{2}$$

Proof. For each k, consider the scenario where each task $j \in L_k \cup \cdots \cup L_K$ releases jobs with execution time $c_j(k)$. □

In the sequel we will call a job *active* if it has been released but has not yet been completed. In the case of a single criticality level and a single machine, it is known that the Earliest Deadline First algorithm, which schedules, preemptively, from among the active jobs the one with earliest deadline first, is optimal [8].

Proposition 2. *A set of tasks with a single criticality level is feasibly scheduled by the Earliest Deadline First algorithm on a single processor of speed σ if and only if $U_1(1) \leq \sigma$.*

3 Single Machine

The scheduling of a collection of independent mixed-criticality jobs on a single processor has already been treated in reference [2], where an algorithm with a speedup bound of 1.619 has been provided. Thus, in this section we focus on task systems. To facilitate understanding we start exposing the case of only 2 criticality levels and then we present the general result.

3.1   Two Criticality Levels

We consider a variant of the Earliest Deadline First algorithm that we call EDF with Virtual Deadlines (EDF-VD).

Algorithm EDF-VD. We distinguish two cases.

Case 1. $U_1(1) + U_2(2) \leq 1$. Apply EDF to the unmodified deadlines of the jobs. As soon as the system reaches level 2, that is a job executes for more than its WCET at level 1, cancel jobs from tasks in L_1.

Case 2. $U_1(1) + \frac{U_2(1)}{1-U_2(2)} \leq 1$ and Case 1 does not hold. Set $\lambda = \frac{U_2(1)}{1-U_1(1)}$. Then, while the system is in level 1: define for task $i \in L_2$, $\hat{p}_i = \lambda p_i$; redefine deadlines by adding $\hat{p}_i$ to the release time of each job of task $i \in L_2$, leaving the deadlines of jobs of tasks in L_1 as they were, and apply EDF to the modified deadlines. As soon as the system reaches level 2: cancel jobs from tasks in L_1; reset the deadlines of each job of task $i \in L_2$, by adding p_i to its release time; apply EDF to these (original) deadlines. $\square$

We give the following sufficient condition for MC-schedulability by EDF-VD.

Theorem 1. *Assume T satisfies*

$$U_1(1) + \min\left(U_2(2), \frac{U_2(1)}{1-U_2(2)}\right) \leq 1.$$

Then T is schedulable by EDF-VD on a unit-speed processor.

Proof. If $U_1(1) + U_2(2) \leq 1$, it is clear that EDF-VD schedules the task system correctly. Therefore assume

$$U_1(1) + \frac{U_2(1)}{1-U_2(2)} \leq 1. \tag{3}$$

First we show that any level-1 scenario is scheduled correctly. The utilization of the task system with the modified deadlines is

$$\sum_{i \in L_1} \frac{c_i(1)}{p_i} + \sum_{i \in L_2} \frac{c_i(1)}{\lambda p_i} = \sum_{i \in L_1} \frac{c_i(1)}{p_i} + \sum_{i \in L_2} \frac{c_i(1)}{\frac{U_2(1)}{1-U_1(1)} p_i}$$

$$= U_1(1) + \frac{1-U_1(1)}{U_2(1)} \sum_{i \in L_2} \frac{c_i(1)}{p_i} = U_1(1) + \frac{1-U_1(1)}{U_2(1)} U_2(1) = 1.$$

Thus, as long as the system is in level-1 it is scheduled correctly by Proposition 2, even satisfying the modified deadlines. Now we have to show that level-2 scenarios are scheduled correctly.

Let t^* denote the time-instant at which the system reaches level 2. Suppose that a job of level-2 task τ_j is released at t^* and is active (i.e., has been released but not yet completed execution). Let r_j denote its arrival time. The real deadline of this job – the time-instant by which it must complete execution – is $d_j := r_j +$

p_i; however, it is EDF-scheduled by EDF-VD assuming a deadline $\hat{d}_j := r_j + \hat{p}_i$. Since the job is active at t^*, it must be that $t^* \leq \hat{d}_j$, since, as we argued above, all jobs would meet their modified deadlines in the schedule for any level-1 scenario. Note that

$$d_j - \hat{d}_j = (r_j + p_i) - (r_j + \hat{p}_i) = p_i - \lambda p_i.$$

This implies that at time t^* for each level-2 job i at least $(1 - \lambda)p_i$ time is left to finish the job in time. We will show that the artificial task system with only level-2 tasks (hence with only a single criticality level) having periods $(1 - \lambda)p_i$ has total utilization less than 1, implying that this system is scheduled correctly by EDF, which implies that the original system can be scheduled correctly from time t^* onwards. From (3) we obtain

$$\frac{U_2(1)}{1 - U_2(2)} \leq 1 - U_1(1) \Rightarrow \lambda = \frac{U_2(1)}{1 - U_1(1)} \leq 1 - U_2(2) \Rightarrow 1 - \lambda \geq U_2(2)$$

Hence, the total utilization of the artificial task system is

$$\sum_{i \in L_2} \frac{c_i}{(1 - \lambda)p_i} \leq \sum_{i \in L_2} \frac{c_i}{U_2(2)p_i} = 1. \qquad \square$$

The above schedulability condition can now be used to obtain a speedup guarantee. Let $\phi = (\sqrt{5} + 1)/2 < 1.619$, the golden ratio.

Theorem 2. *If T satisfies*

$$\max(U_1(1) + U_2(1), U_2(2)) \leq 1,$$

then it is schedulable by EDF-VD on a speed ϕ processor. In particular, if T is MC-schedulable on a unit-speed processor, it is schedulable by EDF-VD on a speed ϕ processor (cf. Proposition 1).

Proof. We show the equivalent claim: if

$$\max(U_1(1) + U_2(1), U_2(2)) \leq 1/\phi, \qquad (4)$$

then it is schedulable by EDF-VD on a speed 1 processor. Let $\Phi := 1/\phi$. We distinguish two cases.

Case $U_2(1) \geq \Phi U_1(1)$. By (4),

$$\Phi \geq U_1(1) + U_2(1) \geq (1 + \Phi)U_1(1).$$

This gives

$$U_1(1) \leq \frac{\Phi}{1 + \Phi} = \Phi^2.$$

So

$$U_1(1) + U_2(2) \leq \Phi^2 + \Phi = 1$$

and by Lemma 1, EDF-VD correctly schedules T on a processor of speed 1. **Case** $U_2(1) \leq \Phi U_1(1)$ Again, by (4),

$$\Phi \geq U_1(1) + U_2(1) \geq \frac{1}{\Phi}U_2(1) + U_2(1) = \frac{\Phi+1}{\Phi}U_2(1) = \frac{1}{\Phi^2}U_2(1).$$

Rewriting gives $U_2(1) \leq \Phi^3$. Then

$$U_1(1) + \frac{U_2(1)}{1-U_2(2)} = U_1(1) + U_2(1) + U_2(1)\frac{U_2(2)}{1-U_2(2)} \leq U_1(1) + U_2(1) + U_2(1)\frac{\Phi}{1-\Phi}.$$

Since $1 - \Phi = \Phi^2$,

$$U_1(1) + \frac{U_2(1)}{1-U_2(2)} \leq U_1(1) + U_2(1) + U_2(1)\frac{\Phi}{\Phi^2} \leq \Phi + \frac{\Phi^3}{\Phi} = 1$$

and by Lemma 1, EDF-VD correctly schedules T on a unit-speed processor. □

3.2 K Criticality Levels

As we have seen, the cases defining the EDF-VD algorithm are directed by the sufficient conditions of Theorem 1. Therefore before defining the algorithm for the K-level problem we first state the sufficient conditions for this case.

For the K-level problem there are $K-1$ conditions, any of which being satisfied is a sufficient condition for schedulability by EDF-VD(K).

Theorem 3. *Assume T satisfies one of the following, for $k = 1, \ldots, K-1$:*

$$\sum_{i=k}^{K-1} U_i(i) + \min\left\{U_K(K), \frac{U_K(K-1)}{1 - \frac{U_K(K)}{\prod_{j=1}^{k}(1-\lambda_j)}}\right\} \leq \prod_{j=1}^{k}(1-\lambda_j), \tag{5}$$

where $\lambda_1 = 0$ and, for $j > 1$,

$$\lambda_j = \frac{\sum_{\ell=j}^{K} U_\ell(j-1)}{\prod_{\ell=1}^{j-1}(1-\lambda_\ell)} \bigg/ \left(1 - \frac{U_{j-1}(j-1)}{\prod_{\ell=1}^{j-1}(1-\lambda_\ell)}\right) \tag{6}$$

Then T is MC-schedulable on a unit-speed processor.

We define the algorithm by presenting the actions if condition k is satisfied for each $k = 1, \ldots, K-1$. In the description of the algorithm we use next to the scaling parameters λ_ℓ defined in (6), scaling parameters μ_k defined by

$$\mu_k = \frac{U_K(K-1)}{\prod_{j=1}^{k}(1-\lambda_j)} \bigg/ \left(1 - \frac{\sum_{i=k}^{K-1} U_i(i)}{\prod_{j=1}^{k}(1-\lambda_j)}\right).$$

Denote by $t_{\ell-1}^*$ the time instant where the system leaves level $\ell-1$ and enters level ℓ. The periods of the tasks are modified in many levels. Once the system switches to level ℓ, an active job of task τ_i can be seen as a task with a new (virtual) period, denoted by $p_i^{(\ell)}$, where $p_i^{(\ell)} = (1 - \lambda_\ell)p_i^{(\ell-1)}$ and $p_i^{(1)} = p_i$.

Algorithm EDF-VD(K). Suppose that condition k of (5) holds and that conditions $1, \ldots, k-1$ do not hold.

Case 1. $U_K(K) \leq \frac{U_K(K-1)}{1 - U_K(K)/\prod_{j=1}^{k}(1-\lambda_j)}$. Then, while the system is in level $\ell \leq k-1$: discard all jobs from tasks in $L_1, \ldots, L_{\ell-1}$; define for task $\tau_i \in L_j$, for $j = \ell+1, \ldots, K$, $\hat{p}_i^{(\ell)} = \lambda_{\ell+1}p_i^{(\ell)}$; for job J_h from task τ_i define a virtual release time $r_h^{(\ell)} = \min\{t_{\ell-1}^*, r_h\}$ and redefine deadlines $\hat{d}_h^{(\ell)} = r_h^{(\ell)} + \hat{p}_i^{(\ell)}$; apply EDF to the modified deadlines.

As soon as the system reaches level k, cancel jobs from tasks in $L_1, \ldots, L_{k-1}$ and reset the deadlines of each job of task $\tau_i \in L_j$, for $j = k, \ldots, K$, by adding p_i to its release time; apply EDF to these (original) deadlines.

Case 2. $U_K(K) > \frac{U_K(K-1)}{1 - U_K(K)/\prod_{j=1}^{k}(1-\lambda_j)}$. Then, while the system is in level $\ell \leq k-1$: discard all jobs from tasks in $L_1, \ldots, L_{\ell-1}$; define for task $\tau_i \in L_j$, for $j = \ell+1, \ldots, K$, $\hat{p}_i^{(\ell)} = \lambda_{\ell+1}p_i^{(\ell)}$; for job J_h from task τ_i define a virtual release time $r_h^{(\ell)} = \min\{t_{\ell-1}^*, r_h\}$ and redefine deadlines $\hat{d}_h^{(\ell)} = r_h^{(\ell)} + \hat{p}_i^{(\ell)}$; apply EDF to the modified deadlines.

As soon as the system reaches level k: cancel jobs from tasks in $L_1, \ldots, L_{k-1}$; reset the deadlines of each job of task $\tau_i \in L_j$, for $j = k, \ldots, K-1$, by adding p_i to its release time; reset the deadlines of each job J_h from task $i \in L_K$ by adding $\mu_k p_i^{(k)}$ to its virtual release time $r_h^{(k)} = t_{k-1}^*$; apply EDF to these deadlines. $\qquad\square$

4   Multiple Identical Machines

4.1   Scheduling a Finite Collection of Independent Jobs

For a single machine, Baruah et al. [2] analyzed the *Own Criticality Based Priority* (OCBP) rule and showed that it guarantees a speedup bound of ϕ on a collection of independent jobs. We show that this approach can be extended to multiple identical machines at the cost of a slightly increased bound.

The OCBP rule consists in determining, before knowing the actual execution times, a fixed priority ordering of the jobs and for each scenario execute at each moment in time the available job with the highest priority.

The priority list is constructed recursively. First the algorithm searches for a job that can be assigned the *lowest* priority: job J_k is a lowest priority job if there is at least $c_k(\chi_k)$ time in $[r_k, d_k]$ assuming that each other job J_j is executed before J_k for $c_j(\chi_k)$ units of time. This procedure is recursively applied to the collection of jobs obtained by excluding the lowest priority job J_k, until all the jobs are ordered or a lowest priority job cannot be found. A collection of jobs is called *OCBP-schedulable* if the algorithm finds a complete ordering of the jobs.

Theorem 4. *Let J be a collection of jobs that is schedulable on m unit-speed processors. Then J is schedulable using OCBP on m processors of speed $\phi + 1 - 1/m$.*

Algorithm 1. Own Criticality Based Priority (OCBP)

---

for $i = 1$ to n **do**

    if $\exists J_k \in J$ such that there is at least $c_k(\chi_k)$ time in $[r_k, d_k]$ assuming each other

    job J_j is executed before J_k for $c_j(\chi_k)$ units of time **then**

        assign priority i to k (higher index denotes higher priority)

        $J \leftarrow J \setminus \{J_k\}$

    else

        return not OCBP-schedulable

    end if

end for

Proof. Let J be a minimal instance that is MC-schedulable on a processor and not OCBP-schedulable on m processors that are s times as fast for some $s > 1$.

Let $\gamma_1 = \sum_{j|\chi_j=1} c_j(1)$ denote the cumulative WCET for jobs with criticality level 1, and let $\gamma_2(1) = \sum_{j|\chi_j=2} c_j(1)$ and $\gamma_2(2) = \sum_{j|\chi_j=2} c_j(2)$ denote the WCETs for jobs with criticality level 2 at level 1 and 2, respectively. Let d_1 and d_2 denote the latest deadlines at level 1 and 2, respectively and let j_1 and j_2 denote the jobs with deadlines d_1 and d_2.

Lemma 1. *The job in J with latest deadline must be of criticality 2. This implies that $d_1 < d_2$.*

Proof. Suppose that $d_1 \geq d_2$. Consider the collection of jobs J' obtained from J by setting the level-2 WCET of criticality-2 jobs to their level-1 WCET. The MC-schedulability of J implies that J' is MC-schedulable. If j_1 cannot be a lowest priority job for J on s-speed processors, then there is not enough available execution time between the release time of j_1 and d_1 if all the other jobs J_j are executed for $c_j(1)$ units of time. Then, level-1 behaviors of J' cannot be scheduled, which is a contradiction. □

A work-conserving schedule is a schedule that never leaves a processor idle if there is a job available. For each $\ell \in \{1,2\}$, we define Λ_ℓ as the set of time intervals on which all the processors are idle before d_ℓ for any work-conserving schedule and λ_ℓ as the the total length of this set of intervals.

Lemma 2. *For each $\ell \in \{1,2\}$, and for each job J_j in J such that $\chi_j \leq \ell$, we have $[r_j, d_j] \cap \Lambda_\ell = \emptyset$. This implies that $\lambda_2 = 0$.*

Proof. Any job J_j in J such that $\chi_j \leq \ell$ and $[r_j, d_j] \cap \Lambda_\ell \neq \emptyset$ would meet its deadline if it were assigned lowest priority. As J is not OCBP-schedulable on a speed-s processor, then the collection of jobs obtained by removing such J_j from J is also not OCBP-schedulable. This contradicts the minimality of J. □

Since J is MC-schedulable on m speed-1 processors then γ_1 cannot exceed $(d_1 - \lambda_1) \cdot m$ in any criticality 1 scenario. Moreover, in scenarios where all jobs executes for their WCET at criticality 1, $\gamma_1 + \gamma_2(1)$ cannot exceed $(d_2 - \lambda_1) \cdot m$

and in scenarios where all jobs execute for their WCET at criticality 2, $\gamma_2(2)$ cannot exceed $(d_2 - \lambda_2) \cdot m$. Hence, the following inequalities hold:

$$\gamma_1 \le (d_1 - \lambda_1)m \tag{7}$$
$$\gamma_1 + \gamma_2(1) \le (d_2 - \lambda_1)m \le d_2 \cdot m \tag{8}$$
$$\gamma_2(2) \le (d_2 - \lambda_2)m = d_2 \cdot m. \tag{9}$$

Since J is not OCBP-schedulable on m speed-s processor, j_1 and j_2 cannot be the lowest priority jobs on such a processor. This implies that

$$\frac{1}{m}(\gamma_1 + \gamma_2(1) - c_{j_1}(1)) + c_{j_1}(1) > (d_1 - \lambda_1)s$$
$$\frac{1}{m}(\gamma_1 + \gamma_2(2) - c_{j_2}(2)) + c_{j_2}(2) > (d_2 - \lambda_2)s = sd_2$$

Hence:

$$\gamma_1 + \gamma_2(1) > sm(d_1 - \lambda_1) - (m-1)c_{j_1}(1) \tag{10}$$
$$\gamma_1 + \gamma_2(2) > smd_2 - (m-1)c_{j_2}(2). \tag{11}$$

Let us define $x = (d_1 - \lambda_1)/d_2$. By inequalities (8) and (10), it follows that

$$d_2 \cdot m > sm(d_1 - \lambda_1) - (m-1)c_{j_1}(1) \Rightarrow \left(1 - \frac{1}{m}\right)c_{j_1}(1) + d_2 > s(d_1 - \lambda_1).$$

By the MC-schedulability, we have: $c_{j_1}(1) \le d_1 - \lambda_1$. Therefore

$$\left(1 - \frac{1}{m}\right)(d_1 - \lambda_1) + d_2 > s(d_1 - \lambda_1), \Rightarrow s < 1 - \frac{1}{m} + \frac{d_2}{d_1 - \lambda_1} = 1 + \frac{1}{x} - \frac{1}{m}.$$

By inequalities (7), (9), (11), we obtain

$$(d_1 - \lambda_1)m + d_2 \cdot m \ge \gamma_1 + \gamma_2(2) > smd_2 - (m-1)c_{j_2}(2).$$

Hence, $d_1 + d_2 + (1 - 1/m)c_{j_2}(2) > sd_2$, By the MC-schedulability, we have $c_{j_2}(2) \le d_2$. Therefore,

$$d_1 - \lambda_1 + \left(2 - \frac{1}{m}\right)d_2 > sd_2, \Rightarrow s < \frac{d_1 - \lambda_1}{d_2} + 2 - \frac{1}{m} = x + 2 - \frac{1}{m}.$$

Hence,

$$s < \min\left\{1 + \frac{1}{x} - \frac{1}{m}, x + 2 - \frac{1}{m}\right\}.$$

As $x + 2 - \frac{1}{m}$ increases and $1 + \frac{1}{x} - \frac{1}{m}$ decreases, with increasing x, then the minimum value of s occurs when $x + 2 - \frac{1}{m} = 1 + \frac{1}{x} - \frac{1}{m}$, that is $x = \phi - 1$ and $s < 1 + \phi - \frac{1}{m}$.   □

4.2   Scheduling a Sporadic Task System

We finally turn to the scheduling of a two-level mixed-criticality sporadic task system on multiple identical machines. We consider *partitioned* algorithms, that is, scheduling algorithms that partition the tasks on the machines and then schedule each machine independently (no migration). Using Theorem 2, the partitioning problem becomes a two-dimensional *vector scheduling* problem [5] where the vectors to be packed are the utilization vectors (u_j) of the tasks. The two-dimensional vector scheduling problem can be approximated in polynomial time within a factor $1+\epsilon$ for any $\epsilon > 0$ [5], and we are able to derive the following theorem.

Theorem 5. *For any $\epsilon > 0$ there is a polynomial-time partitioning algorithm P such that any task system that is MC-schedulable by some partitioned algorithm on m unit-speed processors can be scheduled by the combination of P and EDF-VD on m speed $\phi + \epsilon$ processors.*

References

1. Barhorst, J., Belote, T., Binns, P., Hoffman, J., Paunicka, J., Sarathy, P., Stanfill, J.S.P., Stuart, D., Urzi, R.: White paper: A research agenda for mixed-criticality systems (2009), http://www.cse.wustl.edu/~cdgill/CPSWEEK09_MCAR/
2. Baruah, S.K., Bonifaci, V., D'Angelo, G., Li, H., Marchetti-Spaccamela, A., Megow, N., Stougie, L.: Scheduling real-time mixed-criticality jobs. In: Hliněný, P., Kučera, A. (eds.) MFCS 2010. LNCS, vol. 6281, pp. 90–101. Springer, Heidelberg (2010)
3. Baruah, S.K., Goossens, J.: Scheduling real-time tasks: Algorithms and complexity. In: Leung, J.Y.T. (ed.) Handbook of Scheduling: Algorithms, Models, and Performance Analysis, ch. 28, CRC Press, Boca Raton (2003)
4. Baruah, S.K., Li, H., Stougie, L.: Towards the design of certifiable mixed-criticality systems. In: Proc. 16th IEEE Real-Time and Embedded Technology and Applications Symposium, pp. 13–22. IEEE, Los Alamitos (2010)
5. Chekuri, C., Khanna, S.: On multidimensional packing problems. SIAM Journal on Computing 33(4), 837–851 (2004)
6. Kalyanasundaram, B., Pruhs, K.: Speed is as powerful as clairvoyance. Journal of the ACM 47(4), 617–643 (2000)
7. Li, H., Baruah, S.K.: An algorithm for scheduling certifiable mixed-criticality sporadic task systems. In: Proc. 16th IEEE Real-Time Systems Symp., pp. 183–192. IEEE, Los Alamitos (2010)
8. Liu, C.L., Layland, J.W.: Scheduling algorithms for multiprogramming in a hard-real-time environment. Journal of the ACM 20(1), 46–61 (1973)

Robust Algorithms for Preemptive Scheduling

Leah Epstein[1] and Asaf Levin[2]

[1] Department of Mathematics, University of Haifa, 31905 Haifa, Israel
lea@math.haifa.ac.il.
[2] Faculty of Industrial Engineering and Management, The Technion,
32000 Haifa Israel
levinas@ie.technion.ac.il.

Abstract. Preemptive scheduling problems on parallel machines are classic problems. Given the goal of minimizing the makespan, they are polynomially solvable even for the most general model of unrelated machines. In these problems, a set of jobs is to be assigned to be executed on a set of m machines. A job can be split into parts arbitrarily and these parts are to be assigned to time slots on the machines without parallelism, that is, for every job, at most one of its parts can be processed at each time.

Motivated by sensitivity analysis and online algorithms, we investigate the problem of designing robust algorithms for constructing preemptive schedules. Robust algorithms receive one piece of input at a time. They may change a small portion of the solution as an additional part of the input is revealed. The capacity of change is based on the size of the new input. For scheduling problems, the maximum ratio between the total size of the jobs (or parts of jobs) which may be re-scheduled upon the arrival of a new job j, and the size of j, is called *migration factor*.

We design a strongly optimal algorithm with the migration factor $1 - \frac{1}{m}$ for identical machines. Such algorithms avoid idle time and create solutions where the (non-increasingly) sorted vector of completion times of the machines is minimal lexicographically. In the case of identical machines this results not only in makespan minimization, but the created solution is also optimal with respect to any ℓ_p norm (for $p > 1$). We show that an algorithm of a smaller migration factor cannot be optimal with respect to makespan or any other norm, thus the result is best possible in this sense as well. We further show that neither uniformly related machines nor identical machines with restricted assignment admit an optimal algorithm with a constant migration factor. This lower bound holds both for makespan minimization and for any ℓ_p norm. Finally, we analyze the case of two machines and show that in this case it is still possible to maintain an optimal schedule with a small migration factor in the cases of two uniformly related machines and two identical machines with restricted assignment.

1 Introduction

We study preemptive scheduling on m identical machines with the goals of makespan minimization and minimization of the ℓ_p norm (for $p > 1$) of the

C. Demetrescu and M.M. Halldórsson (Eds.): ESA 2011, LNCS 6942, pp. 567–578, 2011.

machine completion times. There are n jobs, where p_j, the size (or length) of j, is the processing time of the j-th job on a unit speed machine. The load or completion time of a machine is the last time that it processes any job. In makespan minimization, the goal is to assign the jobs for processing on the machines, minimizing the maximum completion time of any machine. In the minimization of the ℓ_p norm, the goal is to minimize the ℓ_p norm of the vector of machine completion times. The set of feasible assignments is defined as follows. In preemptive scheduling, each machine can execute one job at each time, and every job can be run on at most one machine at each time. Idle time is allowed. A job is not necessarily assigned to a single machine, but its processing time can be split among machines. The intervals or time slots in which it is processed do not necessarily have to be consecutive. That is, a job may be split arbitrarily under the constraint that the time intervals, in which it runs on different machines, are disjoint, and their total length is exactly the processing time of the job.

Following a recent interest in problems which possess features of both offline and online scenarios, we study preemptive scheduling with bounded migration, which is also called robust preemptive scheduling. The problem is not an offline problem, in the sense that the input arrives gradually, but it is not a purely online problem either, since some reassignment of the input is allowed. In this variant jobs arrive one by one, and when a new job arrives, its processing time becomes known. The algorithm needs to maintain a schedule at all times, but when a new job j arrives, it is allowed to change the assignment of previously assigned jobs in a very restrictive way. More accurately, the total size of all parts of jobs which are moved to different time slots (or to the same time slot on a different machine) should be upper bounded by a constant factor, called *the migration factor*, times p_j, that is, their total size must be at most a constant multiplicative factor away from the size of the new job. Algorithms which operate in this scenario are called *robust*. We expect a robust algorithm to perform well not only for the entire input, but also for every prefix of the input, comparing the partial output to an optimal solution for this partial input. We stress the fact that in the preemptive variant which we study, at each step the parts of a given job may be scheduled to use different time slots on possibly different machines. When the schedule is being modified, we allow to cut parts of jobs further, and only the total size of part of jobs which are moved either to a different time slot, or to a different machine (or both), counts towards the migration factor.

Next, we review possible machine environments. The most general machine environment is *unrelated machines* [25,24,2]. In this model each job j and machine i have a processing time $p_i(j)$ associated with them, which is the total time that j requires if it is processed on i. In the preemptive model, the job can be split into parts, as long as it is not executed in parallel on different machines. If the total time allocated for the job on machine i is $t_{i,j}$, then in order to complete the job the equality $\sum_{i=1}^m \frac{t_{i,j}}{p_i(j)} = 1$ must be satisfied. In this model the migration factor is not well-defined. We study several important special cases in which the definition of a migration factor is natural. The most basic and natural machine environment is the case of identical machines, for which $p_i(j) = p_j$ for all i, j

[21,27,6]. We study two additional common models which generalize identical machines. In the model of uniformly related machines [22,20,2,4], machine i has a speed $s_i > 0$, and $p_i(j) = \frac{p_j}{s_i}$. In the restricted assignment model [3] each job j has a subset of machines M_j associated with it, also called a *processing set*. Only machines of M_j can process it. We have $p_i(j) = p_j$ if $i \in M_j$ and otherwise $p_i(j) = \infty$.

In the offline scenario of preemptive multiprocessor scheduling [27,28,29,25,26,22,20,23,32,18,7], jobs are given as a set, while in the online problem, jobs arrive one by one to be assigned in this order. A job is assigned without any knowledge regarding future jobs. The exact time slots allocated to the job must be reserved during the process of assignment, and cannot be modified later. Idle time may be created during the process of assignment, and as a result, the final schedule may contain idle time as well. Note that in this variant, unlike non-preemptive scheduling, idle time can be beneficial.

It is known for a while that the offline preemptive scheduling problem, unlike the non-preemptive variant, can be solved optimally even for more general machine models. First, we discuss makespan minimization, on identical machines and on uniformly related machines [27,22,32]. McNaughton [27] designed a simple algorithm for identical machines (see also [28,29] for alternative algorithms). The study of this problem for related machines started with the work of Liu and Yang [26], as well as the work of [25] who introduced bounds on the cost of optimal schedules, which are the average load (that is, the total size of jobs divided by the total speed of all machines), and $m - 1$ bounds resulting from the average load which must be achieved by assigning the k largest jobs to the k fastest machines (for $1 \leq k \leq m-1$). Horvath et al. [22] proved that the optimal cost is indeed the maximum of those m bounds by constructing an algorithm that uses a large (but polynomial) number of preemptions. Gonzalez and Sahni [20] devised an algorithm that outputs an optimal schedule for which the number of preemptions is at most $2(m - 1)$. This number of preemptions was shown to be optimal in the sense that there exist inputs for which every optimal schedule involves at least that many preemptions. This algorithm was later generalized and simplified for jobs of limited splitting constraints by Shachnai et al. [32]. Similar results for a wide class of objective functions, including minimization of the ℓ_p norm, were obtained by Epstein and Tassa [18]. In that work it was shown that while for makespan minimization, the best situation that one can hope for is a flat schedule, where all machines have the same load, this is not necessarily the case for other functions, such as the ℓ_2 norm. Note that due to the option of idle time, the problem of maximizing the minimum load (also known as the *Santa Claus problem*) does not have a natural definition of preemptive schedules, and thus we do not discuss preemptive schedules with respect to this goal. The case of unrelated machines was also shown to admit a polynomial time algorithm. Specifically, Lawler and Labetoulle [23] showed that this problem can be formulated as a linear program that finds the amount of time that each job should spend on each machine, and afterwards, the problem can be solved using an algorithm for open-shop preemptive scheduling.

A number of articles considered the online problem for identical and uniformly related machines [6,17,34,16,12,9,10], including many results where algorithms of optimal (constant) competitive ratio were designed. However, it is impossible to design an online algorithm which outputs an optimal solution with respect to makespan [6,31,17,12]. For restricted assignment (and unrelated machines), the lower bound of $\Omega(\log m)$ given by Azar, Naor and Rom [3] is valid for preemptive scheduling. Thus, it is known for a while that in many of the cases, in order to obtain a robust algorithm which produces an optimal solution, a non-zero migration factor must be used. Another variant of preemptive scheduling which is on the borderline between online and offline algorithms allows to use a reordering buffer of fixed size, where jobs can be stored before they are assigned to time slots on machines [8]. The non-preemptive version of the problem was studied as well (see e.g. [11]).

Robust algorithms were studied in the past for scheduling and bin packing problems. The model was introduced by Sanders, Sivadasan and Skutella [30], where several simple strategies with a small migration factor, but an improved approximation ratio, compared to online algorithms [19], were presented, as well as a robust polynomial time approximation scheme (PTAS). The bin packing problem was shown to admit an asymptotic robust PTAS [13] (and even the problem of packing d-dimensional cubes for any $d \geq 2$ into unit cubes of the same dimension admits an asymptotic robust PTAS [15]). The problem of maximizing the minimum load on identical machines does not admit a robust PTAS, as was shown by Skutella and Verschae [33]. Although the similarity between the makespan objective and minimizing the ℓ_p norm, obtaining a robust PTAS for minimizing the ℓ_p norm of the load vector on identical machines is impossible (a construction similar to the one of [33], using job sizes of an initial batch as in the example of [1], gives a lower bound of 1.00077 for ℓ_2-norm). A similar negative result was shown for bin packing with cardinality constraints [15]. In this version of bin packing an integer parameter t is given, so that in addition to the restriction on the total size of items in a bin, it may contain up to t items. While this problem admits a standard APTAS and an AFPTAS [5,14], it was shown in [15] that it cannot have a robust APTAS.

Our main result is a polynomial time algorithm of migration factor $1-\frac{1}{m}$ which maintains a strongly optimal solution on identical machines. Such a solution is one where the sorted vector of machine completion times is lexicographically minimal. We show that this result is tight in the sense that no optimal robust algorithm for makespan minimization (or for the minimization of some ℓ_p norm) can have a smaller migration factor. Being strongly optimal, the algorithm is optimal with respect to makespan minimization and any ℓ_p norm for $p > 1$. There are several difficulties in obtaining optimal robust algorithms for preemptive scheduling. First, we note that for an algorithm to run in polynomial time, the number of preemptions must be polynomial. A modification to the schedule must reassign parts of jobs very carefully, without introducing a large number of preemptions. For example, if at each step a new preemption is introduced for each part of a job, the resulting number of preemptions would be exponential. Since we are interested in an optimal schedule (or even in a strongly optimal schedule), the structure of the schedule is strict, and the algorithm does not have much freedom. For example, there are jobs that must be assigned to some machine during the entire time that any machine is active. Finally, since no parallelism is allowed, when a part of a job is moved to some time slot, an algorithm must ensure that it is not assigned to any part of this time slot on any of the other $m-1$ machines. We also note that all the known optimal preemptive algorithms for minimizing the makespan on identical machines are not robust.

In addition to the basic case of identical machines, we study the two other, more general, machine models, which are uniformly related machines and identical machines with restricted assignment. We show that in contrast to the result for identical machines, an optimal robust solution cannot be maintained in the last two machine models. Specifically, we show that the migration factor of any robust algorithm for uniformly related machines is at least $m-1$, and the migration factor of any robust algorithm for identical machines with restricted assignment is $\Omega(m)$. Thus, the situation is very different from offline algorithms, where the algorithms for more general machine models are more complex, but are based on similar observations. Note that for identical machines, the set of solutions for (preemptive or non-preemptive) scheduling with the goal function of the ℓ_p norm of machine completion times is the set of strongly optimal schedules with respect to makespan. As stated above, this is not necessarily the case for other machine models [18]. In the full version, we study a relaxation of preemptive scheduling which is fractional assignment. This variant, where jobs can be split arbitrarily among machines (without any restriction on parallelism) is trivial for identical machines and uniformly related machines, but not for restricted assignment, in which case the lower bound of $\Omega(m)$ on the migration factor of an optimal robust algorithm is valid. We design an optimal algorithm for fractional assignment of migration factor $m-1$. In addition, we study preemptive scheduling for the case of two uniformly related machines and the case of two identical machines with restricted assignment, and design robust optimal preemptive algorithms (with a migration factor of 1) for these cases. For uniformly related machines, the algorithm obtains different schedules for the different norms (since there does not exist one schedule which is optimal for all norms), and uses the best possible migration factor for maintaining an optimal schedule for uniformly related machines. For restricted assignment, since no speeds are present, one output can be optimal for all norms if it is strongly optimal for the makespan, and the algorithm produces such an output. The migration factor of 1 is best possible for robust algorithms which maintain a strongly optimal schedule.

2   An Optimal Robust Algorithm for Identical Machines

In this section we present an algorithm which maintains a strongly optimal schedule for identical machines. The algorithm has a migration factor of $1 - \frac{1}{m}$, which is best possible, as shown in Section 3.

Let J denote the sequence of input jobs, and let $n = |J|$. The instance which consists of the first t jobs is denoted by I_t (and so $I_n = J$). Let $P_t = \sum_{i=1}^{t} p_i$ and $p_t^{\max} = \max_{1 \le i \le t} p_i$. We say that a machine i has load L_i which is the total processing times of the fractions of jobs which are assigned to i. Note that if a schedule does not have idle times then the load of a machine is the last time in which it processes a job.

First, we sketch the behavior of our algorithm. At the arrival time of a job, we use a simple algorithm, LOADS which computes the sorted vector of machine loads in an optimal solution. Algorithm LOADS is based on the methods of McNaughton [27] and [22]. The algorithm creates a potential schedule of a simple form (which cannot be used by our algorithm, since it needs to modify its existing schedule rather than creating one from scratch). LOADS is a recursive algorithm which assigns the largest remaining job to run non-preemptively on an empty machine, unless the remaining set of jobs can be scheduled in a balanced way. Thus, the sorted vector of completion times has a parameter k, where the k machines of smallest completion time have the exact same completion time. After finding the output of LOADS, our algorithm finds which machines have an increased completion time, and the new job is assigned. There may be a time slot in which the job can be scheduled on one machine, but typically some parts of jobs need to be moved to make room for the parts of the new job. We carefully move parts of jobs. In the process it is necessary to make sure that no parts of a job are scheduled to run in parallel (not only parts of the new job), and that parts of jobs are not moved unnecessarily, to avoid an increase in the migration factor.

The solution of the static problem for identical machines is relatively simple. The following algorithm was given by McNaughton [27]. The optimal makespan is given by $\text{OPT} = \text{OPT}_n = \max\{\frac{P_n}{m}, p_n^{\max}\}$. The algorithm has an active machine (initialized as the first machine), and keeps assigning jobs there consecutively, starting time zero, until the assignment of a job j would exceed the time OPT_n. In this case only a part of the job is assigned, the algorithm moves on to the next machine as an active machine, and the remainder of j is assigned there, starting time zero. The algorithm does not use idle time.

The algorithms which we design in this section do not use idle time either, but they are in fact strongly optimal. The algorithm of McNaughton [27] is not strongly optimal, but strongly optimal algorithms are known even for uniformly related machines [22]. We start with the description of an algorithm LOADS which computes a strongly optimal solution OPT_t, for the input I_t.

For an input I_t, we define an order as follows. For two jobs i, j we say that i is larger than j if $p_i > p_j$ or if $p_i = p_j$ and $i < j$. When we refer to the largest ℓ jobs of an input we refer to this ordering. The algorithm for an input I_t is recursive, and it outputs a schedule as well as a sequence of loads $L_1^t \le L_2^t \le \ldots \le L_m^t$. We initialize a set of indices $J = \{1, 2, \ldots, t\}$, $P = P_t$, $k = m$.

1. Compute $p^{\max} = \max_{\ell \in J} p_\ell$, and find a job j such that $p_j = p^{\max}$ (ties are broken in favor of a job of the smallest index).

2. If $\frac{P}{k} < p^{\max}$, let $L_k^t = p^{\max}$, assign job j to machine k (non-preemptively, starting at time zero), $J \leftarrow J \setminus \{j\}$, $k \leftarrow k - 1$, $P \leftarrow P - p_j$, and apply the algorithm recursively, that is, go to step 1.

3. Otherwise, let $L_i^t = \frac{P}{k}$ for $1 \le i \le k$, assign the jobs in J using McNaughton's rule to machines $1, 2, \ldots, k$ with a completion time of $\frac{P}{k}$ and halt.

Note that if machine m receives a large job j then the recursive call of the algorithm acts on machines $1, 2, \ldots, m - 1$ and the jobs $I_t - \{j\}$ exactly as it would have acted in the case that this is the complete input.

The property that the algorithm is strongly optimal can be proved by induction. If a large job j is assigned to machine m, there must be at least one machine with a completion time of p_j (since $\mathrm{OPT} \ge p_j$). The strong optimality of the assignment to machines $1, 2, \ldots, m$ follows by induction. If small jobs are assigned to all machines, then all machines have the equal loads, which is clearly strongly optimal. The algorithm can be implemented using a running time of $O(t + m \log m)$.

2.1   Properties of Load Vectors in Strongly Optimal Solutions

Given the strongly optimal solution OPT_t, we say that job j is large in OPT_t and it corresponds to machine k, if in OPT_t the job j is assigned to machine k in step 2. Note that in this case $k > 1$. Otherwise, we say that j is small, and it corresponds to machine κ, which is the final value of k. That is, the last step is a step where L_i^t was defined to be $\frac{P}{\kappa}$ for $1 \le i \le \kappa$ (and $p_j \le \frac{P}{\kappa}$). The following holds due to the construction of the algorithm.

Proposition 1. L_i^t is a monotonically non-decreasing function of i.

In what follows, we refer to L_i^t as the ith completion time of the input I_t.

Lemma 1. L_i^t is a monotonically non-decreasing function of t.

Lemma 2. Let OPT_t and OPT_{t-1} be the solutions obtained by the above algorithm for I_t and I_{t-1}, respectively, and assume that t is a large job in OPT_t which is assigned to machine k. Let $k_{t-1} \le m$ be the maximum index such that $L_i^{t-1} = L_1^{t-1}$ for $1 \le i \le k_{t-1}$ (i.e., the maximum index of a machine that receives small jobs in OPT_{t-1}) and let $k_t \le m$ be the maximum index such that $L_i^t = L_1^t$ for $1 \le i \le k_t$. Then the following properties hold: 1. $k_t < k$ and $k_{t-1} \le k$. 2. $L_i^t = L_i^{t-1}$ for $k < i \le m$. 3. $k_t \ge k_{t-1} - 1$. 4. $L_i^t = L_{i+1}^{t-1}$ for $k_t + 1 \le i \le k - 1$. 5. $L_{k_t+1}^{t-1} \le L_{k_t}^t$.

Lemma 3. Assume that t is a small job in OPT_t. Let k be the machine index to which t corresponds. Let k_{t-1} denote the maximum index j such that $L_1^{t-1} = L_j^{t-1}$. Then $k \ge k_{t-1}$.

2.2   The Procedure ASSIGN

We describe a procedure ASSIGN which will be used by our algorithm. For each new job, ASSIGN is invoked at most once. We will show that the migration factor of this procedure is at most $1 - \frac{1}{m}$ and the claim regarding the migration factor will follow.

The input to ASSIGN consists of the following parameters: a job or a part x of size X of some job, an index of a machine $\ell \geq 1$, a time $\mathcal{L} \geq X$, a preemptive assignment of a set of jobs to machines $1, 2, \ldots, m$, a set of times $L_1 \leq \ldots \leq L_\ell$ for machines $1, 2, \ldots, \ell$, with $L_\ell \leq \mathcal{L}$, such that machine i is free during the time $[L_i, \mathcal{L}]$, and $\ell \mathcal{L} = \sum_{i=1}^{\ell} L_i + X$. Machine ℓ may have a part of the job x assigned starting time $\mathcal{L}$, and no other machine has any idle time. This is the only time during which x is possibly assigned. All other machines of indices $\ell + 1, \ldots, m$ are completely occupied during the time $[0, \mathcal{L}]$. After the application of ASSIGN, no machine will have any idle time.

We start with assigning a part of x, of size $\mathcal{L} - L_1$, to machine 1. Note that $X = \sum_{i=1}^{\ell} (\mathcal{L} - L_i)$, so the remainder of x with size $\tilde{X}$ satisfies $\tilde{X} = X - (\mathcal{L} - L_1) = \sum_{i=2}^{\ell} (\mathcal{L} - L_i)$. Note that no migration was used, and no idle time was introduced. Since $L_1 \leq L_i$ for $1 \leq i \leq \ell$, $\mathcal{L} - L_1 \geq \mathcal{L} - L_i$ for $1 \leq i \leq \ell$, so $X - \tilde{X} \geq \frac{X}{\ell} \geq \frac{X}{m}$.

If $\ell = 1$ then we are done. Assume therefore $\ell \geq 2$, in which case the remainder has a positive size (that is, $\tilde{X} > 0$). We create cells as follows. For $2 \leq i \leq \ell - 1$, if $L_{i+1} > L_i$, there are $i - 1$ cells of the time interval $[L_i, L_{i+1}]$, on machines $2, \ldots, i$. In addition, there are $\ell - 1$ cells of the interval $[L_\ell, \mathcal{L}]$ on machines $2, \ldots, \ell$, if $\mathcal{L} > L_\ell$. The total length of all these intervals is $\sum_{i=2}^{\ell-1} (L_{i+1} - L_i)(i - 1) + (\mathcal{L} - L_\ell)(\ell - 1)$.

Claim. $L_1 \geq \sum_{i=2}^{\ell-1} (L_{i+1} - L_i)(i - 1) + (\mathcal{L} - L_\ell)(\ell - 1) = \tilde{X}$.

We let $\Delta = \sum_{i=2}^{\ell-1} (L_{i+1} - L_i)(i - 1) + (\mathcal{L} - L_\ell)(\ell - 1)$, $\delta_i = L_{i+1} - L_i$ for $2 \leq i \leq \ell - 1$ and $\delta_\ell = \mathcal{L} - L_\ell$. That is, $\Delta = \sum_{i=2}^{\ell} \delta_i (i - 1)$. Since $X - \tilde{X} \geq \frac{X}{\ell}$, we have $\Delta = \tilde{X} \leq (1 - \frac{1}{\ell})X \leq (1 - \frac{1}{m})X$.

We create stripes from time 0 until time Δ. There are $i - 1$ stripes of height δ_i for $i = 2, \ldots, \ell$. The stripes are non-overlapping, and each stripe is created just above the previous one, so that all stripes of one height are consecutive. Consider the stripes in a bottom up manner, and recall that a stripe has a cell associated with it. The total size of parts of jobs that will migrate in one stripe is exactly its height, so the total size of migrating jobs is exactly $\Delta \leq (1 - \frac{1}{m})X$.

We next consider one stripe of height δ_i, of the time interval $[\alpha, \beta]$ ($\beta - \alpha = \delta_i$) and show how to assign a part of x of length δ_i during the time $[\alpha, \beta]$ (possibly preemptively, that is, this part may be split further), while some parts of jobs assigned during this time interval are moved to the cell associated with this stripe, we denote the time slot of this cell by $[\alpha', \beta']$ ($\beta' - \alpha' = \delta_i$) and the machine of this cell by q. We first reassign the parts of jobs of cells $2, 3, \ldots, q - 1$ of the time slot $[\alpha', \beta']$ according to McNaughton's algorithm. Note that this reassignment of parts of jobs which we schedule in the current application of the ASSIGN procedure, does not increase the migration factor of the algorithm since, until the end of the procedure we need not move the jobs, but only declare the necessary changes in the solution. In the resulting assignment to previous cells in this time slot there are at most n times during $(\alpha', \beta']$ in which a machine preempts a job (or finishes a job). Partition the time $(\alpha, \beta]$ into maximum length time intervals $(\alpha + t_i, \alpha + t_{i+1}]$, defined for $0 \leq i \leq r$, where $t_i < t_{i+1}$, such that $t_0 = 0$ and $\beta = \alpha + t_{r+1}$, and there are no preemptions during $(\alpha + t_i, \alpha + t_{i+1}]$ and $(\alpha' + t_i, \alpha' + t_{i+1}]$, i.e., every machine executes the same jobs during all intervals of this form. Thus the number of such intervals is at most n plus the number of maximal consecutive intervals $(\alpha', \beta') \subseteq [\alpha, \beta]$ such that there are no preemptions during (α', β'). For each such interval $[\alpha + t_i, \alpha + t_{i+1}]$, find a machine which executes a job that is not executed on any machine during $[\alpha' + t_i, \alpha' + t_{i+1}]$. Since machine q has an empty cell initially, and the time slot $[\alpha' + t_i, \alpha' + t_{i+1}]$ on it is still empty, there are at most $m - 1$ jobs running during $[\alpha' + t_i, \alpha' + t_{i+1}]$, so there is at least one machine $1 \leq q' \leq m$ running a different job during $[\alpha + t_i, \alpha + t_{i+1}]$, and this part of job is moved from $[\alpha + t_i, \alpha + t_{i+1}]$ on q' to $[\alpha' + t_i, \alpha' + t_{i+1}]$ on q. A part of x is assigned during the time $[\alpha + t_i, \alpha + t_{i+1}]$ on q'.

Corollary 1. *If $\mathcal{L} \geq X$, $L_\ell \leq \mathcal{L}$ and $\ell\mathcal{L} = \sum_{i=1}^{\ell} L_i + X$, procedure* ASSIGN *returns a feasible schedule of the jobs which satisfies the required properties.*

Lemma 4. *The number of preemptions in the solution after n jobs were released, is at most $m^3 n^2$. Moreover, the running time of* ASSIGN *is polynomial in the input size.*

2.3 A Robust Algorithm

We will describe a greedy algorithm for modifying the schedule upon arrival of a job. This algorithm keeps the invariants that there is no idle time, and there are no preemptions after the completion time of the least loaded machine. That is, it receives as an input an arbitrary strongly optimal schedule for I_{t-1}, that has no preemptions after time L_1^{t-1} and converts it into a strongly optimal schedule for I_t, using a migration factor of $1 - \frac{1}{m}$. In order to define the assignment of job t and possible migration of a total size of at most $(1 - \frac{1}{m})p_t$ of other jobs we consider two cases for OPT$_t$.

Case 1. t is a large job in OPT$_t$. Let k be the machine index to which t corresponds. Let $k_{t-1}, k_t \leq m$ be as in Lemma 2. We start with the assignment to machine $k_t + 1$. Let $b = L_{k_t}^t$ and recall that $p_t = L_k^t$. Assign a part of the job t to machine $k_t + 1$ during the time $[L_{k_t}^t, L_k^t]$. After we apply the procedure ASSIGN (see below), the time interval $[L_{k_t+1}^{t-1}, L_{k_t}^t]$ will be occupied as well (so no idle time would be introduced) and we will change the indexing of machines so that machine $k_t + 1$ becomes machine k, and each machine $k_t + 2 \leq i \leq k$ becomes machine $i - 1$ (if $k_t + 1 = k$ then no change of indexing is performed). The new loads of these machines are $(L_{k_t+2}^{t-1}, \ldots, L_k^{t-1}, p_t, L_{k+1}^{t-1}, L_{k+2}^{t-1}, \ldots, L_m^{t-1}) = (L_{k_t+1}^t, \ldots, L_{k-1}^t, L_k^t, L_{k+1}^t, L_{k+2}^t, \ldots, L_m^t)$. Since $k_{t-1} \leq k_t + 1$, these are exactly the sizes of the $m - k_t$ largest jobs in $\{1, 2, \ldots, t\}$. We are left with a part of size b of job t, which needs to be assigned to machines $1, 2, \ldots, k_t + 1$. We invoke the procedure ASSIGN with $\ell = k_t + 1$, $X = b$, $\mathcal{L} = b$, and $L_i = L_i^{t-1}$ for $1 \leq i \leq k_t + 1$. By this assignment, no preemptions are introduced after time

$L_{k_t}^t$. By these definitions, $X = \mathcal{L}$ and $\mathcal{L} \geq L_\ell$. To show $\ell\mathcal{L} = \sum_{i=1}^{\ell} L_i + X$, we recall that $b = \mathcal{L} = \frac{\sum_{i=1}^{k_t+1} L_i}{k_t}$, so $(k_t + 1)b = \sum_{i=1}^{k_t+1} L_i + b$.

Case 2. t is a small job in OPT$_t$. Denote by k the machine index to which t corresponds. We invoke the procedure ASSIGN with $\ell = k$, $X = p_t$, $L_i = L_i^{t-1}$ for $1 \leq i \leq \ell$ and $\mathcal{L} = \frac{Y' + p_t}{\ell}$, where Y' denotes the total size of jobs not assigned to machines $k+1, \ldots, m$ in OPT$_{t-1}$. We need to show $X \leq \mathcal{L}$, $\mathcal{L} \geq L_\ell$, and $\ell\mathcal{L} = \sum_{i=1}^{\ell} L_i + X$. Since job t is associated with machine $\ell = k$ (as a small job), we have $\frac{Y' + p_t}{k} \geq p_t$ so $\mathcal{L} \geq X$. Since $\mathcal{L} = L_k^t$, we get $\mathcal{L} \geq L_k$. Finally, $\ell\mathcal{L} = Y' + p_t = Y' + X$. Since Y' is exactly the total size of jobs assigned to machines $1, 2, \ldots, k$ prior to the assignment of job t, we have $Y' = \sum_{i=1}^{\ell} L_i$. We have proved the following theorem.

Theorem 1. *There exists a polynomial time algorithm of migration factor $1 - \frac{1}{m}$ which maintains a strongly optimal schedule on m identical machines with a polynomial number of preemptions.*

3   Lower Bounds on the Migration Factor

Identical machines. We show that an optimal algorithm (which is not necessarily strongly optimal) must have a migration factor of at least $1 - \frac{1}{m}$. Recall that this is exactly the migration factor of the algorithm given in Section 2. An optimal algorithm which minimizes the ℓ_p norm for some $1 < p < \infty$ on identical machines is simply a strongly optimal with respect to makespan. Thus, we also find that the algorithm of Section 2 has the optimal migration factor for minimization of any ℓ_p norm.

Proposition 2. *Let $\mathcal{A}$ be an optimal robust algorithm for preemptive scheduling on identical machines. The migration factor of $\mathcal{A}$ is no smaller than $1 - \frac{1}{m}$.*

Uniformly related machines and restricted assignment. We show the next lower bounds.

Theorem 2. *Any optimal algorithm for preemptive scheduling on uniformly related machines so as to minimize the makespan or to minimize the ℓ_p norm has a migration factor of at least $m - 1$.*

Theorem 3. *Any optimal algorithm to the problem of preemptive scheduling on parallel machines with restricted assignment so as to minimize the makespan has a migration factor of at least $\frac{m-1}{2}$. The same holds for the problem of minimizing the ℓ_p norm.*

Recall that an algorithm for minimization of the ℓ_p norm for some $1 < p < \infty$ cannot use idle time and must be strongly optimal. Note that the next lower bound does not exclude the option of an algorithm which uses idle time and minimizes the makespan, which has a migration factor of $\frac{1}{2}$ (the lower bounds shown in Proposition 2 and Theorem 3 hold for this case).

Theorem 4. *Any strongly optimal algorithm for preemptive scheduling on two machines with restricted assignment has a migration factor of at least 1.*

Non-preemptive scheduling. Our work is concerned with preemptive scheduling. The case of non-preemptive scheduling was investigated [30], and it was shown that a migration factor of $\Omega(m)$ is required in order to maintain an optimal schedule (even if the computational complexity of the algorithm is not polynomial). We further show that no finite migration factor is possible for non-preemptive scheduling, even for two identical machines.

Consider two identical machines. Let N be a large integer. The first four jobs are of sizes $N + 2, N + 1, N - 1$, and $N - 2$. The unique optimal solution assigns the two jobs of sizes $N + 2$ and $N - 2$ to one machine and the other two jobs to the other machine. The optimal makespan is $2N$. Next, a job of size 6 arrives. The unique optimal solution assigns the three smallest jobs to one machine and the two largest jobs to the other machine, and its makespan is $2N + 3$. To achieve this solution, two of the first four jobs must migrate to another machine. We get a migration factor of at least $\frac{2N-1}{6}$ which can be arbitrarily large. To generalize this example for an arbitrary number of machines m, the initial input is augmented by $m - 2$ jobs of size $2N$. Optimal solutions never combine these jobs with the other jobs. Note that the proof holds both for makespan and for any ℓ_p norm.

References

1. Alon, N., Azar, Y., Woeginger, G.J., Yadid, T.: Approximation schemes for scheduling. In: Proc. 8th Symp. on Discrete Algorithms (SODA), pp. 493–500. ACM/SIAM (1997)
2. Aspnes, J., Azar, Y., Fiat, A., Plotkin, S., Waarts, O.: On-line load balancing with applications to machine scheduling and virtual circuit routing. Journal of the ACM 44(3), 486–504 (1997)
3. Azar, Y., Naor, J., Rom, R.: The competitiveness of on-line assignments. Journal of Algorithms 18(2), 221–237 (1995)
4. Berman, P., Charikar, M., Karpinski, M.: On-line load balancing for related machines. Journal of Algorithms 35, 108–121 (2000)
5. Caprara, A., Kellerer, H., Pferschy, U.: Approximation schemes for ordered vector packing problems. Naval Research Logistics 92, 58–69 (2003)
6. Chen, B., van Vliet, A., Woeginger, G.J.: An optimal algorithm for preemptive on-line scheduling. Operations Research Letters 18, 127–131 (1995)
7. Correa, J.R., Skutella, M., Verschae, J.: The power of preemption on unrelated machines and applications to scheduling orders. In: Dinur, I., Jansen, K., Naor, J., Rolim, J. (eds.) APPROX 2009. LNCS, vol. 5687, pp. 84–97. Springer, Heidelberg (2009)
8. Dósa, G., Epstein, L.: Preemptive online scheduling with reordering. SIAM Journal on Discrete Mathematics 25(1), 21–49 (2011)
9. Ebenlendr, T., Jawor, W., Sgall, J.: Preemptive online scheduling: optimal algorithms for all speeds. Algorithmica 53(4), 504–522 (2009)
10. Ebenlendr, T., Sgall, J.: Optimal and online preemptive scheduling on uniformly related machines. Journal of Scheduling 12(5), 517–527 (2009)
11. Englert, M., Özmen, D., Westermann, M.: The power of reordering for online minimum makespan scheduling. In: Proc. 48th Symp. Foundations of Computer Science (FOCS), pp. 603–612 (2008)

12. Epstein, L.: Optimal preemptive on-line scheduling on uniform processors with non-decreasing speed ratios. Operations Research Letters 29(2), 93–98 (2001)
13. Epstein, L., Levin, A.: A robust APTAS for the classical bin packing problem. Mathemtical Programming 119(1), 33–49 (2009)
14. Epstein, L., Levin, A.: AFPTAS results for common variants of bin packing: A new method for handling the small items. SIAM Journal on Optimization 20(6), 3121–3145 (2010)
15. Epstein, L., Levin, A.: Robust approximation schemes for cube packing (2010)
16. Epstein, L., Noga, J., Seiden, S.S., Sgall, J., Woeginger, G.J.: Randomized online scheduling on two uniform machines. Journal of Scheduling 4(2), 71–92 (2001)
17. Epstein, L., Sgall, J.: A lower bound for on-line scheduling on uniformly related machines. Operations Research Letters 26(1), 17–22 (2000)
18. Epstein, L., Tassa, T.: Optimal preemptive scheduling for general target functions. Journal of Computer and System Sciences 72(1), 132–162 (2006)
19. Fleischer, R., Wahl, M.: Online scheduling revisited. Journal of Scheduling 3(5), 343–353 (2000)
20. Gonzales, T.F., Sahni, S.: Preemptive scheduling of uniform processor systems. Journal of the ACM 25, 92–101 (1978)
21. Graham, R.L.: Bounds for certain multiprocessing anomalies. Bell System Technical Journal 45, 1563–1581 (1966)
22. Horvath, E.C., Lam, S., Sethi, R.: A level algorithm for preemptive scheduling. Journal of the ACM 24(1), 32–43 (1977)
23. Lawler, E.L., Labetoulle, J.: On preemptive scheduling of unrelated parallel processors by linear programming. Journal of the ACM 25(4), 612–619 (1978)
24. Lenstra, J.K., Shmoys, D.B., Tardos, E.: Approximation algorithms for scheduling unrelated parallel machines. Math. Program. 46, 259–271 (1990)
25. Liu, J.W.S., Liu, C.L.: Bounds on scheduling algorithms for heterogeneous computing systems. In: Rosenfeld, J.L. (ed.) Proceedings of IFIP Congress 1974. Information Processing, vol. 74, pp. 349–353 (1974)
26. Liu, J.W.S., Yang, A.T.: Optimal scheduling of independent tasks on heterogeneous computing systems. In: Proceedings of the ACM National Conference, vol. 1, pp. 38–45. ACM, New York (1974)
27. McNaughton, R.: Scheduling with deadlines and loss functions. Management Science 6, 1–12 (1959)
28. Muntz, R.R., Coffman Jr., E.G.: Optimal preemptive scheduling on two-processor systems. IEEE Transactions on Computers 18(11), 1014–1020 (1969)
29. Muntz, R.R., Coffman Jr., E.G.: Preemptive scheduling of real-time tasks on multiprocessor systems. Journal of the ACM 17(2), 324–338 (1970)
30. Sanders, P., Sivadasan, N., Skutella, M.: Online scheduling with bounded migration. Mathematics of Operations Research 34(2), 481–498 (2009)
31. Sgall, J.: A lower bound for randomized on-line multiprocessor scheduling. Information Processing Letters 63(1), 51–55 (1997)
32. Shachnai, H., Tamir, T., Woeginger, G.J.: Minimizing makespan and preemption costs on a system of uniform machines. Algorithmica 42(3-4), 309–334 (2005)
33. Skutella, M., Verschae, J.: A robust PTAS for machine covering and packing. In: de Berg, M., Meyer, U. (eds.) ESA 2010. LNCS, vol. 6346, pp. 36–47. Springer, Heidelberg (2010)
34. Wen, J., Du, D.: Preemptive on-line scheduling for two uniform processors. Operations Research Letters 23, 113–116 (1998)

Approximate Distance Queries
for Weighted Polyhedral Surfaces

Hristo N. Djidjev[1] and Christian Sommer[2]

[1] Los Alamos National Laboratory, Los Alamos, NM 87545
[2] Massachusetts Institute of Technology, Cambridge, MA 02139

Abstract. Let P be a planar polyhedral surface consisting of n triangular faces, each assigned with a positive weight. The weight of a path p on P is defined as the weighted sum of the Euclidean lengths of the portions of p in each face multiplied by the corresponding face weights. We show that, for every $\varepsilon \in (0,1)$, there exists a data structure, termed *distance oracle*, computable in time $O(n\varepsilon^{-2}\log^3(n/\varepsilon)\log^2(1/\varepsilon))$ and of size $O(n\varepsilon^{-3/2}\log^2(n/\varepsilon)\log(1/\varepsilon))$, such that $(1+\varepsilon)$–approximate distance queries in P can be answered in time $O(\varepsilon^{-1}\log(1/\varepsilon) + \log\log n)$. As in previous work (Aleksandrov, Maheshwari, and Sack (*J. ACM 2005*) and others), the big–O notation hides constants depending logarithmically on the ratio of the largest and smallest face weights and reciprocally on the sine of the smallest angle of P. The tradeoff between space and query time of our distance oracle is a significant improvement in terms of n over the previous best tradeoff obtained by a distance oracle of Aleksandrov, Djidjev, Guo, Maheshwari, Nussbaum, and Sack (*Discrete Comput. Geom. 2010*), which requires space roughly quadratic in n for a comparable query time.

1 Introduction

We design an efficient algorithm and a data structure to answer approximate distance queries between points on a weighted planar polyhedral surface. The problems of computing shortest paths arise in numerous application areas and have consistently been among the most active research topics in theoretical computer science.

In many applications, multiple shortest paths have to be computed for the same domain between different pairs of points without knowing the sequence of pairs in advance. In this version of the problem, an appropriate data structure can be computed in a *preprocessing* phase of the algorithm, and a *query* algorithm may utilize the data structure in order to efficiently answer shortest-path or distance queries.

Algorithms for the shortest-path query problem have been developed for planar graphs [6,4,11] and, in general, for graphs with small separators [5]. Algorithms with better space/query-time tradeoffs and faster query times are possible if *approximate* distances and shortest paths are acceptable. A *stretch–α distance oracle* is a data structure that allows computing a distance estimate that does

C. Demetrescu and M.M. Halldórsson (Eds.): ESA 2011, LNCS 6942, pp. 579–590, 2011.
© Springer-Verlag Berlin Heidelberg 2011

not exceed α times the weight of a shortest path [14]. For planar graphs, Thorup [13] and Klein [8] construct, for any $\epsilon > 0$, a $(1 + \epsilon)$–stretch oracle of size $O(n \log(n)/\epsilon)$ in $O(n \log^3(n)/\epsilon^2)$ time that answers distance queries in $O(1/\epsilon)$ time. Their work has been extended to graphs with bounded genus [7] and to minor-free graphs [1].

Modeling a real-world problem as a shortest-path problem often requires the use of non-Euclidean distances. Such types of distances have been intensively studied in recent years and they are the focus of this work. We consider the *weighted region distance* introduced by Mitchell and Papadimitriou [10], where the geometric region is divided into triangles or tetrahedra each assigned an individual weight. The weight or length of a path p is then defined as the weighted sum of the portions of p in each triangle (tetrahedron) multiplied by the corresponding weight. Such distance measures are appropriate, for instance, when computing a route through a terrain with certain terrain properties such as terrain type (e.g. water, sand, or rock), slope, and obstacles, encoded as weights. Another potential application is seismology, where seismic waves follow shortest paths, and the speed of a wave in each layer depends on the type and density of the rock.

Compared to the Euclidean case, computing a shortest path in a weighted region is a more involved problem. While in the interior of each triangle each connected portion of a shortest path is a straight-line segment; when crossing a boundary between two triangles the path locally satisfies *Snell's Law* $w^- \sin(\varphi^-) = w^+ \sin(\varphi^+)$, where w^- and w^+ are the weights of the triangles before and after the boundary and φ^- and φ^+ are the corresponding acute angles between the portions of the paths in the corresponding triangles and the normal to the boundary, respectively. It is believed that an exact algorithm for finding a shortest path between two arbitrary points does not exist for arbitrary values of n.

Accordingly, several algorithms for finding approximate shortest paths have been developed. Mitchell and Papadimitriou [10] first studied the problem and developed an algorithm for computing a $(1 + \varepsilon)$–approximate shortest path between a pair of nodes in $O(n^8 \log(n/\varepsilon))$ time and using $O(n^4)$ space. Their result was subsequently improved in several papers, culminating in the algorithm by Aleksandrov, Maheshwari, and Sack [3] with $O(\frac{n}{\sqrt{\varepsilon}} \log(n/\varepsilon) \log(1/\varepsilon))$ time and $O(n)$ space. The big-O notation hides further dependencies on the maximum and minimum weight ratio and on the angles of the triangulation.

In this paper we study the problem of constructing a distance oracle for answering shortest-path queries between pairs of points on a weighted polyhedral surface. This query version of the shortest-path problem was studied previously by Aleksandrov, Djidjev, Guo, Maheshwari, Nussbaum, and Sack [2]. They prove the following.

Theorem 1 (Aleksandrov et al. [2, Theorem 7]). *Let P be a weighted polyhedral surface of genus g consisting of n triangular faces. Let $\varepsilon \in (0, 1)$ and $q \in (\varepsilon^{-1/2} \log^2(1/\varepsilon), \varepsilon^{-1/2}(g + 1)^{2/3} n^{1/3})$. There exists a data structure for $(1 +$*

ε)–*approximate point-to-point distance queries with space* $O(\frac{(g+1)n^2}{\varepsilon^{3/2}q}\log^4(1/\varepsilon))$, *preprocessing time* $O(\frac{(g+1)n^2}{\varepsilon^{3/2}q}\log(n/\varepsilon)\log^4(1/\varepsilon))$, *and query time* $O(q)$.

The approach of [2] is the following. First, the domain is discretized (see also [3] and Section 2.2) by defining a set of suitably spaced points (called *Steiner points*) on the bisectors of each triangle. Second, a graph is defined with nodes being the Steiner points and edges being added between any pair of nodes whose corresponding Steiner points belong to bisectors of the same or an adjacent triangle. The weight on the edges is equal to the locally computed distance between the corresponding Steiner points. The resulting graph G_ε has $O((n/\sqrt{\varepsilon})\log(1/\varepsilon))$ nodes and $O((n/\varepsilon)\log^2(1/\varepsilon))$ edges.

It is shown that any shortest path in the region between two Steiner points can be approximated by a path in G_ε between the same points with weight at most $1+\varepsilon$ times larger.

In this paper we combine the discretization methodology of Aleksandrov et al. [2,3] with novel algorithms for preprocessing G_ε and querying the resulting data structure. We prove the following result.

Theorem 2. *Let P be a planar polyhedral surface consisting of n triangular faces with a positive weight assigned to each of them. Let $\epsilon \in (0,1)$. There exists a data structure to answer $(1+\epsilon)$–approximate point-to-point distance queries in P with $O(n\epsilon^{-3/2}\log^2(n/\epsilon)\log^2(1/\epsilon))$ space, $O(n\epsilon^{-2}\log^3(n/\epsilon)\log^2(1/\epsilon))$ preprocessing time, and $O(\epsilon^{-1}\log(1/\epsilon) + \log\log n)$ query time.*

The constants in the big-O bounds depend on the geometry of P in the same way as in [3] and [2]. The performance gain (compared to Aleksandrov et al. [2]) has three main reasons: *i)* we only *approximate* distances in G_ε (without any consequences to the quality of the final approximation), *ii)* we use ε–covers on shortest-path separators (as defined in [13], using the scaled version to improve the query time), and *iii)* we work on an implicit representation of a planar version of G_ε to keep the dependency on n almost linear and the dependency on $1/\varepsilon$ quadratic in the preprocessing time, less than quadratic in the space requirements, and linear in the query time.

2   Preliminaries

2.1   Definitions

For the sake of brevity, we assume some familiarity with previous work by Aleksandrov et al. [3,2]. We only include (extracted from [2]) the most important definitions. Let P be a planar polyhedral surface in 3–dimensional Euclidean space consisting of n triangular faces $f_1, \ldots, f_n$. Each face f_i has an associated positive weight w_i, representing the cost of traveling a unit Euclidean distance inside f_i. The cost of traveling along an edge is the minimum of the weights of the triangles incident to that edge. Edges are assumed to be part of the triangle they inherit their weight from. The cost of a path π in P is defined as

$||\pi|| = \sum_{i=1}^{n} w_i |\pi_i|$, where $|\pi_i|$ denotes the Euclidean length of the portion π_i of π in f_i. Path lengths in graphs are denoted by $\mathsf{len}(\cdot)$, distances are denoted by $\mathsf{d}(\cdot, \cdot)$.

2.2   Domain Discretization

The approach of [3] is, given a node s of P and a parameter ε, to first discretize the polyhedral surface P to obtain a graph G_ε. The shortest-path lengths from s in G_ε are $(1 + \varepsilon)$-approximations for the corresponding distances in P [3, Theorem 3.2]. The discretization of P is constructed as follows.

For each triangle, we compute the bisectors for all its angles. Let v denote a node of P, let α denote one of its angles, and let ℓ denote the corresponding bisector. On ℓ, we add Steiner points $p_0, \ldots p_k$ as follows. p_0 is at distance $r(v) := \varepsilon \frac{w_{\min}(v)}{7 w_{\max}(v)} \delta(v)$, where $\delta(v)$ is the minimum Euclidean distance from v to the set of edges incident to triangles around v but *not* incident to v, and $w_{\min}(v)$ and $w_{\max}(v)$ are the minimum and the maximum weight of triangles incident to v, respectively. The remaining Steiner points p_i are chosen as a geometric progression such that $|p_{i-1} p_i| = \sin(\alpha/2) \sqrt{\varepsilon/2} |v p_{i-1}|$ for $i = 1, \ldots, k$. Using these Steiner points, we can compute $(1 + \varepsilon/2)$-approximate distances between any two nodes on the boundary of a triangle.

The number of Steiner points on bisector ℓ of angle α at node v is at most $C(\ell)\varepsilon^{-1/2} \log_2(2/\varepsilon)$, where $C(\ell) < \frac{1.61}{\sin \alpha} \log_2(2 |\ell| / r(v))$ [3, Lemma 2.3]. Following [3,12,2], we assume that $C(\ell)$ is bounded by a constant.

We construct a graph G_ε on $\Theta(n\varepsilon^{-1/2} \log(1/\varepsilon))$ nodes, wherein each node corresponds to either an original node v or to a Steiner point. The edges of G_ε are chosen such that nodes corresponding to Steiner points on neighboring bisectors are connected. Two bisectors are called *neighbors* if the corresponding triangles share at least one edge. Each bisector has nine neighboring bisectors[1] (three within its own triangle and two in each of the adjacent triangles); consequently, the number of edges in G_ε is $\Theta(n\varepsilon^{-1} \log^2(1/\varepsilon))$. The edges have weights defined as the distance in P restricted to the two triangles the corresponding bisectors lie in.

For each shortest path p in P, let $\pi = \pi(p)$ denote the approximating path in G_ε constructed by the algorithm from [3]. We also use the inverse mapping defined by $\pi^{-1}(\pi) = p$. By [3], p and π have the same source and target. We also use the following property that follows from the construction in [3].

Lemma 1. *Let p_1 and p_2 be two non-intersecting shortest paths in P. Then no pair of segments of the paths $\pi(p_1)$ and $\pi(p_2)$ intersect.*

---

[1] Here we use the version of Aleksandrov et al. [2, Section 3.1], wherein each bisector is defined to be a neighbor of all the nine bisectors in adjacent triangles, as opposed to the six bisectors sharing an edge as in [3]. Note that a bisector is a neighbor of itself.

2.3   Approximate Distance Oracles for Planar Graphs

Our preprocessing and query algorithms build on techniques that have been used previously to construct approximate distance oracles for planar graphs [9,13].

The first technique is to *approximately represent shortest paths that intersect a shortest path* [9, Lemma 4]. Let Q be a shortest path of length $O(\delta)$ in a graph G. For any $\epsilon > 0$ there exists a set of nodes $C_\epsilon(Q) \subseteq V(Q)$ (or simply $C(Q)$ if ϵ is clear from the context) termed *cover* of size $O(1/\epsilon)$ such that for those pairs of nodes (u, v) at distance $\delta \leq \mathsf{d}(u, v) \leq 2\delta$ for which all the shortest paths between u and v intersect Q, there is a node q termed *portal* in the cover $q \in C_\epsilon(Q)$ such that

$$\mathsf{d}(u, v) \leq \mathsf{d}(u, q) + \mathsf{d}(q, v) \leq (1 + \epsilon)\mathsf{d}(u, v). \tag{1}$$

The distance oracle involves storing with each node v the portals that cover v with respect to several shortest paths (and the distances associated with these portals).

The second technique is to *recursively separate a planar graph by shortest paths*. Given a triangulated planar graph on N nodes and a rooted spanning tree, Thorup [13] demonstrates how to find a triangle such that the paths from the root of the tree to the corners of the triangle separate the graph into at least two disconnected subgraphs of size at most $2N/3$. The recursive application of this separator theorem yields components that are separated from each other by a constant number of shortest paths.

Note that we cannot directly use the algorithms from [13] due to two main differences between G_ϵ and the graphs considered in [13]: *i)* our graph is *not* planar and *ii)* our weights are not integral but real. Furthermore, a direct adaptation of the algorithms in [13] would result in a rather high dependency on ε, which is considered undesirable in this line of work [3]. We adapt Thorup's algorithms for our scenario with main improvements with respect to the parameter ε, both in the preprocessing and query times.

3   Discretization, Pseudo-planarization, and Separator Decomposition

3.1   Discretizing the Weights

We replace the original weights of G_ε by integral weights in $\{0, 1, \ldots, N\}$, in order to apply the techniques in [13] on G_ε. We determine the value of N and the mapping from the weights of G_ε to $\{0, 1, \ldots, N\}$.

For any $\delta > 0$ consider the mapping i_q defined by the formula $i_q(w) = \lceil \frac{w}{\varepsilon q} \rceil$, where $q = \frac{\varepsilon\delta}{n \log^2(1/\varepsilon)}$. Define $N = N(\delta, q, \varepsilon) = i_q(\delta) = \lceil \frac{\delta}{\varepsilon q} \rceil = O\left(\frac{n \log^2(1/\varepsilon)}{\varepsilon^2}\right)$. Let $\mathsf{w}(e)$ denote the weight of any edge e of G_ε. We have the following.

Lemma 2. *Let G_ε' be the graph with the same nodes and edges as G_ε and weight on each edge e set to $i_q(\mathsf{w}(e))$. Then, for each shortest path p in G_ε' with weight* *at most δ between a pair s, t of nodes, there is a path between s and t in G'_ε whose weight in G_ε is within an additive factor of $O(\varepsilon\delta)$ of the weight of p. The largest edge weight of G'_ε does not exceed N.*

3.2   Pseudo-planarization

One of the main ingredients in Thorup's distance oracle is a *shortest-path sepa-rator* that consists of three shortest paths separating a planar graph into at least two subgraphs of weight at most a third of the size of the original graph. In our case, G_ε is not planar. A set of three shortest paths generally does *not* separate the graph into two edge-disjoint components as each edge of G_ε is intersected (in a geometric sense) by roughly $1/\sqrt{\varepsilon}$ edges. In our construction, we concurrently compute separators and pseudo-planarize the graph. Whenever edges geometri-cally intersect with a shortest-path separator, we split the edge and we add a node to represent this intersection on the separator path.

Note that we cannot use the separator construction algorithm from [13] di-rectly, since our graph is neither planar nor triangulated. The surface P, however, is triangulated, and we use that triangulation for the purpose of the separator construction. We also cannot "planarize" G_ε by adding a node for every intersec-tion of two line segments, since the dependency on ε would increase. While the polynomial dependency on n is of primary importance, in this line of work [3, Table 1], the low dependency on ε^{-1} is also considered relevant.

Construction of $\hat{G}_\varepsilon$ We initialize $\hat{G}_\varepsilon$ as G_ε. For each triangle bisector ℓ, we sort the nodes on ℓ by their distance from the corresponding node of P and for each node on ℓ we store its position. We need these positions to count the number of nodes on the left (right) of a separator path.

Let r be the node in G_ε corresponding to an arbitrary node of a triangle in P. We compute a single-source shortest path tree in G_ε rooted at r. Let T denote that tree and for any node $u \in V(G_\varepsilon)$, let $T(u)$ denote the path from r to u in T. We start with an arbitrary triangle A in P (as opposed to [13], where triangles correspond to faces of the plane graph). Let x, y, z denote the nodes that correspond to the nodes of A, respectively. We remove all edges incident to nodes or intersected by edges of $S(A) = T(x) \cup T(y) \cup T(z)$, as well as all edges intersected by any edge of the triangle A. We say that $S(A)$ defines a *bal-anced* separator if no component of the resulting graph contains more than two thirds of the nodes of the original graph. If $S(A)$ defines a balanced separator, we recurse on each subgraph. Otherwise, we "flip" one node (wlog we flip z) to obtain a neighboring triangle A' such that $T(x) \cup T(y) \cup T(z')$ is a separator with better balance. Once we have found a good separator (corresponding to a triangle in P; let x'', y'', z'' denote the nodes corresponding to its endpoints), we compute, for each edge $e \in T(x'') \cup T(y'') \cup T(z'')$, all the geometric intersec-tions with edges $e' \in E(G_\varepsilon)$. For each intersection a node is added to $V(\hat{G}_\varepsilon)$ and for each such edge e' we add its two parts e'_1, e'_2, each weighted by its length, as edges to $E(\hat{G}_\varepsilon)$. After this step, the union of paths $\hat{T}(x'') \cup \hat{T}(y'') \cup \hat{T}(z'')$ actually separates $\hat{G}_\varepsilon$ into subgraphs. We contract the separator into a new root and continue this procedure recursively in each component as in [13].

Note that $\hat{G}_\varepsilon$ is constructed with respect to a set of shortest-path separators. Even though the graph $\hat{G}_\varepsilon$ is not planar, these paths recursively separate $\hat{G}_\varepsilon$ into subgraphs in a balanced way (note that $\hat{G}_\varepsilon$ is planar in the immediate vicinity of these paths).

The number of edges of $\hat{G}_\varepsilon$ can be estimated as follows. After the corresponding ε-covers and distances are computed, S is contracted to a new node r^S, which is *suppressed* [13]. Suppression means that, while r^S and all its adjacent edges are considered for constructing recursive separators, they are not considered for computing distances/shortest paths. When a separator S of the original graph is constructed, the separator property we use afterwards is that any path between a pair of nodes from different components of $G_\varepsilon \setminus S$ must contain a node from S. Then all new edges are also suppressed. Hence, when recursively constructing separators for the components of $G_\varepsilon \setminus S$, we do not have to add new nodes and edges for intersections between separator edges and suppressed (new) edges. As a result, each edge of G_ε can be divided into two new edges at most once and hence $\left|E(\hat{G}_\varepsilon)\right| \leq 2\left|E(G_\varepsilon)\right|$.

Lemma 3. *The graph $\hat{G}_\varepsilon$ and the shortest-path separators can be computed in time $O(|V| \log^2 |V| + |E| \log |V|)$, where $|V| = O((n/\sqrt{\epsilon}) \log(1/\epsilon))$ and $|E| = O(n\epsilon^{-1} \log^2(1/\epsilon))$. Given two points s and t in $\hat{G}_\varepsilon$, one can find in $O(1)$ time a separator from the shortest-path separator decomposition separating s and t.*

4 Preprocessing Algorithm

We describe the preprocessing algorithm for our approximate distance oracle. We first give a brief overview, next we provide pseudocode, and finally we analyze the algorithm. The algorithm consists of three phases.

1. Discretization: the algorithm discretizes the surface and the weights
2. Pseudo-planarization: the algorithm finds shortest-path separators Q and adds nodes in order to make Q separating (Section 3.2)
3. Data-structure construction: the algorithm computes links to portals on the separator paths Q for increasing scales

Its pseudocode is listed as PREPROCESS in the following.

PREPROCESS (P, ε, δ)
    let G_ε be the graph obtained by the surface discretization in [2,3] and
    the weight discretization described in Section 3.1 with maximum weight N
    compute $\hat{G}_\varepsilon$ as described in the previous section
    compute Nearest Common Ancestor data structure for separator tree
    for each separator path Q
        partition Q into maximal pieces of length $\leq \delta$
        enumerate the subpaths Q_j^δ

for each subpath Q_j^δ compute an ε–cover (equally spaced points) $C(Q_j^\delta)$
enumerate the nodes in all the $C(Q_j^\delta)$
for $r \in \{0, \ldots 4\}$
    let $\mathcal{C}^j$ denote the covers for all subpaths Q_j^δ with $j \mod 5 \equiv r$
    for $i \in \{1, 2, \ldots O(1/\varepsilon)\}$
        let N_i be the set of all nodes i in all the covers in $\mathcal{C}^j$
        compute multiple-source shortest-path tree with all nodes
        in N_i as sources of weighted depth 2δ

Lemma 4 ([9, Lemma 4]). *For any path $u - v$ of length $[\delta, 2\delta]$ in $\hat{G}_\epsilon$ that intersects a shortest-path separator Q_j^δ there is an alternative path $u - q - v$ for a portal $q \in C(Q_j^\delta)$ such that* $\mathsf{len}(u - q - v) \le (1 + \epsilon)\mathsf{len}(u - v)$.

Lemma 5. *For a triangulated surface P with n triangles, given $\varepsilon > 0$ and $\delta > 0$, algorithm* PREPROCESS (P, ε, δ) *computes an $(1+\varepsilon)$–approximate distance oracle for distances of length $\ell \in [\delta, 2\delta]$ in P in time $O(n\epsilon^{-2} \log^2(n/\varepsilon) \log^2(1/\epsilon))$. The space requirement is $O(n\varepsilon^{-3/2} \log(n/\varepsilon) \log(1/\varepsilon))$.*

Proof. Each Steiner point stores ϵ–covers of size $O(\epsilon^{-1})$ per level. There are $O(\log(n/\epsilon) \log(1/\epsilon))$ levels. The space per scale is thus $O(n\varepsilon^{-3/2} \log(n/\varepsilon) \log(1/\varepsilon))$.

Computing the covers requires $O(\epsilon^{-1})$ shortest-path-tree constructions per level. The time per scale is bounded by $O(n\epsilon^{-2} \log^2(n/\varepsilon) \log^2(1/\epsilon))$. $\square$

5 Answering Approximate Shortest-Path Queries

5.1 Overview of the Method

In order to answer approximate distance queries for an arbitrary pair of query points s and t from P, we use an algorithm for answering distance queries in $\hat{G}_\varepsilon$. Let, for any point p in P that is not in G_ε, the neighborhood $\mathcal{N}(p)$ of p be defined as the set of nodes of G_ε contained in the triangle(s) containing p and all adjacent (i.e. sharing an edge) triangles. If p is a node in G_ε, then we define $\mathcal{N}(p) = \{p\}$. Then, the approximate distance between s and t can be computed as

$$\tilde{\mathsf{d}}(s, t) = \min_{p_s \in \mathcal{N}(s), p_t \in \mathcal{N}(t)} \{\mathsf{d}_P(s, p_s) + \mathsf{d}_{\hat{G}_\epsilon}(p_s, p_t) + \mathsf{d}_P(p_t, t)\}. \tag{2}$$

Since points s and p_s are in the same or in neighboring triangles, $\mathsf{d}_P(s, p_s)$ can be computed using an explicit formula based on Snell's law. The same applies for computing $\mathsf{d}_P(p_t, t)$. For computing $\mathsf{d}_{\hat{G}_\epsilon}(p_s, p_t)$, we use the separator decomposition constructed in Lemma 3. Let Q be a shortest-path separator separating p_s and p_t. Then any path between p_s and p_t in $\hat{G}_\varepsilon$ must contain a node in Q and we therefore have $\mathsf{d}_{\hat{G}_\epsilon}(p_s, p_t) = \min_{q \in Q}\{\mathsf{d}_{\hat{G}_\epsilon}(p_s, q) + \mathsf{d}_{\hat{G}_\epsilon}(q, p_t)\}$.

Unless stated otherwise, we assume in this section that $\mathsf{d}_{\hat{G}_\epsilon}(s, t) \in [\delta, 2\delta]$, and that Q is a shortest-path separator in $\hat{G}_\varepsilon$ of length $O(\delta)$ separating s and t, as outlined in the preprocessing algorithm. Under those assumptions, $|C(Q)| = O(1/\varepsilon)$ and one can use Lemma 4 to approximate each distance on the right-hand side of the previous equality, thereby having to look at only $1/\varepsilon$ nodes in $C(Q)$ per distance computation instead at all the nodes in Q, resulting in an $1 + \varepsilon$ approximation of $\mathsf{d}_{\hat{G}_\varepsilon}(p_s, p_t)$. Hence, instead of (2), we can use the equality

$$\hat{d}(s,t) = \min_{p_s \in \mathcal{N}(s), p_t \in \mathcal{N}(t)} \min_{q \in C(Q)} \mathsf{d}_P(s, p_s) + \mathsf{d}_{\hat{G}_\varepsilon}(p_s, q) + \mathsf{d}_{\hat{G}_\varepsilon}(q, p_t) + \mathsf{d}_P(p_t, t). \quad (3)$$

The number of pairs (p_s, p_t) is $|\mathcal{N}(s)| \cdot |\mathcal{N}(t)| = \Theta(\epsilon^{-1} \log^2(1/\epsilon))$ and the size of $|C(Q)|$ is at most $1/\varepsilon$. Hence the total time for answering the approximate distance query based on formula (3) is $O(\epsilon^{-2} \log^2(1/\epsilon))$.

In the following we describe a more efficient divide-and-conquer approach for computing the minimum of formula (3) that avoids looking at all pairs p_s, p_t of nodes in the neighborhoods $\mathcal{N}(s)$ and $\mathcal{N}(t)$ and leads to a query time complexity roughly proportional to $\tilde{O}(\epsilon^{-1})$, ignoring logarithmic factors.

5.2   Divide-and-Conquer Approach

We reduce the problem of computing the minimum from (3) to the problem of finding the distances from all points of $C(Q)$ to s and t. We reduce the problem(s) of finding all distances from (to) $C(Q)$ to a single problem, rather than a sequence of $|C(Q)|$ problems.[2] Once we have those distances, the algorithm requires an additional $O(|C(Q)|) = O(1/\varepsilon)$ time to compute $\hat{d}(s, t)$. We describe the computation of the distances to t as the ones for s are similar.

The idea is the following. There are $|C(Q)|$ nodes from which we want to compute shortest paths distances and each path could possible go through each of the nodes in $\mathcal{N}(t)$. For large $|C(Q)|$ and $|\mathcal{N}(t)|$ many paths intersect each other. It is well-known that in a planar graph, a pair of shortest intersecting paths to the same target can always be replaced by a pair of shortest non-intersecting paths with the same sources and target as the original, as exploited in [6,5]. We could use that property to significantly reduce the search space when computing the shortest paths from $C(Q)$. Unfortunately, our graph $\hat{G}_\varepsilon$ is not planar. We need to prove a similar property for $\hat{G}_\varepsilon$. First we establish several properties of paths in P and $\hat{G}_\varepsilon$ that follow from the definition of $\hat{G}_\varepsilon$.

Lemma 6. *Let p_1 and p_2 be two intersecting shortest paths in P with the same target t. There exists a path p_2^* in P with the same source and target as p_2 that does not intersect p_1 and such that $\mathsf{len}(p_2^*) = \mathsf{len}(p_2)$.*

Lemma 7. *Let π_1 and π_2 be two intersecting paths in $\hat{G}_\varepsilon$ with the same target node t. There exists a path π_2^* with the same source and target as π_2 that does not intersect π_1 and such that $\mathsf{len}(\pi_2^*) \leq (1 + \varepsilon)\mathsf{len}(\pi_2)$.*

---

[2] This is why we compute portals to paths at different scales. By operating at one scale δ, we can use a *single set of portals C per path*, wherein we can efficiently search the best Steiner points for a pair of query points.

The next lemma is instrumental in our divide-and-conquer approach.

Lemma 8. *Let Q be a shortest-path separator, let q', q'' and q be nodes of $C(Q)$ such that q is between q' and q'' on Q. Let π' and π'' be two nonintersecting shortest paths in G_ε from q' and q'' to point t of P and let $\mathcal{N}'(t)$ be the subset of $\mathcal{N}(t)$ that is inside or on the boundary of the cycle determined by Q, π', and π''.*

If π' and π'' contain a node from $\mathcal{N}(t)$, then there is a path π in G_ε from q to t containing a node from $N'(t)$ such that $\mathsf{len}(\pi) \leq (1+\varepsilon)\mathsf{len}(\pi_{opt})$, where π_{opt} is the shortest path from q to t in G_ε.

Proof. If π_{opt} intersects neither π' nor π'', then we set $\pi = \pi_{opt}$. Else, suppose that the first path that π_{opt} intersects is π' and let (u', v') be the edge of π' that intersects an edge (u, v) from π_{opt} with v' between u' and t on π' and v between u and t on π. By Lemma 7, there exists a path π from q to t that does not intersect π' and such that $\mathsf{len}(\pi) \leq (1 + \varepsilon)\mathsf{len}(\pi_{opt})$. Moreover, by the proof of Lemma 7, π coincides with π' in the portion of π' between v' and t and with π_{opt} in the portion between q and u. Since, by assumption, π' and π'' are not intersecting, then π does not intersect π''.   □

The next lemma summarizes the previous results of this section and gives a fast algorithm for answering approximate distance queries in P.

Lemma 9. *Let s, t be two points on P and let Q be a separator path separating s from t and let $\mathcal{C}(Q)$ denote the ϵ-cover of Q of size $1/\varepsilon$. One can find in $O(1/\epsilon)$ time a pair of points $(p_s, p_t) \in \mathcal{N}(s) \times \mathcal{N}(t)$ and $q \in C(Q)$ such that*

$$\hat{\mathsf{d}}(s, t) = \mathsf{d}_P(s, p_s) + \mathsf{d}_{\hat{G}_\epsilon}(p_s, q) + \mathsf{d}_{\hat{G}_\epsilon}(q, p_t) + \mathsf{d}_P(p_t, t) \leq (1 + O(\varepsilon \log(\varepsilon^{-1})))\delta_{\hat{G}_\epsilon}(s, t).$$

Proof. To answer the query, we first compute, for each $q \in C(Q)$, the distance $\hat{\mathsf{d}}(q, s)$ from q to s, then the distance $\hat{\mathsf{d}}(q, t)$ from q to t, and finally compute $\min_{q \in C(Q)} \{\hat{\mathsf{d}}(s, q) + \hat{\mathsf{d}}(q, t)\}$ in $O(|C(Q)|) = O(1/\varepsilon)$ time. Hence we only need to describe how to compute $\hat{\mathsf{d}}(s, q)$ in $O(1/\varepsilon)$ time.

Let q' and q'' be the two outermost cover points of Q. Construct the shortest paths from s to q' and q''. These two paths and q form a cycle c through s. Let $\mathcal{N}(s; q', q'')$ denote the subset of $\mathcal{N}(s)$ that is inside c. Let node $q_m \in C(Q)$ divide $C(Q)$ into to roughly equal subsets with respect to the order on Q. By Lemma 8, there exists a shortest path from q_m to s that contains a point from $\mathcal{N}(s; q', q'')$ that is within $1 + \varepsilon$ factor of the length of the shortest path between q_m and s. Hence the length of such a shortest path can be found in $O(|\mathcal{N}(s; q', q'')|)$ time. That path divides $\mathcal{N}(s; q', q'')$ into two subsets, $\mathcal{N}(s; q', q_m)$ and $\mathcal{N}(s; q_m, q'')$.

By the same method, we find the length of a shortest path from a middle node between q' and q_m to s and a shortest path from a middle node between q_m and q'' to s. The total time to find *both* these lengths is $O(|\mathcal{N}(s; q', q'')|)$. In the next step we find four more shortest-path lengths (approximate distances) from nodes from $C(Q)$ to s in time $O(|\mathcal{N}(s; q', q'')|)$. The time of this algorithm is $O(\varepsilon^{-1})$ and the approximation ratio is $1 + O(\varepsilon \log(\frac{1}{\varepsilon}))$.

In order to achieve stretch $1 + \varepsilon$ in the final algorithm, rather than $1 + O(\varepsilon)$, we need to know an explicit bound on the stretch from Lemma 9, instead of an estimation in terms of big-O asymptotics. The proof of the lemma does not yield the constant, but it does give us that the stretch is not exceeding $(1 + \varepsilon)^{\log(\frac{1}{\sqrt{\varepsilon}})}$. For any specific value of ε, we can estimate the stretch by that formula and use it for computing the parameters for the preprocessing algorithm. On the other hand, by Lemma 9, we know that there exists a constant k independent of ε such that the stretch is bounded by $1 + k\varepsilon \log(\varepsilon^{-1})$.

Lemma 10. *Given two points s and t in P and $\varepsilon \in (0, 1)$, one can find in $O(\epsilon^{-1})$ time a path π in $\hat{G}_\varepsilon$ between s and t such that $\mathsf{len}(\pi) \leq (1 + k\varepsilon \log(\frac{1}{\varepsilon}))\mathsf{d}(s, t)$ for some constant k.*

Proof. By Lemma 3, one can find in $O(\log \log n)$ time a separator path Q of length $O(q)$ separating s and t in $\hat{G}_\varepsilon$. By Lemma 4, there is a set of $C(Q)$ of $O(1/\varepsilon)$ attachment points for Q that can be used to approximate any distance in $[\delta, 2\delta]$ to a point in Q with $1 + \varepsilon$ approximation factor.

Let π be a path between s and t in the approximation graph $\hat{G}_\varepsilon$ such that $\mathsf{len}(\pi) \leq (1+\varepsilon)\mathsf{d}(s, t)$. As Q separates s and t, there exists a node $q \in Q$ such that $\mathsf{len}(\pi) = \mathsf{d}(s, q) + \mathsf{d}(q, t)$. By Lemma 4, there exists a node $q' \in C(Q)$ such that $\mathsf{d}(s, q') \leq (1 + \varepsilon)\mathsf{d}(s, q)$ and $\mathsf{d}(t, q') \leq (1 + \varepsilon)\mathsf{d}(t, q)$. Hence, there is a path π' in $\hat{G}_\varepsilon$ from s to t that contains a node from $C(Q)$ such that $\mathsf{len}(\pi') \leq (1+\varepsilon)^2\mathsf{d}(s, t)$. By Lemma 10, one can find in $O(1/\varepsilon)$ time a path π'' in $\hat{G}_\varepsilon$ that contains a node in $C(Q)$ and is within a $1 + O(\varepsilon \log(\varepsilon^{-1}))$ factor of the distance between s and t in $\hat{G}_\varepsilon$. Hence,

$$\mathsf{len}(\pi'') \leq (1 + O(\varepsilon \log(\varepsilon^{-1})))\delta_{\hat{G}_\varepsilon}(s, t) \leq (1 + O(\varepsilon \log(\varepsilon^{-1})))\mathsf{len}(\pi')$$
$$\leq (1 + O(\varepsilon \log(\varepsilon^{-1})))(1 + \varepsilon)^2\mathsf{d}(s, t) = (1 + O(\varepsilon \log(\varepsilon^{-1})))\mathsf{d}(s, t). \qquad \square$$

Now we are ready to prove the main result of this paper, Theorem 2. Recall that so far in this section we have assumed that Q is a shortest-path separator in $\hat{G}_\varepsilon$ of length $O(\delta)$ separating s and t, and $\mathsf{d}(s, t) \in [\delta, 2\delta]$. We need to compute such δ in our algorithm and to choose an appropriate value for the approximation factor in Lemma 10 so that the computed path is within a $1 + \varepsilon$ approximation factor of the shortest path, as opposed to a $1 + O(\varepsilon \log(\frac{1}{\varepsilon}))$ factor.

5.3   Proof of Theorem 2

Proof (of Theorem 2). In order to find an $O(1)$ approximation for $\mathsf{d}(s, t)$ we do a query using Lemma 10 with $\varepsilon = 1/2$ (strictly speaking, we preprocess another distance oracle for $\varepsilon = 1/2$ and query it here) and do a binary search on the values of δ in $\{1, 2, 4, \ldots, 2^{\lceil \log(nN) \rceil}\}$. We find in $O(\log \log(nN)) = O(\log \log n)$ time the minimum value δ_0 for δ for which the path from Lemma 10 has length $l_{1/2} < \infty$. By Lemma 10, $\mathsf{d}(s, t) \leq l_{1/2} \leq (1 + k/2)\mathsf{d}(s, t)$. By the minimum property of δ, $l_{1/2} > 2\delta_0/2 = \delta_0$ and hence $\delta_0 < l_{1/2} < 2\delta_0$. By combining this with the previous inequality we get $\delta_0/(1 + k/2) < \mathsf{d}(s, t) < 2\delta_0$.

Next, we do a binary search on the values of δ in $\{2^{i_1}, 2^{i_1+1}, \ldots, 2^{i_2}\}$ for $i_1 = \lfloor \log(\delta_0/(1 + k/2)) \rfloor$ and $i_2 = \lfloor \log(2\delta_0) \rfloor$ and ε determined by $\varepsilon \le \epsilon/(k \log(\frac{1}{\varepsilon}))$. Such ε can be determined in $O(\log(\epsilon^{-1}))$ time and the search takes $O(1)$ time. By Lemma 10, the returned distance l satisfies

$$l \le (1 + k\varepsilon \log(\frac{1}{\varepsilon}))\mathsf{d}(s,t) \le (1 + k\epsilon/(k \log(\frac{1}{\varepsilon})) \log(\frac{1}{\varepsilon}))\mathsf{d}(s,t) = (1 + \epsilon)\mathsf{d}(s,t).$$

In order to determine the running time, we use that $\epsilon = \Theta(\varepsilon \log(1/\varepsilon))$ and hence

$$\varepsilon = \Theta(\epsilon/\log(1/\varepsilon)) = O(\epsilon/\log(1/\epsilon))$$

as $\epsilon = o(\varepsilon^2)$. By replacing that estimation for ε in the complexity bounds in Lemmas 5 and 10, we get the claimed complexity bounds.   □

References

1. Abraham, I., Gavoille, C.: Object location using path separators. In: PODC, pp. 188–197 (2006)
2. Aleksandrov, L., Djidjev, H.N., Guo, H., Maheshwari, A., Nussbaum, D., Sack, J.R.: Algorithms for approximate shortest path queries on weighted polyhedral surfaces. Discrete Comput. Geom. 44, 762–801 (2010)
3. Aleksandrov, L., Maheshwari, A., Sack, J.R.: Determining approximate shortest paths on weighted polyhedral surfaces. J. ACM 52(1), 25–53 (2005)
4. Cabello, S.: Many distances in planar graphs. In: SODA, pp. 1213–1220 (2006); a preprint of the Journal version is available in the University of Ljubljana preprint series 47, 1089 (2009)
5. Djidjev, H.: Efficient algorithms for shortest path queries in planar digraphs. Graph-Theoretic Concepts in Computer Science 1197, 151–165 (1997)
6. Fakcharoenphol, J., Rao, S.: Planar graphs, negative weight edges, shortest paths, and near linear time. J. Comput. Syst. Sci. 72(5), 868–889 (2006)
7. Kawarabayashi, K.-i., Klein, P.N., Sommer, C.: Linear-space approximate distance oracles for planar, bounded-genus and minor-free graphs. In: Aceto, L., Henzinger, M., Sgall, J. (eds.) ICALP 2011. LNCS, vol. 6755, pp. 135–146. Springer, Heidelberg (2011)
8. Klein, P.N.: Preprocessing an undirected planar network to enable fast approximate distance queries. In: SODA, pp. 820–827 (2002)
9. Klein, P.N., Subramanian, S.: A fully dynamic approximation scheme for shortest paths in planar graphs. Algorithmica 22(3), 235–249 (1998)
10. Mitchell, J.S.B., Papadimitriou, C.H.: The weighted region problem: finding shortest paths through a weighted planar subdivision. J. ACM 38, 18–73 (1991)
11. Mozes, S., Sommer, C.: Exact distance oracles for planar graphs. CoRR abs/1011.5549 (2010)
12. Sun, Z., Reif, J.H.: On finding approximate optimal paths in weighted regions. J. Algorithms 58(1), 1–32 (2006)
13. Thorup, M.: Compact oracles for reachability and approximate distances in planar digraphs. J. ACM 51(6), 993–1024 (2004)
14. Thorup, M., Zwick, U.: Approximate distance oracles. J. ACM 52(1), 1–24 (2005)

The Union of Probabilistic Boxes: Maintaining the Volume

Hakan Yıldız[1], Luca Foschini[1], John Hershberger[2], and Subhash Suri[1]

[1] University of California, Santa Barbara, USA
[2] Mentor Graphics Corporation, Wilsonville, Oregon, USA

Abstract. Suppose we have a set of n axis-aligned rectangular boxes in d-space, $\{B_1, B_2, \ldots, B_n\}$, where each box B_i is *active* (or present) with an independent probability p_i. We wish to compute the *expected* volume occupied by the union of all the active boxes. Our main result is a data structure for maintaining the expected volume over a *dynamic* family of such probabilistic boxes at an amortized cost of $O(n^{(d-1)/2} \log n)$ time per insert or delete. The core problem turns out to be one-dimensional: we present a new data structure called an *anonymous segment tree*, which allows us to compute the expected length covered by a set of probabilistic segments in logarithmic time per update. Building on this foundation, we then generalize the problem to d dimensions by combining it with the ideas of Overmars and Yap [13]. Surprisingly, while the expected value of the volume can be efficiently maintained, we show that the tail bounds, or the probability distribution, of the volume are intractable— specifically, it is *NP*-hard to compute the probability that the volume of the union exceeds a given value V *even when the dimension is $d = 1$*.

1 Introduction

We consider the problem of estimating the volume covered by a set of overlapping boxes in d-space when the existence of each box is known only with partial certainty. Specifically, we are given a set of n axis-aligned boxes, in which the ith box is known to be present only with probability p_i. (The probabilities of different boxes are independent of each other.) This is a *probabilistic* version of the classical Klee's Measure problem [11]. Besides being a fundamental problem in its own right, it is also a natural framework to model risk and uncertainty in geometric settings. In order to motivate the problem, consider the following scenario.

Suppose that a tract of land has a variety of health hazards, occurring in possibly overlapping regions. The virulence of each hazard is expressed as a survival rate, i.e., the probability that an entity survives after being exposed to the hazard. Assuming independence of the hazards, the probability of survival at a point is the product of the survival probabilities for the different hazards affecting the point. The *average survival rate* within the whole tract is the integral of the survival probabilities of all points in the tract divided by the area the tract. It is easy to see that the integral of concern equals the expected area of

C. Demetrescu and M.M. Halldórsson (Eds.): ESA 2011, LNCS 6942, pp. 591–602, 2011.
© Springer-Verlag Berlin Heidelberg 2011

covered by the hazardous regions, if we treat the survival rates as the probability of *absence* for each region.

Let us now introduce the problem more formally. A d-dimensional rectangular box B, which we call a d-box for convenience, is the Cartesian product of d one-dimensional ranges, namely, $B = \Pi_{i=1}^{d}[a_i, b_i]$. The *volume* of a box B is defined as $vol(B) = \Pi_{i=1}^{d}|b_i - a_i|$. Given a set $\mathcal{B}$ of d-boxes $\{B_1, B_2, \ldots, B_n\}$, its *union* $\bigcup B_i$ is the set of points contained in at least one box of $\mathcal{B}$. In our problem, each box B_i is assumed to be *present* (or active) with an independent probability p_i, and absent otherwise. We wish to compute the *expected value* of the total volume occupied by such a collection of boxes. More generally, we may wish to compute the probability distribution of the volume—for each value V, the probability that the volume of the union is V. In fact, we wish to maintain a collection of such probabilistic boxes so that their volume statistics (expectation or tail bounds) are easily updated as boxes (along with their activation probabilities) are inserted or deleted.

Our problem is a probabilistic and dynamic version of *Klee's measure problem*, which has a long history in computational geometry [3,7,11,12,13]. The fastest algorithm currently known for Klee's problem is due to Chan [7], with worst-case time $O(n^{d/2}2^{O(\log^* n)})$. Despite a long and distinguished history, the computational complexity of the problem has remained largely unresolved for $d \geq 3$ since the breakthrough result of Overmars and Yap [13], with time complexity $O(n^{d/2} \log n)$. Most of the work in the past several years has focused on the conceptually easier case of the *union of cubes* [1,2] or fat boxes [5].

Our main result is a data structure for maintaining the expected volume over a dynamic set of probabilistic boxes in amortized time $O(n^{(d-1)/2} \log n)$ per insert or delete. (Any major improvment in this update complexity will imply a breakthrough on Klee's measure problem because the d-dimensional Klee's problem can be solved by maintaining a $(d - 1)$-dimensional volume over n insertions and deletions [7,13].) The core problem in computing the volume of probabilistic boxes arises already in one dimension, and leads us to a new data structure called an *anonymous segment tree*. This structure allows us to compute the expected length covered by a set of probabilistic segments in logarithmic time per update. Building on this foundation, we then generalize the problem to d dimensions by combining it with the ideas of Overmars and Yap [13]. (The issues underlying this extension are mostly technical, albeit somewhat non-trivial, since the Overmars-Yap scheme uses a space-sweep that requires *a priori* knowledge of the box coordinates, while we assume a fully dynamic setting with no prior knowledge of future boxes.)

Surprisingly, while the expected value of the volume can be efficiently maintained, we show that computing the tail bounds, or the probability distribution, of the volume is intractable—specifically, it is *NP*-hard to compute the *probability* that the volume of the union exceeds a given value V even when the dimension is $d = 1$.

Finally, in order to evaluate the practical usefulness of our, and Overmars-Yap's, scheme in geospatial databases, we implemented the scheme. The results confirm our theoretical bounds in practice and we show that our solution easily outperforms a naïve solution.

2   Probabilistic Volume: Expectation and Tail Bounds

If algorithmic efficiency were not the main concern, then the *expected* volume of the union of n boxes would be easy to compute in polynomial time. A set of n boxes in d-space is defined by $O(dn)$ facets, and the hyperplanes determined by those facets partition the space into $O(n^d)$ rectangular cells, with each cell fully contained in all the boxes that intersect it. For each cell, compute the probability that at least one of its covering boxes is active. By the linearity of expectation, the expected volume is simply the sum of the volumes of these cells weighted by their probability of being covered. This naïve algorithm runs in $O(n^{d+1})$ time, which is polynomial in n for fixed dimension. In Section 4, we present our main result: how to maintain the expected volume much more efficiently under dynamic updates.

On the other hand, we argue below that computing the probability distribution (i.e., the tail bounds) is intractable. In particular, even in one dimension, computing the probability that the union of n probabilistic line segments has volume (length) at least V is *NP*-hard. The following theorem shows that it is hard to compute the probability that the union has length precisely L for some integer L—since we use only integer-valued lengths, computing the tail bound is at least as hard (the probability of length exactly L can be determined from those of lengths at least L and at least $L + 1$.)

Theorem 1. *Given n disjoint line segments on the integer line, where the ith segment is active with probability p_i, it is NP-hard to compute the probability that the union of the active segments has length L.*

Proof. We show a reduction from the well-known *NP*-complete problem SUBSET-SUM [9]. The subset-sum problem takes as input a set of positive integers $A = \{a_1, \ldots, a_n\}$, a *target* integer L, and asks if there is an index subset $I \subseteq \{1, \ldots, n\}$ whose elements sum to the target value L, that is, $\sum_{i \in I} a_i = L$. Given an instance of the subset-sum problem (A, L), we create a set of probabilistic segments $\{s_1, s_2, \ldots, s_n\}$, as follows. The segment s_i begins at point $\sum_{j<i} a_j$ and has length a_i. Each of the n segments occurs with probability $p_i = 1/2$. We observe that because of the uniform probability, each of the 2^n subsets of $\{s_1, s_2, \ldots, s_n\}$ is equally likely, each occurring with probability 2^{-n}.

Since the segments $s_1, \ldots, s_n$ are disjoint, for any index subset $I \subseteq \{1, \ldots, n\}$, the union of the segments $\{s_i | i \in I\}$ has length precisely $\sum_{i \in I} a_i$. Thus, I is a solution to the subset sum problem (A, L) if and only if the union of the segments indexed by I has length L. However, the probability that the union of the active segments has length L is precisely equal to the number of index subsets that are valid solutions of the subset sum problem divided by 2^n. Thus, given the probability that the union of the active segments is L, we can deduce whether the subset sum problem has a solution. This completes the proof.   □

3   Maintaining the Expected Measure in 1D

We begin by describing our data structure in one dimension, and then show how to embed it in an appropriately generalized version of the Overmars-Yap structure for the d-dimensional problem. We describe the data structure, called the *anonymous segment tree*, first without the probabilities, focusing on its form and updates, and then present an *abstraction* that retains all the key elements and yet accommodates probabilistic segments.

3.1   Anonymous Segment Tree

Let S be a dynamic set of n line segments on the number line that undergoes insertions and deletions. Our goal is to maintain the length covered by the union of the segments in S. We simply call this length the *measure* of S. The segments in S split the number line into at most $(2n+1)$ disjoint intervals, called *primitive intervals*. We maintain a balanced binary tree whose keys are the coordinates of the segment endpoints, and whose leaves correspond to the primitive intervals. Each internal node represents the union of all its leaf descendants' intervals. (See Figure 1(a).)

Consider a leaf v and its associated primitive interval $I = (x_1, x_2)$. Let $S_I \subset S$ be the subset of segments that cover I, and define the *coverage count* of v, denoted $cover(v)$, as $|S_I|$. The *measure* of v, denoted $\mu(v)$, is clearly zero if $cover(v) = 0$ and $x_2 - x_1$ otherwise. The measure of S is the sum of $\mu(v)$ over all leaves v. The coverage count is an inefficient mechanism for maintaining the measure when segments are inserted or deleted, so we use a secondary quantity, called $ccover(v)$. (The name *ccover* derives from *complete coverage count*.) The *ccover* values satisfy the two invariants described below.

> SUM INVARIANT: *For any leaf v, $cover(v)$ is the sum of $ccover(a)$ over all ancestors a of v (including v itself).*

A trivial way to achieve the invariant is to set $ccover(v) = cover(v)$ for each leaf v and $ccover(u) = 0$ for each non-leaf node u. But, as we show below, $ccover()$ allows us to support maintenance of the measure through its flexibility. We use $ccover()$ values to maintain the measure as follows, where $L(v)$ is the length of v's interval.

$$\mu(v) = \begin{cases} L(v) & \text{if } ccover(v) > 0 \\ 0 & \text{if } ccover(v) = 0 \ \wedge \ v \text{ is a leaf} \\ \mu(v_l) + \mu(v_r) & \text{if } ccover(v) = 0 \ \wedge \ v \text{ has children } v_l, v_r \end{cases} \qquad (1)$$

The following lemma is easily established.

Lemma 1. *Let a node v be called* exposed *if $ccover(a) = 0$ for all ancestors a of v (excluding v). Then for any exposed node v, $\mu(v)$ is the measure of S restricted to v's interval. In particular, $\mu(root)$ is the measure of S.*

PUSHUP INVARIANT: *For each non-leaf node v with children v_l and v_r, at least one of $ccover(v_l)$ and $ccover(v_r)$ is zero.*

We achieve this invariant by applying the following *push-up* operation at each internal node: Let v_l and v_r be children of v. Decrement $ccover(v_l)$ and $ccover(v_r)$ by $\min(ccover(v_l), ccover(v_r))$ and increment $ccover(v)$ by the same amount. (See Figure 1(b).) This operation propagates the values of $ccover()$ up the tree as much as possible, which in turn allows us to update the $\mu()$ values efficiently.[1]

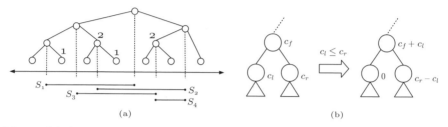

Fig. 1. (a) An anonymous segment tree, positive *ccover* values are shown. (b) the pushup operation, c_i's stand for *ccover* values.

We can maintain the sum and the pushup invariants as segments are inserted or deleted by modifying $O(\log n)$ values in the tree. We briefly outline how this is achieved, omitting standard but technical details. A segment s corresponds to a set of $O(\log n)$ nodes in the tree called *canonical nodes* whose intervals are disjoint but their concatenation equals s. An insertion (deletion) is handled by incrementing (decrementing) the $ccover()$ values of the canonical nodes, maintaining the sum invariant and, thus, the correct μ values in the tree. Due to the changes in primitive intervals, the tree may undergo rebalancing rotations, in which case we temporarily push the $ccover()$ down below the rotating nodes to preserve the sum invariant. Afterwards, we apply the necessary push-ups to restore the pushup invariant. We note that pushing up the $ccover()$ values is necessary otherwise deletions in the tree become inefficient. The push-up invariant guarantees that no $ccover()$ value drops below zero after a deletion.

Lemma 2. *We can maintain the measure of a dynamic set of n segments in $O(\log n)$ time for insert or delete operations, and $O(1)$ time for measure query.*

We next show how to generalize the anonymous segment tree to deal with probabilistic segments. Towards that goal, we introduce an abstract framework that includes the measure of probabilistic segments as a special case.

3.2   An Abstract Anonymous Segment Tree

Let f be a function mapping the segments in $\mathcal{S}$ to some range set G, and let $\oplus$ be a commutative and associative binary operation on G. We consider the

---

[1] The ability to move $ccover()$ values between nodes is the inspiration for the name *anonymous segment tree*: the coverage representation for a node is independent of the covering segments. Coverage of an interval by a single segment is indistinguishable from coverage by an arbitrary number of consecutive short segments.

problem of maintaining the following sum for each primitive interval I of the set $\mathcal{S}$: $F(v) = \sum_{s \in \mathcal{S}_I} f(s)$, where v is the leaf associated with I and the summation uses the $\oplus$ operation.[2]

We compute $F(v)$ indirectly by storing a quantity called $FF(v)$ at each node v of the tree, and maintain the invariant that the sum of $FF(a)$ over all ancestors a of v equals $F(v)$. We require that $\oplus$ is invertible, and there is a total order $\leq_G$ on G such that $A \leq_G B \iff A \oplus C \leq_G B \oplus C$. In other words, $(G, \oplus, \leq_G)$ forms a *totally ordered abelian group*. Finally, we reduce the range of $f()$ from G to G^+, defined as $G^+ = \{g \mid g \in G \wedge e \leq_G g\}$, where e is the identity element of $\oplus$.

The pushup invariant in this abstract setting is that, for each internal node v with children v_l and v_r, at least one of $FF(v_l)$ and $FF(v_r)$ is e and the other is in G^+. Repeated pushup operations in the tree, starting from the leaves, establish this invariant. In particular, let v be an internal node with children v_l and v_r, and without loss of generality assume that $FF(v_l) \leq_G FF(v_r)$. The push-up operation sets $FF(v) = FF(v) \oplus FF(v_l)$, $FF(v_l) = e$ and $FF(v_r) = FF(v_r) \oplus FF(v_l)^{-1}$, where $^{-1}$ denotes the inverse with respect to $\oplus$. We can show that the values $FF()$ can be updated in $O(\log n)$ time as segments are inserted and deleted. (The details are technical, but have no bearing on what follows in the rest of the paper.) We now show below how to use this general framework for maintaining the measure of probabilistic segments.

3.3   Measure of Probabilistic Segments

For the sake of simplicity, we maintain the *complement* of the expected measure: the expected value of the length *not* covered by any active segment.[3] In order to maintain the measure for probabilistic segments, we apply our abstract framework twice. First, for each leaf v, we maintain the number of segments that cover its interval *and* have probability 1. We denote this by $cover(v)$, and use the deterministic coverage count algorithm to maintain it. Second, we maintain the probability that the primitive interval of a leaf v is uncovered by the segments whose probability is strictly less than 1. (The segments with probability 1 are handled separately, and more easily.) We denote this quantity by $prob(v)$, and maintain it using our generalized scheme as follows. We define G as the set of positive reals, $\oplus$ as multiplication, $\leq_G$ as $\geq$, and set $f(s)$ to

$$f(s) = \begin{cases} (1 - p_s) & \text{if } p_s < 1 \\ 1 & \text{if } p_s = 1 \end{cases}$$

Observe that $F(v)$ represents $prob(v)$. For ease of reference we denote the $FF(v)$ values used to maintain $prob()$ by $pprob(v)$. We can define the *uncovered measure* of a node v, denoted $\nu(v)$, recursively as follows:

---

[2] Observe that if G is the set of integers, $\oplus$ is integer addition, and $f(s) = 1$ for every s, then $F(v) = cover(v)$, as in the preceding section.

[3] We assume that all segments are contained in a finite, bounded range, ensuring that the complement is bounded.

$$\nu(v) = \begin{cases} 0 & \text{if } ccover(v) > 0 \\ pprob(v) \cdot L(v) & \text{if } ccover(v) = 0 \ \wedge \ v \text{ is a leaf} \\ pprob(v) \cdot (\nu(v_l) + \nu(v_r)) & \text{if } ccover(v) = 0 \ \wedge \ v \text{ has children } v_l \text{ and } v_r \end{cases} \quad (2)$$

Lemma 3. *Let $\bar{\mu}(v)$ denote the complement of the expected measure of $\mathcal{S}$ restricted to v's interval, let $aprob(v)$ be the product of $pprob(a)$ over all ancestors a of v (excluding v), and let a node v be called* exposed *if $ccover(a) = 0$ at all strict ancestors a of v. Then, for any exposed node v, we have $\bar{\mu}(v) = aprob(v) \cdot \nu(v)$.*

Proof. If an exposed node v has $ccover(v) > 0$, then $\bar{\mu}(v) = 0$. By the first line of (2), $\bar{\mu}(v)$ equals $aprob(v) \cdot \nu(v)$. Now consider an exposed leaf v such that $ccover(v) = 0$. Then $cover(v) = 0$. We write

$$\bar{\mu}(v) = prob(v) \cdot L(v) \ = \ aprob(v) \cdot pprob(v) \cdot L(v)$$

By the second line of (2), this expression equals $aprob(v) \cdot \nu(v)$. Finally, consider an exposed internal node v such that $ccover(v) = 0$. Then v_l and v_r are exposed. By induction, $\bar{\mu}(v_l) = aprob(v_l) \cdot \nu(v_l)$ and $\bar{\mu}(v_r) = aprob(v_r) \cdot \nu(v_r)$. Then

$$\begin{aligned} \bar{\mu}(v) = \bar{\mu}(v_l) + \bar{\mu}(v_r) \ &= \ aprob(v_l) \cdot \nu(v_l) + aprob(v_r) \cdot \nu(v_r) \\ &= aprob(v) \cdot pprob(v) \cdot (\nu(v_l) + \nu(v_r)) \end{aligned}$$

By the third line of (2), the expression equals $aprob(v) \cdot \nu(v)$. $\qquad\square$

Lemma 4. *$\nu(root)$ equals the complement of the expected measure of $\mathcal{S}$.*

By Lemma 4, one can report the expected measure of $\mathcal{S}$ simply by returning the complement of $\nu(root)$. By maintaining $\nu()$ the same way we maintain $\mu()$ in Section 3.1, we end up with a structure that solves the stochastic measure problem:

Theorem 2. *The expected measure of a dynamic set of segments can be maintained in $O(1)$ query time, $O(\log n)$ insertion/deletion time and $O(n)$ space.*

4   Dynamic Probabilistic Volume in d Dimensions

We now show how to maintain the expected volume of the union of a dynamic set of probabilistic boxes in d-space. We adapt the framework of Overmars and Yap's solution to Klee's problem [13]. The first step is to apply Theorem 2 to a very special kind of d-dimensional box arrangement called a *trellis*. A trellis in d dimensions is a rectangular region R and a collection of boxes $\mathcal{B}$ that such that each box in $\mathcal{B}$ forms of an axis-parallel *strip* inside R. In other words, no $(d-2)$-dimensional face (a corner in two dimensions) of a box in $\mathcal{B}$ intersects the interior of R. A two-dimensional example is shown in Figure 2(a), where each box is either a vertical or a horizontal strip.

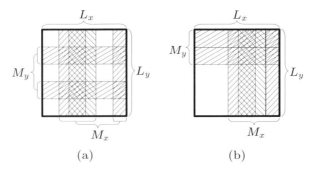

Fig. 2. (a) A two-dimensional trellis formed by 5 boxes. (b) The shape with the same area formed by moving strips.

The volume of a trellis is easy to compute efficiently. First consider the problem in two dimensions. Suppose the horizontal and vertical side lengths of R are L_x and L_y, respectively. Let M_x be the length of the portion of the x-interval of R covered by the vertical strips, and M_y the length of R's y-interval covered by the horizontal strips. Then it is easy to see that the area covered in R is $L_x \times L_y - (L_x - M_x) \times (L_y - M_y)$. (A visual proof is offered in Figure 2(b).)

It follows that computing the area of a trellis reduces to maintaining M_x and M_y separately, i.e., to solving two *one-dimensional* volume problems. In $d > 2$ dimensions, the volume formula for a trellis generalizes easily to

$$\prod_i L_i - \prod_i (L_i - M_i),$$

where the product index ranges from 1 to d, L_i is the side length of R along the ith axis and M_i is the sublength of L_i that is covered by strips orthogonal to the ith axis [13].

To maintain the expected volume within a trellis for stochastic boxes, we use the same formula, except that all the variables in the formula are replaced by their expectations. Specifically, the formula for the d-dimensional case becomes

$$\prod_{1 \leq i \leq d} L_i - \prod_{1 \leq i \leq d} (L_i - E(M_i))$$

where $E(M_i)$ is the expected value of M_i. Note that M_i's are independent. Then, by linearity and multiplicativity of expectation over independent variables, the formula correctly represents the expected volume.

It is clear that the expected volume in a trellis can be maintained in logarithmic time per update by using d instances of anonymous segment tree, each maintaining M_i for $1 \leq i \leq d$. While several efficient solutions are known for maintaining the one-dimensional measure of *non-probabilistic* segments [4,8,10], they all seem quite specialized, focusing on a particular application. It is unclear whether they can be adapted to our probabilistic setting without fairly complicated modifications. The anonymous segment tree, on the other hand, offers a simple, and general, framework suitable for probabilistic measure maintenance.

The second step is to partition the space hierarchically such that the leaves of the partition contain trellis structures. The partition proceeds in d steps. Let us call a face of a box orthogonal to the ith axis an i-*face* and the hyperplane it sits on an i-*bound*. In the first step of the partition, we divide the space into regions called 1-slabs by cutting it with hyperplanes through every $\sqrt{n}$th 1-bound of the boxes along the first axis. Consequently, $O(\sqrt{n})$ 1-slabs are formed, each of which contains $O(\sqrt{n})$ 1-faces. In the second step, each 1-slab is split into 2-slabs by hyperplanes perpendicular to the second coordinate axis. These hyperplanes are introduced as follows: A hyperplane is drawn along every $\sqrt{n}$th 2-bound in $\mathcal{B}$. Additionally, for each box B that has a 1-face inside the 1-slab, two hyperplanes are drawn along both of its 2-bounds. Consequently, each 1-slab is partitioned into $O(\sqrt{n})$ 2-slabs, each of which intersects with $O(\sqrt{n})$ 1-faces and 2-faces of the boxes in $\mathcal{B}$. In the third step, each 2-slab is partitioned into $O(\sqrt{n})$ 3-slabs. This time, the splitting hyperplanes pass along every $\sqrt{n}$th 3-bound and the 3-bounds of each box that has a 1-face or a 2-face intersecting the inside of the 2-slab. This partitioning strategy is continued until the dth step, in which each $(d-1)$-slab is divided into $O(\sqrt{n})$ cells. The following lemma, whose proof can be found in [13], summarizes the key properties of this orthogonal partition.

Lemma 5. *The orthogonal partition contains $O(n^{d/2})$ cells such that each box of $\mathcal{B}$ partially covers $O(n^{(d-1)/2})$ cells, each cell partially intersects $O(\sqrt{n})$ boxes in $\mathcal{B}$, and the boxes partially overlapping a cell form a trellis.*

We can now maintain the uncovered expected volume as follows. For each cell C, we maintain uncovered volume of the boxes that partially intersects C restricted to C. By using the trellis structure we have mentioned, this is doable in logarithmic time per update and constant time per query. Since a box partially intersects $O(n^{(d-1)/2})$ cells, we can update all trellis structures in $O(n^{(d-1)/2} \log n)$ time during a box insertion/deletion.

The cells in a $(d-1)$-slab form a linear sequence. We can track the boxes that completely overlap the cells of a $(d-1)$-slab using a structure called a *slab tree*, which is just an anonymous segment tree with trellises at its leaves. Equation 2 applies at each node of a slab tree, except that if v is a leaf with $ccover(v) = 0$, then $\nu(v) = pprob(v) \cdot \nu_T(v)$, where $\nu_T(v)$ is the uncovered expected measure of the trellis stored at v. The uncovered measure of the whole arrangement is the sum of $\nu(r)$ over all the roots r of the $(d-1)$-slab trees. Updating a slab tree with a box takes logarithmic time. Since there are $O(n^{(d-1)/2})$ $(d-1)$-slabs, updating all slab trees takes $O(n^{(d-1)/2} \log n)$ during a box insertion/deletion. This yields to a total update time of $O(n^{(d-1)/2} \log n)$.

We have a final missing ingredient: a dynamic version of the hierarchical orthogonal partition. Recall that Overmars and Yap's partition is based on making slab cuts at the $\sqrt{n}$th coordinate in each dimension. We relax this constraint by maintaining the sorted sequence of box coordinates in each dimension and cutting along a fairly stable—but not static—set of slab boundaries. The slab boundaries for each dimension partition the corresponding sorted coordinate sequence into *buckets*. We maintain the invariant that each bucket contains at most $2\sqrt{n}$ coordinates, and any two adjacent buckets contain at least $\sqrt{n}$. Whenever one of these invariants is violated, we split or merge the adjacent buckets as necessary by introducing or removing slab boundaries. This invalidates some trellises and slab trees; we restore them by rebuilding. By a potential argument, it can be shown that the amortized cost of all the rebuilding is also $O(n^{(d-1)/2} \log n)$ time per box insertion/deletion, matching the direct cost of data structure updates.

Theorem 3. *The expected volume of a dynamic set of n stochastic boxes in d-space can be maintained in $O(1)$ query time and $O(n^{(d-1)/2} \log n)$ amortized time per update, using an $O(n^{(d+1)/2})$-space data structure.*

5   Discrete Volume

We can also solve the following *discrete* volume version of the problem. *Given a dynamic set $\mathcal{P}$ of points, and a dynamic set $\mathcal{B}$ of d-boxes where each point and box has a probability of being* active, *maintain the expected number of active points in $\mathcal{P}$ that are contained in the union of the active boxes.* The solution idea, in brief, is to represent the points as tiny boxes of size ϵ. We use three structures to maintain the expected volume of the union of: (1) the boxes, (2) the points, and (3) both the boxes and the points. Then, the inclusion-exclusion principle can be used to compute the the expected volume of the intersection between the points and the boxes, which is equal to the expected discrete volume times ϵ. This structure achieves an update time of $O(n^{(d-1)/2} \log n)$, where n is the *total* number of points and boxes. Note that the algorithm has a few technical details. First, one needs to represent ϵ symbolically, so that the algorithm works correctly regardless of the box sizes. Second, the degenerate cases where the points lie on the box boundaries should be handled.

6   Experimental Evaluation

We implemented our algorithm in C++ (available at `http://www.cs.ucsb.edu/~foschini/dynOY/`). The main goal was to evaluate the memory usage and the update time behavior of the data structure. Since those parameters are not affected by the probabilities of the boxes, we performed all our simulations for the deterministic case, namely, $p_i = 1$.

We tested the algorithm using two-dimensional boxes with 64-bit integer coordinates. The experiments were performed on an Intel(R) Core(TM)2 Duo CPU @2.20GHz equipped with 4GB of RAM. To check the correctness of our implementation on large inputs we constructed the union of boxes explicitly using the CGAL [6] library primitives `Boolean_set_operations_2`. This construction is fast when the arrangement of boxes is small, but slows down dramatically when the arrangement size approaches the quadratic worst case. As an example of this, consider a trellis formed by 50 thin vertical and horizontal boxes. Our implementation takes 0.05 seconds to compute the measure of the union, by inserting rectangles one by one, while the CGAL program takes 8.3 seconds. This conforms with the worst case guarantees provided by our analysis.

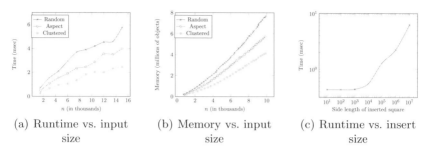

(a) Runtime vs. input size

(b) Memory vs. input size

(c) Runtime vs. insert size

Fig. 3. Experimental Results

Datasets. To test the sensitivity of our algorithm to input distributions, we used three different input configurations: (1) *Random*, a set of 15K rectangles randomly generated, with boundary coordinates uniformly distributed in $[-10^7, 10^7]$, (2) *Aspect*, a set of 15K rectangles with small (< 0.1) aspect ratio and random orientations, with boundary coordinates uniformly distributed in $[-10^7, 10^7]$, and (3) *Clustered*, a set of 15K rectangles with boundary coordinates clustered in ten groups of bounding box $[0, 2 \cdot 10^6]^2$ placed along the diagonal of the $[-10^7, 10^7]^2$ box.

Results. For each dataset, we measured the average time to insert a rectangle as a function of the number of objects n present in the data structure. (Deletion time is about 80% of insertion time.) Since the cost of a single update is predicted to be $O(\sqrt{n} \log n)$ only in an amortized sense, we report the average insertion time of the last 1500 rectangles before each tested n. Results are reported in Figures 3(a) and 3(b). The insertion time is roughly proportional to $\sqrt{n}$, highlighting the fact that the amortized worst case bound is a good indicator of the average case behavior. The memory required (plotted as the number of allocated objects) conforms with the predicted $O(n\sqrt{n})$ worst case bound.

In another experiment we test the sensitivity of the algorithm to the size of the box inserted. In Figure 3(c) we report the average time needed to insert a square into a set of 10K random boxes. the inserted square varying from 20 to 2×10^7 (the size of the full database). The time needed for the insertion increases by a factor of around ten over the range of insertion sizes. This is expected, since a larger square intersects more cells.

The main limitation of the data structure seems to be its memory use. The memory bound of $O(\sqrt{n})$ nodes per box limits the scalability of the algorithm. Especially, if we consider that each box has four edges each affecting $O(\sqrt{n})$ trellises and slab trees, it is easy to see how as few as 15K boxes require roughly 1.5GB of memory in our implementation.

7 Conclusions

In this paper, we considered the problem of maintaining the volume of the union of n boxes in d-space when each box is known to exist with an arbitrary, but independent, probability. We showed that, even in one dimension, computing the probability distribution, namely the probability that the volume exceeds a given value, is NP-hard. On the other hand, we showed that the expected volume of the union can be maintained, nearly as efficiently as in the static and deterministic case. Along the way we introduced a data structure called *anomymous segment tree* that may be of independent interest in dealing with dynamic segment problems with abstract measures. Finally, we also implemented our volume data structure, and showed experimentally that it performs as predicted by theory, and indeed significantly outperforms a naïve solution. Our simulation results also highlight the limitation of a Overmars-Yap type approach: the data structure is memory-intensive, which makes it unsuitable for large data sets. Thus, an interesting future research question is to explore better space-time tradeoffs that might yield scalable solutions to our dynamic and stochastic Klee's problem.

References

1. Agarwal, P.K.: An improved algorithm for computing the volume of the union of cubes. In: Proc. of 26th Symp. on Computational Geometry, pp. 230–239 (2010)
2. Agarwal, P.K., Kaplan, H., Sharir, M.: Computing the volume of the union of cubes. In: Proc. of 23rd Symp. on Computational Geometry, pp. 294–301 (2007)
3. Bentley, J.L.: Solutions to Klee's rectangle problems. Dept. of Comp. Sci., CMU, Pittsburgh PA (1977) (unpublished manuscript)
4. van den Bergen, G., Kaldewaij, A., Dielissen, V.J.: Maintenance of the union of intervals on a line revisited. Computing Science Reports. Eindhoven University of Technology (1998)
5. Bringmann, K.: Klee's measure problem on fat boxes in time $O(n^{(d+2)/3})$. In: Proc. of 26th Symp. on Computational Geometry, pp. 222–229 (2010)
6. CGAL, Computational Geometry Algorithms Library, http://www.cgal.org
7. Chan, T.M.: A (slightly) faster algorithm for Klee's measure problem. Computational Geometry 43(3), 243–250 (2010)
8. Cheng, S.W., Janardan, R.: Efficient maintenance of the union intervals on a line, with applications. In: Proc. of ACM Symp. on Discrete Algorithms, pp. 74–83 (1990)
9. Garey, M., Johnson, D.: Computers and Intractability: A Guide to the Theory of NP-completeness (1979)
10. Gonnet, G.H., Munro, J.I., Wood, D.: Direct dynamic structures for some line segment problems. Computer Vision, Graphics, and Image Processing 23(2), 178–186 (1983)
11. Klee, V.: Can the measure of $\cup[a_i, b_i]$ be computed in less than $O(n \lg n)$ steps? American Mathematical Monthly, 284–285 (1977)
12. van Leeuwen, J., Wood, D.: The measure problem for rectangular ranges in d-space. Journal of Algorithms 2(3), 282–300 (1981)
13. Overmars, M.H., Yap, C.-K.: New upper bounds in Klee's measure problem. SIAM J. Comput. 20(6), 1034–1045 (1991)

Preprocess, Set, *Query!*

Ely Porat and Liam Roditty*

Department of Computer Science, Bar-Ilan University, Ramat-Gan, Israel
{porately,liamr}@macs.biu.ac.il

Abstract. Thorup and Zwick [J. ACM and STOC'01] in their seminal work introduced the notion of *distance oracles*. Given an n-vertex weighted undirected graph with m edges, they show that for any *integer* $k \geq 1$ it is possible to preprocess the graph in $\tilde{O}(mn^{1/k})$ time and generate a compact data structure of size $O(kn^{1+1/k})$. For each pair of vertices, it is then possible to retrieve an estimated distance with multiplicative stretch $2k-1$ in $O(k)$ time. For $k = 2$ this gives an oracle of $O(n^{1.5})$ size that produces in constant time estimated distances with stretch 3. Recently, Pătraşcu and Roditty [FOCS'10] broke the long-standing theoretical status-quo in the field of distance oracles and obtained a distance oracle for sparse unweighted graphs of $O(n^{5/3})$ size that produces in constant time estimated distances with stretch 2.

In this paper we show that it is possible to break the stretch 2 barrier at the price of non-constant query time. We present a data structure that produces estimated distances with $1 + \varepsilon$ stretch. The size of the data structure is $O(nm^{1-\varepsilon'})$ and the query time is $\tilde{O}(m^{1-\varepsilon'})$. Using it for sparse unweighted graphs we can get a data structure of size $O(n^{1.86})$ that can supply in $O(n^{0.86})$ time estimated distances with multiplicative stretch 1.75.

1 Introduction

Thorup and Zwick [15] initiated the theoretical study of data structures capable of representing almost shortest paths efficiently, both in terms of space requirement and query time. Given an n-vertex weighted undirected graph with m edges, they show that for any integer $k \geq 1$ it is possible to preprocess the graph in $\tilde{O}(mn^{1/k})$ time and generate a compact data structure of size $O(kn^{1+1/k})$. For each pair of vertices, it is then possible to retrieve an estimated distance with multiplicative stretch $2k-1$ in $O(k)$ time. An estimated distance has multiplicative (additive) stretch c if for two vertices at distance Δ it is at least Δ and at most $c\Delta$ ($\Delta+c$). We use (α, β) to denote a combination of multiplicative stretch of α and additive stretch of β, that is, the estimated distance is at most $\alpha\Delta + \beta$. Thorup and Zwick called such data structures *distance oracles* as their query time is constant for any constant stretch. Thorup and Zwick [15] showed also that their data structure is optimal for dense graphs. Based on the girth conjecture of Erdős they showed that there exist dense enough graphs which cannot

---

* Work supported by Israel Science Foundation (grant no. 822/10).

be represented by a data structure of size less than $n^{1+1/k}$ without increasing the stretch above $2k - 1$ for any integral k.

Sommer, Verbin, and Yu [13] proved a three-way tradeoff between space, stretch and query time of approximate distance oracles. They show that any distance oracle that can give stretch α answers to distance queries in time $O(t)$ must use $n^{1+\Omega(1/(t\alpha))}/\log n$ space. Their result is obtained by a reduction from lopsided set disjointness to distance oracles, using the framework introduced by Pătrașcu [9]. Any improvement to this lower bound requires a major breakthrough in lower bounds techniques. In particular, it does not imply anything even for slightly non-constant query time as $\Omega(\log n)$ and slightly non-linear space as $n^{1.01}$.

This suggests that upper bounds are currently the only realistic way to attack the Thorup and Zwick space-stretch-query tradeoff. There are several possible ways to obtain a progress:

1. To consider sparse graphs. In particular, graphs with less than $n^{1+1/k}$ edges.
2. To consider also additive error, that is, to get below $2k - 1$ multiplicative stretch with an additional additive stretch.
3. Non-constant query time.

The first way and the second one are closely related. We cannot gain from introducing also additive stretch without getting an improved multiplicative stretch for sparse graphs (i.e., $m = O(n)$). A data structure with size $S(m, n)$ and stretch (α, β) implies a data structure with size $S((\beta+1)m, n + \beta m)$ and multiplicative stretch of α, as if we divide every edge into $\beta+1$ edges then all distances become a multiply of $\beta+1$ and additive stretch of β is useless. For graphs with $m = O(n)$ the size of the data structure is asymptotically the same.

Recently, Pătrașcu and Roditty [10] broke the long-standing theoretical status-quo in the field of distance oracles. They obtained a distance oracle for sparse unweighted graphs of size $O(n^{5/3})$ that can supply in $O(1)$ time an estimated distance with multiplicative stretch 2. For dense graphs the distance oracle has the same size and stretch $(2, 1)$. Pătrașcu and Roditty [10] showed also a conditional lower bound for distance oracle that is based on a conjecture on the hardness of the set intersection problem. They showed that a distance oracle for unweighted graphs with $m = \tilde{O}(n)$ edges, which can distinguish between distances of 2 and 4 in constant time (as multiplicative stretch strictly less than 2 implies) requires $\tilde{\Omega}(n^2)$ space, assuming the conjecture holds. Thus, non-constant query time is essential to get stretch smaller than 2.

In this paper we show that one can gain by allowing non-constant query time. We break the barrier of 2 for multiplicative stretch at the price of non-constant query time. Surprisingly, we show that it is possible to get an arbitrary small multiplicative stretch.

We prove the following:

Theorem 1. *For any unweighted graph and any $\varepsilon > 0$, we can construct in $O(mn)$ time a data structure of size $O(nm^{1-\frac{\varepsilon}{4+2\varepsilon}})$ that given any two vertices s and t returns an estimated distance with multiplicative stretch $1 + \varepsilon$ in*

Table 1. A summary of distance data structures

Reference	Stretch	Query	Space	Input
Trivial	1	$\tilde{O}(m)$	$O(m)$	Weighted digraphs
Trivial	1	$O(1)$	$O(n^2)$	Weighted digraphs
[15]	$2k-1$	$O(k)$	$O(n^{1+1/k})$	Weighted graphs
[8]	$O(k)$	$O(1)$	$O(n^{1+1/k})$	Weighted graphs
[10]	$(2,1)$	$O(1)$	$O(n^{5/3})$	Unweighted graphs
[10]	2	$O(1)$	$O(n^{5/3})$	Unweighted sparse graphs
[1]	$1+\varepsilon$	$O(1)$	$[\varepsilon^{-O(\lambda)} + 2^{O(\lambda \log \lambda)}]n$	Metric space of doubling dimension λ
This paper	$1+\varepsilon$	$\tilde{O}(n^{1-\frac{\varepsilon}{4+2\varepsilon}})$	$O(n^{2-\frac{\varepsilon}{4+2\varepsilon}})$	Unweighted sparse graphs
This paper	1.75	$\tilde{O}(n^{0.86})$	$O(n^{1.86})$	Unweighted sparse graphs

$\tilde{O}(m^{1-\frac{\varepsilon}{4+2\varepsilon}})$ *time. We can provide also an actual path in time proportional to its length.*

When $m = O(n)$ we get a data structure of size $O(n^{2-\frac{\varepsilon}{4+2\varepsilon}})$ with query time $\tilde{O}(n^{1-\frac{\varepsilon}{4+2\varepsilon}})$.

Studying sparse graphs makes a lot of sense not only from theoretical perspective but also from practical point of view. Sparse graphs are the most realistic network scenario in real world applications. Road networks and the Internet are the most popular examples to such scenarios.

Prior to the work of Pătraşcu and Roditty [10], the main focus of the theoretical research on distance oracles was on improving the construction and retrieval times of the distance oracle of Thorup and Zwick. Baswana and Kavitha [2] showed that distance oracles matching the performance of the oracles of Thorup and Zwick can be constructed in $\tilde{O}(n^2)$ time, which is an improvement over the original $O(mn^{1/k})$ construction time for dense enough graphs. Mendel and Naor [8] gave an oracle with $O(1)$ query time (as opposed to $O(k)$ query time of the Thorup-Zwick oracle), at the price of increasing the stretch to $O(k)$. The focus on faster implementations can be partially explained by the fact that it was widely believed that the space/approximation tradeoff of the distance oracle of Thorup and Zwick is optimal. Another line of research considered restricted metric spaces such as Euclidean metric spaces and metric spaces of low doubling dimension. In these restricted cases it is possible to get $1+\varepsilon$ multiplicative stretch with almost linear space. See Bartal *et al.* [1] and reference there in for the state of the art in distance oracles for restricted metric spaces. A summary of the main results appears in Table 1.

Experimental studies on the Thorup-Zwick distance oracle were conducted by Krioukov et al. [7]. Motivated by these experiments, Chen et al. [3] show

---

Algorithm 1. Clusters$(G(V, E), L)$

$E' \leftarrow E$;
$V' \leftarrow V$;
$\mathcal{C} \leftarrow \emptyset$;
while $V' \neq \emptyset$ **do**
$\quad$ **foreach** $u \in V'$ **do**
$\quad\quad \lfloor \; \langle Tree(u), H_u \rangle \leftarrow \text{BFS}(G(V, E'), u, L)$;

$\quad$ Let s be a vertex with minimal fully explored depth;
$\quad$ $\mathcal{C} \leftarrow \mathcal{C} \cup \langle s, Tree(s), r_s, H_s \rangle$;
$\quad$ $E' \leftarrow E' \setminus H_s$;
$\quad$ $V' \leftarrow \{u \mid u$ is in a connected component of $G(V, E')$ with at least L edges $\}$;

return $\mathcal{C}$;

---

that the Thorup-Zwick oracles requires less space on random power-law graphs. Enăchescu et al. [6] obtain similar results for Erdős-Rényi random graphs.

Closely related to the notion of distance oracles is the notion of *spanners*, introduced by Peleg and Schäffer [11]. The main difference between distance oracles and spanners is that spanners are only concerned with the question of how much space is needed to store approximate distances, disregarding the time needed to retrieve approximate distances and paths. Multiplicative-additive spanners were introduced by Elkin and Peleg [5,4]. More constructions of such spanners were obtained by Thorup and Zwick [14] and Pettie [12]. Notice that for spare graphs distance oracles and spanners are very different. If the graph has $O(n)$ edges, a spanner is trivial: just include all the edges. For a distance oracle, such graphs can be hard.

Our algorithm is composed of three stages. In the first stage we preprocess the graph and partition its edges into clusters. It is important to note that we use edge clusters as opposed to the more traditional vertex clusters that are usually used in spanners and distance oracles constructions. In the second stage we set some additional distance information to the clusters. Finally, in the third stage we are ready to answer distance queries. Our algorithm is remarkably simple and can be implemented easily.

The rest of this paper is organized as its title. In the next Section we describe how to preprocess the graph into clusters. In Section 3 we describe the additional information set to the cluster. In Section 4 we present the query algorithm and its analysis. We end in Section 5 with some concluding remarks and open problems.

2   Preprocess

In the preprocessing stage we partition the edges of the graph into clusters as follows. We start to grow in parallel shortest paths trees for each vertex of the graph. As the graph is unweighted each step of the shortest paths computation is a step of Breadth-First-Search (BFS) algorithm and thus equivalent to scanning a single edge. We stop the parallel search when we reach to a certain predefined limit L on the number of edges to be scanned. We refer to the limit parameter

L as the budget of the BFS algorithm, that is, the number of edges it is allowed to scan.

Given a vertex $v \in V$ let H_v be the edges that are scanned by a BFS from v with budget L. (Notice that H_v is the set of *all* edges scanned by the BFS algorithm and thus may contain edges that are not part of the BFS tree.) Let $V(H_v)$ be the vertices that are endpoints of edges from H_v. Let $Tree(v)$ be the shortest paths tree of v in $G(V(H_v), H_v)$. For every $w \in V(H_v)$ we denote with $Tree(v, w)$ the length of the shortest path between v and w in $G(V(H_v), H_v)$. We say that the *fully explored depth* of $Tree(v)$ is its maximal depth minus one and denote it with r_v. We pick the tree with minimal fully explored depth and create a cluster. (If there is more than one we pick one arbitrarily.) Assume that s is the root of the BFS tree. A cluster is represented by the four tuple $\langle s, Tree(s), r_s, H_s \rangle$. We refer to s as the center of this cluster. Notice that a vertex can be the center of more than one cluster. We now proceed to compute clusters in the graph $G(V, E \setminus H_s)$. This process is repeated until every connected component of the graph has less than L edges. The set of clusters is then returned by the algorithm. It is important to note that in each iteration of the algorithm we remove the edges that belong to the new cluster and continue to compute clusters on the remaining graph. As a result of that an edge can be in at most one cluster [1] but a vertex can be in many different clusters. The computation of clusters is presented in Algorithm 1. In the next Lemma we bound the number of clusters.

Lemma 1. *If $\mathcal{C}$ is the set of clusters computed by Algorithm 1 with budget L then $|\mathcal{C}| \leq m/L$.*

Proof. Each cluster contains exactly L edges. When a cluster is formed we removed all its edges from the graph. Thus, the total number of clusters is at most m/L. □

Let $E' = E \setminus \{e \mid \exists \langle u, Tree(u), r_u, H_u \rangle \in \mathcal{C} : r_u \leq \Delta \ \wedge \ e \in H_u\}$, that is, the set of edges that do not belong to clusters of fully explored depth at most Δ. Roughly speaking, in the next Lemma we show that since the graph $G(V, E')$ is relatively sparse it is possible to have a lower bound on the depth of any BFS tree computed from any of its vertices. This useful property will be used later on in the analysis of our query algorithm.

Lemma 2. *Let $\mathcal{C}$ be the set of clusters computed with budget L. For every integral Δ and every $s \in V$ that belongs to a connected component with at least L edges, the fully explored depth of the tree created by BFS($G(V, E'), s, L$) is at least $\Delta + 1$.*

Proof. Clusters are added to $\mathcal{C}$ in a non-decreasing order of their fully explored depth. If we remove all the edges that belong to clusters with fully explored depth of at most Δ then any BFS from any vertex in a connected component of

---

[1] An edge might not be in any cluster as we do not cluster edges in components with less than L edges.

at least L edges with a budget of L must have fully explored depth of at least $\Delta + 1$ as otherwise a cluster of fully explored depth at most Δ is still present in $G(V, E')$, a contradiction.   □

3   Set

We now turn to the second stage. In this stage we set some additional distance information to the clusters that were computed in the first stage. This information is kept in an additional data structure to allow efficient queries. For each cluster we compute the distance to every other vertex of the graph. Let $\delta_E(u, v)$ be the length of the shortest path between u and v in $G(V, E)$. The distance between a cluster $C = \langle s, Tree(s), r_s, H_s \rangle$ and a vertex $u \in V$ is defined as follows:

$$\delta_E(u, C) = \min_{x \in V(H_s)} \delta_E(u, x).$$

It is important to note that the distance between a vertex and a cluster is computed in the original graph $G(V, E)$. Let $p(u, C)$ be a vertex that realizes the minimal distance. If the cluster center s realizes the minimal distance we set $p(u, C)$ to s (the reason to that will be clear later), otherwise, we set $p(u, C)$ to an arbitrary vertex that realizes the minimal distance. This is all the information saved in our data structure. We summarize its main properties in the next Lemma.

Lemma 3. *The data structure described above can be constructed in $O(mn)$ time and has a size of $O(mn/L)$.*

Proof. From Lemma 1 it follows that there are $O(m/L)$ clusters. Thus, the clustering algorithm has $O(m/L)$ iterations and since each iteration takes $O(nL)$ time its total running time is $O(mn)$. We compute also the distances of the graph in $O(mn)$ time. Using the distances and the clusters we compute the additional information. For each cluster we keep the distance to any vertex. This can be done in $O(nL)$ time per cluster. Since the clusters set is of size $O(m/L)$ the total size of the data structure is $O(mn/L)$.   □

4   Query

Let $G(V, E)$ be an unweighted undirected graph and let $s, t \in V$ be two arbitrary vertices. Let $\Delta = \delta_E(s, t)$. In this section we show that for every $\varepsilon > 0$ the query algorithm returns an estimated distance with multiplicative stretch $1 + \varepsilon$ in $\tilde{O}(m^{1 - \frac{\varepsilon}{4 + 2\varepsilon}})$ time. The size of the data structure is $O(nm^{1 - \frac{\varepsilon}{4 + 2\varepsilon}})$.

Roughly speaking, the main idea of the query algorithm is to check whether there is a cluster with a relatively small fully explored depth $(\varepsilon\Delta)$ that intersects the shortest path. If this is the case then we can obtain a good approximation using the information saved in our data structure. If there is no cluster of this scale that intersects the shortest path then the shortest path exists also in a much sparser graph on which a variant of bidirectional BFS can be executed in sub-linear time. Obviously, the full description is much more involved as we do not really know the shortest path or its length. Moreover, the case that $\varepsilon\Delta$ is less than 1 requires a special care.

We now describe more formally the query algorithm. We provide pseudo-code in Algorithm 2. The algorithm is composed of two stages[2]. In the first stage we scan the clusters that were created using the preprocessing algorithm from Section 2. We compute for each cluster $C = \langle u, Tree(u), r_u, H_u \rangle$ the length of the path that is composed from the following three portions, a shortest path between s to $p(s, C)$, a shortest path between $p(s, C)$ and $p(t, C)$ that uses only edges of H_u, and a shortest path between $p(t, C)$ and t. We take the minimal path among all the possible clusters. This stage covers the case in which a shortest path intersects a cluster with a relatively small fully explored depth.

In the second stage we try to improve the path that we have obtained in the first stage by growing shortest paths trees from s and t. This stage is composed of $\log n$ iterations. In the i-th iteration, where $i \in [0, \log n - 1]$ we set $d = 2^i$. The graph that is used in this iteration is $G(V, E')$, where $E' = E \setminus \{e \mid \exists \langle u, Tree(u), r_u, H \rangle \in \mathcal{C} : r_u \leq \varepsilon d \wedge e \in H\}$, that is, only edges that are part of clusters whose fully explored depth is at least $\lceil \varepsilon d \rceil$. We compute two sets of vertices S and T, where $S = \cup_{i=1}^{\lceil \frac{1}{\varepsilon} \rceil} S_i$ and $T = \cup_{i=1}^{\lceil \frac{1}{\varepsilon} \rceil} T_i$. The sets S_1 and T_1 are obtained by executing a BFS with budget L from s and t, respectively. The sets S_{i+1} and T_{i+1} are obtained by executing BFS with budget L for every vertex of S_i and T_i, respectively (see Algorithm 3). We also maintain a distances array h. If $\Delta = \delta_{E'}(s, t)$ then the exact distance is found when $d < \Delta \leq 2d$. Finally, we output the minimal distance that we have obtained. It is important to note that the value of the output corresponds to an actual path in the input graph, thus, we only need to bound the output value from above as it cannot be smaller than Δ.

Next, we show that the algorithm returns a $(1 + \varepsilon)$-approximation of Δ. For the sake of presentation we divide the proof into two Lemmas. In the first Lemma we consider the case that $\varepsilon\Delta \geq 1$ and in the second Lemma we consider the case that $\varepsilon\Delta < 1$.

Lemma 4. *Let $\varepsilon > 0$ and let $\varepsilon\Delta \geq 1$. The algorithm $Query(s, t, \varepsilon)$ returns an estimated distance with $1 + 4\varepsilon$ stretch.*

Proof. Let $P = \{(s, u_1), (u_1, u_2), \dots, (u_\ell, t)\}$ be the edges of a shortest path between s and t. Let $\mathcal{C}' = \{\langle u, Tree(u), r_u, H_u \rangle \mid \langle u, Tree(u), r_u, H_u \rangle \in \mathcal{C} \wedge r_u \leq \varepsilon\Delta\}$. We first consider the case that one of the edges of P belongs to a cluster of $\mathcal{C}'$, that is, there is a cluster $C = \langle u, Tree(u), r_u, H_u \rangle$ in $\mathcal{C}'$, such that $H_u \cap P \neq \emptyset$.

---

[2] There is also a trivial step in which we perform BFS with budget L from the two vertices in the original graph to cover the case that their shortest path is in a component with less than L edges.

Algorithm 2. Query(s, t, ε)

$\hat{\delta}(s, t) \leftarrow \infty$;

// Stage 1

foreach $C = \langle u, Tree(u), r_u, H_u \rangle \in \mathcal{C}$ **do**

$\quad$ **if** $\hat{\delta}(s, t) > \delta_E(s, C) + \delta_{H_u}(p(s, C), p(t, C)) + \delta_E(t, C)$ **then**

$\quad \quad \hat{\delta}(s, t) = \delta_E(s, C) + \delta_{H_u}(p(s, C), p(t, C)) + \delta_E(t, C)$;

// Stage 2

foreach $i \in [0, \log n)$ **do**

$\quad d \leftarrow 2^i$;

$\quad$ **if** $\varepsilon d \geq 1$ **then**

$\quad \quad E' = E \setminus \{e \mid \exists \langle u, Tree(u), r_u, H_u \rangle \in \mathcal{C} : r_u \leq \varepsilon d \ \wedge \ e \in H_u\}$;

$\quad$ **else**

$\quad \quad E' = E \setminus \{e \mid \exists \langle u, Tree(u), r_u, H_u \rangle \in \mathcal{C} : r_u = 0 \wedge e \in H_u \wedge V(H_u) = L+1\}$;

$\quad \langle Tree(s), H_s \rangle \leftarrow \text{BFS}(G(V, E'), s, L)$;

$\quad \langle Tree(t), H_t \rangle \leftarrow \text{BFS}(G(V, E'), t, L)$;

$\quad S \leftarrow S_1 \leftarrow V(H_s)$;

$\quad T \leftarrow T_1 \leftarrow V(H_t)$;

$\quad h(s, \cdot) = Tree(s, \cdot)$;

$\quad h(t, \cdot) = Tree(t, \cdot)$;

$\quad$ **for** $j \leftarrow 1$ **to** $\lceil \frac{1}{\varepsilon} \rceil - 1$ **do**

$\quad \quad S_{j+1} \leftarrow \text{Expand}(s, S_j, h)$;

$\quad \quad T_{j+1} \leftarrow \text{Expand}(t, T_j, h)$;

$\quad \quad S \leftarrow S \cup S_{j+1}$;

$\quad \quad T \leftarrow T \cup T_{j+1}$;

$\quad$ **if** $\hat{\delta}(s, t) > \min_{x \in S \cap T} h(s, x) + h(x, t)$ **then**

$\quad \hat{\delta}(s, t) = \min_{x \in S \cap T} h(s, x) + h(x, t)$

return $\hat{\delta}(s, t)$;

Let $(x, y) \in H_u \cap P$. We show that in this case the first stage of the query algorithm finds a $(1 + \varepsilon)$-approximation of Δ. Recall that in the first stage we consider every cluster and in particular the cluster C. The length of the path between s and t that goes through C is $\delta_E(s, C) + \delta_{H_u}(p(s, C), p(t, C)) + \delta_E(t, C)$. Since $(x, y) \in H_u$ it follows that $\delta_E(s, C) \leq \delta_E(s, x)$ and $\delta_E(t, C) \leq \delta_E(y, t)$. We also know that $\delta_{H_u}(p(s, C), p(t, C)) \leq 2(r_u + 1)$. Since $C \in \mathcal{C}'$ it follows that $r_u \leq \varepsilon\Delta$. Combining this with the fact that $\varepsilon\Delta \geq 1$ it follows that $\delta_{H_u}(p(s, C), p(t, C)) \leq 4\varepsilon\Delta$. We get that $\delta_E(s, C) + \delta_{H_u}(p(s, C), p(t, C)) + \delta_E(t, C) \leq (1 + 4\varepsilon)\Delta$.

We now consider the case in which there is no cluster $C \in \mathcal{C}'$ with edge set H, such that $H \cap P \neq \emptyset$. We show that in this case an exact shortest path is found by the second stage of the query algorithm. In this stage there is an iteration i such that $d = 2^i$ and $d < \Delta \leq 2d$. The graph that is used in this iteration is $G(V, E')$, where, $E' = E \setminus \{e \mid \exists \langle u, Tree(u), r_u, H \rangle \in \mathcal{C} : r_u \leq \varepsilon d \ \wedge \ e \in H\}$. Recall that we compute two sets of vertices S and T, where $S = \cup_{i=1}^{\lceil \frac{1}{\varepsilon} \rceil} S_i$ and $T = \cup_{i=1}^{\lceil \frac{1}{\varepsilon} \rceil} T_i$ using Algorithm 3. We also maintain a distances array h. We now prove that:

Algorithm 3. Expand(u, U, h)

> **foreach** $x \in U$ **do**
> > $\langle Tree(x), H_x \rangle \leftarrow \text{BFS}(G(V, E'), x, L)$;
> > $W \leftarrow W \cup V(H_x)$;
>
> **foreach** $y \in W$ **do**
> > **if** $h(u, y) > h(u, x) + Tree(x, y)$ **then** $h(u, y) = h(u, x) + Tree(x, y)$
>
> **return** W;

Claim 2 *For every* $1 \leq i \leq \lceil \frac{1}{\varepsilon} \rceil$, *if* $\delta_{E'}(s, x) \leq i \lceil \varepsilon d \rceil$ *then* $x \in S_i$ *and* $h(s, x) = \delta_{E'}(s, x)$.

Proof. The proof is by induction on i. For the base of the induction we show that if $\delta_{E'}(s, x) \leq \lceil \varepsilon d \rceil$ then $x \in S_1$ and $h(s, x) = \delta_{E'}(s, x)$. We compute S_1 using a BFS from s with budget L on the graph $G(V, E')$. From Lemma 2 it follows that its fully explored depth is at least $\lceil \varepsilon d \rceil$. Thus, any vertex x for which $\delta_{E'}(s, x) \leq \lceil \varepsilon d \rceil$ is added to S_1 by the algorithm and the value of $h(s, x)$ is set to $\delta_{E'}(s, x)$. For the induction hypothesis we assume that if $\delta_{E'}(s, x) \leq i \lceil \varepsilon d \rceil$ then $x \in S_i$ and $h(s, x) = \delta_{E'}(s, x)$. We prove that if $i \lceil \varepsilon d \rceil < \delta_{E'}(s, x) \leq (i + 1) \lceil \varepsilon d \rceil$ then $x \in S_{i+1}$ and $h(s, x) = \delta_{E'}(s, x)$. Let x' be a vertex on a shortest path between s and x for which $\delta_{E'}(s, x') = i \lceil \varepsilon d \rceil$. From the induction hypothesis it follows that $x' \in S_i$ and $h(s, x') = \delta_{E'}(s, x')$. When Expand is called with S_i, a BFS with budget L is executed from x'. From Lemma 2 it follows that this BFS tree will contain x as its fully explored depth is at least $\lceil \varepsilon d \rceil$ and $\delta_{E'}(x', x) \leq \lceil \varepsilon d \rceil$. As Expand updates $h(s, x)$ with the minimal available distance it also follows that $h(s, x) = \delta_{E'}(s, x)$. □

The same claim holds for the vertex t and the set T with identical proof. We now turn to prove that there is at least one vertex from the shortest path between s and t in $S \cap T$.

Claim 3 *There is a vertex x on the shortest path between s and t such that* $h(s, x) = \delta_{E'}(s, x)$, $h(t, x) = \delta_{E'}(t, x)$ *and* $x \in S \cap T$.

Proof. Recall that $\delta_{E'}(s, t) = \Delta$ and $d < \Delta \leq 2d$. Thus, there is a vertex x from the shortest path between s and t for which $\delta_{E'}(s, x) \leq d$ and $\delta_{E'}(s, x) \leq d$. This implies that $\delta_{E'}(s, x) \leq i \lceil \varepsilon d \rceil$ for some $1 \leq i \leq \lceil \frac{1}{\varepsilon} \rceil$ and $\delta_{E'}(t, x) \leq j \lceil \varepsilon d \rceil$ for some $1 \leq j \leq \lceil \frac{1}{\varepsilon} \rceil$. By applying Claim 2 for s we get that $x \in S_i$ and $h(s, x) = \delta_{E'}(s, x)$. Similarly, by applying Claim 2 for t we get that $x \in T_j$ and $h(t, x) = \delta_{E'}(t, x)$. Hence $x \in S \cap T$. □

From Claim 3 it follows that there is a vertex $x \in S \cap T$ such that $h(s, x) = \delta_{E'}(s, x)$ and $h(t, x) = \delta_{E'}(t, x)$. Thus, the value of $\hat{\delta}(s, t)$ is set by the algorithm to $\delta_{E'}(s, x) + \delta_{E'}(t, x) = \Delta$. □

We now turn to the case that $\varepsilon \Delta < 1$. One possible solution to this case is to divide every edge into a constant number of edges as suggested in the Introduction. This however increases the size of our data structure to $O(m^2/L)$. We show here that by a slight change to the second stage of the query algorithm and a careful analysis it is possible to keep the size of the data structure $O(nm/L)$.

Lemma 5. *Let $\varepsilon > 0$ and let $\varepsilon\Delta < 1$. The algorithm $Query(s, t, \varepsilon)$ returns the exact distance Δ.*

Proof. Let $P = \{(s, u_1), (u_1, u_2), \ldots, (u_\ell, t)\}$ be the set of edges of a shortest path between s and t. Let $\mathcal{C}' = \{\langle u, Tree(u), r_u, H_u \rangle \mid \langle u, Tree(u), r_u, H_u \rangle \in \mathcal{C} \wedge r_u = 0 \wedge V(H_u) = L+1\}$. Notice that every cluster $\langle u, Tree(u), r_u, H_u \rangle \in \mathcal{C}'$ is a star with $L+1$ vertices whose center is u.

We first consider the case that there is a cluster $C = \langle u, Tree(u), r_u, H_u \rangle \in \mathcal{C}'$ such that $H_u \cap P \neq \emptyset$. In the first stage of the query algorithm we consider every cluster and in particular the cluster C. The length of the path between s and t that goes through C is $\delta_E(s, C) + \delta_{H_u}(p(s, C), p(t, C)) + \delta_E(t, C)$.

A shortest path P can intersect a star cluster in two possible ways. One is that $H_u \cap P = \{(x, y)\}$, where $\delta_E(s, x) < \delta_E(s, y)$. The other one is that $H_u \cap P = \{(x, y), (y, z)\}$, where $\delta_E(s, x) < \delta_E(s, y) < \delta_E(s, z)$.

Assume that $H_u \cap P = \{(x, y)\}$, and there is no other shortest path P' such that $H_u \cap P = \{(x, y), (y, z)\}$. The cluster center u must be either x or y. Assume, wlog, that $u = x$. As we assume that there is no other shortest path P' that intersect the cluster in the second way it cannot be that $\delta_E(s, C) < \delta_E(s, u)$. Hence, $\delta_E(s, C) = \delta_E(s, u)$ and $p(s, C) = u$. (Recall that we have set $p(s, C)$ to the cluster center if it realizes the minimal distance.) Moreover, since $p(s, C) = u$ it follows that $\delta_{H_u}(p(s, C), p(t, C)) = 1$. Finally, as $H_u \cap P = \{(x, y)\}$ it follows that $\delta_E(t, C) \leq \delta_E(y, t)$. Adding it all together we get:

$$\delta_E(s, C) + \delta_{H_u}(p(s, C), p(t, C)) + \delta_E(t, C) \leq \delta_E(s, x) + 1 + \delta_E(y, t) = \Delta$$

We now assume that there is at least one shortest path P for which $H_u \cap P = \{(x, y), (y, z)\}$, where $\delta_E(s, x) < \delta_E(s, y) < \delta_E(s, z)$. Since the cluster is a star y must be its center. As $H_u \cap P = \{(x, y), (y, z)\}$ it follows that $\delta_E(s, C) \leq \delta_E(s, x)$ and $\delta_E(t, C) \leq \delta_E(t, z)$. Again as the cluster is a star $\delta_{H_u}(p(s, C), p(t, C)) \leq 2$.

Adding it all together we get:

$$\delta_E(s, C) + \delta_{H_u}(p(s, C), p(t, C)) + \delta_E(t, C) \leq \delta_E(s, x) + 2 + \delta_E(z, t) = \Delta.$$

We now turn to the case in which there is no cluster $C \in \mathcal{C}'$ with edge set H, such that $H \cap P \neq \emptyset$. In the second stage of the query algorithm there exists an iteration i such that $d = 2^i$ and $d < \Delta \leq 2d$. In this iteration $\varepsilon d < \varepsilon\Delta < 1$ and we consider the graph $G(V, E')$, where $E' = E \setminus \{e \mid \exists \langle u, Tree(u), r_u, H_u \rangle \in \mathcal{C} : r_u = 0 \wedge e \in H_u \wedge V(H_u) = L+1\}$, that is, only edges that are not part of a star cluster are used. The next Claim is similar to Claim 2.

Claim 4 *For every $1 \leq i \leq \lceil \frac{1}{\varepsilon} \rceil$, if $\delta_{E'}(s, x) \leq i$ then $x \in S_i$ and $h(s, x) = \delta_{E'}(s, x)$.*

Proof. The proof is by induction on i. For the base of the induction we show that if $\delta_{E'}(s, x) \leq 1$ then $x \in S_1$ and $h(s, x) = \delta_{E'}(s, x)$. We compute S_1 using a BFS from s with budget L on the graph $G(V, E')$. The degree of s is strictly less than L as otherwise $G(V, E')$ contains a star cluster. Hence, x must be reached by a BFS from s with budget L and it is added to S_1. The value of

$h(s,x)$ is set to $\delta_{E'}(s,x)$. We now turn to prove the general case. We assume that if $\delta_{E'}(s,x) \leq i$ then $x \in S_i$ and $h(s,x) = \delta_{E'}(s,x)$. We prove that if $\delta_{E'}(s,x) = i+1$ then $x \in S_{i+1}$ and $h(s,x) = \delta_{E'}(s,x)$. Let x' be a vertex on a shortest path between s and x for which $\delta_{E'}(s,x') = i$. From the induction hypothesis it follows that $x' \in S_i$ and $h(s,x') = \delta_{E'}(s,x')$. When Expand is called with S_i, a BFS with budget L is executed from x'. The degree of x' is strictly less than L as otherwise $G(V,E')$ contains a star cluster. Hence this BFS tree will contain x. As Expand updates $h(s,x)$ with the minimal available distance it also follows that $h(s,x) = \delta_{E'}(s,x)$.   □

The same claim holds for the vertex t and the set T with identical proof. We now show that $S \cap T \neq \emptyset$. Recall that $\delta_{E'}(s,t) = \Delta$ and $d < \Delta \leq 2d$. Thus, there is a vertex x such that $\delta_{E'}(s,x) \leq d$ and $\delta_{E'}(s,x) \leq d$. This implies that $\delta_{E'}(s,x) \leq i$ for some $1 \leq i \leq \lceil \frac{1}{\varepsilon} \rceil$ and $\delta_{E'}(t,x) \leq j$ for some $1 \leq j \leq \lceil \frac{1}{\varepsilon} \rceil$. By applying Claim 4 for s we get that $x \in S_i$ and $h(s,x) = \delta_{E'}(s,x)$. Similarly, by applying Claim 4 for t we get that $x \in T_j$ and $h(t,x) = \delta_{E'}(t,x)$. Hence $x \in S \cap T$ and the value of $\hat{\delta}(s,t)$ is set by the algorithm to $\delta_{E'}(s,x) + \delta_{E'}(t,x) = \Delta$ □

Next, we analyze the running time of the query algorithm and its space usage. The algorithm produces a $(1+4\varepsilon)$ stretch. In the first stage we scan m/L clusters. In the second stage we have $\log n$ iterations. The cost of each iteration is $O(L^{\lceil \frac{1}{\varepsilon} \rceil})$. If we omit logarithmic factor and balance $O(L^{\lceil \frac{1}{\varepsilon} \rceil})$ with m/L we get that $L = m^{\frac{\varepsilon}{1+2\varepsilon}}$ and the cost of a query is $\tilde{O}(m^{1-\frac{\varepsilon}{1+2\varepsilon}})$. In terms of space the size of the data structure is $O(mn/L)$ and for $L = m^{\frac{\varepsilon}{1+2\varepsilon}}$ it is $O(nm^{1-\frac{\varepsilon}{1+2\varepsilon}})$. If we set $\varepsilon' = 4\varepsilon$ we get $1+\varepsilon'$ stretch with query time of $\tilde{O}(m^{1-\frac{\varepsilon'}{4+2\varepsilon'}})$ and space of $O(nm^{1-\frac{\varepsilon'}{4+2\varepsilon'}})$.

5   Concluding Remarks and Open Problems

In this paper we show that it is possible to obtain a multiplicative stretch that is arbitrarily close to 1 with sub-linear space and sub-liner time in sparse graphs. A natural question is whether the tradeoff that we present can be improved further. Another interesting direction is what can we gain from a non-constant query time for stretch 2 and more. Finally, our clustering technique is natural and simple, thus, it will be interesting to see whether it can be used in other closely related problems such as efficient computation of shortest paths approximation.

References

1. Bartal, Y., Gottlieb, L., Kopelowitz, T., Lewenstein, M., Roditty, L.: Fast, precise and dynamic distance queries. In: Proc. of 22th SODA (to appear, 2011)
2. Baswana, S., Kavitha, T.: Faster algorithms for all-pairs approximate shortest paths in undirected graphs. SIAM J. Comput. 39(7), 2865–2896 (2010)
3. Chen, W., Sommer, C., Teng, S.-H., Wang, Y.: Compact routing in power-law graphs. In: Keidar, I. (ed.) DISC 2009. LNCS, vol. 5805, pp. 379–391. Springer, Heidelberg (2009)

4. Elkin, M.: Computing almost shortest paths. ACM Transactions on Algorithms 1(2), 283–323 (2005)
5. Elkin, M., Peleg, D.: (1+epsilon, beta)-spanner constructions for general graphs. SIAM J. Comput. 33(3), 608–631 (2004)
6. Enachescu, M., Wang, M., Goel, A.: Reducing maximum stretch in compact routing. In: INFOCOM, pp. 336–340 (2008)
7. Krioukov, D., Fall, K.R., Yang, X.: Compact routing on internet-like graphs. In: INFOCOM (2004)
8. Mendel, M., Naor, A.: Ramsey partitions and proximity data structures. In: Proc. of 47th FOCS, pp. 109–118 (2006)
9. Pătraşcu, M.: (data) structures. In: Proc. of 49th FOCS, pp. 434–443 (2008)
10. Pătraşcu, M., Roditty, L.: Distance oracles beyond the thorup–zwick bound. In: Proc. of 51st FOCS (2010)
11. Peleg, D., Schäffer, A.A.: Graph spanners. J. Graph Theory, 99–116 (1989)
12. Pettie, S.: Low distortion spanners. ACM Transactions on Algorithms 6(1) (2009)
13. Sommer, C., Verbin, E., Yu, W.: Distance oracles for sparse graphs. In: Proc. of 50th FOCS, pp. 703–712 (2009)
14. Thorup, M., Zwick, U.: Spanners and emulators with sublinear distance errors. In: Proc. of 17th SODA
15. Thorup, M., Zwick, U.: Approximate distance oracles. JACM 52(1), 1–24 (2005)

Cuckoo Hashing with Pages

Martin Dietzfelbinger[1],[*], Michael Mitzenmacher[2],[**], and Michael Rink[1],[*]

[1] Fakultät für Informatik und Automatisierung, Technische Universität Ilmenau
{martin.dietzfelbinger,michael.rink}@tu-ilmenau.de
[2] School of Engineering and Applied Sciences, Harvard University
michaelm@eecs.harvard.edu

Abstract. A downside of cuckoo hashing is that it requires lookups to multiple locations, making it a less compelling alternative when lookups are expensive. One such setting is when memory is arranged in large pages, and the major cost is the number of page accesses. We propose the study of cuckoo hashing with pages, advocating approaches where each key has several possible locations, or cells, on a single page, and additional choices on a second backup page. We show experimentally that with k cell choices on one page and a single backup cell choice, one can achieve nearly the same loads as when each key has $k + 1$ random cells to choose from, with most lookups requiring just one page access, even when keys are placed online using a simple algorithm. While our results are currently experimental, they suggest several interesting new open theoretical questions for cuckoo hashing with pages.

1 Introduction

Standard cuckoo hashing places keys into a hash table by providing each key with k cells determined by hash functions. Each cell can hold one key, and each key must be located in one of its cells. As new keys are inserted, keys may have to move from one alternative to another to make room for the new key. Cuckoo hashing provides high space utilization and worst-case constant-time lookups, making it an attractive hashing variant, with useful applications in both theoretical and practical settings, e.g., [2,8,9,16,17].

Perhaps the most significant downside of cuckoo hashing, however, is that it potentially requires checking multiple cells randomly distributed throughout the table. In many settings, such random access lookups are expensive, making cuckoo hashing a less compelling alternative. As a comparison, standard linear probing works well when memory is split into (not too small) chunks, such as cache lines; the average number of memory accesses is usually close to 1.

In this paper, we consider cuckoo hashing under a setting where memory is arranged into pages, and the primary cost is the number of page accesses. A natural scheme to minimize accesses would be to first hash each key to a page, and then keep a separate cuckoo hash table in each page. This limits the number

[*] Research supported by DFG grant DI 412/10-2.
[**] Research supported by NSF grants IIS-0964473 and CCF-0915922.

C. Demetrescu and M.M. Halldórsson (Eds.): ESA 2011, LNCS 6942, pp. 615–627, 2011.

of pages examined to one, and maintains the constant lookup time once the page is loaded. Such a scheme has been utilized in previous work (e.g., [2]). However, a remaining problem is that the most overloaded page limits the load utilization of the entire table. As we show later, the random fluctuations in the distribution of keys per page can significantly affect the maximum achievable load.

We generalize the above approach by placing most of the cell choices associated with a key on the same primary page. We then allow a backup page to contain secondary choices of possible locations for a key (usually just one). In the worst case we now must access two pages, but we demonstrate experimentally that we can arrange so that for most keys we only access the primary page, leading to close to one page access on average. Intuitively, the secondary page for each key allows overloaded pages to slough off load constructively to underloaded pages, thus distributing the load. We show that we can do this effectively both offline and online by evaluating an algorithm that performs well, even when keys are deleted and inserted into the table.

We note that it is simple to show that using a pure splitting scheme, with no backup page, and page sizes $s = m^\delta$, $0 < \delta < 1$, where m is the number of memory cells, the load thresholds obtained are asymptotically provably the same as for cuckoo hashing without pages. Analysis using such a parametrization does not seem suitable to describe real-world page and memory sizes. While we *conjecture* that the load thresholds obtained using the backup approach, for reasonable parameters for memory and page sizes, match this bound, at this point our *work is entirely experimental*. We believe this work introduces interesting new theoretical problems for cuckoo hashing that merit further study.

Related Work. A number of papers have recently resolved the longstanding issue regarding the load threshold for standard cuckoo hash tables where each key obtains k choices [6,10,11,12,13]. Our work re-opens the issue, as we consider the question of the effect of pages on these thresholds, if the pages are smaller than m^δ, such as for example polylog(m). Practical motivation for this approach includes recent work on real-world implementations of cuckoo hashing [2,17]. In [2], where cuckoo hashing algorithms are implemented on graphical processing units, the question of how to maintain page-level locality for cuckoo hash tables arises. Even though work for lookups can be done in parallel, the overall communication bandwidth can be a limitation in this setting. Ross examines cuckoo hashing on modern processors, showing they can be quite effective by taking advantage of available parallelism for accessing cache lines [17]. Our approach can be seen as attempting to extend this performance, from cache lines to pages, by minimizing the amount of additional parallelism required.

The issue of coping with pages for hash-based data structures is not new. An early reference is the work of Manber and Wu, who consider the effects of pages for Bloom filters [15]. Their approach is the simple splitting approach we described above; they first hash a key to a page, and each page then corresponds to a separate Bloom filter. The deviations in the number of keys hashed to a page yield only small increases in the overall probability of a false positive for the Bloom filter, making this approach effective. As we show below, such deviations have more significant effects on the acceptable load for cuckoo hash tables, leading to our suggested use of backup pages.

Arbitman, Naor, and Segev propose a related scheme [3]. Their structure has two layers. The first layer is a hash table with buckets of size $\Theta\left(\log(1/\varepsilon)/\varepsilon^2\right)$ and one hash function, and the second layer is a cuckoo hash table of size $\varepsilon \cdot n$ for keys that overflow the first layer. If buckets fit on a page, then $1 - \varepsilon$ of the successful searches require only one page access. While we do not have a detailed comparison, given the ε^2 in the denominator of the required bucket size, we expect our approach to be more effective in practice.

Summary of Results. All of our results are experimental. They focus on the setting of four location choices per key. The maximum load factor c_4^* of keys with four hash functions and no paging is known. With small pages and each key confined to one page, we find using an optimal offline algorithm that the maximum achievable load factor is quite low, well below c_4^*. However, if each key is given three choices on a primary page and a fourth on a backup page, the load factor is quite close to c_4^*, even while placing most keys in their primary page, so that most keys require only a single page access. With three primary choices, a single backup choice, and up to 95 percent load, we find that only about 3 percent of keys need to be placed on a backup page (with suitable page sizes). We show that a simple variation of the well-known random walk insertion procedure allows nearly the same level of performance with online, dynamic placement of keys (including scenarios with alternating insertion and deletions). Our experiments consistently yield that at most 5 percent of keys needs to be placed on a backup page with these parameters. This provides a tremendous reduction of the number of page accesses required for successful searches. Regarding unsuccessful searches, spending a little more space for Bloom filters leads to even smaller number of accesses to backup pages. For space reasons not all results are given here; we refer the reader to the online version [7] for more detail.

2   Problem Description

We want to place n keys into $m = n/c$ memory (table) cells where each cell can hold a fixed number of $\ell \geq 1$ keys. The value c is referred to as the load factor. The memory is subdivided into t pages (sub-tables) of equal size $s = m/t$. (Throughout the paper we assume m is divisible by t.) Each key is associated with a *primary page* and a *backup page* distinct from the primary page, as well as a set of k distinct table cells, k_p on the primary page and $k_\mathrm{b} = k - k_\mathrm{p}$ on the backup page. The pages and keys are chosen according to hash functions on the key, and it is useful to think of them as being chosen uniformly at random in each case. For a given assignment let n_p be the number of keys that are placed in their primary page and let n_b be the number of keys that are placed in their backup page. We can state the *cuckoo paging problem* as follows.

Cuckoo Paging Problem: Find a placement of the n keys such that the fraction n_p/n is maximized.

As mentioned, in the standard model with no backup pages and all key locations chosen uniformly at random, there is a threshold load factor $c_{k,\ell}^*$ such that whenever $c < c_{k,\ell}^*$ a placement exists with probability $1 - o(1)$. (See, e.g., [10].)

The aim of this paper is to experimentally investigate the potential for saving access cost by using primary and backup pages. Appropriate algorithms are presented in the next section. For ease of description of the algorithms we also use the following bipartite cuckoo graph model as well as the hashing model.

We consider random bipartite graphs $G = (L \cup R, E)$ with left node set $L = [n]$ and right node set $R = [m]$. The left nodes correspond to the keys, the right nodes correspond to the memory cells of capacity ℓ. The set R is subdivided into t segments $R_0, R_1, \ldots, R_{t-1}$, each of size $s = m/t$, which correspond to the separate pages. Each left node x is incident to $k = k_{\mathrm{p}} + k_{\mathrm{b}}$ edges where its neighborhood $N(x)$ consists of two disjoint sets $N_p(x)$ and $N_b(x)$ determined according to the following scheme (all choices are fully random): choose p from $[t]$ (the index of the primary page); then choose k_{p} different right nodes from R_p to build the set $N_p(x)$; next choose b (the index of the backup page) from $[t] - \{p\}$; and finally choose k_{b} different right nodes from R_b to build the set $N_b(x)$. Let $e = \{x, y\}$ be an edge where $x \in L$ and $y \in R$. We call e a *primary edge* if $y \in N_p(x)$ and call e a *backup edge* if $y \in N_b(x)$.

3 Algorithms

Using the cuckoo graph we can restate the problem of inserting the keys as finding an *orientation* of the edge set of the cuckoo graph G such that the indegree of all left nodes is exactly 1 and the outdegree of all right nodes is at most ℓ. We call such an orientation *legal*. An edge $e = (y, x)$ with x from L and y from R is interpreted as "storing key x in memory cell y." If y is from $N_p(x)$ we call x a *primary key* and otherwise we call x a *backup key*. Each legal orientation which has a maximum number of primary keys is called *optimal*.

In the static case, i.e., if the cuckoo graph G is given in advance, there are well-known efficient (but not linear time) algorithms to find an optimal orientation of G. One possibility is to consider a corresponding *minimum cost matching problem*: Assign a cost of 0 to each primary edge of G and a cost of 1 to each backup edge. Then replace each node y from R with ℓ copies and replace each edge to which y is incident with ℓ copies as well. Initially direct all edges from left to right. Edges from right to left are *matching edges*. The minimum cost matching problem is to find a left-perfect matching (legal orientation) with minimum cost (minimum number of backup keys). In the online version of the paper [7], we describe a variant of the successive shortest path algorithm [1] we used to find such matchings.

In the online scenario the cuckoo graph initially consists only of the right nodes. To begin let us consider the case of insertions only. The keys arrive and are inserted one by one, and with each new key the graph grows by one left node and k edges. To find an appropriate orientation of the edges in each insertion step, we use a random walk algorithm, which is a modification of the common random walk for k-ary cuckoo hashing [9] but with two additional constraints:

1. avoid creating backup keys at the beginning of the insertion process, and
2. keep the number of backup keys below a small fixed fraction.

For the following description of the algorithm we use the hashing model. (The pseudocode can be found in the online version [7].) We refer to a key's k_p cells on its primary page as primary positions, and the k_b cells on its backup page as backup positions. The insertion of an arbitrary key x takes one or more basic steps of the random walk, which can be separated into the following sub-steps. Let x be the key that is currently "nestless", i.e., x is not stored in the memory. First check if one of its primary positions is free. If this is the case store x in such a free cell and stop successfully. Otherwise toss a biased coin to decide whether the insertion of x should be proceed on its primary page or on its backup page.

- If the insertion of x is restricted to the primary page, randomly choose one of its primary positions y. Let x' be the key which is stored in cell y. Store x in y, replace x with x', and start the next step of the random walk.
- If x is to be stored on its backup page, first check if one of the backup positions of x is free. If this is the case store x in such a free cell and stop successfully. Otherwise randomly choose one of the backup positions y on this page and proceed as in the previous case.

The matching procedure is slightly modified to avoid unnecessary back steps. That is, if a key x displaces a key x' and in the next step x' displaces x'' then $x'' = x$ is forbidden as long as x' has another option on this page.

The algorithm uses two parameters.

$\mathfrak{a}$ - the bias of the virtual coin. This influences the fraction of backup keys.
$\mathfrak{b}$ - controls the terminating condition. A global counter is initialized with value $\mathfrak{b} \cdot n$, which is the maximum number of total steps of the random walk summed over all keys. For each basic step the global counter is decremented by one. If the limit is exceeded the algorithm stops with "failure".

Deletions are carried out in a straightforward fashion. To remove a key x, first the primary page is checked for x in its possible cells, and if needed the backup page can then be checked as well. The cell containing x is marked as empty, which can be interpreted as removing the left node x and its k incident edges from G. The global counter is ignored in this setting ($\mathfrak{b} = \infty$).

4   Experiments

For each of the following experiments we consider cuckoo graphs G randomly generated according to some configuration $\kappa = (c, m, s, k_\mathrm{p}, k_\mathrm{b})$ where c is the quotient of left nodes (keys) and right nodes (table cells), m is the total number of right nodes, s is the page size, and $k_\mathrm{p}, k_\mathrm{b}$ are the number of primary and backup edges of each left node. In the implementation the left and right nodes were simply the number sets $[n]$ and $[m]$. All random choices were made via

the pseudo random number generator MT19937 "Mersenne Twister" of the GNU
Scientific Library [14].

If not stated otherwise the total number of cells is $m = 10^6$ and pages are of
size $s = 10^i, i \leq 6$. Our main focus is on situations where $\ell = 1$, i.e., each cell
can hold one key. Moreover we restrict ourselves to the cases $k_{\mathrm{p}} = 3, k_{\mathrm{b}} = 1$ and
(just for comparison) $k_{\mathrm{p}} = 4$ and $k_{\mathrm{b}} = 0$. While we have done experiments with
other parameter values, we believe these settings portray the main points. Also,
while we have computed sample variances, in many cases they are small; this
should be assumed when they are not discussed.

4.1   Static Case

Experimental results for the static case determine the limits of our approach and
serve as a basis of comparison for the dynamic case.

Setup and Measurements. First of all we want to see how far cuckoo hashing with pages carries if there are no backup options at all. Note that for fixed
page size s and larger and larger table size m the fraction of keys that can be
placed decreases. For the case with backup options we try to get an approximation for possible threshold densities. Let $c^-_{s,m}$ and $c^+_{s,m}$ be the loads n/m that
identify the transition from where there is a feasible orientation and where there
is no feasible orientation of G without and with backup option respectively. To
get approximations for $c^-_{s,m}$ and $c^+_{s,m}$ we study different ranges of load factors
$[c^{\mathrm{start}}, c^{\mathrm{end}}]$ and for all c where $c = c^{\mathrm{start}} + i \cdot 10^{-4} \leq c^{\mathrm{end}}$, and $i = 0, 1, 2, \ldots$,
we construct a random graphs and measure the failure rate λ at c. We fit the
sigmoid function

$$f(c; x, y) = \left(1 + \exp(-(c - x)/y)\right)^{-1} \tag{1}$$

to the data points (c, λ) using the method of least squares. The parameter x
(inflection point) is an approximation of $c^-_{s,m}$ and $c^+_{s,m}$ respectively. With $\sum_{\mathrm{res}}$
we denote the sum of squares of the residuals.

Furthermore, for different c and page sizes s, we are interested in the maximum
ratio $r_p = n_{\mathrm{p}}/n$ or load $\alpha_p = n_{\mathrm{p}}/m$ of primary keys, respectively.

For a fixed page p let w be the number of keys that have primary page p but
are inserted on their backup page. Since the number of potential primary keys
for a page follows a binomial distribution, some pages will be lightly loaded and
therefore have a small value of w or even $w = 0$. Some pages will be overloaded
and have to shed load, yielding a large value of w. We want to study the relative
frequency of the values w.

Results. Here we consider results from an optimal placement algorithm.

I. Table 1 gives approximations of the loads where cuckoo hashing with paging
and $k = 4$ hash functions has failure rate $\lambda = 0.5$ in the case of 1 or 0 backup
pages. With no backup pages the number of keys that can be stored decreases
with decreasing page size and the success probability around $c^-_{s,m}$ converges less
rapidly, as demonstrated clearly in Fig. 1. This effect becomes stronger as the
pages get smaller. For this reason the range of load factors $[c^{\mathrm{start}}, c^{\mathrm{end}}]$ of sub-
table (a) grows with decreasing page size. Using only one backup edge per key

Table 1. Approximations of the load factors that are the midpoints of the transition from failure rate 0 to failure rate 1 (without and with backup option) via fitting function (1) to a series of data points. For each data point the failure rate among 100 random graphs was measured. The grey rows correspond to the plots of Fig. 1.

(a) $40 \cdot 2^{6-\log_{10}(s)}+1$ data points, $k_p = 4, k_b = 0$

s	$[c^{\text{start}}, c^{\text{end}}]$	$c^-_{s,m}$	$\sum_{\text{res}}$
10^6	$[0.975, 0.979]$	0.976794	0.014874
10^5	$[0.968, 0.976]$	0.971982	0.096023
10^4	$[0.944, 0.960]$	0.952213	0.299843
10^3	$[0.863, 0.895]$	0.879309	0.653894
10^2	$[0.617, 0.681]$	0.648756	1.382760
10^1	$[0.124, 0.252]$	0.188029	1.809620

(b) 41 data points, $k_p = 3, k_b = 1$

s	$[c^{\text{start}}, c^{\text{end}}]$	$c^+_{s,m}$	$\sum_{\text{res}}$
10^5	$[0.975, 0.979]$	0.976774	0.010523
10^4	”	0.976760	0.014250
10^3	”	0.976765	0.002811
10^2	”	0.976611	0.007172
10^1	$[0.9712, 0.9752]$	0.973178	0.008917

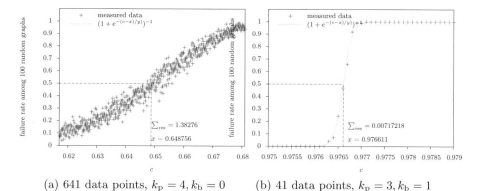

(a) 641 data points, $k_p = 4, k_b = 0$ (b) 41 data points, $k_p = 3, k_b = 1$

Fig. 1. Point of transition (a) without and (b) with backup pages, for $s = 10^2$.

almost eliminates this effect. In this case the values $c^+_{s,m}$ seem to be stable for varying s and are very near to the theoretical threshold of standard 4-ary cuckoo hashing, which is $c^*_4 \approx 0.976770$; only in the case of very small pages $s = 10$ can a minor shift of $c^+_{s,m}$ be observed. The position of $c^+_{s,m}$ as well as the slope of the fitting function appear to be quite stable for all considered page sizes.

II. The average of the maximum fraction of primary keys, allowing one backup option, is shown in Table 2. The fraction decreases with increasing load factor c and decreases with decreasing page size s as well. Interestingly, for several parameters, we found that an optimal algorithm finds placements with more than $c^*_3 \cdot m$ keys sitting in one of their 3 primary positions, where $c^*_3 \approx 0.917935$ is the threshold for standard 3-ary cuckoo-hashing. That is, more keys obtain one of their primary three choices with three primary and one backup choice than what could be reached using just three primary choices even without paging.

III. Figure 2 depicts the relative frequency of the values w among 10^5 pages for selected parameters $(c, s) = (0.95, 10^3)$. In this case about 17 percent of all pages do not need backup pages, i.e., $w = 0$. This is consistent with the idea that pages

Table 2. Average (among 100 random graphs) of the fraction of keys that can be placed on their primary page for different page sizes s and $k_p = 3$, $k_b = 1$. The failure rate is $\lambda = 0$. For $c \geq 0.98$ the random graph did not admit a solution anymore. The entries of the grey cells are larger than c_3^*.

c	$s = 10^5$		$s = 10^4$		$s = 10^3$		$s = 10^2$		$s = 10^1$	
	r_p	α_p	r_p	α_p	r_p	α_p	r_p	α_p	r_p	α_p
0.90	1.000000	0.900000	0.999881	0.899893	0.995650	0.896085	0.975070	0.877563	0.902733	0.812460
0.91	0.999997	0.909997	0.999093	0.909175	0.993008	0.903638	0.971556	0.884116	0.898281	0.817436
0.92	0.998136	0.918286	0.996111	0.916422	0.989452	0.910296	0.967781	0.890358	0.893546	0.822062
0.93	0.990957	0.921510	0.990467	0.921134	0.985015	0.916064	0.963723	0.896263	0.888041	0.825878
0.94	0.983443	0.924436	0.983422	0.924416	0.979730	0.920946	0.959429	0.901863	0.880848	0.827997
0.95	0.975952	0.927154	0.975961	0.927163	0.973744	0.925057	0.954876	0.907132	0.872427	0.828805
0.96	0.968578	0.929835	0.968524	0.929783	0.967224	0.928535	0.947650	0.909744	0.862883	0.828367
0.97	0.961112	0.932279	0.961157	0.932323	0.956892	0.928185	0.935928	0.907850	0.850154	0.824650

with a load below $c_3^* \cdot s$ will generally not need backup pages. The mean $\overline{w}$ is about 2.5 percent of the page size s and for about 87.6 percent of the pages the value w is at most 5 percent of the page size. The relative frequency of w being greater than $0.1 \cdot s$ is very small, about $1.1 \cdot 10^{-3}$.

Summary. We observed that using pages with $(k_p, k_b) = (3, 1)$ we achieve loads very close to the c_4^* threshold ($c_{s,m}^+ \approx c_4^*$). Moreover the load α_p from keys placed on their primary page α_p is quite large, near or even above c_3^*.

Let X be the average (over all keys that have been inserted) number of page requests needed in a search for a key x, where naturally we first check the primary page. If $(k_p, k_b) = (3, 1)$ and a key was equally likely to be in any of its locations, the *expected number of page requests* $\mathrm{E}(X)$ would satisfy $\mathrm{E}(X) = 1.25$. If $(k_p, k_b) = (3, 1)$ and c is near c_4^* then we have roughly $\mathrm{E}(X) \approx c_3^*/c \cdot 1 + (1 - c_3^*/c) \cdot 2$. For example, for $(c, s) = (0.95, 10^3)$, using the values of Table 1 we find $\mathrm{E}(X) \approx 0.974 \cdot 1 + 0.026 \cdot 2 < 1.03$.

Now assume we perform a lookup for a key x not in the table. The disadvantage of using two pages per key is that now we always require two page requests, i.e., $\mathrm{E}(X) = 2$. This can be circumvented by storing an additional set membership data structure, such as a Bloom filter [4], for each page p representing the w many keys that have primary page p but are inserted on their backup page. One can trade off space, computation, and the false positive probability of the Bloom filter as desired. As an example, suppose the Bloom filters use 3 hash functions and their size corresponds to just one bit per page cell. In this case, we can in fact use the same hash functions that map keys to cell locations for our Bloom filter. Bounding the fraction of 1 bits of a Bloom Filter from above via $(k_p \cdot w)/s$, the distribution

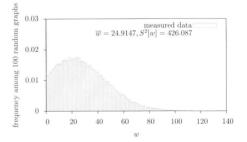

Fig. 2. frequency of $w = 0$ is 0.169, $(c, s) = (0.95, 10^3)$, $a = 10^3$, $\lambda = 0$

of w as in Fig. 2 leads to an average false positive rate of less than 0.15 percent and therefore an expected number of page requests $E(X)$ of less than 1.0015 for unsuccessful searches. By employing more hash functions one could reduce the false positive rate and simultaneously reduce space requirements.

4.2 Dynamic Case

We have seen the effectiveness of optimal offline cuckoo hashing with paging. We now investigate whether similar placements can be found online, by considering the simple random walk algorithm from Sect. 3. We begin with the case of insertions only.

Setup and Measurements. Along with the failure rate λ, the fraction of primary keys r_p and corresponding load α_p, and the distribution of the number of keys w inserted on their backup page, we consider two more performance characteristics:

#st - the average number of steps of the random walk insertion procedure. A step is either storing a key x in a free cell y or replacing an already stored key with the current "nestless" key.

#pr - the average number of page requests over all inserted items. Here each new key x requires at least one page request, and every time we move an item to its backup page, that requires another page request.

We focus on characteristics of the algorithm with loads near $c_{s,m}^+$, varying the number of table cells $m = 10^5, 10^6, 10^7$ and page sizes $s = 10, 10^2, 10^3$. The performance of the algorithm heavily depends on the choice of parameters $\mathfrak{a}$ and $\mathfrak{b}$. Instead of covering the complete parameter space we first set $\mathfrak{b}$ to infinity and use the measurements to give insight into the performance of the algorithm for selected values of $\mathfrak{a}$.

We also study the influence of $\mathfrak{a}$ for a fixed configuration. We vary $\mathfrak{a}$ to see qualitatively how the number of primary keys as well as the number of steps and page requests depend on this parameter. Because of space reasons the (unsurprising) results are omitted but can be found in the online version [7].

It is well known that hashing schemes can perform differently in settings with insertions and deletions rather than insertions alone, so we investigate whether there are substantial differences in this setting. Specifically, we consider the table under a constant load by alternating insertion and deletion steps.

Results. Here we consider results from the random walk algorithm.

I. Tables 3 and 4 show the behavior of the random walk algorithm with loads near $c_{s,m}^+$ for $(c, \mathfrak{a}) = (0.95, 0.97)$ and $(c, \mathfrak{a}) = (0.97, 0.90)$. The number of allowed steps for the insertion of n keys is set to infinity via $\mathfrak{b} = \infty$. The number of trials a per configuration is chosen such that $a \cdot m = 10^9$ (keeping the running time for each configuration approximately constant).

With these parameters the algorithm found a placement for the keys in all experiments; failure did not occur. For fixed page size the sample means are

Table 3. Characteristics of the random walk for $(c, \mathfrak{a}, \mathfrak{b}) = (0.95, 0.97, \infty)$. $\lambda = 0$.

s	m	a	t	$\overline{r_p}$	$\overline{\alpha_p}$	$\overline{\#\text{st}}$	$S^2[\#\text{st}]$	$\overline{\#\text{pr}}$	$S^2[\#\text{pr}]$
10^1	10^5	10^4	10^4	0.860248	0.817236	158.707669	114.072760	10.327258	0.409334
"	10^6	10^3	10^5	0.860219	0.817208	158.618752	11.405092	10.321981	0.040869
"	10^7	10^2	10^6	0.860217	0.817206	158.645056	1.092781	10.323417	0.003914
10^2	10^5	10^4	10^3	0.938431	0.891509	22.807328	1.081478	2.248953	0.003760
"	10^6	10^3	10^4	0.938424	0.891503	22.813986	0.104012	2.249273	0.000366
"	10^7	10^2	10^5	0.938412	0.891491	22.813905	0.010862	2.249201	0.000038
10^3	10^5	10^4	10^2	0.955773	0.907985	16.580150	0.512018	1.892190	0.001779
"	10^6	10^3	10^3	0.955737	0.907950	16.603145	0.052386	1.893515	0.000182
"	10^7	10^2	10^4	0.955730	0.907943	16.598381	0.005534	1.893248	0.000019

Table 4. Characteristics of the random walk for $(c, \mathfrak{a}, \mathfrak{b}) = (0.97, 0.90, \infty)$. $\lambda = 0$.

s	m	a	t	$\overline{r_p}$	$\overline{\alpha_p}$	$\overline{\#\text{st}}$	$S^2[\#\text{st}]$	$\overline{\#\text{pr}}$	$S^2[\#\text{pr}]$
10^1	10^5	10^4	10^4	0.816795	0.792291	158.506335	1222.640379	32.336112	48.892079
"	10^6	10^3	10^5	0.816790	0.792286	153.645339	78.581917	31.363876	3.142566
"	10^7	10^2	10^6	0.816802	0.792298	152.873602	10.759338	31.209210	0.430827
10^2	10^5	10^4	10^3	0.886997	0.860387	23.320507	2.731285	5.361922	0.108700
"	10^6	10^3	10^4	0.886992	0.860382	23.289233	0.256942	5.355625	0.010218
"	10^7	10^2	10^5	0.886985	0.860375	23.268641	0.024796	5.351518	0.000986
10^3	10^5	10^4	10^2	0.898281	0.871332	19.497032	1.550490	4.607751	0.061739
"	10^6	10^3	10^3	0.898232	0.871285	19.486312	0.146267	4.605481	0.005816
"	10^7	10^2	10^4	0.898235	0.871288	19.493215	0.012744	4.606893	0.000507

almost constant; for growing page size the load $\overline{\alpha_p}$ increases, while $\overline{\#\text{st}}$ and $\overline{\#\text{pr}}$ decrease, with a significant drop from page size 10 to 100. For our choices of $\mathfrak{a}$ the random walk insertion procedure missed the maximum fraction of primary keys by up to 2 percent for $c = 0.95$ and by up to 6 percent for $c = 0.97$ and needs roughly the same average number of steps (for fixed page size).

II. To get more practical values for $\mathfrak{b}$ we scaled up the values $\overline{\#\text{st}}$ from Tables 3 and 4 and estimated the failure probability for suitable parameter sets $\pi = (c, s, \mathfrak{a}, \mathfrak{b}) \in \{(0.95, 10^2, 0.97, 30), (0.95, 10^3, 0.97, 25), (0.97, 10^2, 0.90, 30), (0.97, 10^3, 0.90, 25)\}$. For all these parameter sets we observed a failure rate of zero among $a = 10^6$ attempts. We can conclude at a level of significance of at least $1 - e^{-10}$ that for these sets the failure probability of the random walk algorithm is at most 10^{-5}.

III. The results for alternating insertions and deletions for parameters $(c, s) = (0.95, 10^3)$ and $(\mathfrak{a}, \mathfrak{b}) = (0.97, 30)$ are shown in Fig. 3. We measured the current fraction of primary keys r_p and the number of insertion steps with respect to each key $\#\text{st}_{\text{key}}$. Recall that $\#\text{st}$ is the average number of insertion steps concerning all keys.

In the first phase (insertions only) the average number of steps per key grows very slowly at the beginning and is below 10 when reaching a load where about

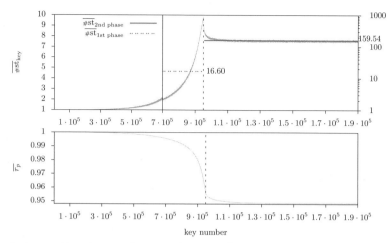

Fig. 3. $(c, s, \mathfrak{a}, \mathfrak{b}) = (0.95, 10^3, 0.97, 30)$, $a = 10^3$, $\lambda = 0$. The ordinate of the right half of the upper plot is in log scale.

1 percent of current keys are backup keys. After that $\overline{\#\mathrm{st}}_{\mathrm{key}}$ grows very fast up to almost the page size. Similarly the sample mean of the fraction of primary keys $\overline{r_p}$ decreases very slowly at the beginning and decreases faster at the end of the first phase. In the second phase (deletions and insertions alternate) $\overline{\#\mathrm{st}}_{\mathrm{key}}$ and $\overline{\#\mathrm{st}}$ decrease and quickly reach a steady state. Since the decrease of $\overline{r_p}$ is marginal but the drop $\overline{\#\mathrm{st}}_{\mathrm{key}}$ is significant we may conclude that the overall behavior is better in steady state than at the end of the insertion only phase. Moreover in an extended experiment with $n = c \cdot m$ insertions and $10 \cdot n$ delete-insert pairs the observed equilibrium remains the same and therefore underpins the conjecture that Fig. 3 really shows a "convergence point" for alternating deletions and insertions.

IV. Figure 4 shows the relative frequency of the values w among 10^5 pages for $(c, s) = (0.95, 10^3)$ and $(\mathfrak{a}, \mathfrak{b}) = (0.97, 30)$ at the end of the insertion only phase, given by Fig. 4 (a), and at the end of the alternation phase, given by Fig. 4 (b). Note that Fig. 4 (a) corresponds to Fig. 2 with respect to the graph parameters.

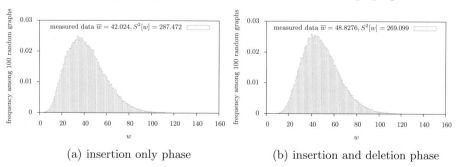

(a) insertion only phase   (b) insertion and deletion phase

Fig. 4. frequency of w, $(c, s) = (0.95, 10^3)$, $a = 10^3$, $\lambda = 0$

The shapes of the distributions differ only slightly, except that in the second phase the number of backup keys is larger. In comparison with the values given by the optimal algorithm in Fig. 2 the distribution of the w values is more skewed and shifted to the right.

Summary. A simple online random-walk algorithm, with appropriately chosen parameters, can perform quite close to the optimal algorithm for cuckoo hashing with paging, even in settings where deletions occur.

With parameters $(c, s) = (0.95, 10^3)$ and $(\mathfrak{a}, \mathfrak{b}) = (0.97, 30)$ the expected number of page requests $\mathrm{E}(X)$ for a successful search is about 1.044, using the values from Table 3. With the Bloom filter approach described in Sect. 4.1 (which can be done only after finishing the insertion of all keys), the distribution from Fig. 4 (a) gives an expected number of page requests for an unsuccessful search of less than 1.0043. Both values are only slightly higher than those resulting from an optimal solution. To improve performance for unsuccessful searches with online insertions and deletions, one can use counting Bloom filters [5] instead, at the cost of more space.

5   Conclusion

Our results suggest that cuckoo hashing with paging may prove useful in a number of settings where the cost of multiple lookups might otherwise prove prohibitive. Perhaps the most interesting aspect for continuing work is to obtain provable performance bounds for cuckoo hashing with pages. Even in the case of offline key distribution with one additional choice on a second page we do not have a formal result proving the threshold behavior we see in experiments.

References

1. Ahuja, R.K., Magnanti, T.L., Orlin, J.B.: Network Flows: Theory, Algorithms, and Applications. Prentice-Hall, Upper Saddle River (1993)
2. Alcantara, D.A., Sharf, A., Abbasinejad, F., Sengupta, S., Mitzenmacher, M., Owens, J.D., Amenta, N.: Real-time parallel hashing on the GPU. ACM Trans. Graph. 28(5) (2009)
3. Arbitman, Y., Naor, M., Segev, G.: Backyard Cuckoo Hashing: Constant Worst-Case Operations with a Succinct Representation. In: Proc. 51st FOCS, pp. 787–796. IEEE, Los Alamitos (2010)
4. Bloom, B.H.: Space/Time Trade-offs in Hash Coding with Allowable Errors. Commun. ACM 13(7), 422–426 (1970)
5. Broder, A., Mitzenmacher, M.: Network applications of Bloom filters: A survey. Internet Mathematics 1(4), 485–509 (2004)
6. Dietzfelbinger, M., Goerdt, A., Mitzenmacher, M., Montanari, A., Pagh, R., Rink, M.: Tight Thresholds for Cuckoo Hashing via XORSAT. In: Abramsky, S., Gavoille, C., Kirchner, C., Meyer auf der Heide, F., Spirakis, P.G. (eds.) ICALP 2010. LNCS, vol. 6198, pp. 213–225. Springer, Heidelberg (2010)
7. Dietzfelbinger, M., Mitzenmacher, M., Rink, M.: Cuckoo Hashing with Pages. CoRR abs/1104.5111 (2011)

8. Erlingsson, Ú., Manasse, M., McSherry, F.: A cool and practical alternative to traditional hash tables. In: Proc. 7th WDAS (2006)

9. Fotakis, D., Pagh, R., Sanders, P., Spirakis, P.G.: Space Efficient Hash Tables with Worst Case Constant Access Time. Theory Comput. Syst. 38(2), 229–248 (2005)

10. Fountoulakis, N., Khosla, M., Panagioutou, K.: The Multiple-orientability Thresholds for Random Hypergraphs. In: Proc. 22nd SODA, pp. 1222–1236. SIAM, Philadelphia (2011)

11. Fountoulakis, N., Panagiotou, K.: Orientability of Random Hypergraphs and the Power of Multiple Choices. In: Abramsky, S., Gavoille, C., Kirchner, C., Meyer auf der Heide, F., Spirakis, P.G. (eds.) ICALP 2010. LNCS, vol. ICALP, pp. 348–359. Springer, Heidelberg (2010)

12. Frieze, A.M., Melsted, P.: Maximum Matchings in Random Bipartite Graphs and the Space Utilization of Cuckoo Hashtables. CoRR abs/0910.5535 (2009)

13. Gao, P., Wormald, N.C.: Load balancing and orientability thresholds for random hypergraphs. In: Proc. 42nd STOC, pp. 97–104. ACM, New York (2010)

14. Gough, B. (ed.): GNU Scientific Library Reference Manual, 3rd edn. Network Theory Ltd (2009), http://www.gnu.org/software/gsl/manual/

15. Manber, U., Wu, S.: An Algorithm for Approximate Membership checking with Application to Password Security. Inf. Process. Lett. 50(4), 191–197 (1994)

16. Pagh, R., Rodler, F.F.: Cuckoo hashing. J. Algorithms 51(2), 122–144 (2004)

17. Ross, K.A.: Efficient Hash Probes on Modern Processors. In: Proc. 23rd ICDE, pp. 1297–1301. IEEE, Los Alamitos (2007)

Approximation Algorithms and Hardness Results for the Joint Replenishment Problem with Constant Demands

Andreas S. Schulz[1] and Claudio Telha[2]

[1] Sloan School of Mgmt. & Operations Research Center, MIT, Cambridge, MA, USA
schulz@mit.edu
[2] Operations Research Center, MIT, Cambridge, MA, USA
ctelha@mit.edu

Abstract. In the Joint Replenishment Problem (JRP), the goal is to coordinate the replenishments of a collection of goods over time so that continuous demands are satisfied with minimum overall ordering and holding costs. We consider the case when demand rates are constant. Our main contribution is the first hardness result for any variant of JRP with constant demands. When replenishments per commodity are required to be periodic and the time horizon is infinite (which corresponds to the so-called general integer model with correction factor), we show that finding an optimal replenishment policy is at least as hard as integer factorization. This result provides the first theoretical evidence that the JRP with constant demands may have no polynomial-time algorithm and that relaxations and heuristics are called for. We then show that a simple modification of an algorithm by Wildeman et al. (1997) for the JRP gives a fully polynomial-time approximation scheme for the general integer model (without correction factor). We also extend their algorithm to the finite horizon case, achieving an approximation guarantee asymptotically equal to $\sqrt{9/8}$.

1 Introduction

In the deterministic Joint Replenishment Problem (JRP) with constant demands, we need to schedule the replenishment times of a collection of commodities in order to fulfill a constant demand rate per commodity. Each commodity incurs fixed ordering costs every time it is replenished and linear holding costs proportional to the amount of the commodity held in storage. Linking all commodities, a joint ordering cost is incurred whenever one or more commodities are ordered. The objective of the JRP is to minimize the sum of ordering and holding costs.

The JRP is a fundamental problem in inventory management. It is a natural extension of the classical economic lot-sizing model that considers the optimal trade-off between ordering costs and holding costs for a single commodity. With multiple commodities, the JRP adds the possibility of saving resources via coordinated replenishments, a common phenomenon in supply chain management.

C. Demetrescu and M.M. Halldórsson (Eds.): ESA 2011, LNCS 6942, pp. 628–639, 2011.

For example, in manufacturing supply chains, a suitable replenishment schedule for raw materials can lead to significant reductions in operational costs. This reduction comes not only from a good trade-off between ordering and holding costs, but also from joint replenishment savings, such as those involving transportation and transactional costs.

Since an arbitrary replenishment schedule may be difficult to implement, it is natural to focus on restricted sets of schedules (often called policies in this context). The *general integer model with correction factor* (GICF) assumes an infinite horizon and constant inter-replenishment time per commodity. The joint ordering cost in the GICF model is a complicated function of the inter-replenishment times, so it is often assumed that joint orders are placed periodically, even if some joint orders are empty. This defines the *general integer model* (GI). In both cases, the time horizon is infinite.

The existence of a polynomial-time optimal algorithm for the JRP with constant demands remains open for all models, regardless of whether the time horizon is finite or not, whether the ordering points are periodic or not, or whether the incurred costs are modeled precisely or not. Given the significant amount of research in this area, it may be a bit surprising that only a few papers mention the lack of a hardness result for the JRP as an issue (the only recent paper we could find is [TB01]). Some papers addressing the JRP with constant demands (e.g. [LY03, MC06]) cite a result by Arkin et al. [AJR89], which proves that the JRP with variable demands is NP-hard. However, it is not clear how to adapt this result to the constant demand case. In fact, we believe that the two problems are completely unrelated from a complexity perspective as their input and output size are incomparable.

In this paper we present the first hardness result for the JRP with constant demands. We show that finding an optimal policy for the general integer model with correction factor is at least as hard as integer factorization, under polynomial-time reductions. Although integer factorization is unlikely to be NP-hard, it is widely believed to be outside P. In fact, this belief supports the hypothesis that RSA and other cryptographic systems are secure [RSA78].

We also give approximation results. An α-approximation algorithm for a minimization problem is an algorithm that produces a feasible solution with cost at most α times the optimal cost. We show that a simple modification of an algorithm by Wildeman et al. [WFD97] gives polynomial-time approximation algorithms with ratios very close to or better than the current best approximations for the JRP with constant demands. We illustrate this in detail for dynamic policies in the finite horizon case and for GI. For the latter model, this yields a fully polynomial-time approximation scheme (FPTAS): for every $\epsilon > 0$, we provide a $(1 + \epsilon)$-approximation algorithm with running time polynomial in n and $1/\epsilon$. Here, n is the number of commodities. To the best of our knowledge, this is the first FPTAS for a model formally known as GI with *variable base*.

In the remainder of this section we formally describe the JRP models we consider in this paper, followed by a review of the literature and a summary of our results.

Mathematical formulation. For all JRP variants considered in this paper the input consists of a finite collection $\mathcal{I} = \{1, \ldots, n\}$ of commodities with constant demand rates $d_i \in \mathbb{Z}_+$, for $i \in \mathcal{I}$. The cost of an order is the sum of the *individual ordering costs* K_i of the commodities involved in the order plus the *joint ordering cost* K_0. The acquired inventory is stored at a *holding cost rate* of h_i per unit of commodity i and per unit of time. The objective is to find an optimal ordering schedule in the time horizon $[0, T)$, where T may be equal to $+\infty$ (the so-called *stationary case*).

If T is finite, a *schedule* S is a finite sequence of joint orders. (An order is called a joint order even if it consists of one commodity only.) If we place N joint orders, the total joint ordering cost is $C_{\text{ord}}^{\text{joint}} \equiv N K_0$. If we replenish commodity i at times $0 = t_1 < t_2 < \ldots < t_{n_i} < T$, its individual ordering cost is $C_{\text{ord}}^{\text{indiv}}(i) \equiv n_i K_i$ and its individual holding cost is $C_{\text{hold}}(i) \equiv \frac{d_i h_i}{2} \sum_{j=1}^{n_i} (t_{j+1} - t_j)^2$, where $t_{n_i+1} = T$. The cost $C[S]$ of the schedule S is the sum of the joint ordering cost $C_{\text{ord}}^{\text{joint}}$, the total individual ordering cost $C_{\text{ord}}^{\text{indiv}} \equiv \sum_{i \in \mathcal{I}} C_{\text{ord}}^{\text{indiv}}(i)$, and the total holding cost $C_{\text{hold}} \equiv \sum_{i \in \mathcal{I}} C_{\text{hold}}(i)$. The objective of the JRP is to minimize $C[S]$.

An arbitrary sequence of joint orders is called a *dynamic* schedule. The structure of an optimal dynamic schedule for the finite horizon case is not known. Potentially, it could be exponential in the size of the input. One can avoid this issue by adding more structure. In the JRP with *general integer policies*, joint orders can be placed only at multiples of a base period p (to be determined), and each commodity $i \in \mathcal{I}$ is periodically replenished every $k_i p$ units of time, for some $k_i \in \mathbb{Z}_+$. The costs are just the time-average version of their counterparts in the finite horizon case. An accurate mathematical description of this scenario is the general integer model with correction factor:

$$
\min \quad \frac{K_0 \Delta(k_1, \ldots, k_{|\mathcal{I}|})}{p} + \sum_{i \in \mathcal{I}} \left(\frac{K_i}{q_i} + \frac{1}{2} h_i d_i q_i \right)
$$

$$
\text{s.t.} \quad q_i = k_i p \tag{GI-CF1}
$$
$$
k_i \in \mathbb{Z}_+
$$
$$
p > 0,
$$

where $\Delta(k_1, \ldots, k_{|\mathcal{I}|})/p$ is the average number of joint orders actually placed in a time interval of length p. With a simple counting argument, it is easy to see that

$$
\Delta(k_1, \ldots, k_{|\mathcal{I}|}) = \sum_{i=1}^{|\mathcal{I}|} (-1)^{i+1} \sum_{I \subseteq \mathcal{I} : |I| = i} \operatorname{lcm}(k_i, i \in I)^{-1},
$$

where $\operatorname{lcm}(\cdot)$ is the least common multiple of its arguments.

The GICF model is complicated to analyze because of the Δ term. Ignoring Δ (i.e. setting the joint ordering cost rate to be K_0/p in GI-CF1) defines the *general integer* model (GI). Note that this change is equivalent to assuming that K_0 is paid at every multiple of the base period.

We defined the GI and GICF formulations in the *variable base model*. Both formulations have a variant where the base p is restricted. The fixed base version of the GI formulation requires p to be multiple of some constant B. The fixed base version of the GICF formulation requires p to be fixed.

Literature Review. We only survey approximation algorithms (see Goyal and Satir [GS89] and Goyal and Khouja [KG08] for a review of other heuristics). For none of the models studied here any hardness results are known.

General integer models. A common approach in this case is to solve the problem for a sequence of values of p and return the best solution found. For instance, Kaspi and Rosenblatt [KR91] approximately solve (GI) for several values of p and pick the solution with minimum cost. They do not specify the number of values of p to test, but they choose them to be equispaced in a range $[p_{\min}, p_{\max}]$ containing any optimal p. For example,

$$p_{\min} = \frac{K_0}{\sqrt{2(K_0 + \sum_{i \in \mathcal{I}} K_i)(\sum_{i \in \mathcal{I}} d_i h_i)}}, \quad p_{\max} = \sqrt{2 \frac{K_0 + \sum_{i \in \mathcal{I}} K_i}{\sum_{i \in \mathcal{I}} d_i h_i}} \quad (1)$$

are a lower and an upper bound for any optimal p. Wildeman et al. [WFD97] transform this idea into a heuristic that converges to an optimal solution. They exactly solve (GI) for certain values of p determined using a Lipschitz optimization procedure. They do not establish a running time guarantee.

A completely different approach uses the rounding of a convex relaxation of the problem. This was introduced by Roundy [Rou85] for the One Warehouse Multi Retailer Problem (OWMR). For the JRP, Jackson et al. [JMM85] find a GI schedule with cost at most $\sqrt{9/8} \approx 1.06$ times the optimal cost for dynamic policies. This approximation is improved to $\frac{1}{\sqrt{2} \log 2} \approx 1.02$ when the base is variable [MR93]. The constants above have been slightly improved by considering better relaxations [TB01].

Using another method, Lu and Posner [Pos94] give an FPTAS for the GI model with fixed base. They note that the objective function is piecewise convex and the problem reduces to querying only a polynomial number of its break points.

General integer policies with correction factor (GICF). No progress has been reported in terms of approximation for this problem, other than the results inherited from the GI model. The incorporation of the correction factor leads to a completely different problem, at least in terms of exact solvability. For example, as the inter-replenishment period goes to 0 the joint ordering cost in the GI model diverges, in contrast to what happens in the GICF model. Porras and Dekker [PD08] show that the inclusion of the correction factor significantly changes the replenishment cycles k_i and the joint inter-replenishment period with respect to those in an optimal GI solution. Moreover, they prove that there is always a solution under this model that outperforms ordering commodities independently, thereby neglecting possible savings from joint orders. This desirable property has not yet been proved for the GI model.

Finite horizon. Most of the heuristics for the finite horizon case assume variable demands and run in time $\Omega(T)$ [LRS04, Jon90]. Some of them can be extended to the constant demand case preserving polynomiality. To our knowledge, the only heuristic with a provable approximation guarantee in this setting is given by Joneja [Jon90]. Their algorithm is designed for variable demands, but for constant demands and $T = \infty$, it achieves an approximation ratio of 11/10.

Summary of results. In Sect. 2 we show that finding an optimal solution for the GICF model in the fixed base case is at least as hard as the integer factorization problem. This is the first hardness result for any of the variants of JRP with constant costs and demands. In Sect. 3 we present, based on [WFD97], a polynomial-time 9/8-approximation algorithm for the JRP with dynamic policies and finite horizon. As the time horizon T increases, the ratio converges to $\gamma \equiv \sqrt{9/8}$. In Sect. 4, we observe that the previous algorithm, extended to the infinity horizon case, is an FPTAS for the class of GI policies (either variable or fixed base model). This result is new for the fixed base case.

2   A Hardness Result for GICF

In this section we prove a hardness result for the JRP in the *fixed base* GICF. In contrast to GI-CF1, this model has p as a parameter:

$$\min \quad \frac{K_0 \Delta(k_1, \ldots, k_{|I|})}{p} + \sum_{i \in \mathcal{I}} \frac{K_i}{q_i} + \frac{1}{2} h_i d_i q_i$$

$$\text{s.t} \quad q_i = k_i p \qquad\qquad\qquad\qquad\qquad \text{(GI-CF2)}$$

$$k_i \in \mathbb{Z}_+$$

Essentially, we prove that if we are able to solve GI-CF2 in polynomial time, then we are able to solve the following problem in polynomial time:

INTEGER-FACTORIZATION: Given an integer M, find an integer d with $1 < d < M$ such that d divides M, or conclude that M is prime.

Reduction. The reduction uses two commodities. The main idea is to set up the costs so that commodity 1 has a constant renewal interval of length M in the optimal solution. Under this assumption, commodity 2 has some incentive to choose an inter-replenishment time q_2 not coprime with M, since this reduces the joint ordering cost with respect to the case when they are coprime. When this happens, and as long as $q_2 < M$, we can find a non-trivial divisor of M by finding the maximum common divisor of M and q, using Euclid's algorithm.

We initially fix $p = 1$, $\frac{1}{2} h_1 d_1 = H_1$, $K_2 = 0$ and $\frac{1}{2} h_2 d_2 = 1$ (here H_1 is a constant we will define later), and therefore GI-CF2 reduces to:

$$\min \quad K_0 \left(\frac{1}{q_1} + \frac{1}{q_2} - \frac{1}{\operatorname{lcm}(q_1, q_2)} \right) + \frac{K_1}{q_1} + H_1 q_1 + q_2$$

$$\text{s.t} \quad q_1, q_2 \in \mathbb{Z}_+ \qquad\qquad\qquad\qquad\qquad\qquad\qquad (2)$$

Note that, except for the term $K_0 / \operatorname{lcm}(q_1, q_2)$, the objective function is the sum of two functions of the form $f(q) = A/q + Bq$. We will frequently use that the minimum of f is equal to $2\sqrt{AB}$, and is attained at $q = \sqrt{A/B}$.

In order to force a replenishment interval $q_1 = M$ in any optimal solution of Program (2), we make commodity 1 "heavy". More precisely, we set $K_1 = 2K_0M^3$, $H_1 = 2K_0M$, so that $\sqrt{K_1/H_1} = M$. Note that if $q_1 = M - 1$ or $q_1 = M + 1$, we get the following relations:

$$\frac{K_1}{M+1} + H_1(M+1) - \left(\frac{K_1}{M} + H_1 M\right) = \frac{H_1}{M+1} > \frac{K_0 M}{M} = K_0$$

$$\frac{K_1}{M-1} + H_1(M-1) - \left(\frac{K_1}{M} + H_1 M\right) = \frac{H_1}{M-1} > K_0,$$

and therefore

$$\frac{K_1}{q_1} + H_1 q_1 + q_2 > \left(\frac{K_1}{M} + H_1 M\right) + K_0 + 1 = K_0 + 4K_0 M^2 + 1 \qquad (3)$$

for $q_1 = M - 1$ or $q_1 = M + 1$. Using that $K_1/q_1 + H_1 q_1$ is convex in $q_1 \in \mathbb{R}^+$ with minimum at $q_1 = M$, we obtain that Eq. (3) holds for any integer $q_1 \neq M$. Since $K_0 + 4K_0 M^2 + 1$ is the objective value of Program (2) when $q_1 = M$ and $q_2 = 1$, we have proven that $q_1 = M$ in any optimal solution.

Now, Program (2) reduces to

$$\min \quad \frac{M^2}{4}\left(\frac{1}{q} - \frac{1}{\operatorname{lcm}(M, q)}\right) + q$$

$$\text{s.t} \quad q \in \mathbb{Z}_+, \qquad\qquad (4)$$

where we eliminated q_1 from the program, renamed q_2 as q, and set $K_0 = M^2/4$. Let us define

$$A(q) = \frac{M^2}{4q} + q, \quad B(q) = \frac{M^2}{4\operatorname{lcm}(M, q)}, \quad F(q) = A(q) - B(q)$$

so the objective value of Program (4) for a given q is equal to $F(q)$.

Let us assume that $M \geq 6$. We now prove that any optimal value q for Program 4 is in $\{2, \ldots, M - 1\}$. First, note that

$$F(1) = \frac{M^2}{4} + 1 - \frac{M}{4} \geq \frac{M^2 - M}{4} \geq M \quad \text{and} \quad F(M) = M.$$

Then, note that for $q \geq M$ we have that $A(q) \geq A(M)$ and $B(q) \leq B(M)$. They follow from the facts that $A(\cdot)$ is convex with minimum $A(M/2) = M$ and that $\operatorname{lcm}(M, q) \geq M$. Therefore, $F(q) \geq M$ for $q \notin \{2, \ldots, M - 1\}$. Finally, since for $M \geq 6$:

$$F(\lfloor M/2 \rfloor) \leq A(\lfloor M/2 \rfloor) - \frac{M}{4\lfloor M/2 \rfloor} \leq A\left(\frac{M}{2}\right) + \frac{1}{M/2 - 1} - \frac{M}{2(M-1)} < M,$$

then any q minimizing Program 4 should be in $\{2, \ldots, M - 1\}$. The second inequality follows from an argument similar to the one used to show that $q_1 = M$. In particular, any such q is either relative prime with M, or it shares a non-trivial divisor of M. The next lemma shows that the latter is always the case when M is odd and composite.

Lemma 1. *Suppose that $M \geq 6$ is an odd composite number. Then every $q' \in \mathbb{Z}_+$ minimizing $F(\cdot)$ satisfies $\gcd(M, q') \neq 1, M$.*

Proof. We already proved that $\gcd(M, q') \neq M$ for any q' minimizing $F(\cdot)$. Suppose M and q' are coprimes. Then $B(q') = \frac{M}{4q'}$, and therefore $F(q') \geq \min_{q \in \mathbb{R}}\{A(q) - \frac{M}{4q}\}$. Let us define L to be this minimum value, and suppose that it is achieved at q^*, then it is easy to see that

$$L = M\sqrt{u}, \quad q^* = \frac{M}{2}\sqrt{u},$$

where $u = 1 - \frac{1}{M}$. To get a contradiction, we will prove that there exists q near to q^* such that $F(q) < L$. To see this, let $3 \leq p \leq \sqrt{M}$ be any non-trivial divisor of M and let $q \in [q^* - p/2, q^* + p/2]$ be any integer divisible by p. Let us write $q = (1 + \epsilon)q^*$, where ϵ may be negative. Using that $\mathrm{lcm}(M, q) \cdot \gcd(M, q) = Mq$, we have that

$$F(q) = \frac{M^2}{4q} + q - \frac{M \gcd(M, q)}{4q} \leq \frac{M^2}{4q} + q - \frac{Mp}{4q}$$

and therefore

$$F(q) - L \leq \frac{M^2 u}{4q} + q - L - \frac{M(p - 1)}{4q}. \tag{5}$$

Using $\frac{M^2 u}{4q^*} = q^*$ we can simplify

$$\frac{M^2 u}{4q} + q - L = \frac{M^2 u}{4(1 + \epsilon)q^*} + (1 + \epsilon)q^* - 2q^* = \frac{q^*}{1 + \epsilon} + \epsilon q^* - q^* = \frac{\epsilon^2}{1 + \epsilon}q^*$$

which, combined with Eq. 5 and $|\epsilon| \leq \left|\frac{p}{2q^*}\right| \leq \frac{1}{\sqrt{Mu}} = \frac{1}{\sqrt{M-1}}$ gives:

$$F(q) - L \leq \frac{1}{1 + \epsilon}\left(\epsilon^2 q^* - \frac{M(p - 1)}{4q^*}\right) \leq \frac{1}{1 + \epsilon}\left(\frac{M\sqrt{u}}{2(M - 1)} - \frac{p - 1}{2\sqrt{u}}\right)$$

Finally, since $1 + \epsilon > 0$ and $\sqrt{u} < 1$, we have that

$$(1 + \epsilon)(F(q) - L) \leq \frac{M\sqrt{u}}{2(M - 1)} - \frac{p - 1}{2\sqrt{u}} < \frac{M}{2(M - 1)} - \frac{p - 1}{2},$$

and it is easy to see that the rightmost expression is negative for $M \geq 3, p \geq 3$, which are true by assumption. Hence, $F(q) < L$, which proves the desired contradiction. □

If 2 is not a divisor of a composite number $M \geq 6$, Lemma 1 guarantees that the greatest common divisor between M and the solution to Program 4 is always a non-trivial divisor. Since we can check if M is prime in polynomial time [AKS04], the following result holds:

Theorem 1. *Suppose that GI-CF2 is polynomial-time solvable. Then* INTEGER-FACTORIZATION *is polynomial-time solvable.*

3 Approximation Algorithm for Finite Horizon

In this section we present a dynamic policy for the finite horizon case. Recall that if we replenish commodity i at times $0 = t_1 < t_2 < \ldots < t_{n_i} < T$, then $C_{\text{ord}}^{\text{indiv}}(i) \equiv n_i K_i$ and $C_{\text{hold}}(i) \equiv \frac{d_i h_i}{2} \sum_{j=1}^{n_i} (t_{j+1} - t_j)^2$, where $t_{n_i+1} = T$. We call the values $t_{j+1} - t_j$ the inter-replenishment lengths.

We temporarily assume that the approximation algorithm has oracle access to N, the total number of joint orders in some optimal solution for JRP. We briefly describe how to remove this assumption in Sect. 4. The description of the algorithm (Alg. 1) is a simple two-step process. In the first step, the algorithm places joint ordering points at every multiple of T/N, starting at $t = 0$. In the second step, each commodity places its orders on a subset of those joint orders in such a way that the individual ordering and holding costs are minimized. Note that this can be carried out separately for each commodity. A similar observation has been used to define an algorithm for GI policies [WFD97].

Algorithm 1

1: **Approx-JRP** (T, h_i, d_i, K_i, K_0)
2:     Guess N, the number of joint orders in an optimal solution.
3:     Set $p = T/N$ to be the joint inter-replenishment length.
4:     Set $J = \{jp : j = 0, \ldots, N - 1\}$, the set of joint order positions.
5:     **for** $i \in \mathcal{I}$ **do**
6:         Choose a subset of J to be the orders of commodity i such that $C_{\text{ord}}^{\text{indiv}}(i) + C_{\text{hold}}(i)$ is minimal.
7:     **return** the schedule obtained.

Running time. We have to be careful in how to execute the algorithm. The set J may have $\Omega(T)$ elements, while the input size is proportional to $\log T$. However, we can explicitly define this set by giving T and N, and it is easy to check that the size of N is polynomial in the input size.

The same difficulty arises in Step 6, but a similar representation can be applied to keep the space polynomial: if the individual schedule for some commodity i minimizes $C_{\text{ord}}^{\text{indiv}}(i) + C_{\text{hold}}(i)$, a simple convexity argument implies that the inter-replenishment lengths can take at most two values and they are consecutive multiples of p. Therefore, we can define this individual schedule by giving the (at most) two inter-replenishment lengths and their frequencies.

It follows that the only step where polynomiality can fail is Step 6. The following lemma establishes its complexity. The proof is omitted due to lack of space.

Lemma 2. *Suppose that commodity i can be ordered only at multiples of some fixed period p. Moreover, assume that T is a multiple of p. Then, it is possible to compute the schedule minimizing $C_{\text{hold}}(i) + C_{\text{ord}}^{\text{indiv}}(i)$ in polynomial time with respect to the input size.*

Approximation analysis. Given an instance of JRP, let OPT be any optimal solution having exactly N joint orders, where N is the value guessed by Alg. 1. For $i \in \mathcal{I}$, let n_i be the total number of individual orders of commodity i in OPT and let **OPT** be the optimal cost. In this section, we may emphasize the dependency on the schedule by including the schedule in brackets. For example, we may write $C[\text{OPT}] = \textbf{OPT}$.

If a commodity is ordered exactly m times, it is easy to show that its holding cost is minimized when the replenishments occur at $\{jT/m : j = 0, \ldots, m-1\}$. We say that m orders are *evenly distributed* when they are placed according to this configuration. This optimality property for the holding cost of evenly distributed orders is the basis for a lower bound on **OPT** we use to prove the approximation guarantee. Our first step in this direction is to define two feasible solutions for the problem:

- The virtual schedule (or VS) places exactly $(1 + \beta_i)n_i$ evenly distributed orders of commodity i, for every $i \in \mathcal{I}$. Each β_i is a parameter to be defined.
- The real schedule (or RS) allows joint orders in $J = \{jp : j = 0, \ldots, N-1\}$. For each commodity i we place exactly $(1+\beta_i)n_i$ orders, that are obtained by shifting each individual order in the virtual schedule to the closest point in J. If there are two closest joint orders, we choose the closest one backwards in time.

Note that both schedules are not defined algorithmically. The real schedule is defined from the virtual schedule, and there is a one-to-one correspondence between their individual orders through the shifting process. We use the term *shifted order* to indicate this correspondence.

Loosely speaking, the cost of the real schedule is closely related to the cost of the schedule output by Alg. 1, while the virtual schedule is related to a lower bound on **OPT**. Both are used as a bridge that relates **OPT** with the cost of the schedule returned by Alg. 1.

Proposition 1. *If $\beta_i \leq 1/8$ for every $i \in \mathcal{I}$, then $C_{\text{hold}}[\text{RS}] \leq \frac{9}{8}C_{\text{hold}}[\text{VS}]$.*

Proof. Consider any commodity $i \in \mathcal{I}$. For simplicity, we omit subindices and write n instead of n_i and β instead of β_i. Let $q = T/(1+\beta)n$ be the inter-replenishment length of the commodity in the virtual schedule. Let $p = T/N$ be the joint inter-replenishment length for the real schedule. Note that $q \geq p/(1+\beta)$.

Suppose first that $p \geq q$. In RS, the commodity is replenished in every joint-order position. Directly evaluating the holding costs gives

$$C_{\text{hold}}[\text{RS}](i) = \frac{T^2 hd}{2N} \leq \frac{T^2 hd}{2n} = (1+\beta)C_{\text{hold}}[\text{VS}](i) \leq \frac{9}{8}C_{\text{hold}}[\text{VS}](i).$$

On the other hand, if $p < q$, let k be the only integer satisfying $kp \leq q < (k+1)p$. Clearly, the inter-replenishment lengths in the real schedule can only take the values kp or $(k+1)p$. Let a be the number of orders of length kp and let b the number of orders of length $(k+1)p$ in the real schedule. We have the relations:

$$a + b = (1+\beta)n \quad \text{and} \quad a(kp) + b(k+1)p = q(1+\beta)n,$$

from where we get, in particular, that $bp = (1 + \beta)n(q - kp)$. Using these three relations, and evaluating the holding cost, we obtain:

$$\frac{C_{\text{hold}}[\text{RS}](i)}{C_{\text{hold}}[\text{VS}](i)} = \frac{a(kp)^2 + b(k+1)^2 p^2}{q^2(1+\beta)n} \leq \frac{(1+\beta)n(kp)^2 + b(2k+1)p^2}{q^2(1+\beta)n}.$$

which can be written after some additional manipulation as

$$\frac{C_{\text{hold}}[\text{RS}](i)}{C_{\text{hold}}[\text{VS}](i)} \leq (-k^2 - k)\left(\frac{p}{q}\right)^2 + (2k+1)\frac{p}{q}.$$

To conclude, note that $-k(k+1)x^2 + (2k+1)x$, as a function of x, has maximum value $\frac{(2k+1)^2}{4k(k+1)}$, which is at most $9/8$ when $k \geq 1$. $\square$

The next proposition shows that the individual ordering and holding costs in RS are within a constant factor of the respective costs in OPT. The proof (not included due to lack of space) uses Prop. 1 and some simple relations among RS, VS and OPT.

Proposition 2. *Let* $\gamma = \sqrt{9/8}$. *Then for every* $\epsilon > 0$ *we can choose* $\{\beta_i\}_{i \in \mathcal{I}}$ *so that the real schedule satisfies the following properties for* T *sufficiently large:*

- $C_{\text{hold}}[\text{RS}] \leq (1+\epsilon)\gamma \cdot C_{\text{hold}}[\text{OPT}]$
- $C_{\text{ord}}^{\text{indiv}}[\text{RS}] \leq (1+\epsilon)\gamma \cdot C_{\text{ord}}^{\text{indiv}}[\text{OPT}]$

Let S be the schedule returned by Alg. 1. Recall that its output is a schedule S that minimizes $C_{\text{ord}}^{\text{indiv}} + C_{\text{hold}}$ restricted to use N evenly distributed joint orders. This and Prop. 2 give the following inequalities for large T:

$$\left(C_{\text{ord}}^{\text{indiv}} + C_{\text{hold}}\right)[\text{S}] \leq \left(C_{\text{ord}}^{\text{indiv}} + C_{\text{hold}}\right)[\text{RS}] \leq (1+\epsilon)\gamma \left(C_{\text{ord}}^{\text{indiv}} + C_{\text{hold}}\right)[\text{OPT}].$$

Since N is the number of joint orders in OPT, then $C_{\text{ord}}^{\text{joint}}[\text{S}] \leq C_{\text{ord}}^{\text{joint}}[\text{OPT}]$. Adding up, we obtain $C[\text{S}] \leq (1+\epsilon)\gamma C[\text{OPT}]$ which is an approximation guarantee asymptotically equal to γ for Alg. 1. For small T, a closer look at our analysis gives an approximation factor of $9/8$.

Theorem 2. *Alg. 1 is a 9/8-approximation algorithm* ($\sqrt{9/8}$ *for large* T) *for dynamic policies in the finite horizon case.*

4   GI Model

We can easily adapt the algorithm described in Sect. 3 to the GI model with variable base (see Alg. 2). We now guess p, the optimal joint inter-replenishment length. Note that Step 5 is simpler, since q_i is always one of the two multiples of p closest to $\sqrt{K_i/h_i}$.

Note that Alg. 2 finds the best value of q_i for the optimal p, and therefore computes the optimal GI policy. Since GI policies approximate unrestricted policies by a factor of $1/(\sqrt{2}\log 2) \approx 1.02$ [MR93, TB01], our algorithm achieves these

---

Algorithm 2. GI model algorithm

1: **Approx-JRP** (T, h_i, d_i, K_i, K_0)
2:     Guess p, the optimal renewal interval in an optimal solution.
3:     Set $J = \{jp : j = 0, \ldots, N-1\}$, the set of joint order positions.
4:     **for** $i \in \mathcal{I}$ **do**
5:         Choose q_i as a multiple of p such that $K_i/q_i + h_i q_i$ is minimum.
6:     **return** the schedule obtained.

---

guarantees. The bound in Sect. 3 (≈ 1.06) is slightly worse since we are not using the powerful machinery available for the stationary case.

From this observation we can obtain a fully polynomial-time approximation scheme for GI policies by exhaustively searching p in powers of $(1 + \epsilon)$. The range of search can be $[p_{\min}, p_{\max}]$, which are the values defined in Eq. (1). The total running time is polynomial in the size of the input and $\frac{1}{\log(1+\epsilon)} = O(1/\epsilon)$. The only thing we need to prove is that choosing p' in the range $p \leq p' \leq p(1 + \epsilon)$ is enough to get a $(1 + \epsilon)$-approximation. This follows from the fact that if $(p, \{k_i\}_{i \in \mathcal{I}})$ defines an optimal schedule with value **OPT**, then $(p/(1 + \epsilon), \{k_i\}_{i \in \mathcal{I}})$ has cost

$$\frac{K_0}{p(1+\epsilon)} + \sum_{i \in \mathcal{I}} \frac{K_i}{(1+\epsilon)k_i p} + \frac{1}{2} h_i d_i k_i (1 + \epsilon) \leq (1 + \epsilon)\mathbf{OPT}.$$

Essentially the same idea can be used to remove the guessing assumption in Alg. 1. We just exhaustively search N in (aproximated) powers of γ.

Finally, Alg. 2 can be extended to the fixed base GI model. The only difference is that we guess p assuming it is a multiple of the base B. The exhaustive search in powers of $(1 + \epsilon)$ has to carefully round the values of p to be multiples of B.

Theorem 3. *Alg. 2 (properly modified) is an FPTAS in the class of GI policies and in the class of fixed base GI policies.*

References

[AJR89]   Arkin, E., Joneja, D., Roundy, R.: Computational complexity of uncapacitated multi-echelon production planning problems. Operations Research Letters 8, 61–66 (1989)

[AKS04]   Agrawal, M., Kayal, N., Saxena, N.: PRIMES is in P. Annals of Mathematics 160(2), 781–793 (2004)

[GS89]   Goyal, S., Satir, A.: Joint replenishment inventory control: Deterministic and stochastic models. European Journal of Operational Research 38(1), 2–13 (1989)

[JMM85]   Jackson, P., Maxwell, W., Muckstadt, J.: The joint replenishment problem with a powers-of-two restriction. IIE Transactions 17(1), 25–32 (1985)

[Jon90]   Joneja, D.: The joint replenishment problem: New heuristics and worst case performance bounds. Operations Research 38(4), 711–723 (1990)

[KG08]   Khouja, M., Goyal, S.: A review of the joint replenishment problem literature: 1989-2005. European Journal of Operational Research 186(1), 1–16 (2008)

[KR91]   Kaspi, M., Rosenblatt, M.: On the economic ordering quantity for jointly replenished items. International Journal of Production Research 29(1), 107–114 (1991)

[LRS04]   Levi, R., Roundy, R., Shmoys, D.: Primal-dual algorithms for deterministic inventory problems. In: Proceedings of the Thirty-Sixth Annual ACM Symposium on Theory of Computing, pp. 353–362. ACM, New York (2004)

[LY03]   Lee, F.C., Yao, M.J.: A global optimum search algorithm for the joint replenishment problem under power-of-two policy. Computers & Operations Research 30(9), 1319–1333 (2003)

[MC06]   Moon, I.K., Cha, B.C.: The joint replenishment problem with resource restriction. European Journal of Operational Research 173(1), 190–198 (2006)

[MR93]   Muckstadt, J., Roundy, R.: Analysis of multistage production systems. In: Rinnooy, A., Zipkin, P. (eds.) Handbooks in Operations Research and Management Science, vol. 4, North-Holland, Amsterdam (1993)

[PD08]   Porras, E., Dekker, R.: Generalized solutions for the joint replenishment problem with correction factor. International Journal of Production Economics 113(2), 834–851 (2008)

[Pos94]   Posner, M.E.: Approximation procedures for the one-warehouse multi-retailer system. Management Science 40(10), 1305–1316 (1994)

[Rou85]   Roundy, R.: 98%-effective integer-ratio lot-sizing for one-warehouse multi-retailer systems. Management Science 31(11), 1416–1430 (1985)

[RSA78]   Rivest, R., Shamir, A., Adleman, L.: A method for obtaining digital signatures and public-key cryptosystems. Communications of the ACM 21(2), 120–126 (1978)

[TB01]   Teo, C., Bertsimas, D.: Multistage lot sizing problems via randomized rounding. Operations Research 49(4), 599–608 (2001)

[WFD97]   Wildeman, R., Frenk, J., Dekker, R.: An efficient optimal solution method for the joint replenishment problem. European Journal of Operational Research 99(2), 433–444 (1997)

Approximation Algorithms for Conflict-Free Vehicle Routing

Kaspar Schüpbach[1] and Rico Zenklusen[2,*]

[1] ETH Zürich, Institute For Operations Research, 8092 Zürich, Switzerland
kaspar.schuepbach@ifor.math.ethz.ch
[2] MIT, Department of Mathematics, Cambridge, MA 02139, USA
ricoz@math.mit.edu

Abstract. We consider a natural basic model for conflict-free routing of a group of k vehicles, a problem frequently encountered in many applications in transportation and logistics. There is a large gap between currently employed routing schemes and the theoretical understanding of the problem. Previous approaches have either essentially no theoretical guarantees, or suffer from high running times, severely limiting their usability. So far, no efficient algorithm is known with a sub-linear (in k) approximation guarantee and without restrictions on the graph topology.

We show that the conflict-free vehicle routing problem is hard to solve to optimality, even on paths. Building on a sequential routing scheme, we present an algorithm for trees with makespan bounded by $O(\mathrm{OPT}) + k$. Combining this result with ideas known from packet routing, we obtain a first efficient algorithm with sub-linear approximation guarantee, namely an $O(\sqrt{k})$-approximation. Additionally, a randomized algorithm leading to a makespan of $O(\mathrm{polylog}(k)) \cdot \mathrm{OPT} + k$ is presented that relies on tree embedding techniques applied to a compacted version of the graph to obtain an approximation guarantee independent of the graph size.

1 Introduction

We investigate the conflict-free routing of vehicles through a network of bidirectional guideways. Conflicts are defined in a natural way, i.e., vehicles cannot occupy the same resource at the same time, hence forbidding crossing and overtaking. The task is to find a routing consisting of a route selection and a schedule for each vehicle, in which they arrive at their destinations as quickly as possible.

Such conflict-free routing algorithms are needed in various applications in logistics and transportation. A prominent example is the routing of Automated Guided Vehicles (AGVs). AGVs are often employed to transport goods in warehouses (for survey papers we recommend [6,23]), or to move containers in large-scale industrial harbors [21]. The guideways can be tracks or any sort of fixed connected and bidirectional lane system. Other related application settings are

* Supported by Swiss National Science Foundation grant PBEZP2-129524, by NSF grants CCF-1115849 and CCF-0829878, and by ONR grants N00014-11-1-0053 and N00014-09-1-0326.

C. Demetrescu and M.M. Halldórsson (Eds.): ESA 2011, LNCS 6942, pp. 640–651, 2011.

the routing of ships in canal systems [16], locomotives in shunting yards [4], or airplanes during ground movement at airports [7,1].

Conflict-free vehicle routing problems can be divided into online problems, where new vehicles with origin-destination pairs are revealed over time, and offline problems, where the vehicles to route are known in advance together with their origin-destination pairs. Here, we concentrate on the offline problem, which is also often a useful building block for designing online algorithms.

Algorithms for conflict-free routings either follow a *sequential* or *simultaneous* routing paradigm. Sequential routing policies consider the vehicles in a given order, and select a route and schedule for each vehicle such that no conflict occurs with previously routed vehicles (see [9,13,10] for sequential routing examples in the context of AGVs). Simultaneous approaches take into account multiple or all vehicles at the same time. Whereas the higher flexibility of those approaches opens up possibilities to obtain stronger routings than the sequential paradigm, they usually lead to very hard optimization problems. Furthermore, they are often difficult to implement in practice. Typically, the routing problem is modeled as an Integer Program which is tackled by commercial solvers [14], column generation methods [4] or heuristics without optimality guarantee [16,7,12].

Sequential algorithms are thus often more useful in practice due to their computational efficiency, but suffer from the difficulty of finding a good sequence to route the vehicles. Furthermore, the theoretical guarantees of these approaches are often weak. The goal of this paper is to address these shortcoming of sequential routing algorithms.

Problem formulation. We consider the following problem setting which captures common structures of many conflict-free vehicle routing problems.

Conflict-Free Vehicle Routing Problem (CFVRP). *Given is an undirected graph $G = (V, E)$, and a set of k vehicles Π with origin-destination pairs (s_π, t_π) $\forall \pi \in \Pi$. Origins and destinations are also called* terminals. *A discretized time setting is considered with vehicles residing on vertices. At each time step, every vehicle can either stay (wait) on its current position or move to a neighboring vertex. Vehicles are forbidden to traverse the same edge at the same time step, also when driving in opposite directions, and no two vehicles are allowed to be on the same node at the same time. A routing not violating the above rules is called* conflict-free. *The goal is to find a conflict-free routing minimizing the* makespan, *i.e. the number of time steps needed until all vehicles reach their destination.*

The CFVRP is a natural first candidate for modeling and analyzing routing problems in a variety of contexts. Clearly, it omits application-specific details and makes further simplifying assumptions.

As a relaxation of the conflict definition above, we assume that vehicles can only be conflicting while in transit, i.e., no conflict is possible before departure and after arrival. The *departure time* of a vehicle is the last time step that the vehicle is still at its origin, and the *arrival time* is the earliest time when vehicle is at its destination. We call this relaxation the *parking assumption*. The parking assumption is natural in many of the listed applications since the terminal node occupations are often managed by separate procedures. In AGV systems the dispatching (task assignment) is usually separated from the routing process and takes care of terminal node occupations. Similarly for airport ground movement problems, where terminals correspond to runways and gates to which airplanes are assigned without conflicts before airplane routing starts. In ship routing in a canal system, the terminals represent harbors having enough space for conflict-free parking of all arriving and departing vessels.

The model setting investigated in [12,19,22] is very similar to the one used here. The differences lie mostly in the modeling of waiting vehicles, which block edges in their setting. In [12], only designated edges can be used for waiting. However, these variations do not significantly change the problem, and our results can easily be transferred. The main reason why we use the CFVRP setting introduced above, is that it leads to a simplified presentation of our algorithms.

CFVRP has many similarities with packet routing [18,15], where the goal is a conflict-free transmission of data packets through cable networks. The crucial difference is that the conflict notion in packet routing is relaxed. It allows for several packets to occupy a node at the same time, as nodes represent network routers with large storage capacity. The concept of edge-conflicts is essentially the same as in our setting and models the limited bandwidth of the transmission links. Hence, the CFVRP setting can as well be seen as a packet routing problem with unit capacities on every node.[1]

Some sequential routing approaches. We briefly discuss some variants of sequential routing schemes, emphasizing on approaches used later in the paper.

The presumably simplest approach is to serially send one vehicle after another on a shortest route to the destination, such that a vehicle departs as soon as the previous one has arrived. The obtained makespan is bounded by $k \cdot L$, where L is used throughout the paper as the maximum origin-destination distance over all vehicles. Since L is clearly a lower bound on OPT, this is a k-approximation. Interestingly, for general graph topologies, no efficient algorithm was known to substantially beat this approach, i.e. with a $o(k)$ approximation guarantee.

Still, several stronger routing paradigms are known and commonly used in practice. An improved sequential routing policy, which we call *greedy sequential routing*, is the following procedure as introduced in [9,13]. Vehicles are considered in a given order, and for each vehicle a route and schedule is determined with earliest arrival time, avoiding conflicts with previously scheduled vehicles. For a fixed ordering of the vehicles, a greedy sequential routing can be obtained efficiently, e.g. by finding shortest paths in a time-expanded graph. Greedy sequential routing is often applied with given origin-destination paths for all vehicles, in which case the task is only to find a schedule for each vehicle that determines how to traverse its origin-destination path over time.

For given origin-destination paths, the following restricted version of sequential routing algorithms, called *direct routing*, often shows to be useful. In direct

---

[1] There are approximation results for packet routing with buffer size 1 in [8]. However, contrary to the CFVRP, they consider bidirectional edges on which two packages can be sent simultaneously in opposite directions.

routing [2], vehicles are not allowed to wait while in transit, i.e., once a vehicle leaves its origin, it has to move to its destination on the given origin-destination path without waiting. An advantage of direct routing is that vehicles only block a very limited number of vertex/time slot combinations.

Combining this concept with the greedy sequential routing, the *greedy direct sequential* algorithm is obtained. Here an ordering of the vehicles is given, as well as a source-destination path for each vehicle. Considering vehicles in the given ordering, the routing of a vehicle is determined by finding the earliest possible departure that allows for advancing non-stop to its destination on the given path, without creating conflicts with previously routed vehicles.

When fixing the origin-destination paths to be shortest paths, greedy sequential routing and its direct variant clearly perform at least as good as the trivial serial algorithm. However, for unfortunate choices of the routing sequence, one can easily observe that the resulting makespan of both approaches can still be a factor of $\Theta(k)$ larger than the optimum (see [22] for details).

Further related results. Spenke [19] showed that the CFVRP is NP-hard on grid graphs. The proof implies that finding the optimal priorities for greedy sequential is also NP-hard.

Polynomial routing policies with approximation quality sublinear in k are known for grid graphs. Spenke introduces a method for choosing a routing sequence with a makespan bounded by $4\text{OPT} + k + 3$. An online version of the problem was investigated by Stenzel [22], again for grid topologies.

Computational results published in [22] indicate that greedy sequential algorithms can have bad performance when the number of route choices of near-shortest-path lengths are limited. For grid topologies, the above-mentioned algorithm of Stenzel [22] takes advantage of the fact that grid graphs contain at least two disjoint routes of almost the same length for each pair of vertices.

Our results. On the negative side we obtain the following result showing that there is not much hope to obtain exact solutions even for seemingly simple settings. The proof is omitted due to space constraints and will appear in a full version of the paper.

Theorem 1. *The CFVRP problem on paths is NP-hard.*

The proof of Theorem 1 considers a problem instance where the optimal routing can be chosen to be a greedy direct routing, thus implying the following result.

Theorem 2. *Choosing an optimal ordering of the vehicles for greedy direct routing is NP-hard, even when the underlying graph is a path.*

The above theorem provides a theoretical explanation for the difficulties encountered in practice when looking for good orderings for sequential routings.

On the positive side, we consider the CFVRP problem on trees, and present a priority ordering of the vehicles leading to a greedy direct routing algorithm with a makespan bounded by $4\text{OPT}+k$. This is achieved by dividing the vehicles into two groups, and showing that each group admits an ordering which leads only to very small delays stemming from vehicles driving in opposite directions.

For general instances, without restrictions on the graph topologies, we show how the tree algorithm can be leveraged to obtain a $O(\sqrt{k})$-approximation, thus leading to the first sublinear approximation guarantee for the CFVRP problem. A crucial step of the algorithm is to *discharge* high-congestion vertices by routing vehicles on a well-chosen set of trees. The purely multiplicative approximation guarantee is obtained despite the $+k$ term in the approximation guarantee for the tree algorithm by exploiting results from the packet routing literature. More precisely, using an approach of Srinivasan and Teo [20], we determine routes for the vehicles with a congestion C bounded by $C = O(\text{OPT})$, and never route more than C vehicles over a given tree.

Additionally, an efficient randomized method with makespan $O(\log^3 k)\text{OPT} + k$ is presented for general graph topologies. This approach relies on obtaining strong tree embeddings in a compacted version of the graph, therefore avoiding a dependency of the approximation guarantee on the size of the graph, which would result by a straightforward application of tree embeddings.

Organization of the paper. In Section 2, we present our approximation algorithm for trees. Section 3 contains the claimed $O(\sqrt{k})$-approximation algorithm for general graph topologies. In Section 4, the randomized algorithm based on tree embeddings and leading to a makespan of $O(\log^3 k)\text{OPT} + k$ is discussed.

2 Tree Approximation

Throughout this section, we assume that the given graph $G = (V, E)$ is a tree and we fix an arbitrary root node $r \in V$. The nodes are numbered as follows: we perform a depth-first search (DFS) on G starting at r, and number the nodes in the order in which they are first visited during the DFS.

The vehicles are partitioned into *increasing vehicles* and *decreasing vehicles*. A vehicle is increasing if the label of its destination is larger than the one of its origin, and decreasing otherwise. We use k^+ and k^- to denote the number of increasing and decreasing vehicles, respectively.

Vehicles will be routed on the unique path from origin to destination. On this path, the node which is closest to the root is called the *bending node* of the vehicle. The labels of the last node before and the first node after the bending node are referred to as *in-label* and *out-label*, respectively. Notice, that the bending node can coincide with the origin or destination node. In this case the in-label or out-label, respectively, is not defined. Increasing vehicles are always guaranteed to have out-labels while decreasing vehicles certainly have in-labels. See Fig. 1 for an illustration. Our tree routing algorithm, TREEROUTING, is a special case of greedy sequential routing, where vehicles are routed in order of decreasing priorities, for any priority list satisfying the following rules:

  i) increasing vehicles have priority over decreasing vehicles,
 ii) among two increasing vehicles, the one with higher out-label has priority,
iii) among two decreasing vehicles, the one with lower in-label has priority,
 iv) ties are broken using an arbitrary fixed vehicle ordering.

Theorem 3. *The makespan obtained by* TREEROUTING *is bounded by* $4L + k$.

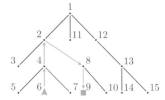

Fig. 1. Tree with nodes labeled in DFS order, starting at the root node with label 1. The route for an increasing vehicle with origin at label 6 and destination at label 9 is indicated. The bending node of the vehicle has label 2, the in-label is 4 and the out-label 8. Note that all increasing vehicles have direction left-to-right or top-to-bottom in this illustration, and the decreasing right-to-left or bottom-to-top, respectively.

Algorithm 1. TREEROUTING

Sort the vehicles in order of decreasing priorities.
Apply greedy sequential routing along the unique paths using this ordering.

Proof of Theorem 3. We start by showing that a particular direct routing exists with the desired makespan, and then deduce Theorem 3 from this result. For the rest of this section, we assume that priorities satisfying the priority rules of TREEROUTING are assigned to the vehicles.

Lemma 1. *There exists a direct routing along the unique origin-destination paths with a makespan of at most $4L + k$, and such that for every node $v \in V$, the vehicles that visit v, do this in order of decreasing priorities.*

Proof. We will show how to construct a direct routing only containing the increasing vehicles with a makespan of $2L + k^+$, such that vehicles visit nodes in order of decreasing priorities. Analogously, a routing with a makespan of $2L + k^-$ can be obtained for the decreasing vehicles[2]. The result then follows by serially applying those two routings.

Notice that direct routing along given paths is fully specified by fixing for each vehicle the passage time at one node on its route. Consider the direct routing that is obtained by fixing for each vehicle the passage time at its bending node as follows: the passage time at the bending node of the highest priority vehicle is set to L, the one with second-highest priority is $L + 1$, and so on, hence leading to a passage time at the bending node of $L + k^+ - 1$ for the increasing vehicle with the lowest priority. Thus, if this routing is conflict-free, then all (increasing) vehicles arrive latest at time step $2L + k^+ - 1$, thus respecting the desired makespan.

We finish the proof by showing that if two vehicles visit the same node v, then the one with lower priority occupies v strictly later than the one with

---

[2] To reduce this case to the increasing case, one can for example swap for every decreasing vehicle the origin and destination, thus turning them into increasing ones. A routing obtained using the procedure for increasing vehicles can then be transformed into a legal routing for the decreasing vehicles by reverting time.

higher priority. This implies as well that the routing is conflict-free. Notice that in particular no edge conflicts are then possible because having an edge conflict between two vehicles would violate the priority ordering at one of its endpoints. Hence, consider two vehicles π and ψ, and assume that ψ has a higher priority than π. We distinguish four cases.

Case 1: π and ψ do not share any node. Here, the claim follows trivially.

Case 2: π and ψ share exactly one common node v. Note that v must be the bending node of at least one of the two vehicles. Let τ_v^ψ and τ_v^π be the passage times of ψ and π at node v, and let τ^ψ and τ^π be the bending times of ψ and π. We show that the higher priority vehicle ψ passes first at v, i.e. $\tau_v^\psi < \tau_v^\pi$, by proving the following chain of inequalities.

$$\tau_v^\psi \leq \tau^\psi < \tau^\pi \leq \tau_v^\pi$$

To check the first inequality assume by sake of contradiction that $\tau_v^\psi > \tau^\psi$, i.e., ψ reaches v after having bent. Hence, v must be the bending node of π, and thus the bending node of π is a descendant of the out-node of ψ (i.e., the out-node of ψ lies on the path between r and the bending node of π). This contradicts that ψ has a higher priority than π. The second, strict inequality holds because of the assignment of smaller bending node passage times to higher priority vehicles. The third inequality holds by a reasoning analogous to the one used for the first inequality.

It remains to discuss cases where π and ψ share more than one node. As G is a tree and routes are paths, these common nodes necessarily form a connected path. It remains to distinguish in which direction this common subpath is traversed by the vehicles.

Case 3: π and ψ use a common subpath in the same direction. Let v denote the common node with the smallest label. It corresponds to the one of the two bending nodes which is closer to the root. The same analysis as in the second case shows that ψ passes v first. As we apply direct routing and ψ and π use the common subpath in the same direction, the lead of ψ over π is the same on all common nodes, and the claim hence follows.

Case 4: π and ψ use a common subpath in opposite directions. The vehicles cannot bend on the common subpath because that would contradict with both vehicles being increasing. One vehicle hence approaches the root while the other one goes away from it. Observe that the approaching vehicle has the larger out-label, and hence must be the higher-priority vehicle ψ. Let v again denote the common node with the smallest label. Using the same reasoning as in the second case, we again obtain that ψ passes v first. It follows that ψ leaves the common path before π enters it, proving the claim.   □

Theorem 3 can now be derived by showing that no vehicle occupies any node later in the TREEROUTING algorithm than in the routing suggested by Lemma 1. A formal proof will appear in the long version of the paper.

3   Hot Spot Routing

Our $O(\sqrt{k})$-approximation for general graph topologies proceeds along the following three steps.

i) Selection of routes for the vehicles with guaranteed upper bounds on both route length and node congestion.

ii) Identification of busy nodes (*hot spots*) and routing of the vehicles going through hot spots with the tree approximation algorithm.

iii) Routing of the remaining vehicles by exploiting the fact that the congestion and hence the conflict potential is limited (*low congestion routing*).

Selection of routes. We will determine for each vehicle $\pi \in \Pi$ an s_π-t_π path $P_\pi \subset E$, satisfying the following properties, where we denote by $\Pi_v \subseteq \Pi$ for $v \in V$, the vehicles whose paths contain v:

i) the *congestion* $C = \max_{v \in V}\{|\Pi_v|\}$ is bounded by $O(\mathrm{OPT})$,

ii) the *dilation* D of the chosen paths, which is the length of the longest path P_π, is bounded by $O(\mathrm{OPT})$.

Notice that both, the congestion C and the dilation D are lower bounds on the minimum makespan that can be achieved with the chosen paths. The problem of finding routes with small congestion and dilation is well-known in packet routing. Using an algorithm of Srinivasan and Teo [20] or a recently improved version presented in [11], a collection of paths with the above properties can be found in polynomial time.

The algorithm. We start by computing origin-destination paths $\{P_\pi\}$ with short congestion and dilation as discussed above. In a first phase, the algorithm goes through the vertices $v \in V$ in any order and checks whether there are more than $\sqrt{k}$ vehicles not routed so far whose origin-destination paths contain v. If this is the case, all those vehicles are routed on a shortest path tree rooted at v using TREEROUTING. Notice that these vehicles are hence not necessarily routed along the paths $\{P_\pi\}$. In a second phase, greedy direct sequential routing (with an arbitrary order) is applied to all vehicles not routed so far. The paths used here are the ones determined at the beginning, i.e., $\{P_\pi\}$. The algorithm is summarized below.

Algorithm 2. HOTSPOT

Generate origin-destination paths $\{P_\pi\}$ with low congestion and dilation

Initialize $\Pi' \leftarrow \Pi$

while There exists a node v with $|\Pi_v \cap \Pi'| > \sqrt{k}$ **do**

    Route the vehicles $\Pi_v \cap \Pi'$ on a shortest path tree with root v, using TREEROUTING

    $\Pi' \leftarrow \Pi' \setminus \Pi_v$

end while

Route the remaining vehicles Π' in an arbitrary order, applying greedy direct sequential routing using the paths $\{P_\pi\}$.

Theorem 4. HOTSPOT *has an approximation quality of* $O(\sqrt{k} \cdot \text{OPT})$.

Proof. The while-loop gets iterated at most $\sqrt{k}$ times, since at each iteration at least $\sqrt{k}$ of the k vehicles are routed. Consider the routing of a group of at least $\sqrt{k}$ vehicles $\Pi_v \cap \Pi'$ during the while-loop. Since each vehicle $\pi \in \Pi_v \cap \Pi'$ is routed along a shortest path tree T rooted at r, and P_π contains v, we have $d_G[T](s_\pi, t_\pi) \leq |P_\pi| \leq D$. Furthermore, the number of vehicles in $\Pi_v \cap \Pi'$ is bounded by C. Thus, the algorithm TREEROUTING routes all vehicles in $\Pi_v \cap \Pi'$ in time $O(C + D)$, and hence, all vehicles routed during the while loop will reach their destination in $O(\sqrt{k}(C + D)) = O(\sqrt{k} \cdot OPT)$ time steps.

Let π be any of the remaining vehicles that are routed during the second phase of the algorithm. Consider all potential departure times for π from its origin, starting after the last vehicle of the first phase has arrived. We want to bound the total number of departure times for π that lead to conflicts due to previously scheduled vehicles during the second phase. If some departure time is not possible, then this must be due to either a node conflict or an edge conflict with another vehicle previously scheduled during the second phase. However, for every node v on the path P_π, at most $\sqrt{k}$ vehicles have previously been routed over v during the second phase. Hence, the occupation of these nodes by other vehicles blocks $O(D\sqrt{k})$ possible departure times. Furthermore, if some departure time t for π is not possible due to some edge conflict with another vehicle ψ routed during the second phase, then either there is also a node conflict for the same departure time (if π and ψ traverse the edge in the same direction), or the departure time $t + 1$ corresponds to a node conflict between π and ψ (if π and ψ traverse the edge in opposite directions). Hence, the total number of departure times that are blocked by edge conflicts is bounded by the total number of departure times blocked by node conflicts which is $O(D\sqrt{k})$. We conclude that π waits at most $O(D\sqrt{k})$ time steps at its origin before directly traveling to its destination in at most D steps. Hence, this second phase is completed in at most $O(D\sqrt{k})$ time steps, thus leading to a total makespan bounded by $O(\sqrt{k}(O+D))$ and proving the claim.   □

4   Low-Stretch Routing

Using tree embeddings to extend algorithms designed for tree topologies to arbitrary graph topologies is a standard approach (see for example [5,17]). However, in our setting, a naïve application of tree embedding would only lead to an approximation guarantee that is polylogarithmic in the number of vertices, whereas we are interested in approximation guarantees independent of the size of the graph. To achieve this goal we will determine a routing by applying tree embedding techniques to a compacted version of the graph G with size of order $O(k^2)$.

The high level idea is to find a collection of $O(\text{polylog}(k))$ trees in G such that for each vehicle $\pi \in \Pi$, there exists a tree T in the collection such that the distance of the s_π-t_π path in T is at most a $O(\text{polylog}(k))$-factor larger than the distance between s_π and t_π in G. Every vehicle π is then assigned to a tree with a short s_π-t_π distance. We then go through the collection of trees in any order and sequentially route first all vehicles assigned to the first tree, then all that are assigned to the second tree and so on. Each group of vehicles that is assigned to the same tree is routed using our tree routing algorithm TREEROUTING.

For any graph $H = (W, F)$ with given edge lengths and two vertices $v, w \in W$, we denote by $d_H(v, w)$ the distance of a shortest path between v and w in H. For $U \subseteq F$, we denote by $H[U]$ the graph (W, U), where U inherits the edge lengths from H. We will apply the following results about low-stretch trees of Dhamdhere et al. [3] to a compacted version of G to find a good collection of spanning trees.

Theorem 5 ([3]). *For any edge-weighted graph $H = (W, F)$, one can draw in polynomial time a spanning tree T of H out of a distribution such that for any $v, w \in W$, the expected stretch is bounded by $O(\log^2 |W|)$, i.e.,*

$$\mathbf{E}\left[d_{H[T]}(v, w)/d_H(v, w)\right] = O\left(\log^2 |W|\right).$$

We transform the unit length network $G = (V, E)$ into a graph $H = (W, F)$ of size $O(k^2)$ with non-negative edge lengths, such that both graphs have the same origin-destination distances for each vehicle. For this purpose, we first compute for each vehicle $\pi \in \Pi$ a shortest path $P_\pi \subseteq E$. To do this, we temporarily perturb the unit edge lengths slightly such that the shortest paths are unique.

Let $W \subseteq V$ be the set of all vertices that are either terminals, or have at least three adjacent edges in $\cup_{\pi \in \Pi} P_\pi$. The graph $H = (W, F)$ is obtained from G by applying the following operations.

  i) Delete all edges and nodes which are not part of any path P_π.
  ii) Every path $P \subseteq G$ between two nodes $v, w \in W$, without any other nodes of W on the path, is replaced by an edge between v and w of length $|P|$.

Notice that H can be interpreted as a compact version of the graph $(V, \cup_{\pi \in \Pi} P_\pi)$. Every edge of H corresponds to a path in G (possibly of length one). More generally, any subset of edges $U \subseteq F$ can be mapped to corresponding edges in G. H has the following properties.

Lemma 2. *The size of $H = (W, F)$ is bounded by $|W| = O(k^2)$, and for each vehicle, the origin-destination distance in H is the same as in G.*

Proof. The claim about origin-destination distances clearly holds, since for every vehicle $\pi \in \Pi$, H contains a compacted version of the path P_π.

To bound the size of H, consider the graph $G' = (V, \cup_{\pi \in \Pi} P_\pi)$. Since H is obtained from G' by eliminating all degree two vertices that are not terminals, each non-terminal vertex in H is a vertex of degree at least three in G'. To prove the claim, it hence suffices to show that G' has at most $O(k^2)$ vertices of degree ≥ 3. If a non-terminal vertex $v \in V$ is of degree at least three in G', then there exist at least two vehicles $\pi, \psi \in \Pi$ such that $|\delta_G(v) \cap (P_\pi \cup P_\psi)| \geq 3$, where $\delta_G(v)$ are the edges adjacent to v in G. We call such a node a *junction* of the two vehicles π, ψ. Observe that for any two vehicles $\pi, \psi \in \Pi$, there are at most two junctions of π and ψ: the paths P_π and P_ψ are unique shortest paths in G w.r.t. the perturbed unit lengths; hence $P_\pi \cap P_\psi$ is a path in G, implying that only the two endpoints of P can be junctions of π and ψ. Since each of the $\binom{k}{2}$ unordered pairs of vehicles leads to at most two junctions, the total number of junctions is bounded by $k(k-1)$, which implies that G' has at most $O(k^2)$ vertices of degree ≥ 3. $\qquad\qquad\qquad\qquad\qquad\qquad\qquad\qquad\qquad\qquad\qquad\qquad\square$

The following lemma now easily follows by standard techniques. Its proof is omitted here due to space constraints.

Lemma 3. *Let $p = 2\log(k)$, and let $U_1, \ldots, U_p \subseteq F$ be random spanning trees of H obtained by applying Theorem 5. Let $T_1, \ldots, T_p$ be the spanning trees in G that correspond to $U_1, \ldots, U_p$. Then, with probability at least $1 - 1/k$, we have that for every vehicle $\pi \in \Pi$, there exists a tree $T \in \{T_1, \ldots, T_p\}$ such that*

$$d_{G[T]}(s_\pi, t_\pi)/d_G(s_\pi, t_\pi) = O(\log^2 k).$$

The algorithm. Our algorithm works as follows. Using Lemma 3 (repeatedly if necessary), we obtain in expected polynomial time a collection of spanning trees $T_1, \ldots, T_p$ with $p = O(\log(k))$ such that we can assign each vehicle π to a tree $T_{i(\pi)}$ satisfying $d_{G[T_{i(\pi)}]}(s_\pi, t_\pi) = O(\log^2 k \cdot d_G(s_\pi, t_\pi))$. Let k_j denote the number of vehicles assigned to T_j. For each tree T_j in the collection, the TREEROUTING algorithm can be used to route all vehicles that are assigned to T_j in time bounded by

$$4\max\{d_{G[T_j]}(s_\pi, t_\pi) \mid \pi \in \Pi, i(\pi) = j\} + k_j = O(\log^2 k)L + k_j,$$

which follows by the guarantee on the stretch provided by Lemma 3. Finally, routing first all vehicles assigned to T_1, then—as soon as all those vehicles arrived at their destinations—route all vehicles assigned to T_2 and so on, thus leads to an algorithm with expected polynomial running time and the claimed bound on the makespan given by

$$O(\log^2 k)pL + \sum_{i=1}^{p} k_j = O(\log^3 k)L + k.$$

Acknowledgments. We are very grateful to Anupam Gupta for discussions related to tree embeddings.

References

1. Atkin, J.A.D., Burke, E.K., Ravizza, S.: The airport ground movement problem: Past and current research and future directions. In: 4th International Conference on Research in Air Transportation, ICRAT(2010)
2. Busch, C., Magdon-Ismail, M., Mavronicolas, M., Spirakis, P.G.: Direct routing: Algorithms and complexity. In: Albers, S., Radzik, T. (eds.) ESA 2004. LNCS, vol. 3221, pp. 134–145. Springer, Heidelberg (2004)

3. Dhamdhere, K., Gupta, A., Räcke, H.: Improved embeddings of graph metrics into random trees. In: Proceedings of the seventeenth annual ACM-SIAM symposium on Discrete algorithm, SODA, pp. 61–69 (2006)
4. Freling, R., Lentink, R.M., Kroon, L.G., Huisman, D.: Shunting of passenger train units in a railway station. Transportation Science 39(2), 261–272 (2005)
5. Galbiati, G., Rizzi, R., Amaldi, E.: On the approximability of the minimum strictly fundamental cycle basis problem. Discrete Applied Mathematics 159(4), 187–200 (2011)
6. Ganesharajah, T., Hall, N.G., Sriskandarajah, C.: Design and operational issues in AGV-served manufacturing systems. Annals of Operations Research 76, 109–154 (1998)
7. Garcia, J., Berlanga, A., Molina, J., Besada, J., Casar, J.: Planning techniques for airport ground operations. In: Proceedings of 21st Digital Avionics Systems Conference., vol. 1, pp. 1D5-1–1D5-12 (2002)
8. auf der Heide, F.M., Scheideler, C.: Routing with bounded buffers and hot-potato routing in vertex-symmetric networks. In: Spirakis, P.G. (ed.) ESA 1995. LNCS, vol. 979, pp. 341–354. Springer, Heidelberg (1995)
9. Kim, C.W., Tanchoco, J.M.A.: Conflict-free shortest-time bidirectional AGV routeing. International Journal of Production Research (1991)
10. Kim, K., Jeon, S., Ryu, K.: Deadlock prevention for automated guided vehicles in automated container terminals. In: Kim, K.H., Günther, H.-O. (eds.) Container Terminals and Cargo Systems, pp. 243–263. Springer, Heidelberg (2007)
11. Koch, R., Peis, B., Skutella, M., Wiese, A.: Real-time message routing and scheduling. In: Dinur, I., Jansen, K., Naor, J., Rolim, J. (eds.) APPROX 2009. LNCS, vol. 5687, pp. 217–230. Springer, Heidelberg (2009)
12. Krishnamurthy, N.N., Batta, R., Karwan, M.H.: Developing conflict-free routes for automated guided vehicles. Operations Research 41(6), 1077–1090 (1993)
13. Möhring, R.H., Köhler, E., Gawrilow, E., Stenzel, B.: Conflict-free real-time AGV routing. In: Proceedings of Operations Research, pp. 18–24 (2005)
14. Oellrich, M.: Minimum-Cost Disjoint Paths Under Arc Dependences - Algorithms for Practice. Ph.D. thesis, Technische Universität Berlin (2008)
15. Peis, B., Skutella, M., Wiese, A.: Packet routing: Complexity and algorithms. In: Bampis, E., Jansen, K. (eds.) WAOA 2009. LNCS, vol. 5893, pp. 217–228. Springer, Heidelberg (2010)
16. Petersen, E.R., Taylor, A.J.: An optimal scheduling system for the welland canal. Transportation Science 22(3), 173 (1988)
17. Sanità, L.: Robust Network Design. Ph.D. thesis, Università Sapienza di Roma (January 2009)
18. Scheideler, C.: Universal Routing Strategies for Interconnection Networks. Springer-Verlag New York, Inc., Secaucus (1998)
19. Spenke, I.: Complexity and Approximation of Static k-splittable Flows and Dynamic Grid Flows. Ph.D. thesis, Technische Universität Berlin (2006)
20. Srinivasan, A., Teo, C.P.: A constant-factor approximation algorithm for packet routing, and balancing local vs. global criteria. In: Proceedings of the ACM Symposium on the Theory of Computing, STOC, pp. 636–643 (1997)
21. Stahlbock, R., Voß, S.: Operations research at container terminals: a literature update. OR Spectrum 30, 1–52 (2008)
22. Stenzel, B.: Online Disjoint Vehicle Routing with Application to AGV Routing. Ph.D. thesis, Technische Universität Berlin (2008)
23. Vis, I.F.: Survey of research in the design and control of automated guided vehicle systems. European Journal of Operational Research 170(3), 677–709 (2006)

A $\frac{3}{2}$ Approximation for a Constrained Forest Problem

Basile Couëtoux

Lamsade, Université Paris-Dauphine

Abstract. We want to find a forest of minimal weight such that each component of this forest is of size at least p. In this paper, we improve the best approximation ratio of $2 - \frac{1}{|V|}$ for this problem obtained by Goemans and Williamson in 1995 to $\frac{3}{2}$ with a greedy algorithm.

1 Introduction

Let $G = (V, E, w)$ be a weighted connected undirected graph, with a node set V, an edge set E and a weight on the edges w, $w(e) \geq 0$. To find a tree or a forest covering a graph with some constraints are network design problems very well studied like the Minimum spanning tree or the Steiner tree problem. The problem we are studying consists in covering a graph such that each tree is of order at least p minimizing the sum of the weight of the edges used. We denote this problem MIN WCF(p). When $p = 2$ we are in the case of the minimal weighted cover and when $p = n$ it becomes the minimal spanning tree problem. These two extrema are can be solved in polynomial time. Yet for any constant p greater or equal to 3 the problem become NP-hard even for a weighted bipartite graphs of maximal degree 3.

About the negative results, Imielinska *et al.* [4] showed that MIN WCF(p) is *NP*-hard for any $p \geq 4$. Monnot and Toulouse [7] proved that the unweighted version of MIN WCF(3) is *NP*-hard even on planar bipartite graphs of maximum degree three. Bazgan *et al.* [1] have studied the difficulty of the approximation, and this problem in *APX*-hard on bipartite graph of maximum degree 3.

About the positive results, a first greedy algorithm is given by Imielinska *et al.* in [4] that achieves a 2-approximation. Interestingly enough, a different algorithm studied by Laszlo and Mukherjee [5] has exactly the same tight worst-case ratio of 2, as well as a common generalization of those two approaches [6]. Applying the method of Goemans and Williamson [2], just a slightly better ratio $2 - \frac{1}{n}$ can be achieved. This method uses the primal dual approach and is eligible for any monotone minimization forest problem. In particular, it works on MIN WCF(p). On the other hand, MIN WCF(p) admits a polynomial time approximation scheme when the graph is planar [1].

A similar problem has been studied by Guttmann-Beck and Hassin in [3], the minimum tree partition problem. Given a complete weighted graph

C. Demetrescu and M.M. Halldórsson (Eds.): ESA 2011, LNCS 6942, pp. 652–663, 2011.

and a set of integer $t_1, \ldots, t_q$, $\sum_1^p t_i = |V|$, they seek a forest of minimal weight with trees $T_1, \ldots T_q$ such that $\forall 1 \leq i \leq q$, $|T_i| = t_i$. This problem is not constant approximable unless P=NP. If the graph verifies the triangular inequality they shown a $2q - 1$ approximation.

For clarity purpose we will say that a tree is *small* if it is of order strictly lesser than p and that it is *big* otherwise. An isolated vertex is a small tree. The initial 2-approximation was a greedy algorithm. Starting with an empty forest, it adds at each step the edge of minimum weight which does not create a cycle or links two big trees together. This algorithm is a really natural greedy algorithm and the analysis is not too complicated.The second algorithm given by Lazlo and Mukherjee is similar to the first one. It starts with a solution containing all the edges and, at each step, it removes the one of maximum weight if it does not create a small tree. Remark that the solution given eventually is acyclic. Using an adaptation of Kruskal's algorithm these algorithms can be implemented in $O(m \log m)$ with $m = |E|$. The ratio of 2 is reached in the instance presented by the Figure 1 for the two algorithms.

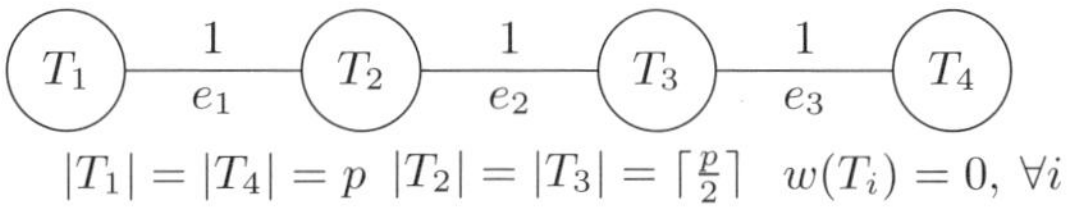

$$|T_1| = |T_4| = p \quad |T_2| = |T_3| = \lceil \tfrac{p}{2} \rceil \quad w(T_i) = 0, \; \forall i$$

Fig. 1. $\{e_1, e_3, T_1, T_2, T_3, T_4\}$ is a solution of weight 2 which can be returned by the algorithms, $\{e_2, T_1, T_2, T_3, T_4\}$ is an optimal solution for this instance of weight 1

We propose here a $\frac{3}{2}$-approximation algorithm. This algorithm is still greedy as the lack of structure on a weighted problem tends to oblige us. The algorithm idea is to choose the edge to add accordingly to the number of small tree its addition suppress. It is easy to see that any edge will act on at most two tree of the partial addition will link two trees forming either a big or a small tree. The choice of the next edge to add come from the following concept. The following concept is used to choose the edge considered by the algorithm:

At each step, a *good* edge to add for F is an edge which links two small trees together such that the resulting tree is big. It decreases the number of small trees by 2. In the following we will refer as a *bad* edge for an edge which is not *good*.

Another formulation of the problem is to seek a forest with no small tree minimizing the weight of the edges used. Therefore we will try to decrease the number of small tree of a partial solution until there is no small tree left. In order to choose the next edge to add to the partial solution, we choose the edge which will decrease the number of small tree with minimal cost. Thus, considering the good and bad edges, we pay up to twice as much for a good edge rather than a bad edge.

2   Algorithm Outline

In the previous greedy algorithms, the approximation ratio is reached when the chosen edges with significant weight in the approximate solution are bad and the ones in the optimal solution are good. The main idea of $\mathcal{AGI}$ is to encourage the selection which will reduce the number of small trees by 2, the good edges, provided that they are not too heavy. Informally, the solution F obtained with $\mathcal{AGI}$ is constructed in the following way: at each iteration an edge e is added such that it minimizes $w(e)/2$ if e is a good edge or $w(e)$ if e is a bad edge. We repeat this operation until a constrained covering forest is obtained. Formally the algorithm is described as follows:

Algorithm 1. $\mathcal{AGI}$

    Data: A graph $G = (V, E)$ and a weight function on the edges w
    Result: A covering forest F containing only big trees
1 $A = E$;
2 $F = \emptyset$;
3 **while** $A \neq \emptyset$ **do**
4   Let e be the lightest of the good edges of A for F and e' the lightest of the other edges of A;
5   **if** $w(e) \leq 2w(e')$ **then**
6         $F \leftarrow F \cup \{e\}$;
7         $A \leftarrow A \setminus \{e\}$;
8   **else**
9   **if** *e′ has an endpoint in a small tree of F and F ∪ {e′} does not contain a cycle* **then**
10            $F \leftarrow F \cup \{e'\}$;
11   **end**
12         $A \leftarrow A \setminus \{e'\}$;
13   **end**
14 **end**
15 Return F ;

The part in $\mathcal{AGI}$ which cost the most is the actualization of the set of good edges. We keep the set of the good edges and the set of the bad edges ordered. An edge is inserted at most once in the good set and once in the bad set. However each time we add an edge to the solution we have to actualize the two sets and therefore to test every incident edges to the two components which are now linked together, this test is bounded by $O(m)$. Since we add at most $n-1$ edges to our solution (otherwise we create a cycle), the complexity of $\mathcal{AGI}$ can be bounded by $O(mn)$.

3   Analysis of the Algorithm

Let F be the forest returned by $\mathcal{AGI}$ and F^* an optimal forest. In first place, we will prove several structural properties which will be needed in the analysis.

After that we will prove an inequality by induction on the step of the algorithm, such that the final step of the algorithm will generate an inequality equivalent to $\frac{3}{2}w(F^*) \geq w(F)$.

We will show in this analysis that if $\mathcal{AGI}$ does not choose a good edge of the optimal solution, then a previous choice of edges in the partial solution forbid this choice and this induce a cost in the optimal solution.

3.1   Notations and Definitions

We order the edges $e_1, \ldots e_m$, by the order they are examined by $\mathcal{AGI}$ (more precisely the order they are removed from A). The inequality will compare the partial forests defined as follows: $F_k^* = F^* \cap \{e_1, \ldots, e_k\}$ and $F_k = F \cap \{e_1, \ldots, e_k\}$. F_k is then the partial solution at the iteration k. For any forest F', we denote $h(F')$ is the number of small trees of the forest F'. An isolated vertex is considered as a small tree. $h(\emptyset) = n$, with n the number of vertices.

Definition 1

$$B = \{e_i \in F \mid h(F_{i-1}) - h(F_i) = 2\}$$
$$B^* = \{e_i \in F^* \mid h(F_{i-1}^*) - h(F_i^*) = 2\}$$

B and B^* are respectively the good edges of F and F^*. Remark that, in the definition, F_{i-1} and F_i are different since $e_i \in F$ but in general F_j may be the same as F_{j-1}. By definition, each tree of F contains exactly one edge of B. Indeed, since F is a forest, each tree contains at most one good edge, and since F is a constrained forest, each tree contains at least one good edge.

We denote c_k the maximum between the weight of the last bad edge for F examined and the half of the weight of the last good edge for F until the k-th iteration. It represents the maximal price that we have paid to decrease the number of trees by one. Formally, $c_k = \max\{\max_{e \in E_k \setminus B} w(e), \max_{e \in E_k \cap B} \frac{w(e)}{2}\}$. Note that every edge, which has not been yet examined at the k-th iteration, is of weight at least c_k.

$T_k(v)$ and $T_k^*(v)$ are the trees of F_k and F_k^* which contains v respectively. We have to keep track of the tree which contains the endpoints of an edge in order to determine whether this edge is good or bad at each step of the algorithm. For any set of edges E' (which will usually be a tree), we denote $V(E')$ the set of the endpoints of edges of E.

Another key concept of the analysis is that the choice of one edge for the solution can *disable* as a good edge for F a good edge of the optimal solution. Intuitively a disabled edge should not link two small trees in a big tree in the execution of the algorithm. Let $1 \leq i \leq m$, i be a step of the algorithm such that $e_i \in F$. We denote T_i the tree of F_i which contains e_i. We will study the necessary conditions to disable an edge of B^*. There is several cases, but we will eventually show that the properties we want holds anyway.

Definition 2. *The set of edges of $B^* \setminus B$ disabled at the iteration i is composed of three kind of edges:*

(α) We disable all the edges e of $B^* \setminus F$ not disabled until now such that $T_i \cup \{e\}$ contains a cycle.

(β) If T_i is big, then we disable all the edges of $B^* \setminus B$ not disabled until now which has exactly one endpoint in $V(T_i)$.

(γ) If T_i is small, we consider the set of edges of B^* not disabled until now which has exactly one endpoint in $V(T_i)$. Within these edges we disable all except one of them which belongs to B if there is one.

(α) and (β) disable are structural. Indeed, (α) comes from the fact that F is a constrained forest. For (β). since T_i is already big, no edge with exactly one endpoint in $V(T_i)$ will be considered as a good edge for F. Eventually (γ) tells that at most one edge of B^* with exactly one endpoint in $V(T_i)$ will be a good edge for F. Moreover, if this edge $e \in B^* \cap B$, it is natural not to disable it. Since there is at most one good edge of F with exactly one endpoint in $V(T_i)$, with (γ)-disable, the edges disabled are in $B^* \setminus B$.

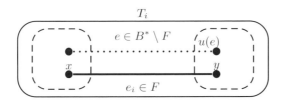

Fig. 2. (α) disable. The addition of the edge $e_i = (x, y)$ links the trees $T_{i-1}(x)$ and $T_{i-1}(y)$, containing respectively the endpoints of e, forming the tree T_i. The edge e, which was not disabled until now, is therefore disabled at the iteration i. The vertex $u(e)$ is chosen arbitrarily between the two endpoints of e.

For an edge $e = (x, y) \in B^* \setminus B$ disabled at the iteration i, we denote $u(e)$, the endpoint of this edge which belongs to $V(T_i)$. If it is a (α)-disable the vertex $u(e)$ is either x, either y indifferently. Remark that $u(e)$ is defined for all edge of $B^* \setminus B$. Figures 2, 3, and 4 illustrate each of the three cases.

Remark that from the Definition 2, for all $k \leq m$, for every tree T of F_k, there is at most one edge of B^* with exactly one endpoint in $V(T)$ which is not disabled at the end of the iteration k and, moreover if T is big there is no edge of B^* with exactly one endpoint in $V(T)$ that is not disabled at the end of the iteration k.

We denote Dis_k the set of edges which are disabled at the iteration k. We denote the set of the edges disabled at the step k which has not been examined yet I_k, formally $I_k = \left(\bigcup_{\ell \leq k} \mathrm{Dis}_\ell \right) \cap \{e_{k+1}, \ldots, e_m\}$. For an iteration $k \in \{1 \ldots m\}$, I_k is the set of edges disabled at an iteration i, $i < k$, which had not been examined yet and $u(I_k) = \{u(e), e \in I_k\}$. I_k is a subset of $\{e_{k+1}, \ldots, e_m\}$.

An edge $e_\ell \in B^* \setminus B$ enters I_k when e_ℓ is disabled and stay there until it is examined at the iteration ℓ of $\mathcal{AGI}$. After this iteration, this edge is not in I_ℓ anymore since it has been examined by the algorithm. We will split $B^* \setminus B$ in

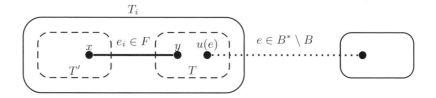

Fig. 3. (β) disable. At the iteration i the edge $e_i = (x, y)$ has been added to F_{i-1}. It forms the tree T_i resulting from $T_{i-1}(x)$ and $T_{i-1}(y)$. T_i is big. Therefore, the dotted edge $e \in B^* \setminus B$, which was not disabled until now, is disabled at the iteration i.

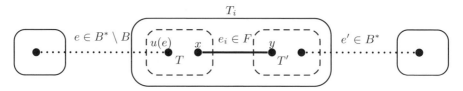

Fig. 4. (γ) disable. The edge $e_i = (x, y)$ link $T_{i-1}(x)$ to $T_{i-1}(y)$ forming T_i. One of the two edges with exactly one endpoint in $V(T_i)$ which was not disabled until now, e, is disabled at the iteration i since it does not belongs to B. If $e' \notin B$ it could have been disabled instead of e. $u(e)$ is the endpoint of e which is in $V(T_i)$.

two sets B_1^* and B_2^*. Informally, $e \in B_1^*$ (resp. $e \in B_2^*$) if and only if, just before its disable, the addition would have linked two small trees in a small (resp. big) one.

Definition 3. $e \in B_i^*$, $i \in \{1, 2\}$, if $e \in B^* \setminus B$ has been disabled at the iteration k and $h(F_{k-1}) - h(F_{k-1} \cup \{e\}) = i$.

3.2 Preliminary Lemmas

The following lemma determines the number of edges which can be disabled at each iteration depending on the nature of the edge examined by the algorithm.

Lemma 1. *For all $k < m$, the following properties about disabling hold.*

(i) *If $e_{k+1} \in B^* \cap B$, then $|Dis_{k+1}| = 0$.*
(ii) *If $e_{k+1} \in F \setminus B$, then $|Dis_{k+1}| \leq 1$.*
(iii) *If $e_{k+1} \in B \setminus B^*$, then $|Dis_{k+1}| \leq 2$.*
(iv) *If $e_{k+1} \notin F$, then $|Dis_{k+1}| = 0$.*

Lemma 2. *For all $k \leq m$, the following assertions hold.*

(i) *Let $e_k \in B^* \setminus B$, then for all $i < k$, $T_i^*(u(e_k))$ is small.*
(ii) *Let v, v' be two vertices belonging to the same tree in F_k^*, formally, $T_k^*(v) = T_k^*(v')$. If $T_k(v) \neq T_k(v')$, then $T_k(v)$ and $T_k(v')$ are big trees.*
(iii) *Let $k \in \{1, \ldots, m\}$. If $T_k(v)$ is a small tree then $V(T_k^*(v)) \subseteq V(T_k(v))$.*
(iv) *Each $e_k \in B^* \setminus B$ is disabled before the iteration k.*

Lemma 3

$$h(F_k^*) \geq h(F_k) + |I_k| \tag{1}$$

Proof. Every vertex $u(e_j)$, with $e_j \in I_k$, is in a small tree $T_k^*(u(e_j))$, Lemma 2 (i). Moreover, each tree T_j^* of F_k^* contains at most one vertex of $u(I_k)$ because each tree of the optimal solution contains exactly one edge of B^*.

Let T_i be a small tree of F_k. From (ii) of the Lemma 2, the forest is formed by the trees $T_k^*(v)$, for $v \in V(T_i)$, covers only the vertices of T_i and contains only small trees. We will show that there exists a tree $T_k^*(v)$, $v \in V(T_i)$ such that it does not cover any vertex of $u(I_k)$. Without loss of generality, we suppose that T_i cover a vertex of $u(I_k)$ (otherwise the property is trivially verified). Let $e \in I_k$ such that $u(e) \in V(T_i) \cap \mathrm{Dis}_j$ with j the largest possible. We will distinguish three cases depending on the way e has been disabled.

- The edge $e = (u(e), v)$ has been (α)-disabled. In this case, e forms a cycle with T_i. $v \in V(T_i)$ and $T_k^*(v)$ does not contains any vertex of $u(I_k)$. Indeed, we know that each tree T^* of F^* contains exactly one edge B^*.
- The edge $e = (u(e), v)$ has been (β)-disabled. This case never happens: $T_j(u(e))$ is big by construction, therefore $T_k(u(e))$ is big; however $T_k(u(e)) = T_i$ that is small by hypothesis.
- The edge $e = (u(e), v)$ has been (γ)-disabled. By construction $v \notin V(T_j(u(e)))$ (the edge e has exactly one endpoint in $V(T_j(u(e)))$). It exists therefore an edge $e' = (x, y)$ with $x \in T_j(u(e))$, $y \notin T_j(u(e))$ and if it had been disabled before the iteration k, $u(e') \notin V(T_i)$. Therefore $T_k^*(x)$ does not contain any vertex of $u(I_k)$ and its vertices are included in $V(T_i)$, from (ii) of the Lemma 2 and because T_i is small.

Every small tree of F_k contains at least one small tree of F_k^* which does not contain any vertex of $u(I_k)$. Therefore $h(F_k)$ is less than the number of small trees of F_k^* which does not contain any vertex of $u(I_k)$. Since there are $h(F_k^*) - |u(I_k)|$ small trees of F_k^* which does not contain any vertex of $u(I_k)$, then

$$h(F_k^*) \geq h(F_k) + |I_k|$$

3.3 Main Inequality

Let w_k the weight of the bad edges e for F^* within $\{e_{k+1}, \ldots, e_m\}$, such that if T^* is the tree of F^* which contains the edge e then the good edge of T^* for F^* has been disabled before the iteration k. Eventually we denote $w(I_k)$ the weight of the edges of I_k.

Lemma 4. *For all $k \in \{1, \ldots, m\}$, the following inequality is verified:*

$$\frac{3}{2}w(F_k^*) \geq w(F_k) - \frac{w_k + w(I_k)}{2} - c_k\left(h(F_k^*) - h(F_k) - |I_k|\right) \tag{2}$$

Proof. To show this lemma we will consider the numerical sequence (g_k) defined as follows:

$$g_k = \frac{3}{2}w(F_k^*) - w(F_k) + \frac{w_k + w(I_k)}{2} + c_k\left(h(F_k^*) - h(F_k) - |I_k|\right)$$

We will show that $g_0 = 0$ and that the sequence $(g_k)_{0 \leq k \leq m}$ is monotone. To show $g_0 = 0$ we just consider the value of parameters at the initialization of the algorithm.

$$w(F_0^*) = w(F_0) = w_0 = w(I_0) = |I_0| = 0$$
$$h(F_0^*) = h(F_0) = n$$

To show that the sequence is monotone we will introduce a half-step g_k' for all $k \in \{0, \ldots, m-1\}$ with:

$$g_k' = \frac{3}{2} w(F_k^*) - w(F_k) + \frac{w_k + w(I_k)}{2} + c_{k+1} \left(h(F_k^*) - h(F_k) - |I_k| \right)$$

Remark that the only modification from g_k is that c_k is replaced by c_{k+1}. However from the Lemma 3, $h(F_k^*) - h(F_k) + |I_k| \geq 0$ and $(c_k)_{0 \leq k \leq m}$ is a monotone sequence (we recall that $c_k = \max\{\max_{e \in E_k \setminus B} w(e), \max_{e \in E_k \cap B} \frac{w(e)}{2}\}$), this implies that $g_k' \geq g_k$. We will consider the difference $g_{k+1} - g_k' = \delta_{k+1}$ and show that this difference is not negative. This difference will come from the variation of the parameters $w(F_k^*), w(F_k), w_k, w(I_k), h(F_k^*), h(F_k), |I_k|$ from which g_k' depends linearly. With this linear dependency we will be able to study a set of events occurring at the iteration $k+1$ independently.

There is three kind of event, the ones related to F the ones related to F^* and the ones related to the disabled edges.

The events $e_{k+1} \in B$ and $e_{k+1} \in F \setminus B$ are the events related to F that track the variations on $w(F_k)$ and $h(F_k)$.

The events $e_{k+1} \in B^* \cap B$, $e_{k+1} \in B^* \setminus B$, $e_{k+1} \in F^* \setminus B^*$ are related to F^* and track the variations on $w(F_k^*), w_k$ and $h(F_k^*)$. Remark that some events overlap with some events related to F but we are only interested in the variation induced by F^*.

Eventually here is a list of the different cases of disabling, which are events related to the value of $w_k, w(I_k)$ and $|I_k|$.

- The event $\mathrm{Dis}_{1,k+1}$ means that an edge of B_1^* is disabled at the iteration $k+1$.
- The event $\mathrm{Dis}_{2,k+1}$ means that an edge of B_2^* is disabled at the iteration $k+1$ with $e_{k+1} \in B$.
- The event $\mathrm{Dis}_{3,k+1}$ means that an edge of B_2^* is disabled at the iteration $k+1$ with $e_{k+1} \in F \setminus B$.

At one iteration at most one event related to F, one event related to F^* and two disabling events can happen.

The set of the disabled edges Dis_{k+1} at the iteration $k+1$ is formed by at most two events within $(\mathrm{Dis}_{i,k+1})_{1 \leq i \leq 3}$, from the Lemma 1. By example if two edges of B_1^* are disabled at the iteration $k+1$ then the event $\mathrm{Dis}_{1,k+1}$ occurs twice. We denote the induced variations by an event A on a parameter by:

$$\delta(A, w(F_k^*)), \delta(A, w(F_k)), \delta(A, w_k), \delta(A, w(I_k)), \delta(A, h(F_k^*)), \delta(A, h(F_k)), \delta(A, |I_k|)$$

Therefore, if by example at the iteration $k+1$ the three events A_1, A_2, A_3 happen, then the variation on the weight of the partial optimal solution $w(F_k^*)$ is given by:

$$w(F_{k+1}^*) - w(F_k^*) = \delta(A_1, w(F_k^*)) + \delta(A_2, w(F_k^*)) + \delta(A_3, w(F_k^*))$$

where $\delta(A, w(F_k^*)) = 0$ if the event A has no effect on the value of $w(F_k^*)$. For every event A, we define δ as follows:

$$\delta(A) = \frac{3}{2}\delta(A, w(F_k^*)) - \delta(A, w(F_k)) + \frac{\delta(A, w_k) + \delta(A, w(I_k))}{2}$$
$$+ c_{k+1}\left(\delta(A, h(F_k^*)) - \delta(A, h(F_k)) - \delta(A, |I_k|)\right)$$

Therefore if by example the events A_1, A_2, A_3 happen at the iteration $k + 1$:

$$g_{k+1} - g_k' = \delta(A_1) + \delta(A_2) + \delta(A_3)$$

We will now determine the δ's associated at the different parameters or at least a lower bound on the variations.

- $e_{k+1} \in B$. In other words the edge e_{k+1} is a good edge for F. The value of the weight of the covering forest increases by $w(e_{k+1})$ and the number of small trees of F_k decreases by 2, i.e.:

$$\delta(e_{k+1} \in B, w(F_k)) = w(e_{k+1})$$
$$\delta(e_{k+1} \in B, h(F_k)) = -2$$
$$\delta(e_{k+1} \in B, X) = 0 \qquad \forall X \neq w(F_k), h(F_k)$$

By reporting in the expression of g_k' we obtain:

$$\delta(e_{k+1} \in B) = 2c_{k+1} - w(e_{k+1})$$

Remark that $w(e_{k+1}) \leq 2c_{k+1}$ ($c_k = \max\{\max_{e \in E_k \setminus B} w(e), \max_{e \in E_k \cap B} \frac{w(e)}{2}\}$). Therefore $\delta(e_{k+1} \in B) \geq 0$

- $e_{k+1} \in F \setminus B$. The edge e_{k+1} is a bad edge for F and therefore we know that $w(e_{k+1}) = c_{k+1}$. The value of the partial forest increases by c_{k+1} and the number of small trees of F_k decreases by 1, i.e.:

$$\delta(e_{k+1} \in F \setminus B, w(F_k)) = c_{k+1}$$
$$\delta(e_{k+1} \in F \setminus B, h(F_k)) = -1$$
$$\delta(e_{k+1} \in F \setminus B, X) = 0 \qquad \forall X \neq w(F_k), h(F_k)$$

We obtain $\delta(e_{k+1} \in F \setminus B) = 0$

- $e_{k+1} \in B^* \cap B$. The number of small trees of F^* decreases by 2 and the value of $w(F_k^*)$ increases by $w(e_{k+1})$:

$$\delta(e_{k+1} \in B^* \cap B, w(F_k^*)) = w(e_{k+1})$$
$$\delta(e_{k+1} \in B^* \cap B, h(F_k^*)) = -2$$
$$\delta(e_{k+1} \in B^* \cap B, X) = 0 \qquad \forall X \neq w(F_k^*), h(F_k^*)$$

We obtain $\delta(e_{k+1} \in B^* \cap B) = \frac{3}{2}w(e_{k+1}) - 2c_{k+1}$. This value may be negative.

- $e_{k+1} \in B^* \setminus B$. Then $e_{k+1} \in I_k$ from (iv) of the Lemma 2 and as $e_{k+1} \notin I_{k+1}$ by definition, because the edge has been examined, we obtain:

$$\delta(e_{k+1} \in B^* \setminus B, w(F_k^*)) = w(e_{k+1})$$
$$\delta(e_{k+1} \in B^* \setminus B, h(F_k^*)) = -2$$
$$\delta(e_{k+1} \in B^* \setminus B, w(I_k)) = -w(e_{k+1})$$
$$\delta(e_{k+1} \in B^* \setminus B, |I_k|) = -1$$
$$\delta(e_{k+1} \in B^* \setminus B, X) = 0 \qquad \forall X \neq w(F_k^*), h(F_k^*), w(I_k), |I_k|$$

Therefore $\delta(e_{k+1} \in B^* \setminus B) = \frac{3}{2}(w(e_{k+1}) - c_{k+1})$. However in this case $w(e_{k+1}) = c_{k+1}$, which enable us to simplify to $\delta(e_{k+1} \in B^* \setminus B) = 0$.

- $e_{k+1} \in F^* \setminus B^*$. w_k decreases by $w(e_{k+1})$ and the number of small trees in the optimal partial solution decreases by 1.

$$\delta(e_{k+1} \in F^* \setminus B^*, w_k) = -w(e_{k+1})$$
$$\delta(e_{k+1} \in F^* \setminus B^*, w(F_k^*)) = w(e_{k+1})$$
$$\delta(e_{k+1} \in F^* \setminus B^*, h(F_k^*)) = -1$$
$$\delta(e_{k+1} \in B^* \setminus B, X) = 0 \qquad \forall X \neq w_k, w(F_k^*), h(F_k^*)$$

We obtain $\delta(e_{k+1} \in F^* \setminus B^*) \geq w(e_{k+1}) - c_{k+1}$. This value is not negative.

- $\mathrm{Dis}_{1,k+1}$. The edge $e = (x, y) \in B_1^*$ has been disabled at the iteration $k+1$. We remind that it means that $e \in B^* \setminus B$ and $h(F_k) - h(F_k \cup \{e\}) = 1$. It implies that $|V(T_k(x))| + |V(T_k(y))| < p$. However $|V(T^*(x))| \geq p$ (F^* is constrained); therefore $V(T^*(x)) \nsubseteq V(T_k(x)) \cup V(T_k(y))$ and there exists an edge $e' \in T^*(x)$ with exactly one endpoint in $V(T_k(x)) \cup V(T_k(y))$. Since this edge e' has exactly one endpoint in $V(T_k(x))$ or in $V(T_k(y))$ and that these two trees are small, it has not been examined by the algorithm at the iteration k. Therefore $T^*(x)$ contains at least one bad edge for F^* which is not examined at the iteration k.

$$\delta(\mathrm{Dis}_{1,k+1}, |I_k|) = 1$$
$$\delta(\mathrm{Dis}_{1,k+1}, w(I_k)) \geq c_{k+1}$$
$$\delta(\mathrm{Dis}_{1,k+1}, w_k) \geq c_{k+1}$$
$$\delta(e_{k+1} \in \mathrm{Dis}_{1,k+1}, X) = 0 \qquad \forall X \neq |I_k|, w(I_k), w_k$$

We obtain that $\delta(\mathrm{Dis}_{1,k+1}) \geq 0$

- $\mathrm{Dis}_{2,k+1}$. The edge $e = (x, y) \in B_2^*$ has been disabled at the iteration $k+1$ by the addition of a good edge for F. Then $w(e) \geq w(e_{k+1})$.

$$\delta(\mathrm{Dis}_{2,k+1}, |I_k|) = 1$$
$$\delta(\mathrm{Dis}_{2,k+1}, w(I_k)) \geq w(e_{k+1})$$
$$\delta(\mathrm{Dis}_{2,k+1}, w_k) \geq 0$$
$$\delta(e_{k+1} \in \mathrm{Dis}_{2,k+1}, X) = 0 \qquad \forall X \neq |I_k|, w(I_k), w_k$$

We obtain $\delta(\mathrm{Dis}_{2,k+1}) \geq w(e_{k+1})/2 - c_{k+1}$. Remark that this value is not necessarily non-negative.

- $\text{Dis}_{3,k+1}$. The edge $e = (x, y) \in B_2^*$ has been disabled at the iteration $k+1$ by the addition of a bad edge for F. Then $w(e) \geq 2w(e_{k+1})$ because otherwise it would have been already examined.

$$\delta(\text{Dis}_{3,k+1}, |I_k|) = 1$$
$$\delta(\text{Dis}_{3,k+1}, w(I_k)) \geq 2w(e_{k+1})$$
$$\delta(\text{Dis}_{3,k+1}, w_k) \geq 0$$
$$\delta(e_{k+1} \in \text{Dis}_{3,k+1}, X) = 0 \qquad \forall X \neq |I_k|, w(I_k), w_k$$

We obtain $\delta(\text{Dis}_{3,k+1}) \geq w(e_{k+1}) - c_{k+1}$. By the way, in the case where e_{k+1} is a bad edge for F, $w(e_{k+1}) = c_{k+1}$ and so we can simplify to $\delta(\text{Dis}_{3,k+1}) \geq 0$.

A selection of δ of interest containing the events where δ can be negative ($e_{k+1} \in B^* \cap B$ and $\text{Dis}_{2,k+1}$) and $e_{k+1} \in B$:

$$\delta(e_{k+1} \in B) = 2c_{k+1} - w(e_{k+1})$$
$$\delta(e_{k+1} \in B^* \cap B) = \tfrac{3}{2}w(e_{k+1}) - 2c_{k+1}$$
$$\delta(\text{Dis}_{2,k+1}) \geq w(e_{k+1})/2 - c_{k+1}$$

Remark that only the events $\text{Dis}_{2,k+1}$ and $e_{k+1} \in B^* \cap B$ can have a δ negative. In these two cases $e_{k+1} \in B$. We will therefore study more precisely the case $e_{k+1} \in B$.

If moreover $e_{k+1} \in B^*$ then no edge has been disabled at the iteration $k+1$ from (i) the Lemma 1, therefore:

$$\delta_{k+1} = \delta(e_{k+1} \in B) + \delta(e_{k+1} \in B^* \cap B) = \frac{w(e_{k+1})}{2} \geq 0$$

Otherwise at most two edges has been disabled from (iii) of the Lemma 1, and these two edges are in B_2^* in the worst case:

$$\delta_{k+1} = \delta(e_{k+1} \in B) + 2\delta(\text{Dis}_{2,k+1}) \geq 0$$

We can eventually deduce that $\delta_k \geq 0$ for all k.

The following theorem is the application of the previous lemma to the last iteration of the algorithm.

Theorem 4. *For every instance I, let F be the forest returned by $\mathcal{AGI}$ and F^* an optimal solution on I. Then:*

$$w(F) \leq \frac{3}{2}w(F^*) \tag{3}$$

Proof. Using the Lemma 4 at the m-th and last iteration we obtain the following inequality:

$$\frac{3}{2}w(F_m^*) \geq w(F_m) - \frac{w_m + w(I_m)}{2} - c_m\left(h(F_m^*) - h(F_m) - |I_m|\right) \tag{4}$$

Since $I_m = \emptyset$ because every edge has been examined, in the same way $w_m = 0$. $F_m = F$ and $F_m^* = F^*$ are two constrained covering forests so $h(F_m) = h(F_m^*) = 0$. We can deduce that

$$w(F) \leq \frac{3}{2}w(F^*) \tag{5}$$

4 Conclusion

We have obtained a better ratio for MIN WCF(p) with a greedy algorithm which run in $O(nm)$. The local search approach might be promising, however a simple k-switch, replacing k edges by $k-1$ of lesser total weight, will not give a better ratio than $\frac{5}{3}$. It would be interesting to study the relaxation of the problem of Guttman-Beck and Hassin, seeking a forest with trees of order at least p_i.

References

1. Bazgan, C., Couëtoux, B., Tuza, Z.: Complexity and approximation of the constrained forest problem. Theoretical Computer Science (to appear, 2011)
2. Goemans, M.X., Williamson, D.P.: A general approximation technique for constrained forest problems. Society for Industrial and Applied Mathematics 1(24), 296–317 (1995)
3. Guttmann-Beck, N., Hassin, R.: Approximation algorithms for minimum tree partition. Discrete Applied Mathematics 87(1-3), 117–137 (1998)
4. Imielinska, C., Kalantari, B., Khachiyan, L.: A greedy heristic for a minmum-weight forest problem. Operations Research Letters 1(14), 65–71 (1993)
5. Laszlo, M., Mukherjee, S.: Another greedy heuristic for the constrained forest problem. Mathematical Programming 1(33), 629–633 (2005)
6. Laszlo, M., Mukherjee, S.: A class of heuristics for the constrained forest problem. Discrete Applied Mathematics 1(154), 6–14 (2006)
7. Monnot, J., Toulouse, S.: The path partition problem and related problems in bipartite graphs. Operations Research Letters 1(35), 677–684 (2007)

External-Memory Network Analysis Algorithms for Naturally Sparse Graphs

Michael T. Goodrich and Paweł Pszona

Dept. of Computer Science
University of California, Irvine

Abstract. In this paper, we present a number of network-analysis algorithms in the external-memory model. We focus on methods for large naturally sparse graphs, that is, n-vertex graphs that have $O(n)$ edges and are structured so that this sparsity property holds for any subgraph of such a graph. We give efficient external-memory algorithms for the following problems for such graphs:

1. Finding an approximate d-degeneracy ordering.
2. Finding a cycle of length exactly c.
3. Enumerating all maximal cliques.

Such problems are of interest, for example, in the analysis of social networks, where they are used to study network cohesion.

1 Introduction

Network analysis studies the structure of relationships between various entities, with those entities represented as vertices in a graph and their relationships represented as edges in that graph (e.g., see [11]). For example, such structural analyses include link-analysis for Web graphs, centrality and cohesion measures in social networks, and network motifs in biological networks. In this paper, we are particularly interested in network analysis algorithms for finding various kinds of small subgraphs and graph partitions in large graphs that are likely to occur in practice. Of course, this begs the question of what kinds of graphs are likely to occur in practice.

1.1 Naturally Sparse Graphs

A network property addressing the concept of a "real world" graph that is gaining in prominence is the *k-core number* [26], which is equivalent to a graph's *width* [16], *linkage* [20], *k-inductivity* [19], and *k-degeneracy* [2,22], and is one less than its Erdős-Hajnal coloring number [14]. A k-core, G', in a graph, G, is a maximal connected subgraph of G such that each vertex in G' has degree at least k. The k-core number of a graph G is the maximum k such that G has a non-empty k-core. We say that a graph G is *naturally sparse* if its k-core number is $O(1)$. This terminology is motivated by the fact that almost every n-vertex graph with $O(n)$ edges has a bounded k-core number, since Pittel *et al.* [24] show that a random graph with n vertices and cn edges (in the Erdős-Rényi model) has k-core number at most $2c + o(c)$, with high probability. Riordan [25] and

C. Demetrescu and M.M. Halldórsson (Eds.): ESA 2011, LNCS 6942, pp. 664–676, 2011.

Fernholz and Ramachandran [15] have also studied k-cores in random graphs. In addition, we also have the following:

- Every s-vertex subgraph of a naturally sparse graph is naturally sparse, hence, has $O(s)$ edges.
- Any planar graph has k-core number at most 5, hence, is naturally sparse.
- Any graph with bounded arboricity is naturally sparse (e.g., see [10]).
- Eppstein and Strash [13] verify experimentally that real-world graphs in four different data repositories all have small k-core numbers relative to their sizes; hence, these real-world graphs give an empirical motivation for naturally sparse graphs.
- Any network generated by the Barabási-Albert [4] preferential attachment process, with $m \in O(1)$, or as in Kleinberg's small-world model [21], is naturally sparse.

Of course, one can artificially define an n-vertex graph, G', with $O(n)$ edges that is not naturally sparse just by creating a clique of $O(n^{1/2})$ vertices in an n-vertex graph, G, having $O(n)$ edges. We would argue, however, that such a graph G' would not arise "naturally." We are interested in algorithms for large, naturally sparse graphs.

1.2 External-Memory Algorithms

One well-recognized way of designing algorithms for processing large data sets is to formulate such algorithms in the *external memory model* (e.g., see the excellent survey by Vitter [28]). In this model, we have a single CPU with main memory capable of storing M items and that computer is connected to D external disks that are capable of storing a much larger amount of data. Initially, we assume the parallel disks are storing an input of size N. A single I/O between one of the external disks and main memory is defined as either reading a block of B consecutively stored items into memory or writing a block of the same size to a disk. Moreover, we assume that this can be done on all D disks in parallel if need be.

Two fundamental primitives of the model are *scanning* and *sorting*. Scanning is the operation of streaming N items stored on D disks through main memory, with I/O complexity

$$scan(N) = \Theta\left(\frac{N}{DB}\right),$$

and sorting N items has I/O complexity

$$sort(N) = \Theta\left(\frac{N}{DB} \log_{M/B} \frac{N}{B}\right),$$

e.g., see Vitter [28].

Since this paper concerns graphs, we assume a problem instance is a graph $G = (V, E)$, with $n = |V|$, $m = |E|$ and $N = |G| = m + n$. If G is d-degenerate, that is, has k-core number, d, then $m \leq dn$ and $N = O(dn) = O(n)$ for $d = O(1)$. We use d to denote the k-core number of an input graph, G, and we use the term "d-degenerate" as a shorthand for "k-core number equal to d."

1.3   Previous Related Work

Several researchers have studied algorithms for graphs with bounded k-core numbers (e.g., see [1,3,12,17,19]). These methods are often based on the fact that the vertices in a graph with k-core number, d, can be ordered by repeatedly removing a vertex of degree at most d, which gives rise to a numbering of the vertices, called a *d-degeneracy ordering* or *Erdős-Hajnal sequence*, such that each vertex has at most d edges to higher-numbered vertices. In the RAM model, this greedy algorithm takes $O(n)$ time (e.g., see [5]). Bauer *et al.* [6] describe methods for generating such graphs and their d-degeneracy orderings at random.

In the internal-memory RAM model, Eppstein *et al.* [12] show how to find all maximal cliques in a d-degenerate graph in $O(d3^{d/3}n)$ time. Alon *et al.* [3] show that one can find a cycle of length exactly c, or show that one does not exist, in a d-degenerate graph in time $O(d^{1-1/k}m^{2-1/k})$, if $c = 4k - 2$, time $O(dm^{2-1/k})$, if $c = 4k - 1$ or $4k$, and time $O(d^{1+1/k}m^{2-1/k})$, if $c = 4k + 1$.

A closely related concept to a d-degeneracy ordering is a *k-core decomposition* of a graph, which is a labeling of each vertex v with the largest k such that v belongs to a k-core. Such a labeling can also be produced by the simple linear-time greedy algorithm that removes a vertex of minimum degree with each iteration. Cheng *et al.* [9] describe recently an external-memory method for constructing a k-core decomposition, but their method is unfortunately fatally flawed[1]. The challenge in producing a k-core decomposition or d-degeneracy ordering in external memory is that the standard greedy method, which works so well in internal memory, can cause a large number of I/Os when implemented in external memory. Thus, new approaches are needed.

1.4   Our Results

In this paper, we present efficient external-memory network analysis algorithms for naturally sparse graphs (i.e., degenerate graphs with small degeneracy). First, we give a simple algorithm for computing a $(2 + \epsilon)d$-degeneracy ordering of a d-degenerate graph $G = (V, E)$, without the need to know the value of d in advance. The I/O complexity of our algorithm is $O(sort(dn))$.

Second, we give an algorithm for determining whether a d-degenerate graph $G = (V, E)$ contains a simple cycle of a fixed length c. This algorithm uses $O\left(d^{1\pm\epsilon} \cdot \left(k \cdot sort(m^{2-\frac{1}{k}}) + (4k)! \cdot scan(m^{2-\frac{1}{k}})\right)\right)$ I/O complexity, where ϵ is a constant depending on $c \in \{4k - 2, \ldots, 4k + 1\}$.

Finally, we present an algorithm for listing all maximal cliques of an undirected d-degenerate graph $G = (V, E)$, with $O(3^{\delta/3}sort(dn))$ I/O complexity, where $\delta = (2 + \epsilon)d$.

One of the key insights to our second and third results is to show that, for the sake of designing efficient external-memory algorithms, using a $(2 + \epsilon)d$-degeneracy ordering is almost as good as a d-degeneracy ordering. In addition to this insight, there are a number of technical details that lead to our results, which we outline in the remainder of this manuscript.

---

[1] We contacted the authors and they confirmed that their method is indeed incorrect.

2 Approximating a d-Degeneracy Ordering

Our method for constructing a $(2 + \epsilon)d$-degeneracy ordering for a d-degenerate graph, $G = (V, E)$, is quite simple and is given below as Algorithm 1. Note that our algorithm does not take into account the value of d, but it assumes we are given a constant $\epsilon > 0$ as part of the input. Also, note that this algorithm destroys G in the process. If one desires to maintain G for other purposes, then one should first create a backup copy of G.

1: $L \leftarrow \emptyset$
2: **while** G is nonempty **do**
3:    $S \leftarrow n\epsilon/(2 + \epsilon)$ vertices of smallest degree in G
4:    $L \leftarrow L|S$ // *append S to the end of L*
5:    remove S from G
6: **end while**
7: **return** L

Algorithm 1. Approximate degeneracy ordering of vertices

Lemma 1. *If G is a d-degenerate graph, then Algorithm 1 computes a $(2+\epsilon)d$-degeneracy ordering of G.*

Proof. Observe that any d-degenerate graph with n vertices has at most $2n/c$ vertices of degree at least cd. Thus, G has at most $2n/(2 + \epsilon)$ vertices of degree at least $(2 + \epsilon)d$. This means that the $n\epsilon/(2 + \epsilon)$ vertices of smallest degree in G each have degree at most $(2 + \epsilon)d$. Therefore, every element of set S created in line 3 has at most $(2+\epsilon)d$ neighbors in (the remaining graph) G. When we add S to L in line 4, we keep the property that every element of L has at most $(2+\epsilon)d$ neighbors in G that are placed behind it in L. Furthermore, note that, after we remove vertices in S (and their incident edges) from G in line 5, G is still at most d-degenerate (every subgraph of a d-degenerate graph is at most d-degenerate); hence, an inductive argument applies to the remainder of the algorithm. $\square$

Note that, after $\lceil \log_{(2+\epsilon)/2}(dn) \rceil = O(\lg n)$ iterations, we must have processed all of G and placed all its vertices on L, which is a $(2 + \epsilon)d$-degeneracy ordering for G and that this property holds even though the algorithm does not take the value of d into account.

The full version of this paper [18] contains proof of the following lemma.

Lemma 2. *An iteration of the* `while` *loop (lines 3-5) of Algorithm 1 can be implemented in $O(sort(dn))$ I/O's in the external-memory model, where n is the number of vertices in G at the beginning of the iteration.*

Thus, we have the following.

Theorem 1. *We can compute a $(2 + \epsilon)d$-degeneracy ordering of a d-degenerate graph, G, in $O(sort(dn))$ I/O's in the external-memory model, without knowing the value of d in advance.*

Proof. Since the number of vertices of G decreases by a factor of $2/(2+\epsilon)$ in each iteration, and each iteration uses $O(sort(dn))$ I/O's, where n is the number of vertices in G at the beginning of the iteration (by Lemma 2), the total number of I/O's, $I(G)$, is bounded by

$$I(G) = O\left(sort(dn) + sort\left((2/(2+\epsilon))dn\right) + sort\left((2/(2+\epsilon))^2 dn\right) + \cdots\right)$$
$$= O\left(sort(dn)\left(1 + \frac{2}{2+\epsilon} + \left(\frac{2}{2+\epsilon}\right)^2 + \cdots\right)\right)$$
$$= O(sort(dn)). \qquad \square$$

This theorem hints at the possibility of effectively using a $(2+\epsilon)d$-degeneracy ordering in place of a d-degeneracy ordering in external-memory algorithms for naturally sparse graphs. As we show in the remainder of this paper, achieving this goal is indeed possible, albeit with some additional alterations from previous internal-memory algorithms.

3   Short Paths and Cycles

In this section, we present external-memory algorithms for finding short cycles in directed or undirected graphs. Our approach is an external-memory adaptation of internal-memory algorithms by Alon *et al.* [3]. We begin with the definition and an example of a *representative* due to Monien [23]. A p-set is a set of size p.

Definition 1 (representative). *Let $\mathcal{F}$ be a collection of p-sets. A sub-collection $\widehat{\mathcal{F}} \subseteq \mathcal{F}$ is a q-representative for $\mathcal{F}$, if for every q-set B, there exists a set $A \in \mathcal{F}$ such that $A \cap B = \emptyset$ if and only if there exists a set $\widehat{A} \in \widehat{\mathcal{F}}$ with this property.*

Every collection of p-sets $\mathcal{F}$ has a q-representative $\widehat{\mathcal{F}}$ of size at most $\binom{p+q}{p}$ (from Bollobás [7]). An optimal representative, however, seems difficult to find. Monien [23] gives a construction of representatives of size at most $O(\sum_{i=1}^{q} p^i)$. It uses a p-ary tree of height $\leq q$ with the following properties.

- Each node is labeled with either a set $A \in \mathcal{F}$ or a special symbol λ.
- If a node is labeled with a set A and its depth is less than q, it has exactly p children, edges to which are labeled with elements from A (one element per edge, every element of A is used to label exactly one edge).
- If a node is labeled with λ or has depth q, it has no children.
- Let $E(v)$ denote the set of all edge labels on the way from the vertex v to the root of the tree. Then, for every v:
  - if v is labeled with A, then $A \cap E(v) = \emptyset$
  - if v is labeled with λ, then there are no $A \in \mathcal{F}$ s.t. $A \cap E(v) = \emptyset$.

Monien shows that if a tree T fulfills the above conditions, defining $\widehat{\mathcal{F}}$ to be the set of all labels of the tree's nodes yields a q-representative for $\mathcal{F}$. As an example,

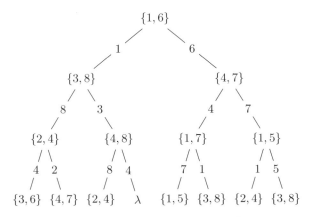

Fig. 1. Tree representation of $\widehat{\mathcal{F}}$

consider a collection of 2-sets, $\mathcal{F} = \{\{2,4\}, \{1,5\}, \{1,6\}, \{1,7\}, \{3,6\}, \{3,8\}, \{4,7\}, \{4,8\}\}$. Fig. 1 presents $\widehat{\mathcal{F}}$, a 3-representative of $\mathcal{F}$ in the tree form.

The main benefit of using representatives in the tree form stems from the fact that their sizes are bounded by a function of only p and q (i.e., maximum size of a representative does not depend on $|\mathcal{F}|$). It gives a way of storing paths of given length between two vertices of a graph in a space-efficient way (see full version of this paper [18] for details).

The algorithm for finding a cycle of given length has two stages. In the first stage, vertices of *high degree* are processed to determine if any of them belongs to a cycle. This is realized using algorithm cycleThrough from Lemma 5. Since there are not many vertices of *high degree*, this can be realized efficiently.

In the second stage, we remove vertices of *high degree* from the graph. Then, we group all simple paths that are half the cycle length long by their endpoints and compute representatives for every such set (see Lemma 3). For each pair of vertices (u, v), we determine (using findDisjoint from Lemma 4) if there are two paths: p from u to v and p' from v to u, such that p and p' do not share any internal vertices. If this is the case, $C = p \cup p'$ is a cycle of required length.

The following representatives-related lemmas are proved in [18].

Lemma 3. *We can compute a q-representative $\widehat{\mathcal{F}}$ for a collection of p-sets $\mathcal{F}$, of size $|\widehat{\mathcal{F}}| \leq \sum_{i=1}^{q} p^i$, in $O\left(\left(\sum_{i=1}^{q+1} p^i\right) \cdot scan(|\mathcal{F}|)\right)$ I/O's.*

Lemma 4. *For a collection of p-sets, $\mathcal{F}$, and a collection of q-sets, $\mathcal{G}$, there is an external-memory method, findDisjoint$(\mathcal{F}, \mathcal{G})$, that returns a pair of sets (A, B) $(A \in \mathcal{F}, B \in \mathcal{G})$ s.t. $A \cap B = \emptyset$ or returns ϵ if there are no such pairs of sets. findDisjoint uses $O\left(\left(\sum_{i=1}^{q+3} p^i + \sum_{i=1}^{p+3} q^i\right) \cdot scan(|\mathcal{F}| + |\mathcal{G}|)\right)$ I/O's.*

Lemma 5. *Let $G = (V, E)$. A cycle of length exactly k that passes through arbitrary $v \in V$, if it exists, can be found by an external-memory algorithm* cycleThrough(G, k, v) in $O\big((k-1)! \cdot scan(m)\big)$ I/O's, where $m = |E|$, via the use of representatives.

Before we present our result for naturally sparse graphs, we first give an external-memory method for general graphs.

Theorem 2. *Let $G = (V, E)$ be a directed or an undirected graph. There is an external-memory algorithm that decides if G contains a cycle of length exactly $c \in \{2k-1, 2k\}$, and finds such cycle if it exists, that takes $O\big(k \cdot sort(m^{2-\frac{1}{k}}) + (2k-1)! \cdot scan(m^{2-\frac{1}{k}})\big)$ I/O's.*

Proof. Algorithm 2 handles the case of general graphs (which are not necessarily *naturally sparse*), and cycles of length $c = 2k$ (the case of $c = 2k-1$ is analogous).

1: $\Delta \leftarrow m^{\frac{1}{k}}$
2: **for all** v – vertex of degree $\geq \Delta$ **do**
3:     $C \leftarrow$ cycleThrough$(G, 2k, v)$
4:     **if** $C \neq \epsilon$ **then**
5:         **return** C
6:     **end if**
7: **end for**
8: remove vertices of degree $\geq \Delta$ from G
9: generate all directed paths of length k in G
10: sort the paths lexicographically, according to their endpoints
11: group all paths $u \rightsquigarrow v$ into collection of $(k-1)$-sets $\mathcal{F}_{uv}$
12: **for all** pairs $(\mathcal{F}_{uv}, \mathcal{F}_{vu})$ **do**
13:     $P \leftarrow$ findDisjoint$(\mathcal{F}_{uv}, \mathcal{F}_{vu})$
14:     **if** $P = (A, B)$ **then**
15:         **return** $C = A \cup B$
16:     **end if**
17: **end for**
18: **return** ϵ

Algorithm 2. Short cycles in general graphs

Since there are at most $m/\Delta = m^{1-\frac{1}{k}}$ vertices of degree at least Δ, and each call to cycleThrough requires $O\big((2k-1)! \cdot scan(m)\big)$ I/O's (by Lemma 5), the first **for** loop (lines 2-7) takes $O\big(m^{1-\frac{1}{k}} \cdot (2k-1)! \cdot scan(m)\big) = O\big((2k-1)! \cdot scan(m^{2-\frac{1}{k}})\big)$ I/O's.

Removing vertices of high degree in line 8 is realized just like line 5 of Algorithm 1, in $O\big(sort(m)\big)$ I/O's. There are at most $m\Delta^{k-1} = m^{2-\frac{1}{k}}$ paths to be generated in line 9. It can be done in $O\big(k \cdot sort(m^{2-\frac{1}{k}})\big)$ I/O's (see [18]). Sorting the paths (line 10) takes $O\big(sort(m^{2-\frac{1}{k}})\big)$ I/O's. After that, creating $\mathcal{F}_{uv}$'s (line 11) requires $O\big(scan(m^{2-\frac{1}{k}})\big)$ I/O's.

The `groupF` procedure groups $\mathcal{F}_{uv}$ and $\mathcal{F}_{vu}$ together. Assume we store $\mathcal{F}_{uv}$'s as tuples (u, v, S), for $S \in \mathcal{F}_{uv}$, in a list F. By $u \prec v$ we denote that u precedes v in an arbitrary ordering of V. For $u \prec v$, tuples $(u, v, 1, S)$ from line 3 mean that $S \in \mathcal{F}_{uv}$, while tuples $(u, v, 2, S)$ from line 5 mean that $S \in \mathcal{F}_{vu}$. The `for` loop (lines 1-7) clearly takes $O\big(scan(m^{2-\frac{1}{k}})\big)$ I/O's. After sorting F (line 8) in $O\big(sort(m^{2-\frac{1}{k}})\big)$ I/O's, tuples for sets from $\mathcal{F}_{uv}$ directly precede those for sets from $\mathcal{F}_{vu}$, allowing us to execute line 9 in $O\big(scan(m^{2-\frac{1}{k}})\big)$ I/O's.

---

proc groupF
1: **for all** (u, v, S) in F **do**
2:     **if** $u \prec v$ **then**
3:         write $(u, v, 1, S)$ back to F
4:     **else**
5:         write $(v, u, 2, S)$ back to F
6:     **end if**
7: **end for**
8: sort F lexicographically
9: scan F to determine pairs $(\mathcal{F}_{uv}, \mathcal{F}_{uv})$

Based on Lemma 4, the total number of I/O's in calls to `findDisjoint` in Algorithm 2, line 13 is

$$O\Big(\sum_{u,v} \big(\sum_{i=1}^{k+2}(k-1)^i \cdot scan(|\mathcal{F}_{uv}| + |\mathcal{F}_{vu}|)\big) \Big)$$
$$= O\Big(\big(\sum_{i=1}^{k+2}(k-1)^i \big) \cdot \sum_{u,v} scan\big(|\mathcal{F}_{uv}| + |\mathcal{F}_{vu}|\big) \Big)$$
$$= O\big((2k-1)! \cdot scan(m^{2-\frac{1}{k}})\big)$$

as we set $p = q = k - 1$ and $\sum_{i=1}^{k+2}(k-1)^i = O\big((k-1)^{k+3}\big) = O\big((2k-1)!\big)$.

Putting it all together, we get that Algorithm 2 runs in $O\big(sort(m^{2-\frac{1}{k}}) + (2k-1)! \cdot scan(m^{2-\frac{1}{k}})\big)$ total I/O's. □

Theorem 3. *Let $G = (V, E)$ be a directed or an undirected graph. There is an external-memory algorithm that, given L – a δ-degeneracy ordering of G (for $\delta = (2 + \epsilon)d$), finds a cycle of length exactly c, or concludes that it does not exist:*

(i) in $O\Big(\delta^{1-\frac{1}{k}} \cdot \big(k \cdot sort(m^{2-\frac{1}{k}}) + (4k)! \cdot scan(m^{2-\frac{1}{k}})\big)\Big)$ I/O's if $c = 4k - 2$

(ii) in $O\Big(\delta \cdot \big(k \cdot sort(m^{2-\frac{1}{k}}) + (4k)! \cdot scan(m^{2-\frac{1}{k}})\big)\Big)$ I/O's if $c = 4k - 1$ or $c = 4k$

(iii) in $O\Big(\delta^{1+\frac{1}{k}} \cdot \big(k \cdot sort(m^{2-\frac{1}{k}}) + (4k)! \cdot scan(m^{2-\frac{1}{k}})\big)\Big)$ I/O's if $c = 4k + 1$

Proof. We describe the algorithm for the case of directed G, with $c = 4k+1$, as other cases are similar (and a little easier). We assume that $\delta < m^{\frac{1}{2k+1}}$, which is obviously the case for *naturally sparse* graphs. Otherwise, running Algorithm 2 on G achieves the advertised complexity.

```
 1: Δ ← m^(1/k)/δ^(1+1/k)
 2: for all v – vertex of degree ≥ Δ do
 3:     C ← cycleThrough(G, 4k + 1, v)
 4:     if C ≠ ε then
 5:         return C
 6:     end if
 7: end for
 8: remove vertices of degree ≥ Δ from G
 9: generate directed paths of length 2k and 2k + 1 in G
10: sort the paths lexicographically, according to their endpoints
11: group all paths u ⤳ v of length 2k into collection of (2k − 1)-sets F_uv
12: group all paths u ⤳ v of length 2k + 1 into collection of (2k)-sets G_uv
13: for all pairs (F_uv, G_uv) do
14:     P ← findDisjoint(F_uv, G_vu)
15:     if P = (A, B) then
16:         return C = A ∪ B
17:     end if
18: end for
19: return ε
```

Algorithm 3. Short cycles in degenerate graphs

Algorithm 3 is remarkably similar to Algorithm 2 and so is its analysis. Differences lie in the value of Δ and in line 9, when only *some* paths of length $2k$ and $2k + 1$ are generated. As explained in [3], it suffices to only consider all $(2k + 1)$-paths that start with two backward-oriented (in L) edges and all $2k$-paths that start with a backward-oriented (in L) edge. The number of these paths is $O(m^{2-\frac{1}{k}}\delta^{1+\frac{1}{k}})$. Since we can generate them in $O\left(k\delta^{1+\frac{1}{k}} \cdot sort(m^{2-\frac{1}{k}})\right)$ I/O's (see [18]), and there are at most $O(m^{1-\frac{1}{k}}\delta^{1+\frac{1}{k}})$ vertices in G of degree $\geq \Delta$, the theorem follows.   □

4   All Maximal Cliques

The Bron-Kerbosch algorithm [8] is often the choice when one needs to list all maximal cliques of an undirected graph $G = (V, E)$. It was initially improved by Tomita *et al.* [27]. We present this improvement as the `BronKerboschPivot` procedure ($\Gamma(v)$ denotes the set of neighbors of vertex v).

---

proc BronKerboschPivot(P, R, X)

1: **if** $P \cup X = \emptyset$ **then**
2:     **output** R //*maximal clique*
3: **end if**
4: $u \leftarrow$ vertex from $P \cup X$ that maximizes $|P \cap \Gamma(u)|$
5: **for all** $v \in P \setminus \Gamma(v)$ **do**
6:     BronKerboschPivot($P \cap \Gamma(v)$, $R \cup \{v\}$, $X \cap \Gamma(v)$)
7:     $P \leftarrow P \setminus \{v\}$
8:     $X \leftarrow X \cup \{v\}$
9: **end for**

The meaning of the arguments to BronKerboschPivot: R is a (possibly non-maximal) clique, P and X are a division of the set of vertices that are neighbors of all vertices in R, s.t. vertices in P are to be considered for adding to R while vertices in X are restricted from the inclusion.

Whereas Tomita *et al.* run the algorithm as BronKerboschPivot($V,\emptyset,\emptyset$), Eppstein *et al.* [12] improved it even further for the case of a d-degenerate G by utilizing its d-degeneracy ordering $L = \{v_1, v_2, \ldots, v_n\}$ and by performing n independent calls to BronKerboschPivot. Algorithm 4 presents their version. It runs in time $O(dn3^{d/3})$ in the RAM model.

1: **for** $i \leftarrow 1 \ldots n$ **do**
2:     $P \leftarrow \Gamma(v_i) \cap \{v_j : j > i\}$
3:     $X \leftarrow \Gamma(v_i) \cap \{v_j : j < i\}$
4:     BronKerboschPivot($P, \{v_i\}, X$)
5: **end for**

Algorithm 4. Maximal cliques in degenerate graph

The idea behind Algorithm 4 is to limit the depth of recursive calls to $|P| \leq d$ and then apply the analysis of Tomita *et al.* [27].

We show how to efficiently implement Algorithm 4 in the external memory model using a $(2 + \epsilon)d$-degeneracy ordering of G. Following [12], we define subgraphs $H_{P,X}$ of G.

Definition 2 (Graphs $H_{P,X}$). *Subgraph $H_{P,X} = (V_{P,X}, E_{P,X})$ of $G = (V, E)$ is defined as follows:*

$$V_{P,X} = P \cup X$$
$$E_{P,X} = \{(u, v) : (u, v) \in E \wedge (u \in P \vee v \in P)\}$$

That is, $H_{P,X}$ contains all edges in G whose endpoints are from $P \cup X$, and at least one of them lies in P. To ensure efficiency, $H_{P,X}$ is passed as an additional argument to every call to BronKerboschPivot with P and X. It is used in determining u at line 4 of BronKerboschPivot (we simply choose a vertex of highest degree in $H_{P,X}$).

Lemmas 6 and 7 are proved in the full version of this paper [18]. In the following, $\delta = (2 + \epsilon)d$.

Lemma 6. *Given a δ-degeneracy ordering L of an undirected d-degenerate graph G, all initial sets P, X, and graphs $H_{P,X}$ that are passed to* BronKerboschPivot *in line 4 of Algorithm 4 can be generated in $O(sort(\delta^2 n))$ I/O's.*

Lemma 7. *Given a δ-degeneracy ordering L of an undirected d-degenerate graph G, in a call to* BronKerboschPivot *that was given $H_{P,X}$, with $|P| = p$ and $|X| = x$, all graphs $H_{P \cap \Gamma(v), X \cap \Gamma(v)}$ that have to be passed to recursive calls in line 6, can be formed in $O(sort(\delta p^2 (p + x)))$ I/O's.*

Theorem 4. *Given a δ-degeneracy ordering L of an undirected d-degenerate graph G, we can list all its maximal cliques in $O(3^{\delta/3} sort(\delta n))$ I/O's.*

Proof. Consider a call to BronKerboschPivot$(P_v, \{v\}, X_v)$, with $|P_v| = p$ and $|X_v| = x$. Define $\widehat{D}(p, x)$ to be the maximum number of I/O's in this call. Based on Lemma 7, $\widehat{D}(p, x)$ satisfies the following recurrence relation:

$$\widehat{D}(p, x) \leq \begin{cases} \max_k \{k \widehat{D}(p - k, x)\} + O\big(sort(\delta p^2 (p + x))\big) & \text{if } p > 0 \\ e & \text{if } p = 0 \end{cases}$$

for constant e greater than zero, which can be rewritten as

$$\widehat{D}(p, x) \leq \begin{cases} \max_k \{k \widehat{D}(p - k, x)\} + c \cdot \frac{\delta p^2 (p + x)}{DB} \log_{M/B}(\delta p^2 (p + x)) & \text{if } p > 0 \\ e & \text{if } p = 0 \end{cases}$$

for a constant $c > 0$. Since $p \leq \delta$ and $p + x \leq n$, we have $\log_{M/B}(\delta p^2 (p + x)) \leq \log_{M/B}(\delta^3 n) = O(\log_{M/B} n)$ for $\delta = O(1)$. Thus, the relation for $\widehat{D}(p, x)$:

$$\widehat{D}(p, x) \leq \begin{cases} \max_k \{k \widehat{D}(p - k, x)\} + \delta p^2 (p + x) \cdot \frac{c' \log_{M/B} n}{DB} & \text{if } p > 0 \\ e & \text{if } p = 0 \end{cases}$$

where c' and e are constants greater than zero. Note that this is the relation for $D(p, x)$ of Eppstein *et. al* [12] (we set $d = \delta$, $c_1 = \frac{c' \log_{M/B} n}{DB}$ and $c_2 = e$). Since the solution for $D(p, x)$ was $D(p, x) = O((d + x)3^{p/3})$, the solution for $\widehat{D}(p, x)$ is

$$\widehat{D}(p, x) = O\Big((\delta + x)3^{p/3} \cdot \frac{c' \log_{M/B} n}{DB}\Big) = O\Big(\frac{\delta + x}{DB} 3^{p/3} \log_{M/B} n\Big)$$

The total size of all sets X_v passed to initial calls to BronKerboschPivot is $O(\delta n)$, and $|P| \leq \delta$. Thus, the total number of I/O's in recursive calls is

$$\sum_v O\Big(\frac{\delta + |X_v|}{DB} 3^{\delta/3} \log_{M/B} n\Big) = O\Big(3^{\delta/3} \frac{\delta n}{DB} \log_{M/B} n\Big) = O\big(3^{\delta/3} sort(\delta n)\big)$$

Combining this with Lemma 6, we get that our external memory version of Algorithm 4 takes $O\big(sort(\delta^2 n) + 3^{\delta/3} sort(\delta n)\big) = O\big(3^{\delta/3} sort(\delta n)\big)$ I/O's. $\square$

References

1. Alon, N., Gutner, S.: Linear time algorithms for finding a dominating set of fixed size in degenerated graphs. Algorithmica 54(4), 544–556 (2009)
2. Alon, N., Kahn, J., Seymour, P.D.: Large induced degenerate subgraphs. Graphs and Combinatorics 3, 203–211 (1987)
3. Alon, N., Yuster, R., Zwick, U.: Finding and counting given length cycles. Algorithmica 17(3), 209–223 (1997)
4. Barabási, A.-L., Albert, R.: Emergence of scaling in random networks. Science 286(5439), 509–512 (1999)
5. Batagelj, V., Zaveršnik, M.: An O(m) algorithm for cores decomposition of networks (2003), http://arxiv.org/abs/cs.DS/0310049
6. Bauer, R., Krug, M., Wagner, D.: Enumerating and generating labeled k-degenerate graphs. In: 7th Workshop on Analytic Algorithmics and Combinatorics (ANALCO), pp. 90–98. SIAM, Philadelphia (2010)
7. Bollobás, B.: On generalized graphs. Acta Mathematica Hungarica 16, 447–452 (1904), doi:10.1007/BF01904851
8. Bron, C., Kerbosch, J.: Algorithm 457: finding all cliques of an undirected graph. Commun. ACM 16(9), 575–577 (1973)
9. Cheng, J., Ke, Y., Chu, S., Ozsu, T.: Efficient core decomposition in massive networks. In: IEEE Int. Conf. on Data Engineering, ICDE (2011)
10. Chrobak, M., Eppstein, D.: Planar orientations with low out-degree and compaction of adjacency matrices. Theor. Comput. Sci. 86(2), 243–266 (1991)
11. Doreian, P., Woodard, K.L.: Defining and locating cores and boundaries of social networks. Social Networks 16(4), 267–293 (1994)
12. Eppstein, D., Löffler, M., Strash, D.: Listing all maximal cliques in sparse graphs in near-optimal time. In: Cheong, O., Chwa, K.-Y., Park, K. (eds.) ISAAC 2010. LNCS, vol. 6506, pp. 403–414. Springer, Heidelberg (2010)
13. Eppstein, D., Strash, D.: Listing all maximal cliques in large sparse real-world graphs. arXiv eprint, 1103.0318 (2011)
14. Erdős, P., Hajnal, A.: On chromatic number of graphs and set-systems. Acta Mathematica Hungarica 17(1-2), 61–99 (1966)
15. Fernholz, D., Ramachandran, V.: The giant k-core of a random graph with a specified degree sequence (2003) (manuscript)
16. Freuder, E.C.: A sufficient condition for backtrack-free search. J. ACM 29, 24–32 (1982)
17. Golovach, P.A., Villanger, Y.: Parameterized complexity for domination problems on degenerate graphs. In: Broersma, H., Erlebach, T., Friedetzky, T., Paulusma, D. (eds.) WG 2008. LNCS, vol. 5344, pp. 195–205. Springer, Heidelberg (2008)
18. Goodrich, M.T., Pszona, P.: External-memory network analysis algorithms for naturally sparse graphs. arXiv eprint, 1106.6336 (2011)
19. Irani, S.: Coloring inductive graphs on-line. Algorithmica 11, 53–72 (1994)
20. Kirousis, L.M., Thilikos, D.M.: The linkage of a graph. SIAM Journal on Computing 25(3), 626–647 (1996)
21. Kleinberg, J.: The small-world phenomenon: an algorithm perspective. In: 32nd ACM Symp. on Theory of Computing (STOC), pp. 163–170 (2000)
22. Lick, D.R., White, A.T.: k-degenerate graphs. Canadian Journal of Mathematics 22, 1082–1096 (1970)
23. Monien, B.: How to find long paths efficiently. Annals of Discrete Mathematics 25, 239–254 (1985)

24. Pittel, B., Spencer, J., Wormald, N.: Sudden emergence of a giant k-core in a random graph. Journal of Combinatorial Theory, Series B 67(1), 111–151 (1996)
25. Riordan, O.: The k-core and branching processes. Probability And Computing 17, 111 (2008)
26. Seidman, S.B.: Network structure and minimum degree. Social Networks 5(3), 269–287 (1983)
27. Tomita, E., Tanaka, A., Takahashi, H.: The worst-case time complexity for generating all maximal cliques and computational experiments. Theor. Comput. Sci. 363(1), 28–42 (2006)
28. Vitter, J.S.: External memory algorithms and data structures: dealing with massive data. ACM Comput. Surv. 33, 209–271 (2001)

Approximate Counting of Cycles in Streams[*]

Madhusudan Manjunath[1], Kurt Mehlhorn[1],
Konstantinos Panagiotou[1], and He Sun[1,2]

[1] Max Planck Institute for Informatics, Saarbrücken, Germany
[2] Fudan University, Shanghai, China

Abstract. We consider the subgraph counting problem in data streams and develop the first non-trivial algorithm for approximately counting cycles of an arbitrary but fixed size. Previous non-trivial algorithms could only approximate the number of occurrences of subgraphs of size up to six. Our algorithm is based on the idea of computing instances of *complex-valued* random variables over the given stream and improves drastically upon the naïve sampling algorithm. In contrast to most existing approaches, our algorithm works in a distributed setting and for the turnstile model, i. e., the input stream is a sequence of edge insertions and deletions.

1 Introduction

Counting the number of occurrences of a graph H in a graph G has wide applications in uncovering important structural characteristics of the underlying network G, revealing information of the most frequent patterns, and so on. We are interested in the situation where G is very large. It is then natural to assume that G is given as a data stream, i.e., the edges of the graph G arrive consecutively and the algorithm uses only limited space to return an approximate value. Exact counting is not an option for massive input graphs. Already counting triangles exactly requires to store the entire graph.

Formally speaking, let $\mathcal{S} = s_1, s_2, \cdots, s_N$ be a stream that represents a graph $G = (V, E)$, where N is the length of the stream and each item s_i is associated with an edge in G. Typical models [12] in this topic include *the Cash Register Model* and *the Turnstile Model*. In the cash register model, each item s_i expresses one edge in G, and in the turnstile model each item s_i is represented by (e_i, sign_i) where e_i is an edge of G and $\mathrm{sign}_i \in \{+, -\}$ indicates that e_i is inserted to or deleted from G. As a generalization of the cash register model, the turnstile model supports the dynamic insertions and deletions of the edges.

In a distributed setting the stream $\mathcal{S}$ is partitioned into sub-streams $S_1, \ldots, S_t$ and each S_i is fed to a different processor. At the end of the computation, the processors collectively estimate the number of occurrences of H with a small amount of communication.

---

[*] The third author was supported by the Alexander von Humboldt-Foundation.

C. Demetrescu and M.M. Halldórsson (Eds.): ESA 2011, LNCS 6942, pp. 677–688, 2011.

Our Results. We present a general framework for counting cycles of arbitrary size in a massive graph. Our algorithm runs in the turnstile model and the distributed setting, and for any constants $0 < \varepsilon, \delta < 1$, our algorithm achieves an (ε, δ)-approximation, i. e. , the returning value Z of the algorithm and the exact value $Z^* = \#C_k$, the number of occurrences of C_k, satisfy $\Pr\left[|Z - Z^*| > \varepsilon \cdot Z^*\right] < \delta$. We also provide an unbiased estimator for general d-regular graphs. This considerably extends the class of graphs that can be counted in the data streaming model and answers partially an open problem proposed by many references, see for example the extensive survey by Muthukrishnan [12] and the 11th open question in the 2006 IITK Workshop on Algorithms for Data Streams [11].

Because the problem of counting the number of cycles of length k, parameterized by k, is $\#\mathbf{W}[1]$-complete [7], our result demonstrates that efficient approximations for $\#\mathbf{W}[1]$-complete problems are possible under certain conditions, even if only a restricted amount of space can be used.

Besides that, we initiate the study of complex-valued hash functions in counting subgraphs. Complex-valued estimators have been successfully applied in other contexts such as approximating the permanent, see [6,10]. In the data streaming setting, Ganguly [8] used a complex-valued sketch to estimate frequency moments. Our main result is as follows:

Theorem 1. *Let G be a graph with n vertices and m edges. For any k, there is an algorithm using S bits of space to (ε, δ)-approximate the number of occurrences of C_k in G provided that $S = \Omega\left(\frac{1}{\varepsilon^2} \cdot \frac{m^k}{(\#C_k)^2} \cdot \log n \cdot \log \frac{1}{\delta}\right)$. The algorithm works in the turnstile model.*

Discussion: A naïve approach for counting the number of occurrences of a k-cycle would either sample independently k vertices (if possible) or k edges from the stream. Since the probability of k vertices (or k edges) forming a cycle is $\#C_k/n^k$ (or $\#C_k/m^k$), this approach needs space $\Omega\left(\frac{n^k \log n}{\#C_k}\right)$ and $\Omega\left(\frac{m^k \log n}{\#C_k}\right)$, respectively. Thus, our algorithm improves upon these two approaches, especially for sparse graphs with many k-cycles, and has the additional benefit that it is applicable in the turnstile model and the distributed setting. Moreover, note that our bound is essentially tight, as there are graphs where the space complexity of the algorithm is $O(\log n)$; consider for example the "extremal graph" with a clique on $\Theta(\sqrt{m})$ vertices, where all other vertices are isolated. Moreover, as a corollary of Theorem 1, when the number of occurrences of C_k is $\Omega\left(m^{k/2-\alpha}\right)$ for $0 \le \alpha < 1/2$, our algorithm with sub-linear space $O\left(\frac{1}{\varepsilon^2} \cdot m^{2\alpha} \cdot \log \frac{1}{\delta}\right)$ suffices to give a good approximation.

Related Work: Counting subgraphs in a data stream was first considered in a seminal paper by Bar-Yossef, Kumar, and Sivakumar [1]. There, the triangle counting problem was reduced to the problem of computing frequency moments. After that, several algorithms for counting triangles have been proposed [2,4,9].

Jowhari and Ghodsi presented three algorithms in [9], one of which is applicable in the turnstile model. Moreover, the problem of counting subgraphs different from triangles has also been investigated in the literature. Bordino, Donato, Gionis, and Leonardi [3] extended the technique of counting triangles [4] to all subgraphs on three and four vertices. Buriol, Fahling, Leonardi and Sohler [5] presented a streaming algorithm for counting $K_{3,3}$, the complete bipartite graph with three vertices in each part. However, except the one presented in [9], most algorithms are based on sampling techniques and do not apply to the turnstile model.

Notation: Let $G = (V, E)$ be an undirected graph without self-loops and multiple edges. The set of vertices and edges are represented by $V[G]$ and $E[G]$ respectively. We will assume that $V[G] = \{1, \cdots, n\}$ and n is known in advance.

Given two directed graphs H_1 and H_2, we say that H_1 and H_2 are *homomorphic* if there is a mapping $i : V[H_1] \to V[H_2]$ such that $(u, v) \in E[H_1]$ if and only if $(i(u), i(v)) \in E[H_2]$. Furthermore, H_1 and H_2 are said to be *isomorphic* if the mapping i is a bijection.

For any graph H, we call a not necessarily induced subgraph H_1 of G an *occurrence* of H, if H_1 is isomorphic to H. We use $\#(H, G)$ to denote the number of occurrences of H in G. When G is the input graph, for simplicity we use $\#H$ to express $\#(H, G)$. Moreover, let C_ℓ be a cycle on ℓ edges.

Organization. Section 2 reviews Jowhari and Ghodsi's algorithm for counting triangles in streams. We generalize Jowhai and Ghodsi's approach in Sect. 3 and get an unbiased estimator for general d-regular graphs. Section 4 discusses the space complexity for counting cycles with arbitrary size. We end this paper with some open problems in Sect. 5.

2   A Review of Jowhari and Ghodsi's Algorithm

We give a brief account of Jowhari and Ghodsi's algorithm [9] in order to prepare the reader for our extension of their approach. Jowhari and Ghodsi estimate the number of triangles in a graph G. Let X be a $\{-1, +1\}$-valued random variable with expectation zero. They associate with every vertex w of G an instance $X(w)$ of X; the $X(w)$'s are 6-wise independent. They compute $Z = \sum_{\{u,v\} \in E[G]} X(u)X(v)$ and output $Z^3/6$ as the estimator for $\#C_3$.

Lemma 1 ([9]). $\mathbb{E}[Z^3] = 6 \cdot \#C_3$.

Proof. For any triple $T \in E^3[G]$ of edges and any vertex w of G, let $\deg_T(w)$ be the number of edges in T incident to w, then $\deg_T(w)$ is an integer no larger than 3. Also

$$\mathbb{E}[Z^3] = \mathbb{E}\left[\left(\sum_{\{u,v\} \in E[G]} X(u)X(v)\right)^3\right]$$

$$= \mathbb{E}\left[\sum_{T=(\{u_1,v_1\},\{u_2,v_2\},\{u_3,v_3\}) \in E^3} X(u_1)X(v_1)X(u_2)X(v_2)X(u_3)X(v_3)\right]$$

Let V_T be the set of vertices that are incident to the edges in T. Then

$$\mathbb{E}[Z^3] = \mathbb{E}\left[\sum_{T \in E^3} \prod_{w \in V_T} X(w)^{\deg_T(w)}\right]$$

By the 6-wise independence of the $X(w), w \in V$, we have

$$\mathbb{E}[Z^3] = \sum_{T \in E^3} \prod_{w \in V_T} \mathbb{E}\left[X(w)^{\deg_T(w)}\right] = \sum_{T \in E^3} \prod_{w \in V_T} \mathbb{E}\left[X^{\deg_T(w)}\right]$$

Since $\mathbb{E}\left[X^{\deg_T(w)}\right] = 1$ if $\deg_T(w)$ is even and $\mathbb{E}\left[X^{\deg_T(w)}\right] = 0$ if $\deg_T(w)$ is odd, we know that $\prod_{w \in V_T} \mathbb{E}\left[X^{\deg_T(w)}\right] = 1$ if and only if the edges in T form a triangle. Since each triangle is counted six times, we have $\mathbb{E}[Z^3] = 6 \cdot \#C_3$. $\quad\square$

The crucial ingredients of the proof are (1) 6-wise independence guarantees that the expectation-operator can be pulled inside, and (2) random variable X is defined such that only vertices with even degree in T have nonzero expectation.

3   Algorithm Framework

We now generalize the algorithm in Section 2 and present an algorithm framework for counting general d-regular graphs. Suppose that H is a d-regular graph with k edges and we want to count the number of occurrences of H in G. The vertices of H are expressed by a, b and c, etc., and the vertices of G are expressed by u, v and w, etc., respectively. We will equip the edges of H with an arbitrary orientation, as this is necessary for the further analysis. Therefore, each edge in H together with its orientation can be expressed as $\overrightarrow{ab}$ for some $a, b \in V[H]$. For simplicity and with slight abuse of notation we will use H to express such an oriented graph.

For each oriented edge $\overrightarrow{ab}$ in H our algorithm maintains a complex-valued variable $Z_{\overrightarrow{ab}}(G)$, which is initialized to zero. The variables are defined in terms of random variables $Y(w)$ and $X_c(w)$, where c is a node of H and w is a node of G. The random variables $Y(w)$ are instances of a random variable Y and the random variables $X_c(w)$ are instances of a random variable X. The range of both random variables is a finite subset of complex numbers. We will realize the random variables by hash functions from $V[G]$ to $\mathbb{C}$; this explains why we indicate the dependence on w by functional brackets. We assume that the variables $X_c(w)$ and $Y(w)$ have sufficient independence as detailed below.

Our algorithm performs two basic steps: First, when an edge $e = \{u, v\} \in E[G]$ arrives, we update each variable $Z_{\overrightarrow{ab}}$ according to

$$Z_{\overrightarrow{ab}}(G) \leftarrow Z_{\overrightarrow{ab}}(G) + \left(X_a(u) \cdot X_b(v) + X_b(u) \cdot X_a(v)\right) \cdot Y(u) \cdot Y(v). \tag{1}$$

Second, when the number of occurrences of a graph H is required, the algorithm returns the real part of $Z/(\alpha \cdot \operatorname{aut}(H))$, where Z is defined via

$$Z := Z_H(G) = \prod_{\overrightarrow{ab} \in E[H]} Z_{\overrightarrow{ab}}(G), \tag{2}$$

α and aut(H) are constant numbers for any given H and will be determined later.

Remark 1. For simplicity, the algorithm above is only for the edge-insertion case. An edge deletion amounts to replacing '+' by '−' in (1).

Remark 2. The first step may be carried out in a distributed fashion, i. e., we have several processors each processing a subset of edges. In the second step the counts of the different processors are combined.

Theorem 2. *Let H be a d-regular graph with k edges. Let us assume that the random variables defined above satisfy the following two properties:*

1. *The random variables $X_c(w)$ and $Y(w)$, where $c \in V[H]$ and $w \in V[G]$, are instances of random variables X and Y, respectively. The random variables are $4k$-wise independent.*
2. *Let Z be any one of $X_c, c \in V[H]$ or Y. Then for any $1 \le i \le 2k$, $\mathbb{E}\left[Z^i\right] \ne 0$ if and only if $i = d$.*

Then $\mathbb{E}[Z_H(G)] = \alpha \cdot \text{aut}(H) \cdot \#(H, G)$, where $\alpha = \left(\mathbb{E}\left[X^d\right] \mathbb{E}\left[Y^d\right]\right)^{2k/d} \in \mathbb{C}$ and aut(H) is the number of permutations and orientations of the edges in H such that the resulting graph is isomorphic to H.

The theorem above shows that $Z_H(G)$ is an unbiased estimator for any d-regular graph H, assuming that there exist random variables $X_c(w)$ and $Y(w)$ with certain properties. We will prove Theorem 2 at first, and then construct such random variables.

Proof (of Theorem 2). We first introduce some notations. For a k-tuple $T = (e_1, \ldots, e_k) \in E^k[G]$, let $G_T = (V_T, E_T)$ be the induced multi-graph, i.e., G_T has edge multi-set $E_T = \{e_1, \ldots, e_k\}$. By definition, we have

$$Z_H(G) = \prod_{\overrightarrow{ab} \in E[H]} Z_{\overrightarrow{ab}}(G)$$

$$= \prod_{\overrightarrow{ab} \in E[H]} \left(\sum_{\{u,v\} \in E[G]} (X_a(u) \cdot X_b(v) + X_a(v) \cdot X_b(u)) \cdot Y(u) \cdot Y(v) \right).$$

Since H has k edges, $Z_H(G)$ is a product of k terms and each term is a sum over all edges of G each with two possible orientations. Thus, in the expansion of $Z_H(G)$, any k-tuple $(e_1, \cdots, e_k) \in E^k[G]$ contributes 2^k different terms to $Z_H(G)$ and each term corresponds to a certain orientation of $(e_1, \cdots, e_k)$. Let $\overrightarrow{T} = (\overrightarrow{e_1}, \cdots, \overrightarrow{e_k})$ be an arbitrary orientation of $(e_1, \cdots, e_k)$, where $\overrightarrow{e_i} = \overrightarrow{u_i v_i}$. So the term in $Z_H(G)$ corresponding to $(\overrightarrow{e_1}, \cdots, \overrightarrow{e_k})$ is

$$\prod_{i=1}^{k} X_{a_i}(u_i) \cdot X_{b_i}(v_i) \cdot Y(u_i) \cdot Y(v_i) , \tag{3}$$

where (a_i, b_i) is the i-th edge of H and $\overrightarrow{u_i v_i}$ is the i-th edge in $\overrightarrow{T}$. We show that (3) is non-zero if and only if the graph induced by $\overrightarrow{T}$ is isomorphic to H (i. e. it also preserves the orientations of the edges).

For a vertex w of G and a vertex c of H, let

$$\theta_{\overrightarrow{T}}(c, w) = \left| \{ i \mid (u_i = w \text{ and } a_i = c) \text{ or } (v_i = w \text{ and } b_i = c) \} \right| . \tag{4}$$

Thus for any $c \in V[H]$, $\sum_{w \in V_T} \theta_{\overrightarrow{T}}(c, w) = d$ since every vertex c of H appears in exactly d edges (a_i, b_i); recall that H is d-regular. Using the definition of $\theta_{\overrightarrow{T}}$, we may rewrite (3) as

$$\left(\prod_{c \in V[H]} \prod_{w \in V_{\overrightarrow{T}}} X_c^{\theta_{\overrightarrow{T}}(c,w)}(w) \right) \cdot \left(\prod_{w \in V_{\overrightarrow{T}}} Y^{\deg_{\overrightarrow{T}}(w)}(w) \right),$$

where $\deg_{\overrightarrow{T}}(w)$ is the number of edges in $\overrightarrow{T}$ incident to w. Therefore

$$Z_H(G)$$

$$= \sum_{\substack{e_1, \cdots, e_k \\ e_i \in E[G]}} \sum_{\overrightarrow{T} = (\overrightarrow{e_1}, \cdots, \overrightarrow{e_k})} \left(\prod_{c \in V[H]} \prod_{w \in V_{\overrightarrow{T}}} X_c^{\theta_{\overrightarrow{T}}(c,w)}(w) \right) \cdot \left(\prod_{w \in V_{\overrightarrow{T}}} Y^{\deg_{\overrightarrow{T}}(w)}(w) \right),$$

where the first summation is over all the k-tuples of edges in $E[G]$ and the second summation is over all their possible orientations. Since each term of Z_H is the product of $4k$ random variables, which by assumption are $4k$-wise independent, we infer by linearity of expectation that

$$\mathbb{E}[Z_H(G)]$$

$$= \mathbb{E} \left[\sum_{\substack{e_1, \cdots, e_k \\ e_i \in E[G]}} \sum_{\overrightarrow{T} = (\overrightarrow{e_1}, \cdots, \overrightarrow{e_k})} \left(\prod_{c \in V[H]} \prod_{w \in V_{\overrightarrow{T}}} X_c^{\theta_{\overrightarrow{T}}(c,w)}(w) \right) \cdot \left(\prod_{w \in V_{\overrightarrow{T}}} Y^{\deg_{\overrightarrow{T}}(w)}(w) \right) \right]$$

$$= \sum_{\substack{e_1, \cdots, e_k \\ e_i \in E[G]}} \sum_{\overrightarrow{T} = (\overrightarrow{e_1}, \cdots, \overrightarrow{e_k})} \prod_{c \in V[H]} \prod_{w \in V_{\overrightarrow{T}}} \mathbb{E}\left[X^{\theta_{\overrightarrow{T}}(c,w)} \right] \cdot \prod_{w \in V_{\overrightarrow{T}}} \mathbb{E}\left[Y^{\deg_{\overrightarrow{T}}(w)} \right] .$$

Let

$$\alpha(\overrightarrow{T}) := \prod_{c \in V[H]} \prod_{w \in V_{\overrightarrow{T}}} \mathbb{E}\left[X^{\theta_{\overrightarrow{T}}(c,w)} \right] \cdot \prod_{w \in V_{\overrightarrow{T}}} \mathbb{E}\left[Y^{\deg_{\overrightarrow{T}}(w)} \right] .$$

We will next show that $\alpha(\overrightarrow{T})$ is either zero or a nonzero constant independent of $\overrightarrow{T}$. The latter is the case if and only if G_T is an occurrence of H in G.

We have $\mathbb{E}\left[X^i \right] \neq 0$ if and only if $i = d$ or $i = 0$. Therefore for any $\overrightarrow{T}$ and $c \in V[H]$, $\prod_{w \in V_{\overrightarrow{T}}} \mathbb{E}[X^{\theta_{\overrightarrow{T}}(c,w)}] \neq 0$ if and only if $\theta_{\overrightarrow{T}}(c, w) \in \{0, d\}$ for all w. Since $\sum_w \theta_{\overrightarrow{T}}(c, w) = \deg_H(c) = d$, there must be a unique vertex $w \in V_{\overrightarrow{T}}$ such that

$\theta_{\overrightarrow{T}}(c, w) = d$. Define $\varphi : V[H] \to V_{\overrightarrow{T}}$ as $\varphi(c) = w$. Then φ is a homomorphism and

$$\prod_{c \in V[H]} \prod_{w \in V_{\overrightarrow{T}}} \mathbb{E}\left[X^{\theta_{\overrightarrow{T}}(c,w)}\right] = \prod_{c \in V[H]} \mathbb{E}\left[X^d\right] = \mathbb{E}\left[X^d\right]^{|V[H]|}.$$

Since $\mathbb{E}[Y^i] \neq 0$ if and only if $i = d$ or $i = 0$, so for any $\overrightarrow{T}$, $\prod_{w \in V_{\overrightarrow{T}}} \mathbb{E}[Y^{\deg_{\overrightarrow{T}}(w)}] \neq 0$ if and only if every vertex $w \in V_{\overrightarrow{T}}$ has degree d in the graph with edge set T. Thus $|V_{\overrightarrow{T}}| = 2k/d = |V[H]|$, which implies that φ is an isomorphism mapping.

We have now shown that $\alpha(\overrightarrow{T})$ is either zero or the nonzero constant

$$\alpha = \left(\mathbb{E}\left[X^d\right] \mathbb{E}\left[Y^d\right]\right)^{2k/d}.$$

The latter is the case if and only if $G_{\overrightarrow{T}}$ is an occurrence of H in G. Let $(G_{\overrightarrow{T}} \equiv H)$ be the indicator expression that is one if $G_{\overrightarrow{T}}$ and H are isomorphic and zero otherwise. Then

$$\mathbb{E}[Z_H(G)] = \sum_{\substack{e_1, \cdots, e_k \\ e_i \in E[G]}} \sum_{\overrightarrow{T} = (\overrightarrow{e_1}, \cdots, \overrightarrow{e_k})} \alpha(\overrightarrow{T}) \cdot (G_{\overrightarrow{T}} \equiv H) = \alpha \cdot \operatorname{aut}(H) \cdot \#(H, G).$$

$\square$

For the case of cycles, we have $\operatorname{aut}(H) = 2k$. We turn to construct hash functions needed in Theorem 2. The basic idea is to choose a $8k$-wise independent hash function $h : D \to \mathbb{C}$ and map the values in D to complex numbers with certain properties. We first show a simple lemma about roots of polynomials of a simple form.

Lemma 2. *For positive interger r, let $P_r(z) = 2 + z^r$ and $z_j = 2^{1/j} \cdot e^{\frac{\pi i}{j}}$. The complex number z_j is a root of the polynomial $P_r(z)$ if and only if $j = r$.*

Proof. We first verify that z_r is a root of the polynomial $P_r(z)$: since $z_r^r = 2 \cdot e^{\pi \cdot i} = -2$, we have $z_r^r + 2 = 0$. To show the converse, we consider z_j^r for $r \neq j$ and verify that $|z_j^r| = \left|2^{r/j} e^{\frac{\pi \cdot i \cdot r}{j}}\right| = 2^{r/j}$. Since $2^{r/j} \neq 2$ if $j \neq r$, the claim follows. $\square$

Let z_j as in Lemma 2 and define random variable H_j as

$$H_j = \begin{cases} 1, & \text{with probability } 2/3, \\ z_j, & \text{with probability } 1/3. \end{cases} \tag{5}$$

Then $\mathbb{E}[H_j^\ell] = \left(2 + z_j^\ell\right)/3 = P_\ell(z_j)/3$ which is nonzero if $j \neq \ell$.

Theorem 3. *For positive integers d and k, let*

$$H = \prod_{1 \leq j \leq 2k, j \neq d} H_j$$

where the H_j are independent. For all integers ℓ between 1 and $2k$, $\mathbb{E}[H^\ell] \neq 0$ if and only if $d = \ell$.

Proof. By independence, $\mathbb{E}[H^\ell] = \prod_{1 \leq j \leq 2k, j \neq d} \mathbb{E}[H_j^\ell]$. This product is nonzero if ℓ is different from all j that are distinct from d, i. e., $\ell = d$. $\square$

4   Proof of the Main Theorem

Now we bound the space of the algorithm for the case of cycles of arbitrary length. The basic idea is to use the second moment method on the complex-valued random variable Z. We first note a couple of lemmas that turn out to be useful: the first lemma is a generalization of Chebyshev's inequality for a complex-valued random variable and the second lemma is an upper bound on the number of closed walks of a given length in terms of the number of edges of the graph. Recall that the conjugate of a complex number $z = a + ib$ is denoted by $\overline{z} := a - ib$.

Lemma 3. *Let X be a complex-valued random variable with finite support and let $t > 0$. We have that*

$$\Pr[|X - \mathbb{E}[X]| \geq t \cdot |\mathbb{E}[X]|] \leq \frac{\mathbb{E}[X\overline{X}] - \mathbb{E}[X]\,\overline{\mathbb{E}[X]}}{t^2 |\mathbb{E}[X]|^2}.$$

Proof. Since $|X - \mathbb{E}[X]|^2 = (X - \mathbb{E}[X])(\overline{X - \mathbb{E}[X]})$ is a positive-valued random variable, we apply Markov's inequality to obtain

$$\Pr[|X - \mathbb{E}[X]| \geq t \cdot |\mathbb{E}[X]|] = \Pr\left[|X - \mathbb{E}[X]|^2 \geq t^2 \cdot |\mathbb{E}[X]|^2\right]$$
$$\leq \frac{\mathbb{E}[(X - \mathbb{E}[X])(\overline{X - \mathbb{E}[X]})]}{t^2 |\mathbb{E}[X]|^2}.$$

Expanding $\mathbb{E}[(X - \mathbb{E}[X])(\overline{X - \mathbb{E}[X]})]$ we obtain that

$$\mathbb{E}[(X - \mathbb{E}[X])(\overline{X - \mathbb{E}[X]})] = \mathbb{E}[X\overline{X}] - \mathbb{E}[X\overline{\mathbb{E}[X]}] - \mathbb{E}[\overline{X}\mathbb{E}[X]] + \overline{\mathbb{E}[X]}\mathbb{E}[X]$$
$$= \mathbb{E}\left[X\overline{X}\right] - \mathbb{E}[X]\overline{\mathbb{E}[X]}.$$

The last equality uses the linearity of expectation and that $\mathbb{E}[\overline{X}] = \overline{\mathbb{E}[X]}$.   □

We now show an upper bound on the number of closed walks of a given length in a graph. This upper bound will control the space requirement of the algorithm.

Lemma 4. *Let G be an undirected graph with n vertices and m edges. Then the number of closed walks W_k with length k in G is at most $\frac{2^{k/2-1}}{k} \cdot m^{k/2}$.*

Proof. Let A be the adjacency matrix of G with eigenvalues $\lambda_1, \cdots, \lambda_n$. Since G is undirected, A is real symmetric and each eigenvalue λ_i is a real number. Then $W_k = \frac{1}{2k} \cdot \sum_{i=1}^n (A^k)_{ii}$ where for a matrix M, M_{ij} is the ij-th entry of the matrix. Because $\sum_{i=1}^n (A^k)_{ii} = \operatorname{tr}(A^k) = \sum_{i=1}^n \lambda_i^k \leq \sum_{i=1}^n |\lambda_i|^k$ and $\left(\sum_{i=1}^n |\lambda_i|^k\right)^{1/k} \leq \left(\sum_{i=1}^n |\lambda_i|^2\right)^{1/2} = (2m)^{1/2}$ for any $k \geq 2$, we have $W_k \leq \frac{1}{2k} \cdot \left(\sum_{i=1}^n |\lambda_i|^2\right)^{k/2} = \frac{2^{k/2-1}}{k} \cdot m^{k/2}$.   □

Corollary 1. *Let G be a graph on m edges and $\mathcal{H}$ be a set of subgraphs of G such that every $H \in \mathcal{H}$ has properties: (1) H has k edges, where k is a constant. (2) Each connected-component of H is an Eulerian circuit. Then $|\mathcal{H}| = O(m^{k/2})$.*

Proof. Fix an integer $r \in \{1, \ldots, k\}$ and consider graphs in $\mathcal{H}$ that have r connected components. By Lemma 4, the number of such graphs is at most

$$\sum_{\substack{k_1, \cdots, k_r \\ k_1 + \cdots + k_r = k}} \prod_{i=1}^{r} W_{k_i} \leq \sum_{\substack{k_1, \cdots, k_r \\ k_1 + \cdots + k_r = k}} \prod_{i=1}^{r} \frac{2^{k_i/2 - 1} \cdot m^{k_i/2}}{k_i} \leq f(k) \cdot (2m)^{k/2},$$

where $f(k)$ is a function of k. Because there are at most k choices of r, we have $|\mathcal{H}| = O(m^{k/2})$. $\qquad\qquad\square$

Observe that the expansion of $\mathbb{E}[Z_H(G)\overline{Z_H(G)}]$ consists of m^{2k} terms and the modulus of each term is upper bounded by a constant. So a naïve upper bound for $\mathbb{E}[Z_H(G)\overline{Z_H(G)}]$ is $O(m^{2k})$. Now we only focus on the case of cycles and use the "cancellation" properties of the random variables to get a better bound for $\mathbb{E}[Z_H(G)\overline{Z_H(G)}]$.

Theorem 4. *Let H be a cycle C_k with an arbitrary orientation and suppose that the following properties are satisfied:*

1. *The random variables $X_c(w)$ and $Y(w)$, where $c \in V[H]$ and $w \in V[G]$ are $8k$-wise independent.*
2. *Let Z be any one of $X_c, c \in V[H]$ or Y. Then for any $1 \leq i \leq 2k$, $\mathbb{E}\left[Z^i\right] \neq 0$ if and only if $i = 2$.*

Then $\mathbb{E}[Z_H(G)\overline{Z_H(G)}] = O(m^k)$.

Proof. By the definition of $Z_H(G)$ we express $Z_H(G)\overline{Z_H(G)}$ as

$$\sum_{\substack{\overrightarrow{T_1} = (\overrightarrow{e_1}, \cdots, \overrightarrow{e_k}) \\ \overrightarrow{T_2} = (\overrightarrow{e_1'}, \cdots, \overrightarrow{e_k'}) \\ e_i, e_i' \in E[G]}} \left(\prod_{\substack{c \in V[H] \\ w \in V_{\overrightarrow{T_1}}}} X_c(w)^{\theta_{\overrightarrow{T_1}}(c,w)} \right) \cdot \left(\prod_{w \in V_{\overrightarrow{T_1}}} Y(w)^{\deg_{\overrightarrow{T_1}}(w)} \right) \cdot$$

$$\left(\prod_{\substack{c \in V[H] \\ w \in V_{\overrightarrow{T_2}}}} \overline{X_c(w)}^{\theta_{\overrightarrow{T_2}}(c,w)} \right) \cdot \left(\prod_{w \in V_{\overrightarrow{T_2}}} \overline{Y(w)}^{\deg_{\overrightarrow{T_2}}(w)} \right),$$

where the function $\theta_{\overrightarrow{T}}(\cdot, \cdot)$ is defined in (4). Using the linearity of expectations and the $8k$-wise independence of the random variables $X_c(w)$ and $Y(w)$, we obtain

$$\mathbb{E}\left[Z_H(G)\overline{Z_H(G)}\right] = \sum_{\substack{\overrightarrow{T_1} = (\overrightarrow{e_1}, \cdots, \overrightarrow{e_k}) \\ \overrightarrow{T_2} = (\overrightarrow{e_1'}, \cdots, \overrightarrow{e_k'}) \\ e_i, e_i' \in E[G]}} Q_{\overrightarrow{T_1}, \overrightarrow{T_2}},$$

where

$$Q_{\overrightarrow{T_1},\overrightarrow{T_2}} = \left(\prod_{c \in V[H]} \prod_{w \in V_{\overrightarrow{T_1}} \cup V_{\overrightarrow{T_2}}} \mathbb{E}\left[X_c(w)^{\theta_{\overrightarrow{T_1}}(c,w)} \overline{X_c(w)}^{\theta_{\overrightarrow{T_2}}(c,w)} \right] \right) \cdot$$
$$\left(\prod_{w \in V_{\overrightarrow{T_1}} \cup V_{\overrightarrow{T_2}}} \mathbb{E}\left[Y(w)^{\deg_{\overrightarrow{T_1}}(w)} \overline{Y(w)}^{\deg_{\overrightarrow{T_2}}(w)} \right] \right) .$$

For any $c \in V[H]$ and $w \in V_{\overrightarrow{T_1}} \cup V_{\overrightarrow{T_2}}$, we write

$$R_{\overrightarrow{T_1},\overrightarrow{T_2}}(c,w) = \mathbb{E}\left[X_c(w)^{\theta_{\overrightarrow{T_1}}(c,w)} \overline{X_c(w)}^{\theta_{\overrightarrow{T_2}}(c,w)} \right].$$

Let $R_{\overrightarrow{T_1},\overrightarrow{T_2}} = \prod_{c \in V[H]} \prod_{w \in V_{\overrightarrow{T_1}} \cup V_{\overrightarrow{T_2}}} R_{\overrightarrow{T_1},\overrightarrow{T_2}}(c,w)$. Then

$$Q_{\overrightarrow{T_1},\overrightarrow{T_2}} = R_{\overrightarrow{T_1},\overrightarrow{T_2}} \cdot \prod_{w \in V_{\overrightarrow{T_1}} \cup V_{\overrightarrow{T_2}}} \mathbb{E}\left[Y(w)^{\deg_{\overrightarrow{T_1}}(w)} \overline{Y(w)}^{\deg_{\overrightarrow{T_2}}(w)} \right].$$

We claim that if the term $Q_{\overrightarrow{T_1},\overrightarrow{T_2}} \neq 0$, then every vertex in $V_{\overrightarrow{T_1}} \cup V_{\overrightarrow{T_2}}$ has even degree in the undirected sense. First, we show that using this claim we can finish the proof of the theorem. Note that $\mathbb{E}[Z_H(G)\overline{Z_H(G)}] = \sum_{G_{\overrightarrow{T_1},\overrightarrow{T_2}} \in \mathcal{E}_{2k}} Q_{\overrightarrow{T_1},\overrightarrow{T_2}}$ where $\mathcal{E}_{2k}$ is the set of directed subgraphs of G on $2k$ edges with every vertex having even degree in the undirected sense. Observing that the undirected graph defined by $G_{\overrightarrow{T_1},\overrightarrow{T_2}}$ is a Eulerian circuit, by Corollary 1 we get $\mathbb{E}[Z_H(G)\overline{Z_H(G)}] \leq \sum_{G_{\overrightarrow{T_1},\overrightarrow{T_2}} \in \mathcal{E}_{2k}} |Q_{\overrightarrow{T_1},\overrightarrow{T_2}}| \leq c \cdot m^k$. Note that an upper bound for the constant c is $\max_{G_{\overrightarrow{T_1},\overrightarrow{T_2}} \in \mathcal{E}_{2k}} |Q_{\overrightarrow{T_1},\overrightarrow{T_2}}|$.

Let us now prove that $Q_{\overrightarrow{T_1},\overrightarrow{T_2}} \neq 0$ implies that every vertex in $V_{\overrightarrow{T_1}} \cup V_{\overrightarrow{T_2}}$ has even degree in the undirected sense. We first make the following observations: For any vertex c of C_k and w in $V_{\overrightarrow{T_1}} \cup V_{\overrightarrow{T_2}}$ we have: $\mathbb{E}\left[X_c^i(w) \right] \neq 0$ if and only if $i = 2$. After expanding $Z_H(G)$ and $\overline{Z_H(G)}$, $X_c(\cdot), c \in V[H]$ appears twice in each term, so we have $\sum_{w \in V_{\overrightarrow{T_1}} \cup V_{\overrightarrow{T_2}}} \theta_{\overrightarrow{T_1}}(c,w) + \theta_{\overrightarrow{T_2}}(c,w) = 4$. Consider a subgraph $G_{\overrightarrow{T_1},\overrightarrow{T_2}}$ on $2k$ edges such that $R_{\overrightarrow{T_1},\overrightarrow{T_2}} \neq 0$. Assume for the sake of contradiction that $G_{\overrightarrow{T_1},\overrightarrow{T_2}}$ has a vertex w of odd degree. This implies that there is a vertex $c \in C_k$ such that $\theta_{\overrightarrow{T_1}}(c,w) + \theta_{\overrightarrow{T_2}}(c,w)$ is either one or three. However $\theta_{\overrightarrow{T_1}}(c,w) + \theta_{\overrightarrow{T_2}}(c,w)$ cannot be one since in this case both $R_{\overrightarrow{T_1},\overrightarrow{T_2}}$ and $Q_{\overrightarrow{T_1},\overrightarrow{T_2}}$ must vanish. Now consider the case where $\theta_{\overrightarrow{T_1}}(c,w) + \theta_{\overrightarrow{T_2}}(c,w) = 3$. This means that $R_{\overrightarrow{T_1},\overrightarrow{T_2}}(c,w)$ is either $\mathbb{E}[X_c^2(w)\overline{X_c(w)}]$ or the symmetric variant $\mathbb{E}\left[X_c(w)\overline{X_c(w)}^2 \right]$. Assume that $R_{\overrightarrow{T_1},\overrightarrow{T_2}}(c,w) = \mathbb{E}[X_c^2(w)\overline{X_c(w)}]$. Since $\sum_{w \in V_{\overrightarrow{T_1}} \cup V_{\overrightarrow{T_2}}} \theta_{\overrightarrow{T_1}}(c,w) + \theta_{\overrightarrow{T_2}}(c,w) = 4$, there must be a vertex $w' \neq w$ in $V_{\overrightarrow{T_1}} \cup V_{\overrightarrow{T_2}}$ such that $R_{\overrightarrow{T_1},\overrightarrow{T_2}}(c,w') = \mathbb{E}[\overline{X_c(w')}]$. This implies that $R_{\overrightarrow{T_1},\overrightarrow{T_2}}$ vanishes and hence $Q_{\overrightarrow{T_1},\overrightarrow{T_2}}$ must also vanish, which leads to a contradiction. □

Now we prove Theorem 1.

Proof (of Theorem 1). First, observe that

$$\frac{\mathbb{E}[Z_H(G)\overline{Z_H(G)}] - \mathbb{E}^2[Z_H(G)]}{|\mathbb{E}[Z_H(G)]|^2} \leq \frac{\mathbb{E}[Z_H(G)\overline{Z_H(G)}]}{|\mathbb{E}[Z_H(G)]|^2}.$$

We run s parallel and independent copies of our estimator, and take the average value $Z^* = \frac{1}{s}\sum_{i=1}^{s} Z_i$, where each Z_i is the output of the i-th instance of the estimator. Therefore $\mathbb{E}[Z^*] = \mathbb{E}[Z_H(G)]$ and

$$\mathbb{E}[Z^*\overline{Z}^*] - |\mathbb{E}[Z^*]|^2 = \frac{1}{s}\left(\mathbb{E}[Z_H(G)\overline{Z_H(G)}] - |\mathbb{E}[Z_H(G)]|^2\right).$$

By Chebyshev's inequality (Lemma 3), we have

$$\Pr\left[|Z^* - \mathbb{E}[Z^*]| \geq \varepsilon \cdot |\mathbb{E}[Z^*]|\right] \leq \frac{\mathbb{E}[Z_H(G)\overline{Z_H(G)}] - \mathbb{E}[Z_H(G)]\overline{\mathbb{E}[Z_H(G)]}}{s \cdot \varepsilon^2 \cdot |\mathbb{E}[Z_H(G)]|^2}.$$

Observe that

$$\mathbb{E}[Z_H(G)\overline{Z_H(G)}] - \mathbb{E}[Z_H(G)]\overline{\mathbb{E}[Z_H(G)]} \leq \mathbb{E}[Z_H(G)\overline{Z_H(G)}] = O(m^k).$$

By choosing $s = O\left(\frac{1}{\varepsilon^2} \cdot \frac{m^k}{(\#C_k)^2}\right)$, we get $\Pr\left[|Z^* - \mathbb{E}[Z^*]| \geq \varepsilon \cdot |\mathbb{E}[Z^*]|\right] \leq 1/3$.

The probability of success can be amplified to $1 - \delta$ by running in parallel $O\left(\log\frac{1}{\delta}\right)$ copies of the algorithm and outputting the median of those values.

Since storing each random variable requires $O(\log n)$ space and the number of random variables used in each trial is $O(1)$, so the overall space complexity is as claimed.   □

5 Conclusions

In this paper we presented an unbiased estimator for counting the number of occurrences of any d-regular graph H in a graph G. For the special case $d = 2$, we proved that the variance of the computed random variables is not too big, thus obtaining an efficient algorithm for computing approximate estimates for the quantities in question. Our work raises a number of challenging open questions.

1. Is it possible to generalize the proposed approach to count other subgraphs, such as for example general cliques? Our results provide an unbiased estimator. However, is there any way of keeping the variance of the underlying random variables small?
2. We used complex-valued hash functions to achieve the desired result. However, there might be other possibilities. Can we use hash functions that take values from other structures, such as Clifford algebras, to obtain better upper bounds for the space complexity of the algorithm?
3. Our algorithm improves significantly upon the naïve sampling algorithms. Unfortunately, it is not clear at all what the optimal memory consumption of an algorithm is. So another fundamental research direction is to obtain lower bounds for counting subgraphs in the turnstile model.

Acknowledgement. The authors would like to thank Divya Gupta for helping them with the implementation of an earlier version of the algorithm. Madhusudan Madhusudan thanks Girish Varma for stimulating discussions.

References

1. Bar-Yossef, Z., Kumar, R., Sivakumar, D.: Reductions in streaming algorithms, with an application to counting triangles in graphs. In: Proceedings of the 13th Annual ACM-SIAM Symposium on Discrete Algorithms, pp. 623–632 (2002)
2. Becchetti, L., Boldi, P., Castillo, C., Gionis, A.: Efficient semi-streaming algorithms for local triangle counting in massive graphs. In: Proceedings of the 14th ACM SIGKDD International Conference on Knowledge Discovery and Data Mining, pp. 16–24 (2008)
3. Bordino, I., Donato, D., Gionis, A., Leonardi, S.: Mining large networks with subgraph counting. In: Proceedings of the 8th IEEE International Conference on Data Mining, pp. 737–742 (2008)
4. Buriol, L.S., Frahling, G., Leonardi, S., Marchetti-Spaccamela, A., Sohler, C.: Counting triangles in data streams. In: Proceedings of the 25th ACM SIGACT-SIGMOD-SIGART Symposium on Principles of Database Systems, pp. 253–262 (2006)
5. Buriol, L.S., Frahling, G., Leonardi, S., Sohler, C.: Estimating clustering indexes in data streams. In: Arge, L., Hoffmann, M., Welzl, E. (eds.) ESA 2007. LNCS, vol. 4698, pp. 618–632. Springer, Heidelberg (2007)
6. Chien, S., Rasmussen, L.E., Sinclair, A.: Clifford algebras and approximating the permanent. Journal of Computer and System Sciences 67(2), 263–290 (2003)
7. Flum, J., Grohe, M.: The parameterized complexity of counting problems. SIAM Journal on Computing 33(4), 892–922 (2004)
8. Ganguly, S.: Estimating frequency moments of data streams using random linear combinations. In: Jansen, K., Khanna, S., Rolim, J.D.P., Ron, D. (eds.) RANDOM 2004 and APPROX 2004. LNCS, vol. 3122, pp. 369–380. Springer, Heidelberg (2004)
9. Jowhari, H., Ghodsi, M.: New streaming algorithms for counting triangles in graphs. In: Wang, L. (ed.) COCOON 2005. LNCS, vol. 3595, pp. 710–716. Springer, Heidelberg (2005)
10. Karmarkar, N., Karp, R., Lipton, R., Lovasz, L., Luby, M.: A Monte-Carlo algorithm for estimating the permanent. SICOMP: SIAM Journal on Computing 22, 284–293 (1993)
11. McGregor, A.: Open Problems in Data Streams and Related Topics. In: IITK Workshop on Algoriths For Data Sreams (2006),
http://www.cse.iitk.ac.in/users/sganguly/data-stream-probs.pdf
12. Muthukrishnan, S.: Data Streams: Algorithms and Applications. Foundations and Trends in Theoretical Computer Science 1(2) (2005)

Algorithms for Solving Rubik's Cubes

Erik D. Demaine[1], Martin L. Demaine[1], Sarah Eisenstat[1],
Anna Lubiw[2], and Andrew Winslow[3]

[1] MIT Computer Science and Artificial Intelligence Laboratory,
Cambridge, MA 02139, USA
{edemaine,mdemaine,seisenst}@mit.edu

[2] David R. Cheriton School of Computer Science,
University of Waterloo, Waterloo, Ontario N2L 3G1, Canada
alubiw@uwaterloo.ca

[3] Department of Computer Science, Tufts University,
Medford, MA 02155, USA
awinslow@cs.tufts.edu

Abstract. The Rubik's Cube is perhaps the world's most famous and iconic puzzle, well-known to have a rich underlying mathematical structure (group theory). In this paper, we show that the Rubik's Cube also has a rich underlying algorithmic structure. Specifically, we show that the $n \times n \times n$ Rubik's Cube, as well as the $n \times n \times 1$ variant, has a "God's Number" (diameter of the configuration space) of $\Theta(n^2/\log n)$. The upper bound comes from effectively parallelizing standard $\Theta(n^2)$ solution algorithms, while the lower bound follows from a counting argument. The upper bound gives an asymptotically optimal algorithm for solving a general Rubik's Cube in the worst case. Given a specific starting state, we show how to find the shortest solution in an $n \times O(1) \times O(1)$ Rubik's Cube. Finally, we show that finding this optimal solution becomes NP-hard in an $n \times n \times 1$ Rubik's Cube when the positions and colors of some cubies are ignored (not used in determining whether the cube is solved).

Keywords: combinatorial puzzles, diameter, God's number, combinatorial optimization.

1   Introduction

A little over thirty years ago, Hungarian architecture professor Ernő Rubik released his "Magic Cube" to the world.[1] What we now all know as the *Rubik's Cube* quickly became a sensation [26]. It is the best-selling puzzle ever, at over 350 million units [15]. It is a tribute to elegant design, being part of the permanent collection of the Museum of Modern Art in New York [18]. It is an icon for difficult puzzles—an intellectual Mount Everest. It is the heart of World Cube Association's speed-cubing competitions, whose current record holders can solve a cube in under 7 seconds (or 31 seconds blindfold) [1]. It is the basis for cube

---

[1] Similar puzzles were invented around the same time in the United States [9][17], the United Kingdom [6], and Japan [10] but did not reach the same level of success.

C. Demetrescu and M.M. Halldórsson (Eds.): ESA 2011, LNCS 6942, pp. 689–700, 2011.
© Springer-Verlag Berlin Heidelberg 2011

art, a form of pop art made from many carefully unsolved Rubik's Cubes. (For example, the recent movie *Exit Through the Gift Shop* features the street cube artist known as Space Invader.) It is the bane of many computers, which spent about 35 CPU years determining in 2010 that the best algorithm to solve the worst configuration requires exactly 20 moves—referred to as *God's Number* [22].

To a mathematician, or a student taking abstract algebra, the Rubik's Cube is a shining example of group theory. The configurations of the Rubik's Cube, or equivalently the transformations from one configuration to another, form a subgroup of a permutation group, generated by the basic twist moves. This perspective makes it easier to prove (and compute) that the configuration space falls into two connected components, according to the parity of the permutation on the cubies (the individual subcubes that make up the puzzle). See [7] for how to compute the number of elements in the group generated by the basic Rubik's Cube moves (or any set of permutations) in polynomial time.

To a theoretical computer scientist, the Rubik's Cube and its many generalizations suggest several natural open problems. What are good algorithms for solving a given Rubik's Cube puzzle? What is an optimal worst-case bound on the number of moves? What is the complexity of optimizing the number of moves required for a given starting configuration? Although God's Number is known to be 20 for the $3 \times 3 \times 3$, the optimal solution of each configuration in this constant-size puzzle still has not been computed [22]. While computing the exact behavior for larger cubes is out of the question, how does the worst-case number of moves and complexity scale with the side lengths of the cube? In parallel with our work, these questions were recently posed by Andy Drucker and Jeff Erickson [4]. Scalability is important given the commercially available $4 \times 4 \times 4$ Rubik's Revenge [25]; $5 \times 5 \times 5$ Professor's Cube [13]; the $6 \times 6 \times 6$ and $7 \times 7 \times 7$ V-CUBEs [27]; Leslie Le's 3D-printed $12 \times 12 \times 12$ [14]; and Oskar van Deventer's $17 \times 17 \times 17$ Over the Top and his $2 \times 2 \times 20$ Overlap Cube, both available from 3D printer *shapeways* [28].

Diameter / God's Number. The diameter of the configuration space of a Rubik's Cube seems difficult to capture using just group theory. In general, a set of permutations (moves) can generate a group with superpolynomial diameter [3]. If we restrict each generator (move) to manipulate only k elements, then the diameter is $O(n^k)$ [16], but this gives very weak (superexponential) upper bounds for $n \times n \times n$ and $n \times n \times 1$ Rubik's Cubes.

Fortunately, we confirm that the general approach taken by folk algorithms for solving Rubik's Cubes of various fixed sizes can be generalized to perform a constant number of moves per cubie, for an upper bound of $O(n^2)$. This result is essentially standard, but we take care to ensure that all cases can be handled.

Surprisingly, this bound is not optimal. Each twist move in the $n \times n \times n$ and $n \times n \times 1$ Rubik's Cubes simultaneously transforms $n^{\Theta(1)}$ cubies (with the exponent depending on the dimensions and whether a move transforms a plane or a half-space). This property offers a form of parallelism for solving multiple cubies at once, to the extent that multiple cubies want the same move to be applied at a particular time. We show that this parallelism can be exploited to reduce the number of moves by a logarithmic factor, to $O(n^2/\log n)$. Furthermore, an easy counting argument shows an average-case lower bound of $\Omega(n^2/\log n)$.

Thus we settle the diameter of the $n \times n \times n$ and $n \times n \times 1$ Rubik's Cubes, up to constant factors. These results are described in Sections 4 and 3, respectively.

n^2-1 *puzzle.* Another puzzle that can be described as a permutation group given by generators corresponding to valid moves is the $n \times n$ generalization of the classic Fifteen Puzzle. This $n^2 - 1$ *puzzle* also has polynomial diameter, though without any form of parallelism, the diameter is simply $\Theta(n^3)$ [20]. Interestingly, computing the shortest solution from a given configuration of the puzzle is NP-hard [21]. More generally, given a set of generator permutations, it is PSPACE-complete to find the shortest sequence of generators whose product is a given target permutation [5,11]. These papers mention the Rubik's Cube as motivation, but neither addresses the natural question: is it NP-hard to solve a given $n \times n \times n$ or $n \times n \times 1$ Rubik's Cube using the fewest possible moves? Although the $n \times n \times n$ problem was posed as early as 1984 [2,21], both questions remain open [12]. We give partial progress toward hardness, as well as a polynomial-time exact algorithm for a particular generalization of the Rubik's Cube.

Optimization algorithms. We give one positive and one negative result about finding the shortest solution from a given configuration of a generalized Rubik's Cube puzzle. On the positive side, we show in Section 6 how to compute the exact optimum for $n \times O(1) \times O(1)$ Rubik's Cubes. Essentially, we prove structural results about how an optimal solution decomposes into moves in the long dimension and the two short dimensions, and use this structure to obtain an algorithm. This result may prove useful for optimally solving configurations of Oskar van Deventer's $2 \times 2 \times 20$ Overlap Cube [28], but it does not apply to the $3 \times 3 \times 3$ Rubik's Cube because we need n to be distinct from the other two side lengths. On the negative side, we prove in Section 5 that it is NP-hard to find an optimal solution to a subset of cubies in an $n \times n \times 1$ Rubik's Cube. Phrased differently, optimally solving a given $n \times n \times 1$ Rubik's Cube configuration is NP-hard when the colors and positions of some cubies are ignored (i.e., they are not considered in determining whether the cube is solved).

2   Common Definitions

We begin with some terminology. An $\ell \times m \times n$ Rubik's Cube is composed of $\ell m n$ cubies, each of which has some position (x, y, z), where $x \in \{0, 1, \ldots, \ell-1\}$, $y \in \{0, 1, \ldots, m-1\}$, and $z \in \{0, 1, \ldots, n-1\}$. Each cubie also has an orientation. Each cubie in a Rubik's Cube has a color on each visible face. There are six colors in total. We say that a Rubik's Cube is *solved* when each face of the cube is the same color, unique for each face.

An *edge cubie* is any cubie which has at least two visible faces which point in perpendicular directions. A *corner cubie* is any cubie which has at least three visible faces which all point in perpendicular directions.

A *slice* of a Rubik's Cube is a set of cubies that match in one coordinate (e.g. all of the cubies such that $y = 1$). A legal move on a Rubik's Cube involves rotating one slice around its perpendicular[2]. To preserve the shape of the cube, there are restrictions on how much the slice can be rotated. If the slice to be rotated is a square, then the slice can be rotated $90°$ in either direction. Otherwise, the slice can only be rotated by $180°$. Finally, note that if one dimension of the cube has length 1, we disallow rotations of the only slice in that dimension. For example, we cannot rotate the slice $z = 0$ in the $n \times n \times 1$ cube.

A *configuration* of a Rubik's Cube is a mapping from each visible face of each cubie to a color. A *reachable configuration* of a Rubik's Cube is a configuration which can be reached from a solved Rubik's Cube via a sequence of legal moves.

For each of the Rubik's Cube variants we consider, we will define the contents of a *cubie cluster*. The cubies which belong in this cubie cluster depend on the problem we are working on; however, they do share some key properties:

1. Each cubie cluster consists of a constant number of cubies.
2. No sequence of legal moves can cause any cubie to move from one cubie cluster into another.

Each cubie cluster has a *cluster configuration* mapping from each visible face of the cubie cluster to its color. Because the number of cubies in a cubie cluster is constant, the number of possible cluster configurations is also constant.

We say that a move *affects* a cubie cluster if the move causes at least one cubie in the cubie cluster to change places. Similarly, we say that a sequence of moves affects a cubie cluster if at least one cubie in the cubie cluster changes position or orientation after the sequence of moves has been performed.

3   Diameter of $n \times n \times 1$ Rubik's Cube

When considering an $n \times n \times 1$ Rubik's Cube we omit the third coordinate of a cubie, which by necessity must be 0. For simplicity, we restrict the set of solutions to those configurations where the top of the cube is orange. We also assume that n is even, and ignore the edge and corner cubies. A more rigorous proof, which handles these details, is available in the full version of this paper.[3]

Consider the set of locations reachable by a cubie at position (x, y). If we flip column x, the cubie will move to position $(x, n - y - 1)$. If we instead flip row y, it will move to position $(n - x - 1, y)$. Hence, there are at most four reachable locations for a cubie that starts at (x, y): (x, y), $(x, n - y - 1)$, $(n - x - 1, y)$, and $(n - x - 1, n - y - 1)$. We call this set of locations the *cubie cluster* (x, y).

We begin by showing that for any reachable cluster configuration, there exists a sequence of moves of constant length which can be used to solve that cluster without affecting any other clusters. Figure 1 gives just such a sequence for each potential cluster configuration.

---

[2] While other definitions of a legal move exist (e.g. rotating a set of contiguous parallel slices), this definition most closely matches the one used in popular move notations.

[3] http://arxiv.org/abs/1106.5736

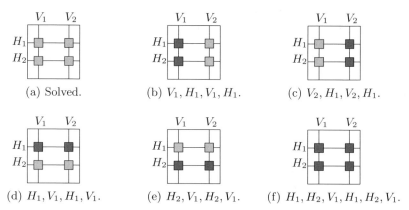

(a) Solved.   (b) V_1, H_1, V_1, H_1.   (c) V_2, H_1, V_2, H_1.

(d) H_1, V_1, H_1, V_1.   (e) H_2, V_1, H_2, V_1.   (f) $H_1, H_2, V_1, H_1, H_2, V_1$.

Fig. 1. The reachable cluster configurations and the move sequences to solve them

In the remainder of Section 3, we use the notation H_1, H_2 and V_1, V_2 to denote the two rows and columns containing cubies from a single cubie cluster. We also use the same symbols to denote single moves affecting these rows and columns. In the special cases of cross and center cubie clusters, we denote the single row or column containing the cluster by H or V, respectively.

3.1 $n \times n \times 1$ Upper Bound

There are n^2 clusters in the $n \times n \times 1$ Rubik's Cube. If we use the move sequences given in Fig. 1 to solve each cluster individually, we have a sequence of $O(n^2)$ moves for solving the entire cube. In this section, we take this sequence of moves and take advantage of parallelism to get a solution with $O(n^2/\log n)$ moves.

Say that we are given columns X and rows Y such that all of the clusters $(x, y) \in X \times Y$ are in the cluster configuration depicted in Fig. 1(b). If we solve each of these clusters individually, the number of moves required is $\Theta(|X| \cdot |Y|)$.

Consider instead what would happen if we first flipped all of the columns $x \in X$, then flipped all of the rows $y \in Y$, then flipped all of the columns $x \in X$ again, and finally flipped all of the rows $y \in Y$ again. What would be the effect of this move sequence on a particular $(x^*, y^*) \in X \times Y$? The only moves affecting that cluster are the column moves x^* and $(n - 1 - x^*)$ and the row moves y^* and $(n - 1 - y^*)$. So the subsequence of moves affecting (x^*, y^*) would consist of the column move x^*, followed by the row move y^*, followed by the column move x^* again, and finally the row move y^* again. Those four moves are exactly the moves needed to solve that cluster.

A generalization of this idea gives us a technique for solving all cubie clusters $(x, y) \in X \times Y$ using only $O(|X| + |Y|)$ moves, if each one of those clusters is in the same configuration. Our goal is to use this technique for a related problem: solving all of the cubie clusters $(x, y) \in X \times Y$ that are in a particular cluster configuration c, leaving the rest of the clusters alone.

For each $y \in Y$, we define $S_y = \{x \in X \mid \text{cluster } (x, y) \text{ is in configuration } c\}$. For each $S \subseteq X$, we define $Y_S = \{y \in Y \mid S_y = S\}$. For each of the $2^{|X|}$ values of

S, we use a single sequence of moves to solve all $(x, y) \in S \times Y_S$. This sequence of moves has length $O(|S| + |Y_S|) = O(|X| + |Y_S|)$. When we sum the lengths up for all Y_S, we find that the number of moves is bounded by

$$O\left(\sum_S (|X| + |Y_S|)\right) = O\left(|X| \cdot 2^{|X|} + \sum_S |Y_S|\right) = O\left(|X| \cdot 2^{|X|} + |Y|\right).$$

To make this technique cost-effective, we partition all $\lfloor n/2 \rfloor$ columns into sets of size $\frac{1}{2} \log n$, and solve each such group individually. This means that we can solve all clusters in a particular configuration c using

$$O\left(\frac{\frac{n}{2}}{\frac{1}{2}\log n} \cdot \left(\frac{1}{2}\log n \cdot 2^{\frac{1}{2}\log n} + \frac{n}{2}\right)\right) = O\left(\frac{n^2}{\log n}\right).$$

moves. When we construct that move sequence for all 6 cluster configurations, we have the following result:

Theorem 1. *Given an $n \times n \times 1$ Rubik's Cube configuration, all cubie clusters can be solved in $O(n^2/\log n)$ moves.*

3.2 $n \times n \times 1$ Lower Bound

Using calculations involving the maximum degree of the graph of the configuration space and the total number of reachable configurations, we have the matching lower bound:

Theorem 2. *Some configurations of an $n \times n \times 1$ Rubik's Cube are $\Omega(n^2/\log n)$ moves away from being solved.*

Omitted proofs may be found in the full version of this paper.

4   Diameter of $n \times n \times n$ Rubik's Cube

For simplicity, we again assume that n is even and ignore all edge and corner cubies. A more rigorous proof, which handles these details, is available in the full version of this paper.

Because the only visible cubies on the $n \times n \times n$ Rubik's Cube are on the surface, we use an alternative coordinate system. Each cubie has a face coordinate $(x, y) \in \{0, 1, \ldots, n-1\} \times \{0, 1, \ldots, n-1\}$. Consider the set of reachable locations for a cubie on the front face with coordinates (x, y). A face rotation of the front face will let it reach the coordinates $(n - y - 1, x)$, $(n - x - 1, n - y - 1)$, and $(y, n - x - 1)$ on the front face. Row or column moves will allow the cubie to move to another face, where it still has to have one of those four coordinates. Hence, it can reach 24 locations in total. We define the *cubie cluster* (x, y) to be those 24 positions that are reachable by the cubie (x, y).

Just as in the case of the $n \times n \times 1$ cube, our goal is to prove that for each cluster configuration, there is a sequence of $O(1)$ moves that can be used to solve the cluster, while not affecting any other clusters. For the $n \times n \times 1$ cube, we wrote these solution sequences using the symbols H_1, H_2, V_1, V_2 to represent a general class of moves, each of which could be mapped to a specific move once the cubie cluster coordinates were known. Here we introduce more formal notation.

Because of the coordinate system we are using, we distinguish two types of legal moves. *Face moves* involve taking a single face and rotating it 90° in either direction. *Row or column moves* involve taking a slice of the cube (not one of its faces) and rotating the cubies in that slice by 90° in either direction. Face moves come in twelve types, two for each face. For our purposes, we will add a thirteenth type which applies the identity function. If a is the type of face move, we write F_a to denote the move itself. Given a particular index $i \in \{1, 2, \ldots, \lfloor n/2 \rfloor - 1\}$, there are twelve types of row and column moves that can be performed — three different axes for the slice, two different indices (i and $n - i - 1$) to pick from, and two directions of rotation. Again, we add a thirteenth type which applies the identity function. If a is the type of row or column move, and i is the index, then we write $RC_{a,i}$ to denote the move itself.

A *cluster move sequence* consists of three type sequences: face types $a_1, \ldots, a_\ell$, row and column types $b_1, \ldots, b_\ell$, and row and column types $c_1, \ldots, c_\ell$. For a cluster (x, y), the sequence of actual moves produced by the cluster move sequence is $F_{a_1}, RC_{b_1,x}, RC_{c_1,y}, \ldots, F_{a_\ell}, RC_{b_\ell,x}, RC_{c_\ell,y}$. A *cluster move solution* for a cluster configuration d is a cluster move sequence with the following properties:

1. For any $(x, y) \in \{1, 2, \ldots, \lfloor n/2 \rfloor - 1\} \times \{1, 2, \ldots, \lfloor n/2 \rfloor - 1\}$, if cluster (x, y) is in configuration d, then it can be solved using the sequence of moves $F_{a_1}, RC_{b_1,x}, RC_{c_1,y}, \ldots, F_{a_\ell}, RC_{b_\ell,x}, RC_{c_\ell,y}$.
2. The move sequence $F_{a_1}, RC_{b_1,x}, RC_{c_1,y}, \ldots, F_{a_\ell}, RC_{b_\ell,x}, RC_{c_\ell,y}$ does not affect cubie cluster (y, x).
3. All three of the following sequences of moves do not affect the configuration of any cubie clusters:

$$F_{a_1}, RC_{b_1,x}, F_{a_2}, RC_{b_1,x}, \ldots, F_{a_\ell}, RC_{b_\ell,x};$$
$$F_{a_1}, RC_{c_1,y}, F_{a_2}, RC_{c_1,y}, \ldots, F_{a_\ell}, RC_{c_\ell,y};$$
$$F_{a_1}, F_{a_2}, \ldots, F_{a_\ell}.$$

Our goal is to construct a cluster move solution for each possible cluster configuration, and then use those solutions to solve multiple cubie clusters in parallel.

In the speed cubing community, there is a well-known technique for solving $n \times n \times n$ Rubik's Cubes in $O(n^2)$ moves, involving a family of constant-length cluster move sequences. These sequences are attributed to Ingo Schütze [24], but due to their popularity in the speed cubing community, their exact origins are unclear. These cluster move sequences can be combined to construct constant-length cluster move solutions for all possible cluster configurations, which is precisely what we wanted. A detailed explanation and proof of correctness for this method can be found in the full version of this paper.

4.1 $n \times n \times n$ Upper Bound

As in the $n \times n \times 1$ case, we wish to solve several clusters in parallel, so that the length of the solution is reduced from $O(n^2)$ to $O(n^2/\log n)$. Say we have a set of columns $X = \{x_1, \ldots, x_\ell\}$ and rows $Y = \{y_1, \ldots, y_k\}$ such that $X \cap Y = \emptyset$ and all cubie clusters $(x, y) \in X \times Y$ have the same cluster configuration d. Solving each cluster individually requires a total of $\Theta(|X| \cdot |Y|)$ moves.

Instead, we will attempt to parallelize. The cluster configuration d must have a constant-length cluster move solution with type sequences $a_1, \ldots, a_m, b_1, \ldots, b_m$, and $c_1, \ldots, c_m$. To construct a parallel move sequence, we use the following sequence of moves as a building block:

$$\text{BULK}_i = F_{a_i}, RC_{b_i, x_1}, RC_{b_i, x_2}, \ldots, RC_{b_i, x_\ell}, RC_{c_i, y_1}, RC_{c_i, y_2}, \ldots, RC_{c_i, y_k}.$$

The full sequence we use is $\text{BULK}_1, \text{BULK}_2, \ldots, \text{BULK}_m$. A careful case-by-case analysis using the properties of cluster move solutions reveals that this sequence of $O(|X| + |Y|)$ moves will solve all clusters $X \times Y$, and that the only other clusters it may affect are the clusters $X \times X$ and $Y \times Y$.

Now say that we are given a cluster configuration d and a set of columns X and rows Y such that $X \cap Y = \emptyset$. Using the same row-grouping technique that we used for the $n \times n \times 1$ case, it is possible to show that there exists a sequence of moves of length $O(|X| \cdot 2^{|X|} + |Y|)$ solving all of the clusters in $X \times Y$ which are in configuration d and limiting the set of other clusters affected to $(X \times X) \cup (Y \times Y)$. By dividing up X into groups of roughly size $\frac{1}{2} \log |Y|$, just as we did for the $n \times n \times 1$ cube, we may show that there exists a sequence of moves with the same properties, but with length $O(|X| \cdot |Y|/\log |Y|)$.

To finish constructing the move sequence for the entire Rubik's Cube, we must account for two differences between this case and the $n \times n \times 1$ case: the requirement that $X \cap Y = \emptyset$ and the potential to affect clusters in $(X \times X) \cup (Y \times Y)$. We handle both cases by taking the initial set of columns $\{1, 2, \ldots, \lfloor n/2 \rfloor - 1\}$ and dividing it into groups $X_1, \ldots, X_j$ of size $\sqrt{n/2}$. We partition the initial set of rows into sets $Y_1, \ldots, Y_j$ in a similar fashion. We then loop through pairs (X_i, Y_j), where $i \neq j$, to solve all clusters in configuration d for all but the clusters $(X_1 \times Y_1) \cup \ldots \cup (X_j \times Y_j)$. Because $j = |X_i| = |Y_i| = \sqrt{n/2}$, the total number of moves required for this step is $O(n^2/\log n)$. To solve the clusters $(X_1 \times Y_1) \cup \ldots \cup (X_j \times Y_j)$, we simply solve each cluster individually, which requires a total of $O(n^{3/2}) < O(n^2/\log n)$ moves. If we add up that cost for each of the $O(1)$ different configurations, the total number of moves is $O(n^2/\log n)$.

Theorem 3. *Given an $n \times n \times n$ Rubik's Cube configuration, all cubie clusters can be solved in $O(n^2/\log n)$ moves.*

4.2 $n \times n \times n$ Lower Bound

Just as we did for the $n \times n \times 1$ lower bound, we can calculate a matching lower bound using the maximum degree of the graph of the configuration space and the total number of reachable configurations:

Theorem 4. *Some configurations of an $n \times n \times n$ Rubik's Cube are $\Omega(n^2 / \log n)$ moves away from being solved.*

5 Optimally Solving a Subset of the $n \times n \times 1$ Rubik's Cube is NP-Hard

In this section, we consider a generalization of the problem of computing the optimal sequence of moves to solve a Rubik's Cube. Say that we are given a configuration of an $n \times n \times 1$ Rubik's Cube and a list of *important* cubies. We wish to find the shortest sequence of moves that solves the important cubies. Note that the solution for the important cubies may cause other cubies to leave the solved state, so this problem is only equivalent to solving an $n \times n \times 1$ Rubik's Cube when all cubies are marked important.

In this section, we prove the NP-hardness of computing the length of this shortest sequence. More precisely, we prove that the following decision problem is NP-hard: is there a sequence of k moves that solves the important cubies of the $n \times n \times 1$ Rubik's Cube? Our reduction ensures that the cubies within a single cluster are either all important or all unimportant, and thus it does not matter whether we aim to solve cubies (which move) or specific cubie positions (which do not move). Therefore the problem remains NP-hard if we aim to solve the puzzle in the sense of unifying the side colors, when we ignore the colors of all unimportant cubies.

Certain properties of the Rubik's Cube configuration can affect the set of potential solutions. For the rest of this section, we will consider only Rubik's Cubes where n is odd and where all edge cubies and cross cubies are both solved and marked important. This restriction ensures that for any cluster, the number of horizontal moves and vertical moves affecting it must both be even. In addition, we will only consider Rubik's Cubes in which all cubie clusters are in the cluster configurations depicted in Figures 1(a), 1(b), and 1(d). This restriction means that the puzzle can always be solved using moves only of types H_1 and V_1. This combination of restrictions ensures that each unsolved cluster must be affected by both vertical and horizontal moves.

Suppose that we are given a configuration and a list of important cubies. Let u_r be the number of rows of index $\leq \lfloor n/2 \rfloor$ that contain at least one important unsolved cubie. Let u_c be the number of columns of index $\leq \lfloor n/2 \rfloor$ that contain at least one important unsolved cubie. Then we say that the *ideal number of moves* for solving the given configuration is $2(u_r + u_c)$. In other words, the ideal number of moves is equal to the smallest possible number of moves that could solve all the important cubies. An *ideal solution* for a subset of the cubies in a particular $n \times n \times 1$ puzzle is a solution for that set of cubies which uses the ideal number of moves. For the types of configurations that we are considering, the ideal solution will contain exactly two of each move, and the only moves that occur will be moves of type H_1 or V_1.

Definition 1. *Let $I_k(m)$ denote the index in the solution of the kth occurrence of move m.*

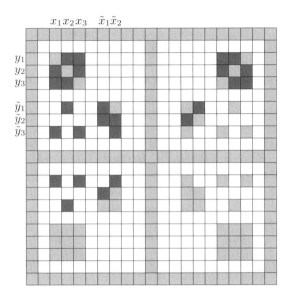

$x_1 x_2 x_3 \quad \tilde{x}_1 \tilde{x}_2$

Fig. 2. A sample of the betweenness gadget from Lemma 1. Important cubies are orange (solved) and blue (unsolved). Unimportant cubies are white. Any ideal solution must either have $I_1(x_1) < I_1(x_2) < I_1(x_3)$ or $I_1(x_3) < I_1(x_2) < I_1(x_1)$.

For our hardness reduction, we develop a gadget (depicted in Fig. 2) which forces a betweenness constraint on the ordering of three different row moves:

Lemma 1. *Given three columns $x_1, x_2, x_3 \leq \lfloor n/2 \rfloor$, there is a gadget using six extra rows and two extra columns ensuring that $I_1(x_2)$ lies between $I_1(x_1)$ and $I_1(x_3)$. This gadget also forces $I_2(x_2) < I_2(x_1)$, $I_2(x_2) < I_2(x_3)$, and*

$$\max_{x \in \{x_1, x_2, x_3\}} I_1(x) < \min_{x \in \{x_1, x_2, x_3\}} I_2(x).$$

The *betweenness problem* is a known NP-hard problem [8,19]. In this problem, we are given a set of triples (a, b, c), and wish to find an ordering on all items such that, for each triple, either $a < b < c$ or $c < b < a$. In other words, for each triple, b should lie between a and c in the overall ordering. Lemma 1 gives us a gadget which would at first seem to be perfectly suited to a reduction from the betweenness problem. However, because the lemma places additional restrictions on the order of all moves, we cannot reduce directly from betweenness.

Instead, we provide a reduction from another known NP-hard problem, Not-All-Equal 3-SAT [8,23]. In this problem, sometimes known as $\neq$-SAT, the input is a 3-CNF formula ϕ and the goal is to determine whether there exists an assignment to the variables of ϕ such that there is at least one true literal and one false literal in every clause. Our reduction from $\neq$-SAT to ideal Rubik solutions closely follows the reduction from hypergraph 2-coloring to betweenness [19].

Theorem 5. *Given a $\neq$-SAT instance ϕ, it is possible to compute in time polynomial in the size of ϕ an $n \times n \times 1$ configuration and a subset of the cubies that has an ideal solution if and only if ϕ has a solution, i.e., belongs to $\neq$-SAT.*

6   Optimally Solving an $O(1) \times O(1) \times n$ Rubik's Cube

For the $c_1 \times c_2 \times n$ Rubik's Cube with $c_1 \neq n \neq c_2$, the asymmetry of the puzzle leads to a few additional definitions. We call a slice *short* if the matching coordinate is z, and *long* otherwise. A *short move* involves rotating a short slice; a *long move* involves rotating a long slice. We define cubie cluster i to be the pair of slices $z = i$ and $z = (n-1) - i$. This definition means that any short move affects the position and orientation of cubies in exactly one cubie cluster.

Each short move affects exactly one cluster. Hence, in the optimal solution, the number of short moves affecting a particular cluster is at most the number of configurations of that cluster. Each cluster has $O(1)$ configurations, so in any optimal solution, any particular short move will be performed $O(1)$ times.

Any sequence of long moves corresponds to an arrangement of $c_1 c_2$ blocks of cubies with dimensions $1 \times 1 \times n$. We call each such arrangement a *long configuration*. There are a constant number of long configurations, so there must exist a *long move tour*: a constant-length sequence of long moves which passes through every long configuration before returning to the initial long configuration.

The effect of a short move depends only on the current long configuration. Hence, to solve a particular $c_1 \times c_2 \times n$ puzzle, it is sufficient to know a sequence of short moves for each cluster, annotated with the long configuration that each such move should be performed in. If we have such a sequence for each cluster, we may construct a full solution to the puzzle by repeatedly performing a long move tour, and inserting short moves into the appropriate places. Then we are guaranteed to be able to perform the kth short move for every cluster during the kth long move tour. Hence, the number of long move tours necessary is bounded by the maximum length of any short move sequence, which is $O(1)$ in any optimal solution. Therefore, any optimal solution contains $O(1)$ long moves.

This bound on the number of long moves allows us to construct an algorithm that does the following:

Theorem 6. *Given any $c_1 \times c_2 \times n$ Rubik's Cube configuration, it is possible to find the optimal solution in time polynomial in n.*

References

1. World Cube Association. Official results (2010),
   http://www.worldcubeassociation.org/results/
2. Cook, S.A.: Can computers routinely discover mathematical proofs? Proceedings of the American Philosophical Society 128(1), 40–43 (1984)
3. Driscoll, J.R., Furst, M.L.: On the diameter of permutation groups. In: Proceedings of the 15th Annual ACM Symposium on Theory of computing, pp. 152–160 (1983)
4. Drucker, A., Erickson, J.: Is optimally solving the $n \times n \times n$ Rubik's Cube NP-hard? Theoretical Computer Science — Stack Exchange post (August-September 2010), http://cstheory.stackexchange.com/questions/783/is-optimally -solving-the-nnn-rubiks-cube-np-hard
5. Even, S., Goldreich, O.: The minimum-length generator sequence problem is NP-hard. Journal of Algorithms 2(3), 311–313 (1981)

6. Fox, F.: Spherical 3x3x3. U.K. Patent 1,344,259 (January 1974)
7. Furst, M., Hopcroft, J., Luks, E.: Polynomial-time algorithms for permutation groups. In: Proceedings of the 21st Annual Symposium on Foundations of Computer Science, pp. 36–41 (1980)
8. Garey, M.R., Johnson, D.S.: Computers and Intractability: A Guide to the Theory of NP-Completeness, 1st edn. Series of Books in the Mathematical Sciences. W. H. Freeman & Co Ltd., New York (1979)
9. Gustafson, W.O.: Manipulatable toy. U.S. Patent 3,081,089 (March 1963)
10. Ishige, T.: Japan Patent 55-8192 (1976)
11. Jerrum, M.R.: The complexity of finding minimum-length generator sequences. Theoretical Computer Science 36(2-3), 265–289 (1985)
12. Kendall, G., Parkes, A., Spoerer, K.: A survey of NP-complete puzzles. International Computer Games Association Journal 31(1), 13–34 (2008)
13. Krell, U.: Three dimensional puzzle. U.S. Patent 4,600,199 (July 1986)
14. Le., L.: The world's first 12x12x12 cube. twistypuzzles.com forum post (November 2009), http://www.twistypuzzles.com/forum/viewtopic.php?f=15&t=15424
15. Seven Towns Ltd. 30 years on... and the Rubik's Cube is as popular as ever. Press brief (May 2010),
    `http://www.rubiks.com/i/company/media_library/pdf/Rubiks%20Cube%20to`
    `%20celebrate%2030th%20Anniversary%20in%20May%202010.pdf`
16. McKenzie, P.: Permutations of bounded degree generate groups of polynomial diameter. Information Processing Letters 19(5), 253–254 (1984)
17. Nichols, L.D.: Pattern forming puzzle and method with pieces rotatable in groups. U.S. Patent 3,655,201 (April 1972)
18. Museum of Modern Art. Rubik's cube,
    http://www.moma.org/collection/browse_results.php?object_id=2908
19. Opatrny, J.: Total ordering problem. SIAM Journal on Computing 8(1), 111–114 (1979)
20. Parberry, I.: A real-time algorithm for the $(n^2 - 1)$-puzzle. Information Processing Letters 56(1), 23–28 (1995)
21. Ratner, D., Warmuth, M.: The $(n^2 - 1)$-puzzle and related relocation problems. Journal of Symbolic Computation 10, 111–137 (1990)
22. Rokicki, T., Kociemba, H., Davidson, M., Dethridge, J.: God's number is 20 (2010), http://cube20.org
23. Schaefer, T.J.: The complexity of satisfiability problems. In: Proceedings of the 10th Annual ACM Symposium on Theory of Computing, San Diego, CA, pp. 216–226 (1978)
24. Schütze, I.: V-cubes Solutions, http://solutions.v-cubes.com/solutions2/
25. Sebesteny, P.: Puzzle-cube. U.S. Patent 4,421,311 (December 1983)
26. Slocum, J.: The Cube: The Ultimate Guide to the World's Bestselling Puzzle — Secrets, Stories, Solutions. Black Dog & Leventhal Publishers (March 2009)
27. V-CUBE. V-cube: the 21st century cube, http://www.v-cubes.com/
28. van Deventer, O.: Overlap cube 2x2x23. shapeways design,
    http://www.shapeways.com/model/96696/overlap_cube_2x2x23.html

Boundary Patrolling by Mobile Agents with Distinct Maximal Speeds

Jurek Czyzowicz[1], Leszek Gąsieniec[2], Adrian Kosowski[3],
and Evangelos Kranakis[4]

[1] Université du Québec en Outaouais, Gatineau, Québec J8X 3X7, Canada
[2] University of Liverpool, Liverpool L69 3BX, UK
[3] INRIA Bordeaux Sud-Ouest, LaBRI, 33405 Talence, France
[4] Carleton University, Ottawa, Ontario K1S 5B6, Canada

Abstract. A set of k mobile agents are placed on the boundary of a simply connected planar object represented by a cycle of unit length. Each agent has its own predefined maximal speed, and is capable of moving around this boundary without exceeding its maximal speed. The agents are required to protect the boundary from an intruder which attempts to penetrate to the interior of the object through a point of the boundary, unknown to the agents. The intruder needs some time interval of length τ to accomplish the intrusion. Will the intruder be able to penetrate into the object, or is there an algorithm allowing the agents to move perpetually along the boundary, so that no point of the boundary remains unprotected for a time period τ? Such a problem may be solved by designing an algorithm which defines the motion of agents so as to minimize the *idle time* I, i.e., the longest time interval during which any fixed boundary point remains unvisited by some agent, with the obvious goal of achieving $I < \tau$.

Depending on the type of the environment, this problem is known as either *boundary patrolling* or *fence patrolling* in the robotics literature. The most common heuristics adopted in the past include the *cyclic strategy*, where agents move in one direction around the cycle covering the environment, and the *partition strategy*, in which the environment is partitioned into sections patrolled separately by individual agents. This paper is, to our knowledge, the first study of the fundamental problem of boundary patrolling by agents with distinct maximal speeds. In this scenario, we give special attention to the performance of the cyclic strategy and the partition strategy. We propose general bounds and methods for analyzing these strategies, obtaining exact results for cases with 2, 3, and 4 agents. We show that there are cases when the cyclic strategy is optimal, cases when the partition strategy is optimal and, perhaps more surprisingly, novel, alternative methods have to be used to achieve optimality.

Keywords: mobile agents, boundary patrolling, fence patrolling, idleness.

C. Demetrescu and M.M. Halldórsson (Eds.): ESA 2011, LNCS 6942, pp. 701–712, 2011.
© Springer-Verlag Berlin Heidelberg 2011

1   Introduction

Consider a Jordan curve $\mathcal{C}$ forming the boundary of a geometric, planar, simply connected object. On the curve are placed k mobile agents, each agent capable of moving at any speed without exceeding its speed limit. The maximal speeds of the agents may be distinct, and the agents are allowed to walk in both directions along $\mathcal{C}$. The main goal is to design the agents' movements so that the maximal time between two consecutive visits to any fixed point of the boundary is minimized. The studied problem has genuine applications. For example, in order to prevent an intruder from penetrating into a protected region, the boundary of the region must be monitored, often with the aid of moving agents such as walking guards, illumination rays, cameras, etc. Since the feasibility of an intrusion likely depends on the time during which the intruder remains undiscovered, it is important to design patrolling protocols which minimize the time during which boundary points are unprotected.

Notation. We assume here that the mobile agents are traversing a continuous, rectifiable curve, where we are interested in the total distance travelled along the curve where some parts can be visited more than once. If the curve is also closed it is called a cycle $\mathcal{C}$. Without loss of generality we may assume that the cycle is represented by a *circle*. The set of *mobile agents* $a_1, a_2, \ldots, a_k$ are moving along $\mathcal{C}$. The speed of each agent a_i may vary during its motion, but its absolute value can never exceed its predefined maximal speed v_i. We assume that a positive speed corresponds to the counterclockwise traversal of the circle and the negative speed to the clockwise movement. Without loss of generality we suppose that the agents are numbered so that $v_1 \geq v_2 \geq \ldots \geq v_k > 0$. Using a scaling argument we assume that the length of the circle is equal to 1 (unit of length), and that in one unit of time an agent using constant speed 1 (one unit of speed) makes exactly one complete counterclockwise tour around the circle.

The position of agent a_i at time $t \in [0, \infty)$ is described by the continuous function $a_i(t)$. Hence respecting the maximal speed v_i of agent a_i means that for each real value $t \geq 0$ and $\epsilon > 0$, s.t., $\epsilon v_i < 1/2$, the following condition is true

$$dist(a_i(t), a_i(t + \epsilon)) \leq v_i \cdot \epsilon \qquad (1)$$

where $dist(a_i(t), a_i(t + \epsilon))$ denotes the distance along the cycle between the positions of agent a_i at times t and $t + \epsilon$.

Definition 1 (Traversal Algorithm). *A traversal algorithm on the cycle for k mobile agents is a k-tuple $\mathcal{A} = (a_1(t), a_2(t), \ldots, a_k(t))$ which satisfies Inequality (1), for all $i = 1, 2, \ldots, k$.*

Definition 2 (Idle time). *Let $\mathcal{A}$ be a traversal algorithm for a system of k mobile agents traversing the perimeter of a circle with the circumference 1.*

1. *The idle time induced by $\mathcal{A}$ at a point x of the circle, denoted by $I_{\mathcal{A}}(x)$, is the infimum over positive reals $T > 0$ such that for each $K \geq 0$ there exists $1 \leq i \leq k$ and $t \in [K, K + T]$ such that $a_i(t) = x$.*

2. *The idle time of the system of k mobile agents induced by $\mathcal{A}$ is defined by $I_{\mathcal{A}} = \sup_{x \in \mathcal{C}} I_{\mathcal{A}}(x)$, the supremum taken over all points of the circle.*
3. *Finally, the idle time, denoted by I_{opt}, of the system of k mobile agents is defined by $I_{opt} = \inf_{\mathcal{A}} I_{\mathcal{A}}$, the infimum taken over all traversal algorithms $\mathcal{A}$.*

Related Work. Patrolling has been intensely studied in robotics, especially in the last 4-5 years (cf. [3,9,10,11,14,16,21]). It is often viewed as a version of *coverage*, a central task in robotics. It is defined as the act of surveillance consisting in walking around an area in order to protect or supervise it. Patrolling is useful, e.g., to determine objects or humans that need to be rescued from a disaster environment. Network administrators may use mobile agent patrols to detect network failures or to discover web pages which need to be indexed by search engines, cf. [16]. Patrolling is usually defined as a perpetual process performed in a static or in a dynamically changing environment.

Notwithstanding several interesting applications and its scientific interest, the problem of boundary and area patrolling has been studied very recently (cf. [2,10,11,19]). On multiple occasions, patrolling has been dealt with using an ad-hoc approach with emphasis on experimental results (e.g. [16]), uncertainty of the model and robustness of the solutions when failures are possible (e.g., [10,11,14]) or non-deterministic solutions (e.g., [2]). In the largely experimental paper [16] one can also find several fundamental theoretical concepts related to patrolling, including models of agents (e.g., visibility or depth of perception), means of communication or motion coordination, as well as measures of algorithm efficiency. In most papers in the domain of patrolling, and also in our paper, algorithm efficiency is measured by its capacity to optimize the frequency of visits to the points of the environment (cf. [3,9,10,11,16]). This criterion was first introduced in [16] under the name of *idleness*. Depending on the approach the idleness is sometimes viewed as the average ([10]), worst-case ([21,5]), probabilistic ([2]) or experimentally verified ([16]) time elapsed since the last visit of the node (see also [3,9]). In some papers the terms of *blanket time* ([21]) or *refresh time* ([19]) are used instead, meaning the similar measure of algorithm efficiency.

Diverse approaches to patrolling based on the idleness criteria were surveyed in [3] — they discussed machine learning methods, paths generated using nego-tiation mechanisms, heuristics based on local idleness, or approximation to the Traveling Salesmen Problem (TSP). In [4] patrolling is studied as a game between patrollers and the intruder. Some papers solved patrolling problem based on swarm or ant-based algorithms ([12,18,21]). In these approaches agents are supposed to be memoryless (or having small memory), decentralized ([18]), i.e., with no explicit communication permitted with other agents or the central station, with local sensing capabilities (e.g., [12]). Ant-like algorithms usually mark the visited nodes of the graph. The authors of [21] present an evolutionary process. They show that a team of memoryless agents, by leaving marks at the nodes while walking through them, after relatively short time stabilizes to the patrolling scheme in which the frequency of the traversed edges is uniform to a factor of two (i.e., the number of traversals of the most often visited edge is at most twice the number of traversal of the least visited one), see also [5].

The author of [9] brings up a theoretical analysis of the approaches to patrolling in graph-based models. The two fundamental methods are referred to as *cyclic strategies*, where a cycle spanning the graph is constructed and the agents consecutively traverse this cycle in the same direction, and *partition-based strategies*, where the region is split into either disjoint or overlapping portions assigned to be patrolled by different agents. The environment and the time considered in the studied models are usually discrete and set in a graph environment. In geometric environments, the skeletonization technique is often applied, where the terrain is first partitioned into cells, and then graph-theoretic methods are used. Usually, cyclic strategies rely either on TSP-related solutions or spanning tree-based approaches. For example, spanning tree coverage, a technique first introduced in [13], was later extended and used in [1,10,14]. This technique is a version of the skeletonization approach where the two-dimensional grid approximating the terrain is constructed and a Hamiltonian path present in the grid is used for patrolling. In the recent paper [19], polynomial-time patrolling solutions for lines and trees are proposed. For the case of cyclic graphs [19] proves the NP-hardness of the problem and a constant-factor approximation is proposed.

Related to our problem is the *lonely runner* conjecture, given this lifelike name in [8], but first stated by J.M. Willis in 1967, [20]. It concerns k runners ($k \geq 2$) running laps on a unit length circular track with constant but pairwise different speeds. It is conjectured that every runner gets at a distance at least $1/k$ along the circular track to every other runner at some time. The conjecture has been proved for up to seven runners. For related work we refer the reader to two recent papers [6] and [7]. Very recently, substantial progress has been announced in an unpublished work [15] using dynamical systems theory. In private communication, the authors point out an equivalence between problems similar to the lonely runner conjecture and certain types of problems from elementary number theory (including Littlewood's, Goldbach's, and Polignac's conjectures).

2   Boundary Patrolling Algorithms

This section contains our results for the patrolling problem with variable-speed agents. The layout of the section is inspired by the categorization from [9] of approaches to patrolling into partition-based strategies (when the environment is partitioned into parts monitored by individual agents) and cyclic strategies (when all agents patrol the environment walking in the same direction along some cycle). In Subsection 2.1 we consider the problem of *fence patrolling*, for which proportional partition-based strategies appear to be the most natural approach. We provide a non-trivial proof that this strategy is indeed optimal for any configuration of speeds for $k = 2$ agents. Next, we consider cyclic strategies for *patrolling a circular boundary*. In Subsection 2.2 we show that such strategies are optimal on the circle for all configurations of speeds with $k \leq 4$ agents, under the additional constraint that all the agents are restricted to motion in the same direction around the boundary. In Subsection 2.3 we show by a technical analysis that the cyclic strategy is optimal for $k = 2$ even for agents which can change direction of motion. Surprisingly, we also show that in this general setting, the cyclic strategy is no longer optimal for $k = 3$ agents, and that a new type of strategy which is neither partition-based nor cyclic achieves a shorter idle time.

Note. Due to space constraints, most proofs (especially the involved proofs of lower bounds) are not present in this version of the paper.

2.1   Fence Patrolling: The Proportional Solution

We first consider the special case with an additional restriction that the boundary contains a special *cutting point* through which no agent is permitted to cross during its movement. This corresponds to the problem of patrolling a segment $S = [0, 1]$ and is known as *fence patrolling* in the robotics literature. We assume that positive speed corresponds to the traversal of the unit segment in the direction from left to right, while negative speed traversal in the opposite direction. We propose the following algorithm:

Algorithm $\mathcal{A}_1$ {for k agents to patrol a segment}
1. Partition the unit segment S into k segments, such that the length of the i-th segment s_i equals $\frac{v_i}{v_1+v_2+\cdots+v_k}$.
2. For each $i = 1, \ldots, k$ place agent a_i at any point of segment s_i.
3. For each $i = 1, \ldots, k$ agent a_i moves perpetually at maximal speed alternately visiting both endpoints of s_i.

Proposition 1. *Traversal algorithm $\mathcal{A}_1$ achieves idle time $I = \frac{2}{v_1+v_2+\cdots+v_k}$.*

Proof. Since each agent covers a non-overlapping segment of the circle (except for its endpoints, which may be visited by two agents) the interior points of each segment s_i are visited by the same agent a_i. The infimum of the frequency of visits of point x inside s_i is achieved for x being its endpoint. Since between two consecutive visits to the endpoint x of s_i agent a_i traverses, using its speed v_i, the segment s_i of length $\frac{v_i}{v_1+v_2+\cdots+v_k}$ twice, the idle time of such a point is $I = 2\frac{v_i}{v_1+v_2+\cdots+v_k}/v_i = \frac{2}{v_1+v_2+\cdots+v_k}$. ∎

We prove below that the algorithm $\mathcal{A}_1$ is optimal for the case of two agents patrolling a segment.

Theorem 1. *The optimal traversal algorithm for two agents patrolling unit segment $S = [0, 1]$ achieves idle time $I_{opt} = \frac{2}{v_1+v_2}$.*

Proof idea. We suppose, by contradiction, that there exists an algorithm $\mathcal{A}$ with an idle time of $I_\mathcal{A} = I_{\mathcal{A}_1} - \epsilon$ for some $\epsilon > 0$. We consider the subsegments S_1 and S_2 forming a decomposition of S, of lengths proportional to the speed bounds of agents a_1 and a_2, respectively ($|S_1| \geq |S_2|$), such that each agent belongs to its subsegment at some specific time. We show that the first agent to visit the common endpoint of S_1 and S_2 has to be the slower agent a_2 (otherwise the other endpoint of S_2 cannot be revisited in time). This forces the faster agent a_1 to visit the other endpoint of S_1, since a_2, in turn, cannot do this in time. As a consequence, a meeting of a_1 and a_2 has to occur. When this meeting happens, we bound from below four values of time, describing the times elapsed from the last visit to each endpoint of S and the times to the next visit to these endpoints. We use for this purpose the information about their speeds and the distance of the point of meeting of the agents from each of the endpoints. We show that the sum of the four considered times is at least equal to twice the idle time of $\mathcal{A}_1$, leading to the conclusion that one of the endpoints must remain unvisited for at least a time of $I_{\mathcal{A}_1}$, a contradiction with $I_{\mathcal{A}} < I_{\mathcal{A}_1}$.   ∎

We conjecture that Theorem 1 extends to the case of any number of agents.

Conjecture 1. The optimal traversal algorithm for n agents patrolling a segment achieves idle time $I_{opt} = \frac{2}{v_1 + v_2 + \ldots + v_k}$.

The approach from Proposition 1 results in a 2-approximation.

Proposition 2. *For any environment (a segment or a circle), the idle time of an optimal traversal algorithm for k agents with maximal speeds $v_1, v_2, \ldots, v_k$ is lower-bounded by $I_{opt} \geq \frac{1}{v_1 + v_2 + \cdots + v_k}$.*

Proof. During any time interval of length I_{opt} all points of the boundary have to be visited by at least one agent, hence the segments of the boundary covered by the corresponding mobile agents must cover the entire boundary. The maximum length of the segment traversed by agent a_i during time I_{opt} equals $I_{opt} v_i$. Since $\sum_{i=1}^{k} I_{opt} v_i \geq 1$ we have $I_{opt} \geq \frac{1}{v_1 + v_2 + \cdots + v_k}$.   ∎

The idea of algorithm $\mathcal{A}_1$ was to balance the work of all agents according to their maximal speeds. Hence the unit segment was partitioned in such a way that the idle time for each sub-segment was equal. The above algorithm seems to imply that we should use all available agents in the patrolling process, i.e., not using some of the agents results in a worse idle time. Indeed, it seems that, if some agent a_i is not being used (i.e., it stays motionless), the sub-segment patrolled by agent a_i in algorithm $\mathcal{A}$ must be covered, entirely or partially, by some other agent a_j. Since a_j must then cover a longer segment using the same maximal speed, this would result in longer idle time for its sub-segment and, consequently, in longer idle time for the algorithm. However the results of the next section indicate that this intuitive observation is not true in the case of patrolling a circle.

Patrolling a segment seems to suffer from an inherent weakness. An agent, after reaching an endpoint of the segment (or its sub-segment), performs a traversal in the opposite direction, first moving through points which were visited very recently, and only revisiting its starting location after two complete traversals of the segment. On the other hand, algorithms on the circle can be designed so that at any time the agent re-visits the location which has been waiting the longest. Consequently, optimal algorithms for a single agent of maximal speed 1 offer idle time 1 for the unit circle and idle time 2 for the unit segment. Therefore, in order to profit from the circle topology it may seem natural to try to traverse the cycle always in the same direction. In the next section we consider the case when all the agents are *required* to do so.

2.2   Patrolling a Unidirectional Boundary

We consider the case of the circle which must be traversed by all agents always in the same direction, say counterclockwise. In another words, every agent must use a strictly positive speed at each period of its movement. We now show an algorithm for which we can prove optimality for a small number of agents. The idea of the algorithm is to use only a subset of r agents having sufficiently high maximal speeds. These agents are spaced on the circle at even distances and ordered to move counterclockwise using the constant speed v_r — the maximal speed of the slowest agent from the subset. The value of r is chosen such that the idle time is minimized.

Algorithm $\mathcal{A}_2$ {for k agents to patrol a unidirectional circle}
1. Let r be such that $\max_{1 \leq i \leq k} iv_i = rv_r$.
2. Place the agents $a_1, a_2 \ldots a_r$ at equal distances of $\frac{1}{r}$ around the circle
3. For each $i = 1, \ldots, r$ agent a_i moves perpetually counterclockwise around the circle at speed v_r.
4. None of the agents $a_{r+1}, a_{r+2} \ldots a_k$ are used by the algorithm.

Theorem 2. *Consider k agents patrolling a unit circle having positive speeds not exceeding the maximal values $v_1 \geq v_2 \geq \ldots \geq v_k > 0$, respectively. Then algorithm $\mathcal{A}_2$ achieves the idle time $I = \frac{1}{\max_{1 \leq i \leq k} iv_i}$.*

Proof. Suppose that i mobile agents spaced at equal distances around the circle walk with speed v_i. Consider any time t. Each agent a_j must visit at some time $t + \Delta$ the point x which was visited at time t by another agent which was predecessor of a_j on the circle. The distance to this point x is equal to $1/i$. Using speed v_i reaching point x takes time $\Delta = \frac{1}{iv_i}$. ∎

Note that the value of $1 \leq r \leq k$ such that $rv_r = \max_{1 \leq i \leq k} iv_i$ is the best possible, since

$$\frac{1}{rv_r} = \frac{1}{\max_{1 \leq i \leq k} iv_i} = \min_{1 \leq i \leq k} \frac{1}{iv_i}$$

We prove that algorithm $\mathcal{A}_2$ is optimal for any setting involving less than 5 mobile agents. For this purpose we introduce first the notion of a *visit pattern*.

Definition 3. *Suppose that k agents $a_1, a_2 \ldots, a_k$ patrol a unit circle according to algorithm $\mathcal{A}$. We say that algorithm $\mathcal{A}$ admits the visit pattern $P = i_1 i_2 \ldots i_p$, where $1 \leq i_j \leq k$ for $j = 1, 2, \ldots, p$, if there exists time t and a point x on the circle, such that starting at time t the following, consecutive visits of point x are made (in this order) by the agents $a_{i_1}, a_{i_2} \ldots a_{i_p}$.*

For example, the visit pattern 131 implies that at some time during the execution of the algorithm, a certain point x on the circle is visited by agent a_1, the next visit to x is made by agent a_3 and the subsequent visit is made again by agent a_1. The notion of the visit pattern may be also extended to the case when more than one agent visits point x at the same time.

We prove below that algorithm $\mathcal{A}_2$ is optimal for the case of $k < 5$ agents. We first make an important observation, which we will use in this proof.

Observation. In order to prove that no algorithm $\mathcal{A}'$ provides a better idle time than algorithm $\mathcal{A}_2$ it is sufficient to focus our attention on a class of algorithms $\mathcal{A}'$ in which the agents $a_1, a_2, \ldots, a_k$ have speeds restricted to, respectively, $v_1 = v, v_2 = v/2, \ldots, v_k = v/k$, for some $v > 0$.

Indeed, assume that the algorithm $\mathcal{A}'$ must run for any sequence of speeds $v_1' \geq v_2' \geq \cdots \geq v_k'$, such that $rv_r' = \max_{1 \leq i \leq k} iv_i'$. Suppose that we change the speeds of the agents to $v_1 = rv_r', v_2 = rv_r'/2, \ldots, v_k = rv_r'/k$. Observe that $v_r = v_r'$ and $v_i \geq v_i'$, for $i \neq r$. Increasing the speeds of the agents may never result in a worse idle time since the algorithm may always choose to use some smaller speeds.

We show below that if some algorithm $\mathcal{A}'$ provides a better idle time than algorithm $\mathcal{A}_2$, then between any two consecutive visits to any point x of the circle by some agent a_c, $1 \leq c \leq k$, there must be at least c visits of point x which are made by other agents.

Lemma 1. *Consider any algorithm $\mathcal{A}$ run for the agents speeds $v_1 = v, v_2 = v/2, \ldots, v_k = v/k$. If algorithm $\mathcal{A}$ admits a visit pattern $ci_1i_2\ldots i_dc$, where $1 \leq d < c \leq k$, then algorithm $\mathcal{A}$ cannot result in a better idle time than $I = \frac{1}{\max_{1 \leq i \leq k} iv_i} = 1/v$ (i.e., the idle time of algorithm $\mathcal{A}_2$).*

Proof. Suppose that algorithm $\mathcal{A}$ admits some visit pattern $ci_1i_2\ldots i_dc$ for some point x of the circle, where $1 \leq d < c \leq k$. Since $v_c = v/c$, the time T between the first and the last visit of point x by a_c is at least

$$T \geq \frac{1}{v_c} = \frac{c}{v}.$$

This time interval is split into $d+1$ sub-intervals by the visits of agents $a_{i_1}, a_{i_2}, \ldots, a_{i_d}$. The largest such sub-interval T' must be at least equal to their average, i.e., since $d + 1 \leq c$,

$$T' \geq \frac{T}{d+1} \geq \frac{c}{v(d+1)} \geq \frac{c}{vc} = \frac{1}{v}.$$

Hence, the idle time of algorithm $\mathcal{A}$ is not less than that of algorithm $\mathcal{A}_2$. ∎

The visit patterns from the statement of Lemma 1, which can never be admitted by any algorithm supposedly offering a better idle time than $\mathcal{A}_2$, will be called *forbidden patterns*.

We now show that for any case of $k < 5$ agents the algorithm $\mathcal{A}_2$ is optimal.

Theorem 3. *Consider the case of $k < 5$ agents patrolling a unit circle having positive speeds not exceeding the maximal values $v_1 \geq v_2 \geq \ldots \geq v_k > 0$, respectively. Algorithm $\mathcal{A}_2$ achieving the idle time $I = \frac{1}{\max_{1 \leq i \leq k} iv_i}$ is optimal for this case.*

Proof idea. The proof is performed for $k = 4$, and implies the result also for $k < 4$. The idea of the proof for $k = 4$ is the following. We show first that, when some particular pairs of agents meet, a forbidden pattern is forced. As a consequence we will show that no agent can overtake agent a_4 more than once. Therefore, the number of agents' visits to any point x of the circle is at most four times the number of visits of this point by agent a_4 (plus a small constant). Hence for any $\epsilon > 0$ no algorithm can offer the idle time smaller than $\frac{1}{4v_4} - \epsilon$ and the idle time of algorithm $\mathcal{A}_2$ cannot be improved.   ∎

We believe that Theorem 3 extends to any number of agents, hence we propose the following conjecture:

Conjecture 2. In the case of k agents patrolling the circle, which have to use positive speeds not exceeding their respective maximal values $v_1 \geq v_2 \geq \ldots \geq v_k > 0$, the algorithm $\mathcal{A}_2$ achieving the idle time $I = \frac{1}{\max_{1 \leq i \leq k} iv_i}$ is optimal.

One can show that the idle time of $\frac{1}{\max_{1 \leq i \leq k} iv_i}$, achieved with positive speeds, is always within a multiplicative factor of $(1 + \ln k)$ away from the theoretical lower bound on idle time of $\frac{1}{\sum_{1 \leq i \leq k} v_i}$ (Proposition 2), which holds even when we allow agents moving in both directions. In this context, it is natural to ask whether using positive speeds by all agents, i.e., traversing the circle in the same direction is always the best strategy. This problem is addressed in the next section.

2.3   Allowing Movement in Both Directions

In this section we consider patrolling of a circle which may be traversed in both, clockwise and counterclockwise directions. It is important to understand whether this additional ability of agents to change directions may be sometimes useful and whether it may lead to a technique better than algorithm $\mathcal{A}_2$ from the previous section. We show that this is not the case for any setting involving $k = 2$ agents. We show, however, that there are settings already for $k = 3$ agents, when using negative speeds by the participating agents leads to a better idle time.

Theorem 4. *Consider two agents patrolling a unit circle with the possibility of movement in both directions. For any pair of maximal speeds $v_1 \geq v_2$ no algorithm $\mathcal{A}$ permitting agents' movement in both directions of the circle can achieve an idle time $I_{\mathcal{A}}$ which is better than $\min\{\frac{1}{v_1}, \frac{1}{2v_2}\}$ (i.e., the idle time provided by algorithm $\mathcal{A}_2$).*

Proof idea. We first show that a pair of agents a_1 and a_2, following algorithm $\mathcal{A}$, may never meet. Then, we prove that at some point of time, the circle is uniquely decomposed into a pair of arcs, each of which contains the set of points which were last visited by the first and by the second agent, respectively. Moreover, the location of each agent is confined to its corresponding arc. The endpoints of the arcs perform a continuous motion in time. We show that one of the endpoints of the arc of agent a_1 is never visited by this agent, whereas the other endpoint is visited regularly for arbitrarily large values of time. Without loss of generality, we will say that the *clockwise orientation* of the circle is the orientation given by traversing the arc of agent a_1 from the endpoint which this agent does not visit, to the other endpoint. We prove that the trajectory of the arc endpoint which is never visited by a_1 is necessarily a monotonous clockwise rotation around the circle (without a limit point). Within every such rotation, we show that each point of the circle is visited by agent a_1 at least twice, and that the distance traversed clockwise in between any two such visits to the same point by agent is at most $\frac{1}{4}$. Taking into account these observations, and the fact that the length of the arc corresponding to agent a_1 is always greater than $\frac{1}{2}$, we perform the main part of the proof which consists in a technical analysis of the trajectory of agent a_1. We finally show that the idle time of algorithm $\mathcal{A}$ cannot be better than $\min\{\frac{1}{v_1}, \frac{1}{2v_2}\}$.   ■

The last theorem implies the following

Corollary 1. *There exists settings when the optimal algorithm solving the boundary patrolling problem does not use some of the agents.*

Indeed, from Theorem 4 it follows that algorithm $\mathcal{A}_2$ is optimal for two agents patrolling a circle. However if agent a_2 has a speed at least twice slower than a_1, patrolling by a_1 (disregarding the behavior of agent a_2), using its maximal speed, results in the optimal idle time.

The next theorem gives an example of the setting for three agents, where using both directions is sometimes necessary to achieve the optimal idle time. This would not be true for every speed setting for three agents (e.g., clearly not for three agents with equal maximal speeds).

Theorem 5. *Consider $k = 3$ agents patrolling a unit circle with the possibility of movement in both directions. There exist settings such that in order to achieve the optimal idle time, some agents need to move in both directions.*

Proof. Consider the setting with $k = 3$ agents having maximal speeds $v_1 = 1, v_2 = v_1/2, v_3 = v_1/3$. Suppose that all agents move in the same counterclockwise direction around the unit circle. By Theorem 3, $\mathcal{A}_2$ is the optimal algorithm for this case and it achieves the idle time $I_{\mathcal{A}_2} = \frac{1}{\max_{1 \le i \le k} i v_i} = 1$. In order to prove the claim of our theorem we need to give an example of an algorithm $\mathcal{A}'$ controlling the movement of the three agents using $v_1, v_2 = v_1/2, v_3 = v_1/3$ as their maximal speeds, such that some agents move in both directions, and such that its idle time $I_{\mathcal{A}'} < 1$. Using the classical concept of distance-time graphs due to E.J. Marey [17] the movement of the agents is described at Fig. 1, where the horizontal axis represents time and the vertical axis refers to the position of the corresponding agent on the circle (with 0 and 1 representing the same point).

A detailed discussion of the construction and its analysis is omitted in this version of the paper. We show that the maximal idle time is equal to $\frac{35}{36} < 1 = I_{\mathcal{A}_2}$, proving the claim of the theorem.   ■

Note that for the speed setting from the example from Figure 1 the algorithm $\mathcal{A}_1$ is not optimal as well since the partition of the circle into segments proportional

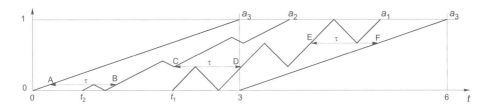

Fig. 1. Example of an algorithm achieving an idle time of $\frac{35}{36}$ for three robots with speeds $v_1 = 1$, $v_2 = 1/2$, $v_3 = 1/3$

to the agents' speeds would result in the idle time of $\frac{2}{1+\frac{1}{2}+\frac{1}{3}} > 1$. Therefore for this speed setting, neither the partition strategy of algorithm $\mathcal{A}_1$, nor the cyclic strategy of algorithm $\mathcal{A}_2$ is the best.

3 Conclusion and Open Problems

The problem of boundary patrolling has been lately intensively studied by the robotics research community. Optimality measures related to idleness — the minimization of the time when a boundary point remained unvisited — are applied in the vast majority of work in the field. This is also the measure of algorithm efficiency adopted in our paper. We have shown that for agents with distinct maximum speeds, in some settings of the problem the decisions made by the optimal algorithms are to some extent counter-intuitive. For example, we showed that it is sometimes advantageous not to make use of all of the agents. We showed that the partition strategy, represented by algorithm $\mathcal{A}_1$, and the cyclic strategy, performed by algorithm $\mathcal{A}_2$ are indeed optimal in certain cases. However, as follows from Theorem 5, in some settings the optimal idle time cannot be obtained by either of these two strategies.

Several problems remain open. The fundamental open problem is to design the optimal strategy for any configuration of speed settings of k agents patrolling a circle with the unit circumference. Two other problems, stated as Conjecture 1 and Conjecture 2, concern the extension of our results to a larger number of agents. Some other important questions include the following. Is it possible that in some setting the optimal algorithm needs agents to overtake (pass) one another? What is the solution to the problem in the case when agents have some radius of visibility (i.e., a point is considered visited when an agent is at some ε neighborhood of the point), potentially different for different agents? Finally, it is interesting to study local coordination scenarios which would allow variable-speed agents to stabilize to an efficient patrolling scheme in a distributed manner.

References

1. Agmon, N., Hazon, N., Kaminka, G.A.: The giving tree: constructing trees for efficient offline and online multi-robot coverage. Ann. Math. Artif. Intell. 52(2-4), 143–168 (2008)

2. Agmon, N., Kraus, S., Kaminka, G.A.: Multi-robot perimeter patrol in adversarial settings. In: ICRA, pp. 2339–2345 (2008)
3. Almeida, A., Ramalho, G., Santana, H., Azevedo Tedesco, P., Menezes, T., Corruble, V., Chevaleyre, Y.: Recent advances on multi-agent patrolling. In: Bazzan, A.L.C., Labidi, S. (eds.) SBIA 2004. LNCS (LNAI), vol. 3171, pp. 474–483. Springer, Heidelberg (2004)
4. Amigoni, F., Basilico, N., Gatti, N., Saporiti, A., Troiani, S.: Moving game theoretical patrolling strategies from theory to practice: An usarsim simulation. In: ICRA, pp. 426–431 (2010)
5. Bampas, E., Gąsieniec, L., Hanusse, N., Ilcinkas, D., Klasing, R., Kosowski, A.: Euler tour lock-in problem in the rotor-router model. In: Keidar, I. (ed.) DISC 2009. LNCS, vol. 5805, pp. 423–435. Springer, Heidelberg (2009)
6. Barajas, J., Serra, O.: Regular chromatic number and the lonely runner problem. Electronic Notes in Discrete Mathematics 29, 479–483 (2007)
7. Barajas, J., Serra, O.: The lonely runner with seven runners. Electron. J. Combin 15(1) (2008)
8. Bienia, W., Goddyn, L., Gvozdjak, P., Sebő, A., Tarsi, M.: Flows, View Obstructions, and the Lonely Runner. Journal of Combinatorial Theory, Series B 72(1), 1–9 (1998)
9. Chevaleyre, Y.: Theoretical analysis of the multi-agent patrolling problem. In: IAT, pp. 302–308 (2004)
10. Elmaliach, Y., Agmon, N., Kaminka, G.A.: Multi-robot area patrol under frequency constraints. Ann. Math. Artif. Intell. 57(3-4), 293–320 (2009)
11. Elmaliach, Y., Shiloni, A., Kaminka, G.A.: A realistic model of frequency-based multi-robot polyline patrolling. In: AAMAS (1), pp. 63–70 (2008)
12. Elor, Y., Bruckstein, A.M.: Autonomous multi-agent cycle based patrolling. In: ANTS Conference, pp. 119–130 (2010)
13. Gabriely, Y., Rimon, E.: Spanning-tree based coverage of continuous areas by a mobile robot. In: ICRA, pp. 1927–1933 (2001)
14. Hazon, N., Kaminka, G.A.: On redundancy, efficiency, and robustness in coverage for multiple robots. Robotics and Autonomous Systems 56(12), 1102–1114 (2008)
15. Horvat, C.H., Stoffregen, M.: A solution to the lonely runner conjecture for almost all points. Technical Report arXiv:1103.1662v1 (2011)
16. Machado, A., Ramalho, G.L., Zucker, J.-D., Drogoul, A.: Multi-agent patrolling: An empirical analysis of alternative architectures. In: Sichman, J.S., Bousquet, F., Davidsson, P. (eds.) MABS 2002. LNCS (LNAI), vol. 2581, pp. 155–170. Springer, Heidelberg (2003)
17. Marey, E.J.: La méthode graphique (1878)
18. Marino, A., Parker, L.E., Antonelli, G., Caccavale, F.: Behavioral control for multi-robot perimeter patrol: A finite state automata approach. In: ICRA, pp. 831–836 (2009)
19. Pasqualetti, F., Franchi, A., Bullo, F.: On optimal cooperative patrolling. In: CDC, pp. 7153–7158 (2010)
20. Wills, J.M.: Zwei Sätze über inhomogene diophantische Approximation von Irrationalzehlen. Monatshefte für Mathematik 71(3), 263–269 (1967)
21. Yanovski, V., Wagner, I.A., Bruckstein, A.M.: A distributed ant algorithm for efficiently patrolling a network. Algorithmica 37(3), 165–186 (2003)

Optimal Discovery Strategies in White Space Networks

Yossi Azar[1,*], Ori Gurel-Gurevich[2], Eyal Lubetzky[3], and Thomas Moscibroda[3]

[1] Tel Aviv University, Tel Aviv, Israel
azar@tau.ac.il
[2] University of British Columbia, Vancouver, BC V6T-1Z2, Canada
origurel@math.ubc.ca
[3] Microsoft Research, Redmond, WA 98052, USA
{eyal,moscitho}@microsoft.com

Abstract. The whitespace-discovery problem describes two parties, Alice and Bob, trying to discovery one another and establish communication over one of a given large segment of communication channels. Subsets of the channels are occupied in each of the local environments surrounding Alice and Bob, as well as in the global environment (Eve). In the absence of a common clock for the two parties, the goal is to devise time-invariant (stationary) strategies minimizing the discovery time.

We model the problem as follows. There are N channels, each of which is open (unoccupied) with probability p_1, p_2, q independently for Alice, Bob and Eve respectively. Further assume that $N \gg 1/(p_1 p_2 q)$ to allow for sufficiently many open channels. Both Alice and Bob can detect which channels are locally open and every time-slot each of them chooses one such channel for an attempted discovery. One aims for strategies that, with high probability over the environments, guarantee a shortest possible expected discovery time depending only on the p_i's and q.

Here we provide a stationary strategy for Alice and Bob with a guaranteed expected discovery time of $O(1/(p_1 p_2 q^2))$ given that each party also has knowledge of p_1, p_2, q. When the parties are oblivious of these probabilities, analogous strategies incur a cost of a poly-log factor, i.e. $\tilde{O}(1/(p_1 p_2 q^2))$. Furthermore, this performance guarantee is essentially optimal as we show that any stationary strategies of Alice and Bob have an expected discovery time of at least $\Omega(1/(p_1 p_2 q^2))$.

1   Introduction

Consider two parties, Alice and Bob, who wish to establish a communication channel in one out of a segment of N possible channels. Subsets of these channels

---

* This work was done during a visit to the Theory Group of Microsoft Research, Redmond.

C. Demetrescu and M.M. Halldórsson (Eds.): ESA 2011, LNCS 6942, pp. 713–722, 2011.

may already be occupied in the local environments of either Alice or Bob, as well as in the global environment in between them whose users are denoted by Eve. Furthermore, the two parties do not share a common clock and hence one does not know for how long (if at all) the other party has already been trying to communicate. Motivated by applications in discovery of wireless devices, the goal is thus to devise time-invariant strategies that ensure fast discovery with high probability (w.h.p.) over the environments.

We formalize the above problem as follows. Transmissions between Alice and Bob go over three environments: local ones around Alice and Bob and an additional global one in between them, Eve. Let A_i, B_i, E_i for $i = 1, \ldots, N$ be the indicators for whether a given channel is open (unoccupied) in the respective environment. Using local diagnostics Alice knows A yet does not know B, E and analogously Bob knows B but is oblivious of A, E. In each time-slot, each party selects a channel to attempt communication on (the environments do not change between time slots). The parties are said to *discover* one another once they select the same channel i that happens to be open in all environments (i.e., $A_i = B_i = E_i = 1$). The objective of Alice and Bob is to devise strategies that would minimize their expected discovery time.

For a concrete setup, let A_i, B_i, E_i be independent Bernoulli variables with probabilities p_1, p_2, q respectively for all i, different channels being independent of each other. (In some applications the two parties have knowledge of the environment densities p_1, p_2, q while in others these are unknown.) Alice and Bob then seek strategies whose expected discovery time over the environments is minimal.

Example. Suppose that $p_1 = p_2 = 1$ (local environments are fully open) and Alice and Bob use the naive strategy of selecting a channel uniformly over $[N]$ and independently every round. If there are $Q \approx qN$ open channels in the global environment Eve then the probability of discovery in a given round is $Q/N^2 \approx q/N$, implying an expected discovery time of about N/q to the very least.

Example. Consider again the naive uniform strategy, yet in this example Alice and Bob examine their local enviroment and each of them selects a channel uniformly over the locally open ones. Suppose for simplicity that $p_1 = p_2 = p$ for some fixed $0 < p < 1$ whereas $q = 1$ (there is no global environment interference), and of-course assume that there *exist* commonly open channels (the probability of not having such channels is exponentially small in N). Then each of Alice and Bob has a total of about pN open channels and the probability that their choice is identical in a given round is about $1/N$. In particular, the uniform strategy has an expected discovery time of about N rounds, diverging with N despite the clear fact that as N grows there are more commonly open channels for Alice and Bob. Our main theorem will show in particular that in the above scenario Alice and Bob can have an $O(1)$ expected discovery time.

In the above framework it could occur that *all* channels are closed, in which case the parties can never discover; as a result, unless this event is excluded the expected discovery time is always infinite. However, since this event has probability at most $(1 - p_1 p_2 q)^N \leq \exp(-N p_1 p_2 q)$ it poses no real problem for applications (described in further details later) where $N \gg 1/(p_1 p_2 q)$. In fact, we aim for performance guarantees that depend only on p_1, p_2, q rather than on N, hence a natural way to resolve this issue is to extend the set of channels to be infinite, i.e. define A_i, B_i, E_i for every $i \in \mathbb{N}$. (Our results can easily be translated to the finite setting with the appropriate exponential error probabilities.)

A *strategy* is a sequence of probability measures $\{\mu^t\}$ over $\mathbb{N}$, corresponding to a randomized choice of channel for each time-slot $t \geq 1$. Suppose that Alice begins the discovery via the strategy μ_a whereas Bob begins the discovery attempt at time s via the strategy μ_b. Let X_t be the indicator for a successful discovery at time t and let X be the first time Alice and Bob discover, that is

$$\mathbb{P}(X_t = 1 \mid A, B, E) = \sum_j \mu_a^t(j) \mu_b^{s+t}(j) A_j B_j E_j , \tag{1}$$

$$X = \min\{t : X_t = 1\} . \tag{2}$$

The choice of μ_a, μ_b aims to minimize $\mathbb{E}X$ where the expectation is over A, B, E as well as the randomness of Alice and Bob in applying the strategies μ_a, μ_b.

Example (*fixed strategies*). Suppose that both Alice and Bob apply the same pair of strategies independently for all rounds, μ_a and μ_b respectively. In this special case, given the environments A, B, E the random variable X is geometric with success probability $\sum_j \mu_a(j) \mu_b(j) A_j B_j E_j$, thus the mappings $A \mapsto \mu_a$ and $B \mapsto \mu_b$ should minimize the value of $\mathbb{E}X = \mathbb{E}\left[\left(\sum_j \mu_a(j) \mu_b(j) A_j B_j E_j\right)^{-1}\right]$.

A crucial fact in our setup is that Alice and Bob have no common clock and no means of telling whether or not their peer is already attempting to communicate (until they eventually discover). As such, they are forced to apply a *stationary* strategy, where the law at each time-slot is identical (i.e. $\mu^t \sim \mu^1$ for all t). For instance, Alice may choose a single μ_a and apply it independently in each step (cf. above example). Alternatively, strategies of different time-slots can be highly dependent, e.g. Bob may apply a periodic policy given by n strategies $\mu_b^1, \ldots, \mu_b^n$ and a uniform initial state $s \in [n]$.

The following argument demonstrates that stationary strategies are essentially optimal when there is no common clock between the parties. Suppose that Alice has some finite (arbitrarily long) sequence of strategies $\{\mu_a^t\}_1^{M_a}$ and similarly Bob has a sequence of strategies $\{\mu_b^t\}_1^{M_b}$. With no feedback until any actual discovery we may assume that the strategies are non-adaptive, i.e. the sequences are determined in advance. Without loss of generality Alice is joining the transmission after Bob has already attempted some β rounds of communication, in which case the expected discovery time is $\mathbb{E}_{0,\beta}X$, where $\mathbb{E}_{\alpha,\beta}X$ denotes the expectation of X as defined in (1),(2) using the strategies $\{\mu_a^{t+\alpha}\}, \{\mu_b^{t+\beta}\}$. Having no common clock implies that in the worst case scenario (over the state of Bob) the expected time to discover is $\max_\beta \mathbb{E}_{0,\beta} X$ and it now follows that Bob is better off modifying his strategy into a stationary one by selecting $\beta \in [M_b]$ uniformly at random, leading to an expected discovery time of $M_b^{-1} \sum_\beta \mathbb{E}_{0,\beta} X$.

1.1 Optimal Discovery Strategies

Our main result is a recipe for Alice and Bob to devise stationary strategies guaranteeing an optimal expected discovery time up to an absolute constant factor, assuming they know the environment densities p_1, p_2, q (otherwise the expected discovery time is optimal up to a poly-log factor).

Theorem 1. *Consider the discovery problem with probabilities p_1, p_2, q for the environments A, B, E respectively and let X denote the expected discovery time. The following then holds:*

(i) There are fixed strategies for Alice and Bob guaranteeing an expected discovery time of $\mathbb{E}X = O(1/(p_1 p_2 q^2))$, namely:
   – *Alice takes $\mu_a \sim \text{Geom}(p_2 q/6)$ over her open channels $\{i : A_i = 1\}$,*
   – *Bob takes $\mu_b \sim \text{Geom}(p_1 q/6)$ over his open channels $\{i : B_i = 1\}$.*
   Furthermore, for any fixed $\varepsilon > 0$ there are fixed strategies for Alice and Bob that do not require knowledge of p_1, p_2, q and guarantee $\mathbb{E}X = O\left(\frac{1}{p_1 p_2 q^2} \log^{2+\varepsilon}\left(\frac{1}{p_1 p_2 q}\right)\right) = \tilde{O}(\frac{1}{p_1 p_2 q^2})$, obtained by taking $\mu_a(j\text{-th open } A \text{ channel}) = \mu_b(j\text{-th open } B \text{ channel}) \propto (j \log^{1+\varepsilon/2} j)^{-1}$.
(ii) The above strategies are essentially optimal as every possible choice of stationary strategies by Alice and Bob satisfies $\mathbb{E}X = \Omega(1/(p_1 p_2 q^2))$.

Remark. The factor $1/6$ in the parameters of the geometric distributions can be fine-tuned to any smaller (or even slightly larger) fixed $\alpha > 0$ affecting the expected discovery time $\mathbb{E}X$ by a multiplicative constant. See Fig. 1 for a numerical evaluation of $\mathbb{E}X$ for various values of α.

Recall that Alice and Bob must apply stationary strategies in the absence of any common clock or external synchronization device shared by them, a restriction which is essential in many of the applications of wireless discovery protocols. However, whenever a common external clock does happen to be available there may be strategies that achieve improved performance. The next theorem, whose short proof appears in the full version of the paper, establishes the optimal strategies in this simpler scenario.

Theorem 2. *Consider the discovery problem with probabilities p_1, p_2, q for the environments A, B, E respectively and let X denote the expected discovery time. If Alice and Bob have access to a common clock then there are non-stationary strategies for them giving $\mathbb{E}X = O(1/(\min\{p_1, p_2\}q))$. Furthermore, this is tight*

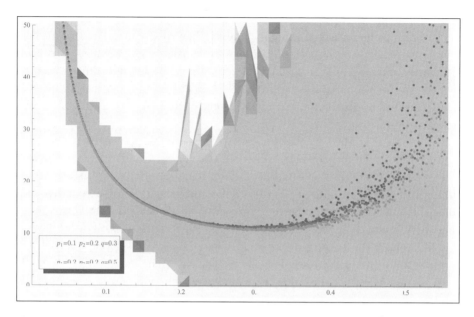

Fig. 1. discovery time $\mathbb{E}X$ as in (2) normalized by a factor of $p_1 p_2 q^2$ for a protocol using geometric distributions with parameters $\alpha p_i q$ for various values of $0 < \alpha < 1$. Markers represent the average of the expected discovery time $\mathbb{E}X$ over 10^5 random environments with $n = 10^4$ channels; surrounding envelopes represent a window of one standard deviation around the mean

as the expected discovery time for any strategies always satisfies $\mathbb{E}X = \Omega(1/(\min\{p_1, p_2\}q))$.

1.2 Applications in Wireless Networking and Related Work

The motivating application for this work comes from recent developments in wireless networking. In late 2008, the FCC issued a historic ruling permitting the unlicensed use of unused portions of the wireless RF spectrum (mainly the part between 512Mhz and 698Mhz, i.e., the UHF spectrum), popularly referred to as "White Spaces" [7]. Due to the potential for substantial bandwidth and long transmission ranges, whitespace networks (which are sometimes also called cognitive radio networks) represent a tremendous opportunity for mobile and wireless communication, and consequently, there has recently been significant interest on white space networking in the networking research community, e.g. [5, 6] as well as industry. One critical rule imposed by the FCC in its ruling is that wireless devices operating over white spaces must not interfere with incumbents, i.e., the current users of this spectrum (specifically, in the UHF bands, these are TV broadcasters as well as licensed wireless microphones). These incumbents are considered "primary users" of the spectrum, while whitespace devices are secondary users and are allowed to use the spectrum only opportunistically, whenever no primary user is using it (The FCC originally mandated whitespace devices to detect the presence of primary users using a combination of sensing techniques and a geo-location database, but in a recent amendment requires only the geo-location database approach [8]). At any given time, each whitespace device thus has a spectrum map on which some parts are blocked off while others are free to use.

The problem studied in this paper captures (and in fact even generalizes) the situation in whitespace networks when two nodes A and B seek to discover one another to establish a connection. Each node knows its own free channels on which it can transmit, but it does not know which of these channels may be available at the other node, too. Furthermore, given the larger transmission range in whitespace networks (up to a mile at Wi-Fi transmission power levels), it is likely that the spectrum maps at A and B are similar yet different. For example, a TV broadcast tower is likely to block off a channel for both A and B, but a wireless microphone — due to its small transmission power — will prevent only one of the nodes from using a channel.

Thus far, the problem of synchronizing/discovery of whitespace nodes has only been addressed when one of the nodes is a fixed access point (AP) and the other node is a client. Namely, in the framework studied in [5] the AP broadcasts on a fixed channel and the client node wishes to scan its local environment and locate this channel efficiently. That setting thus calls for technological solutions (e.g. based on scanning wider channel widths) to allow the client to find the AP channel faster than the approach of searching all possible channels one by one.

To the best of our knowledge, the results in this paper are the first to provide an efficient discovery scheme in the setting where both nodes are remote clients that may broadcast on any given channel in the whitespace region.

1.3   Related Work on Rendezvous Games

From a mathematical standpoint, the discovery problems considered in this paper seem to belong to the field of Rendezvous Search Games. The most familiar problem of this type is known as The Telephone Problem or The Telephone Coordination Game. In the telephone problem each of two players is placed in a distinct room with n telephone lines connecting the rooms. The lines are not labeled and so the players, who wish to communicate with each other, cannot simply use the first line (note that, in comparison, in our setting the channels are labeled and the difficulty in discovery is due to the local and global noise).

The optimal strategy in this case, achieving an expectation of $n/2$, is for the first player to pick a random line and continue using it, whereas the second player picks a uniformly random permutation on the lines and try them one by one. However, this strategy requires the players to determine which is the first and which is the second. It is very plausible that such coordination is not possible, in which case we require both players to employ the same strategy.

The obvious solution is for each of them to pick a random line at each turn, which gives an expectation of n turns. It turns out, however, that there are better solutions: Anderson and Weber [4] give a solution yielding an expectation of $\approx 0.8288497n$ and conjecture it's optimality.

To our knowledge, the two most prominent aspects of our setting, the presence of asymmetric information and the stationarity requirement (stemming from unknown start times) have not been considered in the literature. For example, the Anderson-Weber strategy for the telephone problem is not stationary — it has a period of $n-1$. It would be interesting to see what can be said about the optimal stationary strategies for this and other rendezvous problems. The interested reader is referred to [2, 3] and the references therein for more information on rendezvous search games.

2   Analysis of Discovery Strategies

2.1   Proof of Theorem 1, Upper Bound on the Discovery Time

Let μ_a be geometric with mean $(\alpha p_2 q)^{-1}$ over the open channels for Alice $\{i : A_i = 1\}$ and analogously let μ_b be geometric with mean $(\alpha p_1 q)^{-1}$ over the open channels for Bob $\{i : B_i = 1\}$, where $0 < \alpha < 1$ will be determined later.

Let $J = \min\{j : A_j = B_j = E_j = 1\}$ be the minimal channel open in all three environments. Further let J_a, J_b denote the number of locally open channels prior to channel J for Alice and Bob resp., that is

$$ J_a = \#\{j < J : A_j = 1\}, \quad J_b = \#\{j < J : B_j = 1\}. $$

Finally, for some integer $k \geq 0$ let M_k denote the event

$$ k \leq \max\{J_a p_2 q, \ J_b p_1 q\} < k+1. \tag{3} $$

Notice that, by definition, Alice gives probability $(1 - \alpha p_2 q)^{j-1} \alpha p_2 q$ to her j-th open channel while Bob gives probability $(1 - \alpha p_1 q)^{j-1} \alpha p_1 q$ to his j-th open channel. Therefore, on the event M_k we have that in any specific round, channel J is chosen by both players with probability at least

$$ (1 - \alpha p_1 q)^{\frac{k+1}{p_1 q}} (1 - \alpha p_2 q)^{\frac{k+1}{p_2 q}} \alpha^2 p_1 p_2 q^2 \geq e^{-4\alpha(k+1)} \alpha^2 p_1 p_2 q^2, $$

where in the last inequality we used the fact that $(1 - x) \geq \exp(-2x)$ for all $0 \leq x \leq \frac{1}{2}$, which will be justified by later choosing $\alpha < \frac{1}{2}$. Therefore, if X denotes the expected number of rounds required for discovery, then

$$ \mathbb{E}[X \mid M_k] \leq e^{4\alpha(k+1)} (\alpha^2 p_1 p_2 q^2)^{-1}. \tag{4} $$

On the other hand, J_a is precisely a geometric variable with the rule $\mathbb{P}(J_a = j) = (1 - p_2q)^j p_2q$ and similarly $\mathbb{P}(J_b = j) = (1 - p_1q)^j p_1q$. Hence,

$$\mathbb{P}(M_k) \leq (1 - p_2q)^{k/(p_2q)} + (1 - p_1q)^{k/(p_1q)} \leq 2e^{-k}.$$

Combining this with (4) we deduce that

$$\mathbb{E}X \leq 2\sum_k e^{-k}\mathbb{E}[X \mid M_k] \leq 2e^{4\alpha}(\alpha^2 p_1 p_2 q^2)^{-1}\sum_k e^{(4\alpha-1)k}$$

$$\leq \frac{2e}{\alpha^2\left(e^{1-4\alpha} - 1\right)}\left(p_1 p_2 q^2\right)^{-1} \tag{5}$$

where the last inequality holds for any fixed $\alpha < \frac{1}{4}$. In particular, a choice of $\alpha = \frac{1}{6}$ implies that $\mathbb{E}X \leq 500/\left(p_1 p_2 q^2\right)$, as required. $\qquad\square$

Remark. In the special case where $p_1 = p_2$ (denoting this probability simply by p) one can optimize the choice of constants in the proof above to obtain an upper bound of $\mathbb{E}X \leq 27/(pq)^2$.

Due to space constraints, we postpone the argument establishing discovery strategies oblivious of the environment densities to the full version of the paper.

2.2   Proof of Theorem 1, Lower Bound on the Discovery Time

Theorem 3. *Let* μ_a, μ_b *be the stationary distribution of the strategies of Alice and Bob resp., and let* $R = \sum_j \mu_a(j)\mu_b(j)A_jB_jE_j$ *be the probability of successfully discovering in any specific round. Then there exists some absolute constant* $C > 0$ *such that* $\mathbb{P}(R < Cp_0p_1q^2) \geq \frac{1}{2}$.

Proof. Given the environments A, B define

$$S_k^a = \{j : 2^{-k} < \mu_a(j) \leq 2^{-k+1}\}, \quad S_k^b = \{j : 2^{-k} < \mu_b(j) \leq 2^{-k+1}\}.$$

Notice that the variables S_k^a are a function of the strategy of Alice which in turn depends on her local environment A (an analogous statement holds for S_k^b and B). Further note that clearly $|S_k^a| < 2^k$ and $|S_k^b| < 2^k$ for any k. Let T_k^a denote all the channels where the environments excluding Alice's (i.e., both of the other environments B, E) are open, and similarly let T_k^b denote the analogous quantity for Bob:

$$T_k^a = \{j \in S_k^a : B_j = E_j = 1\}, \quad T_k^b = \{j \in S_k^b : A_j = E_j = 1\}.$$

Obviously, $\mathbb{E}|T_k^a| < 2^k p_2q$ and $\mathbb{E}|T_k^b| < 2^k p_1q$.

Since $\{B_j\}_{j\in\mathbb{N}}$ and $\{E_j\}_{j\in\mathbb{N}}$ are independent of S_k^a (and of each other), for any $\beta > 0$ we can use the Chernoff bound (see, e.g., [9]*Theorem 2.1 and [1]*Appendix A) with a deviation of $t = (\beta - 1)2^k p_2q$ from the expectation to get

$$\mathbb{P}\left(|T_k^a| > \beta 2^k p_2q\right) < \exp\left(-\frac{3}{2}\frac{(\beta - 1)^2}{\beta + 2}2^k p_2q\right),$$

and analogously for Bob we have

$$\mathbb{P}\left(|T_k^b| > \beta 2^k p_1 q\right) < \exp\left(-\frac{3}{2}\frac{(\beta-1)^2}{\beta+2}2^k p_1 q\right).$$

Clearly, setting $K_a = \log_2(1/(p_2 q)) - 3$ and $K_b = \log_2(1/(p_1 q)) - 3$ and taking β large enough (e.g., $\beta = 20$ would suffice) we get

$$\mathbb{P}\left(\bigcup_{k \geq K_a}\left\{|T_k^a| > \beta 2^k p_2 q\right\}\right) \leq 2\mathbb{P}\left(|T_{K_a}^a| > \beta 2^{K_a} p_2 q\right) < \frac{1}{8} \qquad (6)$$

and

$$\mathbb{P}\left(\bigcup_{k \geq K_b}\left\{|T_k^b| > \beta 2^k p_1 q\right\}\right) < \frac{1}{8}. \qquad (7)$$

Also, since $\sum_{k < K_a}|S_k^a| < 2^{K_a} \leq (8p_2 q)^{-1}$ and similarly $\sum_{k < K_b}|S_k^b| < 2^{K_b} \leq (8p_1 q)^{-1}$, we have by Markov's inequality that

$$\mathbb{P}\left(\bigcup_{k < K_a}\left\{|T_k^a| > 0\right\}\right) \leq \sum_{k < K_a}\mathbb{E}|T_k^a| = p_2 q\sum_{k < K_a}\mathbb{E}|S_k^a| < \frac{1}{8} \qquad (8)$$

and similarly

$$\mathbb{P}\left(\bigcup_{k < K_b}\left\{|T_k^b| > 0\right\}\right) < \frac{1}{8}. \qquad (9)$$

Putting together (6),(7),(8),(9), with probability at least $\frac{1}{2}$ the following holds:

$$|T_k^a| \leq \begin{cases} \beta 2^k p_2 q & k \geq K_a \\ 0 & k < K_a \end{cases}, \quad |T_k^b| \leq \begin{cases} \beta 2^k p_1 q & k \geq K_b \\ 0 & k < K_b \end{cases} \quad \text{for all } k. \qquad (10)$$

When (10) holds we can bound R as follows:

$$R = \sum_j \mu_a(j)\mu_b(j)A_j B_j E_j = \sum_k \sum_\ell \sum_{j \in T_k^a \cap T_\ell^b}\mu_a(j)\mu_b(j)$$

$$\leq \sum_k \sum_\ell |T_k^a \cap T_\ell^b|2^{-k+1}2^{-\ell+1}$$

$$\leq \sum_k \sum_\ell \sqrt{|T_k^a||T_\ell^b|}2^{-k+1}2^{-\ell+1} = 4\left(\sum_k \sqrt{|T_k^a|}2^{-k}\right)\left(\sum_\ell \sqrt{|T_\ell^b|}2^{-\ell}\right)$$

$$\leq 4\beta(p_1 p_2)^{1/2}q\left(\sum_{k \geq K_a}2^{-k/2}\right)\left(\sum_{\ell \geq K_b}2^{-\ell/2}\right),$$

where the second inequality used the fact that $|F_1 \cap F_2| \leq \min\{|F_1|,|F_2|\} \leq \sqrt{|F_1||F_2|}$ for any two finite sets F_1, F_2 and the last inequality applied (10). From here the proof is concluded by observing that

$$R \leq 16(p_1 p_2)^{1/2}q2^{-K_a/2}2^{-K_b/2} = 128\beta p_1 p_2 q^2. \qquad \square$$

Corollary 4. *There exists some absolute $c > 0$ such that for any pair of stationary strategies, the expected number of rounds required for a successful discovery is at least $c/(p_1p_2q^2)$.*

Proof. Conditioned on the value of R, the probability of discovery in one of the first $1/(2R)$ rounds is at most $\frac{1}{2}$. Theorem 3 established that with probability at least $\frac{1}{2}$ we have $R < Cp_1p_2q^2$, therefore altogether with probability at least $\frac{1}{4}$ there is no discovery before time $(2Cp_1p_2q^2)^{-1}$. We conclude that the statement of the corollary holds with $c = 1/(8C)$.   □

References

[1] Alon, N., Spencer, J.H.: The probabilistic method, 3rd edn. John Wiley, Chichester (2008)

[2] Alpern, S., Baston, V.J., Essegaier, S.: Rendezvous search on a graph. J. Appl. Probab. 36(1), 223–331 (1999)

[3] Alpern, S., Gal, S.: The theory of search games and rendezvous. International Series in Operations Research & Management Science, vol. 55. Kluwer Academic Publishers, Dordrecht (2003)

[4] Anderson, E.J., Weber, R.R.: The rendezvous problem on discrete locations. J. Appl. Probab. 27(4), 839–851 (1990)

[5] Bahl, P., Chandra, R., Moscibroda, T., Murty, R., Welsh, M.: White space networking with Wi-Fi like connectivity. ACM SIGCOMM Computer Communication Review 39(4), 27–38 (2009)

[6] Deb, S., Srinivasan, V., Maheshwari, R.: Dynamic spectrum access in DTV whitespaces: design rules, architecture and algorithms. In: Proc. of the 15th Annual International Conference on Mobile Computing and Networking, pp. 1–12 (2009)

[7] FCC press release, FCC adopts rules for unlicensed use of television white spaces, Technical Report 08-260, Federal Communications Commision (November 2008)

[8] FCC press release, FCC frees up vacant TV airwaves for Super Wi-Fi technologies and other technologies, Technical Report 10-174, Federal Communications Commision (September 2010)

[9] Janson, S., Luczak Tomasz, A., Rucinsk, a.: Random graphs, p. xii+333 (2000)

Social-Aware Forwarding Improves Routing Performance in Pocket Switched Networks[*]

Josep Díaz[1], Alberto Marchetti-Spaccamela[2], Dieter Mitsche[1],
Paolo Santi[3] and Julinda Stefa[2]

[1] Univ. Politècnica de Catalunya, Barcelona, Spain
[2] Sapienza Università di Roma, Italy
[3] Istituto di Informatica e Telematica del CNR, Pisa, Italy

Abstract. We study and characterize social-aware forwarding protocols in opportunistic networks and we derive bounds on the expected message delivery time for two different routing protocols, which are representatives of social-oblivious and social-aware forwarding. In particular, we consider a recently introduced stateless, social-aware forwarding protocol using interest similarity between individuals, and the well-known BinarySW protocol, which is optimal within a certain class of stateless, social-oblivious forwarding protocols. We compare both from the theoretical and experimental point of view the asymptotic performance of Interest-Based (IB) forwarding and BinarySW under two mobility scenarios, modeling situations in which pairwise meeting rates between nodes are either *independent of* or *correlated to* the similarity of their interests.

1 Introduction

Opportunistic networks, in which occasional communication opportunities between pairs or small groups of nodes are exploited to circulate messages, are expected to play a major role in next generation short range wireless networks [17,18,19]. In particular, pocket-switched networks (PSNs) [9], in which network nodes are individuals carrying around smart devices with direct wireless communication links, are expected to become widespread in a few years. Message exchange in opportunistic networks is ruled by the *store-carry-*and-*forward* mechanism typical of delay-tolerant networks [6]: a node (either the sender, or a relay node) stores the message in its buffer and carries it around, until a communication opportunity with another node arises, upon which the message can be forwarded to another node (the destination, or another relay node).

Given this basic forwarding mechanism, a great deal of attention has been devoted in past years to optimize the forwarding policy of routing protocols. Recently, several authors have proposed optimizing forwarding strategies for PSNs

---

[*] The three first authors were partially supported by the EU through project FRONTS. The work of P. Santi was partially supported by MIUR, program PRIN, Project COGENT.

based on the observation that, being these networks composed of *individuals* characterized by a collection of social relationships, these social relationships can actually be reflected in the meeting patterns between network nodes. Thus, knowledge of the social structure underlying the collection of individuals forming a PSN can be exploited to optimize the routing strategy, e.g., favoring message forwarding towards "socially well connected" nodes. Significant performance improvement of social-aware approaches over social-oblivious approaches has been experimentally demonstrated [3,8,11].

Most existing social-aware forwarding approaches hinge on the ability of *storing* state information that can be used to keep trace of history of past encounters and/or to attempt to predict future meeting opportunities [1,2,3,8,10,11]. On the other hand, socially-oblivious routing protocols such as epidemic [20], two-hops [7] and the class of Spray-and-Wait protocols [18], do not require storing additional information in the node buffers, which are then exclusively used to store the messages circulating in the network. Thus, comparing performance of social-aware vs. social-oblivious forwarding approaches would require modeling node buffers, which renders the resulting network model very complex. If storage capacity on the nodes is not accounted for in the analysis, unfair advantage would be given to social-aware approaches, which extensively use state information.

In [14], a *stateless* approach has been presented; this approach is motivated by the observation that individuals with similar interests meet relatively more often than individuals with diverse interests. The definition of this Interest-Based forwarding approach (IB forwarding in the following) allows a *fair* comparison – i.e., under the same conditions for what concerns usage of storage resources – between social-aware and social-oblivious forwarding approaches in PSNs.

Our contributions. The main goal of this paper is to compare IB and BinarySW forwarding. BinarySW [18] is chosen as a representative element of the class of social-oblivious forwarding protocols, since it is shown in [18] to be optimal within the class of Spray-and-Wait forwarding protocols, and given the extensive simulation-based evidences of its superiority within the class of stateless, social-oblivious approaches. To the best of our knowledge, the notion of interest-based mobility – although empirically verified in [14,16] – has never been formalized in the literature.

Namely, main contributions of this paper are:

1. An asymptotic analysis of IB and BinarySW forwarding performance in two different scenarios: *interest-based* mobility and *social-oblivious* mobility. The first scenario models the situation where node mobility is highly correlated to similarity of individual interests, while the second one models the opposite situation in which node mobility is independent of individual interests. We consider the case when only one relay node can be used to speed up message delivery and we prove, under reasonable probabilistic assumptions, that *IB forwarding provides asymptotic performance benefits compared to BinarySW*: IB forwarding yields *bounded* expected message delivery time and BinarySW yields *unbounded* expected message delivery time. The result that IB forwarding is better than BinarySW forwarding when nodes meet according to an interest based model that favors encounters among similar people might not be surprising. However we remark that such result was not formally proved before; we also observe that while one has constant expected time the other one is unbounded.

2. We confirm the analysis of **1.** through simulations based both on a real-world data trace and a synthetic human mobility model recently introduced in [13].

3. We extend the analysis of **1.** in several ways. First, we consider the case when many relay nodes, more copies of the message, and more hops can be used to speed up message delivery. We show that the expected delivery time of BinarySW is asymptotically the same. We also consider a version of the forwarding algorithm in which the sender knows the ID of the destination, but it does not know its *interest profile* (see next section for a formal definition of interest profile). We show that expected message delivery time with IB forwarding remains bounded even in this more challenging networking scenario if we allow a limited number of relay nodes.

A byproduct of the above analysis is the definition of a simple model of pairwise contact frequency correlating similarity of individual interests with their meeting rate. We believe this model might be useful in studying other social-related properties of PSNs, and we deem such model a contribution in itself.

Due to length constraints, proofs are not reported, and are presented in the full version of the paper [5].

2   The Network Model

We consider a network of $n+2$ nodes, which we denote $\mathcal{N} = \{S, D, R_1, \ldots, R_n\}$: a *source node* S, a *destination node* D, and n potential *relay nodes* $R_1, \ldots, R_n$. Following the model presented in [14], we model each of the $n + 2$ nodes as a point in an m-dimensional *interest space* $[0, 1]^m$, where m is the total number of interests and $m \ll n$. We assume $m = \Theta(1)$. The m-dimensional vector associated with a node defines its *interest profile*, i.e., its degrees of interest in the various dimensions of the interest space. Each node $A \in \mathcal{N}$ is thus assigned an m-dimensional vector $A[a_1, \ldots, a_m]$ in the interest space. As in [14], we use the well-known *cosine similarity* metric [4], which measures similarity between two nodes A and B as $\cos(\angle(AB))$, the cosine of the angle formed by A the origin and B. Since the cosine similarity metric implies that the norm of the vectors is not relevant, we can consider all vectors normalized to have unit norm.

We assume S and D to have orthogonal interests, namely $S[1, 0, \ldots, 0]$, and $D[0, 1, \ldots, 0]$. We call this scenario the *worst-case delivery scenario* since it corresponds to the worst case situation under the interest-based mobility model. Furthermore, in the analysis below, we assume the following concerning the distribution of interest profiles in the interest space: first, the angle α_i between the i-th interest profile and S's interest profile is chosen uniformly at random in $[0, \pi/2]$; then, from all unit vectors in the intersection of the positive orthant of the m-dimensional sphere with that $(m-1)$-dimensional subspace, one vector is chosen uniformly at random.

It is important to observe that, while nodes are assumed to move around according to some mobility model $\mathcal{M}$, node coordinates in the interest space *do not change over time*. This is coherent with what happens in real world, where individual interests change at a much larger time scale (months/years) than needed to exchange messages within the network. Thus, when focusing on a single message delivery session, it is reasonable to assume that node interest profiles correspond to fixed points in the interest space.

In particular, we assume that the mobility metric relevant to our purposes is the *expected meeting time*, which is formally defined as follows:

Definition 1. *Let A and B be nodes in the network, moving in a bounded region R according to a mobility model $\mathcal{M}$. Assume that at time $t = 0$ both A and B are independently distributed in R according to the stationary distribution of $\mathcal{M}$, and that A and B have a fixed transmission range. The* first meeting time T *between A and B is the random variable (r.v.) corresponding to the time interval elapsing between $t = 0$ and the instant of time where A and B first come into each other transmission range. The* expected meeting time *is the expected value of r.v. T.*

Following the literature [17,18,19], we assume that $T_{AB} \sim \exp(\lambda_{AB})$, therefore $\mathbb{E}[T_{AB}] = \frac{1}{\lambda_{AB}}$ (i.e. the meeting time between any pair of nodes A and B is described by a Poisson point process of intensity λ_{AB}).

In the sequel we consider two mobility models and two routing algorithms. The mobility models *social oblivious* and *interest based* are defined as follows:

– *social oblivious mobility*: for any $A, B \in \mathcal{N}$, the meeting rate $\lambda_{AB} = \lambda$ for some $\lambda > 0$ independent on A and B. This corresponds to the situation in which node mobility is not influenced by the social relationships between A and B, and it is the standard model used in opportunistic network analysis [17,18,19].

– *interest-based mobility*: the rate λ_{AB} is defined as $\lambda_{AB} = k \cdot \cos(\alpha_{AB}) + \delta(n)$. Note that the cos term implies higher correlation between nodes with more similar interests while the $\delta(n) > 0$ term accounts for the fact that occasional meetings can occur also between perfect strangers; we are interested in the case $\delta(n) \to 0$ as $n \to \infty$, which corresponds to the fact that as n grows, the probability of a meeting by chance decreases. Finally, $k > 0$ is a parameter modeling the intensity of the interest-based mobility component.

We characterize the performance of *routing algorithms*, i.e. the dynamics related to delivery of a message M from S, to D. We use S, D, or R_i to denote both a node, and its coordinates in the interest space. The dynamics of message delivery is governed by a routing protocol, which determines how many copies of M shall circulate in the network, and the forwarding rules. Namely, the following two routing algorithms are considered:

– *FirstMeeting* (FM) [18]: S generate two copies of M; S always keeps a copy of M for itself. Let R_j be the first node met by S amongst nodes $\{R_1, \ldots, R_n\}$. If R_j is met before node D, the second copy of M is delivered to node R_j. From this point on, no new copy of the message can be created nor transferred to other nodes, and M is delivered to D when the first node among S and R_j gets in touch with D. If node D is met by S before any of the R_i's, M is delivered directly. This protocol is equivalent to Binary SW as defined in [18].

– *InterestBased* [14]: IB(γ) routing is similar to FM, the only difference being that the second copy of M is delivered by S to the first node $R_k \in \{R_1, \ldots, R_n\}$ met by S such that $cos(R_k, D) \geq \gamma$, where $\gamma \in [0, 1]$ is a tunable parameter. Note that IB(0) is equivalent to FM routing. If it happens that after time n still no node in $\{R_1, \ldots, R_n\}$ satisfying the forwarding condition is encountered, then the first relay node meeting S after time n is given the copy of M independently of similarity between interest profiles.

We remark that IB routing is a stateless approach: interest profiles of encountered nodes are stored only for the time needed to locally compute the similarity metrics, and discarded afterwards. Although stateless, IB routing requires storing a limited amount of extra information in the node's memory besides messages: the node's interest profile, and the interest profile of the destination for each stored message. However, note that interest profiles can be compactly represented using a number of bits which is independent of the number n of network nodes, the additional amount of storage requested on the nodes is $O(1)$. On the other hand, stateful approaches such as [1,2,3,8,10,11] require storing an amount of information which is at least proportional to the number of nodes in the network, i.e., it is $O(n)$ (in some cases it is even $O(n^2)$). Thus, comparing IB routing with a socially-oblivious routing protocol can be considered fair (in an asymptotic sense) from the viewpoint of storage capacity. Based on this observation, in the following we will make the standard assumption that node buffers have unlimited capacity [17,18,19], which contributes to simplifying the analysis.

We denote by T_X^{μ} the random variable corresponding to the time at which M is first delivered to D, assuming a routing protocol $X \in \{\text{FM,IB}(\gamma)\}$, where *so* and *ib* represent social-oblivious and interest-based mobility, respectively, under mobility model $\mu \in \{so, ib\}$.

For both algorithms and both mobility models we consider the following random variables: T_1 is the r.v. counting the time it takes for S to meet the first node in the set $\mathcal{R} = \mathcal{N} \setminus \{S\}$; T_2 is 0 if D is the first node in $\mathcal{R}$ met by S; otherwise, if R_j is the relay node, then, T_2 is the r.v. counting the time, starting at T_1, until the first out of S and R_j meets D.

3   Bounds on the Expected Delivery Time

The following proposition states that in the social oblivious mobility scenario $\mathbb{E}[T_{FM}^{so}]$ and $\mathbb{E}[T_{IB(\gamma)}^{so}]$ are asympotically equal. We state the result of IB routing assuming $\gamma := \frac{0.29}{m-1}$ (the extension to other values of $\gamma \in (0, 1)$ is omitted).

Proposition 1. $\mathbb{E}[T_{FM}^{so}] = \mathbb{E}[T_{IB(\gamma)}^{so}] = \frac{1}{2\lambda}(1 + o(1))$ *where* $\gamma := \frac{0.29}{m-1}$.

We now consider the interest base mobility model. Consider first the case when FM routing is used in presence of interest-based mobility. The difficulty in performing the analysis stems from the fact that, under interest-based mobility, the rate parameters of the exponential r.v. representing the first meeting time between S and the nodes in $\mathcal{R}$ are r.v. themselves.

Denote by α_i the r.v. representing $\angle(S, R_i)$ in the interest space, and let $\lambda_i = k \cos \alpha_i + \delta$ be the r.v. corresponding to the meeting rate between S and R_i. Notice that the probability density function for any α_i to have value $x \in [0, \pi/2]$ is $2/\pi$. To compare results for the two mobility cases, we first compute $\mathbb{E}[\lambda_i]$, and set k in such a way that $\mathbb{E}[\lambda_i] = \lambda$. We have

$$\mathbb{E}[\lambda_i] = \int_0^{\pi/2} \frac{2}{\pi} (k \cos(\alpha) + \delta) d\alpha = \frac{2k}{\pi} + \delta, \qquad (1)$$

and thus $k = \frac{\pi}{2}(\lambda - \delta)$. To compute $\mathbb{E}[T_1]$ exactly, we have to consider an n-fold integral taking into account all possible positions of the nodes $R_1, \ldots, R_n$ in the interest space. We will see that T_1 is asymptotically negligible compared with T_2, and hence we can use the trivial upper bound $\mathbb{E}[T_1] \leq \frac{1}{n\delta}$. Computing $\mathbb{E}[T_2]$ exactly also seems difficult. The following theorem gives a lower bound.

Theorem 1. $\mathbb{E}[T_{FM}^{ib}] \geq \min\{\Omega(n/\log n), \Omega(\log(1/\delta))\}$.

The theorem implies that if $\delta = \delta(n) = o(1)$ then $\mathbb{E}[T_{FM}^{ib}] \to \infty$. Theorem 2 analyses IB($\gamma$) routing in presence of interest-based mobility.

Theorem 2. *For some constant $c > 0$ and any $0 < \gamma < 1$, $\mathbb{E}[T_{IB(\gamma)}^{ib}] \leq m\gamma/c$.*

Theorems 1 and 2 establishes asymptotic superiority of IB(γ) over FM routing in case of interest-based mobility.

3.1   Extensions

More copies and more hops. We now discuss how to extend the analysis for interest based mobility to $\ell > 2$ hops and $q > 2$ copies. We consider a variation of the FM routing protocol for interest-based mobility, which we call FM*: we assume that the message M is forwarded from node A to node B only if the interest profile of node B is *more similar* to the destination node than that of node A. If a node has already forwarded M to a set of nodes, then it will forward M only to nodes which are closer to the destination than all the previous ones. We have that, $T_{FM*}^{so} \leq T_{FM}^{so}$, since the first one at least partially accounts for similarity of interest profiles when forwarding messages. Note that the difference between FM* and IB routing is that, while in IB a minimum similarity threshold with D must be satisfied to forward M, in FM* even a tiny improvement of similarity is enough to forward M.

Observe that upper bounds on the asymptotic performance provided by IB routing remain valid also for $\ell, q > 2$. We now show that, even allowing more copies and/or hops and the smarter FM* forwarding strategy, $\mathbb{E}[T_{FM*}^{ib}]$ do not improve asymptotically, with respect to $\ell = 2$ and $q = 2$.

Consider the case of FM* routing with $\ell \geq 2$ ($\ell = \Theta(1)$) hops and exactly 2 copies of M. Let T_1 the r.v. of the first meeting time between S and the first relay node in $\mathcal{N} - \{S\}$, and let T_i be the r.v. of the meeting time between the

$(i-1)$-st relay node and the i-th relay node. Let T_ℓ be the r.v. counting the time it takes for the first relay node among $\mathcal{N} \setminus \{D\}$ to meet D.

The following theorem gives a lower bound on FM* routing with two copies and a constant number of hops.

Theorem 3. $\mathbb{E}[T_{FM^*}^{ib}] = \Omega(\log(1/\delta))$.

Let us consider the situation when we have $q > 2$ copies of M and ℓ hops. Assume wlog $q = 2^w$ for $w \in \mathbb{N}$, and that the copies of M are forwarded at each hop as follows: whenever a node contains 2^s, $s \geq 1$ copies of a message and meets a node different from D, in the independent mobility model it always gives to that node 2^{s-1} copies M – this is the Binary SW strategy of [18]. In the interest-based mobility model it gives to the node 2^{s-1} copies only if the new node is closer to D than the previous hops containing some copies of M. Assume also that all relay nodes keep the last copy for itself and deliver it only if they meet D. Therefore the number of hops is at most $\log_2 q$.

Theorem 4. *For any constant number of copies* $\mathbb{E}[T_{FM^*}^{ib}] = \Omega(\log(1/\delta))$.

Unknown destination. A major limitation of the interest based routing previously considered is that the sender must know the interest profile of the destination i.e., the coordinates $D[a_1, a_2, \ldots, a_m]$ in the interest space. We now relax this assumption assuming that S knows the identity of node D (so delivery of M to D is possible), but not its interest profile and we show that a modified version of the $IB(\gamma)$ routing that uses more than one copy of the message also provides asymptotically the same upper bound.

The idea is that the routing chooses $m-1$ relay nodes with the characteristic that each one the $m-1$ relay nodes will be "almost orthogonal" to the others and to S, and S will pass a copy to each one of them, and keep one. Namely, let $\hat{R}_j$ denote the j-th relay chosen node, $j = 1, 2, \ldots, m-1$. We consider the following routing algorithm Mod-$IB(\gamma)$ to choose relay nodes:

If S meets a node with coordinates $R_i[r_1, r_2, \ldots, r_m]$, the node becomes the j-th relay node $\hat{R}_j$, $j = 1, 2, \ldots, q-1$, if the following conditions are met: $0.05 \leq R_i[1] \leq 0.1$; $\exists k, 2 \leq k \leq m$ s. t. $0.8 \leq R_i[k] \leq 0.85$; $\forall s, 1, \leq s \leq j-1$, $\hat{R}_s[k] < 0.8$

Theorem 5. *For a constant $c > 0$ and $\gamma = \frac{0.29}{m-1}$ we have* $\mathbb{E}[T_{\text{Mod}-IB(\gamma)}^{ib}] \leq m\gamma/c$.

4   Simulations

We have qualitatively verified our asymptotic analysis through simulations, based on both a real world trace collected at the Infocom 2006 conference – the trace used in [14,16] –, and the SWIM mobility model of [13], which is shown to closely resemble fundamental features of human mobility.

4.1   Real-World Trace Based Evaluation

A major difficulty in using real-world traces to validate our theoretical results is that no information about user interests is available, for the vast majority of available traces, making it impossible to realize IB routing. One exception is the Infocom 06 trace [8], which has been collected during the Infocom 2006 conference. This data trace contains, together with contact logs, a set of user profiles containing information such as nationality, residence, affiliation, spoken languages etc. Details on the data trace are summarized in Table 1.

Table 1. Detailed information on the Infocom 06 trace

Experimental data set	Infocom 06
Device	iMote
Network type	Bluetooth
Duration (days)	3
Granularity (sec)	120
Participants with profile	61
Internal contacts number	191,336
Average Contacts/pair/day	6.7

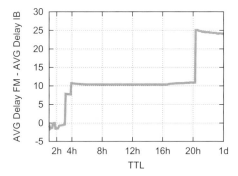

Fig. 1. Difference between average packet delivery delay with FM and IB routing with the Infocom 06 trace as a function of the message TTL

Similarly to [14], we have generated 0/1 interest profiles for each user based on the corresponding user profile. Considering that data have been collected in a conference site, we have removed very short contacts (less than $5min$) from the trace, in order to filter out occasional contacts – which are likely to be several orders of magnitude more frequent than what we can expect in a non-conference scenario. Note that, according to [14], the correlation between meeting frequency of a node pair and similarity of the respective interest profiles in the resulting data trace (containing 53 nodes overall) is 0.57. Thus, the Infocom 06 trace, once properly filtered, can be considered as an instance of interest-based mobility, where we expect IB routing to be superior to FM routing.

In order to validate this claim, we have implemented both FM and IB routing. We recall that in case of FM routing, the source delivers the second copy of its message to the first encountered node, while with IB routing the second copy of the message is delivered by the source to the first node whose interest similarity with respect to the destination node is at least γ. The value of γ has been set to $0.29/(m-1)$ as suggested in the analysis, corresponding to 0.0019 in the Infocom 06 trace. Although this value of the forwarding threshold is very low, it is nevertheless sufficient to ensure a better performance of IB vs. FM routing.

The results obtained simulating sending 5000 messages between randomly chosen source/destination pairs are reported in Figure 1. For each pair, the message is sent with both FM and IB routing, and the corresponding packet delivery time are recorded. Experiments have been repeated using different TTL (TimeToLive) values of the generated message. Figure 1 reports the difference between the average delivery time with FM and IB routing, and shows that a lower average delivery time is consistently observed with IB routing, thus qualitatively confirming the theoretical results derived in the previous section.

4.2   Synthetic Data Simulation

The real-world trace based evaluation presented in the previous section is based on a limited number of nodes (53), and thus it cannot be used to validate FM and IB scaling behavior. For this purpose, we have performed simulations using the SWIM mobility model [13], which has been shown to be able to generate synthetic contact traces whose features very well match those observed in real-world traces. Similarly to [14], the mobility model has been modified to account for different degrees of correlation between meeting rates and interest-similarity. We recall that the SWIM model is based on a notion of "home location" assigned to each node, where node movements are designed so as to resemble a "distance from home" vs. "location popularity" tradeoff. Basically, the idea is that nodes tend to move more often towards nearby locations, unless a far off location is very popular. The "distance from home" vs. "location popularity" tradeoff is tuned in SWIM through a parameter, called α, which essentially gives different weights to the distance and popularity metric when computing the probability distribution used to choose the next destination of a movement. It has been observed in [13] that giving preference to the "distance from home" component of the movement results in highly realistic traces, indicating that users in reality tend to move close to their "home location". This observation can be used to extend SWIM in such a way that different degrees of interest-based mobility can be simulated. In particular, if the mapping between nodes and their home location is random (as in the standard SWIM model), we expect to observe a low correlation between similarity of user interests and their meeting rates, corresponding to a social-oblivious mobility model. On the other hand, if the mapping between nodes and home location is done based on their interests, we expect to observe a high correlation between similarity of user interests and their meeting rates, corresponding to an interest-based mobility model.

Interest profiles have been generated considering four possible interests ($m = 4$), with values chosen uniformly at random in $[0, 1]$. In case of interest-based mobility, the mapping between a node interest profile and its "home location" has been realized by taking as coordinates of the "home location" the first two coordinates of the interest profile. In the following we present simulation results

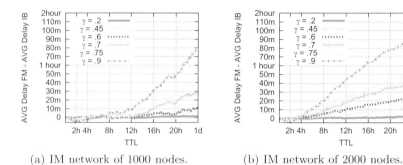

Fig. 2. Difference between average packet delivery delay with FM and IB routing with SWIM mobility in the Interest-based mobility (IM) scenario, as a function of the message TTL

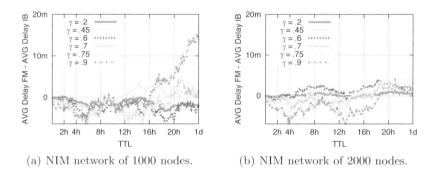

Fig. 3. Difference between average packet delivery delay with FM and IB routing with SWIM mobility in the Non Interest-based mobility (NIM) scenario, as a function of the message TTL

referring to scenarios where correlation between meeting rate and similarity of interest profiles is -0.009 (denoted Non-Interest based Mobility – NIM – in the following), and 0.61 (denoted Interest-based Mobility – IM – in the following), respectively. We have considered networks of size 1000 and 2000 nodes in both scenarios, and sent 10^5 messages between random source/destination pairs. The results are averaged over the successfully delivered messages. In the discussion below we focus only on average delay. However, we want to stress that in both IM and NIM scenarios, the IB routing slightly outperforms FM in terms of delivery rate (number of messages actually relayed): The difference of delivery rates is about 0.015% in favor of IB.

Figure 2 depicts the performance of the protocols for various values of γ on IM mobility. As can be noticed by the figure, the larger the relay threshold γ, the more IB outperforms FM. Moreover, as predicted by the analysis, the performance improvement of IB over FM routing becomes larger for larger networks. Indeed, for $\gamma = .9$ and $TTL = 24h$, message delivery with IB is respectively

$80m$ and $90m$ faster on the network of respectively 1000 nodes (see Figure 2(a)) and 2000 nodes (see Figure 2(a)). This means that, on IM mobility, IB routing delivers more messages with respect to FM, and more quickly.

Notice that the results reported in Figure 2 apparently are in contradiction with Theorem 2, which states an upper bound on the expected delivery time which is directly proportional to γ – i.e., higher values of γ implies a looser upper bound. Instead, results reported in Figure 2 show an increasingly better performance of IB vs. FM routing as γ increases. However, we notice that the bound reported in Theorem 2 is a bound on the *absolute* performance of IB routing, while those reported in Figure 2 are results referring to the *relative* performance of IB vs. FM routing.

The performance of the protocols on NIM mobility is depicted in Figure 3. In this case, the two protocols interchangeably perform better or worse in terms of delay. The negative values in the figure are due to the few more messages that IB delivers to destination whereas FM does not. Some of these messages reach the destination slightly before message TTL, by thus increasing the average delay. However, independently of γ, the values are close to zero. Indeed, note the difference of the $y-$axis between Figures 2 and 3. This indicates that, if mobility is not correlated to interest similarity, as far as the average delay is concerned the selection of the relay node is not important: A node meeting the forwarding criteria in IB routing is encountered on average soon after the first node met by the source.

4.3   Discussion

The Infocom 06 trace is characterized by a moderate correlation between meeting frequency and similarity of interest profiles – the Pearson correlation index is 0.57. However, it is composed of only 53 nodes. Despite the small network size, our simulations have shown that IB routing indeed provides a shorter average message delivery time with respect to FM routing, although the relative improvement is almost negligible (in the order of 0.06%).

To investigate relative FM and IB performance for larger networks, we used SWIM, and simulated both social-oblivious and interest-based mobility scenarios. Once again, the trend of the results qualitatively confirmed the asymptotic analysis: in case of social-oblivious mobility (correlation index is -0.009), the performance of FM and IB routing is virtually indistinguishable for all network sizes; on the other hand, with interest-based mobility (correlation index is 0.61), IB routing provides better performance than FM. It is interesting to observe the trend of performance improvement with increasing network size: performance is improved of about 5.5% with 1000 nodes, and of about 6.25% with 2000 nodes. Although percentage improvements over FM routing are modest, the trend of improvement is clearly increasing with network size, thus confirming the asymptotic analysis. Also, IB forwarding performance improvement over FM forwarding becomes more and more noticeable as the value of γ, which determines selectivity in forwarding the message, becomes higher: with $\gamma = 0.2$ and 2000 nodes, IB improves delivery delay w.r.t. FM forwarding of about 0.1%; with $\gamma = 0.6$ improvement becomes 1.7%, and it raises up to 6.25% when $\gamma = 0.9$.

5   Conclusion

We have formally analyzed and experimentally validated the delivery time under mobility and forwarding scenarios accounting for social relationships between network nodes. The main contribution of this paper is proving that, under fair conditions for what concerns storage resources, social-aware forwarding is asymptotically superior to social-oblivious forwarding in presence of interest-based mobility: its performance is never below, while it is asymptotically superior under some circumstances – orthogonal interests between sender and destination.

We believe several avenues for further research are disclosed by our initial results, such as considering scenarios in which individual interests evolve in a short time scale, or scenarios in which forwarding of messages is probabilistic instead of deterministic.

References

1. Boldrini, C., Conti, M., Passarella, A.: ContentPlace: Social-Aware Data Dissemination in Opportunistic Networks. In: Proc. ACM MSWiM, pp. 203–210 (2008)
2. Costa, P., Mascolo, C., Musolesi, M., Picco, G.P.: Socially-Aware Routing for Publish-Subscribe in Delay-Tolerant Mobile Ad Hoc Networks. IEEE Journal on Selected Areas in Communications 26(5), 748–760 (2008)
3. Daly, E., Haahr, M.: Social Network Analysis for Routing in Disconnected Delay-Tolerant MANETs. In: Proc. ACM MobiHoc, pp. 32–40 (2007)
4. Deza, M.M., Deza, E.: Encyclopedia of Distances. Springer, Berlin (2009)
5. Diaz, J., Marchetti-Spaccamela, A., Mitsche, D., Santi, P., Stefa, J.: Social-Aware Forwarding Improves Routing Performance in Pocket Switched Networks. Tech. Rep. IIT-11-2010, Istituto di Informatica e Telematica del CNR, Pisa (2010)
6. Fall, K.: A Delay-Tolerant Architecture for Challenged Internets. In: Proc. ACM Sigcomm, pp. 27–34 (2003)
7. Grossglauser, M., Tse, D.N.C.: Mobility Increases the Capacity of Ad-Hoc Wireless Networks. In: Proc. IEEE Infocom, pp. 1360–1369 (2001)
8. Hui, P., Crowcroft, J., Yoneki, E.: BUBBLE Rap: Social-Based Forwarding in Delay Tolerant Networks. In: Proc. ACM MobiHoc, pp. 241–250 (2008)
9. Hui, P., Chaintreau, A., Scott, J., Gass, R., Crowcroft, J., Diot, C.: Pocket-Switched Networks and Human Mobility in Conference Environments. In: Proc. ACM Workshop on Delay-Tolerant Networks (WDTN), pp. 244–251 (2005)
10. Ioannidis, S., Chaintreau, A., Massoulie, L.: Optimal and Scalable Distribution of Content Updates over a Mobile Social Networks. In: Proc. IEEE Infocom, pp. 1422–1430 (2009)
11. Li, F., Wu, J.: LocalCom: A Community-Based Epidemic Forwarding Scheme in Disruption-tolerant Networks. In: Proc. IEEE Secon (2009)
12. McPherson, M.: Birds of a feather: Homophily in Social Networks. Annual Review of Sociology 27(1), 415–444 (2001)
13. Mei, A., Stefa, J.: SWIM: A Simple Model to Generate Small Mobile Worlds. In: Proc. IEEE Infocom (2009)
14. Mei, A., Morabito, G., Santi, P., Stefa, J.: Social-Aware Stateless Forwarding in Pocket Switched Networks. In: Proc. IEEE Infocom, MiniConference (2011)
15. Mitzenmacher, M., Upfal, E.: Probability and Computing. Cambridge U.P. Cambridge (2005)

16. Noulas, A., Musolesi, M., Pontil, M., Mascolo, C.: Inferring Interests from Mobility and Social Interactions. In: Proc. ANLG Workshop (2009)
17. Resta, G., Santi, P.: The Effects of Node Cooperation Level on Routing Performance in Delay Tolerant Networks. In: Proc. IEEE Secon (2009)
18. Spyropoulos, T., Psounis, K., Raghavendra, C.S.: Efficient Routing in Intermittently Connected Mobile Networks: The Multi-copy Case. IEEE Trans. on Networking 16(1), 77–90 (2008)
19. Spyropoulos, T., Psounis, K., Raghavendra, C.S.: Efficient Routing in Intermittently Connected Mobile Networks: The Single-copy Case. IEEE Trans. on Networking 16(1), 63–76 (2008)
20. Vahdat, A., Becker, D.: Epidemic Routing for Partially Connected Ad Hoc Networks. Tech. Rep. CS-200006, Duke University (April 2000)

Tolerant Algorithms

Rolf Klein[1], Rainer Penninger[1], Christian Sohler[2], and David P. Woodruff[3]

[1] University of Bonn
[2] TU Dortmund
[3] IBM Research-Almaden

Abstract. Assume we are interested in solving a computational task, e.g., sorting n numbers, and we only have access to an unreliable primitive operation, for example, comparison between two numbers. Suppose that each primitive operation fails with probability at most p and that repeating it is not helpful, as it will result in the same outcome. Can we still approximately solve our task with probability $1 - f(p)$ for a function f that goes to 0 as p goes to 0? While previous work studied sorting in this model, we believe this model is also relevant for other problems. We

- find the maximum of n numbers in $O(n)$ time,
- solve 2D linear programming in $O(n \log n)$ time,
- approximately sort n numbers in $O(n^2)$ time such that each number's position deviates from its true rank by at most $O(\log n)$ positions,
- find an element in a sorted array in $O(\log n \log \log n)$ time.

Our sorting result can be seen as an alternative to a previous result of Braverman and Mossel (SODA, 2008) who employed the same model. While we do not construct the maximum likelihood permutation, we achieve similar accuracy with a substantially faster running time.

1 Introduction

Many algorithms can be designed in such a way that the input data is accessed only by means of certain primitive queries. For example, in comparison-based sorting an algorithm might specify two key indices, i and j, and check whether the relation $q_i < q_j$ holds. Except for such yes/no replies, no other type of information on the key set $\{q_1, \ldots, q_n\}$ is necessary, or this information might not even be available. Similarly, in solving a linear program, given a point and a line, the primitive may return which side of the line the point is on.

Given an oracle that can answer all possible primitive queries on the input data, one can develop an algorithm without worrying about the input data type or numerical issues. This allows for more modular and platform-independent algorithm design. A natural question is what happens if the oracle errs?

There are several ways to model such errors. One natural model is that each time the oracle is queried it returns the wrong answer with a small error probability bounded by some $p > 0$, independent of past queries. In this case, repeating a query can be used to boost the success probability at the cost of additional work. In this paper we shall employ a different model introduced by Braverman and

C. Demetrescu and M.M. Halldórsson (Eds.): ESA 2011, LNCS 6942, pp. 736–747, 2011.

Mossel [3] in the context of sorting. The oracle errs on each query independently with probability at most p, but now each possible primitive query on the input is answered by the oracle *only once*.

There are good reasons to study this model. First, it may not be possible to repeat a query, as observed in [3]. For example, in ranking soccer clubs, or even just determining the best soccer club, the outcome of an individual game may differ from the "true" ordering of the two teams and it is impossible to repeat this game. A different example is ranking items by experts [3], which provides an inherently noisy view of the "true" ranking. Another reason to study this model is that the oracle may err on a particular query due to a technical problem whose consequences are deterministic, so that we would obtain the same wrong answer in any repetition. This is common in computational geometry algorithms due to floating point errors. Geometric algorithms based on primitive queries that additionally work with faulty primitives are more modular and tolerant to errors.

In this model, two natural questions arise: (1) Given the answers to all possible primitive queries, what is the best possible solution one can find if the computational time is unlimited? (2) How good of a solution can be found efficiently, i.e., by using only a subset of the set of oracle answers?

Previous work. Braverman and Mossel [3] consider the sorting problem and provide the following answers to the questions above. With high probability, they construct the maximum likelihood permutation σ with respect to the set of oracle answers. They prove that permutation σ does not place any key more than $O(\log n)$ positions away from its true position. This fact allows the computation of σ using only $O(n \log n)$ key comparisons. The algorithm employs dynamic programming and has running time in $O(n^{3+24c_3})$ for some $c_3 > 0$ that depends on the error probability p.

Feige et al. [5] and Karp and Kleinberg [8] study the model where repeated queries are always independent. Searching games between questioner and a lying responder have been extensively studied in the past; see Pelc [11]. In his terminology, our model allows the responder random lies in response to nonrepetitive comparison questions in an adaptive game. Another related model is studied by Blum, Luby and Rubinfeld [2]. They consider self-testing and self-correction under the assumption that one is given a program P which computes a function f very quickly but possibly not very reliably. The difference with our approach is that we need to work with unreliable *primitives*.

In a series of papers culminating in work by Finocchi *et al* [6], a thorough study of resilient sorting, searching, and dictionaries in a faulty memory RAM model was performed. Here, at most δ memory words can be corrupted, while there is a small set of $O(1)$ memory words that are guaranteed not to be corrupted. The main difference is that only non-corrupted keys need to be sorted correctly. In contrast to this approach, we need to ensure that with high probability every key appears near its true position.

Our results. In this paper we assume the same model as in [3]. That is, primitive queries may fail independently with some probability at most p and noisy answers to all possible queries are pre-computed and available to the algorithm. While [3] only considers the problem of sorting, we initiate the study of a much wider class of problems in this model.

In Section 2 we show how to compute, with probability $1 - f(p)$, the maximum of n elements in $O(n)$ time. Here $f(p)$ is a function independent of n that tends to 0 as p does. There is only a constant factor overhead in the running time for this problem in this model. In Section 3 a sorting algorithm is discussed. Like the noisy sorting algorithm presented by Braverman and Mossel [3], ours guarantees that each key is placed within distance $O(\log n)$ of its true position, even though we do not necessarily obtain the maximum likelihood permutation. Also, we need $O(n^2)$ rather than $O(n \log n)$ key comparisons. However, our algorithm is faster than the algorithm of [3], as the running time is $O(n^2)$, providing a considerable improvement over the $O(n^{3+24c_3})$ time of [3]. Finally, in Section 4, we briefly discuss the noisy search problem. By a random walk argument we show that the true position of a search key can be determined in a correctly sorted list with probability $(1 - 2p)^2$ if all n comparison queries are made. With $O(\log n \log \log n)$ comparisons, we can find the correct position with probability $1 - f(p)$. As an application of our max-finding result, we present in Section 5 an $O(n \log n)$ time algorithm for linear programming in two dimensions. Linear programming is a fundamental geometric problem which could suffer from errors in primitives due, e.g., to floating point errors. Our algorithm is modular, being built from simple point/line side comparisons, and robust, given that it can tolerate faulty primitives. It is based on an old technique of Megiddo [9]. We do not know how to modify more "modern" algorithms for linear programming to fit our error model. The correctness and running time hold with probability $1 - f(p)$.

2   Finding the Maximum of n Numbers

We are given an unordered set $a_1, \ldots, a_n$ of n distinct numbers and we would like to output the maximum using only comparison queries. Each comparison between input elements fails independently with probability p. If the same comparison is made twice, the same answer is given.

Our algorithm LINEARMAX consists of two phases and each phase consists of several stages. In each stage the set of input numbers is pruned by a constant fraction such that with sufficiently high probability the maximum remains in the set. The pruning is done by sampling a set S and comparing each number outside of S with each number inside of S. During the first phase the size of the sample set increases with each stage. The second phase begins once the size of the set of numbers is successfully reduced to below some critical value. Then, we sample fewer elements in each stage so that the probability to accidently sample the maximum remains small. We will say that a stage has an error if either the maximum is removed during the stage or the input is not pruned by a constant fraction. If during a stage the input set is not sufficiently pruned, the algorithm stops and outputs "error". We remark that the purpose of this stopping rule is to simplify the analysis (this way, we make sure that any stage is executed only once and allows us directly to sum up error probabilities of different stages).

If the maximum is removed this will not be detected by the algorithm. In our analysis we derive a bound for the error probability of each stage and then use a union bound to bound the overall error probability.

In the proof we show that for any constant $1/2 > \lambda > 0$ there exists a constant $C = C(\lambda)$ such that for any $p \leq 1/64$, algorithm LINEARMAX succeeds with probability at least $1 - C \cdot p^{\frac{1}{2} - \lambda}$.

LINEARMAX(M, p)

1. $n = |M|$

2.   **while** $|M| \geq \max\{n^{1-\lambda}, \frac{100}{\sqrt{p}}\}$ **do**

3. $j = 1$

4.   Select i such that $n \cdot \left(\frac{15}{16}\right)^{i-1} \geq |M| > n \cdot \left(\frac{15}{16}\right)^{i}$

5.   **while** $j < 100 \cdot i \cdot \log(1/p)$ and $|M| > n \cdot \left(\frac{15}{16}\right)^{i+1}$ **do**

6.   Select a set S of $s_i = 4i$ elements from M uniformly at random

7.   Remove all elements in S from M

8.   Compare all elements from S with M

9.   Let M be the list of elements larger than at least $\frac{3}{4} \cdot s_i$ elements of S

10. $j = j + 1$

11.   **if** $|M| > n \cdot \left(\frac{15}{16}\right)^{i+1}$ **then output** "error" and **exit**

12.   **while** $|M| \geq \frac{100}{\sqrt{p}}$ **do**

13. $j = 1$

14.   Select i such that $\left(\frac{16}{15}\right)^{i} \geq |M| > \left(\frac{16}{15}\right)^{i-1}$

15. $i = i - \lceil \log_{16/15} \frac{100}{\sqrt{p}} \rceil + 1$

16.   **while** $j < 100 \cdot i \cdot \log(1/p)$ and $|M| > \left(\frac{16}{15}\right)^{i-1}$ **do**

17.   Select a set S of $s_i = 4i$ elements from M uniformly at random

18.   Remove all elements in S from M

19.   Compare all elements from S with M

20.   Let M be the list of elements larger than at least $\frac{3}{4} \cdot s_i$ elements of S

21. $j = j + 1$

22.   **if** $|M| > \left(\frac{16}{15}\right)^{i-1}$ **then output** "error" and **exit**

23.   Run any maximum-finding algorithm to determine the maximum m in M

Phase 1 of the algorithm begins at line 2 and Phase 2 begins at line 12. The stages of the phases correspond to the value of i (in Phase 2 this corresponds to the value of i after the subtraction in line 15).

We first analyze the expected running time of the algorithm. We observe that the running time is dominated by the number of comparisons the algorithm performs. In the first phase in stage i in each iteration of the loop the algorithm samples $4i$ elements uniformly at random and compares them to at most $n \cdot \left(\frac{15}{16}\right)^{i}$ elements of M. We will show that the expected number of loop iterations in each stage is $O(1)$. We need the following lemma.

Lemma 2.1. *The probability that in a fixed stage of Phase 1 or 2 the algorithm performs more than $100k$ iterations is at most $\left(\frac{1}{2}\right)^{k}$.*

The proof of the lemma can be found in the full version. From the lemma, it follows immediately that the expected number of loops in a fixed stage is $O(1)$. By linearity of expectation, the overall expected running time for Phase 1 is $O\left(n \cdot \sum_{i=1}^{\infty} \mathbf{E}[\text{iterations in Stage } i] \cdot i \cdot \left(\frac{15}{16}\right)^{i-1}\right) = O(n)$. In order to analyze the second stage, let $i_0 = O(\log n)$ be the maximal value of i, i.e. the first value of i computed in line 15 when the algorithm enters Phase 2. We observe that for any stage $i \leq i_0$ we have $\left(\frac{16}{15}\right)^i \leq n^{1-\lambda}$. It follows that the expected running time of the second stage is $O\left(\sum_{i=1}^{i_0} \mathbf{E}[\text{iterations in Stage } i] \cdot i \cdot \left(\frac{16}{15}\right)^i\right) = O(n)$. The running time of the standard maximum search is also $O(n)$. Hence, the overall expected running time is $O(n)$. We continue by analyzing the error probability of the algorithm. An error happens at any stage if

(a) the maximum is contained in the sample set S,
(b) the maximum is reported to be smaller than $\frac{1}{4} \cdot s_i$ of the elements of S, or
(c) the while loop makes more than $100i \log(1/p)$ iterations.

We start by analyzing (a) during the first phase. The error probability at each loop iteration is $|S|/|M|$. The number of items in M in stage i of phase 1 is at least $n \cdot \left(\frac{15}{16}\right)^i \geq n^{1-\lambda}$. We also have $i = O(\log n)$ and $n \geq \frac{100}{\sqrt{p}}$, which implies $|S|/|M| = O(\log n)/n^{1-\lambda}$. Summing up over the at most $O(i\log(1/p)) \leq O(\log^2 n)$ loop iterations, we obtain that the overall error probability of item (a) in Phase 1 is at most $O(\log^3 n)/n^{1-\lambda} \leq O(1)/n^{1-2\lambda}$. Using that $n \geq 100/\sqrt{p}$ implies $n^{1-2\lambda} \geq (100/p^{1/2})^{1-2\lambda} \geq 1/(p^{1/2-\lambda})$, we obtain that the overall error probability of item (a) in Phase 1 is at most $\frac{C}{5} \cdot p^{1/2-\lambda}$ for a sufficiently large constant $C > 0$. In the second phase, the error probabiliy is at most

$$\sum_{i=1}^{\infty} 4 \cdot i \cdot \left(\frac{15}{16}\right)^{i+\lceil \log_{16/15} \frac{100}{\sqrt{p}} \rceil - 1} \leq \frac{4\sqrt{p}}{100} \cdot \sum_{i=1}^{\infty} i \cdot \left(\frac{15}{16}\right)^{i-1} \leq \frac{C}{5} \cdot \sqrt{p},$$

for sufficently large constant $C > 0$.
    We continue by analyzing (b). Here, an error occurs in stage i if at least $\frac{1}{4} \cdot s_i$ comparisons fail. By the following lemma, this probability is small.

Lemma 2.2. *Let $1 \geq p \geq 0$ be the failure probability of a comparison. Let $k > 0$ be a multiple of 4. The probability that at least $k/4$ out of k comparisons fail is at most $(4ep)^{k/4}$.*

We also prove this lemma in the full version. Since, $p \leq 1/64$, it follows that for $C > 0$ sufficiently large, the probability of failure in each phase is $\sum_{i=1}^{O(\log n)} 100i \cdot \log(1/p) \cdot (4ep)^i \leq \frac{C}{5} \cdot \sqrt{p}$. Next, we analyze (c). By Lemma 2.1 and since $p \leq 1/64$, we have that the error probability for this item is bounded by $\sum_{i=0}^{\infty} \left(\frac{1}{2}\right)^{i \cdot \log \frac{1}{p}} = \sum_{i=0}^{\infty} p^i \leq \frac{C}{5} \cdot \sqrt{p}$, for $C > 0$ a sufficiently large constant. Finally, we consider the error probability of the standard maximum search. A maximum search over a set of n items uses $n - 1$ comparisons. If $n \leq 100/\sqrt{p}$ then the expected number of errors is at most $100p/\sqrt{p} = 100\sqrt{p}$. Let X be the random variable for the number of errors. Since the number of errors is integral, by Markov's inequality we get that the probability of error is

$$\mathbf{Pr}[\text{at least one error occurs}] \leq \mathbf{Pr}[X \geq \frac{1}{100\sqrt{p}} \cdot \mathbf{E}[X]] \leq 100\sqrt{p} \leq \frac{C}{5} \cdot \sqrt{p}$$

for C a sufficiently large constant. Summing up all errors yields an overall error probability of at most $C \cdot \sqrt{p}$.

3   Sorting

In the following section we are given a set S of n distinct keys and the comparison relation $<_E$ with errors. We present an algorihm *SortWithBuckets* that computes an output sequence such that the position of every key in this sequence differs from its rank by at most an additive error of $O(\log n)$. In the following, we use rank(x, R) to refer to the (true) rank of input key x within $R \subseteq S$, i.e., $1 + |\{y \in R : y < x\}|$. We also use rank$_E(x, R)$ to refer to the *virtual* rank of x with respect to a set R, i.e., $1 + |\{y \in R : y <_E x\}|$. Note that there may be more than one key having the same virtual rank.

The algorithm. In the following, we will assume that n is a power of 2. If this is not the case, we can add additional special items which will be assumed to be larger than any input key and run our algorithm on the modified input. Here we may assume no errors in the comparisons as the algorithm can keep track of these items. The algorithm will partition the set $\{1, \ldots, n\}$ into buckets each corresponding to a set of 2^i consecutive numbers for certain i. We call this set the *associated range* of the bucket. At the beginning we will assume that there is a single bucket with associated range $\{1, .., n\}$. In the next step, we will subdivide this bucket into two bucket, with associated ranges $\{1, \ldots, n/2\}$ and $\{n/2 + 1, \ldots, n\}$, respectively. Then the two resulting buckets are further subdivided into four buckets, and so on. The algorithm stops, if the associated ranges contain $O(\log n)$ numbers. Ideally, we would like our algorithm to maintain the invariant that each bucket contains all input numbers whose ranks are in its associated range, e.g., the bucket corresponding to numbers $\{1, \ldots, 2^i\}$ is supposed to contain the 2^i smallest input numbers. Due to the comparison errors, the algorithm cannot exactly maintain this invariant. However, we can almost maintain it in the sense that an item with rank k is either in the bucket whose associated range contains k or it is in one of the neighboring buckets.

This is done as follows. Let us consider a set of buckets B_j with associated ranges of 2^i numbers such that bucket B_j has associated range$\{(j - 1)2^i + 1, \ldots, j2^i\}$. Now assume that the input numbers have been inserted into these buckets in such a way that our relaxed invariant is satisfied. For each bucket B_j let us use S_j to denote the set of input keys inserted into B_j. Now we would like to refine our buckets, i.e., insert the input numbers in buckets B'_j with associated ranges $\{(j - 1)2^{i-1} + 1, j2^{i-1}\}$. This is simply done by inserting an item that is previously in bucket B_j into bucket B'_r, where

$$r = \lfloor \frac{\text{rank}_E(x, \bigcup_{j-2 \le k \le j+2} S_j) + |\bigcup_{k<j-2} S_j|)}{2^{i-1}} \rfloor.$$

Thus, the algorithm computes the cardinality of the buckets up to B_{j-3} and adds to it the virtual rank of x with respect to buckets $B_{j-2}, \ldots, B_{j+2}$. The idea behind this approach is that $\text{rank}(x, S)$ can be written as $|\bigcup_{k<j-3} S_j| + rank(x, \bigcup_{j-2 \le k \le j+2} S_j)$ since, by our relaxed invariant, the keys in buckets $1, \ldots, j-3$ are smaller than x and the keys in $j+3, \ldots$ are larger than x. Now, since we do not have access to $\text{rank}(x, \bigcup_{j-2 \le k \le j+2} S_j)$ we approximate it by $\text{rank}_E(x, \bigcup_{j-2 \le k \le j+2} S_j)$. Since the rank involves fewer elements when the ranges of the buckets decrease, this estimate becomes more and more accurate.

Analysis. Thus, it remains to prove that the relaxed invariant is maintained. The difficulty in analyzing this (and other) algorithms is that there are many dependencies between different stages of the algorithm. In our case, the bucket of an element x is highly dependent on the randomness of earlier iterations. Since analyzing such algorithmic processes is often close to impossible, we use a different approach. We first show that certain properties of the comparison relation $<_E$ hold for certain sets of elements with high probability. Then we show that these properties already suffice to prove that the algorithm sorts with additive error $O(\log n)$. The proof follows using Chernoff bounds and can be found in the full version.

Lemma 3.1. *Let $p \le 1/20$. Let $R \subseteq S$ be a set of $k = 9 \cdot 2^i \ge 100000 \log n$ keys and let $x \in R$. Let $X = |\{y \in R : x < y \text{ and } y <_E x\}| + |\{y \in R : x > y \text{ and } y >_E x\}|$ be the number of false comparisons of x with elements from R. Then $\mathbf{Pr}[X \ge 2^{i-1}] \le n^{-10}$.*

Corollary 3.1. *For $\log(100000 \log n) \le i \le \log n$ and $1 \le j \le n/2^i - 8$ let $S_{ij} = \{x \in S : (j-1) \cdot 2^i + 1 \le \text{rank}(x) \le (j+8) \cdot 2^i\}$. With probability at least $1 - 1/n^8$ we have that every $x \in S_{ij}$ has less than 2^{i-1} comparison errors with elements from S_{ij}.*

Proof: We apply Lemma 3.1 for each S_{ij}. The number of choices for indices i, j is bounded by n^2. Hence by the union bound, the probability that there is an error in any of the set is at most n^{-8}. ☐

We now claim that if the comparison relation $<_E$ satisfies Corollary 3.1 then our algorithm computes a sequence such that any element deviates from its true rank by at most $O(\log n)$. In order to prove this claim, let us assume that our relaxed invariant is maintained for a set of buckets B_j with associated ranges of size 2^i. Let x be an element and j^* be the bucket whose associated range contains x. By our relaxed invariant, x is either in bucket B_{j^*-1}, B_{j^*} or B_{j^*+1}. Hence, to sort x into the next finer bucket, the algorithm inspects (a subset of) buckets $B_{j^*-3}, \ldots, B_{j^*+3}$. By our relaxed invariant, these buckets only contain elements x with $\text{rank}(x) \in S_{i\ell}$ with $\ell = j^* - 4$. Now, in order to sort x into a finer bucket, we compare x with a subset of elements from $S_{i\ell}$. Therefore, the number of errors in this comparison is certainly bounded by the number of errors within $S_{i\ell}$, which is less than 2^{i-1}. Since the next finer buckets have associated ranges of size 2^{i-1}, this implies that our relaxed invariant will be maintained. We summarize our results in the following theorem.

Theorem 3.1. *Let $p \leq 1/20$. There is a tolerant algorithm that given an input set S of n numbers computes in $O(n^2)$ time and with probability at least $1 - 1/n^8$ an output sequence such that the position of every element in this sequence deviates from its rank in S by at most $O(\log n)$.*

4   Searching

In this section we assume that we are given a sorted sequence of keys $a_1 < a_2 < \ldots < a_n < a_{n+1} = \infty$, and a query key q. We know that the sorting is accurate, and that q is different from all a_i. Our task is to determine $\mathrm{rank}(q)$, the smallest index i such that $q < a_i$ holds. A noisy oracle provides us with a table containing answers $q <_E a_i$ or $q >_E a_i$ to all possible key comparisons between q and the numbers a_i. Each of them is wrong independently with probability $p_i \leq p < 1/2$.

Let $\mathrm{conf}(j)$ denote the number of table entries that would be in conflict with $\mathrm{rank}(q) = j$. Our first algorithm, SEARCH, takes $O(n)$ time to read the whole table and to output a position j that minimizes $\mathrm{conf}(j)$; ties are broken arbitrarily. By the same argument as in Braverman and Mossel [3], SEARCH reports a maximum likelihood position for q, given the answer table. As opposed to the sorting problem, we can prove a lower bound to the probability that SEARCH correctly computes the rank of q.

Theorem 4.1. SEARCH *reports $\mathrm{rank}(q)$ with probability at least $(1 - 2p)^2$.*

Proof: We want to argue that positions j to the left or to the right of $\mathrm{rank}(q)$ are less likely to get reported because they have higher conflict numbers. By definition,

$$\mathrm{conf}(j) = \begin{cases} \mathrm{conf}(j-1) + 1, \text{ if } q <_E a_{j-1} \\ \mathrm{conf}(j-1) - 1, \text{ if } q >_E a_{j-1} \end{cases}$$

holds, as can be quickly verified. Let us first consider the indices $j = \mathrm{rank}(q), \ldots, n$ to the right of $\mathrm{rank}(q)$. It makes our task only harder to assume that each oracle answer is wrong with maximum probability p. Thus, the oracle's process of producing these answers, for indices j increasing from $\mathrm{rank}(q)$ to n, corresponds to a *random walk* of the value of $\mathrm{conf}(j)$ through the integers, starting from $\mathrm{conf}(\mathrm{rank}(q))$. With probability $1 - p$, the value of $\mathrm{conf}(j)$ will increase by 1 (since $q < a_j$ holds, by assumption), and with probability p decrease. With probability $\geq (1-p) \cdot (1 - \frac{p}{1-p}) = 1 - 2p$, $\mathrm{conf}(j)$ will increase in the first step *and* never sink below this value again; see the full version for an easy proof of this fact. Thus, with probability $\geq 1 - 2p$, all j to the right of $\mathrm{rank}(q)$ will have values $\mathrm{conf}(j)$ higher than $\mathrm{conf}(\mathrm{rank}(q))$ and, therefore, not get reported. A symmetric claim holds for the indices j to the left of $\mathrm{rank}(q)$. Consequently, SEARCH does with probability $\geq (1 - 2p)^2$ report $\mathrm{rank}(q)$.   ⊡

Theorem 4.1 casts some light on what can be achived utilizing all information available. If sequence $a_1, \ldots, a_n$ is given as a linear list, the $O(n)$ time algorithm SEARCH is of practical interest, too. For a sorted sequence stored in an array, we have a more efficient tolerant binary search algorithm based on SEARCH. The proof of Theorem 4.2 can be found in the full version.

Theorem 4.2. *Given any constant error bound $p \in (0, 1/4)$ we can in time $O(\log n \cdot \log \log n)$ compute the rank of an element in a sorted list of length n with probability at least $1 - f(p)$. Here function $f(p)$ goes to 0 as p goes to 0.*

5   Linear Programming in 2 Dimensions

As an application of our LINEARMAX algorithm, we consider the linear programming problem in two dimensions, namely, the problem of minimizing a linear objective function subject to a family $\mathcal{F}$ of half-plane constraints. We assume our problem is in standard form [4], namely that the problem is to find the lowest (finite) point in the non-empty feasible region, defined by the intersection of a non-degenerate family $\mathcal{F}$ of n half-planes. Define a *floor* to be a half-plane including all points in the plane above a line, whereas a *ceiling* is a half-plane including all points in the plane below a line. We say a half-plane is *vertical* if the line defining it is vertical. In the standard setting, we can assume that vertical lines have been preprocessed and replaced with the constraint $L \leq x \leq R$ for reals L and R, and that no two ceilings and no two floors are parallel. The half-planes are given in a sorted list according to their slope. Our algorithm is based on several basic geometric primitives which can make mistakes.

Error Model: We are given a few black boxes that perform side comparisons.
   The first box is SIDECOMPARATOR, which is given four lines ℓ_A, ℓ_B, ℓ_C, and ℓ_D, and decides if the the intersection of ℓ_A and ℓ_B is to the left of the intersection of ℓ_C and ℓ_D. This description can be simplified to the following: "given two points, is one to the left of the other"? (however, since the input does not contain explicit points, we have chosen to describe the test in this more abstract way).
   The second box is VERTICALCOMPARATOR, which is given four lines ℓ_A, ℓ_B, ℓ_C, and ℓ_D, and returns the line in the set $\{\ell_C, \ell_D\}$ whose signed vertical distance is larger from the intersection point of ℓ_A and ℓ_B. This description can be simplified to the following: "given a point and two lines, which line is closer in vertical distance"? (again, since the input does not contain explicit points, we choose to describe the test in this more abstract way).
   More precisely, if the intersection of ℓ_A and ℓ_B is the point (α, β), and we draw the line $x = \alpha$, then we look at the signed distance from (α, β) to the intersection point of ℓ_C and $x = \alpha$ as well as the intersection point of ℓ_D and $x = \alpha$. Here, by signed, we mean that if (α, β) is above one of these intersection points, then its distance to that point is negative, otherwise it is non-negative. If the signed distances are the same, i.e., the lines ℓ_C and ℓ_D meet $x = \alpha$ at the same point, VERTICALCOMPARATOR reports this.
   Such primitives are basic, and play an essential role in geometric algorithms.

We assume the primitives have a small error probability p of failing. In the case of SIDECOMPARATOR, the box reports the opposite side with probability p. In the case of VERTICALCOMPARATOR, the box reports the further line (in signed vertical distance), or fails to detect if two lines have the same signed distance, with probability p. Multiple queries to the tester give the same answer.

The output of our algorithm is not given explicitly, but rather is specified as the intersection of two half-planes in $\mathcal{F}$ (since this is all that can be determined given abstract access to the input lines).

Theorem 5.1. *There is an algorithm* LP *that terminates in* $O(n \log n)$ *time with probability at least* $1 - 1/n$. *The correcntess probability is* $1 - f(p)$ *where* $f(p)$ *approaches* 0 *as* p *approaches* 0.

We now review Megiddo's algorithm and then describe our main new ideas.

Megiddo's Algorithm: Megiddo's algorithm defines: $g(x) = \max\{a_i x + b_i \mid y \geq a_i x + b_i \text{ is a floor}\}$, and $h(x) = \min\{a_i x + b_i \mid y \leq a_i x + b_i \text{ is a ceiling}\}$. A point x is feasible iff $g(x) \leq h(x)$, that is, if it is above all floors and below all ceilings. The algorithm has $O(\log n)$ stages. The number of remaining constraints in $\mathcal{F}$ in the i-th stage is at most $(7/8)^i \cdot n$. If at any time there is only a single floor g in $\mathcal{F}$, then the algorithm outputs the lowest point on g that is feasible. Otherwise it arbitrarily groups the floors into pairs and the ceilings into pairs, and computes the pair (ℓ_A, ℓ_B) whose intersection point has the median x-coordinate of all intersection points of all pairs.

The algorithm checks if the intersection point (α, β) of ℓ_A and ℓ_B is feasible, i.e., if $g(\alpha) \leq h(\alpha)$. The algorithm then attempts to determine if the optimum is to the right or the left of (α, β). It is guaranteed there is a feasible solution - an invariant maintained throughout the algorithm - and only one side of (α, β) contains a feasible point if (α, β) is infeasible. Megiddo defines the following:

$s_g = \min\{a_i \mid y \geq a_i x + b_i \text{ is a floor and } g(\alpha) = a_i \alpha + b_i\}$,
$S_g = \max\{a_i \mid y \geq a_i x + b_i \text{ is a floor and } g(\alpha) = a_i \alpha + b_i\}$,
$s_h = \min\{a_i \mid y \leq a_i x + b_i \text{ is a ceiling and } h(\alpha) = a_i \alpha + b_i\}$,
$S_h = \max\{a_i \mid y \leq a_i x + b_i \text{ is a ceiling and } h(\alpha) = a_i \alpha + b_i\}$.

Suppose first that (α, β) is infeasible. This means that $g(\alpha) > h(\alpha)$. Then if $s_g > S_h$, any feasible x satisfies $x < \alpha$. Also, if $S_g < s_h$, then any feasible x satisfies $x > \alpha$. The last case is that $s_g - S_h \leq 0 \leq S_g - s_h$, but this implies the LP is infeasible, contradicting the above invariant.

Now suppose that (α, β) is feasible. As Megiddo argues, if $g(\alpha) < h(\alpha)$ then if $s_g > 0$, then the optimal solution is to the left of α. Also, if $S_g < 0$ then the optimal solution is to the right of α. Otherwise $s_g \leq 0 \leq S_g$, and (α, β) is the optimal solution. Finally, if $g(\alpha) = h(\alpha)$, then if (1) $s_g > 0$ and $s_g \geq S_h$, then the optimum is to the left of α, or if (2) $S_g < 0$ and $S_g \leq s_h$, then the optimum is to the right of α. Otherwise (α, β) is the optimum.

Hence, in $O(n)$ time, the algorithm finds the optimum or reduces the solution to the left or right of (α, β). In this case, in each of the pairs of constraints on the other side of α, one of the two constraints can be removed since it cannot participate in defining the optimum. When there are a constant number of constraints left, the algorithm solves the resulting instance by brute force.

Intuition of Our Algorithm: Let Π be a partition of the set $\mathcal{F}$ of input floors and ceilings into pairs. Inspired by our maximum-finding algorithm, instead of computing the median of pairs of intersection points, we randomly sample a set S of $\Theta(\log n)$ pairs from Π. For each pair in Π, we find if the optimum is to the left or right of the intersection point. If (α, β) is the intersection point of a pair of constraints in Π, we use VERTICALCOMPARATOR to find the *lower floor* $\{y \geq a_i x + b_i$ is a floor and $g(\alpha) = a_i \alpha + b_i\}$ as well as the *upper ceiling* $\{y \leq a_i x + b_i$ is a ceiling and $h(\alpha) = a_i \alpha + b_i\}$. By non-degeneracy, each of these sets has size at most 2. We modify our earlier LINEARMAX algorithm to return the maximum two items (i.e., constraints) instead of just the maximum, using VERTICALCOMPARATOR to perform the comparisons. By a union bound, we have the lower floor and upper ceiling with large probability. Using the slope ordering of s_g, S_g, s_h, and S_h we know which side of (α, β) the optimum is on.

To avoid performing the same comparison twice, in any phase each primitive invocation has as input at least one of the lines in our sample set. Since we discard the sample set after a phase, comparisons in different phases are independent. To ensure that comparisons in the same phase are independent, when computing the upper ceiling and lower floor of a sampled intersection point (α, β), we do not include the other sampled pairs of constraints in the comparisons. This does not introduce errors, with high probability, since we have only $\Theta(\log n)$ randomly sampled constraints, while the union of the upper ceiling and lower floor of an intersection point has at most four constraints, and so it is likely the upper ceilings and lower floors of all sampled pairs are disjoint.

Since the sample size is $O(\log n)$, we can show that with high probability we throw away a constant fraction of constraints in each phase, and so after $O(\log n)$ recursive calls the number of remaining constraints is bounded as a function of p alone. The total time is $O(n \log n)$. Our main algorithm is described below.

$\mathrm{LP}(\mathcal{F}, p)$

1. If there are at most $\mathrm{poly}(1/p)$ constraints in $\mathcal{F}$ solve the problem by brute force.
2. If there is at most one floor constraint $f \in \mathcal{F}$, if the slope of f is positive, output the intersection of f and the line $x = L$. If the slope of f is negative, output the intersection of f and the line $x = R$.
3. Otherwise, randomly partition the floors into pairs, as well as the ceilings into pairs (possibly with one unpaired floor and one unpaired ceiling). Let the set of pairs be denoted Π. Draw a set S of $\Theta(\log |\mathcal{F}|)$ pairs of constraints from the pairs in Π uniformly at random, without replacement.
4. Let Φ_F be the set of floors in $\mathcal{F}$, excluding those in S. Let Φ_C be the set of ceilings in $\mathcal{F}$, excluding those in S.
5. For each pair (ℓ_A, ℓ_B) of constraints in S,
   a.   Let $U(\ell_A, \ell_B) = \mathrm{LOWEST}(\ell_A, \ell_B, \Phi_F, p)$.
   b.   Let $L(\ell_A, \ell_B) = \mathrm{HIGHEST}(\ell_A, \ell_B, \Phi_C, p)$.
   c.   Compute $\mathrm{TESTER}(\ell_A, \ell_B, U(\ell_A, \ell_B), L(\ell_A, \ell_B))$.
   d.   Let $T \subseteq S$ be the pairs for which TESTER does not output "fail", and compute the majority output direction dir of the result of TESTER on the pairs in T.
12.   For each pair (ℓ_A, ℓ_B) of constraints in $\Pi \setminus S$,

13.   For each pair $(\ell_C, \ell_D) \in S$, compute SIDECOMPARATOR$(\ell_A, \ell_B, \ell_C, \ell_D)$.
14.   If for at least a 2/3 fraction of pairs in S, the pair (ℓ_A, ℓ_B) is to the right
           (resp. to the left), and if *dir* is to the left (resp. to the right), then
           remove the constraint in the pair (ℓ_A, ℓ_B) from $\mathcal{F}$ that cannot
           participate in the optimum assuming the optimum is really
           to the left (resp. to the right) of the pair (ℓ_A, ℓ_B).
15.   Return $LP(\mathcal{F} \setminus S, p)$.

We defer the analysis to the full version. The subroutines LOWEST, HIGH-EST, and TESTER are also described there. Intuitively, LOWEST finds the upper envelope[1] of a point (that is, the lowest ceilings), and HIGHEST finds the lower envelope (the highest floors). TESTER tests which side of the optimum the point is on based on the slope information in the union of upper and lower envelopes.

References

1. Ajtai, M., Feldman, V., Hassidim, A., Nelson, J.: Sorting and Selection with Imprecise Comparisons. In: Albers, S., Marchetti-Spaccamela, A., Matias, Y., Nikoletseas, S., Thomas, W. (eds.) ICALP 2009. LNCS, vol. 5555, pp. 37–48. Springer, Heidelberg (2009)
2. Blum, M., Luby, M., Rubinfeld, R.: Self-Testing/Correcting with Applications to Numerical Problems. JCSS 47(3), 549–595 (1993)
3. Braverman, M., Mossel, E.: Noisy Sorting Without Resampling. In: Proc. 19th Annual ACM-SIAM Symp. on Discrete Algorithms (SODA 2008), pp. 268–276 (2008)
4. Clarkson, K.L.: Las Vegas Algorithms for Linear and Integer Programming when the Dimension is Small. J. ACM 42(2), 488–499 (1995)
5. Feige, U., Peleg, D., Raghavan, P., Upfal, E.: Computing with Unreliable Information. In: Proceedings 22nd STOC 1990, pp. 128–137 (1990)
6. Finocchi, I., Grandoni, F., Italiano, G.: Resilient Dictionaries. ACM Transactions on Algorithms 6(1), 1–19 (2009)
7. Graham, R.L., Knuth, D.E., Patashnik, O.: Concrete Mathematics, 2nd edn. Addison-Wesley, Reading (1994)
8. Karp, D., Kleinberg, R.: Noisy Binary Search and Applications. In: 18th SODA, pp. 881–890 (2007)
9. Megiddo, N.: Linear Programming in Linear Time When the Dimension Is Fixed. J. ACM 31(1), 114–127 (1984)
10. Mitzenmacher, M., Upfal, E.: Probability and Computing: Randomized Algorithms and Probabilistic Analysis. Cambridge University Press, Cambridge (2005)
11. Pelc, A.: Searching Games with Errors - Fifty Years of Coping with Liars. Theoretical Computer Science 270(1-2), 71–109 (2002)
12. Schirra, S.: Robustness and Precision Issues in Geometric Computation. In: Sack, J.-R., Urrutia, J. (eds.) Handbook of Computational Geometry, pp. 597–632. Elsevier, Amsterdam (2000)
13. Seidel, R.: Small-Dimensional Linear Programming and Convex Hulls Made Easy. Discrete & Computational Geometry (6), 423–434 (1991)

---

[1] Sometimes envelope refers to all boundary lines that touch the convex hull. Here we use it to simply refer to the extremal two constraints of a point.

Alphabet-Independent Compressed Text Indexing*

Djamal Belazzougui[1] and Gonzalo Navarro[2]

[1] LIAFA, Univ. Paris Diderot - Paris 7, France
dbelaz@liafa.jussieu.fr
[2] Department of Computer Science, University of Chile
gnavarro@dcc.uchile.cl

Abstract. Self-indexes can represent a text in asymptotically optimal space under the k-th order entropy model, give access to text substrings, and support indexed pattern searches. Their time complexities are not optimal, however: they always depend on the alphabet size. In this paper we achieve, for the first time, *full alphabet-independence* in the time complexities of self-indexes, while retaining space optimality. We obtain also some relevant byproducts on compressed suffix trees.

1 Introduction

Text indexes, like the suffix tree [1] and the suffix array [18], can *count* the occurrences of a pattern $P[1, m]$ in a text $T[1, n]$ over alphabet $[1, \sigma]$ in time $t_{\mathsf{count}} = O(m)$ or even $t_{\mathsf{count}} = O(m/\lg_\sigma n)$ (suffix trees), or $t_{\mathsf{count}} = O(m + \lg n)$ (suffix arrays). Afterwards, they can *locate* the position of any such occurrence in T in time $t_{\mathsf{locate}} = O(1)$. As the text is available, one can *extract* any substring $T[i, i + \ell - 1]$ in optimal time $t_{\mathsf{extract}} = O(\ell/\lg_\sigma n)$. Yet, their $O(n \lg n)$-bit space complexity renders these structures unapplicable for large text collections.

Compressed text *self-indexes* [21] represent a text $T[1, n]$ over alphabet $[1, \sigma]$ within compressed space and allow not only extracting any substring of T, but also counting and locating the occurrences of patterns.

A popular model to measure text compressibility is the *k-th order empirical entropy* [19], $H_k(T)$. This is a lower bound to the bits per symbol emitted by any statistical compressor that models T considering the context of k symbols that precede (or follow) the symbol to encode. It holds $0 \le H_k(T) \le H_{k-1}(T) \le \lg \sigma$.

Starting with the FM-index [9] and the Compressed Suffix Array [16,23], self-indexes have evolved up to a point where they have reached asymptotically optimal space within the k-th order entropy model, that is, $nH_k(T) + o(n \lg \sigma)$ bits [24,14,10,21,11,3,2]. While remarkable in terms of space, self-indexes have not retained the time complexities of the classical suffix trees and arrays.

---

* Partially funded by the Millennium Institute for Cell Dynamics and Biotechnology (ICDB), Grant ICM P05-001-F, Mideplan, Chile. First author also partially supported by the French ANR-2010-COSI-004 MAPPI Project.

C. Demetrescu and M.M. Halldórsson (Eds.): ESA 2011, LNCS 6942, pp. 748–759, 2011.

Table 1 lists the current space-optimal self-indexes. All follow a model where a sampling step s is chosen (which costs $O((n \lg n)/s)$ bits, so at least we have $s = \omega(\lg_\sigma n)$ for asymptotic space optimality), and then locating an occurrence costs s multiplied by some factor that depends on the alphabet size σ. The time for extracting is linear in $s + \ell$, and is also multiplied by the same factor. There are some recent results [2] where the concept of asymptotic optimality is carried out one step further, achieving $o(nH_k(T)) + o(n) \subseteq o(n \lg \sigma)$ extra space. The only structure achieving locating and extracting times independent of σ is Sadakane's [24], yet its counting time is the worst. Note that a recent FM-index [11] achieves $O(m)$ counting, $O(s)$ locating, and $O(s + \ell)$ extraction time when the alphabet is polylogarithmic in the text size, $\sigma = O(\text{polylog}(n))$.

Only the structures of Grossi et al. [14] escape from this general scheme, however they need to use more than the optimal space in order to achieve alphabet independent times. By using $(2 + \varepsilon)nH_k(T) + o(n \lg \sigma)$ bits, for any $\varepsilon > 0$, they achieve the optimal $O(m/\lg_\sigma n)$ counting time, albeit with an additive polylogarithmic penalty of $p(n) = O(\lg^{(3+\varepsilon)/(1+\varepsilon)} n \lg^2 \sigma)$. They can also achieve sublogarithmic locating time, $O(\lg^{1/(1+\varepsilon)} n)$. Finally the extraction time is also optimal plus the polylogarithmic penalty, $O(\ell/\lg_\sigma n + p(n))$.

Table 1. Current and our new complexities for self-indexes, for the case $\lg \sigma = \omega(\lg \lg n)$. The space results (in bits) hold for any $k \leq \alpha \lg_\sigma(n) - 1$ and constant $0 < \alpha < 1$, and any sampling parameter s. The counting time is for a pattern of length m and the extracting time for ℓ consecutive symbols of T. The space for Sadakane's structure [24] refers to a more recent analysis [21]; see also the clarifications in www.dcc.uchile.cl/gnavarro/fixes/acmcs06.html.

Source	Space $(+O((n \lg n)/s))$	Counting	Locating	Extracting
[14]	$nH_k + o(n \lg \sigma)$	$O(m \lg \sigma + \lg^4 n)$	$O(s \lg \sigma)$	$O((s + \ell) \lg \sigma)$
[24]	$nH_k + o(n \lg \sigma)$	$O(m \lg n)$	$O(s)$	$O(s + \ell)$
[11]	$nH_k + o(n \lg \sigma)$	$O(m \frac{\lg \sigma}{\lg \lg n})$	$O(s \frac{\lg \sigma}{\lg \lg n})$	$O((s + \ell) \frac{\lg \sigma}{\lg \lg n})$
[3]	$nH_k + o(n \lg \sigma)$	$O(m \lg \lg \sigma)$	$O(s \lg \lg \sigma)$	$O((s + \ell) \lg \lg \sigma)$
[2]	$nH_k + o(nH_k) + o(n)$	$O(m \frac{\lg \sigma}{\lg \lg n})$	$O(s \frac{\lg \sigma}{\lg \lg n})$	$O((s + \ell) \frac{\lg \sigma}{\lg \lg n})$
[2]	$nH_k + o(nH_k) + o(n)$	$O(m \lg \lg \sigma)$	$O(s \lg \lg \sigma)$	$O((s + \ell) \lg \lg \sigma)$
Ours	$nH_k + o(nH_k) + O(n)$	$O(m)$	$O(s)$	$O(s + \ell)$

Our main result in this paper is the last row of Table 1. We achieve for the first time *full alphabet independence* for all alphabet sizes, at the price of converting an $o(n)$-bit redundancy into $O(n)$. This is an important step towards leveraging the time penalties incurred by asymptotically optimal space indexes.

We apply various techniques to achieve our result. The general strategy is to find an alternative to the use of *rank* operation on sequences, on which all FM-based indexes build, and for which no constant-time solution is known. We combine FM-indexes with concepts of Compressed Suffix Arrays, monotone minimum perfect hash functions, and compressed suffix trees. As a byproduct we enhance Sadakane's compressed suffix tree [25], which uses $O(n)$ bits on top of an underlying self-index, with a data structure using $O(n \lg \lg \sigma)$ bits that speeds up the important *child* operation; the only one that still depended on the alphabet size and now is also freed from that dependence.

Sections 2 and 3 give the necessary background on self-indexes and monotone minimal perfect hash functions (mmphfs). The latter section finishes with a simple illustration of the power of mmphfs to achieve alphabet independence on locating and extracting time on FM-indexes. This is not in the main path to achieve alphabet independence on counting as well, however, so in Section 4 we reimplement locating and extracting using constant-time *select* operations. Section 5 shows how to use mmphfs to improve the *child* operation on suffix trees, and this is used in Section 6 to reduce the search time on suffix trees. These results are of general interest, but are not used in Section 7, where we use (compressed) suffix trees in a different way to finally achieve linear counting time (in combination with the results of Section 4).

2   Compressed Self-indexes

An important subproblem that arises in self-indexing is that of representing a sequence $S[1, n]$ over an alphabet $[1, \sigma]$, supporting the following operations:

- $access(S, i) = S[i]$, in time t_{access}.
- $rank_c(S, i)$ is the number of times symbol c appears in $S[1, i]$, in time t_{rank}.
- $select_c(S, i)$ is the position in S of the ith occurrence of c, in time t_{select}.

For the particular case of bitmaps, constant-time operations can be achieved using $n + o(n)$ bits [20], or $\lg \binom{n}{m} + O(\lg \lg m) + o(n) = nH_0(S) + O(m) + o(n)$ bits, where m is the number of 1s (or 0s) in S [22]. General sequences can also be represented within asympotically zero-order entropy space $nH_0(S) = \sum_{c \in [1, \sigma]} n_c \lg \frac{n}{n_c}$, where n_c is the number of times c occurs in S. Among the many compressed sequence representations [13,11,3,2,15], we emphasize two results for this paper. The first corresponds to Thm. 1, variant (i), of Barbay et al.'s recent result [2]. The second is obtained by using the same theorem, yet replacing Golynski et al.'s representation [13] for the sequences of similar frequency, by another recent result of Grossi et al. [15] (the scheme compresses itself to $H_k(S) + o(|S| \lg \sigma)$ bits, but with more restrictions; when combining with Barbay et al. we only need that it takes $|S| \lg \sigma + o(|S| \lg \sigma)$ bits).

Lemma 1 ([2,15]). *A sequence $S[1, n]$ over alphabet $[1, \sigma]$ can be represented within $nH_0(S) + o(n(H_0(S) + 1)) + O(\sigma \lg n)$ bits of space, so that the operations are supported in times either (1) $t_{\text{access}} = t_{\text{rank}} = O(\lg \lg \sigma)$ and $t_{\text{select}} = O(1)$, or (2) $t_{\text{select}} = t_{\text{rank}} = O(\lg \lg \sigma)$ and $t_{\text{access}} = O(1)$.*

The FM-index [10] is a compressed self-index built on such sequence representations. In its modern form [11], the index computes the Burrows-Wheeler transform [6] of a text $T[1, n]$, $T^{bwt}[1, n]$, then cuts it into $O(\sigma^k)$ partitions, and represents each partition as a sequence supporting *rank* and *access* operations. From their analysis [11] it follows that if each such sequence S is represented within $|S|H_0(S) + o(|S|H_0(S)) + o(|S|) + O(\sigma \lg n)$ bits of space, then the overall space of the index is $nH_k(T) + o(nH_k(T)) + o(n) + O(\sigma^{k+1} \lg n)$. The latter term is usually removed by assuming $k \leq \alpha \lg_\sigma(n) - 1$ and constant $0 < \alpha < 1$. This is precisely the space Barbay et al. [2] achieve, and the best space reported so far for compressed text indexes under the k-th order entropy model (see Table 1).

A fundamental operation of the FM-index is the so-called *LF-mapping* $LF(i) = C[c] + rank_c(T^{bwt}, i)$, where $c = T^{bwt}[i]$. Here C is a small array storing in $C[c]$ the number of occurrences in T of symbols $< c$. The LF-mapping is used with various purposes. The BWT T^{bwt} is actually aligned with the suffix array [18] $A[1, n]$ of $T[1, n]$, so that $T^{bwt}[i] = T[A[i] - 1]$. The suffix array points to all the suffixes of T in lexicographic order, and thus the occurrences of any pattern $P[1, m]$ in T appear in a range of $A[sp, ep]$. The meaning of the LF-mapping is that, if $A[i] = j$, then $A[LF(i)] = j - 1$, that is, it lets us move virtually backwards in T, while using suffix array positions. The FM-index marks the partitions of the BWT in a sparse bitmap P that is represented within $O(\sigma^k \lg n) + o(n)$ bits and offers constant-time *rank* and *select* [22]. Therefore the time to compute the LF-mapping is $t_{\mathsf{LF}} = O(t_{\mathsf{access}} + t_{\mathsf{rank}})$, where t_{access} and t_{rank} refer to the times in the representation of the partitions.

The time to compute LF impacts all the times of the FM-index. By using a sampling step s, which yields extra space $O((n \lg n)/s)$ bits, any cell $A[i]$ can be computed in time $O(s \cdot t_{\mathsf{LF}})$, and any substring of T of length ℓ can be extracted in time $O((s+\ell) \cdot t_{\mathsf{LF}})$. As no known solution offers $t_{\mathsf{rank}} = O(1)$, we will circumvent the dependence on t_{rank} in order to achieve $t_{\mathsf{LF}} = O(1)$.

The remaining operation offered by the FM-index is *counting*, that is, determining the area $A[sp, ep]$ where pattern P occurs, so that its occurrences can be counted as $ep - sp + 1$ and each occurrence position can be located using $A[i]$, for $sp \leq i \leq ep$. Counting is done via the so-called backward search, which processes the pattern in reverse order. Let $A[sp, ep]$ be the interval for $P[i+1, m]$, then the interval for $P[i, m]$ is $A[sp', ep']$, where $sp' = C[c] + rank_c(T^{bwt}, sp - 1) + 1$ and $ep' = C[c] + rank_c(T^{bwt}, ep)$, where $c = P[i]$. This requires computing $O(m)$ times operation *rank*, yet this *rank* operation is of a more general type than for LF (i.e., it does not hold $T^{bwt}[i] = c$ for $rank_c(T^{bwt}, i)$), and therefore achieving linear time for it will require a more elaborate technique.

The other family of self-indexes are Compressed Suffix Arrays (CSAs) [16,24,14]. Here the main component is function $\Psi(i) = A^{-1}[A[i] + 1]$, which is the inverse of function LF. The array Ψ is represented directly within compressed space and giving constant access time to any value. A sparse bitmap $D[1, n]$ is stored, so that we mark positions $i = 1$ and the positions i such that $T[A[i]] \neq T[A[i-1]]$. In addition, the distinct symbols of T are stored in a string $Q[1, \sigma]$, in lexicographic order. By storing D in compressed form [22], D and Q occupy $O(\sigma \lg n) + o(n)$ bits and we have constant time *rank* and *select* on D. Then we have $T[A[i]] = Q[rank_1(D, i)]$. Moreover, $T[A[i] + k] = T[A[\Psi^k(i)]]$, which gives any string $T[A[i], A[i] + \ell - 1]$ in time $O(\ell)$.

This enables a simple binary-search-based suffix array searching for $P[1, m]$ in time $O(m \lg n)$. By using the same sampling mechanism mentioned for the FM-index, and considering that this time Ψ virtually moves forwards instead of backwards in T, we achieve $O(s)$ locating time and $O(s + \ell)$ extracting time.

For completeness we describe the sampling for the CSA. For locating, sample T regularly every s positions by setting up a bitmap $V[1,n]$ where $V[j] = 1$ iff $A[j] \mod s = 0$ plus an array $S_A[rank_1(V,j)] = A[j]/s$ for those j where $V[j] = 1$. To compute $A[i]$, compute successively $j = \Psi^k(j)$ for $k = 0, 1, \ldots, s-1$ until $V[j] = 1$; then $A[i] = S_A[rank_1(V,j)] \cdot s + k$. For extracting simply store $S_T[j] = A^{-1}[1 + s \cdot j]$ for $j = 0, 1, \ldots, n/s$, then to extract $T[i, i + \ell - 1]$, compute $j = \lfloor (i-1)/s \rfloor$ and extract the longer substring $T[j \cdot s + 1, i + \ell - 1]$. Since the extraction starts from $A[S_T[j]]$ we obtain the first character as $c = T[A[S_T[j]]] = rank_1(D, S_T[j])$, and we use Ψ to find the positions in A pointing to the consecutive characters to extract.

3   Monotone Minimal Perfect Hash Functions

A *monotone minimal perfect hash function (mmphf)* [4,5] $f : [1, u] \to [1, n]$, for $n \le u$, assigns consecutive values $1, 2, \ldots, n$ to domain values $u_1 < u_2 < \ldots < u_n$, and arbitrary values to the rest. Seen another way, it maps the elements of a set $\{u_1, u_2, \ldots, u_n\} \subseteq [1, u]$ into consecutive values in $[1, n]$. Yet a third view is a bitmap $B[1, u]$ with n bits set; then $f(i) = rank_1(B, i)$ where $B[i] = 1$ and $f(i)$ is arbitrary where $B[i] = 0$.

A mmphf on B does not give sufficient information to reconstruct B, and thus it can be stored within less than $\lg \binom{u}{n}$ bits, more precisely $O(n \lg \lg \frac{u}{n} + n)$ bits. This allows using it to speed up operations while adding an extra space that is asymptotically negligible.

As a simple application of mmphfs, we show how to compute the LF-mapping on a sequence $S[1, n]$ within time $O(t_{\mathsf{access}})$, by using additional $O(n(\lg H_0 + 1))$ bits of space. For each character c appearing in the sequence we build a mmphf f_c which records all the positions at which the character c appears in the sequence. This hash function occupies $O(n_c(\lg \lg \frac{n}{n_c} + 1))$ bits, where n_c is the number of occurrences of c in S. Summing up over all characters we get additional space usage $O(n(\lg H_0 + 1))$ bits by using the log-sum inequality[1].

The LF-mapping can now be easily computed in time $O(t_{\mathsf{access}})$ as $LF(i) = C[c] + f_c(i)$, where $c = T^{bwt}[i]$, since we know that f_c is well-defined at c. Therefore the time of the LF function becomes $O(1)$ if we have constant access time to the BWT. Consider now partitioning the BWT as in the FM-index [11]. Our extra space is $O(|S|(\lg H_0(S) + 1))$ within each partition S of the BWT. By the log-sum inequality again[2] we get total space $O(n(\lg H_k(T) + 1))$. We obtain the following result.

Lemma 2. *By adding $O(n(\lg H_k(T) + 1))$ bits to an FM-index built on text $T[1, n]$ over alphabet $[1, \sigma]$, one can compute the LF-mapping in time $t_{\mathsf{LF}} = O(t_{\mathsf{access}})$, where t_{access} is the time needed to access any element in T^{bwt}.*

---

[1] Given n pairs of numbers $a_i, b_i > 0$, it holds $\sum a_i \lg \frac{a_i}{b_i} \ge (\sum a_i) \lg \frac{\sum a_i}{\sum b_i}$. Use $a_i = n_c/n$ and $b_i = -a_i \lg a_i$ to obtain the claim.

[2] This time using $a_i = |S_i|$ and $b_i = |S_i| \lg H_0(S_i)$.

We choose the sequence representation (2) of Lemma 1, so that $t_{\text{access}} = O(1)$. Thus we achieve constant-time LF-mapping (Lemma 2) and, consequently, locate time $O(s)$ and extract time $O(s + \ell)$, at the cost of $O((n \lg n)/s)$ extra bits.

The sequence representation for each partition S takes $|S|H_0(S) + o(|S|H_0(S)) + o(|S|) + O(\sigma \lg n)$ bits. Added over all the partitions [11], this gives the main space term $nH_k(T) + o(nH_k(T)) + o(n) + O(\sigma^{k+1} \lg n)$, as explained. On top of this, Lemma 2 requires $O(n(\lg H_k(T) + 1))$ bits. This is $o(nH_k(T)) + O(n)$ if $H_k(T) = \omega(1)$, and $O(n)$ otherwise.

Theorem 1. *Given a text $T[1, n]$ over alphabet $[1, \sigma]$, one can build an FM-index occupying $nH_k(T) + o(nH_k(T)) + O(n + (n \lg n)/s + \sigma^{k+1} \lg n)$ bits of space for any $k \geq 0$ and $s > 0$, such that counting is supported in time $t_{\text{count}} = O(m \lg \lg \sigma)$, locating is supported in time $t_{\text{locate}} = O(s)$ and extraction of a substring of T of length ℓ in time $t_{\text{extract}} = O(s + \ell)$.*

In order to improve counting time to $O(m)$, however, we will need a much more sophisticated approach that cannot be combined with this first simple result. This is what the rest of the paper is about.

4   Fast Locating and Extracting Using Select

Our strategies for achieving $O(m)$ counting time make use of constant-time *select* operation on the sequences, and therefore will be incompatible with Thm. 1. In this section we develop a new technique that achieves linear locating and extracting time using constant-time *select* operations.

Consider the $O(\sigma^k)$ partitions of T^{bwt}. This time we represent each partition using variant (1) of Lemma 1, so the total space is $nH_k(T) + o(nH_k(T)) + o(n) + O(\sigma^{k+1} \lg n)$ bits. Unlike the case of *access*, the use of bitmap P to mark the beginnings of the partitions and the support for local *select* in the partitions is not sufficient to achieve global *select* on T^{bwt}.

Following Golynski et al.'s idea [13] we set up σ bitmaps B_c, $c \in [1, \sigma]$, of total length $n + o(n)$, as $B_c = 01^{n(c,1)} 01^{n(c,2)} \dots 01^{n(c,\lceil n/b \rceil)}$, where $n(c, i)$ is the number of occurrences of symbol c in partition S_i. So there are overall n 1s and $O(\sigma^{k+1})$ 0s across all the B_c bitmaps, and thus all of them can be represented in compressed form [22] using $O(\sigma^{k+1} \lg n)$ bits, answering *rank* and *select* queries in constant time. Now $q = rank_0(select_1(B_c, j)) = select_1(B_c, j) - j$ tells us the block number where the jth occurrence of c lies in T^{bwt}, and it is the rth occurrence within S_q, where $r = select_1(B_c, j) - select_0(B_c, q)$. Thus we can implement in constant time operation $select_c(T^{bwt}, j) = select_1(P, q) - 1 + select_c(S_q, r)$, since the local *select* operation in S_q takes constant time.

It is known [17] that the Ψ function can be simulated on top of T^{bwt} as $\Psi(i) = select_c(T^{bwt}, j)$, where $c = T[A[i]]$ and i is the j-th suffix in A starting with c. Therefore we can use bitmap D and string Q so as to compute in constant time $r = rank_1(D, i)$, $c = Q[r]$, and $j = i - select_1(D, r) + 1$.

With this representation we have a constant-time simulation of Ψ using an FM-index, and hence we can locate in time $t_{\text{locate}} = O(s)$ and extract a substring of length ℓ of T in time $t_{\mathsf{extract}} = O(s + \ell)$ using $O((n \lg n)/s)$ extra space, as explained in Section 2. This representation is compatible with the linear-time counting data structures that are presented next.

5 Improving Child Operation in Suffix Trees

We now give a result that has independent interest. One of the most important and frequently used operations in compressed suffix trees (CSTs) is also usually the slowest: operation $child(v, c)$ gives the node that descends from node v by symbol c, if it exists. For example, if t_{SA} is the time to compute a cell of the underlying suffix array or of its inverse permutation,[3] then operation $child$ costs time $O(t_{\mathsf{SA}} \lg \sigma)$ in Sadakane's CST [25].

We improve the operation as follows. Given any node of degree d whose d children are labeled with characters $c_1, c_2, \ldots, c_d$, we store all of them in a mmphf f_v occupying $O(d \lg \lg \sigma)$ bits. As the sum of the degrees of all of the nodes in the suffix tree is at most $2n - 1$, the total space usage is $O(n \lg \lg \sigma)$ bits.

To answer $child(v, c)$ we compute $f_v(c) = i$ and verify that the ith child of v, u, descends by symbol c. If so, then $u = child(v, c)$, else v has no child labeled c.

Lemma 3. *Given a suffix tree we can build an additional data structure that occupies $O(n \lg \lg \sigma)$ bits, so as to support operation $child(v, c)$ in the time required by computing the ith child of v, u, for any given i, plus the time to extract the first letter of edge (v, u).*

Sadakane's CST represents the tree topology using balanced parentheses. If we use Sadakane and Navarro's parentheses representation [26], then the ith child of node v is computed in constant time, as well as all the other operations used in Sadakane's CST. Moreoover, computing the first letter of the edge (v, u) takes time $O(t_{\mathsf{SA}})$. Therefore, we reduce the time for operation $child(v, c)$ from $O(t_{\mathsf{SA}} \lg \sigma)$ to $O(t_{\mathsf{SA}})$ at the price of $O(n \lg \lg \sigma)$ extra bits. Sadakane's CST space is $|CSA| + O(n)$ bits, where $|CSA|$ is the size of the underlying self-index. While this new variant raises the space to $|CSA| + O(n \lg \lg \sigma)$, it turns out that, for $\sigma = \omega(1)$, the new extra space is within the usual $o(n \lg \sigma)$ bits of redundancy of most underlying CSAs (though not all of them [2]).

We note that Sadakane [25] also shows how to achieve time complexity $O(t_{\mathsf{SA}})$ for $child$, but at the much heavier expense of using $O(n \lg \sigma)$ extra space.

6 Improving Counting Time in Compressed Suffix Trees

Using the encoding of the $child$ operation as described in the previous section we can find the suffix array interval $A[sp, ep]$ corresponding to a pattern $P[1, m]$ in time $O(m \cdot t_{\mathsf{SA}})$. We show now how to enhance the suffix tree structure with

---

[3] In compressed text indexes it usually holds $t_{\mathsf{SA}} = t_{\mathsf{locate}}$. This holds in particular with the sampling scheme described in Section 2.

$O(n \lg t_{\mathsf{SA}})$ extra bits of space so that this operation requires just $O(m)$ time in addition to that for extracting m symbols from T given its pointer from A.

We use a blind search strategy [8]. We first traverse the trie considering only the characters at branching nodes (moreover we can make mistakes, as seen soon). This returns an interval $A[sp, ep]$ whose correctness is then checked at the end. We store, in addition to the tree topology and to the data structure of Section 5, the number of skipped characters at each node whenever this number is smaller than $t_{\mathsf{SA}} - 1$. If it is larger than that, then we store a special marker. Then, given a pattern P, we traverse the suffix tree top-down and each time we have a branching node and we are at character c in the pattern, we use the result of Section 5 to find the child labeled by c (yet we do not spend time in verifying it) and continue the traversal from that child. For skipping the characters during the top-down traversal, we notice that whenever the skip count of a node is below t_{SA}, we can get it from the node, otherwise we get it in $O(t_{\mathsf{SA}})$ time using Sadakane's CST [25], as the string depth of the node minus that of its parent, $depth(v) - depth(parent(v))$. Note that because we are skipping at least t_{SA} characters, the total time to traverse the trie is $O(m)$ (this is true even if $m < t_{\mathsf{SA}}$ since we know in constant time whether the next skip surpasses the remaining pattern). Finally, after we have finished the traversal, we need to check whether the obtained result was right or not. For that we need to extract the first m characters of any of the suffixes below the node arrived at, and compare them with P. If they match, we return the computed range, otherwise P does not occur in T.

Lemma 4. *Given a text $T[1,n]$ we can add a data structure occupying $O(n \lg t_{\mathsf{SA}})$ bits on top of its CST, so that the suffix array range corresponding to a pattern $P[1,m]$ can be determined within $O(m)$ time plus the time to extract a substring of length m from T whose position in the suffix array is known.*

This gives us a first alphabet-independent FM-index. We can choose any $s = O(\mathrm{polylog}(n))$, so that $\lg t_{\mathsf{SA}} = O(\lg \lg n) \subset o(\lg \sigma)$ whenever $\lg \sigma = \omega(\lg \lg n)$ (recall that the other case is already solved [11]).

Theorem 2. *Given a text $T[1,n]$ over alphabet $[1,\sigma]$, one can build an FM-index occupying $nH_k(T) + o(n \lg \sigma) + O((n \lg n)/s + \sigma^{k+1} \lg n)$ bits of space for any $k \geq 0$ and $0 < s = O(\mathrm{polylog}(n))$, such that counting is supported in time $t_{\mathsf{count}} = O(m)$, locating is supported in time $t_{\mathsf{locate}} = O(s)$ and extraction of a substring of T of length ℓ in time $t_{\mathsf{extract}} = O(s + \ell)$.*
An unsatisfactory aspect of this theorem is that we have increased the redundancy from $o(nH_k(T)) + O(n)$ to $o(n \lg \sigma)$. In the next section we present a more sophisticated approach that recovers the original redundancy.

7   Backward Search in $O(m)$ Time

We can achieve $O(m)$ time and compressed redundancy by using the suffix tree to do backward search instead of descending in the tree. As explained in Section 2, backward search requires carrying out $O(m)$ *rank* operations. We will manage to simulate the backward search with operations *select* instead of *rank*. We will make use of mmphfs to aid in this simulation.

Weiner links. The backward step on the suffix array range for $X = P[i+1, m]$ leads to the suffix array range for $cX = P[i, m]$. When cX corresponds to an explicit (i.e., branching) suffix tree node (and hence that of X is explicit too), this operation corresponds to taking a *Weiner link* [27] on character $c = P[i]$ from the suffix tree node corresponding to $X = P[i+1, m]$. Weiner links are in some sense the inverses of *suffix links*, which lead from the suffix tree node u representing string cX to the node v representing string X, $slink(u) = v$; the Weiner link by c at node v is u, $wlink(v, c) = u$. If cX is not explicit but descends by string aW from its parent u', then X descends by aW from a node v' such that $wlink(v', c) = u'$, and v' is the closest ancestor of v with $wlink(\cdot, c)$ defined.

We use the CST of T [25], so that each node is identified by its preorder value in the parentheses sequence. We use mmphfs to represent the Weiner links. For each symbol $c \in [1, \sigma]$ we create a mmphf w_c and traverse the subtree T_c rooted at $child(root, c)$. As we traverse the nodes of T_c in preorder, the suffix links lead us to suffix tree nodes also in preorder (as the strings remain lexicographically sorted after removing their first c). By storing all those suffix link preorders in function w_c, we have that $w_c(v)$ gives in constant time $wlink(v, c)$ if it exists, and an arbitrary value otherwise. More precisely w_c gives preorder numbers within T_c; it is very easy to convert it to global preorder numbers.

Assume now we are in a suffix tree node v corresponding to suffix array interval $A[sp, ep]$ and pattern suffix $X = P[i+1, m]$. We wish to determine if the Weiner link $wlink(v, c)$ exists for $c = P[i]$. We can compute $w_c(v) = u$, so that *if the Weiner link exists*, then it leads to node u.

We can determine whether u is the correct Weiner link as follows. First, and assuming the preorder of u is within the bounds corresponding to T_c, we use the CST to obtain the range $A[sp', ep']$ corresponding to u [25]. Now we want to determine if the backward step with $P[i]$ from $A[sp, ep]$ leads us to $A[sp', ep']$ or not. Lemma 5 shows how this can be done using four *select* operations.

Lemma 5. *Let $A[sp, ep]$ be the suffix array interval for string X, then $A[sp', ep']$ is the suffix array interval for string cX iff*

$$select_c(T^{bwt}, i-1) < sp \quad \wedge \quad select_c(T^{bwt}, i) \geq sp, \text{ and}$$
$$select_c(T^{bwt}, j) \leq ep \quad \wedge \quad select_c(T^{bwt}, j+1) > ep,$$

where $i = sp' - C[c]$, $j = ep' - C[c]$, $C[c]$ is the number of occurrences of symbols $< c$ in the text T, and T^{bwt} is the BWT of T.

Proof. Note that the range of A for the suffixes that start with symbol c begins at $A[C[c]+1]$. Then $A[sp']$ is the ith suffix starting with c, and $A[ep']$ is the jth. The classical backward search formula (Section 2) for sp' is given next; then we transform it using *rank/select* inequalities. The formula for ep' is similar.

$$sp' = C[c] + rank_c(T^{bwt}, sp-1) + 1 \Leftrightarrow i - 1 = rank_c(T^{bwt}, sp-1)$$
$$\Leftrightarrow select_c(T^{bwt}, i-1) \leq sp-1 \wedge select_c(T^{bwt}, i) \geq sp. \qquad \square$$

Thus we have shown how, given a CST node v, compute $wlink(v, c)$ or determine it does not exists in time $O(t_{\mathsf{select}})$.[4] Now we describe a backward search process on the suffix tree instead of on the suffix array ranges.

The traversal. We start at the tree root with the empty suffix $P[m + 1, m]$. In general, being at tree node v corresponding to suffix $X = P[i + 1, m]$, we look for $u = wlink(v, c)$ for symbol $c = P[i]$. If it exists, then we have found node u corresponding to pattern suffix $cX = P[i, m]$ and we are done for that iteration.

If there is no Weiner link from v, it might be that cX is not a substring of T and the search should terminate. However, as explained, it might also be that there is no explicit suffix tree node for cX, but it falls between node u' representing a prefix Y of cX ($cX = YaW$) and node $u = child(u', a)$ representing string Z, of which cX is a prefix.

Our goal is to find node u, which corresponds to the same suffix array interval of cX. For this sake we consider the parent of v, its parent, and so on, until finding the nearest ancestor v' such that $u' = wlink(v', c)$ exists. If we reach the root without finding a Weiner link, then c is not in T, and neither is P. Once we have found u' we compute $u = child(u', a)$ and we finish.

However, computing *child* would be too slow for our purposes. Instead, we precompute it using a new mmphf w'_c, as follows. For each node $u = child(u', a)$ in T_c, store $v = child(slink(u'), a)$ in w'_c; note each v in T_c is stored exactly once. The preorders of v follow the same order of u, and thus if we call $u' = wlink(v', c)$ (or $v' = slink(u')$), we have the desired child in $w'_c(child(v', a)) = u$.

Now, if $wlink(v, c)$ does not exist, we traverse v and its successive ancestors v' looking for $w'_c(v')$. This will eventually reach node u, so we verify correctness of the mmphf values by comparing (using Lemma 5) the resulting interval directly with the suffix array interval of v. Note this test also establishes that cX is a prefix of Z. Only the suffix tree root cannot be dealt with w'_c, but we can easily precompute the σ nodes $child(root, c)$.

Actually only function w'_c is sufficient. Assume $wlink(v, c) = u$ exists. Then consider u', the parent of u. There will also be a Weiner link from an ancestor v' of v to u'. This ancestor will have a child v'' that points to $w'_c(v'') = u$, and either $v'' = v$ or v'' is an ancestor of v. So we do not check for $wlink(v, c)$ but directly v and its ancestors using w'_c.

Time and space. The total number of steps amortizes to $O(m)$: Each time we go to the parent the depth of our node in the suffix tree decreases. Each time we move by a Weiner link, the depth increases at most by 1, since for any branching node in the path to $u' = wlink(v', c)$ there is a branching node in the path to v'. Since we compute m Weiner links, the total number of operations is $O(m)$. All the operations in the CST tree topology take constant time, and therefore the time t_{select} dominates. Hence the overall time is $O(m \cdot t_{\mathsf{select}})$.

As for the space, the subtree T_c contains n_c leaves and at most $2n_c$ nodes; therefore mmphf w'_c stores at most $2n_c$ values in the range $[1, 2n]$. Therefore it

---

[4] Actually we could by chance get the right range $A[sp', ep']$ from an incorrect node, but this would just speed up the algorithm by finding u ahead of time.

requires space $O(n_c(\lg \lg \frac{n}{n_c} + 1))$ bits, which added over all $c \in [1, \sigma]$ gives a total of $O(n(\lg H_0(T) + 1))$, as in Section 3.

In order to reduce this space we partition the mmphfs according to the $O(\sigma^k)$ partitions of the BWT. Consider all the possible context strings C_i of length k,[5] their suffix tree node v_i, and their corresponding suffix array interval $A[sp_i, ep_i]$. The corresponding BWT partition is thus $S_i = T^{bwt}[sp_i, ep_i]$, of length $n_i = |S_i| = ep_i - sp_i + 1$. We split each function w'_c into $O(\sigma^k)$ subfunctions w^i_c, each of which will only store the suffix tree preorders that correspond to nodes descending from v_i. There are at most $2n_i$ consecutive preorder values below node v_i, thus the universe of the mmphf w^i_c is of size $O(n_i)$. Moreover, the links stored at w^i_c depart from the subtree that descends from string $cC[i]$, whose number of leaves is the number of occurrences of c in S_i, $n(c, i)$. Thus the total space of all the mmphfs is $\sum_{c,i} O(n(c, i)(\lg \lg \frac{n_i}{n(c,i)} + 1)) = O(n(\lg H_k(T) + 1))$ by the log-sum inequality (recall Section 3), as $nH_k(T) = \sum_{c,i} n(c, i) \lg \frac{n_i}{n(c,i)}$.

Note there are $O(\sigma^k)$ nodes with context shorter than k. A simple solution is to make a "partition" for each such node, increasing the space by $O(\sigma^k \lg n)$. It is easy, along our backward search, to know the context C_i we are in, and thus know which mmphf to query.

By combining the results of Section 4, using a sequence representation with $t_{\text{select}} = O(1)$, with our backward counting algorithm, we have the final result.

Theorem 3. *Given a text $T[1, n]$ over alphabet $[1, \sigma]$, one can build an FM-index occupying $nH_k(T) + o(nH_k(T)) + O(n + (n \lg n)/s + \sigma^{k+1} \lg n)$ bits of space for any $k \geq 0$ and $s > 0$, such that counting is supported in time $t_{\text{count}} = O(m)$, locating is supported in time $t_{\text{locate}} = O(s)$ and extraction of a substring of T of length ℓ in time $t_{\text{extract}} = O(s + \ell)$.*

8   Final Remarks

We have achieved alphabet independence on compressed self-indexes. This refers not only to time complexities: Even the space usage is independent of σ. The exception is the extra term $O(\sigma^{k+1} \lg n)$, but it rather limits k and it is essentially unavoidable under the k-th order empirical entropy model [12].

It is open whether we can reduce the $O(n)$ term to $o(n)$, as in the best current space result [2]. More ambitious is to achieve optimal times within optimal space, as already (partially) achieved when using $cnH_k(T)$ bits for $c > 2$ [14].

References

1. Apostolico, A.: The myriad virtues of subword trees. In: Combinatorial Algorithms on Words. NATO ISI Series, pp. 85–96. Springer, Heidelberg (1985)
2. Barbay, J., Gagie, T., Navarro, G., Nekrich, Y.: Alphabet partitioning for compressed rank/Select and applications. In: Cheong, O., Chwa, K.-Y., Park, K. (eds.) ISAAC 2010, Part II. LNCS, vol. 6507, pp. 315–326. Springer, Heidelberg (2010)

---

[5] Actually the compression booster [7] admits a more flexible partition into suffix tree nodes; we choose this way for simplicity of exposition.

3. Barbay, J., He, M., Munro, J.I., Rao, S.S.: Succinct indexes for strings, binary relations and multi-labeled trees. In: SODA, pp. 680–689 (2007)
4. Belazzougui, D., Boldi, P., Pagh, R., Vigna, S.: Monotone minimal perfect hashing: searching a sorted table with $o(1)$ accesses. In: SODA, pp. 785–794 (2009)
5. Belazzougui, D., Boldi, P., Pagh, R., Vigna, S.: Theory and practise of monotone minimal perfect hashing. In: ALENEX (2009)
6. Burrows, M., Wheeler, D.: A block sorting lossless data compression algorithm. Technical Report 124, Digital Equipment Corporation (1994)
7. Ferragina, P., Giancarlo, R., Manzini, G., Sciortino, M.: Boosting textual compression in optimal linear time. J. ACM 52(4), 688–713 (2005)
8. Ferragina, P., Grossi, R.: The string b-tree: A new data structure for string search in external memory and its applications. J. ACM 46(2), 236–280 (1999)
9. Ferragina, P., Manzini, G.: Opportunistic data structures with applications. In: FOCS, pp. 390–398 (2000)
10. Ferragina, P., Manzini, G.: Indexing compressed text. J. ACM 52(4), 552–581 (2005)
11. Ferragina, P., Manzini, G., Mäkinen, V., Navarro, G.: Compressed representations of sequences and full-text indexes. ACM Trans. Alg. 3(2), article 20 (2007)
12. Gagie, T.: Large alphabets and incompressibility. Inf. Proc. Lett. 99(6), 246–251 (2006)
13. Golynski, A., Munro, J.I., Rao, S.S.: Rank/select operations on large alphabets: a tool for text indexing. In: SODA, pp. 368–373 (2006)
14. Grossi, R., Gupta, A., Vitter, J.: High-order entropy-compressed text indexes. In: SODA, pp. 841–850 (2003)
15. Grossi, R., Orlandi, A., Raman, R.: Optimal trade-offs for succinct string indexes. In: Abramsky, S., Gavoille, C., Kirchner, C., Meyer auf der Heide, F., Spirakis, P.G. (eds.) ICALP 2010. LNCS, vol. 6198, pp. 678–689. Springer, Heidelberg (2010)
16. Grossi, R., Vitter, J.: Compressed suffix arrays and suffix trees with applications to text indexing and string matching. In: STOC, pp. 397–406 (2000)
17. Lee, S., Park, K.: Dynamic rank-select structures with applications to run-length encoded texts. In: Ma, B., Zhang, K. (eds.) CPM 2007. LNCS, vol. 4580, pp. 95–106. Springer, Heidelberg (2007)
18. Manber, U., Myers, G.: Suffix arrays: a new method for on-line string searches. SIAM J. Comp. 22(5), 935–948 (1993)
19. Manzini, G.: An analysis of the Burrows-Wheeler transform. J. ACM 48(3), 407–430 (2001)
20. Munro, I.: Tables. In: Chandru, V., Vinay, V. (eds.) FSTTCS 1996. LNCS, vol. 1180, pp. 37–42. Springer, Heidelberg (1996)
21. Navarro, G., Mäkinen, V.: Compressed full-text indexes. ACM Comp. Surv. 39(1), article 2 (2007)
22. Raman, R., Raman, V., Rao, S.: Succinct indexable dictionaries with applications to encoding k-ary trees and multisets. In: SODA, pp. 233–242 (2002)
23. Sadakane, K.: Compressed text databases with efficient query algorithms based on the compressed suffix array. In: Lee, D.T., Teng, S.-H. (eds.) ISAAC 2000. LNCS, vol. 1969, pp. 295–321. Springer, Heidelberg (2000)
24. Sadakane, K.: New text indexing functionalities of the compressed suffix arrays. J. Alg. 48(2), 294–313 (2003)
25. Sadakane, K.: Compressed suffix trees with full functionality. Theo. Comp. Sys. 41(4), 589–607 (2007)
26. Sadakane, K., Navarro, G.: Fully-functional succinct trees. In: SODA, pp. 134–149 (2010)
27. Weiner, P.: Linear pattern matching algorithm. In: Proc. Ann. IEEE Symp. on Switching and Automata Theory, pp. 1–11 (1973)

Distribution-Aware Compressed Full-Text Indexes*

Paolo Ferragina[1], Jouni Sirén[2], and Rossano Venturini[3]

[1] Dept. of Computer Science, Univ. of Pisa
ferragina@di.unipi.it
[2] Dept. of Computer Science, Univ. of Helsinki
jltsiren@cs.helsinki.fi
[3] ISTI-CNR, Pisa
rossano.venturini@isti.cnr.it

Abstract. In this paper we address the problem of building a compressed self-index that, given a distribution for the pattern queries and a bound on the space occupancy, minimizes the expected query-time within that index-space bound. We solve this problem by exploiting a reduction to the problem of finding a minimum weight K-link path in a particular Directed Acyclic Graph. Interestingly enough, our solution is independent of the underlying compressed index in use. Our experiments compare this optimal strategy with several other standard approaches, showing its effectiveness in practice.

1 Introduction

String processing and searching tasks are at the core of modern web search, IR, data base and data mining applications. Most of text manipulations required by these applications involve, sooner or later, *searching* those (long) texts for (short) patterns or *accessing* portions of those texts for subsequent processing/mining tasks. Despite the increase in processing speeds of current CPUs and memories/disks, sequential text searching long ago ceased to be a viable approach, and indexed text searching has became mandatory.

Data compression and indexing seem "opposite approaches" because the former aims at removing data redundancies, whereas the latter introduces extra data in the index to support faster operations. This dichotomy was successfully addressed starting from the year 2000 [4,7], due to various scientific achievements that showed how to relate Information Theory with String-Matching concepts, in a way that index regularities that show up when data is compressible are discovered and exploited to reduce index occupancy without impairing query efficiency (see the surveys [11,3] and references therein). The net result has been

* This work was partially supported by EU-PSP-BPN-250527 (ASSETS), POR-FESR 2007-2013 No 63748 (VISITO Tuscany) projects, MIUR of Italy under project PRIN MadWeb 2008, Yahoo! Research, Finnish Doctoral Programme in Computational Sciences, Academy of Finland (project 1140727), and Nokia Foundation.

C. Demetrescu and M.M. Halldórsson (Eds.): ESA 2011, LNCS 6942, pp. 760–771, 2011.

the design of *compressed data structures* for indexing texts (aka *compressed indexes*, or *compressed and searchable data formats*) that take space close to the *k*th order entropy of the input text, and support the powerful *substring* queries and the *extraction* of arbitrary portions of data. Due to this latter feature, these data structures are sometime called *self*-indexes.

As experimentally shown in [3,5], these self-indexes are very space-efficient (close to best known compressors), and most of them are particularly fast in counting the number of occurrences of the input pattern. Their bottleneck is in the Locate queries, which are roughly between two and three order of magnitude slower than what is achievable with the classic Suffix Array data structure. Also the Extract operation is quite slow compared with other compression methods for sufficiently long decompressed portions. In addition, for Locate and Extract, these indexes need to store some extra information that induces a trade-off between space and time efficiency: the larger is this extra space, the faster is the resulting index. At high level, the extra information is obtained by *sampling* entries of the suffix array at regular distance s_{SA}. This parameter governs the space/time trade-off, because on one hand, each occurrence of the searched pattern is located in at most s_{SA} steps; but on the other hand, the space required is $O(\frac{n \log n}{s_{SA}})$ bits, where n is the length of the indexed text.

Even though the last years have seen a proliferation of different compressed full text indexes [11,3], the above sampling strategy has remained almost unchanged since the very first proposals. This strategy implicitly assumes all text positions to have uniform probability of being located or extracted. But uniform distributions are rare in practice, where we often observe (very) skewed distributions. For example, it is well-known that requests in IR or database systems are drawn accordingly to power law or Zipfian distributions (e.g. see [14] and references therein).

Given these premises we address in this paper the following question: Is it possible to build a distribution-aware compressed self-index that optimizes the expected query-time by occupying a given space? Namely, given the distribution of the subsequent queries and a bound on the space occupancy, the goal is to find a sampling strategy that induces that space bound and minimizes the expected time required for solving Locate/Extract queries drawn accordingly to the input distribution. We solve this problem by exploiting a reduction to the problem of finding a minimum weight K-link path in a particular Directed Acyclic Graph (DAG) (Section 3). Interestingly enough, our solution provides a way to optimally select a set of sampled positions that could be blindly used by mostly known compressed indexes without changing their Locate/Extract algorithms.

In the experimental section (Section 4) we compare our optimal sampling strategy against several other strategies over two large datasets of HTML pages and XML documents. The experiments were performed by using RLCSA, which is an implementation of Compressed Suffix Array (Csa). Although restricted to this single index, our experiments quantify some measures that are independent on the particular implementation of a compressed index, and thus can be adopted to extrapolate conclusions for other indexes as well. Overall we show that our optimal sampling is from 4 to 10 times faster than regular sampling. We also compare our optimal strategy against two heuristic approaches, showing that ours is up to a factor 2.5 faster. One of them is the immediate strategy that "caches" the most probably accessed positions. This heuristic can be poor both in theory and in practice due to the fact that it does not consider interdependencies among sampled positions induced by Locate and Extract algorithms. Roughly speaking, in many circumstances it is more convenient to sample a position whose access probability is not among the top, provided that it is followed by positions having sufficiently high access probabilities. Discovering all these cases is a peculiarity of our optimal solution. These considerations are explained more formally in Section 4, where we quantify also the impact of the various heuristics by performing a significant set of experiments.

2   Background

The large space occupancy of (classical) full-text indexes, like Suffix Tree and Suffix Array, has driven researchers to design the so-called *compressed* full-text indexes. These indexes deploy algorithmic techniques and mathematical tools that lie at the crossing point of three distinct fields: data compression, data structures and databases (see e.g. [4,7,11,3]). Most of these indexes can be classified into two families — namely, FM-indexes (Fmi) and Compressed Suffix Arrays (Csa) — and achieve efficient query times and space close to the one achievable by the best known compressors. In theory, these indexes require $O(nH_k(T)) + o(n \log \sigma)$ bits of space, where $H_k(T)$ is the kth order empirical entropy of text T of length n, and σ is the alphabet size. This bound is appealing because it can be sublinear in $n \log \sigma$ for highly compressible texts. We recall that $nH_k(T)$ is the classic Information-Theoretic lower bound to the storage complexity of T by means of any k-th order compressor (see e.g. [10] for more details). In addition to being compressed, the index is able to efficiently support the following three operations:

- Count($P[1, p]$) returns the number of occurrences of pattern P in the text;
- Locate($P[1, p]$) returns the starting positions of all occurrences of pattern P in the text;
- Extract(l, r) extracts the substring $T[l, r]$.

2.1   The FM-Index Family

These compressed indexes were introduced by Ferragina and Manzini in [4], who devised a way to utilize the relation between the suffix array data structure and the *Burrows-Wheeler Transform* (shortly, Bwt [2]) in efficient time and space. The Bwt is a reversible transformation that permutes the symbols of the input string T into a new string $L = \mathsf{Bwt}(T)$ that is easier to compress. This permutation is the last column of a conceptual matrix $\mathcal{M}(T)$ whose rows are the cyclic rotations of string $T\$$ in lexicographic order.

It is well-known that the original text T can be obtained backwards from L by resorting to a function LF that maps row indexes to row indexes, and is defined as follows [4]: if the Bwt maps $T[j-1]$ to $L[i']$ and $T[j]$ to $L[i]$, then $LF(i) = i'$ (so LF implements a sort of *backward* step over T). Now, since the first row of $\mathcal{M}(T)$ is $\$T$, it can be stated that $T[n] = L[0]$ and, in general, $T[n-i] = L[LF^i(0)]$, for $i = 1, \ldots, n-1$.

Ferragina and Manzini [4] proposed a way to combine the compressibility of the Bwt with the indexing power of the suffix array. In particular, showed that searching operations on T can be reduced to counting queries of *single* symbols in L, now called rank operations. For any symbol $c \in \Sigma$ and position i in L, the query $\mathrm{rank}_c(L, i)$ returns how many times the symbol c appears in $L[1, i]$. An FM-index then consists of three key tools: a compressed representation of $\mathrm{Bwt}(T)$ that supports efficient rank queries, a small array $C[c]$ that tells how many symbols smaller than c appear in T (this takes $O(\sigma \log n)$ bits), and the so called *backward search* algorithm that implements the Count query by using the two structures. More precisely, Fmi searches the pattern $P[1, p]$ backwards in p steps, which eventually identify the interval of text suffixes that are prefixed by P or, equivalently, the interval of rows of $\mathcal{M}(T)$ that are prefixed by P. This is done by maintaining, inductively for $i = p, p-1, \ldots, 1$, the interval $SA[sp_i, ep_i]$ that stores all text suffixes that are prefixed by the pattern suffix $P[i, p]$. The final interval $SA[sp_1, ep_1]$, if any, corresponds to all the suffixes that are prefixed by the pattern $P[1, p]$. Thus, Count(P) can be solved by returning the value $occ = ep_1 - sp_1 + 1$. Since each of the above steps requires the computation of two rank queries over the strings L, $O(p)$ ranks suffice to count the number of occurrences of any pattern P.

In practice, there are various implementations of Fmi, with their main differences in the way the rank-data structure built on $\mathrm{Bwt}(T)$ is compressed. The site Pizza&Chili[1] has several implementations of Fmi that mainly boil down to the following trick: $\mathrm{Bwt}(T)$ is split into blocks (of equal or variable length) and values of rank_c are precomputed for all block beginnings and all symbols $c \in \Sigma$. A query $\mathrm{rank}_c(L, i)$ is solved by summing up the answer available for the beginning of the block that contains $L[i]$, plus the rest of the occurrences of c in that block — they are obtained either by sequentially decompressing the block or by using a proper compressed data structure built on it (e.g. the Wavelet Tree of [6]). The former approach favors compression, the latter favors query speed.

2.2   The CSA Family

These compressed indexes were introduced by Grossi and Vitter [7], who showed how to compactly represent the suffix array SA in $O(n \log \sigma)$ bits and still be able to access any of its entries efficiently. Their solution is based on a function Ψ, which is the inverse of the function LF introduced for Bwt:

$$\Psi(i) = \begin{cases} i' \text{ such that } SA[i'] = SA[i] + 1 & (if \, SA[i] < n) \\ i' \text{ such that } SA[i'] = 1 & (if \, SA[i] = n) \end{cases}$$

In other words, $\Psi(i)$ refers to the position in the suffix array of the text suffix that follows $SA[i]$ in T, namely, the text suffix which is one symbol shorter.

---

[1] http://pizzachili.dcc.uchile.cl/ or http://pizzachili.di.unipi.it/.

Grossi and Vitter show how to hierarchically decompose the suffix array SA in order to obtain its succinct representation that still permits to perform searching operation on it. In their construction they exploit the *piecewise increasing property* of Ψ — namely that $\Psi(i) < \Psi(i+1)$ if $T[SA[i]] = T[SA[i+1]]$ — to represent the suffix array within $O(n \log \sigma)$ bits. The index must keep the original text in a non-compressed form to explicitly compare symbols of the text and the pattern during the searches.

This drawback has been overcome by two subsequent improvements. The first one, due to Sadakane [12], showed that the original text T can be replaced with a binary vector F such that $F[i] = 1$ iff the first symbol of the suffixes $SA[i-1]$ and $SA[i]$ differs. Since the suffixes in SA are lexicographically sorted, one can determine the first symbol of any suffix in constant time by just executing a $rank_1$ query on F. This fact, combined with the retrieval of Ψ's values in constant time, allows to compare any suffix with the searched pattern $P[1,p]$ in $O(p)$ time. Sadakane also provided an improved representation for Ψ achieving $nH_0(T)$ bits. Theoretically, the best variant of Csa is due to Grossi, Gupta and Vitter [6] who used some further structural properties of Ψ to get close to $nH_k(T)$ bits, still preserving the previous time complexity.

In practice, one of the best implementation of the Csa is the one proposed by Sadakane. It does not use the hierarchical decomposition, but orchestrates a compact representation of the function Ψ together with the backward search of the Fmi family.

2.3   Locate and Extract Queries

Even though in the last years we have seen a proliferation of different compressed full text indexes [11,3], Locate and Extract strategies remain almost unchanged since the very first proposals. At a high level, the idea consists in storing the relation between text positions and indexes in the suffix array of some sampled positions of the original text. Recall that Locate(P) requires to return the position $pos(i) = SA[i]$ of any suffix i, while Extract(l, r) extracts the substring $T[l, r]$. Locate is solved by starting from the ith suffix and by going backward or forward in the text by means of LF or Ψ functions. The procedure stops whenever a sampled position is found. Extract(l, r) is solved with a Fmi by starting from the sampled position closest to r and extracting the substring $T[l, r]$ bacwards symbol by symbol. The same strategy is used in Csa, except that we proceed forward starting from the sampled position closest to l.

The Locate algorithm of Fmi and (a practical implementation of) Csa is shown in Fig. 1. This algorithm is used to obtain the position in the text of the suffix that prefixes the ith row of $\mathcal{M}(T)$. As we said, the basic idea is to logically mark a suitable set of rows of $\mathcal{M}(T)$, and keep for each of them their position in T (that is, we store the corresponding SA values). Then, Locate(i) scans the text T backward using the LF-mapping, until a sampled row i' is found, and reports $SA[i'] + t$, where t is the number of backward steps used to find such i'. Csa works by going forward in the text by using Ψ function. To compute the

Algorithm Fmi-Locate(i)	Algorithm Csa-Locate(i)
$i' \leftarrow i, t \leftarrow 0;$	$i' \leftarrow i, t \leftarrow 0;$
while $SA[i']$ is not explicitly stored **do**	**while** $SA[i']$ is not explicitly stored **do**
$\quad i' \leftarrow LF(i');$	$\quad i' \leftarrow \Psi(i');$
$\quad t \leftarrow t + 1;$	$\quad t \leftarrow t + 1;$
return $SA[i'] + t;$	**return** $SA[i'] - t;$

Fig. 1. Algorithms for locating the row with index i in Fmi and Csa

positions of all occurrences of a pattern P, it is thus enough to call Locate(i) for all rows identified by the Count(P) operation.

The sampling rate of $\mathcal{M}(T)$'s rows, hereafter denoted by s_{SA}, is a crucial parameter that trades space for query time. Most Fmi and Csa implementations [3] sample all the $SA[i]$ that are a multiple of s_{SA}. This guarantees that at most s_{SA} steps of LF (or Ψ) suffice for locating the text position of any occurrence. The extra space required to store these positions is $O(\frac{n \log n}{s_{SA}})$ bits. In addition to these positions, we need to store a data structure that is able to, given a row, tell us if the row is sampled and, in that case, return its position in the text. An immediate solution resorts to a bitmap $B[1,n]$ whose ith entry is 1 iff the ith row is sampled. Then, all the sampled $SA[i]$s are stored contiguously in suffix array order, so that if $B[i] = 1$ then one finds the corresponding $SA[i]$ at position $rank_1(B, i)$ in that contiguous storage. In this case the extra space becomes $\frac{n \log n}{s_{SA}} + n + o(n)$ bits. There exist other more space efficient, but probably less practical, solutions. For example, one could resort to Minimal Perfect Hash functions [8]: we create a perfect hash function for the set of marked rows having their positions as satellite data. In this case the extra space is $O(\frac{n \log n}{s_{SA}})$ bits.

For our discussion it is more convenient to sample text positions instead of sampling rows of matrix $\mathcal{M}(T)$. Since there is one-to-one correspondence between $\mathcal{M}(T)$'s rows and text's positions, the problem of sampling positions is exactly the same as the problem of sampling rows.

The algorithm for Extract(l, r) resorts to a similar approach. Each query takes no more than $(r - l + s_{SA} + 1)$ rank queries: at most s_{SA} rank queries are required to reach r starting from the closest sampled position, and $r - l + 1$ queries are required to extract the substring $T[l, r]$ symbol by symbol.

The net result is that the space and time complexities of Fmi and Csa depend on the value s_{SA} and on the performance guaranteed by the data structure used to compute rank queries on the Bwt-string. The extra space required by the best (theoretical) data structures added to support Locate and Extract is bounded by $O((n \log n)/s_{SA})$ bits, which is $o(n)$ whenever s_{SA} is large enough.

3 Optimal Distribution-Aware Locate and Extract

The problem we address in this paper is defined as follows. We assume that, for any position j of the input text T, we know the probability $\Pr(j)$ that the position j will be located (i.e., the probability that we search a pattern P which is a prefix of the jth suffix of T). We have the user defined parameter s_{SA} that specifies the amount of the space that we can use to store information regarding sampled positions. Our aim is that of identifying a optimal set of sampled positions $\mathcal{P}^*$ of size $K = n/s_A$ that allows us to minimize the expected time required to solve Locate queries. The expected time is given by

$$E[\mathcal{P}^*] = \sum_{j=1}^{n} \Pr(j) \cdot c(j, \mathcal{P}^*)$$

where $c(j, \mathcal{P}^*)$ is the cost (e.g., time or number of backward steps) required to reach the first sampled position in $\mathcal{P}^*$, say i, that precedes j in T. We call this problem the *distribution-aware optimal sampling problem*.

We observe that there are several different ways to define $c(j, \mathcal{P})$. For example, by setting $c(j, \mathcal{P}) = j - i$, we are simply counting the number of backward steps required to reach position i from j. This implies that we are implicitly assuming that all the backward steps have the same cost (in terms of CPU usage). Or one could refine the measure by setting $c(j, \mathcal{P})$ to be the sum of the *real cost* of the backward steps required to reach position i from j. To simplify the discussion we will use the first cost type.

We can address the problem of optimally sampling positions for Extract queries by changing the cost function $c()$. In this case, $\Pr(j)$ is the probability of extracting a substring that starts at position i and $c(j, \mathcal{P})$ is the cost of reaching position j starting from the first sampled position in $\mathcal{P}$ that follows j.

The discussion above implicitly assumes that we are dealing with a Fmi. As a Csa scans the text forward in Locate, and starts from the closest sampled position before the substring in Extract, the cost functions are used in the opposite way.

3.1   On Finding a Minimum Weight K-Link Path over a DAG

The Distribution Optimal Sampling Problem can be reduced to the problem of *finding a minimum weight K-link path* [1,13] in a particular Directed Acyclic Graph (DAG) $\mathcal{G}_R$. Given a weighted DAG $\mathcal{G}_R$ and a parameter K, the problem of finding a minimum weight K-link asks to identify a path from v_1 to v_{n+1} consisting of exactly K edges, whose cost is the minimum among all such paths.

In our solution the graph $\mathcal{G}_R$ has a vertex for each text position denoted $v_1, v_2, \ldots, v_n$ plus a dummy vertex v_{n+1} that marks the end of the text. For any pair of positions i and j such that $1 \le i < j \le n+1$, we have an edge (v_i, v_j), whose cost $w(i, j)$ is equal to $\sum_{l=i}^{j-1} \Pr(l) \cdot (l - i)$. Intuitively, $w(i, j)$ accounts the part of expected cost for locating positions between i and $j - 1$, assuming that i is the only sampled position among them.

Efficient solutions for the problem of computing a minimum weight K-link path have been provided in literature [1,13], if the DAG satisfies the so-called *concave Monge condition*.

Definition 1. *A weighted DAG $\mathcal{G}$ satisfies the concave Monge condition if*

$$w(i, j) + w(i + 1, j + 1) \le w(i, j + 1) + w(i + 1, j)$$

holds for all $1 < i + 1 < j < n$.

Lemma 1. *The DAG $\mathcal{G}_R$ satisfies the concave Monge condition.*

The best known solutions for the computation of a minimum weight K-link path on a DAG satisfying the concave Monge condition are summarized in the following Theorems (Proved in [1] and [13]).

Theorem 1. *Given a DAG $\mathcal{G}$ satisfying the concave Monge condition and whose weights are integers, a minimum weight K-link path in $\mathcal{G}$, for any K, can be computed in $O(n \log U)$ time, where U is the maximum absolute value of the weights.*

Theorem 1 provides a weakly polynomial algorithm for the problem, which suffices for most of the interesting cases in practice. In fact, the probabilities of locating positions are typically frequencies derived by observing queries in a query-log of total length, say, m. Thus, we can label the edges of $\mathcal{G}_R$ with integral weights by appropriately multiplying each of these frequencies by m. In this way, the factor $\log U$ in the time complexity of Theorem 1 is $O(\log n + \log m)$. For completeness, we notice that there exists also a solution whose time complexity is independent of the weights.

Theorem 2. *Given a DAG $\mathcal{G}$ satisfying the concave Monge condition, a minimum weight K-link path in $\mathcal{G}$ can be computed in $O(nK^\epsilon)$ time for $K = \Omega(\log n)$ and any fixed ϵ.*

4   Experiments

We implemented our Optimal sampling strategy by resorting to the algorithm of Theorem 1. The algorithm uses binary search to find an adjustment Q, such that $\mathcal{G}_R$ has a minimum weight path from v_1 to v_{n+1} with n/s_{SA} edges, when Q is added to all edge weights. That path is then a minimum-weight n/s_{SA}-link path in $\mathcal{G}_R$. For each candidate of Q, we search for the shortest and the longest minimum-weight paths. If n/s_{SA} falls between the extremes, then a n/s_{SA}-link path can be built by combining the shortest and the longest paths. As we use a simple $O(n \log n)$-time algorithm [9] for finding the minimum-weight paths, the overall time bound is $O(n \log n \log U)$. In practice, the bound is quite pessimistic.

In addition to our Optimal sampling strategy, we implemented three other strategies. Regular sampling is the classical strategy that samples one out of every s_{SA} positions. Greedy sampling selects n/s_{SA} text positions with the largest access probabilities. HalfGreedy first uses regular sampling with rate $2s_{SA}$, and then greedily selects $n/(2s_{SA})$ of the remaining positions.

Before presenting experimental results about these approaches, it is worth to compare the behavior of these strategies for their worst-case distribution with respect to our Optimal strategy. We present these considerations just for Locate, since similar bounds hold for Extract too.

The worst distribution for Regular is clearly the one in which there are n/s_{SA} positions with probability s_{SA}/n, while the others have chance 0 of being located. Each of these positions follows one of the positions that have been sampled by

Regular. Thus, the expected time to solve a locate is $O(s_{SA})$. Clearly, Optimal strategy achieves expected time equal to $O(1)$ by simply sampling all the positions having a positive probability.

Greedy is much worse. Consider the following distribution: each of the first n/s_{SA} positions of the text has probability $\frac{s_{SA}}{n-1}$, while the last n/s_{SA} positions have probability $\frac{s_{SA}}{n+1}$. Greedy wrongly selects the first n/s_{SA} positions, leaving a large part of the text unsampled. Thus its expected time is at least $\Theta(n - n/s_{SA})$.[2] On this distribution Optimal performs much better by sampling every other position with positive probability. In this way, it achieves an expected time of $O(1)$. As far as HalfGreedy is concerned, we observe that its worst expected time is $2s_{SA}$, and this is obtained by using a distribution which is a mixture of the ones used for Regular and Greedy.

The distributions above are specifically designed to highlight the drawbacks of the other strategies. In the remaining part of the section we experimentally compare these strategies on real datasets and with real query-distributions. As we will see, even in this practical setting, Optimal provides a less impressive but yet significant improvement. The different sampling strategies have been plugged in the compressed index RLCSA[3].

The implementation was written in C++ and compiled on g++ version 4.3.3. Experiments were done on a system with 32 gigabytes of memory and two quad-core Intel Xeon E5540 processors running at 2.53 GHz (we used only one core). The system was running Ubuntu 10.04 with Linux kernel 2.6.32. As the optimal sampling requires about $28n$ bytes of memory for a text of length n, we had to use another system with more memory for constructing some of the indexes.

We use two large datasets in the experiments. Html Pages is a 1.24-gigabyte set of web pages obtained by downloading the first 5 Yahoo! search results for all query terms with at least 100 occurrences in a MSN query log. dblp contains the DBLP Computer Science Bibliography[4] in XML format, for a total size of 813 megabytes. Both datasets were downloaded in March 2011.

The set of patterns to be searched for Html Pages was constructed by selecting all terms from the MSN query log and by removing stop words. Each pattern was associated the number of its occurrences in the query log. For dblp, we built a synthetic set of patterns obtained by selecting all author names and all non-stop word terms appearing in paper titles. Each term has associated a number of occurrences that is taken from the previous set of patterns. From the two sets of patterns, we computed the access frequency of each position of the text as follows. For position i, we set its frequency to be the sum of the number of occurrences of those patterns that are prefixes of suffix $T[i, n]$. The frequencies of all positions (suffixes) are plotted in Figure 2 after they have been sorted decreasingly.

---

[2] Notice that at least $n - 2n/s_{SA}$ steps are required to locate each of the last n/s_{SA} positions.

[3] Available at http://www.cs.helsinki.fi/group/suds/rlcsa/.

[4] http://dblp.uni-trier.de/db/

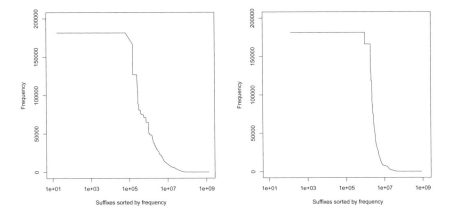

Fig. 2. Distributions of position access frequencies for Html Pages (left) and dblp (right)

Table 1. Average number of LF or Ψ steps required to locate pattern occurrences depending on value of s_{SA} and sampling strategy in use

	Html Pages				dblp			
s_{SA}	Regular	HalfGreedy	Greedy	Optimal	Regular	HalfGreedy	Greedy	Optimal
16	7.5	0.7	0.2	0.1	7.5	0.15	0.005	0.004
32	15.5	4.7	3.0	0.9	15.5	1.2	0.6	0.3
64	31.5	13.9	42.5	4.2	31.5	7.2	4.4	1.9
128	63.5	43.0	104.2	14.7	63.5	26.0	31.4	8.9

For our experiments, we built RLCSA with $s_{SA} = \{16, 32, 64, 128\}$ for both datasets. We searched for $10,000$ patterns randomly selected accordingly to the previously constructed query distributions for a total of about 187.3 million located positions for Html Pages and about 276.6 million positions for dblp. We also extracted snippets of length 16, 32, and 64 from $1,000,000$ randomly selected positions according to position frequencies. In addition to measuring the number of located positions and extracted characters per second (Figure 3), we also determined the average number of LF/Ψ steps required to find a sampled position (see Table 1).

All distribution-aware strategies performed similarly in Locate with low values of s_{SA}, being almost 8 times faster than Regular. This behavior is due to the fact that, for small values of s_{SA}, the distribution-aware strategies are able to sample most of the positions with positive frequencies. With larger s_{SA}, Optimal retained its lead, while Greedy and HalfGreedy became worse. The highest gain of Optimal w.r.t. to Regular is obtained for $s_{SA} = 32$ (factors 8.4 and 10.1 for Html Pages and dblp respectively) while the lowest is obtained for $s_{SA} = 128$ (factors 5.6 and 4.4 respectively). In Extract, Optimal is roughly twice faster than the other strategies. The gain is limited due to the fact that, in any case,

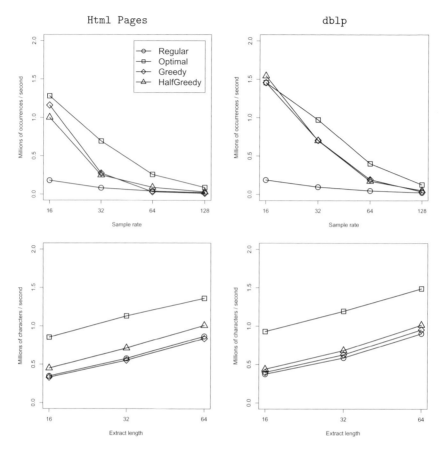

Fig. 3. Experimental results for Html Pages (left) and dblp (right). Locate performance (top) and Extract performance with $s_{SA} = 128$ (bottom).

Extract requires c steps to extract c symbols after finding a sampled position. We observed an odd result for $s_{SA} = 16$ on dblp where HalfGreedy is slightly better than Optimal. This is due to the last step of Locate algorithm that checks if the current row index is sampled or not. Whenever the average number of LF/Ψ steps is close to 0, the cost of this step becomes dominant. This step is usually performed by resorting to rank/select queries over a bit vector. In the current implementation of RLCSA, the time cost of this operation may slightly vary depending on the underlying bit vector. In the final version of the paper, we will investigate the possibility of designing more suitable solutions for this step as well as compare other implementations of compressed indexes.

5   Future Work

In this paper we addressed the problem of designing distribution-aware compressed full-text indexes. We showed that an optimal selection of positions can be computed efficiently in time and space when the distribution of subsequent queries is known beforehand. The advantage at query time is between 4–10 times better than the classical approach to Locate. In case of Extract the advantage is reduced to 2.

An interesting open problem asks for designing distribution-aware compressed indexes that are able to self-adapt themselves to the unknown distribution of queries. We believe that the field of compressed data structures could benefit a lot by following this line of research.

References

1. Aggarwal, A., Schieber, B., Tokuyama, T.: Finding a minimum-weight k-link path graphs with the concae monge property and applications. Discrete & Computational Geometry 12, 263–280 (1994)
2. Burrows, M., Wheeler, D.: A block sorting lossless data compression algorithm. Technical Report 124, Digital Equipment Corporation (1994)
3. Ferragina, P., González, R., Navarro, G., Venturini, R.: Compressed text indexes: From theory to practice. ACM Journal of Experimental Algorithmics, 13 (2008)
4. Ferragina, P., Manzini, G.: Indexing compressed text. J. ACM 52(4), 552–581 (2005)
5. Ferragina, P., Manzini, G.: On compressing the textual web. In: WSDM, pp.391–400 (2010)
6. Grossi, R., Gupta, A., Vitter, J.: High-order entropy-compressed text indexes. In: Proc. 14th ACM-SIAM Symposium on Discrete Algorithms (SODA), pp. 841–850 (2003)
7. Grossi, R., Vitter, J.: Compressed suffix arrays and suffix trees with applications to text indexing and string matching. In: Proc. of the 32nd ACM Symposium on Theory of Computing, pp. 397–406 (2000)
8. Hagerup, T., Tholey, T.: Efficient minimal perfect hashing in nearly minimal space. In: Ferreira, A., Reichel, H. (eds.) STACS 2001. LNCS, vol. 2010, pp. 317–326. Springer, Heidelberg (2001)
9. Hirschberg, D.S., Larmore, L.L.: The least weight subsequence problem. SIAM Journal on Computing 16(4), 628–638 (1987)
10. Manzini, G.: An analysis of the Burrows-Wheeler transform. Journal of the ACM 48(3), 407–430 (2001)
11. Navarro, G., Mäkinen, V.: Compressed full-text indexes. ACM Computing Surveys 39(1) (2007)
12. Sadakane, K.: New text indexing functionalities of the compressed suffix arrays. J. Algorithms 48(2), 294–313 (2003)
13. Schieber, B.: Computing a minimum weight k-link path in graphs with the concave monge property. J. Algorithms 29(2), 204–222 (1998)
14. Silvestri, F.: Mining query logs: Turning search usage data into knowledge. Foundations and Trends in Information Retrieval 4(1-2), 1–174 (2010)

Smoothed Performance Guarantees for Local Search

Tobias Brunsch[1], Heiko Röglin[1], Cyriel Rutten[2], and Tjark Vredeveld[2]

[1] Dept. of Computer Science
University of Bonn, Germany
brunsch@cs.uni-bonn.de, heiko@roeglin.org
[2] Dept. of Quantitative Economics
Maastricht University, The Netherlands
{c.rutten,t.vredeveld}@maastrichtuniversity.nl

Abstract. We study popular local search and greedy algorithms for scheduling. The performance guarantee of these algorithms is well understood, but the worst-case lower bounds seem somewhat contrived and it is questionable if they arise in practical applications. To find out how robust these bounds are, we study the algorithms in the framework of smoothed analysis, in which instances are subject to some degree of random noise.

While the lower bounds for all scheduling variants with restricted machines are rather robust, we find out that the bounds are fragile for unrestricted machines. In particular, we show that the smoothed performance guarantee of the jump and the lex-jump algorithm are (in contrast to the worst case) independent of the number of machines. They are $\Theta(\phi)$ and $\Theta(\log \phi)$, respectively, where $1/\phi$ is a parameter measuring the magnitude of the perturbation. The latter immediately implies that also the smoothed price of anarchy is $\Theta(\log \phi)$ for routing games on parallel links. Additionally we show that for unrestricted machines also the greedy list scheduling algorithm has an approximation guarantee of $\Theta(\log \phi)$.

1 Introduction

The performance guarantee of local search and greedy algorithms for scheduling problems is well studied and understood. For most algorithms, matching upper and lower bounds on their approximation ratio are known. The lower bounds are often somewhat contrived, however, and it is questionable if they resemble typical instances in practical applications. For that reason, we study these algorithms in the framework of smoothed analysis, in which instances are subject to some degree of random noise. By doing so, we find out for which heuristics and scheduling variants the lower bounds are robust and for which they are fragile and not very likely to occur in practical applications. Since pure Nash equilibria can be seen as local optima, our results also imply a new bound on the smoothed price of anarchy, showing that known worst-case results are too pessimistic in the presence of noise.

C. Demetrescu and M.M. Halldórsson (Eds.): ESA 2011, LNCS 6942, pp. 772–783, 2011.

Let us first describe the scheduling problems that we study. We assume that there is a set $J = \{1, \ldots, n\}$ of jobs each of which needs to be processed on one of the machines from the set $M = \{1, \ldots, m\}$. All jobs and machines are available for processing at time 0. The goal is to schedule the jobs on the machines such that the *makespan*, i.e., the time at which the last job is completed, is minimized. Each machine $i \in M$ has a speed s_i and each job $j \in J$ has a processing requirement p_j. We assume w.l.o.g. that $p_j \in [0, 1]$. The time p_{ij} it takes to fully process job j on machine i depends on the machine environment. We consider two machine environments. The first one is the one of *uniform parallel machines*, also known as *related machines*: $p_{ij} = p_j/s_i$. The second machine environment that we consider is the one of *restricted related machines*: a job j is only allowed to be processed on a subset $\mathcal{M}_j \subseteq M$ of the machines. The processing time is therefore $p_{ij} = p_j/s_i$ if $i \in \mathcal{M}_j$ and $p_{ij} = \infty$ if $i \notin \mathcal{M}_j$. An instance I for the scheduling problem consists of the machine speeds $s_1, \ldots, s_m$, the processing requirements $p_1, \ldots, p_n$, and in the restricted case the allowed machine sets $\mathcal{M}_j \subseteq M$ for every job j.

A special case for both machine environments is when all speeds are equal, i.e., $s_i = 1$ for all $i \in M$. In this case, we say that the machines are identical. In the notation of Graham et al. [12] these problems are denoted by $Q||C_{\max}$ and $Q|\mathcal{M}_j|C_{\max}$ for the related machine problems and $P||C_{\max}$ and $P|\mathcal{M}_j|C_{\max}$ in case of identical machines. Since in these problems makespan minimization is equivalent to minimizing the maximum machine finishing time, we may assume that the jobs that are scheduled on a machine i share this processor in such a way that they all finish at the same time.

Even in the case that all speeds are equal, the problems under consideration are known to be strongly NP-hard when m is part of the input (see, e.g., Garey and Johnson [11]). This has motivated a lot of research in the previous decades on approximation algorithms for scheduling problems. Since some of the theoretically best approximation algorithms are rather involved, a lot of research has focused on simple heuristics like *greedy algorithms* and *local search algorithms* which are easy to implement. While greedy algorithms make reasonable ad hoc decisions to obtain a schedule, local search algorithms start with some schedule and iteratively improve the current schedule by performing some kind of local improvements until no such is possible anymore. In this paper we consider the following three algorithms that can be applied to all scheduling variants that we have described above:

– *List scheduling* is a greedy algorithm that starts from an empty schedule and a list of jobs. Then it repeatedly selects the next unscheduled job from the list and assigns it to the machine on which it will be completed the earliest with respect to the current partial schedule. We call any schedule that can be generated by list scheduling a *list schedule*.
– The *jump* and the *lex-jump* algorithms are local search algorithms that start with an arbitrary schedule and iteratively perform a local improvement step. In each improvement step one job is reassigned from a machine i to a different machine i' where it finishes earlier. In the jump algorithm only jobs on *critical*

machines i, i.e., machines that have maximum finishing time, are considered to be improving. In the lex-jump algorithm the jobs can be arbitrary. Note that a local step is lex-jump improving if and only if the sorted vector of machine finishing times decreases lexicographically. A schedule for which there is no jump improvement step or no lex-jump improvement step is called *jump optimal* or *lex-jump optimal*, respectively.

For each of these three algorithms, we are interested in their *performance guarantees*, i.e., the worst case bound on the ratio of the makespan of a schedule to be returned by the algorithm over the makespan of an optimal schedule. The final schedule returned by a local search algorithm is called a *local optimum*. Usually there are multiple local optima for a given scheduling instance both for the jump and the lex-jump algorithm with varying quality. As we do not know which local optimum is found by the local search, we will always bound the quality of the worst local optimum. Since local optima for lex-jump and pure Nash equilibria are the same, this corresponds to bounding the price of anarchy in the scheduling game that is obtained if jobs are selfish agents trying to minimize their own completion time and if the makespan is considered as the welfare function. Similarly, list scheduling can produce different schedules depending on the order in which the jobs are inserted into the list. Also for list scheduling we will bound the quality of the worst schedule that can be obtained.

Notation. Consider an instance I for the scheduling problem and a schedule σ for this instance. By $J_i(\sigma) \subseteq J$ we denote the set of jobs assigned to machine i according to σ. The *processing requirement on a machine* $i \in M$ is defined as $\sum_{j \in J_i(\sigma)} p_j$ and the *load* of a machine is defined by $L_i(I,\sigma) = \sum_{j \in J_i(\sigma)} p_{ij}$. The makespan $C_{\max}(I,\sigma)$ of σ can be written as $C_{\max}(I,\sigma) = \max_{i \in M} L_i(I,\sigma)$. The optimal makespan, i.e., the makespan of an optimal schedule is denoted by $C_{\max}^*(I)$. By Jump(I), Lex(I), and List(I) we denote the set of all feasible jump optimal schedules, lex-jump optimal schedules, and list schedules, respectively, according to instance I.

If the instance I is clear from the context, we simply write $L_i(\sigma)$ instead of $L_i(I,\sigma)$, $C_{\max}(\sigma)$ instead of $C_{\max}(I,\sigma)$, and $C_{\max}^*$ instead of $C_{\max}^*(I)$. If the schedule σ is clear as well, we simplify our notation further to L_i and $C_{\max}$ and we write J_i instead of $J_i(\sigma)$. By appropriate scaling, we may assume w.l.o.g. that $s_{\min} \geq 1$.

Smoothed analysis. As can be seen in Table 1, the performance guarantees of jump and lex-jump is known for all scheduling variants and it is constant only for the simplest case with unrestricted and identical machines. In all other cases it increases with the number m of machines. For list scheduling mainly the case with unrestricted and related machines has been considered. Cho and Sahni [7] and Aspnes et. al. [2] showed that the performance guarantee of list scheduling is $\Theta(\log m)$ in this case.

In order to analyze the robustness of the worst-case bounds, we turn to the framework of smoothed analysis, introduced by Spielman and Teng [20] to explain why certain algorithms perform well in practice in spite of a poor worst-case running time. Smoothed analysis is a hybrid of average-case and worst-case analysis: First, an adversary chooses an instance. Second, this instance is slightly randomly perturbed. The smoothed performance is the expected performance, where the expectation is taken over the random perturbation. The adversary, trying to make the algorithm perform as bad as possible, chooses an instance that maximizes this expected performance. This assumption is made to model that often the input an algorithm gets is subject to imprecise measurements, rounding errors, or numerical imprecision. If the smoothed performance of an algorithm is small, then bad worst-case instances might exist, but one is very unlikely to encounter them if instances are subject to some small amount of random noise.

We follow the more general model of smoothed analysis introduced by Beier and Vöcking [5]. In this model the adversary is even allowed to specify the probability distribution of the random noise. The influence he can exert is described by a parameter $\phi \geq 1$ denoting the maximum density of the noise. Formally we consider the following input model: the adversary chooses the number m of machines and, in the case of non-identical machines, arbitrary machine speeds $s_{\max} := s_1 \geq \ldots \geq s_m =: s_{\min}$. He also chooses the number n of jobs and, in the case of restricted machines, an arbitrary set $\mathcal{M}_j \subseteq M$ for each job $j \in J$. The only perturbed part of the instance are the processing requirements p_j. For each p_j the adversary can choose an arbitrary probability density $f_j : [0,1] \rightarrow [0,\phi]$ according to which it is chosen independently of the processing requirements of the other jobs. We call an instances $\mathcal{I}$ of this kind ϕ-smooth. Formally, a ϕ-smooth instance is not a single instance but a distribution over instances I.

The parameter ϕ specifies how close the analysis is to a worst case analysis. The adversary can, for example, choose for every p_j an interval of length $1/\phi$ from which it is drawn uniformly at random. For $\phi = 1$, every processing requirement is uniformly distributed over $[0,1]$, and hence the input model equals the average case. When ϕ gets larger, the adversary can specify the processing requirements more and more precisely, and for $\phi \rightarrow \infty$ the smoothed analysis approaches a worst-case analysis.

In this paper we analyze the *smoothed performance guarantee* of the jump, the lex-jump, and the list scheduling algorithm. As mentioned above, to define the approximation guarantee of these algorithms on a given instance, we consider the worst local optimum or the worst order in which the jobs are inserted into the list. Now the smoothed performance is defined to be the worst expected approximation guarantee of any ϕ-smooth instance.

Our results. Our results for the jump and lex-jump algorithm are summarized in Table 1. For identical machines the worst case bounds of $\Theta(1)$ trivially hold in the smoothed case as well, and hence are not investigated further. The first remarkable observation is that the smoothed performance guarantees for all variants of restricted machines are robust against random noise. We show that even for large perturbations with constant ϕ, the worst-case lower bounds carry over in

Table 1. Worst-case and smoothed performance guarantees for jump and lex-jump optimal schedules. It is assumed that $s_{\min} \geq 1$ and that $S = \sum_{i=1}^{m} s_i$.

	worst case		ϕ-smooth	
	jump	lex-jump	jump	lex-jump
unrestr. identical	$\Theta(1)$ [10,19]	$\Theta(1)$ [10,19]	$\Theta(1)$	$\Theta(1)$
unrestr. related	$\Theta(\sqrt{m})$ [7,19]	$\Theta\left(\frac{\log m}{\log \log m}\right)$ [8]	$\Theta(\phi)$	$\Theta(\log \phi)$
restricted identical	$\Theta(\sqrt{m})$ [17]	$\Theta\left(\frac{\log m}{\log \log m}\right)$ [3]	$\Theta(\sqrt{m})$	$\Theta\left(\frac{\log m}{\log \log m}\right)$
restricted related	$\Theta\left(\sqrt{m \cdot \frac{s_{\max}}{s_{\min}}}\right)$ [17]	$\Theta\left(\frac{\log S}{\log \log S}\right)$ [17]	$\Theta\left(\sqrt{m \cdot \frac{s_{\max}}{s_{\min}}}\right)$	$\Omega\left(\frac{\log m}{\log \log m}\right)$

asymptotic terms. This can be seen as an indication that neither the jump algorithm nor the lex-jump algorithm yield a good approximation ratio for scheduling with restricted machines in practice.

The situation is much more promising for unrestricted related parallel machines. Here the worst-case bounds are fragile and do not carry over to the smoothed case. Even though both for jump and for lex-jump the worst-case lower bound is not robust, there is a significant difference between these two: while the smoothed approximation ratio for jump grows linearly with the perturbation parameter ϕ, it grows only logarithmically in ϕ for lex-jump optimal schedules. This proves that also in the presence of random noise lex-jump optimal schedules are significantly better than jump optimal schedules. As mentioned earlier this also implies that the smoothed price of anarchy is $\Theta(\log \phi)$. Additionally we show that the smoothed approximation ratio of list scheduling is $\Theta(\log \phi)$ as well. This indicates that both the lex-jump algorithm and the list scheduling algorithm should yield good approximations on practical instances.

Related work. The approximability of $Q||C_{\max}$ is well understood. Cho and Sahni [7] showed that list scheduling has a performance guarantee of at most $1 + \sqrt{2m - 2}/2$ for $m \geq 3$ and that it is at least $\Omega(\log m)$. Aspnes et. al. [2] improved the upper bound to $O(\log m)$ matching the lower bound asymptotically. Hochbaum and Shmoys [13] designed a polynomial time approximation scheme for this problem. For work on restricted related machines we refer to the literature review of Leung and Li [15].

In the last decade, there has been a strong interest in understanding the worst-case behavior of local optima. We refer to the survey [1] and the book [16] for a comprehensive overview of the worst-case analysis and other theoretical aspects of local search. It follows from the work of Cho and Sahni [7] that for the problem on unrestricted related machines the performance guarantee of the jump algorithm is $(1 + \sqrt{4m - 3})/2$ and this bound is tight [19]. For lex-jump optimal schedules, Czumaj and Vöcking [8] showed that the performance guarantee is $\Theta\left(\min\left\{\frac{\log m}{\log \log m}, \log \frac{s_{\max}}{s_{\min}}\right\}\right)$. For the problem on restricted related machines, Recalde et al. [17] showed that the performance guarantee of local optimal schedules with respect to the jump neighborhood is $(1+\sqrt{1 + 4(m - 1)s_{\max}/s_{\min}})/2$ and that this bound is tight up to a constant factor. Moreover, they showed that the performance guarantee of lex-jump optimal schedules is $\Theta\big(\frac{\log S}{\log \log S}\big)$, where $S = \sum_{i=1}^{m} s_i$, assuming that $s_{\min} \geq 1$. When all speeds are equal, Awerbuch et al. [3] showed that the performance guarantee for lex-jump optimal schedules is $\Theta\big(\frac{\log m}{\log \log m}\big)$.

Up to now, smoothed analysis has been mainly applied to running time analysis (see, e.g., [21] for a survey). The first exception is the paper by Becchetti et al. [4] who introduced the concept of smoothed competitive analysis, which is equivalent to smoothed performance guarantees for online algorithms. Schäfer and Sivadasan [18] performed a smoothed competitive analysis for metrical task systems. Englert et al. [9] considered the 2-Opt algorithm for the traveling salesman problem and determined, among others, the smoothed performance guarantee of local optima of the 2-Opt algorithm. Hoefer and Souza [14] presented one of the first average case analyses for the price of anarchy.

The remainder of this paper is organized as follows. In Section 2, we provide asymptotically matching upper and lower bounds on the smoothed performance guarantees of jump optimal schedules, lex-jump optimal schedules, and list schedules in case of unrestricted related machines. In Section 3 we show that smoothing does not help for the setting of restricted machine schedules.

2   Unrestricted Related Machines

Due to space limitations, some proofs have been omitted. They can be found in the full version of this paper [6].

2.1   Jump Optimal Schedules

In this subsection, we show that the smoothed performance guarantee grows linearly with the smoothing parameter ϕ and is independent of the number of jobs and machines. In particular, it is constant if the smoothing parameter is constant.

Theorem 1. *For any ϕ-smooth instance $\mathcal{I}$ with unrestricted and related machines,*

$$\mathop{\mathbf{E}}_{I \sim \mathcal{I}} \left[\max_{\sigma \in Jump(I)} \frac{C_{\max}(I, \sigma)}{C_{\max}^*(I)} \right] < 5.1\phi + 2.5 = O(\phi).$$

Although omitting the proof, we provide some insight. Building on the paper of Cho and Sahni [7], we show that the performance guarantee is negatively correlated with the total processing requirement. However, having a low total processing requirement with respect to the makespan of a jump optimal schedule only happens with a low probability. Therefore, we show that with high probability the performance guarantee of the jump neighborhood cannot be too large.

Corollary 1. *Consider an instance of scheduling with unrestricted and related machines in which the processing requirement of every job is chosen independently and uniformly at random from $[0, 1]$. The expected performance guarantee of the worst jump optimal schedule is $O(1)$.*

Theorem 2. *There is a class of ϕ-smooth instances $\mathcal{I}$ with unrestricted and related machines such that*

$$\mathbf{E}_{I \sim \mathcal{I}} \left[\max_{\sigma \in Jump(I)} \frac{C_{\max}(I, \sigma)}{C_{\max}^*(I)} \right] = \Omega(\phi).$$

2.2 List Schedules and Lex-Jump Optimal Schedules

In this subsection we show that the performance guarantee of lex-jump and list scheduling is $O(\log \phi)$ and that this bound is asymptotically tight.

Theorem 3. *Let α be an arbitrary positive real. For $\phi \geq 2$ and any ϕ-smooth instance $\mathcal{I}$ with unrestricted and related machines*

$$\mathbf{Pr}_{I \sim \mathcal{I}} \left[\max_{\sigma \in Lex(I)} \frac{C_{\max}(I, \sigma)}{C_{\max}^*(I)} \geq \alpha \right] \leq \left(\frac{32\phi}{2^{\alpha/6}} \right)^{n/2}$$

and

$$\mathbf{E}_{I \sim \mathcal{I}} \left[\max_{\sigma \in Lex(I)} \frac{C_{\max}(I, \sigma)}{C_{\max}^*(I)} \right] \leq 18 \log_2 \phi + 30 = O(\log \phi).$$

Both results also hold for $\max_{\sigma \in List(I)} \frac{C_{\max}(I, \sigma)}{C_{\max}^*(I)}$.

Note that the assumption $\phi \geq 2$ in Theorem 3 is no real restriction as for $\phi \in [1, 2)$ any ϕ-smooth instance is a 2-smooth instance. Hence, for these values we can apply all bounds from Theorem 3 when substituting ϕ by 2. In particular, the expected value is a constant.

In the remainder of this section, we will use the following notation. Let $J_{i,j}(\sigma)$ denote the set of all jobs that are scheduled on machine i and have index at most j, i.e., $J_{i,j}(\sigma) = J_i(\sigma) \cap \{1, \ldots, j\}$. If σ is clear from the context, then we just write $J_{i,j}$. We start with observing an essential property both lex-jump optimal schedules and list schedules have in common.

Definition 1. *We call a schedule σ on machines $1, \ldots, m$ with speeds $s_1, \ldots, s_m$ a* near list schedule, *if we can index the jobs in such a way that*

$$L_{i'} + \frac{p_j}{s_{i'}} \geq L_i - \sum_{l \in J_{i,j-1}(\sigma)} \frac{p_l}{s_i} \tag{1}$$

for all machines $i' \neq i$ and all jobs $j \in J_i(\sigma)$. With $NL(I)$ we denote the set of all near list schedules for instance I.

Lemma 1. *For any instance I the relation $Lex(I) \cup List(I) \subseteq NL(I)$ holds.*

Note that in general neither $Lex(I) \subseteq List(I)$ nor $List(I) \subseteq Lex(I)$ holds.

Proof (Lemma 1). For any schedule $\sigma \in \text{Lex}(I)$ we can index the jobs arbitrarily and, by definition, even the stronger inequality $L_{i'} + p_j/s_{i'} \geq L_i$ holds. For $\sigma \in \text{List}(I)$ we can index the jobs in reverse order in which they appear in the list that was used for list scheduling. Consider an arbitrary job $j \in J_i(\sigma)$ and a machine $i' \neq i$. Let $L_i', L_{i'}'$ and $L_i, L_{i'}$ denote the loads of machines i and i' before assigning job j to machine i and the loads of i and i' in the final schedule, respectively. Then, $L_i' + p_j/s_i \leq L_{i'}' + p_j/s_{i'}$ as j is assigned to machine i according to list scheduling. Since $L_i = L_i' + \sum_{l \in J_{i,j}} p_l/s_i$ and $L_{i'} \geq L_{i'}'$, this implies $L_{i'} + p_j/s_{i'} \geq L_i - \sum_{l \in J_{i,j-1}} p_l/s_i$. $\qquad\square$

In the remainder we fix an instance I and consider arbitrary schedules $\sigma \in \text{NL}(I)$ with appropriate indices of the jobs such that Inequality (1) holds.

In our proofs, we adopt some of the notation also used by Czumaj and Vöcking [8]. Given a schedule σ, we set $c = \lfloor C_{\max}(\sigma)/C_{\max}^* \rfloor - 1$. Remember that the machines are ordered such that $s_1 \geq \ldots \geq s_m$. For any integer $k \leq c$ let $H_k = \{1, \ldots, i_k\}$ where $i_k = \max\{i \in M : L_{i'} \geq k \cdot C_{\max}^* \, \forall i' \leq i\}$. Note that $i_k = m$ for all $k \leq 0$ and hence $H_k = M$ for such k. Further, define $\overline{H}_k = H_k \setminus H_{k+1}$ for all $k \in \{0, \ldots, c-1\}$ and $\overline{H}_c = H_c$. Note that this classification always refers to schedule σ even if additionally other schedules are considered. Some properties follow straightforwardly.

Property 1. For each machine $i \in H_k$, $L_i \geq k \cdot C_{\max}^*$.

Property 2. Machine $i_k + 1$, if it exists, is the fastest machine in $M \setminus H_k$.

Property 3. $L_{i_k+1} < k \cdot C_{\max}^*$ for all $k \in \{1, \ldots, c\}$, and $L_1 < (c+2) \cdot C_{\max}^*$.

Let $0 \leq t \leq k \leq c$ be integers. Several times we will consider the first jobs on some machine $i \in H_k$ which contribute at least $t \cdot C_{\max}^*$ to the load of machine i. We denote the set of those jobs by $J_{i,\geq t}$. Formally, $J_{i,\geq t} = J_{i,j_i}$ for $j_i = \min\{j : \sum_{l \in J_{i,j}} p_l/s_i \geq t \cdot C_{\max}^*\}$. on machine i that do not contribute $t \cdot C_{\max}^*$ to the load.

Czumaj and Vöcking [8] show that in a lex-jump optimal schedule the speeds of any two machines which are at least two classes apart differ by a factor of at least 2. The next few lemmata, stated without a proof, show that near list schedules have a similar but slightly weaker property.

Lemma 2. *Let $k \in \{4, \ldots, c\}$. The speed of any machine in class H_k is at least twice the speed of any machine in class $M \setminus H_{k-4}$.*

Let us remark that a similar property has already been shown by Aspnes et. al. [2]. If no machine class would be empty, then the machine speeds would double every 5 classes. Although a machine class can be empty, no two neighboring classes can be empty. Therefore, machine speeds double every 6 classes. To be more formal:

Lemma 3. *Let $0 \leq k_2 \leq k_1 \leq c$ be integers, let i_1 be the first machine of $\overline{H}_{k_1}$ and let $i_2 \in \overline{H}_{k_2}$. Then, $s_{i_2} \leq s_{i_1}/2^{\lfloor \Delta/6 \rfloor}$ where $\Delta = k_1 - k_2$.*

It follows that the slowest machines have exponentially small speeds. Since it can shown that the aggregated total processing requirement assigned to these slow machines in an optimal schedule is large, one finds that many jobs having processing requirements exponentially small in c need to exist.

Lemma 4. *The processing requirement of at least $n/2$ jobs is at most $2^{-c/6+2}$.*

However, having many such small jobs is unlikely when the processing requirements have been smoothed. Therefore, the smoothed performance guarantee cannot be too high, yielding Theorem 3.

Proof (Theorem 3). If $C_{\max}(\sigma)/C_{\max}^* \geq \alpha$, then at least $n/2$ jobs have processing requirement at most $2^{-\alpha/6+3}$ due to Lemma 4 and $c = \lfloor C_{\max}(\sigma)/C_{\max}^* \rfloor - 1 \geq \alpha - 2$. The probability that one specific job is that small is bounded by $\phi \cdot 2^{-\alpha/6+3} = 8\phi \cdot 2^{-\alpha/6}$ in the smoothed input model. Hence, the probability that the processing requirement of at least $n/2$ jobs is at most $2^{-\alpha/6+3}$ is bounded by $2^n \cdot (8\phi \cdot 2^{-\alpha/6})^{n/2} = (32\phi \cdot 2^{-\alpha/6})^{n/2}$. This yields

$$\Pr_{I \sim \mathcal{I}}\left[\max_{\sigma \in \mathrm{NL}(I)} \frac{C_{\max}(I,\sigma)}{C_{\max}^*(I)} \geq \alpha\right] \leq \left(\frac{32\phi}{2^{\alpha/6}}\right)^{n/2}.$$

As for $n = 1$ any schedule $\sigma \in \mathrm{NL}(I)$ is optimal, we just consider the case $n \geq 2$. Let $\alpha_k = \alpha_k(\phi) = 6k \log_2 \phi + 30$, $k \geq 1$. For $\alpha \geq \alpha_k$ we obtain

$$\Pr_{I \sim \mathcal{I}}\left[\max_{\sigma \in \mathrm{NL}(I)} \frac{C_{\max}(I,\sigma)}{C_{\max}^*(I)} \geq \alpha\right] \leq \left(\phi^{1-k}\right)^{n/2} \leq \phi^{1-k} \leq 2^{1-k}$$

as $\phi \geq 2$. Hence,

$$\mathbf{E}_{I \sim \mathcal{I}}\left[\max_{\sigma \in \mathrm{NL}(I)} \frac{C_{\max}(I,\sigma)}{C_{\max}^*(I)}\right] = \int_0^\infty \Pr_{I \sim \mathcal{I}}\left[\max_{\sigma \in \mathrm{NL}(I)} \frac{C_{\max}(I,\sigma)}{C_{\max}^*(I)} \geq \alpha\right] d\alpha$$

$$\leq \alpha_1 + \sum_{k=1}^\infty \int_{\alpha_k}^{\alpha_{k+1}} \Pr_{I \sim \mathcal{I}}\left[\max_{\sigma \in \mathrm{NL}(I)} \frac{C_{\max}(I,\sigma)}{C_{\max}^*(I)} \geq \alpha\right] d\alpha$$

$$\leq \alpha_1 + 6 \log_2 \phi \cdot \sum_{k=1}^\infty 2^{1-k} = 18 \log_2 \phi + 30. \qquad \square$$

There exists a class of ϕ-*smooth* instances and corresponding list and lex-jump optimal schedules which show that the upper bound is asymptotically tight.

Theorem 4. *There is a class of ϕ-smooth instances $\mathcal{I}$ with unrestricted and related machines such that, for any $I \in \mathcal{I}$,*

$$\max_{\sigma \in Lex(I)} \frac{C_{\max}(I,\sigma)}{C_{\max}^*(I)} = \Omega(\log \phi) \quad and \quad \max_{\sigma \in List(I)} \frac{C_{\max}(I,\sigma)}{C_{\max}^*(I)} = \Omega(\log \phi).$$

3   Restricted Machines

In this section we provide lower bound examples showing that the worst-case performance guarantees for all variants of the restricted machines are robust against random noise. Even with large perturbations, that is for a constant ϕ, the worst-case lower bounds still apply. Again, we refer to our technical report for the omitted proofs [6].

3.1   Jump Neighborhood on Restricted Machines

Recalde et al. [17] showed that the makespan of a jump optimal schedule is at most a factor of $1/2 + \sqrt{m - 3/4}$ away from the optimal makespan on restricted identical machines. On restricted related parallel machines they showed the makespan of a jump optimal schedule is no more than a factor of $1/2 + \sqrt{(m - 2) \cdot (s_{\max}/s_{\min}) + 1/4}$ away from the makespan of an optimal schedule. They showed both bounds to be tight up to a constant factor. We show that even on ϕ-smooth instances both bounds are tight up to a constant factor.

Theorem 5. *For every $\phi \geq 2$ there exists a class of ϕ-smooth instances $\mathcal{I}$ on restricted related machines such that*

$$\mathbf{E}_{I \sim \mathcal{I}} \left[\max_{\sigma \in Jump(I)} \frac{C_{\max}(I, \sigma)}{C_{\max}^*(I)} \right] = \Omega \left(\sqrt{m \cdot \frac{s_{\max}}{s_{\min}}} \right).$$

Corollary 2. *For every $\phi \geq 2$ there exists a class of ϕ-smooth instances $\mathcal{I}$ on restricted identical machines such that*

$$\mathbf{E}_{I \sim \mathcal{I}} \left[\max_{\sigma \in Jump(I)} \frac{C_{\max}(I, \sigma)}{C_{\max}^*(I)} \right] = \Omega(\sqrt{m}).$$

3.2   Lex-Jump Optimal Schedules on Restricted Identical Machines

We establish that the bound for the lex-jump neighborhood on unrestricted identical machines is tight even on ϕ-smooth instances. The instance and lex-jump optimal schedule we construct are in the same spirit as the lower bound provided in [3]. Machines are partitioned into classes. When comparing machine class $k + 1$ to the previous class k, the number of machines increases whereas the speeds of machines decreases. Corresponding to each machine class k, a job class is defined whose jobs will be scheduled on machines in class k in a lex-jump optimal schedule whereas they will be scheduled on machines in class $k + 1$ in an optimal schedule. The corresponding processing requirements are picked such that the loads of the machines in class $k + 1$ are slightly less than those of the machines in class k in the lex-jump optimal schedule. However, since processing requirements have been smoothed we need to be more careful. In comparison with [3], the loads between two subsequent machine classes differ less. Also we additionally create a second class of jobs corresponding to each machine class. These jobs all have small processing requirements and are used to balance the loads of machines within a machine class. Working out the details yields the theorem below.

Theorem 6. *For every $\phi \geq 8$ there exists a class of ϕ-smooth instances $\mathcal{I}$ on restricted identical machines such that*

$$\mathbf{E}_{I \sim \mathcal{I}} \left[\max_{\sigma \in Lex(I)} \frac{C_{\max}(I, \sigma)}{C_{\max}^*(I)} \right] = \Omega \left(\frac{\log m}{\log \log m} \right).$$

Remark 1. The worst case upper bound on the performance guarantee for lex-jump optimal schedules on restricted related machines is $O\left(\frac{\log S}{\log \log S}\right)$, where $S = \sum_i s_i/s_m$ [17]. As for identical machines $S = m$, i.e., each machine has speed 1, the upper bound matches the lower bound of Theorem 6 up to a constant factor and smoothing does also not improve the performance guarantee for the worst lex-jump optimal schedules on restricted related machines.

4   Concluding Remarks

We proved that the lower bounds for all scheduling variants with restricted machines are rather robust against random noise. We have also shown that the situation looks much better for unrestricted machines where we obtained performance guarantees of $\Theta(\phi)$ and $\Theta(\log \phi)$ for the jump and lex-jump algorithm, respectively. The latter bound also holds for the price of anarchy of routing on parallel links and for the list scheduling algorithm.

There are several interesting directions of research and we view our results only as a first step towards fully understanding local search and greedy algorithms in the framework of smoothed analysis. For example, we have only perturbed the processing requirements, and it might be the case that the worst-case bounds for the restricted scheduling variants break down if also the sets $\mathcal{M}_j$ are to some degree random. In general it would be interesting to study different perturbation models where the sets $\mathcal{M}_j$ and/or the speeds s_i are perturbed. Lemmas 3 and 4 indicate that there need to exist many machines having exponentially small speeds. We conjecture that if speeds are being smoothed, that the smoothed performance guarantee of near list schedules on restricted related machines is $\Theta(\log \phi)$ as well.

Another interesting question is the following: since we do not know which local optimum is reached, we have always looked at the worst local optimum. It might, however, be the case that the local optima reached in practice are better than the worst local optimum. It would be interesting to study the quality of the local optimum reached under some reasonable assumptions on how exactly the local search algorithms work. An extension in this direction would be to analyze the quality of coordination mechanisms under smoothing.

References

1. Angel, E.: A survey of approximation results for local search algorithms. In: Bampis, E., Jansen, K., Kenyon, C. (eds.) Efficient Approximation and Online Algorithms. LNCS, vol. 3484, pp. 30–73. Springer, Heidelberg (2006)
2. Aspnes, J., Azar, Y., Fiat, A., Plotkin, S.A., Waarts, O.: On-line routing of virtual circuits with applications to load balancing and machine scheduling. Journal of the ACM 44(3), 486–504 (1997)
3. Awerbuch, B., Azar, Y., Richter, Y., Tsur, D.: Tradeoffs in worst-case equilibria. Theoretical Computer Science 361, 200–209 (2006)

4. Becchetti, L., Leonardi, S., Marchetti-Spaccamela, A., Schäfer, G., Vredeveld, T.: Average case and smoothed competitive analysis for the multi-level feedback algorithm. Mathematics of Operations Research 31(3), 85–108 (2006)
5. Beier, R., Vöcking, B.: Random knapsack in expected polynomial time. Journal of Computer and System Sciences 69(3), 306–329 (2004)
6. Brunsch, T., Röglin, H., Rutten, C., Vredeveld, T.: Smoothed Performance Guarantees for Local Search. ArXiv e-prints (May 2011), http://arxiv.org/abs/1105.2686
7. Cho, Y., Sahni, S.: Bounds for list schedules on uniform processors. SIAM Journal on Computing 9, 91–103 (1980)
8. Czumaj, A., Vöcking, B.: Tight bounds for worst-case equilibria. Transactions on Algorithms ACM 3(1) (2007)
9. Englert, M., Röglin, H., Vöcking, B.: Worst case and probabilistic analysis of the 2-opt algorithm for the TSP. In: Proceedings of the 18th ACM-SIAM Symposium on Discrete Algorithms (SODA), pp. 1295–1304 (2007)
10. Finn, G., Horowitz, E.: A linear time approximation algorithm for multiprocessor scheduling. BIT 19, 312–320 (1979)
11. Garey, M.R., Johnson, D.S.: Computers and Intractibility: A Guide to the Theory of NP-Completeness. W.H. Freeman & Co., New York (1979)
12. Graham, R.L., Lawler, E.L., Lenstra, J.K., Rinnooy Kan, A.H.G.: Optimization and approximation in deterministic sequencing and scheduling: a survey. Annals of Discrete Mathematics 5, 287–326 (1979)
13. Hochbaum, D.S., Shmoys, D.B.: A polynomial approximation scheme for machine scheduling on uniform processors: using the dual approximation approach. SIAM Journal on Computing 17, 539–551 (1988)
14. Hoefer, M., Souza, A.: Tradeoffs and average-case equilibria in selfish routing. ACM Transactions on Computation Theory 2(1), article 2 (2010)
15. Leung, J.Y.T., Li, C.L.: Scheduling with processing set restrictions: A survey. International Journal of Production Economics 116, 251–262 (2008)
16. Michiels, W.P.A.J., Aarts, E.H.L., Korst, J.H.M.: Theoretical Aspects of Local Search. Springer, Heidelberg (2007)
17. Recalde, D., Rutten, C., Schuurman, P., Vredeveld, T.: Local search performance guarantees for restricted related parallel machine scheduling. In: López-Ortiz, A. (ed.) LATIN 2010. LNCS, vol. 6034, pp. 108–119. Springer, Heidelberg (2010)
18. Schäfer, G., Sivadasan, N.: Topology matters: Smoothed competitiveness of metrical task systems. Theoretical Computer Science 341(1-3), 3–14 (2005)
19. Schuurman, P., Vredeveld, T.: Performance guarantees of local search for multiprocessor scheduling. Informs Journal on Computing 19(1), 52–63 (2007)
20. Spielman, D.A., Teng, S.H.: Smoothed analysis of algorithms: Why the simplex algorithm usually takes polynomial time. Journal of the ACM 51(3), 385–463 (2004)
21. Spielman, D.A., Teng, S.H.: Smoothed analysis: an attempt to explain the behavior of algorithms in practice. Communications of the ACM 52(10), 76–84 (2009)

Improved Approximations for k-Exchange Systems

(Extended Abstract)

Moran Feldman[1], Joseph (Seffi) Naor[1], Roy Schwartz[1], and Justin Ward[2]

[1] Computer Science Dept., Technion, Haifa, Israel
{moranfe,naor,schwartz}@cs.technion.ac.il
[2] Computer Science Dept., University of Toronto, Toronto, Canada
jward@cs.toronto.edu

Abstract. Submodular maximization and set systems play a major role in combinatorial optimization. It is long known that the greedy algorithm provides a $1/(k + 1)$-approximation for maximizing a monotone submodular function over a k-system. For the special case of k-matroid intersection, a local search approach was recently shown to provide an improved approximation of $1/(k + \delta)$ for arbitrary $\delta > 0$. Unfortunately, many fundamental optimization problems are represented by a k-system which is not a k-intersection. An interesting question is whether the local search approach can be extended to include such problems.

We answer this question affirmatively. Motivated by the *b-matching* and *k-set packing* problems, as well as the more general *matroid k-parity* problem, we introduce a new class of set systems called k-exchange systems, that includes k-set packing, b-matching, matroid k-parity in strongly base orderable matroids, and additional combinatorial optimization problems such as: *independent set in $(k + 1)$-claw free graphs, asymmetric TSP, job interval selection with identical lengths* and *frequency allocation on lines*. We give a natural local search algorithm which improves upon the current greedy approximation, for this new class of independence systems. Unlike known local search algorithms for similar problems, we use counting arguments to bound the performance of our algorithm.

Moreover, we consider additional objective functions and provide improved approximations for them as well. In the case of linear objective functions, we give a non-oblivious local search algorithm, that improves upon existing local search approaches for matroid k-parity.

1 Introduction

The study of combinatorial problems with submodular objective functions has attracted much attention recently, and is motivated by the principle of economy of scale, prevalent in real world applications. Additionally, submodular maximization plays a major role in combinatorial optimization since many optimization problems can be represented as constrained variants of submodular

C. Demetrescu and M.M. Halldórsson (Eds.): ESA 2011, LNCS 6942, pp. 784–798, 2011.

maximization. Often, the feasibility domain of such a variant is defined by a set system. A set system $(\mathcal{N}, \mathcal{I})$ is composed of a ground set $\mathcal{N}$ and a collection $\mathcal{I} \subseteq 2^{\mathcal{N}}$ of independent sets. For a constrained problem defined by $(\mathcal{N}, \mathcal{I})$, $\mathcal{I}$ is the collection of feasible solutions. Here is the known hierarchy of set systems:

$$k\text{-intersection} \subseteq k\text{-circuit bound} \subseteq k\text{-extendible} \subseteq k\text{-system}.$$

A set system belongs to the k-intersection class if it is the intersection of k matroids defined over a common ground set. The class of k-circuit bound contains all set systems in which adding a single element to an independent set creates at most k circuits (i.e., the resulting set contains at most k minimally dependent subsets). Recall that in a matroid, adding an element to an independent set creates at most a single circuit. Therefore, adding an element to an independent set in a k-intersection system closes at most k circuits, one per matroid. The class of k-extendible, intuitively, captures all set systems in which adding an element to an independent set requires throwing away at most k other elements from the set (in order to keep it independent). This generalizes k-circuit bound because in k-circuit bound we need to throw at most one element per circuit closed (i.e., up to k elements). The class of k-system contains all set systems in which for every set, not necessarily independent, the ratio of the sizes of the largest base of the set to the smallest base of the set is at most k (a base is a maximal independent subset).

Motivated by well studied problems such as *matroid k-parity* (which generalizes *matroid k-intersection, k-set packing, b-matching* and *k-dimensional matching*) as well as *independent set in $(k+1)$-claw free graphs*, we propose a new class of set systems which we call k-exchange systems. This class is general enough to capture various well studied combinatorial optimization problems, including matroid k-parity in strongly base orderable matroids, the other problems listed above, and additional problems such as: *job interval selection with identical lengths, asymmetric traveling salesperson* and *frequency allocation on lines*. Note that these last 3 problems, like independent set in $(k+1)$-claw free graphs, do *not* belong to the k-intersection class. On the other hand, we show that this class has a rich enough structure to enable us to present two local search algorithms with provable improved approximation guarantees for submodular and linear functions, respectively. Additionally, we relate the k-exchange class to the notion of strongly base orderable matroids, and show how it relates to the existing set systems hierarchy.

1.1   Our Results

Given a k-exchange system $(\mathcal{N}, \mathcal{I})$ and a function $f : 2^{\mathcal{N}} \to \mathbb{R}^+$, we provide approximation guarantees for the problem of finding an independent set $S \in \mathcal{I}$ maximizing $f(S)$ for several types of objective functions. The main application we consider is *strongly base orderable matroid k-parity*. This problem has two important special cases: *b-matching* and *k-set packing*.

Many interesting applications are k-exchange systems for *small* values of k, For example, the well studied *b-matching* problem is 2-exchange (but not 2-intersection) system. This lets us improve the best approximation of $1/3$ by

[15] to about $1/2$ for the case of a normalized monotone submodular f, and the best approximation of 0.0657 by [17] to about $1/4$ for the case of a general non-negative (not necessarily monotone) submodular f.

The types of objective functions considered in this work are: normalized monotone submodular, general non-negative (not necessarily monotone) submodular, linear and cardinality. Table 1 summarizes these results and the approximation ratio achieved for each application considered in this work.

We present 2 local search algorithms for maximizing submodular and linear objectives, respectively, subject to k-exchange systems. Our first algorithm is a very natural local search algorithm which is essentially identical to the algorithm of Lee et al. [26]. However, unlike the analysis of [26] which uses matroid intersection techniques, our analysis goes through a counting argument applied to an auxiliary graph. Our second algorithm yields improved results for the special case of linear objective functions, and is based on non-oblivious local search techniques employed by Berman [2] for the case of $(k+1)$-claw free graphs. This algorithm is guided by an auxiliary potential function, which considers the sum of the *squared* weights of the elements in the independent set.

It should be noted that as in previous local search algorithms, e.g., [26], the time complexity of our algorithms is exponential in k, thus, k is assumed to be a constant. As mentioned, k is indeed a small constant in our applications (refer to Table 1 for exact values of k).

1.2   Related Work

Extensive work has been conducted in recent years in the area of optimizing submodular functions under various constraints. We mention here the most relevant results. Historically, one of the very first problems examined was maximizing a monotone submodular function over a matroid. Several special cases of matroids and submodular functions were studied in [10,18,19,22,25] using the greedy approach. Recently, the general problem with an arbitrary matroid and an arbitrary submodular function was given a tight approximation of $(1 - 1/e)$ by Calinescu et al. [7]. A matching lower bound is due to [31,32].

The problem of optimizing a normalized, *monotone* submodular function over the intersection of k matroids was considered by Fisher et al. [15] who gave a greedy algorithm with an approximation factor of $1/(k + 1)$, and state that their proof extends to the more general class of k-systems using the outline of Jenkyns [22] (the extended proof is explicitly given by Calinescu et al. [7]). For k-intersection systems, this result was improved by Lee et al. [26] to $1/(k+\delta)$, for any constant $\delta > 0$, while using a local search approach that exploits exchange properties of the underlying combinatorial structure. However, for optimizing a monotone submodular function over k-circuit bound and k-extendible set systems, the current best known approximation is still $1/(k + 1)$ [15].

Maximization of *non-monotone* submodular functions under various constraints has also attracted considerable attention in the last few years. The basic result in this area is an approximation factor of $2/5$, given by Feige et al. [12], for the unconstrained variant of the problem. This was recently improved twice,

Table 1. $k \geq 2$ is a constant, $\delta > 0$ is any given constant and $\beta = 1/(\alpha^{-1}+1)$, where α is the best known approximation for unconstrained maximization of a non-negative submodular function ($\alpha \geq 0.42$ see [13]). f: NMS - normalized monotone submodular, NS - general non-negative submodular, L - linear, C - cardinality.

Maximization Problem	f	k	This Paper	Previous Result
k-exchange	NMS		$1/(k+\delta)$	$1/(k+1)$ [15]
	NS	k	$(k-1)/(k^2+\delta)$	$\beta/(k+2+1/k)$ [17]
	L^{b}		$2/(k+1+\delta)$	$1/k$ [22]
	C^{a}		$2/(k+\delta)$	$1/k$ [22]
Main Applications				
s.b.o.matroid k-parity	NMS		$1/(k+\delta)$	$1/k$ [15]
	NS	k	$(k-1)/(k^2+\delta)$	$\beta/(k+2+1/k)$ [17]
	L^{b}		$2/(k+1+\delta)$	$1/k$ [22]
b-matching	NMS	2	$1/(2+\delta)$	$1/3$ [15]
	NS		$1/(4+\delta)$	$\beta/4.5$ [17]
k-set packing	NMS	k	$1/(k+\delta)$	$1/(k+1)$ [15]
	NS		$(k-1)/(k^2+\delta)$	$\beta/(k+2+1/k)$ [17]
Additional Applications				
independent set in	NMS	k	$1/(k+1+\delta)$	$1/k$ [15]
(k + 1)-claw free graphs	NS		$(k-1)/(k^2+\delta)$	$\beta/(k+2+1/k)$ [17]
job interval selection	NMS	2	$1/(2+\delta)$	$1/3$ [30]
identical lengths	NS	3	$2/(9+\delta)$	$3\beta/16$ [17]
asymmetric traveling	NMS	3	$1/(3+\delta)$	$1/4$ [30]
salesperson	NS		$2/(9+\delta)$	$3\beta/16$ [17]
frequency allocation on	NMS		$1/(3+\delta)$	$1/4$ [15]
lines	NS	3	$2/(9+\delta)$	$3\beta/16$ [17]
	L		$1/(2+\delta)$	$1/3$ [22]
	C		$2/(3+\delta)$	$1-1/e$ [35]

^{a} The result applies for $k \geq 3$.
^{b} For $k = 2$, we also have a PTAS (see Corollary 1).

using a generalization of local search, called simulated annealing, by Gharan and Vondrák [16] and then by Feldman et al. [13]. Chekuri et al. [9] gave a 0.325 approximation for optimization over a matroid. This was very recently improved to roughly $e^{-1} \approx 0.368$ by Feldman et al. [14]. When optimizing over the intersection of k matroids, the current best result is $(k-1)/(k^2+\delta)$, and is due to Lee et al. [26]. A technique for using monotone submodular optimization for non-monotone submodular problems is given by [17]. Gupta et al. [17] use their technique to convert the greedy algorithm into an algorithm achieving an approximation ratio of $1/((\alpha^{-1}+1)(k+2+1/k))$ for k-system, where α is the best know approximation for unconstrained maximization of a non-negative submodular function ($\alpha \geq 0.42$, see [13]).

In the case of maximizing a *linear* objective function over k matroid constraints, an approximation of $1/k$ was given by Jenkyns [22] using a greedy algorithm. This was improved by [26] who gave an approximation of $1/(k-1+\delta)$, for any constant $\delta > 0$, using the same local search techniques as in the monotone submodular case. For the more general k-circuit bounded and k-extendible set systems, the current best known approximation is only $1/k$ [22] (as in the monotone submodular case this result is given for k-system). Hazan et al. [20] give a hardness result of $\Omega(\log k/k)$ that applies to this case.

Among the applications we consider, the most general is matroid k-parity in strongly base orderable matroids. The matroid k-parity problem, described in detail in Definition 3, is related to the matroid k-matching, which is a common generalization of matching and matroid intersection problems. In this problem, we are given a k-uniform hypergraph $H = (V, \mathcal{E})$ and a matroid $\mathcal{M}$ defined on the vertex set V of H. The goal is to find a matching $\mathcal{S}$ in H such that the set of elements covered by the edges in $\mathcal{S}$ are independent in $\mathcal{M}$. The matroid k-parity problem corresponds to the special case in which the edges of H are disjoint. It can be shown that matroid matching in a k-uniform hypergraph is reducible to matroid k-parity as well, and thus the two problems are equivalent [27].

If the matroid $\mathcal{M}$ is given by an independence oracle, there are instances of matroid matching problem (and hence also matroid parity) for which obtaining an optimal solution requires an exponential number of oracle calls, even when $k = 2$ and all weights are 1 [28,23]. These instances can be modified to show that matroid parity is NP-complete (via a reduction from MAXCLIQUE) [34]. In the unweighted case, Lovász [28] obtained a polynomial time algorithm for matroid 2-matching in *linear* matroids. More recently, Lee et al. [27] gave a PTAS for matroid 2-parity in arbitrary matroids, and a $k/2 + \epsilon$ approximation for matroid k-parity in arbitrary matroids.

In the weighted case, it can be shown (see [7]) that the greedy algorithm provides a k-approximation. Although this remains the best known result for general matroids, some improvement has been made in the case of $k = 2$ for restricted classes of matroids. Tong et. al give an exact polynomial time algorithm for weighted matroid 2-parity in gammoids [37]. This result has recently been extended by Soto [36] to a PTAS for the class of all *strongly base orderable* matroids, which strictly includes gammoids. Additionally, Soto shows that matroid 2-matching remains NP-hard even in this restricted case.

An important special case of matroid 2-parity in strongly base orderable matroids is the b-matching problem, in which we are given a maximum degree for each vertex in a graph and seek a collection of edges satisfying all vertices' degree constraints. Many exact algorithms were given for maximum weight linear b-matching problem with an improving dependence of the time complexity on the maximal value of b (see, e.g., [33,29,1]). Mestre [30] gave a linear time approximation algorithm for this problem, and Kalyanasundaram and Pruhs [24] considered an online version of it.

Another important special case of both strongly base orderable matroid k-parity is the k-set packing problem. For linear objective functions, this problem has been considered extensively. For cardinality objective, $2/(k + \delta)$ approximation was already given by Hurkens and Schrijver [21], and this approximation ratio was extended relatively recently to general linear functions by Berman [2], building on an earlier work by Chandra and Halldórsson [8], in the more general context of $(k+1)$-claw free graphs.

Organization. Section 2 contains formal definitions and a description of the relationship of k-exchange systems to existing set systems. In Section 3 we present our main applications. Section 4 contains our local search algorithms and analyses for the case of submodular and linear objective function functions. Due to space limitations, the analysis for the cardinality and non-monotone submodular objective functions is omitted from this extended abstract.

2 Preliminaries

Let $\mathcal{N}$ be a given ground set and $\mathcal{I} \subseteq 2^{\mathcal{N}}$ a non-empty, downward closed collection of subsets of $\mathcal{N}$. We call such a system $(\mathcal{N}, \mathcal{I})$ an independence system. We use the standard terminology for discussing independence systems. Given an independence system $(\mathcal{N}, \mathcal{I})$ we say that a set $S \subseteq \mathcal{N}$ is *independent* if $S \in \mathcal{I}$, and call the inclusion-wise maximal independent sets in $\mathcal{I}$ *bases*.

Algorithmically, an independence system $(\mathcal{N}, \mathcal{I})$ might not be given explicitly since the size of $\mathcal{I}$ might be exponential in the size of the ground set. Therefore, we assume access to an independence oracle that given $S \subseteq \mathcal{N}$ determines whether $S \in \mathcal{I}$. Given an independence system $(\mathcal{N}, \mathcal{I})$, and a function $f : 2^{\mathcal{N}} \to \mathbb{R}^+$, we are concerned with the problem of finding a set $S \in \mathcal{I}$ that maximizes f. In particular, we consider a variety of restricted classes of functions f. In the most restricted setting, $f(S) = |S|$ is simply the *cardinality* of S. A natural generalization is the weighted, or *linear* case in which each element $e \in \mathcal{N}$ is assigned a weight $w(e)$, and $f(S) = \sum_{e \in S} w(e)$. More generally we consider *submodular* function f. A function $f : 2^{\mathcal{N}} \to \mathbb{R}^+$ is *submodular* if for every $A, B \subseteq \mathcal{N}$: $f(A) + f(B) \geq f(A \cap B) + f(A \cup B)$. Equivalently, f is submodular if for every $A \subseteq B \subseteq \mathcal{N}$ and $e \in \mathcal{N}$: $f(A \cup \{e\}) - f(A) \geq f(B \cup \{e\}) - f(B)$. Additionally, a submodular function f is *monotone* if $f(A) \leq f(B)$ for every $A \subseteq B \subseteq \mathcal{N}$, and *normalized* if $f(\emptyset) = 0$.

The description of a submodular function f might be exponential in the size of the ground set. In this paper we assume the *value oracle* model for accessing f, in which an algorithm is given access to an oracle that returns $f(S)$ for a given set $S \subseteq \mathcal{N}$. This is the model commonly used throughout the literature.

The greedy algorithm provides a k-approximation for maximum weighted independent set in all k-systems. Unfortunately, even the more restricted variants of k-systems, such as k-extendible and k-circuit bounded systems, appear too weak to obtain similar (or potentially stronger) results for local search algorithms. With this goal in mind, we propose the following new class of independence systems, which we call k-*exchange systems*:

Definition 1 (k-exchange system). *An independence system $(\mathcal{N}, \mathcal{I})$ is a k-exchange system if, for all S and T in $\mathcal{I}$, there exists a multiset $Y = \{Y_e \subseteq S \setminus T \mid e \in T \setminus S\}$ such that:*

(K1) $|Y_e| \leq k$ *for each* $e \in T \setminus S$.
(K2) *Every* $e' \in T \setminus S$ *appears in at most* k *sets of* Y.
(K3) *For all* $T' \subseteq T \setminus S$, $(S \setminus (\bigcup_{e \in T'} Y_e)) \cup T' \in \mathcal{I}$

The following theorem follows immediately from the definitions of k-extendible and k-exchange systems.

Theorem 1. *Every* k-*exchange system is a* k-*extendible system.*

Mestre shows that the 1-extendible systems are exactly matroids [30]. We can provide a similar motivation for k-exchange systems in terms of *strongly base orderable matroids*, which derive from work by Brualdi and Scrimger on exchange systems [6,4,5].

Definition 2 (strongly base orderable matroid [5]). *A matroid* $\mathcal{M}$ *is strongly base orderable if for all bases* S *and* T *of* $\mathcal{M}$ *there exists a bijection* $\pi : S \to T$ *such that for all* $S' \subseteq S$, $(T \setminus \pi(S')) \cup S'$ *is a base.*

If we restrict S' to be a singleton set in this definition—thus considering only single replacements—we obtain a well-known result of Brualdi [3] that holds for all matroids. In a strongly base orderable matroid, we require additionally that any set of these individual replacements can be performed simultaneously to obtain an independent set. This simultaneous replacement property is exactly what we want for local search, as it allows us to extend the local analysis for single replacements to larger replacements needed by our algorithms.

The following theorem is easily obtained by equating Y_e in Definition 1 with the singleton set $\{\pi(e)\}$, where π is as in Definition 2.

Theorem 2. *An independence system* $(\mathcal{N}, \mathcal{I})$ *is a strongly base orderable matroid if and only if it is a 1-exchange system.*

The cycle matroid on K_4 is not strongly base orderable, and so provides an example of a matroid that is not a 1-exchange system. Conversely, it is possible to find examples of 2-exchange systems that are not 2-circuit bounded.

3   Applications

In this section, we discuss some applications of k-exchange systems. Due to space constraints, we describe only the application for strongly base orderable matroid k-parity. This problem generalizes our other two main applications: k-set packing and b-matching.

Definition 3. *In the matroid* k-*parity problem, we are given a collection* $\mathcal{E}$ *of disjoint* k-*element subsets from a ground set* $\mathcal{G}$ *and a matroid* $(\mathcal{G}, \mathcal{M})$ *defined on the ground set. The goal is to find a collection* $\mathcal{S}$ *of subsets in* $\mathcal{E}$ *maximizing a function* $f : \mathcal{E} \to \mathbb{R}^+$, *subject to the constraint that* $\bigcup \mathcal{S} \in \mathcal{M}$.

We consider matroid k-matching in the special case in which the given matroid is strongly base orderable. For clarity, we use calligraphic letters to denote sets

---

Algorithm 1: LS-k-EXCHANGE$((\mathcal{N}, \mathcal{I}), f, \varepsilon, p)$

---

1 $e \leftarrow \text{argmax}\{f(\{e\}) \mid e \in \mathcal{N}\}$.
2 $S \leftarrow \{e\}$.
3 Let $\mathcal{G} = (\mathcal{I}, \mathcal{E})$ be the p-exchange graph of $(\mathcal{N}, \mathcal{I})$ and $\varepsilon_n = \varepsilon/|\mathcal{N}|$.
4 **while** $\exists (S \rightarrow T) \in \mathcal{E}$ *such that* $f(T) \geq (1 + \varepsilon_n) f(S)$ **do** $S \leftarrow T$.
5 Output S.

---

of sets from the partition $\mathcal{E}$ and capital letters to denote sets of elements from V, (including, in particular, each of the sets in $\mathcal{E}$). Then, matroid k-parity can be expressed as the independence system $(\mathcal{E}, \mathcal{I})$ where $\mathcal{I} = \{S \subseteq \mathcal{E} : \bigcup S \in \mathcal{M}\}$.

Theorem 3. *Strongly base orderable matroid k-parity is a k-exchange system.*

Proof. Consider two solutions $\mathcal{S}, \mathcal{T} \in \mathcal{I}$. We must have $\bigcup \mathcal{S}$ and $\bigcup \mathcal{T}$ in $\mathcal{M}$. Let $\pi : \bigcup \mathcal{S} \rightarrow \bigcup \mathcal{T}$ be the bijection guaranteed by Definition 2, and for any set $E \in \mathcal{S}$ define $Y_E = \{A \in \mathcal{T} : A \cap \pi(E) \neq \emptyset\}$. The sets in $\mathcal{E}$ are disjoint and contain at most k elements. Since π is a bijection, we must therefore have $|Y_E| \leq k$ for all $E \in \mathcal{S}$ and each $A \in \mathcal{T}$ appears in at most k sets Y_E. Thus, Y satisfies Properties (K1) and (K2). Consider a set $\mathcal{C} \subseteq \mathcal{S}$, and let $\mathcal{S}' = (\mathcal{S} \setminus \bigcup \{Y_E : E \in \mathcal{C}\}) \cup \mathcal{C}$. From the definition of π we have $(\bigcup \mathcal{S} \setminus \pi(\bigcup \mathcal{C})) \cup \bigcup \mathcal{C} \in \mathcal{M}$, and $\bigcup \mathcal{S}'$ is a subset of this set, so $\bigcup \mathcal{S}' \in \mathcal{M}$ and hence $\mathcal{S}' \in \mathcal{I}$, showing that (K3) is satisfied.

4 Combinatorial Local Search Approximation Algorithms

Let us define a directed graph representing improvements considered by our local search algorithms:

Definition 4. *Given a k-exchange system $(\mathcal{N}, \mathcal{I})$, $S, T \in \mathcal{I}$, and $p \in \mathbb{N}$, T is p-reachable from S if the following conditions are satisfied:*

1. $|T \setminus S| \leq p$.
2. $|S \setminus T| \leq (k - 1)p + 1$.

Definition 5. *Given a k-exchange system $(\mathcal{N}, \mathcal{I})$, and $p \in \mathbb{N}$, the p-exchange graph of $(\mathcal{N}, \mathcal{I})$ is a directed graph $\mathcal{G} = (\mathcal{I}, \mathcal{E})$ where $(S \rightarrow T) \in \mathcal{E}$ if and only if T is p-reachable from S.*

Our algorithms are local search algorithms starting from a vertex in $\mathcal{G}$ and touring the graph arbitrarily until they find a sink vertex S. The algorithms than output S.

The start point of the algorithms is the singleton of maximum value. It is important to note that this start point is an independent set. Otherwise, the element of the singleton does not belong to any independent set (recall that $(\mathcal{N}, \mathcal{I})$ is monotone), and therefore, can be removed from $(\mathcal{N}, \mathcal{I})$.

Algorithm 2: NON-OBLIVIOUS-LS-k-EXCHANGE$((\mathcal{N}, \mathcal{I}), f, \varepsilon)$

---

1 $e \leftarrow \operatorname{argmax}\{w(e) \mid e \in \mathcal{N}\}$.

2 $S \leftarrow \{e\}$.

3 Round the weights w down to integer multiples of $f(S)\epsilon/n$.

4 Let $\mathcal{G} = (\mathcal{I}, \mathcal{E})$ be the k-exchange graph of $(\mathcal{N}, \mathcal{I})$. **while** $\exists (S \to T) \in \mathcal{E}$ *such that* $w^2(T) > w^2(S)$ **do** $S \leftarrow T$.

5 Output S.

Algorithm 1 attempts to maximizing the objective function itself while touring $\mathcal{G}$. In the case of a linear f, and $k > 2$, we can improve the approximation ratio of algorithm by using the following technique. Since f is linear, we can assign weights $w(e) = f(\{e\})$ for each element $e \in \mathcal{N}$ and express $f(S)$ as $\sum_{e \in S} w(e)$. We construct a new linear objective function $w^2(S) = \sum_{e \in S} w(e)^2$, and use this function to guide our search. Additionally, we ensure convergence by rounding the weights w using the initial solution, as in [2]. Note that in this algorithm, we always search in the k-exchange graph, rather than the p-exchange graph for some given p.

The graph $\mathcal{G}$, like the set $\mathcal{I}$, might be exponential. However, standard techniques can be used to show that the algorithms terminate in polynomial time.

Theorem 4. *For any constants k, $0 < \varepsilon < k$ and $p \in \mathbb{N}$, Algorithms 1 and 2 terminate in polynomial time.*

4.1   Analysis of Algorithm 1

Our analysis of Algorithm 1 proceeds by considering a particular subset of improvements considered by the algorithm in line 4. Let $(\mathcal{N}, \mathcal{I})$ be a k-exchange system, and let $S \in \mathcal{I}$ be the independent set produced by Algorithm 1 and $T \in \mathcal{I}$ be any other independent set. We construct the following bipartite graph $G_{S,T} = (S \setminus T, T \setminus S, E)$, where $E = \{(e, e') \mid e \in T \setminus S, e' \in Y_e\}$. Note that Properties (K1) and (K2) imply that the maximum degree in $G_{S,T}$ is at most k.

The following theorem is a key ingredient in the analysis of Algorithm 1, allowing us to decompose $G_{S,T}$ into a collection of paths, from which we will obtain the improvements considered in our analysis. Like $G_{S,T}$, its use is restricted only to the analysis itself, as no actual construction of $\mathcal{P}(G, k, h)$ is needed.

Theorem 5. *Let G be an undirected graph whose maximum degree is at most $k \geq 2$. Then, for every $h \in \mathbb{N}$ there exists a multiset $\mathcal{P}(G, k, h)$ of simple paths in G and a labeling $\ell : V \times \mathcal{P}(G, k, h) \to \{\emptyset, 1, 2, \ldots, h\}$ such that:*

1. *For every $P \in \mathcal{P}(G, k, h)$, the labeling ℓ of the nodes of P is consecutive and increasing with labels from $\{1, 2, \ldots, h\}$. Vertices not in P receive label $\emptyset$.*
2. *For every $P \in \mathcal{P}(G, k, h)$ and v in P, if $deg_G(v) = k$ and $\ell(v, P) \notin \{1, h\}$, then at least two of the neighbors of v are in P.*
3. *For every $v \in V$ and label $i \in \{1, 2, \ldots, h\}$, there are $n(k, h) = k \cdot (k-1)^{h-2}$ paths $P \in \mathcal{P}(G, k, h)$ for which $\ell(v, P) = i$.*

Note for condition 2, v might be an end vertex of a path P, but still have a label different from 1 and h. This might happen since paths might contain less than h vertices and start with a label different from 1.

Due to space constraints we omit the full proof of Theorem 5, but let us provide some intuition as to why the construction of $\mathcal{P}(G, k, h)$ is possible. Assume that the degree of every vertex in G is exactly k and that G's girth is at least h. Construct the multiset $\mathcal{P}(G, k, h)$ in the following way. ¿From every vertex $u \in V$, choose all possible paths starting at v and containing exactly h vertices. Number the vertices of these paths consecutively, starting from 1 up to h. First, note that all these paths are simple since the girth of G is at least h and all paths contain exactly h vertices. Second, the number of paths starting from u is: $n(k, h) = k \cdot (k-1)^{h-2}$, since all vertices have degree of exactly k. Third, the number of times each label is given in the graph is exactly $n \cdot n(k, h)$. Since the number of vertices at distance i, $1 \le i \le h$, from each vertex u is identical, the labels are distributed equally among the vertices. Thus, the number of paths in which a given vertex u appears with a given label, is exactly $n(k, h)$. This concludes the proof of the theorem in case G has the above properties.

Applying Theorem 5 to $G_{S,T}$ with $h = 2p$, gives a multiset $\mathcal{P}(G_{S,T}, k, 2p)$ of simple paths in $G_{S,T}$ and a labeling $\ell : (S \triangle T) \times \mathcal{P}(G_{S,T}, k, 2p) \to \{\emptyset, 1, 2, \ldots, 2p\}$ with all the properties guaranteed by Theorem 5. We use $\mathcal{P}(G_{S,T}, k, 2p)$ to construct a new multiset $\mathcal{P}'$ of subsets of vertices of $G_{S,T}$. For each path in $P \in \mathcal{P}$ that contains at least one vertex from T [1], we add $(P \cup N(P))$ to $\mathcal{P}'$, where $N(P)$ is the set of all vertices in $G_{S,T}$ neighboring some vertex in P. Intuitively, $\mathcal{P}'$ is the collection of all paths in $\mathcal{P}(G_{S,T}, k, 2p)$ with an extra "padding" of vertices from S that surround P, excluding "paths" composed of a single vertex from S.

Lemma 1. *Every vertex $e \in T \setminus S$ appears in $2p \cdot n(k, 2p)$ sets of $\mathcal{P}'$, and every vertex $e' \in S \setminus T$ appears in at most $2\left((k-1)p + 1\right) \cdot n(k, 2p)$ sets of $\mathcal{P}'$.*

Proof. By property 3 of Theorem 5, every $e \in T \setminus S$ appears in $n(k, 2p)$ paths of $\mathcal{P}(G_{S,T}, k, 2p)$ for every possible label. Since there are $2p$ possible labels, the number of appearances is exactly $2p \cdot n(k, 2p)$. In the creation of $\mathcal{P}'$ from $\mathcal{P}(G_{S,T}, k, 2p)$, no $e \in T \setminus S$ is added or removed from any path $P \in \mathcal{P}(G_{S,T}, k, 2p)$, thus, this is also the number of appearances of every $e \in T \setminus S$ in $\mathcal{P}'$.

Let $e' \in S \setminus T$. By the construction of $\mathcal{P}'$, a set in $\mathcal{P}'$ that contains e' must contain a vertex $e \in T \setminus S$ where $(e, e') \in E$ (e is a neighbor of e' in $G_{S,T}$). Every such neighboring vertex e, by the first part of the lemma, appears in exactly $2p \cdot n(k, 2p)$ sets in $\mathcal{P}'$. Therefore, the number of appearances of e' in sets of $\mathcal{P}'$ is at most: $deg_{G_{S,T}}(e') \cdot 2p \cdot n(k, 2p) \le 2pk \cdot n(k, 2p)$ (recall that the maximum degree of $G_{S,T}$ is at most k). Furthermore, $\ell(e', P) \neq \{1, 2p\}$, by property 2 of Theorem 5, e' has at least two neighbors in $T \setminus S$ which belong to P itself. Hence, $P \cup N(P) \in \mathcal{P}'$ should be counted only once while in the above counting it was counted at least twice. The number of such $P \in \mathcal{P}(G_{S,T}, k, 2p)$ is exactly

---

[1] Note that this implies that this vertex is from $T \setminus S$ since $G_{S,T}$ does not contain vertices from $S \cap T$.

$2(p-1) \cdot n(k, 2p)$ (by property 3 of Theorem 5). Removing the double counting from the bound, we can conclude that for every $e' \in S \backslash T$, the number of sets in $\mathcal{P}'$ it appears in is at most $2pk \cdot n(k, 2p) - 2(p-1) \cdot n(k, 2p) = 2\left((k-1)p+1\right) \cdot n(k, 2p)$.

Note: The proof of Theorem 6 assumes each vertex $e' \in S \backslash T$ appears in *exactly* $2\left((k-1)p+1\right) \cdot n(k, 2p)$ sets in $\mathcal{P}'$. This can be achieved by adding "dummy" sets to $\mathcal{P}'$ containing e' alone.

We are now ready to state our main theorem. In the proof, we use of the following two technical lemmata from [26]:

Lemma 2 (Lemma 1.1 in [26]). *Let f be a non-negative submodular function of $\mathcal{N}$. Let $S' \subseteq S \subseteq \mathcal{N}$ and let $\{T_\ell\}_{\ell=1}^t$ be a collection of subsets of $S \setminus S'$ such that every elements of $S \setminus S'$ appears in exactly k of these subsets. Then, $\sum_{\ell=1}^t [f(S) - f(S \setminus T_\ell)] \le k\left(f(S) - f(S')\right)$.*

Lemma 3 (Lemma 1.2 in [26]). *Let f be a non-negative submodular function of $\mathcal{N}$. Let $S \subseteq \mathcal{N}$, $C \subseteq \mathcal{N}$ and let $\{T_\ell\}_{\ell=1}^t$ be a collection of subsets of $C \setminus S$ such that every elements of $C \setminus S$ appears in exactly k of these subsets. Then, $\sum_{\ell=1}^t [f(S \cup T_\ell) - f(S)] \ge k\left(f(S \cup C) - f(S)\right)$.*

Theorem 6. *For every $T \in \mathcal{I}$ and every submodular f:*

$$f(S \cup T) + \left(k - 1 + \frac{1}{p}\right) \cdot f(S \cap T) \le \left(k + \frac{1}{p} + k\varepsilon\right) \cdot f(S) \ .$$

Proof. Note that by construction, the symmetric difference of $S \triangle P'$ is an independent set, for any $P' \in \mathcal{P}'$ and furthermore $f(S \triangle P')$ p-reachable from S. Since Algorithm 1 terminated with S, it must be the case that S is approximately "locally optimal", and therefore,

$$f(S \triangle P') < (1 + \varepsilon_n) f(S) \ . \tag{1}$$

for all $P' \in \mathcal{P}'$. By submodularity of f, the fact that $S \setminus P' \subseteq S \triangle P'$ and the fact that all vertices in $S \cap P'$ do not belong to either $S \setminus P'$ or $S \triangle P'$, we get:

$$f(S \cup P') - f(S \triangle P') \le f(S) - f(S \setminus P') \ . \tag{2}$$

Adding Inequalities 1 and 2 gives:

$$f(S \cup P') - (1 + \varepsilon_n) f(S) \le f(S) - f(S \setminus P') \ . \tag{3}$$

Inequality 3 holds for every $P' \in \mathcal{P}'$. Summing over all such sets yields:

$$\sum_{P' \in \mathcal{P}'} [f(S \cup P') - f(S)] - \varepsilon_n |\mathcal{P}'| f(S) \le \sum_{P' \in \mathcal{P}'} [f(S) - f(S \setminus P')] \ . \tag{4}$$

Now, we note that any given P' contains only vertices from $S \triangle T$. Thus, Inequality 4 is equivalent to:

$$\sum_{P' \in \mathcal{P}'} [f(S \cup (P' \cap (T \setminus S))) - f(S)] - \varepsilon_n |\mathcal{P}'| f(S)$$

$$\le \sum_{P' \in \mathcal{P}'} [f(S) - f(S \setminus (P' \cap (S \setminus T)))] \ . \tag{5}$$

By Lemma 1 (and the note after it), each vertex in $S \setminus T$ appears in exactly $2\left((k-1)p + 1\right) \cdot n(k, 2p)$ sets in $\mathcal{P}'$, while every vertex of $T \setminus S_{ALG}$ appears in exactly $2p \cdot n(k, 2p)$ sets in $\mathcal{P}'$. Thus, applying Lemma 2 to the right of Inequality 5 and Lemma 3 to the left gives:

$$2p \cdot n(k, 2p)(f(S \cup T) - f(S)) - \varepsilon_n |\mathcal{P}'| f(S) \leq$$
$$2\left((k-1)p + 1\right) n(k, 2p)\left(f(S) - f(S \cap T)\right) .$$

Rearranging terms and using the definition of ε_n we obtain:

$$f(S \cup T) + \left(k - 1 + \frac{1}{p}\right) f(S \cap T) \leq \left(k + \frac{1}{p}\right) f(S) + \frac{\varepsilon |\mathcal{P}'|}{2p \cdot n(k, 2p) |\mathcal{N}|} f(S) .$$
(6)

Finally, we note every set in $P' \in \mathcal{P}'$ contains at least 1 vertex from $G_{S,T}$, and, by Lemma 1, every vertex in $G_{S_{ALG},T}$ appears in exactly $2p \cdot n(k, 2p)$ or $2\left((k-1)p + 1\right) \cdot n(k, 2p)$ sets of $\mathcal{P}'$ (depending on whether the vertex is in $T \setminus S_{ALG}$ or $S_{ALG} \setminus T$). Therefore, in the worst case $|\mathcal{P}'| \leq |S_{ALG} \triangle T| \cdot 2n(k, 2p) \max\{p, (k-1)p + 1\} \leq 2|\mathcal{N}| n(k, 2p)kp$.

By setting $1/p + k\epsilon \leq \delta$ and using basic properties of monotone submodular and linear functions we obtain the following.

Corollary 1. *Given a set function $f : 2^{\mathcal{N}} \to \mathbb{R}^+$ and any $\delta > 0$, Algorithm 1 is a $1/(k + \delta)$ approximation algorithm if f is a normalized monotone submodular function and a $1/(k - 1 + \delta)$ approximation algorithm if f is a linear function.*

4.2 Analysis of Algorithm 2

The analysis of Algorithm 2 closely follows Berman's analysis for $(k + 1)$-claw free graphs [2]. Like the analysis of Algorithm 1, this analysis also considers a subset of the possible improvements considered by the algorithm. Here, however, we consider improvements of the following form. Consider a k-exchange system $(\mathcal{I}, \mathcal{N})$ and the rounded weight function $w : \mathcal{N} \to \mathbb{R}^+$ produced in line 3 of the algorithm (we shall use the rounded weights for the remainder of our analysis). Let S be the solution produced by Algorithm 2, and T be any independent set in $\mathcal{I}$. For each element $x \in S$, let P_x be the set of all elements $e \in T$ such that $x = \arg\max_{y \in Y_e} w(y)$. For elements $e \in S \cap T$, we define $Y_e = \{e\}$, so $P_e = \{e\}$. Then, $\mathcal{P} = \{P_x\}_{x \in S}$ is a partition of T. We consider improvements $P_x \cup N(P_x)$ where $N(P_x) = \bigcup_{e \in P_x} Y_e$. Note that for all $y \in N(P_x)$, we have $w(y) \leq w(x)$. The following theorem from Berman's analysis allows us to relate the value of w^2 to that of w.

Lemma 4. *For all $x \in S$, $e \in P_x$: $w^2(e) - w^2(Y_e \setminus \{x\}) \geq w(x) \cdot (2w(e) - w(Y_e))$.*

We now prove our main theorem regarding Algorithm 2.

Theorem 7. $\frac{k+1}{2} w(S) \geq w(T)$.

Proof. For each element $x \in S$ we consider the improvement $P_x \cup N(P_x)$. By construction, the symmetric difference of $S \triangle (P_x \cup N(P_x))$ is an independent set, for any $x \in S$, and moreover, $S \triangle (P_x \cup N(P_x))$ is k-reachable from S. Since Algorithm 2 terminated, producing solution S, it must be the case that S is locally optimal, and so $w^2(S \triangle (P_x \cup N(P_x))) \leq w^2(S)$ for each $x \in S$. Combining this with the linearity of w^2 and the fact that $x \in Y_e$ for all $e \in P_x$, we have:

$$w^2(P_x) \leq w^2(N(P_x)) \leq w(x)^2 + \sum_{e \in P_x} w^2(Y_e \setminus \{x\}) \ . \tag{7}$$

Rearranging Inequality 7 using $w^2(P_x) = \sum_{e \in P_x} w(e)^2$ we obtain:

$$\sum_{e \in P_x} w(e)^2 - w^2(Y_e \setminus \{x\}) \leq w(x)^2 \ . \tag{8}$$

Applying Lemma 4 to each term on the left of Inequality (8) gives $\sum_{e \in P_x} w(x) \cdot (2w(e) - f(Y_e)) \leq w(x)^2$ Dividing both sides by $w(x)$ we obtain:

$$\sum_{e \in P_x} (2w(e) - w(Y_e)) \leq w(x) \ . \tag{9}$$

Inequality (9) holds for all $x \in S$. Thus, summing over all $x \in S$, we have

$$\sum_{x \in S} w(x) \geq \sum_{x \in S} \sum_{e \in P_x} [2w(e) - w(Y_e)] \ . \tag{10}$$

We note that $\sum_{x \in S} w(x) = w(S)$, and P_x is a partition of T, so Inequality 10 is equivalent to:

$$w(S) \geq \sum_{e \in T} [2w(e) - w(Y_e)] = 2w(T) - \sum_{e \in T} w(Y_e)$$

$$\geq 2w(T) - k \sum_{x \in S} w(x) = 2w(T) - kw(S) \ , \tag{11}$$

where the second inequality follows from (K2).

Corollary 2. *Algorithm 2 is a $2/(k + 1 + \delta)$ approximation algorithm for maximizing a linear function $f : 2^N \to \mathbb{R}^+$ for any $\delta > 0$.*

Proof. It can be shown (as in [2]) that for any ϵ, our rounding operation introduces at most an extra multiplicative error of ϵ with respect to the original weight function, while ensuring that the algorithm makes only $poly(n, \epsilon^{-1})$ improvements.

5   Open Questions

The k-coverable class intersects the k-intersection class, and for both classes the same results are achieved by Algorithm 1, though different insights are used to analysis each case. Finding a common generalization of both classes admitting a uniform analysis is an intriguing question. A more concrete question is whether there is an exact algorithm for maximizing a *linear* function over the 2-coverable class (analogously to Edmonds exact algorithm for 2-intersection [11]). Finally, it would be interesting to see whether Algorithm 2 can be applied to monotone submodular functions. In the submodular case, we no longer have weights to square in the non-oblivious potential function, but one possible approach is to consider the sum of the squared marginals of the submodular function.

Acknowledgments. The last author thanks Julián Mestre and Allan Borodin for providing comments on a preliminary version of the paper.

References

1. Anstee, R.P.: A polynomial algorithm for b-matchings: An alternative approach. Information Processing Letters 24(3), 153–157 (1987)
2. Berman, P.: A d/2 approximation for maximum weight independent set in d-claw free graphs. Nordic J. of Computing 7, 178–184 (2000)
3. Brualdi, R.A.: Comments on bases in dependence structures. Bull. of the Australian Math. Soc. 1(02), 161–167 (1969)
4. Brualdi, R.A.: Common transversals and strong exchange systems. J. of Combinatorial Theory 8(3), 307–329 (1970)
5. Brualdi, R.A.: Induced matroids. Proc. of the American Math. Soc. 29, 213–221 (1971)
6. Brualdi, R.A., Scrimger, E.B.: Exchange systems, matchings, and transversals. J. of Combinatorial Theory 5(3), 244–257 (1968)
7. Calinescu, G., Chekuri, C., Pal, M., Vondrák, J.: Maximizing a submodular set function subject to a matroid constraint. To appear in SIAM Journal on Computing, Special Issue for STOC 2008 (2008)
8. Chandra, B., Halldórsson, M.: Greedy local improvement and weighted set packing approximation. In: SODA, pp. 169–176 (1999)
9. Chekuri, C., Vondrák, J., Zenklusen, R.: Submodular function maximization via the multilinear relaxation and contention resolution schemes. In: STOC, pp. 783–792 (2011)
10. Conforti, M., Cornuèjols, G.: Submodular set functions, matroids and the greedy algorithm: Tight worst-case bounds and some generalizations of the rado-edmonds theorem. Disc. Appl. Math. 7(3), 251–274 (1984)
11. Edmonds, J.: Matroid intersection. In: Hammer, P., Johnson, E., Korte, B. (eds.) Discrete Optimization I, Proceedings of the Advanced Research Institute on Discrete Optimization and Systems Applications of the Systems Science Panel of NATO and of the Discrete Optimization Symposium. Annals of Discrete Mathematics, vol. 4, pp. 39–49. Elsevier, Amsterdam (1979)
12. Feige, U., Mirrokni, V.S., Vondrák, J.: Maximizing non-monotone submodular functions. In: FOCS, pp. 461–471 (2007)
13. Feldman, M., Naor, J.S., Schwartz, R.: Nonmonotone submodular maximization via a structural continuous greedy algorithm. In: Aceto, L., Henzinger, M., Sgall, J. (eds.) ICALP 2011. LNCS, vol. 6755, pp. 342–353. Springer, Heidelberg (2011)
14. Feldman, M., Naor, J.S., Schwartz, R.: A unified continuous greedy algorithm for submodular maximization. To appear in FOCS 2011 (2011)

15. Fisher, M., Nemhauser, G., Wolsey, L.: An analysis of approximations for maximizing submodular set functions - ii. Math. Prog. Study 8, 73–87 (1978)
16. Gharan, S.O., Vondrák, J.: Submodular maximization by simulated annealing. In: SODA, pp. 1098–1116 (2011)
17. Gupta, A., Roth, A., Schoenebeck, G., Talwar, K.: Constrained non-monotone submodular maximization: Offline and secretary algorithms. In: Saberi, A. (ed.) WINE 2010. LNCS, vol. 6484, pp. 246–257. Springer, Heidelberg (2010)
18. Hausmann, D., Korte, B.: K-greedy algorithms for independence systems. Oper. Res. Ser. A-B 22(1), 219–228 (1978)
19. Hausmann, D., Korte, B., Jenkyns, T.: Worst case analysis of greedy type algorithms for independence systems. Math. Prog. Study 12, 120–131 (1980)
20. Hazan, E., Safra, S., Schwartz, O.: On the complexity of approximating k-set packing. Comput. Complex. 15, 20–39 (2006)
21. Hurkens, C.A.J., Schrijver, A.: On the size of systems of sets every t of which have an sdr, with an application to the worst case ratio of heuristics for packing problems. SIAM J. Disc. Math. 2(1), 68–72 (1989)
22. Jenkyns, T.: The efficacy of the greedy algorithm. Cong. Num. 17, 341–350 (1976)
23. Jensen, P.M., Korte, B.: Complexity of matroid property algorithms. SIAM J. Computing 11(1), 184 (1982)
24. Kalyanasundaram, B., Pruhs, K.: An optimal deterministic algorithm for online b-matching. In: Chandru, V., Vinay, V. (eds.) FSTTCS 1996. LNCS, vol. 1180, pp. 193–199. Springer, Heidelberg (1996)
25. Korte, B., Hausmann, D.: An analysis of the greedy heuristic for independence systems. Annals of Discrete Math. 2, 65–74 (1978)
26. Lee, J., Sviridenko, M., Vondrák, J.: Submodular maximization over multiple matroids via generalized exchange properties. To appear in Mathematics of Operations Research (2009)
27. Lee, J., Sviridenko, M., Vondrak, J.: Matroid matching: the power of local search. In: STOC, pp. 369–378 (2010)
28. Lovász, L.: The matroid matching problem. In: Lovász, L., Sós, V.T. (eds.) Algebraic Methods in Graph Theory, Amsterdam (1981)
29. Marsh III., A.B.: Matching algorithms. PhD thesis, The Johns Hopkins University (1979)
30. Mestre, J.: Greedy in approximation algorithms. In: Azar, Y., Erlebach, T. (eds.) ESA 2006. LNCS, vol. 4168, pp. 528–539. Springer, Heidelberg (2006)
31. Nemhauser, G., Wolsey, L.: Best algorithms for approximating the maximum of a submodular set function. Math. Oper. Res. 3(3), 177–188 (1978)
32. Nemhauser, G., Wolsey, L., Fisher, M.: An analysis of approximations for maximizing submodular set functions - i. Math. Prog. 14(1), 265–294 (1978)
33. Pulleyblank, W.: Faces of matching polyhedra. PhD thesis, Deptartment of Combinatorics and Optimization, University of Waterloo (1973)
34. Schrijver, A.: Combinatorial Optimization: Polyhedra and Efficiency. Springer, Heidelberg (2003)
35. Simon, H.: Approximation algorithms for channel assignment in cellular radio networks. In: Csirik, J., Demetrovics, J., Gècseg, F. (eds.) FCT 1989. LNCS, vol. 380, pp. 405–415. Springer, Heidelberg (1989)
36. Soto, J.A.: A simple PTAS for weighted matroid matching on strongly base orderable matroids. To appear in LAGOS (2011)
37. Tong, P., Lawler, E.L., Vazirani, V.V.: Solving the weighted parity problem for gammoids by reduction to graphic matching. Technical Report UCB/CSD-82-103, EECS Department, University of California, Berkeley (April 1982)

Cover-Decomposition
and Polychromatic Numbers

Béla Bollobás[1,*], David Pritchard[2,**], Thomas Rothvoß[3,***], and Alex Scott[4]

[1] University of Memphis, USA and University of Cambridge, UK
[2] EPFL, Lausanne, Switzerland
[3] MIT, Cambridge, USA
[4] Mathematical Institute, University of Oxford, UK

Abstract. A colouring of a hypergraph's vertices is *polychromatic* if every hyperedge contains at least one vertex of each colour; the *polychromatic number* is the maximum number of colours in such a colouring. Its dual, the *cover-decomposition number*, is the maximum number of disjoint hyperedge-covers. In geometric settings, there is extensive work on lower-bounding these numbers in terms of their trivial upper bounds (minimum hyperedge size & degree). Our goal is to get good lower bounds in natural hypergraph families not arising from geometry. We obtain algorithms yielding near-tight bounds for three hypergraph families: those with bounded hyperedge size, those representing paths in trees, and those with bounded VC-dimension. To do this, we link cover-decomposition to iterated relaxation of linear programs via discrepancy theory.

1 Introduction

In a set system on vertex set V, a subsystem is a *set cover* if each vertex of V appears in at least 1 set of the subsystem. Suppose each vertex appears in at least δ sets of the set system, for some large δ; does it follow that we can partition the system into 2 subsystems, such that each subsystem is a set cover?

Many natural families of set systems admit a universal constant δ for which this question has an affirmative answer. Such families are typically called *cover-decomposable*. But the family of *all* set systems is not cover-decomposable, as the following example shows. For any positive integer k, consider a set system which has $2k - 1$ sets, and where every subfamily of k sets contain one mutually common vertex not contained by the other $k - 1$ sets. This system satisfies the hypothesis of the question for $k = \delta$. But it has no set cover consisting of $\leq k - 1$ sets, and it has only $2k - 1$ sets in total; so no partition into two set covers is possible. This example above shows that some sort of restriction on the family is necessary to ensure cover-decomposability.

* Supported in part by NSF grants CNS-0721983, CCF-0728928, DMS-0906634 and CCR-0225610, and ARO grant W911NF-06-1-0076.
** Supported in part by an NSERC PDF.
*** Supported by the Alexander von Humboldt Foundation within the Feodor Lynen program, by ONR grant N00014-11-1-0053, and by NSF contract CCF-0829878.

C. Demetrescu and M.M. Halldórsson (Eds.): ESA 2011, LNCS 6942, pp. 799–810, 2011.

One positive example of cover-decomposition arises if every set has size 2: such hypergraphs are simply graphs. They are cover-decomposable with $\delta = 3$: any graph with minimum degree 3 can have its edges partitioned into two edge covers. More generally, Gupta [1] showed (also [2,3]) that we can partition the edges of any multigraph into $\lfloor \frac{3\delta+1}{4} \rfloor$ edge covers. This bound is tight, even for 3-vertex multigraphs.

Set systems in many geometric settings have been studied with respect to cover-decomposability; many positive and negative examples are known and there is no easy way to distinguish one from the other. In the affirmative case, as with Gupta's theorem, the next natural problem is to find for each $t \geq 2$ the smallest $\delta(t)$ such that when each vertex appears in at least $\delta(t)$ sets, a partition into t set covers is possible. The goal of this paper is to extend the study of cover-decomposition beyond geometric settings. A novel property of our studies is that we use iterated linear programming to find cover-decompositions.

1.1 Terminology and Notation

A *hypergraph* $H = (V, \mathcal{E})$ consists of a ground set V of vertices, together with a collection $\mathcal{E}$ of hyperedges, where each hyperedge $E \in \mathcal{E}$ is a subset of V. Hypergraphs are the same as *set systems*. We will sometimes call hyperedges just *edges* or *sets*. We permit $\mathcal{E}$ to contain multiple copies of the same hyperedge (e.g. to allow us to define "duals" and "shrinking" later), and we also allow hyperedges of cardinality 0 or 1. We only consider hypergraphs that are finite. (In many geometric cases, infinite versions of the problem can be reduced to finite ones, e.g. [4]; see also [5] for work on infinite versions of cover-decomposability.)

To *shrink* a hyperedge E in a hypergraph means to replace it with some $E' \subseteq E$. This operation is useful in several places.

A *polychromatic k-colouring* of a hypergraph is a function from V to a set of k colours so that for every edge, its image contains all colours. (Equivalently, the colour classes partition V into sets which each meet every edge, so-called *vertex covers/transversals*.) The maximum number of colours in a polychromatic colouring of H is called its *polychromatic number*, which we denote by $\mathsf{p}(H)$.

A *cover k-decomposition* of a hypergraph is a partition of $\mathcal{E}$ into k subfamilies $\mathcal{E} = \biguplus_{i=1}^{k} \{\mathcal{E}_i\}$ such that each $\bigcup_{E \in \mathcal{E}_i} E = V$. In other words, each $\mathcal{E}_i$ must be a set cover. The maximum k for which the hypergraph H admits a cover k-decomposition is called its *cover-decomposition number*, which we denote by $\mathsf{p}'(H)$.

The *dual* H^* of a hypergraph H is another hypergraph such that the vertex set of H^* corresponds to the edge set of H, and vice-versa, with incidences preserved. Thus the vertex-edge incidence matrices for H and H^* are transposes of one another. E.g., the standard notation for the example in the introduction is $\left(\binom{[2k-1]}{k} \right)^*$. From the definitions it is easy to see that the polychromatic and cover-decomposition numbers are dual to one another,

$$\mathsf{p}'(H) = \mathsf{p}(H^*).$$

The *degree* of a vertex v in a hypergraph is the number of hyperedges containing v; it is *d-regular* if all vertices have degree d. We denote the minimum degree by δ, and the maximum degree by Δ. We denote the minimum size of any hyperedge by r, and the maximum size of any hyperedge by R. Note that $\Delta(H) = R(H^*)$ and $\delta(H) = r(H^*)$. It is trivial to see that $\mathsf{p} \leq r$ in any hypergraph and dually that $\mathsf{p}' \leq \delta$. So the cover-decomposability question asks if there is a converse to this trivial bound: if δ is large enough, does p' also grow? To write this concisely, for a family $\mathcal{F}$ of hypergraphs, let its extremal *cover-decomposition function* $\overline{\mathsf{p}}'(\mathcal{F}, \delta)$ be

$$\overline{\mathsf{p}}'(\mathcal{F}, \delta) := \min\{\mathsf{p}'(H) \mid H \in \mathcal{F}; \ \forall v \in V(H) : \text{degree}(v) \geq \delta\},$$

i.e. $\overline{\mathsf{p}}'(\mathcal{F}, \delta)$ is the best possible lower bound for p' among hypergraphs in $\mathcal{F}$ with min-degree $\geq \delta$. So to say that $\mathcal{F}$ is cover-decomposable means that $\overline{\mathsf{p}}'(\mathcal{F}, \delta) > 1$ for some constant δ. We also dually define

$$\overline{\mathsf{p}}(\mathcal{F}, r) := \min\{\mathsf{p}(H) \mid H \in \mathcal{F}; \ \forall E \in \mathcal{E}(H) : |E| \geq r\}.$$

In the rest of the paper we focus on computing these functions. When the family $\mathcal{F}$ is clear from context, we write $\overline{\mathsf{p}}'(\delta)$ for $\overline{\mathsf{p}}'(\mathcal{F}, \delta)$ and $\overline{\mathsf{p}}(r)$ for $\overline{\mathsf{p}}(\mathcal{F}, r)$.

1.2   Results

In Section 2 we generalize Gupta's theorem to hypergraphs of bounded edge size. Let $\text{Hyp}(R)$ denote the family of hypergraphs with all edges of size at most R.

Theorem 1. *For all R, δ we have $\overline{\mathsf{p}}'(\text{Hyp}(R), \delta) \geq \max\{1, \delta/(\ln R + O(\ln \ln R))\}$.*

In proving Theorem 1, we first give a simple proof which is weaker by a constant factor, and then we refine the analysis. We use the Lovász Local Lemma (LLL) as well as discrepancy-theoretic results which permit us to partition a large hypergraph into two pieces with roughly-equal degrees. Next we show that Theorem 1 is essentially tight:

Theorem 2. *(a) For a constant C and all $R \geq 2, \delta \geq 1$ we have $\overline{\mathsf{p}}'(\text{Hyp}(R), \delta) \leq \max\{1, C\delta/\ln R\}$. (b) For any sequence $R, \delta \to \infty$ with $\delta = \omega(\ln R)$ we have $\overline{\mathsf{p}}'(\text{Hyp}(R), \delta) \leq (1 + o(1))\delta/\ln(R)$.*

Here (a) uses an explicit construction while (b) uses the probabilistic method.

By plugging Theorem 1 into an approach of [3], one obtains a good bound on the cover-decomposition number of *sparse* hypergraphs.

Corollary 3. *Suppose $H = (V, \mathcal{E})$ satisfies, for all $V' \subseteq V$ and $\mathcal{E}' \subseteq \mathcal{E}$, that the number of incidences between V' and $\mathcal{E}'$ is at most $\alpha|V'| + \beta|\mathcal{E}'|$. Then $\mathsf{p}'(H) \geq \frac{\delta(H) - \alpha}{\ln \beta + O(\ln \ln \beta)}$.*

(Duality yields a similar bound on the polychromatic number.) The proof is analogous to that in [3]: a max-flow min-cut argument shows that in this sparse hypergraph, we can shrink all edges to have size at most β, while keeping the minimum degree at least $\delta(H) - \alpha$.

In Section 3 we consider the following family of hypergraphs: the ground set is the edge set of an undirected tree, and each hyperedge must correspond to the edges lying in some path in the tree. We show that such systems are cover-decomposable:

Theorem 4. *For hypergraphs defined by edges of paths in trees,* $\overline{p}'(\delta) \geq 1 + \lfloor (\delta - 1)/5 \rfloor$.

To prove Theorem 4, we exploit the connection to discrepancy and iterated rounding, using an extreme point structure theorem for paths in trees from [6]. We also determine the extremal polychromatic number for such systems:

Theorem 5. *For hypergraphs defined by edges of paths in trees,* $\overline{p}(r) = \lceil r/2 \rceil$.

This contrasts with a construction of Pach, Tardos and Tóth [7]: if we also allow hyperedges consisting of sets of "siblings," then $\overline{p}(r) = 1$ for all r.

The *VC-dimension* is a prominent measure of set system complexity used frequently in geometry: it is the maximum cardinality of any $S \subseteq V$ such that $\{S \cap E \mid E \in \mathcal{E}\} = 2^S$. It is natural to ask what role the VC-dimension plays in cover-decomposability. We show the following — the proof is deferred to the full version*.

Theorem 6. *For the family of hypergraphs with VC-dimension 1,* $\overline{p}(r) = \lceil r/2 \rceil$ *and* $\overline{p}'(\delta) = \lceil \delta/2 \rceil$.

By duality, the same holds for the family of hypergraphs whose duals have VC-dimension 1. We find Theorem 6 is best possible in a strong sense:

Theorem 7. *For the family of hypergraphs* $\{H \mid VC\text{-}dim(H), VC\text{-}dim(H^*) \leq 2\}$, *we have* $\overline{p}(r) = 1$ *for all* r *and* $\overline{p}'(\delta) = 1$ *for all* δ.

To prove this, we show the construction of [7] has primal and dual VC-dimension at most 2.

All of our lower bounds on $\overline{p}$ and $\overline{p}'$ can be implemented as polynomial-time algorithms. In the case of Theorem 1 this relies on the constructive LLL framework of Moser-Tardos [8]. In the tree setting (Theorem 4) the tree representing the hypergraph does not need to be explicitly given as input, since the structural property used in each iteration (Lemma 15) is easy to identify from the values of the extreme point LP solution. Note: since we also have the trivial bounds $p \leq r, p' \leq \delta$ these give *approximation algorithms* for p and p', e.g. Theorem 1 gives a $(\ln R + O(\ln \ln R))$-approximation for p'.

1.3 Related Work

One practical motive to study cover-decomposition is that the hypergraph can model a collection of sensors [9,10], with each $E \in \mathcal{E}$ corresponding to a sensor

---

* http://arxiv.org/abs/1009.6144

which can monitor the set E of vertices; then monitoring all of V takes a set cover, and $\bar{p}'$ is the maximum "coverage" of V possible if each sensor can only be turned on for a single time unit or monitor a single frequency. Another motive is that if $\bar{p}'(\delta) = \Omega(\delta)$ holds for a family closed under vertex deletion, then the size of a *dual ϵ-net* is bounded by $O(1/\epsilon)$ [11].

A hypergraph is said to be *weakly k-colourable* if we can k-colour its vertex set so that no edge is monochromatic. Weak 2-colourability is also known as *Property B*, and these notions coincide with the property $p \geq 2$. However, weak k-colourability does not imply $p \geq k$ in general.

For a graph $G = (V, E)$, the *(closed) neighbourhood hypergraph* $\mathcal{N}(H)$ is defined to be a hypergraph on ground set V, with one hyperedge $\{v\} \cup \{u \mid \{u, v\} \in E\}$ for each $v \in V$. Then $p(\mathcal{N}(G)) = p'(\mathcal{N}(G))$ equals the *domatic number* of G, i.e. the maximum number of disjoint dominating sets. The paper of Feige, Halldórsson, Kortsarz & Srinivasan [12] obtains upper bounds for the domatic number and their bounds are essentially the same as what we get by applying Theorem 1 to the special case of neighbourhood hypergraphs; compared to our methods they use the LLL but not discrepancy or iterated LP rounding. They give a hardness-of-approximation result which implies that Theorem 1 is tight with respect to the approximation factor, namely for all $\epsilon > 0$, it is hard to approximate p' within a factor better than $(1 - \epsilon) \ln R$, under reasonable complexity assumptions. A generalization of results in [12] to packing polymatroid bases was given in [13]; this implies a weak version of Theorem 1 where the $\log R$ term is replaced by $\log |V|$.

A notable progenitor in geometric literature on cover-decomposition is the following question of Pach [14]. Take a convex set $A \subset \mathbb{R}^2$. Let $\mathbb{R}^2|\text{TRANSLATES}(A)$ denote the family of hypergraphs where the ground set V is a finite subset of $\mathbb{R}^2$, and each hyperedge is the intersection of V with some translate of A. Pach asked if such systems are cover-decomposable, and this question is still open. A state-of-the-art partial answer is due to Gibson & Varadarajan [10], who prove that $\bar{p}(\mathbb{R}^2|\text{TRANSLATES}(A), \delta) = \Omega(\delta)$ when A is an open convex polygon.

The paper of Pach, Tardos and Tóth [7] obtains several negative results with a combinatorial method. They define a family of non-cover-decomposable hypergraphs based on trees and then they "embed" these hypergraphs into geometric settings. By doing this, they prove that the following families are not cover-decomposable: $\mathbb{R}^2|\text{AXIS-ALIGNED-RECTANGLES}$; $\mathbb{R}^2|\text{TRANSLATES}(A)$ when A is a non-convex quadrilateral; and $\mathbb{R}^2|\text{STRIPS}$ and its dual. In contrast to the latter result, it is known that $\bar{p}(\mathbb{R}^2|\text{AXIS-ALIGNED-STRIPS}, r) \geq \lceil r/2 \rceil$ [15]. Recently it was shown [16] that $\mathbb{R}^3|\text{TRANSLATES}(\mathbb{R}^3_+)$ is cover-decomposable, giving cover-decomposability of $\mathbb{R}^2|\text{HOMOTHETS}(T)$ for any triangle T and a new proof (c.f. [17]) for $\mathbb{R}^2|\text{BOTTOMLESS-AXIS-ALIGNED-RECTANGLES}$; the former contrasts with the non-cover-decomposability of $\mathbb{R}^2|\text{HOMOTHETS}(D)$ for D the unit disc [7].

Pálvölgyi [4] poses a fundamental combinatorial question: is there a function f so that in hypergraph families closed under edge deletion and duplication,

$\bar{\mathsf{p}}'(\delta_0) \geq 2$ implies $\bar{\mathsf{p}}'(f(\delta_0)) \geq 3$? This is open for all $\delta_0 \geq 2$ and no counterexamples are known to the conjecture $f(\delta_0) = O(\delta_0)$.

Given a plane graph, define a hypergraph whose vertices are the graph's vertices, and whose hyperedges are the faces. For this family of hypergraphs, it was shown in [3] that $\bar{\mathsf{p}}(\delta) \leq \lfloor (3\delta - 5)/4 \rfloor$ using Gupta's theorem and a sparsity argument. This is the same approach which we exploit to prove Corollary 3.

Several different related colouring notions for paths in trees are considered in [18,19,20].

2   Hypergraphs of Bounded Edge Size

To get good upper bounds on $\bar{\mathsf{p}}'(\mathrm{HYP}(R), \delta)$, we will use the Lovász Local Lemma (LLL):

Lemma 8 (LLL, [21]). *Consider a collection of "bad" events such that each one has probability at most p, and such that each bad event is independent of the other bad events except at most D of them. (We call D the* dependence degree.*) If $p(D+1)\mathrm{e} \leq 1$ then with positive probability, no bad events occur.*

Our first proposition extends a standard argument about Property B [22, Theorem 5.2.1].

Proposition 9. $\bar{\mathsf{p}}'(\mathrm{HYP}(R), \delta) \geq \lfloor \delta/\ln(eR\delta^2) \rfloor$.

I.e. given any hypergraph $H = (V, \mathcal{E})$ where every edge has size at most R and such that each $v \in V$ is covered at least δ times, we must show for $t = \lfloor \delta/\ln(eR\delta^2) \rfloor$ that $\mathsf{p}'(H) \geq t$, i.e. that $\mathcal{E}$ can be decomposed into t disjoint set covers. It will be helpful here and later to make the degree of every vertex *exactly* δ, (this bounds the dependence degree). Indeed this is without loss of generality: else as long as $\deg(v) > \delta$ shrink some $E \ni v$ to $E \backslash \{v\}$ until $\deg(v)$ drops to δ; then observe that if we applying unshrinking to a vertex cover, it is still a vertex cover.

Proof of Proposition 9. Consider the following randomized experiment: for each hyperedge $E \in \mathcal{E}$, assign a random colour between 1 and t to E. If we can show that with positive probability, every vertex is incident with a hyperedge of each colour, then we will be done. In order to get this approach to go through,

For each vertex v define the *bad event* $\mathfrak{E}_v$ to be the event that v is not incident with a hyperedge of each colour. The probability of $\mathfrak{E}_v$ is at most $t(1 - \frac{1}{t})^\delta$, by using a union bound. The event $\mathfrak{E}_v$ only depends on the colours of the hyperedges containing v; therefore the events $\mathfrak{E}_v$ and $\mathfrak{E}_{v'}$ are independent unless v, v' are in a common hyperedge. In particular the dependence degree is less than $R\delta$. It follows by LLL that if

$$R\delta t(1 - \tfrac{1}{t})^\delta \leq 1/\mathrm{e},$$

then with positive probability, no bad events happen and we are done. We can verify that $t = \delta/\ln(eR\delta^2)$ satisfies this bound. $\square$

We will next show that the bound can be raised to $\Omega(\delta/\ln R)$. Intuitively, our strategy is the following. We have that $\delta/\ln(R\delta)$ is already $\Omega(\delta/\ln R)$ unless δ

is superpolynomial in R. For hypergraphs where $\delta \gg R$ we will show that we can partition $\mathcal{E}$ into m parts $\mathcal{E} = \biguplus_{i=1}^{m} \mathcal{E}_i$ so that $\delta(V, \mathcal{E}_i)$ is at least a constant of δ/m, and such that δ/m is polynomial in R. Thus by Proposition 9 we can extract $\Omega((\delta/m)/\ln R)$ set covers from each $(V, \mathcal{E}_i)$, and their union proves $\bar{\mathsf{p}}' \geq \Omega(\delta/\ln R)$.

In fact, it will be enough to consider splitting $\mathcal{E}$ into two parts at a time, recursively. Then ensuring $\delta(V, \mathcal{E}_i) \gtrsim \delta/2$ ($i = 1, 2$) amounts to a discrepancy-theoretic problem: given the incidence matrix with rows for edges and columns for vertices, we must 2-colour the rows by ± 1 so that for each column, the sum of the incident rows' colours is in $[-d, d]$, with the *discrepancy* d as small as possible. To get a short proof of a weaker version of Theorem 1, we can use an approach of Beck and Fiala [23]; moreover it is important to review their proof since we will extend it in Section 3.

Proposition 10 (Beck & Fiala [23]). *In a δ-regular hypergraph $H = (V, \mathcal{E})$ with all edges of size at most R, we can partition the edge set into $\mathcal{E} = \mathcal{E}_1 \uplus \mathcal{E}_2$ such that $\delta(V, \mathcal{E}_i) \geq \delta/2 - R$ for each $i \in \{1, 2\}$.*

Proof. Define a linear program with nonnegative variables $\{x_e, y_e\}_{e \in \mathcal{E}}$ subject to $x_e + y_e = 1$ and for all v, degree constraints $\sum_{e : v \in e} x_e \geq \delta/2$ and $\sum_{e : v \in e} y_e \geq \delta/2$. Note $x \equiv y \equiv \frac{1}{2}$ is a feasible solution. Let us abuse notation and when x or y is 0-1, use them interchangeably with the corresponding subsets of $\mathcal{E}$. So in the LP, a feasible integral x and y would correspond to a discrepancy-zero splitting of $\mathcal{E}$. We'll show that such a solution can be *nearly* found, allowing an additive R violation in the degree constraints. We use the following fact, which follows by double-counting. A constraint is *tight* if it holds with equality.

Lemma 11. *In any extreme-point solution (x, y) to the linear program, there is some tight degree constraint for whom at most R of the variables it involves are strictly between 0 and 1. This holds also if some variables are fixed at integer values and some of the degree constraints have been removed.*

Now we use the following iterated LP rounding algorithm. Each iteration starts with solving the LP and getting an extreme point solution. Then perform two steps: for each variable with an integral value in the solution, fix its value forever; and discard the constraint whose existence is guaranteed by the lemma. Eventually all variables are integral and we terminate.

For each degree constraint, either it was never discarded in which case the final integral solution satisfies it, or else it was discarded in some iteration. Now when the constraint was discarded it had at most R fractional variables, and was tight. So in the sum (say) $\sum_{e : v \in e} x_e = \delta$ there were at least $\delta - R$ variables fixed to 1 on the left-hand side. They ensure $\sum_{e : v \in e} x_e \geq \delta - R$ at termination, proving what we wanted. $\square$

Here is how the Beck-Fiala theorem gives a near-optimal bound on $\bar{\mathsf{p}}'$.

Proposition 12. $\bar{\mathsf{p}}'(\mathrm{HYP}(R), \delta) \geq \delta/O(\ln R)$.

Proof. If $\delta < 4R$ this already follows from Proposition 9. Otherwise apply Proposition 10 to the initial hypergraph, and then use shrinking to make both the resulting $(V, \mathcal{E}_i)$'s into regular hypergraphs. Iterate this process; stop splitting each hypergraph once its degree falls in the range $[R, 4R)$, which is possible since $\delta \geq 4R \Rightarrow \delta/2 - R \geq R$. Let M be the number of hypergraphs at the end.

Observe that in applying the splitting-and-shrinking operation to some $(V, \mathcal{E})$ to get $(V, \mathcal{E}_1)$ and $(V, \mathcal{E}_2)$, the sum of the degrees of $(V, \mathcal{E}_1)$ and $(V, \mathcal{E}_2)$ is at least the degree of $(V, \mathcal{E})$, minus $2R$ "waste". It follows that the total waste is at most $2R(M - 1)$, and we have that $4RM + 2R(M - 1) \geq \delta$. Consequently $M \geq \delta/6R$. As sketched earlier, applying Proposition 9 to the individual hypergraphs, and combining these vertex covers, shows that $\mathsf{p}' \geq M\lfloor R/\ln(eR^3)\rfloor$ which gives the claimed bound. $\square$

Now we get to the better bound with the correct multiplicative constant.

Proof of Theorem 1: $\forall R, \delta,\ \bar{\mathsf{p}}'(\mathrm{HYP}(R), \delta) \geq \max\{1, \delta/(\ln R + O(\ln \ln R))\}$. Now Proposition 9 gives us the desired bound when δ is at most polylogarithmic in R, so we assume otherwise. Due to the crude bound in Proposition 12, we may assume R is sufficiently large when needed. We will need the following well-known discrepancy bound which follows using Chernoff bounds and the LLL; see also the full version.

Proposition 13. *For a constant C_1, in a d-regular hypergraph $H = (V, \mathcal{E})$ with all edges of size at most R, we can partition the edge set into $\mathcal{E} = \mathcal{E}_1 \uplus \mathcal{E}_2$ such that $\delta(V, \mathcal{E}_i) \geq d/2 - C_1\sqrt{d\ln(Rd)}$ $(i = 1, 2)$.*

Let $d_0 = \delta$ and $d_{i+1} = d_i/2 - C_1\sqrt{d_i \ln(Rd_i)}$. Thus each hypergraph obtained after i rounds of splitting has degree at least d_i; evidently $d_i \leq \delta/2^i$. We stop splitting after T rounds, where T will be fixed later to make d_T and $\delta/2^T$ polylogarithmic in R. The total degree loss due to splitting is

$$\delta - 2^T d_T = \sum_{i=0}^{T} 2^i(d_i - 2d_{i+1}) \leq \sum_{i=0}^{T-1} 2^i 2C_1\sqrt{d_i \ln Rd_i} \leq \sum_{i=0}^{T-1} 2^i 2C_1\sqrt{\frac{\delta}{2^i}\ln\frac{R\delta}{2^i}}$$

$$= 2C_1\sqrt{\delta}\sum_{i=0}^{T-1}\sqrt{2^i \ln\frac{R\delta}{2^i}}.$$

This sum is an arithmetic-geometric series dominated by the last term, so that we deduce $\delta - 2^T d_T = O(\sqrt{\delta 2^T \ln(R\delta/2^T)})$. Pick T such that $\delta/2^T$ is within a constant factor of $\ln^3 R$, then we deduce

$$d_T \geq \delta/2^T(1 - O(\sqrt{2^T/\delta \ln(R\delta/2^T)})) \geq \delta/2^T(1 - O(\ln^{-1}(R))).$$

Consequently with Proposition 9 we see that

$$\mathsf{p}' \geq 2^T d_T/(\ln R + O(\ln \ln R)) \geq \delta(1 - O(\ln^{-1}(R)))/(\ln R + O(\ln \ln R))$$

which gives the claimed bound. $\square$

2.1   Lower Bounds

Now we show that the bounds obtained previously are tight.

Proof of Theorem 2(a). We want to show, for a constant C and all $R \geq 2, \delta \geq 1$ we have $\overline{\mathsf{p}}'(\mathrm{HYP}(R), \delta) \leq \max\{1, C\delta/\ln R\}$. Consider the hypergraph $H = \binom{[2k-1]}{k}^*$ in the introduction. It is k-regular, it has $\mathsf{p}'(H) = 1$, and $R(H) = \binom{2k-2}{k-1}$.

Since $\overline{\mathsf{p}}'(\mathrm{HYP}(R), \delta)$ is non-increasing in R, we may reduce R by a constant factor to assume that either $R = 2$, (in which case we are done by Gupta's theorem) or $R(H) = \binom{2k-2}{k-1}$ for some integer k. Note this gives $k = \Theta(\log R)$. Moreover, if $\delta \leq k$ then H proves the theorem, so assume $\delta \geq k$. Again by monotonicity, we may increase δ by a constant factor to make δ a multiple of k. Let $\mu = \delta/k$.

Consider the hypergraph μH obtained by copying each of its edges μ times, for an integer $\mu \geq 1$; note that it is δ-regular. The argument in the introduction shows that any set cover has size at least k and therefore average degree at least $k\binom{2k-2}{k-1}/\binom{2k-1}{k} = k^2/(2k-1) = \Theta(\ln R)$. Thus $\overline{\mathsf{p}}'(\mu H) = O(\delta/\ln R)$ which proves the theorem.   ☐

Proof of Theorem 2(b). We want to show as $R, \delta \to \infty$ with $\delta = \omega(\ln R)$, we have $\overline{\mathsf{p}}'(\mathrm{HYP}(R), \delta) \leq (1 + o(1))\delta/\ln(R)$. We assume an additional hypothesis, that $R \geq \delta$; this will be without loss of generality as we can handle the case $\delta > R$ using the μ-replication trick from the proof of Theorem 2(a), since our argument is again based on lower-bounding the minimum size of an set cover.

Let $\delta' = \delta(1 + o(1))$ and $R' = R(1 - o(1))$ be parameters that will be specified shortly. We construct a random hypergraph with $n = R'^2\delta'$ vertices and $m = R'\delta'^2$ edges, where for each vertex v and each edge E, we have $v \in E$ with independent probability $p = 1/R'\delta'$. Thus each vertex has expected degree δ' and each edge has expected size R'. A standard Chernoff bound together with $np = \omega(\ln m)$ shows the maximum edge size is $(1+o(1))R'$ asymptotically almost surely (a.a.s.); pick R' such that this $(1 + o(1))R'$ equals R. Likewise, since $mp = \omega(\ln n)$ a.a.s. the actual minimum degree is at least $(1 - o(1))\delta'$ which we set equal to δ.

We will show that this random hypergraph has $\mathsf{p}' \geq (1+o(1))\delta/\ln R$ a.a.s. via the following bound, which is based off of an analogous bound for Erdős-Renyi random graphs in [12, §2.5]:

Claim 14. *A.a.s. the minimum set cover size is at least $\frac{1}{p}\ln(pn)(1 - o(1))$.*

The proof is given in the full version. This claim finishes the proof since it implies that the maximum number of disjoint set covers p' is at most $(1 + o(1))mp/\ln(pn) = (1 + o(1))\delta'/\ln(R') = (1 + o(1))\delta/\ln(R)$.   ☐

Aside from the results above, not much else is known about specific values of $\overline{\mathsf{p}}'(\mathrm{HYP}(R), \delta)$ for small R, δ. The Fano plane gives $\overline{\mathsf{p}}(\mathrm{HYP}(3), 3) = 1$: if its seven sets are partitioned into two parts, one part has only three sets, and it is not hard to verify the only covers consisting of three sets are pencils through a point and therefore preclude the remaining sets from forming a cover. Moreover, Thomassen [24] showed that every 4-regular, 4-uniform hypergraph has Property B; together with monotonicity we deduce that $\bar{\mathsf{p}}(\text{HYP}(3), 4) \geq \bar{\mathsf{p}}(\text{HYP}(4), 4) \geq 2$.

3 Paths in Trees

Let TREEEDGES|PATHS denote the following family of hypergraphs: the ground set is the edge set of an undirected tree, and each hyperedge must correspond to the edges lying in some path in the tree. Such systems are cover-decomposable:

Theorem 4. $\bar{\mathsf{p}}'(\text{TREEEDGES}|\text{PATHS}, \delta) \geq 1 + \lfloor (\delta - 1)/5 \rfloor$.

Proof. In other words, given a family of paths covering each edge at least $\delta = 5k + 1$ times, we can partition the family into $k + 1$ covers. We use induction on k; the case $k = 0$ is evidently true.

We will use an iterated LP relaxation algorithm similar to the one used in Proposition 10. However, it is more convenient to get rid of the y variables; it is helpful to think of them implicitly as $y = 1 - x$. Thus our linear program will have one variable $0 \leq x_P \leq 1$ for every path P. Fix integers A, B such that $A + B = \delta$, and the LP will aim to make x the indicator vector of an A-fold cover, and $1 - x$ the indicator vector of a B-fold cover. So for each edge e of the tree, we will have one *covering* constraint $\sum_{P:e \in P} x_e \geq A$ and one *packing* constraint $\sum_{P:e \in P} x_e \leq |P : e \in P| - B$ (corresponding to coverage for y). Note that the linear program has a feasible fractional solution $x \equiv A/\delta$.

As before, the iterated LP relaxation algorithm repeatedly finds an extreme point solution of the linear program, fixes the value of variables whenever they have integral values, and discards certain constraints. We will use the following analogue of Lemma 11, which is an easy adaptation of a similar result for packing in [6]; we give more details in the full version.

Lemma 15. *Suppose some x variables are fixed to 0 or 1, and some covering/packing constraints are discarded. Any extreme point solution x^* has the following property: there is a tight covering or packing constraint involving at most 3 variables which are fractional in x^*.*

When such a constraint arises, we discard it. As before, any non-discarded constraint is satisfied by the integral x at termination. Additionally, consider a discarded constraint, say a covering one $\sum_{P:e \in P} x_e \geq A$ for some P. When it is discarded, it holds with equality, and the left-hand side consists of 0's, 1's, and at most 3 fractional values. Since A is an integer, it follows that there are at least $A - 2$ 1's on the LHS. The final x still has these variables equal to 1; so overall, x is the characteristic vector of an $(A - 2)$-fold cover, and likewise $1 - x$ is the characteristic vector of a $(B - 2)$-fold cover.

Finally, fix $A = 3$ and $B = \delta - 3$. The final integral x covers every edge at least $3 - 2$ times — it is a cover. The final $1 - x$ covers every edge at least $\delta - 5 = 5(k - 1) + 1$ times. Hence we can use induction to continue splitting $1 - x$, giving the theorem. □

For the related settings where we have paths covering nodes, or dipaths covering arcs, more involved combinatorial lemmas [6, full version] give that $\bar{\mathsf{p}}'(\delta)$ is at least $1 + \lfloor (\delta - 1)/13 \rfloor$. We think Theorem 4 is not tight; the best upper bound on $\bar{\mathsf{p}}'$ we know is $\lfloor (3\delta + 1)/4 \rfloor$.

For polychromatic numbers and systems of paths in trees, we have:

Theorem 5. $\bar{\mathsf{p}}(\text{TreeEdges}|\text{Paths}, r) = \lceil r/2 \rceil$.

Proof Sketch. For the lower bound, colour the edges of the tree by giving all edges at level i the colour $i \bmod \lceil r/2 \rceil$.

In the upper bound, it is enough to consider even r. We use a Ramsey-like argument. Take a complete t-ary tree with $\frac{r}{2}$ levels of edges, so each leaf-leaf path has r edges. In any $(\frac{r}{2} + 1)$-colouring, $t/(\frac{r}{2} + 1)$ of the edges incident on the root have the same colour. Iterating the argument, for large enough t, two root-leaf paths have the same sequence of $\frac{r}{2}$ colours, and their union shows the colouring is not polychromatic.   □

4   Future Work

In the *sensor cover* problem (e.g. [10]) each hyperedge has a given duration; we seek to schedule each hyperedge at an offset so that every item in the ground set is covered for the contiguous time interval $[0, T]$ with the *coverage* T as large as possible. Cover-decomposition is the special case where all durations are unit. Clearly T is at most the minimum of the duration-weighted degrees, which we denote by $\bar{\delta}$. Is there always a schedule with $T = \Omega(\bar{\delta}/\ln R)$ if all hyperedges have size at most R? The LLL is viable but splitting does not work and new ideas are needed. In the full version we get a positive result for graphs ($R = 2$):

Theorem 16. *Every instance of sensor cover in graphs has a schedule of coverage at least $\bar{\delta}/8$.*

Acknowledgments. We thank Jessica McDonald, Dömötör Pálvölgyi, and Oliver Schaud for helpful discussions on these topics.

References

1. Gupta, R.P.: On the chromatic index and the cover index of a mulltigraph. In: Alavi, Y., Lick, D.R. (eds.) Theory and Applications of Graphs: Int. Conf. Kalamazoo, May 11-15. Lecture Notes in Mathematics, vol. 642, pp. 204–215. Springer, Heidelberg (1976)
2. Andersen, L.D.: Lower bounds on the cover-index of a graph. Discrete Mathematics 25, 199–210 (1979)
3. Alon, N., Berke, R., Buchin, K., Buchin, M., Csorba, P., Shannigrahi, S., Speckmann, B., Zumstein, P.: Polychromatic colorings of plane graphs. Discrete & Computational Geometry 42(3), 421–442 (2009); Preliminary version appeared in Proc. 24th SOCG, pp. 338–345 (2008)
4. Pálvölgyi, D.: Decomposition of Geometric Set Systems and Graphs. PhD thesis, École Polytechnique Fédérale de Lausanne, arXiv:1009.4641 (2010)

5. Elekes, M., Mátrai, T., Soukup, L.: On splitting infinite-fold covers. Fund. Math. 212, 95–127 (2011); arXiv:0911.2774

6. Könemann, J., Parekh, O., Pritchard, D.: Max-weight integral multicommodity flow in spiders and high-capacity trees. In: Bampis, E., Skutella, M. (eds.) WAOA 2008. LNCS, vol. 5426, pp. 1–14. Springer, Heidelberg (2009)

7. Pach, J., Tardos, G., Tóth, G.: Indecomposable coverings. Canadian Mathematical Bulletin 52, 451-463 (2009); Preliminary version in Proc. 7th CJCDGCGT (2005), pp. 135-148 (2007)

8. Moser, R.A., Tardos, G.: A constructive proof of the general Lovász Local Lemma. J. ACM 57 (2010); Preliminary version in Proc. 41st STOC, pp. 343–350 (2009)

9. Buchsbaum, A.L., Efrat, A., Jain, S., Venkatasubramanian, S., Yi, K.: Restricted strip covering and the sensor cover problem. In: Proc. 18th SODA, pp. 1056–1063 (2007)

10. Gibson, M., Varadarajan, K.: Decomposing coverings and the planar sensor cover problem. In: Proc. 50th FOCS, pp. 159–168 (2009)

11. Pach, J., Tardos, G.: Tight lower bounds for the size of epsilon-nets. In: Proc 27th SoCG, pp. 458–463 (2011) arXiv:1012.1240

12. Feige, U., Halldórsson, M.M., Kortsarz, G., Srinivasan, A.: Approximating the domatic number. SIAM J. Comput. 32, 172–195 (2002); Preliminary version appeared in Proc. 32nd STOC, pp. 134–143 (2000)

13. Călinescu, G., Chekuri, C., Vondrák, J.: Disjoint bases in a polymatroid. Random Struct. Algorithms 35, 418–430 (2009)

14. Pach, J.: Decomposition of multiple packing and covering. In: Kolloquium über Diskrete Geometrie, Salzburg, Inst. Math. U. Salzburg, pp. 169–178 (1980)

15. Aloupis, G., Cardinal, J., Collette, S., Imahori, S., Korman, M., Langerman, S., Schwartz, O., Smorodinsky, S., Taslakian, P.: Colorful strips. In: López-Ortiz, A. (ed.) LATIN 2010. LNCS, vol. 6034, pp. 2–13. Springer, Heidelberg (2010)

16. Keszegh, B., Pálvölgyi, D.: Octants are cover decomposable (2011); arXiv:1101.3773

17. Keszegh, B.: Weak conflict free colorings of point sets and simple regions. In: Proc. 19th CCCG (2007); Extended version "Coloring half-planes and bottomless rectangles" at arXiv:1105.0169

18. Cheriyan, J., Jordán, T., Ravi, R.: On 2-coverings and 2-packings of laminar families. In: Nešetřil, J. (ed.) ESA 1999. LNCS, vol. 1643, pp. 510–520. Springer, Heidelberg (1999)

19. Cheilaris, P., Keszegh, B., Pálvölgyi, D.: Unique-maximum and conflict-free colorings for hypergraphs and tree graphs. In: Proc. 7th Japanese-Hungarian Symp. Disc. Math. Appl., pp. 207–216 (2011); arXiv:1002.4210

20. Erlebach, T., Jansen, K.: The maximum edge-disjoint paths problem in bidirected trees. SIAM J. Discrete Math. 14, 326–355 (2001); Preliminary version in Proc. 9th ISAAC (1998)

21. Erdős, P., Lovász, L.: Problems and results on 3-chromatic hypergraphs and some related questions. In: Hajnal, A., Rado, R., Sós, V.T. (eds.) Infinite and Finite Sets (Coll. Keszthely, 1973). Colloq. Math. Soc. János Bolyai, vol. 10, pp. 609–627. North-Holland, Amsterdam (1975)

22. Alon, N., Spencer, J.H.: The Probabilistic Method, 3rd edn. Wiley, New York (2008)

23. Beck, J., Fiala, T.: "Integer-making" theorems. Discrete Applied Mathematics 3, 1–8 (1981)

24. Thomassen, C.: The even cycle problem for directed graphs. J. Amer. Math. Soc. 5, 217–229 (1992)

Author Index